Design Engineer's Sourcebook

Design Engineer's Sourcebook

Keith L. Richards

CRC Press
Taylor & Francis Group
Boca Raton London New York

CRC Press is an imprint of the
Taylor & Francis Group, an **informa** business

CRC Press
Taylor & Francis Group
6000 Broken Sound Parkway NW, Suite 300
Boca Raton, FL 33487-2742

Printed on acid-free paper

International Standard Book Number-13: 978-1-4987-6341-7 (Hardback)

Library of Congress Cataloging-in-Publication Data

Names: Richards, Keith L., author.
Title: Design engineer's sourcebook / Keith L. Richards.
Description: Boca Raton : Taylor & Francis, CRC Press, 2018. | Includes index.
Identifiers: LCCN 2017021302| ISBN 9781498763417 (hardback) | ISBN 9781315367514 (ebook)
Subjects: LCSH: Mechanical engineering--Handbooks, manuals, etc. | Machine design--Handbooks, manuals, etc. | Mechanical movements--Handbooks, manuals, etc.
Classification: LCC TJ153 .R483 2018 | DDC 620/.0042--dc23
LC record available at https://lccn.loc.gov/2017021302

Visit the Taylor & Francis Web site at
http://www.taylorandfrancis.com

and the CRC Press Web site at
http://www.crcpress.com

Printed and bound in the United States of America by
Edwards Brothers Malloy on sustainably sourced paper

Contents

Section II Mechanics

Section III Design of Mechanical Elements

Author

Keith L. Richards is a retired mechanical design engineer who has worked in the industry for more than 55 years. Initially, he served an engineering apprenticeship with BSA Tools that manufactured a wide range of machine tools including the Acme Gridley, a multi-spindle automatic lathe built under licence, and the BSA single spindle automatic lathes; these were used in Britain and widely exported around the world.

On leaving the BSA, for a number of years he served as a freelance engineering designer in a wide range of industries that also included aluminum rolling mill design, Industrial fork lift trucks, and the Hutton tension leg platform, an offshore oil production platform. His responsibility on this project covered the mooring system components of the platform, and he was answerable to the customer (Conoco) and Lloyds inspectorate for all the engineering aspects. In later years, Keith was more involved in the aerospace industry working on projects covering aircraft undercarriages and environmental control systems for military and commercial aircraft, and also worked on the A380 wing box and trailing edge panels.

Section I

Reference

1

Mathematics

1.1 Trigonometry

1.1.1 Right-Angled Triangle

See Figure 1.1.

$$\text{Area} = \frac{1}{2} a \cdot b \tag{1.1}$$

Pythagoras Theorem

$$c^2 = a^2 + b^2 \tag{1.2}$$

$$A + B + C = 180°, B = 90° - A \tag{1.3}$$

$$\sin(90 - A) = \cos A, \cos(90 - A) = \sin A \tag{1.4}$$

Definitions:

$$\sin A = \frac{\text{opposite}}{\text{hypotenuse}} = \frac{a}{c} \tag{1.5}$$

$$\cos A = \frac{\text{adjacent}}{\text{hypotenuse}} = \frac{b}{c} \tag{1.6}$$

$$\tan A = \frac{\text{opposite}}{\text{adjacent}} = \frac{a}{b} \tag{1.7}$$

$$\text{cosecant } A = \frac{1}{\sin} = \frac{c}{a} \tag{1.8}$$

$$\text{secant } A = \frac{1}{\cos} = \frac{c}{b} \tag{1.9}$$

$$\text{cotangent } A = \frac{1}{\text{tangent}} = \frac{b}{a} \tag{1.10}$$

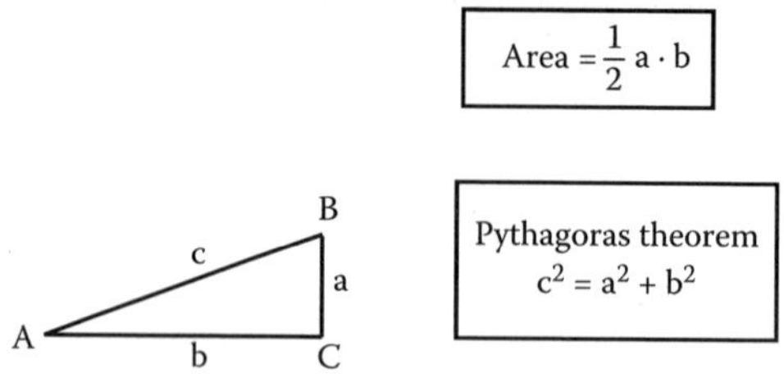

FIGURE 1.1
Right-angled triangle.

1.1.2 Oblique-Angled Triangles

See Figure 1.2.

Cosine rule:

$$a^2 = b^2 + c^2 - 2bc\cos A \tag{1.11}$$

$$b^2 = a^2 + c^2 - 2ac\cos B \tag{1.12}$$

$$c^2 = a^2 + b^2 - 2ab\cos C \tag{1.13}$$

Sine rule:

$$\frac{a}{\sin A} = \frac{b}{\sin B} = \frac{c}{\sin C} \tag{1.14}$$

$$\text{Area} = \sqrt{s(s-a)(s-b)(s-c)} = \frac{1}{2}ab\sin C \tag{1.15}$$

where

$$s = \frac{a+b+c}{2}$$

1.1.3 Trigonometric Relations

$$\sin(A)^2 = \frac{[1-\cos(2A)]}{2} \tag{1.16}$$

$$\sin(A)\cos(A) = \frac{\sin(2A)}{2} \tag{1.17}$$

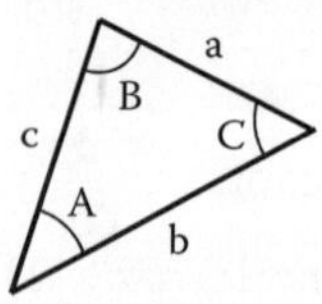

FIGURE 1.2
Oblique-angled triangle.

$$\sin(A+B) = \sin(A)\cos(B) + \cos(A)\sin(B) \tag{1.18}$$

$$\cos(A+B) = \cos(A)\cos(B) - \sin(A)\sin(B) \tag{1.19}$$

$$\cos(A-B) = \cos(A)\cos(B) + \sin(A)\sin(B) \tag{1.20}$$

$$\sin(A)\sin(B) = \frac{[1-\cos(A+B)+\cos(A-B)]}{2} \tag{1.21}$$

$$\cos(-A) = \cos(A) \tag{1.22}$$

$$\sin(A)^2 + \cos A^2 = 1 \tag{1.23}$$

$$\cos(A)^2 = \frac{[1+\cos(2A)]}{2} \tag{1.24}$$

$$1+\cot(A)^2 = \mathrm{cosec}(A)^2 \tag{1.25}$$

1.2 Hyperbolic Functions

Hyperbolic functions involve the exponential functions, e^x and e^{-x}, where e is the base of the Napierian logs (ln). (e=2.7182828…).

Definitions:

$$\text{Hyperbolic sine: } \sinh x = \frac{e^x - e^{-x}}{2} \tag{1.26}$$

$$\text{Hyperbolic cosine: } \cosh x = \frac{e^x + e^{-x}}{2} \tag{1.27}$$

$$\text{Hyperbolic tangent: } \tanh x = \frac{e^x - e^{-x}}{e^x + e^{-x}} = \frac{\sinh x}{\cosh x} \tag{1.28}$$

$$\cosh^2 x - \sinh^2 x = 1 : 1 - \tanh^2 x = \frac{1}{\cosh^2 x} = \mathrm{sech}^2 x \quad \text{(hyperbolic secant)} \tag{1.29}$$

1.2.1 Inverse Hyperbolic Functions

Note:

$\ln = \log_e$

$$y = \sinh^{-1} : \text{'y' equals the inverse hyperbolic sinh of x} \tag{1.30}$$

$$\sinh^{-1} x = \ln(x + \sqrt{x^2+1}) \text{ for all values of x} \tag{1.31}$$

$$\cosh^{-1} x = \pm \ln(x + \sqrt{x^2 - 1}) \quad x \geq 1 \tag{1.32}$$

$$\tanh^{-1} x = \frac{1}{2} \ln \frac{1+x}{1-x} \quad x > 1 \tag{1.33}$$

1.3 Solution of the Quadratic Equation

$$ax^2 + bx + c = 0 \quad a \neq 0, \text{and a, b and c are real} \tag{1.34}$$

The roots are

$$x = \frac{1}{2a}(-b \pm \sqrt{b^2 - 4ac}) \tag{1.35}$$

$ax^2+bx+c=0$, $a\neq0$ and a, b and c are real.

1.4 Solution of Simultaneous Equations (Two Unknowns)

$$a_{11}x_1 + a_{12}x_2 = b_1 \tag{1.36}$$

$$a_{21}x_1 + a_{22}x_2 = b_2 \tag{1.37}$$

The solutions are

$$x_1 = \frac{b_1 a_{22} - b_2 a_{12}}{a_{11}a_{22} - a_{12}a_{21}} \text{ and } x_2 = \frac{b_2 a_{11} - b_1 a_{21}}{a_{11}a_{22} - a_{12}a_{21}} \tag{1.38}$$

1.5 Laws of Exponents

There are only six major laws of exponents that determine all that can be done with exponents and exponential functions:

1. $a^{x+y} = a^x a^y$ (1.39)

2. $a^{x-y} = \frac{a^x}{a^y}$ (1.40)

3. $(a^x)^y = a^{xy}$ (1.41)

4. $(ab)^x = a^x b^x$ (1.42)

5. $a^{x/y} = \sqrt[y]{a^x}$ (1.43)

6. $a^\circ = 1$ (1.44)

1.6 Expansions

$$(a \pm b)^2 = a^2 \pm 2ab + b^2 \tag{1.45}$$

$$(a \pm b)^3 = a^3 \pm 3a^2b + 3ab^2 \pm b^3 \tag{1.46}$$

$$(a^2 - b^2) = (a - b)(a + b) \tag{1.47}$$

$$(a^3 \pm b^3) = (a \pm b)(a^2 - / + ab + b^2) \tag{1.48}$$

$$(a + b)^n = a^n + na^{n-1}b + \frac{n(n-1)}{2!}a^{n-2}b^2 + \cdots \tag{1.49}$$

1.7 Real Root of the Equation f(x) = 0 Using the Newton–Raphson Method

From Figure 1.3, it is seen that the derivative of y = f(x) is continuous and no point of inflexion exists between A and B.

x_0 is the true root (which is unknown) of f(x) = 0, that is when y = 0.

x_1 is a first approximation to x_0: obtained graphically or by trial and error.

A better approximation to x_0 is given by

$$x_2 = x_1 - \frac{f(x_1)}{f'(x_1)} \tag{1.50}$$

This assumes that x_1 is near enough to x_0.

This process may be repeated several times to obtain closer approximations to x_0.

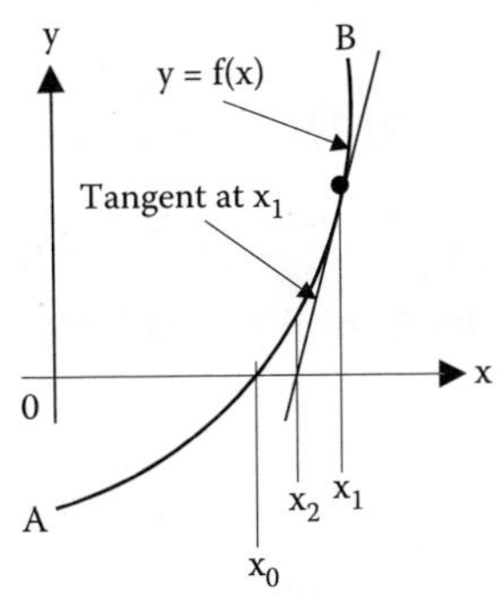

FIGURE 1.3
Real roots of the equation.

1.8 Series

Binomial:

$$(1 \pm x)^n = 1 \pm nx + \frac{n(n-1)}{2!}x^2 \pm \frac{n(n-1)(n-2)}{3!}x^3 + \cdots \tag{1.51}$$

Maclaurin:

$$f(x) = f(0) + xf'(0) + \frac{x^2}{2!}f''(0) + \frac{x^3}{3!}f'''(0) + \cdots \tag{1.52}$$

Taylor:

$$f(a+h) = f(a) + hf'(a) + \frac{h^2}{2!}f''(a) + \frac{h^3}{3!}f'''(a) + \cdots \tag{1.53}$$

1.9 Logarithms

If $y=x^n$ where n is the logarithm of y to the base x, that is, if $100=10^2$, 10 is the base and 2 is the logarithm.

10 is the base for *common* logarithms (log).

e=2.71828… is the base for *natural* (Napierian) logarithms (ln).

$$\log(A \cdot B) = \log(A) + \log(B) \tag{1.54}$$

$$\log\left(\frac{A}{B}\right) = \log(A) - \log(B) \tag{1.55}$$

$$\log(A)^n = n\log(A) \tag{1.56}$$

Also

$$\ln(A \cdot B) = \ln(A) + \ln(B) \tag{1.57}$$

$$\ln\left(\frac{A}{B}\right) = \ln(A) - \ln(B) \tag{1.58}$$

$$\ln(A)^m = m\ln(A) \tag{1.59}$$

Note:

If $x=\log N$ or $x=\ln N$, then $N=10^x$ or $N=e^x$.

N is the antilogarithm of x, that is, $\log(1)=0$ and $\log 10=1$,

$\ln(1)=0$ and $\ln e=1$.

1.10 Differential Calculus

Definitions:

See Figure 1.4.

The value of the derivative at any point 'P' on a curve is equal to the slope of the tangent to the curve at that point.

$$\frac{\Delta y}{\Delta x} = \tan\beta \text{ (slope of tangent at P)} \tag{1.60}$$

as Δx becomes smaller, $\beta \to \alpha$,

that is

$$Q \to P \frac{\Delta y}{\Delta x} \to \frac{dy}{dx} = \tan\alpha \tag{1.61}$$

Note:

The first derivative $\frac{dy}{dx}$ is referred to as $y' = f'(x)$ and the second derivative $\frac{d^2y}{d^2x}, y'' = f''(x), \text{etc.}$

Table 1.1 gives a range of standard derivatives for y or f(x).

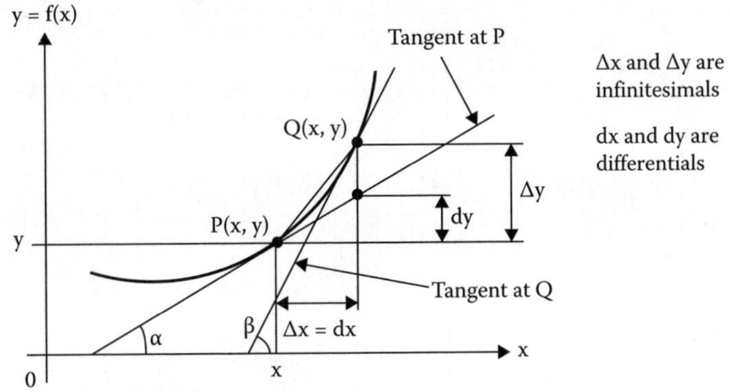

FIGURE 1.4
Definitions for differential calculus.

TABLE 1.1

Table of Standard Derivatives

y or f(x)	dy/dx or f′(x)
1. x	1
2. x^n	nx^{n-1}
3. $\frac{1}{x}$	$\frac{1}{x^{(2)}}$
4. $\sqrt{x}$	$\frac{1}{2\sqrt{x}}$
5. a^x	$a^x \ln a$
6. ax^n	anx^{n-1}
7. e^x	e^x
8. e^{ax}	ae^{ax}
9. enx	nenx
10. log x	$\frac{1}{x}$
11. ln ax	$\frac{1}{x}$
12. x^x	$x^x(\log+1)$
13. sin x	cos x
14. sin ax	a cos ax
15. cos x	–sin x
16. cos ax	–a sin ax
17. tan x	$\sec^2 x$
18. tan ax	$a \sec^2 ax$
19. cot x	$-\text{cosec}^2 x$
20. cot ax	$-a\,\text{cosec}^2 ax$
21. sec x	sec x · tan x
22. sec ax	a sec ax tan ax
23. cosec x	–cosec x · cot x
24. cosec ax	–a cosec ax · cot ax
25. $\sin^{-1} x$	$\frac{1}{\sqrt{1-x^2}}\left(\lvert y\rvert<\frac{\pi}{2}\right)$
26. $\sin^{-1}\frac{x}{a}$	$\frac{1}{\sqrt{a^2-x^2}}\quad(0<y<\pi)$
27. $\cos^{-1} x$	$-\frac{1}{\sqrt{(1-x^2)}}\quad(0<y<\pi)$
28. $\cos^{-1}\frac{x}{a}$	$-\frac{1}{\sqrt{(a^2-x^2)}}$
29. $\tan^{-1} x$	$\frac{1}{1+x^2}\left(\lvert y\rvert<\frac{\pi}{2}\right)$
30. $\tan^{-1}\frac{x}{a}$	$\frac{a}{(a^2+x^2)}$
31. $\sec^{-1} x$	$\frac{1}{x\sqrt{(x^2-1)}}\quad(0<t<\pi x^2<1)$

(Continued)

TABLE 1.1 (*Continued*)
Table of Standard Derivatives

y or f(x)	dy/dx or f′(x)
32. $\sec^{-1}\frac{x}{a}$	$\frac{a}{x\sqrt{(x^2-a^2)}}$
33. $\operatorname{cosec}^{-1} x$	$\frac{1}{x\sqrt{(x^2-1)}}\left(\lvert y\rvert<\frac{\pi}{2}x^2>1\right)$
34. $\operatorname{cosec}^{-1}\frac{x}{a}$	$\frac{1}{x\sqrt{(x^2-a^2)}}$
35. $\cot^{-1} x$	$\frac{1}{1+x^2}\left(\lvert y\rvert<\frac{\pi}{2}\right)$
36. $\cot^{-1}\frac{x}{a}$	$-\frac{a}{a^2+x^2}$
37. sinh x	cosh x
38. sinh ax	a cosh ax
39. cosh x	sinh x
40. tanh x	sech^2 x
41. tanh ax	a sech^2 ax
42. coth x	–cosech^2 x
43. coth ax	–cosech^2 ax
44. sech x	–sech x · tanh x
45. sech ax	–a sech ax
46. cosech x	–cosech x · coth x
47. cosech ax	–cosech ax · cosh ax
48. $\sinh^{-1} x$	$\frac{1}{\sqrt{(1+x^2)}}$
49. $\sinh^{-1}\frac{a}{x}$	$\frac{1}{\sqrt{(a^2+x^2)}}$
50. $\cosh^{-1} x$	$\frac{1}{\sqrt{(x^2-1)}}(y>0, x^2>1)$
51. $\cosh^{-1}\frac{x}{a}$	$\frac{1}{\sqrt{(x^2-a^2)}}$
52. $\tanh^{-1} x$	$\frac{1}{\sqrt{(1-x^2)}}(x^2<1)$
53. $\tanh^{-1}\frac{x}{a}$	$\frac{a}{\sqrt{(a^2-x^2)}}$
54. $\coth^{-1} x$	$\frac{1}{(1-x^2)}(x^2>1)$

(Continued)

TABLE 1.1 (*Continued*)
Table of Standard Derivatives

y or f(x)	dy/dx or f′(x)
55. $\coth^{-1}\frac{x}{a}$	$\frac{a}{(a^2-x^2)}$
56. $\text{sech}^{-1}x$	$-\frac{1}{\sqrt{(1-x^2)}}\ (0<x<1)$
57. $\text{cosech}^{-1}x$	$-\frac{1}{x\sqrt{(x^2+1)}}$
58. $\text{cosech}^{-1}\frac{x}{a}$	$-\frac{a}{x\sqrt{(x^2+a^2)}}$

1.11 Integral Calculus

Definitions:
See Figure 1.5.

1.11.1 Integration Is the Inverse of Differentiation

If

$$F(x)=\int_a^b F(x)dx \quad \text{then} \frac{d}{dx}F(x)=F'(x)=f(x) \tag{1.62}$$

where F(x) is referred to as the primitive of the function f(x).

1.11.2 Indefinite Integrals

All integrals, $\int f(x)dx$ are classed as indefinite integrals and they differ from one another by a constant whose value depends upon the boundary conditions.

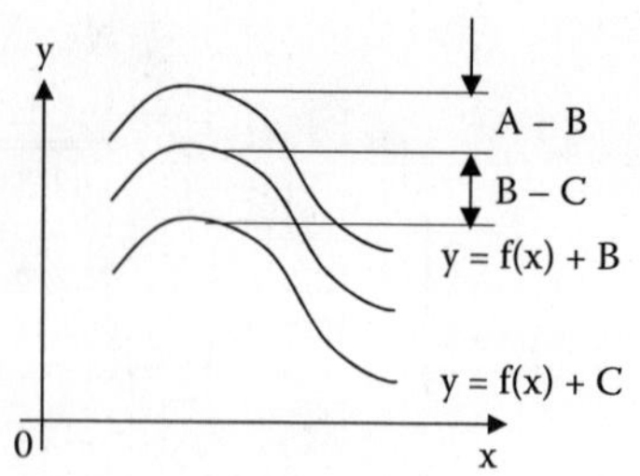

FIGURE 1.5
Definitions for integral calculus.

The following are some common integrals:

f(x)	∫f(x)dx
c (a constant)	cx
x^n	$x^{n+1}/(n+1)$ $(n \neq -1)$
$1/x$	$\ln x$ $(x \neq 0)$
e^{ax}	$e^{ax}/(a)$
$\ln x$	$x \ln x - x$ $(x > 0)$
a^x	$ax/(\ln a)$ $(a > 0, a \neq 1)$
sin ax	$-\cos ax/(a)$
cos ax	$\sin ax/(a)$
tan x	$-\ln \lvert \sec x \rvert$

Note:

The constants of integration have been omitted in the above table.

Table 1.2 gives a more comprehensive list of standard indefinite integrals.

TABLE 1.2

Table of Indefinite Integrals

f(x)	∫f(x) dx
1. $\int dx\ x$	$x+C$
2. $\int x^n\,dx$	$\frac{x^{n+1}}{n+1}+C \quad (n+1 \neq 0)$
3. $\int \frac{1}{x}dx$	$\ln \lvert x \rvert + C$
4. $\int \frac{1}{x \pm a\,dx}$	$\ln \lvert x \pm a \rvert + C$
5. $\int e^x\,dx$	$e^x + C$
6. $\int e^{nx}\,dx$	$\frac{1}{n}e^{nx}+C$
7. $\int a^x\,dx$	$\frac{a^x}{\ln a}+C (a > 0, a \neq 0)$
8. $\int a^{nx}\,dx$	$\frac{a^{nx}}{n \cdot \ln\ a}(a > 0, a \neq 0)$
9. $\int \ln x dx$	$x(\ln x - 1) + C$
10. $\int xe^{nx}\,dx$	$\frac{e^{nx}}{n^2}(nx-1)+C$
11. $\int \sin x\,dx$	$-\cos x + C$
12. $\int \sin ax\,dx$	$-\frac{1}{a}\cos ax + C$
13. $\int \cos x\,dx$	$\sin x + C$
14. $\int \cos ax\,dx$	$\frac{1}{a}\sin ax + C$
15. $\int \tan x\,dx$	$-\ln \lvert \cos x \rvert + C$
16. $\int \cot x\,dx$	$\ln \lvert \sin x \rvert + C$
17. $\int \sin^2 x\,dx$	$\frac{1}{2}x - \frac{1}{4}\sin 2x + C$

(Continued)

TABLE 1.2 (*Continued*)

Table of Indefinite Integrals

f(x)	∫f(x) dx
18. $\int \cos 2x\, dx$	$\frac{1}{2}x + \frac{1}{4}\sin 2x + C$
19. $\int \sec 2ax\, dx$	$\frac{1}{a}\tan ax + C$
20. $\int \text{cosec } 2ax\, dx$	$-\frac{1}{a}\cot ax + C$
21. $\int \frac{1}{\sin x\, dx}$	$\ln \tan \frac{x}{2} + C$
22. $\int \frac{1}{\cos x} dx$	$\ln \tan\left(\frac{\pi}{4} + \frac{x}{2}\right) + C$
23. $\int e^{nx} \sin bx\, dx$	$\frac{1}{n^2 + b^2} e^{nx} (n \cdot bx - b \cdot \cos bx) + C$
24. $\int e^{nx} \cos bx\, dx$	$\frac{1}{n^2 + b^2} e^{nx} (n \cdot \cos bx - b \cdot \sin bx) + C$
25. $\int \sin^{-1} x\, dx$	$x \sin^{-1} x + \sqrt{1 - x^2} + C$
26. $\int \cos^{-1} x\, dx$	$x \cos^{-1} x + \sqrt{1 - x^2} + C$
27. $\int \sinh x\, dx$	$\cosh x + C$
28. $\int \cosh x\, dx$	$\sinh x + C$
29. $\int \tanh x\, dx$	$\ln \cosh x + C$
30. $\int \frac{1}{\sqrt{a^2 - x^2}} dx$	$\sin^{-1} \frac{x}{a} + C \quad (\lvert x \rvert < a)$
31. $\int \frac{1}{\sqrt{a^2 + x^2}} dx$	$\frac{1}{a}\tan^{-1}\frac{x}{a} + C$
32. $\int \frac{1}{\sqrt{x^2 \pm a^2\, dx}}$	$\ln\left(x + \sqrt{x^2 \pm a^2}\right) + C$
33. $\int \sqrt{a^2 - x^2}\, dx$	$\frac{1}{2}\left(x\sqrt{a^2 - x^2} + a^2 \sin^{-1}\frac{x}{a}\right) + C$
34. $\int \frac{1}{x^2 - a^2} dx$	$\frac{1}{2a}\ln\left(\frac{x - a}{x + a}\right) + C \quad (x > a)$

1.11.3 Determination of an Area

Figure 1.6 depicts the area under a known curve f(x) that has been subdivided into vertical rectangular strips.

$$\text{Area A} = \lim_{\delta x_i \to 0} \sum_{i=1}^{i=n} f(x_i)\delta x_i = \int_a^b f(x)dx \tag{1.63}$$

The area is the summation of an infinite number of indefinitely small quantities:

$\int_a^b f(x)dx$ is the definite integral of the function f(x) between x = a and x = b

where a and b are the limits of integration.

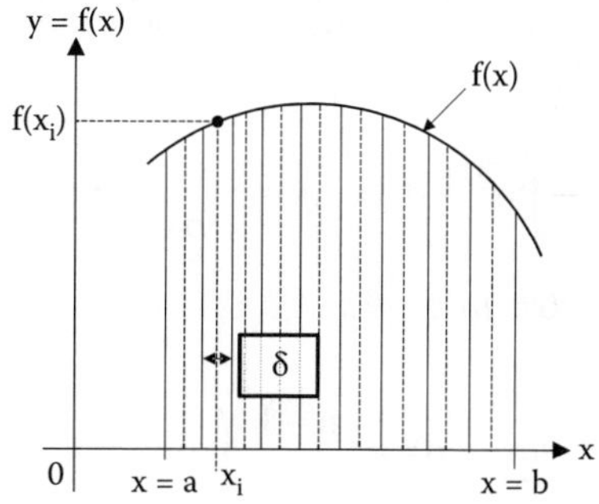

FIGURE 1.6
Determination of an area.

1.11.4 Approximate Integration

In an instance where the curve function is unknown, an approximate value of $\int_a^b f(x)dx$ can be found by two methods:

a. *Simpson's rule:* See Figure 1.7.

$$\frac{h}{3}[y_1 + y_{n+1} + 4(y_2 + y_4 + y_6 + \cdots) + 2(y_3 + y_5 + y_7 + \cdots)] \tag{1.64}$$

where

$$h = (b - a) / n; \quad n = \text{even number of strips}$$

b. *Trapezoidal rule:*

$$h\left[\frac{1}{2}(y_1 + y_{n+1}) + y_2 + y_3 + \cdots + y_n\right] \tag{1.65}$$

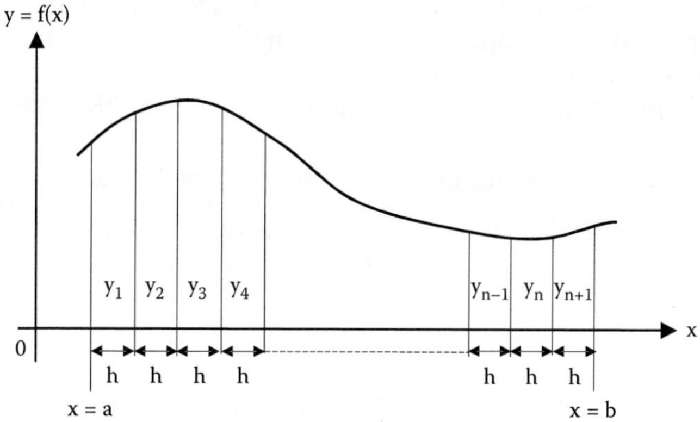

FIGURE 1.7
Method for approximation of an area using Simpson's rule.

In both methods, the accuracy can be improved by increasing the number of strips.

Integration by parts:

$$\int u\frac{dv}{dx}dx = uv - \int \frac{du}{dx}v\,dx \tag{1.66}$$

Reduction formulae for trigonometric integrals:

$$\int_0^{\pi/2} \sin^n x\,dx = \frac{m-1}{m}\int_0^{\pi/2} \sin^{m-2} x\,dx \tag{1.67}$$

$$\int_0^{\pi/2} \cos^m x\,dx = \frac{m-1}{m}\int_0^{\pi/2} \cos^{m-2} x\,dx \tag{1.68}$$

$$\int_0^{\pi/2} \sin^m x\cos^n x\,dx = \frac{m-1}{m+n}\int_0^{\pi/2} \sin^{m-2} x\,\cos^n x\,dx = \frac{n-1}{m+n}\int_0^{\pi/2} \sin^m x\,\cos^{n-2} x\,dx \tag{1.69}$$

where n is an integer.

(These results hold providing that the exponents in the reduced form are greater than –1. There are analogous reduction formulae with other intervals of integration ($1/2k_1\pi$, $1/2k_2\pi$) with k_1, k_2 integral.)

1.12 Laplace Transforms

The Laplace transform is a widely used integral transform used in many applications in engineering, particularly in vibration studies and control engineering. It is denoted by $\mathcal{L}\{f(t)\}$; it is a linear operator of a function f(t) with a real argument t ($t \geq 0$) that transforms f(t) to a function F(s) with complex argument 's'. There is a one-to-one correspondence in the transformation for the majority of practical uses; the most-common pairs of f(t) and F(s) are often given in tables for easy reference. The Laplace transform has the useful property that many relationships and operations over the original f(t) correspond to simpler relationships and operations over its image F(s).

It is named after Pierre-Simon Laplace, who introduced the transform in his work on probability theory.

Standard Laplace transforms are given in Table 1.3. The Laplace transforms of derivatives are as follows.

1.12.1 First Derivative

$$\mathcal{L}\left\{\frac{dy}{dx}\right\} = s\mathcal{L}\{y\} - y(0) \tag{1.70}$$

when y(0) is the value of y at x=0.

TABLE 1.3
Table of Standard Laplace Transforms

Function f(t)	**Laplace Transforms** $\mathcal{L}\{f(t)\}=\int_0^\infty e^{-st}f(t)dt$
1. 1	$\frac{1}{s}$
2. k	$\frac{k}{s}$
3. e^{at}	$\frac{1}{s-a}$
4. sin at	$\frac{a}{s^2+a^2}$ $\mathcal{L}\left\{\frac{d^n y}{dx^n}\right\}=s^n\ L\{y\}-s^{n-1}y(0)-s^{n-2}y'(0)-\cdots-y^{n-1}(0)$
5. cos at	$\frac{s}{s^2+a^2}$
6. t	$\frac{1}{s^2}$
7. t^2	$\frac{2!}{s^3}$
8. t^n (n=positive)	$\frac{n!}{s^{n+1}}$
9. cosh at	$\frac{s}{s^2-a^2}$
10. sinh at	$\frac{a}{s^2-a^2}$
11. $e^{-at}t^n$	$\frac{n!}{(s+a)^{n+1}}$
12. $e^{-at}\sin\omega t$	$\frac{\omega!}{(s+a)^2+\omega^2}$
13. $e^{-at}\cos\omega t$	$\frac{s+a}{(s+a)^2+\omega^2}$
14. $e^{-at}\cosh\omega t$	$\frac{s+a}{(s+a)^2+\omega^2}$
15. $e^{-at}\sinh\omega t$	$\frac{\omega}{(s+a)^2-\omega^2}$

1.12.2 Second Derivative

$$\mathcal{L}\left\{\frac{d^2y}{dx^2}\right\}=s^2\mathcal{L}\{y\}-sy(0)-y'(0) \tag{1.71}$$

when $y'(0)$ is the value of $\frac{dy}{dx}$ at $x=0$.

1.12.3 Higher Derivatives

$$\mathcal{L}\left\{\frac{d^n y}{dx^n}\right\} = s^n \mathcal{L}\{y\} - s^{n-1}y(0) - s^{n-2}y'(0) - \cdots - y^{n-1}(0) \tag{1.72}$$

1.13 Parallel Axis Theorem

The parallel axis theorem may be used to refer the moment of inertia of a rigid body about a given axis to an offset parallel axis which is not necessarily the centre of mass of the body. The theorem is also known as the Huygens–Steiner theorem. The theorem has importance when calculating the sectional properties of a complex section.

1.13.1 Calculation of the Moment of Inertia Using the Parallel Axis Theorem

Consider the Figure 1.8 which shows a rectangular section offset to a specified axis. It is required to calculate the revised moment of inertia with respect to the new axis. In this simple case, the moment of inertia of the rectangle about its own centroid 'x_c:x_c' will be

$$I_c = \frac{b \cdot d^3}{12} \tag{1.73}$$

and the area will be

$$A_c = b \cdot d \tag{1.74}$$

The equation for the theorem is

$$I_{xx} = \sum (I_c + Ah^2) \tag{1.75}$$

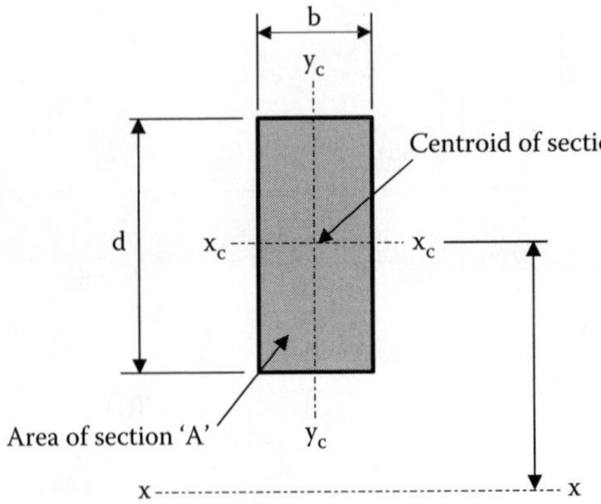

FIGURE 1.8
Parallel axis rule for area of moments.

Example 1.1

Using a practical problem, consider the section shown in Figure 1.9. In this example, the section is broken down into simple rectangular sections. It is a simple matter to construct a table to evaluate the properties of these sections as shown in Table 1.4.

From the table, it is seen that the individual moment of inertia is calculated about each section and the parallel axis theorem is then invoked to determine the new moment of inertia about the neutral axis.

The example shown is known as the *'Area Moment of Inertia'* as it deals with the respective areas of the section.

$$I_{\text{Parallel Axis}} = I_{\text{Centroidal Axis}} + Ah^2 \tag{1.76}$$

where

A is the cross-sectional area.
h is the perpendicular distance between the centroidal axis and the parallel axis.

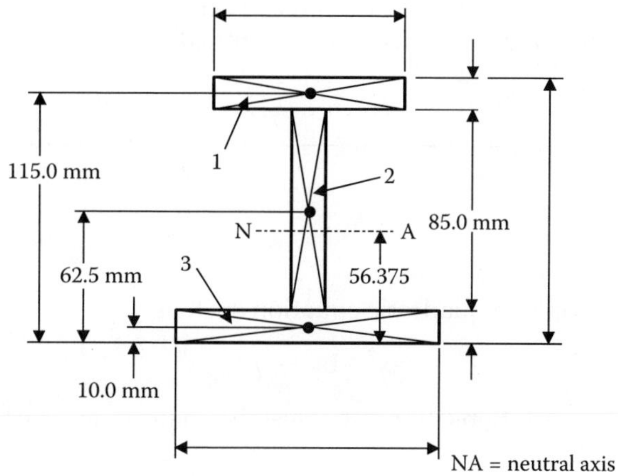

FIGURE 1.9
Example 'I' section.

TABLE 1.4
Sectional Properties for Example 1.1

Section Number	b (mm)	d (mm)	A (mm²)	y (mm)	A·y (mm³)×10³
1	90.0	20.0	1800	115.0	207,000
2	20.0	85.0	1700	62.5	106,250
3	125.0	20.0	2500	10.0	25,000
			ΣA=6000 mm²		ΣAy=338,250 mm³

Position of neutral axis: $\bar{Y} = 56.375\,\text{mm}$

Area Moment of Inertia Calculations

Section Number	I (mm⁴)	h (mm)	I_{xx} + (A·h²) (mm⁴)×10³
1	3000.0	58.6	6189.403
2	12,041.7	6.1	75.818
3	4166.7	46.4	5380.768
	ΣI = 19,208.3 mm⁴		ΣI_{xx} = 11,645.990 mm⁴

The theorem can also be used to determine the *'Mass Moments of Inertia'*

$$I = Mh^2 \tag{1.77}$$

where

M is the mass of the body.
h is the perpendicular distance between the centroidal axis and the parallel axis.

The *'Mass Radius of Gyration'* of a section is a measure of the distribution of the mass in an object from its geometric centre and is a sign of its resistance to rotational motion.

The radius of gyration can also be referred to as an offset parallel axis in a similar manner.

$$k^2 = k_c^2 + h^2 \tag{1.78}$$

where

k is the radius of gyration about an axis parallel to the centroidal axis.
$k_{c'}$ is the radius of gyration about the centroidal axis.
h is the perpendicular distance between the centroidal axis and the parallel axis.

1.14 Complex Numbers

1.14.1 Introduction

When finding the solution to a quadratic equation such as

$$ax^2 + bx + c = 0 \tag{1.79}$$

there will always be two solutions. In most cases, the solution will be straightforward, but in the case of the equation

$$x = 5x^2 - 6x + 5 = 0 \tag{1.80}$$

using the standard formula

$$x = \frac{-b \pm \sqrt{(b^2 - 4ac)}}{2a} \tag{1.81}$$

the solution will be

$$x = \frac{6 \pm \sqrt{(-64)}}{10} \tag{1.82}$$

The solution to this equation relies on solving √(−64).

√(−64) cannot be represented by an ordinary number as there is no real number whose square is a negative quantity.

Now −64 can be written as −1×64, and from this:

$$\begin{aligned} j^7 &= j^4 \times j^3 = -j \\ &= \sqrt{(-1)} \times \sqrt{(64)} \\ &= 8\sqrt{(-1)} \end{aligned} \tag{1.83}$$

This still leaves one to find the solution of √(–1). This cannot be evaluated as a real number, but if √(–1) is represented by j, this will make the workings a lot easier.

The final solution to the quadratic therefore will be

$$x = \frac{6 \pm j8}{10} \quad \therefore x = 0.6 \pm j0.8 \tag{1.84}$$

$$\therefore x = 0.6 + j0.8 \text{ or } x = 0.6 - j0.8 \tag{1.85}$$

Now

$$j = \sqrt{-1} \tag{1.86}$$

$$\therefore j^2 = \sqrt{-1} \times \sqrt{-1} = -1 \tag{1.87}$$

$$j^3 = j^2 \times j = -j \tag{1.88}$$

$$j^4 = j^2 \times j^2 = 1 \tag{1.89}$$

$$j^5 = j^4 \times j^2 = j \tag{1.90}$$

$$j^7 = j^4 \times j^3 = -j \tag{1.91}$$

1.14.2 Argand Diagram

The Argand diagram was devised as a means of representing complex numbers.

Although Jean Robert Argand, a Swiss mathematician in 1806, is credited with developing the diagram, it was in fact described by C. Wessel earlier. The geometric representation of complex numbers as points in a plane made the whole idea of a complex number more acceptable. Indeed, this visualisation helped 'imaginary' and 'complex' numbers become more accepted in mainstream mathematics as a natural extension to negative numbers along the real line.

Figure 1.10 shows the diagram with the complex number z = a + jb plotted as a point with coordinates (a, b). Because the real part of z is plotted on the horizontal axis of the diagram, this is often referred to as the 'real axis'. The imaginary part of z is plotted on the vertical axis and this is then referred to as the 'imaginary axis'.

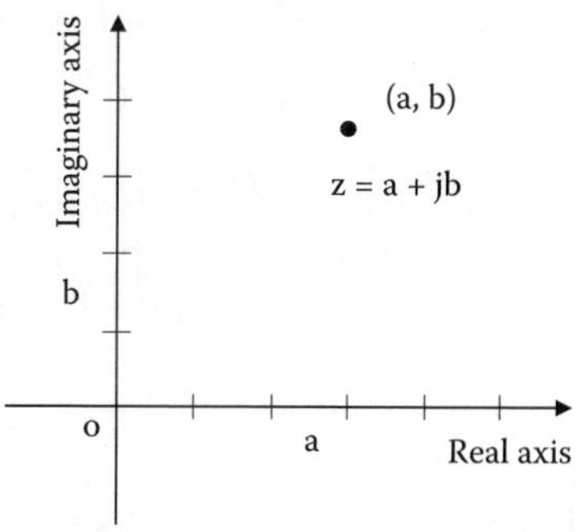

FIGURE 1.10
Argand diagram.

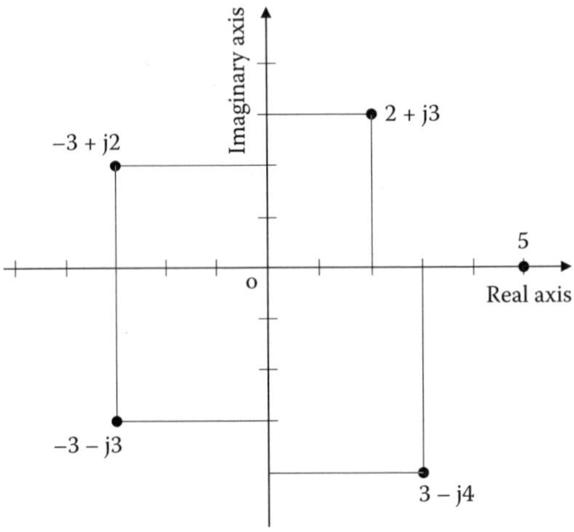

FIGURE 1.11
Argand diagram for Example 1.2.

Example 1.2

Plot the complex numbers 2+j3, −3+j2, −3 − j3, 3 − j4 and 5j on an Argand diagram.

Solution

This is shown in Figure 1.11.

1.14.3 Manipulation of Complex Numbers

1.14.3.1 Addition and Subtraction

The real parts and the imaginary parts of the numbers are added or subtracted separately, for example,

$$(3+j4)+(5+j2)=8+6j \quad (1.92)$$

1.14.3.2 Multiplication

Multiplication is undertaken algebraically; complex multiplication is more difficult to understand from either an algebraic or geometric point of view. Carrying out the algebraic operation first,

$$(3+j2) \text{ and } (1+j4) \quad (1.93)$$

each of these have two terms each and when multiplied together will result in four terms.

$$(3+j2)(1+j4)=3+1j2+j2+j28 \quad (1.94)$$

The 1j2+j2 terms will reduce to j14, which leaves the $j^2 8$ term. Remember that 'j' was introduced as an abbreviation for √(−1), the square root of −1. Therefore, 'j' is something whose square is −1 (see Equation 1.87).

Thus, j28 will equal –8. The product of

$$(3+j2)(1+j4)=(-5+j14)$$

From this example, it is possible to arrive at a general rule for the solution of multiplying complex numbers.

$$(x+jy)(u+jv)=(xu-yv)+j(xv+yu)$$

Remember that (xu – yv) is the real part of the product and is the product of real parts minus the products of the imaginary parts. But (xv +yu), the imaginary part of the product, is the sum of the two products of one real part and the other imaginary part.

1.14.3.3 Division

Consider

$$\frac{2+j3}{3-j4} \tag{1.95}$$

Now

$$\frac{2+j3}{3-j4}=\frac{2+j3}{3-j4}\times\frac{3+j4}{3+j4}=\frac{(2+j3)(3+j4)}{(3^2-4^2)} \tag{1.96}$$

$$=\frac{6+j8+j9-12}{25}$$

$$=\frac{-6+j17}{25}$$

$$=\frac{-6}{25}+\frac{j17}{25}$$

$$=0.24+j0.68 \tag{1.97}$$

1.14.4 Polar Form of a Complex Number

It is sometimes convenient to express a complex number (a+jb) in a different form. From an Argand diagram, OP is a vector a+jb. Let r=length of the vector and θ the angle made with the axis OX (Figure 1.12).

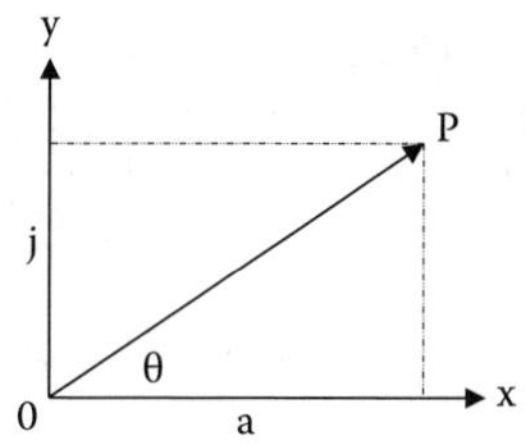

FIGURE 1.12
Polar form of a complex number.

Now

$$r^2 = a^2 + b^2 \quad r = \sqrt{(a^2 + b^2)} \tag{1.98}$$

And

$$\tan\theta = \frac{b}{a}$$

Also

$$a = r\cos\theta \text{ and } b = r\sin\theta \tag{1.99}$$

Since z=a+jb, this can be rewritten

$$z = r\cos\theta + jr\sin\theta \quad \text{that is, } z = r(\cos\theta + j\sin\theta) \tag{1.100}$$

This is called the 'Polar Form' of the complex number a+jb, where

$$r = \sqrt{(a^2 + b^2)} \text{ and } \theta = \tan^{-1}\frac{b}{a}$$

Example 1.3

Express z=4+j3 in polar form.

Solution

From Figure 1.13, it can be seen that

a. $r^2 = 4^2 + 3^2 = 16 + 9$

$= 25$ therefore, $r = 5$

b. $\tan\theta = \frac{3}{4}$

$= 0.75$

$\theta = 36.8667^\circ$

$z = a + jb = r(\cos\theta + j\sin\theta)$

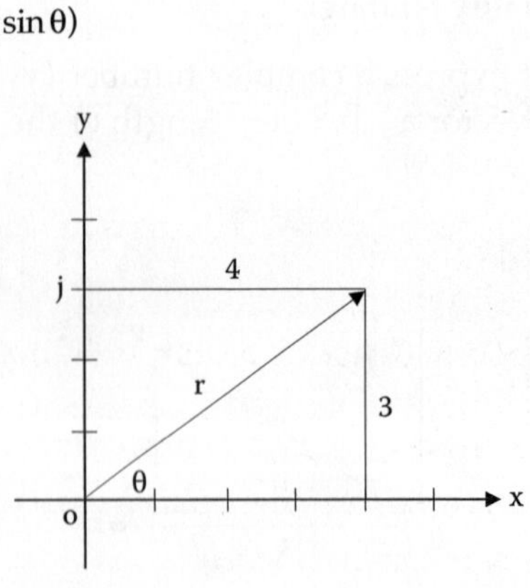

FIGURE 1.13
Argand diagram for Example 1.3.

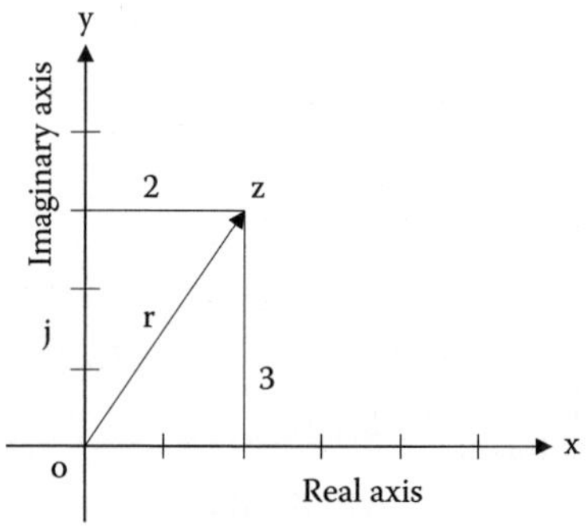

FIGURE 1.14
Argand diagram for Example 1.4.

Therefore, in this case

$$z = 5\,(\cos 36.8667° + j\sin 36.8667°)$$

Example 1.4

Find the polar form of the complex number 2+j3.

Solution

First, construct a sketch diagram (see Figure 1.14).

$z = 2 + j3 = r(\cos\theta + j\sin\theta)$
$r^2 = 4 + 9 = 13$
$r = 3.6056$

$$\tan\theta = \frac{3}{2}$$

$$= 56.310°$$

$z = 3.6056(\cos 56.310° + j\sin 56.310°)$.

1.14.5 Exponential Form of a Complex Number

Thus far, two ways of representing a complex number have been considered, namely:

- Cartesian form ($z = a + jb$) (1.101)
- Polar form ($z = r[\cos\theta + j\sin\theta]$) (1.102)

In this section, a third method of denoting a complex number is introduced:

- Exponential form ($z = r\,.e^{j\theta}$) (1.103)

  The exponential form is derived from the polar form; therefore:

  - The value of 'r' is the same in both cases.
  - The angle is also the same in each case (in the exponential form, the angle has to be in radians).

Example 1.5

Convert the polar form 7(cos 60°+j sin 60°) into the exponential form.

Solution

Converting to exponential form 7(cos 60°+j sin 60°):

'r'=7

'θ'=60° (in radian form $=\frac{\pi}{3}$ radians).

Therefore, in exponential form: $=7e^{j\pi/3}$

If θ is replaced by −θ, the following will result:

$$e^{-j\theta}=\cos(-\theta)+j\sin(-\theta)$$

$$=\cos\theta-\ j\sin\theta$$

Summarising:

$$e^{j\theta}=(\cos\theta+j\sin\theta)$$

$$e^{-j\theta}=(\cos\theta-\ j\sin\theta)$$

1.15 Determinates

1.15.1 Introduction

A determinant is an algebraic operation in which a square matrix is reduced to a numerical value (scalar). It has many uses, particularly in the inversion of matrices.

The notion of determinants predates matrices and linear transformations. Cardano considered determinants towards the end of the sixteenth century and Leibniz also studied them approximately 100 years later.

1.15.2 Description

A determinant is a square array of quantities (elements) to which a numerical value is assigned. Their use is in the analysis and solutions of systems of linear algebraic solutions. In general, the solutions are unmanageable when written out in length. The use of determinants helps simplify the expressions.

As an example of reducing linear algebraic equations to a determinant, consider the following simultaneous equation:

$$a_{11}x_1+a_{12}x_2+a_{13}=0 \tag{1.104}$$

$$a_{22}x_1+a_{22}x_2+a_{23}=0 \tag{1.105}$$

The solution is obtained in the form

$$\frac{x_1}{(a_{12}\cdot a_{23}-a_{22}\cdot a_{13})}=\frac{x_2}{(a_{13}\cdot a_{22}-a_{23}\cdot a_{11})}=\frac{1}{(a_{11}\cdot a_{12}-a_{22}\cdot a_{12})} \tag{1.106}$$

The denominators of this solution can then be expressed in the form of determinants:

$$\begin{vmatrix} a_{12} & a_{13} \\ a_{22} & a_{23} \end{vmatrix} = a_{12} \cdot a_{23} - a_{22} \cdot a_{13} \tag{1.107}$$

$$\begin{vmatrix} a_{13} & a_{11} \\ a_{23} & a_{22} \end{vmatrix} = a_{13} \cdot a_{22} - a_{23} \cdot a_{11} \tag{1.108}$$

$$\begin{vmatrix} a_{11} & a_{12} \\ a_{21} & a_{22} \end{vmatrix} = a_{11} \cdot a_{22} - a_{21} \cdot a_{12} \tag{1.109}$$

1.15.3 Determinant Order

The number of rows/columns in a square matrix determines the order of the determinant, that is

$$\begin{vmatrix} a_{11} & a_{12} \\ a_{21} & a_{22} \end{vmatrix} = \text{Second-order determinant} \tag{1.110}$$

$$\begin{vmatrix} a_{11} & a_{12} & a_{13} \\ a_{21} & a_{22} & a_{23} \\ a_{31} & a_{32} & a_{33} \end{vmatrix} = \text{Third-order determinant} \tag{1.111}$$

$$\begin{vmatrix} a_{11} & a_{12} & a_{13} & a_{14} \\ a_{21} & a_{22} & a_{23} & a_{24} \\ a_{31} & a_{32} & a_{33} & a_{34} \\ a_{41} & a_{42} & a_{43} & a_{44} \end{vmatrix} = \text{Fourth-order determinant} \tag{1.112}$$

Consider a third-order determinant, that is, three rows and three columns:

$$\text{Determinant} = \text{Det}|A| = D = \begin{vmatrix} a_{11} & a_{12} & a_{13} \\ a_{21} & a_{22} & a_{23} \\ a_{31} & a_{32} & a_{33} \end{vmatrix} = \sum_{1}^{3} a_{ij} \tag{1.113}$$

The value of the determinant

$$D = a_{11}(a_{22}a_{33} - a_{23}a_{32}) - a_{12}(a_{21}a_{33} - a_{23}a_{31}) + a_{13}(a_{21}a_{32} - a_{22}a_{31}) \tag{1.114}$$

where the signs are assigned in accordance with the following table:

$$\begin{vmatrix} + & - & + & \cdots \\ - & + & - & \cdots \\ + & - & + & \cdots \end{vmatrix}$$

Signs are associated with determinant element positions.

From the above, it is seen that elements are assigned either positive or negative signs depending on their position within the determinant.

1.15.4 Properties of the Determinant

1. The value of the determinant remains unchanged if rows and columns are interchanged.
2. If two rows of the determinant are interchanged, the numerical value of the determinant is unaltered but its sign is changed.
3. If two rows (or columns) of the determinant are identical, then the value of the determinant is zero.
4. If all the elements of one row (or one column) of the determinant are multiplied by 'k', the new determinant equals kΔ.

1.15.5 Minors and Cofactors

Minor of a determinant element: Consider the fourth-order determinant shown in Equation 1.112. A 'minor' is the determinant of the square matrix formed by deleting one row and one column from some larger square matrix in which the determinant element is required.

$$\begin{vmatrix} a_{11} & a_{12} & a_{13} & a_{14} \\ a_{21} & a_{22} & a_{23} & a_{24} \\ a_{31} & a_{32} & a_{33} & a_{34} \\ a_{41} & a_{42} & a_{43} & a_{44} \end{vmatrix}$$

The minors are calculated as follows:

$$\text{The minor of } a_{11} = \begin{vmatrix} a_{22} & a_{23} & a_{24} \\ a_{32} & a_{33} & a_{34} \\ a_{42} & a_{43} & a_{44} \end{vmatrix}; \quad \text{The minor of } a_{12} = \begin{vmatrix} a_{21} & a_{23} & a_{24} \\ a_{31} & a_{33} & a_{34} \\ a_{41} & a_{43} & a_{44} \end{vmatrix}$$

$$\text{The minor of } a_{14} = \begin{vmatrix} a_{21} & a_{22} & a_{23} \\ a_{31} & a_{32} & a_{33} \\ a_{41} & a_{42} & a_{43} \end{vmatrix}$$

This procedure is then repeated for the remaining elements in the determinant and the cofactor of the determinant is then made up of a matrix representing the minors of the elements. Numerical values are assigned to the elements and the cofactor will be a single numerical value for the determinant.

1.16 Matrices

1.16.1 Introduction

A matrix is an array of mathematical quantities that is used in the solution of linear algebraic equations. Matrices are important in engineering, statistics, physics and so on.

The following is an example of a linear algebraic equation expressed in a matrix form

$$\overbrace{\begin{matrix} y_1 = a_{11}x_1 + a_{12}x_2 + a_{13}x_3 \\ y_2 = a_{21}x_1 + a_{22}x_2 + a_{23}x_3 \\ y_3 = a_{31}x_1 + a_{32}x_2 + a_{33}x_3 \end{matrix}}^{\text{Linear algebraic equation}} = \overbrace{\underset{\text{Column}}{\begin{Bmatrix} y_1 \\ y_2 \\ y_3 \end{Bmatrix}} = \underset{\text{Square}}{\begin{bmatrix} a_{11} & a_{12} & a_{13} \\ a_{21} & a_{22} & a_{23} \\ a_{31} & a_{32} & a_{33} \end{bmatrix}} \underset{\text{Column}}{\begin{Bmatrix} x_1 \\ x_2 \\ x_3 \end{Bmatrix}}}^{\text{Matrices}} \equiv \overbrace{\{y\} = [A]\{x\}}^{\text{Short-hand notation}}$$

From the above, it is seen that (x) and (y) are column matrices (these are often referred to as vectors) and [A] is a square matrix.

1.16.2 Definitions

The following are the basic definitions that are used in matrix algebra.

1.16.2.1 Square Matrix

A square matrix consists of (n) rows and (m) columns and is referred to as a matrix of order $n \times m$.

$$\begin{bmatrix} a_{11} & a_{12} & a_{13} & \cdots & a_{1n} \\ a_{21} & a_{22} & a_{23} & \cdots & a_{2n} \\ a_{31} & a_{32} & a_{33} & \cdots & a_{3n} \\ \vdots & \vdots & \vdots & & \vdots \\ a_{n1} & a_{n2} & a_{n3} & \cdots & a_{nn} \end{bmatrix} = [A] \tag{1.115}$$

The matrix equation can also be written as

$$y_i = \sum_{i=1}^{n} a_{ij} x_i$$

where $i=1, 2, 3$; i is the number of rows and j the number of columns.

1.16.2.2 Row Matrix

A row matrix has one row of numbers as shown below

$$[a_1 \quad a_2 \quad a_3 \quad \cdots \quad a_n] = [A]$$

1.16.2.3 Column Matrix

A column matrix consists of a single column of numbers as shown below

$$\begin{bmatrix} a_1 \\ a_2 \\ \vdots \\ a_n \end{bmatrix} = [A]$$

1.16.2.4 Diagonal Matrix

$$\begin{bmatrix} a_{11} & 0 & 0 \\ 0 & a_{22} & 0 \\ 0 & 0 & a_{33} \end{bmatrix}$$

1.16.2.5 Unit Matrix

$$\begin{bmatrix} 1 & 0 & 0 \\ 0 & 1 & 0 \\ 0 & 0 & 1 \end{bmatrix}$$

1.16.2.6 Symmetric Matrix

$$a_{ij} = a_{ji}\begin{bmatrix} a_{11} & a_{12} & a_{13} \\ a_{12} & a_{22} & a_{23} \\ a_{13} & a_{23} & a_{33} \end{bmatrix}$$

1.16.2.7 Skew Symmetric Matrix—That Is, Anti-Symmetric ($a_{ij}=-a_{ji}$)

$$\begin{bmatrix} 0 & a_{12} & a_{13} \\ -a_{12} & 0 & a_{23} \\ -a_{13} & -a_{23} & 0 \end{bmatrix}$$

1.16.2.8 Null Matrix

$$\begin{bmatrix} 0 & 0 & 0 \\ 0 & 0 & 0 \\ 0 & 0 & 0 \end{bmatrix}$$

where all the elements are zero.

1.16.3 Matrix Algebra

1.16.3.1 Additions of Matrices

Addition of matrices is completed as follows:

$$[A]+[B]=[a_{ij}+b_{ij}]=\begin{bmatrix} (a_{11}+b_{11}) & (a_{12}+b_{12}) & \cdots & \\ . & . & & . \\ . & . & & . \\ . & . & & . \\ . & . & & (a_{mn}+b_{mn}) \end{bmatrix}$$

1.16.3.2 Multiplication of Matrices

Multiplication of matrices is completed as follows:

$$\begin{array}{cc} \begin{bmatrix} & \text{Square} & \end{bmatrix} & \begin{bmatrix} b_{11} & b_{12} & b_{13} & b_{14} \\ b_{21} & b_{22} & b_{23} & b_{24} \\ b_{31} & b_{32} & b_{33} & b_{34} \end{bmatrix} \Leftarrow [B] \\ \begin{bmatrix} a_{11} & a_{12} & a_{13} \\ a_{21} & a_{22} & a_{23} \end{bmatrix} & \begin{bmatrix} c_{11} & c_{12} & c_{13} & c_{14} \\ c_{21} & c_{22} & c_{23} & c_{24} \end{bmatrix} \\ \Uparrow & \Uparrow \\ [A] & [C]=[A][B] \end{array}$$

For example, $c_{11}=a_{11}b_{11}+a_{12}b_{21}+a_{13}b_{31}$.

In general, $[A][B]\neq[B][A]$.

1.16.3.3 Transposition of a Matrix

Transposition of a matrix is where each row element becomes a column element and vice versa:

$$[A]=\begin{bmatrix} a_{11} & a_{12} & a_{13} \\ a_{21} & a_{22} & a_{23} \end{bmatrix} \quad \text{Transpose}: [A]^T=\begin{bmatrix} a_{11} & a_{21} \\ a_{12} & a_{22} \\ a_{13} & a_{23} \end{bmatrix}$$

1.16.3.4 Adjoint of a Matrix

Adjoint of a matrix is where the cofactors (c_{ij}) are transposed:

$$[\text{Adj } A]=[C]^T=\begin{bmatrix} C_{11} & C_{21} & C_{31} \\ C_{12} & C_{22} & C_{32} \\ C_{13} & C_{23} & C_{33} \end{bmatrix}$$

For a definition of cofactors, see Section 1.15.4 on determinants.

1.16.3.5 Inverse of a Square Matrix

$$[A]=[A]^{-1}=\frac{[A]^T}{\text{Det}[A]}=\frac{[A]^T}{|a|}$$

The inverse can only be defined for a square matrix. (*Note*: There are cases where a square matrix cannot be defined.)

1.16.3.6 Transformation from Cylindrical Coordinates to Cartesian Coordinates

An example of the use of matrices is the transformation of cylindrical to Cartesian coordinates in vector algebra. A transformation vector can be determined [A] such that the Cartesian coordinate $\{V\}_{cart}=[A]\{V\}_{cyl}$.

$$[A]=\begin{bmatrix} \cos\theta & -\sin\theta & 0 \\ \sin\theta & \cos\theta & 0 \\ 0 & 0 & 1 \end{bmatrix}$$

$$\begin{bmatrix} V_x \\ V_y \\ V_z \end{bmatrix}=\begin{bmatrix} \cos\theta & -\sin\theta & 0 \\ \sin\theta & \cos\theta & 0 \\ 0 & 0 & 1 \end{bmatrix}\begin{bmatrix} V_R \\ V_0 \\ V_Z \end{bmatrix}$$

$$\{V\}_{cart} = [A]\{V\}_{cyl}$$

2

Introduction to Numerical Methods

2.1 Introduction

Before the advent of electronic digital computers, all computing was accomplished using manual methods.

The aim of this introduction is to present some of the methods used in analysing practical problems arising in mechanical engineering that are not generally solvable using classical methods. These methods were developed prior to the development of the digital computer and have stood the test of time.

The basic mathematics used in numerical analysis is covered in Chapter 1.

The analysis of a physical problem will involve four basic steps:

1. Development of a suitable mathematical model that realistically represents the physical system
2. Derivation of the systems governing equations
3. Solution of the governing equations
4. Interpretation of the results

While an analytical using classical methods will be exact, if it exists, a numerical method will require a number of iterations to generate a solution; this is only an approximation and cannot be considered to be exact by any means.

It is important to understand the errors that arise in numerical analysis, and for this reason a separate section devoted to errors is covered in Section 2.2 and covers the difference between round-off errors and truncation errors.

The numerical methods will be able to solve most complex problems, and one of the advantages is easy programming on a computer using programs such as MATLAB®, MathCAD, Maple and Mathematica. All of these programs are user-friendly.

2.2 Numerical Methods for Integration

There may be a point where the engineer is required to calculate a definite integral that cannot be solved using analytical integration or it is preferred to integrate tabulated data.

The following methods may be used to solve for a definite integral:

1. Manual method
2. Mid-ordinate rule
3. Trapezoidal rule
4. Simpson's rule
5. Quadratic triangulation
6. Romberg integration
7. Gauss quadratic

In this section, the first four methods will be considered as being the most appropriate ones for engineering students to understand.

2.2.1 Manual Method

This method is about the simplest to implement, where the desired curve is traced onto a piece of graph paper. The squares under the curve are counted where they cover more than 50% of the function as in Figure 2.1. The result may give a reasonable estimate of the area under the curve, provided the grid is sufficiently fine.

2.2.2 Mid-Ordinate Rule

The mid-ordinate rule is also referred to as a quadrature or rectangular rule.

In this method, the area to be integrated is sub-divided into rectangles of equal width. The strips may be an even or odd number. In Figure 2.2, the value of $A = \int_a^b f(x)dx$ represents the area under the curve between the values x=a and x=b. As stated above, the area is divided into strips of equal width, $w = (b-a)/n$, where n is the number of strips.

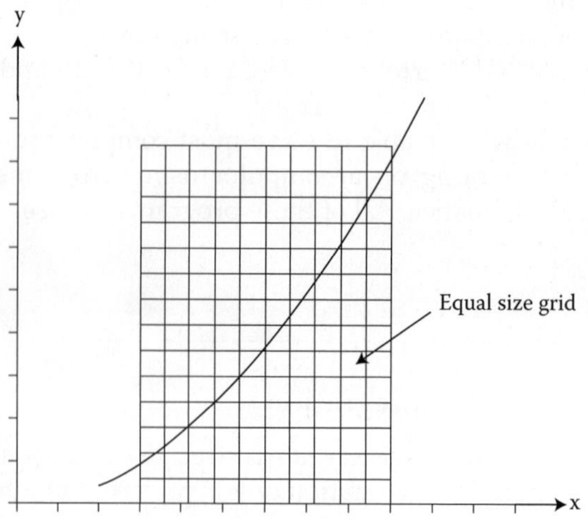

FIGURE 2.1
Manual method of determining the area under a curve.

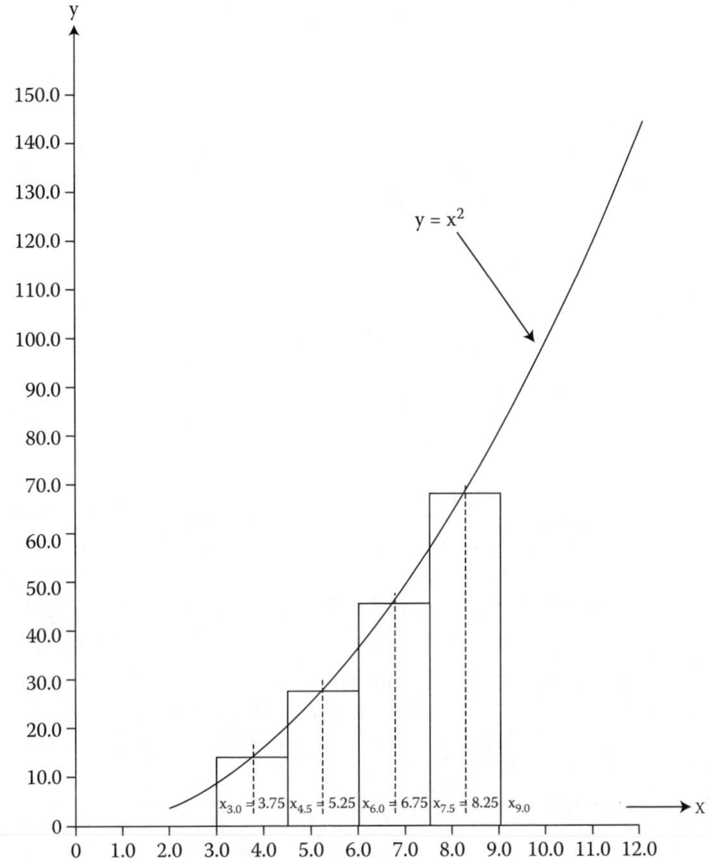

FIGURE 2.2
Mid-ordinate rule.

The height of each rectangle is measured at the point where the mid-point of the strip crosses the curve, that is, halfway between x=0 and x=1, that is x=0.5. The height of the second rectangle will correspond to the 'y'-value for x halfway between x_1 and x_2—namely, at a height of $y_{1.5}$.

This procedure is continued until the mean of the final x-values, x_{n-1} and x_n, is reached and the final y-value, $y_{n-0.5}$.

The formula can be expressed as

$$A = h(y_{0.5} + y_{1.5} + y_{2.5} + y_{n-0.5}) \tag{2.1}$$

where

$$h = \frac{b-a}{n}$$

Purely as an exercise for those readers unfamiliar with the method, it is required to integrate the curve $y=x^2$ between the values of x=3 and x=9. Starting with four rectangular strips, each with a width of four units as in Figure 2.1, it is noted that each strip crosses the curve at its mid-point and the area of the strip is then computed by multiplying the length of the strip by its width. The integral is the sum of the total areas of the strips. Cheating a little, it is already known that the integral uses analytical methods:

$$y = \int_3^9 x^2 \cdot dx$$

$$= \left[\frac{x^3}{3}\right]^9$$

$$= \left[\frac{9^3}{3}\right] - \left[\frac{3^3}{3}\right]$$

$$= 234 \text{ units}$$

Returning to Figure 2.2, tabulating the results of the measured areas, Table 2.1 is constructed.

$$\text{Total area} = x_1(y_1 + y_2 + y_3 + y_4)\,(\text{where } x_1 = x_2 = x_3 = x_4) = 238.35\,\text{mm}^2 \quad (2.2)$$

Increasing the number of strips will increase the accuracy of the integrand.

Table 2.2 tabulates the results of a series of calculations using an increased number of segment widths. It is clearly seen that as the segment strip width is reduced, the values converge.

TABLE 2.1

Strip Dimensions for Figure 2.2

Number of blocks	4
Width of strip	1.5 mm
Height of Strips	
y_1	14.52 mm
y_2	27.90 mm
y_3	45.90 mm
y_4	68.40 mm

TABLE 2.2

Effect on Values of Integrand with the Increase in Number of Strips

Number of Strips	Integral Value
4	238.35
6	237.12
8	236.62
12	236.36
24	236.19

2.2.3 Trapezoidal Rule

A variant of the mid-ordinate rule is the trapezoidal rule which more closely approximates the curve more accurately than the mid-ordinate rule (Figure 2.3). As with the mid-ordinate rule, the number of strips used is immaterial; they may be an even or odd number.

The strip is made up of an element containing both a rectangle and a triangle as shown in Figure 2.4a and b. The area of the trapezoid is obtained by adding the two areas together.

$$A = y_o\Delta x + \frac{1}{2}(y_1 - y_o)\Delta x = \frac{(y_o + y_1)\Delta x}{2} \quad (2.3)$$

Adding the area of n trapezoids, the following approximation is obtained:

$$\int_a^b f(x)dx \approx \frac{(y_o + y_1)\Delta x}{2} + \frac{(y_1 + y_2)\Delta x}{2} + \frac{(y_2 + y_3)\Delta x}{2} + \cdots + \frac{(y_{n-1} + y_n)\Delta x}{2} \quad (2.4)$$

This simplifies to the trapezoidal rule formula

$$\int_a^b f(x)dx \approx \frac{\Delta x}{2}(y_o + 2y_1 + 2y_2 + \cdots + 2y_{n-1} + y_n) \quad (2.5)$$

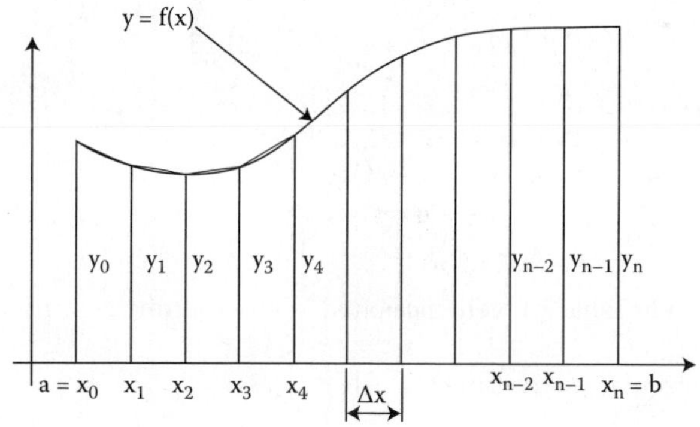

FIGURE 2.3
Trapezoidal rule.

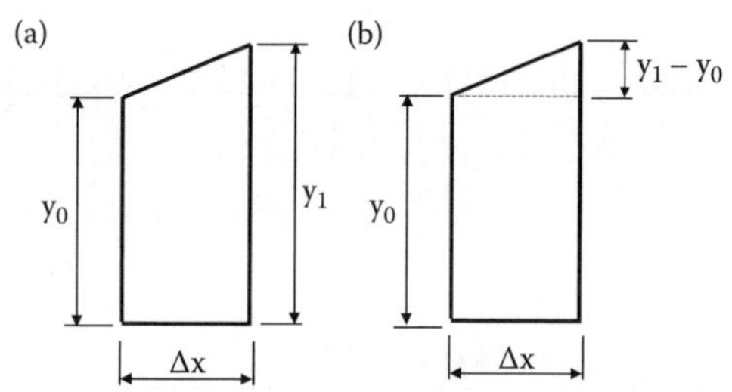

FIGURE 2.4
(a, b) Trapezoidal elements.

TABLE 2.3

Measured Values Used in Example 2.1

x	1	1.5	2	2.5	3	3.5	4	4.5	5
$y=\sqrt{1+x^2}$	$\sqrt{2}$	$\sqrt{3.25}$	$\sqrt{5}$	$\sqrt{7.25}$	$\sqrt{10}$	$\sqrt{13.25}$	$\sqrt{17}$	$\sqrt{21.25}$	$\sqrt{26}$

Example 2.1

Using the trapezoidal rule, estimate the following expression with n=8.

$$\int_1^5 \sqrt{1+x^2}\,dx$$

Solution

For n=8, the strip size Δx will be

$$\Delta x = \frac{5-1}{8} = 0.5$$

The values of $y_o, y_1, y_2, \ldots, y_8$ can now be computed as shown in Table 2.3.

Therefore,

$$\int_1^5 \sqrt{1+x^2}\,dx \approx \frac{0.5}{2}(\sqrt{2}+2\sqrt{3.25}+2\sqrt{5}+2\sqrt{7.25}+2\sqrt{10}$$

$$+2\sqrt{13.25}+2\sqrt{17}+2\sqrt{21.25}+\sqrt{26})$$

which resolves to

$$\approx 12.76\ (\text{Ans})$$

Example 2.2

The results shown in Table 2.4 were measured in an experiment.

Using the trapezoidal rule, estimate $\int_{2.1}^{3.6} y\,dx$

Solution

By inspection, dx=0.3.

Therefore,

$$\int_{2.1}^{3.6} y\,dx \approx \frac{0.3}{2}(3.2+2(2.7)+2(2.9)+2(3.5)+2(4.1)+5.2)$$

$$\approx 5.22\ (\text{Ans})$$

TABLE 2.4

Measured Values Used in Example 2.2

x	2.1	2.4	2.7	3.0	3.3	3.6
y	3.2	2.7	2.9	3.5	4.1	5.2

In the case of the curve in Figure 2.1, the results from the trapezoidal rule will be identical to the mid-ordinate rule. The trapezoidal rule has the advantage where the curve is not continuous as seen in Figure 2.3; with the strip being made up of a rectangle and a triangle, it follows the curve more accurately and reduces the error even with larger strip widths.

2.2.4 Simpson's Rule

Where the trapezoidal rule is more accurate than using the mid-ordinate rule, an even better approximation can be made using Simpson's rule where parabolas are used instead of straight lines to approximate each part of the curve. Simpson's rule is widely used by engineers in preference to most other methods of integration (see Figure 2.5).

As with the previous methods, the area under the required curve is partitioned into an even number of sub-intervals having equal widths such that:

$$\{x_0, x_1, x_2, x_3, x_4, \ldots, x_{n+1}\} \tag{2.6}$$

where n is an even number of strips.

The width of each sub-interval is given by (see Figure 2.5)

$$\Delta x = \frac{b-a}{n}$$

The area under the curve is given by the following formula

$$A \approx \frac{\Delta x}{3}(y_0 + 4y_1 + 2y_2 + 4y_3 + 2y_4 + 4y_5 + \cdots + 4y_{n-1} + y_n) \tag{2.7}$$

By way of an example to the method, consider the definite integral.

Approximate $\int_3^3 [dx / (x+1)]$ using Simpson's rule with n=4.

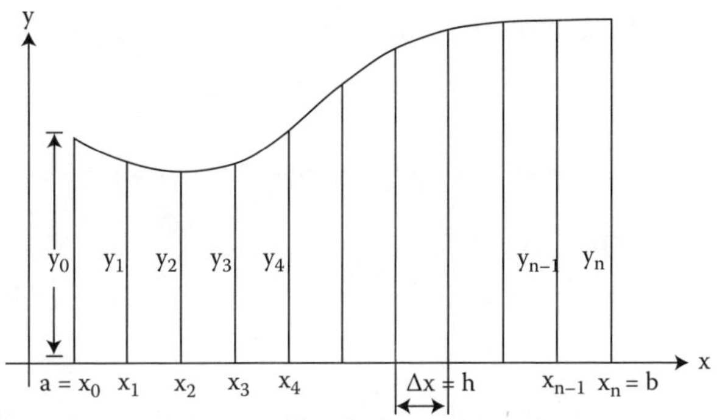

FIGURE 2.5
Simpson's rule.

$$\Delta x = \frac{b-a}{4}$$

$$= 0.25$$

$$y_0 = f(a) = f(2) = 0.333333$$

$$y_1 = f(a+\Delta x) = f(2.25) = \frac{1}{2.25+1} = 0.307692$$

$$y_2 = f(a+2\Delta x) = f(2.5) = \frac{1}{2.5+1} = 0.285714$$

$$y_3 = f(a+3\Delta x) = f(2.75) = \frac{1}{2.75+1} = 0.266667$$

$$y_4 = f(b) = f(3) = \frac{1}{3+1} = 0.25$$

Hence,

$$\text{Area} = \int_a^b (x)dx$$

$$\approx \frac{0.25}{3}[0.333333 + 4(0.307692) + 2(0.285714) + 4(0.266667) + 0.25]$$

$$= 0.287683$$

The actual answer is 0.287682; therefore, the rule is in error by only 0.00036%.

2.3 Evaluation of Errors

From the foregoing section on solving for a definite integral, it was seen how errors arising from the calculation can be reduced by the choice of method, that is, mid-ordinate, triangulation or Simpson's rule, and increasing the number of strips used in the calculation.

2.4 Round-Off and Truncation Errors

Errors arising from the solution of engineering problems can occur due to several factors. The error may be due to

1. Incorrect assumption in the modelling technique, that is that drag force is proportional to the velocity of a vehicle when it is actually proportional to the square of the velocity

2. Errors from mistakes in the programs themselves
3. Errors in the measurement of the physical properties
4. Mixing imperial units with SI units or using incorrect conversion values

In the application of numerical methods, there are two types of error that need to be considered; these are

1. Round-off errors
2. Truncation errors
3. Errors arising from differentiation
4. Integration errors

2.4.1 Round-Off Errors

Round-off errors depend upon the level of precision the engineer is working to.

Consider the number $\frac{1}{4}$. This is represented as 0.25 in decimal format, which is an exact value, but a number such as $\frac{1}{3}$ would be represented as 0.333333 recurring. The round-off error in this case will be $\frac{1}{3} - 0.333333 = 0.0000003\bar{3}$ (the over-bar represents that the last digit is recurring).

There are other numbers that cannot be represented exactly, such as π or $\sqrt{2}$; these are known as transcendental numbers. They continue to infinity without repeating and therefore need to be approximated in calculations.

When there are calculations involving a number of decimal values that have been either rounded up or down, then the errors in the final calculation can become significant, and these errors need to be considered carefully.

2.4.2 Truncation Errors

A truncation error is defined as an error occurring when a mathematical procedure is prematurely terminated.

By way of an example, consider the Maclaurin series for e^x

$$e^x \approx 1 + x + \frac{x^2}{2!} + \frac{x^3}{3!} + \frac{x^4}{4!} + \frac{x^5}{5!} + \cdots \tag{2.8}$$

This series has an infinite number of terms but when using it to calculate e^x, only a finite number of terms is actually considered to calculate e^x.

In this example, if only the first three terms are used to calculate e^x, such that:

$$e^x \approx 1 + x + \frac{x^2}{2!} \tag{2.9}$$

A truncation error for this approximation will occur as the following terms will not be considered in the calculation.

The question is raised, how can this truncation error be controlled?

The concept of relative approximate error can be used to see how many terms need to be considered to minimise the error. It will not be solved unless an infinite number of terms are considered; therefore, the engineer has to be realistic and determine the minimum number of terms that have to be considered to give the level of accuracy that can be acceptable.

As an example, calculate $e^{1.6}$ using the Maclaurin series:

$$f(x)=f(0)+xf'(0)+\frac{x^2}{2!}f''(0)+\frac{x^3}{3!}f'''(0)+\cdots \tag{2.10}$$

$$e^{1.6}=1+1.6+\frac{1.6^2}{2!}+\frac{1.6^3}{3!}+\cdots \tag{2.11}$$

Using a hand calculator, the inbuilt function gives $e^{1.6}=4.9530324244$ to 10 decimal places.

Table 2.5 tabulates the errors resulting from the series after a certain number of elements, and Figure 2.6 depicts the convergence of the series $e^{1.6}$ towards a finite value.

From the table, it will be seen that the error reduces as the number of terms are increased, and the series at element number 7 will have an error <1.0%.

TABLE 2.5
Truncation Errors for $e^{1.6}$ Using Maclaurin Series

Number of Terms Considered	$e^{1.6}$	Relative Approximate Error	Absolute Approximation Error
1	1.00000	0.00000	—
2	2.60000	1.60000	61.53846
3	3.88000	1.28000	32.98969
4	4.56267	0.68267	14.96201
5	4.83573	0.27307	5.64685
6	4.92311	0.08738	1.77492
7	4.94642	0.02330	0.47108
8	4.95174	0.00533	0.10756
9	4.95281	0.00107	0.02151
10	4.95300	0.00019	0.00382

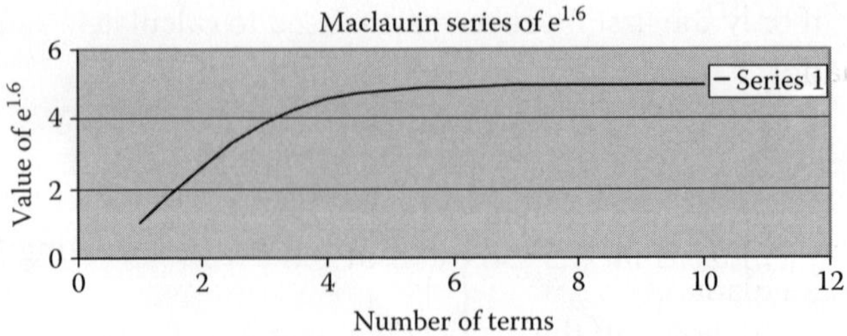

FIGURE 2.6
Showing the convergence of $e^{1.6}$ with an increase in the number of terms used.

2.5 Errors Arising from Differentiation

The majority of errors that arise from differentiation occur due to the equation being 'ill conditioned'.

As an example of ill conditioning, consider the function

$$f(x) = \frac{1}{(x-1)}$$

when

$$f(1.1),\ f(x) = 10$$

$$f(1.01),\ f(x) = 100$$

and

$$f(1.001), f(x) = 1000$$

It is clearly seen that a small change in 'x' of <0.1 will result in a rapid change in 'f(x)' approaching 1000. Therefore, evaluating f(x) as 'x' approaches 1.00, f(x) will tend to ∞ making this type of equation very difficult to solve and will introduce errors in its solution.

Consider the equation $f(x) = \sqrt{x}$. When

$$f(1.1), f(x) = 1.04880$$

$$f(1.01), f(x) = 1.00498$$

and

$$f(1.001), f(x) = 1.000499$$

This equation is considered well-conditioned and will yield sensible results with minimum errors.

2.6 Integration Errors

Consider the equation

$$y = x^2 \tag{2.12}$$

It is required to determine the area under the curve between the values x=3 and x=9.

Using classical calculus for the solution, consider the integral

$$y = \int_3^9 x^2 \cdot dx \tag{2.13}$$

Using classical integral calculus for the solution:

$$\left[\frac{9^3}{3}\right]-\left[\frac{3^3}{3}\right]$$

$$\text{Area} = 234 \text{ units}$$

2.7 Series

The study of infinite series is important in the understanding of numerical analysis. Of particular, importance is the 'power series' for two reasons:

1. The evaluation of functions and definite integrals. The subroutines used in computers and pocket calculators are based on series expansions for evaluating functions such as sin x and e^x.
2. Many numerical methods including the solution of differential equations are based on the Taylor series expansion of an arbitrary function.

An infinite series, in general, is an expression in the form

$$u_1 + u_2 + u_3 + u_4 + \cdots + u_n \tag{2.14}$$

where the u_n terms are formed in accordance with a definite rule. The partial sum of the first 'n' terms:

$$s_n = \sum_{i=1}^{n} u_i \tag{2.15}$$

If the limit of s_n as n approaches infinity exists and is finite, the series is said to converge. Conversely, if the limit exists but is infinite, the series diverges. Converging series is the current concern.

A series is said to be absolutely convergent when it remains convergent where all the terms are made positive. Cauchy ratio test provides a simple convergence check. The series will converge if the ratio:

$$R = \lim_{n \to \infty} \left| \frac{u_n + 1}{u_n} \right| \tag{2.16}$$

is <1.0 and diverges if the ratio is >1.0. Additional information regarding convergent series will be found in any standard calculus text.

2.8 Newton–Raphson Method

It often becomes necessary to solve algebraic equations of the third or higher orders as well as transcendental equations. To find the roots of these equations often involves a very inefficient trial and error process. The method being described in this section is often known as the Newton–Raphson method, after I. Newton (1666) and J.R. Raphson (1690) who improved on the method.

The Newton–Raphson method is a root-finding process using the first few terms of the Taylor series of a function f(x) in the vicinity of a suspected root and uses an iteration procedure requiring an initial approximation to the solution; the approximation is improved following each iteration. It is a characteristic of iterative methods that minor errors at any step will usually be corrected in the following steps.

2.8.1 Demonstration of the Method

Consider a non-linear equation

$$f(x) = 0 \tag{2.17}$$

The Taylor series representation of f(x) in the vicinity of a point x_0 is

$$f(x) = f(x_0) + (x - x_0)f'(x) + \frac{1}{2}(x - x_0)^2 f''(x_0) + \cdots \tag{2.18}$$

If $x = x_1$ is a solution of Equation 2.17, then $f(x_1) = 0$ and Equation 2.18 becomes:

$$0 = f(x_0) + (x_1 - x_0)f'(x_0) + \frac{1}{2}(x_1 - x_0)^2 f'(x_0) + \cdots \tag{2.19}$$

When the term $(x_1 - x_0)$ becomes sufficiently small, powers of $(x_1 - x_0)$ higher than the first can then be ignored and Equation 2.19 reduces to

$$0 = f(x_0) + (x_1 - x_0)f'(x_0) \tag{2.20}$$

The solution of this equation for x_1 gives the relationship which is then used to find the roots of Equation 2.17:

$$x_1 = x_0 - \frac{f(x_0)}{f'(x_0)} \tag{2.21}$$

Using this expression as a recurrent relationship for determining improved values of x_1, each based on a previous approximation x_0, it is possible to evaluate the real roots of Equation 2.17 to any required accuracy.

When Equation 2.21 is used in this way, it is written as

$$x_{n-1} = x_n - \frac{f(x_n)}{f'(x_n)}, \quad n = 0, 1, 2, 3, \ldots \tag{2.22}$$

The steps used in the application of Newton's method are summarised as follows:

1. Tabulate or graph f(x) versus x to find an approximate value x_0 of a real root.
2. This starting point can be improved by calculating a corrected value x_1 from Equation 2.21.
3. Successive corrections are made in the same manner using Equation 2.22.
4. The solution is complete when the term $f(x_n)/f'(x_n)$ has no effect on the last decimal place to be retained in x_{n+1}.

Example 2.3

The equation $x^3 - 3x - 4 = 0$ is known to have a root at approximately $x = 2$. It is required to find a more accurate root.

Solution

Now, $f(x_0) = x^3 - 3x - 4$; therefore, $f'(x) = 3x^{2-}3$.

A first approximation will be $x_0 = 2$; then

$$f(x_0) = f(2) = -2 \text{ and } f'(x_o) = f''(2) = 9$$

A more improved approximation (x_1) will be

$$x_1 = x_0 - \frac{f(x_0)}{f'(x_0)}$$

$$x_1 = 2 - \frac{(-2)}{9}$$

$$= 2.22222$$

Therefore,

$$x_0 = 2.00000; x_1 = 2.22222$$

For the next iteration starting from x_1

$$x_2 = 2.22222 - \frac{0.30724}{11.81479}$$

$$x_2 = 2.19622$$

A further iteration produces

$$x_3 = 2.19622 - \frac{0.00455}{11.47015}$$

$$x_3 = 2.19582$$

and a further iteration results in

$$x_4 = 2.19622 - \frac{0.0000010361}{11.46492}$$

$$x_4 = 2.19582$$

this result is similar to 5 decimal places as x_4.

Hence, the required is 2.19582 to 5 decimal places.

It is seen that the method is simple but effective and can be repeated with each repetition or iteration giving a result closer to the required root.

The method lends itself to using a spreadsheet.

2.9 Iterative Methods for Solving Linear Equations

2.9.1 Gauss Elimination Method

One of the problems with using the Gauss elimination method is that it is sensitive to rounding errors and the method does not offer any refinement to reduce these errors. Also, when the number of simultaneous equations becomes large, that is, 10 or more, the process then becomes laborious.

There are two iterative techniques currently used by engineers:

1. Jacobi iterative method
2. Gauss–Seidel method

2.9.2 Jacobi Iterative Method

Carl Gustav Jacob Jacobi (1804–1851) developed an iterative method for matrices which will be studied in this section.

This method makes two assumptions:

1. That the system of equations given by him has a unique solution.

$$\begin{aligned} a_{11}x_1 + a_{12}x_2 + a_{13}x_3 + \cdots + a_{1n}x_n &= b_1 \\ a_{21}x_1 + a_{22}x_2 + a_{23}x_3 + \cdots + a_{2n}x_n &= b_2 \\ &\;. \\ &\;. \\ a_{n1}x_1 + a_{n2}x_2 + a_{n3}x_3 + \cdots + a_{nn}x_n &= b_n a \end{aligned} \tag{2.23}$$

2. And the coefficient matrix 'A' has no zeros on the main diagonal. If any of the diagonal values $a_{11}, a_{22}, a_{33}, \ldots, a_{nn}$ are zero, then the rows or columns will need to be interchanged to obtain a coefficient matrix that has nonzero values on the main diagonal.

To review the Jacobi method, consider the Equation 2.23, solve the first equation for x_1, the second equation for x_2 and so on, to x_n as follows:

$$\begin{aligned} x_1 &= \frac{1}{a_{11}}(b_1 - a_{12}x_2 - a_{13}x_3 - \cdots - a_{1n}x_n) \\ x_2 &= \frac{1}{a_{22}}(b_2 - a_{21}x_1 - a_{23}x_3 - \cdots - a_{2n}x_n) \\ &\vdots \\ &\vdots \\ x_n &= \frac{1}{a_{nn}}(b_n - a_{n1}x_1 - a_{n2}x_2 - \cdots - a_{nn}x_{n-1}) \end{aligned} \tag{2.24}$$

The next step is to make an initial approximation of the solution,

$$(x_1, x_2, x_3, \ldots, x_n), \tag{2.25}$$

these values of x_1 are substituted into the right-hand side of the rewritten equations to obtain the first approximations. After this procedure has been completed, one iteration is achieved. The second iteration is performed in exactly the same way, but this time substituting the recalculated values for x_1, x_2 x_3, …, x_n into the right-hand side of the equation.

This procedure is then repeated forming a sequence of approximations until they begin to converge to the actual. The procedure is illustrated in Example 2.4.

Example 2.4

Use the Jacobi method to approximate the solution of the following system of linear equations.

$$12x_1 - 3x_2 + x_3 + 4x_4 = 25$$

$$x_1 + 15x_2 - 7x_3 - x_4 = 6$$

$$4x_1 + x_2 - 20x_3 + 5x_4 = -34$$

$$2x_1 - 8x_2 + x_3 + 10x_4 = 29$$

Solution

To begin, write the above equations in the following form using the rules of algebra;

$$x_1 = \left(\frac{1}{12}\right)(25 + 3x_2 - x_3 + 4x_4)$$

$$x_2 = \left(\frac{1}{15}\right)(6 - x_1 + 7x_3 + x_4)$$

$$x_3 = \left(\frac{1}{20}\right)(34 + 4x_1 + x_2 + 5x_4)$$

$$x_4 = \left(\frac{1}{10}\right)(29 - 2x_1 + 8x_2 - x_3)$$

Because the actual solution is unknown, choose as a convenient initial approximation:

$$x_1 = 0, x_2 = 0, x_3 = 0, x_4 = 0$$

The first approximation is

$$x_1 = \left(\frac{1}{12}\right)[25 + 3(0) - (0) + 4(0)] = 2.08333$$

$$x_2 = \left(\frac{1}{15}\right)[6 - (0) + 7(0) + (0)] = 0.400$$

$$x_3 = \left(\frac{1}{20}\right)[34 + 4(0) + (0) + 5(0)] = 1.700$$

$$x_4 = \left(\frac{1}{10}\right)[29 - 2(0) + 8(0) - (0)] = 2.900$$

These values are now substituted into the right-hand side of the equations to get

$$x_1 = \left(\frac{1}{12}\right)[25 + 3(0.400) - 1(700) + 4(2.900)]$$

$$x_2 = \left(\frac{1}{15}\right)[6 - (0.400) + 7(1.700) + (2.900)]$$

$$x_3 = \left(\frac{1}{20}\right)[34 + 4(2.0833) + (0.400) + 5(2.900)]$$

$$x_4 = \left(\frac{1}{10}\right)[29 - 2(2.0833) + 8(0.400) - (1.700)]$$

This procedure is continued until it is clearly seen that the values are converging. Table 2.6 shows the results for this example after 13 iterations, where it is clearly seen that the values have converged after 11 iterations with a very inaccurate start value. If more accurate start values had been chosen, then the convergence would have been quicker.

As the last two columns in Table 2.6 are almost identical, therefore it can be concluded that substituting the values for x_1, x_2, x_3 and x_4 back into the original equations will result in

$$\text{Line } 1 = 24.99885, \text{Line } 2 = 5.99914, \text{Line } 3 = -34.00203, \text{Line } 4 = 28.99875$$

If the iteration had been terminated at the tenth iteration, the results would have been

$$\text{Line } 1 = 24.98414, \text{Line } 2 = 6.01237, \text{Line } 3 = -33.99865, \text{Line } 4 = 28.9612$$

It can be seen that increasing the number of iterations results in a more refined answer. A choice has to be made on how many iterations will be needed depending upon the degree of accuracy required.

Although these calculations have been completed by hand to five decimal places, it is possible that the procedure can be automated in Excel or similar programs such as MathCad.

2.9.3 Gauss–Seidel Method

Carl Friedrich Gauss (1777–1855) and Philipp L. Seidel (1821–1896) proposed a modification to the Jacobi method that carries their names. This method is considered no more difficult to use than the Jacobi method, and it often requires fewer iterations to obtain the same degree of accuracy.

With the Gauss–Seidel method, when the value for x_1 is determined from the first equation, its value is then used in the second equation to obtain the new x_2. These new values are then used in the third equation obtaining the new x_3; this procedure is then repeated for the remaining equations.

TABLE 2.6

Results of Example 2.4

Start Value	1	2	3	4	5	6	7	8	9	10	11	12	13	14	15	16
0.00000	2.08333	1.07500	1.27903	1.19117	1.00068	1.05351	1.01851	0.99870	1.00914	1.00085	1.00002	1.00141	0.99984	1.00010	1.00019	0.99994
0.00000	0.40000	1.24778	1.83933	1.77120	1.92935	1.92114	1.96652	1.99548	1.99667	1.99545	2.00012	1.99928	1.99947	2.00010	1.99985	1.99996
0.00000	1.70000	2.86167	2.63572	2.89704	2.98982	2.94386	2.99585	2.99698	2.99200	3.00090	2.99902	2.99904	3.00029	2.99975	2.99992	3.00005
0.00000	2.90000	2.63333	3.39706	3.85209	3.78902	3.94436	3.97982	3.96993	3.99695	3.99631	3.99610	4.00190	3.99924	3.99958	4.00009	3.99985

Example 2.5

Use the Jacobi method to approximate the solution of the following system of linear equations.

$$12x_1 - 3x_2 + x_3 + 4x_4 = 25 \quad (1)$$

$$x_1 + 15x_2 - 7x_3 - x_4 = 6 \quad (2)$$

$$4x_1 + x_2 - 20x_3 + 5x_4 = -34 \quad (3)$$

$$2x_1 - 8x_2 + x_3 + 10x_4 = 29 \quad (4)$$

As with Example 2.4, the equations are rearranged as follows:

$$x_1 = \left(\frac{1}{12}\right)(25 + 3x_2 - x_3 + 4x_4) \quad (5)$$

$$x_2 = \left(\frac{1}{15}\right)(6 - x_1 + 7x_3 + x_4) \quad (6)$$

$$x_3 = \left(\frac{1}{20}\right)(34 + 4x_1 + x_2 + 5x_4) \quad (7)$$

$$x_4 = \left(\frac{1}{10}\right)(29 - 2x_1 + 8x_2 - x_3) \quad (8)$$

The procedure used for the Jacobi iteration is similar but with a small change.

In this example taking x_2=0, x_3=0 and x_4=0, a value for x_1=2.08333 is found from substitution in Equation 5.

In Equation 6 to calculate x_2, substituting the value for x_1=2.08333 together with x_3=0 and x_4=0, a value for x_2=0.26111 is obtained.

To calculate x_3, substituting the revised value for x_1=2.14826 and x_2=0.26111 into Equation 7, the value for x_3=2.12972 is obtained.

To calculate x_4, after substituting the above values for x_1, x_2 and x_3 into Equation 8, the value for x_4=2.46626 is obtained.

From this point onwards, the iteration procedure is identical to that for the Jacobi method.

The results are shown in Table 2.7, and it will be noted that in this method the values quickly begin to converge on the solution. This does not always happen so quickly unless, as in this example, the coefficients on the main diagonal of the matrix are the largest in each row.

The results also show that using the same number of iterations as the Jacobi method, the results are more accurate.

TABLE 2.7
Results of Example 2.5

	Gauss–Seidel Method															
Start Value	**1**	**2**	**3**	**4**	**5**	**6**	**7**	**8**	**9**	**10**	**11**	**12**	**13**	**14**	**15**	**16**
2.08333	2.08333	2.14826	1.97113	1.39535	1.11312	1.13571	1.03553	1.01530	1.01876	1.00212	1.00266	1.00236	0.99994	1.00051	1.00025	1.00009
0.00000	0.26111	0.26111	1.25065	1.43294	1.82066	1.89952	1.92759	1.98364	1.98516	1.99146	1.99867	1.99768	1.99914	1.99988	1.99964	1.99998
0.00000	0.00000	2.12972	2.14271	2.77332	2.87372	2.88614	2.98375	2.97945	2.98639	2.99949	2.99643	2.99867	3.00005	2.99940	2.99993	2.99991
0.00000	0.00000	0.00000	2.46626	3.29202	3.48995	3.84653	3.90386	3.93659	3.98591	3.98574	3.99280	3.99876	3.99781	3.99932	3.99986	3.99996

2.10 Non-Linear Equations

Section 2.8 introduced and demonstrated two methods for obtaining the real roots of linear equations. In this section, the procedure is extended to that of finding the roots of non-linear equations using two iterative procedures. The first iteration technique uses a variant of Newton's method and the second is a direct iteration method.

There are a number of physical examples demonstrating non-linearity behaviour, including the motion of a pendulum or the electrical AC waveform, and in control theory, the behaviour of an aerial dish when subjected to wind gusts.

Consider the following two equations:

$$F(x, y) = 0 \tag{2.26a}$$

$$G(x, y) = 0 \tag{2.26b}$$

These two equations define a curve in the x–y plane. Let the problem be to determine the point or points of intersection of these two curves.

Initially as a rough guide to help understand the problem, a graph of the two equations could be drawn to identify the approximate co-ordinates of the points, and once they have been found, the accuracy may be improved using either of the two techniques discussed in this section.

2.10.1 Newton's Method

In Section 2.7, Newton's method was introduced for the solution of one variable; this method can be adopted to a system of equations comprising several variables. As discussed in that section, the method is based on the Taylor series expansion of a function.

A function of two variables, F(x, y), may be expanded within the neighbourhood of a point, (x_0, y_0).

$$\begin{aligned} F(x, y) &= F(x_0 + \alpha \cdot y_0 + \beta) \\ &= F(x_0, y_0) + \alpha F_x(x_0, y_0) + \beta F_y(x_0, y_0) + \cdots \end{aligned} \tag{2.27}$$

where α and β are small increments applied to x and y. The subscripts x and y indicate partial differentiation with respect to x and y. Hence, the notation $F_x(x_0, y_0)$ indicates the partial derivative of F(x, y) with respect to x evaluated at $x = x_0$ and $y = y_0$.

Following the expansion of the original equations in this form, only the linear terms in α and β are retained,

$$\begin{aligned} F(x, y) &= F(x_o, y_o) + \alpha F_x(x_o, y_o) + \beta F_y(x_o, y_o) \\ G(x, y) &= G(x_o, y_o) + \beta G_x(x_o, y_o) + \beta G_y(x_o, y_o) \end{aligned} \tag{2.28}$$

From Equation 2.26a,

$$F(x, y) = 0$$

Therefore, Equations 2.26a, 2.26b and 2.27 become:

$$\alpha F_x + \beta F_y = -F \tag{2.29}$$

$$\alpha G_x + \beta G_y = -G \tag{2.30}$$

where all the functions are evaluated at (x_o, y_o).

If (x_o, y_o) is an initial approximation to the, Equation 2.23 can be used to improve the approximation by solving for α and β, then calculating:

$$\begin{aligned} x_1 &= x_o + \alpha \\ y_1 &= y_o + \beta \end{aligned} \tag{2.31}$$

as an improved approximation. This process can be continued until successive approximations only differ by a very small amount.

As in this example, only two equations are considered, Cramer's rule can be applied to solve for α and β at each step.

$$\alpha = \frac{\begin{vmatrix} -F & F_y \\ -G & G_y \end{vmatrix}}{D} \quad b = \frac{\begin{vmatrix} F_x & -F \\ G_x & -G \end{vmatrix}}{D} \quad \text{where } D = \begin{vmatrix} F_x & F_y \\ G_x & G_y \end{vmatrix} \tag{2.32}$$

When D disappears in the neighbourhood of a solution, there may be multiple solutions, several solutions close together or no solution at all.

Example 2.6

Consider the following set of equations to demonstrate the method:

$$\begin{aligned} y &= \sin x \\ x^2 + y^2 &= 1 \end{aligned} \tag{2.33}$$

Solution

For these equations

$$\begin{array}{lll} F = y - \sin x & F_x = -\cos x & F_y = 1 \\ G = x^2 + y^2 - 1 & G_x = 2x & G_y = 2y \end{array}$$

A plot of these equations indicate that an intersection occurs in the first quadrant close to x=0.7 and y=0.7.

$$\begin{array}{lll} F = 0.056 & F_x = -0.765 & F_y = 1.000 \\ G = -0.020 & G_x = 1.400 & G_y = 1.400 \end{array}$$

Substituting in Equation 2.25

$$D = -2.471, \ \alpha = \frac{-0.0984}{-2.471} = 0.040, \ \beta = \frac{0.0631}{-2.471} = -0.026$$

From Equation 2.24, the improved approximation will be

$$x_1 = 0.700 + 0.040 = 0.740$$

$$y_1 = 0.700 - 0.026 = 0.674$$

Continuing with this procedure using x_1 and y_1,

$$F = 0.000 \quad F_x = -0.738 \quad F_y = 1.000$$

$$G = 0.002 \quad G_x = 1.480 \quad G_y = 1.348$$

and

$$D = -2.475, \ \alpha = \frac{0.002}{-2.475} = -0.001, \ \beta = \frac{0.0015}{-2.475} = 0.001$$

hence,

$$x_2 = 0.740 - 0.001 = 0.739$$

$$y_2 = 0.674 - 0.001 = 0.673$$

A further application of the procedure gives

$$x_3 = 0.739$$

$$y_3 = 0.674$$

For these values, F=0.000 and G=0.000. Therefore, to three decimal places, the solution to the problem is

$$x = 0.739$$
$$y = 0.674$$

3

Probability

3.1 Introduction

Probability is the measure of events that owe their occurrence to chance. The probability, or likelihood, of an event occurring can be gauged from knowledge of the circumstances of the event and then expressed as a simple figure. A simple example is that of tossing a coin, which has 50% chance of landing with the "head" facing upward.

3.2 Numerical Value of Probability

Probability is expressed on a scale between 0 for an impossible event to occur and 1 for an event that is certain to occur. Other events have probabilities within this range, that is, when the coin discussed in the introduction is tossed, the probability that the "head" landing face up will occur is 0.5.

A horse with a 5–4 odds on chance of winning its race is 5/6 or 0.556, therefore the probability it will lose will therefore be 4/5 or 0.444.

A further example, the probability of attaining a six in a throw of an unbiased dice or die will be 1/6 or 0.167. There are six possible outcomes of throwing the dice, all of which are equally likely; hence, the chance of throwing a particular number is 1/6.

3.3 Calculation of Probability

It is important to understand some of the terms that are used in discussing probability before proceeding further.

Events are *equally likely or equally probable* if considering all the relevant evidence that one event cannot be expected to occur more often any another. As an example, when throwing an unbiased dice, the chance of throwing a "three" is equal to the chance of throwing a "six."

Events are *mutually exclusive* if any two of them cannot occur simultaneously; that is, the occurrence of one event precludes the simultaneous occurrence of any of the other events. For example, it is impossible for one die to simultaneously show two numbers at the same time.

A set of events is an *exhaustive set* when all the possible events that can occur in an experiment are included. Going back to the example of one die, the six possible outcomes are 1, 2, 3, 4, 5, and 6, and this forms the exhaustive set of possibilities.

3.3.1 Proportion of Cases Favorable in an Exhaustive Set

Consider an exhaustive set containing 'n' equally and likely and mutually exclusive events, and in 'm' of these the event 'A' occurs, the probability that the event 'A' will occur is the ratio m/n, which is the ratio of

$$\frac{\text{number of cases favorable to A}}{\text{number of cases in exhaustive set}}$$

This can best be explained by considering some examples:

Example 3.1

Find the probability of scoring an even number in a throw of one die.

There are three ways of throwing an even number, that is, by throwing either 2, 4, or 6.

The total number of ways of throwing the die is 6, that is, the number of events in the exhaustive set is 6.

Thus, the probability of throwing an even number is

$$P = \frac{m}{n} = \frac{\text{number of favorable cases}}{\text{number in exhausted set}} = \frac{3}{6} = 0.5$$

Example 3.2

In a normal pack of 52 playing cards, determine the probability of picking a "club" at random.

There are 13 clubs in the pack and therefore, m = number of favorable ways = 13.

The pack contains 52 possible cards that can be chosen, therefore 'n' = the number of ways of choosing a card = 52.

$$\text{Thus, } P = \frac{m}{n} = \frac{13}{52} = \frac{1}{4} = 0.25$$

Example 3.3

What is the probability of picking at random a red card with a value greater than nine from a normal pack of playing cards?

There are 10 cards (5 diamonds and 5 hearts) that satisfy this requirement.

$$\text{Thus, } P = \frac{m}{n} = \frac{10}{52} = 0.192$$

Example 3.4

A box contains a selection of balls consisting of 4 black, 5 yellow, and 3 white balls. Find the probability of choosing at random:

(3.4.1) a yellow ball
(3.4.2) a nonwhite ball
(3.4.3) a white ball

(3.4.1) The number of yellow balls is 5 and there are five ways of choosing a yellow ball. The total number of balls is 12, and therefore the exhaustive set contains 12 possible events, hence the probability of choosing a yellow ball is:

$$P_y = \frac{4}{12} = 0.333$$

(3.4.2) The total number of nonwhite balls is 9, therefore the probability of picking a nonwhite ball will be

$$P_{nw} = \frac{9}{12} = 0.75$$

(3.4.3) The probability of choosing a white ball is

$$P_w = \frac{3}{12} = 0.25$$

From the above examples, it will be deduced that the probability of choosing a nonwhite ball is (1 − probability of choosing a white ball). This simple relationship, that the probability of an event occurring is (1 − the probability of the event not happening) this often enters into probability and statistical calculations and therefore, should be remembered.

3.3.2 Addition of Probabilities

The probability of any one of the mutually exclusive events A_1, A_2, A_3 …. A_n occurring is the sum of the probabilities of these events.

Where the probability of A_1 occurring is p_1
Where the probability of A_2 occurring is p_2
Where the probability of A_3 occurring is p_3
Where the probability of A_n occurring is p_n

The probability that A_1 or A_2 or A_3 or A_n will occur is $P = p_1 + p_2 + p_3 + \cdots + p_n$.

This can be seen by considering 'N' to be the total number of all possible equally likely and mutually exclusive events that can occur, and m_1 to be the number of the favorable to A_1, m_2 the number favorable to A_2, etc.

Hence, there are $m_1 + m_2 + m_3 + \cdots + m_n$ cases favorable to the occurrence of A_1 or A_2 or A_3 or … A_n, and the probability that A_1 or A_2 or A_3 or … A_n will occur is

$$P = \frac{m_1 + m_2 + m_3 + \cdots m_n}{N} = \frac{m_1}{N} + \frac{m_2}{N} + \frac{m_3}{N} + \cdots + \frac{m_n}{N}$$

$$= p_1 + p_2 + p_3 + \cdots + p_n$$

Two examples follow to demonstrate this principle:

Example 3.5

Determine the probability of choosing a yellow or a black ball, as in Example 6.4.

Probability of choosing a yellow ball is $P_y=0.333$
Probability of choosing a black ball is $P_b=5/12=0.417$
Therefore, the probability of choosing a yellow or a black ball will be

$$P = P_y + P_b = 0.333 + 0.417 = 0.75$$

This is confirmed with the answer in Example (3.4.2).

Example 3.6

Determine the probability of obtaining a number '7' or '11' at a throw of two dice.

(3.6.1) A '7' can be 'thrown' in the following ways:

$$(1+6), (2+5), (3+4), (5+2), (6+1)$$

Hence, there are six ways of 'throwing' a '7.'

As there are six possible ways of 'throwing' the first die and for each of these six ways of 'throwing' the second, the total number of ways of 'throwing' the two dice is 36; therefore, the probability of throwing a '7' is

$$\frac{6}{36} = \frac{1}{6}$$

(3.6.2) The '11' can be 'thrown' in only two ways:

$$(5+6) \text{ and } (6+5)$$

and the probability of obtaining '11' is

$$\frac{2}{36} = \frac{1}{18}$$

(3.6.3) The probability of obtaining a '7' or '11' is

$$\frac{1}{6} + \frac{1}{18} = \frac{4}{18} = 0.2\dot{2}\dot{2}$$

This is slightly more than once in every five attempts.

3.4 Permutations and Combinations

3.4.1 Permutations

Consider a number of items that are arranged among themselves in different orders, then each order or arrangement is a permutation of the objects, then the total number of permutations possible for 'n' items is

$$^{n}P_{n}$$

where $^{n}P_{n} = n(n-1)(n-2)$

This is seen by considering that for the first place in each arrangement there is a choice of n items, for each of those, once the first place has been allocated, there is a choice of (n−1) items for the second place, and so on.

n(n − 1)(n − 2)… is known as factorial 'n' and is usually written n! or n.

If a number 'm' of the items are alike, then the number of possible arrangements of the items is reduced by the number of permutations of the 'm' items among themselves, as each of these permutations will be identical.

Hence, the number of permutations or arrangements of 'n' items, 'm' of which are alike, is $\frac{n!}{m!}$.

The number of permutations of 'n' different items taken 'r' at a time is

$$^{n}P_{r} = \frac{n!}{(n-r)!}$$

It can be seen that the first place can be filled in 'n' ways, the second place filled in (n−1) ways, until the 'r' place is filled in (n−r+1) ways.

$$\therefore\ ^{n}P_{r} = n(n-1)(n-2)\cdots(n-r-1)$$

$$= \frac{n(n-1)(n-2)\cdots(n-r+1)(n-r)(n-r-1)\cdots}{(n-r)(n-r-1)\cdots}$$

$$= \frac{n!}{(n-r)!}$$

3.4.2 Combinations

When groups of 'r' items are selected from 'n' items without due attention being paid to the order of the 'r' items among themselves, each group is referred to as a combination and the number of groups is called the number of combinations of 'n' taken 'r' at a time. This can be written as

$$^{n}C_{r} \text{ or } \binom{n}{r}$$

where $\binom{n}{r} = \frac{n!}{(n-r)!r!}$

It will be seen that by considering in any one of the combinations, the 'r' items can be arranged in r! different orders. If this is carried out to all the combinations, then they will have been permutated.

$$\therefore \binom{n}{r} \times r! = {}^{n}P_{r} = \frac{n!}{(n-r)!}$$

$$\therefore \binom{n}{r} = \frac{n!}{(n-r)!r!} = \binom{n}{n-r}$$

When calculating combinations, it is found to be convenient to write n(n−1)(n−2) … to r(or n−r) factors, whichever is less, in the numerator and 1.2.3….r (or n−r) in the denominator

$$\text{hence, } \binom{7}{4} = \frac{7.6.5}{3.2.1} = 35 = \binom{7}{3}$$

The following examples will illustrate the use of combinations.

Example 3.7

A container contains four faulty items and seven good items. Calculate the probability of selecting at random a pair of faulty items.

The number of ways of choosing two faulty items from four is

$$\binom{4}{2} = \frac{4\times3}{2\times1} = 6$$

The total number of ways of selecting 2 items from 11 is

$$\binom{11}{2} = \frac{11\times10}{2\times1} = 55$$

The possibility that the pair of items selected will both be faulty is

$$\frac{6}{55} = 0.109$$

Example 3.8

From the same container, a group of five items are chosen at random. Calculate the probability that the group will contain two faulty and three good items.

Following the Example (3.7), the number of ways two faulty items from four is

$$\binom{4}{2} = 6$$

The number of ways of choosing three good items from seven is

$$\binom{7}{3} = \frac{7\times6\times5}{3\times2\times1} = 35$$

Any one of the pairs of faulty items can be associated with any one of the groups of three good items; therefore, the number of possible combinations of two faulty and three good items will be

$$\binom{4}{2}\times\binom{7}{3} = 6\times35 = 210$$

The total number of ways of selecting 5 items from 11 is

$$\binom{11}{5} = \frac{11\times10\times9\times8\times7}{5\times4\times3\times2\times1} = 462$$

Hence, the required probability is

$$\frac{210}{462} = 0.\dot{4}5\dot{4}$$

Example 3.9

Consider a hand of 13 playing cards. Find the probability that the hand contains:

(3.9.1) Two aces.
(3.9.2) Four aces.

In this example, there are $\binom{52}{13}$ possible different hands, all mutually exclusive and equally likely.

(3.9.1) The number of ways of selecting two aces from the four in the pack will be

$$\binom{4}{2} = 6$$

For each of these, the number of ways of selecting the other cards from the remaining 48 cards is $\binom{48}{11}$

The probability of choosing 2 aces in 13 cards is

$$P_2 = \frac{\binom{48}{11}\binom{4}{2}}{\binom{52}{13}}$$

$$= \frac{\frac{4\times3}{2\times1}\times\frac{48\times47\times46\times\cdots\times39\times38}{11\times10\ \times\cdots\times2\times1}}{\frac{52\times51\times\cdots\times41\times40}{13\times12\times11\times\cdots\times2\times1}} \lim_{x\to\infty}$$

$$P_2 = 0.216$$

(3.9.2) There is only one way that four aces can be selected from the four in this pack,

$$\text{i.e.,} \binom{4}{4} = 1$$

The number of ways of selecting the remaining 9 cards from the 48 cards that remain is $\binom{48}{9}$

Thus, the possibility of 4 aces in 13 cards is

$$P_4 = \frac{\binom{4}{4}\binom{48}{9}}{\binom{52}{13}}$$

$$= \frac{\dfrac{1\times 48\times 47\times\cdots\times 41\times 40}{9\times 8\times\cdots\times\ 2\times 1}}{\dfrac{52\times 51\times\cdots\times 41\times 40}{13\times 12\times 11\times\cdots\times 2\times 1}}$$

$$P_4 = \frac{1}{375},\ \text{or } 0.00267$$

3.4.3 Compound Probability and Multiplication of Probabilities

3.4.3.1 Compound Events and Dependence

When an event 'A' is considered that is in association with an event 'B,' the occurrence of the combined (or compound) event 'AB' consists of the simultaneous occurrences of 'A' and 'B.'

Before the probability of occurrence of this combined event can be calculated, it is necessary to first determine whether or not 'A' and 'B' are independent.

Taking as an example the two dice introduced in (3.6), if 'AB' is the occurrence of a 12 when throwing two dice, this will require scoring a 6 on both of them, the first one 'A' scoring a 6 will in no way affect the probability of the second dice 'B' showing a 6. In this case 'A' and 'B' are considered *independent*.

In a second example, if 'AB' represents the drawing of two red cards in succession without replacement from a pack of 52 playing cards, the event that the first card is red will affect the probability of drawing a second red card, as the number of red cards remaining has been reduced. Therefore, in this case 'A' and 'B' are considered *dependent*.

In the instance where each card is replaced after drawing the individual card, in this case 'A' and 'B' will then be considered to be *independent*.

Example 3.10

A box contains eight unused valve stems and seven used ones. Two are drawn, one at a time, without replacement. Find the probability of drawing two unused valve stems.

Probability that the first valve stem is unused is

$$P_A = \frac{8}{15}$$

Probability that the second selection is unused is

$$P_{B,A} = \frac{7}{14}$$

The probability that both are unused is

$$P_{A,B} = \frac{8}{15}\times\frac{7}{14} = \frac{4}{15} = 0.267$$

This example could also have been calculated using the method shown in Example 3.7, considering that the items being drawn together, this gives the probability as

$$P = \frac{\begin{pmatrix} 8 \\ 2 \end{pmatrix}}{\begin{pmatrix} 15 \\ 2 \end{pmatrix}} = \frac{\frac{8 \times 7}{2 \times 1}}{\frac{15 \times 14}{2 \times 1}} = 0.267$$

Example 3.11

Returning to the pack of playing cards, determine the probability of obtaining a "flush," that is, five cards of all the same suit, in a deal of five cards.

The required probability of obtaining five cards of any particular suit is

$$P = \frac{13}{52} \times \frac{12}{51} \times \frac{11}{50} \times \frac{10}{49} \times \frac{9}{48} = 0.000495$$

There are four different suits from which to choose, by the "addition of probabilities," the probability of a flush will be $4 \times 0.000495 = 0.00198$.

This answer could have been obtained using the method outlined in Example 3.7.

3.5 Binomial, Poisson, and Multinomial Laws of Probability

3.5.1 Series of Trials

A *trial* is defined as an attempt to produce, under a set of given conditions, an event 'E' that is not certain to occur. When trials are repeated, they form a set or series of trials. They will be *independent* if the result of a trial does not depend on the outcome of previous trials.

An example of a series of independent trials could be likened to tossing a coin a number of times. To find the probability of obtaining any particular combination of results, the probability of each sequence of results, in association with the number of possible arrangements of these results, must be considered.

Considering the probability of obtaining four heads and one tail in a series of five trials, the probability of any one sequence or arrangement, such as HTHHH, will be $\left(\frac{1}{2}\right)^5$. This must be multiplied by the number of possible arrangements or orders of occurrence of the four heads and one tail in the five trials.

The number of orders is

$$\frac{5!}{4!\ 1!} = \begin{pmatrix} 5 \\ 4 \end{pmatrix} = \begin{pmatrix} 5 \\ 1 \end{pmatrix}$$

That is the number of ways of arranging five objects, that is, four alike of one kind and another of a different kind.

Thus, the probability of four heads and one tail will be

$$\begin{pmatrix} 5 \\ 1 \end{pmatrix} \left(\frac{1}{2}\right)^5$$

Similarly, the probability of three heads and two tails is

$$\binom{5}{2}\binom{1}{2}^5$$

The probability of two heads and three tails is

$$\binom{5}{3}\binom{1}{2}^5 = \binom{5}{2}\binom{1}{2}^5$$

The probability of one head and four tails will be

$$\binom{5}{4}\binom{1}{2}^5 = \binom{5}{1}\binom{1}{2}^5$$

And, for completeness, the probability of no heads and five tails is

$$\binom{5}{0}\binom{1}{2}^5 = \binom{5}{2}\binom{1}{2}^5$$

where $\binom{5}{0} = 1$

The successive probabilities can be seen to follow a pattern; this is the sequence of terms of the binomial expansion of $\left(\frac{1}{2}+\frac{1}{2}\right)^5$.

3.5.2 Binomial Distribution

This is a particular case of the *'Binomial Law of Probability,'* which states that in a series of 's' independent trials in each of which the known probability of success of an event 'E' is 'p' and p+q=1, the probability that 'E' will occur 'm' times is

$$P = \binom{s}{m} p^m q^{s-m}$$

This can be proved by considering a box containing white and black balls where a ball is chosen at random, 'p' is the proportion of white balls and 'q' is the proportion of black balls in the box.

After each trial the ball chosen is replaced in the box so that the proportion of balls remains unchanged. The probability of the first 'm' trials giving white balls and the remaining (s−m) giving black balls is $p^m q^{s-m}$.

The probability of any other order is the same and the number of possible orders is

$$\frac{s!}{m!(s-m)!} = \binom{s}{m}$$

Hence, the probability of 'm' white is $\binom{s}{m} p^m q^{s-m}$

For various values of 'm,' the corresponding probabilities are the successive terms of the binomial expansion of $(q-p)^s$

$$=q^s+\binom{s}{1}q^{s-1}p+\binom{s}{2}q^{s-2}p^2+\cdots+p^s$$

Example 3.12

Five dice are thrown simultaneously (or one die is thrown five times in succession). Determine the probability of two "sixes" in the five results:

$$P=\binom{5}{2}\left(\frac{5}{6}\right)^3\left(\frac{1}{6}\right)^2=\frac{1250}{216\times 36}=0.1608$$

Example 3.13

Six coins are tossed:

(3.12.1) Find the probability of 2, 3, or 4 heads.
(3.12.2) Find the probability of 0, 1, 5, or 6 heads.
(3.12.1) The probability of 2, 3, or 4 heads is

$$P=\binom{6}{2}\left(\frac{1}{2}\right)^2\left(\frac{1}{2}\right)^4+\binom{6}{3}\left(\frac{1}{2}\right)^3\left(\frac{1}{2}\right)^3+\binom{6}{4}\left(\frac{1}{2}\right)^4\left(\frac{1}{2}\right)^3$$

$$=\frac{6\times 5}{2\times 1}\left(\frac{1}{2}\right)^6+\frac{6\times 5\times 4}{3\times 2\times 1}\left(\frac{1}{2}\right)^6+\frac{6\times 5}{2\times 1}\left(\frac{1}{2}\right)^6$$

$$=\frac{15}{64}+\frac{20}{64}+\frac{15}{64}$$

$$=0.781$$

(3.12.2) The probability of 0, 5, or 6 heads is

$$P=\binom{6}{0}\left(\frac{1}{2}\right)^6+\binom{6}{1}\left(\frac{1}{2}\right)^6+\binom{6}{5}\left(\frac{1}{2}\right)^6+\binom{6}{6}\left(\frac{1}{2}\right)^6$$

As $\binom{6}{0}$ is the number of ways of obtaining no heads (i.e., six tails) and there is only one way of achieving this, $\binom{6}{0}=1$; and similarly $\binom{6}{6}$ the number of ways of obtaining six heads is 1.

$$P=\frac{1}{64}+\frac{6}{64}+\frac{6}{64}+\frac{1}{64}$$

$$=\frac{14}{64}$$

$$=0.219$$

This result would have been achieved in (3.12.1) above as (1–0.781).

3.5.3 The Poisson Distribution

This distribution is used in the same manner as the binomial distribution, though there are subtle differences between them.

The binomial distribution deals with events where both the probability of the event occurring and the event not occurring in any one trial are real and known. Both of these outcomes are real distinct happenings that do occur in varying proportions when the series of trials are made.

The Poisson distribution deals with the occurrence of isolated events occurring at random and at unpredictable moments in time. For example, the probable number of flashes of lightning occurring in a storm, or road accidents in a day, or goals in a soccer match cannot be assessed by the binomial distribution, as there is no way of accounting for the number of times the lightning did not flash or vehicles did not crash or a goal was not scored. These are intangible events and there is no way of making an estimate of their effects in a binomial distribution.

In brief, the value of 's' in the fundamental expression (q+p)s is not known.

It was found, largely by accident, in the course of experiments, that where 'x' is the expected or average number of occurrences of an event, in a series of trials, or in a definite interval of time (where the average number of occurrences with time are known), the respective terms of the series:

$$1+x+\frac{x^2}{2!}+\frac{x^2}{3!}+\cdots,\text{ etc.}$$

were proportional to the probabilities of obtaining the occurrences of the event 0, 1, 2, 3 …, etc., times in the series of trials.

To use this as a probability distribution in the same way as the binomial distribution, the sum of all the terms must be unity, which means that each term is the actual numerical value of probability of that number of events occurring in any one series of trials.

In the binomial distribution $(q+p)=(q+p)^s$ for all values of 'x.'

$$\text{Since } 1+x+\frac{x^2}{2!}+\frac{x^2}{3!}+\cdots\text{etc.}=e^x$$

(e=2.7183 to four decimal place)

$$\text{and } e^{-x}\times e^x=1=e^{-x}\left(1+x+\frac{x^2}{2!}+\frac{x^3}{3!}+\cdots,\text{ etc.}\right)$$

This expression satisfies the requirements of a distribution and is called the *Poisson probability distribution.*

The Poisson distribution is used extensively in medical and biological research, as well as in industrial quality control and with engineering reliability prediction, which is covered in Chapter 8.

One of its first uses was used in the Crimea War for calculating the number of cavalrymen killed per year by horse kicks, and this showed that the accidents were the result of chance. It was also used during the latter parts of WWII when the UK Government was trying to discover the accuracy of the Vl flying bombs that were hitting London at a frightening rate. Recording and analyzing all the landing sites, it was established that the mechanism for cutting off the fuel to the engine of the bombs was not very accurate and the landing happened purely by chance.

3.5.4 The Multinomial Law of Probability

The Multinomial Law of Probability is similar to the binomial law but is more applicable to repeated trials where there are more than two possible outcomes of each trial. The law states that in a series of 's' independent trials in each of which the probability of an event 'E_1' is p_1, of an event E_2 is p_2 ... of an event and E_m is p_m, where $p_1+p_2+\ldots p_m=1$, the probability that E_1 will occur exactly l_1 times, E_2 occur l_2 times, ... E_m occurs l_m times is given by

$$\frac{s!}{l1!\ l2!\ l3.....lm!}\pi_1^{l_1}\ \pi_2^{l_2}\ \pi_3^{l_3}........\pi_m^{l_m}$$

The probabilities are obtained from the successive terms of the expansion of $(p_1+p_2+\ldots. p_m)^s$ for different values of l_1, l_2, etc.

Example 3.14

A box contains 5 white, 2 black, 8 blue, and 7 red balls. If balls are selected one at a time with replacement of each ball, find the probability of drawing 2 white, 4 blue, 3 red, and 1 black in 10 trials.

$$P_w=\frac{5}{22},\ P_{bl}=\frac{8}{22}, P_r=\frac{7}{22},\ P_{bk}=\frac{2}{22}$$

The required probability is

$$P=\frac{10!}{2!\ 4!\ 3!\ 1!}\left(\frac{5}{22}\right)^2\left(\frac{2}{22}\right)\left(\frac{8}{22}\right)^4\left(\frac{7}{22}\right)^3$$

$$=12600\times0.0516\times0.0909\times0.0175\times0.0322$$

$$=12600\times0.00000262$$

$$=0.033,\ (\text{or one chance in }30)$$

Example 3.15

Consider the box containing valve stems.

Box 'A' contains 8 new and 7 old valve stems, and box 'B' contains 4 new and 8 old valve stems. A box is chosen at random and a valve stem is picked from it. Determine the probability that the valve stem chosen is new.

Probability of choosing box 'A' $=\frac{1}{2}=$ probability of choosing box 'B'

Probability of selecting a new valve stem from box 'A' is $\frac{8}{15}$

Probability of selecting a new valve stem from box 'B' is $\frac{4}{12}$

Hence, the required probability is $\left(\frac{1}{2}\times\frac{8}{15}\times\frac{1}{2}\times\frac{4}{12}\right)=\frac{13}{30}=0.433$

4

Statistics

4.1 Introduction

In recent years, the role of statistics has gained more importance in engineering, particularly in quality control. Statistical work can generally be divided into the following categories:

4.1.1 Collection of data
4.1.2 Tabulation of data
4.1.3 Analysis
4.1.4 Interpretation

4.1.1 Collection of Data

Collection of data implies observing and studying the subject under investigation; it may be just counting the type and number of automobiles passing a fixed point in a highway or taking measurements of machined parts coming off an automatic machine line.

There are two types of data, namely

a. Qualitative data:

These are generally nonnumerical data such as recording the hair color of hair, etc.

b. Quantitative data:

These are data that are measurable, such as the length of bars off a cutting machine or the diameter of pins coming off a grinding machine.

4.1.2 Tabulation of Data

The collection and tabulation of data is generally combined activities and in practice are treated together. Tabulation of the collected data is necessary to enable mathematical exercises to be performed on this data later. In some instances, nothing further is required to be undertaken on the data, whereas in other instances considerable calculation is required so that conclusions may be eventually reached that may not have been readily seen in the initial glance at the figures.

It should be established that the very act of obtaining and measuring the data could cause an error in the data itself. As an example, consider a flow meter fitted within a hydraulic circuit, the very presence of the flow meter may affect the pressure drop within the piping and thus alter the flow while measuring it.

Statistics is the study of a group or groups of *individuals*, each of which possess one or more characteristic capable of measurement. The *individuals* (also referred to as a population) may be people, machined components, medicines, etc.

The characteristics may be height, length, concentricity, reliability, etc. A characteristic such as height or weight can be measured quantitatively, that is, numerically, but a characteristic such as hair color must be measure qualitatively or nonnumerically. The tabulation of data depends on whether the measurement of the characteristic is either qualitative or quantitative.

4.1.2.1 Sample Values

The samples are taken from a population, and the samples form a *set* within this population. As an example, a set of 100 machined components might be taken from a population of 100,000 components. The samples must be truly random, meaning the sample does not bias the results to one extreme or the other. If the set is large enough and truly random, then it can be expected that any information will apply equally well to a larger population.

The values within the set are referred to as *variables* and can include length, weight, size, etc. The measured value could be a decimal number depending on the accuracy of the measurement. In the extreme, one sample would be obtained for each exact decimal value. This would result in an unwieldy table, which will be no use at all. In this situation, the values could be rounded off so that those values that are close together could be counted as the same, or values that fall within a specific range with say 'x' being the value at the center of the range.

4.1.2.2 Raw Data

The original data collected and initially tabulated prior to any analysis being performed is known as *'raw data'* an example could be the individual failure stresses measured in a series of fatigue tests, or the examination marks for 200 students, in this latter case the individual marks are shown in the examples in Table 4.1.

4.1.2.3 Bands

To reduce the size of the "Raw Data" table, the data are *banded* (or grouped). Table 4.2 shows an example where in this case the table shows groups/bands of marks between 0–10, 10–19, 20–29, and so forth. The table is then populated with the number of marks that occur within each group, for example, in the first group, there are only three marks, and in the second group, there are four marks. The third group has 10 marks and the fourth group 24 marks, and so on.

This table is known as a *"Frequency Distribution table,"* and if the values are plotted against the midpoint of each band, then a graph similar to that of Figure 4.1 will result. This graph is known as the "Frequency Distribution" graph and can be drawn in a number of ways, where the marks are plotted vertically and the frequency horizontally.

Figure 4.1a shows the graph in polygon form. Figure 4.1b is drawn as a bar graph, and Figure 4.1c shows it in a histogram form. Notice that in (a) and (b) the points are drawn in the center of the band, and the rectangles that form the histogram are drawn between the limits of each band.

TABLE 4.1
Examination Marks

6	71	30	71	21	70	60	71	41	53
60	58	61	57	72	42	59	71	53	35
59	60	13	51	61	36	62	52	43	32
44	90	63	57	44	61	55	72	62	53
51	70	54	31	45	38	69	33	54	15
63	52	64	74	23	59	42	55	33	46
51	34	91	47	64	75	58	65	48	37
49	70	56	66	40	62	17	65	50	24
25	67	52	70	77	46	59	66	42	35
49	91	58	47	68	34	57	42	67	50
59	68	16	77	86	68	51	7	69	48
38	76	57	47	69	58	77	27	43	57
43	73	67	59	36	64	46	75	58	47
58	92	49	72	48	48	74	86	39	51
75	88	75	9	60	73	48	50	73	40
53	95	38	82	44	72	59	37	61	45
44	62	71	63	63	42	29	59	66	54
30	70	41	74	64	65	85	55	32	21
41	85	50	31	22	81	35	56	40	45
87	33	80	37	62	33	52	26	89	28

TABLE 4.2
Frequency Distribution Table

Sample range	0–9	10–19	20–29	30–39	40–49	50–59	60–69	70–79	80–89	90–100
Marks	3	4	10	24	36	44	36	28	10	5

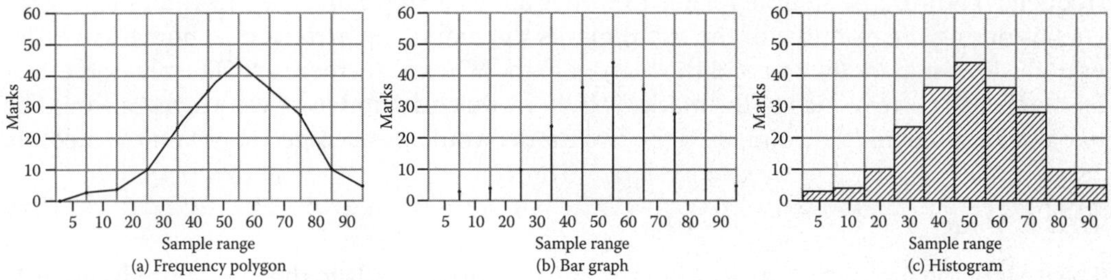

FIGURE 4.1
Frequency distribution graph in (a) polygon form, (b) bar graph form, and (c) histogram form.

4.1.2.4 Mean

The 'mean' of the data is the easiest to compute. All that is required is for all the values in a set of data to be added together and the result to be divided by the number of values in the dataset. Take Table 4.1 as an example:

TABLE 4.3
Mean Value of Example

'x' Midpoint	4.5	14.5	24.5	34.5	44.5	54.5	64.5	74.5	84.5	94.5	Total
f	3	4	10	24	36	44	36	28	10	5	200
fx	13.5	58	245	828	1602	2398	2322	2086	845	472.5	10870

Summing all the marks obtained by the students gives a total figure of 10,837. Dividing this figure by the number of students (200) gives a mean value of

$$\text{Mean } \bar{x} = \frac{10837}{200} = 54.185$$

Thus, the mean is 54.185. This approach can be simplified by banding the data, as discussed in 4.1.2.3.

This results in the Table 4.2 Frequency distribution table. Reducing the sample range to the midpoint of the range will give Table 4.3, where the sample number is 'x' and the number of marks within the range is 'f.' Multiplying 'x' by 'f' gives 'fx,' and summing this row gives a value of 10,870. Dividing this value by the number of students will give

$$\text{Mean } \bar{x} = \frac{10830}{200} = 54.35$$

This mean value is not quite as accurate as the previous mean value, as the values have been taken at the midpoint of each band, but for the purposes of future calculations is generally considered accurate enough.

4.1.2.5 Median

The *median* of the data is the value of the variate where half the number of items (or total frequency) within the sample relates to values above it, and half relate to values below it.

Considering the example of the exam marks the marks are arranged in numerical order with the lowest mark (6) first and the highest mark (95) last. As there are 200 values, the center point (100) corresponds to the mark 55. If there are odd number of values, there would be two marks at the middle point, then the two marks would be averaged to obtain the median.

4.1.2.6 Mode

The *mode* of the data is the value of the variate that corresponds to the maximum frequency. In Table 4.3, this would correspond to the class with a mark of 54.5, as there are 44 marks within this group.

It is possible for the maximum figure to occur more than once within the distribution.

4.1.2.7 Ogive and Quartiles

Consider Table 4.3, and add a further row enabling the accumulative frequency to be calculated; the date is then plotted against the resulting graph, called an *"Ogive"*. The Ogive makes it easier to determine the median of the data set as shown in Table 4.4.

TABLE 4.4

Cumulative Frequency for Table 4.3

x											
Midpoint	4.5	14.5	24.5	34.5	44.5	54.5	64.5	74.5	84.5	94.5	Total
f	3	4	10	24	36	44	36	28	10	5	200
fx	13.5	58	245	828	1602	2398	2322	2086	845	472.5	10870
Cum 'f'	3	7	17	41	77	121	157	185	195	200	

Figure 4.1 shows the resulting graph for Table 4.3.

For the purposes of the graph, the cumulative frequency is converted into a percentage figure and is plotted as a vertical scale.

The median of the data corresponds to the 50% level.

A quartile is one of four equal groups in a set. The graph in Figure 4.1 is shown in three sections: (a) Frequency Polygon, (b) A bar graph and (c) A histogram. The section from 0% to 25% is referred to as the "*Lower quartile.*" The section from 75% to 100% is referred to as the "*Upper quartile,*" and the section between 25% and 75% is called the "*Interquartile.*"

The interquartile can be divided into two parts, and these are called the '*semi-interquartile*' and are a measure of how the samples are distributed around the median. Any range that corresponds to a change of 1% is called a '*percentile.*'

An improved method of showing how the samples are distributed about the median is to use the "*Standard Deviation*".

Example 4.1

A company manufactures pneumatic cylinders with a nominal bore of 20.00 mm. A selection of 20 cylinders are chosen at random for the purposes of quality control, and the bores are carefully measured. The results are shown in Table 4.5.

1. Produce a frequency distribution table having bands of 0.1 mm.
2. Calculate the mean of the sample cylinders.
3. Draw a histogram.
4. Plot the Ogive.
5. Calculate the median, mode, and upper and lower quartiles together with the semi-interquartile.

Solution

Total = 398.3
Total samples 'n' = 20
Mean = 398.3/20 = 19.915.

Ranked order:
The median is the middle value of the samples when they have been ranked in order of size, and this is the ninth sample, hence the median is 19.9 mm.

TABLE 4.5

For Example 4.1

Sample no.	**1**	**2**	**3**	**4**	**5**	**6**	**7**	**8**	**9**	**10**
Diameter	19.9	19.8	20.1	19.9	19.7	20.1	20.0	19.6	19.7	20.1
Sample no.	**11**	**12**	**13**	**14**	**15**	**16**	**17**	**18**	**19**	**20**
Diameter	20.2	20.0	19.9	19.8	20.1	20.0	19.7	19.6	19.9	20.2

4.2 Standard Deviation

Before discussing standard deviation, the following points must be reviewed.

4.2.1 Dispersion

Important information regarding the general mass of the observations recorded is covered by the mean, mode, and median. Information about the spread (dispersion) of the values around a central value will not be gained from these statistics.

Consider the following sets:

Set 1: 17, 18, 19, 20, 21 has a mean of 19.

Set 2: 4, 13, 18, 23, 37 has a mean of 19.

These two sets have the same mean value but it is clearly seen that Set 1 is more closely arranged around the mean compared to Set 2. It would be useful to have a measure that indicates the spread of the observed values around the mean.

4.2.2 Range

Studying Set 1, the spread or *range* of the values 21−17=4, and for Set 2, the spread is 37−4=33. The disadvantage of using the range is that only the extreme values of the set is dealt with and it gives no information about the spread of the intermediate values.

4.2.3 Variance

The *variance* is a numerical value used to indicate how widely individuals in a group vary. If individual observations vary greatly from the group mean, then the variance can be quite large.

It is important to distinguish between the variance of a population and the variance of a sample. They have different notation, and they are computed differently. The variance of a population is denoted by σ^2; and the variance of a sample, by s^2.

4.2.3.1 Variance of a Population

The variance of a *population* is defined by the following formula:

$$\sigma^2 = \frac{\Sigma(x_i - x)^2}{n} \tag{4.1}$$

where σ^2 is the population variance, 'x' is the population mean, 'x_i' is the *i*th element from the population, and 'n' is the number of elements in the population.

Example 4.2

Variance of a Population

Considering the exam marks previously shown in Table 4.1, the following values are deduced.

Standard deviation will be discussed in Section 4.2.2 but is included here for comparison with the examples deduced from Example 4.3, Variance of a Sample.

4.2.3.2 Variance of a Sample

The variance of a *sample* is defined by a slightly different formula:

$$s^2 = \frac{\Sigma(x_i - x)^2}{(n-1)} \tag{4.2}$$

where s^2 is the sample variance, 'x' is the sample mean, 'x_i' is the *i*th element from the sample, and n is the number of elements in the sample. Using this formula, the variance of the sample is an unbiased estimate of the variance of the population.

The variance is equal to the square of the standard deviation.

Example 4.3

Variance of a Sample

From Table 4.1 for the 22 sets of marks, a random set of marks are selected three times.

Table 4.6 shows that the errors of making a small selection of the marks from Table 4.1 results in a small error.

BEWARE: If the original set of marks had been small, say less than 20, then the sample variance calculated using Equation 4.2 will result in a large error (Table 4.7).

4.2.4 Standard Deviation

The *standard deviation* is the most important measure of dispersion.

The standard deviation (σ) is the square root of the variance and is a statistic that indicates how tightly the date is spread around the mean in a set of the data.

Hence, σ is:

$$\sigma = \sqrt{\text{variance}}$$
$$= \text{standard deviation}$$

One standard deviation either side of the mean, as shown in Figure 4.2 (2σ), accounts for approximately 68% of the samples (Table 4.8).

Two standard deviations either side of the mean (4σ) accounts for roughly 95% of the samples, and three deviations (6σ) accounts for nearly 99% of the samples. If the curve had been flatter and more spread out, then the standard deviation would have to be larger in order to account for the 68%. This explains why the standard deviation is able to show the spread of the examples within the set from the mean.

TABLE 4.6

Ranked Order for Table 4.1

Sample	**1**	**2**	**3**	**4**	**5**	**6**	**7**	**8**	**9**	**10**
Diameter	19.6	19.6	19.7	19.7	19.7	19.8	19.8	19.9	19.9	19.9
Sample	**11**	**12**	**13**	**14**	**15**	**16**	**17**	**18**	**19**	**20**
Diameter	19.9	20.0	20.0	20.0	20.1	20.1	20.1	20.1	20.2	20.2

TABLE 4.7

Random Set of Marks Selected from the Marks Shown in Table 4.1

Ident number	Selection 1		Selection 2		Selection 3	
	Mark	Mean	Mark	Mean	Mark	Mean
1	4	1709.823	17	642.6225	5	1548.423
2	7	1395.023	30	374.4225	10	1046.523
3	13	861.4225	36	301.0225	18	592.9225
4	16	694.3225	45	178.2225	24	455.8225
5	27	455.8225	46	178.2225	33	336.7225
6	44	205.9225	53	128.8225	33	336.7225
7	45	178.2225	75	28.6225	34	336.7225
8	49	152.5225	84	11.2225	41	235.6225
9	55	128.8225	97	0.1225	46	178.2225
10	57	107.1225	100	0.4225	47	178.2225
11	65	69.7225	104	7.0225	57	107.1225
12	74	40.3225	106	7.0225	78	18.9225
13	86	11.2225	115	21.6225	85	11.2225
14	106	7.0225	148	135.7225	87	5.5225
15	107	7.0225	160	244.9225	115	21.6225
16	120	21.6225	164	277.2225	143	113.4225
17	126	44.2225	165	277.2225	147	135.7225
18	148	135.7225	167	277.2225	159	244.9225
19	149	160.0225	175	386.1225	161	244.9225
20	176	386.1225	180	426.4225	165	277.2225
21	183	513.0225	194	1132.323	180	426.4225
22	187	710.2225	198	1343.223	187	710.2225
Mean		6647.3		7995.3		6379.8
Variance		316.54		380.73		303.80
SD		17.79		19.51		17.43

Consider a process line is manufacturing pistons and the diameter is being machined using a cylindrical grinding machine, assuming the acceptance standard for the diameter is set at one standard deviation either side of a mean diameter, then 68% of the pistons produced will be acceptable and anything larger will be rejected, i.e., the larger the value of 'σ,' the more rejects. Now if the mean of the sample begins to move as production continues, samples will then be rejected more on one side than the other, and this would be indicative that the grinding wheel is beginning to wear and will require replacing.

TABLE 4.8

Averaging the Mean, Variance, and Standard Deviation of the Values Shown in Table 4.7

Mean	7146.40	% Error
Variance	340.30	3.80%
Standard deviation (SD)	18.43	1.85%

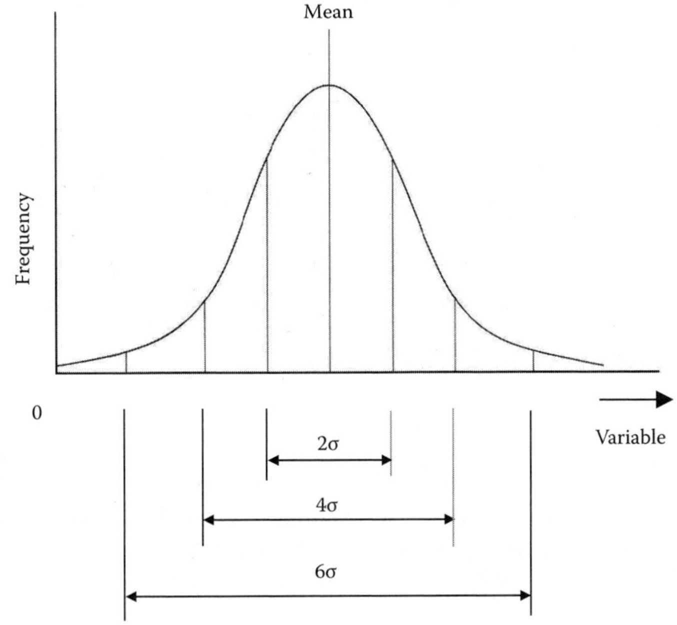

FIGURE 4.2
Standard deviations.

4.3 Fitting Functions to Experimental Data

When undertaking experimental work that requires logging data being read from instruments, etc., it is understandable that errors will be inherent within the data, such as errors within the recording instruments or errors in reading these instruments, particularly if it is analogue. If the unit is digital, then there could be thermal problems within the instruments circuit, giving small false readings or other errors not accounted for. Substantial statistical work has been undertaken to improve the accuracy of fitting the data to minimize the errors between the plotted function and the experimental data.

The most widely used method is that of *"least squares"* also referred to as the *linear regression method.*

A simple example illustrates the method:

Consider an experiment to establish a law relating frictional force against the reaction between the two sliding surfaces.

Table 4.9 tabulates the readings taken during the experiment to measure the coefficient of friction.

From our understanding of friction, it is expected that the relationship between applied force and sliding force will be linear. It is expected that there will be small errors in measuring the weights and forces together with small variations of the surface texture.

TABLE 4.9

Readings Taken from Experiment for Coefficient of Friction

F (N)	0	6	8	12	12.2	18	20	24.2	24.2	32	34
R (N)	0	100	200	300	400	500	600	700	800	900	1000

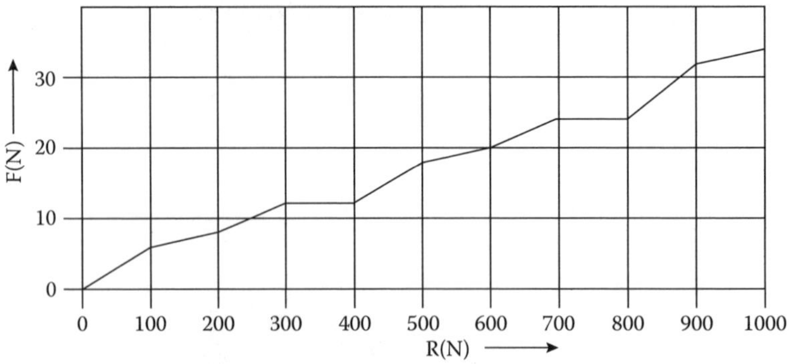

FIGURE 4.3
Graph for Table 4.9.

The graph shown in Figure 4.3 shows that joining the plotted points together does not produce a satisfactory graph, but it also shows that it would be possible to draw a straight line through an average of the points. The question is how to draw the best line straight line that minimizes the error between the points and the line.

Example 4.4

Fit the best straight line to the following data:

x	y
0	81
6	63
12	64
18	62
24	58
30	60
36	41
42	47
48	38

	x	y	x^2	xy
	0	81	0	0
	6	63	36	378
	12	64	144	768
	18	62	324	1116
	24	58	576	1392
	30	60	900	1800
	36	41	1296	1476
	42	47	1764	1974
	48	38	2304	1824
Σ				
	216	514	7344	10728

Centroid of data:
Mean 'x' = 24.00
Mean 'y' = 57.11

$$\bar{y} = a + b.\bar{x} \quad 57.11 = a + 24.b \tag{4.3}$$

$$\sum xy = a\sum x + b.x^2 \quad 10728 = 216.a + 7344.b \tag{4.4}$$

Multiply Equation 4.3×216 and subtract from Equation 4.4

$$12335.976 = 216.a + 5184.b \tag{4.5}$$

$$10728.000 = 216.a + 7344.b \tag{4.6}$$

$$1607.976 = 2160.b \tag{4.7}$$

$$\therefore b = 0.7444 \tag{4.8}$$

Back substitute Equation 4.8 into Equation 4.3 and evaluate for 'a':

$$57.11 = a \times 24 \times 0.7444$$

$$57.11 = a \times 17.866$$

$$a = 17.866 + 57.11$$

$$a = 74.976 \tag{4.9}$$

Trend line is $y = 74.976 - 0.7444x$

Table of residuals $(y_c - y_o)$

i.e., compare the original value for 'x' against the calculated value (Figure 4.4).

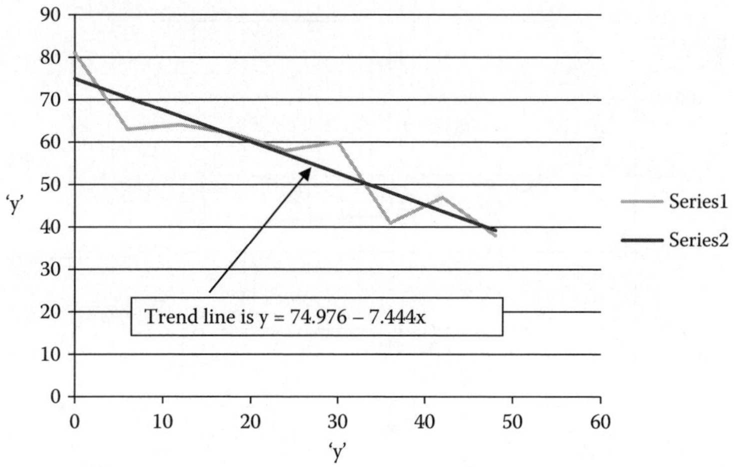

FIGURE 4.4
Trend line for Example 4.4.

x	0.7444x	74.976−0.7444x	y_o	Residuals
		y_c		(y_o-y_c)
0	0	74.976	81	−6.024
6	4.466	70.510	63	7.510
12	8.933	66.043	64	2.043
18	13.399	61.577	62	−0.423
24	17.866	57.111	58	−0.890
30	22.332	52.644	60	−7.356
36	26.798	48.178	41	7.178
42	31.265	43.711	47	−3.289
48	35.731	39.245	38	1.245
			Σ residuals=−0.0064	

Example 4.5

In an experiment to evaluate tool life when cutting identical materials at various cutting speeds, the following results were obtained (Figure 4.5).

Speed 'v' (m/min)	120	130	170	200
Life 'T' (min)	62	25	7.2	2.8

Assuming the law relating 'v' and 'T' is $v = aT^k$

Determine the constants a and k to give the best fit for the data.

$$v = aT^k \quad \log v = k \log T + \log a$$

$$Y = mX + C$$

X=log T	1.792	1.398	0.857	0.447
Y=log v	2.079	2.114	2.230	2.301

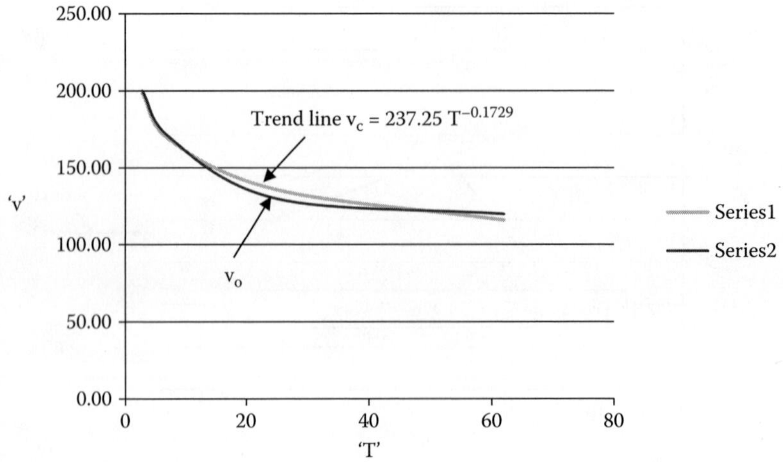

FIGURE 4.5
Trend line for Example 4.5.

$$S = \sum_{1}^{n} (Y_i - mX_i - c)^2$$

$$\frac{\partial s}{\partial m} = -2\sum_{1}^{n} X(Y - mX - c) = 0$$

$$\therefore \sum xy - m\sum x^2 - c\sum x = 0$$

$$\frac{\partial s}{\partial c} = -2\sum_{1}^{n} (Y - mX - c) = 0$$

$$\therefore \sum \left(Y - m\sum X - nc\right) = 0$$

	X	Y	XY	X²
	1.792	2.079	3.726	3.211
	1.398	2.114	2.955	1.954
	0.857	2.230	1.912	0.735
	0.447	2.301	1.020	0.200
Σ	4.495	8.724	9.622	6.101

The normal equation gives

$$9.622 - 6.101m - 4.495c = 0 \tag{4.10}$$

and

$$8.724 - 4.495m - 4c = 0 \tag{4.11}$$

Dividing both equations by the coefficient of m:

$$m + 0.7396c = 1.5774 \tag{4.12}$$

$$m + 0.8899c = 1.9408 \tag{4.13}$$

$$\text{Therefore, } 0.1530c = 0.3634$$

$$c = 2.375$$

Substituting for 'm' in Equation 4.12

$$m = 1.5774 - 0.7369(2.375) = 1.5774 - 1.7503$$

$$m = -0.1729$$

$$\text{Therefore, } Y = -0.1729X + 2.375$$

$$m = n \rightarrow n = 0.1729$$

$$c = \log a \quad \log a = 2.375$$

$$a = 237.25$$

Hence, the law is

$$v = 237.25\ T^{-0.1729}$$

Check:

	Corrected	Original	Residuals
T	v_c	v_o	$v_o - v_c$
62	116.06	120	3.94
25	135.80	130	–5.80
7.2	168.43	170	1.57
2.8	198.33	200	1.67
		Σ residuals = 1.38	

5

Properties of Sections and Figures

The figures in this chapter list the properties for both plane shapes and solid shapes that are most commonly used in engineering design.

The figures are in two parts:

Figure 5.1 considers the properties of plane areas.

Figure 5.2 lists the properties of solid shapes.

The equations shown are for the following:

5.1 Centroid C_x, C_y, C_z

The centroid of an area is the point which is the average distance from all the points on the surface of the figure act.

C_x, C_y and C_z are the centroid co-ordinates.

5.2 Moment of Inertia/Second Moment of Area

This is a geometric property of the shape, usually associated with the cross section of a beam. It is an indication of the beam's ability to resist bending: the smaller the value of the moment of inertia, the more the beam will deflect; and conversely, the greater the value, the less the beam will deflect.

The moment of inertia will pass through the centroid of the area and this will be the datum axis for the section. If the axis does not pass through this datum axis, it is possible for the revised moment of inertia to be recalculated for the new axis position.

5.3 Polar Moment of Inertia of a Plane Area

The polar moment of inertia will relate to an axis that is perpendicular to the plane of the area.

If all of the area is considered to comprise an infinitely small number of small areas (da), then the polar moment of inertia will be the sum of all these areas multiplied by r^2, where r is the radius of da from the perpendicular axis.

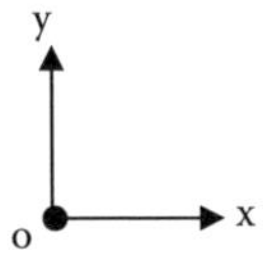	**Area = A**	C_x C_y	I_{xx}	I_{yy}
Rectangle 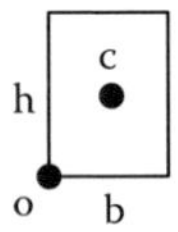	$b \cdot h$	b/h $h/2$	$A \cdot h^2/12$	$A \cdot b^2/12$
Parallelogram 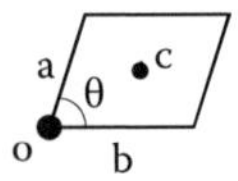	$a \cdot b \cdot \sin\theta$	$(b + a\cos\theta)/2$ $a \cdot \sin\theta/2$	$A(a \cdot \sin\theta)^2/12$	$A(b^2 + a^2 \cos 2\theta)/12$
Triangle 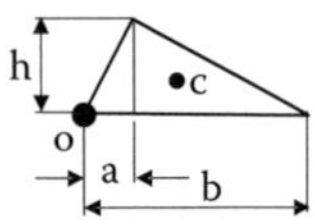	$b \cdot h/2$	$(a + b)/3$ $h/3$	$A \cdot h^2/18$	$A(b^2 - ab + a^2)/18$
Trapezium 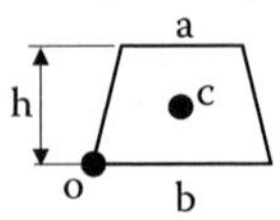	$h(a + b)/2$	$C_y = \dfrac{h(2a + b)}{3(a + b)}$	$\dfrac{A \cdot h^2(a^2 + 4^2ab + b^2)}{18(a + b)^2}$	—
Hexagon 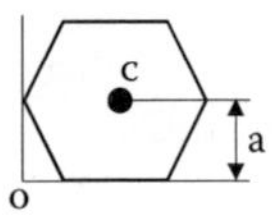	$6a^2 \cdot \tan(30°)$ $= 3.464\, a^2$	$a/\cos(30°)$ $= 1.156\, a$	$\dfrac{A}{12}\left(\dfrac{a^2[1 + 2\cos^2(30°)]}{\cos^2(30°)}\right)$ $= 0.962\, a^4$	$\dfrac{A}{12}\left(\dfrac{a^2[1 + 2\cos^2(30°)]}{\cos^2(30°)}\right)$ $= 0.962\, a^4$
Circle 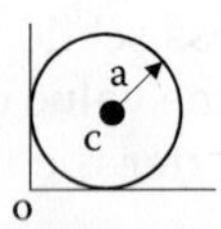	πa^2	$C_x = a$ $C_y = a$	$\dfrac{A \cdot a^2}{4}$	$\dfrac{A \cdot a^2}{4}$

FIGURE 5.1
Areas, volumes, centroids and moments of inertia for plane figures.

(Continued)

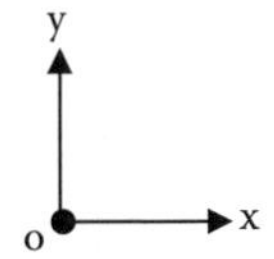	Area = A	C_x C_y	I_{xx}	I_{yy}
Annulus 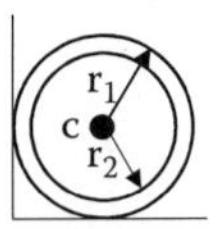	$\pi(r_1^2 - r_2^2)$	$C_x = r_1$ $C_y = r_1$	$\frac{\pi(r_1^4 - r_2^4)}{4}$	$\frac{\pi(r_1^4 - r_2^4)}{4}$
Semicircle 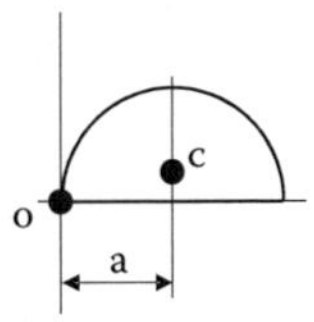	$\frac{\pi a^2}{2}$	$C_x = a$ $C_y = \frac{4a}{3 \cdot \pi}$	$\frac{Aa^2 (9\pi^2 - 64)}{36\pi^2}$ $= 0.1098\ a^4$	$\frac{Aa^2}{8}$
Sector of a circle 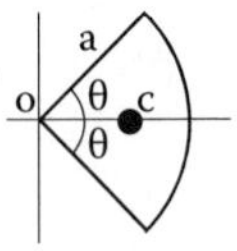	$a^2 \cdot \theta$	$C_x = \frac{2 \cdot a \cdot \sin\theta}{3 \cdot \theta}$ $C_y = 0$	$\frac{Aa^2}{4\theta}(\theta - \sin(\theta) \cdot \cos(\theta))$	$\frac{Aa^2}{4\theta}\left[\theta + \sin(\theta) \cdot \cos(\theta) - \frac{16\sin^2\theta}{9\theta}\right]$
Rectangle 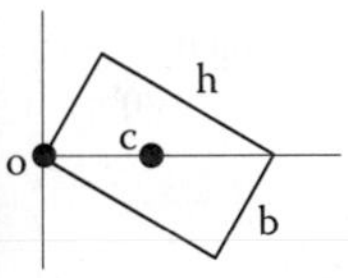	$b \cdot h$	$C_x = \frac{\sqrt{(b^2 + h^2)}}{2}$ $C_y = 0$	$\frac{Ab^2 \cdot h^2}{6(h^2 + b^2)}$	$\frac{A(h^4 \cdot b^4)}{12(h^2 + b^2)}$
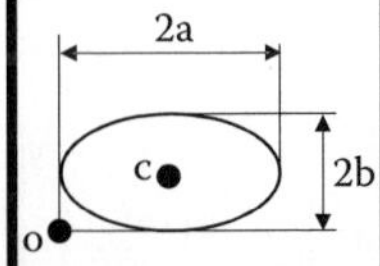	$\pi \cdot a \cdot b$	$C_x = a$ $C_y = b$	$\frac{A \cdot b^2}{4}$	$\frac{A \cdot a^2}{4}$
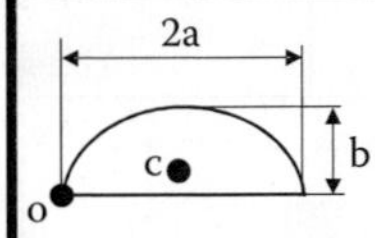	$\frac{A \cdot b}{2}$	$C_x = a$ $C_y = \frac{4 \cdot b}{3 \cdot \pi}$	$\frac{A \cdot b^2(9\pi^2 - 64)}{36\pi^2}$	$\frac{A \cdot a^2}{4}$

FIGURE 5.1 (CONTINUED)
Areas, volumes, centroids and moments of inertia for plane figures.

Solid 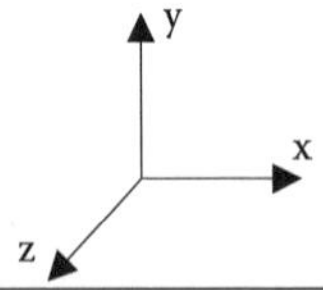	Area volume	C_x C_y C_z	I_{xx}	I_{yy}	I_{zz}
Uniform rod 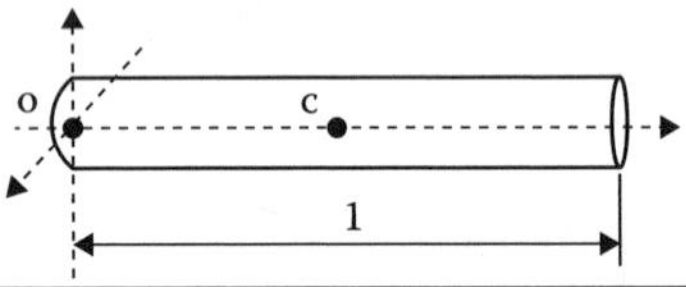	$\frac{\pi \cdot a^2}{4}$ $\frac{\pi \cdot a^2}{4} . 1$	1 0 0	0	$\frac{m \cdot l^2}{12}$	$\frac{m \cdot l^2}{12}$
Rectangular prism 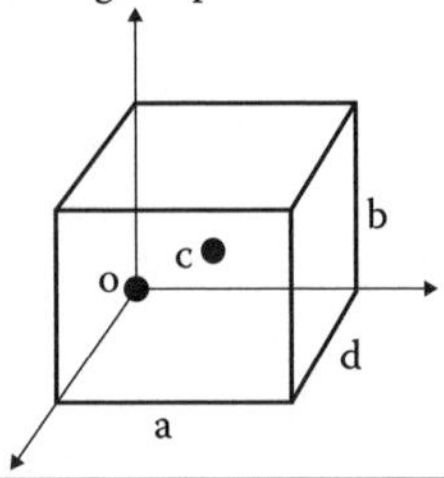	$2(a \cdot b + a \cdot c + b \cdot c)$ $a \cdot b \cdot c$	$\frac{a}{2}$ $\frac{b}{2}$ $\frac{c}{2}$	$\frac{m(b^2 + c^2)}{12}$	$\frac{m(c^2 + a^2)}{12}$	$\frac{m(a^2 + b^2)}{12}$
Right rectangular pyramid 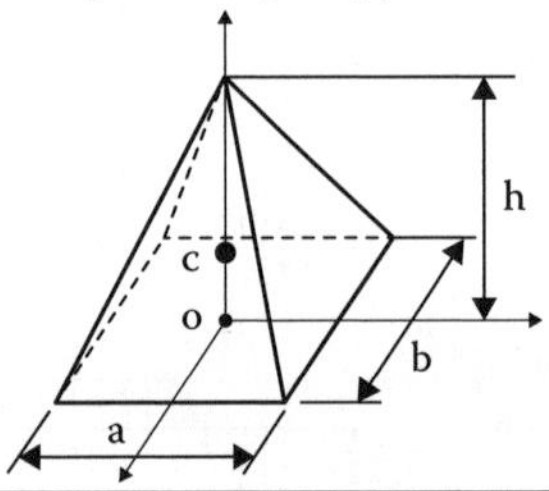	$\frac{2 \cdot \pi \cdot a \cdot b \cdot c}{3}$ $\frac{a \cdot b \cdot h}{3}$	0 $\frac{h}{4}$ 0	$\frac{m(4b^2 + 3h^2)}{80}$	$\frac{m(a^2 + b^2)}{20}$	$\frac{m(4b^2 + 3h^2)}{80}$
Uniform hoop 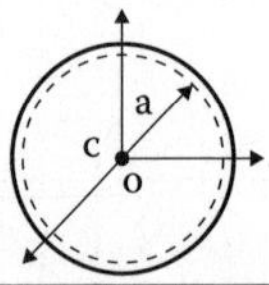	0 0	0 0 0	$\frac{m \cdot a^2}{2}$	$\frac{m \cdot a^2}{2}$	$m \cdot a^2$
Spherical shell 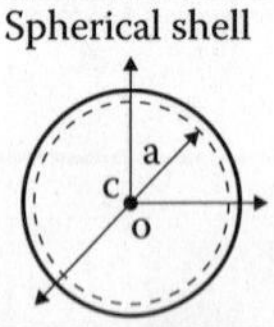	$\frac{4 \cdot \pi \cdot a^2}{0}$	0 0 0	$\frac{2 \cdot m \cdot a^2}{3}$		
Hollow sphere 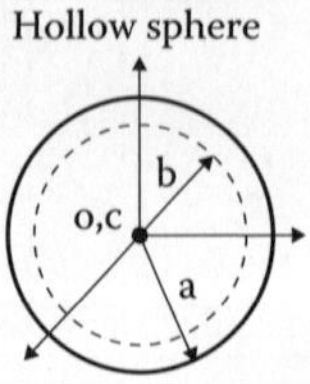	$4 \cdot \pi \cdot a^2$ $\frac{4 \cdot \pi(a^3 - b^3)}{3}$	0 0 0	$\frac{2 \cdot m(a^5 - b^5)}{5(a^3 - b^3)}$		

FIGURE 5.2
Areas, volumes, centroids and moments of inertia for solid figures.

(Continued)

Solid 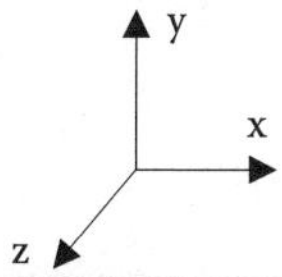	Area volume	C_x C_y C_z	I_{xx}	I_{yy}	I_{zz}
Sphere 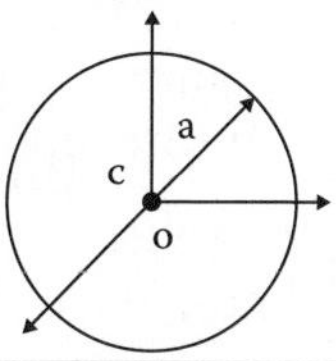	$4 \cdot \pi \cdot a^2$ $\frac{4 \cdot \pi \cdot a^3}{3}$	0 0 0		$\frac{2 \cdot m \cdot a^2}{5}$	
Hemisphere 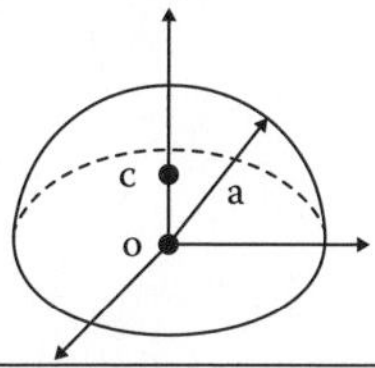	$2 \cdot \pi \cdot a^2$ $\frac{2 \cdot \pi \cdot a^3}{3}$	0 $\frac{3 \cdot a}{8}$ 0	—	$\frac{2 \cdot m \cdot a^2}{5}$	—
Right circular cylinder 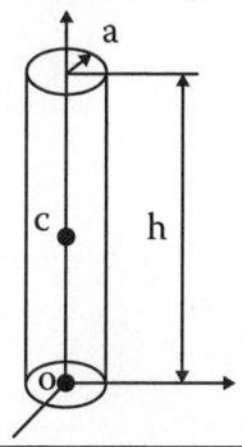	$2 \cdot \pi \cdot a(h + a)$ $\pi \cdot a^2 \cdot h$	0 $\frac{h}{4}$ 0	$\frac{3 \cdot m(4 \cdot a^2 + h^2)}{80}$	$\frac{3 \cdot m \cdot a^2}{10}$	$\frac{3 \cdot m(4 \cdot a^2 + h^2)}{80}$
Ellipsoid 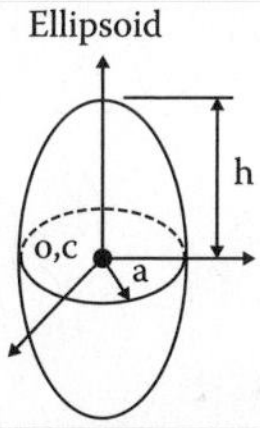	— $\frac{2 \cdot \pi \cdot a \cdot b \cdot c}{3}$	0 0 0	$\frac{m \cdot (b^2 + c^2)}{5}$	$\frac{m \cdot (c^2 + a^2)}{5}$	$\frac{m \cdot (a^2 + b^2)}{5}$
Segment of spherical shell 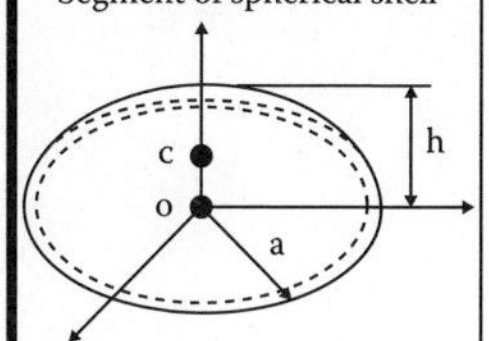	$2 \cdot \pi \cdot r \cdot h$	0 $\frac{h}{2}$ 0	—	—	—
Segment of sphere 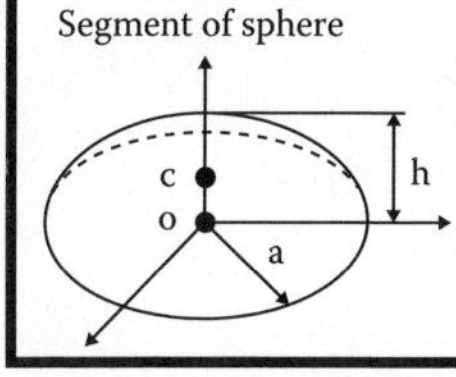	$2 \cdot \pi \cdot r \cdot h$ $\pi \cdot h^2 \left(r - \frac{h}{3}\right)$	0 $\frac{h(4a - h)}{4(3 \cdot a - h)}$ 0	$\frac{mh}{80}\left(\frac{A}{(3a - h)^2}\right)$ $A = 240a^3 - 220 \cdot a^2 \cdot h + 72a \cdot h^2 - 9h^3$	$m\left(a^2 - \frac{3ah}{4} + \frac{3h^2}{20}\right) \times \frac{2h}{3a - h}$	$\frac{mh}{80}\left(\frac{A}{(3a - h)^2}\right)$ $A = 240a^3 - 220 \cdot a^2 \cdot h + 72a \cdot h^2 - 9h^3$

FIGURE 5.2 (CONTINUED)
Areas, volumes, centroids and moments of inertia for solid figures.

The polar moment of inertia is the sum of any two moments of inertia about an axis acting at 90° to each other in the case of a plane area:

$$J = I_{xx} + I_{yy}$$

where I_{xx} and I_{yy} are the moments of inertia for axes through the centroid in directions x and y.

For a plane solid, it will be the sum of any three moments of inertia, for example, where I_{xx}, I_{yy} and I_{zz} are the moments of inertia for axes through the centroid 'c' in directions x, y and z.

$$J = I_{xx} + I_{yy} + I_{zz}$$

6

Properties of Engineering Materials

6.1 Introduction

Advances in materials research and technology offer great promise for the future. Materials science forms the foundation for engineers in engineering and product development, because the structures, components, and devices that engineers design are limited by the properties of the materials available and the techniques that can be used for fabrication.

Selecting the most appropriate material for a particular project or application is usually a difficult task, requiring compromises between different material properties including the following:

6.1 General physical properties
6.2 Mechanical properties
6.3 Thermal properties

6.2 General Physical Properties

6.2.1 Density

Density is one of the most fundamental physical properties of any material. It is defined as the ratio of an object's mass to its volume. Because most designs are limited by either size and or weight, density is an important consideration in many calculations.

Density is a function of the mass of the atoms making up the materials and the distance between them. Massive, closely packed atoms characterize high-density materials, such as tungsten or neptunium. In contrast light, relatively distant atoms compose low-density materials such as beryllium or aluminum. Density at a macroscopic level is also a function of the microscopic structure of a material. A relatively dense material may be capable of forming a cellular structure, such as a honeycomb or foam, which can be nearly as strong and much less dense than the bulk material. Composites, including natural constituents such as wood and bone, for example, generally rely on microscopic structure to achieve densities far lower than common monolithic materials.

6.2.2 Availability/Manufacturability

Availability and manufacturability requirements are often unseen limiting factors in materials selection. The importance of a material being available is obvious. Materials that are not available cannot be used. The importance of processing is not always so obvious.

Any other desirable qualities are useless if a material cannot be processed into the shape required to perform its function. Most engineering materials in use today have well-known substitutes that would perform better and often at lower cost, but processes for forming, cutting, machining, joining, etc., are not available or commercially viable. There is often a period of time after a new material is introduced during which its application is severely limited while processing techniques are developed to facilitate its use.

6.2.3 Cost

A material's cost is also generally a limiting factor. While cost is universally recognized and perhaps the easiest of all properties to understand, there are specific cost considerations for materials selection. Just as materials and their processing go hand in hand, so too do material costs and processing costs. Understanding the entire processing sequence is critical to accurately evaluating the true cost of a material.

6.2.4 Appearance

Because the appearance of many mechanical components seems fairly trivial, it is also easy to overlook its importance in the marketing and commercial success of a product.

6.3 Mechanical Properties

The mechanical properties of a material describe how it will react to physical forces. Mechanical properties occur as a result of the physical properties inherent to each material and are determined through a series of standardized mechanical tests.

6.3.1 Strength

Strength has several definitions, depending on the material type and application. Before choosing a material based on its published or measured strength, it is important to understand the manner in which strength is defined and how it is measured. When designing for strength, material class and mode of loading are important considerations.

For metals, the most common measure of strength is the yield strength. For most polymers, it is more convenient to measure the failure strength, the stress at the point where the stress–strain curve becomes obviously nonlinear. Strength for ceramics, however, is more difficult to define. Failure in ceramics is highly dependent on the mode of loading. The typical failure strength in compression is 15 times the failure strength in tension. The more common reported value is the *compressive failure* strength.

6.3.2 Elastic Limit

The *elastic limit* is the highest stress at which all deformation strains are fully recoverable. For most materials and applications, this can be considered the practical limit to the maximum stress a component can withstand and still function as designed. Beyond the elastic limit, permanent strains are likely to deform the material to the point where its function is impaired.

6.3.3 Proportional Limit

The *proportional limit* is the highest stress at which stress is linearly proportional to strain. This is the same as the elastic limit for most materials. Some materials may show a slight deviation from proportionality while still under recoverable strain. In these cases, the proportional limit is preferred as a maximum stress level because deformation becomes less predictable above it.

6.3.4 Yield Strength

Yield strength is the minimum stress that produces permanent plastic deformation. This is perhaps the most common material property reported for structural materials, because of the ease and relative accuracy of its measurement. The yield strength is usually defined at a specific amount of plastic strain, or offset, which may vary by material and or specification. The *offset* is the amount that the stress–strain curve deviates from the linear elastic line. The most common offset for structural metals is 0.2%.

6.3.5 Ultimate Tensile Strength

Ultimate tensile strength is an engineering value calculated by dividing the maximum load on a material experienced during a tensile test by the initial cross section of the test sample. When viewed in light of the other tensile test data, the ultimate tensile strength helps to provide a good indication of a material's toughness but is not by itself a useful design limit. Conversely, this can be construed as the minimum stress that is necessary to ensure the failure of a material.

6.3.6 True Fracture Strength

True fracture strength is the load at fracture divided by the cross-sectional area of the sample. Like ultimate tensile strength, true fracture strength can help an engineer to predict the behavior of the material but is not itself a practical strength limit. Because the tensile test seeks to standardize variables such as specimen geometry, strain rate, and uniformity of stress, it can be considered a kind of best-case scenario of failure.

6.3.7 Ductility

Ductility is a measure of how much deformation or strain a material can withstand before breaking. The most common measure of ductility is the percentage of change in length of a tensile sample after breaking. This is generally reported as % El or percent elongation. The R.A. or reduction of area of the sample also gives some indication of ductility.

6.3.8 Toughness

Toughness describes a material's resistance to fracture. It is often expressed in terms of the amount of energy a material can absorb before fracture. Tough materials can absorb a considerable amount of energy before fracture, whereas brittle materials absorb very little. Neither strong materials (such as glass) nor very ductile materials (such as lead) can absorb large amounts of energy before failure. Toughness is not a single property, but rather is a combination of strength and ductility.

The toughness of a material can be related to the total area under its stress–strain curve. A comparison of the relative magnitudes of the yield strength, ultimate tensile strength, and percent elongation of different material will give a good indication of their relative toughness. Materials with high yield strength and high ductility have high toughness. Integrated stress–strain data is not readily available for most materials, so other test methods have been devised to help quantify toughness. The most common test for toughness is the Charpy impact test.

In crystalline materials, the toughness is strongly dependent on crystal structure. Face-centered cubic materials are typically ductile, whereas hexagonal close-packed materials tend to be brittle. Body-centered cubic materials often display dramatic variation in the mode of failure with temperature. In many materials, the toughness is temperature dependent. Generally, materials are more brittle at lower temperatures and more ductile at higher temperatures. The temperature at which the transition takes place is known as the *Ductile to Brittle Transition Temperature* (DBTT). The DBTT is measured by performing a series of Charpy impact tests at various temperatures to determine the ranges of brittle and ductile behavior. Use of alloys below their transition temperature is to be avoided, due to the risk of a catastrophic failure.

6.3.9 Fatigue Ratio

The dimensionless *fatigue ratio* f is the ratio of the stress required to cause failure after a specific number of cycles to the yield stress of a material. Fatigue tests are generally run through 10^7 or 10^8 cycles. A high fatigue ratio indicates materials that are more susceptible to crack growth during cyclic loading.

6.3.10 Loss Coefficient

The *loss coefficient* is another important material parameter in cyclic loading. It is the fraction of mechanical energy lost in a stress–strain cycle. The loss coefficient for each material is a function of the frequency of the cycle. A high-loss coefficient can be desirable for damping vibrations, whereas a low-loss coefficient transmits energy more efficiently. The loss coefficient is also an important factor in resisting fatigue failure. If the loss coefficient is too high, then cyclic loading will dissipate energy into the material, leading to fatigue failure.

6.4 Thermal Properties

6.4.1 Thermal Conductivity

Thermal conductivity is the rate of heat transfer through a material in steady state. It is not easily measured, especially for materials with low conductivity, but reliable data is readily available for most common materials.

6.4.2 Thermal Diffusivity

Thermal diffusivity is a measure of the transient heat flow through a material.

6.4.3 Specific Heat

Specific heat is a measure of the amount of energy required to change the temperature of a given mass of material. Specific heat is measured by calorimetry techniques and is usually reported both as C_V, the specific heat measured at constant pressure, or C_P, the specific heat measured at constant pressure.

6.4.4 Melting Point

The *melting point* is the temperature at which a material goes from the solid to the liquid state at one atmosphere. The melting temperature is not usually a design criteria, but it offers important clues to other material properties.

6.4.5 Glass Transition Temperature

The *glass transition temperature*, or T_g, is an important property of polymers. The glass transition temperature is a temperature range that marks a change in mechanical behavior. Above the glass transition temperature, a polymer will behave like a ductile solid or highly viscous liquid. Below T_g, the material will behave as a brittle solid. Depending on the desired properties, materials may be used both above and below their glass transition temperature.

6.4.6 Thermal Coefficient of Expansion

The *thermal coefficient of expansion* is the amount a material will change in dimension with a change in temperature. It is the amount of strain due to thermal expansion per degree Kelvin, expressed in units of K^{-1}. For isotropic materials, is the same in all directions, anisotropic materials have different values depending on the direction.

6.4.7 Thermal Shock Resistance

Thermal shock resistance is a measure of how large a change in temperature a material can withstand without damage. Thermal shock resistance is important to most high-temperature designs. Measurements of thermal shock resistance are highly subjective, because it is extremely process dependent. Thermal shock resistance is a complicated function of heat transfer, geometry, and material properties. The temperature range and the shape of the part both play a key role in the material's ability to withstand thermal shock. Tests must be carefully designed to mimic anticipated service conditions to accurately assess the thermal shock resistance of a material.

6.4.8 Creep Resistance

Creep is slow, temperature-aided, time-dependent deformation. Creep is typically a factor in materials above one-third of their absolute melting temperature or two-thirds of their glass transition temperature. *Creep resistance* is an important material property in high-temperature design, but it is difficult to quantify with a single value. Creep response is a function of many material and external variables, including stress and temperature. Often, other environmental factors, such as oxidation or corrosion, play a role in the fracture process.

Creep is plotted as strain vs. time. A typical creep curve shows three basic regimes. During stage I, the primary or transient stage, the curve begins at the initial strain, with a relatively high slope or strain rate, which decreases throughout stage I until a steady state is reached. Stage II, the steady state stage, is generally the longest stage and represents most of the response. The strain rate again begins to increase in stage III, and rupture generally follows quickly (see Figure 6.1).

Different applications call for different creep responses. In situations where long life is desired, minimum creep rate is the most important material consideration. Testing through stage II should be sufficient for determining minimum creep rate. It is not necessary to proceed all the way to rupture. For this type of test, the longer the test, the more accurate the creep rate will be. Unfortunately, practicality limits most creep tests to times shorter than would be desirable for high accuracy.

For short-lived applications, such as rocket nozzles, the time to failure may be the only consideration. The main issue is whether or not the component fails, not the amount of deformation it may undergo. For this application, creep tests may be run to completion but without recording any data other than the time to rupture. In this case, temperatures may be elevated above expected conditions, to provide a margin of safety.

The main objective of a creep test is to study the effects of temperature and stress on the minimum creep rate and the time to rupture. Creep testing is usually run by placing a sample under a constant load at a fixed temperature. The data provided from a complete creep test at a specific temperature, T, and stress includes three creep constants: the dimensionless creep exponent, n; the activation energy Q; and A, a kinetic factor.

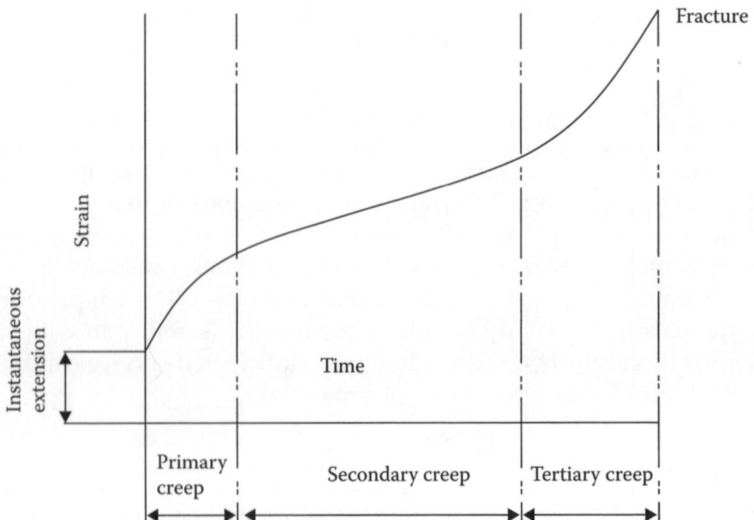

FIGURE 6.1
Creep curve.

6.5 Materials: Ferrous Metals

As the most abundant of all commercial metals, alloys of iron and steel continue to cover a broad range of structural applications. Iron ore is readily available, constituting about 5% of the earth's crust, and is easy to convert to a useful form. Iron is obtained by fusing the ore to drive off oxygen, sulfur, and other impurities. The ore is melted in a furnace in direct contact with the fuel using limestone as a flux. The limestone combines with impurities and forms a slag, which is easily removed.

6.5.1 Cast Iron

Cast iron is defined as an iron alloy with more than 2% carbon as the main alloying element. In addition to carbon, cast irons must also contain from 1 to 3% silicon, which combined with the carbon give them excellent castability. Cast iron has a much lower melting temperature than steel and is more fluid and less reactive with molding materials. However, they do not have enough ductility to be rolled or forged.

The precipitation of carbon (as graphite) during solidification is the key to cast iron's distinctive properties. The graphite provides excellent machinability (even at wear-resisting hardness levels), dampens vibration, and aids lubrication on wearing surfaces (even under borderline lubrication conditions), which makes it an excellent choice for machine tool slideways.

Steels and cast irons are both primarily iron with carbon as the main alloying element. Steels contain <2% and usually <1% carbon, while all cast irons contain more than 2% carbon. About 2% is the maximum carbon content at which iron can solidify as a single-phase alloy, with all of the carbon in solution in austenite. Thus, cast irons by definition solidify as heterogeneous alloys and always have more than one constituent in their microstructure.

In addition to carbon, cast irons also must contain appreciable silicon (Si), usually from 1%–3%, and thus they are actually iron-carbon-silicon alloys. The high carbon content and the Si in cast irons make them excellent casting alloys. Table 6.1 gives Percentage Composition of Typical Unalloyed Cast Irons

6.5.2 Carbon Steel

Carbon steel is a malleable, iron-based metal containing <2% carbon (usually <1%), small amounts of manganese, and other trace elements. Steels can either be cast to shape or wrought into various mill forms from which finished parts are formed, machined, forged,

TABLE 6.1

Percentage Composition of Typical Unalloyed Cast Iron

	Carbon (%)	Silicon (%)	Manganese (%)	Sulfur (%)
Grey	2.5–4.0	1.0–3.0	0.2–1.0	0.02–0.25
Ductile	3.0–4.0	1.8–2.8	0.1–1.0	0.01–0.03
Compacted graphite	2.5–4.0	1.0–3.0	0.2–1.0	0.01–0.03
Malleable (cast white)	2.0–2.9	0.9–1.9	0.15–1.2	0.02–0.20
White	1.8–3.6	0.5–1.9	0.25–0.8	0.06–0.30

stamped, or otherwise shaped. Carbon steels are specified by chemical composition, mechanical properties, method of deoxidation, or thermal treatment.

6.5.3 Alloy Steel

Steels that contain specified amounts of alloying elements—other than carbon and the commonly accepted amounts of manganese, copper, silicon, sulfur, and phosphorus—are known as alloy steels. Alloying elements are added to change mechanical or physical properties. A steel is considered to be an alloy when the maximum of the range given for the content of alloying elements exceeds one or more of these limits: 1.65% Mn, 0.60% Si, or 0.60% Cu; or when a definite range or minimum amount of any of the following elements is specified or required within the limits recognized for constructional alloy steels: aluminum, chromium (to 3.99%), cobalt, columbium, molybdenum, nickel, titanium, tungsten, vanadium, zirconium, or other element added to obtain an alloying effect. Technically, then, tool and stainless steels are alloy steels.

6.5.4 Stainless Steel

Stainless steel is the generic name for a number of different steels used primarily for their resistance to corrosion. The one key element they all share is a certain minimum percentage (by mass) of chromium: 10.5%. Although other elements, particularly nickel and molybdenum, are added to improve corrosion resistance, chromium is always the deciding factor. The vast majority of steel produced in the world is carbon and alloy steel, with the more expensive stainless steels representing a small, but valuable niche market.

Stainless steels are commonly divided into five groups:

1. Martensitic stainless steels
2. Ferritic stainless steels
3. Austenitic stainless steels
4. Duplex (ferritic-austenitic) stainless steels
5. Precipitation-hardening stainless steels

Martensitic stainless steels, typified by types 410/420/440, containing about 12Cr and 0.1C wt% as the basic composition. They are not as corrosion resistant as the other classes, but are extremely strong and tough, as well as highly machineable, and can be hardened by heat treatment. They contain 11.5%–18% chromium and significant amounts of carbon. Some grades include additional alloying elements in small quantities.

Ferritic stainless steels contain larger amounts of Cr, which stabilizes the ferritic phase. Ferritic stainless steels are highly corrosion resistant, but are far less durable than austenitic grades and cannot be hardened by heat treatment. They contain between 10.5% and 27% chromium and very little nickel, if any. Typical applications may include appliances, and automotive and architectural trim (i.e., decorative purposes), as the cheapest stainless steels are found in this family (type 409).

Austenitic stainless steels, such as type 304, typically contain 18Cr and 8Ni wt% (aka 18/8 stainless). Austenitic stainless steels comprise over 70% of total stainless steel production. They contain a maximum of 0.15% carbon, a minimum of 16% chromium, and sufficient nickel and/or manganese to retain an austenitic structure at all temperatures from the cryogenic region to the melting point of the alloy. Other standard grades have different preferred applications; for example, type 316, which contains up to 3 wt% Mo, offers an improved general and pitting corrosion resistance, making it the material of choice for marine applications and coastal environments.

Duplex stainless steels are two-phase alloys based on the Fe–Cr–Ni system. The specific advantages offered by duplex stainless steels over conventional 300 series stainless steels are strength (approximately twice that of austenitic stainless steels), improved toughness and ductility (compared to ferritic grades), and a superior chloride SCC resistance and pitting resistance. The high yield strength offers designers the use of thin-wall material (which can lead to major reductions in weight) with adequate pressure-containing and load-bearing capacity. Duplex stainless steels have found widespread use in a range of industries, particularly the oil and gas, petrochemical, and pulp and paper industries.

Specialist grades include the precipitation hardened or oxide dispersion strengthened alloys.

6.5.5 Tool Steels

The defining properties of *tool steels* include resistance to wear, stability during heat treatment, strength at high temperatures, and toughness. To develop these properties, tool steels are always heat treated. Because the parts may distort during heat treatment, precision parts should be semi-finished, heat treated, then finished. Tool steels are classified into several broad groups, some of which are further divided into subgroups according to alloy composition, hardenability, or mechanical similarities.

- Type W—Water-hardening, or carbon, tool steels rely on carbon content for their useful properties.
- Type S—Shock-resisting tool steels are strong and tough, but are not as wear resistant as many other tool steels.
- Types O, A, and D—Cold-work tool steels including oil- and air-hardened types are often more costly but can be quenched less drastically than water-hardening types. Type O steels are oil hardening; Types A and D steels are air hardening (the least severe quench) and are best suited for applications such as machine ways, brick mold liners, and fuel-injector nozzles. The air-hardening types are specified for thin parts or parts with severe changes in cross section—parts that are prone to crack or distort during hardening. Hardened parts from these steels have a high surface hardness; however, these steels should not be specified for service at elevated temperatures.
- Type H—Hot-work steels serve well at elevated temperatures.
- Types T (tungsten alloy) and M (molybdenum alloy)—High-speed tool steels make good cutting tools because they resist softening and maintain a sharp cutting edge at high service temperatures.
- Type L-A—Special-purpose, low-cost, low-alloy, tool steel often specified for machine parts when wear resistance combined with toughness is important.
- Type F—Carbon-tungsten alloys (Type F) are shallow hardening and wear resistant but not suited for high temperatures or for shock service.
- Type P-A—Mold steels are designed specifically for plastic-molding and zinc die-casting dies.

6.6 Materials: Nonferrous Metals

Non-ferrous metals are metals that do not contain iron. There are two groups of metals; ferrous and nonferrous. *Ferrous metals* contain iron, for example carbon steel, stainless steel (both alloys; mixtures of metals) and wrought iron. *Nonferrous metals* don't contain iron, for example aluminum, brass, copper (which can be remembered as ABC), and titanium. You can also get nonferrous metals as alloys; for example, brass is an alloy of copper and zinc.

Nonferrous metals are specified for structural applications requiring reduced weight, higher strength, nonmagnetic properties, higher melting points, or resistance to chemical and atmospheric corrosion. They are also specified for electrical and electronic applications.

Nonferrous metals are discussed in the following sections.

6.6.1 Aluminum

Pure aluminum is a silvery-white metal with many desirable characteristics. It is light, nontoxic (as the metal), nonmagnetic, and nonsparking. It is easily formed, machined, and cast. Pure aluminum is soft and lacks strength, but alloys with small amounts of copper, magnesium, silicon, manganese, and other elements have very useful properties. Aluminum is an abundant element in the earth's crust, but it is not found free in nature. The Bayer process is used to refine aluminum from bauxite, an aluminum ore. Because of aluminum's mechanical and physical properties, it is an extremely convenient and widely used metal.

6.6.1.1 Some Common Uses

Building and construction industry

- Door and window frames
- Wall cladding, roofing, awnings

Manufacture of electrical products

- High-tension power lines, wires, cables, busbars
- Components for television, radios, refrigerators, and air-conditioners

Packaging and containers

- Beverage cans, bottle tops
- Foil wrap, foil semi-rigid containers

Cooking utensils

- Kettles and saucepans

Aeronautical, aviation, and automotive industries

- Propellers
- Airplane and vehicle body sheet
- Gearboxes, motor parts

Leisure goods

- Tennis racquets, softball bats
- Indoor and outdoor furniture

6.6.1.2 Properties

- Very lightweight (about 1/3 the mass of an equivalent volume of steel or copper) but with alloying can become very strong
- Excellent thermal conductor
- Excellent electrical conductor (on a weight-for-mass basis, aluminum will conduct more than twice as much electricity as copper)
- Highly reflective to radiant energy in the electromagnetic spectrum
- Highly corrosion resistant in air and water (including sea water)
- Highly workable and can be formed into almost any structural shape
- Nonmagnetic
- Nontoxic

6.6.2 Copper

Copper provides a diverse range of properties: good thermal and electrical conductivity, corrosion resistance, ease of forming, ease of joining, and color. However, copper and its alloys have relatively low strength-to-weight ratios, and low strengths at elevated temperatures. Some copper alloys are also susceptible to stress-corrosion cracking unless they are stress relieved. Next to silver, copper is the next best electrical conductor. It is a yellowish red metal that polishes to a bright metallic luster. It is tough, ductile, and malleable. Copper has a disagreeable taste and a peculiar smell. Copper is resistant to corrosion in most atmospheres, including marine and industrial environments. It is corroded by oxidizing acids, halogens, sulfides, and ammonia-based solutions.

Copper and its alloys—the brasses and bronzes—are available in rod, plate, strip, sheet, tube shapes, forgings, wire, and castings.

6.6.3 Nickel

Nickel suits many applications that require specific corrosion resistance or elevated temperature strength. Some nickel alloys are among the toughest structural materials known. When compared to steel, other nickel alloys have ultrahigh strength, high proportional limits, and high moduli of elasticity. Commercially pure nickel has good electrical, magnetic, and magnetostrictive properties.

6.6.4 Titanium

There are three structural types of titanium alloys:

- *Alpha Alloys* are non-heat treatable and are generally very weldable. They have low to medium strength, good notch toughness, and reasonably good ductility, and possess excellent mechanical properties at cryogenic temperatures. The more highly alloyed alpha and near-alpha alloys offer optimum high temperature creep strength and oxidation resistance, as well.

- *Alpha-Beta Alloys* are heat treatable and most are weldable. Their strength levels are medium to high. Their hot-forming qualities are good, but the high temperature creep strength is not as good as in most alpha alloys.
- *Beta or near-beta Alloys* are readily heat treatable, generally weldable, capable of high strengths and good creep resistance to intermediate temperatures. Excellent formability can be expected of the beta alloys in the solution treated condition. Beta-type alloys have good combinations of properties in sheet, heavy sections, fasteners, and spring applications.

6.6.5 Zinc

Zinc is a silvery blue-grey metal with a relatively low melting point (419.5°C) and boiling point (907°C). When unalloyed, its strength and hardness is greater than that of tin or lead, but is appreciably less than that of aluminum or copper. The pure metal cannot be used in stressed applications due to low creep resistance. For these reasons, most uses of zinc are after alloying with small amounts of other metals or as a protective coating for steel.

6.6.5.1 Uses

One of the most useful characteristics of zinc is its resistance to atmospheric corrosion, and just over half of its use is for the protection of steelwork. In addition to its metal and alloy

TABLE 6.2

Mechanical and Physical Properties—Zinc

Tensile strength (cast): 28 MN/m² (4,000 psi)
(rolled—with grain)
(99.95% zinc soft temper): 126 MN/m² (18,000 psi)
(98.0% zinc hard temper): 246 MN/m² (35,000 psi)
Elongation:
(rolled—with grain)
(99.95% zinc soft temper): 65%
(98.0% zinc hard temper): 5%
Modulus of elasticity: 7×104 MN/m² (1×107 psi)
Brinell hardness, 500 kg load:
for 30 s: 30
Impact resistance:
(pressed zinc, elongation=30%): 6.5–9 J/cm² (26–35 ft-1bs/in.²)
Surface tension—liquid (450°C): 0.755 N/m
Surface tension—liquid (419.5°C): 0.782 N/m
Viscosity-liquid (419.5°C): 0.00385 N/m
Velocity of sound (20°C): 3.67 km/s
Coefficient of friction:
(rolled zinc versus rolled zinc): 0.21
Hardness: 2.5 Mohs

forms, zinc also extends the life of other materials, such as steel (by hot dipping or electrogalvanizing), rubber and plastics (as an aging inhibitor), and wood (in paints). Zinc is also used to make brass, bronze, and die-casting alloys in plate, strip, and coil; foundry alloys; superplastic zinc; and activators and stabilizers for plastics. Table 6.2 lists some important properties for zinc.

Section II

Mechanics

7

Statics

Statics is the study of rigid bodies that are in equilibrium and concerned with the analysis of loads such as a force or moment/torque acting on a physical body or system, that is, it is in a state where the relative positions of sub-structures do not vary over time or where the structures or sub-structures are moving at a constant velocity without any variation in that velocity.

In statics, it is assumed that the bodies are perfectly rigid without any deformation. In truth, this is not correct. Any physical object will deform or deflect a small amount depending on the level of physical force or load being applied.

These deformations are considered to be insignificant and will not affect the conditions of equilibrium or motion and therefore can be disregarded.

A body can be considered rigid when the relative movement between its constituent parts is negligible.

The basic concepts of statics are

Space: The geometric region occupied by the body where its position can be described by linear and angular measurements relative to a prescribed co-ordinate system.

Time: This is the measure of the succession of events relative to a datum event measured in seconds, minutes or days.

Mass: The measure of the inertia of a body, which is resistant to a change in motion. This is measured in kilogrammes in the SI system of measurement.

Force: The action of an external force acting on a body such as gravitational acceleration or the impact of another body resulting from kinetic energy transfer.

The sections of this chapter will consider:

1. Force, mass and moments
2. Structures and frameworks
3. Vectors and vector analysis

7.1 Force, Mass and Moments

Newton developed the fundamentals of mechanics in which the concepts of space, time and mass are absolute and independent of each other.

> *Newton's first law*: A particle remains at rest or continues to move in a straight line with a constant speed if there is no unbalanced force acting on it (resultant force=0).

This implies that the net force or net torque on every part of the structure is zero. Therefore, quantities such as stress or pressure can be derived. The net force equalling zero is known as the first condition for equilibrium and the net torque equalling zero is known as the second condition for equilibrium.

For completeness, the two remaining laws are as follows:

> *Newton's second law:* The acceleration of a particle is proportional to the resulting force acting on it and is in the direction of this force.

that is

$$\text{Force} = \text{mass} \times \text{acceleration}$$
$$F = ma \tag{7.1}$$

> *Newton's third law:* The forces of action and reaction between interacting bodies are equal in magnitude and opposite in direction and act along the same line of action.

7.1.1 System of Units

The units of measurement used in statics are as follows:

Measurement system	Force	Length	Mass	Time
SI units	Newton (N)	Metre (m)	Kilogram (kg)	Second (s)

$$1\,\text{N} = (1\,\text{kg})(1\,\text{m/s}^2)$$

Hence, 1 N is the force required to give a mass of 1 kg an acceleration of 1 m/s^2.

Weight is a force.

Consider a weight of 1 kg mass:

W = mg (where g is the gravitational constant).
W = (1 kg)(9.81 m/s^2).
W = 9.81 N.

7.1.2 Free-Body Diagrams

A free-body diagram is a very useful method of depicting the interaction of all forces and moments acting on a body at a given situation. It is essentially a sketch of a body that is in equilibrium and entirely separated from its surroundings. Figure 7.1 shows some examples of free-body diagrams.

7.1.3 Forces and Moments

7.1.3.1 Force

A force results from one body acting on another and tends to move it in the direction of its action. A force can be characterised by a localised vector defined by its magnitude, direction and point of application. A single resultant force using the principles of vector addition as shown in Section 7.3.1 can replace a number of forces acting at any point.

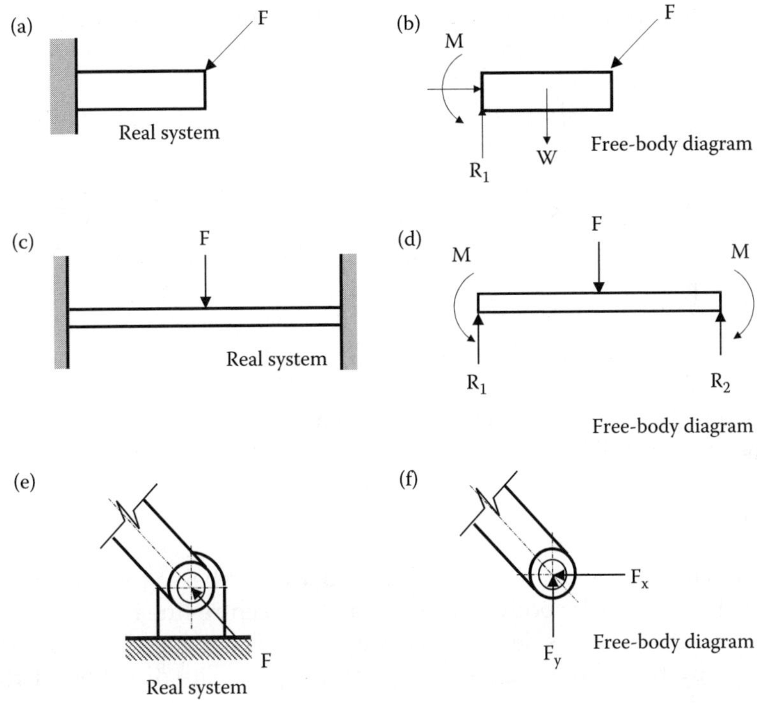

FIGURE 7.1
Examples of free-body diagrams: A cantilever beam (a and b); a beam in encastrè (c and d); frictionless pivot (e and f).

Considering Newton's first law (above), it is clear that a body will remain at rest if the resultant force acting on the body is zero. This is referred to as the equilibrium condition and using the Cartesian co-ordinate system can be stated as follows:

$$R = \sum F = 0 \tag{7.2}$$

or

$$\left(\sum F_x\right) \cdot i + \left(\sum F_y\right) \cdot j + \left(\sum F_z\right) \cdot k = 0 \tag{7.3}$$

This can be simplified by only considering the scalar identities….

$$\sum F_x = 0 \quad \sum F_y = 0 \quad \sum F_z = 0 \tag{7.4}$$

As there are three equations of equilibrium for the three-dimensional case, there can only be three unknowns that can be determined from these equations.

7.1.3.2 Moments

In addition, for the force having the tendency to move a body in the direction of its application, a force can also cause the body to rotate about an axis. The axis of the force may be any line which neither intersects nor is parallel to the line of action of the force. This rotational tendency is referred to as the moment of the force (M) and is the product of the perpendicular distance (d) of the line of action from the point of application of the force (F) multiplied by the magnitude of the force.

The moment of a force is defined in a vector form using determinant algebra (in two-and three-dimensional cases) as

$$M = Fd = \begin{vmatrix} i & j & k \\ d_x & d_y & d_z \\ F_x & F_y & F_z \end{vmatrix} = (d_yF_z - d_zF_y)\cdot i + (d_zF_x - d_xF_z)\cdot j + (d_xF_y - d_yF_x)\cdot k \quad (7.5)$$

$$M_o = d(F_1 + F_2 + F_3 \cdots) = d\cdot F_1 + d\cdot F_2 + D\cdot F_3 \cdots \quad (7.6)$$

Where a number of concurrent forces are applied to a point 'P', the combined moment caused by these forces will be equal to the sum of the moments of these individual forces.

7.1.3.3 Couples

When two forces have an equal magnitude 'F' and are parallel to the line of action but act on either side of the body, the body will rotate about a centre line in either a clockwise or anticlockwise direction. This is defined as a couple and is the magnitude of the individual force multiplied by the distance between the two forces. The direction of the couple is identified by the right-hand rule:

$$\text{Couple} = e\cdot F \quad (7.7)$$

It can also be proved that the moment of the couple is the same magnitude at any location.

7.1.3.4 Rigid-Body Equilibrium

The static equilibrium of a particle is an important concept in the study of statics. A particle is in equilibrium only if all the resultant forces acting upon it are equal to zero and the resultant moments are also equal to zero.

$$\sum F = 0 \quad \sum M = 0 \quad (7.8)$$

Therefore,

$$\sum F_x = 0 \quad \sum F_y = 0 \quad \sum F_z = 0 \quad \sum M_x = 0 \quad \sum M_y = 0 \quad \sum M_z = 0 \quad (7.9)$$

There are six independent equations of equilibrium. The six equations are derived from the free-body diagrams that show all of the applied forces and moments together with all the resulting reaction forces and moments. From these six equations, it is possible to solve for six unknowns. If there are more than six unknowns, then the system is statically indeterminate.

In the special case of a two-dimensional equilibrium, which is often applied to beams and simple structures, the equations for plane static equilibrium are shown below

$$\sum F_x = 0 \quad \sum F_y = 0 \quad \sum M_z = 0 \quad (7.10)$$

7.2 Structures

A structure refers to an assembly of materials whose function is to support a load or loads, be they concrete, brick, metal linkages and so on. The term structure can be applied to an aircraft wing, bridge, building or any form where a load or force is being resisted. The component parts of a loaded structure are in a state of stress that is tensile, compressive or shear. The distribution of these loads or forces is analysed to determine which materials are most appropriate to safely support the load or forces.

Structures are classified into two basic groups:

1. Framed structures
2. Mass structures

The former is manufactured from either separate linkages or plates welded or riveted together to form a lattice. The latter will depend upon the mass of material to provide a resistance to the load, such as a masonry dam to withstand the water pressure in a reservoir.

For the purposes of this discussion, only the former (framed structures) will be considered as the latter is more in the civil engineering domain.

In the notes that follow, the framework or trusses are considered to be manufactured from linkages connected to each other at pin joints that do not transmit moment forces. The connecting linkages can only transmit either tensile forced (ties) or compressive forces (struts). For the purposes of this discussion, it is assumed that the ties and struts do not experience any axial and transverse deflections. Figure 7.2 shows examples of pin-jointed truss structures.

Structures can be classified as either 2-dimensional (plane frames) or 3-dimensional (space frames).

For a 2-dimensional frame, the number of linkages (N) required with J joints will be

$$N = 2 \cdot J - 3 \tag{7.11}$$

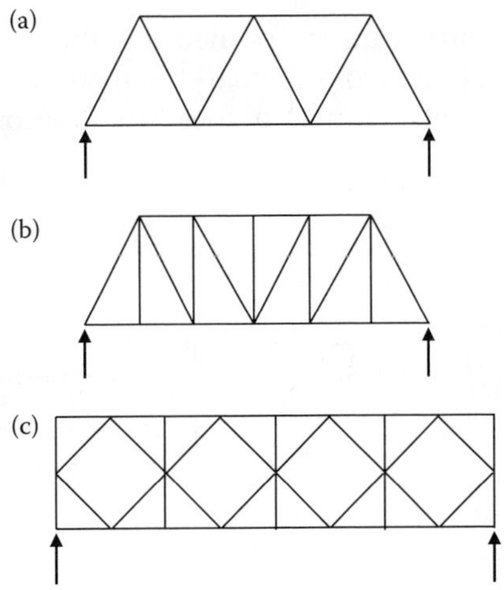

FIGURE 7.2
Types of truss structures. Warren truss (a); Pratt truss (b); Bailey truss (c).

And for a 3-dimensional frame, the number of linkages (N) required with J joints will be

$$N = 3 \cdot J - 6 \tag{7.12}$$

Note: It will be clear to the reader that in practice most frame joints are not pinned. The principles described here can be used with reasonable accuracy for a majority of frames where the ratio of the length of the linkage to its depth (L/d) is in excess of 10. In the case of frames that have relatively rigid joints and short linkages, the analysis is statically indeterminate and therefore outside the scope of this discussion.

7.2.1 Pin Joint

A pin joint allows the joined members to swivel as opposed to a rigid joint which does not allow any rotation. A rigid joint may be welded, whereas a pinned joint may be a bolt, rivet or any other form of swivel pin.

Figure 7.3 shows two methods of providing a pin joint.

Two important points about a pin joint are

1. The connected members are free to rotate.
2. The force in the member can only carry axial forces or loads.

7.2.1.1 Struts and Ties

Consider a member depicted in Figure 7.4a,b with a pin joint at each end. A pin joint cannot transmit rotation from one member to another as each can only push or pull on the joint along the direction of its length. A member subject to a tension load is known as a 'TIE' and is shown with arrows pointing inwards at each end as in Figure 7.4a. A member in compression is called a 'STRUT' and is shown with arrows pointing outwards at each end as shown in Figure 7.4b.

7.2.1.2 Bow's Notation

When several members of a structure are pinned together and the joints are in equilibrium, the resultant force must be zero. When all the forces are added up as vectors, they must form a closed polygon. If one or more of these are unknown, then it must be the vector that closes the polygon.

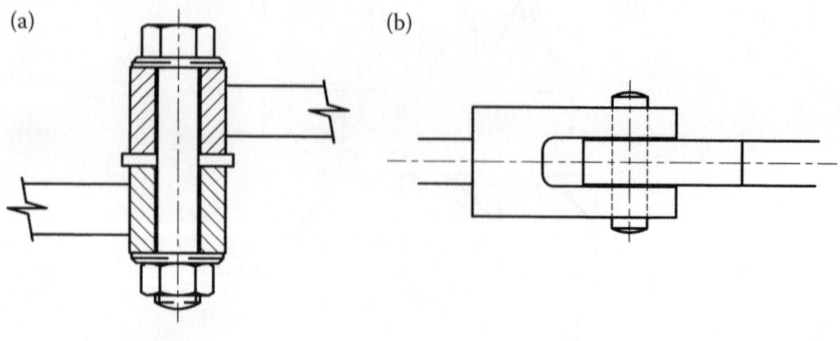

FIGURE 7.3
(a) and (b): Types of pin joints.

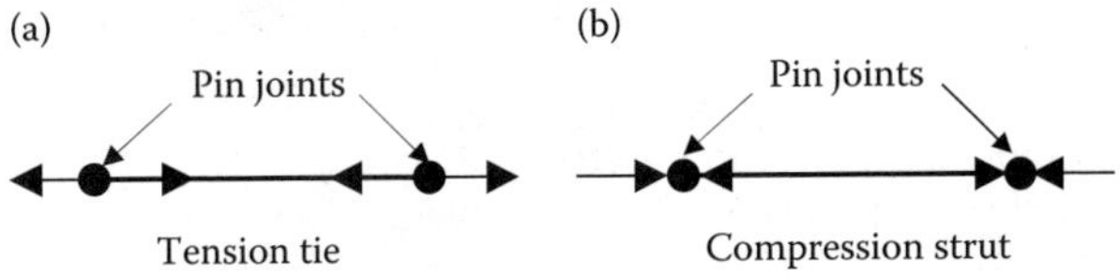

FIGURE 7.4
Ties (a) and struts (b).

Consider three members joined by a pin as shown in Figure 7.5. Only one of these forces is known. A method known as 'Bow's notation' is used to help identify and label each member enabling a polygon (or in this case) a triangle of forces to be drawn. The process is as follows:

1. Label the spaces between each member as shown in Figure 7.5. This diagram is also known as a 'space diagram'.
2. Starting at any space, say 'A', identify each member individually by moving clockwise around the joint, so that the first member is identified as a–b, the next member as b–c and the last as c–a (in this case only).
3. Draw the known vector a–b in the same direction as the space diagram. The next vector b–c starts at 'b', but at the moment the length is not known; therefore, draw a line representing b–c. When all the vectors are added together, they must form a closed triangle; hence, c–a will end at 'a'. Complete the line c–a and where the two 'c' lines cross, this will be the point 'c' as shown in Figure 7.6.
4. Finally, transferring the arrows back to the space diagram in the same direction as shown on the triangle of forces (Figure 7.7) where they push onto the pin joint, the member will be in compression and therefore will be a strut. If the arrow pulls away from the pin joint, the member will be in tension and is therefore a tie.

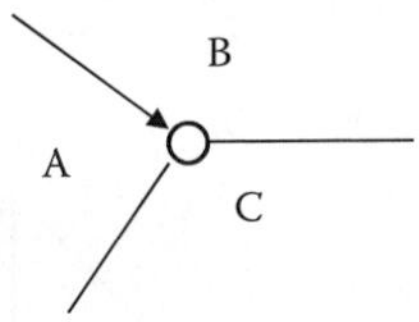

FIGURE 7.5
Bow's notation.

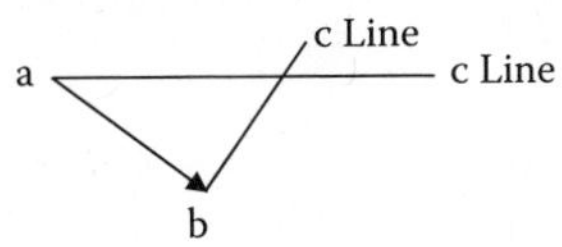

FIGURE 7.6
Vector diagram.

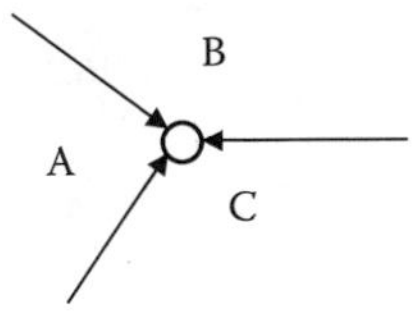

FIGURE 7.7
Triangle of forces.

Example 7.1

A column (strut) is held vertically by two guy ropes as shown in Figure 7.8a. The maximum allowable compressive force in the strut is 25 kN. Calculate the forces in each rope bearing in mind that the ropes can only be in tension.

Solution

Following the procedure as outlined in Section 7.2.1.2, first draw the space diagram (Figure 7.8b) and then complete the triangle of forces (Figure 7.8c).

The forces in each of the ropes will be b–c and c–a. These forces can be found by either drawing to scale or by calculation.

First, ensure that the internal angles of the triangle of forces add up to 180°. In this instance, the angle between vectors ac and bc will be

$$180° - (30° + 45°) = 105°$$

$$\frac{bc}{\sin 45°} = \frac{25}{\sin 105°}$$

$$bc = 18.30 \text{ kN(Tension)}$$

and

$$\frac{ca}{\sin 30°} = \frac{25}{\sin 105°}$$

$$ca = 12.94 \text{ kN(Tension).}$$

Figure 7.9 summarises these forces.

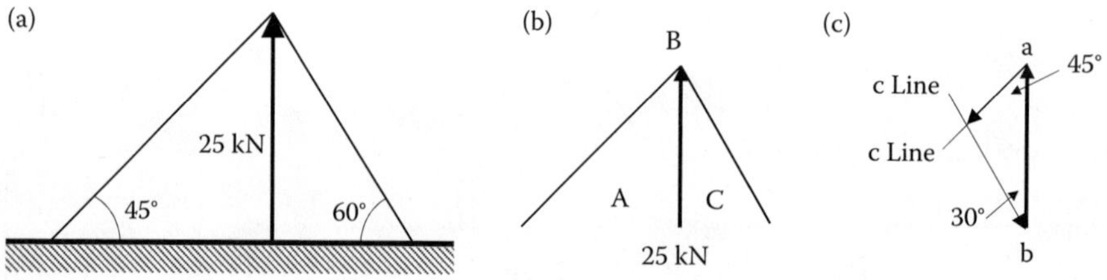

FIGURE 7.8
(a) Example 7.1, (b) Bow's notation, and (c) vector diagram for Example 7.1.

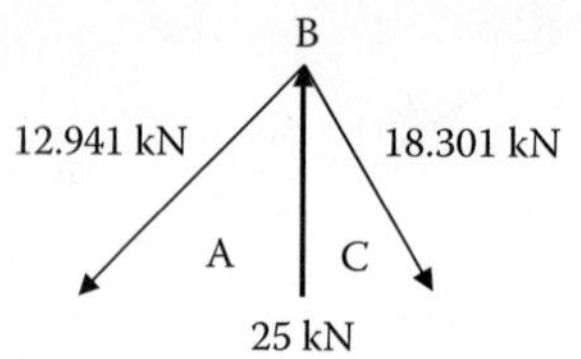

FIGURE 7.9
Summary of forces for Example 7.1.

7.2.2 Solving Forces in Pin-Jointed Frames

There are many examples of lattice work frames that are made up of a number of pin-jointed members such as

- Roof trusses
- Crane jibs
- Bridges
- Electrical pylons

Many of these structures use riveted or bolted joints and are not entirely free to rotate at the joint, but the theory for pin-jointed frames appears to work quite well. Bow's notation is applied to each joint in turn and solving the forces in each member. Transferring the force directions back from the polygon to the framework diagram, it can then be deduced which members are struts or ties. The force direction at the other end of the member can then be determined, and this is what is needed to help solve the forces in the other pin joints.

Example 7.2

A pin jointed framework has a load of 250 kN applied to the joint as shown in Figure 7.10a. Solve the forces and reactions for the framework shown.

Solution

1. Draw the space diagram and then label the spaces as shown in Figure 7.10a using Bow's notation.
2. Solve the known joint using either drawing or trigonometry (Figure 7.10b.)

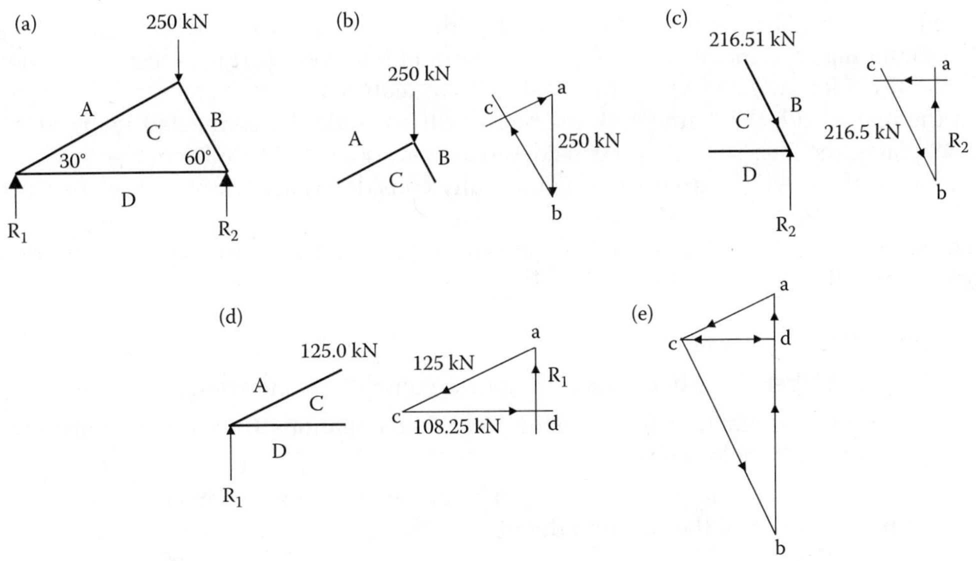

FIGURE 7.10
(a) Example 7.2, (b) joint ABC, (c) joint CBD, (d) joint ACD and (e) combined diagram.

$$ac = 250\text{ kN} \times \sin 30° = 125.0\text{ kN (strut)}$$

$$bc = 250\text{ kN} \times \cos 30° = 216.5\text{ kN (strut)}$$

3. Solve for the other joint (Figure 7.10c)

$$bd = 216.5 \times \cos 30° = 187.5\text{ kN } (R_2)$$

$$cd = 216.5 \times \sin 30° = 108.25\text{ kN}$$

4.

$$ad = 125\text{ kN} \times \sin 30° = 62.50\text{ kN } (R_1)$$

$$cd = 125\text{ kN} \times \cos 30° = 108.25\text{ kN}$$

The reaction R_2 is easily deduced as the total upward force is 250 kN. Hence, $R_2 = (250\text{ kN} - 187.5\text{ kN}) = 62.50$ kN. The solution for the remaining joint is not needed except for completeness (see Figure 7.10d).

When the reader becomes proficient with this work, it will be found to be more convenient to complete this work in one diagram (see Figure 4.10e).

There are three methods for analysing frameworks:

1. Method of joints
2. Graphical methods
3. Method of sections

7.2.3 Method of Joints

The method of joints is a technique for finding the internal forces within a framework and it works on the assumption that the framework members are pin connected, making them two force members. Equations of static equilibrium can then be written for each pin joint and the set of equations can be solved simultaneously to determine the forces acting in the framework members.

The biggest problem with the 'method of joints' is the amount of work that has to go into calculating each member force. It may be difficult to solve just for these forces alone; instead, the solution may require more forces than required.

In the real world, the framework members will normally be connected using welded gusset plates; the idealised joint is considered as that connected by a frictionless pin.

Although there will be an error, it is normally considered acceptable when the framework members are long and slender, that is, $L/d > 10$.

The steps in applying the method of joints as applied to the analysis of a plane framework are as follows:

1. Label all the pin joints (A, B, C, … etc.).
2. Draw and label a free-body diagram for the complete framework.
3. Determine the reaction forces using the three equations for static equilibrium applied to the whole framework.
4. Draw a free-body diagram for each pin assembly. This may be one free-body diagram that shows all the pin joint details.
5. If possible, begin solving the equilibrium equations at a joint where only two unknown reactions exist. Work from joint to joint using the criterion of two unknown reactions.

6. Apply the two equations of static equilibrium relating to F_x and F_y, (moments are zero) at each joint to identify the forces in the attached members.

Example 7.3

A truss as shown in Figure 7.11 has a force acting at the apex of the truss. Draw the free-body diagram and determine the forces acting within the individual links.

Solution

See Table 7.1 for the link calculations.

7.2.4 Graphical Methods as Applied to a 2-Dimensional Framework

The graphical method for the analysis of frameworks is a convenient method for simple frames and trusses, but as the framework becomes more complex, the method then becomes more complicated and cannot be used for 3-dimensional space frames.

Consider the frame in Figure 7.12.

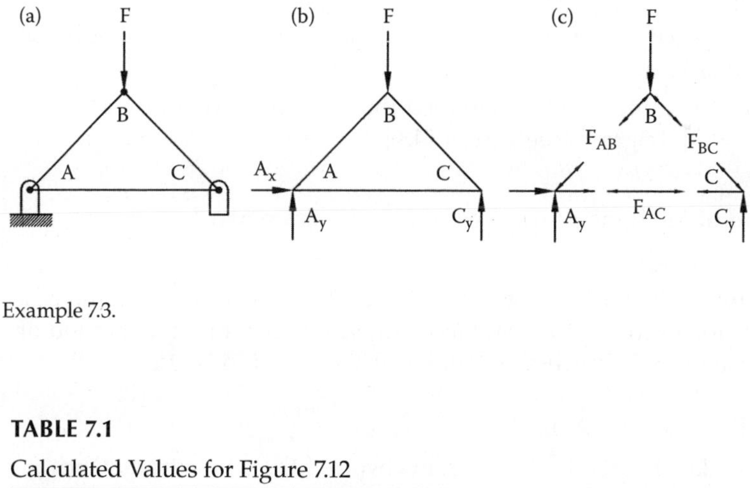

FIGURE 7.11
(a), (b), and (c): Example 7.3.

TABLE 7.1

Calculated Values for Figure 7.12

Figure 7.11a	Figure 7.11b	Figure 7.11c
Angle A & C=45°	$\Sigma M_z=0$	*Joint A*
AC=1000 mm	*Moments about A*	$F_{AB}=707/1000=A_y$
AB=1000_x sin 45°	C_{yx} 1000=1000 N×500	$F_{AB}=A_{yx}$ 1000/707
=707 mm	C_y=500 N	=707 N
F=1000 N	$\Sigma F_x=0$	$F_{ac}=F_{ab}\times 707/1000$
	$A_x=0$	=707 N
	$\Sigma F_y=0$	*Joint B*
	A_y=1000 N−C_y=500 N	$F_{BC\,x}$ 707/1000
		=1000 N−F_{AB}
		707/1000 N
		F_{BC}=707 N
		Joint C
		F_{AC}=707 N
		F_{BC}=707 N

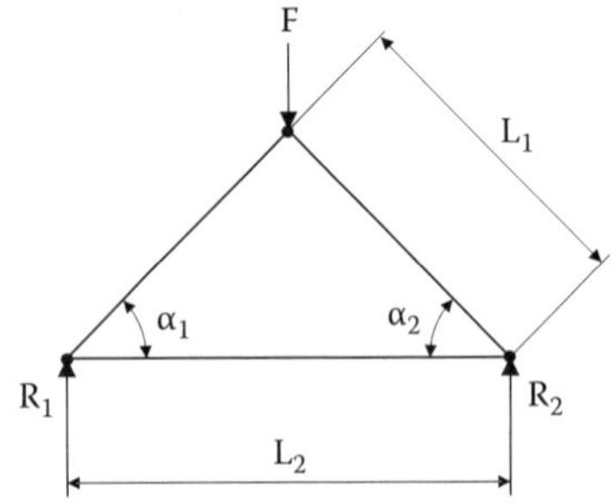

FIGURE 7.12
Basic truss frame.

As with the method of joints procedure, the initial steps is to

- Label all the pin joints.
- Draw and label a free-body diagram for the whole framework.
- Establish the reaction forces.
- Draw a free-body diagram for each individual joint; this may be combined onto a single free-body diagram.
- Produce a force vector polygon for each joint starting with one having no more than two unknowns.
- The force diagrams can be combined into one single diagram [this is also known as a Maxwell diagram (see Figure 7.13)].

7.2.5 Method of Sections as Applied to a Plane Framework

The graphical method demonstrated in Section 7.2.4 can be expanded to cover more complicated figures such as that shown in Figure 7.14. It also can, in some cases, be used to determine the forces in selected bars more rapidly than using the method of joints.

The steps used in this process are similar to that used in Section 7.2.4:

1. Label all pin joints—A, B, C, etc.
2. Draw and label a free-body diagram covering the whole framework.
3. Determine the reaction forces using the three equations for static equilibrium applied to the whole framework.
4. Draw a free-body diagram for a selected part of the frame, which can include two joints breaking the bars under consideration.
5. Calculate the forces in the bars to one side of the 'cut', again using the three equations for static equilibrium.

Figure 7.14 shows an example of a Warren-type girder carrying a variety of loads. It is required for the purposes of this exercise to determine the forces in members AB and JK.

Start at a point on the frame where the forces are known (see Figure 7.15).

A cutting line is drawn through the members under consideration and a free-body diagram is constructed for the part of the framework to the left of the cutting line (see Figure 7.15).

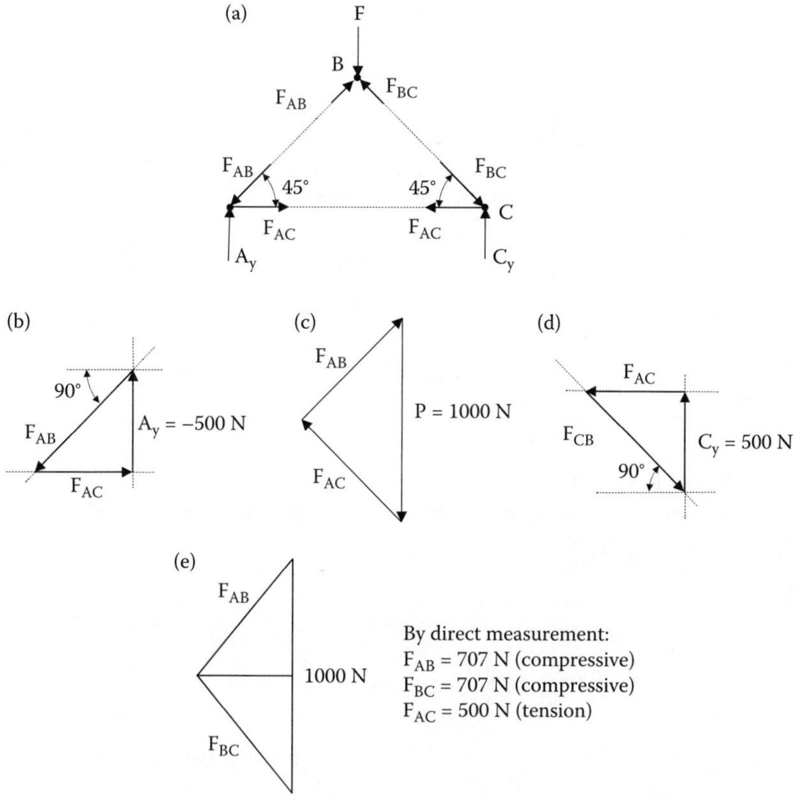

FIGURE 7.13
(a)–(e): Graphical method of joint analysis.

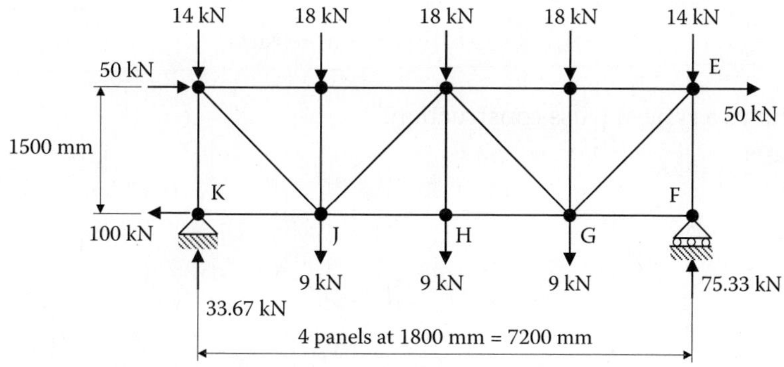

FIGURE 7.14
Warren girder.

The unknown forces are shown in tension.
Taking moments about joint A will eliminate forces F_{AJ} and F_{AB} leaving force $F_{KJ.}$
Sum the moments about A to zero:

$$0 = 100(1.5) - F_{KJ}(1.5) \tag{7.13}$$

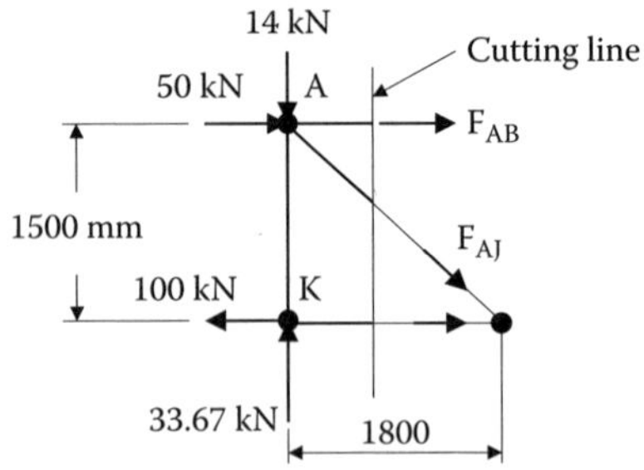

FIGURE 7.15
Free-body diagram for Figure 7.14.

therefore, $F_{KJ} = 100$ kN (tension).

Now taking moments about joint J which is outside the free-body diagram, this will eliminate F_{AJ} and F_{KJ} leaving only F_{AB}.

Summing moments about J to zero:

$$0 = 50(1.5) - 14(1.8) + 33.67(1.8) + F_{AB}(1.5)$$

$$= -73.60 \text{ kN (tension)}$$

Summing moments vertically to zero, this will eliminate both F_{KJ} and force F_{AB} leaving F_{AJ}.

$$0 = 33.67 - 14 - F_{AJ}(5/7.81)$$

$$F_{AJ} = +30.72 \text{ kN (tension)}$$

$$F_{KA} = -33.67 \text{ kN (compression)}$$

Figure 7.16 shows a typical truss construction.

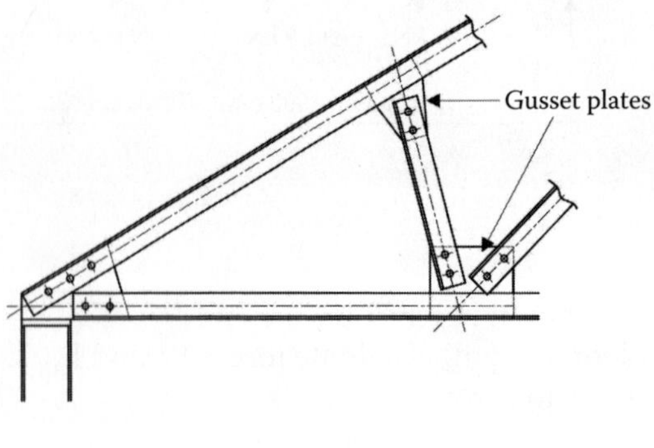

FIGURE 7.16
Typical truss construction.

7.3 Vectors and Vector Analysis

Vector algebra is particularly important when dealing with problems involving forces, displacements, velocities, accelerations, moments and so on in either two or three dimensions. These quantities have magnitude and direction that define them.

Quantities such as mass, volume, power and so on, however, have magnitude, direction is not involved and these are referred to as scalars.

Vectors do not obey the normal rules of addition and subtraction; they are added or subtracted using the Parallelogram rule.

7.3.1 Vector Addition

Consider two vectors $\vec{R}$ and $\vec{Q}$ acting on a single point: From Figure 7.17a through c, it can be seen that these can be replaced by a single vector $\vec{R}$. By drawing lines parallel to the vectors to complete the parallelogram, the resultant 'R' is found.

In Figure 7.17c, the vector Q has been moved parallel to itself and its tail has been placed to the head of P, and the resultant 'R' is as shown. This is known as 'triangular construction'.

The triangular rule may be applied when it is more convenient to use the parallelogram rule.

Analytically, the magnitude R of the vector R is given by

$$R = \sqrt{P^2 + Q^2 - 2PQ\cos(180 - q)} \text{ (cosine rule)} \tag{7.14}$$

7.3.2 Vector Subtraction

Vectors may be subtracted, that is $\vec{P} - \vec{Q}$, by applying the rule thus: Draw $\vec{P}$ and $-\vec{Q}$ and add either the parallelogram or triangular rules to obtain $\vec{R}$ as shown in Figure 7.18b.

7.3.3 Resolving a Vector into Components

Consider a vector $\vec{P}$ as in Figure 7.19 and determine its components.

In this example, there is a requirement to convert the vector P into its components. P_x and P_y are the rectangular components along the Cartesian axes o_x and o_y, and their magnitudes will be

$$P_x = P\cos\theta \text{ and } P_y = P\sin\theta \tag{7.15}$$

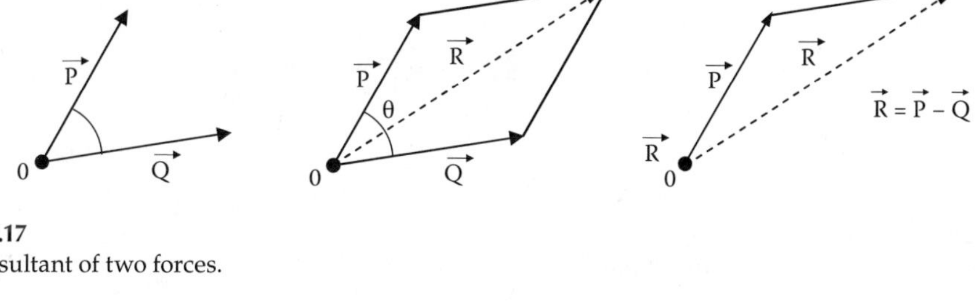

FIGURE 7.17
(a)–(c): Resultant of two forces.

$$P \cdot Q = PQ \cos \phi$$

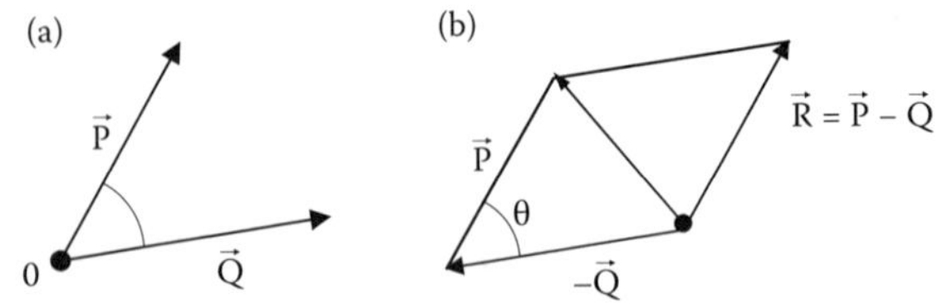

FIGURE 7.18
(a) and (b): Subtracting vectors.

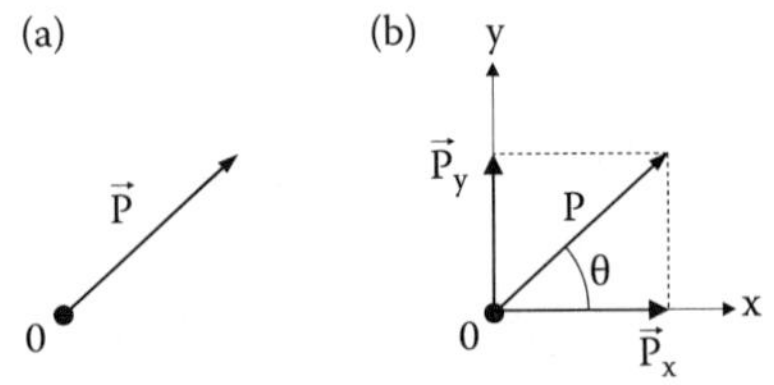

FIGURE 7.19
(a) and (b): Resolving co-ordinates.

It also follows that:

$$P = \sqrt{P_x^2 + P_y^2} \text{ in magnitude} \tag{7.16}$$

Next, consider the case where the co-ordinates are not rectangular:

Figure 7.20 shows a space diagram of a force vector and it is required to determine the respective forces given the directions.

7.3.4 Analytical Determination of the Components of the Vector

Figure 7.21 shows a space diagram giving the force and component directions for the following example:

Let

$$\gamma = 180 - (\alpha + \beta) \tag{7.17}$$

From the sine rule:

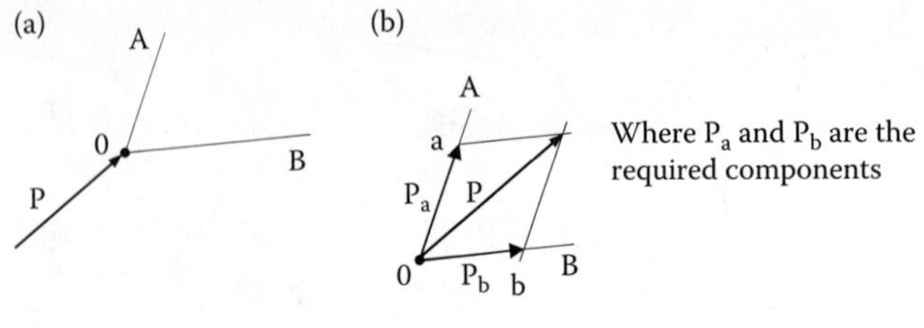

FIGURE 7.20
Space diagram for given force and direction.

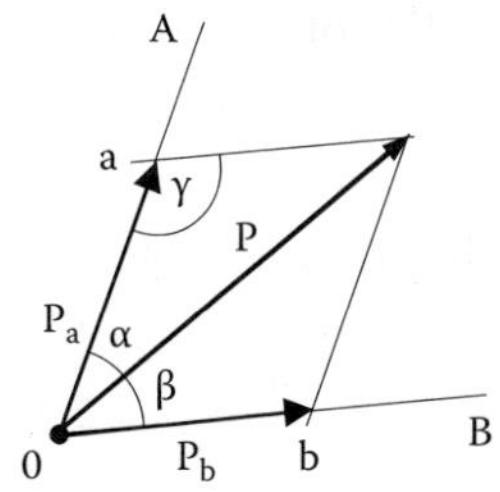

FIGURE 7.21
Space diagram for given force and component directions.

$$\frac{P}{\sin\gamma} = \frac{P_a}{\sin\beta} = \frac{P_b}{\sin\alpha} \tag{7.18}$$

From which:

$$P_a = \frac{\sin\beta}{\sin(\alpha+\beta)} P \tag{7.19}$$

$$P_b = \frac{\sin\alpha}{\sin(\alpha+\beta)} P \tag{7.20}$$

If the co-ordinates are rectangular as in Figure 7.19, then

$$P_a = P\sin\beta(\text{vertical}) \text{ and } P_b = P\cos\beta(\text{horizontal}) \tag{7.21}$$

Note: In the majority of cases the directions 0A and 0B are perpendicular with each other in which case the angles α and β will be required.

7.3.5 Resultant of a Number of Coplanar Vectors (More than Two Vectors)

Note: For vectors to be concurrent their lines of action must meet at a point.

The procedure for constructing a vector polygon is as follows (Figure 7.22):

1. Establish an origin '0' and from this draw P_1 parallel to the given vector 'P_1' and drawing the magnitude to scale.

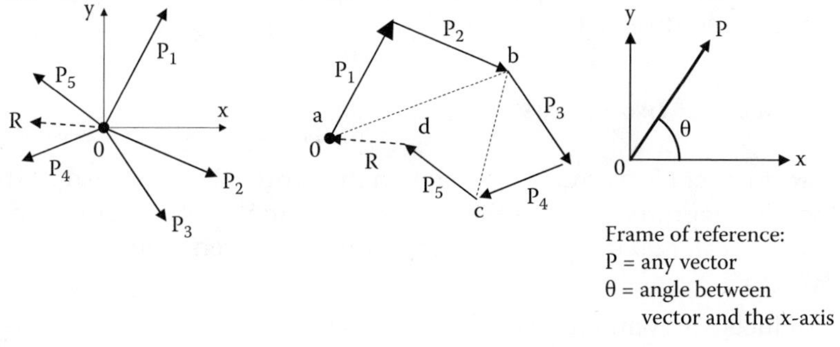

FIGURE 7.22
Construction of vector polygon. Given vectors (a); vector polygon (b).

2. From the head of 'P_1', place the tail of 'P_2' and draw the vector 'P_2' parallel to the given vector 'P_2' and in this instance applying a triangular construction.
3. Repeat this procedure for the remaining vectors.
4. The *resultant* (R) will be given by the vector ad.

7.3.6 Analytical Solution to Figure 7.22

Resultant 'R'

1. Magnitude of vector:

$$R = \sqrt{\left(\sum P\sin\theta\right)^2 + \left(\sum P\cos\theta\right)^2} \tag{7.22}$$

Where

$$\sum P\sin\theta = P_1\sin\theta_1 + P_2\sin\theta_2 + \cdots + P_5\sin\theta_5$$

$$\sum P\cos\theta = P_1\cos\theta_1 + P_2\cos\theta_2 + \cdots + P_5\cos\theta_5$$

2. Direction relative to the 'x'-axis
   Let

$$R_x = \sum P\cos\alpha \text{ and } R_y = \sum P\sin\alpha$$
$$\alpha = \arctan\left(\frac{R_y}{R_x}\right) = \arctan\left(\frac{\Sigma P\sin\alpha}{\Sigma P\cos\alpha}\right) \tag{7.23}$$

7.3.7 Product of Vectors

7.3.7.1 *Multiplication of a Vector 'P' by a Scalar 'K'*

The product of a vector with a magnitude of 'P' and direction 'θ' and a scalar is a vector increased in magnitude. Thus, a vector P multiplied by a scalar K will give a vector with a magnitude of KP and the direction remains unaltered at 'θ'. If K is –ve, the sense of the vector is in the reversed direction.

7.3.7.2 *Scalar Product of Two Vectors*

1. The scalar product of two vectors, as shown in Figure 7.23, is defined as the product of the two magnitudes times the cosine of the angle between the two vectors. The result is a scalar quantity. The scalar product is written with a dot between the symbols, thus:
2. Scalar products are commutative:

$$A \cdot B = B \cdot A$$

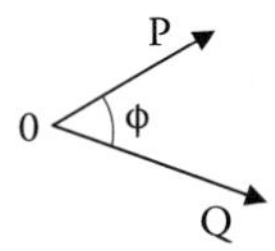

FIGURE 7.23
Product of two vectors.

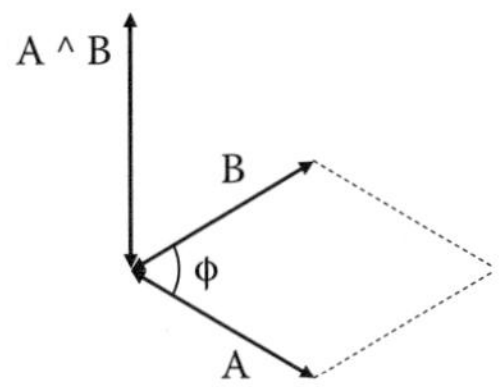

FIGURE 7.24
Vector product.

3. Scalar products are distributive:

$$A \cdot (B + C) = A \cdot B + A \cdot C$$

4. If A · B=0, then A=0 or B=0 or A and B are at right angles.

7.3.8 Vector (or Cross) Product

1. The vector product of two vectors (Figure 7.24) is defined as a vector whose magnitude is equal to the product of magnitudes times the sine of the angle between the vectors. The direction is perpendicular to the plane containing the two vectors and forms a right-handed system.
   The symbol for a vector product is ^.
   Thus, A ^ B=(AB sin ϕ)n, where AB sin ϕ is the magnitude and n is the unit normal vector to the plane containing A and B (see Figure 7.24).
2. Vector products are not commutative, that is, A ^ B≠B ^ A.
   However,

$$A \wedge B = -B \wedge A$$

3. Vector products are distributive provided the order of the vectors is maintained.

$$A \wedge (B + C) = A \wedge B + A \wedge C$$

4. If A ^ B=0, then A=0 or B=0 or A and B are parallel.

8

Dynamics

8.1 Kinematics

Kinematic analysis is the study of the motion of a particle or body without regard to the forces that may be generating the motion. The motion may be either rectilinear or rotary and in mechanical engineering this includes the motion of linkages in mechanisms including that of robotic arms and so on. The definition of a mechanism in this context is rigid bodies connected by joints and is applied to the combination of geometric bodies that constitute a machine or part of a machine. They move with definite relative motions with respect to each other and are often referred to as the geometry of motion.

8.2 Nomenclature

s	linear displacement (m)
v	average velocity (m/s)
a	acceleration (m/s^2)
t	time (s)
u	initial velocity (m/s)
v	final velocity (m/s)
θ	rotation angle (radians)
ω_1	initial angular velocity (rad/s)
ω_2	final angular velocity (rad/s)
η	angular speed (rev/min)
α	angular acceleration (rad/s^2)

8.3 Newton's Laws of Motion (Constant Acceleration)

In Chapter 7 (Statics) the reader was introduced to Newton's three laws of motion. In this chapter, the second law is considered as a set of equations. There are four equations and these describe linear and angular motions.

8.3.1 Linear Motion Equations

$$s = ut + \frac{at^2}{2} \tag{8.1}$$

$$v = u + at \tag{8.2}$$

$$v^2 = u^2 + 2as \tag{8.3}$$

$$s = \left(\frac{u+v}{2}\right)t \tag{8.4}$$

$$v = \frac{ds}{dt} \tag{8.5}$$

$$a = \frac{dv}{dt} = v\frac{dv}{ds} \tag{8.6}$$

8.3.2 Angular Motion Equations

$$\theta = \omega_i t + \frac{\alpha t^2}{2} \tag{8.7}$$

$$\omega_f = \omega_i + \alpha_t \tag{8.8}$$

$$\omega_f^2 = \omega_i^2 + 2\alpha\theta \tag{8.9}$$

$$\theta = \left(\frac{\omega_i + \omega_f}{2}\right)t \tag{8.10}$$

$$\omega = \frac{d\theta}{dt} \tag{8.11}$$

$$\alpha = \frac{d\omega}{dt} = \omega\frac{d\omega}{d\theta} \tag{8.12}$$

These laws will be expanded in the following sections.

8.4 Rectilinear Motions

Rectilinear motion (also known as linear motion) is motion in one direction only, that is, along a straight line and can be described mathematically as having one spatial dimension. The motion can be of two types:

- Uniform linear motion with a constant velocity (or zero acceleration)
- Non-linear motion with a variable velocity (or non-zero acceleration)

The motion of a particle along a straight line can be described by its position 'x', which varies with time 't'. A good example of linear motion is a machine tool slide on a lathe.

8.4.1 Uniform Linear Motion

Of all the motions, linear motion is the most basic and according to Newton's first law of motion, providing an object does not experience any net force and will continue to carry on moving in a straight line at a constant velocity until it is subjected to an external force. External forces such as gravity and friction can cause an object to change the direction of its motion such that the motion therefore cannot be described as linear.

In Figure 8.1, the object travels a distance 's' over a time 't' at a uniform velocity.

8.4.2 Non-Uniform Linear Motion

In linear motion where the object has a non-uniform velocity, the object's velocity is continuously changing, in which case the acceleration will also be changing in a non-linear relationship. An example is if a body is falling to earth, it will fall with an increasing acceleration, and hence the object's velocity will be varying in a non-linear manner.

Figure 8.2 shows the displacement/time curve of such a situation where the displacement 's' is changing in accordance with a specific law and the distance travelled at a particular point can be determined by drawing a tangent to the curve at a time $t=t_1$ and measuring the distance. Knowing the relationship between the distance 's' travelled and the time 't', differentiating 's' with respect to 't' will result with the velocity of the object at that point.

8.4.3 Variable Velocity

Figure 8.3 shows a curve where the linear or angular velocity is changing with respect to time. Differentiating with respect to time will give the linear or angular acceleration.

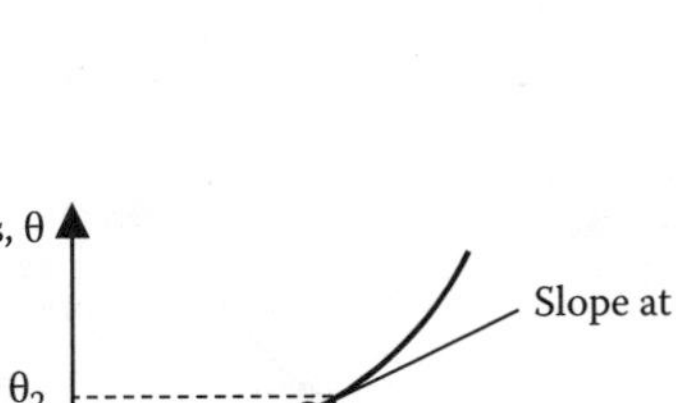

FIGURE 8.1
Linear displacement–time.

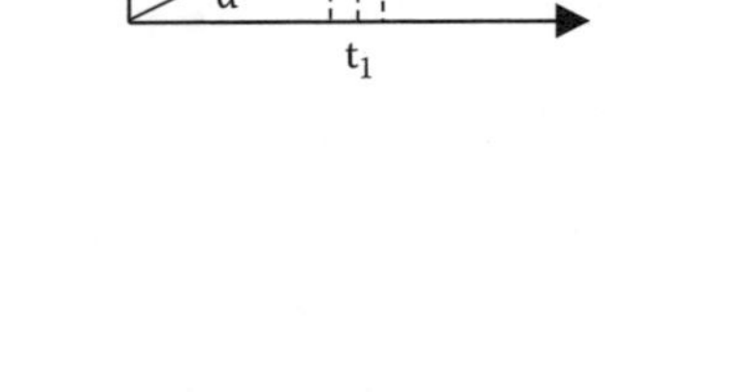

FIGURE 8.2
Variable displacement–time.

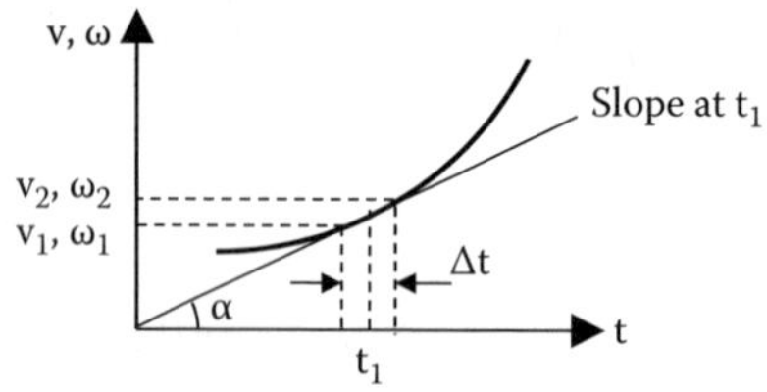

FIGURE 8.3
Velocity–time.

8.4.4 Variable Acceleration

Consider the case where the acceleration of a body is increasing in a non-linear relationship:

Figure 8.4 depicts the relationship between acceleration 'a' and 'a' with respect to time.

Linear velocity:

$$v = \frac{ds}{dt} = \dot{s} = \tan\alpha \tag{8.13}$$

Angular velocity:

$$\omega = \frac{d\theta}{dt} = \dot{\theta} = \tan\alpha \tag{8.14}$$

Linear acceleration:

$$a_l = \frac{dv}{dt} = \dot{v} = \tan\beta \tag{8.15}$$

Angular acceleration:

$$a_r = \frac{d\omega}{dt} = \dot{\omega} = \tan\beta \tag{8.16}$$

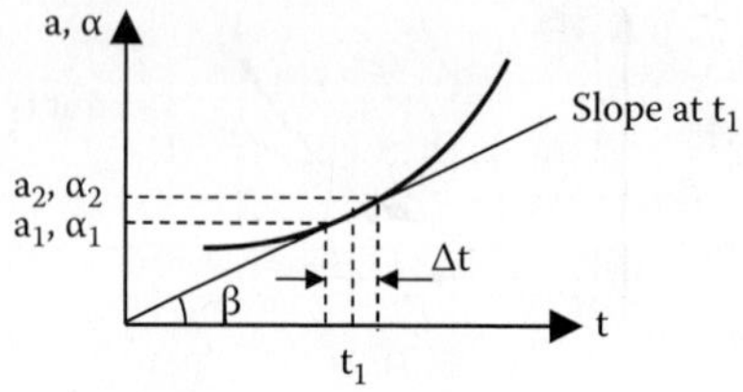

FIGURE 8.4
Acceleration–time.

8.5 Circular Motion

8.5.1 Motion on a Circular Path

Figure 8.5 describes the motion of a particle in a circular path with 'ω' given:

$$v = r\omega \tag{8.17}$$

$$a = r\omega^2 = \frac{v^2}{r} \tag{8.18}$$

The acceleration 'a' is towards the centre 'C' and it is referred to as the centripetal acceleration.

8.5.2 Rolling Wheel

Figure 8.6a, b shows a rolling wheel.

With 'v' and 'a' given:

$$\omega = \frac{v}{r} \tag{8.19}$$

$$\alpha = \frac{a}{r} \tag{8.20}$$

At any point 'P' on the wheel rim:

$$v_p = \omega AP \tag{8.21}$$

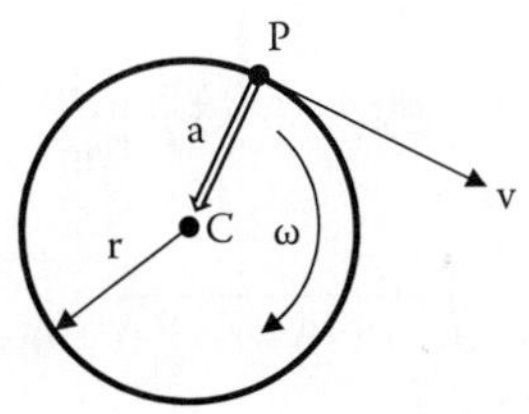

FIGURE 8.5
Motion in a circular path.

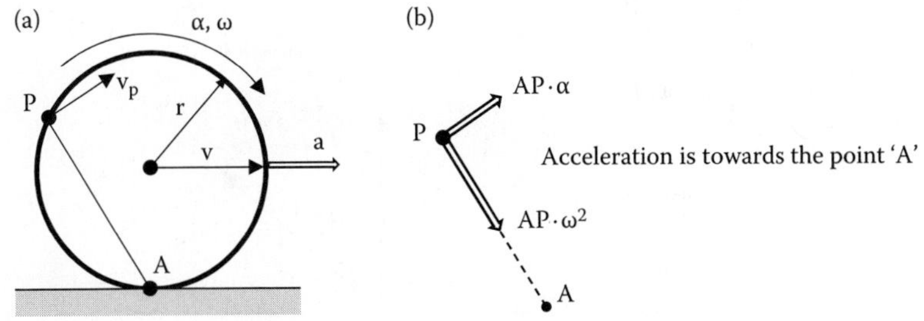

FIGURE 8.6
(a) A particle on a rolling wheel and (b) the vector representation of that point.

At the top of the wheel:

$$\text{Velocity} = 2 \cdot v \tag{8.22}$$

8.6 Absolute and Relative Motion

An example of absolute motion can be explained by considering a frame of reference based on the earth, an individual sitting in either a motor vehicle or train carriage; the motion of the individual relative to the earth can then be considered absolute against other individuals or objects within the same frame of reference.

If the reference frame is now changed and is based upon the train carriage, when the individual in the train carriage decides to move along the carriage, the individual's motion will be considered relative against the local reference frame.

8.7 Rotating Unit Vector

Consider a rotating unit vector e....

See Figure 8.7a:

$$e = i\cos\theta + j\sin\theta \tag{8.23}$$

$$\frac{di}{dt} = \frac{dj}{dt} = 0 \text{ (i and j are unit vectors in x and y directions)} \tag{8.24}$$

$$\dot{e} = \dot{\theta}(-i\cos\theta + j\sin\theta) \tag{8.25}$$

See Figure 8.7b:

$$|\dot{e}| = \dot{\theta}\sqrt{\sin^2\theta + \cos^2\theta} = \theta = \omega \tag{8.26}$$

$$e = \dot{\theta}k \ldots (\omega \text{ is in the z direction}) \tag{8.27}$$

$$\dot{e} = \omega x e \ldots (kxj = -i, kxi = j) \tag{8.28}$$

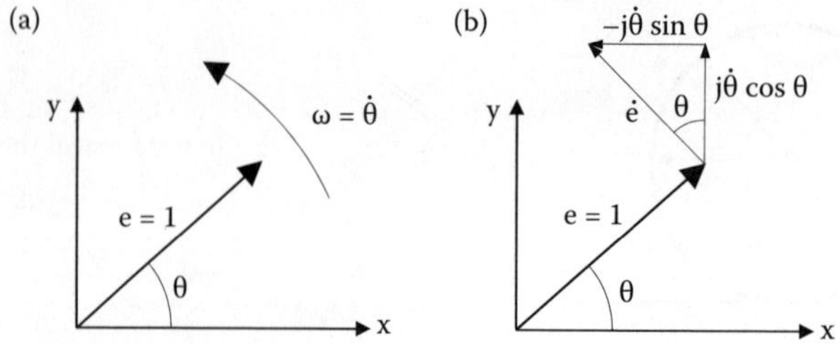

FIGURE 8.7
(a) Rotating unit vector and (b) the equivalent vector representation.

8.8 Vector of Point in a Rotating Reference Frame

In Figure 8.8, a point 'P' having a vector position 'r' is translating in the xy plane. 'e_1' and 'e_2' identify the position of the vector in respect to a reference frame that is rotating with an angular velocity 'ω'. 'r_1' and 'r_2' are components of the motion with respect to the 'e_1' and 'e_2' directions.

The position of the vector is expressed as follows:

$$r = r_1 e_1 + r_2 e_2 \tag{8.29}$$

The derivative with respect to time:

$$\dot{r} = \dot{r}_1 e_1 + r_1 \dot{e}_1 + \dot{r}_2 e_2 + r_2 \dot{e}_2 \tag{8.30}$$

$$r_1 \dot{e}_1 = r_1 \omega x e_1 = \omega x r_1 e \tag{8.31}$$

$$r_2 \dot{e}_2 = r_2 \omega x e_2 = \omega x r_2 e \tag{8.32}$$

From Equation 8.28

$$\frac{de}{dt} = \omega x e,$$

Therefore,

$$\dot{r} = \dot{r}_1 \dot{e}_1 + \dot{r}_2 e_2 + \omega x r_1 e_1 + \omega x r_2 e_2 \tag{8.33}$$

$$= \dot{r}_1 + \omega x (r_1 e_1 + r_2 e_2) \tag{8.34}$$

$$= \dot{r}_1 + \omega x r \tag{8.35}$$

(r_1 = rate of change of 'r' as measured relative to e_1, e_2)

$$\dot{r}_1 = \dot{r}_1 e_1 + \dot{r}_2 e_2 \tag{8.36}$$

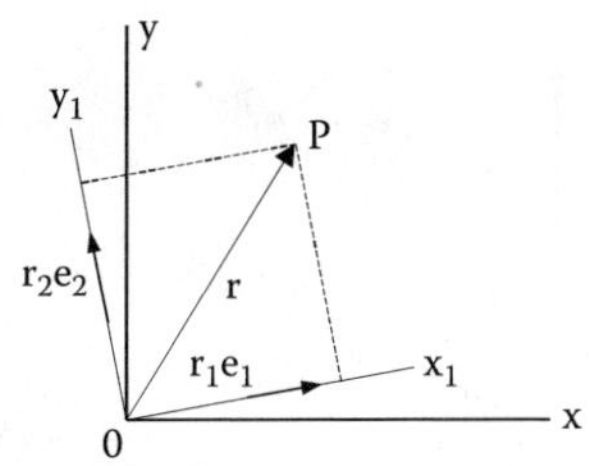

FIGURE 8.8
Vector of a point in rotating the reference frame.

8.9 Velocity of a Point in a Moving Reference Frame

Consider a point 'P' moving within a reference frame 'xy' as shown in Figure 8.9 having a position vector of

$$r_p = r_{o'} + r \tag{8.37}$$

r_o' locates o′ relative to the fixed reference frame xy axes and 'r' locates the point 'P' relative to axes (x_1, x_2).

$$r_p = r_{o'} + r \tag{8.38}$$

$$v_p = \dot{r}_p = \dot{r}_{o'} + \dot{r} \tag{8.39}$$

$$\dot{r} = \dot{r}_r + \omega x r \tag{8.40}$$

$$\dot{r}_p = \dot{r}_{o'} + \dot{r}_r + w x r \tag{8.41}$$

The moving frame is not rotating, $\omega=0$

$$v_p = \dot{r}_p = \dot{r}_{o'} + r_r \tag{8.42}$$

8.10 Acceleration of a Particle

Differentiating (with respect to time) the for velocity 'v' results in the for acceleration 'a'

$$a_p = \ddot{r}_p = \ddot{r}_{o'} + \ddot{r} \tag{8.43}$$

Following the principles outlined above

$$r = r_r + \omega x r \tag{8.44}$$

$$\ddot{r} = \frac{d}{dt}(\dot{r}_r + \omega x r) = \frac{d\dot{r}_r}{dt} + \frac{d}{dt}(\omega x r) \tag{8.45}$$

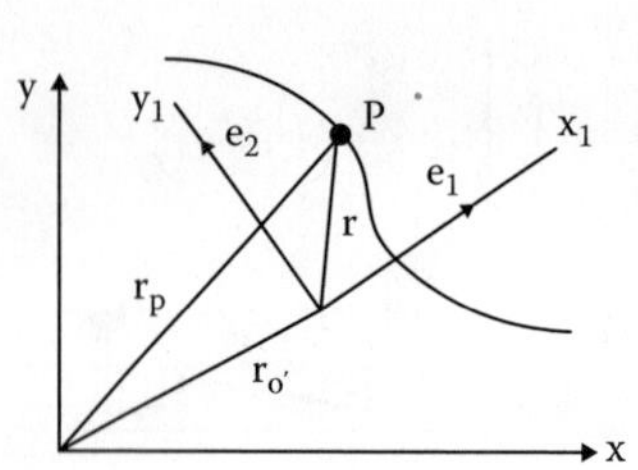

FIGURE 8.9
Vector of a point.

$$\frac{d\dot{r}_r}{dt} = \ddot{r}_r + \omega x \dot{r}_r \tag{8.46}$$

$$\frac{d}{dt}(\omega x r) = \dot{\omega} x r + \omega x(\dot{r}_r + \omega x r) \tag{8.47}$$

$$= \dot{\omega} x r + \omega x \dot{r}_r + \omega x(\omega x r) \tag{8.48}$$

Therefore,

$$r = \ddot{r}_r + \dot{\omega} x r + \omega x(\omega x r) + 2\omega x \dot{r}_r \tag{8.49}$$

$$a_p = \ddot{r}_o + \ddot{r}_r + \dot{\omega} x r + \omega x(\omega x r) + 2\omega x \dot{r}_r \tag{8.50}$$

$$a_p = \ddot{r}_o + \ddot{r}_r \tag{8.51}$$

8.11 Kinematics of Rigid Bodies in One Plane

The definition of a 'rigid body' is where all the internal points are fixed relative to each other. There are three co-ordinates required to identify the position and orientation of the body in plane motion. A rigid body will have three degrees of freedom in plane motion.

Further definitions:

- Rectilinear translation is where all the points comprising the rigid body move in a straight line.
- Curvilinear translation is where all the fixed points in the rigid body maintain the same orientation and remain fixed while the body moves along a curved path.
- Rotation about a fixed line, where all the points in a rigid body move in a circular motion around a fixed line.
- Plane motion, each point in a rigid body remains fixed to each other while moving in a path parallel to a fixed plane.

The equations for motion for a particle can be used to develop the plane motion of a rigid body. These are the same equations for the motion of a point, but as all the points are fixed relative to each other, the terms involving velocity and acceleration of these points relative to each other will be zero.

The equations for velocity and acceleration where the reference axes 'x' and 'y' are fixed to a rigid body are as follows:

From Equation 8.37:

$$r_p = r_{o'} + r$$

Point velocity:

$$v_p = v_{o'} + w \cdot r \text{ (as above but terms involving } \dot{r}_r \text{ are zero).}$$

Point acceleration:

$$a_p = a_{0'} + w \cdot r + w(w \cdot r) \text{ (as above but terms including } \dot{r}_r \text{ and } \ddot{r}_r \text{ are zero).}$$

If the position of the moving reference frame is fixed such that r_o' is constant, then

$$v_p = \omega \cdot r \tag{8.52}$$

$$a_p = a \cdot r + \omega(\omega \cdot r) \tag{8.53}$$

8.12 Instantaneous Centre of Rotation

The instantaneous centre of rotation is also known as the 'instant centre of rotation'.

For a body moving in space, the motion can be defined from its position, velocity and acceleration of any point on the body. There is a point for which the instantaneous translational velocity is zero, which is only the rotation of the body about the point where it occurs. This is known as the 'instantaneous centre of rotation'. The relative values of the linear and angular velocities will determine its location. When the angular velocities are near zero (i.e., near translation motion), the location of the instantaneous centre of rotation will be near infinity.

When the location of a point and its associated velocity and angular velocity are known: (see Figure 8.10a, b). The point 'C' will be on a line passing through 'A', which is normal to the direction of the velocity as shown in Figure 8.10a:

$$r_A = \frac{v_A}{\omega} \tag{8.54}$$

If the location of two points and their respective velocities are known: where 'C' is the intersection of the lines drawn normal to the velocities of each of these points.

Where the lines are coincident, the radius can be established using geometry as in Figure 8.10b:

$$AC = \frac{AB(v_A)}{(v_A - v_B)} \tag{8.55}$$

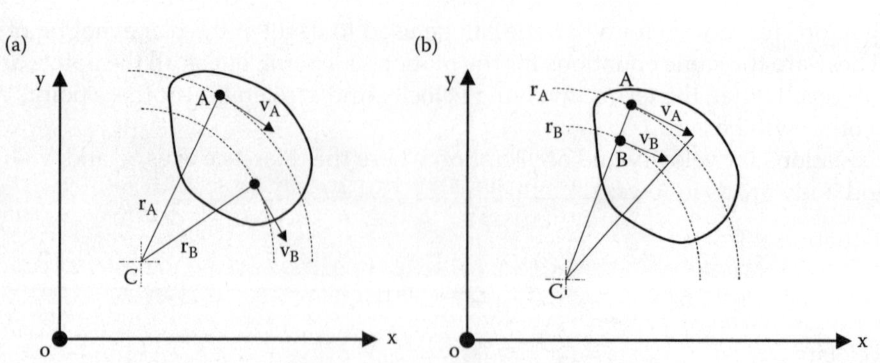

FIGURE 8.10
Instantaneous centre of rotation. (a) The location of two separate points and (b) coincident lines.

8.13 Kinematics of Rigid Bodies in Three Dimensions

Three points are required to specify the position of a particle relative to the selected co-ordinate system and are identified as having three degrees of freedom. For a rigid body, the location of three separate points relative to a selected co-ordinate system is also required. As these points are relative to one another, only six independent co-ordinates are required to locate the body in a three-dimensional space and an unrestrained rigid body is therefore said to have six degrees of freedom. As an example, a rigid body can be positioned by locating the position of one of the bodies (three co-ordinates), then positioning a line on the body (two co-ordinates) and finally identifying a rotation about the line (one co-ordinate). This will sum up to six co-ordinates.

The motion of a rigid body existing in three dimensions can have a number of modes:

- *Rectilinear translation:* All points on a rigid body move in a straight line.
- *Curvilinear translation:* The relationship of all points in a rigid body remains fixed when the body moves along a curved path.
- *Rotation about a fixed axis:* The relationship of all points in a rigid body remains fixed when the body moves about a fixed axis.
- *Rotation about a fixed point:* All points in a rigid body maintain their fixed relationship with each other when the body moves in a circular motion about a fixed point.
- *General motion:* The rigid body motion includes translation and rotation.

Notes:

- If a rigid body rotates about the 'x' axis from position 'A' through an angle of $\pi/2$ and then proceeds to rotate about the 'y' axis, it will have moved to a certain position 'B'.
- If the same rigid body is rotated from position 'A' through an angle of $\pi/2$ about the 'y' axis and further rotated $\pi/2$ about the 'x' axis, it will have moved to position 'C'.
- Position 'C' will not be the same as position 'B'. Finite rotation does not obey the same laws as vector addition.

8.14 Theorems

1. *Euler's Theorem:* Two component rotations about different axes passing through a point are equivalent to a single resultant rotation about an axis passing through the point.
2. *Chasles's Theorem:* Any displacement of a rigid body may be compounded from a single rotation about any selected point plus a translation of that point.
3. *Poinsot's Central Axis Theorem:* Any finite displacement of a rigid body may be reduced to a single rotation about an axis plus translation parallel to the same axis. This theorem only relates to the displacement of the rigid body and not to paths taken by the points.

These theorems can be applied to angular velocities.

Any rotation of a rigid body can be described by a single angular velocity plus a translational velocity parallel to the angular velocity vector.

Any motion of a body about a point may be represented by a single velocity about an axis through that point.

Any motion of a rigid body may be represented by the velocity of a point plus the angular velocity about an axis passing through the points.

8.15 Translation Motion

Rectilinear translation and curvilinear translation: all points on a body will move in parallel straight curved lines. There is no relative velocity or acceleration between any points on the body (see Figure 8.11).

Position:

$$r_P = r_Q + r_{PQ} \tag{8.56}$$

Velocity:

$$v_P = v_Q + v_{PQ} \tag{8.57}$$

Acceleration:

$$a_P = a_Q + a_{PQ} \tag{8.58}$$

8.16 Rotation about a Fixed Axis

Rotation of a rigid body about a fixed axis is shown in Figure 8.12. The vector of the rotary motion has sense and direction in accordance with the right-hand rule when this is aligned with the direction of the axis. There is zero velocity due to the rotation of the axis.

Velocity:

$$v_P = w \cdot r_P \tag{8.59}$$

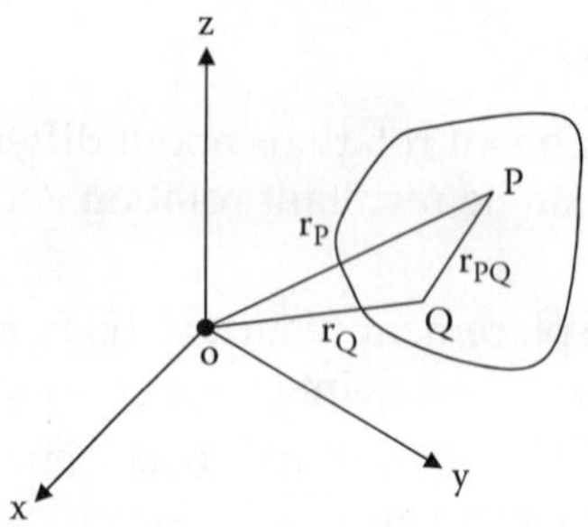

FIGURE 8.11
Rectilinear, translation and curvilinear.

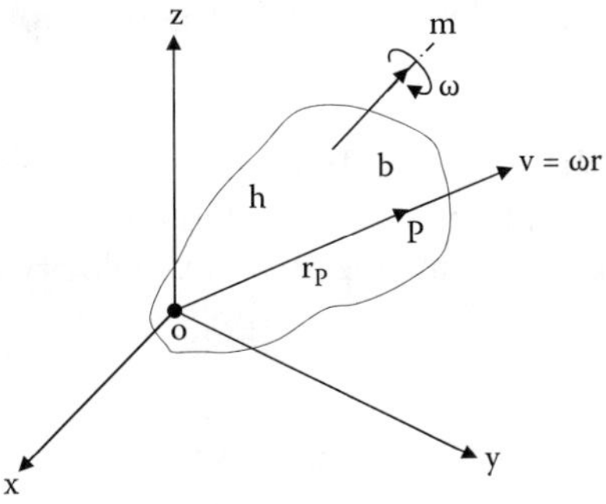

FIGURE 8.12
Rotation of a rigid body about a fixed axis.

Acceleration:

$$a_P = \dot{v}_P = \dot{\omega} \cdot r_P + \omega(\omega \cdot r_P) \tag{8.60}$$

8.17 Rotation about a Fixed Point

Figure 8.13 shows that the rotation of a rigid body about a fixed point can always be reduced to rotation of a body about an instantaneous axis of rotation.

Considering, for example, a rigid body (say a cone) which is rotating about a horizontal axis (say a shaft), angular velocity=ω_1 and the shaft is itself rotating about a vertical axis, with an angular velocity=ω_2.

If ω_2=0, then the axis of rotation is the centre line of the shaft and the velocity of any point on the shaft is proportional to the radius (max=r) from the shaft. If the rotation about the vertical axis is increased to a certain value, the velocity of the shaft is proportional to ω_2x the shaft length (max=l) from the axis radius. If the two angular velocities are the same and the radius of the cone (r=l), then at any instant the velocity of the top surface of the cone is zero, that is, the top surface is then the instantaneous axis of rotation.

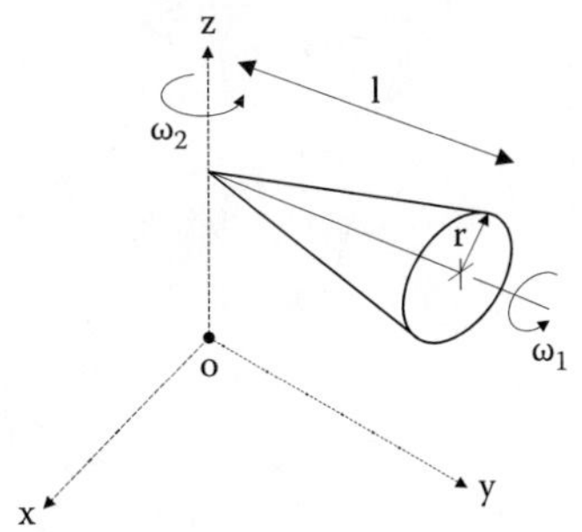

FIGURE 8.13
Rotation about a fixed axis.

This very simplified representation illustrates the principle; if ω_2 is increased, the cone representing the path of the instantaneous axis of rotation will be larger than the actual surface of the cone.

For the model illustrated, there is also a space cone, which is the path the instantaneous axis of rotation follows in space, that is, an inverted cone centred on the vertical axis.

For the example above, the instantaneous angular velocity of the cone will be

$$w = w_1 + w_2 \tag{8.61}$$

$$= w_1 i + w_2 j \tag{8.62}$$

In the real world, the motions and shapes are not the same as shown but the interactions between the body rotation on its axis, and the angular motion in space still result in an instantaneous axis of rotation between an instantaneous body cone and an instantaneous space cone. The body cone may rotate outside or inside the space cone (see Figure 8.14).

When determining the instantaneous angular velocity, the angular acceleration 'α' is tangential to the contact point of the two cones as shown in Figure 8.14. The velocity and acceleration at any point are simply determined as below.

Velocity:

$$v_P = \omega \cdot r \tag{8.63}$$

Acceleration:

$$a = \dot{v} = \dot{\omega} \cdot r + \omega(\omega \cdot r) \tag{8.64}$$

$$= \alpha \cdot r + \omega(\omega r) \tag{8.65}$$

8.18 General Motion

The general case of 3D motion can be reduced to translation+rotation around a fixed axis. This is basically a generalisation of the theorems described above.

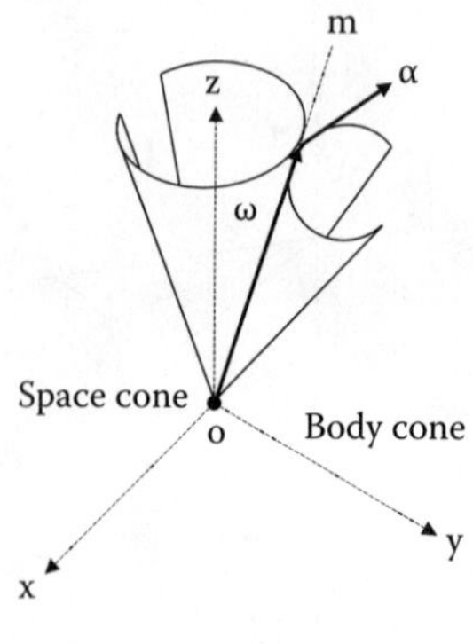

FIGURE 8.14
Rotation about a fixed point.

For a body possessing linear and angular motion, it is often not possible to have an instantaneous axis of rotation as all the points may have non-zero velocities. The most convenient method of kinematic analysis of rigid bodies in 3D space is to use the principles of relative motion.

The following reference system may be either a translating or a rotating one.

A translating reference system (see Figure 8.15):

This motion simply develops the motion already studied in previous sections for particle motions.

The basic motion equations are

$$r_P = r_Q + r \tag{8.66}$$

$$v_P = v_Q + v_{PQ} \tag{8.67}$$

$$a_P = a_Q + a_{PQ} \tag{8.68}$$

The above equations are sufficient if the angular velocity is zero but if this is not the case, the following equations for velocity and acceleration will be found to be more definitive.

Velocity:

$$v = v_Q + \omega \cdot r_{PQ} \tag{8.69}$$

Acceleration:

$$a_P = v_P = v_Q + \omega \cdot r_{PQ} + \omega(\omega \cdot r_{PQ}) \tag{8.70}$$

$$= a_Q + \alpha \cdot r_{PQ} + \omega(\omega \cdot r_{PQ}) \tag{8.71}$$

A rotating reference system (see Figure 8.16).

A more general form of the relative reference axis method using the rotating reference axes. The reference axes x, y, z are rotating with an angular velocity of Ω. The rigid body will have a rotation velocity 'ω' as shown before.

The basic motion equations are as shown above. The expressions for velocity and acceleration of point 'P' as shown before.

The derivation is a simple extension of that provided above for 2D motion with a rotating relative axis with the third dimension (z) added.

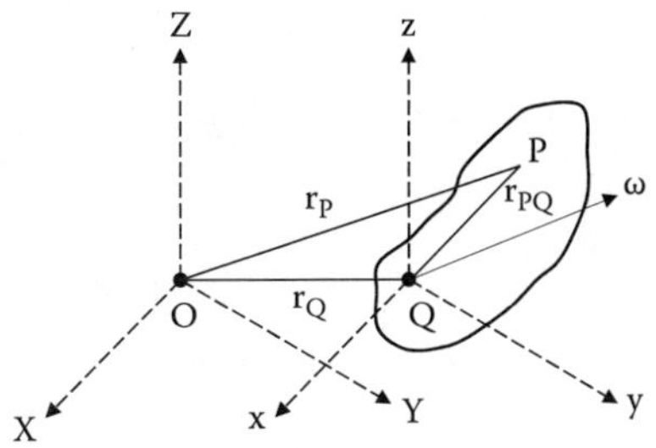

FIGURE 8.15
Translating a reference system.

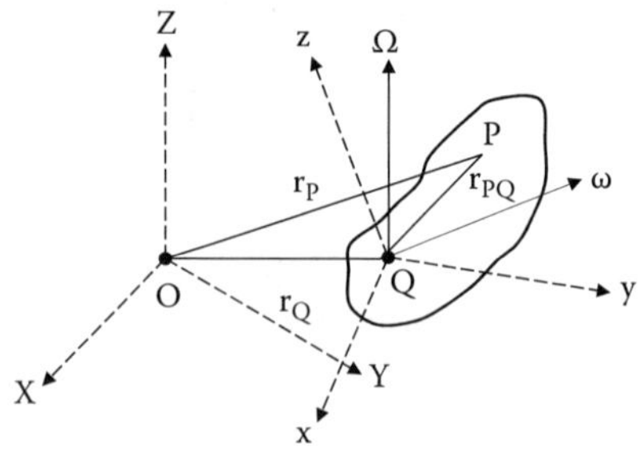

FIGURE 8.16
Rotating a reference system.

Velocity:

$$v_P = \dot{r}_Q + \Omega \cdot r_{PQ} + \dot{r}_r \tag{8.72}$$

$$= v_Q + \Omega \cdot r_{PQ} + v_r \tag{8.73}$$

Acceleration:

$$a_P = \dot{v}_P \tag{8.74}$$

$$= \dot{a}_Q + \dot{\Omega} \cdot r_{PQ} + \Omega(\Omega \cdot r_{PQ}) + 2\Omega \cdot v_r + a_r \tag{8.75}$$

v_r and a_r = the velocity and acceleration of 'P' relative to the rotating x, y, z axis.

Note: Equations 8.72 through 8.75 are based on the general case where the angular velocity of the rigid body (ω) is different from the angular velocity of the axis (Ω). If the reference axis is fixed to the body, then ω will be equal to Ω and v_r and a_r will be equal to zero. For this case, the formula will be the same as for the translating reference axis as above.

9

Mechanical Vibrations

9.1 Introduction

Mechanical vibrations are defined as oscillations in mechanical dynamic systems and are the motions of a particle or a body or system of connected bodies that have been displaced from a position of equilibrium. The majority of vibrations are undesirable in machines or structures as they can result in increased stresses, causing increased wear such as fretting and increased bearing loads. Mechanical fatigue can also result from vibrations and rotating machine parts including aero engine parts and will need careful balancing in order to prevent any damage resulting from vibrations.

Although most vibration problems are undesirable, such as the Tacoma Narrow Bridge failure in the United States in the 1940, and innumerable airframe failures resulting from vibration-induced fatigue, some mechanical systems such as the Beale free-piston Stirling engine rely on the vibration characteristics of the system to function correctly. In the mining and quarrying industries, these rely on sifting different sized particles using vibrating screed beds. In the manufacturing industry, vibration conveyors are used to convey components from one machining process to another.

Vibrations can be classified into four basic categories:

1. Free
2. Forced
3. Self-excited
4. Random

Free vibration of a system is vibration that occurs in the absence of any external force.

External force acting on a system will cause forced vibrations; in this instance, the exciting force is continuously supplying energy to the system. These types of vibrations may be either deterministic or random (see Figure 9.1).

Self-excited vibrations are periodic and deterministic oscillations. Under certain conditions, the equilibrium state becomes unstable and any disturbances will cause the perturbations to grow until some effect limits any further growth. This is in contrast to force vibrations, where the exiting force is independent of the vibrations and can still persist even when the system is prevented from vibrating.

This chapter will concentrate on most of the aspects of vibration ranging from simple harmonic free vibrations to forced vibrations.

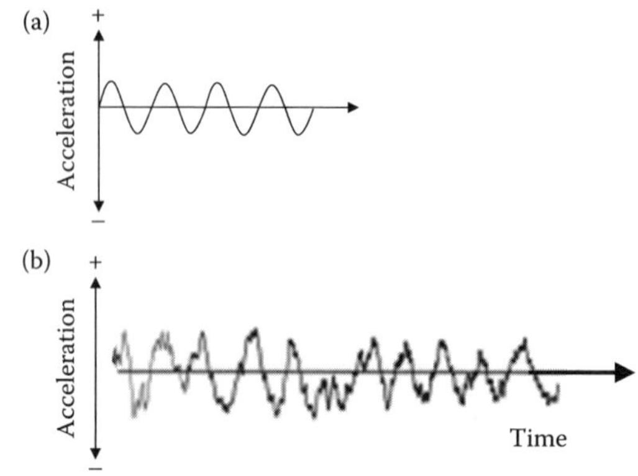

FIGURE 9.1
(a) Deterministic and (b) random vibrations.

9.2 Single Degree of Freedom: Free Vibrations

9.2.1 Free Natural Vibrations

A free vibration is one that occurs naturally without any energy being added to the vibrating system. The vibration is started by an input of energy but the vibrations die away over time as the energy is dissipated. In each case, when the body is moved away from the rest position, there is a natural force that will try to restore the body back to its rest position. This can be better demonstrated by considering first a pendulum which can be considered as one degree of freedom, where the pendulum is displaced from its rest position and allowed to swing backwards and forwards with its amplitude gradually diminishing as the pendulum loses energy under the influence of gravity. This can be further demonstrated when considering a mass suspended on a tension spring hanging vertically. With the mass displaced vertically downwards and the spring 'stretched', when the mass is released it will continue to rise and fall until again the energy in the spring is dissipated and the mass will come to rest in its equilibrium position (see Figure 9.2a and b).

The motion that these two examples display is known as 'simple harmonic motion' or generally referred to as 'SHM'.

This motion is characterised by the fact that when the displacement is plotted against time, the resulting graph is basically sinusoidal. The displacement can be either angular (e.g., the angle moved by the simple pendulum) or linear (e.g., the displacement of the mass on the spring). Although this study is concerned with natural vibrations, it may help to understand the nature of SHM if the forced vibration produced by a mechanism such as the Scotch yoke is considered.

9.2.2 Simple Harmonic Motion

Consider Figure 9.3 which shows a typical Scotch yoke arrangement. With the crank rotating at a constant ω (rad/s), a pin fitted in the crank slides in the slot of the yoke and point 'P' on the yoke will oscillate up and down as it is constrained to move only in the vertical direction by the bearing through which it is allowed to slide. The motion of point 'P' is considered to be sinusoidal when the displacement of the yoke against the angular

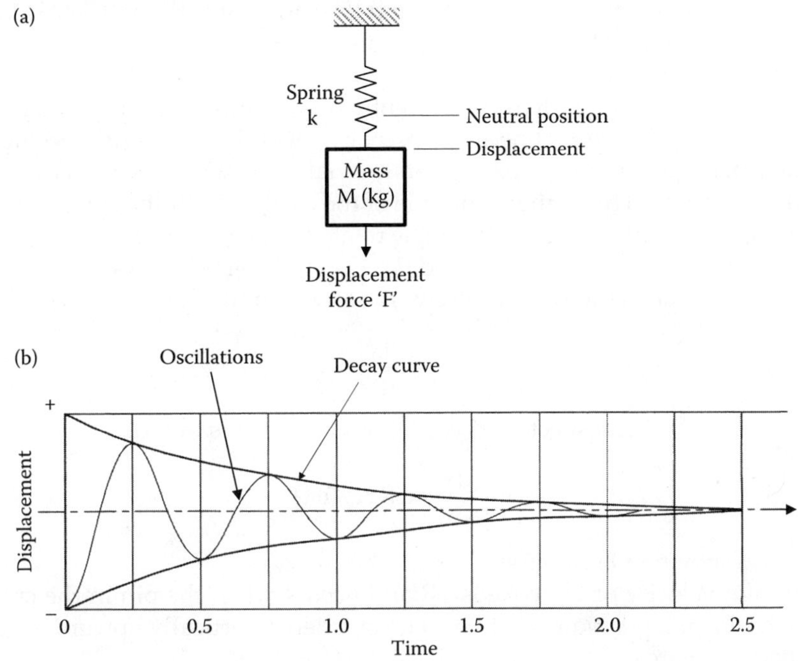

FIGURE 9.2
Deflected mass on a spring (a) and resultant vibration curve (b).

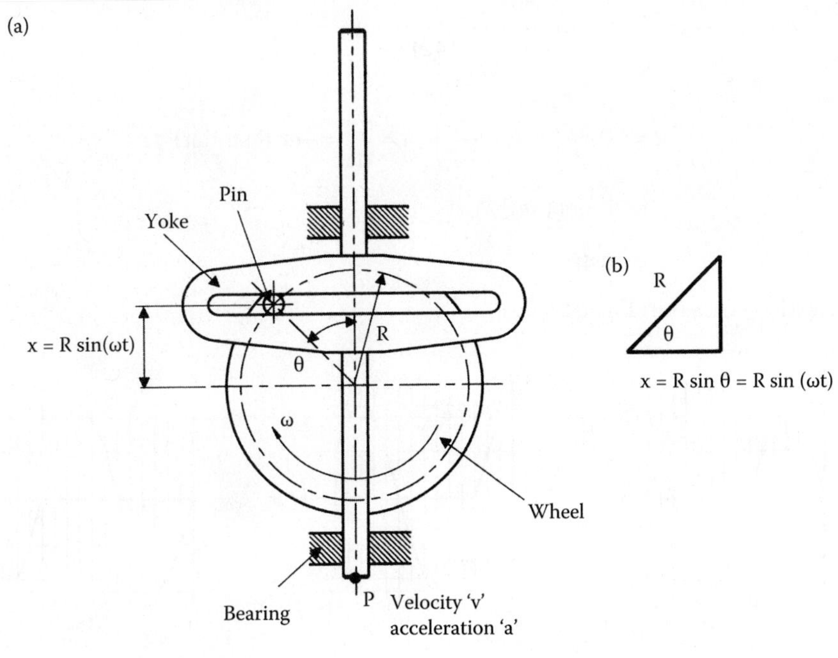

FIGURE 9.3
(a, b): The Scotch yoke arrangement.

position of the crank is plotted against time and therefore by definition is SHM. The point 'P' moving up and down will at any instantaneous point have a displacement 'x', a velocity 'v' and an acceleration 'a'.

With the pin located at radius 'R' from the centre of the crank, the vertical displacement of the pin from the horizontal centreline at any point is 'x'. This is also the displacement of point 'P'. The yoke reaches a maximum displacement equal to 'R' when the pin is at the top and '–R' when at the bottom. This is the amplitude of the oscillation. If the crank is rotating at a constant ω (rad/s), then after time t (s) the angle rotated will be $\theta = \omega t$ radians. From the right-angled triangle (Figure 9.3b), $x = R\sin(\omega t)$ and the graph of x against θ is shown in Figure 9.4a.

Velocity is the rate of change of distance with respect to time and in calculus form

$$v = \frac{dx}{dt} \tag{9.1}$$

If 'x' is differentiated with respect to 't', the following result is found:

$$v = \frac{dx}{dt} = \omega R\cos(\omega t) \tag{9.2}$$

This plot is also shown in Figure 9.4b.

The maximum velocity of the yoke is ωR and occurs when the pin in the crank passes through the horizontal position. Positive (+) is considered vertically upwards and negative (–) vertically downwards.

Acceleration is the rate of change of velocity with respect to time and again in calculus form:

$$a = \frac{dv}{dt} \tag{9.3}$$

Differentiating velocity 'v', the following is obtained

$$x = 20\sin\left((50 \times 0.3142) + \frac{\pi}{8}\right)$$

$$x = 20\sin\left(1.571 + \frac{\pi}{8}\right) \quad a = \frac{dv}{dt} = -\omega^2 R\sin(\omega t) \tag{9.4}$$

$$= 20\sin(1.9637)$$

$$= 18.481$$

This plot is also shown in Figure 9.4c.

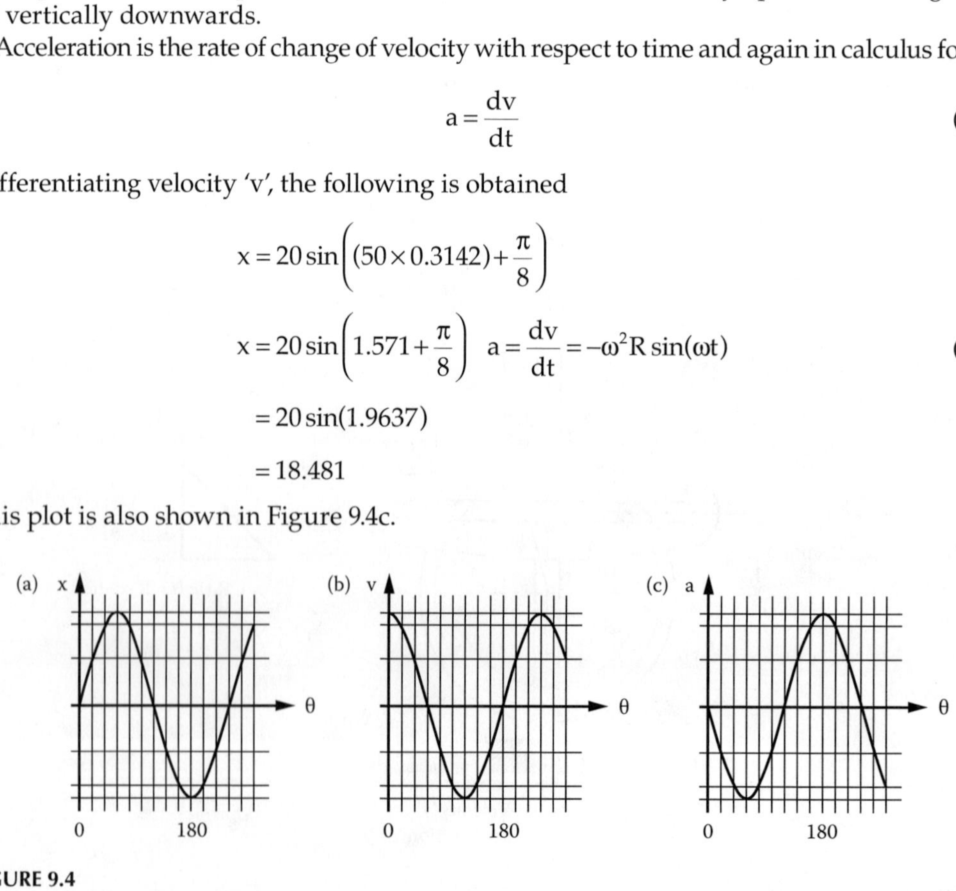

FIGURE 9.4
Graph of the Scotch yoke rotation. (a) Displacement, (b) velocity and (c) acceleration.

The amplitude is $\omega^2 R$ and this is positive at the bottom of the crank and negative at the top (when the yoke is about to change direction).

Now, since $R \sin(\omega t) = x$, substituting for x

$$a = -\omega^2 x \tag{9.5}$$

This is the usual definition of SHM and the equation states that any body that performs sinusoidal motion must have an acceleration that is directly proportional to the displacement and will always be directed to the point of zero displacement. The constant of proportionality is ω^2.

Any vibrating body that has a motion that can be described in this way must vibrate with SHM and have the same equations for displacement, velocity and acceleration.

9.2.2.1 *Angular Frequency, Frequency and Periodic Time*

The angular velocity of the crank is 'ω' but in any vibration problem such as the mass on a spring, this is referred to as the angular frequency as no physical crank exists.

The frequency of the crank in revolutions/s is equivalent to the frequency of the vibration. If the crank is rotating at 2 rev/s, the time for one revolution is 1/2 s. If the crank is rotating at 5 rev/s, the time for one revolution will be 1/5 s. Hence, if the crank is rotating at f (rev/s), then the time for one complete revolution will be 1/f. This relationship is important and gives the periodic time.

Periodic time 'T' is the time required to complete one cycle.

Where f is the frequency or number of cycles per second, it follows that:

$$T = \frac{1}{f} \tag{9.6}$$

and

$$f = \frac{1}{T} \tag{9.7}$$

It can be seen that each cycle of an oscillation will be equivalent to one rotation of the crank and one revolution is an angle of 2π radians.

Therefore, when $\theta = 2\pi$, $t = T$. Hence, it follows that since $\theta = \omega t$, $2\pi = \omega T$. Rearranging this will result in $\omega = 2\pi/T$. Substituting $T = 1/f$, therefore,

$$\omega = 2\pi f \tag{9.8}$$

9.2.2.2 *Equations for SHM*

From the three equations derived previously

$$\text{Displacement: } x = R\sin(\omega t) \tag{9.9}$$

$$\text{Velocity: } v = \frac{dx}{dt} = \omega R\cos(\omega t) \tag{9.10}$$

$$\text{Acceleration: } a = \frac{dv}{dt} = -\omega^2 R\sin(\omega t) \tag{9.11}$$

The plots of x, v and a against the angle θ are shown in Figure 9.4a. In the calculations made so far, the measured angle from the horizontal position 'θ' was decided and concluded that the time was zero at this point. If the timing had been started after the angle had reached a value of 'φ' from this point, in this case 'φ' would be called the phase angle and be measured in radians.

The resulting equations for displacement, velocity and acceleration can then be rewritten as follows to take account of the phase angle

$$\text{Displacement: } x = R\sin(\omega t + \varphi) \tag{9.12}$$

$$\text{Velocity: } v = \frac{dx}{dt} = \omega R\cos(\omega t + \varphi) \tag{9.13}$$

$$\text{Acceleration: } a = \frac{dv}{dt} = -\omega^2 R\sin(\omega t + \varphi) \tag{9.14}$$

From Figure 9.4a and b, the plots of x, v and a are the same but from Figure 9.4b it will be noted that the vertical axis has been displaced by 'φ'. The point to note between Figure 9.4a and b is that the velocity curve has been displaced by 1/4 cycle (90°) to the left, and the acceleration curve has been displaced a further 1/4 cycle, making it 1/2 cycle out of phase with 'x'.

Example 9.1

The displacement of a body subject to an SHM is described by the following equations.

$$x = R\sin(\omega t + \varphi)\ (\text{from Equation 9.12})$$

where
R = amplitude
ω = natural frequency
φ = phase angle

Given that R = 20 mm, ω = 50 rad/s and φ = π/8.
Calculate the following:

1. Frequency
2. Periodic time
3. Displacement, velocity and acceleration when t = T/4

Frequency:

$$f = \frac{\omega}{2\pi} = \frac{50}{2\pi} = 7.9577\ \text{Hz}$$

Periodic time:

$$T = \frac{1}{f} = 0.12566\ \text{s}$$

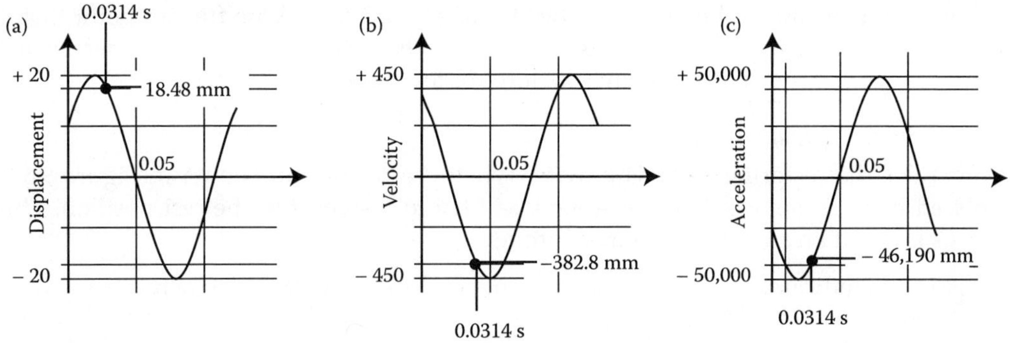

FIGURE 9.5
Figure for Example 9.1. (a) Displacement, (b) velocity and (c) acceleration.

Time (t):

$$t = \frac{T}{4}$$
$$= 0.03142 \text{ s}$$

Displacement (solve for t = 0.03142 s):

$$x = 20 \sin\left((50 \times 0.3142) + \frac{\pi}{8}\right)$$

$$x = 20 \sin\left(1.571 + \frac{\pi}{8}\right)$$
$$= 20 \sin(1.9637)$$
$$= 18.481 \text{ mm}$$

The equations for velocity 'v' and acceleration 'a':

$$x = 20 \sin(\omega t + \varphi)$$
$$v = 20\omega \cos(\omega t + \varphi)$$
$$= 20 \times 50 \times \cos(1.9637)$$
$$= -382.872 \text{ mm/s}$$

$$a = -20\omega^2 \sin(\omega t + j)$$
$$= -20 \times 50^2 \sin(1.9637)$$
$$= -46{,}190.061 \text{ mm/s}^2$$

These results are confirmed in Figure 9.5.

9.2.2.3 Free Natural Vibrations of a Single-Degree-of-Freedom System

In this section, it will be shown that some simple cases of natural vibrations are extended examples of SHM. One important point common to all these cases is that there must be a natural force that causes the body to move to the rest position. One further point that is common to all the following examples is that the body must possess a mass (inertia) and in order to accelerate this mass an inertial force or torque must be present.

9.2.2.3.1 Simple Pendulum

In this case, the restoring force is gravity. When the pendulum shown in Figure 9.6 is displaced through an angle 'θ', the weight will try to restore it to the rest position. This analysis is based on a moment of force (torque).

Note: Mass is indicated as 'M'. This should not be confused with 'm', which is being used as an abbreviation for metre.

Restoring force:

$$\text{Weight} = Mg$$
$$\text{Torque} = T \cdot g = \text{weight} \times Mg(L \sin \theta)$$

Inertia torque:

As the pendulum has an angular acceleration 'α' as the pendulum slows down and then speeds up, it requires an inertial force to produce this effect. T_j denotes this torque.

From Newton's second law for angular motion, $T_j = I\alpha$, where α is the angular acceleration and I is the moment of inertia.

It is assumed that the mass is concentrated at radius 'L' from the pivot point and the moment of inertia is then $I = ML^2$.

Balance of moments:

If there is no applied torque from any external source acting on the mass, then the total torque acting on the body must equal zero, that is,

$$T = \frac{1}{f}$$
$$= 0.0498 \text{ s}$$

The sine of small angles is very similar to the angle itself in radians. The smaller the angle, the more accurate it becomes. In such cases, sin(θ) = θ radians; therefore, this expression can be simplified to

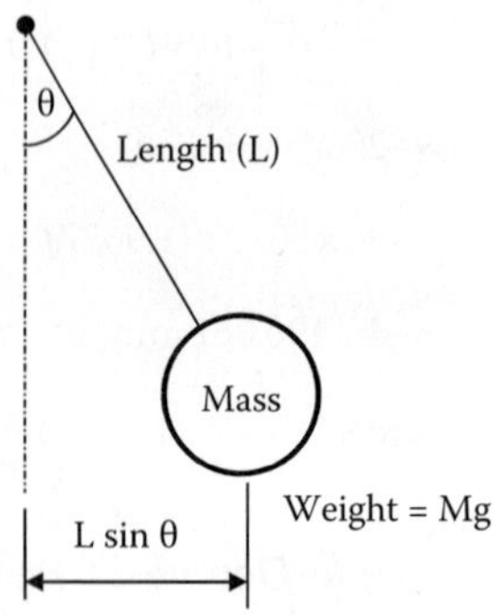

FIGURE 9.6
Simple pendulum.

$$g\alpha = -L\alpha$$

$$\alpha = -\left(\frac{g}{L}\right)\theta \tag{9.15}$$

This expression meets the requirements for SHM since the acceleration 'α' is directly proportional to the displacement 'θ' and the minus sign indicates that the mass is always accelerating towards the rest position. It follows that the constant of proportionality is (g/L), that is,

$$\omega^2 = \frac{g}{L}, \quad \omega = \left(\frac{g}{L}\right)^{1/2} \tag{9.16}$$

If this displacement 'θ' is plotted against time 't', similar graphs as shown in Figure 9.4a and b will result. The displacement in this example is angle and is not to be confused with the angle on the Scotch yoke. The frequency of oscillation is obtained from

$$f = \frac{\omega}{2\pi}$$

Hence

$$f = \frac{1}{2\pi}\sqrt{\frac{g}{L}} \tag{9.17}$$

It will be noted that mass does not enter into the equation for frequency. On the earth, the only way to alter the frequency is to alter the length 'L'.

Please remember that the above equations are valid only if the pendulum swings through a small angle. If the angle is large, the motion will not be a perfect SHM.

Example 9.2

A mass is suspended by a string 60 mm long. It is nudged so that it makes a small swinging oscillation. Determine the frequency and the periodic time of the swing.

$$f = \left(\frac{1}{2}\pi\right)\left(\frac{g}{L}\right)^{1/2}$$

$$f = \left(\frac{1}{2}\pi\right)\left(\frac{9.81}{0.06}\right)^{1/2}$$

$$= 20.085 \text{ Hz}$$

$$T = \frac{1}{f}$$

$$= 0.0498 \text{ s}$$

9.2.2.4 *Elementary Parts of a Vibrating System*

Up to now, a simplified introduction to SHM has been considered. Taking a more academic approach to a vibrating system that consists of a spring (a means of storing potential

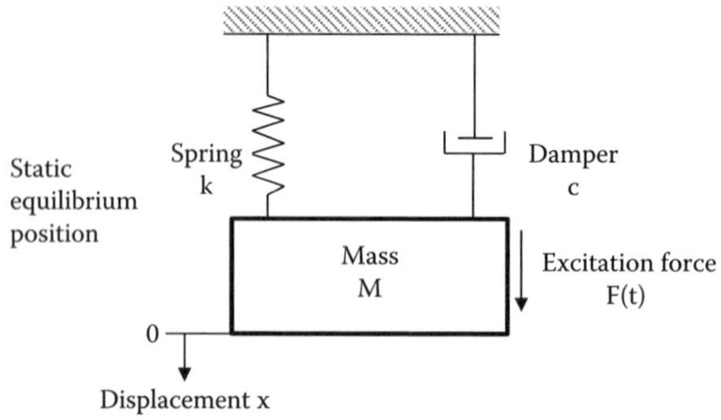

FIGURE 9.7
Elementary parts of a vibrating system.

energy), a mass or inertia (a means of storing kinetic energy) and a damper (a means by which energy is gradually lost from the system), this is as shown in Figure 9.7. An undamped vibrating system will involve the transfer of potential energy to kinetic energy and kinetic energy back to potential energy.

In a damped vibrating system, some energy is dissipated in each cycle of vibration and will require to be replaced by an external force if a steady state of vibration is to be maintained.

9.2.2.5 Linear Elastic Oscillations

The majority of natural oscillations will occur due to the restoring force due to a spring. The spring can be considered to be an elastic body supporting a mass. The spring can be any structural member of a discrete length and does not have to be a helical spring (either tension or compression) in the literal sense. Although a helical spring is shown diagrammatically, it represents any elastic member that has a spring stiffness 'k' and will either extend or compress, twist or bend under a specific load.

Consider a mass suspended on a spring as shown in Figure 9.8. The mass is subject to a force 'F' which displaces the mass downwards. The spring is extended by a distance 'x_o' and this is called the initial displacement.

If the force is removed, the mass will begin to oscillate up and down with an SHM.

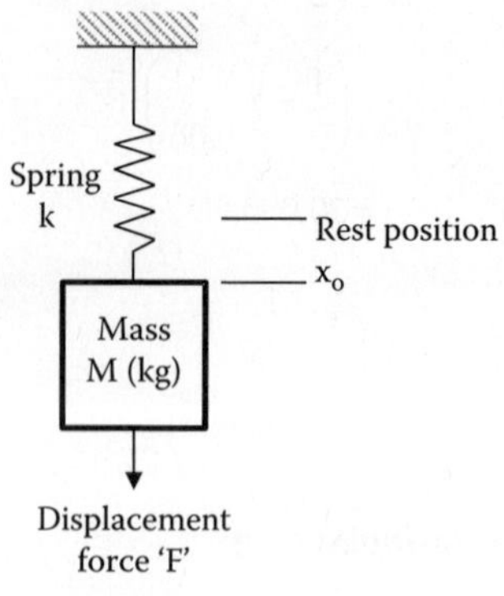

FIGURE 9.8
Mass suspended on a spring.

Let F_s = spring force that tries to return the mass to its original rest position. From spring theory

$$F_s = kx$$

where
k is spring rate
x is displacement

As the motion of the mass has an acceleration when released, an inertial force 'F_i' is present.
From Newton's second law of motion

$$F_i = Ma \text{ (mass} \times \text{acceleration)}$$

Therefore, balancing the forces acting on the mass gives

$$F = F_i + F_s = Ma + kx$$

If the mass is physically disturbed and released so that it is oscillating freely, the applied force is obviously zero and this is the requirement for oscillation to be free and natural; hence,

$$0 = Ma + kx,$$

rearranging the equation for 'a':

$$a = -\left(\frac{k}{M}\right) \cdot x \tag{9.18}$$

Thus, this is the equation for SHM where the acceleration is directly proportional to the displacement and is directed towards the rest position of the mass.

If the displacement 'x' is plotted against time 't', a sinusoidal graph would result. The constant of proportionality k/M is the square of the angular frequency so

$$\omega = \sqrt{\frac{k}{M}} \tag{9.19}$$

The frequency of oscillation will be

$$\begin{aligned} f &= \frac{\omega}{2\pi} \\ &= \frac{1}{2\pi}\sqrt{\frac{k}{M}} \end{aligned} \tag{9.20}$$

As this is a natural oscillation, the frequency is generally denoted as ω_n and f_n. This equation is true for all elastic oscillations.

Example 9.3

A spring having a stiffness of 20 kN/m supports a mass of 4 kg. The mass is displaced 8 mm vertically downwards and then released to produce linear oscillations.

Calculate the frequency and the periodic time and the displacement, velocity and acceleration 0.05 s after the mass has been released.

$$\omega_n = \sqrt{\frac{k}{M}}$$

$$= \sqrt{\frac{20\times 10^3\ \text{N/m}}{4\ \text{kg}}}$$

$$= 70.71\ \text{rad/s}$$

$$f_n = \frac{\omega_n}{2\pi}$$

$$= 11.25\ \text{Hz}$$

$$T = \frac{1}{f_n}$$

$$= 0.089\ \text{s}$$

The oscillations start at the bottom of the cycle so $x_o = -8.0$ mm and the resulting graph of x against time will therefore be a negative cosine curve with an amplitude of 8.0 mm.

The equations that describe the motion are as follows:

$$x = x_o \cos\omega t\ (\text{when } t = 0.05\ \text{s}, x = -8.0\cos(70.71\times 0.05))$$

$$x = 7.387\ \text{mm (Note: Angles are expressed in radians)}$$

To obtain the expression for velocity, the expression displacement is differentiated once

$$v = -\omega\cdot x_o\cdot \sin\omega t$$

$$v = -70.71(-8.0)\sin(70.71\times 0.05)$$

$$= -217\ \text{mm/s}$$

Differentiating the displacement equation a second time will give the equation for acceleration, that is,

$$a = -\omega^2 x_o \cos\omega t \quad \text{since } x = x_o\cos\omega t\ a = -\omega^2 x$$

$$a = -70.712\times 7.387$$

$$= -36{,}934.3\ \text{mm/s}^2$$

Figure 9.9a through c confirms these results.

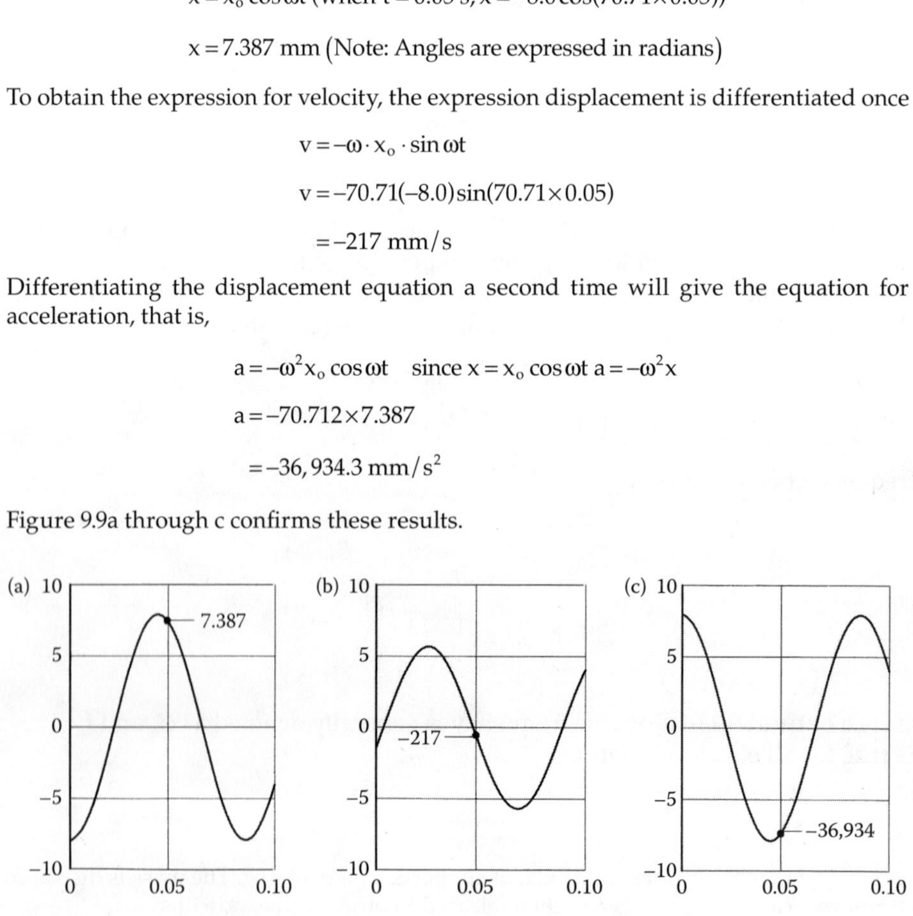

FIGURE 9.9
(a–c): Example 9.3.

9.2.2.6 *Transverse Vibrations*

A transverse vibration is where the motion is normal to the length. This generally occurs in beams and shafts and may be solved by either an angular or linear motion. In this section, a linear motion is used.

9.2.2.6.1 Cantilever

Consider a mass at the end of a cantilever beam as in Figure 9.10. If the mass is displaced at right angles to the beam, the beam will bend and using the beam deflection equations, it will be found that the force is directly proportional to displacement for small beam deflections. If the beam deflections become large, then the constant of proportionality will fail.

The cantilever can be considered as a simple transverse spring.

The stiffness of the beam can be calculated using the beam theory where the deflection of the cantilever due to a point load acting at the end of the beam is given by

$$y = \frac{Fl^3}{3EI} \tag{9.21}$$

In a beam equation, 'y' is used to denote the deflection of the beam since 'x' has already been used. In the studies, thus far 'x' has been used to denote the deflection; therefore, 'x' will be used for future deflections.

Therefore, Equation 9.21 will be rewritten as

$$x = \frac{Fl^3}{3EI} \tag{9.22}$$

Hence, the stiffness will be

$$\begin{aligned} k &= \frac{F}{x} \\ &= \frac{3EI}{l^3} \end{aligned} \tag{9.23}$$

It can be shown that the theory is the same as for a mass on the end of a spring:

$$f_n = \frac{1}{2\pi}\sqrt{\frac{k}{M}} \tag{9.24}$$

$$= \frac{1}{2\pi}\sqrt{\frac{3EI}{Ml^3}} \tag{9.25}$$

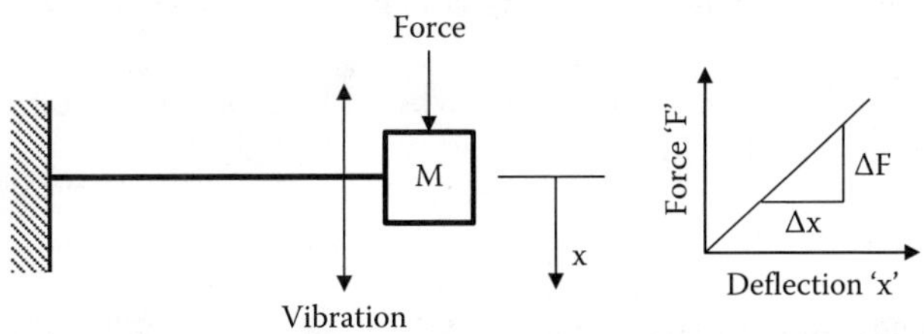

FIGURE 9.10
Cantilever beam with a mass at the free end.

9.2.2.6.2 Simply Supported Beam

Evaluating the deflection of a simply supported beam with a point load acting mid-span, the central deflection will be (see Figure 9.10)

$$x = \frac{Fl^3}{48EI} \tag{9.26}$$

and it follows that:

$$k = \frac{48EI}{l^3} \tag{9.27}$$

and

$$f_n = \frac{1}{2\pi}\sqrt{\frac{48EI}{Ml^3}} \tag{9.28}$$

9.2.2.6.3 Static Deflection

The stiffness can also be found by measuring the static deflection of the beam. Assuming that the mass deflects a small distance 'x_s' under its own weight, the force in this case is the weight; hence,

$$F = Mg \tag{9.29}$$

It follows that:

$$k = \frac{Mg}{x_s} \tag{9.30}$$

$$f_n = \frac{1}{2\pi}\sqrt{\frac{k}{M}} \tag{9.31}$$

and

$$= \frac{1}{2\pi}\sqrt{\frac{g}{x_s}} \tag{9.32}$$

This formula will work both for a cantilever and a simply supported beam. It may be noted that none of the foregoing calculations have considered the mass of the beam or shaft and provided this is small in comparison to the concentrated mass, the formula will give an accurate answer.

Later work will consider beams and shafts where the mass of the beam and shaft is significant in comparison with the load.

Example 9.4

Consider a rod 20.0 mm in diameter and 1200 mm long is rigidly fixed at one end and having a mass of 2 kg acting at the free end. Ignoring the weight of the rod, calculate the frequency of transverse vibrations. Take E = 200 GPa.

Second moment of area for the circular section

$$I = \frac{\pi \cdot D^4}{64}$$

$$= \frac{\pi \times 0.024}{64}$$

$$= 7.854 \times 10^{-9}\ m^4 \tag{9.33}$$

$$f_n = \frac{1}{2\pi}\sqrt{\frac{3EI}{Ml^3}}$$

$$= \frac{1}{2\pi}\sqrt{\frac{3 \times 200 \times 10^9 \times 7.854 \times 10^{-9}}{2 \times 1.23}}$$

$$= 5.88\ Hz \tag{9.34}$$

Example 9.5

A horizontal shaft sitting in simple bearings has a mass of 40 kg placed at mid-span. There is a deflection of 1.00 mm due to the mass. Ignoring the mass of the shaft, determine the frequency of transverse oscillation ($g = 9.81\ m/s^2$).

$$f_n = \frac{1}{2\pi}\sqrt{\frac{g}{x_n}}$$

$$= \frac{1}{2\pi}\sqrt{\frac{9.81}{0.001}}$$

$$= 15.76\ Hz$$

9.2.2.7 Energy Methods (Rayleigh)

A method for solving complex oscillations for a mass–spring system was developed by Rayleigh and is based on that. During an oscillation, the maximum kinetic energy of the oscillating mass is equal to the maximum strain energy. This is best illustrated by referring to Figure 9.11.

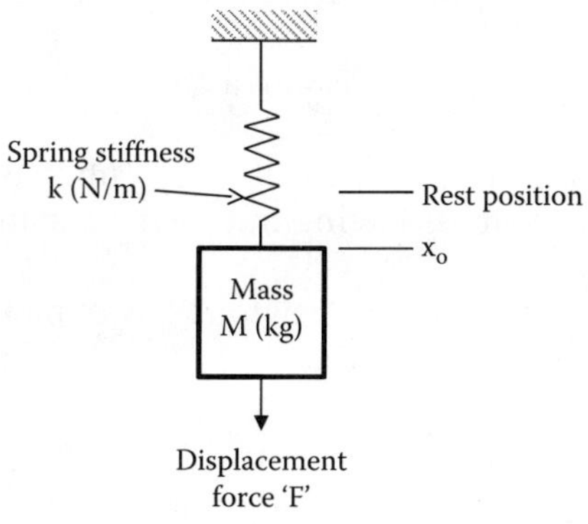

FIGURE 9.11
Rayleigh's method for transverse oscillations.

Let the maximum deflection of the mass be x_o.
The spring force will be $F=kx_o$.

$$\text{The work done} = \frac{1}{2}Fx_o$$

$$= \frac{1}{2}kx_o^2 \tag{9.35}$$

Let the displacement at any time be

$$x = x_o \sin \omega t \tag{9.36}$$

The velocity will be

$$v = \omega x_o \cos \omega t \tag{9.37}$$

The maximum velocity will be $\omega \cdot x_o$.

$$\text{The maximum kinetic energy} = \frac{1}{2}Mv^2 \tag{9.38}$$

$$KE_{max} = \frac{1}{2}M\omega^2 x_o^2 \tag{9.39}$$

Equating energy and $(1/2)M\omega^2 x_o^2$

$$\omega^2 = \frac{k}{M} \tag{9.40}$$

It is important to point out that this result is independent of the deflection (x_o). This method is particularly useful for determining the frequency of transverse oscillations.

9.2.2.8 Kinetic Energy

If a mass 'm' kg is at rest and is then accelerated 'a' to a velocity 'v' m/s. a force 'F' is required to provide the acceleration. This force is given by Newton' second law of motion:

$$F = m \cdot a \tag{9.41}$$

Consider a mass at rest having an initial velocity of 'V_o' (m/s) is then accelerated to a velocity 'V_t' (m/s) in a time 't' (s), the mass having travelled a distance of 'x' (m) with a uniform acceleration of 'a' (m/s^2)

These two quantities are related by the Newton's law of motion:

$$a = \frac{v}{t} \tag{9.42}$$

$$x = \frac{(V_o - V_t)}{2} \times t \tag{9.43}$$

The work done in accelerating the mass:

$$P = \frac{F}{a} \tag{9.44}$$

Energy has been expended in doing this work and has therefore been stored in the mass it will be carried along with it. This energy is called "Kinetic Energy"

$$K.E = \frac{m \cdot v^2}{2} \tag{9.45}$$

Example 9.6

Consider the shaft depicted in Figure 9.12.

It is required to establish the natural frequency of the shaft, ignoring the mass of the shaft.

Take the flexural stiffness as $EI = 20{,}300\ N \cdot m^2$.

The solution to this problem can be derived using two different methods:

1. Static deflection method
2. Strain energy method

In this instance, the strain energy method will be used for the solution.

Determine the reactions R_A and R_B.

Taking moments about R_A:

$1.5 \times 196.2 = 1.0 \times R_C$

$R_C = 294.3\,N$ (down).

Taking moments about R_C:

$1.00\ R_A = 0.50 \times 196.2$

$R_C = -98.1\,N$ (up).

Strain energy for section 'A' to 'C':

$$\begin{aligned} U &= \frac{1}{2EI} \int M^2\, dx \\ &= \frac{1}{2 \times 20{,}300} \int_0^1 (-98.1x)^2\, dx \\ &= \frac{98.1^2}{2 \times 20{,}300} \left[\frac{x^3}{3}\right]_0^1 \\ &= 0.079\ J \end{aligned} \tag{9.46}$$

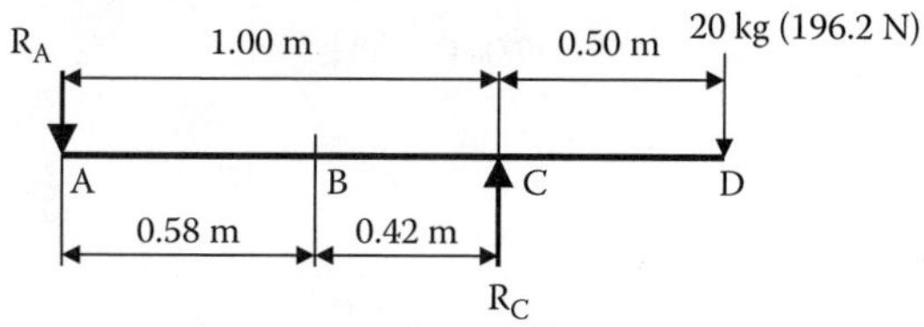

FIGURE 9.12
Shaft in Example 9.6.

Strain energy for section 'C' to 'B':

This will be simplified if 'x' is measured from the free end where M=196.2x.

$$U = \frac{1}{2EI}\int M^2\,dx$$

$$= \frac{1}{2 \times 20{,}300}\int_0^{0.51} (-196.2x)^2\,dx$$

$$= \frac{196.2^2}{2 \times 20{,}300}\left[\frac{x^3}{3}\right]_0^{0.5}$$

$$= 0.0395\,J$$

The total strain energy 'U'=0.079+0.0395=0.1185 J.

Let the deflection produced at the free end be y_m.

The strain energy

$$U = \frac{1}{2}Fy_m$$

$$F = Mg \qquad (9.47)$$

$$= 20\text{ kg} \times g$$

$$= 196.2\text{ N}$$

Hence,

$$y_m = 2 \times \frac{0.1185}{196.2}$$

$$= 0.0012\text{ m}$$

Now if the mass is oscillating up and down sinusoidally

$$y = y_m \sin \omega t \qquad (9.48)$$

The velocity of the oscillation will be

$$v = \omega y_m \cos(\omega t) \qquad (9.49)$$

The kinetic energy will be

$$KE = \frac{1}{2}\frac{Mv^2}{2} \qquad (9.50)$$

$$= \frac{1}{2} M\{\omega y_m \cos(\omega t)\}^2 \tag{9.51}$$

The maximum value of KE can be calculated as follows:
Now y_m is the deflection at the free end of the shaft.

$$= 0.0021\ \text{m}, M = 20\ \text{kg}$$

Equating energies

$$\frac{1}{2} M\{\omega y_m\}^2 = 0.1185$$

$$10\omega^2(0.0021^2) = 0.1185$$

$$\omega^2 = \frac{0.1185}{\{10(0.0021^2)\}}$$

$$= 8229$$

$$\omega = 90.71\ \text{rad/s}$$

$$f_n = \frac{90.71}{(2\pi)}$$

$$= 14.43\ \text{Hz}$$

The static deflection of the shaft could have been determined using the beam theory.

9.3 Damped Vibrations

This section covers the theory of natural vibrations with damping. In Section 9.2, it was intimated that once a system was set to vibrate naturally, it would carry on vibrating as the energy that had been put into the system by the initial disturbance has no way of escaping the system. In practice, the vibrations will gradually reduce in amplitude and die away over a period of time. Figure 9.13 shows the response of a system to varying degrees of damping following an initial displacement 'A'.

The definition of damping is where the free vibration of a system is controlled. A system that is critically damped returns to its stable reference position as quickly as possible without any under- or overshoot. A good example is the machine gun when the recoil mechanism of a gun is designed with critical damping such that the system returns to its firing position in the quickest time without any overshoot. A further example is the automotive suspension system where the vehicle passes over an obstacle and

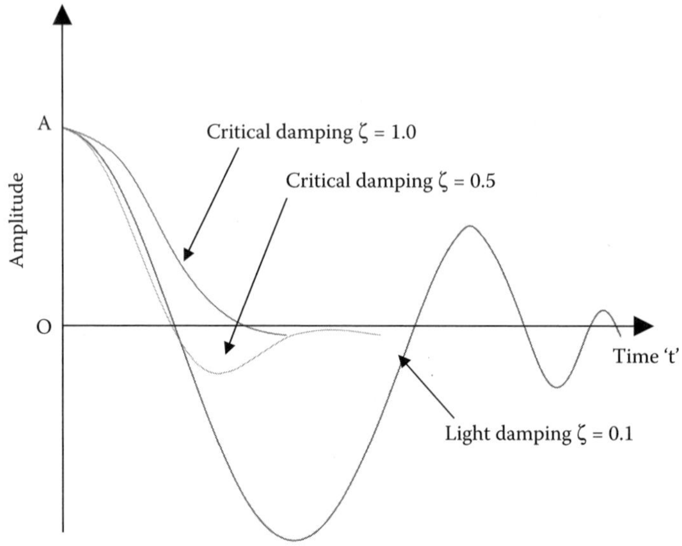

FIGURE 9.13
Damping following an initial displacement 'A'.

returns the vehicle to its correct ride height as quickly as possible without any undesirable undershoot.

Damping can be accomplished by a number of methods including

1. Viscous damping
2. Coulomb damping (dry friction)
3. Inertial damping
4. Internal damping

The basic nomenclature used in damping is covered in Table 9.1. Table 9.2 tabulates the natural frequencies and coefficients 'K' for various modes.

9.3.1 Viscous Damping

Damping that uses a fluid to provide the damping force is known as viscous damping.

Figure 9.14a and b shows two types of viscous dampers. These kinds of dampers are also known as dashpots. In Figure 9.14a, the dashpot uses a fluid (usually oil) and the piston has a number of small orifices passing through the piston. These holes are sized so that the fluid is restricted when it passes from one side to the other. Because of this restriction, a force is required to move the piston within the dashpot and the force can be made to oppose any external force generated by the vibration.

In Figure 9.14b, air is the fluid and as the piston moves within the dashpot, it will suck or force air through the orifice in the bottom of the dashpot. Again, the orifice is sized to provide a resistance to the air as it passes through the orifice and this will provide a force opposing the motion of the piston.

TABLE 9.1

Nomenclature Used in Forced Vibrations

Symbol	Description	Units
L=	Length	m
m=	Mass per unit length	kg/m
I=	Area moment of inertia	m^4
g=	Acceleration due to gravity	($9.81\,m/s^2$)
E=	Modulus of elasticity	N/m^2
G=	Torsional modulus	N/m^2
f=	Frequency of vibration	Hz
y=	Deflection	m
k=	Radius of gyration	m
J=	Shaft polar moment of inertia	m^4
s=	Shaft torsional stiffness (GJ/L)	Nm·rad

TABLE 9.2

Natural Frequencies and Coefficients 'K' for Various Modes

Case	Graphic	Equation	Harmonic Mode				
1	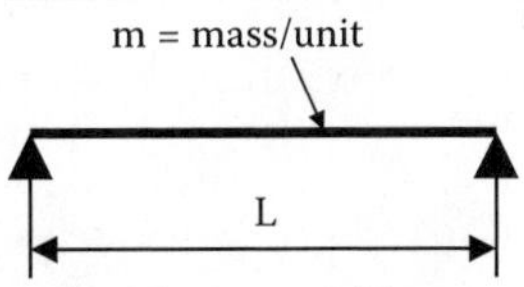 Simply supported beam	$f = \frac{K}{2\cdot\pi\cdot L^2}\sqrt{\frac{E\cdot I}{m}}$	Mode	1	2	3	4
			K	9.87	39.5	88.8	158
2	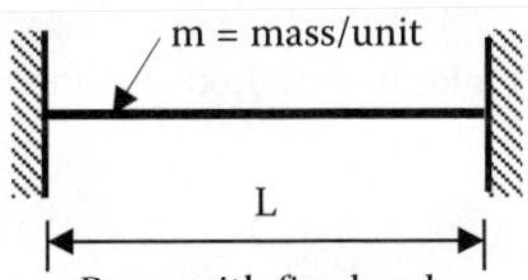 Beam with fixed ends	$f = \frac{K}{2\cdot\pi\cdot L^2}\sqrt{\frac{E\cdot I}{m}}$	Mode	1	2	3	4
			K	22.4	61.7	121.0	200.0
3	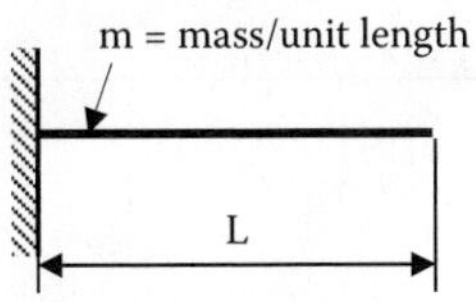 Cantilever subject to transverse vibrations	$f = \frac{K}{2\cdot\pi\cdot L^2}\sqrt{\frac{E\cdot I}{m}}$	Mode	1	2	3	4
			K	3.52	22.0	61.7	121
4	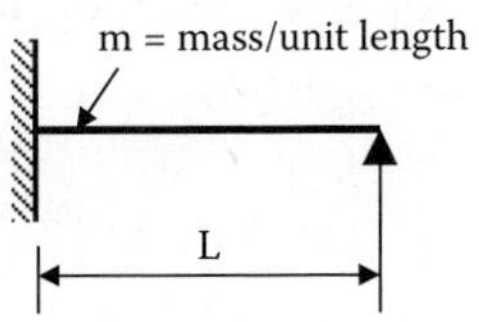 Cantilever one end fixed and simply supported at the free end	$f = \frac{K}{2\cdot\pi\cdot L^2}\sqrt{\frac{E\cdot I}{m}}$	Mode	1	2	3	4
			K	15.4	50.0	104.0	178.0

(*Continued*)

TABLE 9.2 (*Continued*)

Natural Frequencies and Coefficients 'K' for Various Modes

Case	Graphic	Equation	Harmonic Mode
5	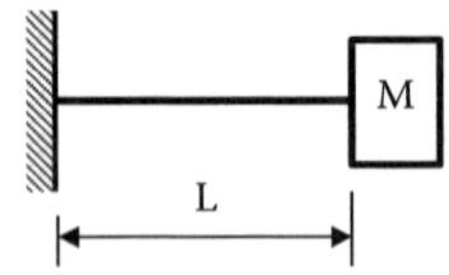 Cantilever with mass at the free end (assume the beam has a negligible mass)	$f = \frac{1}{2\cdot\pi}\sqrt{\frac{3\cdot E\cdot I}{M\cdot L^3}}$	
6	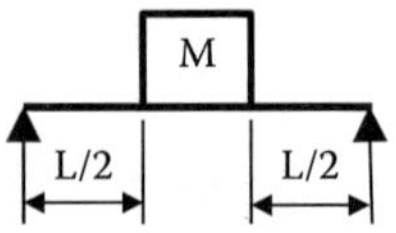 Central mass on simply supported beam (assume the beam has a negligible mass)	$f = \frac{1}{2\cdot\pi}\sqrt{\frac{48\cdot E\cdot I}{M\cdot L^3}}$	
7	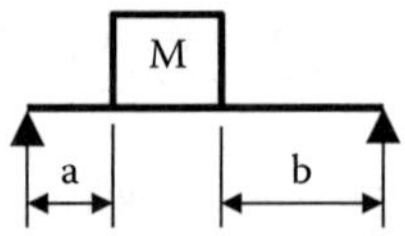 Off-centre mass on a simply supported beam (assume the beam has a negligible mass)	$f = \frac{1}{2\cdot\pi}\sqrt{\frac{3\cdot E\cdot I\cdot L}{M\cdot a^2\cdot b^2}}$	L=a+b
8	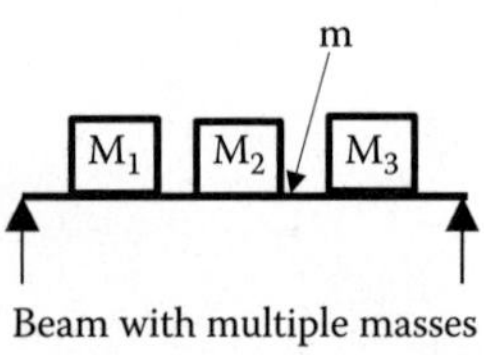 Beam with multiple masses	$\frac{1}{f} = \frac{1}{f^2} + \frac{1}{f_1^2} + \frac{1}{f_2^2} + \frac{1}{f_3^2}$	Using Dunkerly's method of combined loading

It can be shown that for both systems the force opposing the motion (the damping force) is proportional to the velocity of the piston within the dashpot. The equation for this force is

$$F_{dashpot} = \text{constant} \times \text{velocity}$$
$$= c\frac{dx}{dt} \quad (9.52)$$

The constant of proportionality 'c' is called the damping coefficient and has units of Ns/m and is determined by the sizes of the orifices in the piston or the dashpot cylinder.

The damped system is characterised as in Figure 9.15.

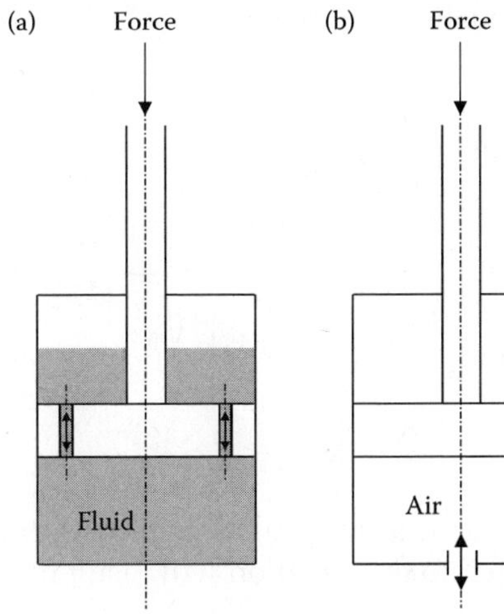

FIGURE 9.14
Dashpot designs. (a) Fluid dashpot and (b) pneumatic dashpot.

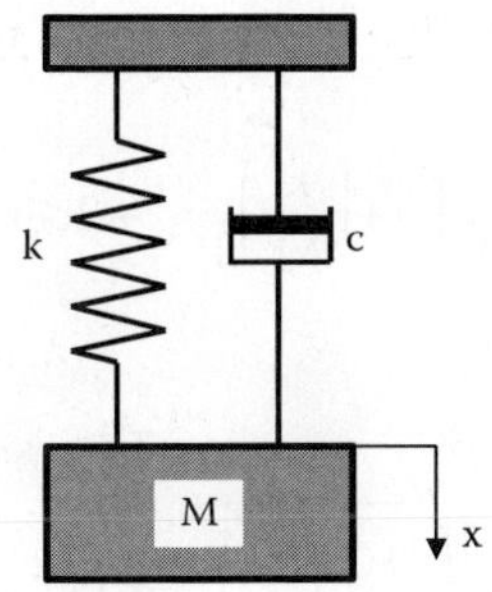

FIGURE 9.15
Spring–mass–viscous damper system.

For a free natural vibration with damping, the equations of motion for the above damper/spring/mass system with damping will be

$$M \cdot \frac{d^2x}{dt^2} = -k \cdot x - c \cdot \frac{dx}{dt} \tag{9.53}$$

Dividing Equation 9.53 through by M gives

$$\frac{d^2x}{dt^2} + \left(\frac{c}{M}\right) \cdot \frac{dx}{dt} + \left(\frac{k}{M}\right) \cdot x = 0 \tag{9.54}$$

Substituting $\omega_n{}^2$ for k/M, making

$$\begin{aligned} \delta &= \frac{c}{2 \cdot M \cdot \omega_n} \\ &= \frac{c}{2 \cdot M \cdot \sqrt{k / M}} \end{aligned} \tag{9.55}$$

The following version of the equation will result:

$$\frac{d^2x}{dt^2} + 2 \cdot \delta \cdot \omega_n \cdot \frac{dx}{dt} + \omega_n{}^2 \cdot x = 0 \tag{9.56}$$

Now

$$\begin{aligned} 2\delta\omega_n &= 2 \cdot \frac{c}{2\sqrt{M \cdot k}} \sqrt{\frac{k}{M}} \\ &= \frac{c}{M} \end{aligned} \tag{9.57}$$

This equation may be solved by assuming a solution in the form $x = e^{s \cdot t}$ (s=constant).
Substituting this relationship into the equation will result in the following formula

$$\left[s^2 + \left(\frac{c}{M}\right) \cdot s + \left(\frac{k}{M}\right)\right] e^{st} = (s^2 + 2\delta\omega_n s + \omega_n^2) e^{st} = 0 \tag{9.58}$$

Hence,

$$s^2 + \left(\frac{c}{M}\right) \cdot s + \left(\frac{k}{M}\right) = s^2 + 2\delta\omega_n s + \omega_n^2 \tag{9.59}$$

There are two roots to this equation

$$\begin{aligned} s_{1.2} &= -\frac{c}{2M} \pm \sqrt{\left(\frac{c}{2M}\right)^2 - \frac{k}{M}} = -\delta\omega_n \pm \sqrt{\delta^2\omega_n^2 - \omega_n^2} \\ &= \omega_n(-\delta \pm \sqrt{\delta^2 - 1}) \end{aligned} \tag{9.60}$$

The solution of this equation in general forms

$$x = Ae^{s1}t + Be^{s2}t \tag{9.61}$$

A and B are constants which can be evaluated from initial values of 'x' and dx/dt. The substitution of the roots into the general form will result in the following equation

$$x = e^{-\left(\frac{c}{2M}\right)t} \left(Ae^{\left(\sqrt{\left(\frac{c}{2M}\right)2 - \frac{k}{M}}\right)^t} + Be^{-\left(\sqrt{\left(\frac{c}{2M}\right)2 - \frac{k}{M}}\right)^t} \right) \tag{9.62}$$

Alternatively,

$$x = e^{-\delta\omega_n t} \left(Ae^{+\omega_n\left(\sqrt{\delta^2 - 1}\right)t} + Be^{-\omega_n\left(\sqrt{\delta^2 - 1}\right)t} \right) \tag{9.63}$$

The term $e^{-(c/2M)t}$ represents an exponential decaying factor. There are three general results arising from the expression within the brackets that will have a significant effect on the results, that is,

1. If $(c/2M)^2=(k/M)\cdots(\delta^2=1)$, the factor within the bracket is 0 and the solution will be considered to be critically damped.
2. If $(c/2M)^2 < (k/M)\cdots(\delta^2<1)$, the factor inside the bracket is negative; therefore, the solution will be underdamped.
3. If $(c/2M)^2 > (k/M)\cdots(\delta^2>1)$, the factor inside the bracket is positive; therefore, the solution will be overdamped.
4. If $(\chi/2M)^2=0\cdots(\delta^2>0)$, the factor inside the bracket shows that the system is undamped.

9.3.2 Coulomb Damping

When a machine slide is accelerated along a slideway, friction will offer a resistance to the slide's movement. This resistance is referred to as Coulomb damping and is proportional to the friction in the system (see Figures 9.16 and 9.17).

Static friction occurs when the two objects are stationary or undergoing no relative motion. The frictional force 'F' exerted between the two surfaces having no relative movement cannot exceed a value that is proportional to the product of the normal force 'N' and the coefficient of static friction μ_s.

$$F_s = \mu_s \cdot N \tag{9.64}$$

Kinetic friction occurs when the two contacting surfaces are undergoing relative motion and sliding against each other. In this case, the friction force 'F' between the two surfaces is proportional to the product of the normal force 'N' and the coefficient of kinetic friction μ_k.

$$F_k = \mu_k \cdot N \tag{9.65}$$

In both these cases, the frictional force will always oppose the direction of movement between the surfaces. The normal force is perpendicular to the direction of motion and is equal to the weight (Mg) of the sliding object, where the damping force $F_d=\mu_{s\cdot k}\cdot N$ and is independent of velocity.

The system is given an initial displacement 'A' and is then released as shown in Figure 9.13.

$$\frac{2\pi}{\omega_n}$$

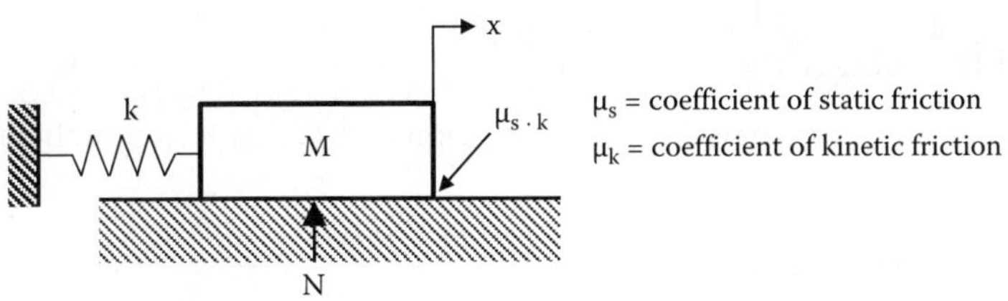

FIGURE 9.16
Coulomb damping.

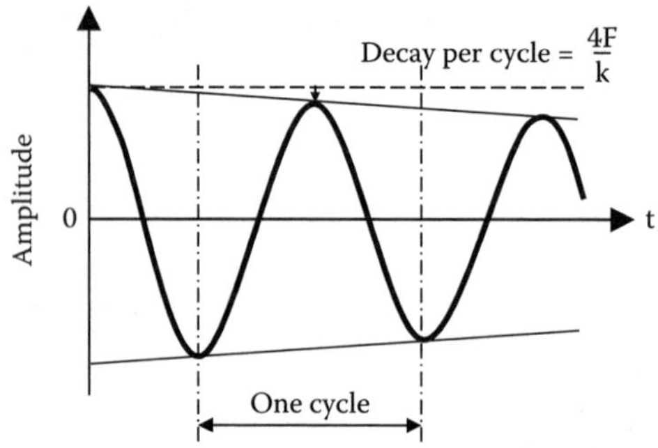

FIGURE 9.17
Coulomb diagram.

The equations of motion for each direction

$$M\ddot{x} + kx \pm F_d = 0$$

$$x = \left(A \pm \frac{F}{k}\right)\cos\omega_n t \mp \frac{F}{k} \quad \mp \text{Since 'F' opposes motion}$$

9.3.3 Inertial Damping

A significant mass is attached to the vibrating system. This has the effect of absorbing the vibration energy; it does not eliminate the vibration but introduces a phase lag such that when the vibration peak passes, the inertia of the mass absorbs that energy and this then reduces the instantaneous energy. An example is the counterweights fitted to the crankshaft of an engine as they even out the vibrations created within the engine.

A further example is the ballast fitted to boats where they damp out the rolling action created by the sea waves.

Internal damping (also known as hysteretic and viscoelastic damping) is mainly concerned with the support of a mass using an isolator between the mass and supporting structure such as the ground.

There are two methods of isolation:

1. Active isolation
2. Passive isolation

Figure 9.18a and b is a schematic diagram of a mass subjected to a vertical periodic force.

9.3.4 Internal Damping

In general, rubber mounts are used to isolate vibrations under items such as machine tools and so on.

D=Specific damping energy
 = energy dissipated per cycle per unit volume of material
 = area of stress–strain loop (Hysteresis loop)

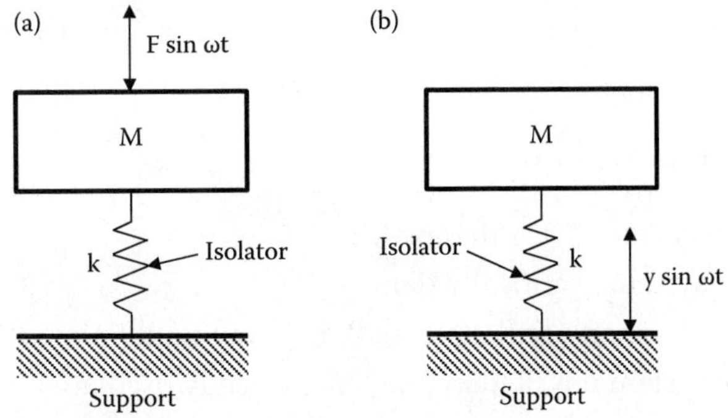

FIGURE 9.18
Mass subject to a vertical periodic force. (a) Active isolation and (b) passive isolation.

U=Maximum energy stored per unit volume during loading
ψ=Specific damping capacity
α=Loss angle, the phase difference between stress σ and strain ε
tan ϕ=Loss factor
$D=\pi\sigma\varepsilon \sin\phi$
$U=1/2\ \sigma\varepsilon\cos\phi$
$\psi=D/(2U)=\pi\tan\phi$

For a mass supported by rubber isolators

$$\delta=\frac{1}{2}\tan\alpha$$

The value of α for natural rubber will vary between 4° and 11°.

For neoprene, α=9° and for butyl α=15°.

The equations of motion will be

$$x=A^{e-\delta\omega_n t}\sin(\omega_d t+\psi)\quad x+\tan\alpha\omega_n x+\omega_n^2 x=0$$

with the solution

$$x=A^{e-\delta\omega_n t}\sin(\omega_d t+\psi)$$

9.4 Single Degree of Freedom: Forced Vibrations

In the previous section covering damped oscillations, it was shown that a free vibration will eventually die away over a period of time as the energy is dissipated by the damping. It was shown that the equation for the displacement of a damped oscillation is given by

$$x=Ce^{-\delta\omega_n t}\cos(\omega_n t) \tag{9.66}$$

where

δ is the damping factor

ω_n is the natural frequency

The following cases are described:

1. When $\delta > 1$, the system is overdamped.
2. When $\delta = 1$, the system is critically damped.
3. When $\delta < 1$, the damped oscillations will die away with time.
4. When $\delta = 0$, the system has no damping and a steady oscillation will occur.

Figure 9.19 illustrates the effect of the damping ratio.

If the damping factor 'δ' is less than zero, that is, negative, this would be the opposite to damping and would reverse the energy flow in that energy would be put back into the system instead of removing it. As the energy is put back into the system, it is possible for the oscillations to increase, the energy being supplied by an external force, and such oscillations are called forced oscillations.

An excellent example is that of a child sitting on a swing. If the child begins to swing but does nothing, the swing will gradually come to a halt. If someone gives the swing a small push at the start of each swing, energy is being added to the system and the swing (oscillations) begins to increase, that is, gets higher and higher. This phenomenon is called excitation.

Many engineering structures are prone to vibration when being excited at or near their natural frequency. If the excitation is close to the natural frequency, then the oscillations may become out of control. An example is the wind blowing around cooling towers, chimney stacks and suspended cables. The phenomenon is known as vortex shedding and is a major problem for the designers of these structures as the oscillations may increase and lead to a catastrophic failure.

At a lower level, if an automotive is travelling over a corrugated surface such as Belgian pave, the disturbance may be close to the natural frequency of the suspension causing the vehicle to possibly bounce out of control. Most vehicle-testing laboratories will usually have a section of pave as part of the suspension test.

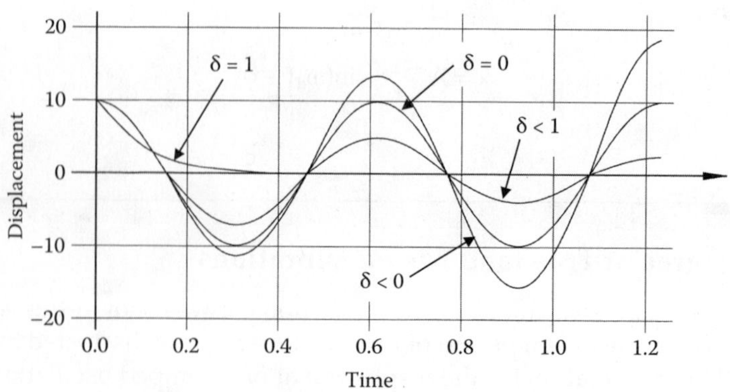

FIGURE 9.19
Comparison of various damping ratios.

9.4.1 Forced Vibrations

Two types of forced vibrations will be considered:

1. When a mass has a disturbing force acting on it.
2. When the spring support is disturbed harmonically.

9.4.1.1 Disturbing Force Acting on Mass

Figure 9.20 shows a system where a mass sitting on a spring and fitted with a damper has a fixed support. Located on the mass is a small rotating machine that is out of balance. This is equivalent to a small mass 'm' that is rotating at a radius 'r' and is producing the out-of-balance force. The magnitude of the force is $F_o = mr\omega^2$. The main mass is restrained within guides and is restricted to be only able to move vertically (one degree of freedom). At the position shown with the small mass rotated through an angle of 'θ', the component of F_o acting vertically is $F' = F_o \sin\theta$.

Any force applied to the main mass must overcome the inertia, damping force and spring force.

The applied force is

$$F = F_i + F_d + F_s \tag{9.67}$$

$$F = M\frac{d^2x}{dt^2} + c\frac{dx}{dt} + kx \tag{9.68}$$

In this case, as the mass is restricted to move vertically, the only force applied to the mass is the vertical component of the centrifugal force.

$$F_o \sin(\omega t) = M\frac{d^2x}{dt^2} + c\frac{dx}{dt} + kx \tag{9.69}$$

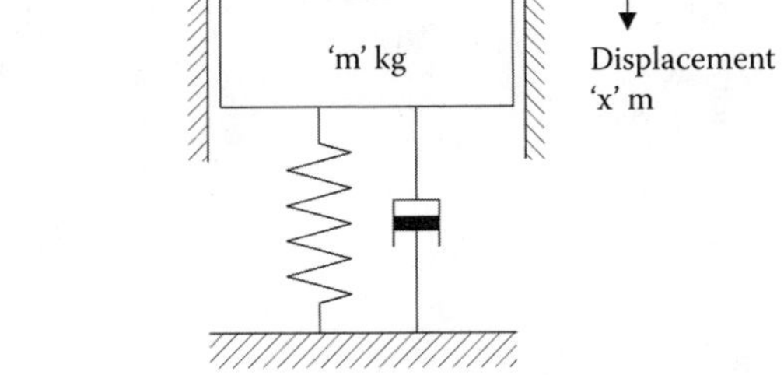

FIGURE 9.20
Mass subject to rotor excitation.

9.4.1.2 Phasor Representation

It is assumed that the mass oscillates vertically with a sinusoidal oscillation 'A'. Assume that the oscillation is going to start when it passes through the rest position. The displacement is given by $x = A\sin \omega t$ where A is the amplitude.

The velocity is $v = (dx/dt) = A\omega \cos \omega t$ where $A\omega$ is the amplitude.

The acceleration will be $a = (dv/dt) = -A\omega^2 \sin \omega t$ where $A\omega^2$ is the amplitude.

The displacement 'x', velocity 'v' and acceleration 'a' when plotted against time 't' will result in Figure 9.21. Each graph may be generated by a vector rotating at ω (rad/s) and with a length equivalent to the amplitude. Such vectors are referred to as phasors. At a given point in time, the tip of each vector is projected across to the appropriate point as indicated.

It is clearly seen that in order to produce the results, the velocity vector is 90° in front of the displacement vector and the acceleration vector is 90° in front of the velocity vector.

Considering Figure 9.20, it is obvious that the spring force is directly proportional to the displacement 'x', so therefore it must be in phase with 'x'. The damping force is then directly proportional to the velocity 'v' and is in phase with 'v'; and finally, the inertial force is directly proportional to the acceleration 'a'; hence, it follows that the inertia is in phase with 'a'.

All three forces can then be represented by phasors rotating at an angular velocity ω (rad/s).

The spring force is in phase with the movement; therefore, the displacement vector can be drawn horizontally and the velocity vector and acceleration vector can be drawn 90° and 180° ahead, respectively as shown in Figure 9.22.

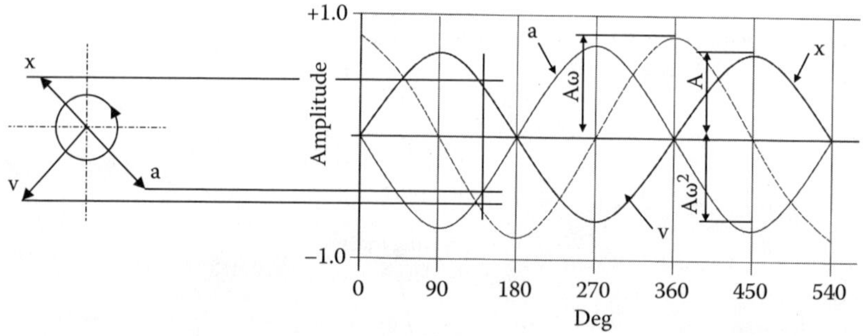

FIGURE 9.21
Phasor representation.

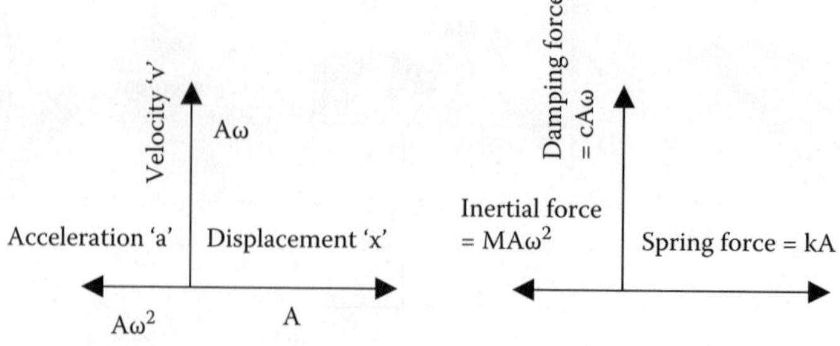

FIGURE 9.22
Representation of vectors.

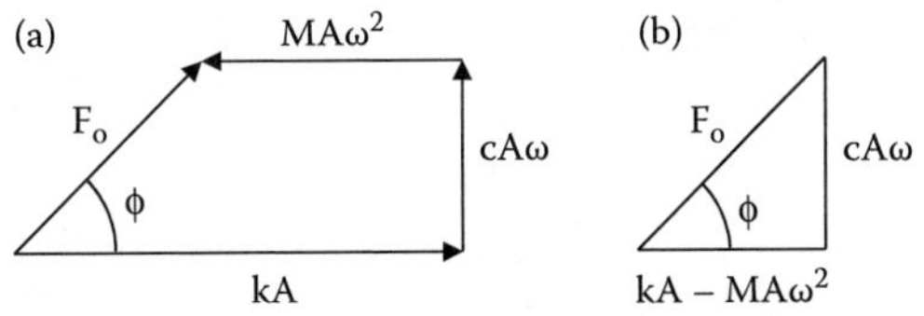

FIGURE 9.23
(a, b): Addition of vectors.

The sum of these three forces can be represented by a force F_o and by adding these three forces a typical vector diagram will result as shown in Figure 9.23.

From the diagram, it is seen that the applied force F_o is at an angle ϕ to the horizontal axis; therefore, it must be displaced by a phase angle 'ϕ' relative to 'x'.

$$F_o^2 = (kA - MA\omega^2)^2 + (cA\omega)^2 \tag{9.70}$$

$$F_o{}^2 = A^2(k - M\omega^2)^2 + A^2(c\omega)^2 \tag{9.71}$$

$$F_o{}^2 = A^2\left[(k - M\omega^2)^2 + (c\omega)^2\right] \tag{9.72}$$

Dividing through by M^2

$$\frac{F_o{}^2}{M^2} = A^2\left[(k - \omega^2)^2 + \left(\frac{c\omega}{M}\right)^2\right] \tag{9.73}$$

From Equations 9.55 and 9.57, it was shown that $\omega_n^2 = k/m$ and $c/M = 2\delta\omega_n$.

$$A^2 = \left(\frac{F_o}{M}\right)^2 \left\{\frac{1}{(\omega_n^2 - \omega^2)^2 + (2\delta\omega\omega_n)^2}\right\} \tag{9.74}$$

From the triangle, the phase angle 'ϕ' is obtained.

$$\tan\phi = \frac{2\delta\omega\omega_n}{\omega_n^2 - \omega^2} \tag{9.75}$$

Plotting displacement (x) and the applied force (F_o) against time gives a graph similar to Figure 9.24.

9.5 Natural Frequency of Beams and Shafts

This section discusses the natural frequency of transverse vibrations.

9.5.1 Degrees of Freedom

In Section 9.3, vibration in one degree of freedom was discussed. In this section, this discussion will be extended to three degrees of freedom.

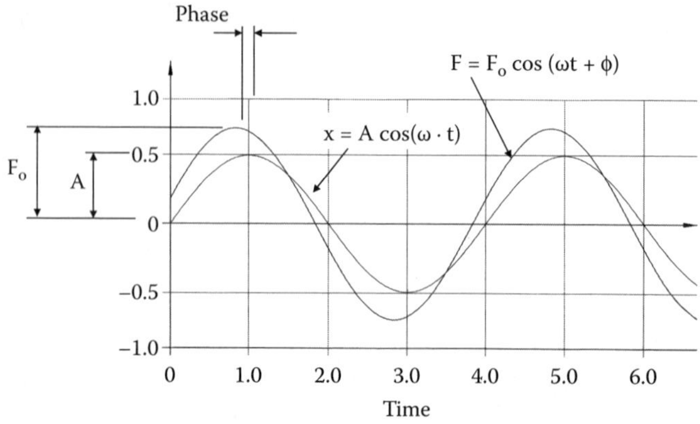

FIGURE 9.24
Frequency versus time.

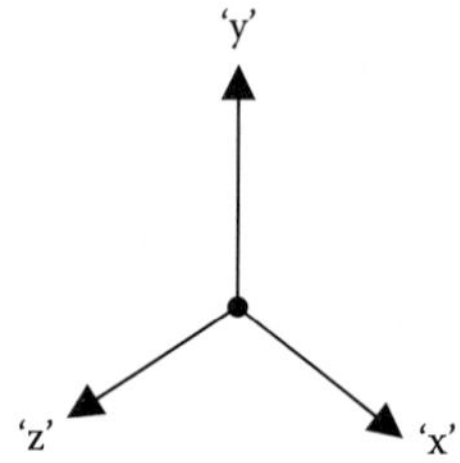

FIGURE 9.25
Three degrees of freedom axes.

There are three translation axes as shown in Figure 9.25; they are in the 'x', 'y' and 'z' directions in which an object can move unrestrained.

There are three other degrees of freedom which cover pitch, roll and yaw. These are superimposed on the translation axes as shown in Figure 9.26.

In the study of free vibrations, only one degree of freedom will be considered.

9.5.2 Beams Subject to Transverse Vibrations

A beam or cantilever when subject to transverse vibrations will exhibit various modal behaviour characteristics. These behaviours will be dependent upon the natural frequency of the beam or cantilever.

9.5.3 Simply Supported Beam Subject to Transverse Vibration

Consider a beam as shown in Figure 9.27a which is simply supported at each end and subjected to a transverse vibration. This is referred to as a mode-1 vibration.

If the frequency of the vibration is increased, the second harmonic is reached (see Figure 9.27b). And if the frequency is increased even further, a third harmonic will be reached resulting in Figure 9.27c.

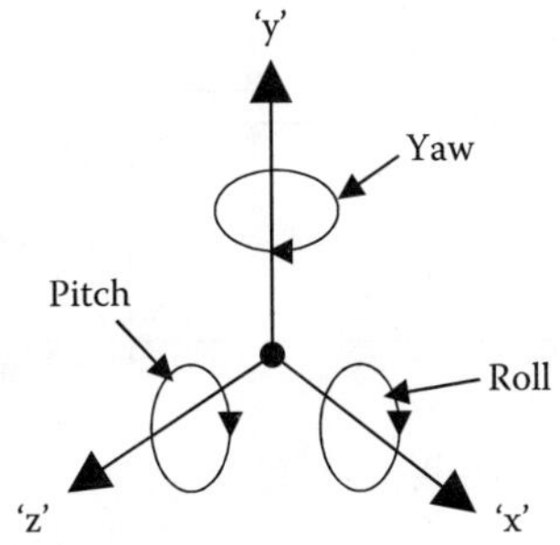

FIGURE 9.26
Six degrees of freedom.

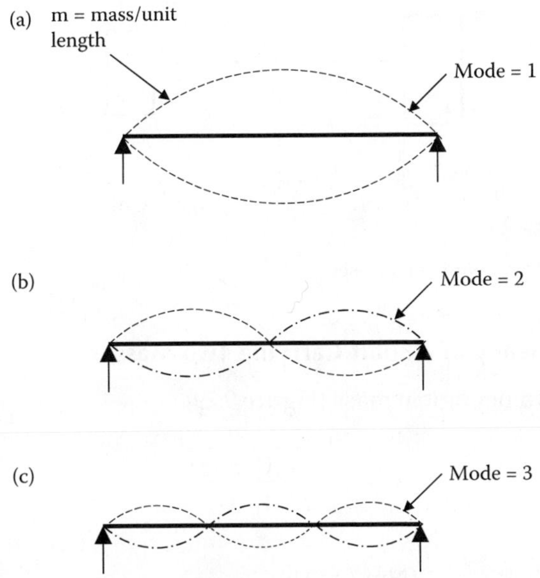

FIGURE 9.27
Transverse harmonics. (a) First harmonic, (b) second harmonic and (c) third harmonic.

9.5.4 Torsional Frequency of a Cantilevered Shaft Carrying a Mass at the Free End

Assume the beam has a negligible mass (Figure 9.28).

$$f_n = \frac{1}{2\pi}\sqrt{\frac{k}{J_m}}\,\text{Hz} \tag{9.76}$$

where

$$k = \text{torsional stiffness of shaft} = \frac{G \cdot J}{L}$$

$$J = \text{polar moment of inertia} = \frac{\pi}{32}D^4$$

$$J_m = \text{polar moment of inertia of mass} = Mk^2$$

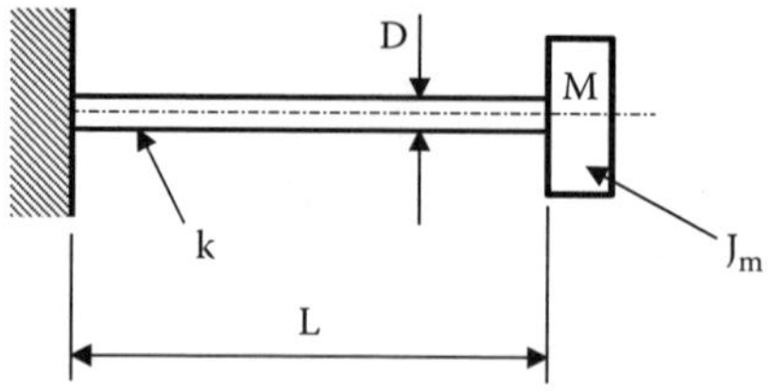

FIGURE 9.28
Torsional frequency of a cantilever shaft.

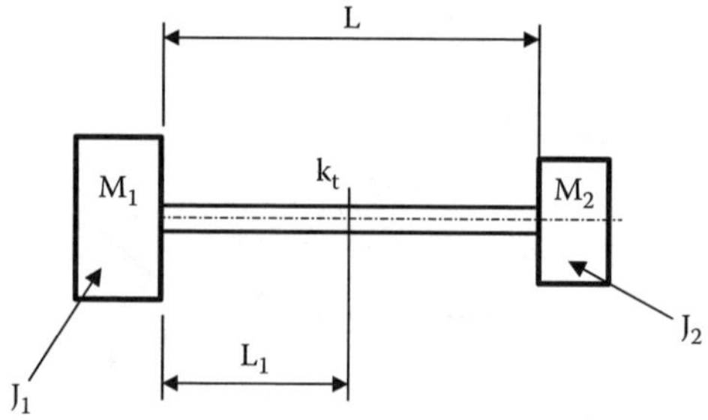

FIGURE 9.29
Torsional frequency of shaft carrying two masses.

9.5.5 Torsional Frequency of a Shaft Carrying Two Masses

Assume the beam has a negligible mass (Figure 9.29).

$$f_n = \frac{1}{2\pi}\sqrt{\frac{k_t(J_1 + J_2)}{(J_1 \cdot J_2)}}\,\text{Hz} \tag{9.77}$$

Torsional stiffness of shaft $= \frac{G \cdot J}{L}$ (Nm / rad).

G=shear modulus (N/m^2)
θ=twist in the shaft (rad)
Polar moment of inertia of masses=$M_1k^2+M_2k^2$
Position of node $= L_1 = L / (1+(J_{M_1} / J_{M_2}))$
Ratio of the twist in the shaft

$$\frac{\theta_1}{\theta_2} = \frac{J_2}{J_1} \tag{9.78}$$

Figure 9.30 shows the position of the node.

9.5.6 Torsionally Equivalent Shafts

In the previous two examples, a uniform diameter shaft was considered. In actual practice, shafts with variable diameters and lengths are more likely to be used.

In such cases, a shaft may theoretically be replaced by an equivalent shaft that has a uniform diameter.

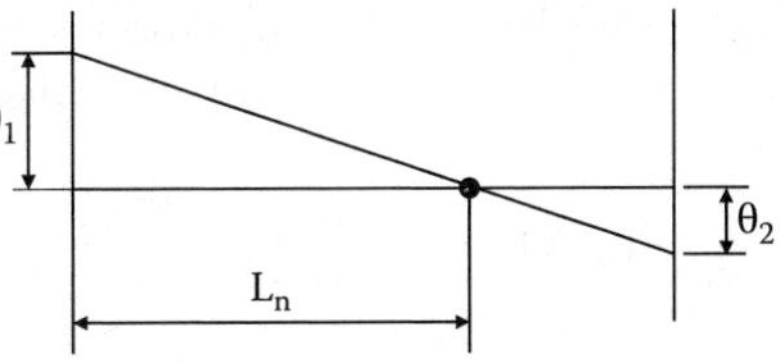

FIGURE 9.30
Natural frequency of a two mass system and node position.

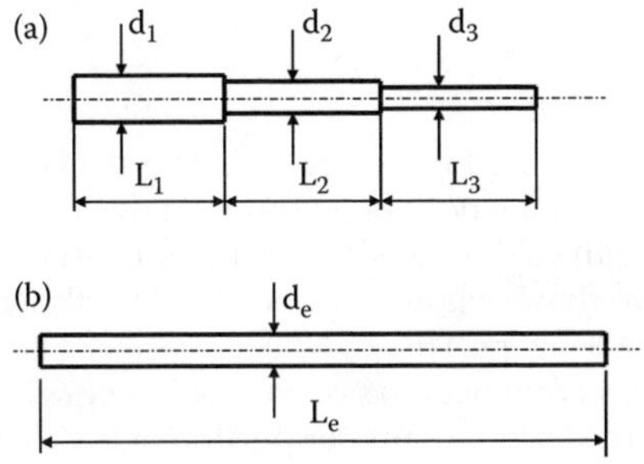

FIGURE 9.31
(a, b): Three-stepped equivalent shaft.

Figure 9.31a depicts a shaft having three steps with diameters d_1, d_2 and d_3 with lengths L_1, L_2 and L_3, respectively. Consider that this shaft is replaced by an equivalent shaft having a uniform diameter 'd_e' and length 'L_e' as shown in Figure 9.31b. The equivalent shaft has to have the same angle of twist as the original shaft when opposing torques 'T' are applied from both ends of the shaft.

Let

d_1, d_2 and d_3=diameters for the corresponding lengths 'L'.

θ_1, θ_2 and θ_3=angle of twist for lengths L_1, L_2 and L_3, respectively.

θ_T=total angle of twist.

J_1, J_2 and J_3=polar moments of inertia for the shafts having diameters d_1, d_2 and d_3, respectively.

As the total angle of twist of the equivalent shaft is equal to the sum of the individual angles of twists of the different lengths, therefore

$$\theta_1 = \frac{T \cdot L_1}{G \cdot J_1}, \quad \theta_2 = \frac{T \cdot L_2}{G \cdot J_2}, \quad \theta_3 = \frac{T \cdot L_3}{G \cdot J_3} \tag{9.79}$$

Hence

$$\frac{T \cdot L_e}{G \cdot J_e} = \frac{T \cdot L_1}{G \cdot J_1} + \frac{T \cdot L_2}{G \cdot J_2} + \frac{T \cdot L_3}{G \cdot J_3} \tag{9.80}$$

Extracting T/G which is common throughout:

$$\frac{L_e}{J_e} = \frac{L_1}{J_1} + \frac{L_2}{J_2} + \frac{L_3}{J_3} \tag{9.81}$$

Now

$$J_e = \frac{\pi \cdot d_e^4}{32}, J_1 = \frac{\pi \cdot d_1^4}{32}, J_2 = \frac{\pi \cdot d_2^4}{32} \text{ and } J_3 = \frac{\pi \cdot d_3^4}{32} \tag{9.82}$$

Replacing 'J' and multiplying throughout with d_1^4 where the diameter of the equivalent shaft diameter 'd' is taken as d_1, the length of the equivalent shaft is

$$L_e = L_1 + L_2 \frac{d_1^4}{d_2^4} + L_3 \frac{d_1^4}{d_3^4} \tag{9.83}$$

Whenever confronted with a stepped shaft in torsional vibration problems, always convert it into an equivalent shaft of uniform diameter and then determine the torsional frequency as discussed for the two rotor example.

Example 9.7

Consider a steel shaft as illustrated in Figure 9.32. The steel shaft is 1500 mm long. It is stepped at 95 mm diameter for a length of 600 mm, followed by a diameter of 60 mm for a length of 500 mm and finally a diameter of 50 mm for the remaining length of 400 mm. The shaft is fitted with two flywheels at each end. The first flywheel has a mass of 900 kg and a radius of gyration of 850 mm. This flywheel is located on the 95 mm diameter portion of the shaft. The second flywheel has a mass of 700 kg with a radius of gyration of 550 mm and is located at the 50 mm diameter portion of the shaft.

Calculate the position of the node together with the natural frequency of the free torsional vibration of the shaft. The modulus of rigidity (G) of the shaft material may be taken as 7.2 GPa.

Solution

Given that

$L = 1500$ mm

$L_1 = 600$ mm	$d_1 = 95$ mm
$L_2 = 500$ mm	$d_2 = 60$ mm
$L_3 = 400$ mm	$d_3 = 50$ mm
$M_a = 900$ kg	$k_a = 850$ mm
$M_b = 700$ kg	$k_b = 550$ mm
$G = 7.2$ GPa	

Determine the length of the equivalent shaft assuming its diameter as $d_1 = 95$ mm.

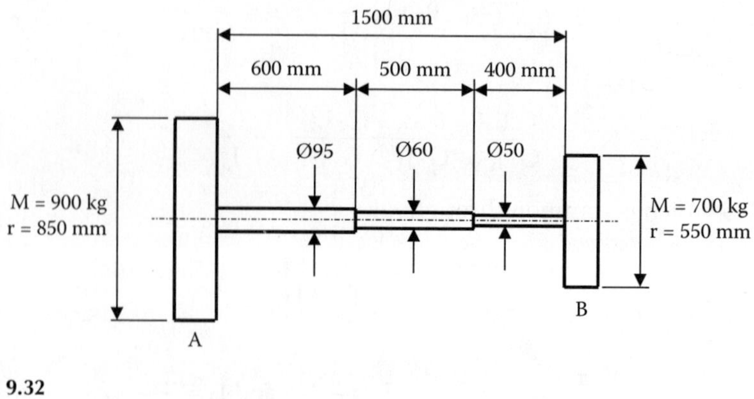

FIGURE 9.32
Stepped shaft carrying two flywheels.

From Equation 9.83, the length of the equivalent shaft is calculated as

$$L_e = L_1 + L_2 \frac{{d_1}^4}{{d_2}^4} + L_3 \frac{{d_1}^4}{{d_3}^4}$$

$$L_e = 600\,\text{mm} + 500\,\text{mm} \times \frac{95\,\text{mm}^4}{60\,\text{mm}^4} + 400\,\text{mm} \times \frac{95\,\text{mm}^4}{50\,\text{mm}^4}$$

$$L_e = 8955\,\text{mm}\,(8.955\,\text{m})$$

Location of the Node

Assume the position of the node of the equivalent shaft lies at 'N' as shown in Figure 9.33.

Let

$$L_A = \text{distance of the node from the flywheel 'A'}$$

$$L_B = \text{distance of the node from the flywheel 'B'}$$

Mass moment of inertia of flywheel 'A':

$$J_A = M_a \times {k_a}^2$$

$$J_A = 900\text{ kg} \times (850\text{ mm})^2$$

$$J_A = 650.25\text{ kg}\cdot\text{m}^2$$

Mass moment of inertia of flywheel 'B':

$$J_B = M_b \times {k_b}^2$$

$$J_B = 700\,\text{kg} \times (550\text{ mm})^2$$

$$J_B = 211.75\,\text{kg}\cdot\text{m}^2$$

Hence,

$$L_A \times J_A = L_B \times J_B$$

$$L_A = \frac{L_B \times J_B}{J_A}$$

$$L_A = \frac{L_B \times 211.75\text{ kg}\cdot\text{m}^2}{650\text{ kg}\cdot\text{m}^2}$$

$$L_A = 0.326\ L_B$$

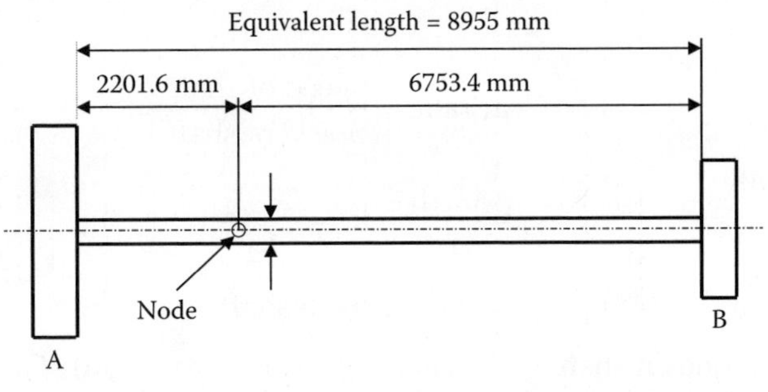

FIGURE 9.33
Mode position for Example 9.7.

Now

$$L_A + L_B = L = 8955 \text{ mm}$$

$$L_A = 0.326\ L_B + L_B$$

$$L_A = 1.326\ L_B$$

$$L_B = \frac{8955 \text{ mm}}{1.326}$$

$$\therefore\quad L_B = 6753.4 \text{ mm}$$

$$L_A = 8955 - 6753.4 \text{ mm}$$

$$L_A = 2201.6 \text{ mm}$$

Hence, the node lies at 2201.6 mm from flywheel 'A' and 6753.4 mm from flywheel 'B'.

Natural Frequency of Free Torsional Vibrations

The polar moment of inertia of the equivalent shaft:

$$J_e = \frac{\pi \cdot d_1^4}{32}$$

$$J_e = \frac{\pi \times (0.095 \text{ m})^4}{32}$$

$$J_e = 8.0 \times 10^{-6} \text{ m}$$

$$f_n = \frac{1}{2\pi}\sqrt{\frac{G \cdot J_e}{L_A \cdot J_B}}$$

$$f_n = \frac{1}{2\pi}\sqrt{\frac{72 \times 10^9 \text{N/m}^2 \times 8.0 \times 10^{-6} \text{m}^4}{2.2 \text{ m} \times 650 \text{ kg} \cdot \text{m}^2}}$$

$$f_n = 3.194 \text{ Hz}$$

9.5.7 Torsional Frequency of a Geared Shaft Carrying Two Masses

Note: The shafts are assumed to be light (Figure 9.34).

$$f_n = \frac{1}{2\pi}\sqrt{\frac{k_1 k_2}{G_R^2 k_2 + k_1}\left(\frac{J_1 + G_R^2 J_2}{J_1 \cdot J_2}\right)} \text{Hz (Neglecting the inertias of the gears)} \tag{9.84}$$

where

$$G_R = \text{Gear ratio} = \frac{\text{Speed of shaft 2}}{\text{Speed of shaft 1}} \tag{9.85}$$

$$k_1 = \frac{G_{R1} I_1}{L_1} \quad k_2 = \frac{G_{R2} I_2}{L_2} \tag{9.86}$$

Note: Torque and inertia are referred to the driven shaft.

$$\text{Torque on shaft 2} = \text{torque on shaft 1} \times (\text{gear ratio})\ (G_R).$$

$$\text{Inertia of shaft 2} = \text{inertia of shaft 1} \times (\text{gear ratio})^2\ (G_R)^2.$$

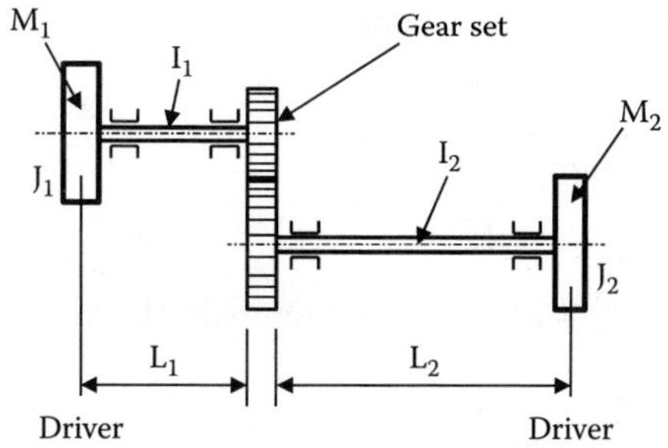

FIGURE 9.34
Torsional frequency of a geared shaft with two masses.

Example 9.8

A steel shaft 100 mm in diameter and 1000 mm long carries a flywheel at its end measuring 1000 mm in diameter and weighing 1000 kg.

Calculate the torsional frequency of this single rotor system (see Figure 9.35).

Solution

Given that the diameter of the shaft = 100 mm, length of the shaft = 1000 mm, modulus of rigidity (G) = 72×10^9 Pa and density (ρ) = 7.8 kg/m^3.

Polar moment of inertia of the shaft:

$$J_{shaft} = \frac{\pi\times(\text{dia' shaft})^4}{32}$$

$$J_{shaft} = \frac{\pi\times100\ \text{mm}^4}{32} = 9.817\times10^{-6}\ \text{m}^4$$

Stiffness of the shaft:

$$S = \frac{J\cdot G}{L}$$

$$S = \frac{9.817\times10^{-6}\text{m}^4\times72\times10^9\ \text{Pa}}{1000\ \text{mm}} = 7.069\times10^5\ \text{N}\cdot\text{m}$$

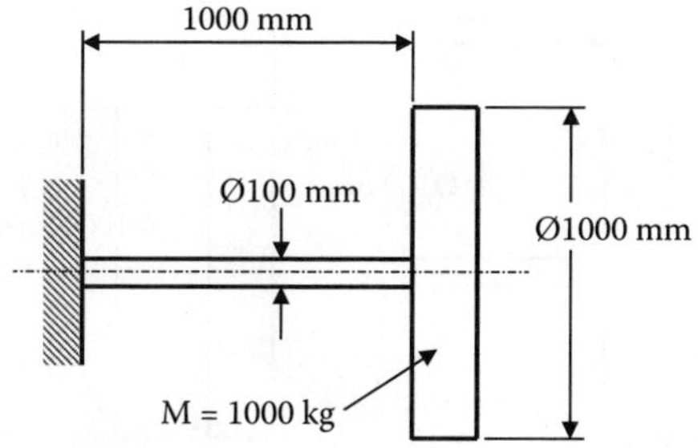

FIGURE 9.35
Torsional frequency of a plane shaft with one flywheel.

Polar Moment of Inertia of the Flywheel

Mass moment of inertia:

Mass = 1000 kg

Radius of gyration = 500 mm

$$J_m = M \cdot k^2$$

$$J_m = 1000\ \text{kg} \times 500^2\ \text{mm} = 250\ \text{kg} \cdot \text{m}^2$$

$$f_n = \frac{1}{2\pi} \cdot \sqrt{\frac{S}{J_m}}$$

$$f_n = \frac{1}{2\pi} \cdot \sqrt{\frac{7.069 \times 10^5\ \text{Nm}}{250\ \text{kg} \cdot \text{m}^2}}$$

$$f_n = 8.463\ \text{Hz}$$

Example 9.9

A steel shaft 100 mm in diameter and 1000 mm long carries two identical flywheels at each end measuring 1000 mm in diameter and weighing 1000 kg each.

Calculate the torsional frequency of this two rotor system and determine the node position. (Refer to Figure 9.36).

Solution

Given that the diameter of the shaft = 100 mm, length of the shaft = 1000 mm, modulus of rigidity (G) = 72×10^9 Pa and density (ρ) = 7.8 kg/m³.

Polar Moment of Inertia of Shaft

$$J_{shaft} = \frac{\pi \times (\text{dia' shaft})^4}{32}$$

$$J_{shaft} = \frac{\pi \times 100\ \text{mm}^4}{32} = 9.817 \times 10^{-6}\ \text{m}^4$$

Stiffness of the shaft

$$S = \frac{J \cdot G}{L}$$

$$S = \frac{9.817 \times 10^{-6}\ \text{m}^4 \times 72 \times 10^9\ \text{Pa}}{1000\ \text{mm}} = 7.069 \times 10^5\ \text{N} \cdot \text{m}$$

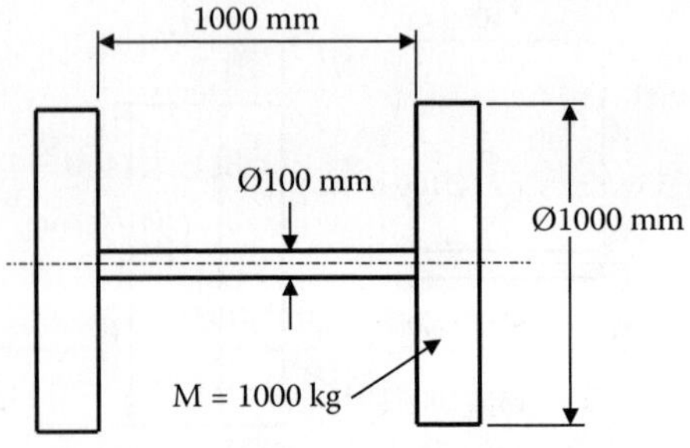

FIGURE 9.36
Torsional frequency of a plane shaft with two flywheels.

Polar Moment of Inertia of Flywheels

Mass moment of inertia:
Mass = 1000 kg
Radius of gyration = 500 mm

$$J = M \cdot k^2$$

$$J = 1000\ \text{kg} \times 500^2\ \text{mm} = 250\ \text{kg} \cdot \text{m}^2$$

Torsional Frequency

$$f_n = \frac{1}{2\pi}\sqrt{\frac{S \cdot (2J)}{J^2}}$$

$$f_n = \frac{1}{2p}\sqrt{\frac{7.069 \times 10^5 \times (2 \times 250\ \text{kg} \cdot \text{m}^2)}{(250\ \text{kg} \cdot \text{m}^2)^2}} = 11.968\ \text{Hz}$$

Position of Node

$$L_n = \frac{1000\ \text{mm}}{(1 + (250\ \text{kg} \cdot \text{m}^2 / 250\ \text{kg} \cdot \text{m}^2))}$$

$$L_n = 500\ \text{mm (from either end)}$$

Example 9.10

In this exercise, a shaft connects a motor to a gear set and a further shaft then connects to a pump.

The pump speed is one-third of that of the motor. The diameter of the shaft from the motor to the pinion of the gear set is 60 mm and the length of the shaft is 300 mm. The moment of inertia of the motor is 400 kg·m². The pump shaft is 100 mm in diameter and is 600 mm long. The moment of inertia of the pump is 1500 kg·m². Neglecting the inertia of the shafts and gears, determine the frequency of torsional vibration of the system. The modulus of rigidity of the shaft material is 72 GPa.

Solution

Given that

$G_r = N_a/N_b = 3{:}1$ (G_r = gear ratio), $d_1 = 60$ mm, $L_1 = 300$ mm, $d_2 = 100$ mm, $L_2 = 600$ mm, $J_a = 400$ kg·m², $J_b = 1500$ kg·m² and $G = 72 \times 10^9$ N/m².

The first action is to establish the moment of inertia of the equivalent rotor 'B' and the additional length of the equivalent shaft assuming its diameter is $d_1 = 60$ mm.

The Moment of Inertia of the Equivalent Rotor 'B'

$$J_b = \frac{J_b}{Gr^2}$$

$$J_b = \frac{1500\ \text{kg} \cdot \text{m}^2}{3^2}$$

$$J_b = 166.667\ \text{kg} \cdot \text{m}^2$$

The Additional Length of the Equivalent Shaft

$$L_3 = G_r^{\ 2} \cdot L_2 \cdot \left(\frac{d_1}{d_2}\right)^4$$

$$L_3 = 3^2 \times 600 \text{ mm} \times \left(\frac{60 \text{ mm}}{100 \text{ mm}}\right)^4$$

$$L_3 = 700 \text{ mm}$$

The Total Length of the Equivalent Shaft

$$L_e = L_1 + L_3$$

$$L_e = 1000 \text{ mm}$$

Let the node of the equivalent system lie at position 'N' as shown in Figure 9.37a and b.

$$L_a \times J_a = L_b \times J_b$$

$$L_a = \frac{L_b \cdot J_b}{J_a}$$

$$L_a = \frac{700 \text{ mm} \times 166.67 \text{ kg} \cdot \text{m}^2}{400 \text{ kg} \cdot \text{m}^2}$$

$$L_a = 291.67 \text{ mm}$$

The Polar Moment of Inertia of the Equivalent Shaft

$$J_e = \frac{\pi}{32} d_1^{\ 4}$$

$$J_e = \frac{\pi}{32} \times (60 \text{ mm})^4$$

$$J_e = 1.27 \times 10^6 \text{ mm}^4$$

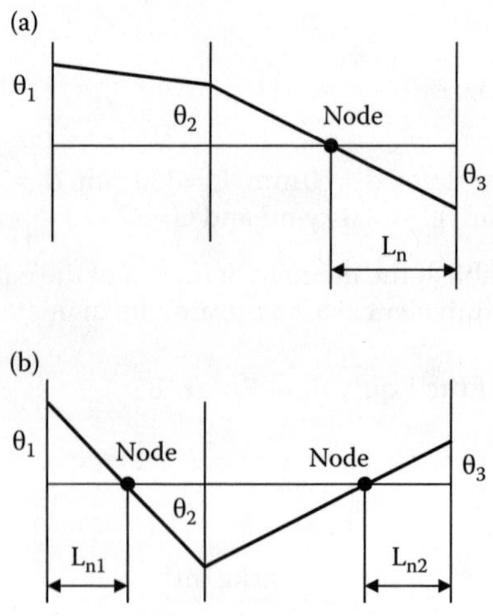

FIGURE 9.37
(a, b): First and second modes for Example 9.10.

Torsional Frequency of the Equivalent Shaft

$$L_e = L_1 + L_2 \frac{d_1^{\,4}}{d_2^{\,4}} + L_3 \frac{d_1^{\,4}}{d_3^{\,4}}$$

$$f_n = \frac{1}{2\pi}\sqrt{\frac{G \cdot J_e}{L_a \cdot J_a}}$$

$$f_n = \frac{1}{2\pi}\sqrt{\frac{72 \times 10^9 \, P_a \times 1.272 \times 10^6 \text{ mm}^4}{291.67 \text{ mm} \times 400 \text{ kg} \cdot \text{m}^2}}$$

$$f_n = 4.46 \text{ Hz}$$

9.5.8 Torsional Frequency of a Shaft Carrying Three Masses (Figure 9.38)

Note: The shaft is assumed to be light.

First mode (low frequency):

$$F_{n1} = \frac{1}{2\pi}\sqrt{\frac{1}{2}\left[(c_1 + c_2) - \sqrt{(c_1 - c_2)^2 + 4\frac{k_1 \cdot k_2}{J_2^{\,2}}}\right]} \tag{9.87}$$

where

$$k_1 = \frac{G_1 \cdot I_1}{L_1} \text{ and } k_2 = \frac{G_2 \cdot I_2}{L_2}$$

Ratio of twists in the shaft:

$$\frac{\theta_2}{\theta_3} = \frac{k_2 - J_3\omega_1^2}{k_2}; \quad \frac{\theta_3}{\theta_1} = \frac{k_2(k_1 - J_1\omega_1^2)}{k_1(k_2 - J_3\omega_1^2)}; \quad \frac{\theta_2}{\theta_1} = \frac{k_1 - J_1\omega_1^2}{k_1} \tag{9.88}$$

Position of node:

$$L_n = \frac{L_2}{(\theta_2 / \theta_1) + 1} \tag{9.89}$$

$$F_{n2} = \frac{1}{2\pi}\sqrt{\frac{1}{2}\left[(c_1 + c_2) + \sqrt{(c_1 - c_2)^2 + 4\frac{k_1 \cdot k_2}{J_2^{\,2}}}\right]} \tag{9.90}$$

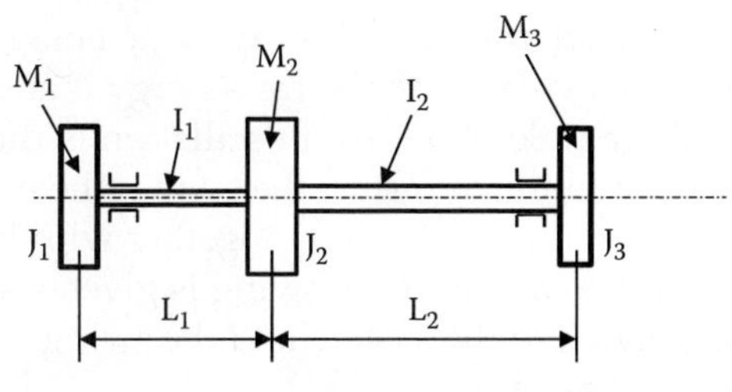

FIGURE 9.38
Torsional frequency of a shaft carrying three masses.

Ratio of twists in the shaft:

$$\frac{\theta_2}{\theta_3}=\frac{k_2-J_3\omega_2^2}{k_2};\quad \frac{\theta_3}{\theta_1}=\frac{k_2(k_1-J_1\omega_2^2)}{k_1(k_2-J_3\omega_2^2)};\quad \frac{\theta_2}{\theta_1}=\frac{k_1-J_1\omega_2^2}{k_1} \tag{9.91}$$

$$k_1=\frac{G_1\cdot I_1}{L_1};\quad k_2=\frac{G_2\cdot I_2}{L_2} \tag{9.92}$$

$$c_1=k_1\left(\frac{1}{J_1}+\frac{1}{J_2}\right);\quad c_2=k_2\left(\frac{1}{J_2}+\frac{1}{J_3}\right) \tag{9.93}$$

Position of nodes:

$$L_{n1}=\frac{L_1}{(\theta_2/\theta_1)+1};\quad L_{n2}=\frac{L_2}{(\theta_2/\theta_3)+1} \tag{9.94}$$

9.6 Forced Vibrations

9.6.1 Overview

In Section 9.3, which covers damped vibrations, it was demonstrated that a free vibration will diminish over time as the energy contained in the vibrating system is dissipated by the damping. The displacement in a damped oscillation is given as

$$x=Ce^{-\delta\omega_n t}\cos(\omega t) \tag{9.95}$$

where

δ=damping ratio and ω_n=natural angular frequency

1. When $\delta > 1$, the system is overdamped.
2. When $\delta = 1$, the system is critically damped.
3. When $\delta < 1$, the damped oscillation will gradually diminish in time.
4. When $\delta = 0$, the system is undamped and steady oscillation will occur.

Figure 9.19 compares the effects of various damping factors on oscillations.

When the damping ratio (δ) is less than zero, that is, negative, instead of energy being taken out of the system, there is an external force inputting energy into the system. As the energy is added, the amplitude increases. When this energy is added, such oscillations are said to be 'forced'. A classic example of a forced oscillation is that of a child on a swing. When the child is given an initial push to start the swing, the oscillations will gradually decrease due to air resistance on the child's body together with friction at the swing supports and will eventually come to a halt. If the swing is given a slight push at the start of each swing, energy is being added to the system and the swing will go higher and higher. This phenomenon is known as excitation.

Two further examples of forced vibrations are when an automobile is being driven over Belgian pave. In some cases, the vehicle's suspension is not able to cope with the rough ride and there is excessive vibration and the suspension oscillations will increase uncontrollably. The second example happens when a vehicle's wheels are out of balance and that creates excessive vibration of the wheel.

There are many other examples in engineering structures that vibrate at or near the natural frequency of the structure such as a pump and motor set on a suspended floor, suspension bridges and chimney columns subject to a wind blowing around them causing vortex shedding.

There are three types of forcing mechanisms that will be considered and these will be applied to a spring–mass system as shown in Figure 9.39a through c. The masses are constrained in the horizontal plane and will be able to move only in the vertical plane; hence, they are subjected to one degree-of-freedom.

9.6.2 External Forcing

Figure 9.39a models the behaviour of a system that has a time-varying force acting on it such as a structure being subject to wind loading.

Base excitation: This system model represents the behaviour of a vibration isolation system where the base of the spring is given a prescribed motion, causing the mass to vibrate. The arrangement is shown in Figure 9.39b. Examples include a vehicle suspension system or the earthquake response of a structure.

Rotor excitation: Consider a motor attached to a large mass (M) and fitted with an out-of-balance mass (m_o) that rotates at a constant speed at radius 'r' as depicted in Figure 9.39c. This causes the main mass 'M' to oscillate. In this case, the disturbing force will be harmonic, that is, sinusoidal.

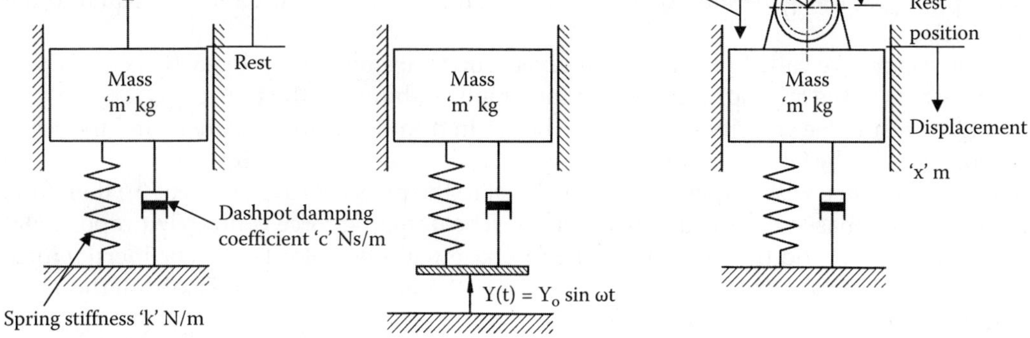

FIGURE 9.39
External forcing examples. (a) External forcing, (b) base excitation and (c) rotor excitation.

Any force that is applied to the mass to make it move has to overcome the inertia (F_i) of the mass, spring force (F_s) from the supporting structure and damping (F_d). The applied force (F_a) will be

$$F_a = F_i + F_s + F_d \tag{9.96}$$

$$F = M\frac{d^2x}{dt^2} + c\frac{dx}{dt} + kx \tag{9.97}$$

In the cases shown, the mass can only move vertically; therefore, the only force applied to it will be the vertical component of the centrifugal force:

$$F_o \sin(\omega t) = M\frac{d^2x}{dt^2} + c\frac{dx}{dt} + kx \tag{9.98}$$

When the mass is subjected to a harmonic disturbing force, it will oscillate vertically with a sinusoidal motion having amplitude 'A'. Consider that the timing of the oscillation begins when the oscillation passes through the rest position. The displacement is given by

$$x = A \sin \omega t \tag{9.99}$$

where A is the amplitude.

The velocity will be

$$v = \frac{dx}{dt} = A\omega \cos \omega t \tag{9.100}$$

where Aω is the amplitude.

The acceleration will be

$$a = \frac{dv}{dt} = -A\omega^2 \sin \omega t \tag{9.101}$$

where $A\omega^2$ is the amplitude.

In Figure 9.21, the displacement 'x', velocity 'v' and acceleration 'a' are plotted against time. Each graph is generated by a vector rotating at ω rad/s with a length equal to the amplitude.

Such vectors are called phasors. At a given point in time, the tip of each vector is projected across to the appropriate point in the graph as shown in the figure.

For this result, the velocity vector has to be 90° in front of the displacement and the acceleration needs to be 90° in advance of the velocity.

The spring force will be directly proportional to the physical displacement 'x'; therefore, it has to be in phase with the displacement 'x'. The damping force is directly proportional to the velocity 'v' and therefore has to be in phase with the velocity 'v'. The inertial force is directly proportional to the acceleration 'a' and therefore will be in phase with the acceleration 'a'.

The three forces can be represented by phasors rotating at an angular velocity ω (rad/s), choosing a time when the displacement is horizontal as shown in Figure 9.22a.

The spring force being in phase with the displacement is drawn horizontally. The other vectors will be 90° and 180° ahead, respectively, as depicted in Figure 9.22b.

The sum of these three vectors is F_o; adding these together, a typical vector diagram as shown in Figure 9.23 will be obtained.

It will be seen from the figure that the applied force F_o is at an angle 'θ' to the horizontal; hence, it must be displaced by the phase angle 'θ' relative to 'x'.

From Pythagoras's theorem

$$F_o^2 = A^2(k - M\cdot\omega^2)^2 + A^2(c\cdot\omega)^2 \text{ (Extracting } A^2) \tag{9.102}$$

$$F_o^2 = A^2[(k - M\cdot\omega^2)^2 + (c\cdot\omega)^2] \text{ (Simplifying)} \tag{9.103}$$

$$\frac{F_o^2}{M^2} = A^2\left[\left(\frac{k}{M} - \omega^2\right)^2 + \left(\frac{c\cdot\omega}{M}\right)^2\right] \text{(Dividing each term by } M^2) \tag{9.104}$$

Now

$$w_n^2 = \frac{k}{M} \text{ and } \frac{c}{M} = 2\delta\omega_n$$

$$A^2 = \left(\frac{F_o}{M}\right)^2\left[\left(\frac{1}{(\omega_n^2 - \omega^2)^2 + (2\delta\omega\cdot\omega n)^2}\right)\right] \tag{9.105}$$

From Figure 9.24, it is possible to deduce the phase angle

$$\tan\phi = \frac{2\delta\cdot\omega\cdot\omega_n}{\omega_n^2 - \omega^2} \tag{9.106}$$

9.6.3 Frequency Response Diagrams

Consider the spring–mass–damper model of Figure 9.39a with a harmonic force externally applied.

This type of force could, for example, be generated by a rotating imbalance in, say, a motor or a wheel on a vehicle. If the speed of rotation (ω) is gradually increased from zero and taking a value of $\omega_n = 10$, plotting 'θ' against 'ω' for various values of 'δ', a graph like Figure 9.40 will result.

The graph shows that the phase angle 'θ' starts at zero and reaches 90° when $\omega = \omega_n$. As the speed of rotation is increased, the phase angle approaches 180°.

Plotting the amplitude 'A' against frequency 'ω' for various values of 'δ' will result in the graph shown in Figure 9.41.

Equation 9.104 is analysed at three frequencies.

When

1. $\omega = 0$
2. $\omega = \omega_n$
3. $\omega > \omega_n$

$$A^2 = \left(\frac{F_o}{M}\right)^2\left[\left(\frac{1}{(\omega_n^2 - \omega^2)^2 + (2\delta\omega\cdot\omega n)^2}\right)\right] \tag{9.107}$$

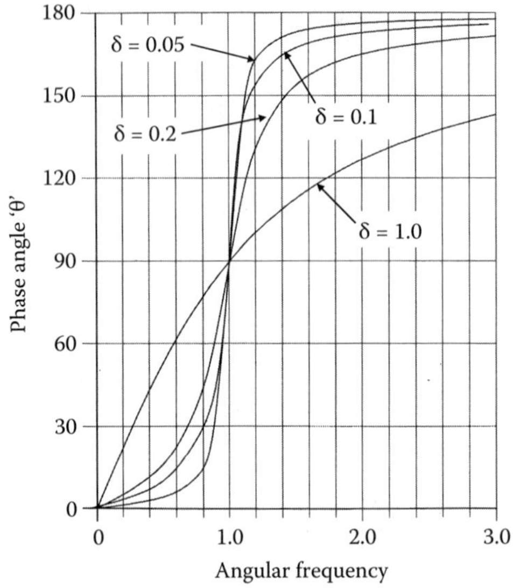

FIGURE 9.40
Phase angle with respect to damping.

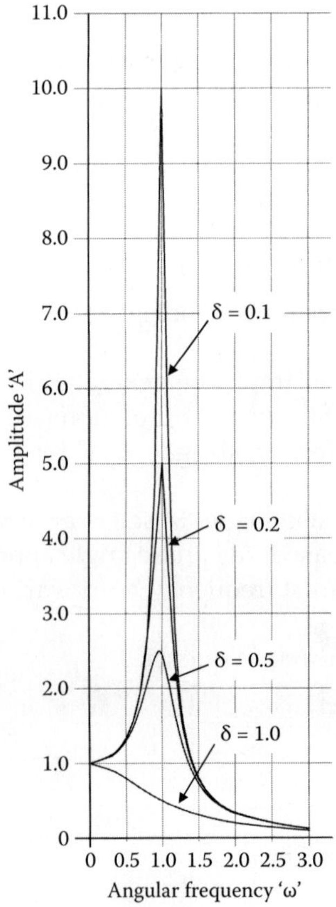

FIGURE 9.41
Amplitude against angular frequency.

1. $\omega = 0$: The equation reduces to

$$A = \left(\frac{F_o}{M}\right)\left\{\frac{1}{\omega_n^2}\right\} \tag{9.108}$$

   This will have a finite value (1 in the Figure 9.41) and will be the same starting point for all values of 'δ'.

2. $\omega = \omega_n$: At this point, the amplitude will become

$$A = \left(\frac{F_o}{M}\right)\left\{\frac{1}{(2\delta\omega^2)}\right\}$$

   The value of the amplitude will depend upon the value of 'δ'. The smaller the value of damping (δ), the larger the peak value of 'A'. If the damping factor 'δ' is reduced to zero, then theoretically 'A' $\rightarrow \infty$.

3. $\omega > \omega_n$: As the frequency increases past 'ω_n', the amplitude will begin to diminish down to zero for all values of 'δ'.

   It can be concluded that when an out-of-balance machine rotates at a speed very much greater than 'ω_n', there will be very little disturbance to the system, but when the speed of rotation is reduced and approaches ω_n, the amplitude will increase and could possibly become very large. In the past, it has led to catastrophic failures, so care has to be exercised. As an aside, very large armatures as used in electrical power stations are designed to have a low resonant frequency and when starting up have to pass through all the harmonic frequencies. Before it settles down at its operating speed, the armature will remain at this speed for many months to minimise any possible damage when slowing down. It should be noted that the frequency at which the amplitude peaks is known as the resonant frequency, which is not quite the same as the natural frequency of the system.

Example 9.11

A mass–spring–damper system is subjected to a harmonic disturbing force given by the equation F = 400 sin (30 t) N as shown in Figure 9.42. Calculate the amplitude of the mass and the phase angle.

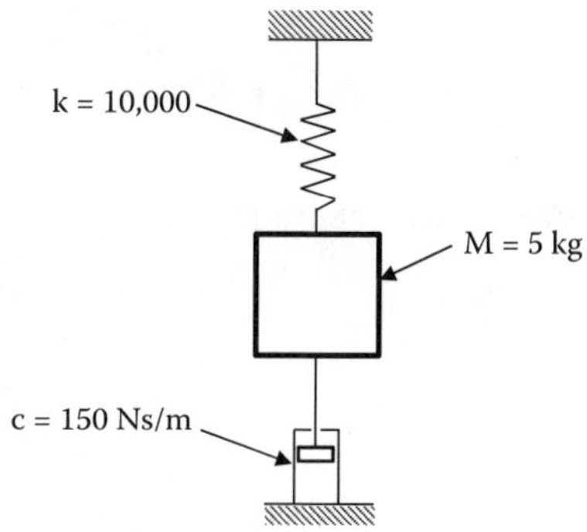

FIGURE 9.42
Example 9.11.

Solution

Given that k=10,000 N/m, M=5 kg and c=150 N·s/m.

$$\omega_n = \sqrt{\frac{k}{M}}$$

$$\omega_n = \sqrt{\frac{10{,}000\ \text{N/m}}{5\ \text{kg}}}$$

$$\omega_n = 44.271\ \text{Hz.}$$

$$c_c = \sqrt{4\ \text{Mk}}$$

$$c_c = \sqrt{4 \times 5\ \text{kg} \times 10{,}000\ \text{N/m}}$$

$$c_c = 447.214\ \text{kg/s}$$

$$\delta = \frac{c}{c_c}$$

$$\delta = \frac{150\ \text{kg/s}}{447.214\ \text{kg/s}}$$

$$\delta = 0.335$$

From the above equation:

$$(F = 400 \sin(30t)\text{N})$$

$$F_o = 400\ \text{N}$$

$$\omega = 30\ \text{rad/s}$$

$$A^2 = \left(\frac{F_o}{M}\right)^2 \left\{\frac{1}{(\omega_n^2 - \omega^2)^2 + (2\delta\omega \cdot \omega_n)^2}\right\}$$

$$A^2 = \left(\frac{400\ \text{N}}{5\ \text{kg}}\right)^2 \left\{\frac{1}{((44.72\ \text{rad/s})^2 - (30\ \text{rad/s})^2)^2 + (2 \times 0.335 \times 30\ \text{rad/s} \times 44.71\ \text{rad/s})^2}\right\}$$

$$A^2 = 0.003168\ \text{m}^2$$

$$A = 0.056\ \text{m}\ (56.0\ \text{mm})$$

$$\tan\theta = \frac{2\delta\omega \cdot \omega_n}{\omega_n^2 - \omega^2}$$

$$\tan\theta = \frac{2 \times 0.335 \times 30\ \text{rad/s} \times 44.721\ \text{rad/s}}{44.721\ \text{rad/s}^2 - 30\ \text{rad/s}^2}$$

$$\tan\theta = 0.818$$

$$\theta = 39.283°$$

9.6.4 Harmonic Movement of the Support

Figure 9.39b shows a mass–spring–damper system subject to a base excitation. The mass is restrained to move vertically only and the foundation is subject to a motion described by the equation $y = a\sin(\omega t)$.

It is assumed that the mass will move harmoniously having an amplitude 'A', but it cannot be assumed that the motion of the mass will be in phase with that of the support. Hence, the equation of motion will be amended to $x = A\sin(\omega t + \theta)$ where θ is the phase angle.

During the cycle, the spring will be stretched or shortened by an amount $(x-y)$ at any time. The spring force will be

$$F = k(x - y) \tag{9.109}$$

The three forces acting on the mass are

$$\text{Spring force} = k(x - y) \tag{9.110}$$

$$\text{Damping force} = c\,dx/dt \tag{9.111}$$

$$\text{Inertial force} = Md^2y/dt^2 \tag{9.112}$$

In this instance, there is no external force being directly applied; so balancing the forces gives

$$\theta_t = \theta - \tan^{-1}\left(\frac{F_d}{F_s}\right) \tag{9.113}$$

$$0 = M\frac{d^2x}{dt^2} + c\frac{dx}{dt} + k(x - y) \tag{9.114}$$

$$0 = M\frac{d^2x}{dt^2} + c\frac{dx}{dt} + kx - ky \tag{9.115}$$

$$ky = M\frac{d^2x}{dt^2} + c\frac{dx}{dt} + kx \tag{9.116}$$

$$k(a\sin\omega t) = M\frac{d^2x}{dt^2} + c\frac{dx}{dt} + kx \tag{9.117}$$

Comparing this result with Equation 9.104

$$F_o\sin(\omega t) = M\frac{d^2x}{dt^2} + c\frac{dx}{dt} + kx \tag{9.118}$$

It will be seen that the equations are similar except that the term 'ka' replaces the term F_o. It follows that solutions will be the same with this substitution.

$$A^2 = \left(\frac{ka}{M}\right)^2\left\{\frac{1}{(\omega_n^{\;2} - \omega^2)^2 + (2\delta\omega\omega_n)^2}\right\} \tag{9.119}$$

$$\tan\theta = \frac{2\delta\omega\omega_n}{\omega_n^{\;2} - \omega^2} \tag{9.120}$$

9.6.5 Magnification Factor

The ratio A/a is also known as the magnification ratio and applies when the support is excited. Equation 9.106 can be rearranged into the following form.

$$\text{M.F.} = \frac{A}{a} = \left(\frac{k}{M}\right)\sqrt{\left\{\frac{1}{(\omega_n^2 - \omega^2)^2 + (2\delta\omega\omega_n)^2}\right\}} \quad (9.121)$$

As $k/M = \omega_n^2$,

$$\text{M.F.} = \sqrt{\left\{\frac{(\omega_n^2)^2}{(\omega_n^2 - \omega^2)^2 + (2\delta\omega\omega_n)^2}\right\}} \quad (9.122)$$

$$\text{M.F.} = \sqrt{\left\{\frac{1}{\left(1 - \frac{\omega^2}{\omega_n^2}\right)^2 + \left(2\delta\frac{\omega}{\omega_n}\right)^2}\right\}} \quad (9.123)$$

Equation 9.123 also applies to the case when a harmonic disturbing force is applied as $k_a = F_o$; therefore, it follows that the M.F. will equal the maximum force in the spring F_o.

Figure 9.43 shows the response graph and will be similar to both cases. At low values of 'ω', the support and the mass will move together in synchrony. As the speed is increased and 'ω' approaches 'ω_n', the amplitude 'A/a' increases and the phase angle

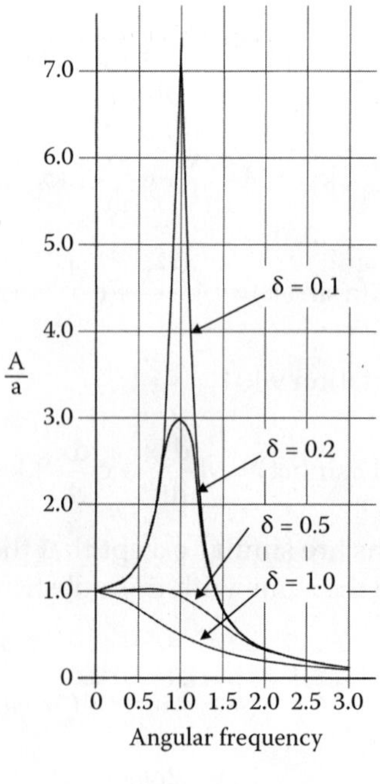

FIGURE 9.43
Response graph.

approaches 90°. As the speed passes through resonance and there is any further increase in speed, the amplitude 'A/a' will reduce and eventually become almost static. The phase angle will tend to 180° at higher speeds. The magnification ratio will be largest at the resonance frequency. As stated before, do not confuse the resonance frequency with the natural frequency.

The maximum magnification factor occurs when

$$\left(1-\frac{\omega^2}{\omega_n{}^2}\right)^2+\left(2\delta\frac{\omega}{\omega_n}\right)^2 \tag{9.124}$$

is at a maximum. Using the max and min theory, this expression can be simplified to

$$(1-r^2)^2+(2\delta r)^2 \tag{9.125}$$

Differentiating Equation 9.125 w.r.t. r

$$\frac{d\{(1-r^2)^2+(2\delta r)^2\}}{dr}=2(1-r^2)(-2r)+8\delta^2 r \tag{9.126}$$

Equating to zero

$$2(1-r^2)(-2r)+8\delta^2 r=0$$

$$r=\surd(1-2\delta^2) \tag{9.127}$$

Hence, mean peak will occur when

$$\omega=\omega_n\ \surd(1-2\delta^2) \tag{9.128}$$

Example 9.12

Figure 9.44 shows a mass–spring–damper system where the support is subject to a motion of y = 6 sin(40 t) mm.

Determine the maximum amplitude of the mass and corresponding phase angle.

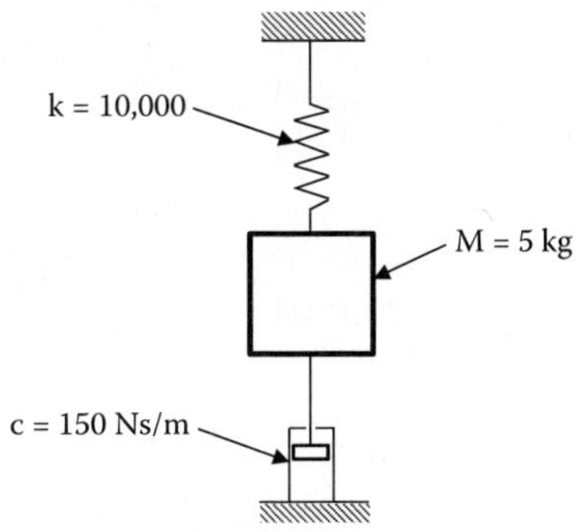

FIGURE 9.44
Example 9.12, mass–spring–damper system.

Solution

Given that k=10,000 N/m, M=5 kg and c=150 N·s/m.

$$\omega_n = \sqrt{\frac{k}{M}}$$

$$\omega_n = \sqrt{\frac{10{,}000\ \text{N/m}}{5\ \text{kg}}}$$

$$\omega_n = 44.721\ \text{Hz}$$

$$c_c = \sqrt{4\ \text{Mk}}$$

$$c_c = \sqrt{4 \times 5\ \text{kg} \times 10{,}000\ \text{N/ma}}$$

$$c_c = 447.214\ \text{kg/s}$$

$$\delta = \frac{c}{c_c}$$

$$\delta = \frac{150\ \text{kg/s}}{447.214\ \text{kg/s}}$$

$$\delta = 0.335$$

From the equation of motion where a=6 mm and 40 rad/s.

$$\frac{A}{a} = \left(\frac{k}{M}\right)\sqrt{\frac{1}{(\omega_n^2 - \omega^2)^2 + (2\delta\omega \cdot \omega_n)^2}}$$

$$\frac{A}{a} = \left(\frac{10{,}000}{5}\right)\sqrt{\frac{1}{(44.721^2 - 40^2)^2 + (2 \times 335 \times 40 \times 44.721)^2}}$$

$$\frac{A}{a} = 1.581$$

$$A = 1.581 \times a$$

Therefore,

$$A = 9.487\ \text{mm}$$

The phase angle

$$\tan\theta = \frac{2 \times \delta \times \omega \times \omega_n}{\omega_n^2 - \omega^2}$$

$$\tan\theta = \frac{2 \times 0.335 \times 40 \times 44.721}{44.721^2 - 40^2}$$

$$\tan\theta = 2.997$$

$$\theta = 71.545°$$

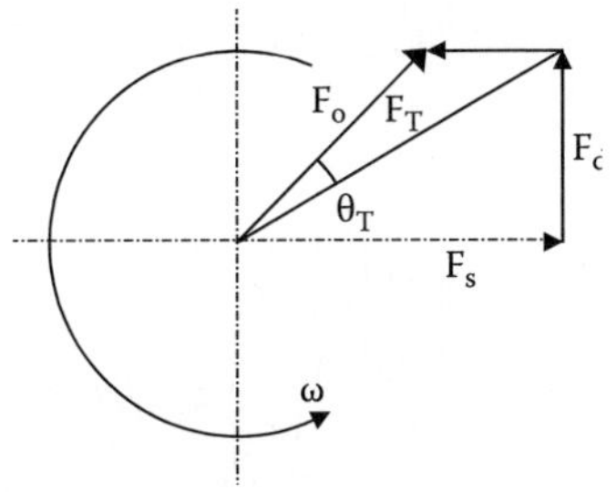

FIGURE 9.45
Vector diagram.

9.6.6 Transmissibility

When a mass is vibrating on an elastic support, a force is transmitted through the spring and damper the supporting frame or ground. This force will be the sum of the spring and damping force. Figure 9.45 depicts the corresponding vector diagram.

From the vector diagram, it is deduced that the transmitted force is

$$F_t = \sqrt{(F_s{}^2 + F_d{}^2)} \tag{9.129}$$

Now

$$F_s = k \cdot A$$

$$F_d = c \cdot A \cdot \omega$$

Hence,

$$F_T = \sqrt{[(k \cdot A)^2 + (c \cdot A \cdot \omega)]} \tag{9.130}$$

The transmissibility ratio is defined by F_T/F_o.

The phase angle between the transmitted force and the applied force i

$$\theta_T = \theta - a\tan\left(\frac{F_d}{F_s}\right) \tag{9.131}$$

If the substitution F_o=ka is made to the above work, then it will apply to both harmonic disturbing forces and harmonic motion of the support.

Example 9.13

Calculate the transmitted force and the phase angle for Example 9.12.

Solution

Given that k=10, 000 N/m, M=5 kg, c=150 N·s/m, ω= 30 rad/s, ω_n= 44.721 rad/s and A=9.487 mm, from the equations of motion:

F_s=k·A
F_s=10,000 N×9.487 mm
F_s=94.87 N
F_d=c×A×ω
F_d=150 N·s/m×9.487 mm×30 rad/s
F_d=42.692 N

$F_T = \sqrt{(F_s^2 + F_d^2)}$
$F_T = \sqrt{[(94.87\,N)^2 + (42.692\,N)^2]}$
$F_T = 104.033\,N.$

$$\theta_T = \theta - a\tan\left(\frac{F_d}{F_s}\right)$$

$$\theta_T = 39.283^\circ - a\tan\left(\frac{42.692\text{ N}}{94.87\text{ N}}\right)$$

$$\theta_T = 15.055^\circ$$

9.6.7 Using Forced Vibration Response to Measure the Properties of a Structure

The natural frequency and damping coefficient of a structure or component can be measured a number of ways. The simplest one is to attach an accelerometer to a small structure or component and 'striking' it with a 'calibrated' hammer. The hammer is a piece of laboratory equipment and is fitted with an accelerometer in its head. It is used with a 'modal analysis testing equipment'. The accelerometer on the structure is used to measure its response. Another method is to attach the component or small structure it with a vibrating table which inputs a forced vibration into the item. Again, an accelerometer attached to the part measures via the 'modal analyser' and the part's response is measured as the frequency and amplitude of the vibration are increased.

A third method, which is used for very large structures such as bridges, uses an 'exciter' which is essentially an electric motor fitted with an out-of-balance mass. By adjusting the position of the mass on a crank attached to the motor, the vibration amplitude is adjustable and the speed of the motor will vary the frequency.

In the case of measuring the frequency response of the bridge, the exciter is placed at a point mid-span. A number of accelerometers are then placed at various points along the structure at measured distances from the position of the exciter with one accelerometer positioned close to the exciter. Measurements from the accelerometers are recorded for various frequencies and amplitudes.

A graph similar to the one shown in Figure 9.46 is plotted from the results. The maximum response 'X_{max}' is measured and a line is drawn at the amplitude $X_{max} / \sqrt{2}$. The frequencies ω_1 and ω_2 together with ω_{max} are then measured as shown in the figure.

The bandwidth of the response ($\Delta\omega$) is defined as

$$\Delta\omega = \omega_2 - \omega_1 \tag{9.132}$$

$$A_o = \frac{1\times 10^{-4}\times 500}{\left[\left(1-\frac{[\pi^2]}{2.236^2}\right)^2 + \left(2\times 0.224\times\frac{\pi}{2.236}\right)^2\right]^{0.5}}$$

Like the logarithmic decrement, the bandwidth is a measure of the damping that exists in the system.

The natural frequency and damping coefficient can be estimated using the following formula

$$\delta \approx \frac{\Delta\omega}{2\omega_{max}} \tag{9.133}$$

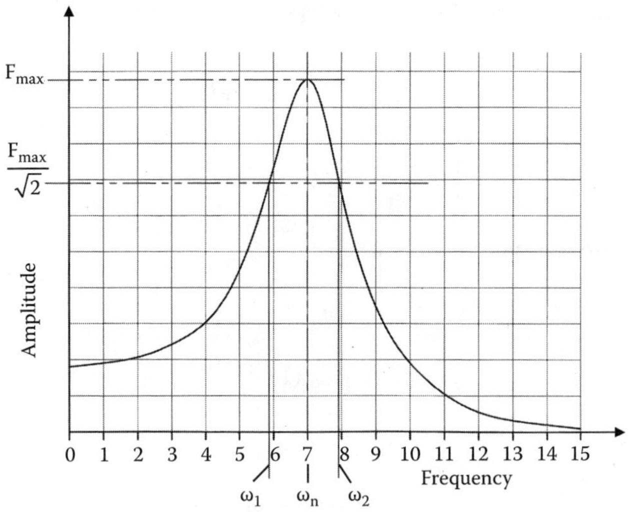

FIGURE 9.46
Bandwidth.

This formula is accurate for small δ – say $\delta > 0.2$.

From the study of steady state response, it can be shown that

$$\omega_{max} = \omega_n \sqrt{1 - 2\delta^2} \tag{9.134}$$

and for small δ

$$\omega_{max} \approx \omega_n \tag{9.135}$$

The next stage is to develop an expression relating bandwidth $\Delta\omega$ to δ. The frequencies ω_1 and ω_2 are calculated next. The maximum amplitude of vibration is calculated by setting $\omega = \omega_n$, which gives

$$A_{max} = \frac{K \cdot F_o}{2\delta\sqrt{1-\delta^2}} \tag{9.136}$$

where $K = 1/k$.

At the two frequencies of interest, it is known that $A_o = A_{max}/\sqrt{2}$; hence, ω_1 and ω_2 have to be the solutions of the equation:

$$\frac{KF_o}{2\delta\sqrt{1-\delta^2}} \cdot \frac{1}{\sqrt{2}} - \frac{KF_o}{\sqrt{\left(1-\frac{\omega^2}{\omega_n^2}\right)^2 - (2\delta\omega\omega_n)^2}} \tag{9.137}$$

Rearranging:

$$\omega^4 \frac{1}{2} 2\omega^2\omega_n^2(1-2\delta^2) - \omega_n^4 - 8\delta^2\omega_n^4(1-\delta^2) = 0 \tag{9.138}$$

This produces a quadratic equation for ω^2 and has the solutions

$$\omega_1 = \left\{\omega_n^{\ 2}(1-2\delta^2) - 2\omega_n 2\delta\sqrt{1-\delta^2}\right\}^{1/2} \tag{9.139}$$

$$\omega_2 = \left\{\omega_n^{\ 2}(1-2\delta^2) + 2\omega_n 2\delta\sqrt{1-\delta^2}\right\}^{1/2} \tag{9.140}$$

Expanding both expressions in a Taylor series

$$\omega_1 \approx \omega_n(1-\delta) \tag{9.141}$$

$$\omega_2 \approx \omega_n(1+\delta) \tag{9.142}$$

Finally,

$$\Delta\omega = \omega_2 - \omega_1 \tag{9.143}$$

$$\Delta\omega = 2\delta\omega_n \tag{9.144}$$

Example 9.14

Consider an idealised spring–mass–damper structure having a stiffness of 10 kN/m, a mass of 2000 kg and a damping coefficient of 2 kN·s. The structure is subject to a harmonic force of 500 N at a frequency of 0.5 Hz.

Calculate the steady state amplitude of vibration.

Solution

Given that k=10 kN/m, M=2000 kg, c=2 kN·s/m, F=500 N, frequency=0.5 Hz, A=500 N, $\omega=0.5\times 2\pi$, $\omega=\pi$ rad/s:

$$\omega_n = \sqrt{\frac{k}{M}}$$

$$\omega_n = \sqrt{\frac{10\ \text{kN/m}}{2000\ \text{kg}}}$$

$$\omega_n = 2.236\ \text{rad/s}$$

$$\delta = \frac{c}{2\cdot\sqrt{k\cdot M}}$$

$$\delta = \frac{24\ \text{kN m}}{2\cdot\sqrt{10\ \text{kN/m}\times 2000\ \text{kg}}}$$

$$\delta = 0.224$$

$$K = \frac{1}{k}$$

Steady state amplitude:

$$A_o = \frac{K \times F}{[(1-(\omega^2/\omega_n{}^2))^2 + (2\cdot\delta\cdot(\omega/\omega_n))^2]^{0.5}}$$

$$A_o = \frac{1\times 10^{-4} \times 500}{\left[\left(1-\frac{[\pi^2]}{2.236^2}\right)^2 + \left(2\times 0.224 \times \frac{\pi}{2.236}\right)^2\right]^{0.5}}$$

$$A_o = 43.14 \text{ mm}$$

10

Control Systems

10.1 Introduction to Control Systems Modelling

10.1.1 Introduction

Automatic control of machines and processes is a fundamental to the successful performance of modern industry.

Although an automatic control system was first developed by the Egyptians in the third century B.C with the Ktesibios water clock in Alexandria using a form of a feedback control device this field was largely left alone until 1620 using a closed loop feedback control for a furnace which is attributed to Drebbel and 1788 with the design of a centrifugal flyball governor developed by James Watt to regulate the speed of steam engines being built by Boulton and Watt (Figure 10.1.1).

In 1868 J.C. Maxwell in his paper 'On Governors' studied the instability of the flyball governor using differential equation. This was one of the first papers to use mathematics to describe a control system and demonstrate the importance of mathematical models to explain the complex phenomena. This signalled the beginning of mathematical control and systems theory. Parts of the control theory had appeared earlier but Maxwell's paper was the first to bring it all together.

Significant developments were made in control theory in the following one hundred years in developing new mathematical techniques and made it then possible to control more complex dynamic processes. These developments included optimal control methods in the 1970's and 1980's which has led to safer and more efficient aircraft travel and automotive engines together with more efficient (and safer) chemical processes, the list is expanding.

Modern day control engineering (also known as control systems engineering) is a relatively new field of study that has gained significant attention in other areas than mechanical and electrical engineering such as financial and biological processes being amenable to control techniques.

The function of any control system is to automatically regulate the output of a system and maintain it at a desired value. The desired value is the input to the system. If the input is changed the output must respond to the new set value.

There are some basic properties and terminology that is used to model a control system and these will be discussed more fully later in this chapter.

10.1.1.1 Basics of Control Theory

In this section the reader will be introduced to the basic underlying theory of control systems and the terminology used.

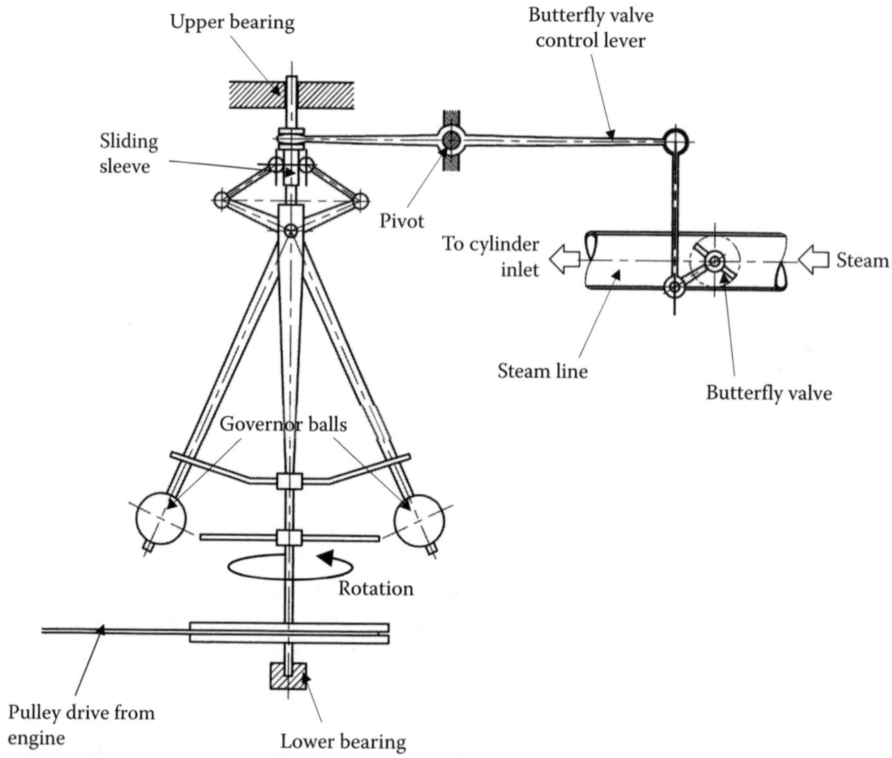

FIGURE 10.1.1
Centrifugal flyball governor.

To start, there are two general classifications of control systems:

1. Open Loop systems where the control action is independent of the output from the system.
2. Closed loop systems in which the control action is dependent on the output.

10.1.1.2 Open Loop Control System

An open loop system requires an independent external action to ensure the required output. As an example: Human intervention is needed when a saucepan is boiling on a heating ring switching it off to prevent it boiling dry.

Open loop systems have two important features:

- They depend on their calibration for accurate operation.
- Open loop systems are not normally affected by stability problems, there is little risk that the input will result in an unexpected output.

Figure 10.1.2 gives an example of an open loop system.

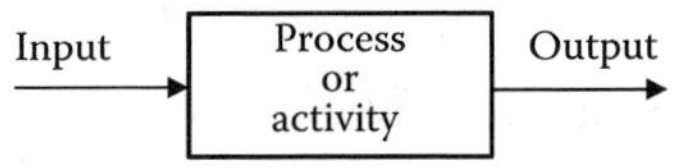

FIGURE 10.1.2
Example of an open-loop system.

10.1.1.3 Closed Loop Control System

Closed loop systems are also referred to as "Feed Back" systems where the output is compared against the input into the system. These types of systems are complicated requiring the use of differential equations for their solution.

Figure 10.1.3 illustrates a typical closed loop control system using a feedback loop.

10.1.1.4 Control System Definitions

The following are some basic control system definitions:

10.1.1.4.1 System

A system is a collection of entities that form and act as a single unit.

10.1.1.4.2 Input

A signal supplied from an external source to produce a specified response from the control system.

10.1.1.4.3 Output

The output is the actual response from the control system, it may not be equal to the response implied by the input signal.

10.1.1.4.4 Open Loop

An open loop control system is where the control action is independent of the output.

10.1.1.4.5 Closed Loop

A closed loop control system is one where the control action is related to the input.

10.1.1.4.6 Feedback

Feedback is the property of a closed loop control system where the output signal is compared with the input signal to enable an appropriate control action may be completed in accordance with the requirements of the control system.

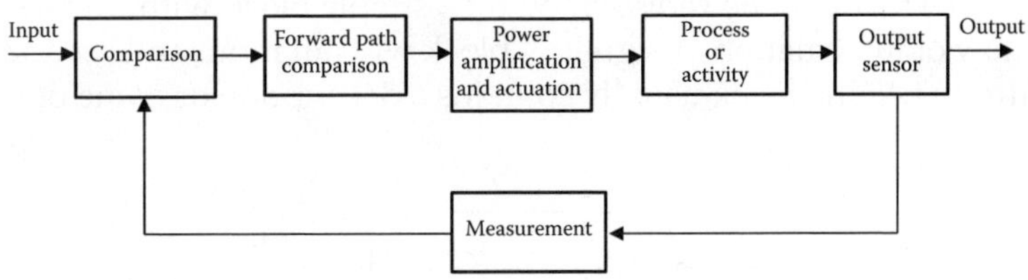

FIGURE 10.1.3
Example of a closed-loop system.

10.1.1.4.7 Servomechanism

This is a power amplifying feedback control system in which the controlled variable is either a mechanical position or a time derivative of position such as velocity or acceleration.

10.1.1.4.8 Regulator

A feedback control system where the reference input is fixed over the operating time period. The primary function of the regulator is to maintain a constant output signal.

10.1.1.5 Feedback Characteristics

A feedback in a control system results in the following advantages:

10.1.1.5.1 Increased accuracy, the output can be made to reproduce the input.

10.1.1.5.2 Reduced sensitivity to system characteristics.

10.1.1.5.3 Reduction in the effects of non-linearities.

10.1.1.5.4 Increased bandwidth, the system can be made to respond to a wider range of frequencies.

10.1.1.5.5 The major disadvantage of using feedback is the increased risk of instability and the cost of implementation.

10.1.1.6 Control Models

The study of control systems require a good working knowledge of:

10.1.1.6.1 Differential equations and other mathematical techniques.

10.1.1.6.2 Block diagrams and transfer functions.

10.1.1.6.3 Signal flow diagrams.

Block diagrams and signal flow paths are shorthand representations used in the construction of schematic diagrams representing the physical system, or the set of mathematical equations that characterise the individual component parts of the control system.

10.1.1.7 Block Diagrams and Transfer Functions

Any item in any system may be represented by a simple block with arrows representing the direction flow of information or signal. A block with an input and associated output is shown in Figure 10.1.4. The block usually contains a description or name of the element or

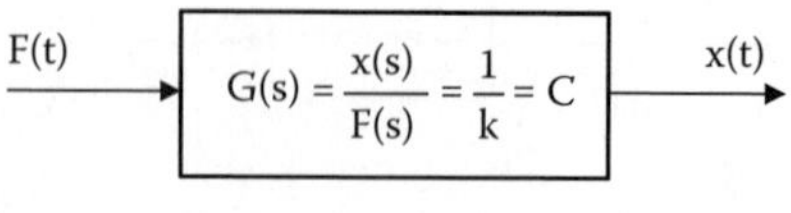

FIGURE 10.1.4
Block diagram.

the mathematical operation to be accomplished. In this example the block represents the function:

$$G(s) = \frac{x(s)}{F(s)} = \frac{1}{k} = C \tag{10.1.1}$$

In general terms the input is designated as 'θ_i' and the output as 'θ_o'. It may also be seen as 'θ_1' and 'θ_2' respectively. The ratio of the output to the input is mostly shown as $G=\theta_o/\theta_i$. When the model is a differential equation, the Laplace transform is used which introduces the complex operator 's'. In this case the 'G' is called the "transform function" and is written as:

$$G(s) = \frac{\theta_o(s)}{\theta_i(s)} \tag{10.1.2}$$

A generalised feedback control system is shown in Figure 10.1.5.

10.1.2 Engineering System Models

When studying control systems, it will be surprising to see the similarity there is between the various branches of engineering. The reader will need to have a broad base of knowledge of engineering science to understand the various elements and see how many of them are mathematically similar to each other.

In this section different kinds of systems will be reviewed and how they conform to similar laws with clear analogies between them including:

- Mechanical.
- Electrical.
- Thermal.
- Fluids.

the fundamental laws which are used most include:

- Resistance (R).
- Capacitance (C).

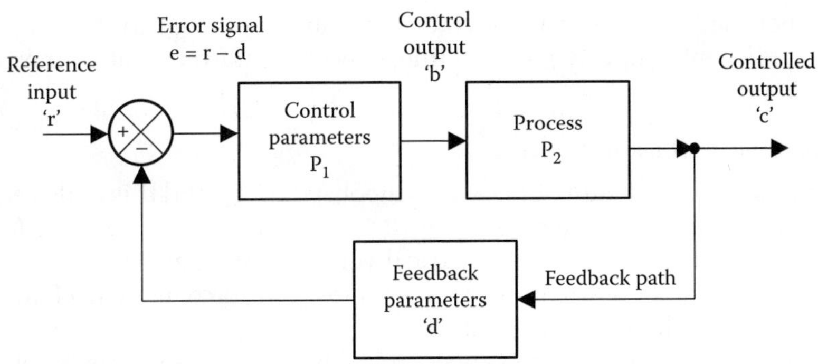

FIGURE 10.1.5
A generalised feedback control system.

TABLE 10.1.1

Comparison of Various Quantities between Systems

Mechanical	Fluid	Thermal	Electrical
Spring:	Fluid capacitor:	Thermal capacitor:	Electrical capacitor:
x=C·F=(1/k)·F	M=C·Δp	Q=CΔT	Q=C·V
Damper:		Heat transfer laws:	Ohm's law
Force=kd×velocity F=kd dx/dy Torque=kd×angular velocity	Fluid friction laws do not conform	ΔT=RΦ ΔT=R dQ/dt	V=I·R V=R dQ/dt
Newtons second law of motion:	Fluid interance: $\Delta p = L\, d^2v/dt^2$	No equivalence	Law of inductors: $V = L\, d^2q/dt^2$
Force=mass×acceleration			
$F = Md^2x/dt^2$			
D'Alembert's principles: ΣForce=0 ΣMoment=0	Laws of conservation of mass: ΣMass=constant	Laws of conservation of energy: ΣEnergy=constant	Kirchoff's law: ΣCurrent=0

- Inductance (L).
- Conservation laws.

Table 10.1.1 shows the equivalent components between the various systems. It will be useful to note that capacitance is a zero-order differential equation and resistance will be a first-order differential equation, where inertia and inductance are second-order differential equations.

10.1.2.1 Similarities of Elements between Systems

10.1.2.1.1 Capacitance

The symbol 'C' is used in electrical, thermal and fluid capacitance. Capacitance in mechanical systems is equivalent to 1/k where k is the spring stiffness.

10.1.2.1.2 Resistance

In electrical and thermal systems the symbol 'R' is used for resistance.

10.1.2.1.3 Inductance, Inertia and Inertance

'L' is the symbol used for electrical inductance and fluid inertance. In mechanical systems, mass is an equivalent property used in linear motion and moment of inertia used for angular motion.

10.1.2.1.4 Other Symbols Used

Electrical charge and the quantity of heat is symbolised by 'Q' and this is also equivalent to displacement in mechanical systems such as distance (usually 'x') and angle (usually 'θ').

'V' is recognised as the symbol for electrical voltage (potential difference or e.m.f) and has its equivalent to temperature in thermal systems. The force 'F' in mechanical systems and 'p' for pressure in fluid systems is used.

In electrical systems 'I' or 'i' is the symbols used in electrical systems for electrical current, where the symbol 'Φ' is used to symbolise heat flow rate and 'v' or 'u' are the symbols used for velocity in mechanical systems.

10.1.2.2 Laplace Transforms

Laplace transforms have been covered in some detail in Chapter 1, Mathematics and will be touched upon in later sections in this chapter.

To recap: the use of Laplace transforms is to allow differential equations be converted into a normal algebraic equation where the quantity 's' is a normal algebraic quantity. It is considered a shorthand method for writing differential coefficients, i.e.,

$$\frac{d\theta}{dt} \text{ can be written as } s\theta$$

$$\frac{d^2\theta}{dt^2} \text{ becomes } s^2\theta$$

and

$$\frac{d^n\theta}{dt^n} \text{ becomes } s^n\theta$$

10.1.2.3 Transfer Functions

Models of systems can often be written in the form of the ratio of output/input and if the model is turned into a function of 's' it is called a transfer function and will usually be represented as G(s).

i.e.,

$$G(s) = \frac{x}{F}s = \frac{1/k}{s^2(M/k) + s(k_d/k) + 1}$$

In the next section mathematical models of some basic mechanical systems will be considered.

10.1.2.4 Linear Mechanical Systems

10.1.2.4.1 Spring

The basic law of a mechanical spring (either helical or leaf) is Force ∞ change in length. Figure 10.1.6 shows the model with the mechanical symbols and with it depicted as a block diagram Figure 10.1.7.

This relationship has no derivatives in that it may be written as a function of either 't' or 's' with no transform involved.

As a function of time it can be written F(s) = kx(s).

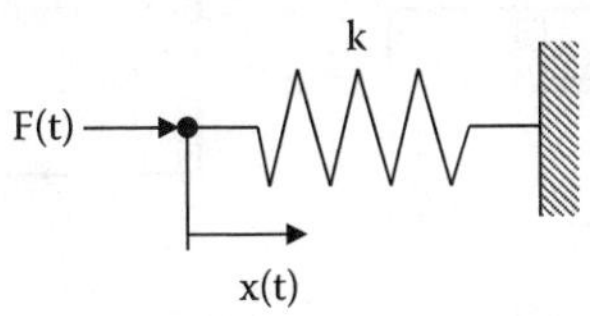

FIGURE 10.1.6
Model of a spring.

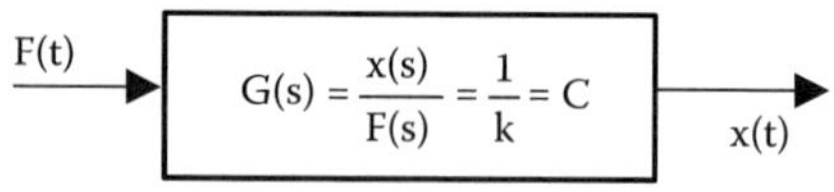

FIGURE 10.1.7
Block diagram representation for a spring.

This equation can be re-arranged as a transfer function such that $\frac{x}{F}(s) = \frac{1}{k} = C$ where 'C' is the reciprocal of stiffness and is referred to as the mechanical capacitance.

The use of 'k' is the preferred symbol in mechanics but 'C' is mostly used as it is directly analogous to electrical capacitance.

10.1.2.4.2 Damper or Dashpot

A damper may be characterised as a piston within a cylinder and within a viscous fluid, the force is directly proportional to the velocity of the piston.

$F \propto v$ where v is the first derivative of distance.

This equation can be written as:

$$F \propto \frac{dx}{dt}$$

The basic law of a damper is:

$$F_t = k_d \frac{dx}{dt} \text{ where } k_d \text{ is the damping coefficient.}$$

When changed into Laplace form:

$$F = k_d s x$$

Rearranged into a transfer function:

$$\frac{x}{F}(s) = \frac{1}{k_d s}$$

k_d is the damping coefficient which has units of Force/Velocity or N s/m. Figure 10.1.8 shows the model as a mechanical symbol and its equivalent control block.

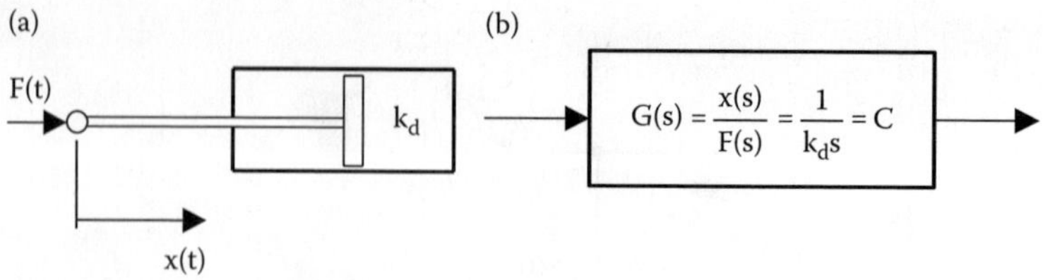

FIGURE 10.1.8
Damper symbol. (a) Schematic for a damper system and (b) the block model for a damper.

10.1.2.4.3 Mass

Newton's second law of motion states that a force is required to accelerate a mass and is written as:

$$\text{Force} = \text{mass} \times \text{acceleration}.$$

Acceleration is the second derivative of 'x' with time.

The basic law is:

$$F(t) = M\frac{d^2x}{dt^2}$$

And when changed into Laplace form:

$$F = Ms^2x$$

and rearranged into a transfer function:

$$\frac{x}{F}(s) = \frac{1}{Ms^2}$$

Figure 10.1.9 depicts the mass under the influence of a force and its equivalent control block.

10.1.2.4.4 Mass–Spring System

In Chapter 9 Mechanical Vibrations the spring–mass system was introduced in which motion only occurs in one direction hence the system has only one degree of freedom. It is normal for the direction of motion to be expressed as the 'x' direction regardless of the actual direction.

Figure 10.1.10a depicts the spring–mass system and the free body diagram Figure 10.1.10b. The input is a disturbing force 'F' which will be a function of time F(t). This disturbing force could be a sinusoidal force. The output is a motion 'x' which will be a function of time x(t).

Now, let 'x' be a positive direction vertically, the input force is opposed by the spring force and inertia force which will always oppose any changes to motion as stated in Newton's third law of motion.

Hence:

$$\text{Spring force} = kx$$

$$\text{Inertia force} = M\frac{d^2x}{dt^2}$$

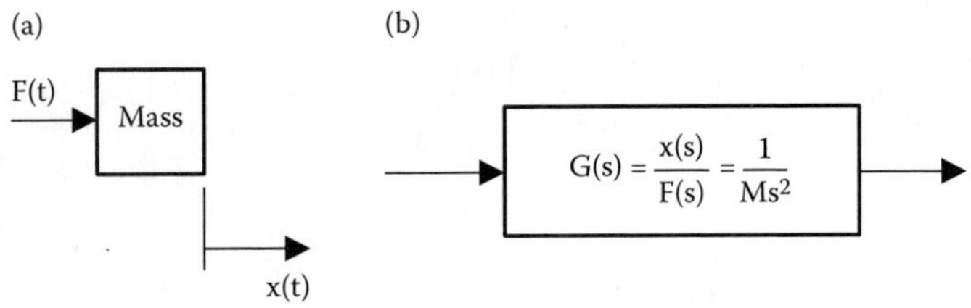

FIGURE 10.1.9
Block diagram of a mass subject to acceleration. (a) The free body diagram for a mass subject to acceleration and (b) the equivalent block diagram.

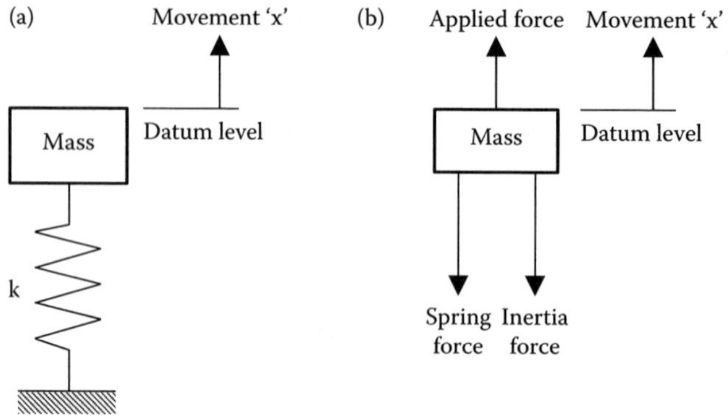

FIGURE 10.1.10
(a) Physical system and (b) free body diagram for a mass–spring system.

D'Alembert's principle states that all forces and moments on a body should equate to zero which in this case means:

$$F(t) - kx(t) - M\frac{d^2x}{dt^2(t)} = 0$$

or

$$F(t) = M\frac{d^2x}{dt^2(t)} + kx(t)$$

Changing to a function of 's':

$$\begin{aligned} F(t) &= Ms^2x + kx \\ &= x\left[Ms^2 + k\right] \end{aligned}$$

$$\begin{aligned} x(s) &= \frac{F}{Ms^2 + k} \\ &= \frac{F(1/M)}{s^2 + k/M} \end{aligned}$$

This may be shown as a transfer function:

$$\begin{aligned} G(s) &= \frac{x(s)}{F(s)} \\ &= \frac{1/M}{s^2 + k/M} \end{aligned}$$

Figure 10.1.11 shows the block diagram.

10.1.2.4.5 Spring–Damper System

Consider Figure 10.1.12a:

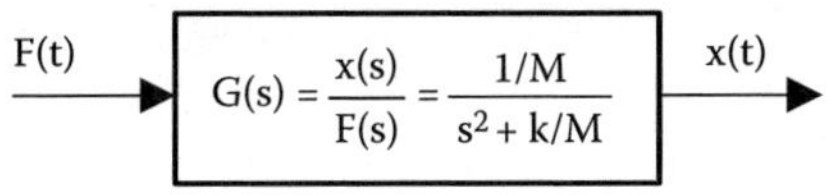

FIGURE 10.1.11
Block diagram for a mass–spring system.

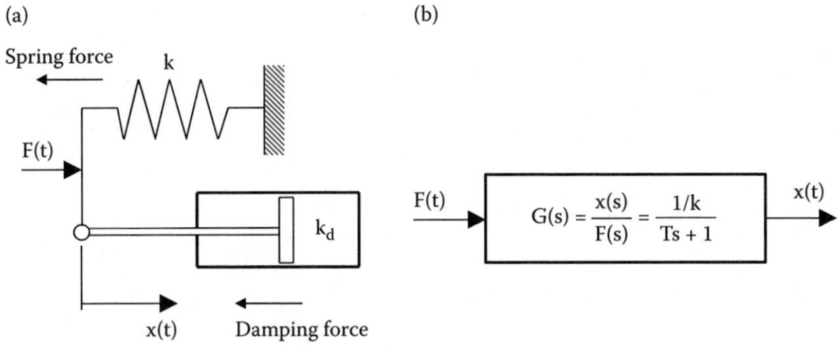

FIGURE 10.1.12
System (a) and block model (b) for a spring–damper.

Force balance as a function of time:

$$F(t) = kx + kd\frac{dx}{dt}$$

Force balance as a function of 's'

$$F(s) = kx + k_d \cdot sx$$

Rearranging as a transfer function:

$$\frac{x}{F}(s) = \frac{1/k}{(k_d/k)s + 1}$$

The units of k_d/k are seconds and this is the time constant for the damped system:

$$T = \frac{k_d}{k}$$

The standard first-order equation:

$$\frac{x}{F}(s) = \frac{1/k}{Ts + 1}$$

This result will be studied a number of times in the following sections.
The block diagram is shown in Figure 10.1.12b.

10.1.2.4.6 Mass–Spring–Damper System

Figure 10.1.13a depicts a mass–spring–damper system in which a force 'F' is applied to the system resulting in a displacement 'x', both being a function of time 't'

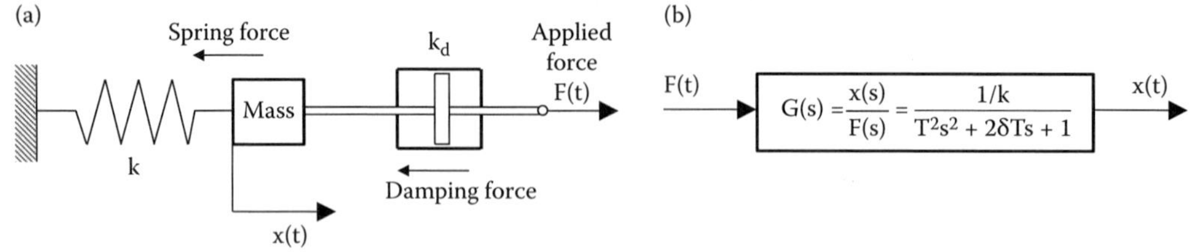

FIGURE 10.1.13
(a) Schematic for a spring–mass–damper. (b) Block diagram for a spring–mass–damper.

Hence:

$$\text{Spring force Fs} = kx$$

$$\text{Damping force } F_d = k_d\frac{dx}{dt}$$

$$\text{Inertia force } F_i = M\frac{d^2x}{dt^2}$$

These three forces oppose the motion of the system, so if the total force acting on the system is zero then:

$$F = F_i + F_d + F_s$$

$$F(t) = M\frac{d^2x}{dt^2} + k_d\frac{dx}{dt}kx$$

$$F(s) = Ms^2x + k_dsx + kx$$

$$G(s) = \frac{x}{F}s = \frac{1/k}{s^2(M/k) + s(k_d/k) + 1}$$

An examination of the units of $(M/k)^{0.5}$ is in seconds and this is the second-order time constant with the symbol 'T'. Therefore the transfer function can be written as:

$$G(s) = \frac{x}{F}(s)$$

$$= \frac{1/k}{T^2s^2 + s\delta Ts + 1}$$

where δ is the damping ratio which is defined as:

$$\delta = \frac{k_d}{C_c}T$$

where C_c is the critical damping ratio defined as $(4Mk)^{0.5}$.

The term 2δT is

$$2\frac{k_d}{C_c}T = \frac{2k_d\sqrt{M/k}}{\sqrt{4Mk}}$$

$$= \frac{2k_d\sqrt{M}}{2\sqrt{M}\sqrt{k}\sqrt{k}}$$

$$= \frac{k_d}{k}$$

Therefore the forgoing is correct.

Figure 10.1.13b depicts the block diagram for the transfer function for this type of system. It is a second-order order transfer function and an analysis of it will be dealt in more detail later.

Example 10.1.1

A spring–mass–damper system has the following values:
Stiffness (k) = 1000.0 N, Mass (M) = 5.0 kg, Damping coefficient (k_d) = 30.0 Ns/m

1. Calculate:
   (a) the time constant,
   (b) the critical damping coefficient,
   (c) the damping ratio.
2. Derive the equation for the force required when the mass is accelerating.
3. Using the derived equation evaluate the static deflection when the force 'F' = 15.0 N.
4. Using the equation evaluate the force required to accelerate the mass at 6.0 m/s² when the velocity is 0.70 m/s.

Solution

k = 1000.0 N
M = 5.0 kg
k_d = 30.0 Ns/m

1. $T = \sqrt{\frac{M}{k}} = \sqrt{\frac{5.0}{1000.0}} = 0.0707\ \text{s}$

$$C_c = \sqrt{4Mk} = \sqrt{4.5.1000} = 141.421\ \text{Ns/m}$$

$$\delta = \frac{k_d}{C_c} = \frac{30}{141.421} = 0.212$$

2. For a constant acceleration $S^2X = a$ (acceleration) and $SX = V$ (velocity)

$$F = k \cdot X(T^2S_2 + 2\delta T \cdot S + 1)$$

$$F = 1000 \times X(0.0707^2S^2 + 2.0 \times 212.0 \times 0707S + 1)$$

$$F = X(5.1409S^2 + 29.9768S + 1000)$$

$$F = 5.1409\ S^2X + 29.9768\ SX + 1000X$$

$$F = 5.1409a + 29.9768V + 1000X$$

3. For a constant force and a static position there is neither acceleration or velocity, therefore the S^2 and S terms are zero.
   Consider the deflection 'X' when the force 'F' is 15.0 N

$$\frac{F}{X} = 1000\,\text{N},\; X = \frac{15}{1000} = 0.015\,\text{m (15.0 mm)}$$

4. Force needed to accelerate the mass at 6.0 m/s^2 when the velocity is 0.7 m/s:
   a = 6.0 m/s^2; V = 0.70 m/s.

$$F = (5.1409 \times 6.0) + (29.9768 \times 0.70) + 1000X$$

$$F = 30.8454 + 20.9838 + 1000X$$

$$F = 51.8292 + 1000X$$

The deflection 'X' will need to be evaluated from $x = \frac{V^2}{2a}$ which in this case yields 0.0408 m (4.08 mm)

$$F = 92.663\,\text{N}$$

10.1.2.5 Rotary Mechanical Systems

The following work is essentially the rotary equivalent of the previous work.

10.1.2.5.1 The Torsion Bar

A metal rod which is restrained at one end and supporting a mass at the other is equivalent to the mass and spring (Figure 10.1.14). If a torque is applied to the rod a twist will result in the rod. The applied torque is directly proportional to the angle of twist and the ratio T/θ is the torsional stiffness of the rod and is denoted as 'k' as in the linear systems.

The nomenclature used is:

T	Torque	Nm
θ	Angle of twist	Radians
k	Torsional stiffness	Nm/rad

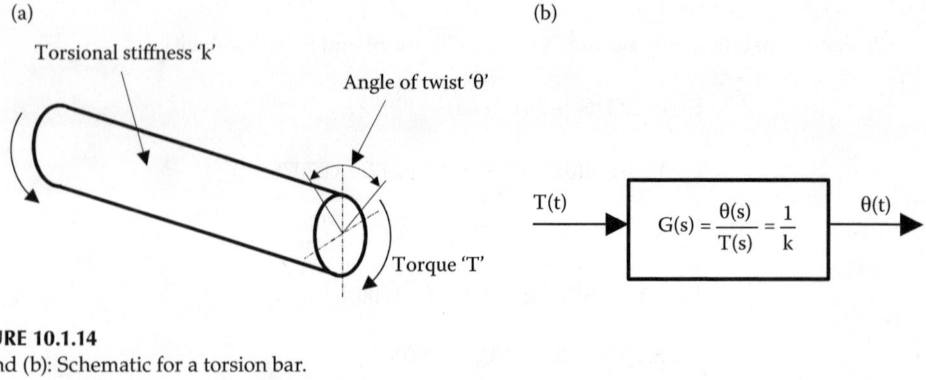

FIGURE 10.1.14
(a) and (b): Schematic for a torsion bar.

Equating the torques: $T(t)=k\theta\ (t)$
Change to Laplace form $T(s)=k\theta\ (s)$
Write as a transfer function $\frac{\theta}{T}(s)=\frac{1}{k}$.

10.1.2.5.2 Torsion Damper

Rather like the dampers used in the linear systems which characterises a piston operating in an enclosed tube filled with a viscous fluid, the rotary dampers may be idealised using vanes also rotating in a viscous fluid where the torque required to rotate the vanes is directly proportional to the angular velocity. k_d is the torsion damping coefficient with values of Nm s/radian.

Hence:

$$T(t)=k_d\frac{d\theta}{dt},\quad T(s)=k_d s\theta,\quad G(s)=\frac{\theta}{T}(s)=\frac{1}{k_d s}$$

Figure 10.1.15a depicts a representation of the torsional damper and Figure 10.1.15b shows the equivalent block diagram.

10.1.2.5.3 Moment of Inertia

A rotating mass will oppose the change to rotary motion and Newton's second law for rotating masses may be represented by $T=I\frac{d^2\theta}{dt^2}$ where I is the moment of inertia in kg m².

$$T=I\frac{d^2\theta}{dt^2}\quad T(s)=Is^2\theta\quad G(s)=\frac{q}{T}(s)=\frac{1}{Is^2}$$

Note: The symbol 'J' is also used to represent the moment of inertia.

Figure 10.1.16a depicts a torsion bar supporting a mass subject to an applied torque and Figure 10.1.16b shows the equivalent block diagram.

10.1.2.5.4 Geared Systems

Geared systems is a common element that is found in many mechanical systems where an inertial mass is driven through a gear system via a drive motor, the effect of the inertia is significantly altered when the inertia is referred back to the motor.

Consider a motor coupled to a load through a set of gears as shown in Figure 10.1.17. The two bearings provide damping in the system (viscous friction).

θ_i is the input rotation provided in this case by the motor and θ_o is the output rotation.

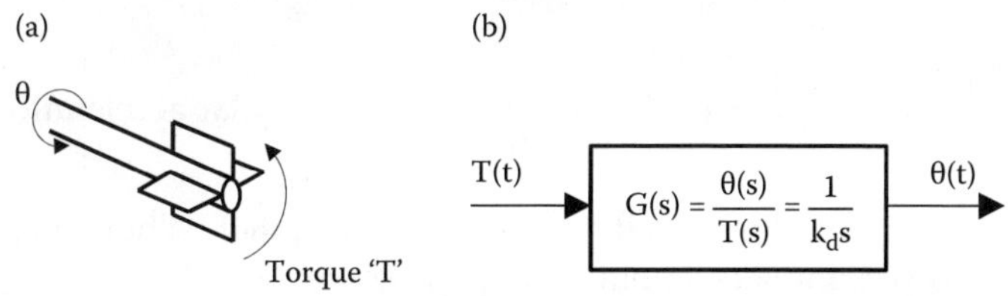

FIGURE 10.1.15
(a) and (b): Schematic for a torsion damper.

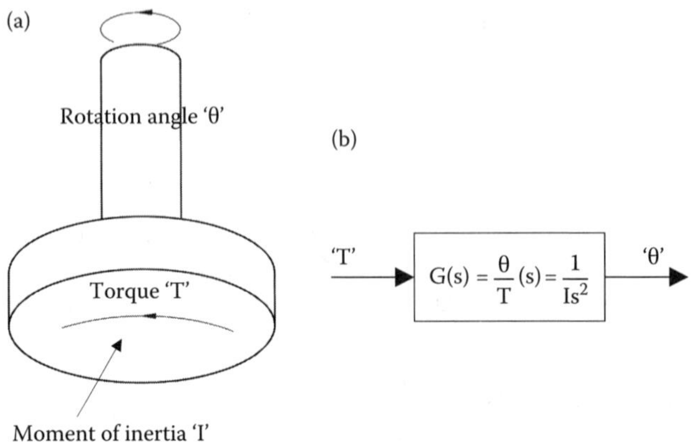

FIGURE 10.1.16
(a) and (b): Torsion bar supporting a mass subject to an applied torque.

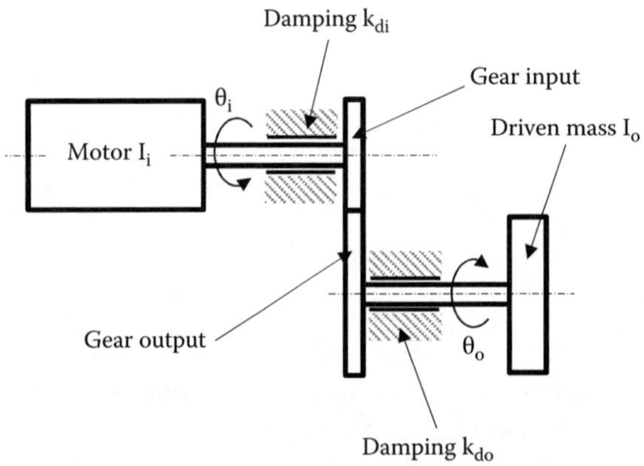

FIGURE 10.1.17
Geared system.

The gear ratio is $G_r = \dfrac{\theta_o}{\theta_i}$. As this is a fixed ratio and is not a function of time, the speed and acceleration will also be in the same ratio.

Therefore:

$$\frac{d\theta_i}{dt} = \omega_i \quad \frac{d\theta_o}{dt} = \omega_o \quad Gr = \frac{\omega_o}{\omega_i} \quad \omega \text{ is the angular velocity}$$

$$\frac{d^2\theta_i}{dt^2} = \alpha_i \quad \frac{d^2\theta_o}{dt^2} = \alpha_o \quad Gr = \frac{\alpha_o}{\alpha_i} \quad \alpha \text{ is the angular acceleration}$$

The power that is transmitted by a shaft is given by Power $=\omega T$. Where no power is lost, the output and input powers must be equal hence, it follows:

$$\omega_i T_i = \omega_o T_o \text{ therefore}$$

$$T_i = \frac{\omega_o T_o}{\omega_i}$$

$$= G_r T_o$$

In practice friction will have a significant effect on torque.

The torque due to the inertia on the output shaft I_o:

$$T_o = I_o \alpha_o$$

$$= I_o \alpha_o G_r$$

Torque on the input shaft:

$$T_i = T_o \cdot G_r$$

$$= I_o \alpha_i \cdot G_r^2$$

The damping torque on the output shaft:

$$T_o = k_{do} \cdot \omega_o$$

$$= k_{do} \cdot \omega_i \cdot G_r$$

The damping torque on the input shaft:

$$T_i = T_o \cdot G_r$$

$$= k_{do} \cdot \omega_i \cdot G_r^2$$

Now considering that there is an inertia and damping torque on the input shaft and on the output shaft, the total torque produced on the input shaft:

$$T_i = I_i \cdot \alpha_i + k_{di} \cdot \omega_i + G_r \cdot T_o$$

$$T_o = I_o \cdot \alpha_o + k_{do} \omega_o$$

$$T_i = I_i \cdot \alpha_i + k_{di} \cdot \omega_i + G_r \{I_o \cdot \alpha_o + k_{do} \cdot \omega_o\}$$

$$T_i = I_i \cdot \alpha_i + k_{di} \cdot \omega_i + G_r^2 \cdot I_o \cdot a_i + G_r^2 k_{do} \cdot \omega_i$$

$$T_i = \alpha_i \cdot \left(I_i + G_r^2 \cdot I_o\right) + \omega_i \cdot \left(k_{di} + G_r^2 \cdot k_{do}\right)$$

Now:

$\left(I_i + G_r^2 \cdot I_o\right)$ is the effective moment of inertia I_e, $\left(k_{di} + G_r^2 \cdot k_{do}\right)$ is the effective damping coefficient k_{de}.

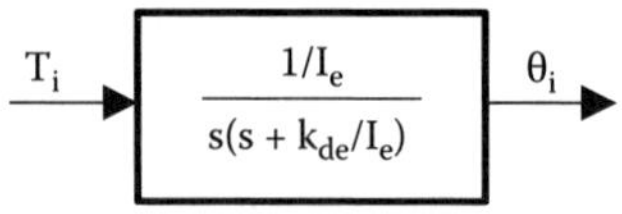

FIGURE 10.1.18
Block diagram for a geared example.

The equation may be written:

$$T_i = \alpha_i(I_e) + \omega_i(k_{de})$$

and in calculus form this can be written as:

$$T_i = \frac{d^2\theta}{dt^2}(I_e) + \frac{d\theta}{dt}(k_{de})$$

Changing this equation into a function of 's':

$$T_i(s) = s^2\theta(I_e) + s_\theta(k_{de})$$
$$= s\theta(sI_e + k_{de})$$

The output is the angular rotation of the input shaft (motor) and the input is the input torque, the geared system can be represented as a transfer function:

$$\frac{\theta(s)}{T_i(s)} = \frac{1/I_e}{s}\left(s + \frac{k_{de}}{I_e}\right)$$

Figure 10.1.18 represents the above transfer function as a block diagram.

Example 10.1.2

It is required to rotate a Radar aerial through a geared reduction using a DC servo motor.

The system parameters are:

- Gear reduction ratio is 10:1.
- The servo motor has a moment of inertia of 0.5 kg m^2.
- The driven mass has a moment of inertia of 1.2 kg m^2.
- The damping on the motor is 0.1 Nm·s/rad.
- The damping on the aerial bearings is 0.05 Nm·s/rad.

Determine the transfer function θ/T_m in its simplest form and calculate the motor torque required to:

1. Turn the aerial at a constant rate of 0.02 rad/s
2. Accelerate the aerial at 0.005 rad/s^2 when $\omega = 0$

Solution

1.

$$
\begin{aligned}
I_e &= \left(I_i + G_r{}^2 \cdot I_o\right) \\
&= \left(0.5 + 10^2 \times 1.2\right) \\
&= 120.5\ \mathrm{kg \cdot m^2}.
\end{aligned}
$$

$$
\begin{aligned}
k_{de} &= \left(k_{di} + G_r{}^2 \cdot k_{do}\right) \\
&= \left(0.1 + 10^2 \times 0.05\right) \\
&= 5.1\ \mathrm{Nm\ s/rad}
\end{aligned}
$$

$$\frac{\theta(s)}{T_i(s)} = \frac{(1/I_e)}{s}\left\{s + \frac{k_{de}}{I_e}\right\}$$

$$\frac{\theta}{T_i}(s) = \frac{1/I_e}{s(s + k_{de}/I_e)}$$

$$T_i = I_e \cdot \alpha + k_{de} \cdot \omega$$

$$T_i = 120.5\alpha + 5.1\omega$$

If the aerial is moving at a constant speed α is zero.
Therefore:

$$
\begin{aligned}
T_i &= 5.1\omega \\
&= 5.1 \times 0.02 \\
&= 0.102\ \mathrm{Nm}
\end{aligned}
$$

2. When the aerial is accelerating at 0.005 rad/s^2 the input (motor) acceleration will be 10 times larger (0.005 rad/s^2 × 10) = 0.05 rad/s^2

$$
\begin{aligned}
Ti &= 120.5\alpha + 5.1\omega \\
&= 60.25\ \mathrm{Nm}\ \text{when}\ \omega = 0.
\end{aligned}
$$

10.1.2.6 Thermal Systems

10.1.2.6.1 Heating and Cooling

If a mass (M) is heated by immersing in a bath of hot fluid (as shown in Figure 10.1.19), the quantity of heat (Q) absorbed by the mass will be a function of the specific heat capacity (C) multiplied by the change in temperature from ambient (T_1) to the final temperature of the mass (T_2)

This can be expressed in algebraic form:

$$Q = Mc(T_2 - T_1)$$

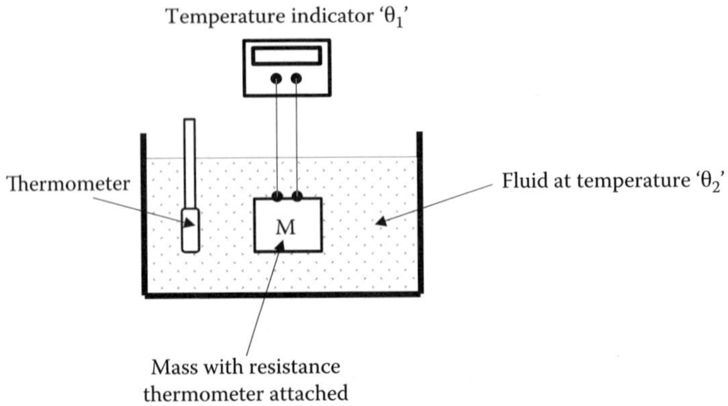

FIGURE 10.1.19
Diagram for an immersed mass.

From the laws of heat transference:

$$\begin{aligned} dQ &= Mc\,d\theta_1 \\ &= C \cdot d\theta_1 \end{aligned}$$

where C=Mc and is the thermal capacitance expressed in Joules/Kelvin.

Dividing both sides by dt:

$$\frac{dQ}{dt} = \Phi = C\frac{d\theta_1}{dt}$$

The rate of heat transfer into the mass is $\Phi = C\frac{d\theta_1}{dt}$ and the rate of transfer is governed by the thermal resistance between the liquid and the mass. Ohm's law is similar so that:

$$\Phi = \frac{(\theta_2 - \theta_1)}{R}$$

where 'R' is the thermal resistance in Kelvin per Watt.

Equating for Φ:

$$C\frac{d\theta_1}{dt} = \frac{\theta_1 - \theta_2}{R}$$

$$\frac{d\theta_1}{dt} = \frac{\theta_1 - \theta_2}{RC}$$

$$\frac{d\theta_1}{dt} + \frac{\theta_1}{RC} = \frac{\theta_2}{RC}$$

In mechanical, electrical, thermal and fluid systems the product of resistance and capacitance is time constant 'T' therefore:

$$\frac{d\theta_1}{dt} + \frac{\theta_1}{T} = \frac{\theta_2}{T}$$

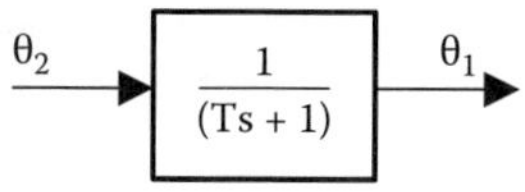

FIGURE 10.1.20
Block diagram for an immersed mass transfer function.

Converting from a function of time to a function of 's':

$$s\theta_1 + \frac{\theta_1}{T} = \frac{\theta_2}{T}$$

$$\theta_1(Ts + 1) = \theta_2$$

$$\frac{\theta_1}{\theta_2}(s) = \frac{1}{(Ts + 1)}$$

Figure 10.1.20 depicts the block diagram for this transfer function.

Note: The transfer function is the same standard first-order equations that was derived for the spring–damper system and the thermal capacitance 'C' will be equivalent to 1/k and resistance will be equivalent to k_d.

10.1.2.6.2 Process Heating System

In the chemical process industry, precise control of heating being applied to process tanks is essential, particularly when dealing with volatile solutions such as oil where if overheating occurs would lead to a catastrophic event. In dangerous situations pneumatics are used in preference to electronic controllers to reduce the risk of any fire that may occur.

Figure 10.1.21 shows a schematic diagram of an industrial heater using pneumatics for the control of the temperature of a tank of liquid that is heated by steam. The tank is fitted with a temperature sensor that sends a signal (within a range of 0.2 to 1.0 bar) to a temperature controller. The controller has an input temperature set by adjusting the control. A pressure sensing device produces another pneumatic signal between 0.2 and 1.0 bar depending upon the error that exist between the heating tank sensor and the set temperature on the controller. This error signal is then sent to a pneumatically controlled flow control valve and this controls the flow of steam being sent to the heating coils within the tank. Generally these tanks are fitted with stirrers to ensure the temperature of the heated fluid is consistent throughout the tank. Hence if the temperature sensor within the tank senses the fluid is cooling additional steam is introduced to the heating coils and when any overheating is detected the steam flow is reduced.

The model for this system will not be derived here as it is more complicated than simple:

$$\frac{\theta_1}{\theta_2}(s) = \frac{1}{(Ts + 1)}$$

Generally these type of controllers are known as three term control (proportional, integration and differentiation) and will be covered later in the chapter.

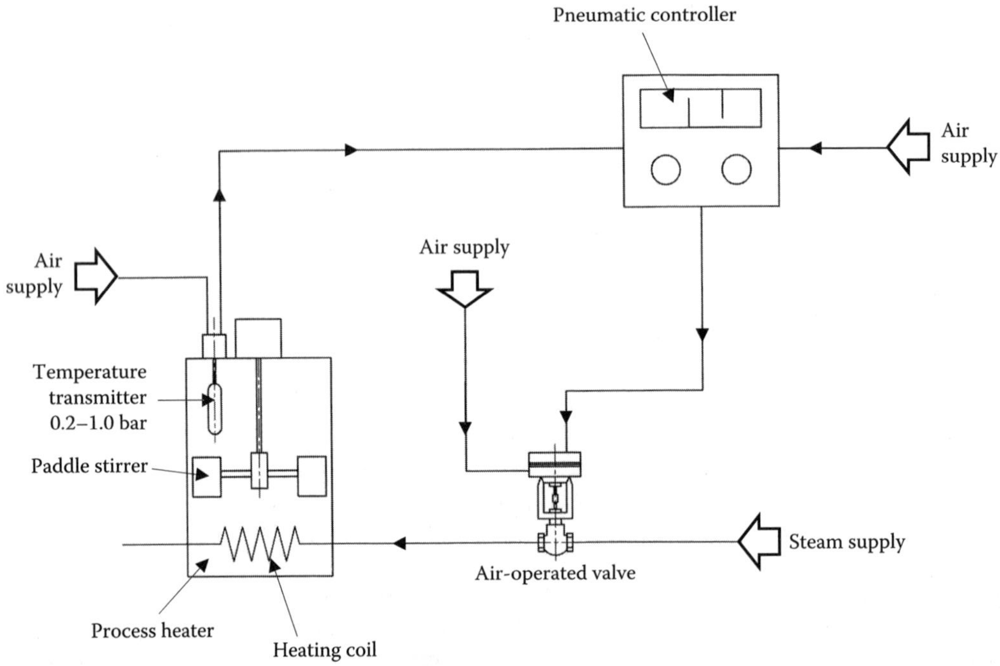

FIGURE 10.1.21
Typical pneumatic process heater control circuit.

Example 10.1.3

Consider a simple thermal heating system that has a transfer function $\frac{\theta_1}{\theta_2}(s) = \frac{1}{(Ts+1)}$.

The temperature of the system is θ_o and this is at ambient temperature (20°C), When the set temperature is changed from 20°C to 100°C, the time constant 'T' is 4s. Deduce a formula that shows how the system temperature changes with time.

Solution

$$\frac{\theta_o}{\theta_i} = \frac{1}{T(s)+1} \quad \therefore \theta_i = Ts\theta_o + \theta_o$$

Let θ_i be a constant (100°C) at all values of time after t=0 (the start of the change).

$$\theta_i - \theta_o = Ts\theta_o$$

$$\theta_i - \theta_o = T\frac{d\theta_o}{dt}$$

Let

$$\theta_i - \theta_o = x$$

Differentiating and $-d\theta_o = dx$

The equation will become:

$$x = T\frac{d\theta_o}{dt}$$

$$= -T\frac{dx}{dt}$$

Integrating (without limits):

$$\theta_o{=}100 - 80e^{-t/4} - \frac{t}{T} = \ln(x) + A$$

Substituting for 'x'

$$-\frac{t}{T} = \ln(\theta_i - \theta_o) + A$$

Now: when t=0, (θ_o is the starting temperature and is equal to θ_i)

$$-\frac{t}{T} = 0 = \ln(\theta_i - \theta_i) + A$$

$$A = -\ln(\theta_i - \theta_i) \ \ (\theta_i - \theta_i = \text{the change in temperature } \Delta\theta)$$

Substituting for 'A'

$$-\frac{t}{T} = \ln(\theta_i - \theta_o) - \ln(\Delta\theta)$$

$$= \ln\frac{(0_i - 0_o)}{\Delta\theta}$$

Taking anti-logs:

$$e^{-t/T} = \frac{(\theta_i - \theta_o)}{\Delta\theta}$$

$$\Delta\theta e^{-t/T} = (\theta_i - \theta_o)$$

$$\theta_o = \theta_i - \Delta\theta e^{-t/T}$$

Substituting the value T=4:

$$\Delta\theta = 100 - 20$$

$$= 80$$

therefore θ_i=100.

Hence the law is:

$$\theta_o = 100 - 80e^{-t/4}$$

Figure 10.1.22 evaluates and shows the resultant plot.

t (s)	θ (°C)
0	20.000
2	51.478
4	70.570
6	82.150
8	89.173
10	93.433
12	96.017
14	97.584
16	98.533
18	99.111
20	99.461

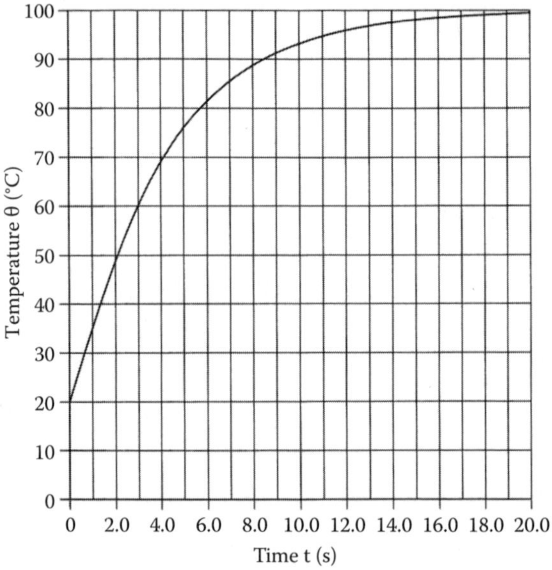

FIGURE 10.1.22
Resultant plot for Example 10.1.3.

10.1.2.7 Hydraulic System

10.1.2.7.1 Hydraulic Motor

Figure 10.1.23 is a representation of a hydraulic motor in which an inlet is supplied with pressurised fluid at a flow rate Q (m^3/s), at a pressure of p (N/m^2). This will result in an output speed ω (rad/s) at a torque T (Nm). It is assumed that the motor is 100% efficient and this will result in the outlet pressure being zero.

The following derivation of a model for use in control theory.

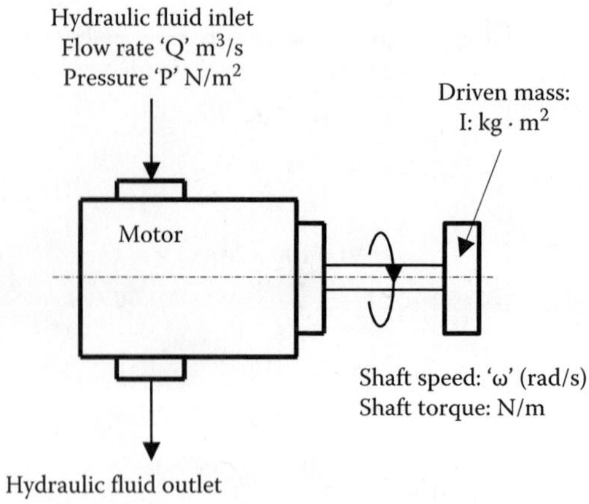

FIGURE 10.1.23
Hydraulic motor.

The formula relates flow rate 'Q' and the speed of rotation 'ω':

$$Q = k_q \omega$$
$$= k_q \frac{d\theta}{dt}$$

k_q is a constant known as the nominal displacement and has units of m^3 per radian. θ is the angle of rotation specified in radians.

Written as a function of 's' this is expressed as:

$$Q = k_q s\theta$$

If the flow rate is taken as the input and the angle of rotation is taken as the output the transfer function will be:

$$G(s) = \frac{\theta}{Q}$$
$$= \frac{1}{k_q s}$$

The following formula relates the system pressure 'p' to the output torque 'T':

$$T = k_q \cdot p$$

The input pressure and output torque can be related as:

$$G(s) = \frac{T}{p} = k_q$$

This a further definition of the constant k_q.

Example 10.1.4

If a hydraulic motor has a nominal displacement of 10 cm^3/radian, calculate the torque the motor will produce at a pressure of 110 bar.

Solution

$$T = p \cdot k_q$$
$$= 110 \times 10^5 (N/m^2) \times 10 \times 10^{-6} (m^3/rad)$$
$$= 110\,N \cdot m$$

10.1.2.7.2 Hydraulic Cylinder

Where the hydraulic motor converts pressure into rotary motion, the hydraulic cylinder converts pressure into a linear motion. Figure 10.1.24a shows the basic elements of a double acting cylinder.

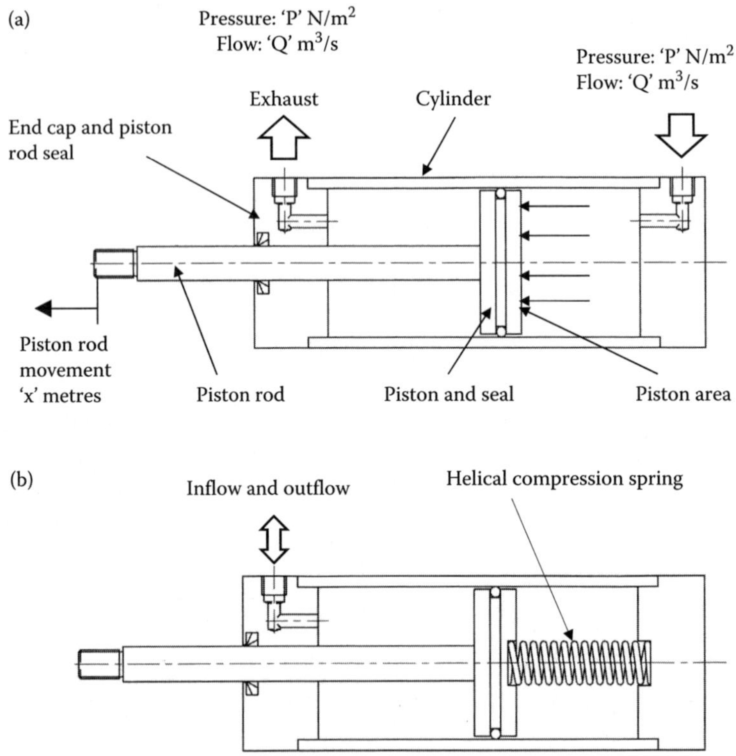

FIGURE 10.1.24
(a) and (b): Elements of double- and single-acting cylinders.

A single acting cylinder is similar but hydraulic pressure is only applied to one side of the cylinder and a spring returns the piston to its original position as shown in Figure 10.1.24b. In some instance the spring may be replaced by a chamber of air.

The following discussion will be restricted to the double acting cylinder.

The flow rate and piston movement are related by the law:

$$G(s) = \frac{x}{Q}$$

$$= \frac{1}{As}$$

This can be expressed as a transfer function with 'x' as the output and 'Q' the input:

$$Q = A\frac{dx}{dt}$$

Pressure and force are related by the law:

$$F = \frac{p}{A}$$

and the corresponding transfer function with 'p' as the input and 'F' as the output will be:

$$G(s) = \frac{x_o}{x_i} = \frac{1}{T^2s^2 + 2\delta Ts} \qquad G(s) = \frac{F}{p} = A$$

10.1.2.7.3 Directional Valve and Actuator

For a hydraulic cylinder (also referred to as an actuator) to function it requires a special type of valve known as a directional control valve, this controls the fluid flow from one side of the piston to the other.

Figure 10.1.25 represents such a valve connected to a double acting cylinder.

The input to the system is the x_i and this allows a flow of fluid into the cylinder of Q (m^3/s) this will force the piston rod to move a distance 'x_o'.

If an assumption is made that for a constant supply pressure, the flow rate is directly proportional to the valve position, it can be stated that:

$$Q = k_v \cdot x_i$$

k_v is the valve constant and the units are m^2/s.

Considering Figures 10.1.24a and 10.1.25, the area of the piston is A m^2.

The velocity of the piston is:

$$v = \frac{dx_o}{dt}$$

This is related to the fluid flow and the piston area by the laws of continuity such that:

$$\begin{aligned} Q &= k_v \cdot x_i \\ &= A\frac{dx_o}{dt} \end{aligned}$$

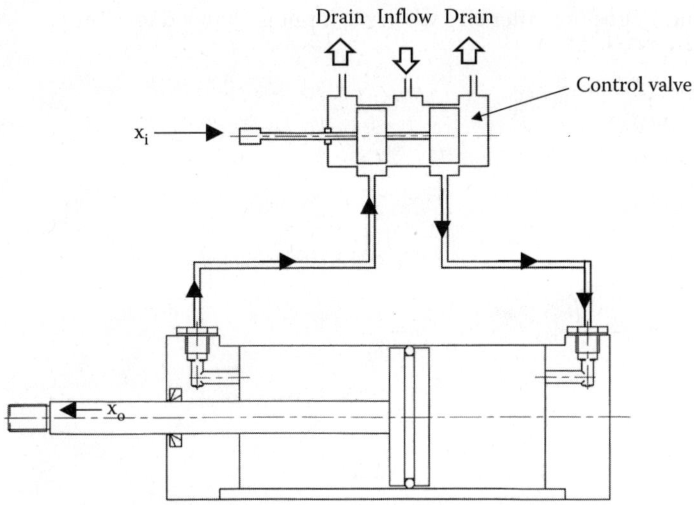

FIGURE 10.1.25
Directional control valve and actuator.

Changing this to a function of 's' this will become:

$$k_v x_i = As \cdot x_o$$

Expressed as a transfer function:

$$G(s) = \frac{x_o}{x_i}(s)$$

$$= \frac{1}{(A / k_v)s}$$

The units of A/k_v are seconds and it can be deduced that this is a time constant 'T'.

$$G(s) = \frac{x_o}{x_i}(s) = \frac{1}{Ts}$$

Note:

This equation is not quite the standard first-order equation $\frac{1}{T_s + 1}$, the difference is the output is variable for a given input, unlike the previous examples where a limit is imposed on the output.

If the cylinder was replaced by a hydraulic motor the equation would be similar where the output would be measured as an angle instead of linear motion.

Example 10.1.5

A hydraulic cylinder has a bore of 100 mm and is controlled using a directional valve with a constant of 0.3 m^2/s.

1. Estimate the time constant 'T'
2. Given that x_i and x_o are zero when t=0, calculate the velocity of the piston and its output position after 0.1 s when the input is changed to 5.0 mm.

Solution

Area of the piston:

$$A = \frac{\pi D^2}{4}$$

$$= 7.854 \times 10^{-3}\ m^2$$

$$T = \frac{A}{k_v}$$

$$= \frac{7.854 \times 10^{-3}}{0.3}$$

$$= 0.026\ s$$

$$G(s) = \frac{x_o}{x_i}(s) = \frac{1}{Ts}$$

now:

$$Ts \cdot x_o = x_i$$

$$T\frac{dx_o}{dt} = x_i$$

$$\frac{dx_o}{dt} = \text{velocity}$$

$$\text{velocity} = \frac{x_i}{T}$$

$$\text{velocity} = \frac{0.005\ \text{m}}{0.026}$$

$$= 0.191\ \text{m/s}$$

$$\text{velocity} = \frac{\text{distance}}{\text{time}}$$

$$\text{distance} = x_o = v \cdot t$$

$$= 0.191 \times 0.1$$

$$= 0.0191\ \text{m}$$

10.1.2.7.4 Directional Control Valve and Actuator Connected to a Mass

Considering the previous example with a directional control valve is connected to the input and output of a double acting actuator, this next example discusses when a mass is connected to the actuator and the mass is subject to a damping force.

Figure 10.1.26 shows the arrangement with the actuator subject to a hydraulic pressure.

The applied force is due to the pressure F_d and is determined by the pressure acting on the area 'A' such that:

$$F_p = pA$$

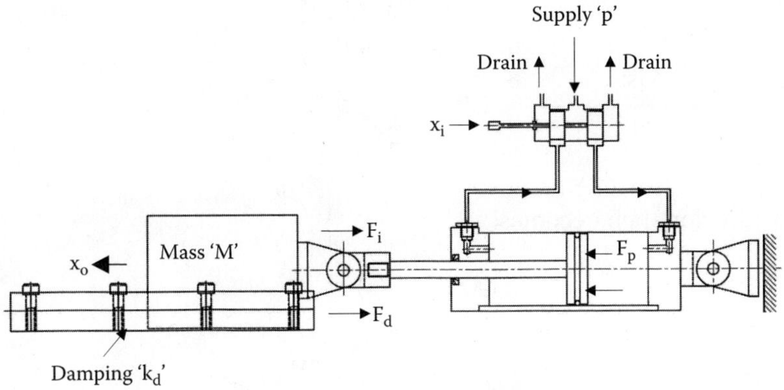

FIGURE 10.1.26
Double-acting cylinder attached to a mass.

The applied force is opposed by the inertia force F_i and the damping force F_d.

$$F_i = M\frac{d^2x_o}{dt^2}$$

and

$$F_d = k_d\frac{dx_o}{dt}$$

Balancing these forces will give:

$$pA = M\frac{d^2x_o}{dt^2} + k_d\frac{dx_o}{dt}$$

Substituting $p=k_v \cdot x_i$

$$k_v \cdot x_i \cdot A = M\frac{d^2x_o}{dt^2} + k_d\frac{dx_o}{dt}$$

In Laplace form:

$$k_v \cdot x_i \cdot A = Ms^2x_o + k_d s x_o$$

Rearranging into a transform function:

$$G(s) = \frac{x_o}{x_i}(s) = \frac{1}{(M / Ak_v)s^2 + (k_d / Ak_v)s}$$

Examining the units it will be found that:

$$\frac{M}{Ak_v} = T^2$$

where T is a time constant.

The critical damping coefficient is:

$$C_c = \sqrt{(4MAk_v)}$$

and the damping ratio is:

$$\frac{\delta k_d}{C_c}$$

The transfer function then becomes:

$$G(s) = \frac{x_o}{x_i}(s)$$

$$= \frac{1}{T^2s^2 + 2\delta Ts}$$

The block diagram for this transfer function is seen in Figure 10.1.27.

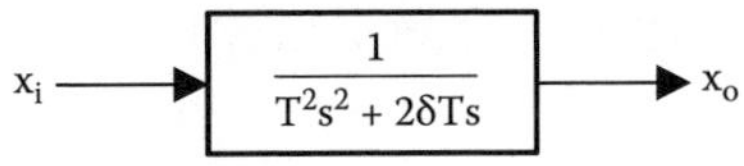

FIGURE 10.1.27
Block diagram for Figure 10.1.26.

There is a similarity with the standard 2nd order equation $\frac{1}{T^2s^2 + 2\delta Ts + 1}$. The difference is due to there being no limitation on the output.

Example 10.1.6

Consider a double acting hydraulic actuator with a bore of 100.0 mm and rod diameter of 25.0 mm and moves a mass of 60 kg. It is fitted with a directional control valve with a constant of k_v=200000 Pa/m There is a damping coefficient of 160 Ns/m.

Determine the time constant T, C_c and δ.
Given that x_i and x_o are zero when t=0, calculate the initial acceleration of the mass when the output is changed suddenly to 6.0 mm.
Calculate the acceleration when the velocity reaches 2 mm/s.
Calculate the velocity when the acceleration is zero.

Solution

The piston area (A):

$$A = \frac{\pi\left(D^2 - d^2\right)}{4}$$

$$A = \frac{\pi\left(100.0^2 - 25.0^2\right)}{4}$$

$$= 7.363 \times 10^{-3}\ m^2$$

$$T = \sqrt{\left(\frac{M}{Ak_v}\right)}$$

$$= \sqrt{\left(\frac{60}{20000 \times 7.363 \times 10^{-3}}\right)}$$

$$= 0.638\ s$$

$$C_c = \sqrt{\left(4MAk_v\right)}$$

$$= 187.997\ N \cdot s/m$$

$$\delta = \frac{k_d}{C_c}$$

$$\delta = 0.851$$

$$G(s) = \frac{x_o}{x_i} = \frac{1}{T^2s^2 + 2\delta Ts}$$

Or in terms of time

$$x_i = \left(T^2 \times \text{acceleration}\right) + \left(2\delta T \times \text{velocity}\right)$$

The initial velocity is zero.

$$0.005 = T^2 a + 0$$

$$0.005 = 0.638^2 a + 0$$

Therefore,

$$a = 6.939 \times 10^{-3}\ \text{m/s}^2$$

Now when v = 0.002

$$0.005 = T^2 a + (2\delta Tv)$$
$$0.005 = 0.638^2 a + (2 \times 0.851 \times 0.638 \times 0.002)$$
$$a = 6.939 \times 10^{-3}\ \text{m/s}^2$$

The system initially accelerates and will eventually settle down to a constant velocity with no acceleration.

Putting 'a' = 0

$$0.005 = 0 + (2 \times 0.851 \times 0.638) \times \text{velocity}$$

Hence,

$$\text{Velocity} = 0.004602\ \text{m/s or } 4.602\ \text{mm/s}$$

10.1.2.8 Electrical System Models

10.1.2.8.1 Resistance

When applying Ohm's Law:

$$V = I.R \text{ or } \frac{V}{I} = R \quad G(s) = \frac{V}{I}(s) = sL \text{ or } s^2Q$$

This equation may be a function of time or 's'.

The equation may be expressed in term of the electrical charge 'Q'.

Now:

$$\text{Since } I = \frac{dQ}{dt} \quad I(s) = sQ$$

Hence:

$$G(s) = \frac{V}{Q}(s) = sR$$

It will be seen that this expression is similar to that of the damper.

Figure 10.1.28 illustrates the symbol for a resistor.

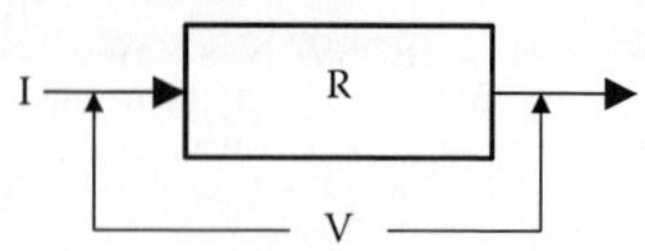

FIGURE 10.1.28
Schematic for a resistor.

10.1.2.8.2 Capacitance

The law for a capacitor is:

$$Q = C.V \quad \frac{V}{Q} = \frac{I}{C}$$

This expression is similar to that for a spring.

Differentiating with respect to time:

$$\frac{dQ}{dt} = C\frac{dV}{dt}$$

where $\frac{dQ}{dt}$ is the current 'I' hence the equation can be expressed as:

$$I = C\frac{dV}{dt}$$

and expressed as a function of 's' this then becomes:

$$I(s) = CsV$$

The transfer function is:

$$G(s) = \frac{V}{I}(s) = \frac{1}{sC}$$

Figure 10.1.29 illustrates the symbol for a capacitor.

10.1.2.8.3 Inductance

From Faraday's Law: $V = L\frac{dI}{dt} = L\frac{d^2Q}{dt^2}$

This expression is similar to the model for mass and can be expressed as either a first-order or second-order equation.

As a function of 's':

$$V(s) = L\,sI \text{ or } Ls^2Q$$

$$G(s) = \frac{V}{I}(s) = sL \text{ or } s^2Q$$

Figure 10.1.30 is the circuit symbol for an inductor.

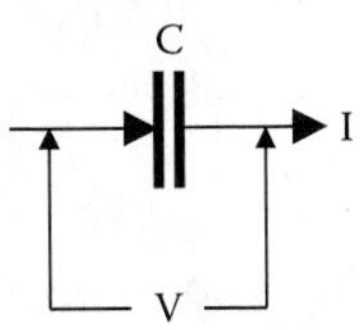

FIGURE 10.1.29
Schematic for a capacitor.

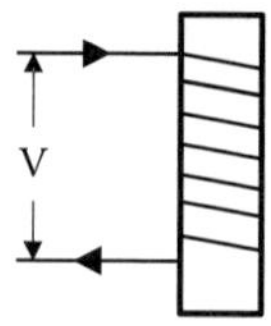

FIGURE 10.1.30
Schematic for an inductor.

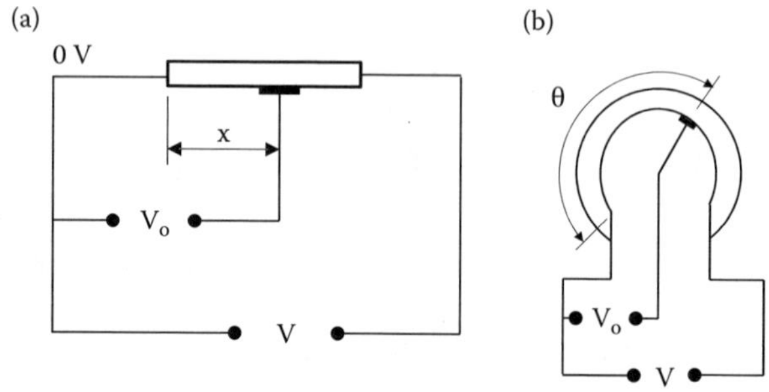

FIGURE 10.1.31
Schematics for linear (a) and angular (b) potentiometers.

10.1.2.8.4 Potentiometer

Figure 10.1.31 illustrates two versions of a potentiometer:

Figure 10.1.31a is a linear potentiometer.

Figure 10.1.31b is and angular version.

In a potentiometer, if the supply voltage is constant and the current is very small or negligible, the output voltage 'V' is directly proportional to the position 'x' or angle 'θ'.

In this case a simple transfer function is easily obtained:

$$V_o = V_i - \Delta V e^{-t/T} \quad G = \frac{V_o}{x}(s) = \text{constant} = k_p \text{ (linear)}$$

$$G(s) = \frac{V_o}{\theta}(s) = \text{constant} = k_p \text{ (angular)}$$

10.1.2.8.5 R–C Series Circuit

Figure 10.1.32 illustrates a typical R–C series circuit in which the sum of the voltage across the resistor and the capacitor is:

$$V_i = IR + \frac{I}{Cs}$$

$$V_i = I\left(R + \frac{1}{Cs}\right)$$

The output will be the voltage across the capacitor so:

$$V_o = \frac{I}{Cs}$$

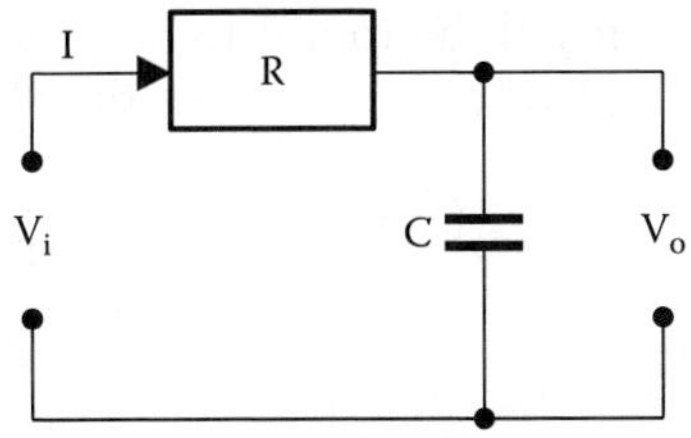

FIGURE 10.1.32
Schematic for an R–C circuit.

The transfer function is then:

$$G(s) = \frac{V_o}{V_i}(s) = \frac{I/Cs}{I(R + 1/Cs)} = \frac{1}{RCs + 1}$$

The units of RC will be seconds and this is another electrical time constant 'T'. The transfer function can be written as:

$$G(s) = \frac{V_o}{V_i}(s) = \frac{1}{RCs + 1}$$

This is a standard first-order equation and is the same as the spring and damper also the thermal example.

10.1.2.8.6 L–C–R in Series

This is three sub-systems in series. Figure 10.1.33 depicts such a circuit, in this case the output is the voltage on the capacitor and the input is the voltage across the series circuit.

The input voltage is the sum of all three voltages and is found by summing them:

$$V_i = IR + IsL + \frac{I}{sC}$$

The output voltage is $V_o = \frac{I}{sC}$. The transfer function is:

$$G(s) = \frac{V_o}{V_i}(s) = \frac{I/Cs}{I(R + sL + 1/Cs)} = \frac{1}{RCs + CLs^2 + 1}$$

$$G(s) = \frac{V_o}{V_i}(s) = \frac{1}{s^2CL + sCR + 1}$$

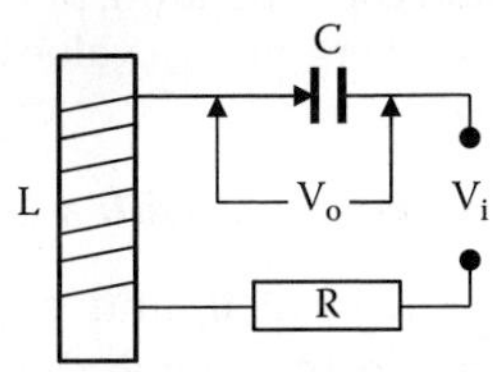

FIGURE 10.1.33
Schematic for L–C–R in series.

If the units of CL are examined it will be found to be seconds^2 and this is another time constant which is defined as T^2=CL, the equation can then be rewritten as:

$$G(s) = \frac{V_o}{V_i}(s) = \frac{1}{T^2s^2 + 2\delta Ts + 1}$$

The damping ratio δ is defined as:

$$\delta + \frac{R}{\sqrt{4\frac{L}{C}}}.$$

The expression $\sqrt{4\frac{L}{C}}$ is called the critical damping value.

Note: This is a standard second-order equation and is identical to the mass–spring–damper system.

Example 10.1.7

A capacitance of 300 μF is connected in series with a resistor of 30 kΩ as shown in Figure 10.1.32.

The transfer function for this circuit is:

$$\frac{\theta_o}{\theta_i}(s) = \frac{1}{(Ts+1)}$$

There is a step voltage change from 3 to 10 V across the resistor.

Calculate the time constant 'T' and derive a formula for how the voltage across the capacitor varies with time.

Solution

$$T = RC = 20 \times 10^3 \times 200 \times 10^{-6} = 9\text{ s}$$

The derivation follows that used for Example 10.1.3 where 'V' replaces 'θ' and the following will result:

$$V_o = V_i - \Delta Ve^{-t/T}$$

Substituting the values T=9

$$\Delta V = 10 - 3 = 7 \quad V_i = 10$$

$$V_o = 10 - 7e^{-t/9}$$

Figure 10.1.34 shows the results and corresponding plot for this equation. From the plot it is clearly seen that the curve is converging to a value of V_o = 10 units.

10.1.2.9 Closed Loop System Transfer Function with a Unity Feedback

Consider a simple system having an input θ_i and an output θ_o. For the system to be controlled where the output is to change and match the value of the input which will have a set value, the input will need to reflect the error θ_e instead of the set value. This error will be obtained by comparing the output value against the input value using a summing

t (s)	V_o
0	3.000
2	4.395
4	5.512
6	6.406
8	7.122
10	7.696
12	8.155
14	8.522
16	8.817
18	9.053
20	9.241
22	9.393
24	9.514
26	9.611
28	9.688
30	9.750

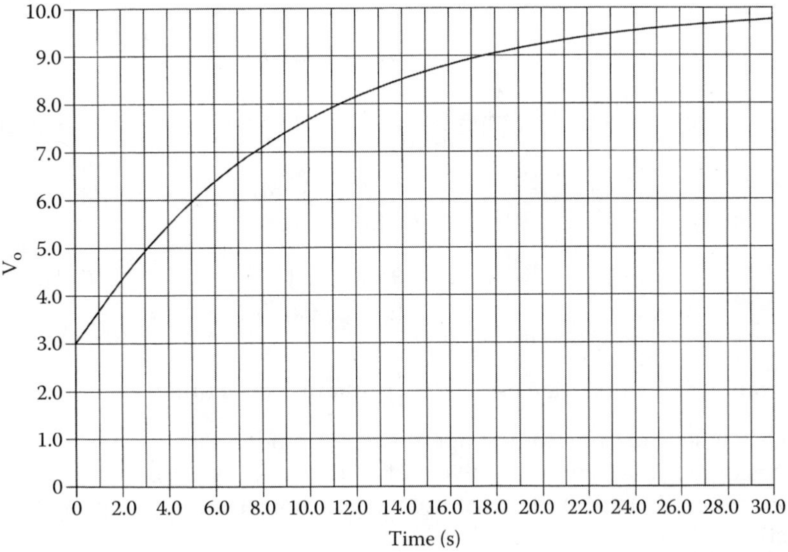

FIGURE 10.1.34
Results for Example 10.1.7.

junction (see Section 10.1.3.3 for a full description). This will produce the result $\theta_e = \theta_i - \theta_o$ and as θ_o is subtracted from θ_i, this is referred to as a "Negative Feedback". The block diagram shows that the signal is returned around a closed loop to the summing junction hence the name "Closed Loop System".

Figure 10.1.35 shows the representative block diagram for such a system. This particular system is known as a "Unitary Feedback System" as the feedback loop does not contain any processing element shown in the feedback loop (this will be discussed in more detail in Section 10.1.3.4).

$$G(s) = \frac{\theta_o}{\theta_e}$$

Substitute $\theta_e = \theta_i - \theta_o$.

$$G(s) = \frac{\theta_o}{\theta_i - \theta_o}$$

Divide the bottom line by θ_o.

$$G(s) = \frac{1}{\frac{\theta_i}{\theta_o} - 1}$$

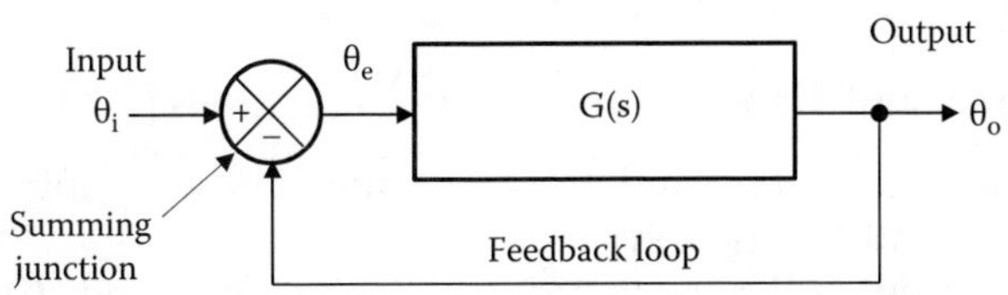

FIGURE 10.1.35
Closed-loop system with unity feedback.

Let
$$\frac{\theta_i}{\theta_o} - 1 = \frac{1}{G(s)}$$

$$\frac{\theta_i}{\theta_o} = \frac{1}{G(s)} + 1 = \frac{1+G(s)}{G(s)} \text{ and invert}$$

$$\frac{\theta_o}{\theta_i} = \frac{G(s)}{1+G(s)}$$

or
$$\frac{\theta_o}{\theta_i} = \frac{1}{1/G(s)+1}$$

The transfer function for a closed loop system with a unitary feedback is $\frac{\theta_o}{\theta_i} = \frac{1}{1/G(s)+1}$

The transfer function for the open loop system is G(s).

Considering the hydraulic valve and actuator as discussed in Section 10.1.2.7.3.

When the actuator and valve are turned into a closed loop system the two transfer functions become:

$G_{cl} = \frac{1}{Ts+1}$ for the first-order version and $G_{cl} = \frac{1}{T^2s^2} + 2\delta Ts + 1$ for the second-order version.

It will be seen that these two models are mathematically similar to the transfer function for the mass–spring–damper and L–C–R circuits.

Example 10.1.8

An open loop circuit has a transfer function $G(s) = \frac{2}{s^2+2s+1}$. Determine the closed loop transfer function when the feedback is unity.

Solution

$$G_{cl} = \frac{1}{\frac{1}{G(s)+1}}$$

$$= \frac{1}{\frac{s^2+2s+1}{2}+1}$$

$$= \frac{2}{s^2+2s+1+2+1}$$

$$= \frac{2}{s^2+2s+1+3}$$

10.1.3 Block Diagram and Transfer Function Manipulations

In Section 10.1.2 block diagrams and transfer functions were introduced for various mechanical, hydraulic and electrical systems.

In this section the reader will be introduced to "Open Loop" and "Closed Loop" control systems and the use of transfer functions in the study of these systems, but first a distinction needs to be made between these two types of systems.

To start, there are two general classifications of control systems:

1. Open Loop systems where the control action is independent of the output from the system.
2. Closed loop systems in which the control action is dependent on the output.

10.1.3.1 Open Loop Control System

An open loop system requires an independent external action to ensure the required output. As an example: Human intervention is needed when a saucepan is boiling on a heating ring switching it off to prevent it boiling dry.

Open loop systems have two important features:

- They depend on their calibration for accurate operation.
- Open loop systems are not normally affected by stability problems, there is little risk that the input will result in an unexpected output.

Figure 10.1.36 gives an example of an open loop system.

10.1.3.2 Closed Loop Control System

Closed loop systems are also referred to as "Feed Back" systems where the output is compared against the input into the system. These types of systems are complicated requiring the use of differential equations for their solution.

Figure 10.1.37 illustrates a typical closed loop control system using a feedback loop together with its corresponding block Figure 10.1.38.

10.1.3.3 Summing Junctions

In order for a control system to be regulated, the error between the output and input needs to be determined.

The summing junction is used to represent the addition or subtraction of signals.

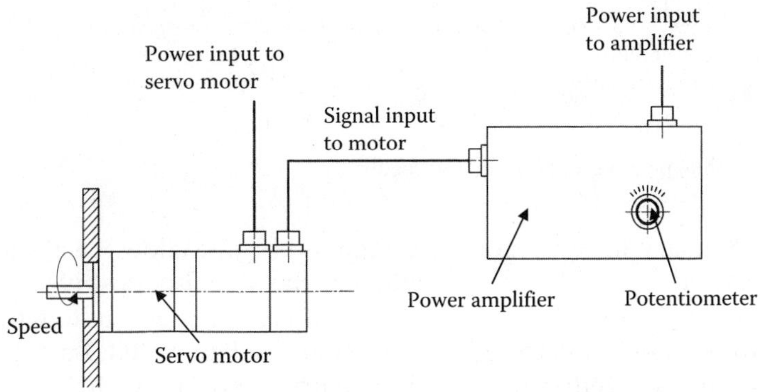

FIGURE 10.1.36
Open-loop control circuit.

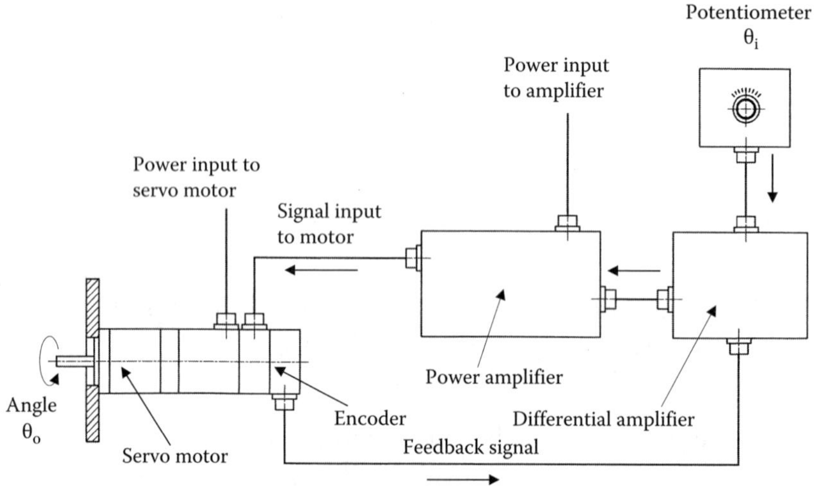

FIGURE 10.1.37
Closed-loop control circuit.

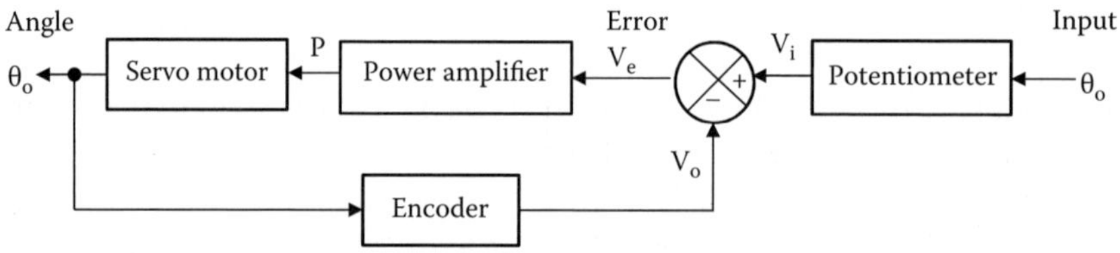

FIGURE 10.1.38
Bloch diagram for the closed-loop circuit in Figure 10.1.37.

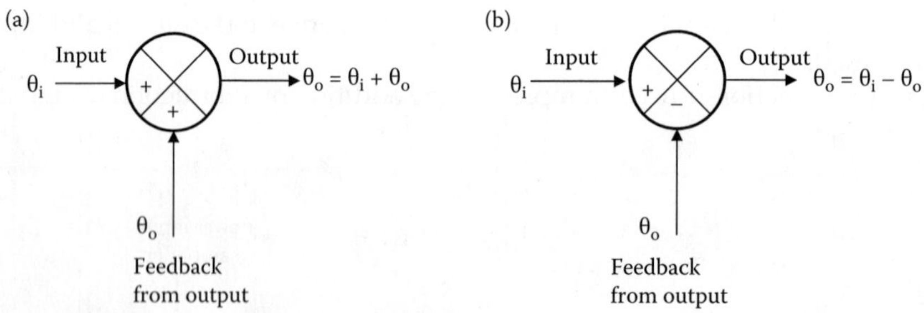

FIGURE 10.1.39
Summing junctions for (a) positive and (b) negative feedback.

Arrows are necessary to represent the direction of signal flow. Each incoming arrow has a sign associated with it to indicate whether the signal is positive or negative. The summing junction generally has two or more inputs and one output to which the sum of the inputs are routed as shown in Figure 10.1.39a and b. Figure 10.1.39a depicts a junction adding the two inputs together for a positive output and Figure 10.1.39b is subtracting the two inputs. In general the junction shown in (b) is the most common junction used in negative feedback systems. Very occasionally the junction (a) may be found in positive feedback systems but these will not be discussed in this chapter as they are specialized control techniques.

10.1.3.4 Closed Loop System Transfer Functions

The basic block diagram for a closed loop system with unitary feedback is shown in Figure 10.1.40. The main block is an open loop system with a transfer function G_{ol}. This relates the output and the error so that $G_{ol}=\theta_o/\theta_i$.

The transfer function for a closed loop system is G_{cl}. The input θ_i is related to the output θ_o.

By comparing the output value with the input value the error is obtained in the summing junction. This will produce the result $\theta_e=\theta_i-\theta_o$ and as θ_o is subtracted this is referred to as "Negative Feedback". The block diagram shows that the signal is passed around a closed loop and this is known as a "Closed Loop" system.

The system shown in Figure 10.1.40 is said to have a "Unitary Feedback" as there is no processing shown in the feedback path.

The following is an important result for all closed loop systems and will be used later.

$$G(s)=\frac{k}{(T^2s^2+2\delta Ts)}$$

$$\theta_e=\theta_i-\theta_o$$

$$G_{ol}=\frac{\theta_o}{\theta_i-\theta_o}$$

$$G_{ol}(\theta_i-\theta_o)=\theta_o$$

$$G_{ol}\theta_i-G_{ol}\theta_o=\theta_o$$

$$G_{ol}\theta_i=\theta_o+G_{ol}\theta_o$$

$$G_{ol}\theta_i=\theta_o(1+G_{ol})$$

$$G_{cl}=\frac{\theta_o}{\theta_i}$$

$$G_{cl}=\frac{G_{ol}}{1+G_{ol}}$$

$$G_{cl}=\frac{1}{(1/G_{ol})+1}$$

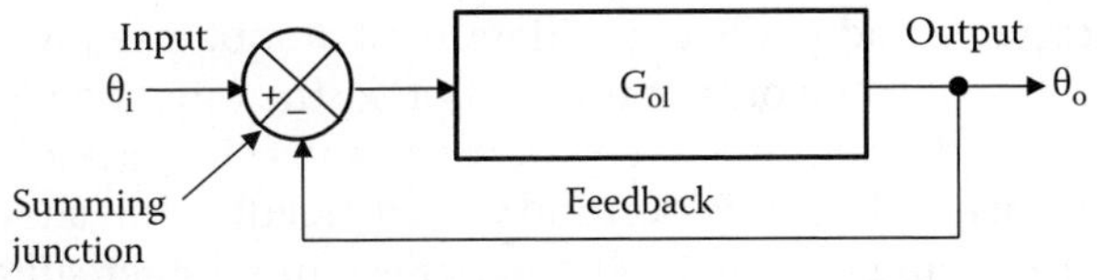

FIGURE 10.1.40
Basic block diagram for a closed loop with unity feedback.

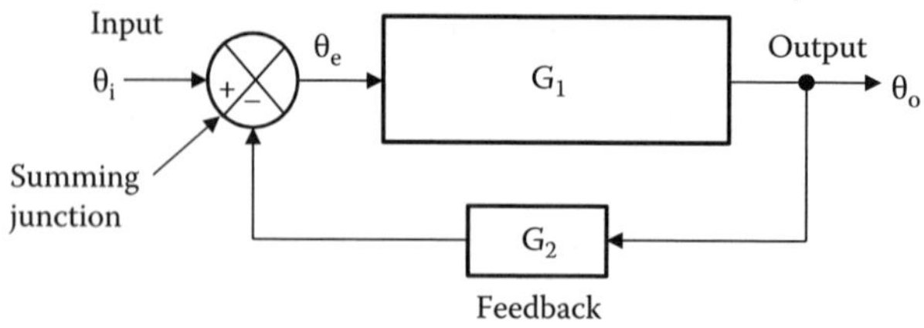

FIGURE 10.1.41
Basic block diagram for a closed loop with processor in the feedback loop

This is the transfer function for a closed loop unity feedback system although in practice there will be a transducer to measure the output, therefore a block in the feedback loop should be shown to represent the transducer as shown in Figure 10.1.41. There may be other signal processing instrumentation in the path such as amplifiers or signal conditioners and these will also need to be shown. In the figure, G_1 has been substituted in place of G_{ol}, this is the open loop transfer function and is in the forward path and G_2 is in the feedback path.

The open loop transfer function can be related to G_1 and G_2 as follows:

$$G = \frac{\theta_o}{\theta_e}$$

$$\theta_e = \theta_i - G_2\theta_o$$

$$G_1 = \frac{\theta_o}{\theta_i - G_2\theta_o}$$

$$G_1(\theta_i - G_2\theta_o) = \theta_o$$

$$G_1\theta_i - G_1G_2\theta_o = \theta_o$$

$$G_1\theta_i = G_1G_2\theta_o + \theta_o$$

$$G_1\theta_i = \theta_o(1 + G_1G_2)$$

$$G = \frac{\theta_o}{\theta_i} = \frac{G_1}{1 + G_1G_2}$$

This is the transfer function for the closed system.

10.1.3.5 Velocity Feedback

When a system is unstable and tends to oscillate, a technique used to stabilise the system is to use velocity feedback. If the output is a motion x_o the rate of change dx_o/dt will be a true velocity.

A typical servo system uses both position and velocity feedback as shown in Figure 10.1.42. The velocity signal is generated from a tacho-generator or another suitable speed measuring device. The resulting signal is compared with the input and output and the resulting error signal will be:

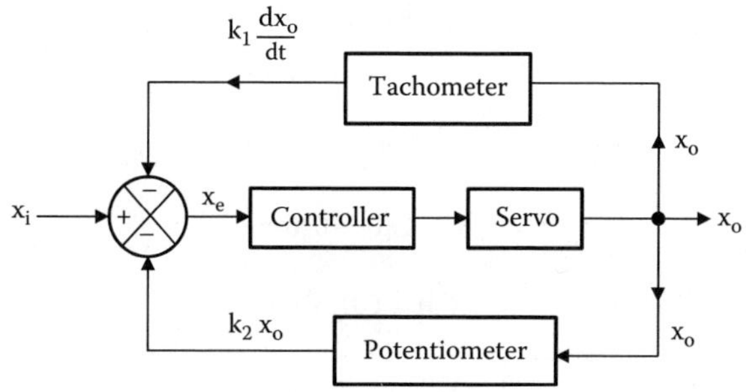

FIGURE 10.1.42
Block diagram for a closed-loop system with velocity feedback.

$$x_e = x_i - k_2 x_o - k_1 \frac{dx_o}{dt}$$

When a step change is made to the input, the error will be a maximum so the output will change very rapidly. The velocity feedback will be greatest at the start. The effect of the feedback will be to reduce this error directly proportional to the velocity. When the output is static the feedback will be zero and therefor no error will result. The feedback will have the same effect as damping and if an analysis is completed it will be seen that there will be control over k_1 giving control over damping. This method is very useful for stabilising an oscillating system.

Example 10.1.9

Figure 10.1.43 depicts a closed loop control system having both velocity and negative feedback. The transfer functions for the system is $G(s) = \frac{k}{(T^2s^2 + 2\delta Ts)}$. It is required to derive the closed loop transfer function.

Solution

The open loop transfer function is given as:

$$G(s) = \frac{k}{(T^2s^2 + 2\delta Ts)}$$

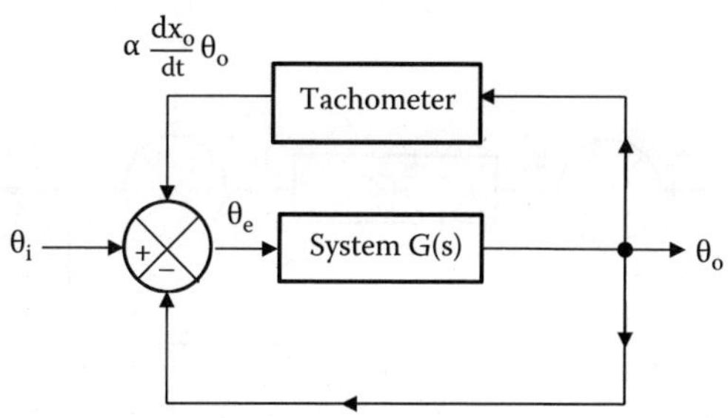

FIGURE 10.1.43
Closed-loop circuit for Example 10.1.8.

$$\frac{\theta_o}{\theta_i} = \frac{k}{T^2s^2 + s(2\delta T + k\alpha) + k}$$

$$\theta_e = \theta_i - \theta_o - (\alpha s + 1)$$

$$\theta_o = G\theta_e = G[\theta_i - \theta_o(\alpha s + 1)]$$

$$\theta_o = G\theta_i - G\theta_o(\alpha s + 1)$$

$$\theta_o + G\theta_o(\alpha s + 1) = G\theta_i$$

$$\theta_o[1 + G(\alpha s + 1)] = G\theta_i$$

$$\frac{\theta_o}{\theta_i} = \frac{G}{1 + G(\alpha s + 1)}$$

$$\frac{\theta_o}{\theta_i} = \frac{1}{1/G + (\alpha s + 1)}$$

Substituting $G = \dfrac{k}{(T^2s^2 + 2\delta Ts)}$

$$\frac{\theta_o}{\theta_i} = \frac{1}{\frac{[T^2s^2 + 2\delta Ts]}{k} + (\alpha s + 1)}$$

$$\frac{\theta_o}{\theta_i} = \frac{k}{T^2s^2 + 2\delta Ts + k(\alpha s + 1)}$$

$$\frac{\theta_o}{\theta_i} = \frac{k}{T^2s^2 + s(2\delta T + k\alpha) + k}$$

This is the closed loop transfer function.

The term with 's' is the effective damping term and is affected by the value of $k\alpha$.

10.1.3.6 Disturbance

Figure 10.1.44 characterises the effect on the output signal by a disturbance.

The disturbance 'd' is added to the output 'θ' to produce a new output θ_o.

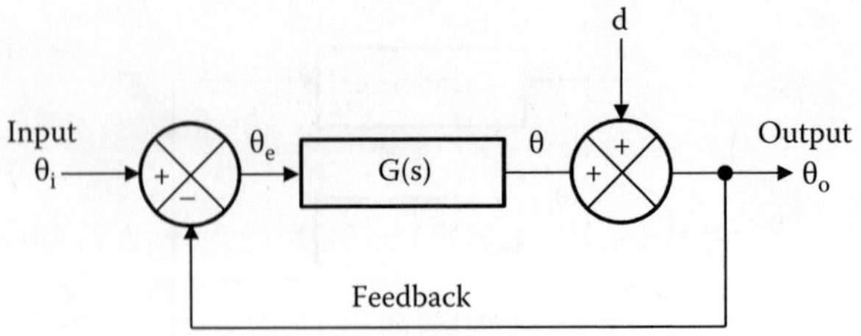

FIGURE 10.1.44
Closed-loop circuit subject to a disturbance.

G is the forward path transfer function:

$$\theta_e = \theta_i - \theta_o.$$

$$\theta_o = \theta + d$$

$$\theta = G\theta_e.$$

$$\theta_o = G\theta_e$$

$$\theta_o = G\theta_e + d$$

$$\theta_o = G(\theta_i - \theta_o) + d$$

$$\theta_o = G\theta_i - \theta_o + d$$

$$\theta_o(1+G) = G\theta_i + d$$

$$\theta_o = \frac{(G\theta_i + d)}{(1+G)}$$

Example 10.1.10

A simple closed loop circuit as shown in Figure 10.1.44 consists of an amplifier with a gain of 10. For an input of 4 mA, show that the effect of a disturbance when added to the output with a magnitude of:

1. 0.
2. 2.

Solution

1. Let G=10, θ_i=4, d=0.

$$\theta_o = \frac{(10\times 4+0)}{(1+10)}$$

$$\theta_o = \frac{40}{11}$$

2. Let G=10, θ_i=4, d=2

$$\theta_o = \frac{(G\theta_i + d)}{(1+G)}$$

$$\theta_o = \frac{(10\times 4+2)}{1+10}$$

$$\theta_o = \frac{42}{11}$$

This result shows that a disturbance of 2 will produce an output error of 2/11.

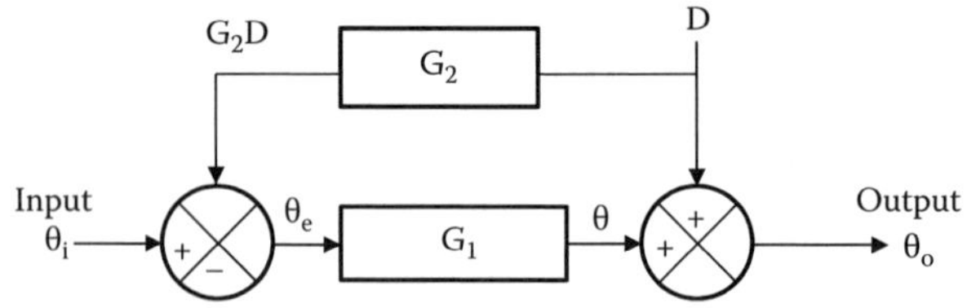

FIGURE 10.1.45
Closed-loop circuit elimination of disturbance.

10.1.3.6.1 To Eliminate the Effect of a Disturbance

To reduce or eliminate the effect of a disturbance acting on the output, it is possible to introduce a special feedback path. Figure 10.1.45 shows the idealised system.

The disturbance 'D' in this case is processed through a transfer function 'G_2' and this is added to the input. With G_1 as the forward path transfer function, G_2 is the feedback path transfer function.

$$\theta = G_1\theta_e = G_1(\theta_i - G_2D) \quad \theta_o = \theta + D$$

$$\theta_o = G_1(\theta_i - G_2D) + D \quad \theta_o = G_1\theta_i - G_1G_2D + D$$

From the last line it will be seen that if $G_1G_2D = D$ then $\theta = G_1\theta_i$

The effect of the disturbance will be completely removed when $G_1 = 1/G_2$.

Example 10.1.11

For the system described in Section 10.1.3.6.1 the forward path transfer function is

$$G = \frac{12}{4s^3 + 7s^2 + 60s^2} = \frac{12}{4s^3 + 60s^2} \quad G_1(s) = \frac{4}{(s+1)}$$

Determine the transfer function for the feedback path to eliminate the effects of the disturbance acting on the input:

Solution

$$G_2(s) = \frac{1}{G_1(s)} = \frac{(s+1)}{4}$$

10.1.3.7 Proportional and Differential Control

A transfer function used in process engineering is the P+D transfer function (Proportional plus Differentiation. Figure 10.1.46 depicts the block diagram for this function.

For a proportional system the output is directly proportional to the input. It can be considered as an amplifier or attenuator with k_1 as a simple ratio.

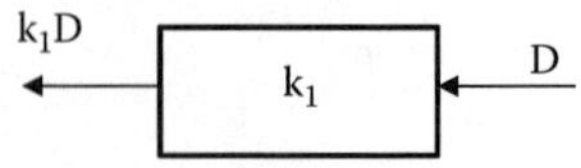

FIGURE 10.1.46
Feedback diagram for proportional control.

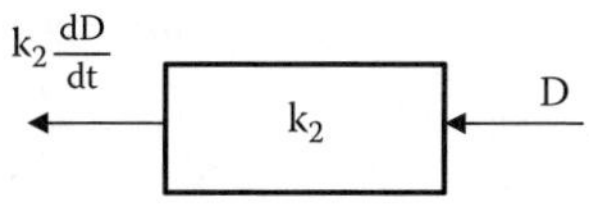

FIGURE 10.1.47
Feedback block diagram for differential control.

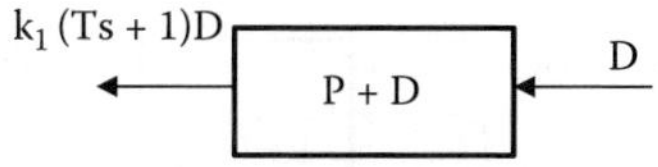

FIGURE 10.1.48
Feedback block diagram for proportional and differential control.

Figure 10.1.47 shows a differential block in which the output is proportional to the rate of change of the input with time. Fitting a tachometer to the output is such an example. In Laplace form the output is k_2sD.

The units of k_2 is in seconds and can be written as $k_2 = Tk_1$ where 'T' is the time derivative.

Proportional control can be combined with differential control and Figure 10.1.48 illustrates such a block. For P+D it follows that the transfer function is k_1 (Ts+1).

Example 10.1.12

Given that the open path transfer function is $G_1 = \frac{5}{(s+2)}$ Determine the value of the time derivative 'T' and the constant which will eliminate the disturbance 'D'.

Figure 10.1.49 illustrates the block diagram.

Solution

$$G_2 = \frac{1}{G_1} = \frac{(s+2)}{5}$$

$$G_2 = k_1(Ts+1)$$

Rearranging so that the forms match,

$$0.2(s+2) = k_1(Ts+1)$$

$$0.4(0.5s+1) = k_1(Ts+1)$$

Hence by comparison $k_1 = 0.4$ and $T = 0.5$.

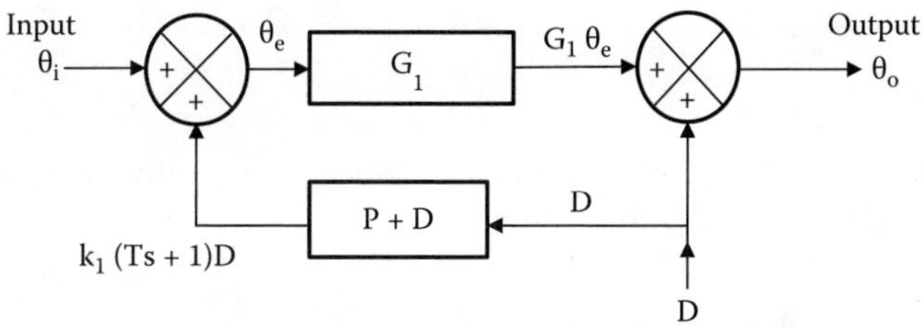

FIGURE 10.1.49
Closed-loop circuit for Example 10.1.11.

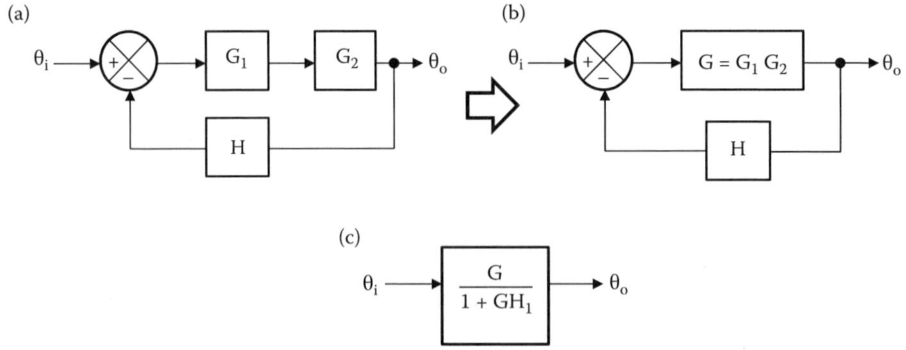

FIGURE 10.1.50
(a)–(c): Reducing a block diagram, complexity (1).

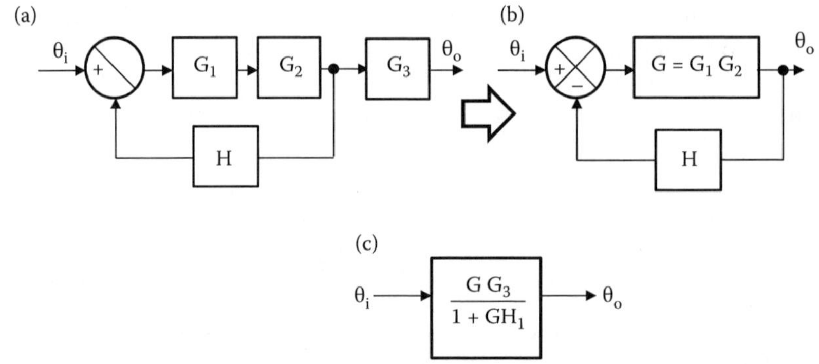

FIGURE 10.1.51
(a)–(c): Reducing a block diagram, complexity (2).

10.1.3.8 Simplifying Complex Systems

The symbol 'H' is commonly used for the feedback transfer function but any appropriate letter can be used to help simplify complex circuits.

Figures 10.1.50 through 10.1.52 show the stages in simplifying a block diagram to a single block with one transfer function.

Example 10.1.13

Consider the block diagram shown in Figure 10.1.53a and derive the transfer function with

$$G_1 = \frac{3}{s},\ G_2 = \frac{1}{(4s+5)},\ G_3 = 4,\ G_4 = \frac{1}{s},\ H_1 = 5,\ H_2 = 0.5.$$

Solution

Figure 10.1.53c shows the simplification of the block diagram that is shown in Figure 10.1.53a to a single block and transfer function.

Substituting the above data results in

$$D_1 = \frac{\frac{4}{4s+5}}{1+\frac{4\times 0.54}{4s+5}} = \frac{4}{(4s+5)\left(1+\frac{2}{4s+5}\right)} = \frac{4}{4s+7}$$

$$G = \frac{\left(\frac{3}{s}\right)\left(\frac{4}{4s+7}\right)(D_1)}{1+\left(\frac{3}{s}\right)(5)(D_1)} = \frac{\frac{3}{s^2}\left(\frac{4}{4s+7}\right)}{1+\frac{15}{s}\left(\frac{4}{4s+7}\right)} = \frac{\frac{12}{s^2(4s+7)}}{1+\frac{60}{4s^2+7s}} = \frac{12}{s^2(4s+7)+s^2(4s+7)\left(\frac{60}{4s+7}\right)}$$

$$G = \frac{12}{4s^3+7s^2+60s^2} = \frac{12}{4s^3+67s^2}$$

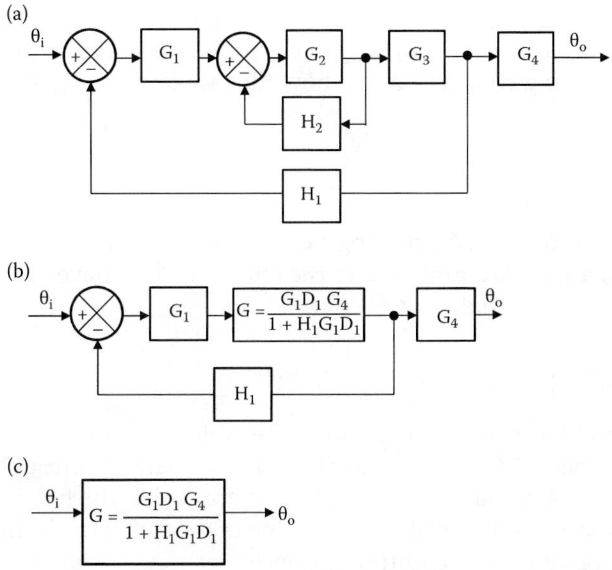

FIGURE 10.1.52
(a)–(c): Reducing a block diagram, complexity (3).

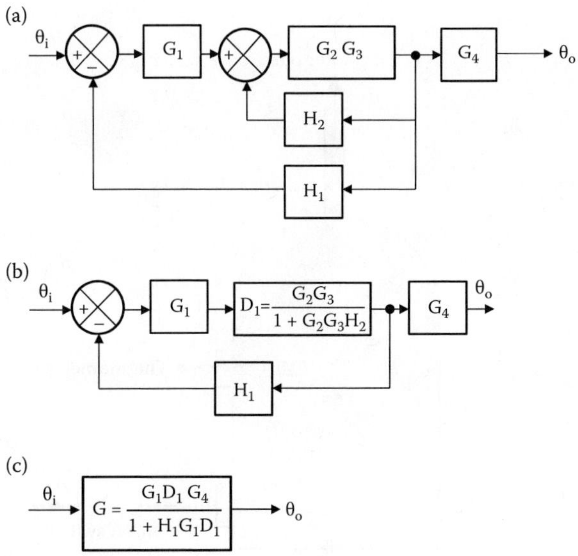

FIGURE 10.1.53
(a)–(c): Reducing a block diagram for Example 10.1.13.

10.2 Analysis of Control Systems

10.2.1 Introduction

In Section 10.1, the reader was introduced to the basics of control theory and how systems are modeled.

In this Section 10.2, the methods that are used to analyze the performance of the system are discussed.

Section 10.1 introduced the concept of transfer functions and how they represented the model of the system being considered and how they will be used to study the behavior of the system.

10.2.1.1 System Response

The primary function of the Control Engineer is to determine how a control system is going to respond to a particular input, and the effect this will have on the output. To start the discussion, consider a simple on/off control.

10.2.2 On/Off Control

In a typical domestic hot water storage tank, the water may be heated using an electric heating element connected to a power supply that is operated through a simple switch. When the switch is closed, the electrical current is applied to the heating element and the water is then heated. In this instance, the operation of the switch is the input action and the resultant water temperature is the output action.

The water heater shown in Figure 10.2.1 is comprised of a cold water feed into the tank and a hot water takeoff. The tank is fitted with a thermometer and the heating element (an on/off switch to control the electrical power to the heater) completes the assembly.

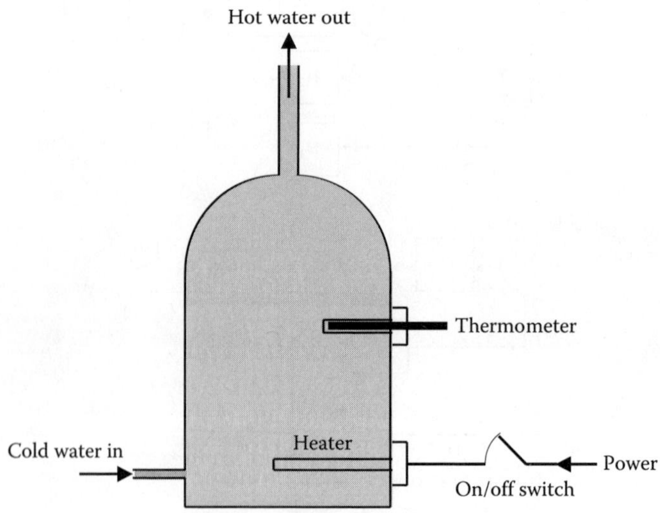

FIGURE 10.2.1
Control of hot water in storage tank.

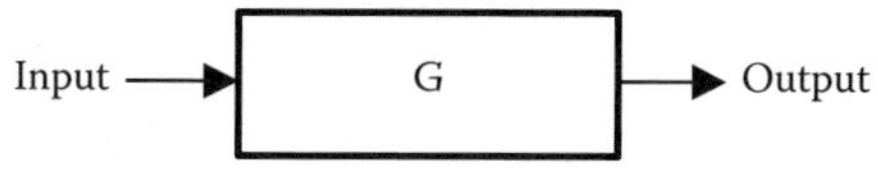

FIGURE 10.2.2
Block diagram showing open-loop system.

This arrangement is represented by a simple block diagram, as shown in Figure 10.2.2, from which it can be clearly seen that this is an "open-loop system" and there is no direct control over the water temperature, thus relying on human intervention to turn the switch on or off to control the water temperature.

An improvement will be to replace the thermometer with a thermocouple and replace the on/off switch with a thermostat. The thermostat is essentially an on/off switch that is directly controlled from the thermocouple, as shown in Figure 10.2.3. The thermostat has two adjust temperature settings, the lower temperature setting is when the switch is closed and the upper setting is when the switch opens.

The operation is to set the desired temperature 'θ_i' (input) on the thermostat and the desired water temperature 'θ_o' (output). The thermostat then compares the measured temperature with the set temperature. When the water temperature is too cold, the thermostat remains closed and heat is provided to the water. When the water temperature reaches the desired temperature, the thermostat is then switched open.

A *step change* is when the water in the storage tank is at ambient temperature and the switch is closed, or when the set temperature is suddenly increased.

Ideally, the thermostat should switch off the power to the heating element when the set temperature is reached. In practice, this will not occur, for as the element is cooling down residual heat from the element will still continue to heat the water for a short time, resulting in a small "overshoot" of the required temperature. Conversely, as the water in the tank begins to cool down at a rate that will be dependent upon the amount of insulation surrounding the tank, when the water temperature reaches the low temperature limit, the thermostat will not switch on immediately until it has detected an error between

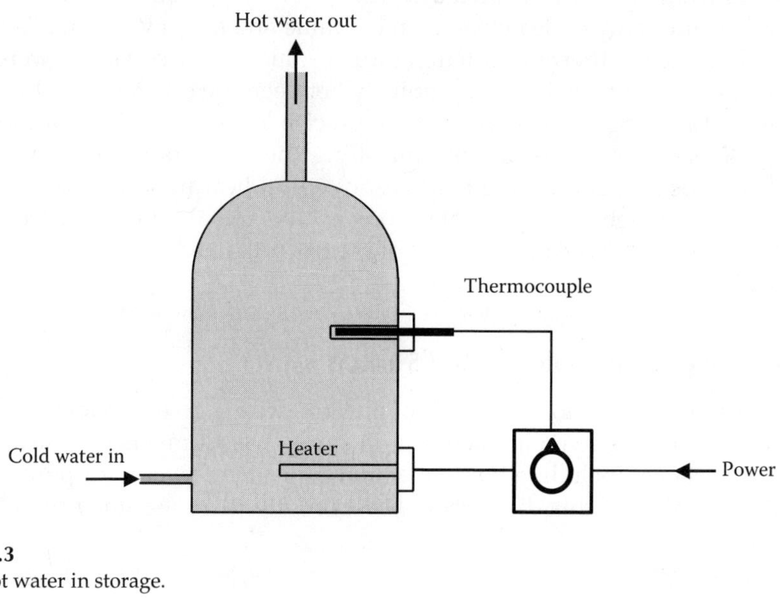

FIGURE 10.2.3
Control of hot water in storage.

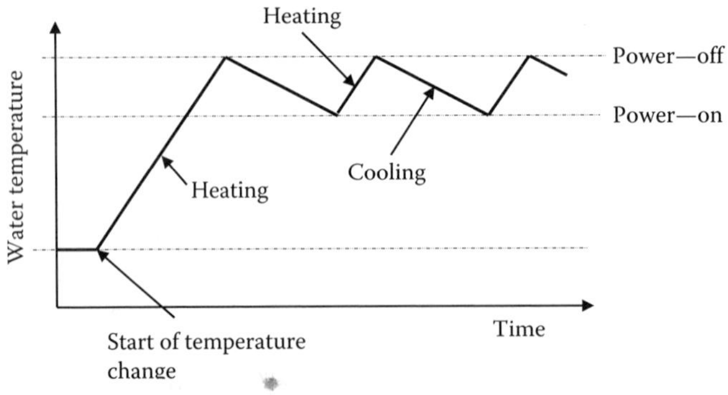

FIGURE 10.2.4
System time response diagram.

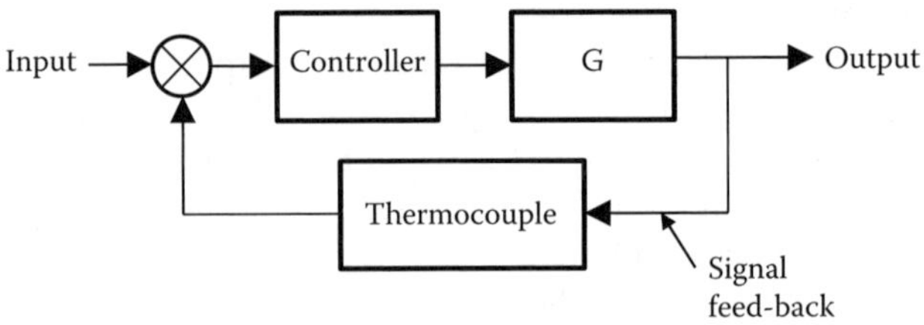

FIGURE 10.2.5
Block diagram showing open-loop system.

the required temperature and the set temperature on the thermostat. This procedure will continue indefinitely, resulting in a "Temperature-Time" graph, as shown in Figure 10.2.4. This graph is known as a "System-Time Response" graph.

The block diagram depicted in Figure 10.2.5 shows that the introduction of the thermostat into the switching circuit has converted the system into a "closed-loop system."

This type of on/off control is known as a "Bang/Bang" system, for it can be seen that the output will continually cycle between an off value and an on value that is fractionally above and below the set thermostat temperature values. If these values were set closer together, the system would switch more quickly between the on and off values. With this type of system, it is difficult to obtain a more precise level of accuracy, but for domestic purposes it is more than adequate for controlling the temperature of the water in the storage tank and also for controlling a domestic central heating system where the actual temperatures are relatively unimportant so long as they remain within the comfort level.

Systems that require a more precise level of control will need a more advanced method of regulation.

10.2.3 Response of Continuous Control Systems

This section discusses how to set about solving the way a given system performs with respect to time. It should be mentioned that in some cases 'time' is not always the main variable; other variables including how a machine tool slide moves in response to an input could equally be valid. In this discussion, however, it will be assumed that 'time' is the main variable.

10.2.3.1 Standard Models

The initial step is to derive a mathematical model along the lines shown in Section 10.1, where it was shown that the following models applied to a number of different systems that were analogs to each other. These are the standard first-order and second-order equations.

Standard first-order equation:

$$\theta_i = T\frac{d\theta_o}{dt} + \theta_o \tag{10.2.1}$$

Standard second-order equation:

$$\theta_i = T^2\frac{d^2\theta_o}{dt^2} + 2\delta T\frac{d\theta_o}{dt} + \theta_o \tag{10.2.2}$$

where

'T' is a time constant, and

'δ' is a damping ratio.

'T' and 'δ' of system parameters or constants and will be discussed later in this section.

10.2.4 Standard Inputs

10.2.4.1 Impulse (Figure 10.2.6a)

An impulse is an instantaneous change in 'θ_i' lasting for a zero length of time and returning to its initial value. This type of event is mostly confined to digital systems where instantaneous values are sampled by digital-to-analog converters. Impulse response is widely used as a standard input to a control system to measure and compare the responses of different systems to the same type of event.

10.2.4.2 Step Change (Figure 10.2.6b)

A *step change* is an instantaneous change in the input value, which then remains at the new value.

$$\theta_i = H \text{ for all values after } t = 0$$

where 'H' is the change in value or the height of the step.

10.2.4.3 Ramp or Velocity Change (Figure 10.2.6c)

In a *ramp change*, the input changes at a constant rate. This is also called a *velocity input*.

$$\frac{d\theta_i}{dt} = c \text{ or } \theta_i = ct$$

where 'c' is the rate of change (velocity).

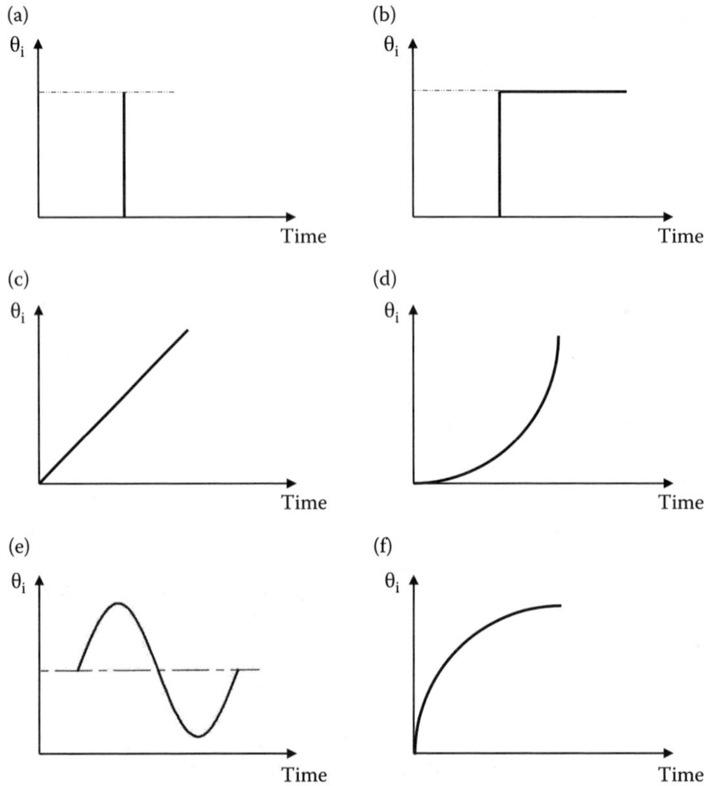

FIGURE 10.2.6
(a) Impulse Input (b) a step change (c) a ramp or velocity change (d) parabolic or acceleration change (e) a sinusoidal change (f) and an exponential change.

10.2.4.4 Parabolic or Acceleration Change (Figure 10.2.6d)

In a *parabolic change,* the rate of change is exponential, i.e., $\frac{d^2\theta_i}{dt^2} = a$ or $\theta_i = \frac{at^2}{2}$.

This change is also known as an *acceleration change,* since the rate of change is a constant 'a.'

10.2.4.5 Sinusoidal Change (Figure 10.2.6e)

In *sinusoidal change,* the input changes sinusoidally with time.

$$\theta_i = A\sin(\omega t + \phi)$$

where A is the amplitude and ϕ is the phase angle.

10.2.4.6 Exponential Change (Figure 10.2.6f)

Exponential change is when the input changes exponentially with time.

$$\theta_i = A(1 - e^{-at})$$

When an equation has been developed to represent how the input to a system will change with time, it is then necessary to establish the output changes with time. From this, it will be possible to see how accurate and how fast the response of the system to the input will be.

The following subsections provide further definitions and terminologies used in the description of types of systems.

10.2.5 Linear System

An important property of a linear system is that of *superposition*, that is, if one input generates a given output and a further input generates a second output, then combining the two inputs together will result in the sum of the two outputs.

A further property of a linear system is that of *homogeneity*, that is, if an input generates a given output, then multiplying the input will also multiply an output by the same value.

10.2.6 Linear Time-Invariant Systems

A pressure flow meter is an example of a linear time-invariant system. The mathematical model for a differential pressure flow meter is shown to be

$$Q = C(\Delta p)^n$$

where

'C' is a constant for the meter

'n' is a parameter

'Q' and 'p' in the equation are variables

If the constants and parameters remain unchanged over a period of time, then the system is known as a *linear time-invariant system*.

10.2.7 Laplace Transforms

To allow the solution of θ_o as a function of time, a mathematical transformation known as *Laplace transforms* is used. This allows the equation to be changed from a differential equation to an algebraic equation, where it can be manipulated and converted back into a differential equation by using inverse transforms. Section 10.1.2.2 briefly introduces Laplace transforms, together with Table 10.2.1 showing standard Laplace transforms.

Using Laplace transforms, the differential equation 'd/dt' is replaced with the symbol 's':

$$\frac{d\theta_o}{dt} \text{ becomes } s\theta_o \text{ and } \frac{d^2\theta_o}{dt^2} \text{ becomes } s^2\theta_o$$

It may be appreciated that θ_o on its own becomes $s^0\theta_o$ and that the integral of θ_o becomes $s^{-1}\theta_o$, and so forth.

Referring back to Equations 10.2.1 and 10.2.2 in the standard models (Section 10.2.3.1) and incorporating these changes produces the following algebraic equations:

$$\theta_i = Ts\,\theta_o + \theta_o$$

$$\theta_i = \theta_o(Ts + 1) \qquad (10.2.3)$$

TABLE 10.2.1

Table of Further Common Laplace Transforms

Ref	Time Domain f(t)	Frequency Domain f(s)	Comments
1	$e^{-at} f(t)\,dt$	$f(s+a)$	
2	δt	1	Unit impulse
3	H	$\frac{H}{s}$	Step 'H'
4	ct	$\frac{c}{s^2}$	Ramp
5	$H(t-T)$	$H\frac{e^{-sT}}{s}$	Delayed step
6		$\frac{1-e^{-sT}}{s}$	Rectangular pulse
7	ke^{-at}	$\frac{k}{s+a}$	Exponential
8	$kt\,e^{-at}$	$\frac{k}{(s+a)^2}$	
9	$K(e^{-at}-e^{-bt})$	$\frac{k(b-a)}{(s+a)(s+b)}$	
10	$k\sin(\omega t)$	$\frac{k\omega}{s^2+\omega^2}$	Sinusoidal
11	$k\omega t\sin(\omega t)$	$\frac{2k\omega^2 s}{(s^2+\omega^2)^2}$	
12	$k\cos(\omega t)$	$\frac{ks}{s^2+\omega^2}$	Cosinusoidal
13	$ke^{-at}\sin(\omega t)$	$\frac{k\omega}{(s+a)^2+\omega^2}$	Damped sinusoidal
14	$k\,e^{-at}\cos(\omega t)$	$\frac{s+a}{(s+a)^2+\omega^2}$	Damped cosinusoidal
15	$k\left\{1-e^{-\frac{t}{T}}\right\}$	$\frac{ka}{s(s+a)}\quad a=\frac{1}{T}$	Exponential growth
16	$k\left\{t-\left(1-e^{-t/T}\right)\right\}$	$\frac{ka}{s^2(s+a)}$	
17	$k(1-\cos\omega t)$	$\frac{k\omega^2}{s(s^2+\omega^2)}$	
18	$k\sin(\omega t+\phi)$	$\frac{k\{\omega\cos\varphi+s\sin\varphi\}}{s^2+\omega^2}$	
19	$1-e^{-at}\left(\cos bt+\frac{a}{b}\sin bt\right)$	$\frac{a^2+b^2}{s\left[(s+a)^2+b^2\right]}$	Standard second-order system step response

and:

$$\theta_i = T^2s^2\theta_o + 2\delta Ts\theta_o + \theta_o$$

$$\theta_i = \theta_o(T^2s^2 + s\delta Ts + 1) \tag{10.2.4}$$

The equations can be written as a ratio to produce a *transfer function*:

$$\frac{\theta_o}{\theta_i} = \frac{1}{Ts+1} \tag{10.2.5}$$

$$\frac{\theta_o}{\theta_i} = \frac{1}{T^2s^2 + 2\delta Ts + 1} \tag{10.2.6}$$

Here, the transfer functions are expressed in terms of s f(s) instead of functions of time (ft).

Prior to the solution of the transfer function, the input θ_i is required to be changed from a function of time f(t) into a function of s f(s). The use of standard s (Example 10.2.1) will aid the main transformation instead of trying to transform it mathematically.

Following are two examples.

Example 10.2.1

Determine the Laplace transform when h_1 is a constant (a step change).

Solution

Hence for a unit step h=1 and the Laplace transform is $\frac{1}{s}$.

Example 10.2.2

Find the Laplace transform of e^{-at}:

Solution

$$L\left(e^{-at}\right) = \int_0^\infty f(t)dt = \int_0^\infty e^{-st}e^{-at}dt = \int_0^\infty {}^{-(s+a)t}dt$$

$$L\left(e^{-at}\right) = -\left[\frac{\left(e^{-(s+a)}t\right)}{s+a}\right]_0^\infty$$

$$L\left(e^{-at}\right) = -\left[\frac{0}{s+a} - \frac{1}{s+a}\right]$$

$$L\left(e^{-at}\right) = f(s) = \frac{1}{s+a}$$

10.2.8 The Standard First-Order System

10.2.8.1 Response to an Impulse and Step Input

Example 10.2.3

Compare the time response of a system with the transfer function $\frac{\theta_o}{\theta_i} = \frac{1}{Ts+1}$ to a unit impulse and a unit impulse input.

Solution

Consider a *unit impulse*:

$$\theta_o(s) = \frac{1\theta_i}{Ts+1}, \quad \text{substitute } \theta_i(s) = 1 \text{ for an impulse input}$$

$$\theta_o(t) = \text{inverse Laplace transform of } \frac{1(1)}{Ts+1}$$

rearrange into a recognizable transform and $\frac{1/T}{s+1/T}$ $\theta_o(t) = (1/T)e^{-t/T}$.

Unit step input

$$\theta_o(s) = \frac{1\theta_i}{Ts+1}, \quad \text{substitute } \theta_i(s) = \frac{1}{s} \text{ for a step input}$$

$$\theta_o(t) = \text{inverse Laplace transform of } \frac{1}{s(Ts+1)}$$

rearrange into a recognizable transform and $\frac{1/T}{s(s+1/T)}$ $\theta_o(t) = 1 - e^{-t/T}$.

Figure 10.2.7 compares the two responses.

10.2.9 First-Order Response Graph to a Step Input

For any standard first-order system, the response to a step input, the output response is $\theta_o = H(1 - e^{-t/T})$. If the input and output is plotted against time, then the result would be as shown in Figure 10.2.8.

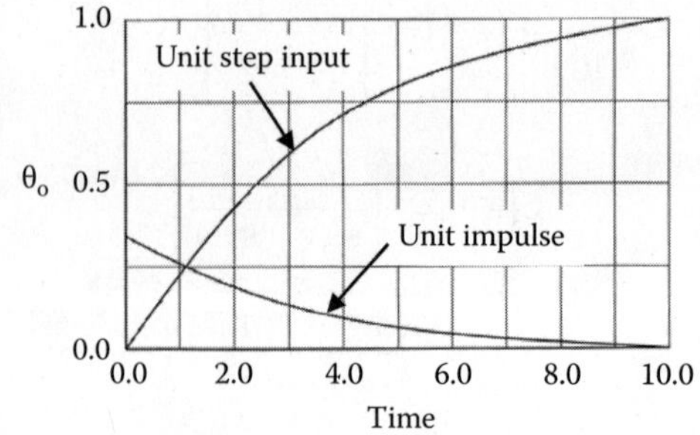

FIGURE 10.2.7
Comparison between a unit impulse and a unit step.

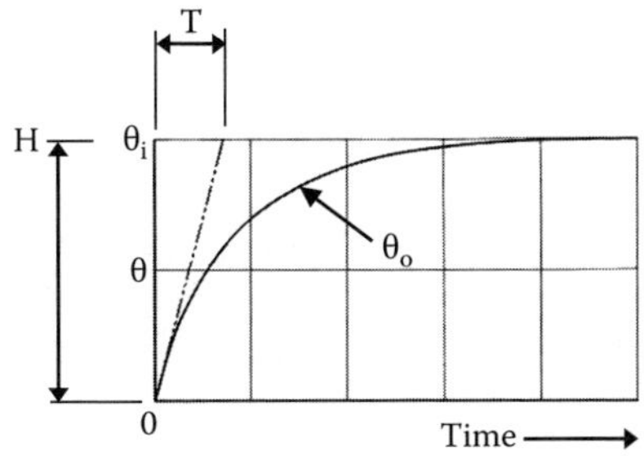

FIGURE 10.2.8
First-order response graph.

10.2.9.1 First-Order Time Constant

The time constant 'T' has a real meaning. Referring to the response graph in Figure 10.2.8, the rate of change of θ_o is

$$\frac{d\theta_o}{dt} = H\left(\frac{1}{T}\right)e^{-t/T}$$

At the start of the change t=0, therefore the initial rate of change is

$$\frac{d\theta_o}{dt} = H\left(\frac{1}{T}\right)e^{-0} = \frac{H}{T}$$

The initial gradient (rate of change) is H/T. If the initial gradient is projected as shown in Figure 10.2.8, it will intercept θ_i at a time of t=T. This will lead to a definition of 'T' as the time it would take the output to match the input if it maintained changing at the same initial rate.

A further method of defining the time constant is as follows:

At time t=T, the value of the output is

$$\theta_o = H\left\{1 - e^{-1}\right\} = 0.6321H$$

This answer is saying the output will reach 63.2% of the desired value after 'T' seconds.

When t=5T, the output will be

$$\theta_o = H\left\{1 - e^{-1}\right\} = H\left\{1 - e^{-5}\right\} = 0.9933H$$

When t=14T,

$$\theta_o = H\left\{1 - e^{-1}\right\} = H\left\{1 - e^{-14}\right\} = 0.99999H$$

It can be seen that after t=5 seconds, the system has reached 99.33% of the correct value, whereas it will not reach 100% of the correct value for a further 9 seconds (i.e., when t=14 seconds).

10.2.10 Response of a Standard First-Order System to a Ramp Input

A ramp or velocity input occurs when

$$\theta_i(t) = ct \text{ and } \theta_i(s) = \frac{c}{s^2}$$

$$\frac{\theta_o}{\theta_i}(s) = \frac{1}{(Ts+1)}$$

$$\theta_o(s) = \frac{\theta_i(s)}{(Ts+1)} = \frac{c}{s^2(Ts+1)} = \frac{c(1/T)}{s^2(s+1/T)}$$

If 'c' is replaced with 'k' and 1/T is replaced with 'a,' then the equation can be rewritten as

$$\theta_o(s) = \frac{ca}{s^2(s+a)}$$

The results can be found in the table of transforms. The inverse transform gives the following solution:

$$\theta_o(t) = c\left\{t - T\left(1 - e^{-t/T}\right)\right\}$$

Plotting the output 'θ_o' against time 't,' the term $\left(e^{-t/T}\right)$ becomes negligibly small and can then be ignored. The equation reduces to

$$\theta_o(t) = c\{t - T(1)\} = c(t - T)$$

The error becomes

$$\theta_e(t) = \theta_i - \theta_o = ct + cT$$

$$\theta_e(t) = cT$$

A constant error 'cT' will result after an initial transient stage and this is the steady-state error, as shown in Figure 10.2.9.

10.2.11 Response of a Standard First-Order System to a Sinusoidal Input

The standard first-order system transfer function for a sinusoidal input is

$$\frac{\theta_o}{\theta_i}(s) = \frac{k}{Ts+1}$$

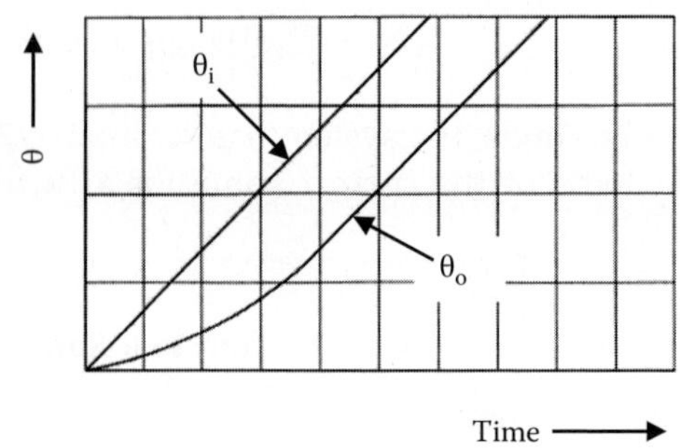

FIGURE 10.2.9
First-order response to a ramp input.

In the steady state, 's=0'

$$\frac{\theta_o}{\theta_i}(s) = \frac{k}{t} = k$$

It follows, therefore, that 'k' is the steady-state system gain.

For a sinusoidal input, let $\theta_i(t) = A\sin(\omega t)$.

The Laplace transform is

$$\theta_i(s) = \frac{\omega}{\omega^2 + s^2}$$

The output will be

$$\theta_i(s) = \frac{\theta_i}{(Ts+1)}$$

$$= \frac{k}{(Ts+1)}\frac{\omega}{\left(\omega^2 + s^2\right)}$$

The inverse transform will give the output as a function of time

$$\theta_o(t) = AK\frac{\sin(\omega t)}{\left(1+T^2\omega^2\right)} - AK\omega T\frac{\cos(\omega t)}{1+T^2\omega^2} + AK\omega T^2\frac{\exp\left(\frac{-1}{T}\right)t}{\left(T+T^2\omega^2\right)}$$

Plotting the input and output gives a typical result as shown in Figure 10.2.10 with A=1 and K=1.

This figure shows that after an initial transient stage (as in the ramp input), the output quickly settles down into a steady-state response with a sinusoidal form but having a different amplitude and lagging the input with a phase shift. The transient stage will only last one cycle.

10.2.12 Gain of a First-Order System

DC gain is an electrical term and is the gain when there is no variation of the input in the time domain, that is, when there is no frequency in the input (0 Hz), the input is constant and all the derivative values of the transfer functions are zero, hence the DC is the magnitude of G(s) when s=0.

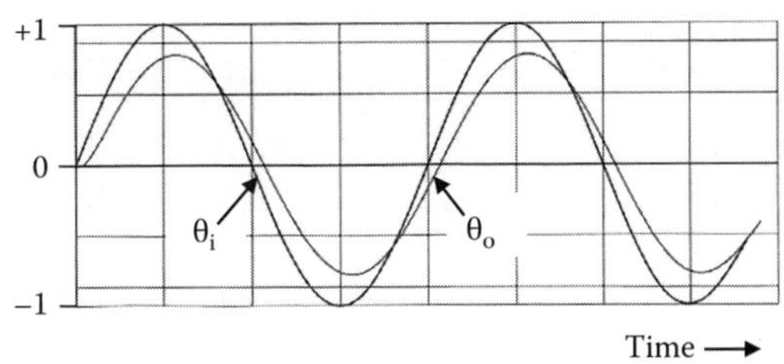

FIGURE 10.2.10
First-order response to sinusoidal input.

The gain of the standard first-order closed-loop transfer function $\frac{\theta_o}{\theta_i}(s) = \frac{1}{Ts+1}$ is unity.

When the transfer function is of the form $\frac{\theta_o}{\theta_i}(s) = \frac{K}{Ts+1}$, the DC gain is 'K.'

It is not always obvious what the DC gain is, for example, consider the case where $\frac{\theta_o}{\theta_i}(s) = \frac{A}{Ts+B}$

When s=0, the DC gain will be A/B, and the transfer function can be rearranged to

$$\frac{\theta_o}{\theta_i}(s) = \frac{A/B}{(T/B)s+1}$$

This equation is now in the form:

$$\frac{\theta_o}{\theta_i}(s) = \frac{K}{T_2 s+1}$$

As a result of this change, T_2 is a new time constant. In a steady state 's'=0, therefore, 'K' is the steady-state gain of the system.

$$\text{DC gain} = A/B \quad T_2 = T/B$$

The DC gain will also have an effect on the magnitude of a dynamic response.

Following a step change 'H,' the response will settle down at 'KH' (see Figure 10.2.11a) and the response to a ramp input will not settle at a fixed value, but to $\theta_o(t) = Kc\left\{t - T\left(1 - e^{-t/T}\right)\right\}$. The effect of introducing gain to a ramp input ($\theta_i = ct$) is shown in Figure 10.2.11b, where the output passes and exceeds the input increasingly as time progresses.

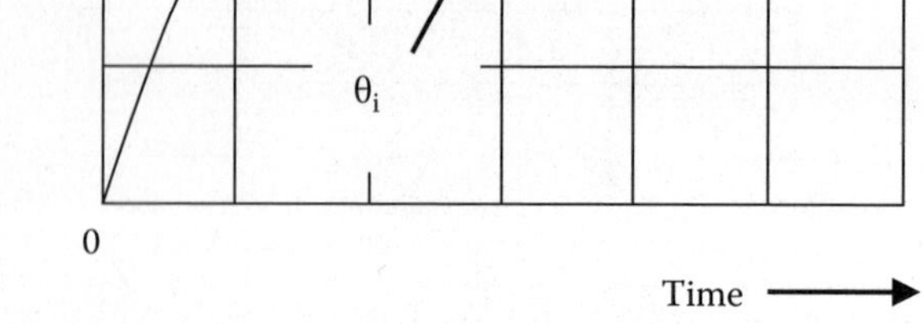

FIGURE 10.2.11
Gain of a first-order system to a step input.

10.2.13 The Use of Partial Fractions to Solve Responses

Partial fractions can be used to solve most types of responses. Following are two examples using the standard first-order transfer function:

$$\frac{\theta_o}{\theta_i} = G(s) = \frac{1}{(Ts+1)} = \frac{1/T}{s+1/T}$$

Put $1/T = p$ and $\frac{\theta_o}{\theta_i}(s) = \frac{p}{s+p}$

10.2.13.1 Unit Step Input

A unit step input in Laplace transform form is $\theta_o(s)=1/s$:

$$\theta_o(s) = \frac{p}{s(s+p)}$$

Converting into partial fractions:

$$\theta_o(s) = \frac{A_o}{s} + \frac{A_1}{s+p}$$

Resolving A_o and A_1 in the normal manner for partial fractions:

$$\frac{A_o}{s} + \frac{A_1}{s+p} = \frac{p}{s(s+p)}$$

$$\frac{A_o(s+p) + A_1 s}{s(s+p)} = \frac{p}{s(s+p)}$$

Therefore, the numerators must equate:

$$A_o(s+p) + A_1 s = p$$

and this will be true for any value of 's':

$$\text{Put } s = -p \text{ and } A_o(0) - A_1 p \quad \text{so, } A_1 = -1$$

$$\text{Now, put } s = 0 \quad A_o(p) + 0 \quad \text{so, } A_o = +1$$

Substitute into the solution:

$$\theta_o(s) = \frac{1}{s} - \frac{1}{(s+p)}$$

The inverse Laplace transform will be

$$\theta_o(t) = 1 - e^{-tp} = \left(1 - e^{-tp}\right)$$

Putting p=1/T

$$\theta_o(t) = 1 - e^{-t/T}$$

This solution will be found by other methods covered previously.

10.2.13.2 Unit Ramp

$$\frac{\theta_o}{\theta_i}(s) = \frac{p}{s+p}$$

Putting into partial fractions:

$$\theta_o(s) = \frac{B_o}{s} + \frac{B_1}{s^2} + \frac{B_2}{s+p} = \frac{p}{s^2(s+p)}$$

$$\theta_o(s) = \frac{s(s+p)B_o + (s+p)B_1 + s^2B_2}{s^2(s+p)} = \frac{p}{s^2(s+p)}$$

$$s(s+p)B_o + (s+p)B_1 + s^2B_2 = p$$

Equating the numerator and first let p=–s:

$$s(0)B_0 + (0)B_1 + p^2B_2 = p$$

$$\text{put } s = -p \text{ and } B_2 = \frac{1}{p}$$

Putting s=0 and B_1=1.

Next, put s=1 and substituting for B_1 and B_2 to solve for B_0 and find B_0=–1/p:

$$\theta_o(t) = -\frac{1}{p} + t + \frac{1}{p}e^{-pt} \quad \text{Let } \frac{1}{T} = p$$

The solution is

$$\theta_o(t) = -T + t + Te^{-t/T}$$

$$\theta_o(t) = t - T\left(1 - e^{-t/T}\right)$$

This solution will have been found using other previous methods.

Example 10.2.4

A system has a transfer function $\frac{\theta_o}{\theta_i}(s) = \frac{1}{(s+1)(s+2)}$. Establish the time response to a unit step input.

Solution

$$\theta_o(s) = \frac{\theta_i}{(s+1)(s+2)}$$

$$\theta_o(s) = \frac{1}{s(s+1)(s+2)}$$

Converting to partial fractions:

$$\theta_o(s) = \frac{1}{2s} - \frac{1}{s+1} + \frac{1}{2(s+2)}$$

Inverse transform:

$$\theta_o(t) = \frac{1}{2} - e^{-t} + \frac{1}{2}e^{-2t}$$

The response curve is illustrated in Figure 10.2.12.

Example 10.2.5

A system has a transfer function $\frac{\theta_o}{\theta_i}(s) = \frac{\theta_i}{(s+2)(s+3)}$. Determine the output time response to a unit step input.

Solution

$$\theta_o(s) = \frac{\theta_i}{(s+2)(s+3)}$$

$$\theta_o(s) = \frac{1}{(s+2)(s+3)}$$

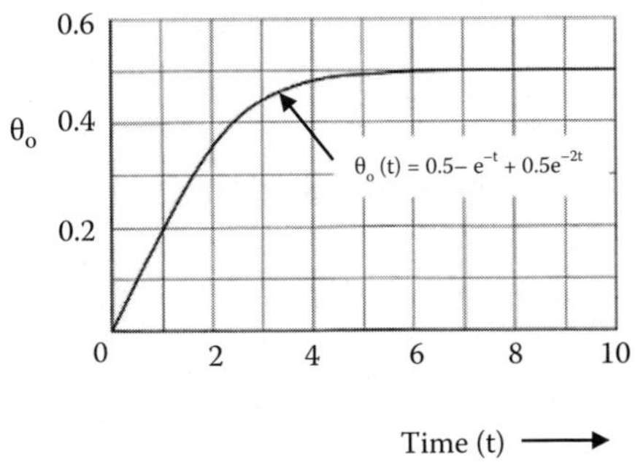

t	θ_o
0	0
1	0.200
2	0.374
3	0.451
4	0.482
5	0.493
6	0.498
7	0.499
8	0.500
9	0.500
10	0.500

FIGURE 10.2.12
Example 1: Response curve.

Convert to partial fractions:

$$\theta_o(s) = \frac{1}{6s} - \frac{1}{2(s+2)} + \frac{1}{3(s+2)}$$

Inverse transform:

$$\theta_o(t) = \frac{1}{6} - \frac{1}{2}e^{-2t} + \frac{1}{3}e^{-3t}$$

10.2.14 Standard Second-Order System

10.2.14.1 Forms of the Standard Transfer Functions

Second-order transfer functions can be stated in a number of ways. The following are a number of the most useful forms.

1. In terms of the time constant 'T'

$$\frac{\theta_o}{\theta_i}(s) = \frac{k}{T^2s^2 + 2T\delta s + 1}$$

2. In terms of the natural frequency 'ω_n'

$$\frac{\theta_o}{\theta_i}(s) = \frac{\omega_n{}^2 k}{s^2 + 2\delta\omega_n s + \omega_n{}^2}$$

3. In terms of the poles (polynomial)

$$\frac{\theta_o}{\theta_i}(s) = \frac{\omega_n{}^2 k}{(s - p_1)(s - p_2)}$$

4. In terms of two parameters σ and ω_r

$$\frac{\theta_o}{\theta_i}(s) = \frac{\omega_n{}^2 k}{(s-\sigma)^2 + \omega_r{}^2}$$

$$= \frac{\omega_r{}^2 + \sigma\omega_r{}^2}{(s-\sigma)^2 + \omega_r{}^2}$$

Note: 'δ' is the damping factor but 'ζ' may be sometimes encountered as a symbol for the damping factor.

10.2.14.2 Time Constant Form

'T' is the second-order time constant of the system. This results from various parameters of a particular system:

$$\frac{\theta_o}{\theta_i}(s) = \frac{k}{T^2s^2 + 2T\delta s + 1}$$

This is the form that generally results from the analysis of a real system.

10.2.14.3 Natural Frequency Form

The system's natural frequency is defined as $\omega_n = \frac{1}{T}$. Substituting $T = \frac{1}{\omega_n}$ and then rearranging the form to obtain:

$$\frac{\theta_o}{\theta_i}(s) = \frac{{\omega_n}^2 k}{s^2 + 2\delta\omega_n s + {\omega_n}^2}$$

10.2.14.4 Polynomial Form

In a polynomial form, the denominator of the transfer function is a quadratic equation and should factorize to give

$$\frac{\theta_o}{\theta_i}(s) = \frac{{\omega_n}^2 k}{(s+x)(s+y)}$$

yielding

$$\frac{\theta_o}{\theta_i}(s) = \frac{{\omega_n}^2 k}{s^2 + s(x+y) + xy}$$

Comparing the denominator $(s^2 + s(x+y) + xy)$ with the denominator from the natural frequency form $(s^2 + 2\delta\omega_n s + {\omega_n}^2)$ and it will be apparent that $x + y = 2\delta\omega_n$ and $xy = {\omega_n}^2$.

$$x = \frac{\omega_n^2}{y} \qquad \frac{\omega_n^2}{y+y} = 2\delta\omega_n \qquad \omega_n^2 + y^2 = y2\delta\omega_n \qquad y^2 - y2\delta\omega_n + \omega_n^2 = 0$$

or

$$y = \frac{\omega_n^2}{x} \qquad x + \frac{\omega_n^2}{x} = 2\delta\omega_n \qquad x^2\omega_n^2 + x^2 = x2\delta\omega_n \qquad x^2 - x2\delta\omega_n + \omega_n^2 = 0$$

As the solution of the quadratic is the same for both, x and y will be the same and therefore will be the roots of the denominator.

Solve the following quadratic equation:

$$y = \frac{-b \mp \sqrt{b^2 - 4ac}}{2a}$$

Solve for when $a = 1$, $b = -2\delta\omega n$, $c = \omega_n^2$

Hence:

$$\begin{aligned} y = x &= \frac{2\delta\omega_n \mp \sqrt{(2\delta\omega_n)^2 - 4\omega_n^2}}{2} \\ &= \delta\omega_n \mp \sqrt{(\delta\omega_n)^2 - \omega_n^2} \\ &= \delta\omega_n \mp \omega_n\sqrt{\delta^2 - 1} \\ &= \delta\omega_n \mp j\omega_n\sqrt{1 - \delta^2} \end{aligned}$$

$$\frac{\theta_o}{\theta_i}(s) = \frac{\omega_n^2 k}{\left\{s + \delta\omega_n + j\omega_n\sqrt{1-\delta^2}\right\}\left\{s + \delta\omega_n - j\omega_n\sqrt{1-\delta^2}\right\}}$$

10.2.14.5 σ and ω_r Form

Substituting $\sigma = \delta\omega_n$ and $\omega_r = \omega_n\sqrt{1-\delta^2}$ in the last equation, where ω_r is the damped resonant frequency, the output value is maximum. The transfer function is

$$\frac{\theta_o}{\theta_i}(s) = \frac{\omega_n{}^2 k}{\{s + \sigma + j\omega_r\}\{s + \sigma - j\omega_r\}}$$

This will further simplify to

$$\frac{\theta_o}{\theta_i}(s) = \frac{\omega_n{}^2 k}{(s+\sigma)^2 + \omega_r{}^2}$$

This equation can be further manipulated to a form where it can be used for inverse Laplace transforms.

From the definitions

$$\sigma = \delta\omega_n \text{ and } \omega_r = \omega_n\sqrt{\left(1-\delta^2\right)}$$

The following can be determined:

$$\omega_r{}^2 = \omega_n{}^2\left(1-\delta^2\right)$$

$$\omega_r{}^2 = \omega_n{}^2 - \delta^2\omega_n{}^2$$

$$\omega_r{}^2 = \omega_n{}^2 - \sigma^2$$

$$\omega_n{}^2 = \omega_r{}^2 + \sigma^2\omega_r{}^2$$

$$\frac{\theta_o}{\theta_i}(s) = \frac{\omega_r{}^2 + \sigma\omega_r{}^2}{(s+\sigma)^2 + \omega_r{}^2}$$

10.2.14.6 Poles

These are the values of 's' that make the denominator zero. The denominator is in a quadratic form in 's' that in general will satisfy the solution of 's' in the equation $\left(s^2 + 2\delta\omega_n s + \omega_n{}^2\right) = k$ are the roots of the equation. The poles are the roots when k=0. The equation $\left(s^2 + 2\delta\omega_n s + \omega_n{}^2\right) = 0$ requires to be solved to find the poles. Using the quadratic formula to solve the equation:

$$s = \frac{-b \mp \sqrt{b^2 - 4ac}}{2a}$$

where $a=1$, $b=-2\delta\omega_n$, $c=\omega_n^2$.

Therefore,

$$s=\frac{-2\delta\omega_n \mp \sqrt{(2\delta\omega_n)^2-4\omega_n^2}}{2}$$

$$s=-\delta\omega_n \mp \sqrt{(\delta\omega_n)^2-\omega_n^2}$$

$$s=-\delta\omega_n \mp \omega_n\sqrt{\delta^2-1}$$

$$s=-\delta\omega_n \mp j\omega_n\sqrt{1-\delta^2}$$

This is usually written as

$$s=-\sigma \mp j\omega_r.$$

From the above, it will be apparent that the poles are the negative values of 'x' and 'y.'

J is the complex operator (√–1) and these roots may be a complex number that can be plotted using the Argand diagram (see Section 1.14).

If the poles are denoted 'p_1' and 'p_2,' it will be seen that $p_1=-\sigma+j\omega_r$ and $p_2=-\sigma-j\omega_r$.

Expressing the transfer function in polynomial form gives

$$\frac{\theta_o}{\theta_i}(s)=\frac{\omega_n^2 k}{(s-p_1)(s-p_2)}$$

10.2.14.7 Response to a Step Input

Many real systems are represented by standard second-order transfer functions.

Consider the time response of a standard second-order system to a step input when $\delta=0$ (this is an undamped system):

$$\frac{\theta_o}{\theta_i}=\frac{1}{T^2s^2+2\delta Ts+1}$$

Putting $\delta=0$ and making θ_o the subject:

$$\theta_o=\frac{\theta_i}{T^2s^2+1}$$

For a step input where $\theta_i=H$ and $\theta_i(s)=H/s$:

$$\theta_o(s)=\frac{H/s}{T^2s^2+1}$$

$$\theta_o(s)=\frac{H}{s(T^2s^2+1)}$$

Now, rearranging the equation into a form that can be recognized from the form above:

$$\theta_o(s)=\frac{H/T^2}{s(s^2+1/T^2)}$$

Substituting $1/T=\omega$, the following will be recognized from the form above:

$$\theta_o(s) = \frac{H\omega^2}{s(s^2+\omega^2)} \qquad \text{(10.1.3 case 17).}$$

From the manipulation of transforms it will be found that the inverse transform that converts the output into a function of time:

$$\theta_o(t) = H(1-\cos\omega t)$$

From this result, it is deduced that the output is a negative cosine curve, as depicted in Figure 10.2.13 where the amplitude is 'H' and the frequency of oscillation is $\omega=1/T$ (rad/s).

In the previous example, the step input had no damping. In the following example, various values of damping will be examined and the effect on the output will be considered.

The standard second-order form will be used:

$$\frac{\theta_o}{\theta_i}(s) = \frac{\omega_n{}^2}{(s+\sigma)^2+\omega_r{}^2}$$

For a unit step $\theta_i(s)=\frac{1}{s}$ hence,

$$\theta_o(s) = \frac{1}{s}\frac{\omega_r{}^2+\sigma\omega_r{}^2}{(s+\sigma)^2+\omega_r{}^2}$$

From the table of Laplace transforms, the inverse is

$$\theta_o(t) = 1-e^{-\sigma t}\left(\cos\omega_r t+\frac{\sigma}{\omega_r}\sin\omega_r t\right) \qquad \text{(10.2.1 case 14 (Damped Co-sinusoidal)).}$$

This equation may be plotted for different values of damping ratios (δ) to produce a normalized graph. The settling level is 1.0. ω_n is the natural angular frequency of the system. The natural frequency is

$$f_n = \frac{\omega_n}{2\pi}$$

This is the frequency at which the system would oscillate if there were no damping.

A more-detailed analysis of the step input response will be given later in this section, but the following will be useful at this stage.

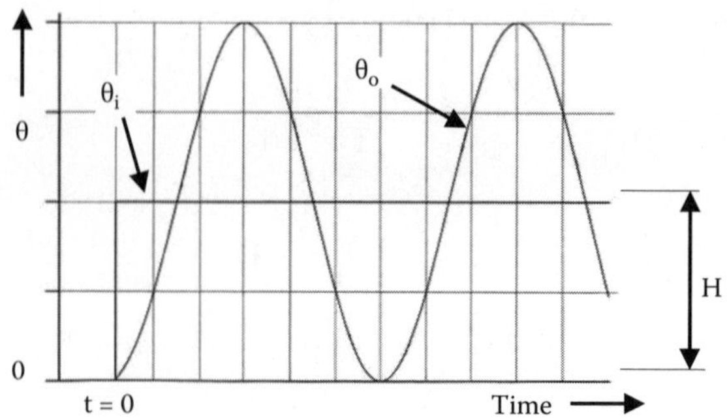

FIGURE 10.2.13
Response to a step.

It was shown earlier through the use of the polynomial form that the step response is

$$\theta_o(t) = 1 - e^{-\sigma t}\left(\cos\omega_r t + \frac{\sigma}{\omega_r}\sin\omega_r t\right)$$

where $\sigma = \delta\omega_n$ and $\omega_{r=}\omega_n\sqrt{(1-\delta^2)}$.

As this is a unit step input, it will have an amplitude of $\theta_A(t) = -e^{-\sigma t}\left(\cos\omega_r t + \frac{\sigma}{\omega_r}\sin\omega_r t\right)$.

Figure 10.2.14 compares the amplitude and response plots for the unit step.

10.2.14.8 Amplitude Reduction Factor

Consider two oscillations with amplitudes θ_1 and θ_2, respectively, one occurring T_p seconds after the first oscillation.

The first oscillation

$$\theta_1 = -e^{-\sigma t}\left(\cos\omega_r t + \frac{\sigma}{\omega_r}\sin\omega_r t\right)$$

The second amplitude occurs 'T_p' seconds later, and $T_p = 2p/\omega_r$ is the periodic time:

$$\theta_2 = -e^{-\sigma(t+T_p)}\left\{\cos\omega_r\left(t+T_p\right) + \frac{\sigma}{\omega_r}\sin\omega_r\left(t+T_p\right)\right\}$$

$$\frac{\theta_1}{\theta_2} = \frac{-e^{-\sigma t}\left\{\cos\omega_r(t) + \frac{\sigma}{\omega_r}\sin\omega_r(t)\right\}}{-e^{-\sigma(t+T_p)}\left\{\cos\omega_r\left(t+T_p\right) + \frac{\sigma}{\omega_r}\sin\omega_r\left(t+T_p\right)\right\}} = 1$$

It may be shown that

$$\frac{\left\{\cos\omega_r(t) + \frac{\sigma}{\omega_r}\sin\omega_r(t)\right\}}{\left\{\cos\omega_r\left(t+T_p\right) + \frac{\sigma}{\omega_r}\sin\omega_r\left(t+T_p\right)\right\}} = 1$$

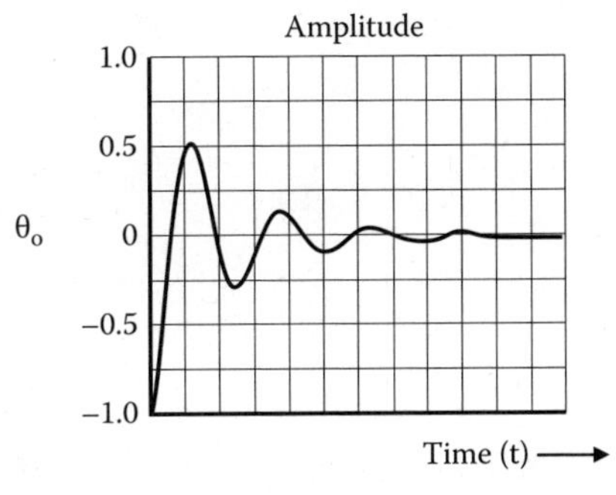

FIGURE 10.2.14
Amplitude plot for a unit step input.

Therefore, the amplitude ratio of the two successive oscillations will be

$$\frac{\theta_1}{\theta_2} = \frac{-e^{-\sigma t}}{-e^{-\sigma(t+T_p)}}$$

$$\frac{\theta_1}{\theta_2} = -e^{-\sigma t} - e^{-\sigma(t+T_p)}$$

$$\frac{\theta_1}{\theta_2} = -e^{-\sigma T_p}$$

Hence,

$$\ln\left(\frac{\theta_1}{\theta_2}\right) = \sigma T_p$$

This expression is usually referred to as the *logarithmic decrement*.

Substituting for σ and T_p:

$$\ln\left(\frac{\theta_1}{\theta_2}\right) = \frac{2\pi\delta}{\sqrt{1-\delta^2}}$$

This expression is called the *amplitude reduction factor.*

10.2.14.9 Ramp Input

Considering the standard transfer function:

$$\frac{\theta_o}{\theta_i}(s) = \frac{1}{T^2s^2 + 2\delta Ts + 1}$$

In this section, the input is a ramp, and therefore

$$\theta_i(s) = \frac{c}{s^2}$$

Substitute in the above equation:

$$\theta_o(s) = \frac{c}{s^2\left(T^2s^2 + 2\delta Ts + 1\right)}$$

The inverse transform is

$$\theta_o(t) = -2c\delta T + tc - cT\exp\left(\frac{-\delta}{T}t\right)\frac{\sinh\left(\frac{1}{T}\sqrt{-1+\delta^2 t}\right)}{\sqrt{-1+\delta^2}}$$

$$-2c\delta^2 T\exp\left(\frac{-\delta}{T}t\right)\frac{\sinh\left(\frac{1}{T}\sqrt{-1+\delta^2 t}\right)}{\sqrt{-1+\delta^2}} + 2cTd\exp\left(\frac{-\delta}{T}t\right)\cosh\left(\frac{1}{T}\sqrt{-1+\delta^2 t}\right)$$

The plot of the function is shown in Figure 10.2.15 and shows that the output is oscillatory and that the level of oscillation will depend on the damping ratio 'δ.' When $\delta = 1$, the result will be similar to that of a first-order system.

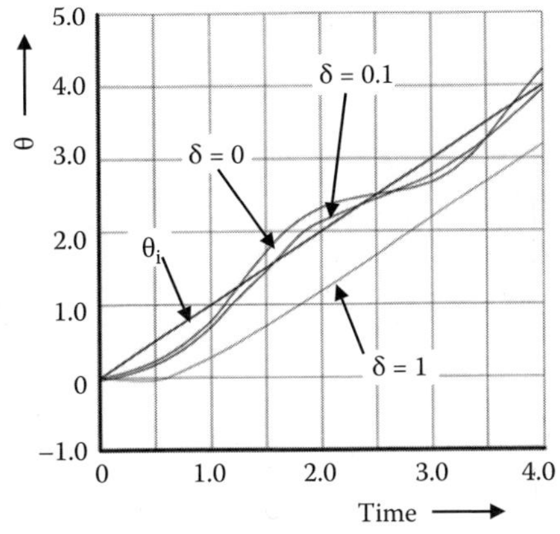

FIGURE 10.2.15
Ramp input subject to various damping ratios.

10.2.14.10 Sinusoidal Input

Considering the standard transfer function,

$$\frac{\theta_o}{\theta_i}(s) = \frac{K}{T^2s^2 + 2\delta Ts + 1}$$

'K' is the gain and 'δ' is the damping ratio.

In this instance, the input is sinusoidal, therefore

$$\theta_i(t) = A\sin(\omega t) \quad \text{where 'A' is the amplitude.}$$

The Laplace transform is

$$\theta_i(s) = \frac{\omega}{\omega^2 + s^2}$$

Substitute

$$\theta_o(s) = \frac{K}{\left(T^2s^2 + 2\delta Ts + 1\right)} \frac{A\omega}{\omega^2 + s^2}$$

The inverse transfer when calculated using the software program Mathcad produces a very long and complicated expression. It is considered too long to reproduce here. Plotting the results with

$\omega < 1/T$ results in Figure 10.2.16. Reviewing the plot, it is apparent that during the transient stage, the output grows but settles down to a constant amplitude that is out of phase with the input.

If $\delta = 1$ (critical damping), it will reach the steady-state condition more quickly.

When the plot is repeated with a higher frequency, with $\omega = 1/T$, a condition called *resonance* is reached, as shown in Figure 10.2.17.

A lightly damped response will grow rapidly to a much larger amplitude than the input and the phase shift will be 1/4 cycle. If the damping is small, the output amplitude would be huge. If the frequency is raised further so that $\omega > 1/T$, then the result will be very different, as shown in Figure 10.2.18. It will be observed that the output is much reduced and the phase shift will be approaching a half cycle.

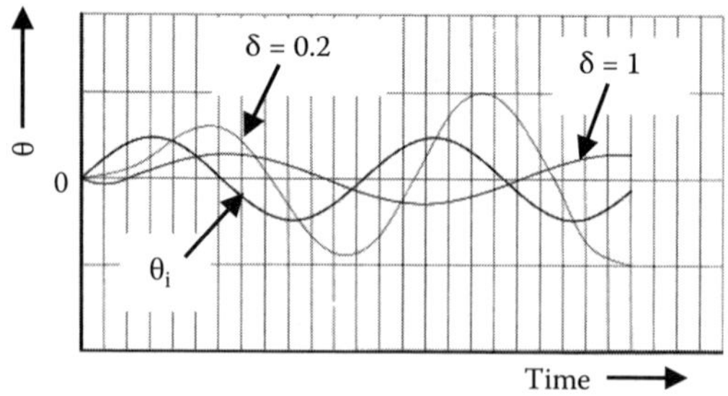

FIGURE 10.2.16
Sinusoidal input plot.

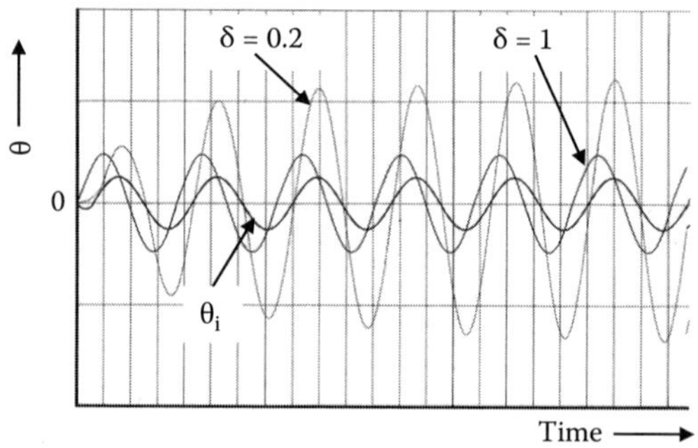

FIGURE 10.2.17
Sinusoidal input-repeated pattern.

10.2.14.11 DC Gain of Second-Order Systems

As with a first-order system, the DC gain is the magnitude of the transfer function when it contains no dynamic terms, i.e., 's' = 0:

$$\frac{\theta_o}{\theta_i}(s) = \frac{K}{\left(T^2s^2 + 2\delta Ts + 1\right)}$$

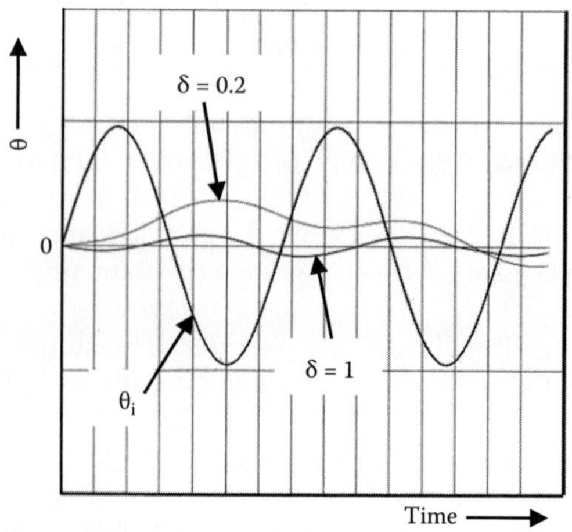

FIGURE 10.2.18
Sinusoidal input lightly.

When the last figure on the bottom line is plus 1, the system has a negative feedback and 'K' is the DC gain of the system.

Example 10.2.6

A system has the following transfer function: $G(s)=\left(\frac{1}{4s^2+8s+3}\right)$. Calculate the DC gain.

Solution

$$G(s)=G(s)=\left(\frac{1}{4s^2+8s+3}\right)$$

put s=0, and the gain is 1/3=0.333, which is an attenuation.

Steady-State Error

The steady-state error that is produced by a system may be found by applying the *Final Value theorem*. This states that the final value of a time function G(t) as $t \to 0$ is given by the value of s G(s) as $s \to 0$.

Example 10.2.7

Find the final value of $G(s)=\frac{1}{s+4}$.

Solution

The final value in the time domain as $1 \to \infty$ is

$$G(t)=s\frac{1}{s+4}=0 \text{ when } s \to 0$$

The final value of G(t) is 0.

Example 10.2.8

Find the final value of $G(s)=\frac{1}{s(s+4)}$.

Solution

The final value in the time domain is

$$G(t)=s\frac{1}{s(s+4)}=\frac{1}{4}=0.25 \text{ when } s \to 0$$

The final value of G(t) is 0.25.

The steady-state error of a system may be found by applying the theorem to the transfer function representing the error.

Consider a system with a closed-loop transfer function $G(s)_{c1}$. The error is

$$\theta_e=\theta_i-\theta_o$$

$$\theta_o = G_{cl}\theta_i$$

$$\theta_e = \theta_i - G_{cl}\theta_i = \theta_i\left(1 - G_{cl}\right)$$

The steady-state error

$$\theta_e(t) = s\theta_e(s) = s\theta_i(s)\left(1 - G_{cl}\right) \text{ as } s \to 0$$

Now consider a system that has an open-loop transfer function $G(s)_{cl}$ and a unit feedback. The closed-loop transfer function is

$$G(s)_{cl} = \frac{\theta_o}{\theta_i} = \frac{G_{ol}}{1 + G_{ol}}$$

The steady-state error is

$$\theta_e(t) = s\theta_e(s) = \frac{s\theta_i(s)}{1 + G_{ol}} \text{ as } s \to 0$$

Example 10.2.9

Determine the steady-state error when a unit step input is applied to a system having an open-loop transfer function of $G(s) = \frac{1}{(s+4)}$ with a unit feedback.

Solution

$$\theta_e(t)(\text{steadystate}) = \frac{s\theta_i(s)}{1 + G}$$

For a step input $\theta_i(s) = \frac{1}{s}$

$$\text{so } \theta_e(t) = \frac{1}{1 + G}$$

$$\theta_e(t) = \frac{1}{1 + \left(\frac{1}{s+4}\right)} \text{ and as } s \to 0 \text{ this becomes} = \frac{1}{1 + 1/4 = 0.8}$$

10.2.15 The s-Plane, Poles, and Zeros of the Second-Order System

This section is an introduction to the concept of *root locus plots*. Root locus analysis is a graphical method for examining how the roots of a system change with variation of a certain system parameter, commonly a gain within a feedback system. This section applies to a standard second-order system.

The form of the transient function $\frac{\theta_o}{\theta_i}(s) = \frac{\omega_n{}^2 k}{(s - p_1)(s - p_2)}$ has been previously examined in Section 10.2.14.1.

Here, p_1 and p_2 are poles where the value of 's' makes the denominator zero. It was shown that the poles are given by $p_1 = -\sigma + j\omega_r$ and $p_2 = -\sigma - j\omega_r$ and that these are complex numbers that can be plotted on an Argand diagram (see Figure 10.2.19). This diagram is now referred to as the s-plane, and the coordinates are $-\sigma$ and $j\omega_r$.

When the poles are plotted for a range of values of 'δ,' the diagram shown in Figure 10.2.19 is obtained. This is a root locus plot with respect to 'δ' and takes the form of a circle with a radius of ω_n.

For values of 'δ' between 1 and 0, a typical pair p_1 and p_2 will be found.

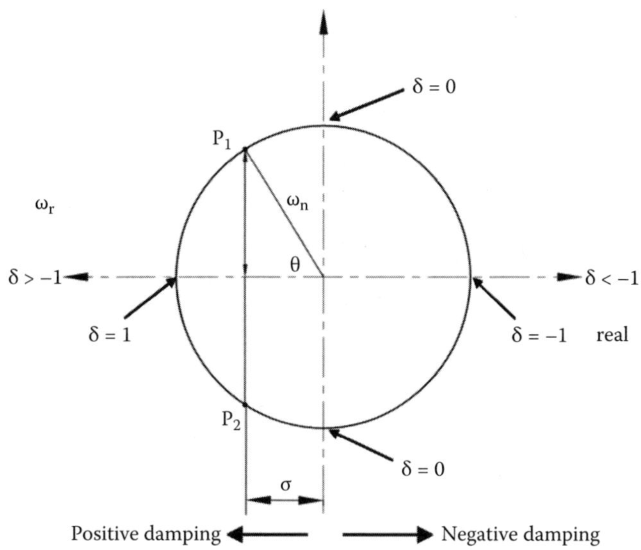

FIGURE 10.2.19
Position of poles.

From Figure 10.2.19, consider the angle 'θ':

$$\cos\theta = \frac{\sigma}{\omega_n} = \frac{\delta\omega_n}{\omega_n} = \delta$$

Consider the radius:

$$\text{radius} = \sqrt{\left(\sigma^2 + \omega_r^{\,2}\right)}$$

$$= \sqrt{\left[\left(\delta.\omega_n\right)^2 + \omega_n^{\,2}\left(1-\delta^2\right)\right]} = \omega_n$$

It follows that the position of the poles will be dependent upon the values of 'δ' and 'ω_n.' The complex pole can be represented in terms of damping ratio 'δ' and the natural frequency of the system 'ω_n.'

It should be noted that the smaller the value of 'δ,' the more oscillatory the response to a step input becomes and the closer the resonant frequency is to the natural frequency.

The solution to the time response to a unit step input was shown earlier as

$$\theta_o(t) = 1 - e^{-at}\left(\cos\omega_r t + \frac{\sigma}{\omega_r}\sin\omega_r t\right)$$

It should be noted that positive values of damping lead to s systems and for negative values, the system will be unstable. The diagram shown in Figure 10.2.20 depicts the response of the system to a unit step input for various pole positions. The diagram shows that the greatest stability occurs when the poles are on the negative real axis, and the maximum instability occurs when the pole is on the positive real axis. The response will be more oscillatory when the pole has a large imaginary (j) term. This is an important point to consider when this is studied in more detail later in the chapter.

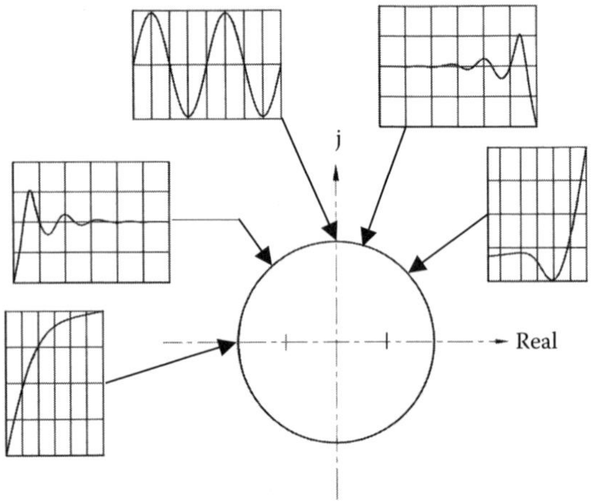

FIGURE 10.2.20
Phasors.

Example 10.2.10

It is found that a second-order system has a natural frequency of 120 rad/s together with a damping ratio of 0.4. Calculate the position of the poles on the 's' plane.

Solution

$$\theta = \cos^{-1}\delta = \cos^{-1} 0.4 = 66.422°$$

$$\sigma = \omega_n \delta = 120 \times 0.4 = 48$$

$$\omega_r = \omega_n \sin\theta = 120 \times \sin 66.422° = 109.98 \text{ rad/s}$$

The poles will be located at −48 ± j110.

Example 10.2.11

A standard second-order system has a natural frequency of 50 rad/s together with a damping ratio of 0.3. Establish the position of the poles on the 's' plane.

Solution

$$\theta = \cos^{-1}\delta = \cos^{-1} 0.3 = 72.542°$$

$$\sigma = \omega_n \delta = 50 \times 0.3 = 15$$

$$\omega_r = \omega_n \sin\theta = 50 \sin 72.542° = 47.7 \text{ rad/s}$$

The poles are located at −47.7 ± j50.

10.2.16 Sinusoidal Response

To recap,

The standard first-order transfer function is $\frac{\theta_o}{\theta_i}(s) = \frac{k}{Ts+1}$.

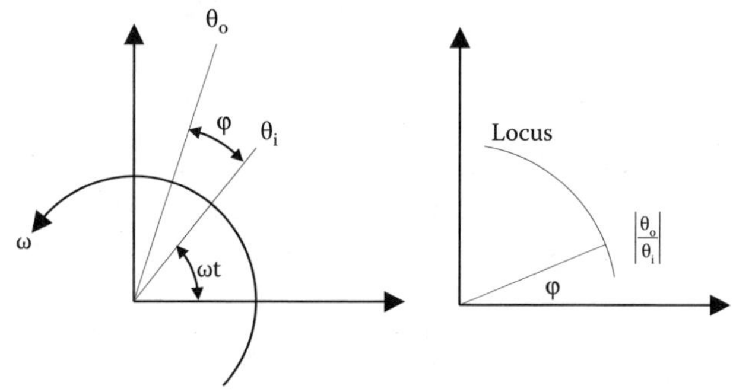

FIGURE 10.2.21
Sinusoidal frequency response.

The standard second-order transfer function is $\frac{\theta_o}{\theta_i}(s) = \frac{k}{T^2s^2 + 2T\delta s + 1}$.

Applying a sinusoidal signal to the input will produce an output that is also sinusoidal, having a different amplitude and phase angle. Both signals can be represented as phasors, as shown in Figure 10.2.20. With the input having unit amplitude and held at θ_o, the output phasor will have a phase angle ϕ to it, and the length of the phasor will be the ratio of the amplitudes.

Plotting the vectors for all frequencies between $\omega=0$ and ∞ will produce a polar plot where the amplitude ratio and phase angle can be seen for any frequency.

A form of a frequency response diagram is shown in Figure 10.2.21.

In this plot, the transfer function of 'jω' by replacing 's' with 'jω.'

The standard first-order transfer function becomes $\frac{\theta_o}{\theta_i}(j\omega) = \frac{k}{j\omega T + 1}$.

The standard second-order transfer function becomes $\frac{\theta_o}{\theta_i}(j\omega) = \frac{k}{T^2(j\omega)^2 + 2T\delta j\omega + 1}$.

This is now in a *Fourier transform* form and may be made into a complex number in the form:

$$G = A + jB$$

This may now be plotted as a vector on the complex plane.

It should be noted that in the steady state when $\omega=0$, $G(j\omega)=k$, and this is the DC gain. 'T' is a time constant and 'δ' is the damping ratio.

The polar plot can be made for either an open-loop or closed-loop system. It is more usual for the open loop to be plotted, as this is the easiest. There is an important reason for doing this when stability is studied later in this chapter.

At this point in the discussion, the polar plots of some basic blocks will be reviewed.

10.2.16.1 Integrator (Figure 10.2.22a and b)

Consider a control system element where the input is integrated:

$$\theta_o(t) = \int \theta_i \, dt \quad \theta_o(s) = s^{-1}\theta_i \quad \frac{\theta_o}{\theta_i}(j\omega) = \frac{1}{s}$$

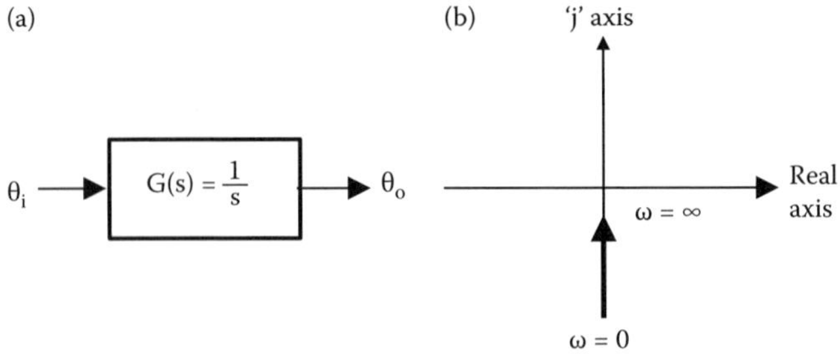

FIGURE 10.2.22
(a) Integrator and (b) differentiator.

Substituting s=jω:

$$\frac{\theta_o}{\theta_i}(j\omega) = \frac{1}{j\omega}$$

Multiply by the conjugate $\frac{\theta_o}{\theta_i}(j\omega) = \frac{1}{j\omega} \times \frac{-j\omega}{-j\omega} = \frac{-1}{\omega}$.

The polar plot would be a line running from –∞ to 0 on the –j axis, as shown in Figure 10.2.22. The radius is 1/ω and the angle will be –90° for all values of 'ω.'

10.2.16.2 Differentiator (Figure 10.2.23a and b)

Consider a control system element where the input is differentiated:

$$\theta_o(t) = \frac{d\theta_i}{dt} \quad \theta_o(s) = s\theta_i \quad \frac{\theta_o}{\theta_i} = s$$

Substituting s=jω:

$$\frac{\theta_o}{\theta_i}(j\omega) = \frac{1}{j\omega}$$

The polar plot would be a line running from 0 to ∞ on the +j axis, as shown in Figure 10.2.23b. The radius is 'ω' and the angle is +90° for all values of 'ω.'

10.2.16.3 Exponential Delay (Figure 10.2.24a and b)

The exponential delay can be expressed in the following form.

$$\theta_o(t) = e^{-t}\theta_i dt, \quad \frac{\theta_o}{\theta_i}(s) = \frac{1}{1+s}$$

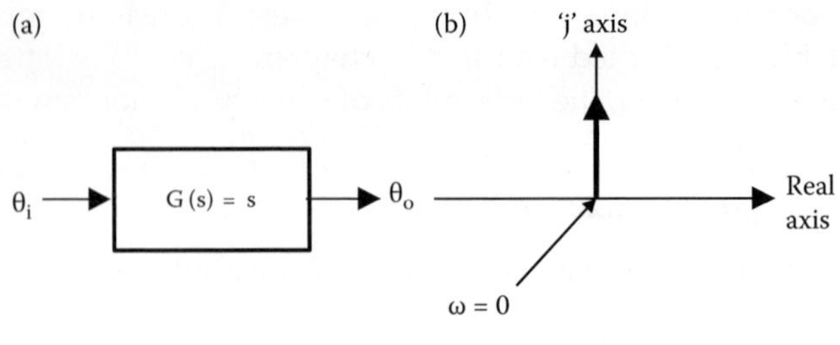

FIGURE 10.2.23
(a) and (b) Differentiator.

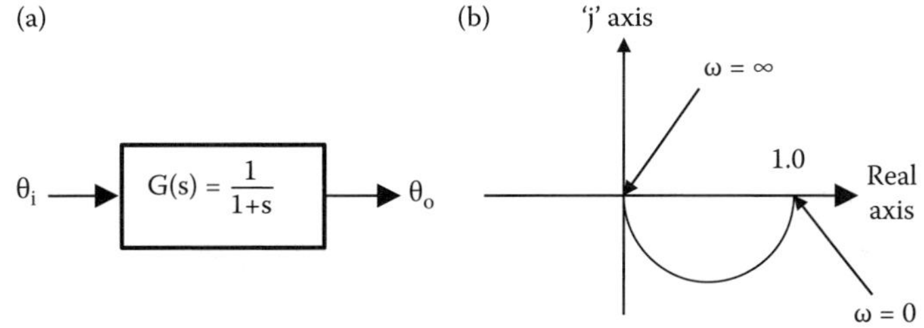

FIGURE 10.2.24
(a) and (b) Exponential delay.

In the jω form, the transfer function becomes $\frac{\theta_o}{\theta_i}(j\omega) = \frac{1}{1+j\omega}$.

Convert this transfer function into a complex number by multiplying the top and bottom of the equation by the conjugate number:

$$\frac{\theta_o}{\theta_i}(j\omega) = \frac{(1-j\omega)}{(1+j\omega)(1-j\omega)} = \frac{(1-j\omega)}{(1+j\omega^2)} = \frac{1}{1+\omega^2} - j\frac{\omega}{1+\omega^2}$$

$$\frac{\theta_o}{\theta_i}(j\omega) = A - jB \quad A = \frac{1}{\left(1+\omega^2\right)} \quad B = \frac{\omega}{\left(1+\omega^2\right)}$$

The polar plot is shown and formed by plotting the coordinates 'A' and 'B' for all values of 'ω.'

The radius is $\frac{1}{\sqrt{1+\omega^2}}$ and the angle is $-\tan^{-1}(\omega)$.

10.2.17 The Standard First-Order Transfer Function

Writing the transfer function in jω form gives $\frac{\theta_o}{\theta_i}(j\omega) = \frac{1}{1+j\omega T}$.

Convert this equation into a complex number by multiplying the top and bottom rows by the conjugate number:

$$\frac{\theta_o}{\theta_i}(j\omega) = \frac{\left(1-j\omega T\right)}{\left(1+j\omega T\right)\left(1-j\omega T\right)} = \frac{\left(1-j\omega T\right)}{\left(1+j\omega^2 T^2\right)} = \frac{1}{1+\omega^2 T^2} - j\frac{\omega}{1+\omega^2 T^2}$$

$$\frac{\theta_o}{\theta_i}(j\omega) = A - jB \quad A = \frac{1}{\left(1+\omega^2 T^2\right)} \quad B = \frac{\omega T}{\left(1+\omega^2 T^2\right)}$$

The polar plot is obtained by plotting 'A' horizontally and 'B' vertically. This is completed for all frequencies from '0' to '∞,' and the resulting locus is called a *frequency response diagram.*

The plot takes the form of a semicircle and shows that at 'ω' = 0, the output and input have identical amplitude and the phase angles are zero. Increasing the frequency results in the amplitude of the output reducing to zero, and the phase angle tending to 90°.

The geometry shown in Figure 10.2.25a and b shows that

$$\tan(-\phi) = \frac{B}{A} = \omega T. \quad \text{Therefore, } -\phi = \tan^{-1}\omega T$$

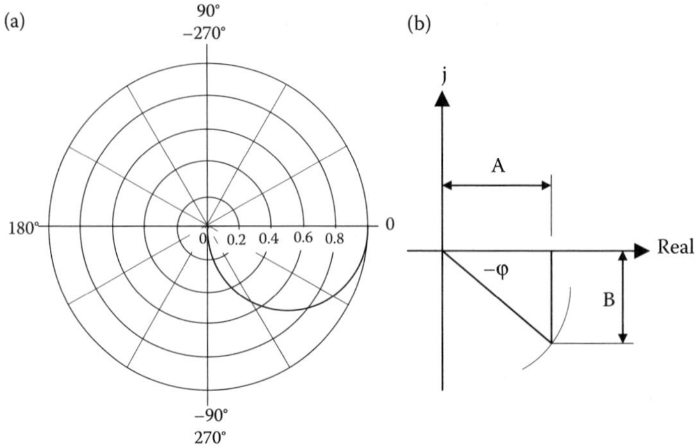

FIGURE 10.2.25
(a) Frequency response diagram and (b) polar plot.

The ratio of the output amplitude to the amplitude of the input signal is

$$\frac{a_2}{a_1} = \sqrt{A^2 + B^2} = \frac{1}{\sqrt{\left(1+\omega^2 T^2\right)}}$$

This is the length of the vector. For any given frequency, the steady-state plot θ_o and θ_i versus time is typically as shown in Figure 10.2.26a and b.

If the transfer function was presented in the form $\frac{\theta_o}{\theta_i}(s) = \frac{K}{n + T's}$ and it had been changed into standard form: $\frac{\theta_o}{\theta_i}(s) = \frac{k}{1 + Ts}$, then the DC gain will be 'k'=K/n and the primary time constant will be T=T′/n. Repeating the above process will reveal that the resulting radius and phase angle would be

$$\text{Radius} = \frac{k}{\sqrt{1+T^2\omega^2}} = \frac{K/n}{\sqrt{1+\frac{T'^2\omega^2}{n^2}}}$$

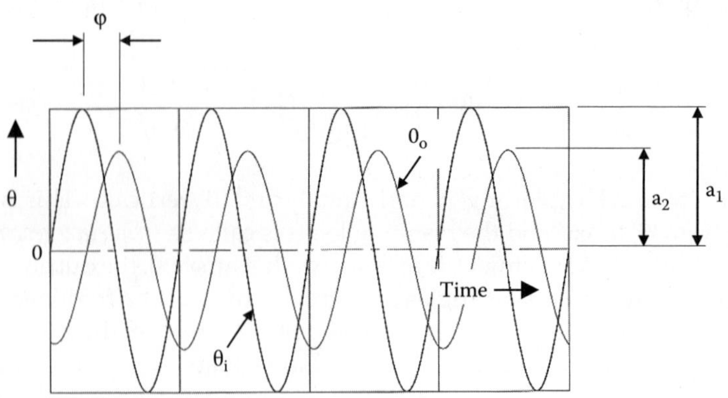

FIGURE 10.2.26
Steady-state sinusoidal input.

TABLE 10.2.2

Results for Locus $\frac{\theta_o}{\theta_i}(j\omega)$

ω (rad/s)		0	1	5	10	20	50	70	100
Mod (θo/θo)	$k/\sqrt{1+T^2\times\omega^2}$	1	0.93	0.45	0.24	0.12	0.05	0.04	0.03
ϕ (degrees)	$-\tan^{-1}(w\times T)$	0	−21.8	−63.4	−76.0	−82.9	−87.1	−88.0	−88.6

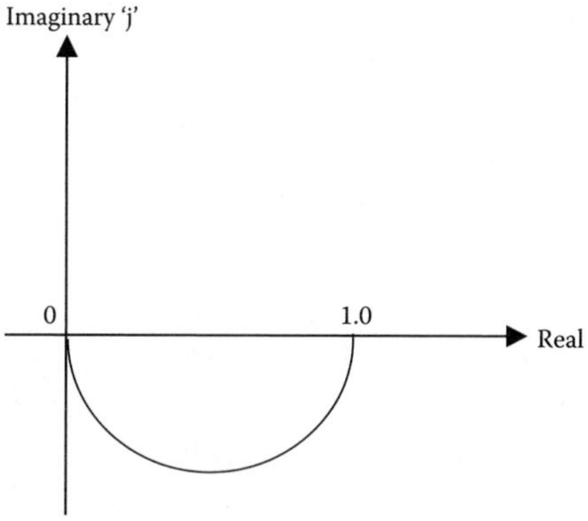

FIGURE 10.2.27
Polar plot for Example 10.2.6.

and the phase angle will be $\phi = -\tan^{-1}(\omega T) = -\tan^{-1}\left(\frac{\omega T'}{n}\right)$.

Example 10.2.12

In a simple first-order system, the transfer function is $\frac{\theta_o}{\theta_i}(s) = \frac{1}{(Ts+1)}$.

The input is a sinusoidal signal and the system time constant is 0.4 s. Produce a polar plot of the locus of $\frac{\theta_o}{\theta_i}(j\omega)$ covering frequencies from '0' to 100 rad/s.

Solution

T = 0.4 s.

k = 1

Figure 10.2.27 displays the polar plot for the Table 10.2.2.

10.2.18 The Standard Second-Order Transfer Function

To recap,

The standard second-order transfer function is $\frac{\theta_o}{\theta_i}(s) = \frac{k}{T^2s^2 + 2T\delta s + 1}$.
This may be written as

$$G(s) = \frac{1}{T^2s^2 + 2T\delta s + 1}$$

Written in jω form

$$G(jw) = \frac{1}{T^2(jw)^2 + 2T\delta jw + 1}$$

$$G(jw) = \frac{1}{A + jB}$$

where $A = (1 - T^2\omega^2)$ and $B = 2T\delta\omega$.

Multiplying the top and bottom lines by the conjugate number $A - jB$

$$G(jw) = \frac{A - jB}{(A + jB)(A - jB)}$$

$$G(jw) = \frac{A - jB}{(A^2 + B^2)} = C - jD$$

$$C = \frac{A}{A^2 + B^2} = \frac{\left(1 - \omega^2 T^2\right)}{\left\{\left(1 - \omega^2 T^2\right) + \left(2\delta\omega T\right)^2\right\}}$$

$$D = \frac{B}{A^2 + B^2} = \frac{(2\delta T\omega)}{\left\{\left(1 - T^2\omega^2\right) + \left(2\delta\omega T\right)^2\right\}}$$

The polar plot is drawn by plotting 'C' horizontally and 'D' vertically. The plot for all frequencies between $\omega = 0$ and ∞ is shown in Figure 10.2.28a and b.

The phase angle is found from $\phi = \tan^{-1} D/C$. The length of the vector is found from

$$G(s) = \sqrt{\left(C^2 + D^2\right)}$$

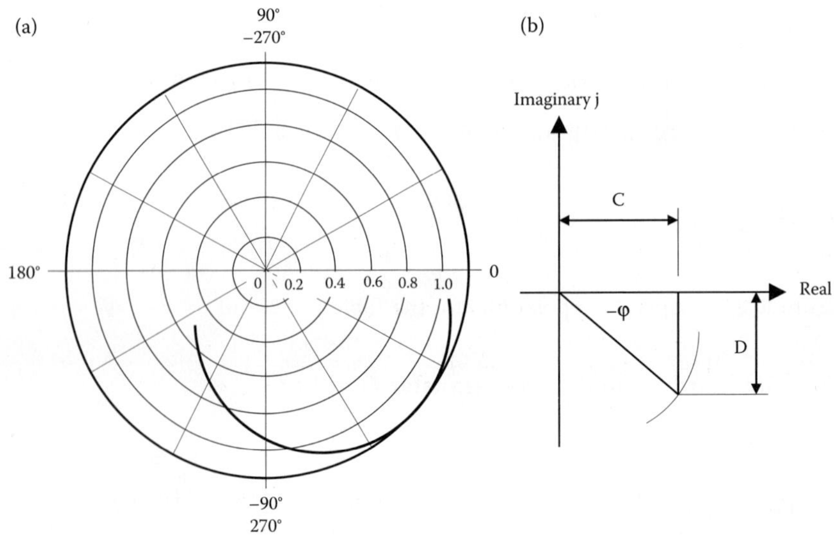

FIGURE 10.2.28
(a) Frequency response diagram and (b) polar plot.

This is the amplitude ratio of the output to the input signals. The overall result shows that at ω=0, the amplitudes are the same and the phase angle are also the same. As the frequency increases (ω), the amplitude of the output increases [depending upon the value of the damping ratio (δ)] and then reduces to zero as the phase angle increases to 180°.

Example 10.2.13

A hydraulic cylinder is controlled by a system having a transfer function $\frac{\theta_o}{\theta_i} = \frac{1}{\left(T^2s^2 + 2\delta Ts + 1\right)}$.

The time constant 'T' for the system is 0.02 s, and the damping ratio 'δ' is 0.5.

The input is varied harmonically as $x_i = 20\sin(\omega t)$ at 15 rad/s. Calculate the phase angle and the amplitude of the output.

Solution

From the question, ω=15 rad/s, δ=0.5, t=0.02 s:

$$A = \left(1 - T^2\omega^2\right) = 1 - 0.02^2 \times 15^2 = 0.91$$

$$B = (2T\delta\omega) = j2 \times 0.02 \times 0.5 \times 15 = 0.3$$

$$C = \frac{A}{A^2 + B^2} = \frac{0.91}{0.91^2 + 0.3^2} = 0.991$$

$$D = \frac{B}{A^2 + B^2} = \frac{0.3}{0.91^2 + 0.3^2} = 0.327$$

$$\phi = \tan^{-1}\frac{D}{C} = \tan^{-1}\frac{0.327}{0.991} = -18.25^\circ$$

$$\left|\frac{\theta_o}{\theta_i}\right| = \sqrt{\left(C^2 + D^2\right)} = \sqrt{\left(0.991^2 + 0.327^2\right)} = 1.043$$

It follows that the output amplitude is 1.043×20=20.87 and the phase angle is −18.25°:

$$\theta_o(t) = 20.87\sin\left(\omega t - 18.25^\circ\right)$$

10.2.19 Stability Analysis

10.2.19.1 Introduction

The stability of a control system is extremely important and is a safety issue for the control system in accurately performing its intended task. As an example, consider an automatic pilot for a commercial aircraft carrying a large number of passengers. Instability of the control system could lead to an unimaginable catastrophic accident.

10.2.19.2 Definition

A working definition for the stability of a control system relates its response to its input or an external disturbance. A system that remains in a constant state unless affected by an external action and returns to a constant state when the external disturbance is removed can be considered to be stable.

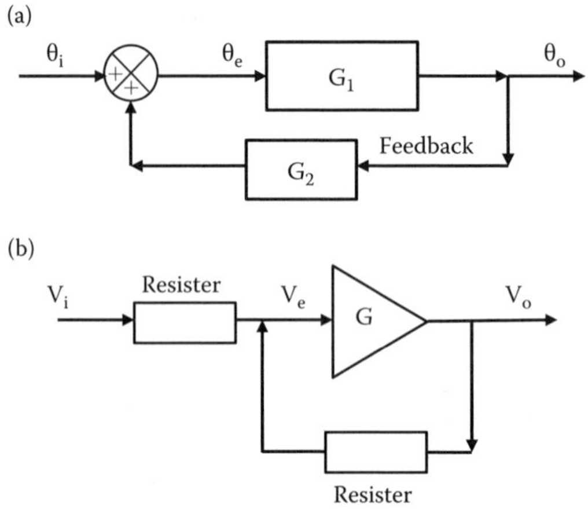

FIGURE 10.2.29
(a) Block diagram and (b) circuit diagram.

10.2.19.3 Stability of a Linear Control System

Stability is a condition where the system output has become unattenuated, that is, in a control system (Figure 10.2.29a and b) that increases the signal amplitude in a feedback loop to the summer, for example, the gain has been increased.

A system designed for negative feedback, with a summer that subtracts the feedback signal from the input signal, should be s, but there are occasions when the signal varies, such as with a sinusoidal signal, where it is possible for the negative feedback to become converted into a positive feedback. This would occur when the phase shift of the feedback becomes 180° to the input signal and the gain is one or more. When this happens, the system becomes unstable.

An example of positive feedback occurs when a microphone is placed in front of a loudspeaker and a loud oscillation occurs. This also occurs at events when the gain of the amplifier is turned up to high, and when the announcer speaks a loud, piercing oscillation occurs that continues to increase in volume.

The closed-loop transfer functions for Figure 10.2.29a may be derived from Table 10.2.3.

10.2.19.4 Analysis of Circuit Stability

There are a number of criteria that covers the stability of control systems, and they include but are not restricted to the following:

- 10.2.19.4.1 The Hurwitz criterion
- 10.2.19.4.2 Routh criterion

TABLE 10.2.3

Closed-Loop Transfer Functions

$G_1 = \frac{\theta_o}{\theta_e}$	$\theta_e = \theta_i + G_2\theta_o$	$G_1 = \frac{\theta_o}{\theta_i + G_2\theta_o}$	$G_1(\theta_i + G_2\theta_o) = \theta_o$
$G_1\theta_i + G_1G_2\theta_o = \theta_o$	$G_1\theta_i = \theta_o - G_1G_2\theta_o$	$G_1\theta_i = \theta_o(1 - G_1G_2)$	$G_{cl} = \frac{\theta_o}{\theta_i} = \frac{G_1}{1 - G_1G_2}$

- 10.2.19.4.3 Nyquist criterion
- 10.2.19.4.4 Nyquist criterion using Nyquist plots
- 10.2.19.4.5 Simplified forms of the Nyquist criterion
- 10.2.19.4.6 The Nyquist criterion using Bode plots
- 10.2.19.4.7 The Root Locus method

In this section, the concentration will be on the Nyquist and Bode pots.

10.2.19.5 Nyquist Diagrams

A negative feedback circuit will become unstable when the feedback signal arriving at the summer is larger than the input signal and has also shifted 180° relative to the input. Consider the block diagram of the closed-loop circuit shown in Figure 10.2.29. If a sinusoidal voltage is applied to the input into the circuit via a signal generator and the feedback is subtracted from the incoming signal using the summer to produce the error voltage. Due to the time delay the feedback signal may be 180° out of phase to the input signal, when the signals are summed the result could be an error signal significantly larger than the input signal. This results in an instability, the output voltage will then continue to increase exponentially.

A practical method of establishing whether this will happen is to disconnect the feedback loop from the summer circuit and monitor the feedback voltage over a wide range of frequencies.

For control systems being initially designed, the Nyquist diagram is one method of identifying the stability of the system. This is based on the open-loop transfer function where the locus is plotted on the complex plane (Figure 10.2.30). In the case of the system being inherently unstable, the Nyquist diagram (The Nyquist plot is a parametric plot of a frequency response used in automatic control and signal processing. The most common use of Nyquist plots is for assessing the stability of a system with feedback. In Cartesian coordinates, the real part of the transfer function is plotted on the X axis.) will enclose the

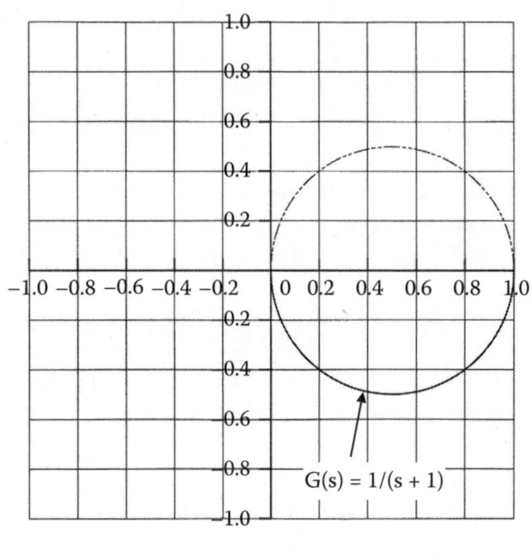

FIGURE 10.2.30
Block diagram.

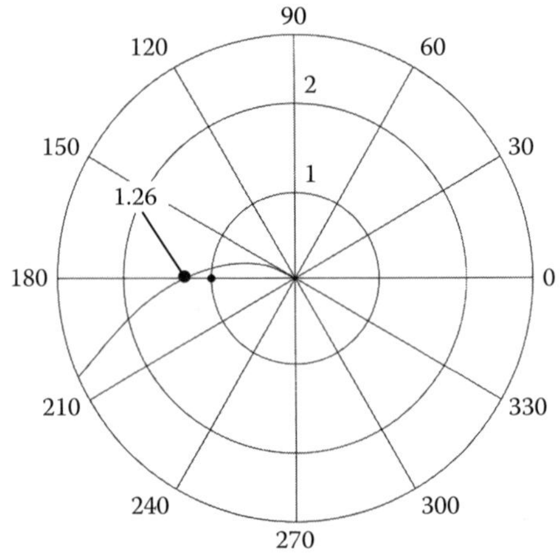

FIGURE 10.2.31
Phase and gain margins for Example 10.2.14.

point-1 (this is the point where the phase angle is 180° and having a unity gain), as shown in Figure 10.2.31.

By way of an example, consider the following:

Example 10.2.14

A control system has the following open-loop transfer function:

$$G(s) = \frac{80}{\{(s+1)(s+2)(s+4)\}}$$

Draw a polar plot for $\omega=1$ to $\omega=10$.

Estimate the phase and gain margins.

Solution

Evaluate the polar coordinates for $\frac{80}{s+1}, \frac{1}{s+2}, \frac{1}{s+4}$.

The radius may be found using $\frac{\frac{K}{n}}{\sqrt{1+\frac{T^2\omega^2}{n^2}}}$ and the angle $-\tan^{-1}\left(\frac{\omega T}{n}\right)$

Table 10.2.4 tabulates the values of the radius (R) against the angle (φ).

Figure 10.2.32 is a plot of the Table 10.2.4.

As shown in Figure 10.2.32, the gain margin is established to be 0.12 and the phase margin is 4.0°.

10.2.20 Bode Plots

A *Bode plot* is where the magnitude (radius of the polar plot) and phase angle of the transfer function is plotted on a logarithmic scale. It is first necessary to be able to express the gain in decibels.

To recap, 'G' is a power gain and is the ratio of power out and power in:

$$G = \frac{\text{Power out}}{\text{Power in}}$$

TABLE 10.2.4

(Page 10.2.39) Open-Loop Transfer Function

$$G(s) = \frac{80}{(s+1)(s+2)(s+4)}$$

Radius (R)

$$R = \frac{\left(\frac{K}{n}\right)}{\sqrt{1+\frac{T^2 \cdot \omega^2}{n^2}}}$$

Angle (ϕ)

$$\phi = -a\tan\left(\frac{\omega \cdot T}{n}\right)$$

$R_1 = \frac{80}{1+s}$	$R_2 = \frac{1}{2+s}$	$R_3 = \frac{1}{4+s}$	$\phi_1 = -a\tan\left(\frac{\omega \cdot T}{n}\right)$	$\phi_2 = -a\tan\left(\frac{\omega \cdot T}{n}\right)$	$\phi_3 = -a\tan\left(\frac{\omega \cdot T}{n}\right)$
T=1	T=1	T=1	T=1	T=1	T=1
K=80	K=1	K=1	K=80	K=1	K=1
n=1	n=2	n=4	n=1	n=2	n−1

ω	**R_1**	**R_2**	**R_3**	**R**	**ϕ_1**	**ϕ_2**	**ϕ_3**	**ϕ**
1	56.569	0.447	0.243	6.145	−45.000	−26.565	−14.036	−85.601
1.5	44.376	0.400	0.234	4.154	−56.310	−36.870	−20.556	−113.736
2	35.777	0.354	0.224	2.837	−63.435	−45.000	−26.565	−135.000
2.5	29.711	0.312	0.212	1.965	−68.199	−51.340	−32.005	−151.544
3	25.298	0.277	0.200	1.402	−71.565	−56.310	−36.870	−164.745
3.5	21.978	0.248	0.188	1.025	−74.055	−60.255	−41.186	−175.496
4	19.403	0.224	0.177	0.769	−75.964	−63.435	−45.000	−184.399
4.5	17.354	0.203	0.166	0.585	−77.471	−66.038	−48.366	−191.875
5	15.689	0.186	0.156	0.455	−78.690	—68.199	−51.340	−198.229
5.5	14.311	0.171	0.147	0.360	−79.695	−70.017	−53.973	−203.685
6	13.152	0.158	0.139	0.289	−80.538	−71.565	−56.310	−208.413
6.5	12.165	0.147	0.131	0.234	−81.254	−72.897	−58.392	−212.543
7	11.314	0.137	0.124	0.192	−81.870	−74.055	−60.255	−216.180
7.5	10.573	0.129	0.118	0.161	−82.405	−75.069	−61.928	−219.402
8	9.923	0.121	0.112	0.134	−82.875	−75.964	−63.435	−222.274
8.5	9.347	0.115	0.106	0.114	−83.290	−76.759	−64.799	−224.848
9	8.835	0.108	0.102	0.097	−83.660	−77.471	−66.038	−227.169
9.5	8.375	0.103	0.097	0.084	−83.991	−78.111	−67.166	−229.268
10	7.960	0.098	0.093	0.073	−84.289	−78.690	−68.199	−231.178

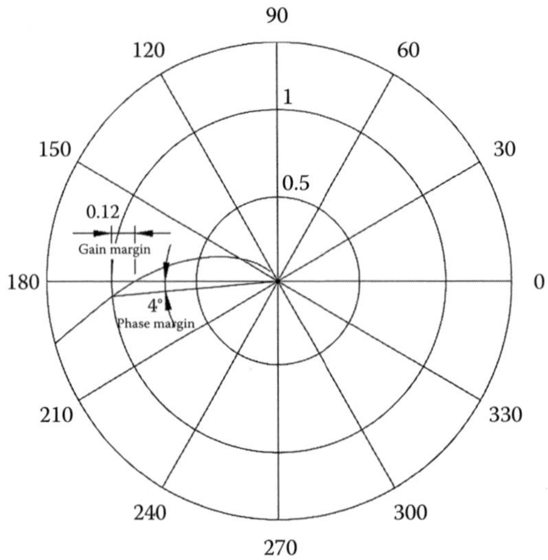

FIGURE 10.2.32
Phase and gain margins for Example 10.2.14.

If the power in and out is considered in electrical terms, it can be stated as

$$G = \frac{V_{out} I_{out}}{V_{in} I_{in}}$$

With the same value of resistance at the input and output and using Ohm's law, this can be expressed as

$$G = \frac{V_o^2}{V_i^2} \text{ or } \frac{I_o^2}{I_i^2}$$

Expressing 'G' in decibels

$$G(db) = 10 \log \frac{V_o^2}{V_i^2}$$

or

$$G(db) = 20 \log \frac{V_o}{V_i}$$

or

$$G(db) = 20 \log \frac{I_o}{I_i}$$

Hence, it usual to express the modulus of 'G' as

$$|G| = 20 \log \left| \left(\frac{\theta_o}{\theta_i} \right) \right|$$

Note that the gain in db is the 20 log R where 'R' is the radius of the polar plot from previous section.

Consider the transfer function $G(s) = \frac{1}{s}$

$$G(j\omega) = \frac{1}{j\omega} = \frac{-1}{\omega}$$

$$|G| = \left|\frac{-j}{\omega}\right| = \left(\frac{1}{\omega}\right)$$

$$G(db) = 20\ \log\left(\frac{1}{\omega}\right) = -20\ \log\omega$$

Figure 10.2.33 shows this equation when plotted in semilog graph form It shows a straight line that passes through 0 db at $\omega=1$ and has a gradient of –20 db per decade. The phase angle is –90° at all values of 'ω.'

Consider the following transfer function

$$G(s) = \frac{1}{Ts+1} \quad |G| = \left|\frac{1}{j\omega T+1}\right| = \frac{1}{|j\omega T+1|}$$

$$G(j\omega) = \frac{1}{1+\omega^2 T^2} - j\frac{\omega T}{1+\omega^2 T^2}$$

The radius of the polar plot coordinate is

$$\frac{1}{\left(T^2 w^2 + 1\right)^{0.5}}$$

and this is the gain.

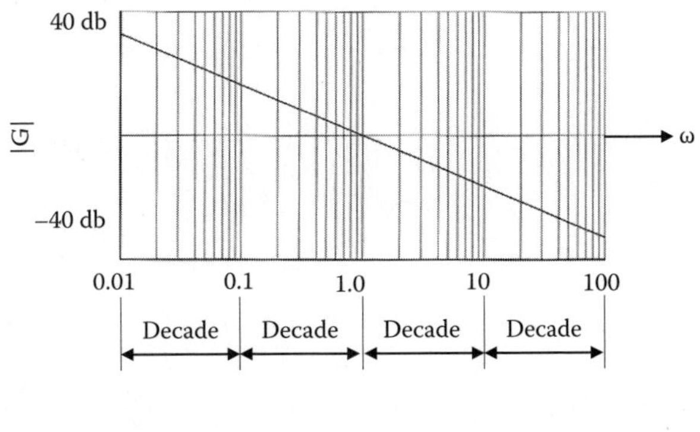

FIGURE 10.2.33
Plot for equation G(s) = 1/s.

The gain in db will be

$$G(db) = 20\left[\log\left(\frac{1}{\left(T^2\omega^2+1\right)^{0.5}}\right)\right]$$

$$= -20\left[\frac{1}{2}\log\left(T^2\omega^2+1\right)\right]$$

$$= -10\log\left(T^2\omega^2+1\right)$$

The phase angle is –tan(ωT).

As an example and putting T=1, when the result is plotted, the product is two distinct straight lines that are shown in Figure 10.2.34a. The horizontal line is produced by very small values of 'ω' and is referred to as the *low-frequency asymptote*. The sloping line occurs at higher values of 'ω' and is called the *high-frequency asymptote* and has a gradient

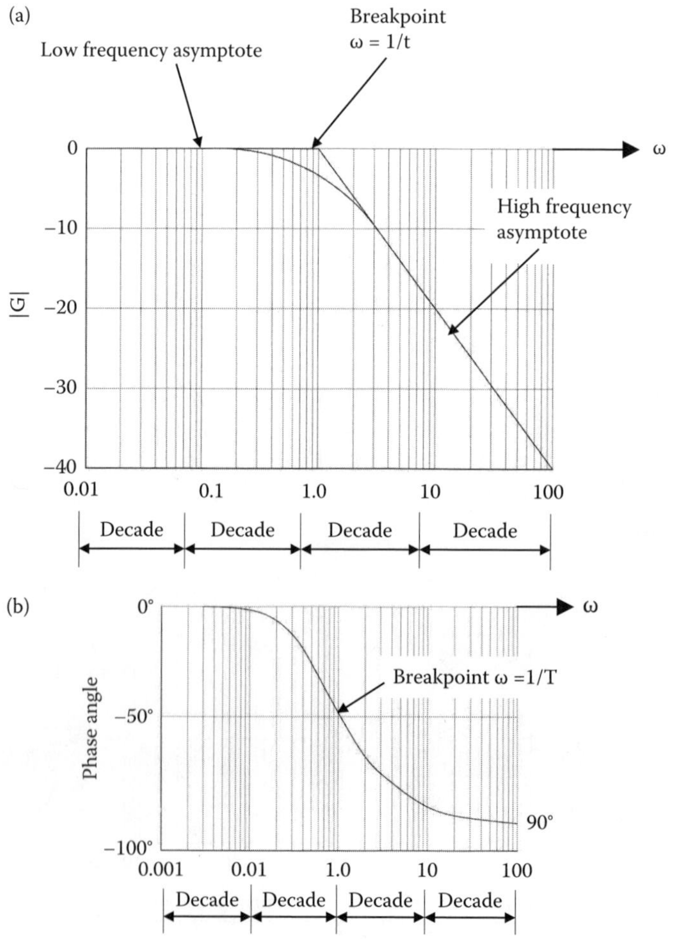

FIGURE 10.2.34
(a) Low- and high-frequency asymptotes and (b) phase angle.

of –20 db per decade. The two lines meet at the *breakpoint frequency* or natural frequency and in this is given by

$$\omega_n = \frac{1}{T} = 1$$

Figure 10.2.34b shows the phase angle plotted against 'ω' between the values of '0' to –90°. The 45° point occurs at the breakpoint frequency.

As a further example, consider the following transfer function, which is a standard first-order response to a step input:

$$G(s) = \frac{K}{s(1+sT)} \quad G(j\omega) = \frac{K}{j\omega(1+j\omega T)}$$

An easier way to determine $|G|$ is to separate the two parts and establish the modulus of each part separately, i.e.,

$$|G| = K\frac{1}{|j\omega|}\frac{1}{\left|\left(1+\omega^2T^2\right)\right|}$$

$$\left|\frac{1}{j\omega}\right| = \frac{1}{|j\omega|} = \frac{1}{\omega}$$

$$\left|\frac{1}{1+j\omega T}\right| = \frac{1}{|1+j\omega T|} = \frac{1}{\left(1+\omega^2T^2\right)^{0.5}}$$

$$|G| = K\frac{1}{|\omega|}\frac{1}{\left|\left(1+\omega^2T^2\right)^{0.5}\right|} = K\left(\frac{1}{\omega}\right)\left(\frac{1}{\left(1+\omega^2T^2\right)^{0.5}}\right)$$

Taking logs, the following is obtained:

$$\log|G| = \log K + \log\left(\frac{1}{\omega}\right) + \log\left(\frac{1}{\left(1+\omega^2T^2\right)^{0.5}}\right)$$

$$\log|G| = \log K - \log\omega - \frac{1}{2}\log\left(1+\omega^2T^2\right)$$

$$\log|G|\,\text{db} = 20\left[\log K - \log\omega - \frac{1}{2}\log\left(1+\omega^2T^2\right)\right]$$

There are three components to this equation and they may be plotted separately, as shown in Figure 10.2.35a. The graph for the complete equation is the sum of all three components. This results in two distinctive slopes of –20 db per decade and –40 db per decade. (In this case, K and T have been taken arbitrarily as 10, resulting in the break-point of $\omega = 1/T = 0.1$.)

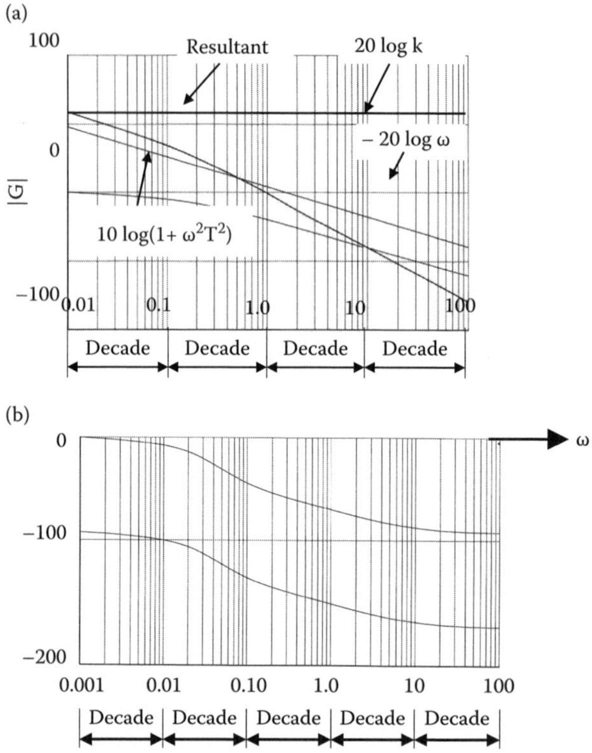

FIGURE 10.2.35
(a) Standard first-order response to a step and (b) plot of phase angle.

Plotting the phase angle against frequency on the logarithmic scale shows that the phase angle changes by 90° every time it passes through a breakpoint frequency. The plot for the case under examination is shown in Figure 10.2.35b.

A good approximation will be obtained by sketching the asymptotes for each and adding them together.

Example 10.2.15

A system has a transfer function $G(s) = \dfrac{1}{s(Ts+1)}$ where the time constant is 0.5 seconds.

Plot the Bode diagram for the gain and phase angle and determine the low-frequency gain per decade, together with the high-frequency gain per decade and the breakpoint frequency.

Solution

$$G(s) = \frac{1}{s(Ts+1)}, \quad G(j\omega) = \frac{1}{j\omega}\left(\frac{1}{j\omega T+1}\right), \quad |G| = \left|\frac{1}{j\omega}\right|\left|\frac{1}{j\omega T+1}\right| = \left(\frac{1}{w}\right)\left[\frac{1}{\left(\omega^2T^2+1\right)^{0.5}}\right]$$

$$|G|(db) = 20\left[-\log\omega - \frac{1}{2}\log\left(\omega^2T^2+1\right)\right] \quad \phi = -90 - \tan^{-1}\left(\frac{1}{\omega T}\right)$$

TABLE 10.2.5
Values for Example 10.2.15

ω	0.001	0.01	0.1	1.0	10	100	1000
–log ω	3	2	1	–1	–1	–2	–3
$-1/2 \log(\omega^2T^2+1)$	Small	Small	Small	–0.048	–0.707	–1.707	–2.707
Total gain (units)	3	2	1	–0.48	–1.707	–3.707	–5.707
Total gain (db)	60	40	20	–0.96	–34.14	–74.14	–114.14
ϕ degrees	–90	–90.3	–93.9	–117	–169	179	–180

Table 10.2.5 shows that the gain drops by 20 db per decade at the low frequencies and by 40 db per decade for the high frequencies. When these values are plotted on semilogarithmic paper, it is revealed that a breakpoint frequency of 2 rad/s occurs. This value can also be found from

$$\omega_n = \frac{1}{T} = \frac{1}{0.5} = 2$$

Make a quick way of drawing an approximate Bode plot to evaluate the gain in db at the breakpoint frequency, and then draw the asymptotes (see Figure 10.2.36a), first with a slope of –20 db per decade prior to the breakpoint and –40 db per decade following the breakpoint. The phase angle can be found by adding the two components:

$$G(s) = \frac{1}{s(Ts+1)}, \quad G(j\omega) = \frac{1}{j\omega} + \frac{1}{j\omega T+1}$$

The phase angle for the first part will be the angle of a vector at position $-\frac{1}{\omega}$ on the 'j' axis corresponding to –90° (Figure 10.2.36b).

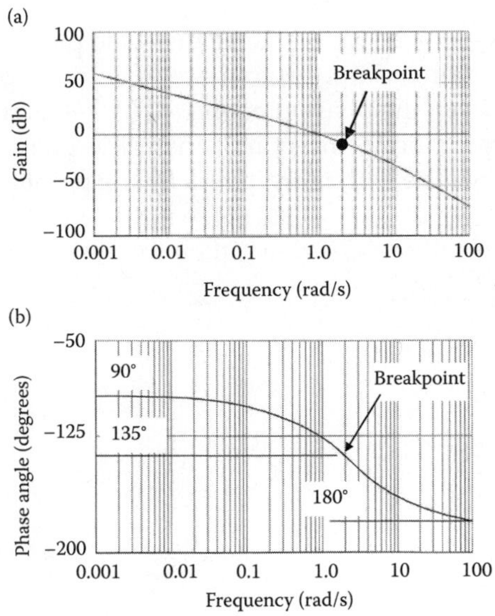

FIGURE 10.2.36
(a) Bode plot for example and (b) example phase plot.

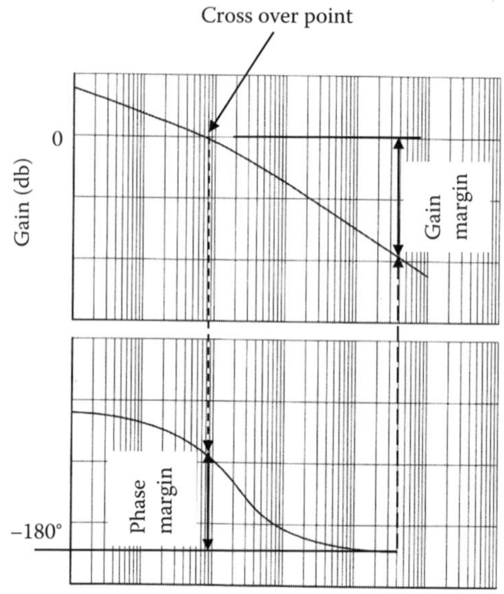

FIGURE 10.2.37
Gain and phase margins.

The phase angle for the second part of the angle of a vector at –1 on the real axis and –ωT on the 'j' axis. The two phase angles may be added together to produce the overall result.

The Bode phase plot may be drawn quickly by noting the breakpoint frequency that occurs at the midpoint of the phase shift (–135° in this) and drawing the asymptotes such that they change by 90° at each breakpoint frequency.

10.2.21 Gain and Phase Margins Using the Bode Plot

The gain and phase angles may be found using the Bode plot by locating the point where the gain is zero db (unity gain) and projecting down onto the phase diagram. The phase margin is the margin between the phase plot and –180° (see Figure 10.2.37).

To locate the gain margin, locate the point where the phase angle reaches –180° and project back to the gain plot. The gain margin will be the point between this point and the zero db level.

If the gain is increased until it reaches zero, then the system will then become unstable.

11

Physics

11.1 Heat

Matter is made up of molecules that are continuously in a state of movement, the amount of movement dependent on how much energy the matter possesses. As more energy added, the more active the molecules become, moving around rapidly, growing gradually hotter as a result.

As energy is removed from the matter or body, the molecules become less active and their speed reduces, and the matter then becomes cooler.

Heat energy may be added to a body by a number of means, including the direct application of a flame, electrical resistance, friction, etc.

11.1.1 Temperature

The heat of a body may be measured using a thermometer. The most common thermometers used consist of a liquid (usually alcohol or mercury) within a glass tube that is closed at both ends and evacuated of air. These liquid thermometers are based on the principle of thermal expansion. As a substance is heated, it expands to a greater volume. Nearly all substances exhibit this behavior of thermal expansion and are the basis of the design and operation of all liquid thermometers. There are other types of thermometers that give a digital readout of the temperature, and these rely on the change of electrical resistance of a wire, usually platinum.

Another type of thermometer depends upon the electromagnetic force change with temperature of two distinct metals (copper and iron) that are joined together using solder. The ends of the wires are connected to a galvanometer and as the junction is heated, a current flows within the circuit, causing the galvanometer to deflect.

All thermometers must be calibrated so that they will measure absolute temperatures. This is undertaken by measuring two distinct temperatures. First, the thermometer (or junction) is immersed in freezing liquid that is on the point of thawing and the temperature of the liquid is measured. The thermometer is then immersed in boiling water and the second measurement is then made. These measurements are made at an atmospheric pressure of 1 atm.

By convention, the freezing water is marked at 0 and the boiling water is marked at 100. The scale is then divided into 100 equal divisions between the marks. Accurate measurements can then be made of the temperature of any object within the temperature range for which the thermometer has been calibrated.

11.1.1.1 Temperature Scales

The thermometer calibration process described above will result in what is known as a *centigrade* thermometer. A centigrade thermometer has 100 divisions between the normal freezing point of water and the normal boiling point of water. The centigrade scale is known as the *Celsius scale*, after the Swedish astronomer Anders Celsius, who is credited with its development. The Celsius scale has been accepted worldwide with the exception of the United States of America, which still uses the Fahrenheit scale in a similar manner to that described above, based on freezing water measures at 32°F and boiling water measured at 212°F with 180 divisions between.

Temperatures measured in the Fahrenheit scale can be converted to the equivalent Celsius scale using the following formula:

$$°C = \frac{(°F - 32)}{1.8} \tag{11.1.1}$$

Conversely, temperatures measured in the Celsius scale are easily converted to the Fahrenheit scale using the following formula:

$$°F = 1.8 \times °C + 32 \tag{11.1.2}$$

11.1.1.2 Kelvin Scale

While the Celsius and Fahrenheit scales are widely used today, there have been other scales developed have included the Rankine scale, the Newton scale, and the Romer scale. All of these are rarely used today. A further scale, the *Kelvin temperature scale*, has been adopted as the standard metric system of temperature measurement and is possibly the widest-used scale by engineers and scientists. The Kelvin scale is similar to the Celsius scale in that there are 100 divisions between the freezing and boiling points of water. However, the zero-degree mark on the Kelvin scale is 273.15 units cooler than it is on the Celsius scale. Hence, the temperature of 0 K is equivalent to a temperature of –273.15°C. Note that the degree symbol is not used on the Kelvin scale, so that a temperature of 250 units above 0 K is referred to as 250 K and not as 250° 0 K; the temperature is abbreviated as 250 K. Conversions between Celsius temperatures and Kelvin temperatures can be performed using one of the two following equations:

$$°C = K - 273.15° \tag{11.1.3}$$

$$K = °C + 273.15° \tag{11.1.4}$$

The zero point on the Kelvin scale is referred to as *absolute zero* and it is the temperature where all movements of molecules making up matter become stationary. Currently, engineers and scientists have been able to lower the temperature to just above the absolute zero point, but have not quite yet reached it.

In the process of reducing the temperature to absolute zero, a number of unusual properties have been discovered. These properties include the following:

- *Superconductivity*
  In a superconductor, the resistance drops abruptly to zero when the material is cooled below its critical temperature. An electric current flowing in a loop of superconducting wire can persist indefinitely with no power source.

- *Superfluidity*
  Superfluidity is a state of matter in which the matter behaves like a fluid with zero viscosity. This phenomenon was first discovered in helium, where as long as it remains superfluid, it creeps up the wall of a container as a thin film.

11.1.1.3 Effects of Temperature

Most matter will expand when heated; this increase in size is called *thermal expansion*. This expansion can be useful in the case of the alcohol in the thermometer or gases in most heat engines. But such expansion can also have a detrimental effect on bridges and rail lines. It is for this reason that large gaps are left in structures to accommodate this expansion, otherwise large forces will be generated if the expansion is resisted, sufficient to buckle rail lines.

11.1.1.3.1 Linear Expansion

In gases, the expansion due to the application of heat is used for good effect in increasing pressure that is used in most forms of heat engines, including the two- and four-stroke vehicle engines.

Besides causing an increase in size or pressure, a change of temperature will also cause a change of state. As an example, solid ice will change to a liquid (water) and further application of heat will change the liquid into a gas (steam). It will be noted that when there is a change in state, the temperature remains constant until all the matter (in the case of ice) becomes completely converted to the next phase (in the case of the ice example, to water). The liquid phase will completely change to a gas only after all the liquid has boiled, and the gas will then increase in temperature until the gas has the same temperature as the liquid. This phase is known as *saturated gas* as there will be liquid droplets suspended in the gas. These droplets then become completely evaporated and the temperature of the gas will begin to rise. Figure 11.1.1 depicts the change in phases from solid–liquid–gas. The heat required to allow a change in state is known as *sensible heat*.

This process is completely reversible, in that as the gas begins to cool it will revert to being a liquid, and subject to continuous cooling then returns to a solid. The body that had absorbed heat in the process of becoming a liquid then a gas will then release all its heat on cooling. This heat is referred to as *latent heat*.

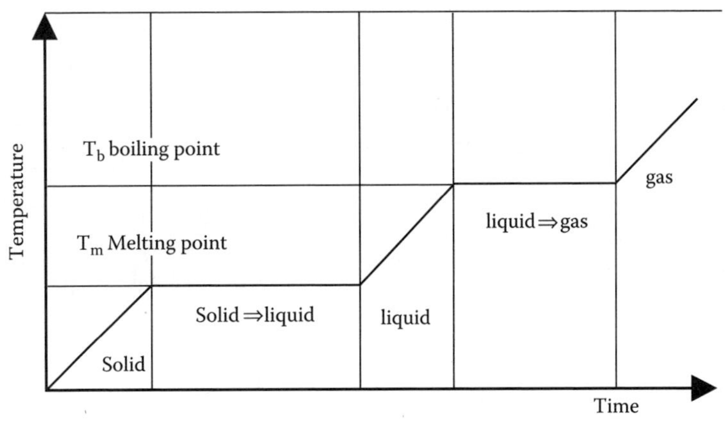

FIGURE 11.1.1
Changes of state.

Materials subject to temperature will increase in size. A direct example of this phenomenon is in a thermometer, where either the mercury or alcohol increases in length within the thermometer tube, giving an indication of the temperature. When the heat is removed, the material will return to its original size. The amount of increase depends upon three factors:

1. Original size of the component
   It would be expected that the increase in length of 1 km of a railway line would be greater (480 mm) than 100 mm (0.048 mm) of the same line for a temperature rise of 40°C, all other conditions being equal.
2. Temperature rise
   Considering the example of the thermometer, the change in length of the mercury column is much greater for a 100°C temperature rise than for a 10°C temperature rise.
3. The material itself
   It will be found that that the increase in a 250 mm length of the mercury column will be 1.525 mm for a 100°C rise in temperature, while a corresponding increase in the same length of steel will be 0.30 mm, only 20% of the amount for mercury.

Table 11.1.1 tabulates values of thermal expansion coefficients for various materials.

The equation for the calculation for the change in length for a specified temperature rise is

$$\Delta L = L \times \alpha \times \Delta T \tag{11.1.5}$$

where

ΔL = change in length
L = original length
α = coefficient of linear expansion
ΔT = change in temperature

TABLE 11.1.1

Linear Expansion Coefficients for Various Materials

Material	Linear α (10^{-6} K^{-1})	Material	Linear α (10^{-6} K^{-1})
Aluminum	23.1	Silver	18.9
Brass	20.3	Solder (lead–tin)	25.0
Carbon (graphite)	6.5	Steel, stainless	17.3
Chromium	4.9	Steel, structural	12.0
Copper	16.5	Tin	22.0
Gold	14.2	Titanium	8.5
Invar (64% Fe, 36% Ni)	1.2	Tungsten	4.5
Iron	11.8	Uranium	13.9
Lead	28.9	Water, ice (0°C)	51.0
Nickel	13.3	Zinc	30.2

11.1.1.4 Newton's Law of Cooling

Newton's law of cooling states that the hotter an object is the faster it will cool, and the rate of cooling is proportional to the temperature difference between the object and the ambient temperature. For example, if a measure of water that is at the boiling point is allowed to cool naturally in an uninsulated container and the temperature of the water is monitored and documented continuously, it will result in the graph shown in Figure 11.1.2. This curve is a classic equation of exponential decay over time.

The rate of loss from the water is directly proportional to the temperature difference between the water and its surroundings, provided the difference is small.

$$\text{i.e.,} \quad \frac{dT}{dt} \propto \left(T_t - T_s\right)$$

Newton's law of cooling is given by

$$\frac{dT}{dt} = k\left(T_t - T_s\right)$$

where

T_t is the temperature at a given time
T_s is the temperature of the surroundings
k is a constant
Newton's law of cooling formula is given by

$$T(t) = T_s + \left(T_o - T_s\right)e^{-kt}$$

where

T(t) is the temperature of the given body at time 't'
T_s is the temperature of the surroundings

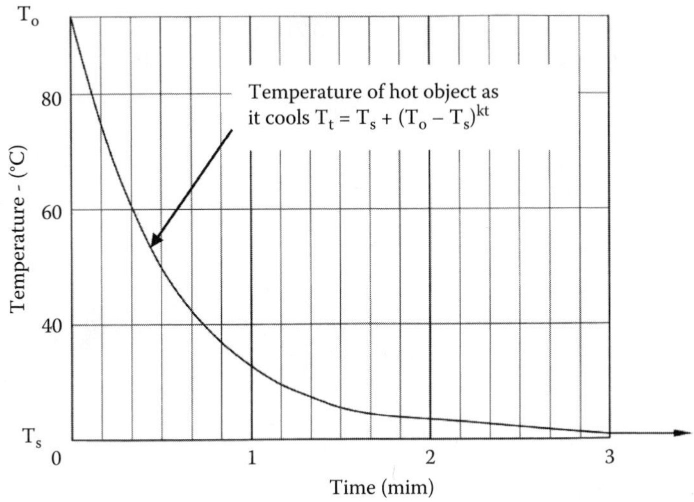

FIGURE 11.1.2
Newton's law of cooling.

T_o is the initial temperature of the body
t is the time taken
k is the constant

The greater the difference in temperature between the body and its surroundings, the quicker the body temperature changes.

The parameter 'k' can be derived either from the curve or derived mathematically and is specific for this type of problem with cooling in air naturally in an uninsulated measuring container.

Example

Oil is heated to 60°C and allowed to cool down to 50°C after 6 min. Calculate the time it would take the oil to cool from 50°C to 40°C with the surrounding temperature at 25°C.

Solution

Initial temperature of the oil = 60°C
At t = 6 min, the temperature is 50°C.
$T_o = 60°C$
$T_s = 25°C$
t = 6 min
Newton's law of cooling is given by $T(t) = T_s + (T_o - T_s)e^{-kt}$

$$\frac{T(t) - T_s}{T_o - T_s} = e^{-kt}$$

$$-kt \ln e = \ln \frac{T(t) - T_s}{T_o - T_s}$$

$$-kt \ln e = \ln \frac{50 - 25}{60 - 25}$$

$$= \ln 71428$$

$$k = -\frac{0.71428}{6}$$

$$= -\frac{\ln(0.71428)}{6}$$

$$k = -0.0561$$

The time taken for the oil to cool down to 40°C:
From the equation,

$$-kt \ln e = \ln \frac{T(t) - T_s}{T_o - T_s}$$

$$-(0.0561)t = \ln\left(\frac{40 - 25}{60 - 25}\right)$$

$$-0.056 = -0.8473$$

$$t = 15.13 \text{ min}$$

11.1.2 Heat Capacity

The total heat capacity (C) of a body is the amount of heat energy required to raise the temperature of the body by 1°C. It is expressed in joules per degree K (J K^{-1}).

The specific heat capacity (c) of a substance is the heat required to raise a unit mass (m) through 1°C, it is the heat capacity per unit mass of the substance and its units are expressed in either joule per gram per degree K (J g^{-1} K^{-1}) or joule per kilogram per degree K (J kg^{-1} K^{-1}). The specific heat of water (c_w) is approximately 4200 J kg^{-1} K^{-1}.

The *kilocalorie* is the amount of heat required to raise 1 kg of water by 1°C. In S.I. units the calorie is equivalent to 4.184 J. These reversible conversions of work and heat energy are called the *mechanical equivalent of heat*.

From the definition of specific heat capacity, it follows that

$$\text{Heat capacity, C} = \text{mass} \times \text{specific heat capacity.}$$

As an example, the specific heat capacity of copper is 0.39 J g^{-1} K^{-1} or 390 J kg^{-1} K^{-1}.

$$\begin{aligned}\text{The heat capacity of 5 kg of copper} &= 5 \times 390 \\ &= 1950\ \text{J K}^{-1} \\ &= 1.95\ \text{kJ K}^{-1}.\end{aligned}$$

Table 11.1.2 gives the specific heat capacities for a range of materials.

11.1.3 Heat Transfer

Heat is transferred from one location to another by one of three methods:

- Conduction
- Convection
- Radiation

TABLE 11.1.2

Specific Heat Capacities for Various Materials

Material	Sp Ht: J kg^{-1} $K^{-1} \times 10^3$	Material	Sp Ht: J kg^{-1} $K^{-1} \times 10^3$
Aluminum	0.91	Solder	0.18
Brass	0.38	Glass	0.70
Copper	0.39	Ice	2.10
Iron	0.47	Rubber	1.70
Steel	0.45	Stone	0.90
Lead	0.13	Wood	1.70
Mercury	0.14	Alcohol	2.50
Nickel	0.46	Ether	2.40
Platinum	0.13	Paraffin oil	2.10
Silver	0.24	Turpentine	1.76

11.1.3.1 Conduction

Consider an iron bar that is partially placed in a furnace to heat an area of the bar for a blacksmith to work on.

The heat from the furnace will travel along the bar gradually before it becomes too hot to hold the bar comfortably.

The transfer rate (H) is the ratio of the amount of heat per time for the heat to transfer from one location to another. This may be expressed as

$$H = \frac{Q}{\Delta t} \tag{11.1.6}$$

where

H has units of J/s or W
Q is in J
t is in seconds

There must be a temperature difference for conduction to take place, as heat will only flow from a hot body to a cold body and not the other way around.

Considering the iron bar has a length (L) and a cross-sectional area (A), the heat conduction along the length of the bar is given by

$$H = \frac{Q}{\Delta t} = kA\frac{(T_2 - T_1)}{L} \tag{11.1.7}$$

where heat flows from position T_2 to T_1 and $T_2 > T_1$, as shown in Figure 11.1.3.

The constant (k) is called *thermal conductivity*. Table 11.1.3 gives values of thermal conductivities for a range of materials.

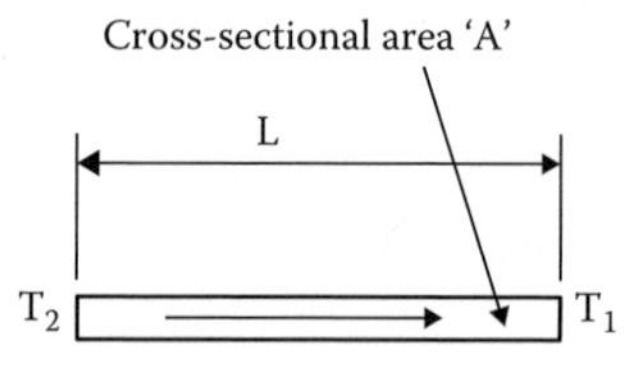

FIGURE 11.1.3
Heat conduction along an iron bar.

TABLE 11.1.3
Thermal Conductivities for Various Materials

Material	'k' W m^{-1} K^{-1}	Material	'k' W m^{-1} K^{-1}
Aluminum	210	Platinum	0.13
Brass	109	Silver	420
Copper	380	Solder	55 (~)
Iron (pure)	76	Glass	1.1
Iron (wrought)	59	Ice	2.10
Steel	46	Rubber	1.70
Lead	35	Wood	0.21
Mercury	9.2	Alcohol	0.18
Nickel	87	Paraffin oil	0.13

It will be noted that the metal materials have a far higher conductivity constant than do liquids or other nonmetal materials.

11.1.3.2 Convection

Liquids and gases are fluids. The particles in these fluids can move from place to place. Convection occurs when particles with a lot of heat energy in a liquid or gas move and take the place of particles with less heat energy. Heat energy is transferred from hot places to cooler places by *convection.*

Liquids and gases expand when they are heated. This is because the particles in liquids and gases move faster when they are heated than they do when they are cold. As a result, the particles take up more volume. This is because the gap between particles widens, whereas the particles themselves stay the same size.

The liquid or gas in hot areas is less dense than the liquid or gas in cold areas, so it rises into the cold areas. The denser cold liquid or gas falls into the warm areas. This sets up convection currents that transfer heat from place to place.

11.1.3.3 Radiation

Thermal radiation is a process by which energy, in the form of electromagnetic radiation is emitted by a heated surface in all directions and travels directly to its point of absorption at the speed of light; thermal radiation does not require an intervening medium to carry it.

A fire in an open grate is an excellent example of thermal radiation, as the heat is felt directly without any intervening fluid between the radiant (fire) and the recipient.

Thermal radiation ranges in wavelength from the longest infrared rays through the visible-light spectrum to the shortest ultraviolet rays. The intensity and distribution of radiant energy within this range is governed by the temperature of the emitting surface. The total radiant heat energy emitted by a surface is proportional to the fourth power of its absolute temperature (the Stefan–Boltzmann law). The Stefan–Boltzmann law (also Stefan's constant) is a physical constant denoted by the Greek letter σ (sigma) and is the constant of proportionality. "The total intensity radiated over all wavelengths increases as the temperature increases", of a black body which is proportional to the fourth power of the thermodynamic temperature.

$$E = \sigma T^4 \tag{11.1.8}$$

where

E = radiant heat energy emitted from a unit area in 1 s

σ = constant of proportionality (the Stefan–Boltzmann constant)

T = the absolute temperature (in degrees Kelvin)

The constant has a value of 5.6704×10^{-8} W/m^2 K^4

The law only applies to blackbodies, theoretical surfaces that absorb all incident heat radiation.

The rate at which a body radiates (or absorbs) thermal radiation depends on the nature of the surface, as well. Objects that are good emitters are also good absorbers (Kirchhoff's radiation law). A blackened surface is an excellent emitter as well as an excellent absorber. If the same surface is silvered, it becomes a poor emitter and a poor absorber. A blackbody (sometimes spelled "black body") is a theoretically ideal radiator and absorber of energy

TABLE 11.1.4

Comparisons between Two Radiator Surfaces

Color	Finish	Ability to Emit Thermal Radiation	Ability to Absorb Thermal Radiation
Dark	Dull or matte	Good	Good
Light	Shiny	Poor	Poor

at all electromagnetic wavelengths. The term comes from the fact that a cold blackbody appears visually black.

Table 11.1.4 gives a comparison between two radiant surfaces.

A common misconception in home heating is the use of the word *radiator.* A radiator is a thin panel (often painted white) that is fed using hot water. They would be better at radiating heat if they were painted with black matte paint, but, in fact, despite their name, radiators transfer most of their heat to a room by convection, not by thermal radiation.

11.1.4 Calorimetry

Calorimetry is the science of measuring heat and is derived from the Latin *calor* meaning heat and the Greek *metry* meaning to measure. All calorimetric techniques are based on the measurement of heat that may be generated (exothermic process), consumed (endothermic process), or simply dissipated by the sample. There are many methods to measure such heat and since its advent in the late eighteenth century, a wide range of techniques have been developed.

A calorimeter is a device that is used to measure the heat of reaction and to measure the heats of reaction they are enclosed in a calorimeter, initiate the reaction and then measure the temperature difference before and after the reaction. The temperature difference enables the heat released in the reaction to be made.

The method for determining the calorific value will depend upon the type of fuel or substance being tested. For gaseous fuels, these are tested in a "Boyes" calorimeter under a constant pressure and under steady flow conditions. Solid and liquid fuels are tested under constant volume in a bomb calorimeter.

Most mechanical engineering students particularly those studying thermodynamics will be familiar with the bomb calorimeter and this will now be described in a little more detail.

The bomb is a small stainless-steel vessel measuring 100 mm diameter by 150 mm long in which a small mass of the fuel is held in a crucible within the vessel (see Figure 11.1.4). If the fuel is solid, it will be usually crushed, passed through a sieve, and pressed into the form of pellets. A piece of platinum or nichrome wire is contained within the pellet and forms part of the ignition circuit. For liquid fuels, a known mass of the fuel is placed in the crucible which is also in contact with the ignition wire. The bomb is then filled with oxygen to aid ignition and burning of the sample.

The bomb is then immersed in a sealed insulated tank called a Dewar filled with water that is continuously stirred before, during and after the test. The Dewar is also fitted with a sensitive thermometer for accurate measurement of the water temperature during the test. Figure 11.1.5 shows the results of a typical test to determine the higher calorific value of a sample of fuel oil to be used in an oil engine trial. Table 11.1.5 gives details of the masses used in the test.

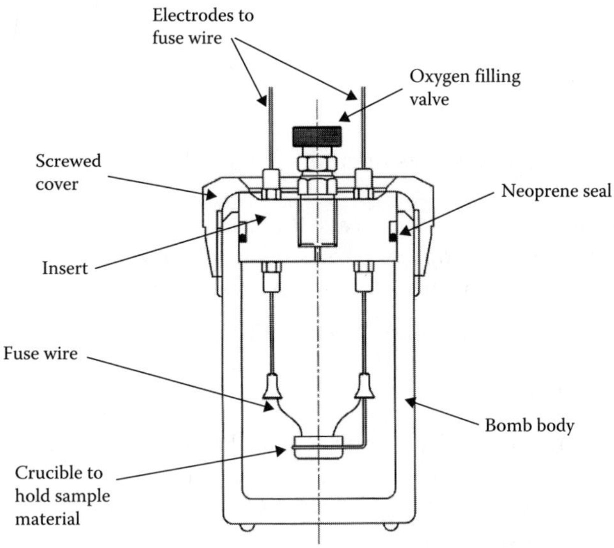

FIGURE 11.1.4
Bomb calorimeter.

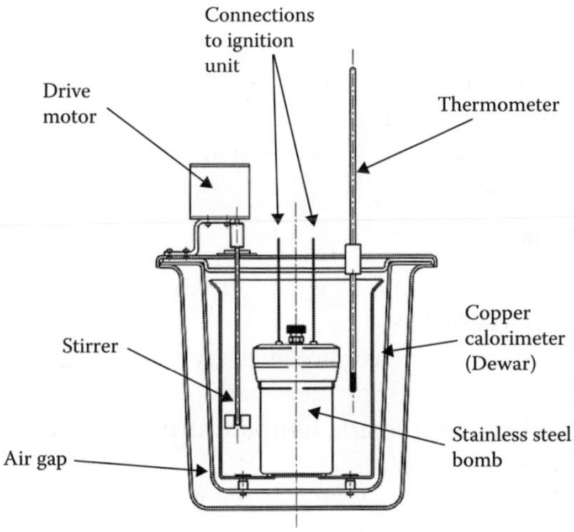

FIGURE 11.1.5
Calorimeter.

TABLE 11.1.5
Individual Masses of Items Used in the Calorimeter

Description	Mass (g)
Water equivalent of the apparatus	485
Mass of water in the inner calorimeter	2500
Mass of crucible	4.1
Mass of crucible and fuel oil	5.1
Mass of fuel oil fired	1.0

11.2 Light

11.2.1 Fundamentals

Light waves, radio waves, and X-rays (including gamma rays) are all types of electromagnetic waves.

Light is a transverse, electromagnetic wave that can be seen by humans. The wave nature of light was first illustrated through experiments on diffraction and interference. Like all electromagnetic waves, light can travel through a vacuum. The transverse nature of light can be demonstrated through polarization.

11.2.2 Production of Light

Light is produced by one of the two methods as follows:

Incandescence is the emission of light from 'hot' matter ($T \gtrsim 800$ K).

Luminescence is the emission of light when excited electrons fall to lower energy levels.

11.2.3 Velocity

The velocity of light in a vacuum is a universal constant and in a vacuum is fixed at 299,792,458 m/s. The velocity of all electromagnetic waves is the same in a vacuum.

The velocity of light in a medium is always slower than the speed of light in a vacuum and is dependent upon the medium through which it travels.

11.2.4 Other Characteristics

The amplitude of a light wave is related to its intensity, which is the absolute measure of a light wave's power density (W/m^2).

Brightness is the relative intensity as perceived by the average human eye and is called its *luminous intensity* (I). The unit of brightness is the *candela* (cd). The candela is defined as the light emitted by a black at the freezing temperature of platinum (1773°C). The intensity of a light source is called its *candlepower*.

The amount of light that falls on a given surface is called luminous flux (F). The lumen (lm) is the unit of luminous flux. One lumen is equal to the luminous flux that falls on each m^2 of a 1 metre radius sphere when a 1cd isotropic light source is at the center of the sphere.

The frequency of a light wave is related to its color. Monochromatic light is described by only one frequency and laser light is effectively monochromatic.

There are six simple, named colors in English (and many other languages) associated with a band of monochromatic light. They are shown in Table 11.2.1 in order of increasing frequency.

Light is sometimes also known as visible light to differentiate it from ultraviolet light and infrared light, neither of which are visible to the unaided human eye.

Other forms of electromagnetic radiation that are not visible to humans are sometimes also referred to as light.

Figure 11.2.1 shows a section of the electromagnetic spectrum and illustrates where the light spectrum is in relation to the other areas of radiation, including the radio and TV spectrum.

Polychromatic light is described by many different frequencies and nearly every light source is polychromatic. White light is polychromatic as it is a mixture of the principle colors.

TABLE 11.2.1

Wavelength and Frequencies of Visible Light

Color	Wavelength (nm)	Frequency (THz)
Violet	380–450	668–789
Blue	450–495	606–668
Green	495–570	526–606
Yellow	570–590	508–526
Orange	590–620	484–508
Red	620–750	400–484

A graph of relative intensity versus frequency is called a *spectrum*, the plural is spectra. Although this term is frequently associated with light, the term can be applied to any wave phenomena, including vibration. A *continuous* spectrum is one in which every frequency is present within some range.

Blackbody radiators emit a continuous spectrum.

A *discrete* spectrum is one in which only a well-defined set of isolated frequencies is present. (A discrete spectrum is a finite collection of monochromatic light waves.)

The excited electrons in a gas emit a discrete spectrum.

The wavelength of a light wave is inversely proportional to its frequency and is often described by its wavelength in a vacuum. Light ranges in wavelength from 400 nm on the violet end to 700 nm on the red end of the visible spectrum. Phase differences between light waves can produce visible interference effects such as Figure 11.1.1 Newton's rings.

11.2.5 Polarization

Light is an electromagnetic wave that vibrates at very high speed back and forth as it moves. Individual light waves have their own wavelength as well as angle of vibration, which can range through a full 360°. The angle of vibration of a light wave is defined by its electric and magnetic components (vectors) that are perpendicular to each other and to the direction of the wave propagation. Ordinary visible light is unpolarized, that is, it is

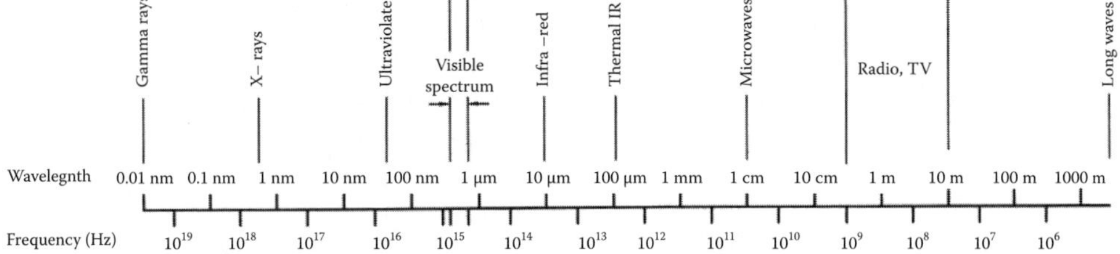

FIGURE 11.2.1
Electromagnetic spectrum.

composed of many different waves vibrating in all directions. But light waves that vibrate the same way (for example, horizontally) are called *polarized*.

11.2.6 Classification of Polarization

There are three classifications of polarized light—namely

- Linear polarization
- Circular polarization
- Elliptical polarization

Polarized light is used extensively in the study of stresses and strains in mechanical components using models constructed from birefringent materials that shows the existence of high stress areas.

11.2.6.1 Linear Polarization

When the electric fields of light waves constituting a light beam oscillate in one plane, it results in *linearly polarized* light. A plane electromagnetic wave is said to be linearly polarized. The transverse electric field wave has an accompanying magnetic wave as illustrated in Figure 11.2.2a.

11.2.6.2 Circular Polarization

When the plane of oscillation rotates around the propagation axis, light is defined as *circularly polarized*. Circularly polarized light consists of two perpendicular electromagnetic waves having an equal amplitude with a phase difference of 90°. Figure 11.2.2b illustrates right circularly polarized light.

11.2.6.3 Elliptical Polarization

Elliptically polarized light consists of two perpendicular waves having unequal amplitudes differing in phase by 90°. Figure 11.2.2c shows right elliptically polarized light.

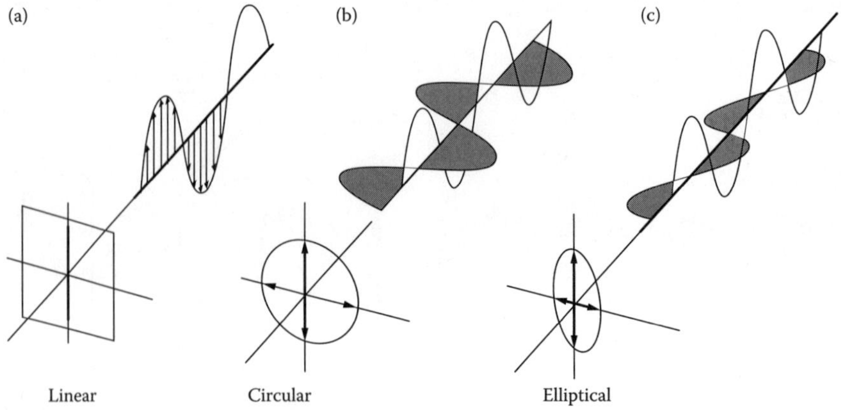

FIGURE 11.2.2
(a) Linear, (b) circular and (c) elliptical polarised light.

11.2.7 Optics

Optics is the study of the nature and behavior of light and can be divided into subdisciplines based on the type of model used to describe light:

In physical optics, light is assumed to behave like a classical wave.

In quantum optics, light is assumed to have both wave and particle properties.

In geometric optics, light is assumed to travel in a definite direction with relatively little diffraction. This behavior is known as *rectilinear propagation*.

Particles of light are called *photons*.

11.2.7.1 Propagation

The path of propagation of a light wave is a geometric ray. The principle of reversibility states that light will follow exactly the same path if its direction of travel is reversed.

11.2.7.2 Rays

Rays are

Straight lines in empty Euclidean space

Geodesics in general relativistic space–time

The eye can see something only if a ray from the object reaches the eye.

A ray is the path of least action connecting two points in space and is also:

The path of least time (the quickest path)

The path of least distance (the shortest path)

Unique and therefore reversible

The rays of geometric optics:

Are perpendicular to the wave fronts of physical optics

Indicate the most probable path of the photons of quantum optics

11.2.7.3 *Interface*

An interface is the boundary between:

Two different media

Two regions of a medium with different characteristics such as:

- Density (which is often related to temperature)
- Concentration of solute (salinity, for example)
- Mechanical stress

When an incident ray meets an interface, it will be partially:

Reflected

Reflected rays obey the law of reflection

Transmitted

Transmitted rays obey Snell's law

Absorbed

Absorbed rays obey the law of conservation of energy (the energy of the ray is not destroyed, but changes form).

11.2.7.4 Angles

Angles in geometric optics are measured with respect to a line normal to the interface:

The angle of incidence is the angle between the incident ray and the normal.

The angle of reflection is the angle between the reflected ray and the normal.

The angle of refraction is the angle between the transmitted ray and the normal.

11.2.7.5 Reflection

The law of reflection states that the angle of incidence equals the angle of reflection.

The law of reflection can be derived from the principle of least action.

11.2.8 Radiometry and Photometry

There are two methods used for measuring light:

Radiometry

Photometry

11.2.8.1 Radiometry

Radiometry is the science of measuring optical radiation within the frequency range between 3×10^{11} and 3×10^{16} Hz. This range corresponds to wavelengths between 0.01 and 1000 micrometers and covers the spectra called ultraviolet, visible, and infrared.

Table 11.2.2 shows the units of radiometry.

TABLE 11.2.2

Radiometric Units

Quantity	Radiometric
Power	Watt (W)
Power per unit area	W/m^2
Power per unit solid angle	W/sr
Power per area per solid angle	W/mm^2-sr

TABLE 11.2.3
Photometric Units

Quantity	Radiometric
Power	Watt (W)
Power per unit area	W/m^2
Power per unit solid angle	W/sr
Power per area per solid angle	W/mm^2-sr

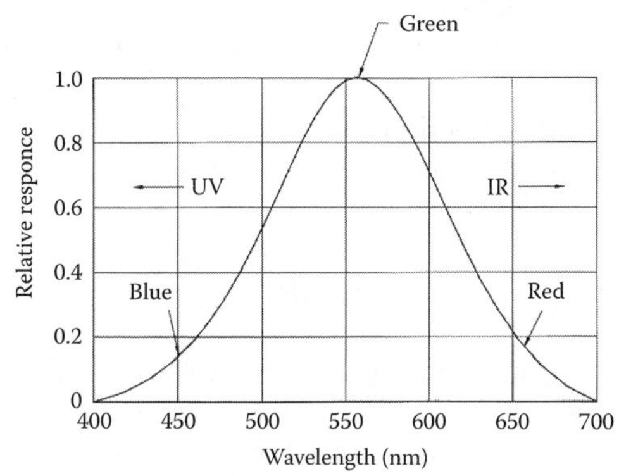

FIGURE 11.2.3
Photopic curve.

11.2.8.2 Photometry

Photometry is the measurement of light that is defined as electromagnetic radiation that is detectable with the human eye. It is restricted to the range of wavelengths covering 360–830 nm. Photometry is like radiometry, except that everything is weighted by the spectral response of the eye.

Visual photometry uses the eye as a comparison a detector, whereas physical photometry uses optical detectors that are constructed to imitate the spectral response of the eye.

Table 11.2.3 shows the units of photometry.

Figure 11.2.3 shows a curve that indicates the average response of the eye to different wavelengths of light under carefully controlled conditions. The curve is called the *photopic curve.*

The only real difference between radiometry and photometry is that the former covers the entire optical radiation spectrum and the latter is confined to the visible spectrum, as defined by the response of the eye.

11.2.9 Holography

In recent years, holography has become an important research tool, particularly in the study of vibration. One specific area that has met with considerable success is the dynamics of gas turbine blades, which have been investigated using the holographic technique to help reduce blade failure rates resulting from resonance frequencies. Reducing the torsional vibrations of the blades has also helped to substantially improve the thermal efficiency of the engines, particularly in the compression section of the engine, correctly directing the air flow through to the subsequent compression stages.

Holography is a technique that enables a light field, which is generally the product of a light source scattered off objects to be recorded and later reconstructed when the original light field is no longer present due to the absence of the original objects. Holography can be thought of as somewhat similar to sound recording, whereby a sound field created by vibrating matter like musical instruments or vocal cords is encoded in such a way that it can be reproduced later, without the presence of the original vibrating matter.

The holographic recording itself is not an image; it consists of an apparently random structure of varying intensity, density, or profile.

Holography enables three-dimensional images to be made. It involves the use of a laser, interference, diffraction, light intensity recording, and suitable illumination of the recording. The image changes as the position and orientation of the viewing system changes in exactly the same way as if the object were still present, thus making the image appear three-dimensional.

11.2.9.1 Principle of Producing a Holographic Image

Figure 11.2.4 shows an arrangement for recording a hologram. This requires the laser beam to be aimed through a series of elements. The first element is a beam splitter that divides the beam into two identical beams, each aimed in two different directions:

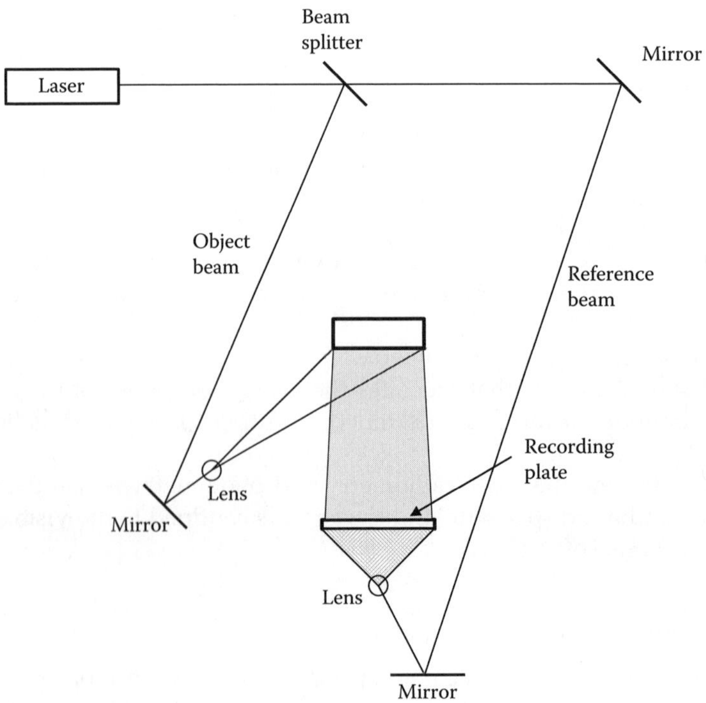

FIGURE 11.2.4
Basic layout for creating a hologram.

The first beam (known as the *illumination* or *object beam*) is spread using lenses and is directed onto the scene using a mirror. This reflected light then falls onto the recording medium.

The second beam passes through the beam splitter and is then reflected to an additional mirror and spread using a further lens, but is directed so that it does not come in contact with the scene, and instead travels directly onto the recording medium and is referred to as the reference beam.

Several different materials can be used as the recording medium. One of the most common is a film similar to photographic film (silver halide photographic emulsion), but with a much higher concentration of light-reactive grains, making it capable of the much higher resolution that holograms require. A layer of this recording medium (e.g., silver halide) is attached to a transparent substrate, which is commonly glass, but may also be plastic.

When the two laser beams reach the recording medium, their light waves intersect and interfere with each other. It is this interference pattern that is imprinted on the recording medium. The pattern itself is seemingly random, as it represents the way in which the scene's light interfered with the original light source—but not the original light source itself. The interference pattern can be considered an encoded version of the scene requiring a particular key—the original light source—in order to view its contents.

This missing key is provided later by shining a laser identical to the one used to record the hologram onto the developed film. When this beam illuminates the hologram, it is diffracted by the hologram's surface pattern. This produces a light field identical to the one originally produced by the scene and scattered onto the hologram. The image this effect produces in a person's retina is known as a *virtual image*.

Holograms can also be produced for viewing in ordinary light and not requiring a laser light for viewing.

11.3 Sound

11.3.1 Sound

Sound waves are produced by a vibrating body. The vibrating object moves in one direction and compresses the air directly in front of it. As the vibrating object moves in the opposite direction, the pressure on the air is reduced so that an expansion or rarefaction, of air molecules occurs. One compression and one rarefaction make up one *longitudinal wave*. The vibrating air molecules move back and forth parallel to the direction of motion of the wave, receiving energy from adjacent molecules nearer the source and passing the energy to adjacent molecules farther away from the source. As the wave progresses there is a gradual reduction in the pressure. Sound is the reception of such waves at the ears and their perception by the brain.

11.3.2 The Decibel

Musicians will be familiar with the fact that when the frequency of a note is doubled, its pitch will raise an octave. An octave is a series of eight notes occupying the interval between (and including) two notes, one having twice or half the initial frequency. The increase in pitch will sound the same to the human ear if the frequency is doubled from 100 to 200 Hz as from 500 to 1000 Hz. Similarly, increase in the loudness depending upon the ratio of the intensities and not on the absolute differences in intensity of the sound level.

If the power of a source of sound is increased from say 0.1 to 0.2 Watt and similarly from 0.2 to 0.4 W, the loudness of the source to the ear increases in equal steps. This equality is dependent upon the equality of the ratio of the powers and not on their difference. In commercial practice, the increase in loudness is calculated by taking the logarithm of the ratio of the powers to the base 10, which is $\log_{10}2$, or in this case 0.3.

The relative intensities or powers are expressed in *bels*, named after Graham Bell (the inventor of the telephone). If the power of a source of sound changes from P_1 to P_2

$$\text{Number of bels} = \log_{10}\left(\frac{P_2}{P_1}\right) \tag{11.3.1}$$

The bel is far too large a unit of measurement to be practical and the *decibel (dB)* has been adopted. A decibel is one-tenth of a bel and in the above case,

$$\text{Number of decibels} = 10\log_{10}\left(\frac{P_2}{P_1}\right) \tag{11.3.2}$$

The minimum change in power that the ear can detect is approximately 1 dB. This corresponds to an increase in power of approximately 25%.

Example 11.3.1

A loudspeaker of an amplifier is producing 50 milliwatts of power, the volume control is increased to produce 500 mW of power.

The increase in power is given by:

$$\begin{aligned}\text{Number of decibels} &= 10\log_{10}\left(\frac{P_2}{P_1}\right)\\ &= 10\times\log\left(\frac{500}{50}\right)\\ &= 10\text{ dB}\end{aligned}$$

If the volume control is now increased until the power is increased to 1000 mW, then the increase in power compared to the original value is

$$\begin{aligned}&= 10\times\log_{10}\left(\frac{1000}{50}\right)\\ &= 10\times\log_{10}20\\ &= 13\text{ dB}\end{aligned}$$

When the volume control is turned down such that the power is decreased from 1000 to 200 mW, the change in power is now

$$\begin{aligned}&= 10\times\log_{10}\left(\frac{200}{1000}\right)\\ &= 10\log_{10}2 - 10\log_{10}10\\ &= -7\text{ dB}\end{aligned}$$

The negative sign indicates there is a decrease in power.

Besides its use in acoustics, the decibel is used extensively in electronics and control systems.

11.3.3 Intensity and Pitch

Pitch in sound is analogous to color in light; the pitch of a sound depends upon the frequency of the tone that the ear receives. High notes are produced by an object that is vibrating at a higher frequency than for a low note, which only vibrates at a lower frequency.

The *intensity* of a sound is the amount of energy crossing a unit area in unit time or the power flowing through the unit area. The SI unit is watts per meter^2. The loudness of the sound depends upon the subjective effect of intensity of sound waves on the human ear. In general, a more intense sound is also louder, but the ear does not respond linearly to all frequencies, so two tones of the same intensity but with different pitches may appear to have different loudness.

The intensity of the threshold of hearing (I_o), which is the intensity that can be barely heard by a normal person is about 0–12 (W/m^2) when measured by acoustical devices. The relation between loudness and intensity is nearly logarithmic. The intensity level of sound is measured in decibels and is given by the equation $\beta = 10 \log I/I_o$, where 'β' is the recognized symbol for intensity in decibels, 'I' is the sound intensity, and 'I_o' is the intensity of the threshold of hearing. For example, normal conversation is about 60 dB, and a power saw is about 110 dB.

11.3.4 Resonance

The tuning fork is a useful instrument for investigating sound because it vibrates at only one frequency in contrast to most musical instruments that produce several different frequencies simultaneously. A struck tuning fork vibrates at a natural frequency dictated by the fork's characteristics—the dimensions and the material from which it is made. If the stem of a vibrating tuning fork is set on a table top, the tone becomes louder because the fork forces the table top to vibrate. Because the table top has a larger vibrating area, the sound is more intense. This principle of forced vibrations is applied in most musical instruments by using a part of the instrument, such as the piano sounding board to intensify the sound.

Consider two matched tuning forks with the same frequency mounted on separate sounding boxes. As shown in Figure 11.3.1, the vibrating air column set up by one tuning fork will cause the other tuning fork to vibrate weakly. This action is called *resonance* or *sympathetic vibration*. Resonance occurs when the natural vibration rates of two objects are the same or when one has a natural vibration rate that is a multiple of the other.

11.3.4.1 Beats

When two sound waves having a slight difference in frequencies merge a slight variation in the volume of the sound will be perceived, giving rise to a *beat frequency*. Examples such as two tuning forks that are a few frequencies apart and a twin engine aircraft flying overhead will have a similar phenomenon, creating a fluctuating droning sound. The beat frequency occurs when the two frequencies are added together and the resultant amplitude will vary or oscillate at a rate that will be the difference between the frequencies. The beat frequency will create a beat envelope around the original sine wave as shown in Figure 11.3.2.

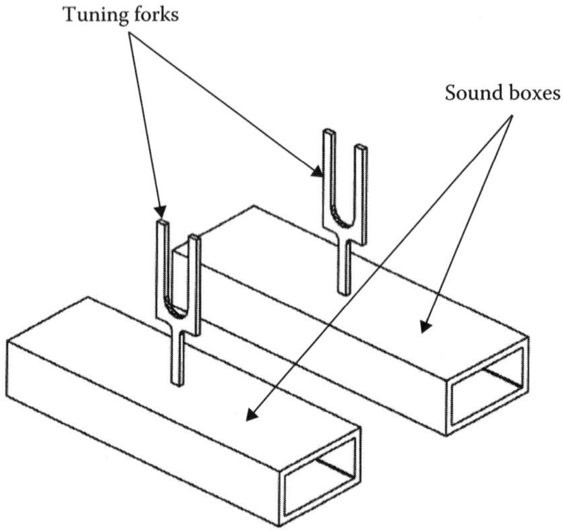

FIGURE 11.3.1
Demonstration of resonance.

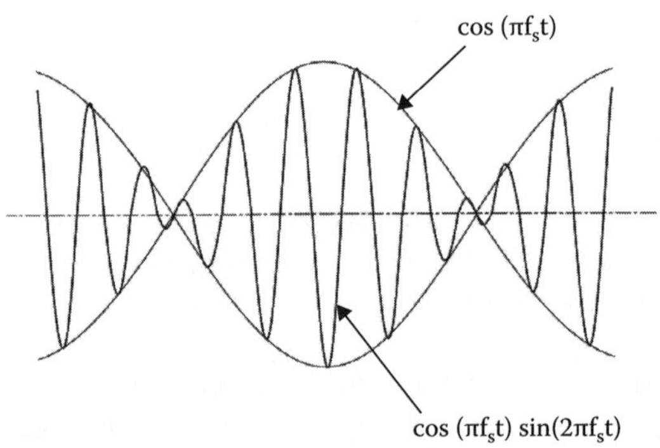

FIGURE 11.3.2
Beat frequency.

If the frequencies of the two sounds are close to each other, a sound that is the average of the two will be heard, but the modulation of the amplitude will also be heard as a beat frequency. This is the difference between the two initial frequencies:

$$f_b = |f_1 - f_2|$$

where
fb is the beat frequency
f_1 and f_2 are the two sound frequencies
$|f_1 - f_2|$ is the absolute value or positive (+) value of the difference

A practical application of using the beat frequency is piano tuning. When tuning an instrument accurately, the tuner listens carefully to the sound and adjusts the instrument to eliminate beats between the instrument and a given pitch.

TABLE 11.3.1

Decibel (Loudness) Comparison Chart

Common Sound	Sound Level (dB)	Effect
Rocket launch pad	180	Short-term exposure—will cause permanent hearing loss
Jet engine at 100 ft	140	Painfully loud
Pneumatic riveter at 4 ft	125	Extremely loud
Rock concert	120	Extremely loud
City traffic	90	Very loud
Hair dryer	80	Annoying
Business office	70	Intrusive
Air conditioning unit	60	Noticeable
Conversational speech	50	Quiet
Bedroom	40	Moderately quiet
Library	30	Very quiet
Soft whisper at 15 ft	20	Audible
Broadcast studio	20	Just audible
Weakest sound heard	0	Hearing begins

11.3.5 Comparison of Common Sounds

Table 11.3.1 lists the volume levels of various common (and not so common) sounds and their effect on hearing.

11.4 Atomic Physics

11.4.1 Atomic Theory

Rutherford suggested that the atom has the same construction as the solar system. It consists of a nucleus, which carries the greater part of the mass, with electrons clustering around in circular orbits. In order to avoid the difficulty that revolving electrons should continuously give off energy in the form of electromagnetic radiation, Danish physicist Niels Bohr suggested that only certain orbits were possible, and that when in one of these orbits no radiation takes place and if an electron changed orbit, then a small amount of energy is released, detectable by a spectrograph.

The *nucleus* was considered to be made up of *protons* of unit mass and carrying a unit positive charge. The *electron* was thought to have a negative mass but carried a comparative unit negative charge. To explain the difference between observed molecular weight and theoretical molecular weight, it is necessary to postulate the existence within the nucleus of particles other than protons. These were called *neutrons*, which have a unit mass but no charge.

If the atomic number is called 'z,' then the number of protons and orbiting electrons must all equal 'z.' The number of neutrons must equal 'A' – 'z' where 'A' is the atomic weight.

The neutrons can be used to explain the discrepancy that arises with fractional atom weights. This is achieved by the use of isotopes.

An isotope is a form of a chemical element whose atomic nucleus contains a specific number of neutrons, in addition to the number of protons that uniquely defines the element. All isotopes of a given element have the same number of protons in each atom.

Take for example chlorine (Cl). Naturally occurring chlorine is a mixture of the isotopes Ci-37 and Cl-35.

In Cl-37 there are 17 protons and 20 neutrons and Cl-35 there are 17 protons and 18 neutrons.

The number of protons is the atomic number (the atomic number of Cl is 17). The number of neutrons in Cl-35 is: 35 – 17 = 18 and in Cl-37 is 37 – 17 = 20.

The isotopes have two different mass numbers (Cl-35 has a mass number of 35 and Cl-37 has a mass number of 37). The number of protons in each is the same (the atomic number is still 17) but the number of neutrons is different.

Protons	17	17
Neutrons	20	18
Atomic weight	37	35

In the later developments of the atomic theory, it was realized that the classical laws of particle dynamics could not be applied to these fundamental particles.

For example, in classical dynamics it was necessary to specify the position and momentum of a particle, but in atomic dynamics if either the position or the momentum is known, the other products cannot be determined. This is *Heisenberg's uncertainty principle.* The consequence of Heisenberg is that there is no longer reason to think of the electron revolving in a fixed orbit, but rather as a wave function. This function only specifies the probability of a particular electron having a particular position in the space surrounding the nucleus; consequently, instead of specifying the motion of an electron by a single number or integer 'n,' it is now necessary to use four quantum numbers to specify the electron state:

n	Principle quantum
l	Azimuthal (orbital)
m	Magnetic quantum (inner)
s	Spin

The energies of electrons having the same values of 'n' and 'l' but different values of 'm' are the same, provided there is no magnetic field present (Zeeman effect, splitting of the spectral lines).

The spin quantum number 's' for the electron having the same value of 'n,' 'l,' and 'm' can have one of the two values +1/2 and –1/2 (spins in opposite directions); again, only in a magnetic field will the energies of the two electrons with opposite spin be different:

For example:

n = 1	n = 2	n = 3
l = 0	l = 0, 1.	l = 0, 1, 2.
m = 0	m = +1, 0, –1.	
s = ±1/2	s ± 1/2	

11.4.2 Nomenclature of the Electronic States

Since the values of the energies are determined by 'n' and 'l,' it is only necessary to consider these in the nomenclature. The principal quantum number describes the size of the orbital of an electron. It is denoted by n. The higher the value of n, the larger is the size of orbital. Value of n can never be zero, but the azimuthal quantum number describes the shape and type of orbital. It is denoted by *t* and its value ranges from 0 to n–1. Azimuthal quantum number is of 4 types: s, p, d and f.

For example, when the principle quantum number is 1, 'l' must be equal to zero and an electron in that state would be designated 1s; such a state can only have a single value of the inner quantum number, for example, 'm' = 0 but can have the values of 's' ±1/2.

Another basic principle of modern quantum theory is that no two electrons can have the same four quantum numbers (the Pauli exclusion principle) on the same atom.

To understand the periodic table in terms of electron states, it is necessary to consider the significance of these four quantum numbers. The most important number is 'n,' since this is mostly responsible for determining the energy of the electron. It can have integral values beginning with 1, which is the lowest energy state and therefore the most stable. The electrons having a principle quantum number 'n' can take up integral values of the orbital quantum number 'l' between 0 and n – 1. The *orbital quantum number* is associated with the angular momentum of the electron (shape of orbit):

- For a given value of 'n,' the electron having the lowest value of 'l' will have the lowest energy level.
- The orbital quantum number is associated with the angular momentum of the electron (shape of orbit). For a given value of 'n,' the electron having the lowest value of 'l' will have the lowest energy level.

The two quantum numbers 'm' and 's' are concerned with the orientation of the electron orbit around the nucleus and the direction of the spin of the electron, respectively.

- For a given value of 'l,' the electron may have an integral value of 'm' from '+l' through zero to '–l.'
- Two electrons in any one atom can be in 1s state once the 1s states have been filled. The next lowest energy state must be 'n' = 2, therefore electrons can exist in either 2s or 2p state, since the energy of the 2s state is lower than the 2p state, the 2s will be filled first. Again, there can only be two electrons in the 2s state.
- The electrons in 'p' state can have the values of 'm' +1, 0, and –1, and the electrons with these values of 'm' can have s values of +1/2 or –1/2, for example, six electrons in any 2p state, for 'n' = 3 this gives 2, 3s states; 6, 3p states, and 10, 3d states; similarly, when 'n' = 4: 2, 4s states; 6, 4p states, 10, 4d states, and 14, 4f states.

11.4.3 Periodic Table

The periodic table orders the information about chemical elements. All the elements are arranged in order of increasing atomic number. Figure 11.4.1 depicts the periodic table. The rows are called "periods" and the columns are called "groups".

Elements within the same group have similar chemical properties. The table covers both metals and non-metals. The main groups are numbered 1 to 7 reading from left to right,

1 **H** 1.008																	2 **He** 4.00
3 **Li** 6.94	4 **Be** 9.01											5 **B** 10.81	6 **C** 12.01	7 **N** 14.01	8 **O** 16.00	9 **F** 19.00	10 **Ne** 20.18
11 **Na** 22.99	12 **Mg** 24.31											13 **Al** 26.98	14 **Si** 28.09	15 **P** 30.97	16 **S** 32.07	17 **Cl** 35.45	18 **Ar** 39.95
19 **K** 39.20	20 **Ca** 40.08	21 **Sc** 44.96	22 **Ti** 47.88	23 **V** 50.94	24 **Cr** 52.00	25 **Mn** 54.94	26 **Fe** 55.85	27 **Co** 58.93	28 **Ni** 58.69	29 **Cu** 63.55	30 **Zn** 65.39	31 **Ca** 69.72	32 **Ge** 72.61	33 **As** 74.92	34 **Se** 78.96	35 **Br** 79.90	36 **Kr** 83.80
37 **Rb** 85.47	38 **Sr** 87.62	39 **Y** 88.91	40 **Zr** 91.22	41 **Nb** 92.91	42 **Mo** 95.94	43 **Tc** (98)	44 **Ru** 101.0	45 **Rh** 102.9	46 **Pd** 106.4	47 **Ag** 107.8	48 **Cd** 112.4	49 **In** 114.8	50 **Sn** 118.7	51 **Sb** 121.7	52 **Te** 127.6	53 **I** 126.9	54 **Xe** 131.2
55 **Cs** 132.9	56 **Ba** 137.3	57 **La** 138.9	72 **Hf** 178.5	73 **Ta** 180.1	74 **W** 183.9	75 **Re** 186.2	76 **Os** 190.2	77 **Ir** 192.2	78 **Pt** 195.1	79 **Au** 197.0	80 **Hg** 200.6	81 **Tl** 204.4	82 **Pb** 207.2	83 **Bi** 209.0	84 **Po** (209)	85 **At** (210)	86 **Rn** (222)
87 **Fr** 223.0	88 **Ra** 226.0	89 **Ac** 227.0	104 **Rf** (261)	105 **Db** (262)	106 **Sg** (263)	107 **Bh** (262)	108 **Hs** (265)	109 **Mt** (266)	110 **Ds** (281)	111 **Rg** (272)	112 **Uub** (285)	113 **Uut** (284)	114 **Uuq** (289)	115 **Uup** (288)	116 **Uuh** (292)		

58 **Ce** 140.1	59 **Pr** 141.0	60 **Nd** 144.2	61 **Pm** (145)	62 **Sm** 150.4	63 **Eu** 153.0	64 **Gd** 157.3	65 **Tb** 158.9	66 **Dy** 162.5	67 **Ho** 164.9	68 **Er** 167.3	69 **Tm** 168.9	70 **Yb** 173.0	71 **Lu** 175.0
90 **Th** 232.4	91 **Pa** 231.4	92 **U** 238.0	93 **Np** (237)	94 **Pu** (240)	95 **Am** (243)	96 **Cm** (247)	97 **Bk** (248)	98 **Cf** (251)	99 **Es** (252)	100 **Fm** (257)	101 **Md** (257)	102 **No** (259)	103 **Lr** (262)

FIGURE 11.4.1
Periodic table of the elements.

and the last group on the far right is group 0. The transitional metals are elements that are found between groups 2 and 3.

Well known examples include iron, copper and gold. These elements are generally quite dense and many form brightly coloured compounds.

Once the set of states corresponding to a given principle quantum number is filled and the electrons in these states are said to fill a closed shell, it is a consequence of quantum mechanics that once the shell is filled, the energy of that shell falls to a low value and the electrons are in a very stable state.

The electrons surrounding an atom are located in regions around the nucleus called "energy levels".

An energy level represents the 3-dimensional space surrounding the nucleus where electrons are most likely to be.

The first energy level is closest to the nucleus.

The second energy level is a little farther away than the first. The third is a little farther away than the second, and so on.

Each energy level can accommodate or "hold" a different number of electrons before additional electrons begin to go into the next level.

When the first energy level has two electrons, the next electrons go into the second energy level until the second level has eight electrons.

When the second energy level has eight electrons, the next electrons go into the third energy level until the third level has eight electrons.

When the third energy level has eight electrons, the next two electrons go into the fourth energy level. The electrons in the energy level farthest from the nucleus are called valence electrons.

Atoms in the same column (group) in the periodic table have the same number of valence electrons.

11.4.4 Van der Waals Forces

Relatively weak electric forces that attract neutral molecules to one another in gases, in liquefied and solidified gases, and in almost all organic liquids and solids. The forces are named for the Dutch physicist Johannes van der Waals, who in 1873 first postulated these intermolecular forces in developing a theory to account for the properties of real gases. Solids that are held together by van der Waals forces characteristically have lower melting points and are softer than those held together by the stronger ionic, covalent, and metallic bonds.

Van der Waals forces may arise from three sources as follows:

- First, the molecules of some materials, although electrically neutral, may be permanent electric dipoles. Because of fixed distortion in the distribution of electric charge in the very structure of some molecules, one side of a molecule is always positive, and the opposite side moderately negative. The tendency of such permanent dipoles to align with each other results in a net attractive force.
- Second, in the presence of molecules that are permanent dipoles, this temporarily distorts the electron charge in other nearby polar or nonpolar molecules, hence inducing further polarization. An additional attractive force will result from the interaction of a permanent dipole with a neighboring induced dipole.
- Third, even though no molecules of a material are permanent dipoles (e.g., in the noble gas argon or the organic liquid benzene), a force of attraction exists between the molecules, accounting for condensing to the liquid state at sufficiently low temperatures.

11.4.5 Metallic Bonding

In metals, the electrostatic structure of the atom is such that the gain, loss, or sharing of electrons as described previously does not take place, and thus, the valence atoms are easily detached so that they form with other atoms an electron cloud. A bond then develops between the cloud of negatively charged electrons and positively charged ions.

11.4.6 Structure of Metals

Metals in their solid form are crystalline, that is, their atoms (ions) are arranged so as to produce a regular pattern. In metals, there are three common arrangements as follows.

11.4.6.1 Face-Centered Cubic Structure

In a face-centered cubic (FCC) structure, the number of atoms per unit cell is four; that is there are eight corner atoms shared by eight unit cells giving the equivalent of one atom per unit cell. There are six face atoms each shared by two unit cells giving the equivalent of three atoms per unit cell, that is a total of four atoms. If it is assumed the atoms to be solid spheres, then it is possible to derive a relationship between 'r' (the atomic radius) and 'a' (the lattice parameter) which is the length of the unit cell.

An example of face-centered cubic metals include: aluminum, copper, lead, gold, and silver.

11.4.6.2 Body-Centered Cubic Structure

In the body-centered cubic (BCC) arrangement, there is a single atom in the center of the unit cell. The number of atoms per unit cell is two.

Examples of BCC include tungsten, chromium, vanadium, and iron (at room temperature).

11.4.6.3 Hexagonal Close-Packed Structure

In the hexagonal close-packed (HCP) structure, there are six atoms per unit cell. This structure is denser than the two previous structures. It is characterized by the fact that each atom in one layer is located directly above or below the interstices formed by the three atoms in the adjacent layers. Examples of metals with this type of crystal structure include: zinc, magnesium, cobalt, and cadmium. These metals tend to be more brittle than metals with FCC and BCC structures.

11.4.7 Atomic Weight and Atomic Number

It is possible to calculate densities provided that the structure is known together with the atomic weight and the lattice parameters. The weight of an element is called the *atomic weight* and ranges from 1.008 for hydrogen (which has only one proton) to nearly 250 for some of the more unstable transuranium elements. Atomic weights are expressed in grams per gram-atom. One gram-atom always contains 6.02×10^{23} atoms and is known as Avogadro's number).

Thus,

$$\text{Weight of atom} = \frac{\text{Atomic weight}}{6.02 \times 10^{23}} \tag{11.4.1}$$

11.4.8 Atomic Coordination

The coordination number in any lattice system is the number of nearest neighbors surrounding any atom. For BCC metals, this will correspond to eight; in FCC and HCP metals, the coordination number will be 12.

11.4.9 Stacking Sequences

The FCC and HCP structures are equally dense and made by the stacking of close-packed layers, the difference being one of stacking sequence.

In FCC metals, the packing arrangement is in layers ABC, ABC, ABC, etc., whereas in HCP metals, the packing sequence is AB, AB, AB, etc.

In the case of HCP metals, it is not possible to completely define the lattice by a single parameter, therefore it is necessary to state HCP using two lattice parameters, 'a' and 'c'. Ideally, the ratio 'c/a' assumes close-packed and equal spheres = 1.633. In practice, the 'c/a' ratio might be greater or less than the ideal.

11.4.10 Crystal Planes

Crystal planes play an important part in deformation processes. Consequently, it is vital that they are understood and identified. In any metallic lattice, the number of atomic planes is unlimited. Planes where the atomic density is high are termed *crystallographic planes*, and often the description of a crystal is defined in terms of these planes.

The study of metallic crystals is called *crystallographic analysis* and is an important tool for stress analysis, as it helps analyze the deformation of the crystal due to an applied stress and identify any residual stress locked in the metallic structure.

11.4.11 Indices Planes

The various planes and directions within a crystal are specified in terms of the *Miller index notation.* A plane is defined by the length of its intercepts on the three (or four, with HCP) crystal axes (the three edges of the unit cell) measured from the origin of the coordinates; the intercepts are expressed in terms of the unit cell, which are the unit distances along the three axes. The reciprocals of these three intercepts reduced to the three smallest integers having the same notation are known as the *Miller indices.*

There are certain conventions used when expressing crystallographic directions using the Miller system. Directions are enclosed in square brackets [ ].

Negative directions are represented by putting a bar over the appropriate integer.

$$\text{For example, } \left[\bar{1}\bar{1}\bar{1}\right] \text{ is the opposite of } \left[111\right].$$

The choice of negative directions is arbitrary, but it is essential to maintain consistency of use.

11.4.12 Miller–Bravais System for HCP

In the notation to describe the crystallographic planes in a hexagon lattice, such as magnesium, four axes are used to define a plane and direction.

The advantage of four axes for hexagonal crystals is that crystallographically similar planes and directions give the same form of similar indices. The four axes system bring sixfold symmetry of the hexagonal crystal lattice.

11.4.13 Allotropy

Allotropy is the property possessed by certain elements to exist in two or more distinct forms that are chemically identical but have different physical properties. In the case of iron, the crystal structure has one form at room temperature and another at high temperature. When heated between 910°C and 1400°C, the atomic structure changes from body-centered cubic to face-centered cubic; from 1400° to its melting point at 1535°C, it reverts to a body-centered cubic. This change is associated with a volumetric and property change in the metal.

This process is reversible and when the metal is cooled it reverts to its original condition. The allotropy of iron modifies the solubility of carbon, and it is because of this that steel can be hardened.

11.5 Nuclear Physics

11.5.1 Introduction

More than 99% of the mass of visible matter in the universe is nuclear matter. Protons and neutrons are the building blocks of atomic nuclei. Exotic forms of nuclear matter were present in the early universe and continue to exist today in neutron stars. Nuclear fusion processes at the core of the sun are the source of the vast energy flow that sustains life on Earth. Nuclear fusions in stars and nuclear processes at the end of stellar life have formed the rich spectrum of elements, which is observed in nature.

Nuclear physics studies the structure of nuclei—their formation, stability, and decay. It aims to understand the fundamental nuclear forces in nature, their symmetries, and the resulting complex interactions between protons and neutrons in nuclei and among quarks inside hadrons, including the proton.

11.5.2 Atom Nucleus

In Section (11.4), it was shown that the classical atom is comprised of protons, neutrons, and electrons and forms the building blocks of the physical universe. Recapping, the atom has a number of protons and neutrons bonded together using nuclear forces in the central nucleus.

It was shown that the atomic number of an element is the number of protons within the nucleus of its atom. An element will always have the same number of protons but can have a different number of neutrons. The atomic weight of the element is the sum of the protons and neutrons. *Isotope* is the name given to an element having different number of neutrons, which results in the element having dissimilar weights.

The masses of atoms and molecules are measured with atomic mass units (u).

One atomic unit $=1\ u=1.660\times10^{-27}$ kg. (this mass is exactly the same as 1/12 the mass of a stable carbon atom with 12 nucleons).

The mass of a proton $m_p = 1.673 \times 10^{-27}$ kg = 1.007277 u.
The mass of a neutron $m_n = 1.675 \times 10^{-27}$ kg = 1.008665 u.

11.5.3 Binding Energy

When the masses of the constituent particles of a nucleus are added together, the sum is less than the nucleus itself. There is a mass deficiency. The energy equivalent of the missing mass is called the *binding energy*. The higher the binding energy, the more stable the atom.

11.5.4 Radioactivity

Some nuclei are unstable and spontaneously emit radiation, which is called *radioactivity*. The radiation is of three types:

- Alpha decay, in which the emitted particles are helium nuclei of two protons and two neutrons
- Beta decay, in which the emitted particles are electrons
- Gamma decay, in which high energy photons are emitted

The original nucleus is called the *parent nucleus*, and the nucleus remaining after the decay is called the *daughter nucleus*. The process of one element changing into another through radioactivity is called *transmutation*.

If a nucleus emits an alpha particle, it loses two protons and two neutrons; therefore, the daughter nucleus has an atomic mass of four less and an atomic number of two less than the parent nucleus. An example of alpha decay of uranium is 92238U →90234 Th +22He.

If a nucleus emits a beta particle, it loses an electron. Since the mass of the electron is so small compared to that of a proton and a neutron, the atomic mass of the daughter nucleus is the same as the parent nucleus, but the atomic number of the daughter nucleus is one greater than that of the parent nucleus. An example of beta decay of bismuth is 83212Bi → 84212P0 ±0 e.

Frequently, the daughter nucleus is left in an excited state after either alpha or beta decay. Then the nucleus can give up excess energy by emission of gamma radiation. The following example shows a typical situation where gamma decay occurs: 512B → 612C* ±10 e; then, 612C* →612C+γ, where the asterisks indicate an excited nucleus.

The rules for radioactive decay are based on conservation laws. Examination of the preceding examples reveals that the number of nucleons and the electric charge are conserved; that is, the total on one side of the equation equals the total on the other side of the equation. Other conservation laws that must be observed are those of energy, momentum, and angular momentum.

11.5.5 Half-Life

The decay rate (R) or the activity of a sample of radioactive material is defined as the number of decays per second, given by $R = -\lambda N$, where N is the number of radioactive nuclei at some instant and λ is the decay constant.

The half-life (T) is defined as the time required for half of a given number of radioactive nuclei to decay. It is different for each type of radioactive element:

- The general decay curve for a radioactive sample relating the number of nuclei present at a given time to the original number of nuclei is exponential. The expression is 01 n +92235U → 56241Ba +3692Kr+301 n.
- The total rest mass of the products is less than the original rest mass of the original uranium by 220 Mega electron volts (MeV). This is an enormous amount of energy compared to energy releases in chemical processes and when considering that a relatively modest piece of uranium has so many nuclei.
- Nuclear fusion occurs when light nuclei are combined to form a heavier nucleus. The sun is powered by nuclear fusion.

The binding energy is related to stability. When the mass energy of the parent nucleus is greater than the total mass energy of the decay products, spontaneous decay will take place. If the decay products have a greater total mass energy than the parent nucleus, additional energy is necessary for the reaction to occur. Energy is released when light nuclei combine (fusion) and when heavy nuclei split (fission) (Table 11.5.1).

TABLE 11.5.1

Types of Nuclear Decay

Type of Decay	Gamma Decay	Alpha Decay	Beta Decay	Electron Capture	Positron Capture
Decay event	Emission of gamma ray reduces energy of nucleus	Emission of alpha particle reduces size of nucleus	Emission of electron by proton changes it to a neutron	Capture of electron by proton changes it to a neutron	Emission of Positron by a proton changes it to a neutron
Daughter product	Reduction in energy level	Atomic number change	Atomic number change	Atomic number change	Atomic number change
Reason for instability	Nucleus has excess energy	Nucleus is to large	Nucleus has too many neutrons relative to protons	Nucleus has too many protons	Nucleus has too many protons

12

Thermodynamic Basics

12.1 Introduction

12.1.1 What Is Thermodynamics?

Thermodynamics is the branch of physics that studies the effects of temperature and heat on physical systems at the macroscopic scale. In addition, it also studies the relationship that exists between heat, work and energy.

Thermal energy is found in many forms in today's society including power generation of electricity using gas, coal or nuclear, heating water by gas or electric, rocket propulsion, astronomy and cosmology, amongst many others.

12.1.2 Brief History

Engineering thermodynamics has its beginnings in the 1800s at the start of the industrial revolution in the UK when Thomas Newcomen (1712) designed and manufactured the first atmospheric steam engine, later improved by James Watt around 1764; this work laid the foundations for the development of the steam engine.

Although Thomas Savery (1650–1715) is acknowledged as the true inventor of the steam engine in 1698, he built the world's first commercially useful steam engine. It was originally intended for pumping water out of deep mines. The machine did not have a piston and relied on the opening and closing of various valves to direct the steam into the standing water, forcing it out in the exhaust.

In 1824, Sadi Carnot (1796–1832) published a paper entitled 'Reflections on the Motive Power of Fire' and this was cited to be the starting point for thermodynamics as a modern science. James Joule (1818–1889) together with Herman Helmholtz (1821–1894) independently formulated the relationship between heat and mechanical work. This discovery led to the theory of conservation of energy resulting in the first law of thermodynamics.

Josiah Willard Gibbs (1839–1903) is considered to be one of the founders of modern thermodynamics and a pioneer in graphical methods for thermal sciences and gives his name to the 'Gibbs Phase Rule'.

In 1862, James Clerk Maxwell put forward the philosophical argument that both light and radiant heat were forms of electromagnetic waves and this then led to the start of the quantitative analysis of thermal radiation. Jožef Stefan further observed that the total radiant flux from a black body is proportional to the fourth power of its temperature. This work was derived theoretically in 1884 by Ludwig Boltzmann and is known as the Stefan–Boltzmann law.

In later years, many workers have developed the theory of thermodynamics in a much wider field covering other sciences including astronomy and medical research.

12.2 Basic Thermodynamics

12.2.1 Basic Concepts

Properties are characteristics of a system which include mass, volume, energy, pressure and temperature. Thermodynamics also considers other quantities that are not physical properties, such as mass flow rates and energy transfers by work and heat. Properties are also considered to be either intensive or extensive.

Table 12.1 lists various quantities used in thermodynamics together with their units.

12.2.2 Extensive

An extensive property is one that is divisible in size; for example, volume when divided will become smaller. Mass and energy are other examples.

12.2.3 Intensive

Intensive properties are those that are independent of the size of the system; for example, temperature, pressure and density of a substance will remain unchanged when divided into smaller masses.

12.2.4 Specific and Total Quantities

These relate only to extensive properties.

A *specific quantity* represents in the case of, say, mass, the quantity per kilogram and is obtained by dividing the property by the mass. These properties are designated by lower case letters, such as 'v' for the specific volume (m^3/kg) and 'h' for specific enthalpy (J/kg).

Total quantity is always designated by higher case letters, say 'V' for volume (m^3) and 'H' for enthalpy (J).

Specific volume is principally used for gases and vapours. The inverse of a specific volume is density (ρ) (kg/m^3) and this property is chiefly used for solids and liquids but can also be used for gases ($\rho=1/v$).

As the same letter is often used to designate more than one property, quite often alternative letters are used. As an example, 'v' for specific volume may occur in the same work as velocity, and 'u' and 'c' are sometimes used for velocity to avoid confusion in the case of initial velocity and final velocity.

12.2.5 Energy Forms

Fluids and solids can possess several forms of energy. All fluids possess energy due to their temperature and this is referred to as 'internal energy'. They will also possess 'gravitational or potential energy' (PE) due to distance (z) above a datum level and if the fluid is moving at a velocity (v), it will also possess 'kinetic energy' (KE). If the fluid is pressurised, it will possess 'flow energy' (FE).

Pressure and temperature are the two governing factors and internal energy can be added to FE to produce a single property called 'enthalpy'.

In the following paragraphs, each of these terms will be considered in more detail.

TABLE 12.1
Symbols Used in Thermodynamics and Heat Transfer

Quantity	Units	Derived Units	S.I. Symbol
Absolute temperature	K		T
Adiabatic index			γ
Area	m^2		A
Celsius temperature	°C		θ
Characteristic gas constant	N m/kg K	J/kg K	R
Density	kg/m^3		ρ
Dynamic viscosity	$N\ s/m^2$	Pa s	η or μ
Energy	N m/kg K	Joule	
Enthalpy	N m	Joule	H
Entropy	J/K		S
Force	$kg\ m/s^2$	N	F
Heat transfer	N m	Joule	Q
Heat transfer rate	N m/s	Watt	Φ
Internal energy	N m	Joule	U
Kinematic viscosity	m^2/s		v
Length	m		Various
Mass	kg		m
Mass flow rate	kg/s		
Polytropic index			n
Pressure	N/m^2	Pascal	p
Pressure head	m		h
Specific enthalpy	N m/kg	J/kg	h
Specific entropy	J/kg K		s
Specific heat capacity	m/s	Joule/kg K	c
Specific internal energy	N m/kg	J/kg	u
Specific volume	m^3/kg		ν
Time	s		t
Universal gas constant	J/kmol K		R_o
Velocity	m/s^2		v or u
Volume	m^3		V or Q
Volumetric flow rate	m^3/s		
Weight	$kg\ m/s^2$	N	W
Work	N m	Joule	W
Work rate (power)	N/m/s	Watt	P

12.2.6 Internal Energy

The molecules of a fluid possess both kinetic KE and PE relative to an internal datum. Generally, this is regarded simply as the energy due to its temperature and the change in internal energy in a fluid that undergoes a temperature change is given by

$$\Delta U = mc\Delta T \tag{12.1}$$

The total internal energy is denoted by the symbol 'U', which has values of joule, kilojoule or megajoule; also the specific internal energy 'u' has the values of kilojoule per kilogram.

Note: The change in temperature is in either degrees Celsius or Kelvin. The law which states internal energy is a function of temperature only; this is known as *'Joule's Law'* and is independent of its volume and pressure.

12.2.7 Gravitational or Potential Energy

When a mass 'm' kilogram is raised to a height of 'z' metres above a datum level, a lifting force is required. This force obviously has to be greater than the mass being lifted.

The work done in raising the mass is force×distance moved, so

$$\text{Work} = mgz \text{ where g is the gravity constant} \tag{12.2}$$

As energy has been used to do this work and cannot be destroyed, it will follow that energy has been stored in the mass. This energy is called the 'gravitational energy' or 'potential energy' (PE).

$$PE = mgz \tag{12.3}$$

12.2.8 Kinetic Energy

If a mass 'm' kilogram is at rest and is then accelerated 'a' to a velocity 'v' (m/s), a force 'F' is required to provide the acceleration. This force is given by Newton's second law of motion:

$$F = m \cdot a \tag{12.4}$$

After a time 't' (s), the mass will have travelled a distance of 'x' metres and have reached a velocity of 'v' (m/s). These two quantities are related by the laws

$$a = \frac{v}{t} \tag{12.5}$$

$$x = \frac{v \cdot t}{2} \tag{12.6}$$

The work done in accelerating the mass

$$P = \frac{F}{A} \tag{12.7}$$

Energy has been expended in doing this work and has therefore been stored in the mass and will be carried along with it. This energy is called 'kinetic energy'.

$$KE = \frac{m \cdot v^2}{2} \tag{12.8}$$

12.2.9 Flow Energy

When fluid is pumped under pressure along a pipe or conduit, energy has been used to do the pumping. This energy is contained within the fluid and may be recovered as an example using a hydraulic cylinder.

Consider a piston pushing fluid within a cylinder. Fluid pressure is 'P' (N/m^2) and the force required by the piston is

$$F = P \cdot A \tag{12.9}$$

where

P = fluid pressure

A = internal area of the pipe or conduit

When the piston has moved a distance of 'x' metres, the work done is

$$W = F \cdot x = P \cdot A \cdot x \tag{12.10}$$

Since $A \cdot x = V_{ol}$ and is the volume pumped in the cylinder, the work done is

$$W = PV_{ol} \tag{12.11}$$

Energy has been used in doing this work and it is stored in the fluid as 'flow energy':

$$FE = P \cdot V_{ol} \tag{12.12}$$

12.2.10 Enthalpy

As stated in Section 12.2.5, *Enthalpy* (H) requires both pressure and temperature; it therefore must possess both flow (FE) and internal energy (U). These two energies are added together

$$H = FE + U \tag{12.13}$$

The units are Joule (J) for the total enthalpy or kJ/kg for the specific enthalpy.

12.2.11 Gas Laws

In this section, a *'perfect gas'* will be discussed and the following will be considered:

- Derive the basic gas laws.
- Derive the characteristic gas law.
- Examine the universal gas law.
- Define the 'mol'.

12.2.12 Theory

A gas is made up of molecules which are consistently moving around in a random manner. In a perfect gas, these molecules may collide but will have no tendency to join or repel each other. In reality, there is a small force of attraction between the gas molecules but as this force is so small, gas laws formulated for a perfect gas work quite well for a practical gas.

Each molecule possesses an instantaneous velocity and hence has KE. The sum of this energy is the internal energy 'U'. The velocity of the molecule will depend upon the temperature. When there is a change in the temperature, there will be a corresponding change in the internal energy. The internal energy is for all intents and purposes zero at –273°C. This temperature is known as absolute zero or 0 to K.

(Remember when converting from °Celsius to Kelvin to add 273 to the Celsius temperature.)

12.2.13 Pressure

When a fluid or a gas is compressed as a function of the applied force and area in the following relationship:

$$P = \frac{F}{A} \tag{12.14}$$

where

F = force

A = area

The unit of pressure in SI units is N/m^2 or Pascals.

12.2.14 A Perfect Gas

A perfect gas is one in which its working temperature is well above its critical temperature and obeys the following:

- Boyle's law.
- Charles's law.
- Joule's law of internal energy.
- Dalton's law of partial pressures.
- Its specific heat is constant.

To obey these laws, the gas would not change its state even at absolute zero.

The distance between its molecules needs to be so far apart as not to have any intermolecular forces or collisions.

At normal pressures and temperatures, the permanent gases such as oxygen, nitrogen, helium and so on closely obey these laws and are known as 'semi-perfect' gases.

12.2.15 Boyle's Law

Boyle's law states that provided the temperature 'T' of a perfect gas remains constant, the volume 'V' of a given mass of gas is inversely proportional to its pressure 'P' of the gas, that is, $P \propto 1/V$ (as shown in Figure 12.1) or $P \times V = \text{constant}$ if the temperature remains constant.

If the gas experiences a change in state during an isothermal process, then

$$P_1V_1 = P_2V_2 = \text{constant} \tag{12.15}$$

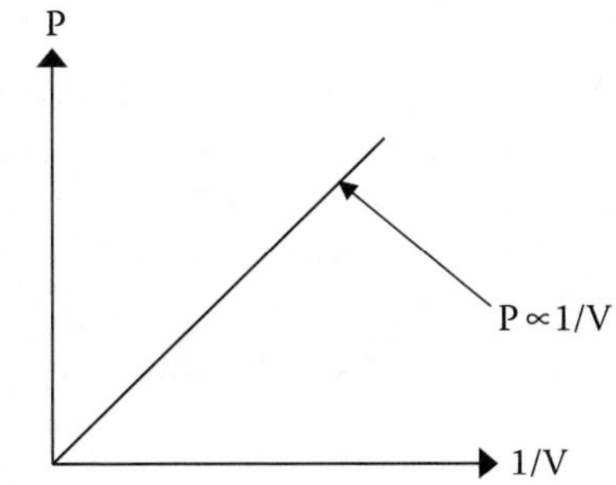

FIGURE 12.1
Graph of $P \propto 1/V$.

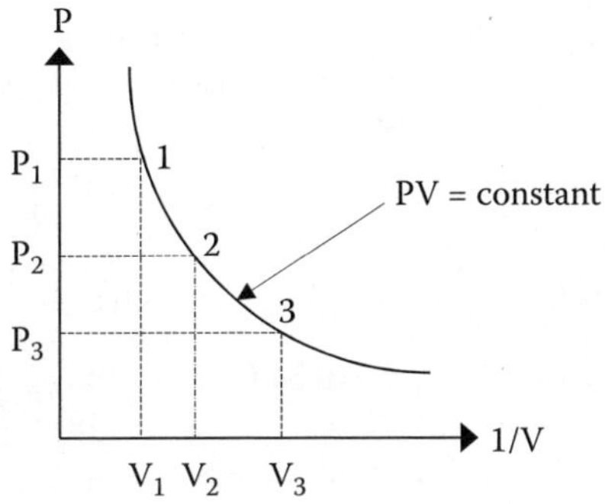

FIGURE 12.2
Graph of PV=constant.

Representing the process on a graph having axes of pressure 'P' and volume 'V', the result will be as shown in Figure 12.2. The curve is known as a rectangular hyperbola and has the mathematical equation x·y=constant.

Example 12.1

A certain perfect gas is heated at a constant temperature from an initial state of 0.22 m^3 and 325 kN/m^2 to a final state of 170 kN/m^2. Determine the final pressure of the gas.

Solution

State 1: $P_1 = 325$ kN/m^2 and $V_1 = 0.22\,m^3$.
State 2: $P_2 = 170$ kN/m^2 and V_2 =?
From the equation, $P_1 \cdot V_1 = P_2 \cdot V_2$

$$V_2 = V_1 \times \frac{P_1}{P_2}$$

$$\frac{V_1}{T_1} = \frac{V_2}{T_2} = \text{constant}$$

$$V_2 = 0.22\ m^3 \times \frac{325\ kN/m^2}{170\ kN/m^2}$$

$$V_2 = 0.4206\ m^3$$

12.2.16 Charles's Law

Charles's law states that provided the pressure 'P' of a given mass of gas remains constant, the volume 'V' of the gas will be directly proportional to the absolute temperature 'T' of the gas, that is, $V \propto T$, or $V = \text{constant} \times$'T'. Therefore, $V/T = \text{constant}$ for a constant pressure 'P'.

If the gas experiences a change in state during a constant pressure process, then

$$\frac{V_1}{T_1} = \frac{V_2}{T_2} = \text{constant} \tag{12.16}$$

This process is represented on a graph as shown in Figure 12.3.

Example 12.2

A quantity of gas is subjected to a constant pressure process causing the volume of gas to reduce from 0.54 m³ at a temperature of 345°C to 0.32 m³. Calculate the final temperature of the gas at the end of this process.

Solution

From the question

$$V_1 = 0.54\ \text{m}^3$$

$$T_1 = 345 + 273\ \text{K}$$

$$V_2 = 0.32\ \text{m}^3$$

Now

$$\frac{V_1}{T_1} = \frac{V_2}{T_2} = \text{constant}$$

$$T_2 = T_1 \times \frac{V_2}{V_1}$$

$$= 618\ \text{K}\left(\frac{0.32\ \text{m}^3}{0.54\ \text{m}^3}\right)$$

$$T_2 = 366.22\ \text{K}$$

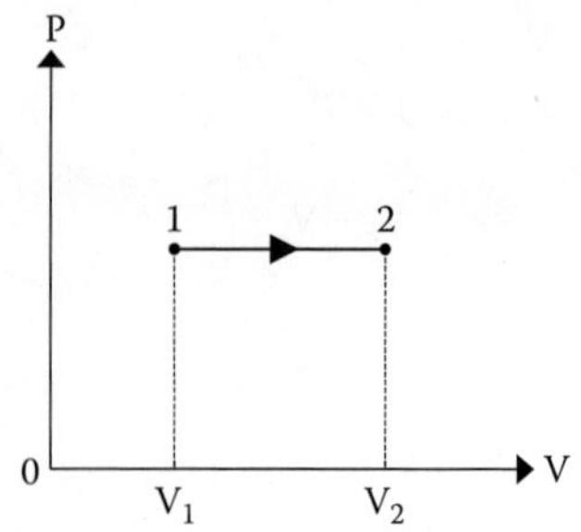

FIGURE 12.3
P–V graph of constant pressure.

12.2.17 Universal Gas Law

The universal gas equation combines pressure, volume and temperature and the relation between Boyle's and Charles's laws is expressed in Equation 12.17

$$\frac{P.\,V}{T} = \text{constant} = R \tag{12.17}$$

where R is known as the universal gas constant. That is

$$\frac{P_1V_1}{T_1} = \frac{P_2V_2}{T_2} \tag{12.18}$$

Example 12.3

A volume of gas at a pressure of 325 kN/m² and temperature of 618 K is compressed to a volume 0.16 m³ and a pressure of 380 kM/m³. Determine the final temperature of the gas.

Solution

State 1: P_1=325 kN/m³, V_1=0.22 m³ and T_1=618 K.
State 2: P_2=380 kN/m³, V_2=0.16 m³ and T_2 =?

From Equation 12.18:

$$\frac{P_1V_1}{T_1} = \frac{P_2V_2}{T_2}$$

$$T_2 = \frac{618\text{ K} \times 380\text{ kN/m}^2 \times 0.16\text{ m}^3}{325\text{ kN/m}^2 \times 0.22\text{ m}^3}$$

$$T_2 = 525.52\text{ K}$$

Gases, in practice, do not obey this law rigidly but many do tend towards it. This law is also referred to as the *characteristic equation of state of a perfect gas.*

The constant 'R' is called the gas constant and its units are Nm/kg K or J/kg K. Each perfect gas will have a different gas constant.

The characteristic gas equation is usually written as

$$PV = RT \tag{12.19}$$

or for 'm' kg, occupying V m³

$$PV = mRT \tag{12.20}$$

The characteristic gas equation can be expressed in a different form which is derived using the kilogramme-mole as a unit. The kilogramme-mole is defined as a quantity of gas equivalent to meter kilogram of the gas, where M is the molecular weight of the gas. Oxygen has a molecular weight of 32; hence, 1 kg mol of oxygen is equivalent to 32 kg of oxygen.

From the definition of the kilogramme-mole, for 'm' (kg) of the gas

$$m = nM \tag{12.21}$$

where n is the number of moles.

Note well: As the standard of mass in SI units is the kilogramme (kg), the kilogramme-mole will be written as 'mole'.

Substituting for m from Equation 12.21 in Equation 12.20

$$P \cdot V = nMRT \text{ or } MR = \frac{P \cdot V}{nT} \tag{12.22}$$

Avogadro's hypothesis states that the volume of 1 mole of any gas will be the same as the volume of 1 mole of any other gas when the gases are at the same temperature and pressure. Therefore, V/n is the same for all gases at the same values of P and T.

The quantity PV/nT is a constant for all gases. This constant is called the 'universal gas constant' and has the symbol 'R_o'.

$$MR = R_o = \frac{PV}{nT} \tag{12.23}$$

or

$$PV = nR_oT \tag{12.24}$$

As $MR = R_o$ then

$$R = \frac{R_o}{M} \tag{12.25}$$

It has been shown by experiment that the volume of 1 mole of any perfect gas at a pressure of one bar and at a temperature of 1°C is approximately 22.71 m^3.

From Equation 12.24:

$$\begin{aligned} R_o &= \frac{PV}{nT} \\ &= \frac{1 \times 10^5 \times 22.71}{1 \times 273.15} \\ &= 8314.11 \text{ J/mole K} \end{aligned}$$

The gas constant can be found for any gas when the molecular weight is known using the Equation 12.25. The molecular weight for oxygen is 32; hence, the gas constant is

$$\begin{aligned} R &= \frac{R_o}{M} \\ &= \frac{8314.11}{32} \\ &= 259.81 \text{ J/kg K} \end{aligned}$$

Example 12.4

The volume of a gas at 0.046 m^3 is contained in a sealed cylinder under a pressure of 300 kN/m^2 at a temperature of 45°C. The gas is then compressed until a pressure of 1.27 MN/m^2 at a temperature of 83°C is reached. Assuming the gas is a perfect gas, given R=0.29 kJ/kg K, calculate

1. The mass of the gas (kg)
2. The final volume of the gas (m^3)

Solution

State 1: $V_1=0.046\,m^3$, $P_1=300\ kN/m^2$ and $T_1=273+45.0\ K=318.0\ K$.
State 2: $P_2=1.27\ MN/m^2$, $T_2=273+83.0\ K=356.0\ K$ and $R=0.29\ kJ/kg\ K$.

1. From Equation 12.20:

$$PV = mRT$$

$$m = \frac{P_1V_1}{RT_1}$$

$$= \frac{300\times 0.046}{0.29\times 318}$$

$$= 0.1496\ kg$$

2. From Equation 12.16:

$$\frac{V_1}{T_1} = \frac{V_2}{T_2}$$

$$V_2 = V_1\left(\frac{T_2}{T_1}\right)$$

$$V_2 = 0.046\ m^3\left(\frac{356\ K}{318\ K}\right)$$

$$V_2 = 0.051\ m^3$$

12.2.18 Specific Heat Capacity

The specific heat capacity of any substance is defined as the amount of energy required to raise a unit mass through one degree temperature rise. In thermodynamics, there are two specified conditions used:

1. Constant volume (C_v)
2. Constant pressure (C_p)

The two specific heat capacities do not have the same value, and it is very important to distinguish between them.

12.2.19 Specific Heat Capacity at Constant Volume (C_v)

Consider 1 kg of a gas supplied with an amount of heat energy sufficient to raise its temperature by 1 K while the volume of the gas constant, the amount of heat energy supplied, is known as the specific heat capacity at constant volume and is denoted by C_v. The basic unit of C_v is J/kg K.

For a reversible non-flow process at constant volume:

$$dQ = mC_v dT \tag{12.26}$$

For a perfect gas, the value of C_v will be constant for any one gas at all pressures and temperatures. Equation 12.26 can be expanded as follows.

Heat flow in a constant volume process between two states:

$$Q_{12} = mC_v(T_2 - T_1) \tag{12.27}$$

From the non-flow energy equation:

$$Q - W = (U_2 - U_1)$$

$$mC_v(T_2 - T_1) - 0 = (U_2 - U_1)$$

$$(U_2 - U_1) = mC_v(T_2 - T_1) \tag{12.28}$$

that is

$$dU = Q \tag{12.29}$$

where 'U' is the internal energy

Note: In a reversible constant volume process, there will be no work energy transfer as the piston will be unable to move; therefore, $W = 0$.

Figure 12.4 shows the P–V diagram for the reversible constant volume process.

Example 12.5

A quantity of 4.5 kg of gas is heated at a constant volume of 1.5 m^3 and temperature 20°C until the temperature rose to 150°C. If the gas is assumed to be perfect, determine

1. The heat flow during the process
2. The pressure at the beginning of the cycle
3. The final pressure

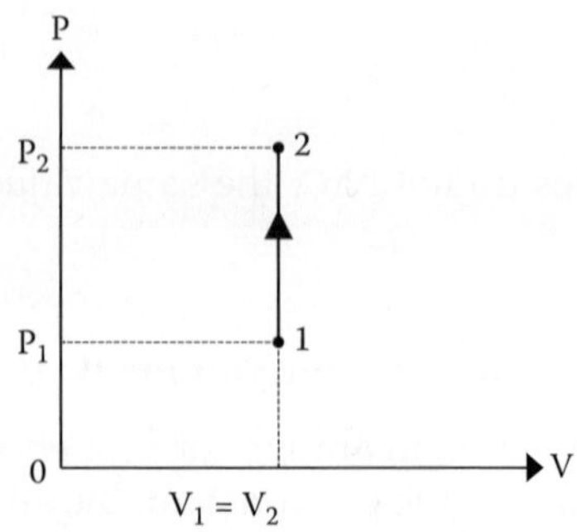

FIGURE 12.4
P–V graph for the reversible constant volume process.

Given:

$m =$	4.5	kg
$V_1 =$	1.5	m^3
$V_2 =$	1.5	m^3
$T_1 =$	20 + 273 = 293	K
$T_2 =$	150 + 273 = 423	K
$C_v =$	0.72	kJ/kg K
$R =$	0.287	kJ/kg K

$$C_v = 0.72 \text{ kJ/kg K and } R = 0.287 \text{ kJ/kg K}$$

Solution

1. From Equation 12.27:

$$\begin{aligned} Q_{12} &= mC_v(T_2 - T_1) \\ &= 4.5 \text{ kg} \times 0.72 \text{ kJ/kg K} \times (423\text{–}293) \text{ K} \\ &= 421.2 \text{ kJ} \end{aligned}$$

2. From the Equation 12.20, PV = mRT

For state 1:

$$P_1V_1 = mRT$$

$$P_1 = \frac{mRT_1}{V_1}$$

$$P_1 = \frac{4.5 \text{ kg} \times 0.287 \text{ kJ/kg K} \times 293 \text{ K}}{1.5 \text{ m}^3}$$

$$P_1 = 252.27 \text{ kN/m}^2$$

3. For state 2:

$$P_2V_2 = mRT$$

$$P_2 = \frac{mRT}{V_2}$$

$$P_2 = \frac{4.5 \text{ kg} \times 0.287 \text{ kJ/kg K} \times 423 \text{ K}}{1.5 \text{ m}^3}$$

$$P_2 = 364.20 \text{ kN/m}^2$$

12.2.20 Specific Heat Capacity at Constant Pressure (C_p)

When 1 kg of a gas is supplied with an amount of heat energy sufficient to raise the temperature by 1 K while the pressure of the gas remains constant, the amount of heat energy that is supplied is known as the specific heat capacity at constant pressure and is denoted by C_p. The unit of C_p is J/kg K.

For a reversible non-flow process at constant pressure:

$$dQ = mC_p dT \tag{12.30}$$

For a perfect gas, the value of C_p is constant for any one gas at all pressures and temperatures. Equation 12.20 can be expanded as follows:

In a reversible constant pressure process, the heat flow

$$Q = mC_p(T_2 - T_1) \tag{12.31}$$

12.2.21 Relationship between the Specific Heats

Consider a perfect gas being heated at constant pressure from T_1 to T_2. Referring to the non-flow equation, $Q=U_2-U_1+W$ and the equation for a perfect gas, $U_2-U_1=mC_v(T_2-T_1)$, combining will give

$$Q = mC_v(T_2 - T_1) + W$$

In a constant pressure process, the work done by the fluid is given by

$$W = P \cdot \Delta V$$

that is

$$W = P(V_2 - V_1) \tag{12.32}$$

Using the equation $PV=mRT$:

$$W = mR(T_2 - T_1) \tag{12.33}$$

Substituting:

$$\begin{aligned} Q &= mC_v(T_2 - T_1) + mR(T_2 - T_1) \\ &= m(C_v + R)(T_2 - T_1) \end{aligned} \tag{12.34}$$

Equating Equations 12.31 and 12.34 for the heat flow 'Q':

$$mC_p(T_2 - T_1) = m(C_v + R)(T_2 - T_1)$$

Therefore,

$$C_p = C_v + R \tag{12.35}$$

This equation may also be written as

$$R = C_p - C_v \tag{12.36}$$

12.2.22 Specific Heat Ratio 'γ'

The ratio of specific heat at constant pressure to the specific heat at constant volume is given by the symbol 'Υ' (gamma).

$$\gamma = \frac{C_p}{C_v} \tag{12.37}$$

From Equation 12.25, it is clear that C_p has to be greater than C_v for a perfect gas. It follows, therefore, that the ratio $C_p/C_v = \gamma$ is always greater than unity. In general, 'γ' is approximately 1.4 for diatomic gases such as carbon monoxide (CO), hydrogen (H_2), nitrogen (N_2) and oxygen (O_2). For monatomic gases such as argon (A) and helium (He), γ is approximately 1.6 and for triatomic gases including carbon dioxide (CO_2) and sulphur dioxide (SO_2), γ is about 1.3.

Some useful relationships between C_p, C_v, R and γ can be derived from Equations 12.36 and 12.37.

$$C_p - C_v = R$$

Dividing through by 'C_v':

$$\frac{C_p}{C_v} - 1 = \frac{R}{C_v}$$

Now from Equation 12.57:

$$\gamma = \frac{C_p}{C_v}$$

$$\gamma - 1 = \frac{R}{C_v} \tag{12.38}$$

$$C_v = \frac{R}{(\gamma - 1)}$$

Also from Equation 12.37, $C_p = \gamma C_v$ and substituting in Equation 12.38:

$$C_p = \gamma C_v = \frac{\gamma R}{(\gamma - 1)}$$

$$C_p = \frac{\gamma R}{(\gamma - 1)} \tag{12.39}$$

Example 12.6

A particular perfect gas has a specific heat as follows:

$$C_p = 0.846 \text{ kJ/kg K and } C_v = 0.657 \text{ kJ/kg K}$$

Determine the gas constant and the molecular weight of the gas.

Solution

From Equation 12.36

$$R = C_p - C_v$$

that is

$$R = 0.846 - 0.652$$
$$= 0.189 \text{ kJ/kg K}$$
$$\text{or } R = 189 \text{ Nm/kg K}$$

From Equation 12.25

$$M = \frac{R_o}{R}$$

that is

$$M = \frac{8314.4}{189}$$
$$= 44.0 \text{ kg/kmol}$$

12.3 Laws of Thermodynamics

12.3.1 Conservation of Energy

The law of conservation of energy was first formulated in the nineteenth century.

This law states that the total energy of an isolated system remains constant regardless of changes within the system. An alternative way of stating this is that energy is a quantity that can be converted from one form to another, but cannot be created nor destroyed.

It was Gottfried Wilhelm Leibniz during the period from 1676 to 1689 who first attempted a mathematical formulation of the kind of energy which is connected with motion (KE). Leibniz noticed that in many mechanical systems (of several masses, m_i each with velocity v_i), $\sum_i m_i v_i^2$ was conserved so long as the masses did not interact. He called this quantity the vis viva or living force of the system. The principle represents an accurate statement of the approximate conservation of KE in situations where there is no friction.

12.3.2 First Law of Thermodynamics

From the Equation 12.40, it is seen that in terms of heat flow and work output into and out of the engine, the system is in a state of balance. This state is known as the 'steady state' and can be explained by the *'First Law of Thermodynamics'* which states that the first law is a consequence of energy and requires that a system may exchange energy with its surroundings strictly by heat flow or work.

Thus,

$$\Delta E = \Delta Q - \Delta W \tag{12.40}$$

where

ΔE is the change in energy
ΔQ is the change in heat
ΔW is the change in work

12.3.3 Steady Flow Process

The process which changes the state of a thermodynamic system can be divided into two main types depending upon whether the system is an open or closed system.

12.3.4 Flow Process

A flow process is known as an open system where work energy, heat energy and the working fluid may be transferred across the boundary (Figure 12.5).

General energy equation

$$gZ_1 + U_1 + p_1V_1 + \frac{C_1^2}{2} + Q = gZ^2 + U^2 + p_2V_2 + \frac{C_2^2}{2} + W \tag{12.41}$$

where

gZ_1 = potential energy

U = internal energy

$\frac{C^2}{2}$ = kinetic energy

Equation 12.40 may be written as

$$gZ_1 + h_1 + \frac{C_1^2}{2} + Q = gZ_2 + h_2 + \frac{C_2^2}{2} + W \tag{12.42}$$

where U + pV = h = *enthalpy*.

Enthalpy is a measure of the total energy of a thermodynamic system. It includes the internal energy, which is the energy required to create a system, and the amount of energy required to make room for it by displacing its environment and establishing its volume and pressure.

Enthalpy is a thermodynamic potential. It is a state function and an extensive quantity. The unit of measurement for enthalpy in the International System of Units (SI) is the joule.

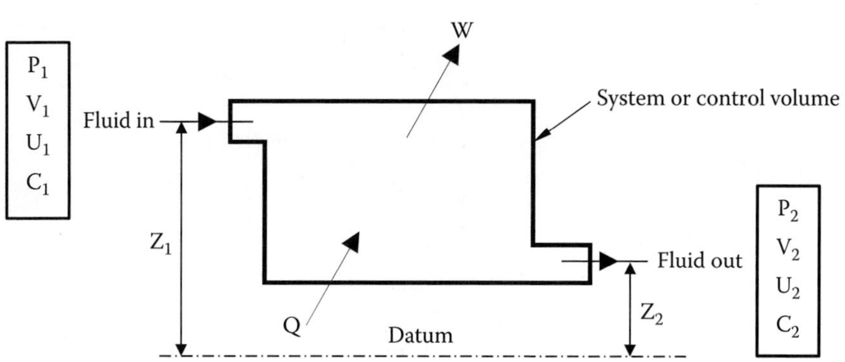

FIGURE 12.5
Thermodynamic process.

12.3.5 Consider a Boiler at Constant Pressure

Figure 12.6 depicts a boiler supplying steam with water and heat being supplied.

Applying the general energy equation:

PE is small when compared with the heat energy being supplied and therefore can be neglected.

KE is also very small and is neglected.

As there are no working parts, 'W' = 0.

The equation can now be written in the following manner

$$h_1 + Q = h_2 \tag{12.43}$$

$$Q = h_2 - h_1 \text{ (for M kg flow of fluid)}$$

$$\therefore Q = M(h_2 - h_1) \tag{12.44}$$

Example 12.7

A boiler operating at a constant pressure of 1.5 MPa evaporates fluid at the rate of 1000 kg/h. At the boiler inlet, the entering fluid has an enthalpy of 165 kJ/kg. At the exit to the boiler, the enthalpy of the exiting steam is 2200 kJ/kg. Calculate the heat energy required by the boiler.

Solution

From the general energy equation

$$Q - W = M\left[(h_2 - h_1) + \left(\frac{C_2^2 - C_1^2}{2}\right)\right]$$

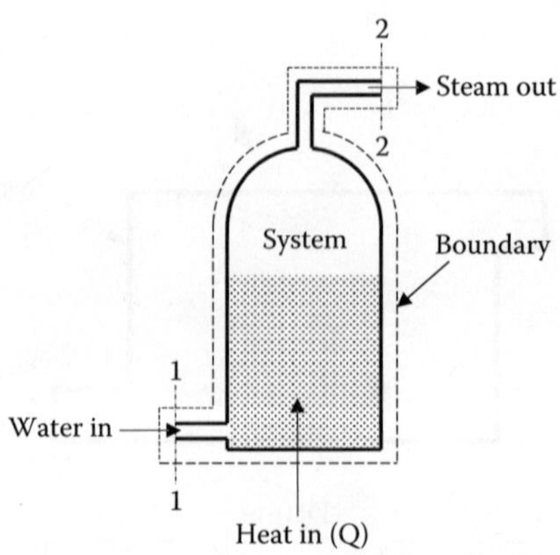

FIGURE 12.6
Boiler supplying steam.

1. For the boiler, $(C_2^2 - C_1^2)/2$ is assumed to be zero.
2. No working or moving parts; therefore, W=0.

where

Q = heat energy per hour required by the boiler
W = work energy leaving the system = 0
m = fluid flow rate = 1000 kg/h
h_1 = enthalpy at inlet = 165 kJ/kg
h_2 = enthalpy at outlet = 2200 kJ/kg

Hence,

$$Q = 1000(2200 - 165)\ \text{kJ/h}$$

$$Q = 2.035 \times 10^6\ \text{kJ/h}$$

If 60% of the heat energy supplied to the boiler is used in evaporating the fluid, calculate the rate of fuel consumption required to maintain the rate of evaporation when 1 kg of fuel produces 30,000 kJ of heat energy.

$$\text{Heat energy required per hour} = \frac{2.035 \times 10^6}{0.60}$$

$$= 3.392 \times 10^6\ \text{kJ/h}$$

The heat energy available from the fuel

$$= 30{,}000\ \text{kJ/h}$$

Hence

$$\text{Fuel required} = \frac{3.392 \times 10^6}{30{,}000}\frac{\text{kJ}}{\text{h}} \times \frac{\text{kg}}{\text{kJ}}$$

$$= 113.1\ \text{kg/h}$$

Example 12.8

A condenser has fluid flowing through it at the rate of 35 kg/min. The fluid enters with a specific enthalpy of 2500 kJ/kg and leaves with a specific enthalpy of 250 kJ/kg. Calculate the rate of heat energy loss from the system.

Solution

The steady FE equation gives

$$Q - W = \dot{m}\left[(h_2 - h_1) + \left(\frac{C_2^2 - C_1^2}{2}\right)\right]$$

For a condenser, W=0 and the terms representing the change in KE and PE may be neglected. The equation will reduce to

$$Q = \dot{m}(h_2 - h_1)$$

Hence,

$$Q = 35\ \text{kg/min}(250 - 2500)\ \text{kJ/kg}$$

$$Q = -78{,}750\ \text{kJ/min}$$

12.3.6 Nozzle

A nozzle is an efficient way to convert thermal energy into KE.

Figure 12.7 depicts a convergent nozzle with a mass flow of fluid passing through it. The steady FE equation gives

$$Q - W = \dot{m}\left[(h_2 - h_1) + \left(\frac{C_2^2 - C_1^2}{2}\right) + (gZ_2 - gZ_1)\right]$$

The average velocity of the fluid through the nozzle will be high and only spends a short duration within the nozzle. As a result, it is considered that there is not sufficient time for any heat energy to flow either into or out of the fluid; as a result, it is considered that 'Q' = 0.

As a nozzle has no moving parts, hence no work energy will be transferred to or from the fluid as it passes through the nozzle; therefore, 'W' = 0.

PE is generally small and can be neglected.

The equation becomes

$$0 = \dot{m}\left[(h_2 - h_1) + \left(\frac{C_2^2 - C_1^2}{2}\right)\right] \tag{12.45}$$

Often C_1 is negligible compared with C_2, and in this case the equation will reduce to

$$0 = \dot{m}\left[(h_2 - h_1) + \left(\frac{C_2^2}{2}\right)\right] \tag{12.46}$$

or

$$\frac{C_2^2}{2} = (h_1 - h_2) \tag{12.47}$$

Equation 12.47 can also be rewritten as

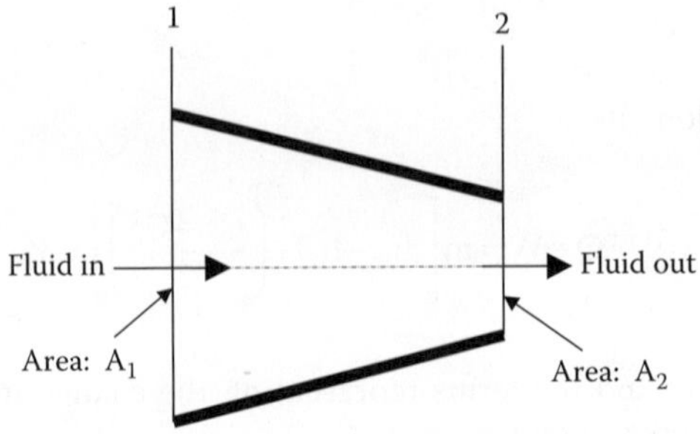

FIGURE 12.7
Mass flow through a convergent nozzle.

$$C_2 = \sqrt{2(h_1 - h_2)} \tag{12.48}$$

Example 12.9

Fluid flowing through a horizontal nozzle has the following properties:

1. At the inlet to the nozzle, the fluid has a specific enthalpy of 2800 kJ/kg with negligible velocity.
2. At its outlet, the fluid has a specific enthalpy of 2250 kJ/kg and its specific volume is 1.25 m^3/kg.

Calculate the required area of the nozzle assuming the flow is adiabatic.

Solution

From the steady FE equation:

$$Q - W = \dot{m}\left[(h_2 - h_1) + \left(\frac{C_2^2 - C_1^2}{2}\right) + (gZ_2 - gZ_1)\right]$$

When this equation is applied to the nozzle, it becomes

$$0 = \dot{m}\left[(h_2 - h_1) + \left(\frac{C_2^2 - C_1^2}{2}\right)\right]$$

As the inlet C_1 is negligible, the equation can be rewritten as

$$C_2 = \sqrt{2(h_1 - h_2)}$$

$$= \sqrt{[2(2800 - 2250)]}$$

$$= 33.166\ \text{m/s}$$

Applying the equation of continuity at the outlet to the nozzle gives

$$\dot{m} = \frac{A_2 C_2}{v_2}$$

Rearranging to solve for 'A_2':

$$A_2 = \frac{\dot{m} \cdot v_2}{C_2}$$

$$A_2 = \frac{14\ \text{kg/s} \times 1.25\ \text{m}^3/\text{kg}}{33.166\ \text{m/s}}$$

$$A_2 = 0.5276\ \text{m}^2 \quad \text{(Area of exit of convergent nozzle)}$$

The respective diameter will be

$$d_2 = \sqrt{\frac{4 \cdot A_2}{\pi}}$$

$$d_2 = 0.8196\ \text{m}$$

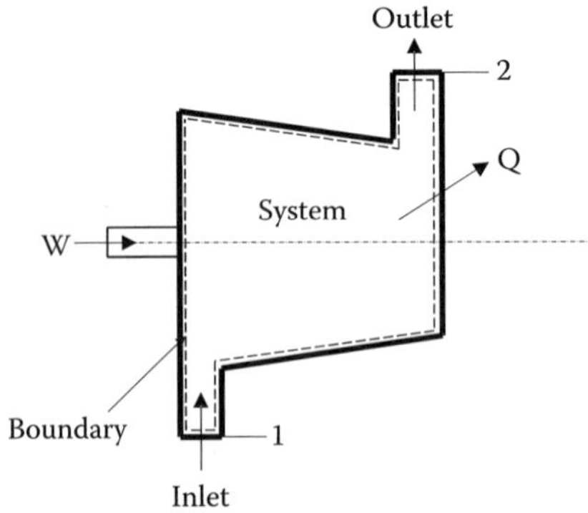

FIGURE 12.8
Pump.

12.3.7 Pump

A pump is a device that uses external work energy to produce a pressure rise in a flowing fluid.

Figure 12.8 portrays a pump defining its boundary with its input and output together with its losses. The steady FE equation gives

$$Q - W = \dot{m}\left[(h_2 - h_1) + \left(\frac{C_2^2 - C_1^2}{2}\right) + (gZ_2 - gZ_1)\right]$$

Although the velocities of the fluid will be high, the differences between the input and output velocities through the pump will not be large. The term representing the change in KE can be disregarded.

PE will be generally small and can also be neglected.

'W' is the amount of work energy/second required to drive the pump.

The steady state FE equation becomes

$$-W = \dot{m}(h_2 - h_1) \tag{12.49}$$

as $h_2 > h_1$, 'W' will be found to be negative.

Example 12.10

Fluid is being pumped at the rate of 45 kg/min. The specific enthalpy of the fluid at the inlet to the pump is 46 kJ/kg and at the outlet of the pump it is 175 kJ/kg. Assuming 105 kJ/kg of heat energy is lost to its surroundings by the pump, calculate the power required to drive the pump assuming the pump efficiency is 85%.

Solution

The fluid flow rate, '$\dot{m}$' = 45 kg/min

$$= 0.75\ \text{kg/s}$$

Heat loss 'Q' = –105 kJ/min

$$= -1.75 \text{ kJ/s}$$

Specific enthalpy (h_1) at inlet=46 kJ/kg and specific enthalpy (h_2) at exhaust=1.27 kJ/kg.

The kinetic and potential energies are neglected. Substituting the above data into the steady FE equation

$$-1.75 - W = 0.75 \times (175 - 46) \text{ kJ/s}$$

Hence,

$$W = -1.75 - (0.75 \times 129)$$

$$= -98.5 \text{ kJ/s}$$

$$= -98.5 \text{ kW (the negative sign indicating this is work energy required to drive the pump)}$$

As the pump efficiency is 85%, power required to drive the pump

$$= 98.5 / 0.85$$

$$= 115.88 \text{ kW}$$

12.3.8 Turbine

A turbine is opposite to that of a pump in that a pressure drop across the turbine results in work energy being produced (see Figure 12.9).

The steady FE equation gives

$$Q - W = \dot{m}\left[(h_2 - h_1) + \left(\frac{C_2^2 - C_1^2}{2}\right) + (gZ_2 - gZ_1)\right]$$

- The average velocity flow of fluid through a turbine is normally high and passes quickly through the turbine. It can be assumed that heat energy does not have time to flow either into or out of the fluid, and hence Q = 0.

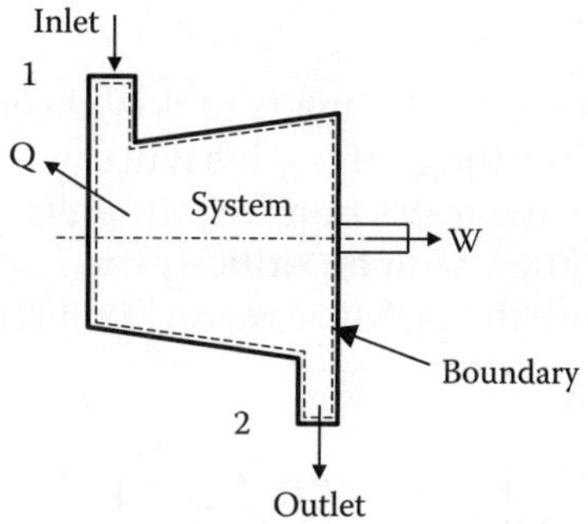

FIGURE 12.9
Turbine.

- The velocities through the turbine are high and the differences between the inlet velocity and outlet velocity are not large; therefore, the term representing the change in KE can be neglected.
- PE is generally small and can be neglected.
- 'W' is the amount of external work energy produced per second.

The steady FE equation can therefore be reduced to

$$-W = \dot{m}(h_2 - h_1) \tag{12.50}$$

or

$$W = \dot{m}(h_1 - h_2) \tag{12.51}$$

Example 12.11

In a turbine, fluid flows through it at the rate of 45 kg/min. The specific enthalpy drop of the fluid is 580 kJ/kg. The turbine loses 2100 kJ/min in the form of heat energy. Calculate the power produced by the turbine assuming the changes in potential and KE are small and can be neglected.

Solution

From the question, Q=heat energy flow into the system=–2100 kJ/min, W=work energy flow from the system (kJ/min), m=fluid flow rate=45 kg/min, h_2-h_1=–580 kJ/kg, C_1 and C_2=neglected and Z_1 and Z_2=neglected.

The steady FE equation becomes

$$-2100\text{ kJ/min} - W = 45\text{ kg/min} \times -580\text{ kJ/kg}$$

$$W = (26{,}100 - 2100)\text{ kJ/min}$$

$$= 24{,}000\text{ kJ/min}$$

$$= 400\text{ kJ/s}$$

$$= 400\text{ kW}$$

12.3.9 Throttling

A throttling process is where the fluid is made to flow through a restriction; for example, an orifice or a partially open throttling valve. This will cause a substantial pressure drop in the fluid within the vicinity of the restriction. Throttling is a non-reversible flow process.

Figure 12.10 shows a pipe fitted with an orifice plate. One reason for fitting an orifice plate in a pipeline is to measure the pressure within a fluid passing through the pipe.

The general steady FE equation

$$Q - W = M\left[(h_2 - h_1) + \left(\frac{C_2^2 - C_1^2}{2}\right) + (gZ_2 - gZ_1)\right]$$

PE can be considered to be small enough to be neglected.

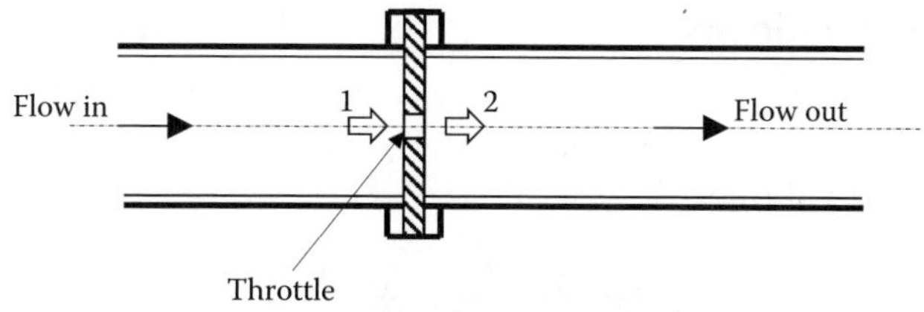

FIGURE 12.10
Throttling process.

There are no working parts; no energy can be transferred in the form of work energy; therefore, W = 0.

The throttling takes place over a very small distance and the available area through which heat energy can flow either into or from the fluid is very small, so it can be assumed that very little energy is lost by heat transfer; hence, Q = 0.

Any difference that exists between C_1 and C_2 will be small, and as a result any change in the KE is normally not considered.

The steady FE equation becomes

$$0 = \dot{m}(h_2 - h_1) \tag{12.52}$$

that is, during a throttling process, the enthalpy will remain constant.

Example 12.12

A fluid flowing through a pipeline undergoes a throttling process where the pressure is reduced from 10 bar to 1 bar when passing through an orifice plate. Prior to throttling, the specific volume of the fluid is 0.3 m^3/kg, and after throttling it is 1.8 m^3/kg. Calculate the change in specific internal energy during this throttling process.

Solution

For a throttling process, the steady state FE equation can be written as

$$0 = \dot{m}(h_2 - h_1) \text{ or } h_2 - h_1$$

But $h_2 = u_2 + P_2v_2$ and $h_1 = u_1 + P_1v_1$.

The change in specific internal energy

$$= u_2 - u_1$$

$$= (h_2 - P_2v_2) - (h_1 - P_1v_1)$$

$$= (h_2 - h_1) - (P_2v_2 - P_1v_1)$$

$$= 0 - (1 \times 1.8 - 10 \times 0.3) \text{ bar m}^3/\text{kg}$$

$$= 120 \times 10 \text{ Nm/kg}$$

$$= 120 \text{ kJ/kg}$$

12.3.10 Equation of Continuity

Consider any section of the cross-sectional area 'A', where the fluid velocity is 'C'; then the rate of volume flow past the section is 'CA'. Also, since the mass flow is volume flow divided by the specific volume

$$\text{Mass flow rate, } \dot{m} = \frac{CA}{\nu}$$

where ν=specific volume at the particular section

With reference to Figure 12.11

$$\dot{m} = \frac{C_1 \cdot A_1}{\nu_1} = \frac{C_2 \cdot A_2}{\nu_2} \tag{12.53}$$

12.3.11 Non-Flow Processes

In the previous section, the discussion centred on a fluid flowing through a system. In this section, the discussion will be where the fluid undergoes a series of processes where it does not flow. An example of this is the cylinder of an internal combustion engine. Referring to Figure 12.12 in the suction stroke, the working fluid is drawn into the cylinder where it is temporarily sealed. Whilst the cylinder is sealed, the fluid is compressed by the rising piston. Thermal energy is then introduced which produces sufficient energy forcing the piston back down the cylinder, thereby creating external work. On completion of the working stroke, the exhaust valve is opened and the fluid is forced out of the cylinder back

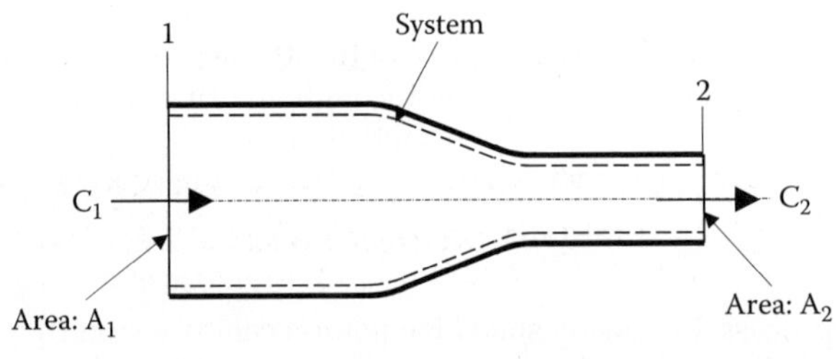

FIGURE 12.11
Mass flow through a system.

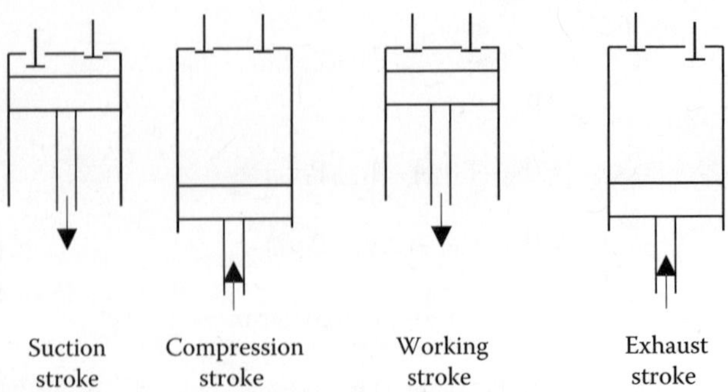

FIGURE 12.12
Internal combustion engine cycle.

into its surroundings. The processes which are subject to action where the working fluid cannot cross the system boundary are called *non-flow processes*. This process occurred during the compression and working strokes in the previous example.

Referring to the general energy equation in Equation 12.41

$$gZ_1 + U_1 + P_1V_1 + \frac{C_1^2}{2} + Q = gZ^2 + U^2 + P_2V_2 + \frac{C_2^2}{2} + W$$

If the fluid undergoes a non-flow process from state (1) to state (2), the terms from the general energy equation for p_1V_1 and p_2V_2 will be zero as the fluid is already in the system and will still be in the system at the end of the process. For similar reasons, the changes in kinetic and potential energies of the fluid will also be zero. The general energy equation will reduce to

$$U_1 + Q = U_2 + W \tag{12.54}$$

or

$$U_2 - U_1 = Q - W \tag{12.55}$$

that is, in a non-flow process, the change in internal energy of the fluid is equal to the net amount of heat energy supplied to the fluid minus the net amount of work energy flowing from the fluid.

This equation is known as the *non-flow energy equation* and in the following discussion it will be shown how this is applied to the various non-flow processes.

12.3.12 Constant Temperature (Isothermal Process) (pV–C)

In a process where the change in temperature is very small, that process can be approximated to an isothermal process. As an example, consider the slow expansion or compression of a working fluid in a cylinder that is perfectly cooled by water; this process may be analysed assuming the temperature remains constant.

Referring to Figure 12.13a and b which shows a constant temperature (isothermal) process, the general relationship of the properties between the initial and final states of a perfect gas will be as follows:

$$\frac{p_1V_1}{T_1} = \frac{p_2V_2}{T_2} \tag{12.56}$$

If the temperature remains constant throughout the process, that is, $T_1 = T_2$, then the above relationship becomes

$$p_1V_1 = p_2V_2 \tag{12.57}$$

From this equation, it will be seen that an increase in the volume will result in a decrease in the pressure. In other words, in an isothermal process, the pressure is inversely proportional to its volume.

12.3.12.1 Work Transfer

Referring to the process represented on the p–V diagram in Figure 12.13, it is noted that the volume increases during the process, that is, the fluid is expanding and the expansion work is given by

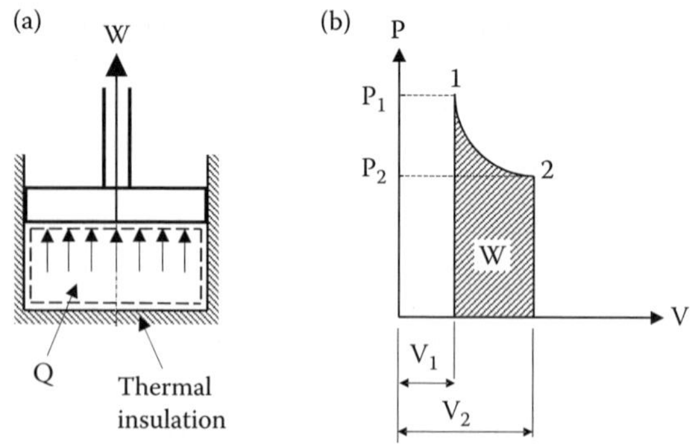

FIGURE 12.13
(a,b): Constant temperature (isothermal) process.

$$
\begin{aligned}
W &= \int_1^2 p\,dV \\
&= \int_1^2 \frac{c}{V}\,dV \quad (pV = c\text{, a constant}) \\
&= c\int_1^2 \frac{dV}{V} \\
&= p_1V_1 \ln\frac{V_2}{V_1} \\
&= mRT_1 \ln\frac{V_2}{V_1} \quad (p_1V_1 = mRT) \\
&= mRT_1 \ln\frac{p_1}{p_2} \quad \left(\text{since } \frac{V_2}{V_1} = \frac{p_1}{p_2}\right)
\end{aligned}
\tag{12.58}
$$

On the p–V diagram shown in Figure 12.13b, the shaded area under the curve represents the amount of work being transferred from the system.

As this is an expansion process (that is an increasing volume), the work is being done by the system, that is, the system produces work output and this is shown by the direction of the arrow representing 'W'.

12.3.12.2 Heat Transfer

Applying the heat balance to this case

$$U_1 + Q = U_2 + W \quad \text{(from Equation 12.54)}$$

For a perfect gas:

$$U_1 = mc_vT_1 \text{ and } U_2 = mc_vT_2$$

As the temperature is constant:

$$U_1 = U_2$$

Substituting in the energy balance equation:

$$Q = W \tag{12.59}$$

Note that the heat flow is equivalent to the work done in an isothermal process *for a perfect gas only*. It should also be pointed out that on the p–V diagram when the process takes place from right to left, the work done by the fluid is negative. That is, work is being done on the fluid.

Example 12.13

A process compresses 0.4 kg of oxygen isothermally from 1.01 bar to 5.5 bar at 22°C. Establish the work done during the compression and the heat transferred during the process. Assume that the oxygen is a perfect gas and the molecular weight of the oxygen is M = 32 kg/kmol.

Solution

From the question, $m = 0.4$ kg, $p_1 = 1.01$ bar, $T_1 = 22°C$, $p_2 = 5.5$ bar and $M = 32$ kg/kmol.

From the equation $R = \frac{R_o}{M}$

$$= \frac{8314}{32}$$

$$= 0.260 \text{ kJ/kg K}$$

For an isothermal process, work input

$$W = mRT_1 \ln\frac{p_2}{p_1} \quad \text{(from Equation 12.58)}$$

$$= 0.4 \times 0.260 \times (22 + 273) \ln\frac{5.5}{1.01}$$

$$= 52 \text{ kJ}$$

As this is an isothermal process, all the work input is rejected as heat; hence, Q = W = 52 kJ.

12.3.13 Adiabatic Process (Q = 0)

An adiabatic process is one in which there is no heat transferred to or from the fluid during the process. Such a system is thermally isolated and the process within such a system may be idealised as an adiabatic process. Such a process can be either reversible or irreversible. Only the reversible process will be considered here.

Steam engine cylinders and gas turbine casings are generally well insulated to minimise any heat loss from the system. The fluid expansion process in such cases can be assumed to be adiabatic.

Figure 12.14a and b depicts an adiabatic process (zero heat transfer).

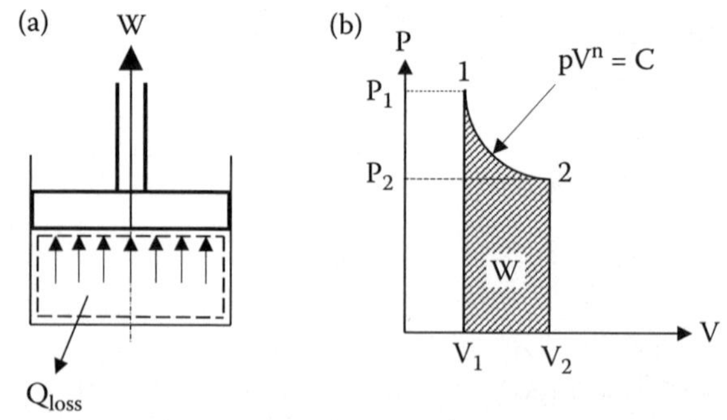

FIGURE 12.14
(a,b): Polytropic process.

For a perfect gas, the equation for an adiabatic process is

$$pV\gamma = C \tag{12.60}$$

where γ = ratio of specific heat = Cp/Cv.

Equation 12.60 applies to states 1 and 2 as

$$p_1 V_1^{\gamma} = p_2 V_2^{\gamma} \tag{12.61}$$

Equation 12.61 can be rewritten as

$$\frac{p_2}{p_1} = \left[\frac{V_1}{V_2}\right]^{\gamma} \tag{12.61a}$$

By manipulating Equations 12.56 and 12.61, the following relationship can be determined:

$$\frac{T_2}{T_1} = \left[\frac{p_2}{p_1}\right]^{\gamma^{-1}/\gamma} = \left[\frac{V_1}{V_2}\right]^{\gamma^{-1}} \tag{12.62}$$

Examining Equations 12.61 and 12.62 for adiabatic processes on a perfect gas, the following conclusions can be made.

- Any increase in pressure will result in an increase in temperature.
- Any increase in volume will result in a decrease in pressure.
- Any increase in volume will result in a decrease in temperature

12.3.13.1 Work Transfer

A reversible adiabatic process for a perfect gas is shown on the p–V diagram in Figure 9.14b and it is noted that the volume increases during the process. The work done is given by the shaded area and this area can be evaluated by integration—that is

$$W = \int_1^2 p\, dV \tag{12.63}$$

Since $pV^{\gamma}=c$, a constant, then

$$= \int_1^2 \frac{c}{V^{\gamma}} dV$$

$$= c \int_1^2 \frac{dV}{V^{\gamma}}$$

$$W = \frac{p_1V_1 - p_2V_2}{\gamma - 1} \tag{12.64}$$

Note that after expansion, p_2 is smaller than p_1.

As this is an expansion process, the work is done by the system, that is, the system produces work output and this is shown by the direction of the arrow representing 'W'.

12.3.13.2 Heat Transfer

As stated in an adiabatic process, Q=0.

Applying an energy balance to this case (Figure 12.13b):

$$U_1 - W = U_2$$

$$W = U_1 - U_2 \tag{12.65}$$

In an adiabatic expansion, the work output will be equal to the decrease in internal energy, that is, because of the work output, the internal energy of the system will decrease by a corresponding amount.

Hence, for a perfect gas:

$$U_1 = mc_vT_1$$

$$U_2 = mc_vT_2$$

Substituting into Equation 12.65:

$$W = mc_v(T_1 - T_2) \tag{12.66}$$

Now

$$c_p - c_v = R$$

or

$$c_v = \frac{R}{\gamma - 1} \tag{12.67}$$

Substituting in Equation 12.66

$$W = \frac{mR(T_1 - T_2)}{\gamma - 1} \tag{12.68}$$

But $mRT_1 = p_1V_1$ and $mRT_2 = p_2V_2$

Substituting in Equation 12.68, the expression for the expansion becomes

$$W = \frac{p_1 V_1 - p_2 V_2}{\gamma - 1} \tag{12.69}$$

Example 12.15

A thermally insulated reciprocating compressor compresses air at 0.98 bar and 20°C to one-sixth of its original volume. Calculate the final pressure and volume of the air after compression.

If the mass of air within the cylinder is 0.05 kg, determine the required work input. For air, take $\gamma = 1.4$ and $c_v = 0.718$ kJ/kg K.

Solution

From the question, $p_1 = 0.98$ bar, $T_1 = 20 + 273 = 293$ K, $V_2/V_1 = 1/6$, $m = 0.05$ kg and $cv = 0.718$ kJ/kg.

As the cylinder is thermally insulated, it is considered that heat transfer is negligible and that the process is treated as adiabatic and considers the air as a perfect gas.

From Equation 12.61:

$$\frac{p_2}{p_1} = \left[\frac{V_1}{V_2}\right]^{\gamma}$$

Now

$$p_2 = 0.98 \times 6^{1.4}$$

$$= 12.04 \text{ bar}$$

From Equation 12.62:

$$\frac{T_2}{T_1} = \left[\frac{V_1}{V_2}\right]^{\gamma - 1}$$

$$T_2 = 293 \times 6^{0.4}$$

$$= 599.97 \text{ K} \quad \text{say } 600 \text{ K}$$

$$= 327°\text{C}$$

Rewriting Equation 12.66:

$$W = mc_v (T_1 - T_2)$$

$$W = 0.05 \times 0.718 (600 - 293)$$

$$= 11.02 \text{ kJ}$$

12.3.14 Polytropic Process ($pV^n = C$)

In practice, it is found that many processes approximate to a reversible law in the form

$$pV^n = \text{constant} \tag{12.70}$$

where n is a constant. Vapours and perfect gases obey this type of law closely in many non-flow processes. Such processes are internally reversible.

If a piston in a cylinder is cooled perfectly and a compression or expansion is carried out slowly, the process will be isothermal where $n=1$.

If a compression is carried out rapidly and again the piston and cylinder are perfectly insulated, the process will be adiabatic and in this case the constant $n=\gamma$.

When a compression or expansion is carried out at a moderate speed and the piston and cylinder assembly are cooled to some degree, the process will be somewhere between the two conditions described above. This will be the general situation in many engineering applications where the index 'n' will take some value that is between 1 and γ dependent upon the degree of cooling.

Some examples include

Compression in a water-cooled air compressor: $n=1.1$

Compression in a fan-cooled air compressor: $n=1.2$

Compression in an air-cooled air compressor: $n=1.3$

Figure 12.14a and b depicts the polytropic process.

Equation 12.70 applies to the states 1 and 2 as

$$p_1 V_1^n = p_2 V_2^n \tag{12.71}$$

or

$$\frac{p_2}{p_1} = \left[\frac{V_1}{V_2}\right]^n \tag{12.72}$$

For a perfect gas, the general property relationship between the two states:

$$\frac{p_1 V_1}{T_1} = \frac{p_2 V_2}{T_2} \tag{12.73}$$

Manipulation of Equations 12.71 and 12.72 results in the following relationship:

$$\frac{T_2}{T_1} = \left[\frac{p_2}{p_1}\right]^{\frac{n-1}{n}} = \left[\frac{V_1}{V_2}\right]^{n-1} \tag{12.74}$$

When examining Equations 12.72 and 12.74, the following conclusions can be made:

- Any increase in volume will result in a decrease in pressure.
- Any increase in volume will result in a decrease in temperature.
- Any increase in pressure will result in an increase in temperature.

12.3.14.1 Work Transfer

Referring to Figure 12.14b, there is an increase in volume during the process and as the fluid expands, the expansion work is given by

$$W = \int_1^2 p\,dV$$

$$= \int_1^2 \frac{c}{V^n} dV \quad (pV^n = c, \text{a constant})$$

$$= c\int_1^2 \frac{dV}{V^n} \tag{12.75}$$

$$W = \frac{p_1V_1 - p_2V_2}{n-1} \tag{12.76}$$

Note that after expansion, p_2 is smaller than p_1.

Following the expansion process, the work is done by the system and this is shown by the direction of the arrow that represents work 'W' as shown in Figure 12.14b.

12.3.14.2 Heat Transfer

The energy balance is applied to this case as

$$\begin{aligned} U_1 - Q_{loss} - W &= U_2 \\ Q_{loss} &= (U_1 - U_2) - W \end{aligned} \tag{12.77}$$

or

$$W = (U_1 - U_2) - Q_{loss} \tag{12.78}$$

In a polytropic expansion, the work output is reduced due to the heat losses.

Examining Equation 12.78 and Figure 12.14b, it will be noted that during the polytropic process, as the volume increases there is a corresponding reduction in pressure. For a perfect gas, Equation 12.74 also shows that a decrease in pressure will also result in a drop in the system temperature.

Example 12.16

Following compression, the combustion gases in a petrol engine are at 35 bar and 900°C. The gases then expand through a volume ratio (V_2/V_1) of 8.5/1 and occupy $0.51\times10^{-3}m^3$ after expansion. The polytropic expansion index n=1.15 when the engine is air cooled. Calculate the temperature and pressure of the gas after expansion and establish what the work output will be?

Solution

From the question, p_1=30 bar, t_1=900°C (T_1=900+273=1173 K), $V_2=0.51\times10^{-3}m^3$, n=1.15 and V_2/V_1=8.5.

Treating the air as a perfect gas, for a polytropic process the property relationship is given by Equation 12.74

$$\frac{T_2}{T_1} = \left[\frac{V_1}{V_2}\right]^{n-1}$$

$$= 1173 \times \left[\frac{1}{8.5}\right]^{1.15-1}$$

$$= 850.9\ \text{K}$$

$$= 577.9°\text{C}$$

From Equation 12.72:

$$p_2 = p_1 \left[\frac{V_1}{V_2}\right]^{n}$$

$$= 35 \times \left[\frac{1}{8.5}\right]^{1.15}$$

$$= 2.99\ \text{bar}$$

and

$$\frac{V_2}{V_1} = 8.5$$

Then

$$V_1 = \frac{0.510 \times 10^{-3}}{8.5}$$

$$= 60 \times 10^{-6}\ \text{m}^3$$

Equation 12.76 gives the work output during the polytropic expansion:

$$W = \frac{p_1 V_1 - p_2 V_2}{n-1}$$

$$= \frac{(35 \times 10^5)(60 \times 10^{-6}) - (2.99 \times 10^5)(0.510 \times 10^{-3})}{1.15 - 1}$$

$$= 529.6\ \text{J}$$

$$= 0.530\ \text{kJ (Ans)}$$

12.3.15 Constant Volume Process

In certain chemical processes, fluids are held in fixed volume rigid-walled vessels whilst the fluid or gas is either heated or cooled as shown in Figure 12.15a and b. In this case, the process is considered a constant volume process as the vessel has a fixed volume.

The general property relation between the initial and final states of a perfect gas is applied as

$$\frac{p_1 V_1}{T_1} = \frac{p_2 V_2}{T_2}$$

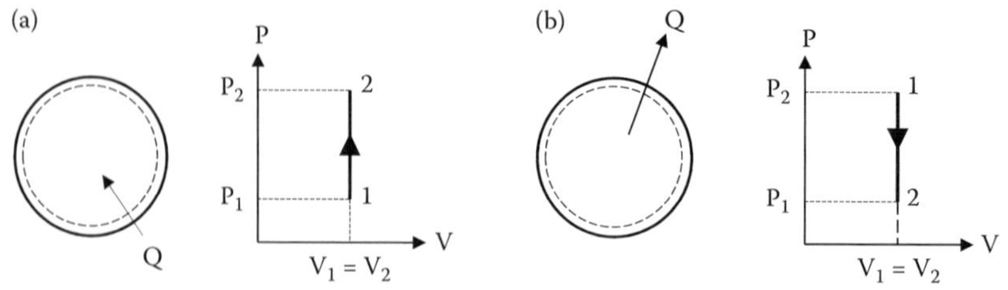

FIGURE 12.15
Constant volume process ($V_1 = V_2$). (a) Heating and (b) cooling.

As the volume remains constant during the process, $V_2 = V_1$ and the above equation then reduces to

$$\frac{p_1}{T_1} = \frac{p_2}{T_2} \tag{12.79}$$

or

$$\frac{T_2}{T_1} = \frac{p_2}{P_1} \tag{12.80}$$

From this equation, it is clearly seen that an increase in temperature will result in a corresponding increase in pressure. Hence, the temperature is proportional to the pressure.

12.3.15.1 Work Transfer

Work transfer (p dV) will be zero as the change in volume (dV) during the process will also be zero. Some process vessels (reactors) usually have a stirrer or paddlewheel installed to assist in mixing the fluid, in which case some work will be transferred.

12.3.15.2 Heat Transfer

Applying the non-flow energy equation from Equation 12.55

$$Q - W = U_2 - U_1$$

This will give

$$Q - 0 = U_2 - U_1$$

that is,

$$Q = U_2 - U_1 \tag{12.81}$$

Note: This result is important and shows that the net amount of heat energy supplied to or taken from the fluid during a constant volume process is equal to the change in the internal energy of the fluid.

Example 12.17

During a constant volume process, the specific internal energy of a fluid is increased from 120 to 180 kJ/kg. Calculate the amount of heat energy supplied to 2 kg of fluid to increase the internal energy.

Solution

From the non-flow energy equation:

$$Q - W = U_2 - U_1$$

For a constant volume process:

$$W = 0$$

Therefore, the equation becomes

$$\begin{aligned} Q &= U_2 - U_1 \\ &= (180 - 120)\ \text{kJ/kg} \\ &= 60\ \text{kJ/kg} \end{aligned}$$

For a mass of 2 kg of fluid:

$$\begin{aligned} Q &= 60 \times 2 \\ &= 120\ \text{kJ/kg} \end{aligned}$$

that is, 120 kJ/kg of heat energy will be required.

12.3.16 Constant Pressure Process

Intensifiers are used in hydraulic and gas supply systems to maintain a constant pressure within the system, even though the flow may be varying due to usage. These consist of having a cylinder fitted with a piston that has a constant load applied to it as depicted in Figure 12.16a and b. This is an example of a constant pressure process.

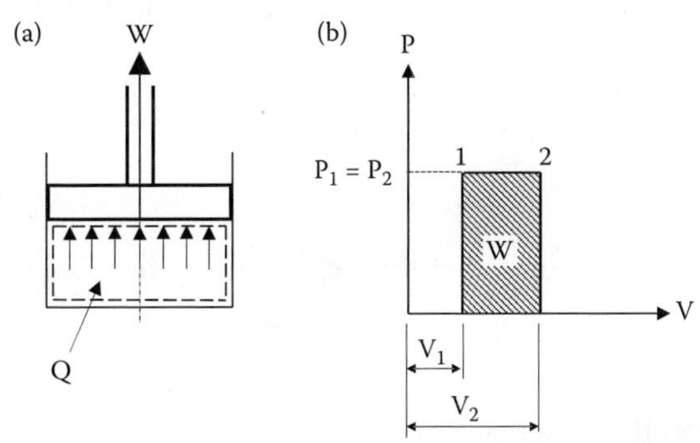

FIGURE 12.16
(a,b): Constant pressure process.

The general property relation between the initial and final states of a perfect gas is applied as

$$\frac{p_1 V_1}{T_1} = \frac{p_2 V_2}{T_2}$$

When the pressure remains constant during the process, $p_2 = p_1$, the above relation then becomes

$$\frac{V_1}{T_1} = \frac{V_2}{T_2}$$

or

$$\frac{T_2}{T_1} = \frac{V_2}{V_1} \tag{12.82}$$

It is obvious from this equation that any increase in volume will result in a corresponding increase in temperature in the fluid. Therefore, the temperature is proportional to the volume.

12.3.16.1 Work Transfer

Referring to the p–V diagram in Figure 12.16b, it will be noted that the increase in volume expands during the process. This expansion work is given by (similar to Equation 12.63)

$$W = \int_1^2 p\, dV$$

$$= p \int_1^2 dV \quad \text{(p is constant)}$$

$$= p(V_2 - V_1) \tag{12.83}$$

On a p–V diagram, the area under the process line represents the amount of work transfer. From Figure 12.16b

$$W = \text{Area of the shaded rectangle}$$

$$= \text{Height} \times \text{width}$$

$$= p(V_2 - V_1)$$ (It will be noted that this expression is identical to Equation 12.83.)

12.3.16.2 Heat Transfer

Again applying the non-flow energy equation

$$Q - W = U_2 - U_1$$

or

$$Q = (U_2 - U_1) + W \tag{12.84}$$

Part of the heat supplied is converted into work energy and the remainder is utilised in increasing the internal energy of the system.

Now, substituting for 'W' in Equation 12.84:

$$\begin{aligned} Q &= (U_2 - U_1) + p(V_2 - V_1) \\ &= U_2 - U_1 + p_2V_2 - p_1V_1 \quad (\text{since } p_2 = p_1) \\ &= (U_2 + p_2V_2) - (U_1 + p_1V_1) \end{aligned}$$

Now

$$H = U + pV$$

Hence,

$$Q = H_2 - H_1 \tag{12.85}$$

Referring to Figure 12.16b, it will be noted that heating will increase the volume, that is, the fluid will expand. For a perfect gas, the equation suggests that an increase in volume will result in a corresponding increase in temperature.

Example 12.18

A cylinder contains a fluid with a volume of $0.1\,m^3$ at a constant pressure of 7 bar and having a specific enthalpy of 210 kJ/kg. The volume expands to $0.2\,m^3$ following the application of heat energy to the fluid and the specific enthalpy increases to 280 kJ/kg.

Determine:

1. The quantity of heat energy supplied to the fluid.
2. The change in the internal energy of the fluid.

Solution

From the question, $p = 7.0$ bar, $V_1 = 0.1\,m^3$ and $V_2 = 0.2\,m^3$.

1. Heat energy supplied = change in enthalpy of the fluid:

$$\begin{aligned} Q &= H_2 - H_1 \\ &= m(h_2 - h_1) \\ &= 2.25(280 - 210) \\ &= 157.5\text{ kJ} \end{aligned}$$

2. For a constant pressure process:

$$\begin{aligned} W &= P(V_2 - V_1) \\ &= 7 \times 10^5 \times (0.2 - 0.1) \\ &= 7 \times 10^4\text{ J} \\ &= 70\text{ kJ} \end{aligned}$$

Applying the non-flow energy equation

$$Q - W = U_2 - U_1$$

gives

$$U_2 - U_1 = 157.5 - 70$$
$$= 87.5 \text{ kJ}$$

13

Fluid Mechanics

13.1 Fluid Properties

The term fluid covers both liquids and gasses. It should be noted that there is a degree of commonality in Chapter 12, where a number of terms used are also used in fluid mechanics. Table 13.1 lists some of these terms used together with their units.

A liquid is hard to compress and takes the shape of the container it is in. It has a fixed volume and an upper-level surface. However, although liquids are generally considered to be incompressible, they are compressible only when they are highly pressurised.

Gasses are very easy to compress and expand to fill their container and unlike liquids, they do not have a free surface.

The important characteristics of fluids from the point of view of fluid mechanics include density, pressure, viscosity and compressibility.

13.1.1 Density

There are three forms of density and a distinction must be carefully made between them.

1. *Mass density* (ρ). This is the mass per unit volume with the unit of kilogram per cubic meter. The density of gas changes according to its pressure, but that of a liquid may be considered constant unless the relevant pressures are very high.
2. *Specific weight* (g). In fluid mechanics, specific weight represents the force exerted by gravity on a unit volume of a fluid. For this reason, the unit is expressed as force per unit volume (N/m^3). Specific weight can be used as a characteristic property of a fluid.
3. *Specific gravity (sg) or relative density (RD).* This is a dimensionless quantity being the ratio of the density (mass of a unit volume) of a substance to the density of a given reference material. Sg usually means relative density with respect to water. The term 'relative density' is often preferred in modern scientific usage.

The sg for liquids and solids is generally identified as the ratio of the material density relative to water under the same conditions, say 4°C and 1.013 bar (ambient pressure). For gasses, the sg is the ratio of the gas density to the density of air, both under the same conditions as identified above.

For a liquid:

$$RD = \frac{\rho}{\rho_w} \tag{13.1}$$

TABLE 13.1

Symbols Used in Fluid Mechanics

Symbol	Description	Units	Symbol	Description	Units
a	Acceleration	m/s^2	R	Gas constant	J/(kg·K)
A	Area	m^2	R_o	Universal gas constant	J/(kg·mol·K)
F	Force	N	ρ	Fluid density	$kg{\cdot}m^3$
g	Acceleration due to gravity	m/s^2	ρ	Density	kg/m^3
h	Fluid head	m	ρ_r	Density	kg/m^3
K	Bulk modulus	MPa	s	Specific volume	m^3/kg
m	Mass	kg	u	Fluid velocity	m/s
M	Molecular weight		v	Fluid velocity	m/s
p	Fluid pressure	N/m^2	x	Depth of centroid	m
P_{abs}	Absolute pressure	N/m^2	β	Compressibility	1/MPa
P_{gauge}	Gauge pressure	N/m^2	θ	Slope	Radians
P_{atm}	Atmospheric pressure	N/m^2	τ	Shear stress	N/m^2
P_s	Surface pressure	N/m^2	μ	Viscosity	Pa·s
Q	Volumetric flow	m^3/s	v	Kinematic viscosity	m^2/s
q	Heat transfer/unit mass	J/kg	υ	Specific volume	m^3/kg
			γ	Ratio of specific heats	

For a gas:

$$D = \frac{\rho}{\rho_a} \tag{13.2}$$

The density of gasses and vapours is greatly affected by pressure. For 'perfect gasses', the density can be calculated from the formula:

$$\rho = \frac{P}{RT} \tag{13.3}$$

$$R = \frac{R_o}{M} \tag{13.4}$$

'R_o' is the universal gas constant=8314 J/kg K and 'M' is the molecular weight of the material. Hence,

$$R = \frac{8314}{M} \frac{J}{kg\ K} \tag{13.5}$$

The reciprocal of density, that is, the volume per unit mass is called the specific volume (υ)

$$\upsilon = \frac{1}{\rho} \tag{13.6}$$

The dimensionless unit for density is ML^{-3} and the corresponding unit for specific volume is $M^{-1} L^3$ (see Section 13.5).

13.1.2 Pressure

On planet Earth, a fluid is always subject to pressure and is the force per unit area at a point. The absence of pressure can only occur in a complete vacuum. A complete vacuum is really a theoretical concept as even in deep space, there is a partial pressure.

The normal pressure occurring on the surface of the Earth is called the 'atmospheric pressure'. Pressure is measured in two ways:

- Absolute pressure, where the pressure is measured relative to that of a perfect vacuum.
- Gauge pressure, where the pressure is measured relative to the local atmospheric pressure. These measured pressures are referred to as gauge pressures.

Figure 13.1 shows the relationship between the gauge and the atmospheric pressure. The figure shows two measurements:

1. A pressure less than atmospheric pressure
2. A pressure greater than atmospheric pressure

The SI unit for pressure is Pascal (Pa) with the unit of N/m^2.

13.1.3 Static Pressure and Head

When considering fluid pressure, it has been found more convenient in hydrostatics and fluid dynamics to use a pressure head as a means of measuring pressures. Considering Figure 13.2, a volume of fluid in an open container is subject to atmospheric pressure acting on the surface of the fluid. Consider a vertical tube that is connected to the bottom of the container and terminates at a distance above the surface of the fluid into a vessel held at a pressure of absolute vacuum. It will be found that the fluid will be forced up the tube to a point where gravity acting on the fluid in the tube balances the force due to the pressure at the bottom of the head of the fluid. Assuming that the area of the tube is A_t and the density of the fluid is ρ, the pressure at the top of the tube is zero. The force of the fluid acting on the section 'x:x' as shown in the figure will be

$$F_{xx} = 0 + hA\rho g \tag{13.7}$$

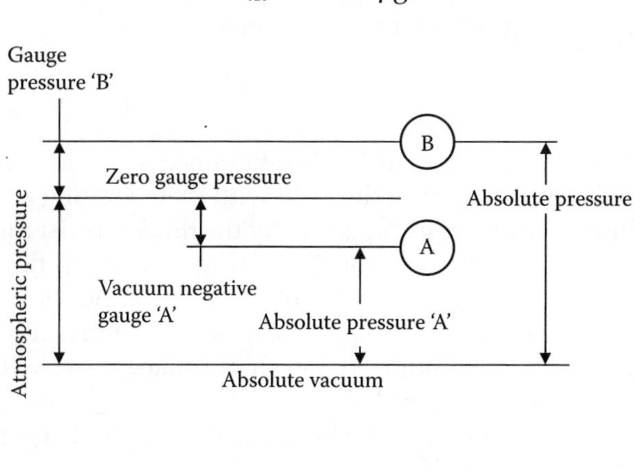

FIGURE 13.1
Atmospheric scale.

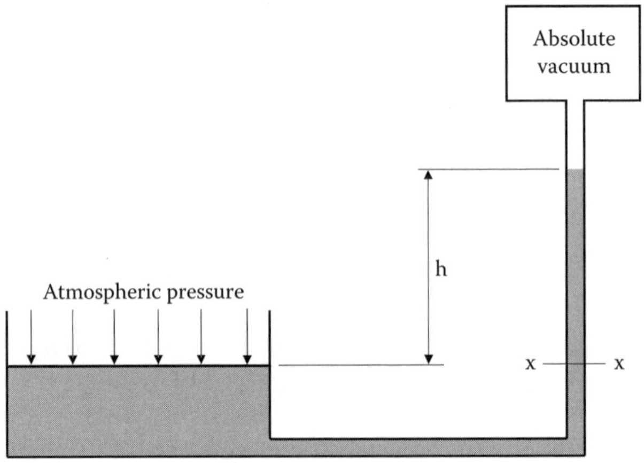

FIGURE 13.2
Atmospheric pressure.

The pressure at 'x:x' is equal to

$$P_{xx} = \frac{0 + hA\rho g}{h\rho g}$$

For a fluid having a known fixed density, the height 'h' can be conveniently used to identify the pressure. For water, the atmospheric pressure is approximately 10.5 m. In practice, water will vaporise into the vacuum at the top of the tube, thereby reducing the column height by approximately 180 mm.

Therefore, mercury is used for measuring pressure. With an atmospheric pressure of 1 bar, this will support a column height of 750.06 mm of mercury.

Mercury is used as it has a low-vapour pressure and the vacuum will only be reduced by approximately 0.16 Pa. This is very small compared to an atmospheric pressure of 10^5 Pa (1 bar).

From this description, it is clear that gauge and vacuum pressures can be easily determined using this simple method. As a point of interest, barometers measure the local atmospheric pressure readings in millimeters of mercury (Hg).

13.1.4 Viscosity

A fluid at rest cannot resist any shearing forces, but once the fluid is in motion, shearing forces are then set up between layers of the fluid with them moving at different velocities. The viscosity of a fluid is a measure of the ability of the fluid to resist these shearing forces.

Perfect fluids cannot, in theory, transmit shear stresses. All real fluids will resist shear flow to some degree or another. The viscosity of the fluid defines the degree of resistance to flow it possesses. Consider Figure 13.3a that shows a concentric tube located on a shaft and separated from the shaft by a fluid. As the shaft rotates, there will be a tendency for the tube to rotate at an angular velocity (ω).

The velocity distribution is shown in Figure 13.3b. The torque required to rotate the tube is an indication of the viscosity of the fluid.

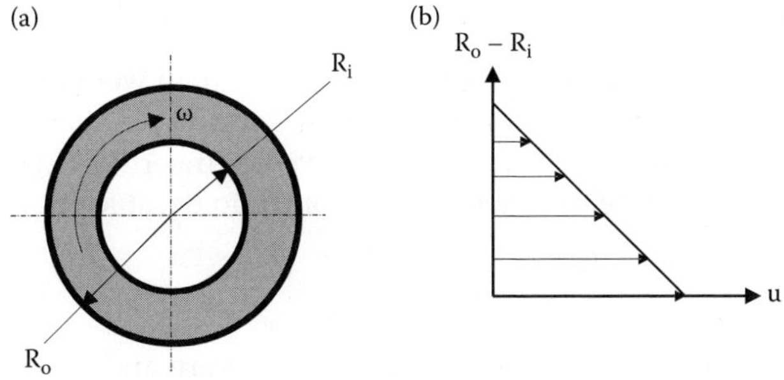

FIGURE 13.3
(a, b) Viscosity of a fluid.

13.1.4.1 Coefficient of Dynamic Viscosity

The coefficient of dynamic viscosity (η) is defined as the shear force per unit area that is required to drag one layer of a fluid with a unit velocity past another layer a unit distance away from it in the fluid. The SI unit of measurement for dynamic viscosity is

$$\text{Coefficient of dynamic viscosity} = \frac{\text{N} \cdot \text{s}}{\text{m}^2} \text{ or } \frac{\text{kg}}{\text{m} \cdot \text{s}} \text{ or Pa} \cdot \text{s}$$

The old unit for dynamic viscosity (η) was dyn·s/cm^2 and this was called a 'poise' (after Poiseuille) and the SI unit is related to the Poise:

$$10 \text{ Poise} = 1\frac{\text{Ns}}{\text{m}^2} \quad \text{(this is not considered an acceptable multiple)}$$

Since

$$1 \text{ centipoise (1 cP)} = 0.001\frac{\text{N} \cdot \text{s}}{\text{m}^2} \quad \text{cP is an accepted SI unit}$$

13.1.4.2 Kinematic Viscosity

Kinematic viscosity (v) is the ratio of the dynamic viscosity to the mass density:

$$v = \frac{\eta}{\rho} \tag{13.8}$$

The base unit is m^2/s. The previous metric unit was cm^2/s and this was called the 'Stoke' after a British scientist. The SI unit is related to the Stoke.

$$1 \text{ Stoke (St)} = 0.0001\frac{\text{m}^2}{\text{s}} \quad \text{(this is not considered acceptable as an SI unit)}$$

Since

$$1 \text{ CentiStoke (cSt)} = 0.000001\frac{\text{m}^2}{\text{s}} \quad \text{and this is an acceptable multiple}$$

$$1 \text{ cSt} = 0.00000\frac{\text{m}^2}{\text{s}} = 1\frac{\text{mm}^2}{\text{s}}$$

13.1.4.3 Other Units

Other units of viscosity will be encountered by the way they were originally derived. As an example, Redwood seconds came from the name of the Redwood viscometer. Other units include Engler's degrees, Society of Automotive Engineers (SAE) numbers and so on. Conversion charts are available to enable conversions into useable engineering or SI units.

13.1.5 Compressibility

For liquids, the relationship between the change in pressure and the change in volume is given by the bulk modulus 'K'

$$\text{Bulk modulus} = \frac{\text{Change in pressure intensity}}{\text{Volumetric strain}} \tag{13.9}$$

$$= \frac{\text{Change in pressure intensity}}{(\text{Change in volume/original volume})} \tag{13.10}$$

The relationship between pressure and volume for a gas can be derived from the gas laws.

For all perfect gases:

$$pV = RT \tag{13.11}$$

where

p represents absolute pressure
V represents specific volume=1/w
T represents absolute temperature
R represents gas constant

If any change occurs isothermally (at constant temperature):

$$pV = \text{constant} \tag{13.12}$$

If any change occurs adiabatically (no heat gain or loss):

$$pV^{\gamma} = \text{constant} \tag{13.13}$$

where γ represents the ratio of the specific heat at constant pressure to the specific heat at constant volume.

13.2 Fluid Flow

When a fluid is at rest, there is no shear force developed but when the fluid is in motion, shear forces set up due to viscosity and turbulence, opposing the motion and producing frictional effects.

This section will study the patterns of flow, both inside a conduit and outside it.

13.2.1 Patterns of Flow

A fluid may be considered as consisting of a large number of individual particles moving in the general direction of flow but usually, are not parallel with each other. The velocity of any particle is a vector quantity having a magnitude and direction that varies from moment to moment. The path followed by a particle is called a 'streamline' or 'path line'.

When considering the flow of a large body of fluid, it is sometimes convenient to consider a small section. If streamlines are drawn through every point on the circumference of a small area, a 'stream tube' is formed (see Figure 13.4). Fluid particles can only flow along a streamline and particles cannot flow across the streamline; the fluid inside a stream tube can only enter or leave at its ends.

13.2.2 Types of Flow

There are two types of flow.

13.2.2.1 Internal Flow

Internal flow is within the boundary walls and these include all types of flow within pipes, channels, airflow within ducts and so on.

13.2.2.2 External Flow

External flow covers the flow that is outside a boundary, body or conduit. Examples of these types of flow include immersed bodies, airflow around buildings and flow over aircraft wings and suspension bridges and around automobile vehicles.

Two distinct types of flow can occur; these are as follows.

13.2.2.3 Laminar Flow

This type of flow is also known as viscous flow or streamline in which the particles move in an orderly manner and retain the same relative positions in successive cross sections.

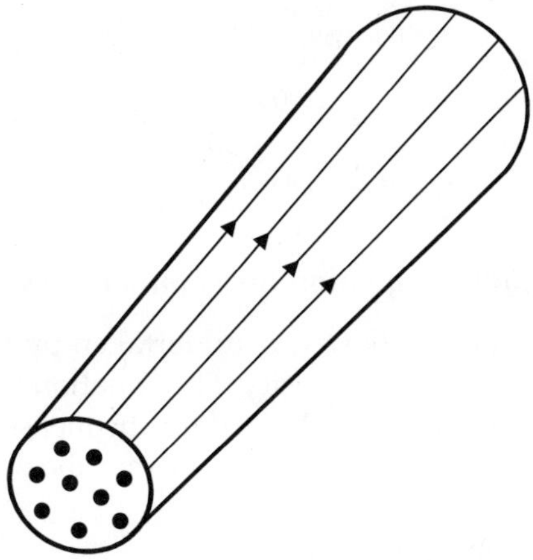

FIGURE 13.4
Stream tube.

13.2.2.4 Turbulent Flow

In this type of flow, it may be visualised that the particles of fluid move in a disorderly manner if they occupy different relative positions in successive cross sections.

13.2.3 Laminar Flow

Following a study of laminar flow, Osborne Reynolds concluded that this type of flow can be defined by the velocity, density and viscosity of the fluid together with the dimensional size of the conduit and depends on the value of

$$\frac{\rho v d}{\eta} \tag{13.14}$$

This relationship is known as 'Reynolds number', in which
ρ represents the mass density of the fluid
v represents velocity
d represents a typical dimension for the size of the pipe
η represents the viscosity of the fluid

When the Reynolds number is <2100 approximately, the flow will always be viscous. If the relevant Reynolds number is >4000, the flow will be turbulent. When the flow in the transition region is termed 'critical', it may be either laminar or turbulent or a mixture of both.

Example 13.1

A fluid with a density of 860 kg/m³ has a kinematic viscosity of 40 cSt. Determine the critical velocity when it is flowing through a pipe having a bore of 50 mm.

Solution

$$R_e = \frac{u_m d}{v}$$

$$u_m = \frac{R_e v}{d}$$

$$= \frac{2100 \times 40 \times 10^{-6}}{0.05}$$

$$= 1.68\ \text{m/s}$$

13.2.4 Derivation of Poiseuille's Equation for Laminar Flow

The original derivation for laminar flow was undertaken by Poiseuille and relates pressure loss in a pipe to the velocity and viscosity. The equation is the basis for the measurement of viscosity; hence, his name has been used for the measurement of viscosity.

Consider a pipe having a laminar flow in it. A stream tube of length DL, a radius 'r' and thickness dr is shown in Figure 13.5.

Let 'y' be the distance from the pipe wall:

$$y = R - r \tag{13.15}$$

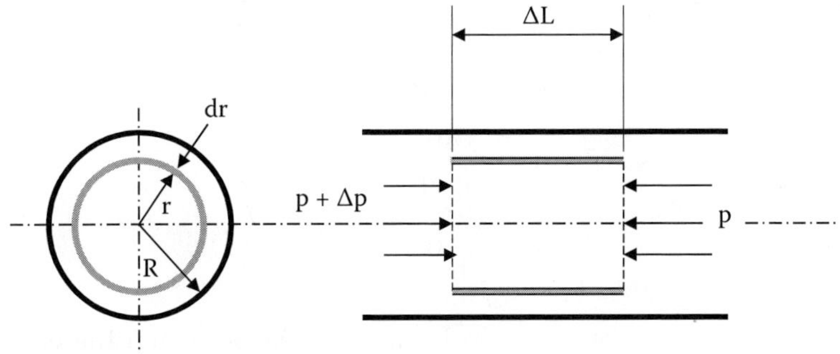

FIGURE 13.5
Derivation for laminar flow.

$$dy = -dr \frac{du}{dy} = -\frac{du}{dr} \tag{13.16}$$

The shear stress on the outside of the stream tube is 'τ'. The force (F_s) acting from right to left is due to the shear stress and is obtained by multiplying 'τ' by the surface area.

Hence,

$$F_s = \tau \times 2\pi r \Delta L \tag{13.17}$$

For a Newtonian fluid:

$$\tau = \mu \frac{du}{dy} = -\mu \frac{du}{dr} \tag{13.18}$$

Substituting for 'τ', the following equation is obtained:

$$F_s = -2\pi r \Delta L \mu \frac{du}{dr} \tag{13.19}$$

The pressure difference between the left end and the right end of the section is Δp. The force due to F_p is Δp × circular area of radius 'r'.

$$F_p = \Delta p \times pr^2$$

Equating forces:

$$-2\pi r \mu \Delta L \frac{du}{dr} = \Delta p \pi r^2$$
$$du = \frac{\Delta p}{2\mu \Delta L} r dr \tag{13.20}$$

Integrate Equation 13.20 to obtain the velocity of the streamline at any radius 'r' between the limits u=0 when r=R and u=u when r=r.

$$\int_0^u du = -\frac{\Delta p}{2\mu\Delta L}\int_R^r r\,dr$$

$$u = -\frac{\Delta p}{4\mu\Delta L}(r^2 - R^2) \tag{13.21}$$

$$u = \frac{\Delta p}{4\mu L}(r^2 - R^2)$$

It is recognised that this is the equation for a parabola; so, when the equation is plotted to show the boundary layer, it extends from zero to a maximum value at the centre (Figure 13.6).

For the maximum velocity, substitute r=0:

$$u_1 = \frac{\Delta p R^2}{4\mu\Delta L} \tag{13.22}$$

The average height of a parabola is the maximum value/2; so, the average velocity

$$u_m = \frac{\Delta p R^2}{8\mu\Delta L} \tag{13.23}$$

It may be required to calculate the pressure drop in terms of diameter 'd'. Therefore, substitute R=d/2 and rearrange the equation:

$$\Delta p = \frac{32\mu\Delta L\mu_m}{d^2} \tag{13.24}$$

The volumetric flow rate is

Average velocity × cross-sectional area

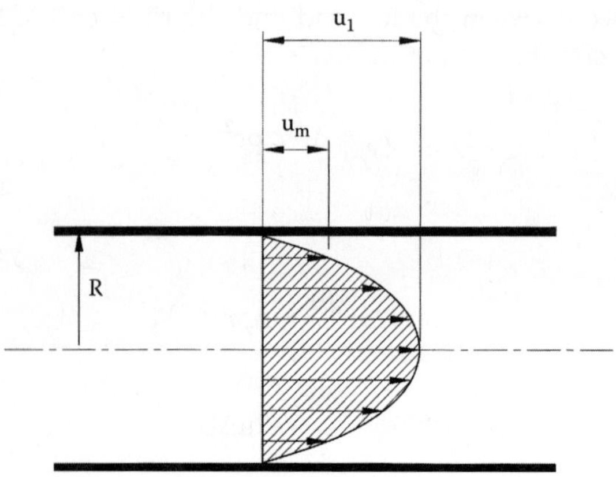

FIGURE 13.6
Laminar flow in a pipe.

Hence,

$$Q = \frac{\pi R^2 \Delta p R^2}{8\mu \Delta L}$$

$$= \frac{\pi R^4 \Delta p}{8\mu \Delta L} \qquad (13.25)$$

$$= \frac{\pi d^4 \Delta p}{128\mu \Delta L}$$

This equation can be rearranged for the pressure head; the friction head for a length 'L' is derived from

$$h_f = \frac{\Delta p}{\rho g} \qquad (13.26)$$

Hence,

$$h_f = \frac{32\mu L \mu_m}{\rho g d^2} \qquad (13.27)$$

This is Poiseuille's equation and applies only to laminar flow.

Example 13.2

A capillary tube is 30 mm long and has a 1.0 mm bore. To obtain a flow rate of 8 mm^3/s requires a head of 30 mm. The density of the fluid is 800 kg/m^3.

Calculate the dynamic and kinematic viscosity of the fluid.

Solution

Rearranging Poiseuille's equation:

$$\mu = \frac{hf \rho g d^2}{32 L u_m}$$

Now

$$A = \frac{\pi d^2}{4}$$

$$= \frac{\pi \times 1.0^2}{4}$$

$$= 0.7854\ mm^2$$

$$u_m = \frac{Q}{A}$$

$$= \frac{8}{0.7854}$$

$$= 10.18\ mm/s$$

Substituting values:

$$\begin{aligned}\mu &= \frac{0.03 \times 800 \times 9.81 \times 0.001^2}{32 \times 0.03 \times 0.01018} \\ &= 0.02409\ \text{N} \cdot \text{s/m} \quad \text{or} \\ &= 241\ \text{cP}\end{aligned}$$

$$\begin{aligned}v &= \frac{\mu}{\rho} \\ &= \frac{0.02409}{800} \\ &= 30.114 \times 10^{-6}\ \text{m}^2/\text{s} \quad \text{or} \\ &= 30.11\ \text{cSt}\end{aligned}$$

Example 13.3

A fluid is flowing through a pipe of 150 mm bore having a Reynolds number of 250. The dynamic viscosity is 0.018 N·s/m² and the density is 900 kg/m³.

Calculate the pressure drop per metre length together with the velocity and radius at which it occurs.

Solution

$$R_e = \frac{\rho \mu_m d}{\mu}$$

$$\mu_m = \frac{R_e \cdot \mu}{\rho d}$$

$$\begin{aligned}\mu_m &= \frac{(250 \times 0.018)}{(900 \times 0.15)} \\ &= 0.0333\ \text{m/s}\end{aligned}$$

$$\begin{aligned}\Delta p &= \frac{32 \mu L \mu_m}{d^2} \\ &= \frac{32 \times 0.018 \times 1.0 \times 0.05}{0.15^2} \\ &= 1.28\ \text{Pa}\end{aligned}$$

$$u = \left(\frac{\Delta p}{4L\mu}\right)(R^2 - r^2)$$

The above equation is rearranged to solve for 'r':

$$\begin{aligned}r &= \sqrt{R^2 - \frac{4L\mu u}{\Delta p}} \\ &= 0.0692\ \text{m} \\ r &= 69.2\ \text{mm}\end{aligned}$$

13.2.5 Turbulent Flow

In the previous section, it was shown that Poiseuille's equation applies only to laminar flow. At some critical velocity, the flow will begin to become turbulent forming eddies and showing other chaotic behaviour that do not contribute to the volumetric flow rate. This turbulence increases the resistance dramatically so that large increases in pressure will be required to further increase the volume flow rate.

In a circular pipe, turbulence is generated by the friction existing on the internal surface of the pipe. If the internal surface was completely smooth without any surface roughness, then the flow would remain laminar at higher Reynolds numbers into the critical zone. Where there is any surface roughness, this will initiate turbulence at lower Reynolds numbers. Consider Figure 13.7 that shows the internal surface roughness of a pipe. Relative roughness is expressed as D/ε.

That is, less rough pipes will have a high D/ε whereas more rough pipes will have lower D/ε.

Table 13.2 gives some comparative roughness values for materials commonly used in fluid transport.

To begin the discussion of turbulent flow in circular pipes, it is necessary to define the friction coefficient. It is defined as follows:

$$\begin{aligned} C_f &= \frac{\text{wall shear stress}}{\text{dynamic pressure}} \\ &= \frac{2D\Delta p}{4L\rho\mu_m^2} \end{aligned} \tag{13.28}$$

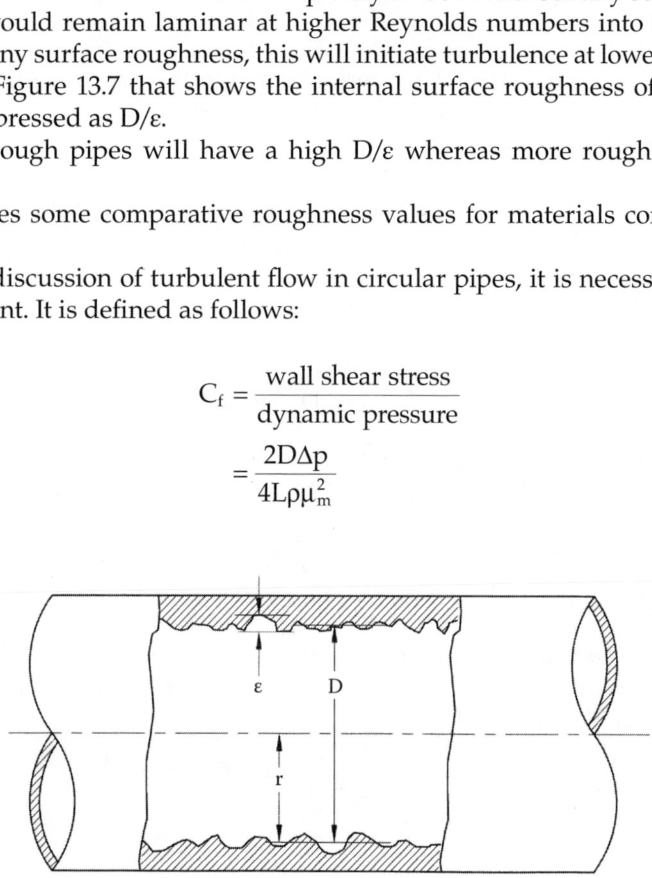

FIGURE 13.7
Surface roughness in a pipe.

TABLE 13.2
Comparative Roughness Values for a Range of Materials

Material	Surface Roughness ε (m)
Plastic	3.0×10^{-7}
Steel plain	4.6×10^{-5}
Steel galvanised	1.5×10^{-4}
Concrete	1.2×10^{-4}

Rearranging Equation 13.28 to equate to Δp:

$$\Delta p = \frac{4C_f L \rho \mu_m^2}{2D} \tag{13.29}$$

This expression is frequently used to state the friction head h_f.

$$h_f = \frac{\Delta p}{\rho g} = \frac{4C_f L u_m^2}{2gD} \tag{13.30}$$

This expression is known as the 'Darcy formula' and in the case of laminar flow, Darcy's formula must give the same result as Poiseuille's equation.

Hence, equating the two expressions:

$$\frac{4C_f L u_m^2}{2gD} = \frac{32\mu L u_m}{\rho g D^2} \tag{13.31}$$

Hence

$$C_f = \frac{16\mu}{\rho u_m D} = \frac{16}{R_n} \tag{13.32}$$

This gives the same result as for laminar flow.

In pipes where the Reynolds number exceeds 3000, it can be safely assumed that turbulent flow will exist.

13.2.6 Fluid Resistance

An alternative approach to solving problems that involve losses is to use fluid resistance. The above equations can be expressed in terms of the flow rate 'Q' by substituting $u=Q/A$.

From Equation 13.30:

$$h_f = \frac{4C_f L u_m^2}{2gD} \tag{13.33}$$

$$= \frac{4C_f L Q^2}{2gDA^2} \tag{13.34}$$

Substituting $A=\pi D^2/4$, the following equation is derived:

$$h_f = \frac{32C_f L Q^2}{g\pi^2 D^5} = RQ^2 \tag{13.35}$$

where R is the fluid resistance; hence

$$R = \frac{32C_f L}{g\pi^2 D^5} \tag{13.36}$$

Note: This equation contains the friction coefficient (C_f) and this will vary with changes in velocity and surface roughness; therefore, 'R' should be used with caution.

13.2.7 Moody's Diagram

An American engineer L. F. Moody (1944) conducted a series of experiments that resulted in a chart known as the Moody chart (see Figure 13.8). This shows a graph in non-dimensional form that relates friction factor, Reynolds number and the relative roughness in circular pipes.

The diagram shows that the friction coefficient decreases with the Reynolds number but at a certain point, it becomes constant. When this point is reached, the flow is said to be a fully developed turbulent. This point occurs at lower Reynolds numbers for rougher pipes.

A formula that gives an estimated value for any surface roughness was proposed by Professor Haaland of the Norwegian Institute of Technologies in 1984.

$$\frac{1}{\sqrt{C_f}} = -3.6\log_{10}\left\{\frac{6.9}{R_n} + \left(\frac{e}{3.71}\right)^{1.11}\right\} \tag{13.37}$$

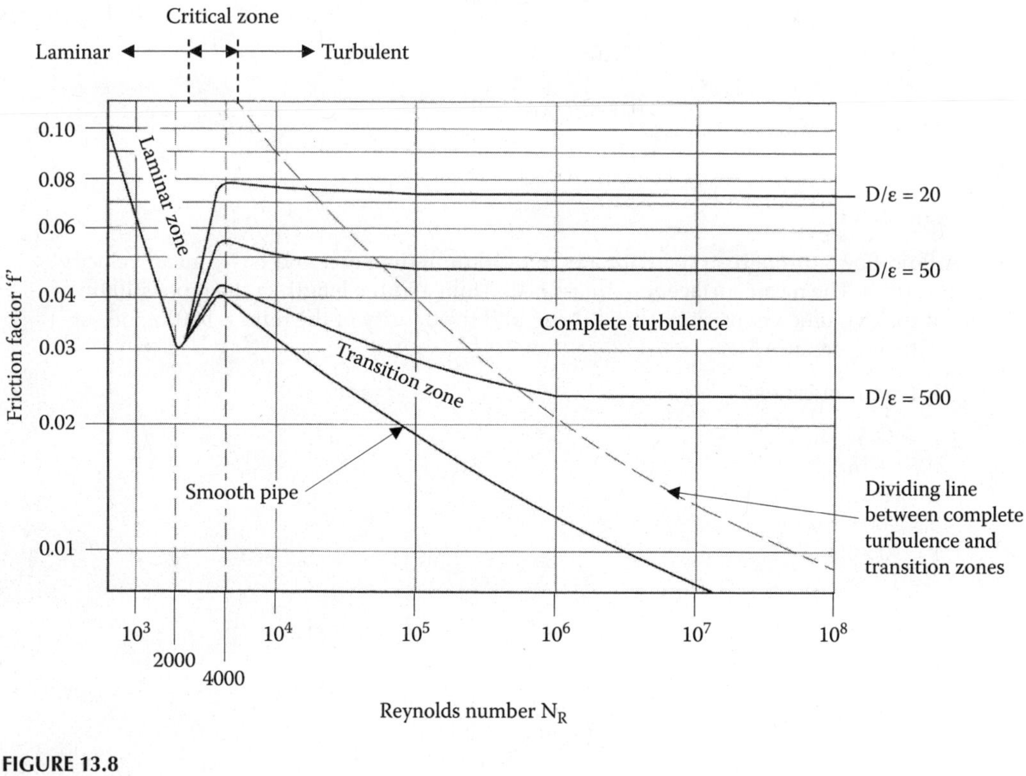

FIGURE 13.8
The Moody diagram.

Example 13.4

A circular pipe with a 100-mm-diameter bore, having a mean surface roughness of 0.06 mm, has a fluid flowing through it with a Reynolds number of 20,000.

Calculate the friction coefficient.

Solution

The mean surface roughness:

$$\begin{aligned}\varepsilon &= k / d \\ &= \frac{0.06}{100} \\ &= 0.006\end{aligned}$$

From the Moody diagram (Figure 13.8), locate the line for $\varepsilon = k/d$.

Trace this line till it meets the vertical line at $R_n = 20{,}000$. Read off the value for C_f on the left-hand axis.

The answer will be $C_f = 0.0067$.

This answer can be checked using the Haaland formula:

$$\begin{aligned}\frac{1}{\sqrt{C_f}} &= -3.6\log_{10}\left\{\frac{6.9}{R_n} + \left(\frac{e}{3.71}\right)^{1.11}\right\} \\ &= -3.6\log_{10}\left\{\frac{6.9}{20{,}000} + \left(\frac{0.0006}{3.71}\right)^{1.11}\right\} \\ \frac{1}{\sqrt{C_f}} &= 12.206 \\ C_f &= 0.0067 \text{ (Ans)}\end{aligned}$$

Example 13.5

A fluid flows through a pipe with a bore of 100 mm diameter and has a mean velocity of 6.0 m/s. The mean surface roughness is 0.03 mm and the length of the pipe is 100 m.

If the dynamic viscosity is 0.005 N·s/m^2 and the density of the fluid is 900 kg/m^3, calculate the pressure loss.

From the question:

$\rho = 900 \text{ kg/m}^3$
$u = 6.0 \text{ m/s}$
$d = 0.10 \text{ m}$
$\eta = 0.005 \text{ N·s/m}^2$
$k = 30 \times 10^{-6} \text{m}$
$L = 100 \text{ m}$

$$\begin{aligned}R_n &= \frac{\rho u d}{\eta} \\ &= \frac{900 \times 6.0 \times 0.1}{0.005} \\ &= 108{,}000\end{aligned}$$

$$\varepsilon = \frac{k}{d}$$
$$= \frac{30 \times 10^{-6}}{0.1}$$
$$= 0.0003$$

Using Haaland's formula and checking with Moody's chart:

$$\frac{1}{\sqrt{C_f}} = -3.6 \log_{10}\left\{\frac{6.9}{R_n} + \left(\frac{\varepsilon}{3.71}\right)^{1.11}\right\}$$
$$= -3.6 \log_{10}\left\{\frac{6.9}{108{,}000} + \left(\frac{0.0003}{3.71}\right)^{1.11}\right\}$$
$$= 14.521$$
$$C_f = 0.00474$$

$$h_f = \frac{4 \cdot C_f \cdot L \cdot u^2}{2 \cdot d \cdot g}$$
$$= \frac{4 \times 0.00474 \times 100 \times 6.0^2}{2 \times 0.10 \times 9.81}$$
$$= 34.821 \text{ m}$$

$$\Delta p = \rho \cdot g \cdot h_f$$
$$= 900 \times 9.81 \times 34.821$$

Pressure drop due to the length of the pipe:

$$= 307.326 \text{ kPa (Ans)}$$

13.3 Continuity Equation

The continuity equation is a mathematical version of the principle of the conservation of mass as applied to fluid flow. There are two laws that apply:

- The law of conservation of mass (continuity equation)
- The law of conservation of energy (Bernoulli's equation)

13.3.1 Conservation of Mass

Consider a fluid flowing at a constant rate in a pipe or conduit; the mass flow rate must be the same in all sections along the length. Figure 13.9 shows the length of piping connected to a pump filling a storage tank.

The mass flow rate in any section will be

$$m = \rho A u_m \tag{13.38}$$

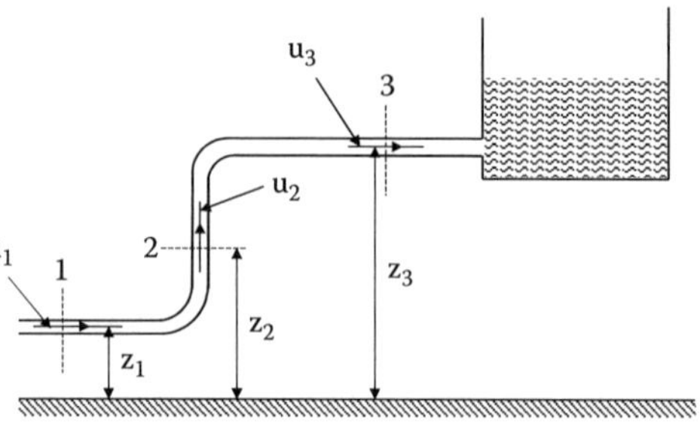

FIGURE 13.9
Conservation of mass.

where
ρ represents fluid density (kg/m³)
A represents cross-sectional area (m²)
u_m represents mean fluid velocity (m/s)

For the system shown in Figure 13.9, the mass fluid flow rate must be the same in sections (1), (2), and (3); hence

$$\rho_1 A_1 u_1 = \rho_2 A_2 u_2 = \rho_3 A_3 u_3 \tag{13.39}$$

When the liquid is a fluid, the density will be equal in all sections; therefore, it cancels out, so:

$$A_1 u_1 = A_2 u_2 = A_3 u_3 = \text{constant} \tag{13.40}$$

13.3.2 Conservation of Energy

Forms of energy.

13.3.2.1 Flow Energy

This is the energy a fluid possesses due to its pressure.

$$\text{Flow energy (FE)} = p \cdot Q \text{ (joules)}$$

p represents the pressure in the fluid (Pascal)
Q represents volumetric flow rate (m^3)

13.3.2.2 Potential Energy

The energy a fluid possesses due to its height relative to a datum level.

$$\text{Potential energy (PE)} = mgz \text{ (joules)} \tag{13.41}$$

where
m represents mass (kg)
g represents gravitational constant (m/s^2)
z represents the height above the datum (m)

13.3.2.3 Kinetic Energy

This is the energy a fluid possesses due to its velocity.

$$\text{Kinetic energy (KE)} = \frac{1}{2}\mu_m^2 \text{ (joules)} \quad (13.42)$$

where μ represents the mean velocity (m/s).

13.3.2.4 Specific Energy

Specific energy is the energy per kilogram.

There are three forms of specific energy:

1. $$\frac{\text{Flow energy}}{m} = \frac{p \cdot Q}{m} = \frac{p}{\rho} \text{ (joules/kg)} \quad (13.43)$$

2. $$\frac{\text{Potential energy}}{m} = g \cdot z \text{ (joules/kg)} \quad (13.44)$$

3. $$\frac{\text{Kinetic energy}}{m} = \frac{1}{2}u^2 \text{ (joules/kg)} \quad (13.45)$$

13.3.2.5 Energy Head

When the energy terms are divided by the weight (m·g), energy per Newton will be the result. Studying the units closely:

$$\frac{J}{N} = \frac{N \cdot m}{N} = m \text{ (metres)}$$

This form of energy is normally referred to as the 'energy head' and the three forms of energy expressed in this way are as follows:

$$\frac{FE}{mg} = \frac{p}{\rho g} = h$$

$$\frac{PE}{mg} = z$$

$$\frac{KE}{mg} = \frac{u^2}{2g}$$

The term FE is referred to as the pressure head; previously, it was shown that $(p/\rho g) = h$.

This is the height a liquid will rise to in a vertical pipe that is connected to the system. The potential energy term is the actual height relative to a datum level. The term $u^2/2g$ is called the kinetic head; this is the pressure head that results from converting the velocity into pressure.

13.3.3 Bernoulli's Equation

Bernoulli's equation is an expression of the conservation of energy. If no energy is added to the system such as work or heat, then the total energy of the fluid is conserved. Initially, the only forces considered are gravity, pressure and inertia forces. The viscosity forces are assumed to be negligible and the fluid is assumed to be a perfect inviscid fluid under steady flow conditions.

Referring to Figure 13.8, the total energy (E_t) at sections (1) and (2) must be equal to each other. So

$$E_t = p_1 Q_1 + mgz_1 + m\frac{u_1^2}{2} = p_2 Q_2 + mgz_2 + m\frac{u_2^2}{2} \tag{13.46}$$

Dividing by the mass gives the specific energy form:

$$\frac{E_t}{m} = \frac{p_1}{\rho_1} + gz_1 + \frac{u_1^2}{2} = \frac{p_2}{\rho_2} + gz_2 + \frac{u_2^2}{2} \tag{13.47}$$

Dividing by 'g' will give the energy terms per unit weight:

$$\frac{E_t}{mg} = \frac{p_1}{g\rho_1} + z_1 + \frac{u_1^2}{2g} = \frac{p_2}{g\rho_2} + z_2 + \frac{u_2^2}{2g} \tag{13.48}$$

Now, since $p/\rho g$ = pressure head 'h', the total head is given by the following equation:

$$h_t = h_1 + z_1 + \frac{u_1^2}{2g} = h_2 + z_2 + \frac{u_2^2}{2g} \tag{13.49}$$

This is Bernoulli's equation expressed in the form of 'h', in which each term is an energy head in metres. 'z' is the potential (or gravitational) head and $u^2/2g$ is the kinetic or velocity head.

When considering an actual system, there will be friction within the pipe and elsewhere. Heat is produced but will be absorbed by the fluid resulting in a rise in the internal energy and hence temperature. This temperature rise will be very small and if the pipe is long, the energy might be lost as heat transfer to the surroundings. As the equations do not include internal energy, the balance will be lost and an extra term is required on the right-hand side of the equation maintaining the balance. This term is either the head lost due to friction (h_L) or the pressure loss (p_L)

$$h_1 + z_1 + \frac{u_1^2}{2g} = h_2 + z_2 + \frac{u_2^2}{2g} + h_L \tag{13.50}$$

Expressing the equation in the form of pressure:

$$p_1 + \rho g z_1 + \frac{\rho u_1^2}{2g} = p_2 + \rho g z_2 + \frac{\rho u_2^2}{2g} + p_L \tag{13.51}$$

Example 13.6

Figure 13.10 depicts a tank being filled via a pump delivering water through a pipe of 30 mm bore.

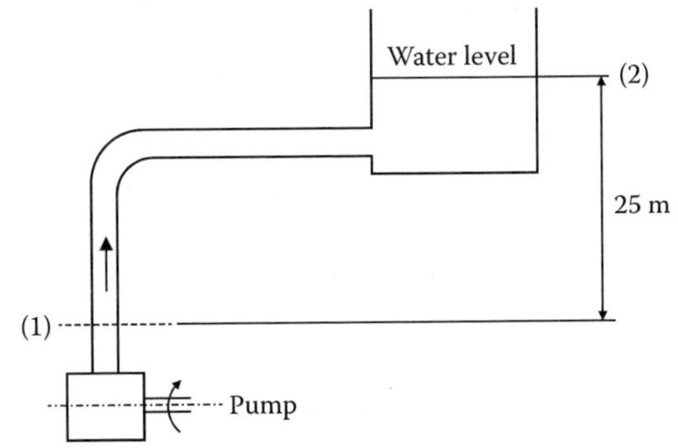

FIGURE 13.10
Example 13.6.

Determine the pressure at position '1' when the flow rate is 0.0014 m^3/s. The density of water is 1000 kg/m^3 and the pressure loss due to friction is 50 kPa.

Solution

Internal area of the pipe:

$$A = \frac{\pi d^2}{4} = \frac{\pi \times 0.03^2}{4} = 706.86 \times 10^{-6}\ m^2$$

Flow rate (Q):

$$= 0.0014\ m^3/s$$

Mean velocity of water (u_m):

$$= \frac{Q}{A}\ m^2 = \frac{0.0014}{706.86 \times 10^{-6}} = 1.981\ m/s$$

Applying Bernoulli's equation between position '1' and the water level in the tank:

$$p_1 + \rho g z_1 + \frac{\rho u_1^2}{2} = p_2 + \rho g z_2 + \frac{\rho u_2^2}{2} + p_L$$

Make position '1' the low level of the datum; therefore, 'z_1'=0 and z_2= 25 m.

The pressure at the water level will be zero-gauge pressure

$$p_L = 50{,}000\ Pa$$

The velocity of water at position '1' is 1.98 m/s and at the surface of the water level, the velocity is zero.

Therefore,

$$p_1 + 0 + \frac{1000 \times 1.98^2}{2} = 0 + 1000 \times 9.854 \times 25.0 + 0 + 50{,}000$$

Hence,

$$p_1 = 294.39 \text{ kPa (Ans)}$$

Example 13.7

A horizontal nozzle is discharging water into an open vessel at atmospheric pressure. The inlet of the nozzle has a bore area of 600 mm^2 and the discharge has a bore area of 200 mm^2. Figure13.11 shows the diagram of the nozzle. Calculate the water flow rate with the inlet pressure that is 400 Pa. Assume there are no energy losses.

Solution

Applying Bernoulli's equation between positions (1) and (2):

$$p_1 + \rho g z_1 + \frac{\rho u_1^2}{2} = p_2 + \rho g z_2 + \frac{\rho u_2^2}{2} + p_L$$

Using the gauge pressure, p_2=0 and as the venture is horizontal, the potential terms cancel out. The loss term is also zero; hence, the equation simplifies to the following:

$$p_1 + \frac{\rho u_1^2}{2} = \frac{\rho u_2^3}{2}$$

From the continuity equation:

$$u_1 = \frac{Q}{A_1} = \frac{4Q}{\pi \times 0.6^2} = 3.537Q$$

$$u_2 = \frac{Q}{A_2} = \frac{4Q}{\pi \times 0.2^2} = 31.831Q$$

Substituting these values into Bernoulli's equation:

$$400 + 1000 \times \frac{(3.537Q)^2}{2} = 1000 \times \frac{(31.831Q)^2}{2}$$

$$400 + 6255.185Q^2 = 506606.3Q^2$$

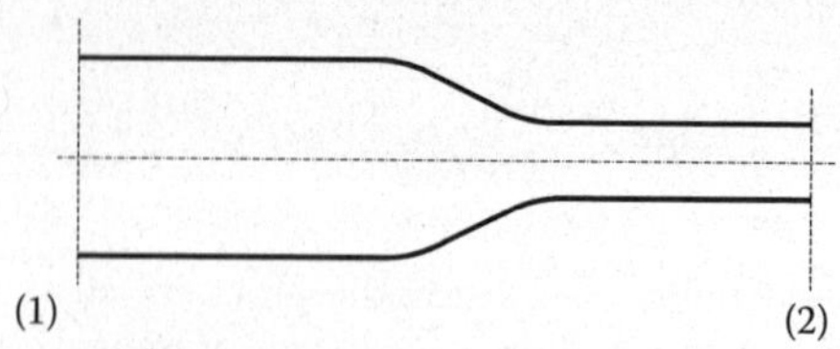

FIGURE 13.11
Horizontal venturi. (1) Inlet and (2) discharge.

Therefore,

$$400 = (506{,}606.3 - 6255.185)Q^2$$

$$\frac{400}{500{,}351.11} = Q^2$$

$$Q = 0.0283\ \text{m}^3/\text{s (Ans)}$$

13.3.4 Stagnation Point

Consider a circular bar located within a flowing fluid whether it is a liquid or gas. The fluid will hit the part and divide to go to either side of the body as shown in Figure 13.12.

At the point where the fluid parts, there will be an area in which the fluid will have zero velocity. Stagnation points exist at the surfaces of objects within the flow field where the fluid is brought to rest by the object. Wings on an aircraft are a further example, where the airstream parts at the leading edge to either flow above or below the wing surface, thereby providing lift.

In the flow stream, Bernoulli's equation shows that the energy comprising pressure + velocity + head is constant and as a result at the stagnation point, the pressure is increased from 'p' to $p+(\rho u^2/2)$ as the velocity energy is converted into pressure energy. For a uniform density fluid, the value of $p+(\rho u^2/2)$ is known as the stagnation pressure of the streamline.

A manometer connected to the point 'X' (Figure 13.12) would indicate the stagnation pressure $((p/\rho g)+(u^2/2g))$. If the static head $(p/\rho g)$ is known, then by subtraction, the velocity head and hence the velocity can be easily calculated. This is the basis of operation for a Pitot tube to measure airspeed.

13.4 Hydrostatics

The definitions of hydrostatics include the following:

The pressure exerted by a fluid at equilibrium at a given point within the fluid, due to the force of gravity. Hydrostatic pressure increases in proportion to the depth measured

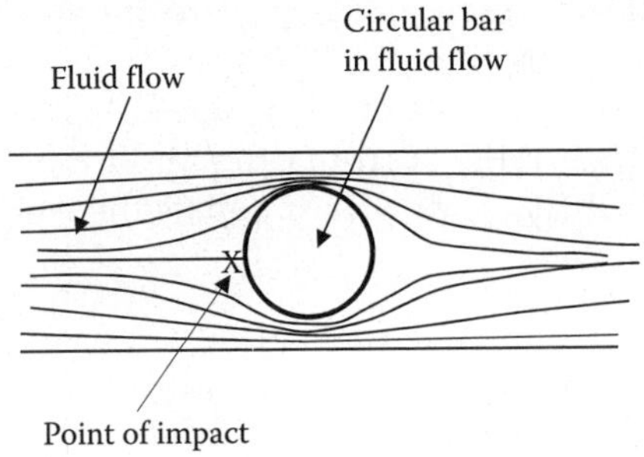

FIGURE 13.12
Circular bar in a fluid flow.

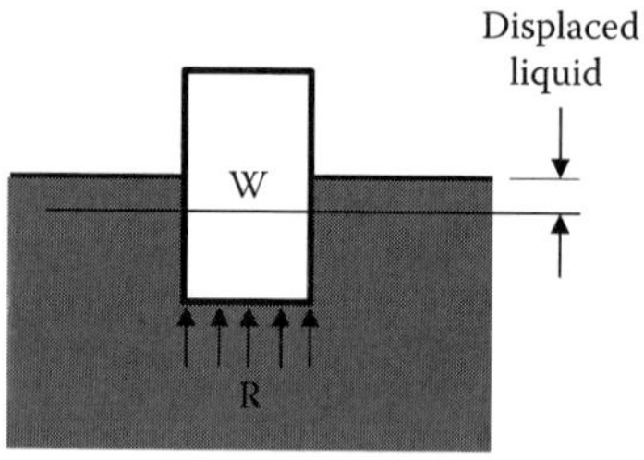

FIGURE 13.13
Archimedes' principle.

from the surface because of the increasing weight of the fluid exerting downward force from above.

In essence, the study of hydrostatics covers fluid bodies that are at rest or moving sufficiently slowly; so, there is no relative motion between the adjacent parts of the body.

There are only pressure forces acting perpendicular to any surface and there are no shear stresses.

13.4.1 Buoyancy

The principles of Archimedes state that the upthrust on a body fully or partly immersed in a fluid at rest is equal to the amount of fluid displaced.

A floating body will be in equilibrium under the action of the weight 'W' acting vertically downwards at its centre of gravity 'G' as shown in Figure 13.13. The upthrust 'R' acts vertically upwards through the centre of buoyancy 'B' for equilibrium.

Hence,

$$\text{Weight of body 'W'} = \text{Upthrust 'R'}$$

$$= \text{Weight of liquid displaced}$$

'R' and 'W' must act in the same vertical straight line.

The surrounded fluid acting on an immersed body generates an upward thrust and this is known as the force of buoyancy. This thrust acts through the centroid of the displaced volume, and this is referred to as the centre of buoyancy. The centre of buoyancy is not the same as the centre of gravity that relates to the distributed weight within the body. If the body is solid with a uniform density identical to water and the body is fully immersed in water, the force of the buoyancy will be exactly equal to the weight, and in this case, the centre of buoyancy will be the same as the centre of gravity. The body will then be in equilibrium with the surrounding fluid.

This principle also applies to gasses as well as liquids. A balloon, filled with hydrogen having a lower density to that of the surrounding air at the sea level, will then rise to a height where the weight of air that is displaced equals the weight of gas in the balloon.

Example 13.8

It is required to lay a steel pipeline for conveying gas across a stretch of water. The pipeline will be completely immersed in water and will be anchored at intervals of 3.0 m along its length. The details of the pipeline are

Internal diameter = 1.20 m
External diameter = 1.25 m

Calculate the buoyancy force in newton per metre run and the upward force acting on each anchorage. The density of steel is 7900 kg/m^3, and that of water is 1000 kg/m^3.

Solution

$$\text{Buoyancy force/metre run} = \text{upthrust/metre run}:$$

$$= \text{Weight of water displaced/metre run}$$

$$= (1000 \times 9.81) \times \frac{\pi \times 1.25^2}{4}$$

$$= 12.039 \text{ kN/m}$$

The anchorages are 3.0 m apart; therefore

$$\text{Upward force on anchorage} = (\text{buoyancy force} - \text{weight})$$

For 3.0 m length of pipe:

$$\text{The weight of 3 m length of pipe} = 3.0 \times (7900 \times 9.81) \times \frac{\pi(1.25 - 1.20)^2}{4}$$

$$= 22.369 \text{ kN}$$

Buoyancy force on 3.0 m length of pipe:

$$= 3.0 \times 12.039$$

$$= 36.117 \text{ kN}$$

Hence, upward force acting on the anchorage:

$$= (36.117 - 22.369) \text{ kN}$$

$$= 13.748 \text{ kN (Ans)}$$

In a situation where the centre of gravity (G) does not have the same location as the centre of buoyancy (B), the body will orientate itself so that the centre of gravity is below the centre of buoyancy (see Figure 13.14). The figure considers a hollow vessel that has a heavy mass in a fixed position within the vessel. In this case, the centre of buoyancy is well above the centre of gravity for the vessel and as such, the vessel is in a stable equilibrium position.

If the vessel had a centre of buoyancy that was below the centre of gravity but on the same centreline as in Figure 13.15a, the vessel would, in theory, be in a stable position. If the vessel was displaced by a small angle, the vessel would then be in an unstable position; a couple would be generated to rotate the vessel to an equilibrium position where the centre of buoyancy will be above the centre of gravity as in Figure 13.15b.

The definition for stable and unstable equilibrium can be simply stated as when an immersed body, initially at rest, is displaced so that the force of buoyancy and the centre of gravity are not on the same vertical line:

1. The body is stable if the resulting couple tends to bring the body back to its original position.
2. The body is unstable if the resulting couple moves the body away from its original position.

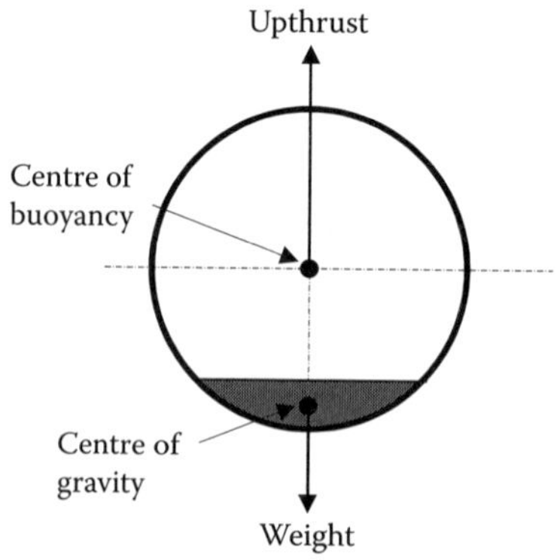

FIGURE 13.14
Centre of gravity.

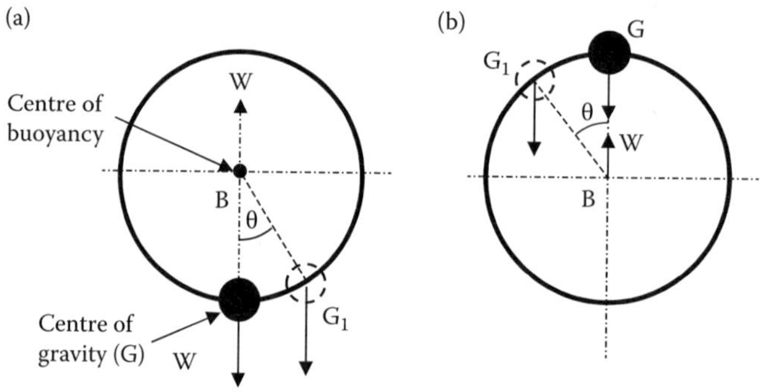

FIGURE 13.15
Centre of gravity. (a) Stable and (b) unstable.

13.4.2 Metacentre and Metacentre Height

In Figure 13.16a, a rectangular vessel is immersed in water where the centre of buoyancy is at position 'B', the centre of gravity is at 'G' and the waterline is at 'S:S'. Consider the vessel that is now tilted to the position shown in Figure 13.16b where the waterline is now at 'S:S'. The centre of buoyancy has now moved to position 'B_1'. There is an upthrust 'W' due to the buoyancy at 'B_1' and the weight of the vessel 'W' is acting down at 'G' (the centre of gravity). A couple 'W·a' is acting to restore the vessel back to its original equilibrium position.

Drawing a line vertically to intersect the line joining the centre of buoyancy and the centre of gravity will give the position of the metacentre for the position of the vessel shown. Heeling the vessel further will produce different metacentre positions. Provided that the metacentre remains above the centre of gravity, the vessel will remain stable and will return to its equilibrium position.

The metacentre height is the distance between the metacentre and the centre of gravity.

If the vessel is tilted far enough where the metacentre is below the centre of gravity, then the vessel will become unstable and could result in capsizing.

From the above discussion, the following conclusions can be made:

- A floating vessel is stable provided that the metacentre lies above the centre of gravity 'G'.

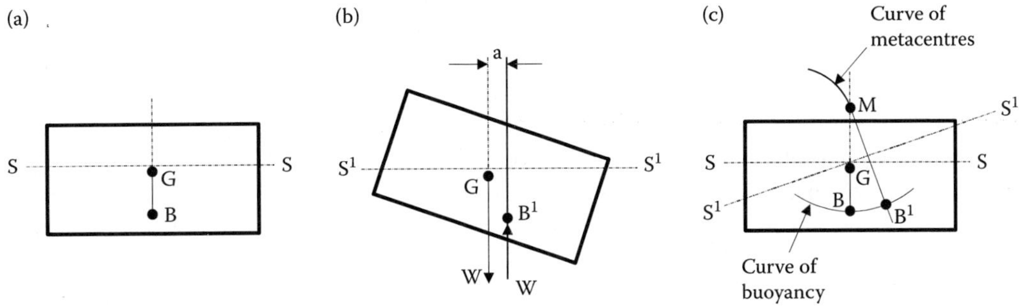

FIGURE 13.16
Metacentre height.

- A floating vessel is in neutral equilibrium if the metacentre lies on the centre of gravity 'G'.
- A floating vessel is unstable if the metacentre lies below the centre of gravity 'G'.

Example 13.9

A vessel has a displacement of 2.5×10^6kg in freshwater. A mass of 20×10^3kg is moved across the deck of the vessel causing a tilt of 4.4°.

Calculate the transverse metacentre height.

Solution

The overturning moment created when a mass of 20×10^3kg is moved 9 m across the deck

$$= (20\times10^3\times9.81)\times9\ \text{N/m}$$

$$= 1.7658\times10^6\ \text{N/m}$$

The restoring moment:

$= W \cdot x$

$= W \cdot GM \cdot \theta$ (GM is the distance between the centre of gravity and the metacentre.)

Since $W = 2.5\times10^6\times9.81$ N and $\theta = 4.4°$.

The restoring moment

$$= 2.5\times106\times9.81\times4.4°\times GM$$

The vessel is in equilibrium while in the tilted position.

The overturning moment represents the righting moment:

$$20\times10^3\times9.81\times9 = 2.5\times10^6\times4.4°\times GM$$

Metacentric height (GM):

$$= \frac{20\times9}{2500\times0.07667}$$
$$= 0.939\ \text{m (Ans)}$$

13.4.3 Pressure in Liquids

A perfect fluid cannot oppose or apply any shear force and is defined as non-viscous (or inviscid) under all conditions. The intensity of normal forces is called the pressure and will be positive if the fluid is compressed.

Consider a small element of fluid of a uniform width that is subject to pressures p_x, p_y and p_z as shown in Figure 13.17. It is assumed that the element is so small that the pressures are assumed to be uniform across the faces of the element (gravity is not considered).

Equating forces:

$$P_z \cdot A \cdot \sin\theta = A \sin\theta P_y \tag{13.52}$$

$$P_z \cdot A \cdot \cos\theta = A \cos\theta P_x \tag{13.53}$$

This example illustrates a perfect liquid and to a large extent, for real fluids, the pressure at a point will be the same in all directions (or the element would move in the direction of least pressure). If no other forces are acting on the body of the fluid, the pressure must be the same at all neighbouring points. Therefore, in this case, the pressure will be the same throughout the fluid and will be the same in any direction at a point (Pascal's principle). In SI units, pressure is expressed in N/m^2 (Pascal).

A practical application of the above principle is seen in the humble car jack where a small piston can exert sufficient force to raise an automobile off the road. Figure 13.18 illustrates the principle behind this.

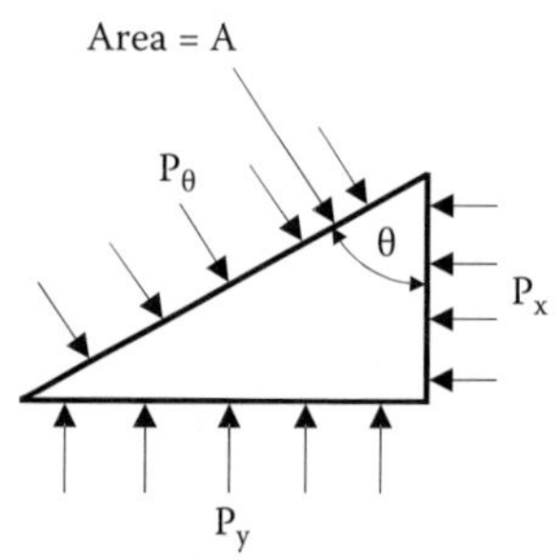

FIGURE 13.17
Pressure in liquids.

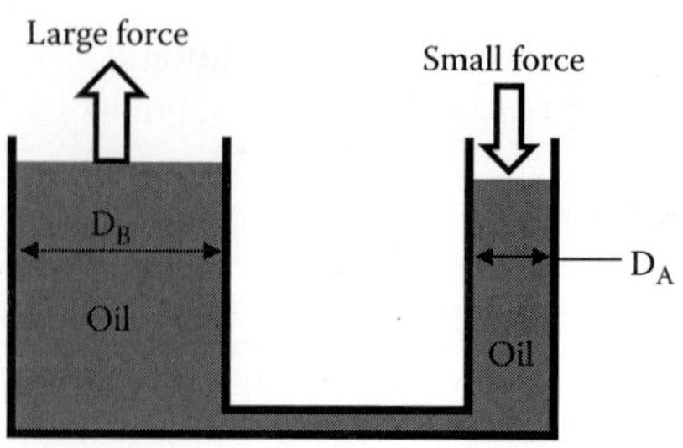

FIGURE 13.18
Force amplification.

In the figure, it is seen that a small piston and cylinder (A) are connected to a much larger piston and cylinder (B). When a force is applied to the piston (A), the pressure in the fluid (A) will be

$$P_A = \frac{F_A}{A_A} \tag{13.54}$$

where
P_A is the pressure in cylinder (A)
F_A is the force from piston (A)
A_A is the area of cylinder (A)

As the two cylinders are interlinked, the fluid pressure in cylinder (B) will be equal to that of cylinder (A) according to Pascal's law.

That is

$$P_A = P_B \tag{13.55}$$

Therefore, the force generated by piston (B) will be

$$F_B = A_B \times P_A \tag{13.56}$$

or

$$F_B = \frac{A_B}{A_A} F_A \tag{13.57}$$

Distance moved:
The distance 'h' moved by piston (A) will displace a volume of fluid:

$$\Delta V = h_A \cdot A_A \tag{13.58}$$

The piston (B) will move a distance:

$$h_B = \frac{A_A}{A_B} h_A \tag{13.59}$$

Example 13.10

It is required to lift a part of a vehicle to facilitate changing a wheel. The vehicle weight is 1.5 ton.

Design a hydraulic jack that will lift a mass of 560 kg using a handle of 300 mm long with a hand force of approximately 40 N (4.07 kg).

Figure 13.19 depicts the arrangement of the jack that is fitted with a non-return valve to enable a suitable height to be obtained with steady pumping.

Solution

Vehicle weight = 1.5 ton.

Assuming the jack carries a quarter of this weight and allowing a safety factor of 2 for the proof load:

$$\text{Jack proof load} = \frac{1500\ \text{kg}}{4} \times 2$$
$$= 750\ \text{kg}$$

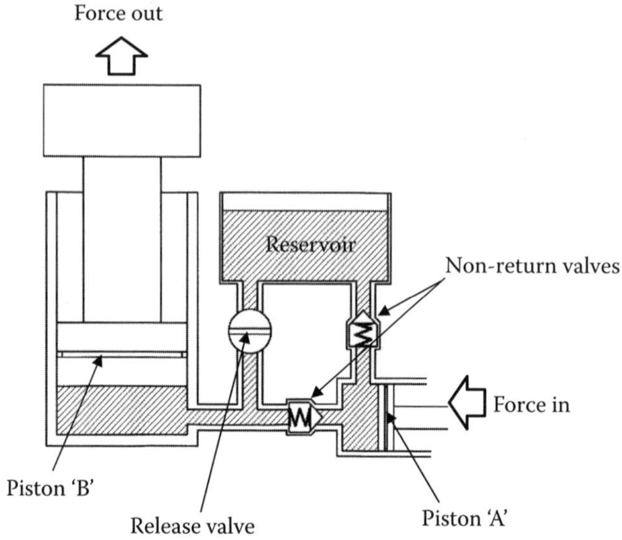

FIGURE 13.19
Example 13.10.

The force to be exerted by the lifting piston:

$$= 751 \times 9.81\ \text{N}$$

$$= 7354.98\ \text{N}$$

The diameter of the lifting cylinder:

$$= 50.0 \times 10^{-3}\ \text{m}$$

$$\text{Area}\,(A_L) = \frac{\pi \times (50.0 \times 10^{-3})^2}{4}$$

$$= 1.9635 \times 10^{-3}\ \text{m}^2$$

The pressure in the lifting cylinder (P_L):

$$P_L = \frac{\text{lifting force}}{\text{area of lifting cylinder}}$$

$$= \frac{7354.98\ \text{N}}{1.9635 \times 10^{-3}\ \text{m}^2}$$

$$P_L = 3.746 \times 10^6\ \text{Pa}$$

The diameter of the pumping cylinder (Dia'_p)

$$\text{Dia}'_p = 12.0 \times 10^{-3}\ \text{m}$$

The area of the pumping cylinder (A_p):

$$A_P = \frac{\pi \times (12.0 \times 10^{-3})^2}{4}$$
$$= 113.097 \times 10^{-6}\ m^2$$

The force required on the pumping cylinder (F_p):

$$F_P = \text{pressure} \times A_P$$
$$= 3.746 \times 10^6\ Pa \times 113.097 \times 10^{-6}\ m^2$$
$$= 423.647\ N$$

Pumping force F_A: Taking moments about a pivot point offset 25.0 mm from the centre of piston 'A' (Figure 13.19):

$$F_A = \frac{25.0 \times 10^{-3}\ mm \times 423.647\ N}{300 \times 10^{-3}\ mm}$$
$$F_A = 35.3\ N\ (3.6\ kg)$$

The stroke of the pumping cylinder (h_p) will be approximately 12 mm and the stroke of the lifting cylinder (h_L) will be

$$h_L = \frac{area_P}{area_L} \times h_P$$
$$= \frac{113.097 \times 10^{-6}}{1.9635 \times 10^{-3}} \times 12.0 \times 10^{-3}$$
$$= 0.691 \times 10^{-3}\ m \quad \text{say } 0.690\ mm$$

13.4.4 Pressure due to the Weight of a Liquid

If a tank with a cross-sectional area 'A' and a depth 'h' is open to the atmosphere, the surface of the liquid will be at atmospheric pressure. If pressure measurements are made at various depths, the pressure will be found to increase linearly as shown in Figure 13.20.

The volume of the liquid is

$$V = Ah\ m^3 \tag{13.60}$$

The mass of the liquid will be

$$m = \rho V$$
$$= \rho A h\ kg$$

Multiplying the mass by the gravitational constant 'g':

$$W = mg$$
$$= \rho A h g\ \text{Newton}$$

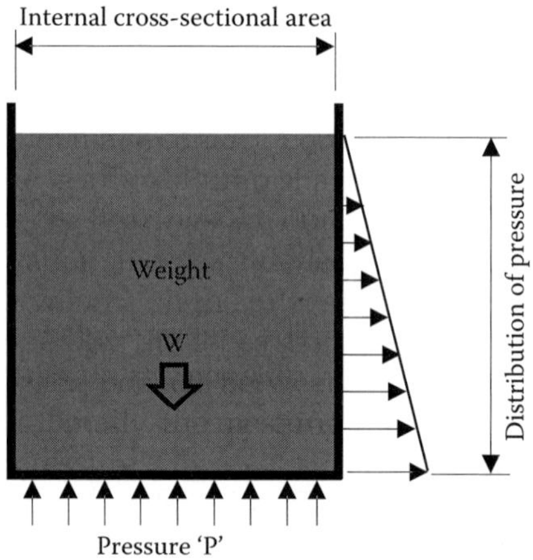

FIGURE 13.20
Pressure due to the weight of a fluid.

It follows that the pressure at the bottom of the tank will be

$$p = \frac{W}{A}\ N/m^2$$

The pressure at a depth 'h' in a liquid is given by the following equation:

$$p = \rho g h\ N/m^2\ (\text{pascal}) \qquad (13.61)$$

Example 13.11

A water tank is fitted with a circular inspection hatch of 0.75 m diameter and is located on the bottom of the tank at a depth of 5 m. The density of water is 1000 kg/m^3.

Calculate the force acting on the hatch due to the water pressure.

Solution

The pressure at the bottom of the tank:

$$p = \rho g h$$

where

$\rho = 1000\ kg/m^3$
$g = 9.81\ m/s^2$
$h = 5\ m$

$$p = 1000 \times 9.81 \times 5$$
$$= 49.05\ kPa$$

The force acting on the hatch is a product of the hatch area and pressure.

$$A = \frac{\pi d^2}{4}$$
$$= \frac{\pi \times 0.75^2}{4}$$
$$= 0.4418\ m^2$$

The force 'F' = p · A:

$$= 49.05 \times 10^3 \times 0.4418$$

$$= 21.67 \times 10^3 \text{ N}$$

$$= 21.67 \text{ kN}$$

13.4.5 Forces on Submerged Surfaces

A section submerged vertically below the surface of a liquid will have variable force acting on it due to the varying pressure of the liquid with the increase in depth.

Figure 13.21 shows a vertical rectangular section fully immersed in a liquid.

From Section 13.4.4, the pressure at a depth 'h' in a liquid is given by Equation 13.61:

$$p = \rho g h$$

In this example, 'h' is being replaced by 'y' to denote depth for reasons that will become obvious.

Hence,

$$p = \rho g y \tag{13.62}$$

The force acting on the elemental strip due to this pressure is

$$dF = p\,dA = \rho B g y\,dy \tag{13.63}$$

The total force acting on the surface due to the pressure is denoted by 'F' and is obtained by integrating this expression between the limits y_1 and y_2.

$$F = \rho g B\left(\frac{y_2^2 - y_1^2}{2}\right)$$

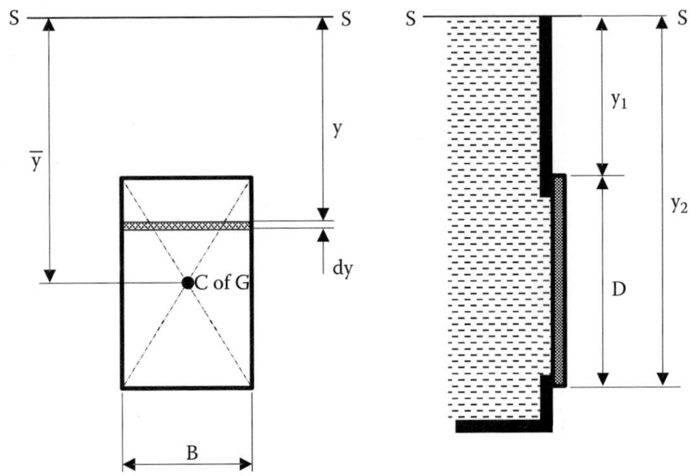

FIGURE 13.21
Submerged section.

This expression can then be factored:

$$F = \rho g B \left[\frac{(y_2 - y_1)(y_2 + y_1)}{2} \right]$$

Now, $(y_2-y_1)=D$ (the depth of the surface); so, $B(y_2-y_1)=B{\cdot}D$=total area of the surface. $(y_2-y_1)/2$ is the distance from the free surface to the centroid $\bar{y}$.

It follows that the total force is given by the expression:

$$F = \rho g A\bar{y} \tag{13.64}$$

The term 'Ay' is the first moment of area and the total force on the submerged surface is

$$F = \rho g \times \text{first moment of area about the free surface}$$

Note: The first moment of area is the area×the distance from the centroid of the area to an axis, in this case, the free surface 'S:S'.

13.4.6 Centre of Pressure

The centre of pressure is that point on which it is considered that the total force is assumed to act on a submerged section. Referring to Figure 13.21, the force acting on the strip is 'dF'. This force produces a turning moment on the section with respect to the axis 'S:S'. The turning moment due to dF is

$$dM = y\ dF = \rho g B y^2\ dy \tag{13.65}$$

The total turning moment about the axis 'S:S' due to the pressure is found by integrating Equation 13.65 between the limits y_1 and y_2.

Hence

$$M = \int_{y_1}^{y_2} \rho g B y^2\ dy$$

$$= \rho g B \int_{y_1}^{y_2} y^2\ dy$$

By definition:

$$I_{ss} = B \int_{y_1}^{y_2} y^2\ dy$$

Therefore

$$M = \rho g I_{ss}$$

This moment should also equal to that obtained from the total force 'F' multiplied by the distance '$\bar{h}$'. The position at the depth 'h' is called the 'centre of pressure'.

$\bar{h}$ is determined by equating the moments:

$$\begin{aligned} M &= \bar{h} \cdot F \\ &= \bar{h}\rho g A \bar{y} \\ &= \rho g I_{SS} \end{aligned}$$

$$\begin{aligned} \bar{h} &= \frac{\rho g I_{SS}}{\rho g A \bar{y}} \\ &= \frac{I_{SS}}{A\bar{y}} \end{aligned}$$

$$\bar{h} = \frac{\text{second moment of area}}{\text{first moment of area}} \text{about 'S : S'} \tag{13.66}$$

To resolve this expression requires a degree of experience with the parallel axis theorem that is required to resolve the second moment of area about the free surface.

The parallel axis theorem is as follows:

$$I_{SS} = I_{CG} + A\bar{y}^2$$

where

I_{SS} is the second moment about the free surface S:S

TABLE 13.3

Geometric Properties for Some Common Figures

		Area "A"	Second Moment of Area I_{gg} about Axis G:G through the Centroid
Rectangle	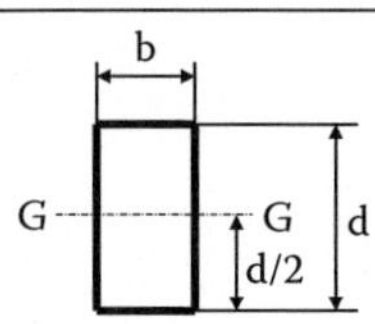	bd	$\frac{bd^3}{12}$
Triangle	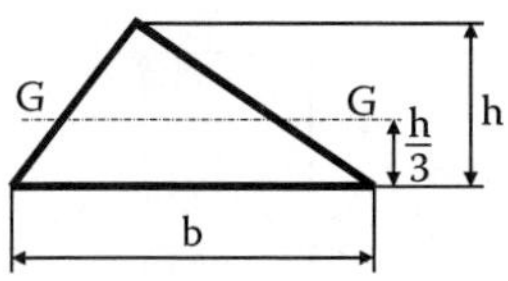	$\frac{b \cdot h}{4}$	$\frac{b \cdot h^3}{36}$
Circle	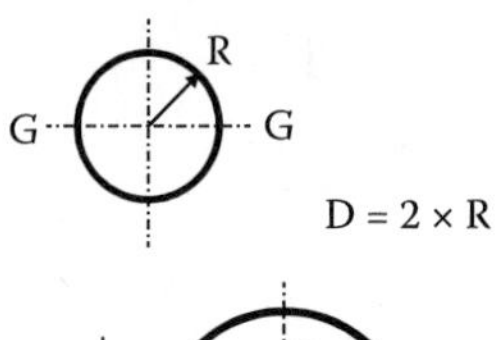	$\frac{\pi D^2}{4}$ >	$\frac{\pi D^4}{64}$
Semi-circle	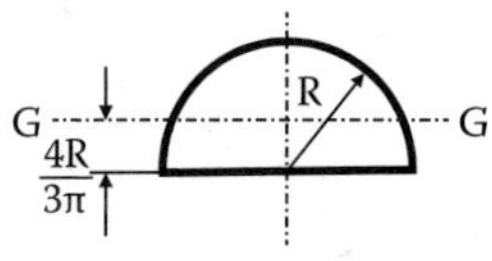	$\frac{\pi R^2}{2}$	$0.1102\ R^4$

I_{CG} is the second moment of area about the centroid C of G
Table 13.3 gives some geometric properties of common figures.

Example 13.12

A rectangular inspection plate is fitted onto a steel water-storage tank as shown in Figure 13.22.

Calculate the total force acting on the inspection plate and the position of the centre of pressure. Also, calculate the total moment about the bottom edge of the plate. Assume the water density is 1000 kg/m³.

Solution

Total force $F = \rho g A\bar{y}$

For the plate:

$$\bar{y} = 0.75 + 0.5\ (\text{m})$$
$$= 1.25\ \text{m}$$
$$\text{Area} = 1.0 \times 0.75\ (\text{m}^2)$$
$$= 0.75\ \text{m}^2$$
$$F = 1000 \times 9.81 \times 0.75 \times 1.25$$
$$= 9.197\ \text{kN}$$

$$\bar{h} = \frac{\text{second moment of area}}{\text{first moment of area}}$$

First moment of area:

$$= A\bar{y} = 0.75 \times 2 = 1.5\ \text{m}^3$$

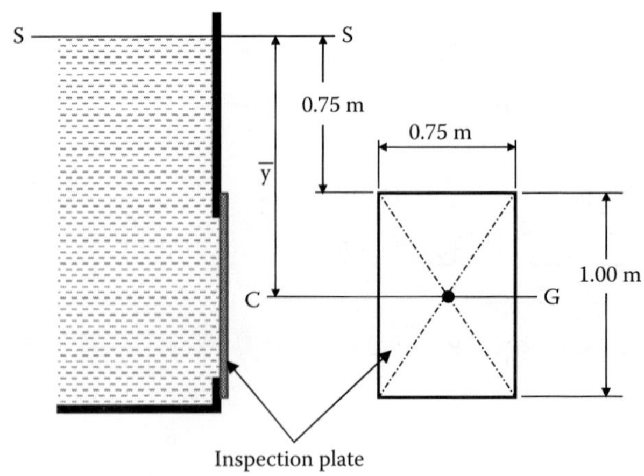

FIGURE 13.22
Example 13.12.

Second moment of area:

$$I_{SS} = \left(\frac{BD^3}{12}\right) + Ay^2$$
$$= \left(\frac{0.75 \times 1.0^3}{12}\right) + (0.75 \times 1.25^2)$$
$$= 1.234\ m^4$$

$$\bar{h} = \frac{I_{SS}}{A\bar{y}}$$
$$= 0.823\ m$$

The distance from the bottom edge:

$$x = 2.5 - 0.823$$
$$= 1.677\ m$$

The moment about the bottom edge:

$$= Fx$$
$$= 9.197 \times 10^3 \times 1.677$$
$$= 15.42\ kN\ m$$

13.5 Dimension Analysis

The United Kingdom has adopted the system of metric units known as the Système International d'Units, abbreviated to SI. In some old designs imperial units are still being used, these are being phased out as these designs are updated.

The SI has six basic units; these are arbitrarily defined as

Length	Metre (m)
Mass	Kilogram (kg)
Time	Second (s)
Temperature	Degree (θ)
Electric current	Ampere (A)
Luminous intensity	Candela (cd)

The last two items are not used in fluid mechanics, temperature is only used infrequently.

All other units are derivations of these four fundamental units, M, L, T and θ. Other systems of units could be used but staying with M, L, T, θ removes any constraints to a particular system of measurement.

13.5.1 Dimensions

In mechanics, all quantities can be expressed in terms of these fundamental dimensions of mass M, length L and time T.

Hence

$$\text{Acceleration} = \frac{\text{Distance}}{(\text{Time})^2}$$

So that

$$\text{Dimensions of acceleration} = \frac{\text{Dimension of distance}}{(\text{Dimension of time})^2} = \frac{L}{T^2}$$

So that

$$\text{Dimension of force} = \text{Dimension of mass} \times \text{dimension of acceleration}$$

$$= \frac{ML}{T^2}$$

And when written down in dimensional form:

$$= MLT^{-2}$$

Table 13.4 shows the dimensions of common quantities.

13.5.2 Dimensional Equations

When an equation is to represent a physical or real quantity, the terms on both sides must be of the same sort. For example, all forces as well as both sides should be numerically equal, otherwise, the equation will be meaningless. Every term must have the same dimensions, so that like is compared with like.

As an example, consider the equation $v^2 = u^2 + 2as$.

This equation calculates the final velocity 'v' of a mass which starts with an initial velocity 'u' and then receives an acceleration of 'a' for a distance of 's'. When dimensions are substituted for the quantities, each term must have the same dimensions if the equation is to be true.

The dimensions of the quantities are

$$v = LT^{-1}$$

$$u = LT^{-1}$$

$$a = LT^{-2}$$

$$s = L$$

The dimensions of v^2 are $(LT^{-1})^2 = L^2T^{-2}$.

The dimensions of u^2 are $(LT^{-1})^2 = L^2T^{-2}$.

The dimensions of 2 as are $(LT^{-2}) \times L = L^2T^{-2}$.

TABLE 13.4

Dimensions of Some Common Quantities

Quantity	Description	Dimension
Length, size	Include all linear measurements	L
Area	Length × length	L^2
Volume	Area × length	L^3
First moment of area	Area × length	L^3
Second moment of area	Area × length^2	L^4
Angle	A ratio; arc/radius	1
Strain	A ratio	1
Time		T
Velocity	Distance/time	LT^{-1}
Angular velocity	Angle/time	T^{-1}
Acceleration	Velocity/time	LT^{-2}
Angular acceleration	Angular velocity/time	T^{-2}
Kinematic viscosity	Dynamic viscosity/mass density	L^2T^{-1}
Mass		M
Force	Mass × acceleration	MLT^{-2}
Weight	Force	MLT^{-2}
Mass density	Mass/volume	ML^{-3}
Specific weight	Weight/volume	$ML^{-2}T^{-2}$
Pressure (intensity)	Force/area	$ML^{-1}T^{-2}$
Shear stress	Force/area	$ML^{-1}T^{-2}$
Elastic modulus	Stress/strain	$ML^{-1}T^{-2}$
Impulse	Force/time	MLT^{-1}
Momentum	Mass × velocity	MLT^{-1}
Work, energy	Force × distance	ML^2T^{-2}
Power	Work/time	ML^2T^{-3}
Moment of a force	Force × distance	ML^2T^{-2}
Dynamic viscosity	Shear stress/velocity gradient	$ML^{-1}T^{-1}$

All three dimensions have the same form and therefore the equation is dimensionally correct and therefore could represent a real event.

There are other quantities that require more careful consideration when writing in the basic MLTθ form. Force is one such important unit. In SI, the unit of force is the 'N'. Engineers have agreed to define force as the need to accelerate 1 kg mass at an acceleration of 1 m/s^2. The Newton is a derived unit that is equal to 1 kg m/s^2. In a dimensional form, the dimensions of force becomes MLT^{-2} and this has to be considered when writing dimensions containing force.

Example 13.13

Deduce the basic dimensions for pressure.

Solution

The definition of pressure is

$$p = \frac{\text{force}}{\text{area}}$$

Pascal is the SI unit for pressure and the units are N/m^2.

The dimensions for force is MLT^{-2} and for area is L^2. Therefore, the basic dimensions for pressure is

$$p = \frac{MLT^{-2}}{M^2}$$

This can be written in standard form:

$$p = ML^{-1}T^{-2} \text{ (Ans)}$$

13.6 Fluid Drag

If an object is placed in a flowing fluid, a resistance to the flow will be generated causing the object to be pulled by the fluid in the direction of the fluid flow. This resistance is known as 'drag' and is a function of the form of the object and skin friction between the object and the fluid.

1. Form drag (or pressure drag) is based on the pressure difference between the upstream and downstream surfaces of the object.
2. Skin friction resulting from the viscous shear of the fluid flowing over the object's surfaces.

13.6.1 Form Drag

Form or pressure drag is applicable to bodies that are tall in comparison with their length in the direction of the flow. Figure 13.23a depicts such a body in a flowing fluid and these bodies are known as 'bluff bodies'. Figure 13.23b shows a streamlined body in a flowing fluid and as is seen there is very little disturbance to the flow around the body.

Examples of bluff bodies would be the valve plate in a shut-off valve and the leg of a pier or bridge support in a river. In this instance, the fluid locally speeds up around the leading

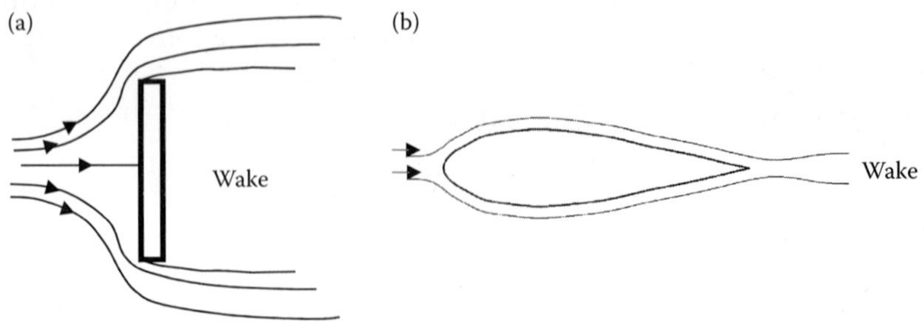

FIGURE 13.23
(a) Bluff and (b) streamlined bodies.

edge of the body and the boundary layer quickly breaks away from the surface. The fluid is sucked in from behind the body in the opposite direction. This creates a wake behind the body and this in turn creates a lower pressure region resulting in a significant form drag.

If the body was shaped as in Figure 13.23b, the amount of drag produced by the section would be appreciably reduced, the low-pressure region will be reduced, and the wake produced will also be smaller. If a road vehicle is fitted with a streamlined body, it is able to attain a higher speed than without it, showing the significant reduction in form drag.

13.6.2 Skin Friction Drag

If a flat plate is immersed in a flowing fluid with the plate orientated to be parallel to the flow, when the flow hits the leading edge of the plate the flow is retarded and a thin layer is formed by the fluid flowing along one side of the flat plate as shown in Figure 13.24a and b. The flow close to the surface of the plate slows down and the flow in this region is considered laminar. This slowing of the fluid is due to the viscous nature of the fluid.

The layers above the surface are moving therefore shearing is taking place between the layers of the fluid. The shear stress acting between the surface and the first moving layer is called 'the wall shear stress' and is denoted by the symbol 'τ_w'.

The velocity of the fluid is increasing from zero at the surface of the plate to a maximum 'u_o' at a distance 'δ' above it. This layer is called the 'Boundary Layer' and 'δ' is the boundary layer thickness. Figure 13.24b shows how the velocity 'u' varies with the thickness of the layer 'y'.

Refer to Figure 13.24a again. At the leading edge of the plate, as the flow is retarded due to the viscous resistance and the boundary layer is originated, more flow is slowed down due to the viscous forces and the boundary layer then begins to thicken. At some point, the laminar flow loses stability and the flow becomes less even. This point where the laminar flow deteriorates is known as the 'transition point' and is at the start of the region called the transition region where the laminar flow changes to turbulent flow. At the plate surface, a thin laminar sub-layer will remain below the turbulent boundary layer.

The boundary layer thickness 'δ' grows in thickness with distance from the leading edge. At some distance from the leading edge, the layer reaches a constant thickness and then it is called a 'fully developed boundary layer'. The Reynolds number in these cases is

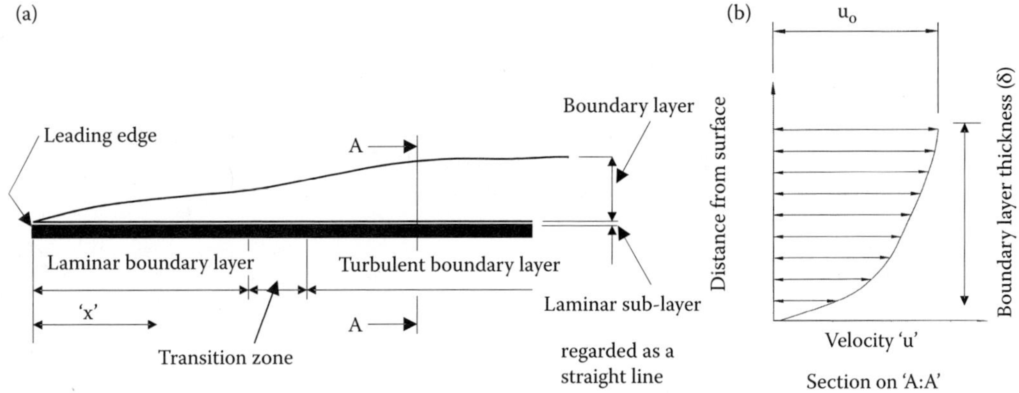

FIGURE 13.24
(a) and (b) Boundary layer.

$$(R_e)_x = \frac{\rho u_o x}{\mu} \tag{13.67}$$

where x is the distance from the leading edge.

At low Reynolds numbers, the boundary layer may be laminar throughout the entire thickness of the layer. At higher Reynolds numbers, the flow is turbulent meaning at some point from the leading edge the flow within the boundary layer becomes turbulent.

A turbulent boundary layer is very unstable where the streamlines do not remain parallel to each other. The shape of the boundary layer represents an average of the velocity of the fluid at any height.

Between the laminar and turbulent section, there is a transition. The turbulent boundary layer exists on top of a thin laminar known as the 'laminar sub-layer'. The velocity gradient within this layer is linear. Further analysis would show that, for long surfaces, the boundary layer will be turbulent over most of its length. A number of equations have been developed to describe the shape of the laminar and turbulent boundary layers.

13.6.3 Estimating Skin Drag

Skin drag is a result of the wall shear stress 'τ_w' and acts on the wetted surface area. The drag force is

$$R = \tau_w \times \text{wetted area} \tag{13.68}$$

The dynamic pressure is the pressure that results from the conversion of the kinetic energy of the fluid into pressure energy and is defined by the expression $\rho u_o^2 / 2$. The drag coefficient is defined as

$$\begin{aligned} C_{Df} &= \frac{\text{Drag force}}{\text{Dynamic pressure} \times \text{wetted area}} \\ &= \frac{2R}{\rho u_o^2 \times \text{wetted area}} \\ &= \frac{2\tau_w}{\rho u_o^2} \end{aligned} \tag{13.69}$$

Example 13.14

Calculate the drag force on both sides of a thin smooth rectangular plate 1.0 m wide × 2.0 m long where the 2.0 m length is parallel to the flow of fluid having a velocity of 40 m/s. The density of the fluid is 1000 kg/m³ and its dynamic viscosity is 8 cP.

Solution

$$\begin{aligned} (R_e)_x &= \frac{\rho u_o L}{\mu} \\ &= \frac{1000 \times 40 \times 2}{0.008} \\ &= 10 \times 10^6 \end{aligned}$$

$$C_{Df} = 0.074 \times (10 \times 10^6)^{-1/5}$$
$$= 0.00295$$

$$\text{Dynamic pressure} = \frac{\rho u_o^2}{2}$$

$$= \frac{1000 \times 40^2}{2}$$
$$= 800 \text{ kPa}$$

$$t_w = \text{CDf} \times \text{dynamic pressure}$$
$$= 0.00295 \times 800 \times 10^3$$
$$= 2360 \text{ Pa}$$
$$R = t_w \times \text{wetted area}$$
$$= 2360 \times 2 \times 1.0$$
$$= 4720.0 \text{ N (Ans)}$$

13.6.4 General Notes on Drag Coefficients

In fluid dynamics, the drag coefficient is commonly denoted by c_d and is a dimensionless quantity. This coefficient is used to quantify the resistance of an object in a flowing fluid and is a helpful guide in understanding the importance of the shape of an object in drag resistance. It is also useful for comparing various shapes.

Table 13.5 shows the drag coefficients for some common shapes.

A general approach to solving form drag which is due to pressure changes only and the drag coefficient due to the pressure only is represented by C_{dp} and is defined as

$$C_{dp} = \frac{\text{Drag force}}{\text{Dynamic pressure} \times \text{projected area}} = \frac{2R}{\rho u_o^2 \times \text{projected area}} \tag{13.70}$$

The projected area is the outline of the shape that is directly facing the fluid flow and the pressure that is acting on any point of the surface of the shape is 'p'. The force that is being exerted by the pressure on a small surface area will be p dA. If the shape is inclined at an angle 'θ' with the fluid flow, the force will be

$$p = \cos\theta \, dA \tag{13.71}$$

The total force can be found by integrating the projected surface.

$$R = \oint p \cos\theta \, dA \tag{13.72}$$

TABLE 13.5

Measured Drag Coefficients for Some Common Shapes

Shape		Figure	Drag Coefficient
Sphere	→	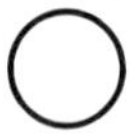	0.47
Half-sphere	→		0.42
Cone	→		0.50
Cube	→		1.05
Angled cube	→		0.80
Long cylinder	→	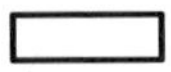	0.82
Short cylinder	→	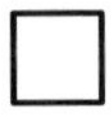	1.15
Streamlined body	→		0.04
Streamlined half-body	→		0.09

The pressure distribution over the surface is expressed in the form of a pressure coefficient defined as follows:

$$C_p = \frac{2(p - p_o)}{\rho u_o^2} \tag{13.73}$$

where

p_o is the static pressure of the undisturbed flowing fluid
u_o is the velocity of the undisturbed flowing fluid
$\rho u_o^2 / 2$ is the dynamic pressure of the stream

If any streamline that is affected by the surface is considered, by applying Bernoulli's theorem between an undisturbed point and any point on the surface, the following will result:

$$p_o + \frac{\rho u_o^2}{2} = p + \frac{\rho u^2}{2}$$

$$p - p_o = \frac{r}{2}\left(u_o^2 - u^2\right) \tag{13.74}$$

$$C_p = \frac{2(p - p_o)}{\rho u_o^2}$$

$$= \frac{2\left(\frac{p}{2}\left(u_o^2 - u^2\right)\right)}{\rho u_o^2}$$

$$= \frac{\left(u_o^2 - u^2\right)}{u_o^2}$$

$$= 1 - \frac{u^2}{u_o^2} \tag{13.75}$$

The equation shows that when $u < u_o$, the pressure will be positive and when $u > u_o$, the pressure will be negative.

13.6.5 Total Drag

As explained previously, drag is made up of two components:

1. Skin friction drag
2. Form drag

This applies to all vehicles including aircraft and automobiles as well as bluff objects such as squares, cylinders and spheres. Calculating the drag forces on an object using theoretical methods is difficult. The majority of the data is captured using experimental methods including wind tunnel analysis and the concept of a drag coefficient is widely used.

The 'drag coefficient' is denoted by C and is defined by

$$C_d = \frac{\text{Resistance force}}{\text{Dynamic pressure} \times \text{projected area}}$$

$$C_d = \frac{2R}{\rho u_o^2 \times \text{projected area}} \tag{13.76}$$

Example 13.15

A cylinder 100 mm diameter × 350 mm long is in a fluid stream moving at 0.75 m/s. The axis of the cylinder is normal to the direction of the fluid flow. The density of the fluid is 1000 kg/m^3. The drag force is found to be 75 N. At a point on the surface of the body, the pressure is measured at 90 Pa above the ambient pressure.

Calculate

1. The drag coefficient.
2. The velocity at the specified point on the surface.

Solution

$$\text{Projected area} = 0.10 \times 0.35\ \text{m}^2$$

$$= 0.035\ \text{m}^2$$

$R = 75\ \text{N}$

$u_o = 0.75\ \text{m/s}$

$\rho = 1000\ \text{kg/m}^3$

Dynamic pressure (P_d):

$$P_d = \frac{\rho \cdot u^2}{2}$$
$$= \frac{1000 \times 0.75^2}{2}$$
$$P_d = 281.25 \text{ Pa}$$

$$C_d = \frac{\text{Resistance force}}{P_d \times \text{projected area}}$$
$$= \frac{75}{281.25 \times 0.035}$$
$$C_d = 7.619$$

$$p - p_o = \frac{\rho}{2}\left(u_o^2 - u^2\right)$$
$$90 = \frac{1000}{2}(0.75^2 - u^2)$$

Rearranging the equation to solve for 'u':

$$u = \sqrt{u_o^2 - \left(\frac{2(p - p_o)}{r}\right)}$$
$$= 0.618 \text{ m/s}$$

13.6.6 Drag on a Cylinder

From Equation 13.76, the drag coefficient is defined as

$$C_d = \frac{2R}{\rho u_o^2 \times \text{projected area}}$$

where the projected area is the length×diameter of the cylinder.

The flow pattern in the wake of the cylinder is dependent upon the Reynolds number and is illustrated in Figure 13.25a through d. The figure considers an infinitely long cylinder in a fluid flow that provides a two-dimensional flow pattern.

1. At very low velocities, where the Reynolds number is $R_e < 0.5$, the inertial forces are very low compared with the viscous forces and the streamlines return to the flow pattern behind the cylinder and the drag is roughly proportional to u_n. The drag is inversely proportional to C_d.

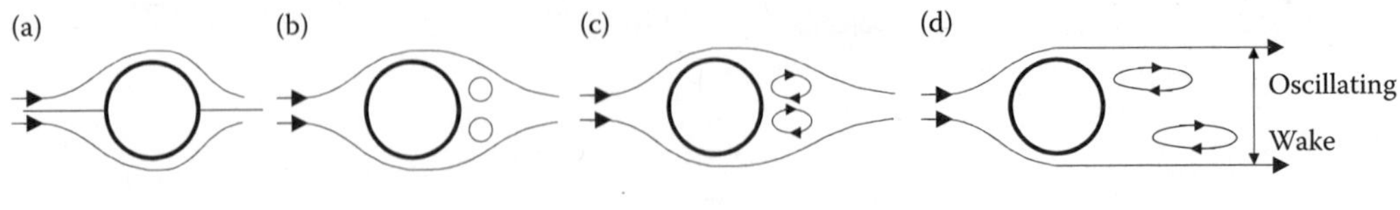

FIGURE 13.25
Fluid flow around a cylinder. (a) $R_e < 0.5$, (b) $R_e = 2–30$, (c) $R_e = 40–70$ and (d) $R_e > 90$.

2. Between the Reynolds numbers 2–30, a wake begins to form behind the cylinder. The stream lines come together behind the cylinder but small eddies are formed which rotate in opposite directions.
3. With the Reynolds numbers between 40 and 70, the eddies elongate and the wake flow becomes unstable. As the Reynolds number increases to 70, the eddies elongate and begin to break away alternatively from each side of the cylinder.
4. At the higher levels of the Reynolds numbers ($R_e > 90$), the eddies begin to form vortices downstream from the cylinder. It is this unsymmetrical behaviour of the flow pattern that gives rise to vortex shedding and is also known as the Karman vortex street.

The drag coefficient (C_d) reaches a minimum of approximately 0.9 at R_e=2000. It then increases slightly due to the increasing turbulence of the wake and this position of separation gradually moves upstream. The profile drag at this stage is nearly all due to the pressure (form) drag.

As the Reynolds number reaches about 20×10^5, the laminar boundary layer becomes more turbulent before separation. The turbulent boundary layer has a higher kinetic energy than the laminar-based layer and as such is better able to withstand the adverse pressure gradient.

There is a sudden drop in the drag dropping to about a C_d of 0.3.

With a further increase in the Reynolds number, there is a consequent increase in the drag coefficient to approximately 0.7. For the Reynolds numbers above 4×10^6, the drag becomes independent of the Reynolds number.

Figure 13.26 shows an approximate relationship between C_d and R_e for an infinitely long cylinder.

It needs to be borne in mind that C_d is not an absolute constant for a given shape. It will vary with the speed of the fluid flow being a function of the Reynolds number (R_e). As an example, the C_d of a smooth sphere will vary from a high value for a laminar flow to 0.47 when in a turbulent fluid flow. Figure 13.27 shows the variation in the C_d factor for a sphere with respect to the increasing Reynolds number.

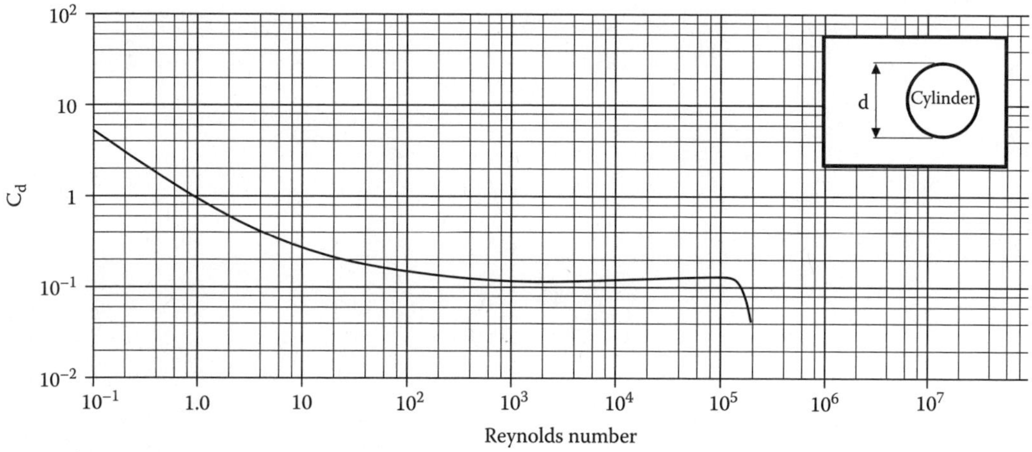

FIGURE 13.26
Drag factor (C_d) for a cylinder.

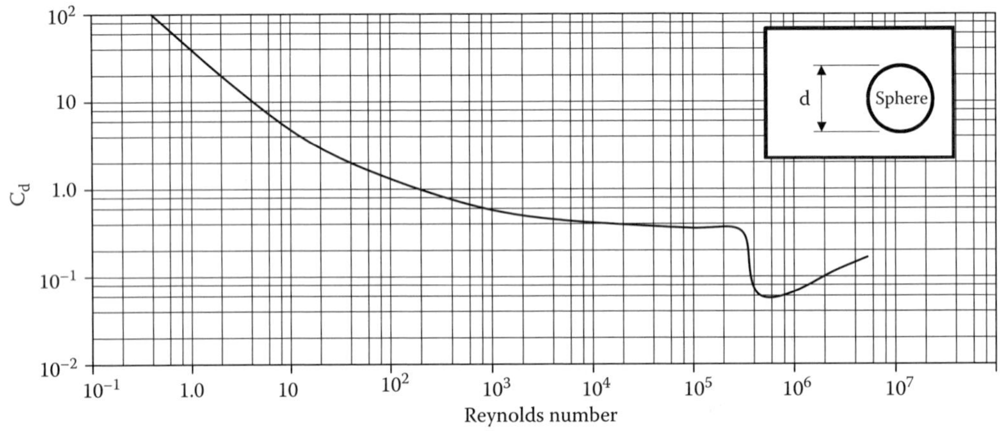

FIGURE 13.27
Drag factor (C_d) for a sphere.

An empirical formula that covers the Reynolds number in the range R_e=0.2 to 10^5 is as follows:

$$C_d = \frac{24}{R_e} + \frac{6}{1+\sqrt{R_e}} + 0.4 \tag{13.77}$$

Example 13.16

A 40.0 mm diameter sphere is suspended in a flowing fluid having a density of 750 kg/m^3 and a dynamic viscosity of 50 cP. The fluid is flowing at 0.6 m/s. (1 cP = 0.001 N · s/m^2.)

Calculate the drag force acting on the sphere.

Solution

$$\begin{aligned} R_e &= \frac{\rho u d}{\mu} \\ &= \frac{750 \times 0.6 \times 0.04}{0.05} \\ &= 360 \end{aligned}$$

From Equation 13.77:

$$\begin{aligned} C_d &= \frac{24}{R_e} + \frac{6}{1+\sqrt{R_e}} + 0.4 \\ &= \frac{24}{360} + \frac{6}{\sqrt{360}} + 0.4 \\ C_d &= 0.7671 \end{aligned}$$

Now

$$\text{Projected area} = \frac{\pi d^2}{4}$$

$$= \frac{\pi \times 0.04^2}{4}$$

$$= 1.2566 \times 10^{-3}\ \text{m}^2$$

$$C_d = \frac{2R}{\rho u^2 \times \text{projected area}}$$

Rearranging the equation to solve for 'R':

$$R = \frac{C_d \rho u^2 \cdot A}{2}$$

$$= \frac{0.8 \times 750 \times 0.6^2 \times 1.2566 \times 10^{-3}}{2}$$

$$= 0.136\ \text{N (Ans)}$$

13.7 Properties of Water

Water is described as a colourless, transparent, odourless, and tasteless compound of oxygen and hydrogen and is the most abundant compound found on the planet. It covers approximately 70% of the Earth's surface. Water exists in all three states: solid, liquid and gas. It is in dynamic equilibrium between liquid and gas states.

Table 13.6 lists the chemical and physical properties of water and Table 13.7 gives the properties of water over a range of varying temperatures and saturation pressures. Except where noted otherwise, data are given for water in their standard state (at 25°C, 100 kPa).

In line with a number of substances, water can take three basic forms:

Solid: The solid phase of water is known as ice and commonly takes the structure of hard, amalgamated crystals, such as ice cubes or loosely accumulated granulated crystals such as snow.

Liquid: Water is mostly in the liquid phase at the Earth's surface at normal temperature and pressure.

Gas: When heat is applied to water, it transforms to a gaseous phase and is known as water vapour or steam, which is essentially minute droplets suspended in the air and forming a transparent cloud.

With an increase in temperature and pressure, the water enters a fourth state which is a supercritical fluid. This state is much less common than the other three; it only very rarely occurs in nature and happens in deep water hydrothermal vents where the water is heated to the supercritical temperature by the volcanic magna and the critical pressure is achieved by the crushing weight of the ocean at the extreme depths where the vents are located.

In power generation water boilers, where a supercritical phase can be achieved, the boiler will be operating at its most efficient rate where the water is immediately converted into supercritical steam without the vapour phase occurring.

TABLE 13.6
Chemical and Physical Properties of Water

Some Common Thermal Properties of Water	
Maximum density at 4°C	1000 kg/m^3
Specific weight at 4°C	9.807 kN/m^3
Freezing temperature	0°C
Boiling temperature	100°C
Latent heat of melting	334 kJ/kg
Latent heat of evaporation	2.270 kJ/kg
Critical temperature	380–386°C
Critical pressure	22.1 MPa (MN/m^2)
Specific heat—water	4.187 kJ/kg K
Specific heat—ice	2.108 kJ/kg K
Specific heat water vaporisation	2.996 kJ/kg K
Thermal expansion from 4°C to 100°C	0.42×10^{-3} K^{-1}
Bulk modulus elasticity	2.15×109 Pa (N/m^2)
Identifiers	
CAS number	7732-18-5
PubChem	962
ChemSpider	937
UNII	059QF0KO0R
ChEBI	CHEBI:15377
ChEMBL	CHEMBL1098659
RTECS number	
Properties	
Molecular formula	H_2O
Molar mass	18.01528(33) g/mol
Density	1000 kg/m^3
Melting point	0°C, 32°F, 273.15 K
Boiling point	99.98°C, 211.97°F, 373.13 K
Acidity (pK_a)	15.74~35–36
Basicity (pK_b)	15.74
Refractive index (nD)	1.3330
Viscosity	0.001 Pa s at 20°C
Structure	
Crystal structure	Hexagonal
Molecular shape	Bent
Dipole moment	1.85 D

13.7.1 Specific Heat Capacity of Water

Water has a very high specific heat capacity (the second highest after ammonia) together with a high heat of vaporisation (2257 kJ/kg at normal boiling point). This is a result of extensive bonding of hydrogen between its molecules.

The specific heat is defined as the heat required to raise the temperature of a 1.0 kg of a substance by 1.0°C (K). Its units are J/kg K (or J/kg °C).

The designation for specific heat at constant pressure is 'c'.

TABLE 13.7

Properties of Water at Varying Temperatures and Saturation Pressures

Temp	Saturation Pressure	Density	Density	Specific Volume	Specific Volume	Viscosity	Viscosity
t	p_{sat}	ρ_{lig}	ρ_{vap}	v_{liq}	v_{vap}	μ_{lig}	μ_{vap}
	Pressure	Liquid	Vapour	Liquid	Vapour	Liquid	Vapour
°C	MPa	kg/m³	kg/m³	m³/kg	m³/kg	μPa·s	μPa·s
0.01	0.0006112	999.8	0.004851	0.001	206.1	1792	9.216
10	0.001228	999.7	0.009407	0.001	106.3	1306	9.461
20	0.002339	998.2	0.01731	0.001002	57.76	1002	9.727
30	0.004247	995.6	0.03042	0.001004	32.88	797.4	10.01
40	0.007385	992.2	0.05124	0.001008	19.52	653.0	10.31
50	0.01235	988.0	0.08315	0.001012	12.03	546.8	10.62
60	0.01995	983.2	0.1304	0.001017	7.667	466.4	10.93
70	0.0312	977.7	0.1984	0.001023	5.040	403.9	11.26
80	0.4741	971.8	0.2937	0.001029	3.405	354.3	11.59
90	0.07018	965.3	0.4239	0.001036	2.359	314.4	11.93
100	0.1014	958.3	0.5982	0.001043	1.672	281.7	12.27
110	0.1434	950.9	0.8269	0.001052	1.209	254.7	12.61
120	0.1987	943.1	1.1220	0.001060	0.8912	232.1	12.96
130	0.2703	934.8	1.4970	0.001070	0.6680	212.9	13.30
140	0.3615	926.1	1.9670	0.001080	0.5085	196.5	13.65
150	0.4762	917.0	2.548	0.001091	0.3925	182.5	13.99
160	0.6182	907.4	3.260	0.001102	0.3068	170.2	14.34
170	0.7922	897.5	4.122	0.001114	0.2426	159.6	14.68
180	1.003	887.0	5.159	0.001127	0.1938	150.1	15.03
190	1.255	876.1	6.395	0.001141	0.1564	141.8	15.37
200	1.555	864.7	7.861	0.001157	0.1272	134.3	25.71
210	1.908	852.7	9.588	0.001173	0.1043	127.6	16.06
220	2.320	840.2	11.62	0.001190	0.6609	121.5	16.41
230	2.797	827.1	13.99	0.001209	0.0715	116.0	16.76
240	3.347	813.4	16.75	0.001229	0.0597	110.9	17.12
250	3.976	798.9	19.97	0.001252	0.05008	106.1	17.49
260	4.692	783.6	23.71	0.001276	0.04217	101.7	17.88
270	5.503	767.5	28.07	0.001303	0.03562	97.50	18.28
280	6.417	750.3	33.16	0.001333	0.03015	93.51	18.70
290	7.442	731.9	39.13	0.001366	0.02555	89.66	19.15
300	8.588	712.1	46.17	0.001404	0.02166	85.90	19.65
310	9.865	690.7	54.54	0.001448	0.01833	82.17	20.21
320	11.280	667.1	64.64	0.001499	0.01547	78.41	20.85
330	12.860	640.8	77.05	0.001561	0.01296	74.54	21.61
340	14.600	610.7	92.76	0.001638	0.01078	70.43	22.55
350	16.530	574.7	113.6	0.001740	0.008802	65.88	23.82
360	18.67	527.6	143.9	0.001895	0.006949	60.33	25.72

When heat is transferred to a liquid, the temperature of the liquid will rise and is directly proportional to the heat transferred (Q):

$$Q = mc\Delta T,$$

where
m is the mass (kg)
c is the specific heat index
ΔT is the rise in temperature (°C or K)

The specific heat 'c' is reasonably constant but will change significantly when the pressure or temperature change is very large.

Example 13.17

Calculate the rise in temperature of 5 kg of water when 84 kJ of heat energy is applied.

Solution

Specific heat capacity of water = 4.187 kJ/kg K.
Heat energy input:

$$Q = mc\Delta T$$
$$\Delta T = \frac{Q}{mc}$$
$$= \frac{84 \times 10^3}{5 \times 4.187 \times 10^3}$$
$$= 4.012\ \text{K (Ans)}$$

13.7.2 Enthalpy of Fusion

The enthalpy of fusion or heat of fusion is the change in enthalpy resulting from heating a given quantity of a substance to change its state from a solid to a liquid. The temperature at which this occurs is called the melting point.

The 'enthalpy' of fusion is a latent heat, because during the change in state, the introduction of heat cannot be observed as a temperature change, as the temperature remains constant during the process. The latent heat of fusion is the enthalpy change of any amount of substance when it melts. When the heat of fusion is referenced to a unit of mass, it is usually called the specific heat of fusion, while the molar heat of fusion refers to the enthalpy change per amount of substance in moles.

The liquid phase has a higher internal energy than the solid phase. This means energy must be supplied to a solid in order to melt it and energy is released from a liquid when it freezes, because the molecules in the liquid experience weaker intermolecular forces and have a larger potential energy.

When liquid water is cooled, its temperature falls steadily until it drops just below the freezing point at 0°C. The temperature then remains constant at the freezing point while the water crystallises. Once the water is completely frozen, its temperature continues to fall.

The specific enthalpy of fusion of water is 333.55 kJ/kg at 0°C. Of common substances, only that of ammonia is higher.

13.7.3 Enthalpy of Vaporisation

The enthalpy of vaporisation, H_V, also known as the heat of vaporisation or heat of evaporation, is the energy required to transform a given quantity of a substance from a liquid into a gas at a given pressure (often atmospheric pressure).

It is often measured at the normal boiling point of a substance; although tabulated values are usually corrected to 298 K, the correction is often smaller than the uncertainty in the measured value.

The heat of vaporisation is temperature dependent (Table 13.8), though a constant heat of vaporisation can be assumed for small temperature ranges and for $Tr \ll 1.0$. The heat of vaporisation diminishes with increasing temperature and it vanishes completely at the critical temperature ($T_r = 1$) because above the critical temperature, the liquid and vapour phases no longer co-exist.

Table 13.9 gives the constant pressure heat capacity (C_p) for a range of temperatures in joule per gram kelvin at a pressure of 100 kPa.

13.8 Channel Flow

Open-channel flow, a branch of hydraulics, is a type of liquid flow within a conduit with a free surface, known as a channel. The other type of flow within a conduit is pipe flow. These two types of flow are similar in many ways, but differ in one important respect: the free surface. Open-channel flow has a free surface, whereas most pipe flows do not.

TABLE 13.8

Heat of Vaporisation

Temperature	Hv (kJ/mol (°C))
0	45.054
25	43.99
40	43.35
60	42.482
80	41.585
100	40.657
120	39.684
140	38.643
160	37.518
180	36.304
200	34.962
220	33.468
240	31.809
260	29.93
280	27.795
300	25.3
320	22.297
340	18.502
360	12.966
374	2.066

TABLE 13.9

Constant-Pressure Heat Capacity

Temperature (°C)	Cp (J/(g K) at 100 kPa)
0	4.2176
10	4.1921
20	4.1818
25	4.1814
30	4.1784
40	4.1785
50	4.1806
60	4.1843
70	4.1895
80	4.1963
90	4.205
100	4.2159

Open channels are found in a variety of situations; namely, drainage systems and irrigation systems. They are also found when diverting water from a natural reservoir in nearby hills to a water treatment plant for supplying clean drinkable water to towns and cities.

13.8.1 Channel Flow

In the above examples, the water flows down an open channel with its top surface open to the air. Engineers have to ensure that these channels have sufficient capacity to carry the anticipated flow of water.

Figure 13.28a through d shows various cross-sections of typical open channel shapes. Types of open channel flow:

- Steady flow—when the flow rate (Q) does not change with time.
- Uniform flow—when the depth of fluid does not change for a selected length or section of the channel.
- Uniform steady flow—when the discharge does not change with time and the depth remains constant for a selected section, cross-section should remain constant.
- Varied steady flow—when the depth changes but the discharge remains the same.
- Varied unsteady flow—when both the depth and discharge changes along the length of the channel being measured.
- Rapidly varying flow—depth change is rapid, and the cross-section remains constant.
- Gradually varying flow—the depth change is gradual with a constant cross-section remaining constant.

13.8.2 Hydraulic Radius

A parameter that is used to express the ratio of the flow cross-sectional area (A) of the channel and the wetted perimeter (WP):

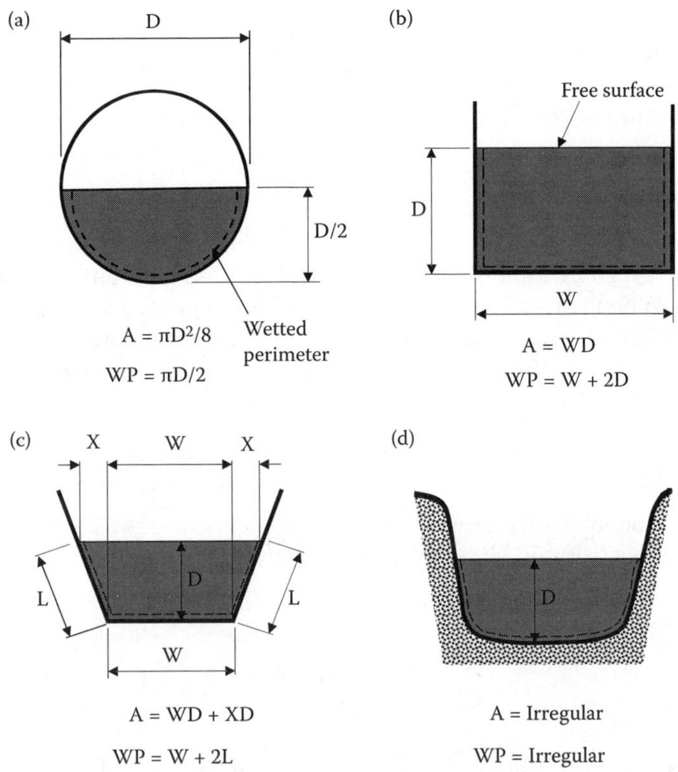

FIGURE 13.28
Typical open channels. (a) Circular pipe running half full, (b) rectangular channel, (c) trapezoidal channel and (d) natural channel.

$$R = \frac{A}{WP} \tag{13.78}$$

13.8.3 Flow Rate

Flow rate (Q) is the rate at which the fluid, in terms of volume, flows along a channel. Its units are m^3/s.

If it is assumed that the flow in the channel is uniform, then the flow rate is a function of

- The area of the channel (m^2)
- The hydraulic radius (m)
- The slope of the channel (in the direction of the flow)
- The roughness of the channel material that is in direct contact with the fluid

The above assumptions can be represented using the 'Manning' Equation:

$$Q = \frac{A}{n} \times R^{2/3} \times S^{1/2} \tag{13.79}$$

where

A represents cross-sectional area of the channel (m^2)
R represents hydraulic radius (m)
S represents channel slope in the direction of flow (m/m)
n represents Manning coefficient of roughness

13.8.4 Roughness

Roughness is dependent upon the nature of the channel material. Some values of Manning's values for 'n' are given in Table 13.10. From the table, it is seen that most materials have a range of values. When considering concrete as a material for construction, the smoothest concrete with a high class finish has a normal value of approximately 0.011, with a range of 0.010–0.013. Laboratory channels with glass side, used for research purposes and modelling, have a value of 0.010.

Example 13.18

Estimate the normal discharge for a 200 mm inside diameter pipe, lined with a common clay tile, running half full. The slope drops 1.0 m over 1000 m.

Solution

$$A = \frac{\pi D^2}{8} = \frac{\pi \times (200 \times 10^{-3})^2}{8} = 0.0157\ m^2$$

$$WP = \frac{\pi D}{2} = \frac{\pi \times 200 \times 10^{-3}}{2} = 0.3142\ m$$

$$R = \frac{A}{WP} = \frac{0.0157}{0.3142} = 0.050\ m$$

From Table 13.10, Manning's coefficient for clay tile is 0.013.
Substituting these values into Equation 13.79:

TABLE 13.10
Various Values of "n" for the Most Common Channel Surfaces

Channel Surface Description	n
Glass, copper, plastic or other smooth surface	0.010
Smooth, unpainted steel, planed wood	0.012
Painted steel or coated cast iron	0.013
Smooth asphalt, common clay drainage tile, trowel finished concrete, glazed brick	0.013
Uncoated cast iron, black wrought iron pipe, vitrified clay sewer pipe	0.014
Brick in cement mortar, float finished concrete, concrete pipe	0.015
Formed unfinished concrete, spiral steel pipe	0.017
Smooth earth	0.018
Clean excavated earth	0.022
Corrugated metal storm drain	0.024
Natural channel with stones and weeds	0.030
Natural channel with light brush	0.050
Natural channel with tall grasses and reeds	0.060
Natural channel with heavy brush	0.100

$$Q = \frac{A}{n} \times R^{2/3} \times S^{1/2}$$
$$= \frac{0.0157}{0.013} \times 0.05^{2/3} \times 0.001^{1/2}$$
$$Q = 5.183 \times 10^{-3}\ m^3/s\ (Ans)$$

Example 13.19

Calculate the slope of the trapezoidal channel shown in Figure 13.30 when the flow rate is 1.42 m³/s. It is formed from unfinished concrete.

Solution

$$A = WD + XD$$
$$= (2.0 \times 0.50) + (0.5 \times 0.5)$$
$$= 1.25\ m^2$$
$$WP = W + 2L$$
$$= 2.0 + (2 \times 0.7071)$$
$$= 3.414\ m$$
$$R = \frac{A}{WP}$$
$$= \frac{1.25}{3.414}$$
$$= 0.366\ m$$

Substitute values into Equation 13.79:

$$Q = \frac{A}{n} \times R^{2/3} \times S^{1/2}$$
$$1.42 = \frac{1.25}{0.017} \times 0.366^{2/3} \times S^{1/2}$$

Rearranging:

$$S = \left(\frac{n \cdot Q}{A \cdot R^{2/3}}\right)^2$$
$$= 1.445\ m\ per\ 1000\ m\ (Ans)$$

13.9 Orifice Plate

An orifice plate is an artificial obstruction placed in a pipeline that is used for the measurement of flow rate. Volumetric or mass flow rates may then be determined, this being dependent upon the calculation method that is associated with the specific type of orifice plate used.

Bernoulli's principle is used in the measurement of the fluid flow, where there is a relationship between the pressure of the fluid and its velocity, that is, when the velocity increases there is a corresponding drop in the fluid pressure and vice versa.

13.9.1 Description

An orifice plate is a thin plate having a hole carefully machined in the centre of the plate. The plate is fixed in the pipeline and is mounted between two coupling flanges in the pipe. The fluid is forced to go through the small orifice; there is a point of maximum convergence that occurs a short distance downstream of the plate. This convergence is known as the vena contracta (see Figure 13.29). The velocity of the fluid is increased when going through the orifice and there is a corresponding reduction in the fluid pressure. Beyond the vena contracta, the fluid expands back to the pipeline diameter and the velocity and pressure return to their original values. Pressure tappings are provided, one upstream of the plate and the other just downstream.

There are three locations for the tappings:

1. *Flange location:* The tap locations are 25 mm upstream and 24 mm downstream from the face of the orifice plate.
2. *Vena contracta location:* The tap locations are one internal pipe diameter upstream and 0.3–0.8 pipe diameter downstream from the face of the orifice plate.
3. *Pipe location:* The tap location 2.5 internal pipe diameter upstream, and 8 pipe diameter downstream from the orifice plate.

It should be noted here that an orifice plate will only work well with a fully developed flow profile. This is achieved by using a long upstream pipe usually between 20 and 40 pipe diameter before the orifice plate depending upon the Reynolds number or by the use of a flow conditioner. Orifice plates are inexpensive and small but the pressure drop is not so well recovered as well as a venturi nozzle. If there is sufficient space available, a venturi nozzle is more efficient than an orifice plate.

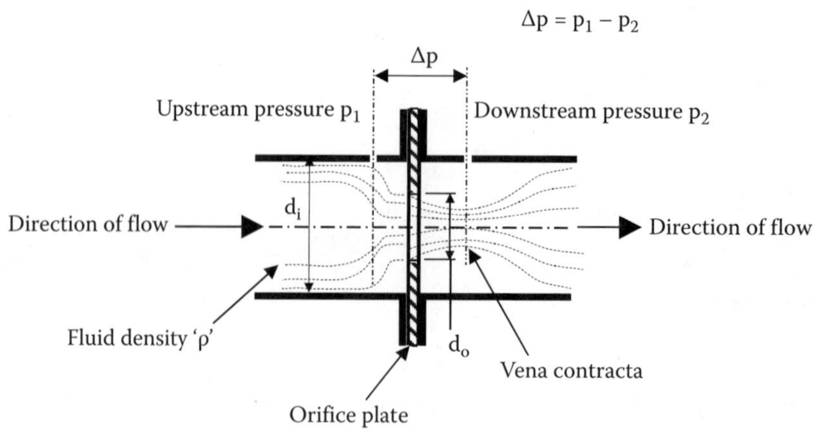

FIGURE 13.29
Orifice plate.

13.9.2 Measurement

When the fluid flows through the orifice plate, there are frictional energy losses which together with the contraction in the diameter of the flow affect the value of 'C_d' (coefficient of discharge). The shape of the orifice lip is also important; this may be either sharp or flat.

If the Bernoulli equation is applied between positions 1 and 2 in Figure 13.29, the following is obtained:

$$h_1 + \frac{u_1^2}{2g} = h_2 + \frac{u_2^2}{2g} \tag{13.80}$$

Neglecting u_1^2:

$$h_1 = h_2 + \frac{u_2^2}{2g} \tag{13.81}$$

$$2g(h_1 - h_2) = u_2^2$$

$$u_2 = \sqrt{2g(h_1 - h_2)}$$

or

$$u_2 = \sqrt{\frac{2\Delta p}{\rho}} \tag{13.82}$$

now

$$Q = u_2 \cdot A_o$$

Introducing 'C_d' to allow for friction and other factors:

$$Q = C_d \cdot A_o\sqrt{2g(h_1 - h_2)}$$

which can be expressed as

$$Q = C_d A_o\sqrt{\frac{2\Delta p}{\rho}} \tag{13.83}$$

where

- Q represents volumetric flow rate (m/s^3)
- A_o represents area of orifice diameter (m)
- Δp represents differential pressure (Pa)
- ρ represents fluid density (kg/m^3)

The majority of commercial orifice plate calculations only consider the diameter of the orifice to calculate the volumetric flow rate.

Example 13.20

A horizontal pipe with an internal diameter of 25 mm is fitted with an orifice plate having a 20 mm diameter hole. The differential pressure across the plate is measured at 750 kPa. It can be assumed that the coefficient of discharge is 0.64 and the density of water is 1000 kg/m^3.

Estimate the volumetric flow rate in cubic meter per second.

Solution

$$A_o = \frac{\pi d_o^2}{4} = \frac{\pi \times 0.02^2}{4} = 0.314 \times 10^{-3}\ m^2$$

$$Q = C_d A_o \sqrt{\frac{2\Delta p}{\rho}}$$

$$= 0.64 \times 0.314 \times 10^{-3} \times \sqrt{\frac{2 \times 750{,}000}{1000}}$$

$$= 0.007783\ m^3/s\ (Ans)$$

13.10 Fluid Machines

A fluid machine is a device that converts the kinetic or potential energy held in a fluid into mechanical energy, or it is reversible and converts mechanical energy into fluid energy.

Essentially, there are two main types of fluid machines, namely:

1. Positive displacement machines
2. Rotodynamic machines

13.10.1 Positive Displacement Machines

These come in three forms:

1. Single rotor
2. Double rotor
3. Reciprocating piston

13.10.1.1 Single Rotor

Sliding vane: In this design the vanes are moved by a rotor, drawing fluid into and then forcing the fluid out from the pumping chamber that is formed within the body of the pump (see Figure 13.30). Carefully selected vane materials make these pumps well-suited for handling low-viscosity, non-lubricating liquids including solvents, fuel oils, petrol and liquidified gas. Fluid viscosities range from 0.5 to 220,000 cSt.

Axial piston: This is a variation in the piston pump described above; here the individual pistons are arranged in a circular manner within a cylinder block and are driven by a cam plate fitted on the drive shaft. In this type of pump, the fluid is drawn in and forced out as the inclined cam plate is rotated about the centreline of the pump. This type of pump has the advantage that the flow can be variable; the angle of the cam plate is varied from a zero angle where there will be no piston displacement, up to a maximum angle where the pistons will have a maximum stroke. These pumps can also be supplied with a fixed cam plate.

Axial piston pumps have a relatively low flow rate of approximately 70 m^3/h (300 galls/min) but are capable of generating pressures up to 25 MPa (3626 $lbf/in.^2$). Figure 13.31 depicts an axial piston pump.

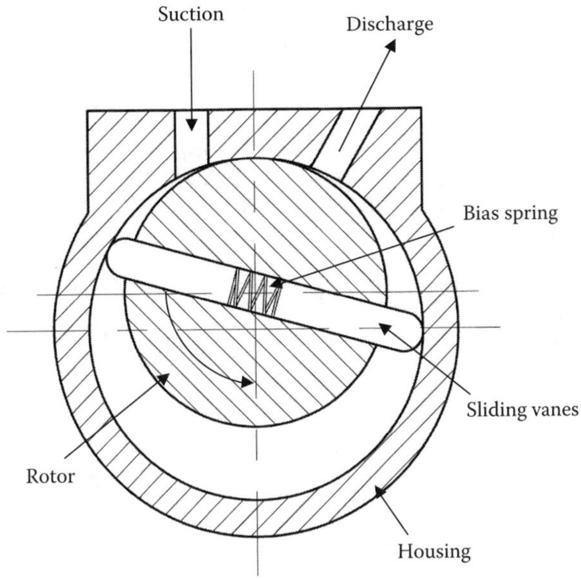

FIGURE 13.30
Sliding vane pump.

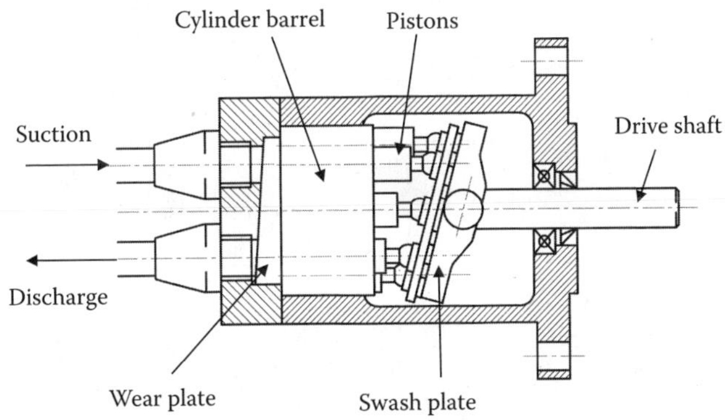

FIGURE 13.31
Axial piston pump.

Reciprocating piston pump: Fluid is drawn in and out of the pump chamber by a piston that reciprocates within a cylinder. The flow of the fluid is controlled by port valves attached to the cylinder (see Figure 13.32).

Peristaltic pump: A peristaltic pump is a type of positive displacement pump used for pumping a variety of fluids. The fluid is contained within a flexible tube fitted inside a circular pump casing (though linear peristaltic pumps have been made). A rotor with a number of 'rollers', 'shoes', 'wipers' or 'lobes' attached to the external circumference of the rotor compresses the flexible tube. As the rotor turns, the part of the tube under compression is pinched closed (or 'occludes'), thus forcing the fluid to be pumped to move through the tube (Figure 13.33). Additionally, as the tube opens to its natural state after passing the cam (restitution or resilience), fluid flow is induced into the pump. This process is called peristalsis. Typically, there will be two or more rollers or wipers, occluding the tube, trapping between them a body of fluid. The

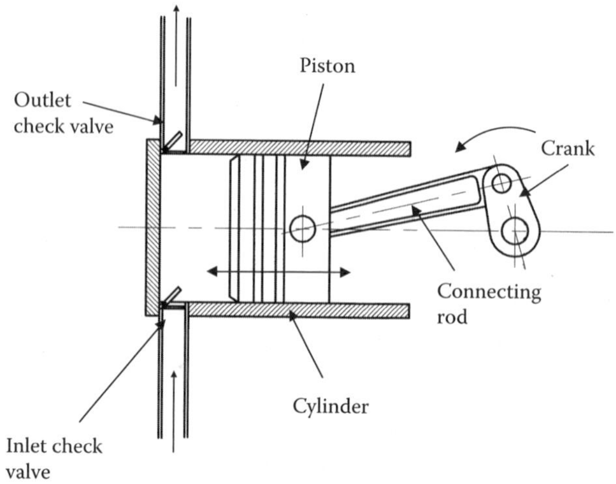

FIGURE 13.32
Reciprocating pump.

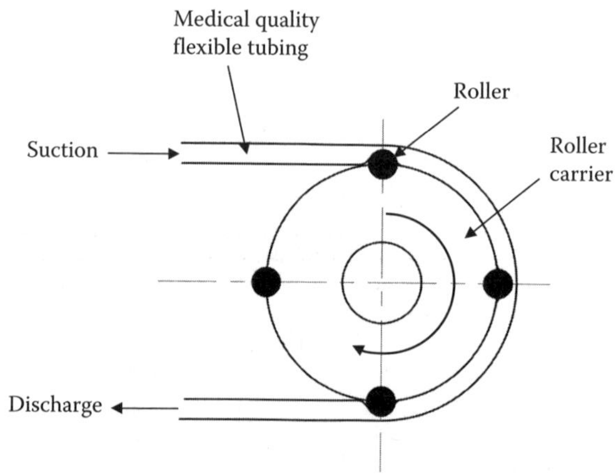

FIGURE 13.33
Peristaltic pump.

body of fluid is then transported, at ambient pressure, towards the pump outlet. These pumps are able to pump fluids with solids in suspension and low-to-medium viscosity levels. Peristaltic pumps are available with flow rates up to 1.4 m^3/min (307 gallons/min) with differential pressures up to 1.6 MPa (232 lbf/in.^{2}).

Smaller models generally operate at speeds up to 200 rev/min and the larger models are limited to speeds below 100 rev/min.

13.10.1.2 Double Rotor

Circumferential pump: This type of pump as shown in Figure 13.34 is similar to the lobe pump but here there are only two 'teeth' per shaft and the fluid is carried around in the space between the piston surfaces and the fluid chamber. Like the

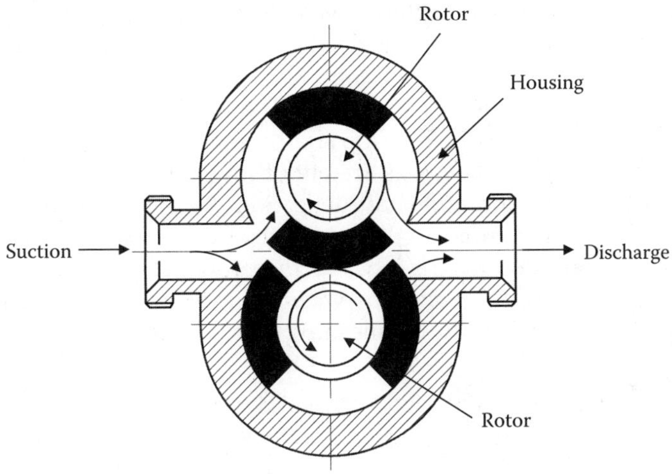

FIGURE 13.34
Circumferential piston.

lobe pump, there is no sealing contact between the piston surfaces, clearances between the pistons are maintained by external timing gears. These pumps have proven to be very reliable. As the circumferential pump rotates, the expanding volume draws the fluid in and is then forced out the discharge port by the reducing volume on the discharge side.

Circumferential pumps are available in a range of sizes capable of flow rates up to 2.3 m^3/min (506 galls/min) and discharge pressures up to 3.1 MPa (450 lbf/in.^{2}) covering a viscosity range from 50 to 1,000,000 cSt. They are able to pump any product that can be moved being able to handle rather large solids and shear-sensitive fluids. They are also suitable to run dry for extended periods of time.

Gear pump: (Figure 13.35). Fluid is carried between the gear teeth within a tight fitting circular cavity and is expelled by the meshing teeth. The inner surface of the pump housing provides a continuous sealing and either rotor is capable of driving the other.

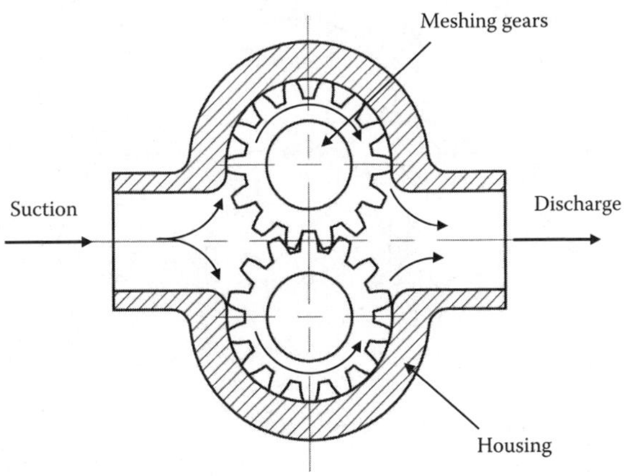

FIGURE 13.35
Gear pump.

Lobe: Lobe pumps are similar to external gear pumps in operation in that the fluid flows around the interior of the casing (Figure 13.36). There are three lobes per shaft leaving a cavity between the lobe features. Unlike external gear pumps, however, the lobes do not make contact. Lobe contact is prevented by external timing gears located in the gearbox. Pump shaft support bearings are located within the gearbox, and since the bearings are out of the pumped liquid, the pressure is limited by bearing location and shaft deflection. As the lobes come out of mesh, they create expanding volume on the inlet side of the pump. The liquid flows into the cavity and is trapped by the lobes as they rotate. Liquid travels around the interior of the casing in the pockets between the lobes and the casing—it does not pass between the lobes. Finally, the meshing of the lobes forces the liquid through the outlet port under pressure.

Single-screw pump: The fluid is carried between the rotor screw threads which are a close fit within a cylinder block. The Archimedes screw is one such example of this machine (see Figure 13.37). They can handle fluids containing abrasive and solid particles up to 90 mm (3.50 in.) and can also handle multiphase fluids containing up to 99% gas. These types of pumps are self-priming and can lift fluids up to 8.5 m (28 ft) and are able to handle any fluid that is compatible with the materials of construction.

Double-screw pump: Double-screw pumps have twin external screw rotors that intermesh as shown in Figure 13.38. The fluid enters at the central inlet and then splits axially into two end sections and as the rotating screws intermesh, chambers are formed and convey the fluid to the end receiving chambers formed in the housing and returned to a separate central port.

As with the single screw pumps, they can handle a wide range of multiphase fluids and are often found in oil production/pipeline applications.

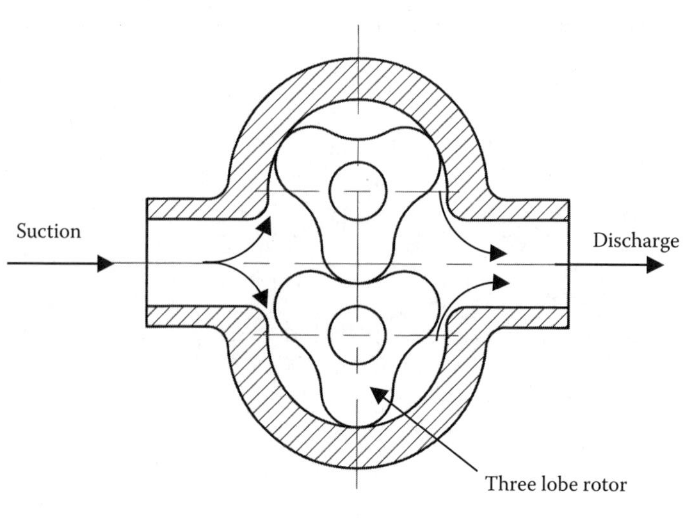

FIGURE 13.36
Lobe pump.

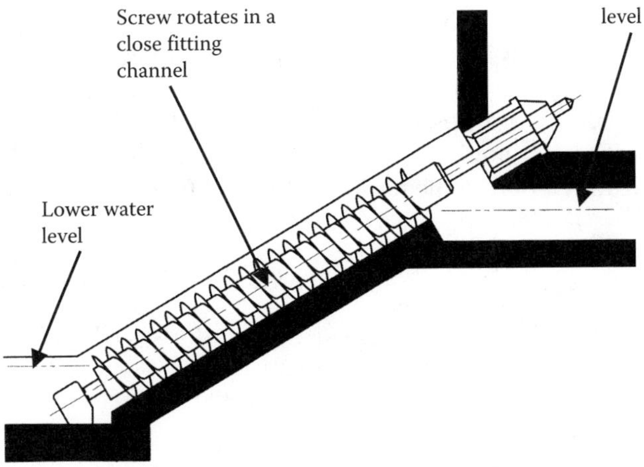

FIGURE 13.37
Single-screw pump.

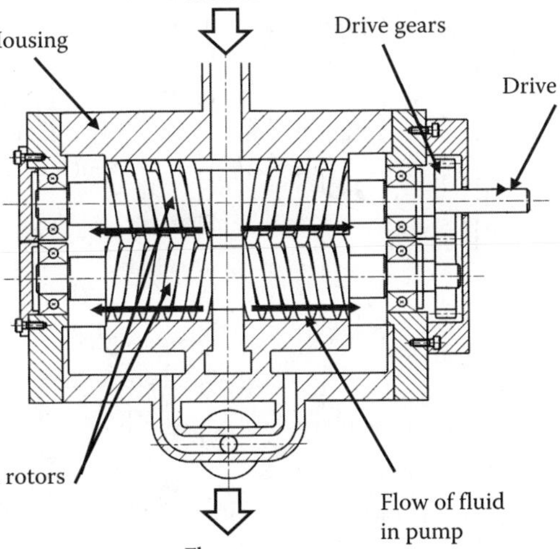

FIGURE 13.38
Double-screw pump.

Rotodynamic machines: Within this class of machines, there are three basic pump categories:

1. Radial flow
2. Mixed flow
3. Axial flow

Radial flow pumps—(Figure 13.39): Often simply referred to as centrifugal pumps. The fluid enters along the axial plane, is accelerated by the impeller and exits at right angles to the shaft (radially). Radial-flow pumps operate at higher pressures and lower flow rates than axial and mixed-flow pumps.

Mixed-flow pumps—(Figure 13.40): Mixed-flow pumps function as a compromise between radial and axial-flow pumps. The fluid experiences both radial acceleration and lift and exits the impeller somewhere between 0° and 90° from the axial direction. As a consequence, mixed-flow pumps operate at higher pressures than axial-flow pumps while delivering higher discharges than radial-flow pumps. The exit angle of the flow dictates the pressure head-discharge characteristic in relation to the radial and mixed flow.

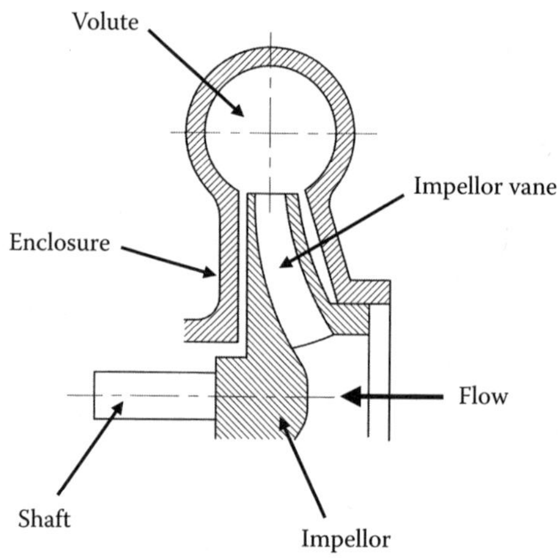

FIGURE 13.39
Radial flow compressor.

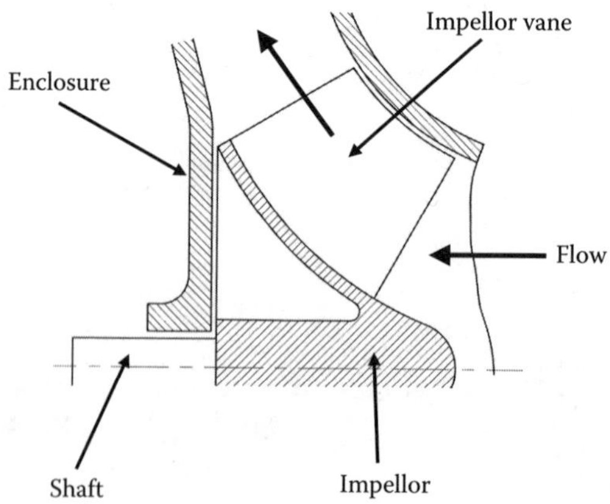

FIGURE 13.40
Mixed flow compressor.

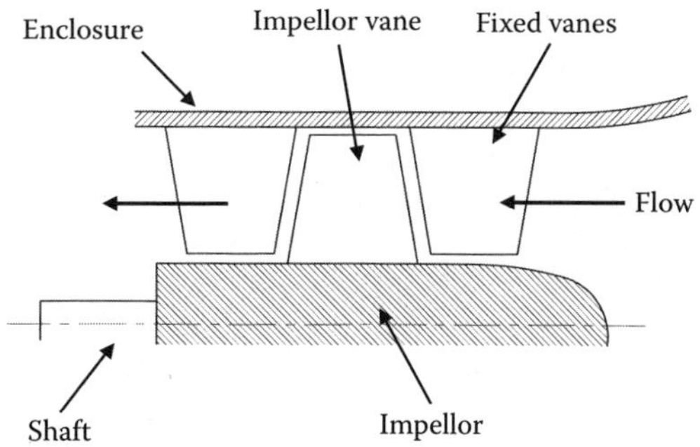

FIGURE 13.41
Axial flow compressor.

Axial flow pumps—(Figure 13.41): Axial-flow pumps differ from radial flow pumps in that the fluid enters and exits along the same direction parallel to the rotating shaft. The fluid is not accelerated but instead 'lifted' by the action of the impeller. They may be likened to a propeller spinning in a length of a tube. Axial-flow pumps operate at much lower pressures and higher flow rates than radial-flow pumps.

14

Introduction to Analytical Stress Analysis and the Use of Mohr's Circle

14.1 Introduction

Of all the graphical methods used by an engineer for visualizing problems, there is little doubt that Mohr's circle is the best known; in its various applications it is an aid to visualizing a stress or strain problem.

The object of this chapter is to explore with a reasonable thoroughness three important uses of Morh's circle. These applications confront the engineer and designer in many branches of mechanical and structural engineering.

In the development of the circle construction, it has been considered desirable to present a concise account of the underlying theory. It is possible for the reader more interested in applications to accept these results and go straight to those sections dealing with the constructions and applications.

Nearly all textbooks, of necessity, deal with the circle in too cursory a manner and leave many detailed questions unanswered. It is hoped that this chapter devoted solely to the circle will answer more of these questions. In illustrating the text with examples of a practical nature, other topics of strength of materials will be quoted, the background of which can be found in the most strength of materials textbooks.

14.2 Notation

A = Area

a = Circle constant

B = Breadth (of beam cross section)

D = Depth (of beam-section)

E = Modulus of elasticity

e = Strain (usually with suffix to indicate direction)

F = Force

h = Distance

I = Second moment of area (with suffix of the type xx or cg)

J = Polar moment (with suffix of the type xy)
M = Moment (with suffix to indicate axis)
n = Factor of safety (FoS)
O = Origin of graph
P = Force, or pole point
$P_{1,2}$ = Pole points
R = Radius of circle, radius of curvature
T = Torque (usually with suffix to indicate axis)
y = Distance from neutral axis of a beam to a given point
β = Angle
γ = Poisson's ratio, angle
δ = Deflection
θ = Angle of plane
θ_p = Angle of principal plane
σ = Stress (with suffix of the type x to indicate plane)
σ_y = Yield stress, or normal stress in y direction
σ_w = Working stress
τ = Shear stress (with suffix of the type xy)
φ = Shear strain (with suffix of the type xy)

14.3 Two-Dimensional Stress Analysis

Consider the rectangle ABCD shown in Figure 14.1 of unit thickness and subject to direct stresses σ_x and σ_y, together with shearing stresses τ_{xy}.

The stress normal to the plane of the diagram σ_z is equal to zero. This is the condition of plane stress.

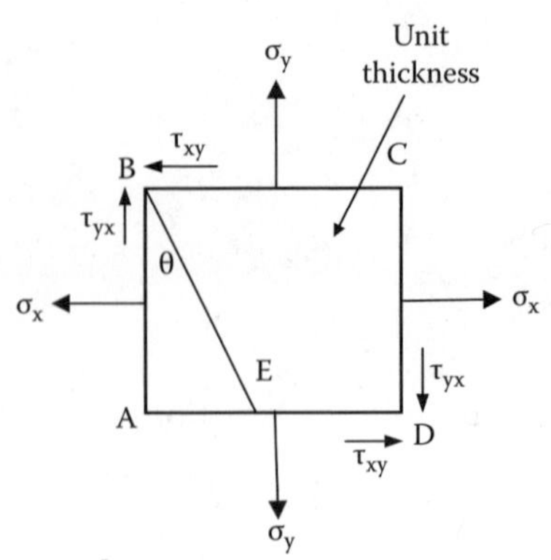

FIGURE 14.1
A unit thickness element subject to a biaxial stress field.

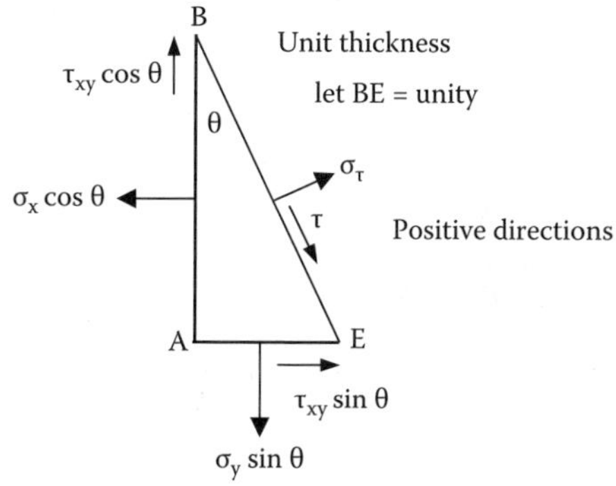

FIGURE 14.2
Section of element in Figure 14.1.

It is necessary to investigate the stress conditions on any plane making an angle θ with the side AB and the consequence that follows from the existence of such stresses.

A triangular element is isolated from the rectangle, the stresses on the original planes AB and AE being known and the sloping side being an unknown normal stress of σ_n and an unknown shearing force τ are postulated. The problem is to obtain expressions for these unknown stresses by considering *forces* (not stresses, please note) acting on this triangular element of unit thickness. The length BE is assumed to be one unit, as shown in Figure 14.2. The forces are then resolved in normal and tangential directions to the slope BE.

For equilibrium, the sum of all the forces in a given direction must equal zero.

Thus, resolving normally to BE in accordance with the sign convention accompanying Figure 14.2,

$$\Sigma P_n = -\sigma_x \cos\theta\cos\theta + \tau_{xy}\cos\theta\sin\theta + \tau_{xy}\cos\theta\sin\theta - \sigma_y\sin\theta\sin\theta + \sigma_n = 0$$

$$\sigma_n = \sigma_x\cos^2\theta + \sigma_y\sin^2\theta - 2\tau_{xy}\sin\theta\cos\theta$$

and in terms of double angles,

$$\sigma_n = 1/2(\sigma x + \sigma y) + 1/2(\sigma x - \sigma y)\cos 2\theta - \tau_{xy}\sin 2\theta \quad (14.1)$$

Resolving tangentially along BE and again carefully following the sign convention,

$$\Sigma P_\theta = -\sigma_x\cos\theta\sin\theta - \tau_{xy}\cos^2\theta + \sigma_y\sin\theta\cos\theta + \tau_{xy}\sin^2\theta + \tau = 0$$
$$\tau = 1/2(\sigma_x - \sigma_y)\sin 2\theta + \tau_{xy}\cos 2\theta \quad (14.2)$$

14.4 Principal Stresses and Principal Planes

σ_n is clearly a quantity that varies with the angle θ. By differentiating with respect to θ and equating to zero, the conditions for maximum and minimum normal stress can be found from Equation 14.1:

$$\frac{d\sigma_n}{d\theta} = -(\sigma_x - \sigma_y)\sin 2\theta - 2\tau_{xy}\cos 2\theta = 0 \quad (14.3)$$

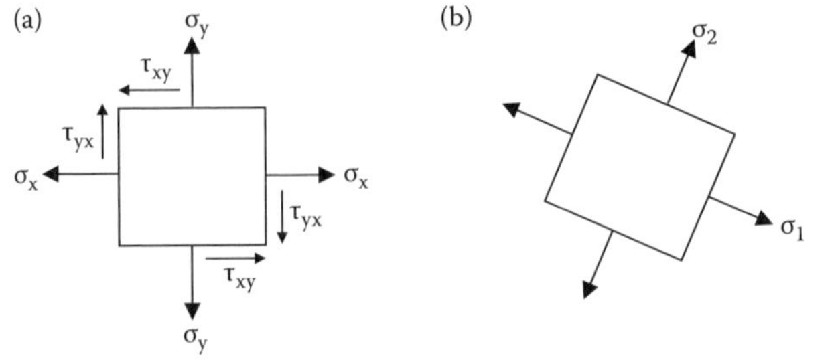

FIGURE 14.3
Principal stresses and planes: complex stress system.

Thus,

$$\tan 2\theta = -\frac{2\tau_{xy}}{(\sigma_x - \sigma_y)} \tag{14.4}$$

A comparison of Equations 14.2 and 14.3 shows that τ is zero on the planes where the maximum or minimum normal stresses are found.

This is an extremely important finding: the conditions for σ_{max} or σ_{min} have been derived, and on the planes where these stresses occur the shear stresses are zero. There are two of these planes because there are two values of θ, 90° apart, which satisfy Equation 14.4.

These planes are called *principal planes*, and the normal stresses σ_{max} and σ_{min} acting on these planes are called *principal stresses*. On all the other planes, shear stresses will be acting. It is also possible to show that in a three-dimensional system there are three mutually perpendicular planes that possess no shear stresses.

Figure 14.3a and b shows, respectively, the complex stress situation and the equivalent principal stress pattern. It must be clearly understood that these are equivalent systems. In the solution of a problem, the engineer chooses the system that best fits the purpose.

Generally, the stresses shown in Figure 14.3a are obtained by some preliminary calculations and the principal stress system is then deduced. Frequently, this second system is the one more easily assessed. Expressions for the principal stresses, now to be called σ_1 and σ_2, can be obtained from Equations 14.1 and 14.3, giving

$$\sigma_{1,2} = 0.5(\sigma_x + \sigma_y) + \frac{0.5(\sigma_x - \sigma_y)\cdot 0.5(\sigma_x - \sigma_y) + \tau_{xy}{}^2}{\sqrt{\left(\frac{\sigma_x - \sigma_y}{2}\right)^2 + \tau_{xy}{}^2}}$$

$$\sigma_{1,2} = 0.5(\sigma_x + \sigma_y) \pm \sqrt{\left(\frac{\sigma_x - \sigma_y}{2}\right)^2 + \tau_{xy}{}^2} \tag{14.5}$$

14.4.1 Maximum Shear Stress

If now Equation 14.2 is differentiated and equated to zero, then the condition for maximum shear stress within the element can be obtained:

$$\frac{d\tau}{d\theta} = (\sigma_x - \sigma_y)\cos 2\theta - 2\tau_{xy}\sin 2\theta = 0$$

Hence,

$$\tan 2\theta = \left(\frac{\sigma_x - \sigma_y}{2 \cdot \tau_{xy}} \right) \quad (14.6)$$

This is the negative reciprocal of Equation 14.3. Consideration of Equations 14.3 and 14.5 shows the values of 2θ defined by each differ by 90°. Thus, θ differs by 45° in each case. This means that the maximum shear stress occurs on planes 45° to the planes of principal stress. This is shown in Figure 14.4a and b. It is worth noting that the maximum shears are accompanied by normal stresses.

Equation 14.6 can be used to obtain τ_{max} from Equation 14.2:

$$\begin{aligned} \tau_{max} &= \frac{0.5(\sigma_x - \sigma_y)0.5(\sigma_x - \sigma_y) + \tau_{xy}{}^2}{\sqrt{\left(\frac{\sigma_x - \sigma_y}{2}\right)^2 + \tau_{xy}{}^2}} \\ &= \pm\sqrt{\left(\frac{\sigma_x - \sigma_y}{2}\right)^2 + \tau_{xy^2}} \end{aligned} \quad (14.7)$$

Expressions 14.5 and 14.7 lead to two simple and important relations, that is,

$$\tau_{max} = \text{half the principal stress difference}$$

and

$$\sigma_1 + \sigma_2 = \sigma_x + \sigma_y = \text{a constant value}$$

Thus, the sum of the normal stresses on any pair of perpendicular faces is invariant:

$$\tau_{max} = \frac{1}{2}(\sigma_1 - \sigma_2)$$

This small number of equations forms the basis of stress analysis. The following sections will consider the geometric interpretation of these equations.

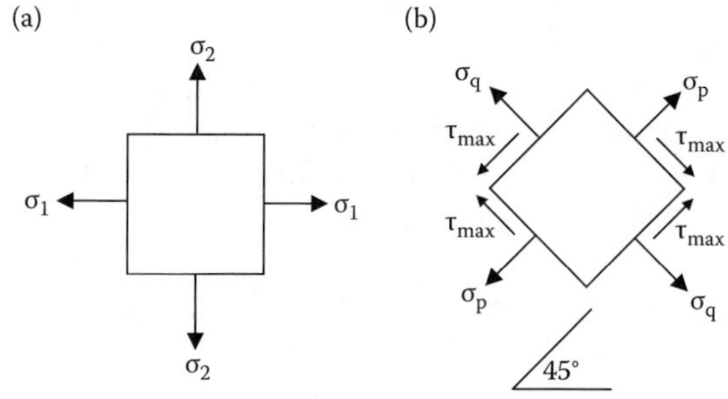

FIGURE 14.4
Principal stresses and planes: maximum shear stresses.

14.4.2 Geometric Interpretation

For the present purpose, Equations 14.1 and 14.2 are rearranged as follows:

$$\sigma_n - \frac{1}{2}(\sigma_x - \sigma_y) = \frac{1}{2}(\sigma_x - \sigma_y) \cdot \cos 2\theta - \tau_{xy} \cdot \sin 2\theta$$

$$\tau = \frac{1}{2}(\sigma_x - \sigma_y) \cdot \sin 2\theta + \tau_{xy} \cdot \cos 2\theta$$

Squaring each side, these become

$$\left[\sigma_n - \frac{1}{2}(\sigma_x - \sigma_y)\right]^2 = \frac{1}{4}(\sigma_x - \sigma_y)^2 \cdot \cos^2 2\theta - (\sigma_x - \sigma_y) \cdot \tau_{xy} \cdot \sin 2\theta \cdot \cos 2\theta + \tau_{xy}{}^2 \sin^2 2\theta$$

$$\tau^2 = \frac{1}{4}(\sigma_x - \sigma_y)^2 \sin^2 2\theta + (\sigma_x - \sigma_y) \cdot \tau_{xy} \cdot \sin 2\theta \cdot \cos 2\theta + \tau_{xy}{}^2 \cos^2 2\theta$$

and adding the two equations together results in

$$\left[\sigma_n - \frac{1}{2}(\sigma_x + \sigma_y)\right]^2 + \tau^2 = \frac{1}{4}(\sigma_x - \sigma_y)^2 \cdot \tau_{xy}{}^2 \tag{14.8}$$

Note the right-hand side of Equation 14.8 is a constant quantity for a given system of applied stresses. On inspection, the equation can be seen to be of the type

$$(x - a)^2 + y^2 = R^2$$

and is thus the equation of a radius of a circle.

This circle is the graphical representation of Equations 14.5 and 14.7 and, thus, defines all possible stress conditions.

14.5 Construction of Mohr's Circle

The starting point of this discussion is any two-dimensional stress system, as shown in Figure 14.3a. In some cases, some stresses may be zero. The Mohr's circle is fundamentally a graph, and as such it requires horizontal and vertical axes.

The horizontal axis is used to represent direct or axial stresses, and the vertical axis is reserved for shear stresses. If a circle is to be produced, then both axis scales must be identical; otherwise, an ellipse will be generated.

The stresses σ_x and σ_y are marked off along the horizontal axis, and from these points ordinates are projected to equal τ_{xy} and its complementary τ_{yx}. This is shown in Figure 14.6, where these coordinates define the points A and B.

A consistent sign convention must be adopted for the stresses, and here it is assumed that σ_x and σ_y will be positive and drawn to the right-hand side of the origin when they represent tensile stresses.

Shear stress systems give rise to a surprising number of sign convention possibilities, and in this instance shear stresses producing a clockwise couple on an element will be taken as positive (see Figure 14.5).

Returning to the construction, the points A and B just obtained are joined up. The joining line cuts through the base line at M. Now draw a circle, at center M and radius AM.

Examination of Figure 14.6 shows that

$$OM = \frac{\sigma_x + \sigma_y}{2}$$

and that

$$ME = \frac{\sigma_x - \sigma_y}{2}$$

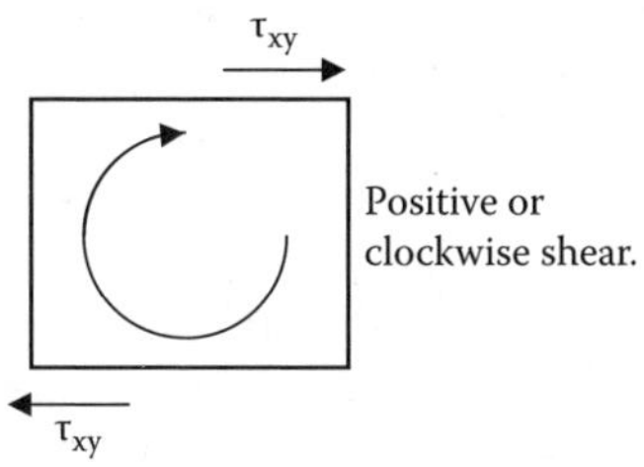

FIGURE 14.5
Sign conventions.

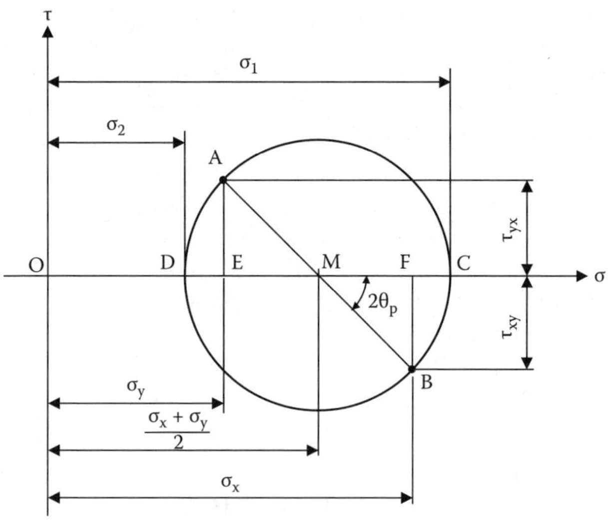

FIGURE 14.6
Mohr's circle diagram.

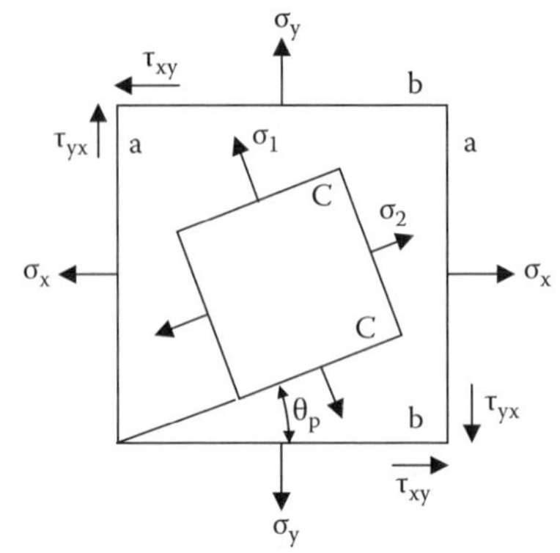

FIGURE 14.7
Stresses acting on the element.

The radius is

$$MA = \sqrt{\left(\frac{\sigma_x - \sigma_y}{2}\right)^2 + \tau_{xy}^{\ 2}}$$

Thus, the diagram is analogous to the various related formulas derived earlier.

It is important to note (and remember) that the angle CMB=$2\theta_p$ on the circle diagram corresponds to the angle θ_p on the original stresses element.

14.5.1 Conclusions and Deductions

The construction of the circle can quickly become an automatic activity. There is no need to consciously use the algebraic equations, because the circle provides an alternative.

The two principal stresses σ_1 and σ_2 can be measured immediately. Figure 14.7 shows that there is no shear stress associated with the principal stress. The principal stresses are also clearly seen to be the maximum and minimum values of the normal stress. The half-height of the circle gives the maximum shear stress (accompanied, in general, by an axial stress).

Note that the principal stresses, on the circle, are 180° apart, and also that the maximum shear stress is 90° away from the principal stresses. This signifies that the corresponding angles in the element are 90° and 45°, respectively.

Point C on the diagram represents the maximum principal stress σ_1 and is separated from B by a counterclockwise rotation of $2\theta_p$. On the element, the principal plane upon which σ_1 acts is therefore at an angle θ_p counterclockwise from the planes b (Figure 14.7). The plane for the minimum principal stress σ_2 is obtained by turning through a further 90°, since D on the circle is 180° from C.

14.6 Relationship between Direct and Shear Stress

At this point in the discussion it may be of interest to the reader to show the graphical relationship between the direct stress σ and shear stress τ.

Consider a bar of cross-sectional area A subjected to an axial tensile load P, and it is required to investigate the nature of the stresses on any plane x:x making an angle θ with the normal cross section (Figure 14.8).

Let σ_θ and τ_θ be the normal, that is, direct and tangential (shear), components of the resultant stress on the interface, the directions being assumed to be as shown (Figure 14.9).

Stress on the normal cross section: $\sigma = P/A$

Area of section x:x: $= A \sec\theta = A/\cos\theta$

Resolving forces perpendicular to x:x: $\sigma_\theta \times A/\cos\theta = P\cos\theta$.

Therefore, $\sigma_\theta = (P/A)\cos^2\theta$.

Hence, $\sigma_\theta = \sigma\cos^2\theta$.

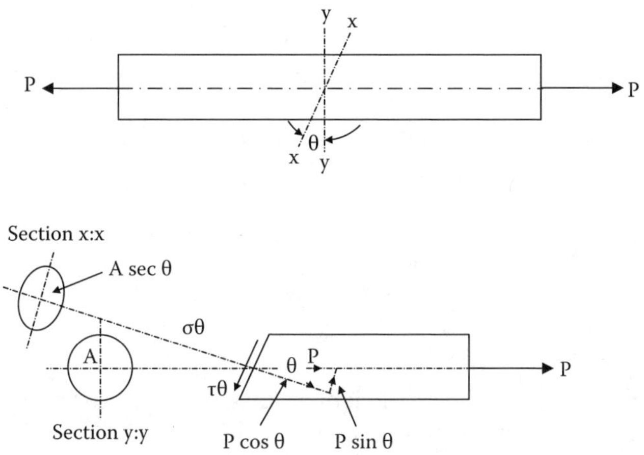

FIGURE 14.8
Section through a prismatic bar.

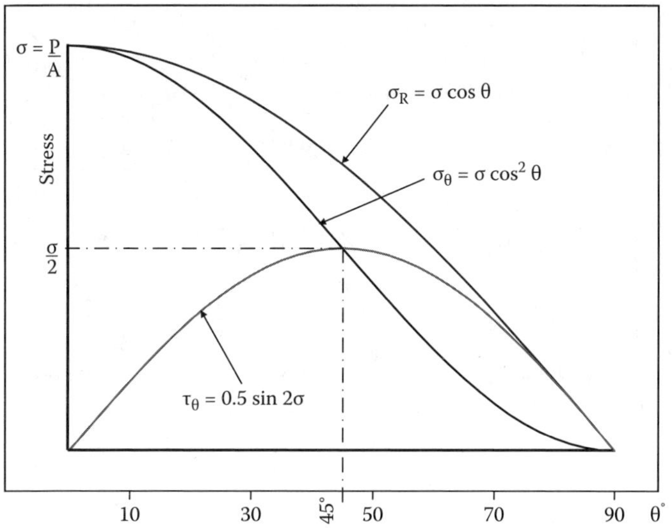

FIGURE 14.9
Distribution of stress from Figure 9.8.

Since $\cos^2\theta$ is positive for all values of θ, σ_θ can never be negative, that is, compressive.

The maximum value of σ_θ occurs when $\cos^2\theta$ is a maximum, that is, when $\theta=0°$, then $\sigma_\theta=\sigma$.

This is the stress on y:y.

Resolving forces parallel to x:x:

$\tau_\theta \times A/\cos\theta = P\sin\theta$.

Therefore,

$\tau_\theta = P/A \sin\theta\cos\theta$.

Hence:

$\tau_\theta = 1/2\ \sigma\sin 2\theta$.

For values of θ greater than 90°, the above expression becomes negative and the direction of τ_θ is the reverse of that assumed.

The maximum value of τ_θ occurs when $\sin 2\theta$ is a maximum, that is, when $\sin 2\theta=1$. Then $2\theta=90°$, that is, $\theta=45°$. Then $\tau_\theta=\sigma/2$.

14.7 The Pole of Mohr's Circle

The method just outlined would be sufficient to answer any two-dimensional stress problem. However, with a very simple addition the circle can be made to indicate directly the axes or planes of the stresses. The circle diagram of Figure 14.6 can be extended by drawing through point B a line parallel to the τ axis and cutting the circle at P_1, which is termed

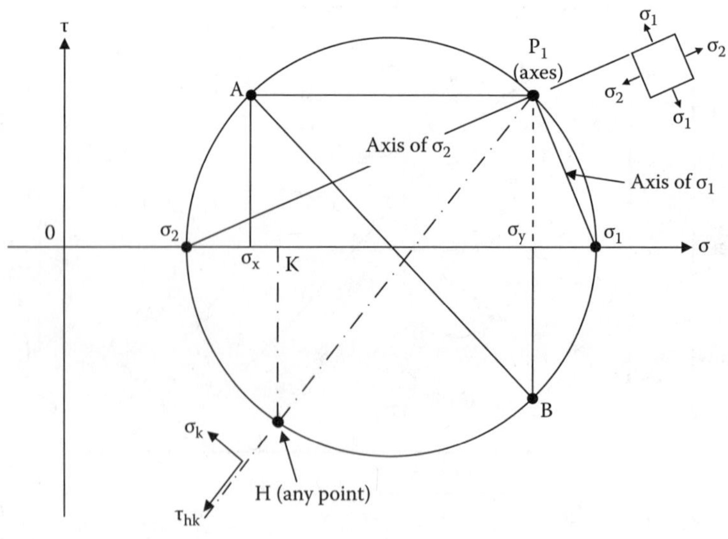

FIGURE 14.10
The pole of the Mohr's circle (a).

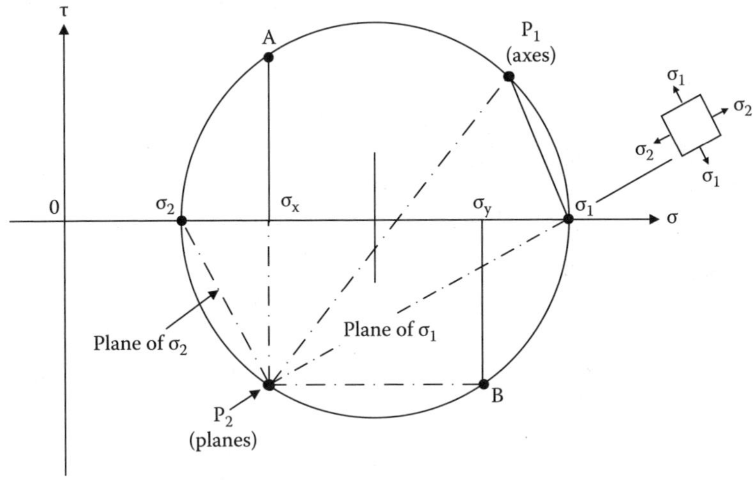

FIGURE 14.11
The pole of the Mohr's circle (b).

the *pole*. This point P_1 is the pole for the axes of direct stresses. If a line is drawn from any point on the circle to pass through the pole, then the direction of this line gives the axis for the stress component represented by the point on the circle from which the line originated. This is shown in Figure 14.10.

If instead of drawing lines through A and B parallel to the stress axes, you draw lines parallel to the planes upon which the stresses act, then the pole point P_2 will be formed from which planes of any stress condition can be found. This is shown in Figure 14.11. The two pole points are diametrically opposite to each other, that is, 180° apart on the circle and 90° on the element, as one would expect for planes and axes. These simple ideas will now be used for a number of different examples.

14.7.1 A Few Special Cases

Before moving on to typical practical examples, it is useful to examine some simple stress situations, both for the practice of visualizing the conditions as represented by the circle and for the useful place certain of these cases possess as working concepts in stress analysis.

The following thumbnail sketches in Figure 14.12 will be considered in turn:

- a,b The single stress shown is, of course, a principal stress of zero. Note that the maximum shear stress shown by the circle is $\tau_{max}=1/2\sigma_y$.
- c In the example of pure torque, say a shaft, the counterclockwise, and hence negative, shearing stresses shown by the vertical arrows are balanced by the clockwise, therefore positive, complementary shear stresses. These are marked off the vertical axis of the diagram and a circle; center O is drawn to touch them. The circle demonstrates that this shear system gives rise to tensile and compressive stresses equal in magnitude and, as shown by the pole construction, at 45° to the shear planes.
- d This example illustrates the so-called two-dimensional hydrostatic stress system where equal compressive stresses act in the x and y directions. The circle now shrinks to a point showing there is no shear. The case of equal tensile stresses is similar and has important consequences in certain theories of elastic failure.

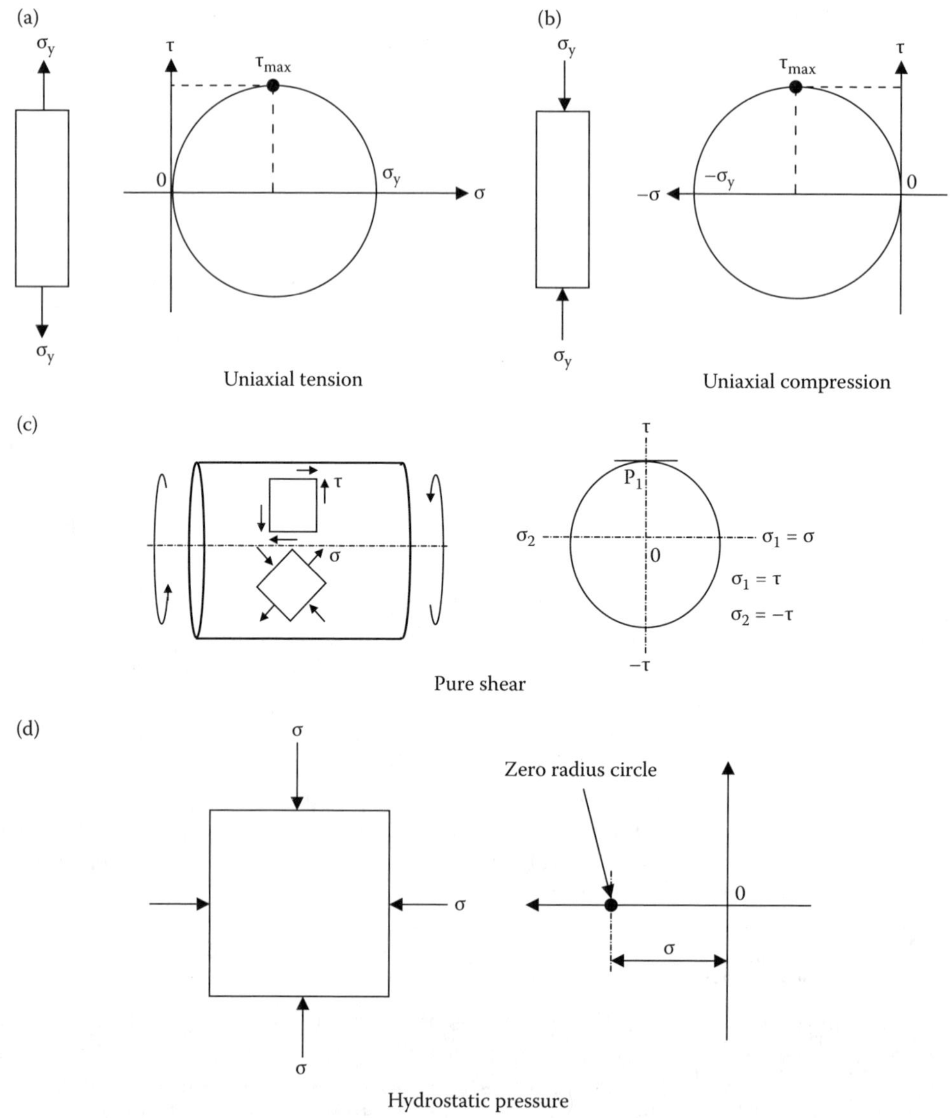

FIGURE 14.12
Types of loading that is treated using the Mohr's circle.

14.8 Examples

In this section, the ideas so far developed will be used in examples of a more realistic nature. It is hoped that the ease of the method will become apparent and that the readers will be able to make similar analysis for their own problems. The first two examples are not taken directly from a practical problem, but indicate the form into which a practical problem must be put before it can be analyzed by the circle, or indeed, by any other method.

Example 14.1

Given a complex stress situation, find the principal stresses and their inclination.

Figure 14.13a shows the complex stress situation. Choose a suitable scale for the stress and draw horizontal and vertical axes. Mark off the 90 MPa in the positive direction, and from this position draw in vertically the shear stress of 80 MPa. Note that the shears on the vertical plane are counterclockwise and therefore negative, so that the resulting point B lies below the σ axis.

Similar treatment of the stresses on the horizontal planes gives the point A. AB is a diameter of the basic Mohr's circle, which may therefore be constructed with center M and radius MA. The values of the maximum and minimum principal stresses can be scaled to give +118 and −147 MPa.

A line is drawn from point A parallel to the axis of the 120 MPa stress cutting the circle, and as a check, another line from B parallel to the axis of the 90 MPa stress; then these will meet on the circle at P_1, which is the pole point for the axes.

If the principal stress points C and D are joined to P_1, CP_1, and DP_1, give the directions of the axes of σ_1 and σ_2, respectively. The angle DCP_1 gives the inclination of the axis of maximum principal stress to that of the 90 MPa stress in the original system and is found by measurement to be approximately 18°.

If it had been necessary, a pole point P_2 for planes could have been drawn just as easily. It would appear at the opposite end of the diameter to P_1.

Example 14.2

Find the stress components on any plane when the principal planes and stresses are given.

Figure 14.14a shows two principal stresses. Investigate the value of the stresses on a given line or plane, 40° from the vertical plane.

On the σ axis measure off the two principal stresses. In this example, note that these provide the circle extremities because the shear stress component is zero. Find the

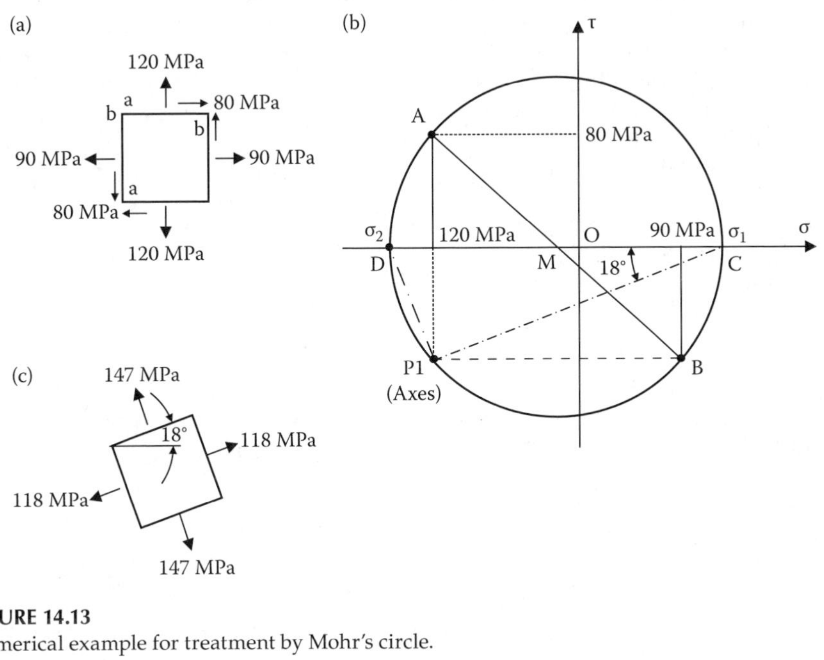

FIGURE 14.13
Numerical example for treatment by Mohr's circle.

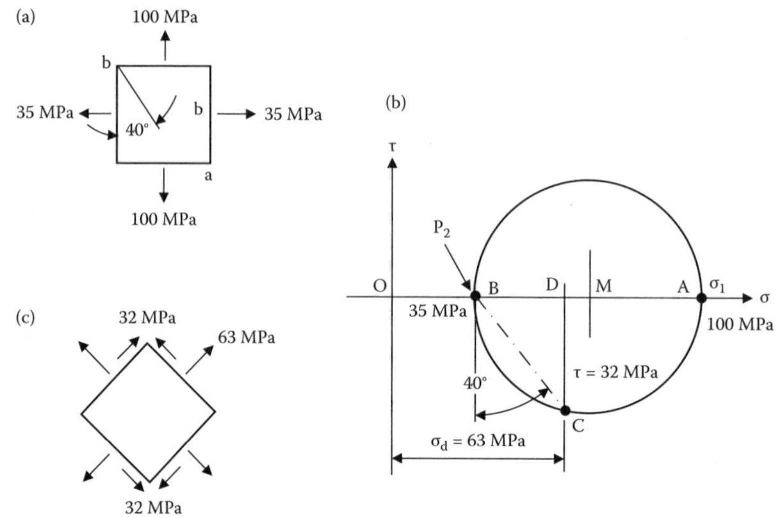

FIGURE 14.14
Example 14.2.

midpoint of the distance AB and draw in a circle. The construction for the pole of the planes P_2, as indicated in Figure 14.10, shows that in this case P_2 coincides with B. Draw a line through P_2 parallel to the required plane.

This cuts the circle at C, which has coordinates OD and DC. OD is the normal stress on the plane and is found to be 63 MPa, positive and hence tensile. DC is the shear stress and is –32 MPa; consequently, it acts counterclockwise, as shown in Figure 14.14c.

Example 14.3

The loading on a shaft is such that it causes simultaneous bending and twisting. Calculate the stresses at the surface of the shaft due to these moments acting separately and then using the Mohr's circle to find the principal stresses and planes for the combined loading (Figure 14.15).

Finally, to calculate the FoS based on a given yield stress, use the von Mises shear stress–strain energy criterion.

Data

Bending moment M = 1600 Nm
Twisting moment T = 850 Nm
Yield in pure tension = 300 MPa
Diameter d = 65 mm
Factor of safety (FoS) = To be determined
Material = Mild steel

In the following calculations, the simple theory of bending and torsion will be assumed. The appendix briefly describes the main theories of elastic failure.

Calculation of the bending stress

$$I_{xx} = \frac{\pi \cdot d^4}{64}$$

$$I_{xx} = 876240.51\ \text{mm}^4$$

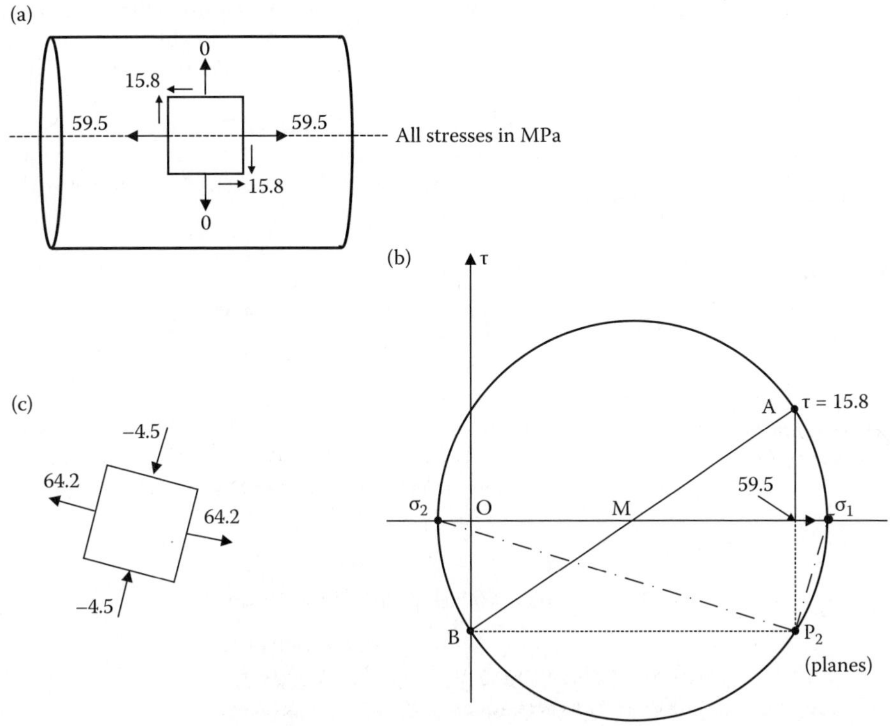

FIGURE 14.15
Example 14.3: Loading on a shaft.

Now applying the beam theory

$$\sigma_x = \frac{M \cdot d}{I_{xx} \cdot 2}$$

$$\sigma_x = 514.34 \text{ MPa}$$

Calculation for the shear stress due to torsion

$$J = \frac{\pi \cdot d^4}{32}$$

$$J = 1752481.01 \text{ mm}^4$$

Now applying the torsion theory

$$T_{xy} = \frac{\tau \cdot d}{J \cdot 2}$$

$$T_{xy} = 15.76 \text{ MPa}$$

Figure 14.15a shows the system of maximum stresses at the shaft surface, in which it should be noted that the circumferential direct stress is zero and that the bending stress is taken as tensile. An identical system will exist at the opposite end of the shaft, except the bending stress will be compressive.

The above values are now used to construct the circle in Figure 14.15b; from the points A and B lines are drawn parallel to the appropriate planes to obtain a pole point P_2. The principal stresses are measured and found to be 64.2 MPa tensile and 4.5 MPa compressive. The principal stress points on the circle are joined up to the pole P_2 and the principal plane diagram is drawn in Figure 14.15c. Carefully note that this is a pole for planes, and the axial stresses are at right angles to the plane. It would help to fix ideas if the reader would sketch in a pole P_1 for the axes of the direct stresses and show that it produces an identical result.

To find the FoS

The von Mises theory for a two-dimensional stress system is given by

$$\sigma_y^2 = \sigma_1^2 + \sigma_2^2 - \sigma_1 \cdot \sigma_2$$

where σ_1 and σ_2 are the principal stresses at which failure will occur if the yield stress in pure tension is σ_y.

In this case, there is a FoS=σ_y/σ_w, where σ_w is the working stress given by

$$\sigma_w^2 = \sigma_1^2 + \sigma_2^2 - \sigma_1 \cdot \sigma_2$$

$$\sigma_w = 66.564 \text{ MPa}$$

$$\text{FoS} = \frac{\sigma_y}{\sigma_w}$$

$$\text{FoS} = 4.507$$

If the circle construction were applied at the opposite end of the shaft, where the bending stress is compressive, the principal stresses would be numerically identical but opposite in sign, so that σ_w and consequently FoS would have the same value as above.

Example 14.4

Figure 14.16a shows a cross section of a universal beam 416 mm×154 mm×74 kg/m³. The other dimensions relative to the problem are shown on the diagram.

A vertical shear force of 140 kN is applied to the section, together with a bending moment of 100 kNm, which acts simultaneously with the shear force. The second moment of area is 270×10⁶ mm⁴.

In the first instance, estimate the maximum principal stress, which would occur just above the flange in the web.

Second, if the maximum principal stress is limited to 100 MPa, find how much the shear stress and force could be increased if the bending moment is maintained constant.

Figure 14.16b shows the element under consideration. There will be a tensile stress due to bending and shearing stresses due to the vertical shear force. These are calculated in the following section.

The stress due to bending can be calculated from the beam theory:

$$M = 100 \times 10^3 \text{ Nm}$$

$$y = 190 \text{ mm}$$

$$I = 270 \times 10^6 \text{ mm}^4$$

$$\sigma_b = \frac{My}{I}$$

$$\sigma_b = 70.37 \text{ N/mm}^2$$

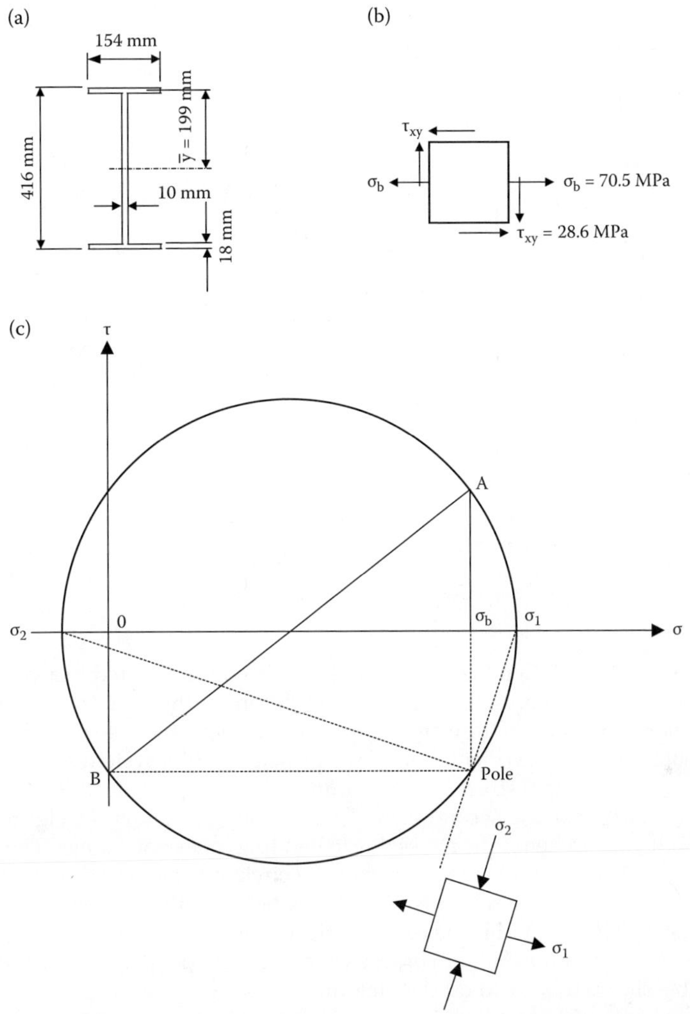

FIGURE 14.16
Example 14.4: Universal beam part 1.

For the shear stress, at the required point, the following relationship is used:

$Q = 140 \times 10^3$ N(vertical shear force)

$I = 270 \times 10^6$ mm^4 – second moment of area

$A = 2772$ mm^2(area of section above the line upon which shear stress is required)

$y = 199$ mm(vertical distance of centroid of this area from neutral axis line)

$b_o = 10$ mm(width of section at the point where shear stress is required)

$$\tau_{xy} = \frac{Q}{I \cdot b_o}(Ay)$$

$$\tau_{xy} = \frac{140 \times 10^3 \text{ N} \times 154 \text{ mm} \times 18 \text{ m} \times 199 \text{ m}}{270 \times 10^6 \text{ mm}^4 \times 10 \text{ mm}}$$

$$\tau_{xy} = 28.6 \text{ N/mm}^2$$

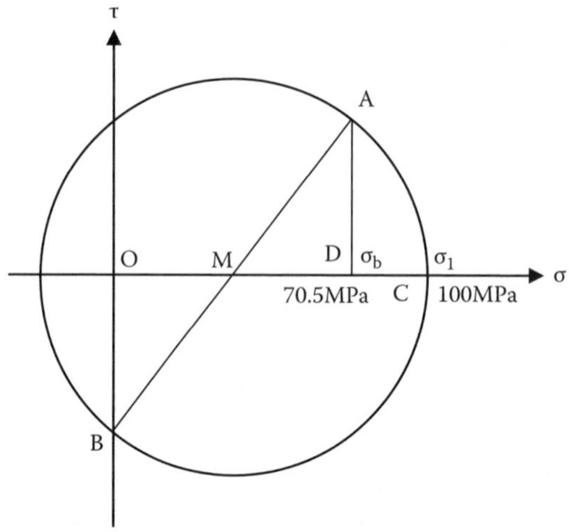

FIGURE 14.17
Example 14.4: Universal beam part 2.

For the stress system, shown the Mohr's circle has been constructed in accordance with Figure 14.16c. Point A represents the stress condition on the vertical face and point B that on the horizontal face of the element. For completeness, the pole point for the planes has been obtained and the orientation of the principal stress system shown. The value of the maximum principal stress is 80.6 N/mm^2.

The Mohr's circle for the second part of the problem is shown in Figure 14.17. The bending moment, and hence the stress due to bending, does not change. The maximum principal stress is given so the extremity of the circle can be placed on the horizontal axis at C. The center of the circle must lie halfway between the origin O and the point D representing σ_b. The point M is the center of the circle.

Thus, the circle of radius MC can be drawn. Points A and B can now be obtained and the allowable shear stress read off the diagram.

New shear stress = 53.9 MPa.

Hence, the new value of shear force can be obtained, pro rata, from the original value:

$$Q = \frac{140 \times 53.9}{28.6}$$
$$= 264 \text{ kN}$$

14.9 The Analysis of Strain

Previous sections have shown the basis for the use of the Mohr's circle when solving complex stress problems. However, in some areas of analysis the starting point may be data recorded as strains generally obtained from electrical resistance strain gauges of one kind or another.

Ultimately this information must be processed and converted into information about stresses. Therefore, the next section will derive a set of equations relating to strains.

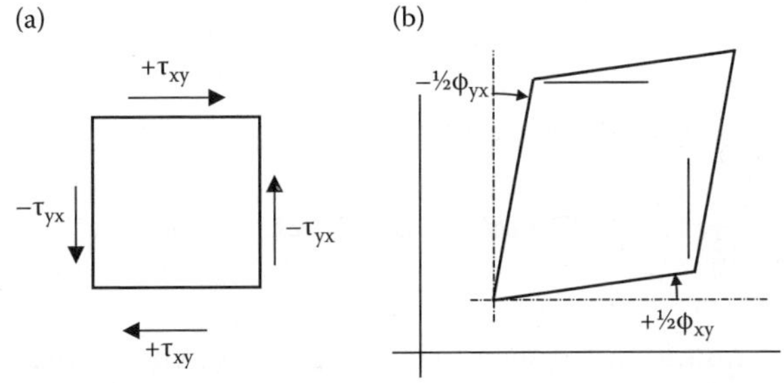

FIGURE 14.18
Sign conventions for strain.

14.9.1 Sign Conventions for Strains

For linear strain, extensions are considered positive. For the case of shear strain it is desirable to choose a system that is simple to use and is fully consistent with the stress conventions already adopted.

Figure 14.18a and b shows a pure shear stress system with the signs added in accordance with the conventions to be used here. The deformation of the element is shown and the signs are chosen for the semi-shear strains to correspond to those used for the stresses.

From the early sections the reader will recall that the first letter in the double suffix notation refers to the plane and the second letter gives the direction of the normal to the plane on which the stress acts.

Note that now τ_{xy}, which is positive, accompanies $1/2\varphi_{xy}$, which is a counterclockwise rotation and is also taken as positive. This has the added advantage of agreeing with the standard mathematical usage. The other rotation of $1/2\varphi_{yx}$ is negative and corresponds with the stress conditions.

It can be shown that the equations for transforming the strains are as follows:

$$\varepsilon_x^1 = \frac{\varepsilon_x + \varepsilon_y}{2} + \frac{\varepsilon_x - \varepsilon_y}{2}\cos 2\theta + \frac{\gamma_{xy}}{2}\sin 2\theta \tag{14.9}$$

$$\varepsilon_y^1 = \frac{\varepsilon_x + \varepsilon_y}{2} + \frac{\varepsilon_x - \varepsilon_y}{2}\cos 2\theta - \frac{\gamma_{xy}}{2}\sin 2\theta \tag{14.10}$$

$$\gamma_{xy}^1 = \frac{-(\varepsilon_x - \varepsilon_y)}{2}\sin 2\theta + \frac{\gamma_{xy}}{2}\cos 2\theta \tag{14.11}$$

14.10 Comparison of Stress and Strain Equations

The following equations compare the stress and strain used in the construction of the stress circle.

$$\begin{aligned} \sigma_n &= \frac{1}{2}(\sigma_x + \sigma_y) + \frac{1}{2}(\sigma_x - \sigma_y)\cos 2\Theta - t_{xy}\sin\Theta \quad \text{From Equation 14.1} \\ \varepsilon_a &= \frac{1}{2}(\varepsilon_x + \varepsilon_y) + \frac{1}{2}(\varepsilon_x - \varepsilon_y)\cos 2\Theta - 2\varphi_{xy}\sin 2\Theta \end{aligned} \tag{14.12}$$

and

$$\tau = \frac{1}{2}(\sigma_x - \sigma_y)\sin 2\Theta + t_{xy}\cos 2\Theta \quad \text{From Equation 14.2}$$
$$\frac{1}{2\varphi_{ab}} = \frac{1}{2}(\varepsilon_x - \varepsilon_y)\sin 2\Theta - \frac{1}{2}\varphi_{xy}\cos 2\Theta \tag{14.13}$$

The stress and strain equations are seen to be completely analogous, except that the strain quantity corresponding to τ_{xy} is $1/2\varphi_{xy}$ and not φ_{xy}.

Provided that the vertical axis is represented by $1/2\varphi$, the circle can be represented by a circle diagram.

The construction of the strain circle will be identical to the stress circle if the linear and shear strains are part of the data.

14.10.1 The Strain Rosette

One of the most common methods of measuring strain on a component is to use an electrical resistance strain gauge. They can only measure tensile or compressive strains, but are not able to measure shearing strains. When the direction of the principal strains is not known, it is usual practice to fit a strain rosette consisting of three gauges, which are displaced with an angular relationship to each other. Figure 14.19 shows some arrangements of strain gauge rosettes.

14.10.2 Construction

1. Select suitable scales for the linear strain axis (abscissa) and half the shear strain (ordinate).
2. Considering Figure 14.19c. The points representing the strains ε_1 and ε_3 must lie on the circumference of the circle and be 180° apart (since grids 1 and 3 are 90° apart). To establish the center of the circle, algebraically add ε_1 and ε_3, divide by 2, and plot the result, observing the sign on the abscissa.
3. To establish the radius, the magnitude of $\varphi_{1.3}$ is determined by $\varepsilon_2-(\varepsilon_1+\varepsilon_3)/2$; the value is laid off from ε_1 (observing the sign) parallel to the ordinate. The circle is then drawn.
4. Values for ε_{max} and ε_{min} occur where the circle crosses the ε axis (since the shearing strains are zero at these points). γ_{max} is the diameter of the circle on the $\gamma/2$

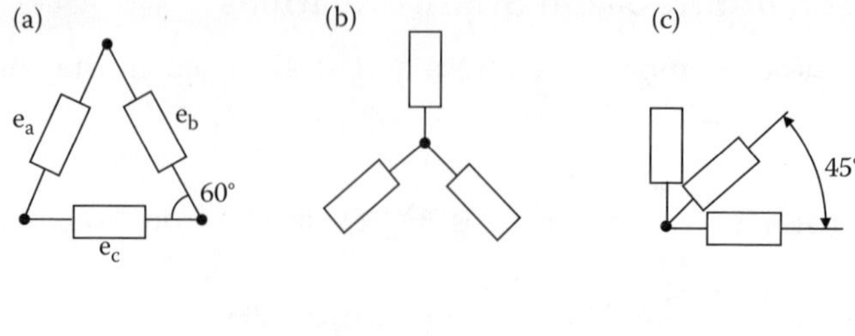

FIGURE 14.19
The strain rosette.

scale (i.e., twice the radius). The angle of the maximum principal strain is that subtended by the abscissa and the radius (where $\gamma_{1,3}$ meets the circle), divided by 2, the sign according to the code illustrated.

The Mohr's strain circle can also be constructed for strains measured at angles other than 45°.

The method is due to F. A. McClintock.

Let the measured strains be orientated as shown in Figures 14.20 through 14.26.

The strains (ε) can have any value including the angles (θ).

To Draw the Mohr's Strain Circle

1. The rosette axes are arranged by extending, if necessary, so that they are
   - a Arranged in sequence, that is, in order of ascending or descending order of strain magnitudes
   - b The included angles between the axes of the maximum and minimum strains are less than 180°.

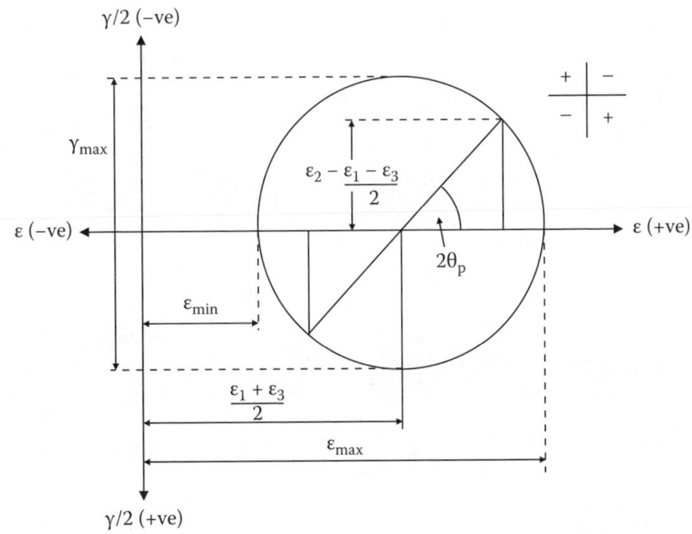

FIGURE 14.20
Mohr's strain circle.

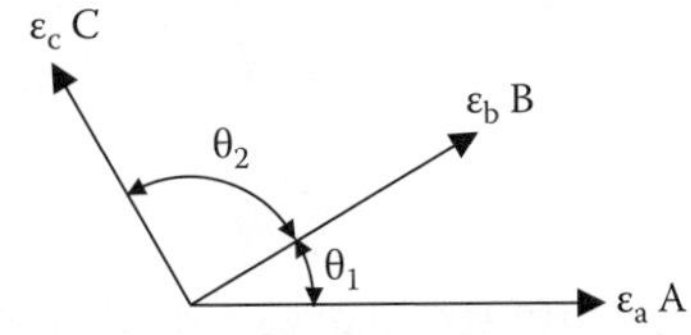

FIGURE 14.21
Rosette strains.

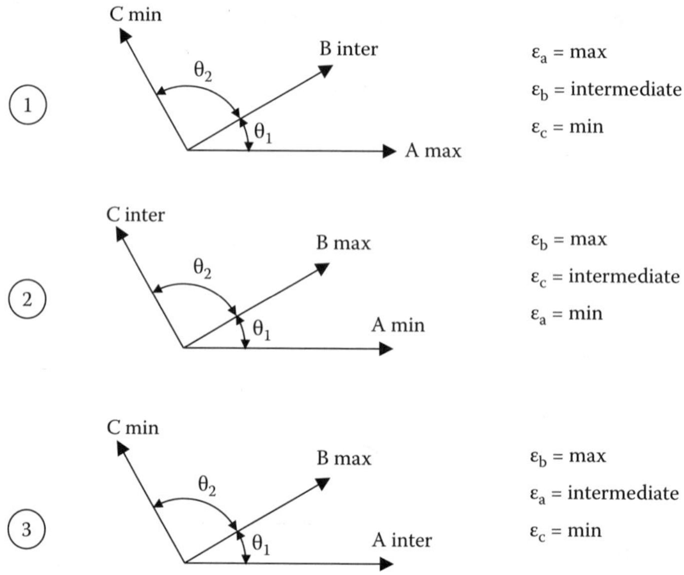

FIGURE 14.22
Rosette axes.

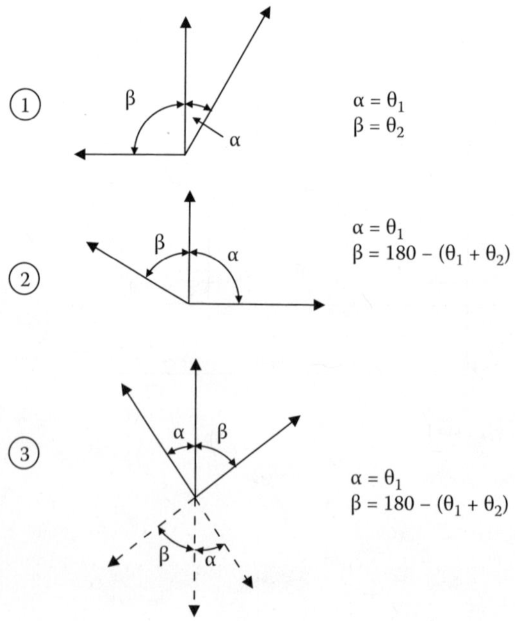

FIGURE 14.23
Rearrangement of rosette axes.

If α=angle between the max and strain axes, and β=angle between the intermediate and min strain axes, rearrange the layouts such that the intermediate strain is in the vertical position.

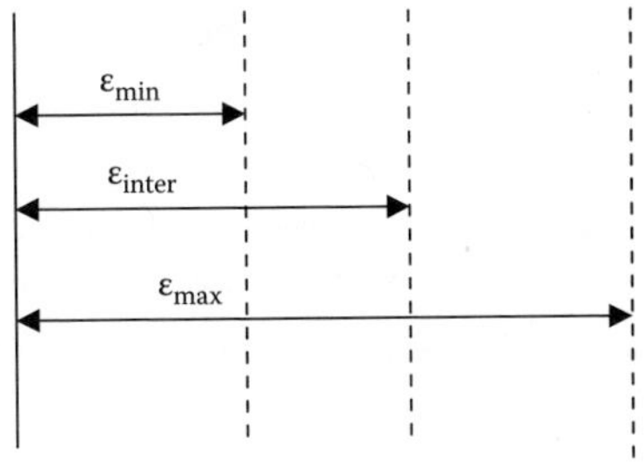

FIGURE 14.24
Strain plot.

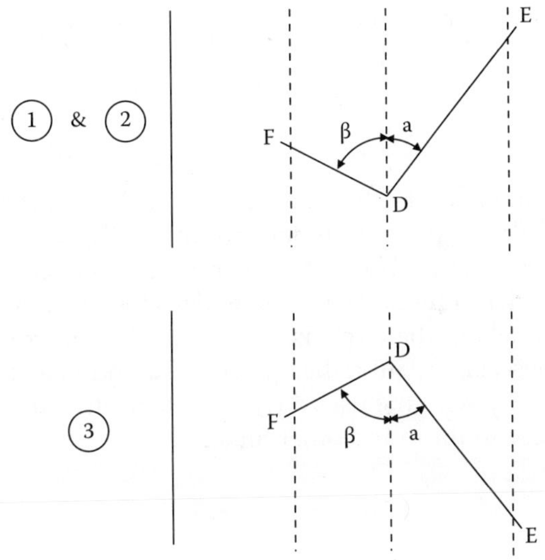

FIGURE 14.25
Construction of Mohr's strain circle.

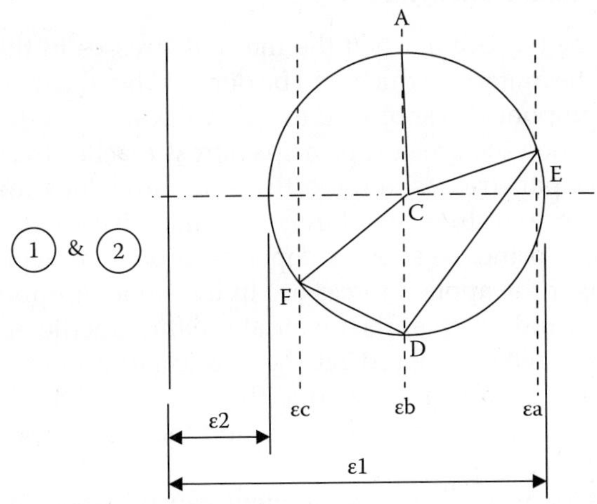

FIGURE 14.26
Construction of Mohr's strain circle (continued).

In Case 3, the axis of max strain falls to the left of the intermediate, and hence further extensions will be necessary.

2. Plot the strains as shown without inserting an x axis.
3. Construct a circle through points DEF. This is Mohr's strain circle. Now draw X axis.

For the case where $\varepsilon_a > \varepsilon_b > \varepsilon_c$, ε_1 and ε_2, and the principal strains, the points E, A, and F represent ε_a, ε_b, and ε_c, respectively. They must be chosen to satisfy the following requirements:

1. The magnitudes of ε_a, ε_b, and ε_c.
2. The rotational sequence must correspond to the rosette layout.

Principal stresses are given by

$$\sigma_1 = \frac{E(\varepsilon_1 + \nu\varepsilon_2)}{1-\nu^2}, \quad \sigma_2 + \frac{E(\varepsilon_2 + \nu\varepsilon_1)}{1-\nu^2}$$

14.10.3 Conclusion

This concludes the introduction to the construction of the Mohr's circle, and the author hopes that it has given the student an understanding of the construction and use of the circle in solving complex stress problems, together with the use of the circle in transposing strains derived from electrical resistance strain gauges to principal stresses.

There are other areas where the circle is useful: notably in certain flat-plate problems and soil mechanics. These topics are rather specialized and have not been treated in this text. However, a familiarity with the uses advocated here should make it a comparatively easy task to extend its use in these and other topics.

14.11 Theories of Elastic Failure

14.11.1 Steady Load Failure Theories

Elastic failure is assumed to occur when the internal stresses in the test piece reach the elastic limit stress for the material, which will be denoted by σ_o when subjected to a simple tension test. When a component is subjected to a complex stress system, elastic failure may not necessarily occur when the greatest principal stress reaches the elastic limit stress σ_o. The other (lesser) principal stresses in the orthogonal directions may affect the limiting value of the greatest principal stress at failure, which may then be greater or lesser than σ_o.

The effect of the lesser principal stresses depends upon a number of factors and applications, that is, such as if the applied forces are in the same or opposite directions to the greatest principal stress and whether the material is either ductile or brittle.

Various theories have been proposed on the elastic failure of a material under complex stress, and these are usually associated with the name of the originator. Substantial research has been undertaken on these theories, and the most important applications are given after each theory.

The system of stresses applied to a component can be resolved into three principal stresses, as shown in Figure 14.27 and in the following. It is assumed that $\sigma_x > \sigma_y > \sigma_z$ and that they are all the same sign.

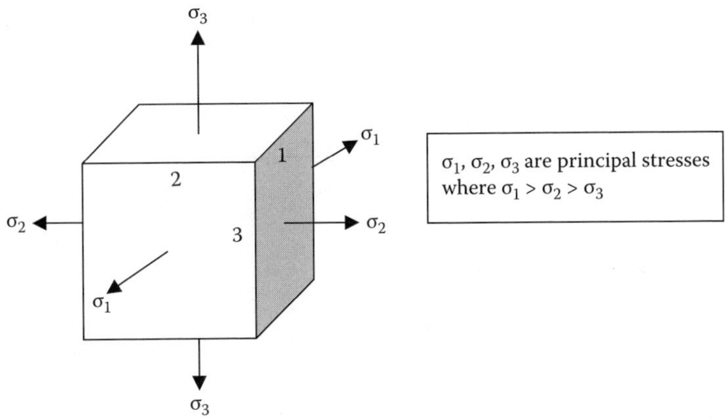

FIGURE 14.27
Principle orthogonal principal stresses.

14.11.2 Maximum Principal Stress (Rankine's) Theory

Failure will be considered to have occurred when one of the principal stresses reaches the elastic limit stress in a simple tensile stress test, irrespective of the other principal stresses (see Figure 14.28).

Failure occurs when

$$\sigma_1 = \sigma_{t1} \quad \sigma_{t1} = \text{yield strength in tension}$$

$$\sigma_3 = \sigma_{c1} \quad \sigma_{c1} = \text{yield strength in compression}$$

In a biaxial stress condition, the theory implies that if the stress state falls within the boundary depicted in Figure 14.28, the material will not fail.

The theory has been found to be approximately true for brittle materials, but not for ductile materials. Rupture will usually take place on a plane inclined to the plane of greatest direct stress, indicating that failure is due to shear.

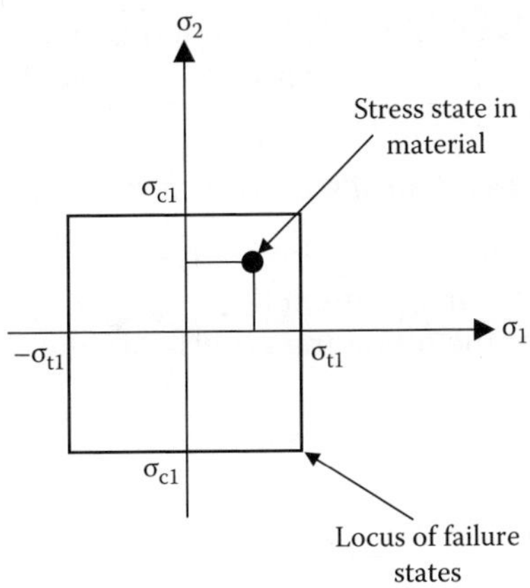

FIGURE 14.28
Maximum principal stress (Rankine's) theory.

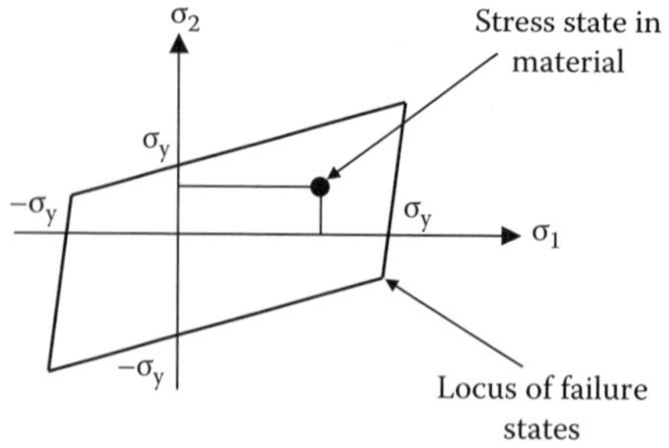

FIGURE 14.29
Maximum principal strain (Saint Venant's) theory.

14.11.3 Maximum Principal Strain (Saint Venant's) Theory

This theory predicts that failure will occur when the greatest of the three principal strains becomes equal to the strain corresponding to the yield strength (see Figure 14.29).

That is,

$$E\varepsilon_1 = \sigma_1 - \nu(\sigma_2 + \sigma_3) = \pm\sigma_y$$
$$E\varepsilon_2 = \sigma_2 - \nu(\sigma_1 + \sigma_3) = \pm\sigma_y$$
$$E\varepsilon_3 = \sigma_3 - \nu(\sigma_1 + \sigma_2) = \pm\sigma_y$$

E=Modulus of elasticity
ν=Poisson's ratio

For a biaxial condition (plane stress)

That is,

$$\sigma_3 = 0$$
$$\sigma_1 \nu \sigma_2 = \pm\sigma_y$$
$$\sigma_2 \nu \sigma_1 = \pm\sigma_y$$

As long as the stress state falls within the polygon, the material will not yield.

This theory is not substantiated by experiment and finds little general support.

14.11.4 Maximum Shear Stress (Guest's or Tresca's) Theory

Yielding begins whenever the maximum shear stress in a part becomes equal to the maximum shear stress in a tension test specimen that begins to yield

Yielding will occur if any of the following criteria are met (Figure 14.30):

$$\pm\sigma_y = \sigma_1 - \sigma_2$$
$$\pm\sigma_y = \sigma_2 - \sigma_3$$
$$\pm\sigma_y = \sigma_1 - \sigma_3$$

For the biaxial case, (plane strain) $\sigma_3=0$:

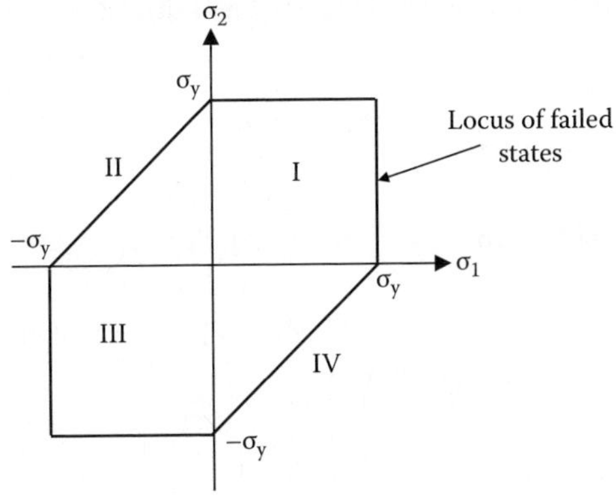

FIGURE 14.30
Maximum shear stress (Guest's or Tresca's) theory. *Note:* In quadrants I and III, the maximum principal stress theory and the maximum shear stress theory are the same for the biaxial case.

$$\pm\sigma_y = \sigma_1 - \sigma_2$$
$$\pm\sigma_y = \sigma_2$$
$$\pm\sigma_y = \sigma_1$$

Note: In quadrants I and III, the Maximum Principal Stress Theory and the Maximum Shear Stress Theory are the same for the biaxial case. This theory gives good correlation with experimental results obtained from ductile material.

14.11.5 Distortion Energy Theory

14.11.5.1 Strain Energy (Haigh's) Theory

Yielding will occur when the distortion energy per unit volume equals the distortion energy per unit volume in a uniaxial tension specimen stressed to its yield strength (see Figure 14.31).

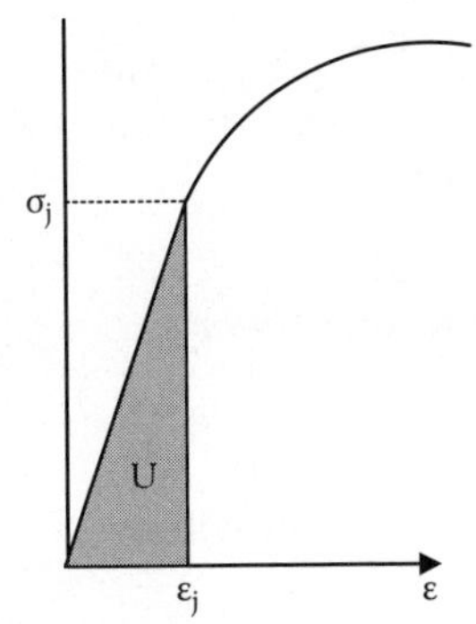

FIGURE 14.31
Strain energy (Haigh's) theory.

The strain energy per unit volume is given by the equation

$$U = \frac{1}{2}\sigma_1 \cdot \varepsilon_1 + \frac{1}{2}\sigma_2 \cdot \varepsilon_2 + \sigma_3 \cdot \varepsilon_3$$

$$\text{Units}:$$

$$[U] = [N/mm^2]*[mm/mm] = [N \cdot mm/mm^3]$$

An expression for the strain energy per unit volume in terms of stress only can be obtained by making use of the stress–strain relationship.

In algebraic format:

$$\varepsilon_1 = \frac{1}{E}(\sigma_1 - \nu\sigma_2 - \nu\sigma_3)$$

$$\varepsilon_2 = \frac{1}{E}(\sigma_2 - \nu\sigma_1 - \nu\sigma_3)$$

$$\varepsilon_3 = \frac{1}{E}(\sigma_3 - \nu\sigma_1\nu\sigma_2)$$

$$U = \frac{1}{2}\cdot\sigma_1\cdot\varepsilon_1 + \frac{1}{2}\cdot\sigma_2\cdot\varepsilon_2 + \frac{1}{2}\cdot\sigma_3\cdot\varepsilon_3$$

$$= \frac{1}{2}\sigma_1\left[\frac{1}{E}\cdot(\sigma_1 - \nu\sigma_2 - \nu\sigma_3)\right] + \frac{1}{2}\sigma_2\left[\frac{1}{E}\cdot(\sigma_2 - \nu\sigma_1 - \nu\sigma_3)\right] + \frac{1}{2}\sigma_3\left[\frac{1}{E}\cdot(\sigma_3 - \nu\sigma_1 - \nu\sigma_2)\right]$$

$$U = \frac{1}{2E}(\sigma_1^2 + \sigma_2^2 + \sigma_3^2 - 2\nu(\sigma_1\cdot\sigma_2 + \sigma_2\cdot\sigma_3 + \sigma_3\cdot\sigma_1))$$

In matrix format: $$\begin{Bmatrix} \varepsilon_1 \\ \varepsilon_2 \\ \varepsilon_3 \end{Bmatrix} = \begin{pmatrix} 1 & -\nu & -\nu \\ -\nu & 1 & -\nu \\ -\nu & -\nu & 1 \end{pmatrix} \begin{Bmatrix} \sigma_1 \\ \sigma_2 \\ \sigma_3 \end{Bmatrix}$$

This theory is well supported by experimental results on ductile materials, particularly with thick cylinders. It breaks down in the case of hydrostatic pressure ($\sigma_1 = \sigma_2 = \sigma_3$).

It predicts failure when

$$\sigma = \frac{\sigma_o}{\sqrt{3\cdot(1-2\nu)}}$$

when in fact no failure would occur.

14.11.5.2 Shear Strain Energy (von Mises) Theory

This theory is known by a number of names:

- Maximum shear strain energy per unit volume theory
- Distortion energy theory
- von Mises–Hencky theory
- Maximum octahedral shear stress theory

This theory is commonly known as the von Mises theory.

Failure is predicted to occur when the shear strain energy stored per unit volume in a strained material reaches the shear strain energy per unit volume at the elastic limit in a simple tension test. This is similar to the previous theory, but it assumes that the volumetric strain energy plays no part in producing elastic failure.

$$\text{Thus,}\quad \frac{1}{6G}\left[\sigma_x^2+\sigma_y^2+\sigma_z^2-(\sigma_x\cdot\sigma_y+\sigma_y\cdot\sigma_z+\sigma_z\cdot\sigma_x)\right]=\frac{\sigma_o}{6G}$$

$$\text{or}\quad \sigma_x^2+\sigma_y^2+\sigma_z^2-(\sigma_x\cdot\sigma_y+\sigma_y\cdot\sigma_z+\sigma_z\cdot\sigma_x)=\sigma_o^2$$

This theory has received considerable verification in practice and is widely regarded as the most reliable basis for design.

14.11.6 Conclusions

It has been found that both the distortion energy theory and the maximum shear stress theory provide reasonable estimates for the onset of yielding in the case of static loading for ductile, homogeneous, isotropic materials whose compression and tensile strengths are the same.

Both distortion energy theory and the maximum shear stress theory predict the onset of yielding and are independent of hydrostatic stress. This agrees reasonably well with experimental data for moderate hydrostatic pressures.

Both the distortion energy theory and the maximum shear stress theory underpredict the strength of brittle materials loaded in compression. Brittle materials often have much higher compressive strengths than tensile strengths.

The distortion energy theory is slightly more accurate than the maximum shear stress theory. The distortion energy theory is the yield criteria most often used in the study of plasticity. Its continuous nature makes it more mathematically amenable.

14.12 Interaction Curves, Stress Ratios Margins of Safety, and Factors of Safety

14.12.1 Interaction: Stress Ratio

Yielding or failure of a structural member subjected to combined stresses may be predicted without using principal stresses by using what is known as the interactive method. The method represents applied and allowable stress conditions on a structural member by stress ratios.

Stress ratios are nondimensional coefficients R given by

$$R=\frac{\text{applied stress}}{\text{allowable stress}}$$

Load ratios may also be used instead of the stress ratios, if this is more convenient.

The method involves determining the allowable stress or loading for each separate failure mode, such as tension, compression, bending, buckling, and shear. Stress ratio R for each separate failure condition is calculated and combined using interactive equations, if these loads act simultaneously on the member.

The equation generally takes the following form:

$$R_1^x R_2^y + R_3^z \ldots = 1.0$$

where x, y, and z are exponents defining interactive relationships.

Failure is when the sum of the stress ratios is greater than 1.0.

14.12.2 Interactive Curve

Interactive curves are used, for convenience, to show the relationship given by the interaction equations (see Figure 14.32).

For example, point A is located with the coordinates R_1 and R_2.

The margin of safety (MS) assuming each load increases proportionately until failure occurs at point B is

$$MS = \frac{OB}{OA} - 1$$

The MS assuming R_1 remains constant with R_2 increasing until failure occurs at point C is

$$MS = \frac{EC}{EA} - 1$$

The MS assuming R_2 remains constant with R_1 increasing until failure occurs at point D is

$$MS = \frac{FD}{FA} - 1$$

14.12.3 Interaction, Stress Ratios, Yield Conditions

To investigate yielding on a compact structure (crippling and buckling not pertinent) under the action of two-dimensional combined stresses, the following method may be used, as an alternative to calculating principal stresses.

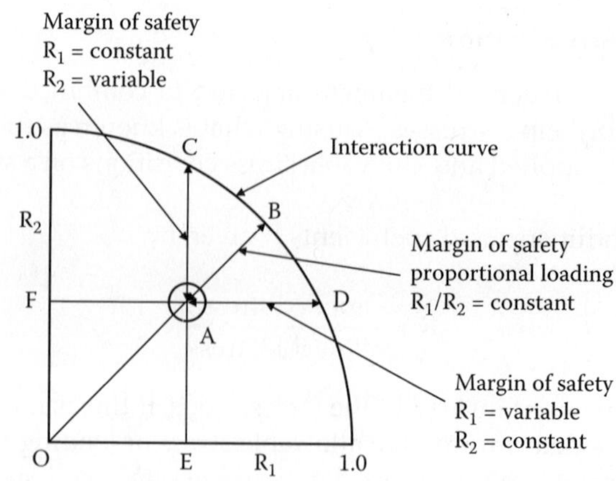

FIGURE 14.32
Interaction curve.

A maximum equivalent Stress Ratio $\bar{R}$ is computed and the MS (yield) obtained there from:

$$\text{MS (yield)} = \frac{1}{\bar{R}} - 1$$

R is given by the following formula (which is based on the maximum shear stress energy per unit volume theory):

$$R = \left[R_{n1}^2 + R_{n2} - R_{n1}R_{n2} + R_s^2\right]^{0.5}$$

where $R_n = R_t + R_b$ or $R_c + R_b$ and subscripts 1 and 2 indicates mutually perpendicular directions.

Stress ratios due to direct stress are positive for tension and negative for compression. The effect of bending stress may be added to the above, ensuring consistency of signs; that is, bending stresses are positive for tension and negative for compression.

$$R_t = \frac{f_t}{\bar{F}_{ty}}$$

$$R_c = \frac{f_c}{\bar{F}_{cy}}$$

$$R_s = \text{shear stress ratio} = R_s = \frac{f_s}{\bar{F}_{sy}}$$

where $\bar{F}_{sy}$ = may be taken as $0.577 \cdot \bar{F}_{ty}$.

14.12.4 Interaction Equations: Yield Conditions

Table 14.1 gives the interaction equations for the yield conditions.

14.12.5 Interaction Equations: Failure Conditions

14.12.5.1 Compact Structures: No Bending

The following interaction equations for failure conditions are similar to those for the yield conditions, as shown, except for the following:

where

$$R_t = \frac{f_t}{\bar{F}_{tu}}$$

$$R_c = \frac{f_c}{\bar{F}_{cu}}$$

$$R_s = \frac{f_s}{\bar{F}_{su}}$$

where

$\bar{F}_{tu}$ = ultimate allowable tensile strength of the material
$\bar{F}_{cu}$ = ultimate allowable compressive strength of the material
$\bar{F}_{su}$ = ultimate allowable shear strength of the material

TABLE 14.1

Interaction Equations—Yield Conditions

Type of Combined Loading	Tension Stresses +ve Compression Stresses −ve
Uniaxial tension and shear 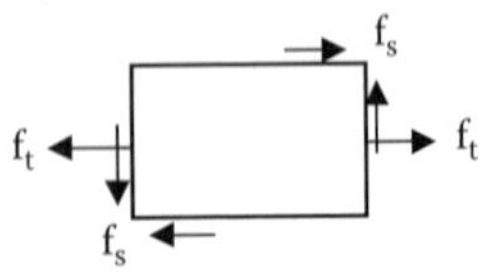	$R_t = \frac{f_t}{\bar{F}_{ty}} \quad R_s = \frac{f_s}{\bar{F}_{sy}}$ Interaction formula $= R_t^2 + R_s^2 = 1$ $\text{M.S.} = \frac{1}{(R_t^2 + R_s^2)^{0.5}} - 1$
Uniaxial compression and shear 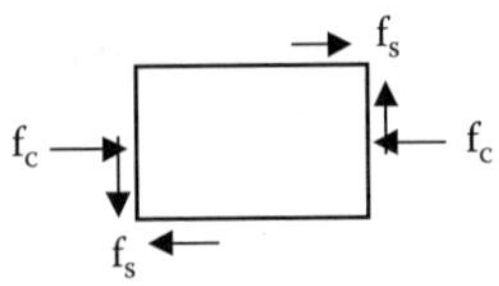	$R_c = \frac{f_c}{\bar{F}_{cu}} \quad R_s = \frac{f_s}{\bar{F}_{sy}}$ where $\bar{F}_{cu} = 2\bar{F}_{sy}$ Interaction formula $= R_c^2 + R_s^2 = 1$ $\text{M.S.} = \frac{1}{(R_c^2 + R_s^2)^{0.5}} - 1$
Biaxial tension 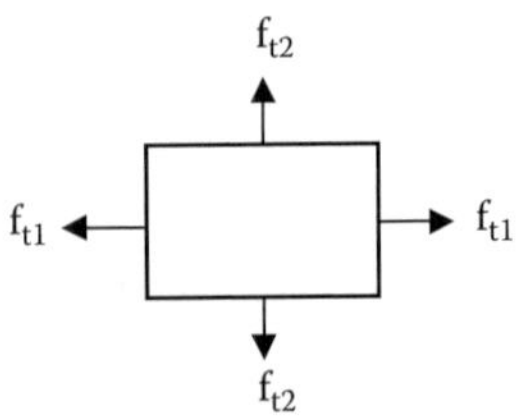	No interaction between axes $\text{M.S.} = \frac{1}{R} - 1$ R is greater of R_{t1} and R_{t2} $R_{t1} = \frac{f_{t1}}{\bar{F}_{ty1}} \quad R_{t2} = \frac{f_{t2}}{\bar{F}_{ty2}}$
Biaxial compression 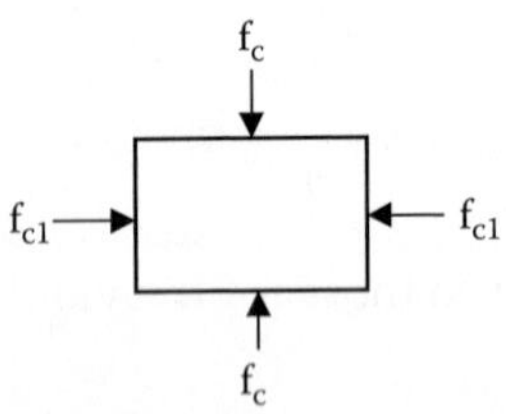	No interaction between axes $\text{M.S.} = \frac{1}{R} - 1$ R is greater of R_{c1} and R_{c2} $R_{c1} = \frac{f_{c1}}{\bar{F}_{cy1}} \quad R_{c2} = \frac{f_{c2}}{\bar{F}_{cy2}}$
Biaxial normal stresses and shear 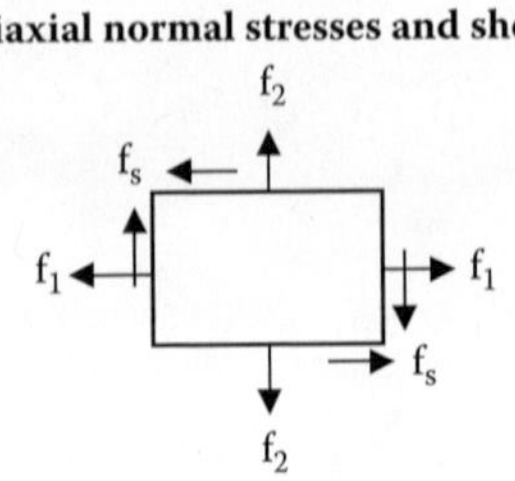	$R_{n1} = R_{t1}$ or $R_{c1} \qquad R_{n2} = R_{t2}$ or R_{c2} Interaction formula $= R_{n1}^2 + R_{n2}^2 - R_{n1}^2 R_{n2}^2 + R_s^2 = 1$ $\text{M.S.} = \frac{1}{\left[R_{n1}^2 + R_{n2}^2 - R_{n1}^2 R_{n2}^2 + R_s^2\right]^{0.5}} - 1$

14.12.6 Compact Structures: Bending

1. Bending within the elastic limit. The stresses resulting from bending; that is, f_t and f_c, and the resulting stress ratios should be treated as per the yield conditions.
2. Bending in the plastic range. Pure bending in the plastic range is covered in Chapter 1, where the allowable bending moment $\bar{M}_b$ is evaluated:

$$R_s = \frac{M_b}{\bar{M}_b}$$

   where
   M_b is the applied bending moment
   $\bar{M}_b$ is the allowable bending moment.
3. Plastic bending and flexural shear:

$$R_s = \frac{f_s}{f_{su}}$$

   where
   f_s is the applied bending moment
   $\underline{F}_{su}$ is the allowable bending moment.

Interaction equation:

$$\approx R_b^2 + R_s^2 = 1$$

$$M.S. = \frac{1}{\left[R_b^2 + R_s^2\right]0.5} - 1$$

14.12.7 General Interaction Relationships (Figures 14.33 through 14.36)

Example 14.5

Consider a structural member subject to a combined axial compression and bending. In this case, the axial compression will cause additional deflection, which in turn increases the moment of the bending load. This increase can easily be taken care of by an amplification factor k:

$$k = \frac{1}{1 - \frac{P}{P_{cr}}}$$

where

$$P_{cr} = \frac{\pi^2 EI}{L^2}$$

$$k = \frac{1}{\cos\left(\frac{\pi}{2}\sqrt{\frac{P}{P_{cr}}}\right)}$$

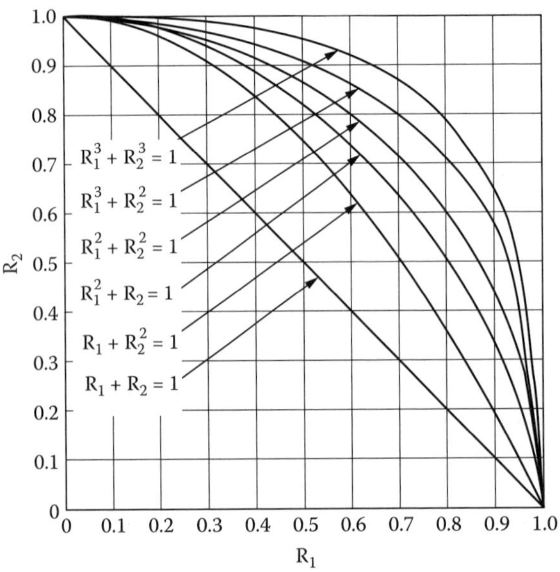

FIGURE 14.33
General interactive relationships.

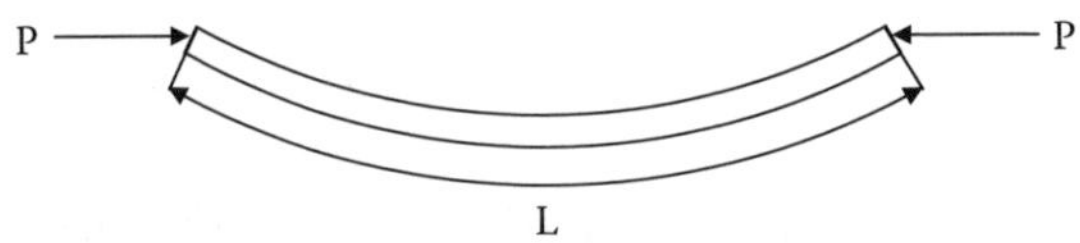

FIGURE 14.34
Direct compression and bending.

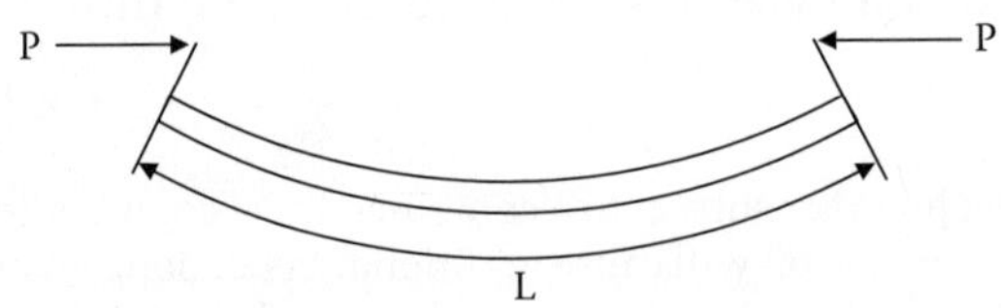

FIGURE 14.35
Offset compression and bending.

The bending moment applied to the structural member (chosen at the maximum cross section) is then multiplied by the amplification factor k, and this value is then used as the applied moment M in the ratio:

$$R_b = \frac{M}{M_u}$$

The chart in Figure 14.36 is then used to determine the amplification factor k for the bending moment applied to a beam when it is also subject to axial compression.

The resulting combined stress is then found from the following:

$$\sigma = \frac{P}{A} \pm \frac{kM_c}{I}$$

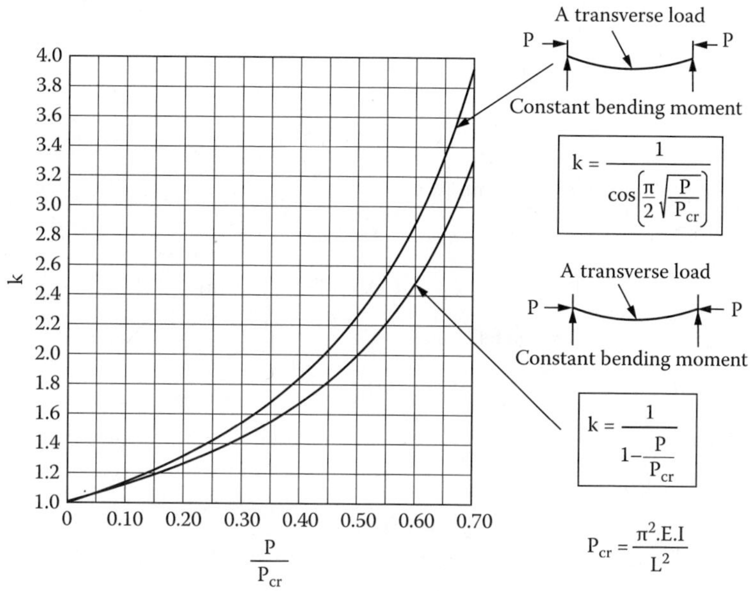

FIGURE 14.36
Amplification factor "k" for bending moment.

Example 14.6

A loading platform is to be fabricated from a 10 mm thick top plate and an under frame manufactured from 25 mm equal angle. The whole structure is in the form of a truss (Figure 14.37).

Determine the compound stress (axial compression and bending) in the top compression panel.

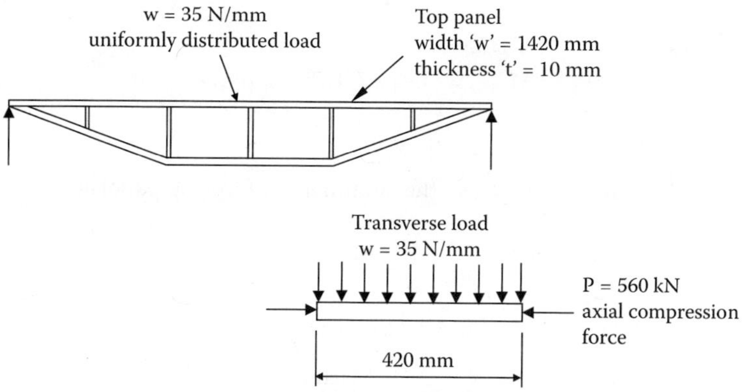

FIGURE 14.37
Truss diagram.

Summary

$$L = 420.0\ \text{mm} \qquad I = \frac{W \cdot t^2}{12}$$

$$W = 1420.0\ \text{mm} \qquad I = 118.333 \times 10^3\ \text{mm}^4$$

$$t = 10.0\ \text{mm}$$

$$w = 35.0\ \text{N/mm} \qquad A = W \cdot t$$

$$P = 560.0\ \text{kN} \qquad A = 14.2 \times 10^3\ \text{mm}^2$$

$$E = 206.842\ \text{GPa}$$

Critical load (P_{cr}):

$$P_{cr} = \frac{p^2 \cdot E \cdot I}{L^2}$$

$$P_{cr} = 1.369 \times 10^6\ \text{N}$$

Ratio:

$$\frac{P}{P_{cr}} = 0.409$$

Bending moment:

$$M = \frac{w \cdot L^2}{8}$$

$$M = 771.75\ \text{kN} \cdot \text{mm}$$

Obtaining the amplification factor k for the sinusoidal bending moment from the curve (Figure 14.36):

$$k = 1.692$$

The actual applied moment due to the extra deflection is found to be

$$M_{actual} = k \cdot M$$
$$M_{actual} = 1.692 \times 771.75\ \text{kN} \cdot \text{mm}$$
$$= 1.306 \times 10^6\ \text{N} \cdot \text{mm}$$

The resulting combined stress above the neutral axis of the top panel is

$$\sigma_c = \frac{P}{A} + \frac{M_{actual} \cdot t}{I \cdot 2}$$

$$\sigma_c = 94.611\ \text{MPa}$$

$$\sigma_L = \frac{P}{A} - \frac{Mactual \cdot t}{I \cdot 2}$$

$$\sigma_L = -15.738\ \text{MPa}$$

14.12.8 Determination of Safety Factors

The ultimate load values for this structural member are in compression alone.

For the values for bending alone, that is, without the combined compression, the following is to be used.

Assuming material properties, aluminum plate to BS 1470 5154A-H4,

$$\sigma_{0.2} = 225 \text{ MPa}$$

For compression alone:

$$\text{Since } \frac{L}{r} = 150 \quad \text{(where L = length, r = radius of gyration)}$$

$$\text{(assume)} P_u = P_{cr}$$

For bending alone:

The plastic or ultimate bending moment is

$$Mu = \left(W \cdot \sigma_{0.2} \cdot \frac{t}{2} \right) \frac{t}{2}$$

$$Mu = 7.987 \times 10^6 \text{ N} \cdot \text{mm}$$

These ultimate values are represented on the interactive curve shown in Figure 14.38.

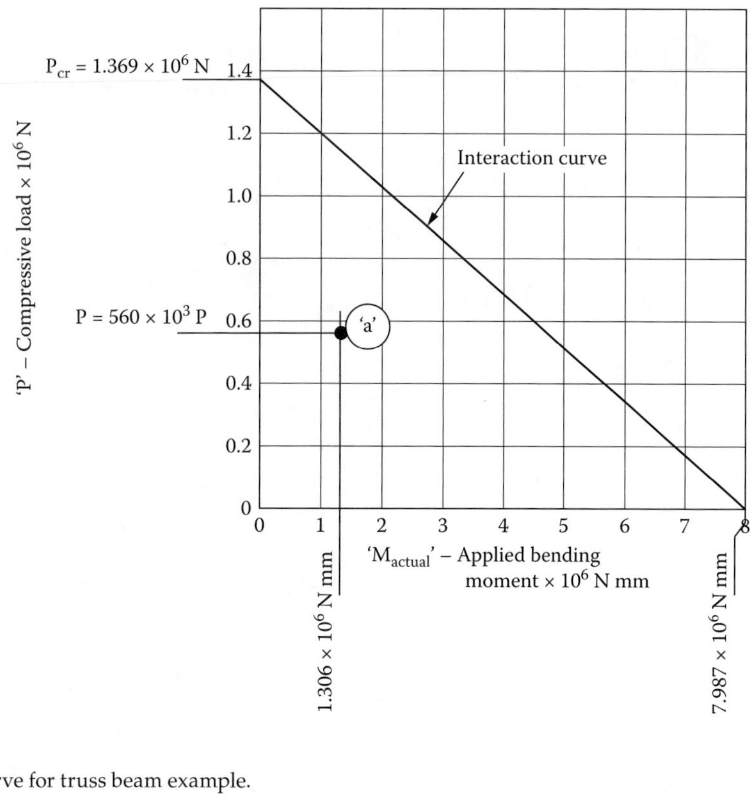

FIGURE 14.38
Interaction curve for truss beam example.

Plotting the present load values indicates that there is an approximately 2.4:1 FoS before the top panel will begin to buckle.

If this had been an extremely short plate (very low L/r ratio), the critical value (P_{cr}) could be much higher than the actual ultimate value (P_u).

15

Introduction to Experimental Stress Analysis

It is important that design calculations are verified, particularly when the design is a complex structure, such as an air vehicle wing box or fuselage. A number of factors are used in the analytical analysis, essentially to compensate for lack of specific knowledge generally associated with materials. These factors are referred to as *factors of ignorance* or "fudge" factors. These factors are used essentially to enable the design to proceed with the analysis being revisited when there is better information available.

With the enormous improvements in the use of computers in engineering analysis (CAE) and design (CAD), many designs are bought to the marketplace without any preliminary models being produced. The manufacturers rely on the analytical calculations to identify any high-stressed areas that may lead to a design failure. In most cases, a mechanical failure in something like a hair dryer will only cause an inconvenience to the owner, but when a failure occurs in the aforementioned wing box or fuselage, then the outcome will be a catastrophic accident with a potential for loss of life.

In this instance, the manufacturer will produce a full or scale model of the part or assembly and apply appropriate loading to it and monitor any high-stressed areas to see if the part will reach or exceed its calculated design life. The analysis will be undertaken using a variety of instrumentation, including but not restricted to the following:

- Photoelastic coatings
- Brittle lacquer coatings
- Electrical resistance strain gauges

15.1 Photoelasticity

For many years, photoelastic analysis was at the forefront of experimental stress analysis where solid models were manufactured from photoelastic material. These were loaded in a manner similar to the actual part, then heated and allowed to carefully cool in a low-temperature oven over several hours. When the model was cured, it was carefully cut up and the sections were analyzed using a polariscope; the technique relied on the use of polarized light, and changes in the stress patterns in the material created visible fringes. These fringes gave an indication of the internal stresses locked into the part, and using special calibration techniques, accurate stresses could be assigned to these fringe patterns. This chapter discusses the principles behind both three-dimensional and two-dimensional photoelastic models.

15.1.1 The Principles of Photoelasticity

The underlying principles will be outlined covering the basic optical instrument, the polariscope, and its different variations are described together with the techniques used for the determination of separating the principal stress values from the photoelastic data.

This chapter will not cover the detailed optical theory of the photoelastic effect and the various forms of the polariscope; this material is given in standard textbooks covering this subject.

Photoelastic stress analysis is an extremely powerful technique that enables a "whole field" analysis to be carried out on a model of the component, compared with using strain gauges, which only consider the strain at a single point on the component.

This work, when carried out at the design stage, will verify the results of a finite element (FE) analysis. In some case studies, using the photoelastic method, stresses as much as 20% higher were found than were predicted in the FE model. The accuracy of the FE model is totally dependent on the mesh density. With modern high-speed computers, it is possible to increase the mesh density to improve accuracy, but at a cost in computing power.

The technique of photoelastic analysis involves the making of a scale model, either full-size or to scale, manufactured from a suitable material (usually Araldite CT 200), and then placing the model in a beam of polarized light using a polariscope.

If the model is loaded in a similar direction to that of the actual component, then an interference pattern will be visible in the model. The interference pattern will consist of colored fringes called *isochromatics*. These fringe patterns can be interpreted to give information on the magnitudes of the principal stresses that are present in the model.

Using a simple formula, the stresses in the actual part can then be determined even though the component material may be either steel, aluminum, cast iron, ceramic, or even plastic itself.

A second set of interference fringes is also visible under certain conditions; these fringes are termed *isoclinics*. Isoclinics allow determination of the directions of the principal stresses at any point in the model, and from these, stress trajectories or stress flow lines can then be drawn for the actual component under similar loading conditions.

15.1.2 Principles

The process is based on the property of birefringence, which is exhibited by certain transparent materials. *Birefringence* is the property of double refraction, wherein a ray of light passing through a birefringent material experiences two refractive indices. Many optical crystals exhibit the property of birefringence or double refraction. But photoelastic materials show this property of birefringence only when the model, manufactured from this material, has a stress applied to it and the magnitude of the refractive indices at each point in the model is directly related to the state of stress existing at that point. Thus, the state of the stress in the model will be similar to the state of the stress in the actual structural component.

When a ray of plane-polarized light is passed through a photoelastic material, it will be resolved along the two principal stress directions, and each of these components experiences different refractive indices. The difference in the refractive indices leads to a relative phase retardation between the two component waves. The *stress optic law* gives the magnitude of the relative retardation:

$$R = CT(\sigma_{11} - \sigma_{22})$$

where R is the induced retardation, C is the stress optic coefficient, T is the specimen thickness, σ_{11} is the first principal stress, and σ_{22} is the second principal stress.

The two waves are bought together in a polariscope. The phenomenon of optical interference then takes place and a fringe pattern is produced, depending on the relative retardation. From the study of the fringe pattern, the state of stress at various points in the material can then be determined.

15.1.3 Isoclinics and Isochromatics

Two types of fringe patterns can be observed in photoelasticity: isoclinic and isochromatic fringes are viewed when the source light is monochromatic. These fringes appear as light and dark fringes, whereas with white light illumination, colored fringes are observed. The difference in principal stresses is related to the birefringence, and hence the fringe color through the stress optic law.

The definition of an isoclinic is that it is the locus of points at which there is a constant inclination of principal stress directions, that is, wherever either principal stress direction coincides with the axis of polarization of the polarizer.

Isoclinics cannot give information on the magnitude of the principal stress differences, but give valuable information on the principal stress directions.

Isochromatics are the locus of the points along which the difference in the first and second principal stresses will remain the same; hence, they are the lines that join the points with equal maximum magnitudes of shear stress. When viewed in white light, the fringes are seen in brilliant colored bands and have the same colored sequence as is observed with a film of oil on water and with Newton's rings.

15.1.4 Plane Polariscope

In this setup, the plane polariscope (Figure 15.1) consists of two linear polarizers and a light source. The light source can emit either monochromatic or white light, depending on the investigation being carried out.

First, the light passes through the first polarizer, which converts the light into plane-polarized light. The apparatus is set up in such a way that this plane-polarized light then passes through the stressed test specimen. This light then follows, at each point of the

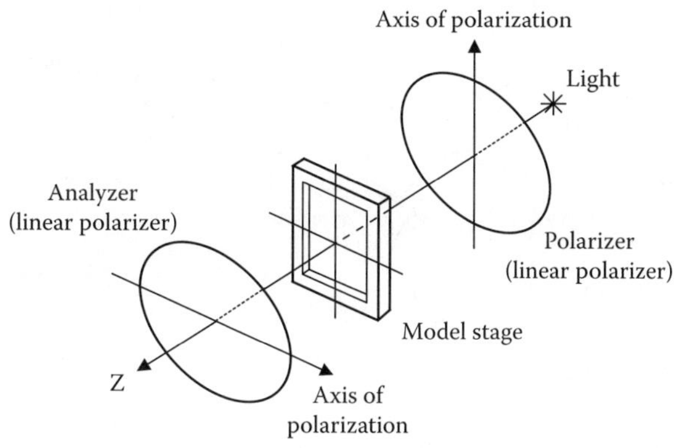

FIGURE 15.1
Schematic of the working components for a plane polariscope.

specimen, the direction of principal stress at that point. The light is then made to pass through the analyzer, and the fringe pattern can then be observed.

The fringe pattern in a plane polariscope setup consists of both the isoclinics and isochromatics. The isoclinics change with the orientation of the polariscope, whereas there is no change in the isochromatics.

15.1.5 Circular Polariscope

Two quarter-wave plates are added to the experimental setup of the plane polariscope. The first quarter-wave plate is placed in between the polarizer and the test specimen, and the second quarter-wave plate is placed between the specimen and the analyzer. The effect of adding the quarter-wave plates is to generate circularly polarized light.

The essential advantage of a circular polariscope (Figure 15.2) over a plane polariscope is that in a circular polariscope setup, only the isochromatics will be seen. The isoclinics are eliminated. This will eliminate the problem of trying to differentiate between the isoclinics and isochromatics.

15.1.6 Two-Dimensional and Three-Dimensional Photoelasticity

Photoelasticity can be applied to either two- or three-dimensional analysis.

Two-dimensional photoelasticity is carried out on test specimens manufactured from thin material shaped to the basic profile under investigation, and when mounted in a special test rig, allows loads to be applied to the specimen and the change in the fringes to be observed in real time.

Three-dimensional photoelasticity is more complex. It requires a three-dimensional model to be manufactured and to be fitted in a special heated oven. A steady load is applied to the model prior to a heating program being applied. The model is then heated

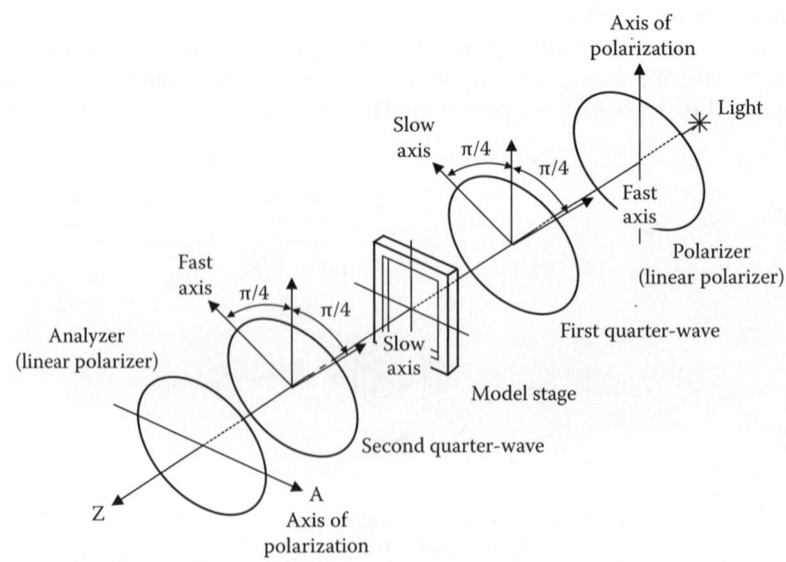

FIGURE 15.2
Schematic of the working components for a circular polariscope.

and cooled under a carefully controlled heating program. This allows stresses generated due to the loading arrangement to be locked into the model on withdrawal from the oven.

This is known as the *stress freezing technique.*

The model is then sliced up in the areas of specific interest. The samples can then be observed and analyzed as two-dimensional specimens using either the plane or the circular polariscopes, depending on the experimental setup.

15.1.7 Further Development

Photoelastic analysis has been used on a wide range of stress analyses and was the preferred experimental stress analysis method before the advent of finite element analysis (FEA).

When using FEA, it is crucial to assess the accuracy of the numerical model. This can only be assessed by using experimental verification. As an example, a threaded joint experiences nonuniform contact. This is difficult to incorporate accurately into the FE model. Idealized models tend to significantly underestimate the actual stress concentrations at the root of the thread.

Photoelastic analysis still remains a major tool in the modern stress analyst's arsenal, validating mathematical models and identifying the accurate placing and orientation of electrical resistance strain gauges.

15.2 Photoelastic Coatings

In certain instances, it may be desirable to carry out a dynamic or static examination of a physical part, such as a landing gear structure or a gearbox housing, to determine the surface strains. With the use of photoelastic coating it is possible to mold a special strain-sensitive coating around the part under examination. The technique is "full field" identical to the two- and three-dimensional methods.

Photoelastic coating has a history of successful applications in a wide range of situations covering manufacturing and construction where stress analysis is used, including aerospace, automotive, pressure vessels, etc. Other applications are found in medical engineering, including the determination of the strength of femur prostheses.

15.2.1 Preparation of the Coating

A liquid plastic is cast on a flat-plate mold and allowed to partially polymerize. While it is still in a pliable state, the sheet is removed from the mold and formed by hand to the contours of the test part. When fully cured, the plastic coating is bonded in place with special reflective cement and the part is then ready for testing.

For plane surfaces, premanufactured flat sheets are available and can be cut to size and bonded directly to the test part.

15.2.2 Analysis of the Coating

The coating is illuminated by polarized light from a reflection polariscope (Figure 15.3). When viewed through the polariscope, the coating displays the strains in identical patterns to the plane and circular polariscopes used on the two-dimensional models.

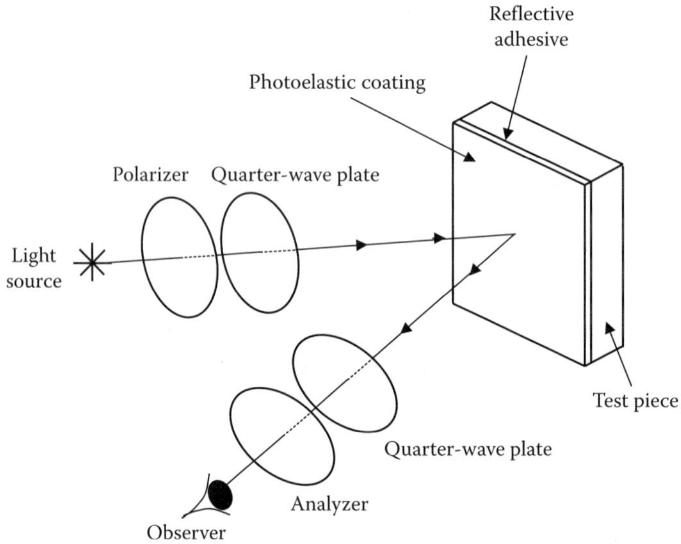

FIGURE 15.3
Schematic representation of a reflection polariscope.

The polariscope is similar to the circular polariscope (Figure 15.2) and is able to be handheld or mounted on a tripod.

The polariscope can be fitted with an optical transducer (compensator) where quantitative stress analysis can then be quickly performed. The use of photography or video recording can maintain a permanent record of the overall strain distributions, particularly in dynamic analysis.

Although the coating is not permanent, the coating will have a reasonable life for fairly long-term examinations to be carried out. Unlike brittle lacquer coatings, the photoelastic coating method may be used over a wide range of loading spectra. Providing the coating is protected, the results are not subject to environmental effects such as humidity or temperature.

The light source can either be a standard light source for taking static measurements or be replaced with a stroboscopic light accessory for cyclic dynamic measurements.

15.2.3 Coating Materials

A wide selection of coatings is available from specialist suppliers in either flat sheet and liquid form for applications to metals, concrete, rubber, and most other materials. Also available are specially designed application kits for the installation on most test parts.

15.2.4 Full-Field Interpretation of Strain Distribution

Photoelastic coatings allow strain measurement to be undertaken in either the laboratory or on-site for in situ measurements during construction or manufacture. The advantage of the photoelastic coatings is that it is full field. Areas of high stresses can be easily identified together with the principal stress direction, so that the coating can be locally cut away and an electrical resistance strain gauge fitted, allowing for a quantitative measurement to be taken to confirm the strain reading. It has to be borne in mind that the strain gauge reading may not match that of the photoelastic coating; this is due to the length of the gauge chosen. Standard length gauges will have a gauge length of 12.5 mm, and therefore the strain reading will be integrated over this length, resulting in a lower strain reading. If a shorter gauge length is chosen, then the reading will be higher as the integration length is reduced.

The interpretation of the fringe patterns is identical to the two- and three-dimensional photoelastic analysis technique. Photoelastic fringes have their own characteristic patterns, which is very helpful in interpreting the fringe patterns. As an example, the fringes are ordinarily continuous bands, forming either curved lines or closed loops, like the isobars on a weather map. The black zero-order fringes are usually isolated spots, lines, or areas surrounded by higher-order fringes. The fringes never intersect or otherwise lose their identity. The fringe order, and therefore the strain level, is uniform at every point on a fringe. Also, the fringes always exist in a continuous sequence by both number and color; that is, if the first and third fringes are identifiable, the second-order fringe will lie between them. The color sequence in any direction establishes if the fringe order and strain level are increasing or decreasing in that direction.

If there is a zero-order fringe within the field of view, it will be obvious by the black color. Assuming that the coated test piece has a free square corner or pointed projection, the stress in this area will always be zero, and a zero-order fringe will exist at this position, irrespective of the load magnitude. The zero-order fringe may shrink slightly as the load magnitude is increased.

From Table 15.1 note the fringe order counting from the zero-order fringe. It is important to note that the analyst should have good color vision; otherwise misinterpretation of the color fringes will occur. Most people tend to have a problem with the color green, and this could lead to reading the stresses a magnitude higher. To overcome this problem, it is possible to view the model in monochromatic light and physically count the fringes from the zero-order fringe to the area of interest.

Generally, when viewing the photoelastic coating without any load applied, it is possible to note the direction of the fringes as the load is incrementally applied. Without any load being applied the color of the coating will be dominated by the zero-order fringe (black),

TABLE 15.1

Isochromatic Fringe Characteristics

Color	Fringe Order
Black	0.00
Pale yellow	0.60
Dull red	0.90
Red/blue transition	1.00
Blue-green	1.22
Yellow	1.39
Rose red	1.82
Red/green transition	2.00
Green	2.35
Yellow	2.50
Red	2.65
Red/green transition	3.00
Green	3.10

and as the load is progressively applied, additional colored fringes in the correct color sequence will begin to appear and spread across the test piece. The analyst will note the appearance of the fringes; closely spaced fine fringes will denote steep strain gradients. Areas where almost uniformly black or gray exist usually indicate a significantly understressed region.

15.3 Introduction to Brittle Lacquer Coatings

The advantages offered for using brittle lacquer is it can be applied to a wide range of materials, including metals, many plastics, glass, ceramics, and wood products. It is inexpensive and requires no instrumentation to analyze the results.

The coating is sprayed on to the surface and then allowed to cure in either air or heat to attain the brittle properties. As the component is loaded, the coating will begin to crack when the threshold strain or strain sensitivity is exceeded.

15.3.1 Loading and Testing Techniques

The brittle lacquer coating fractures in response to the surface strain underneath it. Analysis of the coating indicates the direction and magnitude of stress within the elastic limit of the test piece material. The brittle lacquer coating gives a graphic picture of the distribution, direction, location, sequence, and magnitude of the tensile strains. The coating cracking is a predetermined value, which is determined by a simple calibration procedure.

Under normal conditions, the coating will crack in the range from 5.08×10^{-6} to 76.2×10^{-6} m/m. In terms of stress, these values represent between 41 and 620 MPa.

The qualitative picture of the stress distribution gives an immediate visual indication for immediate design improvements. With very careful handling, it is possible to obtain quantitative results with a ±10% accuracy.

Flaking of the brittle lacquer coating is an indication of local yielding in the test piece.

Cracking occurs where the strain is highest, giving an immediate indication of the presence of stress concentrations. The cracks also indicate the directions of maximum strain at these areas, as they always align at right angles to the direction of the maximum principal tensile strain. The method has been in use for many years, principally to provide test engineers with quick and reliable information about the strain response of the material to which the coating has been bonded. In addition, as it provides information on the principal strain directions, this information is of great use in the selection, location, and orientation of strain gauges for the accurate measurement of peak stresses.

15.3.2 Effects of Change in Relative Humidity and Temperature

The lacquers generally have a nominal threshold sensitivity of 500 microstrain at specified temperatures and humidity conditions specified by the lacquer manufacture in a coating selection chart. The selection of a suitable coating depends on the environmental conditions occurring at the time of the test.

The coating begins to lose its brittleness and sensitivity begins to shift as the relative humidity rises above 20%. This only begins to become noticeable over large changes in the relative humidity.

Other disadvantages include that it is affected by humidity and temperature, that is, requiring the test temperature to be maintained so that an approximate strain level can be determined by comparison with calibration test pieces.

Strain sensitivity changes by approximately 4 microstrain per 1% change in relative humidity.

The effect of temperature on the coating is approximately 70 microstrain per degree Celsius change. As the temperature increases, the coating begins to soften and becomes more plastic, and therefore loses its sensitivity. The opposite applies as the temperature falls. According to StressKote (a major supplier of brittle lacquer coatings), a simple ready calculation is that a 5°C change will cause a sensitivity shift to 360 microstrain, whereas at a constant temperature a 25% change in relative humidity will cause a shift of 100 microstrain.

The grade of coating selected for the test will be dependent on the test temperature and relative humidity.

Prior to the test program, a calibration test strip is prepared and loaded as per the manufacturer's instructions. The calibration is carried out at the test temperature and relative humidity. This allows the microstrain sensitivity of the coating to be read from the calibration test strip.

The calibration bar is simply a rectangular bar that fits into a calibration jig that supports the bar at one end. The bar is then essentially a cantilever. An eccentric cam is located at the free end of the cantilever, and this produces a known strain distribution along the cantilever length. When the lacquer on the bar is fully cured, the cam is rotated to depress the bar by a known amount. The position at which the cracking of the lacquer begins provides the strain sensitivity when compared with a marked strain scale. Quantitative measurements of strain levels can then be made on the component under test, as, for example, if the calibration sensitivity is shown to be 1000 microstrain (strain $\times 10^{-6}$), then the strain at the point on the component at which cracks first appear will also be 1000 microstrain.

15.3.3 Measuring Strain under Static Loading

The technique required is to incrementally load the component, holding the load for a few minutes and releasing it to zero prior to the application of the next increment, noting the crack patterns at each load application. It is recommended to mark key information with a wax pencil. Photographs can be taken with a camera fitted with a flash. Care should be taken when using floodlights, for if they are too close to the test piece, the heat may soften the coating and cause the cracks to heal.

15.4 Introduction to Strain Gauges

The most widely used experimental stress analysis technique in industry today is the strain gauge. The gauges come in different types, but the most common are vibrating wire strain gauges and electrical resistance strain gauges.

15.4.1 Vibrating Wire Strain Gauges

Vibrating wire stain gauges are mostly used in civil engineering for monitoring bridge and tunnel loads. They generally have a large gauge length and are not unduly affected

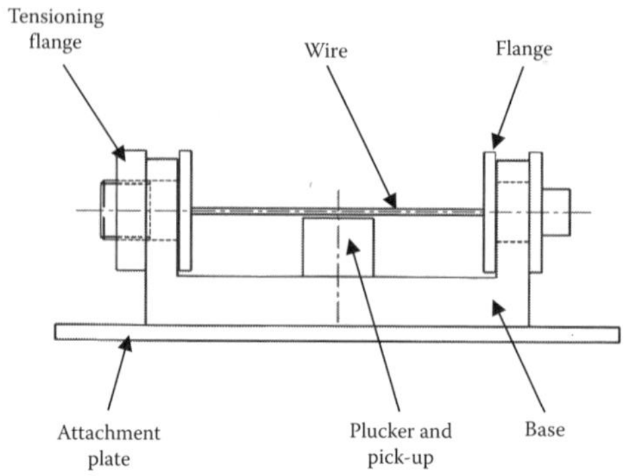

FIGURE 15.4
Schematic diagram of a vibrating wire strain gauge.

by the size of the aggregate used in concrete. The gauge comes in a variety of fittings. It may be embedded in concrete, able to be spot welded to steel beams or columns among a variety of alternative attachments.

The basic construction is that of a fine wire supported under tension between two supports (see Figure 15.4). As the structure to which the gauge is secured moves, this will cause tension on the wire and in turn affect the frequency of the wire. When a measurement is required, the wire is electrically "plucked" and a pick-up then measures the frequency of the vibrating wire. Instrumentation converts this frequency into a strain reading.

The advantage of the vibrating wire gauge is that it is able to lie dormant for several hours or days without any power being applied to it, and when required it can be switched on at will to give an accurate reading without any further calibration required. Indeed, the connection to the gauge can be severed and, provided the gauge position is not disturbed, a calibrated signal will be available when the signal cable is reconnected. Other advantages include being weatherproof and able to operate in any environmental conditions, including corrosive atmospheres provided adequate protection is provided.

Because of its large gauge length (it can be supplied in shorter gauge lengths), the vibrating wire strain gauge has little application in engineering measurements outside of the field of civil engineering.

15.4.2 Electrical Resistance Strain Gauges

In 1843, Charles Wheatstone published his work on the bridge circuit he had invented. He had noticed the change in resistance in an electrical conductor due to the effects of mechanical stress.

William Thomson went further with some work published in 1956.

In 1938, two workers in the United States, Edward E. Simmons and Arthur C. Ruge, developed the basics of the modern strain gauge.

The electrical resistance strain gauge is simply a length of wire or foil formed into a shape of a continuous grid, as shown in Figure 15.5. This grid is cemented to a nonconductive backing.

The gauge can then be bonded securely to the surface of the component under investigation. Any strain experienced in the surface of the component will be transmitted to the gauge itself.

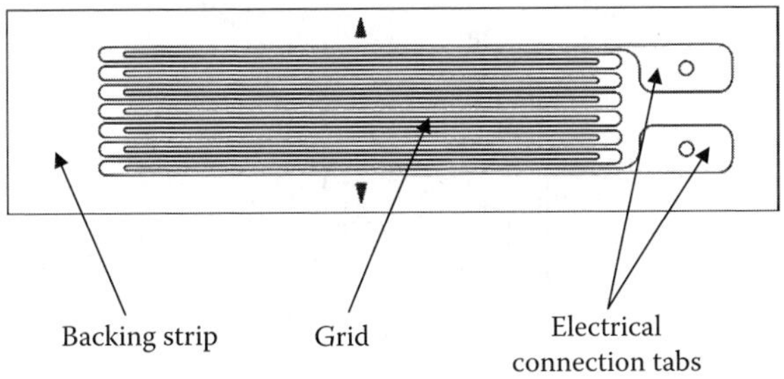

FIGURE 15.5
Elements of a linear electrical resistance strain gauge.

The fundamental equation for the electrical resistance R of a length of the conductor is

$$R = \frac{\rho L}{A} \tag{15.1}$$

where L is the length of the conductor, A is the cross-sectional area, and ρ is the specific resistance or resistivity. Any change in the length of the gauge, and hence sectional area, will result in a change of resistance. Measurement of this resistance change using suitably calibrated equipment enables a direct reading of the linear strain to be obtained. The relationship that exists for a number of alloys over a substantial strain range between the change in resistance and strain may be expressed as follows:

$$\frac{\Delta R}{R} = K \times \frac{\Delta L}{L} \tag{15.2}$$

where ΔL and ΔR are the change in length and resistance, respectively, with K being termed the gauge factor.

Thus,

$$\text{Gauge factor } K = \frac{\Delta R/R}{\Delta L/L} = \frac{\Delta R/R}{\varepsilon} \tag{15.3}$$

where ε is the strain. The strain gauge manufacturer will always supply the value of the gauge factor and can be checked using simple calibration procedures if required. Typical values of K for most conventional foil strain gauges will lie in the region of 2.0 and 2.2, and most modern strain gauge instruments, with the value of K set accordingly, allow the strain values to be recorded directly.

Changes in resistance produced by normal strain levels that are experienced in engineering components are very small, and as a consequence, sensitive instrumentation is required to measure these changes. Strain gauge instruments are based on the Wheatstone bridge networks as shown in Figure 15.6.

The condition of balance for this network is

$$R_1 \times R_3 = R_2 \times R_4 \tag{15.4}$$

In its simplest half-bridge wiring system, gauge R_1 is the active gauge, that is, the gauge actually being strained. Gauge R_2 is called a dummy gauge and is bonded to a piece of material identical to that which R_1 is bonded to, but in this case, it is unstrained. The purpose for this is to cancel out in gauge R_1 any resistance change due to temperature

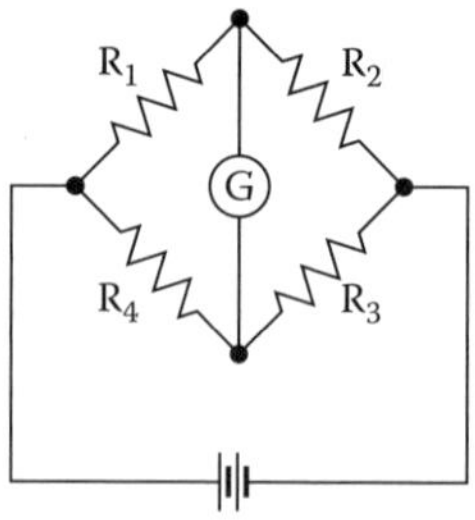

FIGURE 15.6
Wheatstone bridge circuit.

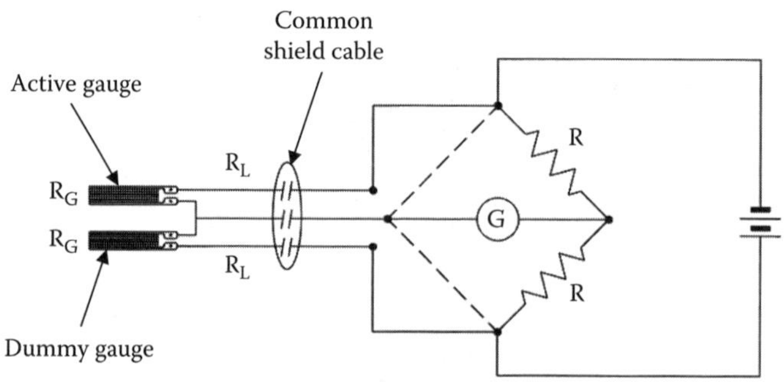

FIGURE 15.7
Three lead-wire system for half-bridge (dummy-active) setup.

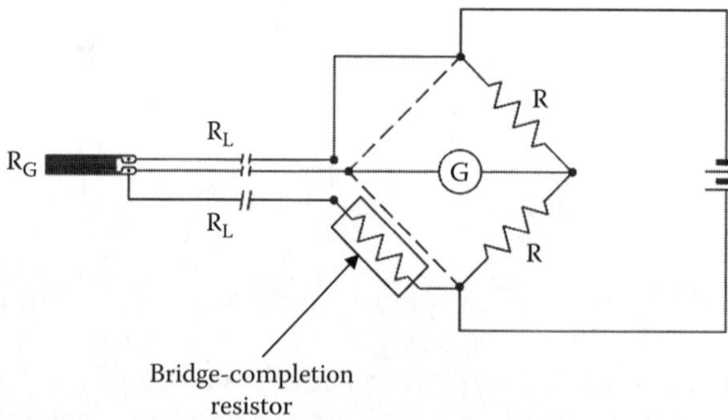

FIGURE 15.8
Three lead-wire system for quarter-bridge (dummy-active) setup with a single self-temperature compensated gauge.

changes. Gauges R_1 and R_2 represent the working half of the network—hence the name *half-bridge system* (Figure 15.7). Gauges R_3 and R_4 are standard resistors built into the measuring instrument. Alternative wiring systems utilize one (quarter bridge) or all four (full bridge) (Figure 15.8) resistors of the bridge resistance arms.

15.4.3 Unbalanced Bridge Circuit

With the bridge initially balanced to zero, any strain that occurs at gauge R_1 will cause the galvanometer needle to deflect. The deflection can be calibrated to read the strain in R_1, as stated above, and the circuit can have an arrangement where the gauge factor can be adjusted. Strain readings can, therefore, be taken with the pointer off the zero position and the bridge is thus unbalanced.

15.4.4 Null Balance or Balanced Bridge Circuit

An alternative measurement procedure is to make use of a variable resistance in one arm of the bridge. This will cancel out any deflection of the galvanometer needle. This adjustment can be calibrated directly as strain, and the readings are taken with the pointer on zero, that is, in the balanced position.

15.4.5 Installation Procedures

The quality and success of any strain gauge installation are influenced by the consideration and choice of the installation procedure together with the choice of the adhesive. The apparently mundane procedure of actually cementing the gauge onto the component is a critical step in the operation. Every precaution must be taken to ensure a scrupulously clean surface if perfect adhesion is to be attained. J. Pople covers full details of the surface preparation and installation techniques in the BSSM (British Society for Strain Measurement) *Strain Measurement Reference Book*.

Following the successful attachment of the gauge to the surface, lead wires are then attached to the electrical connection tabs using solder. Extreme care is required in this operation, as it is very easy to generate a dry joint with its associated noise. An insulation check must be carried out to confirm the installation; the installation resistance to ground (RTG) should be a minimum of 104 MΩ at 30 V.

The sequence for successfully installing an electrical strain gauge is as follows:

1. Determine the measurement required. What feature do you want to measure? Bending, torsion, tension, compression, etc.?
2. Select the most appropriate strain gauge for that measurement.
3. What is the specimen material? Select the appropriate adhesive for the particular application and compatible with the gauge backing.
4. Select the area where the gauge is being fitted and ensure there is adequate access for fitting the gauge.
5. Follow the installation procedure carefully, ensuring that the gauge, if clamped, is equal over its length and the pressure is between 34 and 138 kPa. Unequal clamping will result in an unequal glue line.
6. Select the style of terminal pad and bond adjacent to the strain gauge.
7. Select the correct type of cable between the gauge and bridge. This is of vital importance, as the cable in a quarter-bridge configuration forms part of the measuring circuit. Any change in the cable resistance will affect the accuracy of the strain measurement.
8. Select the correct type of solder and flux for the application: only use the solder and flux that has been approved by the gauge manufacturer. Many types of commercial solders and fluxes are unsuitable for strain gauge applications.
9. Carefully solder the jumper leads between the strain gauge and the terminal pad and the lead wire from the terminal pad to the bridge.

10. The lead wire may have to terminate in a plug and socket at an appropriate point if the measuring point is a distance from the instrumentation. It is essential that the resistance remains consistent during the measurement phase and the resistance value is as low as possible.
11. Depending on the application, a choice will have to be made if the strain gauge requires protection. This will depend on the environmental conditions the gauge is expected to operate in. If the gauge is operating in clean-air conditions within laboratory conditions and the test is expected to last a very short time, the gauge can in this circumstance be painted with a paint compatible with the lead insulation to prevent any oxidation at the gauge–structure interface. But if the test is being undertaken in workshop conditions, then a minimal degree of protection will be required. Encapsulating the gauge in an insulating layer of polysulfide rubber (Bostik 2114–5) will provide this. This will give protection for up to 6 months against any moisture ingress.
12. In situations where the gauge is expected to operate in the most adverse conditions, a suitable metal surround will be required, particularly if the gauge has to face sea water conditions such as on oil rigs, together with the extremes of temperature. Layers of polysulfide epoxy further protect the polysulfide rubber encapsulation. The mechanical housing has an extension to it providing a totally leak-proof connection for the signal cable. This is also fitted with an adequate insulation.
13. At several points in the installation of the gauge, the insulation will have to be tested using a gauge insulation tester to ensure that the insulation resistance is in excess of 20,000 MΩ for foil strain gauges when installed under laboratory conditions. A value of 10,000 MΩ should be considered a minimum. Readings below this generally indicate trapped foreign matter, moisture, residual flux, or backing damage due to soldering, as well as incomplete solvent evaporation from the overcoating.

15.5 Extensometers

Extensometers were developed for use on tensile testing machines for measuring strains; they measure the very small extensions or contractions in the test pieces when under increasing load. They are not precluded from use in measuring strains in other applications. The extensometer was invented by Charles Huston in the 1870s (Figure 15.9).

Modern extensometers utilize electronics in their construction, using a very sensitive displacement transducer. Special clip-on fasteners enable the extensometer to be quickly attached to the test specimen. These instruments are extremely accurate, having a resolution of <0.3 μm. These instruments require a conditioning module for their operation.

These extensometers are known as *contact extensometers.*

15.5.1 Contact Extensometers

Figure 15.10 shows an extensometer for measuring strains in plastic test pieces.

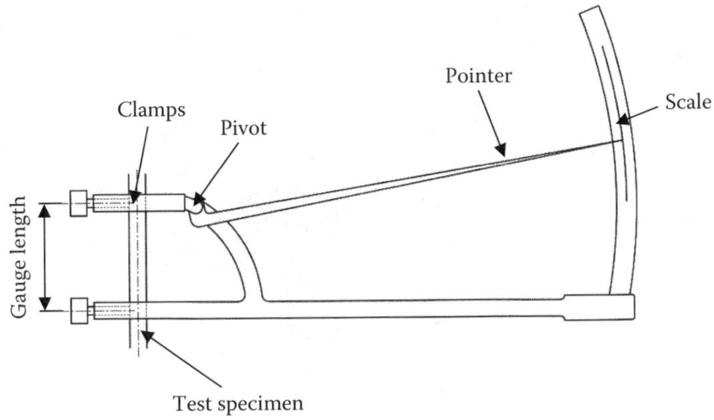

FIGURE 15.9
Huston's extensometer.

FIGURE 15.10
Extensometer for plastic test coons.

15.5.2 Noncontact Extensometers

Another class of extensometers, known as *noncontact*, utilize lasers for performing the strain or elongation measurements. The principle utilizes the reflection of a laser light from the surface of the test specimen via a charge-coupled interface digital (CCD) camera. Resolutions of typically 0.1 μm can be attained (Figure 15.11).

Targets are attached to the test specimen at a known distance apart, and a laser unit scans the targets and measures the distance apart. Although costly compared to the contact type, they are very easy to set up. The laser light is low power, and, therefore, providing the operator does not stare into the light, they are safe to use.

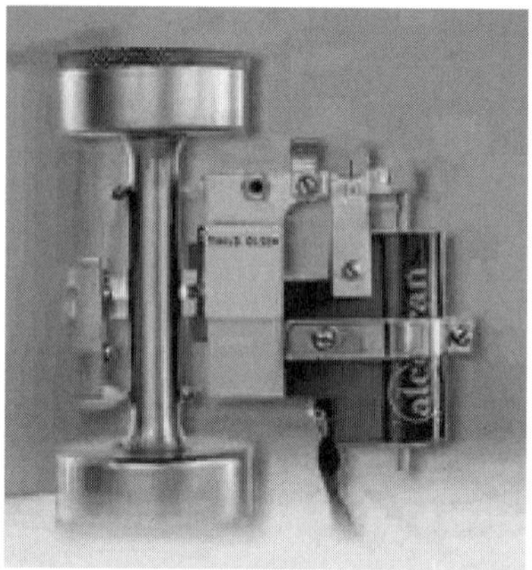

FIGURE 15.11
Extensometer for metal test coons.

15.5.2.1 General Notes

The big disadvantage with these types of extensometers is the "gauge length," which tends to be rather large for very local strain measurements. As with the electrical resistance strain gauge as well as the vibrating wire gauges, the strain is integrated over a larger length than those of the electrical resistance strain gauges, and hence are very local strains or not identified. Having said that, they are on a par with the vibrating wire strain gauges, where these do have quite large gauge lengths.

In civil engineering, where most of the structures are made using concrete, the size of the aggregate will dictate the gauge length. In some instances, the size of the aggregate can be as large as 25 mm, in which case the gauge length will need to be >100 mm to be able to yield any meaningful results.

The disadvantage with these, as with the electrical resistance strain gauges, is that the datum reading will be lost if there is a power failure. Therefore, it is essential to have a reliable battery backup.

15.5.3 Applications

As with the vibrating wire strain gauges, these extensometers have found a use in the civil engineering and mining industries for monitoring walls, etc. Plotting displacements against time, geotechnical engineers can predict if a failure is imminent.

16

Introduction to Fatigue and Fracture

16.1 Introduction and Background to the History of Fatigue

The earliest investigation into metallic fatigue was undertaken by August Wöhler between 1852 and 1870, who had been commissioned by the French Railway Company SNCF. It had experienced a number of axle failures and wanted to know and understand the reasons for them.

Wöhler constructed a rotating bending test rig (see Figure 16.1) to emulate and investigate the loading on a full size railway axle. These axles were subjected to a range of rotation speeds and loads.

Figure 16.2 shows a representation of the rotating cantilever test rig identifying its major component part.

He constructed a curve (Figure 16.3) comparing the applied load on the axle to the number of cycles to failure. From this curve he observed that the curve exhibited an endurance limit where stresses below that limit did not display any further damage.

From these results it was established that metals had a limiting stress where failure could be estimated with a reasonable degree of accuracy under the conditions prevailing at that time.

The next breakthrough in understanding the mechanics of fatigue was observed by Bauschinger in 1886. His experiments showed that the limit of elasticity on the first compression loading is less than the initial tension loading; this is called the Bauschinger effect. Figure 16.4 shows that if point C is extrapolated, C′, using the B-C slope, then $\Delta\varepsilon_b$ will show the permanent cyclic strain softening or cyclic strain hardening. The second part of Figure 16.4 is drawn to an absolute scale for stress-strain.

When strain hardening occurs, as in cast aluminum, this effect becomes more apparent. These plots are generated during a fatigue test showing the Bauschinger effect and are often referred to as hysteresis loops.

The first fatigue tests were conducted using a fully reversible cycle, i.e., alternating loads between compression and tensile with the mean stress at 0.

This situation very rarely occurs in real life, and a way was found to offset the normal Wöhler type curve. Goodman, Soderberg, and Gerber undertook this work. Their work is based on either the ultimate tensile stress σ_{uts} or yield stress σ_{yield}, and the endurance limit stress σ_e or the fatigue strength for a given number of cycles σ_{am} (for zero mean stress) as the baseline for a safe design and is drawn on two axes, with the x axis representing the mean stress and the alternating stress plotted on the y axis. Figure 16.5 shows these three most widely used empirical relations. The straight line joining the alternating fatigue stress to the tensile stress is the modified Goodman law. Goodman's original law included the assumption that the alternating fatigue limit was equal to one-third of

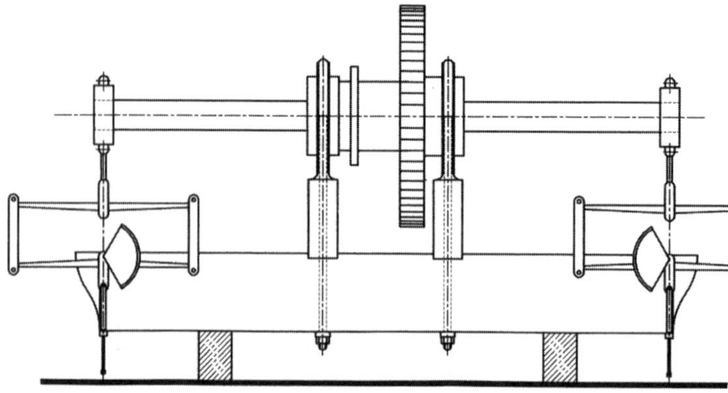

FIGURE 16.1
Wöhler's original railway axle.

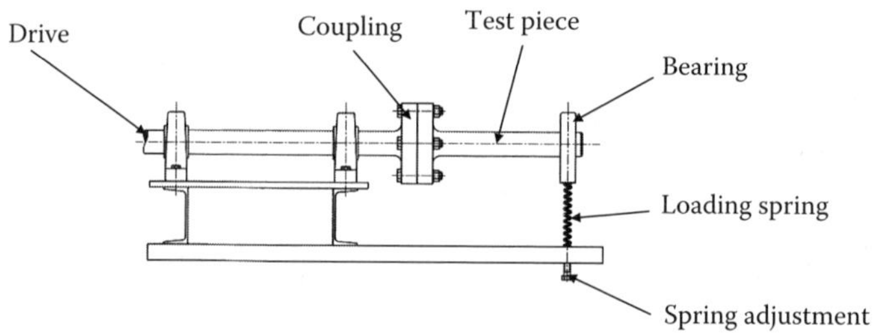

FIGURE 16.2
Representation of a rotating cantilever test rig.

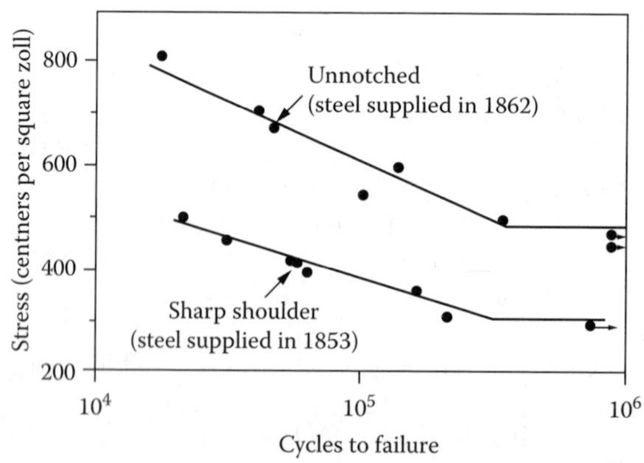

FIGURE 16.3
The S-N curve as described by Wöhler. (1 çentners per zoll^2=0.75 Mpa.)

the tensile stress, but this has been modified to the relation shown, using the alternating fatigue stress determined experimentally. The original law is no longer used and has been replaced by the modified law; this is referred to as the Goodman law.

Gerber deduced that the early results found by Wöhler fitted closely to a parabolic relation, and this is often referred to as Gerber's parabola. A straight line connecting

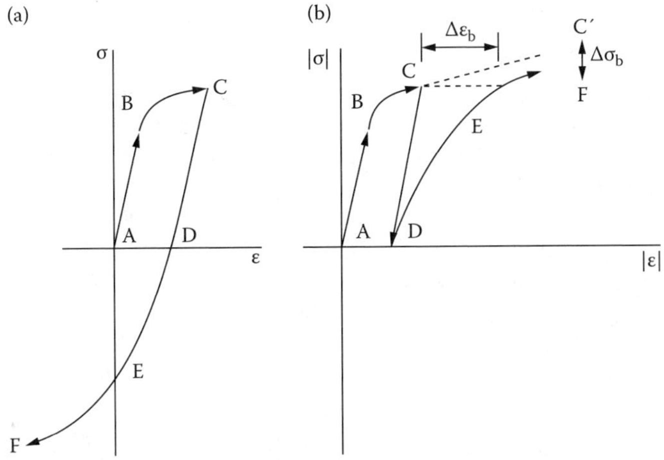

FIGURE 16.4
The stress and strain curve for cycliс loading.

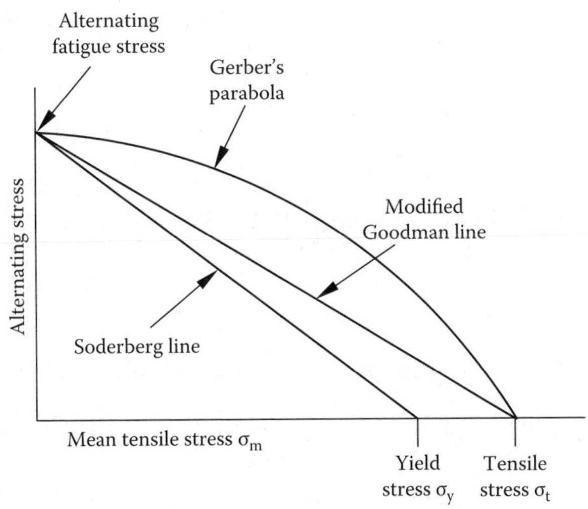

FIGURE 16.5
Fatigue stress-static stress diagram.

the alternating fatigue stress to the static yield stress gives the third relation, known as Soderberg's law. For many purposes it is essential that the yield stress is not exceeded, and this relation is intended fulfill the condition that neither fatigue failure nor yielding shall occur.

The Goodman curve gives good results for brittle materials and conservative results for ductile materials. The Gerber relation will give good results for ductile materials.

It was found that a compressive mean stress improved the fatigue life of a component and decreased with a tensile mean stress. Gerber's parabolic relationship may therefore produce erroneous results to the conservative side in the compressive mean stress region.

16.1.1 Later Developments

Old notions regarding the fatigue mechanism based on the crystallization theory were proved incorrect by Ewing and Rosenheim in 1900, closely followed by Ewing and Humfrey in 1903, showing that slip bands developed in many grains of the polycrystalline material. These slip bands broaden as cyclic deformation continues, leading to extrusions and intrusions on the surface of the component, and then one dominant flaw will then lead to failure.

The empirical relationship of the stress-cycle or S-N curve was proposed by Basquin in 1910 using a log-log scale showing the linear relationship (see Equation 16.1).

$$\frac{\Delta\sigma}{2} = \sigma_a \sigma_f' \cdot (2 \cdot N_f)^b \tag{16.1}$$

where $\Delta\sigma$ is the stress range for one fully reversed cycle, σ_f is the fatigue strength coefficient, b is the fatigue stress exponent, and N_f is one fully reversed cycle.

One fully reversed cycle will be in the following sequence (see Figure 16.4a):

$$A \rightarrow B \rightarrow C \rightarrow D \rightarrow E \rightarrow F \text{ back to A}$$

Palmgren in 1924 and Miner in 1945 also noticed the accumulation of damage and presented a linear cumulative damage rule that excluded any sequence of loading. The equation is shown in Equation 16.2:

$$\sum_{i=1}^{m} \left(\frac{n_f}{N_f} \right) = 1 \tag{16.2}$$

where n_i is the number of cycles corresponding to the amplitude σ_{ai}, which is the stress amplitude, and N_{fi} is the number of cycles to failure σ_{ai}.

Notch effects at the tip of a notch had been found by Neuber to follow the well-known rule as in Equation 16.3.

$$K_t = \sqrt{K_\sigma \cdot K_\varepsilon} \tag{16.3}$$

where K_σ is the ratio of the maximum local stress to the nominal stress, and K_ε is the ratio of the maximum local strain to the nominal strain.

With strain being the concentration now, this will also address the plastic deformation at the tip of the notch. The same can also be performed for the fatigue notch factor by replacing K_t with K_f.

K_t can be viewed as the macroscopic stress-strain notch factor, and K_f as the stress-strain notch factor associated with the microscopic properties, such as material, surface finish, and inclusions.

16.1.2 Recent Developments

Further development was independently undertaken by Coffin and Manson, where they proposed that plastic strain was responsible for fatigue damage. They also proposed an empirical relationship between the number of load reversals to failure and plastic strain.

$$\frac{\Delta\varepsilon_P}{2} = \varepsilon_f' \cdot (2 \cdot N_f)^c \tag{16.4}$$

where $\Delta\varepsilon_p$ is the strain range for one fully reversed cycle, ε_f' can be set equal to ε_f, which is the fracture ductility in a simple monotonic test, and c is in the range of –0.5 to –0.7 for most metals.

Fracture mechanics has its origins in the stress analysis of Inglis and the energy method of Griffith. In their analysis of brittle solids and later work by Irwin it was shown that the amplitude of the stress singularity ahead of the crack tip can be expressed by the stress intensity factor, K.

The stress fields at the crack tip can be derived for three major modes of loading, each involving different crack surface displacements (Figure 16.6):

- Mode I: Opening or tensile mode. Crack surfaces moving directly apart.
- Mode II: Sliding or in plane shear mode. Crack surfaces moving relative to one another, normal to the crack front.
- Mode III: Tearing or antiplane shear mode. Crack surfaces moving relative to one another parallel to the crack front.

Most engineering situations involving cracked components are covered by mode I loading. Mode III occurs in pure shear situations. Mode II is rarely encountered except in rolling contact bearing races. Mixed modes are encountered in biaxial loading situations.

Equation 16.5 gives a relationship for the stress intensity factor:

$$K_1 = F \cdot \sigma \cdot \sqrt{\pi \cdot a} \leq \tag{16.5}$$

with I generally indicating mode I, Fa geometrical factor that will refer to the crack size and loading condition, all in relation to the specimen, and "a" the crack size in depth. Further work by Paris, Gromez, and Anderson showed that the range in stress intensity factor could be related to the change in fatigue crack length per load cycle. Equation 16.6 shows this law.

$$\frac{da}{dN} = C \cdot \Delta K^m \tag{16.6}$$

where

$$\Delta K = K_{max} - K_{min}$$

where C and m are material constants.

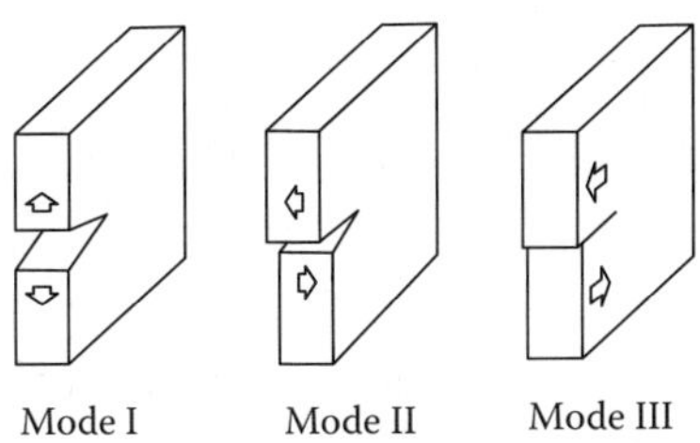

FIGURE 16.6
Various modes of failure.

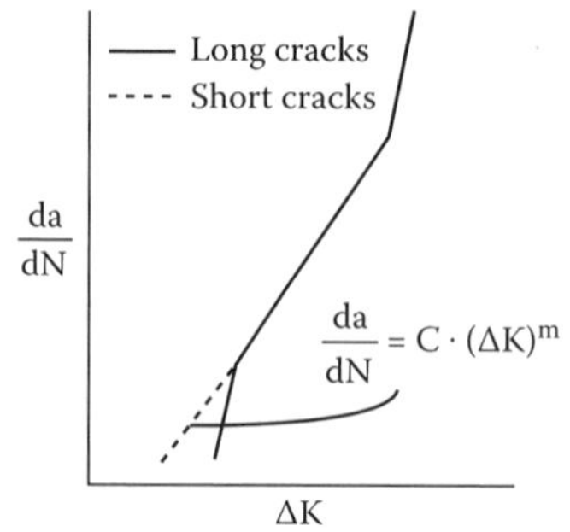

FIGURE 16.7
Crack-growth rate curve.

The knowledge of crack initiation and subsequent slip bands that formed to initiation cracks has been increased with the development and use of the electron and optical microscopes. This has led to the discovery of so-called persistence slip bands (PSBs) by Zappfe and Worden. The term *persistence* refers to the phenomenon that even after a layer of surface material is removed where the slip bands have previously occurred, the slip bands will occur in the same place.

The next area of interest is that the stress intensity factor can change as a result of the crack advancement. In other words, the ΔK can be influenced by the details of crack closure or fracture surface contact in the wake of the advancing fatigue crack tip. Elber showed that under certain conditions cracks will stay closed in a cyclic tensile load. The fatigue crack growth is therefore not only influenced by the values of ΔK alone, but also by the prior history and crack size. The so-called short crack phenomenon was identified by Pearson, who showed that small fatigue flaws, typically an order smaller than other fatigue cracks, will exhibit faster crack growth when the crack lengthens. Figure 16.7 shows a typical crack growth rate curve.

This has been a brief introduction into the mechanics of fatigue and fracture. The following sections in this chapter will give more detail on this subject.

16.1.3 Basic Definitions

The following are definitions of some terms frequently used in the discussion of fatigue analysis (please refer to Figure 16.8):

Stress cycle: The smallest division of the stress-time function that is repeated.

Nominal stress: Obtained from the simple theory in tension, bending, and torsion, neglecting geometric discontinuities.

Maximum stress: The largest or highest algebraic value of a stress in a stress cycle. Positive for tension (S_{max}).

Minimum stress: The smallest or lowest algebraic value of a stress cycle. Positive for tension (S_{min}).

Mean stress: The algebraic mean of the maximum and minimum stress in one cycle (S_{mean}).

Stress range: The algebraic difference between the maximum and minimum stresses in one cycle (S_r).

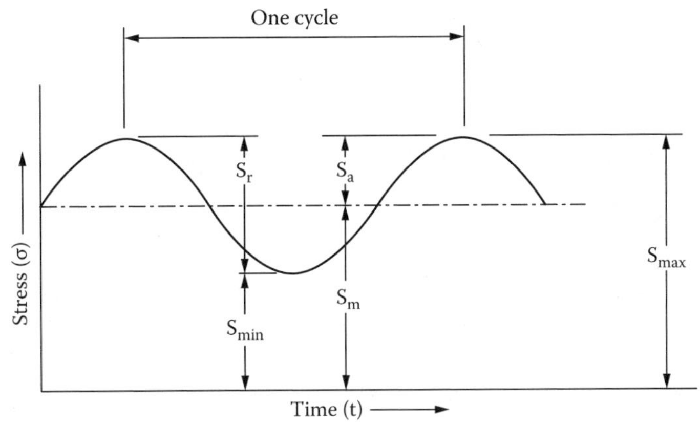

FIGURE 16.8
Stress versus time curve.

Stress amplitude: Half the value of the algebraic difference between the maximum and minimum stresses in one cycle, or half the value of the stress range (S_a).

Stress ratio: The ratio of minimum stress to maximum stress.

Fatigue life: The number of stress cycles that can be sustained for a given test condition.

Fatigue strength: The greatest number of stress cycles that can be sustained by a member for a given number of stress cycles without fracture.

Fatigue limit: The highest stress level that a member can withstand for an infinite number of load cycles without failure.

Fatigue life for P percent survival, N_p: The fatigue life for which P percent of the sample has a longer life; for example, N90 is the fatigue life for which 90% will be expected to survive and 10% will fail.

16.2 The Fatigue Process

From the introduction it is seen that from the tests undertaken by Wöhler, the catastrophic failures of the rail axles were due to fatigue. The axles had failed due to the repeated application of a sinusoidal cyclic load, each of a magnitude, well below the yield stress of the material used. Clearly it can be seen that at any point on the axle, it would be subjected to a fully reversed stress cycle each single revolution of the axle. See Figure 16.9 for a schematic arrangement.

Modern fatigue testing very rarely subjects the complete component or assembly to a fatigue test, but usually relies on subjecting a standardized test specimen to the test loads. This has the advantage that the test is subjected to identically similar geometric components, and a detailed database can be developed comparing the results loading, frequency, and materials.

Modern fatigue testing subjects the standardized specimens to a wide range of types of testing strategies, including push-pull, two-and three-point bending, fluctuating torsion, and other modes of stressing besides rotating bending.

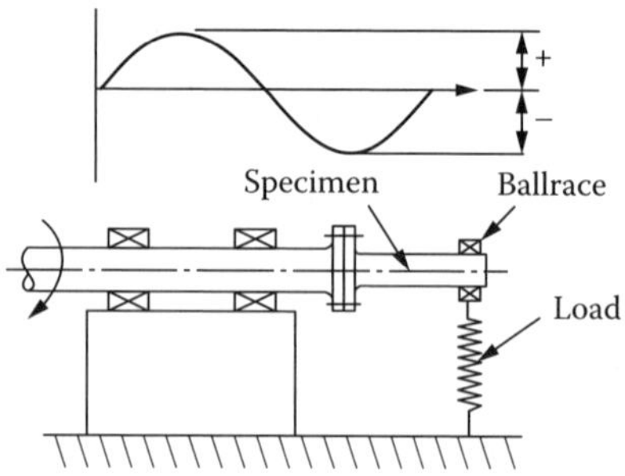

FIGURE 16.9
Schematic arrangement of a Wöhler fatigue test.

The nature of the testing has to be carefully considered in light of the nature of the failure mode being investigated. In Wöhler's case he subjected his test pieces to a rotating bending test. It would not be appropriate to compare the results from a push-pull test with those of a rotating bending test. They are essentially totally dissimilar tests.

The simple rotating cantilever test imposes a sinusoidal loading cycle about a zero mean stress. More complex fatigue tests may superimpose a fluctuating stress on a mean stress that will not be zero or use load cycles that may not be sinusoidal.

In carrying out fatigue tests, a batch of nominally identical test specimens is prepared and a single specimen is subject to a load W and the number of cycle to failure N_f is counted and logged. The remaining specimens are tested at other loads, W_1, W_2, W_3, …, W_n, to determine the lives N_{f1}, N_{f2}, N_{f3}, …, N_{fn}.

Stress amplitudes S corresponding to the full load cycle are calculated according to the specimen size and geometry, and the results are presented as a plot of S vs. log N_f. A typical plot is shown diagrammatically in Figure 16.10, and these graphs are referred to as S/N curves.

There are two types of S/N curves:

a. This curve shows a definite horizontal portion or "knee" at long lives.
b. Here the curve becomes asymptotic to the N axis at very high values of N.

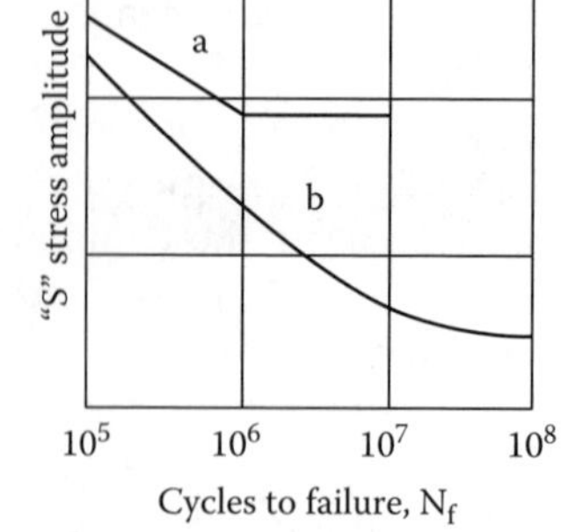

FIGURE 16.10
Typical fatigue curves.

For curve a, the horizontal portion of the curve defines a value of S below which failure is not likely to occur, however large the value of N becomes. This value of S is known as the fatigue limit of the material (or endurance limit) when expressed in units of stress. This type of behavior is typical of ferrite steels (and also some aluminum alloys). For most materials this value, as a rough guide, is about 40% of the uniaxial tensile strength.

In contrast, most nonferrous alloys and polymeric materials exhibit an S-N curve, which continues to fall as the stress amplitude is lowered. Curve b in Figure 16.10 is a typical example, and the endurance limit is defined as the stress amplitude necessary to cause fatigue failure after a specified number of stress cycles (for example, $N = 10^7$ cycles).

Fatigue testing is not limited to the rotating bending mode. Other tests include push-pull, plane bending, fluctuating torsion, and other modes of stressing. The control of load amplitude may be replaced by other parameters, such as total strain, plastic strain amplitude, or displacement.

The simple rotating cantilever test imposes a sinusoidal load cycle acting about a zero mean stress on the specimen. More complex tests may superimpose a fluctuating stress on a mean stress that is not zero, or use a load cycle that is not sinusoidal and employs load sequences that attempt to simulate service conditions (Figure 16.11). At its most complex, a whole structure such as a vehicle chassis or aircraft structure may be subjected to computer-controlled cyclic loading that more approximately represents those that will be experienced in service.

Fatigue testing is essential for the acquisition of data to help predict a fatigue failure in a component. This is performed in a variety of ways, depending upon the stage of the design or production phase. The following main types of tests can be identified:

- Stress-life testing of small specimens
- Strain-life testing of small specimens
- Crack growth testing
- S-N tests of components
- Prototype testing for design validation

The first three tests are generally idealized tests using standard specimens; these are used for producing information on the material response. The use of the results from these tests in life prediction of components and structures requires additional knowledge of the influencing factors related to size, geometry, surface conditions, and any corrosive environment.

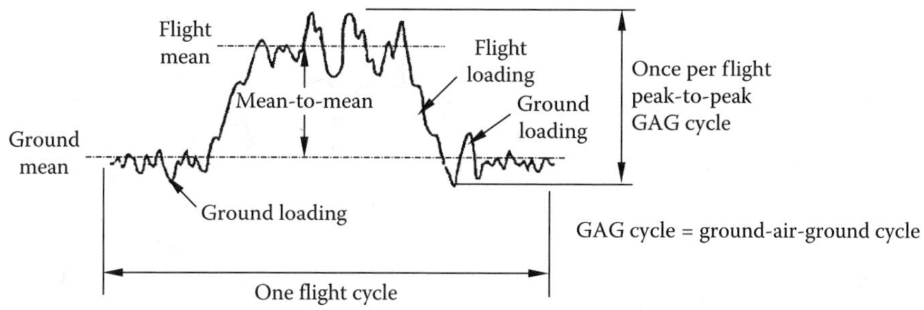

FIGURE 16.11
Example of a typical flight cycle.

Life prediction tests are normally carried out using standardized specimens for producing more accurate S-N curves that are not influenced by the presence of notches and other surface conditions.

Rotating bending machines have been used in the past for generating large amounts of test data in a relatively inexpensive way. These are now being replaced by the use of closed-loop electrohydraulic test machines; these are fitted with hydraulic grips that facilitate the insertion and removal of the test specimens.

Figure 16.12 depicts an axial (push-pull) electrohydraulic servo fatigue test machine that is widely used in the majority of material test laboratories. These machines are capable of precise control of almost any type of stress-time, strain-time, or any load patterns.

Special equipment such as environmental chambers allow the test specimens to be tested at high or low temperatures in addition to providing corrosive atmospheres around the specimen and evaluate the fatigue characteristics in these atmospheres.

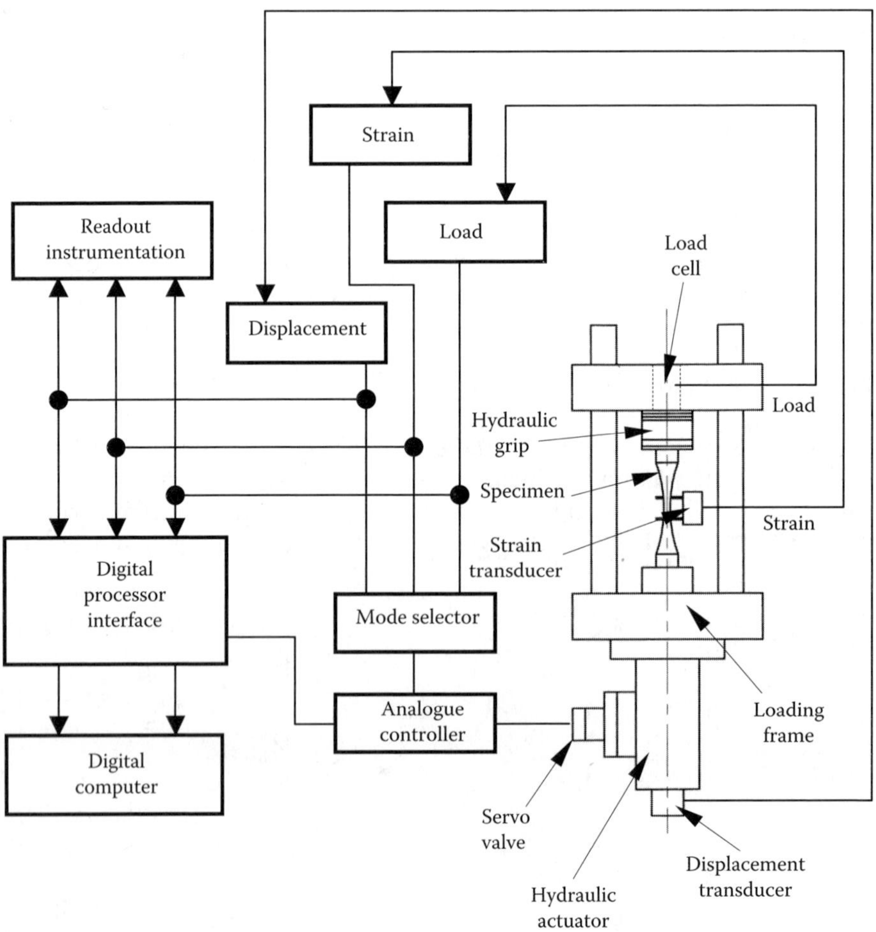

FIGURE 16.12
Electrohydraulic fatigue test machine.

16.3 Initiation of Fatigue Cracks

Fatigue failure in metals starts with the initiation of a crack invariably at a free surface. Then follows a period where the crack begins to nucleate with other local cracks and starts a slow growth until the remaining section cannot carry the load and a catastrophic separation occurs.

There are two distinct fatigue processes involved: initiation and propagation.

The initial crack in engineering alloys will often occur at a site other than a slip band. Grain boundaries, second-phase particles, particle-matrix interfaces, corrosion pits, and machining marks have all been observed at sites of crack initiation. The common feature of these sites is that they must be capable of localizing plastic deformation that occurs in directions that intersect a free surface.

If a crack or crack-like defect preexists, then there will be no initiation process required and the whole life of the specimen is then spent in crack growth.

The cyclic slip process (shear controlled) has been widely held to be responsible for crack initiation on the surface of uniform polished specimens. It is believed that slip will occur between crystals orientated in the direction of the maximum shear stresses, which is roughly at an angle of 45° to the applied cyclic load.

Several equivalent models have been proposed to try and explain the initiation of fatigue cracks by local deformation. Wood has proposed a model depicted in Figure 16.13. During the rising load part of the cycle, slip occurs on a favorably orientated slip plane. In the falling load part slip takes place in the reverse direction on a parallel slip plane.

Since slip on the first plane is inhibited, strain hardening is by oxidation of the newly created free surface. This first cyclic slip can give rise to an extrusion or intrusion in the metal surface. An intrusion can grow into a crack by continuing plastic flow during subsequent cycles.

An intrusion-induced crack or imperfection appears on the metal surface, and once started (stage I crack growth) can also grow by a mechanism of reversed slip, leading to microcracking, which in turn will lead to a macrocrack (stage II crack growth).

Many components live most of their lives with microcracks preexisting within them. It is obvious that for many components it is the ability to propagate cracks rather than initiate them that is important.

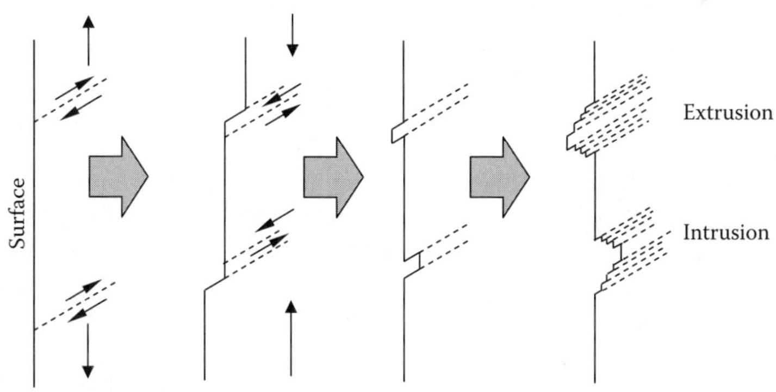

FIGURE 16.13
Wood's model for fatigue crack initiation.

Stage I initiation (slip band growth) may be completely absent in practical cases where cracks originate from highly stressed, sharply notched. Stage I growth can account for up to 90% of the total fatigue life of smooth ductile specimens at low stresses.

Stage II propagation is the later stages of the crack growth and represents the remaining 10% of the growth to failure. The fracture surfaces may show the well-known macroscopic progression marks known as beach marks.

These marks are often curved, with the center of curvature at the origin. They serve as a useful guide to direct the investigator to the fracture initiation site.

16.4 Factors Affecting Fatigue Life

It would be perfect if the effect of any single variable on the separate processes of stage I and stage II crack growth was fully understood. Unfortunately, many investigators have only studied the effects of particular variables on the total number of cycles to failure (N_f). The effect is further complicated, as a single variable may have different effects in low-stress/high-cycle fatigue from those of high-stress/low-cycle fatigue. Variables may also interact; as an example, frequency of cycling is apparently a single variable, but altering it will usually also change strain rate and the time available for environmental effects at the crack tip opened by the tensile component of the load cycle.

16.4.1 Stress Amplitude

The effect of stress amplitude on the cycles to failure is large, as indicated by the S/log N curves, and it is generally accepted that the tensile component is the most damaging. The effect of periodic large amplitude loading is complex, and it may be either beneficial or damaging. A large load early in the life may induce some work hardening or blunt a notch or groove by plastic deformation; either event will have beneficial results. A tensile overload during stage II growth may induce compressive residual stresses at the crack tip, slowing down crack growth during the immediate or immediately subsequent cycles. A sufficiently large tensile overload during stage II growth may produce a burst of overload failure that will extend the crack a considerable distance. This effect is referred to as crack jumping.

Summarizing, increasing the stress amplitude will result in early crack initiation and an increase in the rate of both modes of crack growth.

16.4.2 Mean Stress

It is unusual for engineering components to be subject to stress cycling about a zero mean stress. It is more usual for an alternating stress to be superimposed on a mean stress that is either compressive or tensile (see Figure 16.14).

In the case where the stress amplitude is expressed as a maximum and minimum, as in Figure 16.15, the mean stress is conveniently expressed by a dimensionless parameter R, where:

$$R = \frac{\sigma_{min}}{\sigma_{max}}$$

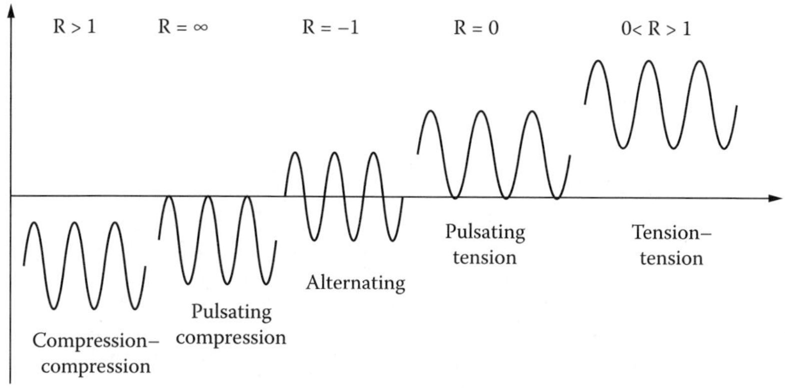

FIGURE 16.14
Stress cycles with different mean stresses and "R"-ratios.

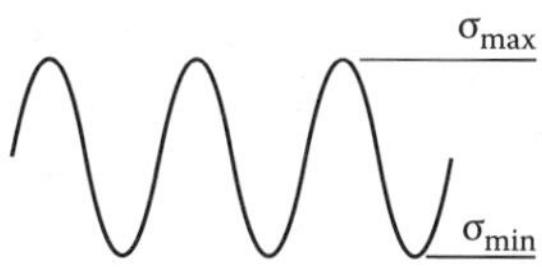

FIGURE 16.15
Amplitudes of the stress cycle.

It is well established that tensile mean stress reduces fatigue life, and plotting the stress amplitude S against mean stress shows the relationship σ_{mean} for a particular value of N_f (see Figure 16.16).

As described in Section 16.1, a number of models have been proposed for the analytical and graphical representations of the relationship between the mean life and the alternating stress for a given fatigue life. The most well-known among these models are those proposed by Gerber, Goodman, and Soderberg. The life plots are displayed in Figure 16.17 and are described by the following expressions:

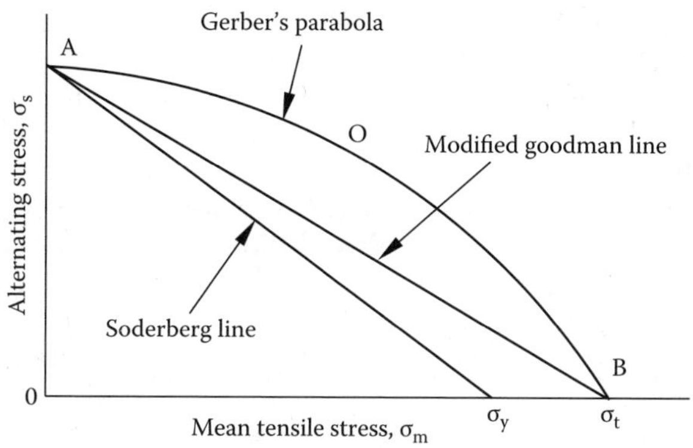

FIGURE 16.16
Comparison of Gerber, Soderberg, and Goodman laws viewed against stress amplitude and mean tensile stress.

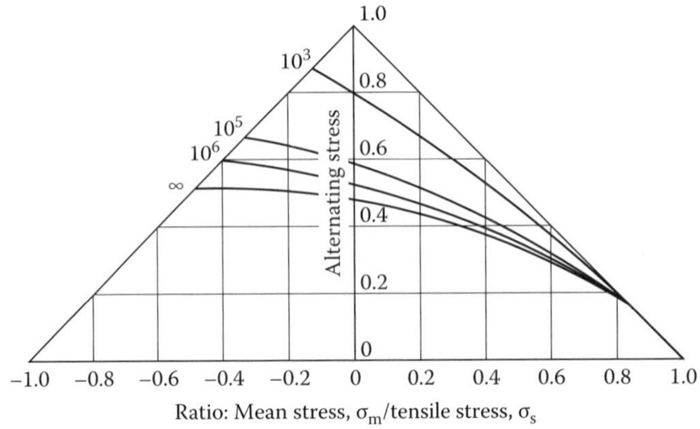

FIGURE 16.17
Typical master diagram.

$$\text{Gerber parabola}: \frac{\sigma_a}{\sigma_e} + \left(\frac{\sigma_m}{\sigma_{ult}}\right)^2 = 1$$

$$\text{Goodman line}: \frac{\sigma_a}{\sigma_e} + \frac{\sigma_m}{\sigma_{ult}} = 1$$

$$\text{Soderberg line}: \frac{\sigma_a}{\sigma_e} + \frac{\sigma_m}{\sigma_{ys}} = 1$$

where

σ_a = Fatigue strength in terms of stress amplitude when $\sigma_m \neq 0$

σ_e = Fatigue strength (for a fixed life) in terms of stress amplitude for a fully reversed loading ($\sigma_m = 0$ and $R = -1$) in some references $\sigma_e = \sigma_{ao}$)

σ_{ult} = Ultimate tensile strength of material

σ_{ys} = Material yield strength

Most metals and alloys give results that lie between the Goodman's line and the Gerber's parabola, with the Goodman's line matching experimental data quite closely for brittle materials but is conservative for ductile alloys.

The Soderberg line is more conservative than the Goodman line but is physically more meaningful in that it puts the maximum possible mean stress as equal to the yield stress.

Gerber's parabola is generally good for ductile alloys.

Whole families of these diagrams are needed to describe the behavior of a material over a range of N_f values, and it is simpler to show the effect of R on a so-called master diagram of the sort shown in Figure 16.17. It is usual in such diagrams to normalize alternating stress (plotted on the y axis) and the mean stress (plotted on the x axis) by dividing the values by the tensile stress. A single diagram can thus be used as a guide to a whole range, such as steel or aluminum alloys.

Such diagrams may be plotted from experimental data but are frequently derived from empirical formulas. Heywood uses the following formula to plot a master diagram giving the fatigue life of unnotched steels as a function of mean stress and alternating stress amplitude:

$$\frac{\sigma_a}{\sigma_t} = \left(1 - \frac{\sigma_m}{\sigma_t}\right)[A_o + \gamma(1 - A_o)]$$

where

$$A_o = \frac{1+0.0038N^4}{1+0.008N^4}$$

and

$$\gamma = \frac{\sigma_m \left(2+\frac{\sigma_m}{\sigma_t}\right)}{3\sigma_t}$$

where
- σ_m = mean stress
- σ_a = alternating stress
- σ_t = ultimate tensile stress
- N = cycles to failure

16.5 Stress Concentrations

Any discussion on fatigue analysis has to include a discussion on the effects stress concentrations have on the life of a component.

Consider a component in Figure 16.18 manufactured with a hole in the middle of the strip and subject to a tensile force P.

At position 1, the stress is uniform across the section and equal to the load divided by the cross-sectional area, i.e.:

$$\sigma_1 = \frac{P}{A_1} \tag{16.7}$$

At position 2, it is clearly seen that the stress is significantly higher at the edge of the hole and then drops off rapidly with the transverse distance from the hole. At a distance of about one hole diameter, the stress is less than P/A_2; this is due to reasons of equilibrium. That is:

$$\int_{A_2} \sigma_2 dA + P \tag{16.8}$$

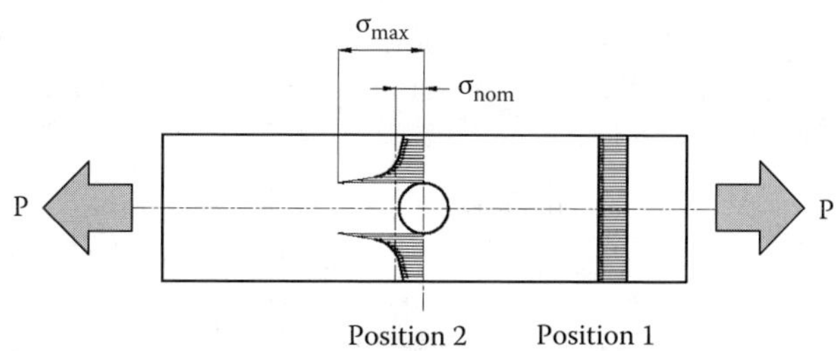

FIGURE 16.18
Effect of a stress concentration.

The integral of the normal stress over the whole area must equal the applied force, since the average normal stress times the total area balances the applied force. Therefore, if the stress is greater than P/A_2 on one portion of the cross, it has to be less than P/A_2 elsewhere.

16.5.1 The Elastic Stress Concentration Factor

The index of the severity of the stress concentration is referred to as the stress concentration factor, and the commonly accepted symbol is K_t for statically loaded parts.

The stress concentration factor is defined as the ratio of the maximum stress to the nominal stress at the section containing the discontinuity, or as in Figure 16.18.

$$K_t = \frac{\sigma_{2max}}{\sigma_{2nom}} = \frac{\sigma_{2max}}{\frac{P}{A_2}} \tag{16.9}$$

By way of an example:

If $P = 20{,}000\,N$ and $A_2 = 120\ mm^2$,

$$\sigma_{2nom} = 167\ MPa$$

In this instance, if the maximum stress was found to be 420 MPa,

$$\sigma_{2max} = 420\ MPa$$

Then

$$K_t = \frac{420\ MPa}{167\ MPa} = 2.51$$

In this instance a safety factor as high as 2, based on the nominal stress at the critical section, would still be insufficient to prevent a failure. This is a case where a safety factor should not be used to allow for the presence of a stress concentration.

16.5.2 The Fatigue Stress Concentration Factor

The presence of any surface discontinuities, either external or internal, as discussed previously, will have a significant effect on the endurance limit of the component.

The static stress concentration factor will not be applicable to the stress concentration factor used for fatigue analysis. The stress concentration factor for fatigue is denoted K_f. This is influenced by the presence of notches from which, due to the high stress concentration, cracks may start to grow, depending upon the severity of the notch. The effect of the notches in fatigue K_f is defined as the unnotched to notched fatigue strength, obtained from fatigue tests (see Figure 16.19).

$$K_f = \frac{\text{fatigue strength of unnotched specimen}}{\text{fatigue strength of notched specimen}}$$

Both material strengths are measured at the same value of N_f.

The difference in magnitude of K_f and K_t for some materials permits the definition of a notch sensitivity factor q where

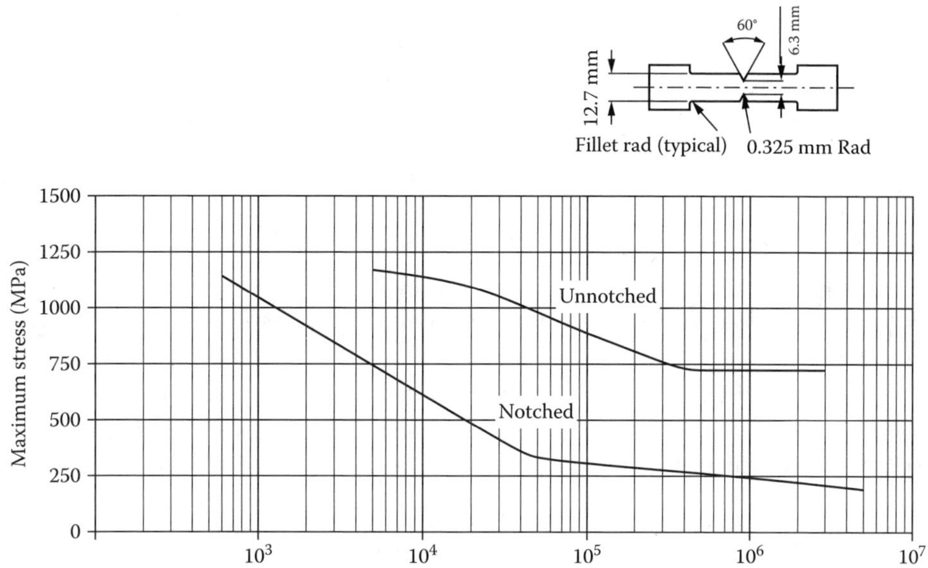

FIGURE 16.19
Experimental S-N curves for a notched and un-notched specimen.

$$q = \frac{K_f - 1}{K_t - 1} \tag{16.10}$$

For fully notch-sensitive materials $K_f = K_t$ and $q = 1$. For totally notch-insensitive materials $K_f = 1$ and $q = 0$. Ductile materials have lower notch sensitivity than stronger, more brittle materials, so that little benefit is usually gained in changing from a low-strength steel in attempting to improve notch fatigue life. Heywood gives excellent and detailed accounts of notch fatigue strength and extensive data on many materials.

16.6 Structural Life Estimations

Estimation of fatigue life is a fairly simple process but one beset with several complexities. Obviously load magnitude and sequence are very important elements of the process. Although a number of techniques have been developed to satisfy specific conditions, the simplest and most widely used, as well as practical, is the Palmgren-Miner hypothesis. As discussed in Section 16.1, Palmgren devised the technique in 1924 relative to calculating the life of ball bearings. In 1945 Miner presented a paper reporting its application to structural elements. The Palmgren-Miner method merely proposes that the fraction of the fatigue life used up in service is the ratio of the applied number of load cycles at a given level divided by the allowable number of load cycles to failure at the same variable stress level. If several levels of variable stresses are applied to a part, then the sum of the respective cycle ratios is the fraction of fatigue life used up.

This damage is usually referred to as the cycle ratio or cumulative damage ratio. If the repeated loads are continued at the same level until failure occurs, the cycle ratio will be equal to unity. When fatigue loading involves a number of stress amplitudes, the total damage is the sum of the different cycle ratios and failure should still occur when the cycle ratio sum equals unity.

where

$$D = \sum_{i=1}^{k} \left(\frac{n_i}{N_i} \right) = 1 \tag{16.11}$$

where

n_i = Number of loading cycles at the ith stress level

N_i = Number of loading cycles to failure for the ith stress level based on a constant amplitude

k = Number of stress levels to be considered in the analysis

Fatigue crack initiation is assumed to occur when the D value is equal to unity. There are three parameters, which affect the magnitude of the summation of the cycle ratios.

1. The order of the load applications. Consider, for example, two different stress levels, f_1 and f_2, and their respective cycle lives, N_1 and N_2. Consider also that f_1 is greater than f_2; if f_1 is applied prior to f_2 the life will be shorter than if f_2 is applied first.
2. Summation of cycle ratios. The second effect on the summation of cycle ratios is due to the amount of damage caused by continuous loading at the same level. The summation of cycle ratios for different stress levels is accurate only if the number of continuous cycles at each stress level is small. For most applications, the loading is random and the stress level is constantly changing. The number of continuous cycles at a particular level should be small and the summation of cycle ratios is considered fairly accurate.
3. Notched and unnotched features. This parameter affects the summation of the cycle ratios and whether or not the fatigued part is notched (such as fastener holes, etc.) or unnotched. The unnotched part generally gives a summation less than unity, while the notched part gives a summation greater than unity. Since most of the structural fatigue failures originate in some form or other from a notch, it indicates that a good average value of 1.5 should be used for the cycle ratio to predict failure of, say, an airframe major structural member, such as a wing box. For simpler structural members a cycle ratio of 1.0 should be used.

16.7 Introduction to Linear Elastic Fracture Mechanics

Notation:

a = Crack half-length for a crack free to extend at both ends

a_c = Critical crack length at failure

a_0 = Initial crack length

2_b = Characteristic dimension of component width

C = Coefficient in simple crack growth rate law defined by $da/dN = C(\Delta K)^n$

E = Modulus of elasticity

G = Release rate of potential energy/unit area
K = Stress intensity factor (mode I)
$\mathbf{K_c}$ = Critical value of stress intensity factor (also plane stress fracture toughness)
$\mathbf{K_{Ic}}$ = Plane strain fracture toughness in mode I
ΔK = Stress intensity range
N = Number of cycles
R = Stress ratio, S_{min}/S_{max}
r = Polar coordinate of stress in component
$\mathbf{S_c}$ = Critical stress at failure
T = Thickness of component in the region of the crack tip
A = Geometric factor in stress intensity factor
$\boldsymbol{\sigma_y}$ = Yield (or 0.2% proof) stress of material
$\boldsymbol{\sigma_r}$ = Stress along axis of polar coordinate *r*
$\boldsymbol{\theta}$ = Angle of polar coordinate r from the *x* axis
$\boldsymbol{\nu}$ = Poisson's ratio

16.7.1 Preamble

The first major study into the fracture phenomenon in cracked bodies was undertaken by A.A. Griffith, who presented the paper "The Phenomena of Rupture and Flow in Solids" in 1921 following his work on glass specimens. His basic premise was that unstable propagation of a crack occurs if any increment of crack growth results in more stored energy being released than being absorbed by the creation of the new crack surface. From further testing he developed a constant failure value relating to the applied stress field and crack length over a range of crack lengths.

During World War II there were a series of catastrophic failures involving mass-produced Liberty cargo ships that fractured into two pieces on or immediately after being launched. There followed further spectacular failures immediately after the war, notably the Comet aircraft in the 1950s, and there was a resurgence of activity in developing the concept of fatigue and failure analysis. G.R. Irwin evolved an alternative interpretation of fracture phenomena known as the stress intensity factor approach in the 1950s. This work then focused attention on the mechanics near the crack tip and is now widely used for solving both the residual strength and fatigue life calculations.

16.7.2 Comparison of Fatigue and Fracture Mechanics

Similarities and dissimilarities between fatigue and fracture mechanics are summarized in Table 16.1. Both fatigue and fracture mechanics rely on results from testing; however, the fracture mechanics concept makes it possible to handle fracture considerations in a quantitative manner and has shown greater applicability to fatigue crack propagation.

16.7.3 The Difference between Classical Fatigue Analysis and Fracture Mechanics

The fundamental difference between fatigue analysis and fracture mechanics is that in fatigue analysis, the analysis is based on the rate of crack growth. No consideration is given to the preexistence of flaws or cracks that would eventually lead to the generation of a starting crack, or the time spent in the stage I phase. The S-N curves indicate the average number of cycles taken to a failure for a given stress value, the specimens either precracked or being manufactured with a standard groove in the test coupon. This groove

TABLE 16.1

Similarities and Dissimilarities between Fatigue and Fracture Mechanics

Fatigue Characteristics or Considerations	Fracture Mechanics Characteristics or Considerations
Considers no initial material flaws, e.g., voids, inclusions, etc.	Assumes pre-existence of flaws, inhomogeneities and discontinuities in a material.
Data presented in the form of a plot of stress versus number of cycles to failure, S.N. curve.	Data presented in the form of stress intensity factor versus cycles to failure or flaw growth rates.
Life prediction utilises cumulative damage theories.	Life prediction is based on minimum flaw growth potential, i.e., the growth of an initial flaw to critical value.
Analysis carried out in two steps: 1. Relating repeated loads to stress. 2. Evaluating stresses using the cumulative damage theory to predict structural life.	Imposes limits on non-destructive inspections and procedures.
Does not consider sustained loading.	Predicts fatigue behaviour such as those stemming from stress corrosion or fatigue.
A purely analytical fatigue design method is not yet available.	Considers sustained loading.
The scatter inherent in fatigue behaviour and in service conditions would require that results be interpreted statistically.	Considers sequence of operational load. Has shown greater applicability to fatigue crack propagation because conditions for fatigue less than critical.
Considers fractures for relatively large numbers of cycles only (10×10^3 and over).	Considers fractures for relative small numbers of cycles ($0 < \text{cycles} < 10 \times 10^3$)

creates a stress concentration that then leads to the starting crack in a small number of cycles. The original fatigue test results for a large number of materials, including aluminum alloys and steels, have been lost; therefore it is not possible to subject these results to any modern statistical analysis.

The underlying premise of fracture mechanics is that all parts contain minute sharp-edge cracks. These cracks grow in a stable manner under either static or cyclic stress, and when a crack has reached a critical length, the growth rate then becomes unstable, leading to fast or brittle fracture on the next load cycle.

Referring to cracks as defects does not imply that the material is defective; it implies that it is inevitable that engineering size cracks and other discontinuities such as voids, cold-shuts, and incomplete weld penetrations will exist.

16.7.4 Stress Intensity

Fracture mechanics gives a numerical description of the elastic stress field near the crack tip, called the stress intensity factor, and characterizes the behavior of the stress field just beyond the small plastic zone at the crack tip in the presence of small-scale yielding.

The stress intensity factor relates the level and mode of the loading, as well as geometry, to the stress state at the crack tip. It has to be emphasized that the factor is not to be considered as a stress concentration factor in the geometric sense of the term. However, having said that, the square root of the crack length used in the stress intensity factor corresponds to the square root of the notch radius used to calculate the stress concentration factors.

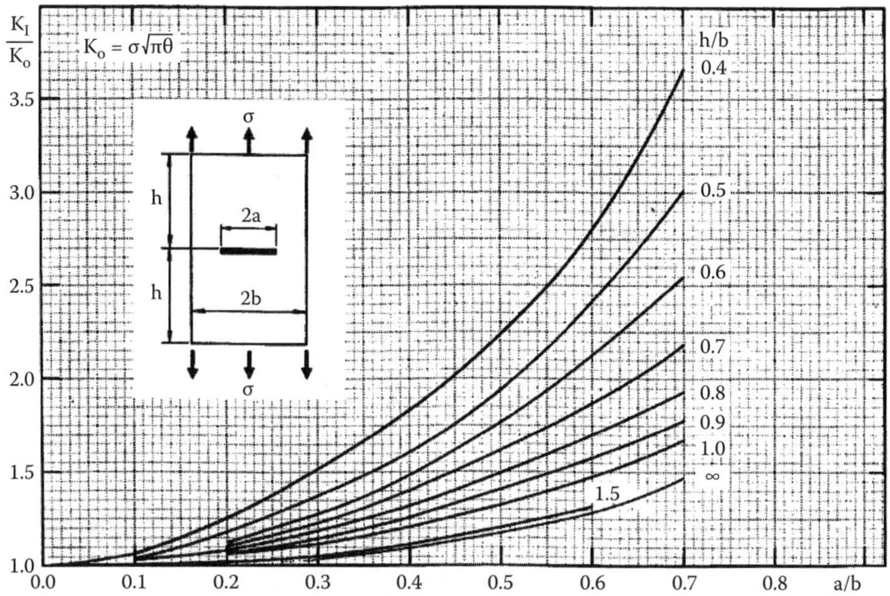

FIGURE 16.20
A typical set of 'K' curves for a rectangular plate with a central crack subject to a uniform uniaxial tensile stress.

There is no single equation that describes the general relationship between the applied stress and stress intensity. The relationship for a specific case is dependent on the mode of loading and geometry (Figure 16.20).

For a constant stress amplitude and through thickness cracks, the stress intensity factor is given by

$$K_1 = \sigma\sqrt{\pi a} \tag{16.12}$$

16.7.4.1 General Stress Intensity Solution

Irwin used classical stress analysis methods to investigate the detailed stress distributions near the crack tip. Based on the complex stress function approach of Wesergaard, Irwin showed that the elastic stress field in the neighborhood of the crack tip (see Figure 16.21) was given by

$$\sigma_x = \frac{K}{\sqrt{2\pi \cdot r}}\cos\left(\frac{\theta}{2}\right)\left[1 - \sin\left(\frac{\theta}{2}\right)\sin\left(\frac{3\theta}{2}\right)\right] \tag{16.13}$$

$$\sigma_y = \frac{K}{\sqrt{2\pi \cdot r}}\cos\left(\frac{\theta}{2}\right)\left[1 + \sin\left(\frac{\theta}{2}\right)\sin\left(\frac{3\theta}{2}\right)\right] \tag{16.14}$$

$$\tau_{xy} = \frac{K}{\sqrt{2\pi r}}\cos\left(\frac{\theta}{2}\right)\sin\left(\frac{\theta}{2}\right)\sin\left(\frac{3\theta}{2}\right) \tag{16.15}$$

It should be noted that these stress distributions are inversely proportional to the square root of the distance from the crack tip. At the crack tip itself (r=0), the stress distributions predict infinite stresses, but this is an idealized situation, known as a stress singularity, resulting

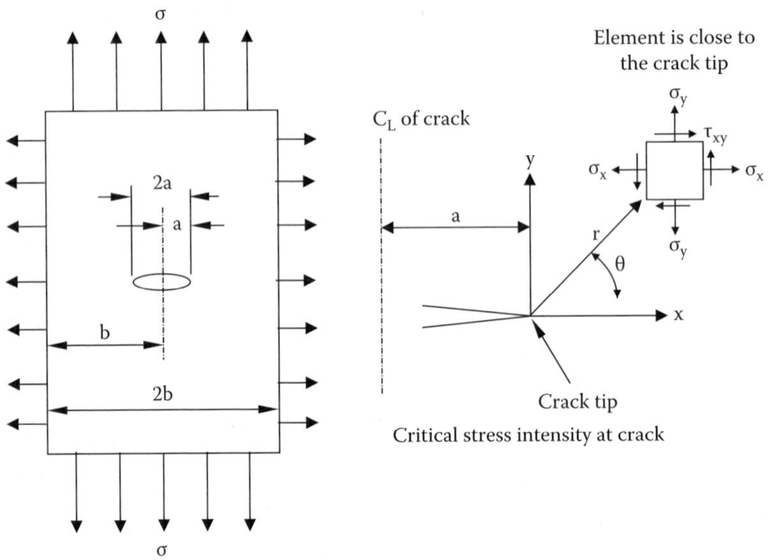

FIGURE 16.21
Mode I crack under a biaxial stress field.

from the assumption of elastic behavior without any limiting criterion. On the plane of the crack (θ=0, y=0), the shear stress is zero and the direct stress components are given by

$$\sigma_x = \sigma_y = \frac{K}{\sqrt{2\pi r}} \tag{16.16}$$

The term $\sigma\sqrt{(\pi r)}$ is dependent only on the applied stress and crack size and defines the gradient of stress with inverse square root of the distance away from the singularity at the crack tip. The term $\sigma\sqrt{(\pi r)}$ was defined by Irwin as the stress intensity factor and given the symbol K. It should be noted that K is not a stress concentration factor and has dimensions with units of stress × √(length). Although the definition of the stress intensity factor as $K=\sigma\sqrt{(\pi r)}$ is the one generally used, in the case of a central crack in an infinite plate subject to remote tension, there are some papers in the literature where an alternative definition is used without the π, namely, $K=\sigma\sqrt{(a)}$. Care must be taken to check which definition is being used in any particular case. In these notes the Irwin definition of $K=\sigma\sqrt{(\pi r)}$ is used throughout.

It is important to recognize that the stress singularity and the stress intensity factor, which dominate the stress field at the crack tip, are features of tension loading. These arise because tension forces cannot be transferred across the free surface of the crack and are redistributed around the edges of the crack in a nonuniform manner. When compression loading is applied to the cracked plate and if the crack surfaces are in contact, forces can be transmitted directly through the crack, so there is no requirement for redistribution, and hence no stress singularity, and the stress intensity factor is zero. This has important consequences when fatigue loading is applied to a cracked component.

16.7.5 Fracture Toughness and Crack Growth

Fracture toughness is the second premise after the stress intensity factor of fracture mechanics, and it characterizes the crack growth of a material containing a defect, either surface breaking or subterrarium, and is a material property found by experiment (usually performed using the Charpy test).

Fracture toughness is the ability of a part containing a crack or defect to sustain a load without catastrophic failure. As long as the stress intensity factor K stays below a critical value of K_{1c} (the parameter representing a critical value for fracture toughness in mode I) (see Figure 16.6), the crack is considered stable. If K reaches or exceeds K_{1c}, the crack will propagate and lead to sudden failure. Propagation rates can reach speeds over 1500 m/s.

Analytical techniques involving the application of fracture toughness data have become indispensable tools for the design of fail-safe structures, particularly those involving high-strength materials (Figure 16.22).

From the work of Paris and Erdogan it has been established that there is a simple relationship between crack growth rate and the range of stress intensity factor during the loading cycle.

$$\frac{da}{dN} = A(\Delta K)^m \tag{16.17}$$

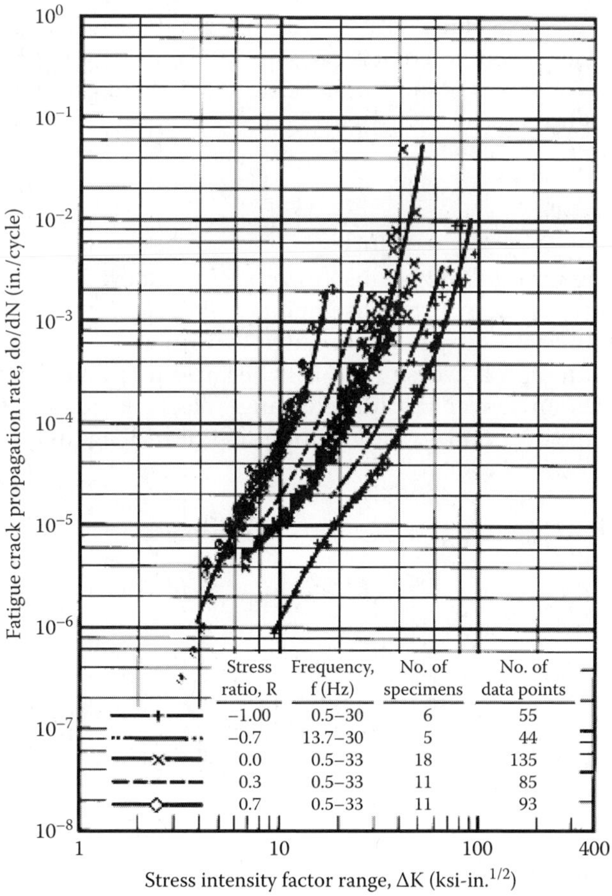

FIGURE 16.22
A typical set of stress intensity curves.

This is known as the Paris-Erdogan law, where A and m are constants and determined experimentally.

If the crack propagation law for the material is known, it is possible to calculate by integration the number of cycles required for the crack to propagate from one length to another. Further, if the fracture toughness value for the material is known, then it is possible with the value of the maximum design stress to calculate the critical value of crack length at which fast fracture will occur—hence, by integration of the Paris law, the total life in cycles of the cracked component.

During fatigue crack growth, $\Delta K = \Delta\sigma a^{1/2} \cdot \alpha$, where α is the compliance factor for the given geometry. The crack growth law gives $da/dN = A(\Delta K)^m = A(\Delta\sigma \cdot \alpha c^{1/2})^m$.

Integrating this expression for the number of cycles required for the crack to grow from an initial length a_0 to the critical length for fast fracture gives:

$$\int_{a_i}^{a_c} a^{(-m/2)}\,da = A\alpha^m \Delta\sigma^m \int_{N=i}^{N=N_f} dN$$

$$= \frac{2}{2-m}\left[a_i^{(1-m/2)} - a_f^{(1-m/2)}\right] = A\alpha^m \Delta\sigma^m N_f$$

Note that this expression fails at $m=2$.

Early literature reported values of m mostly lying between 2 and 4, although values much higher have been encountered. It is becoming clear that for the purpose of the above calculation appropriate values of m lie between 2 and 3, the higher values being found in materials of low toughness. Thus, when $m=3$,

$$N_f = \frac{2}{A^3(\Delta\sigma)^3}\left(\frac{1}{\sqrt{a_i}} - \frac{1}{\sqrt{a_f}}\right) \tag{16.18}$$

Although integrating for total life is a useful procedure, more information is obtained if the calculation is carried out in a stepwise manner. This demonstrates that growth of the crack accelerates with respect to life, as shown in Figure 16.23.

In fact, the real situation is worse than this because it is now known that the Paris-Erdogan law applies only over the middle range of crack growth rates, i.e., between rates of about 10^{-6} and 10^{-4} mm/cycle. A plot of log da/dN against log ΔK shows three

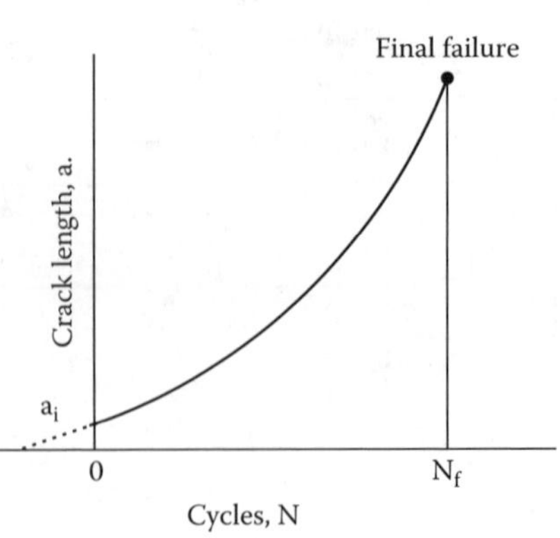

FIGURE 16.23
Fatigue crack as a function of life measured in cycles of stress.

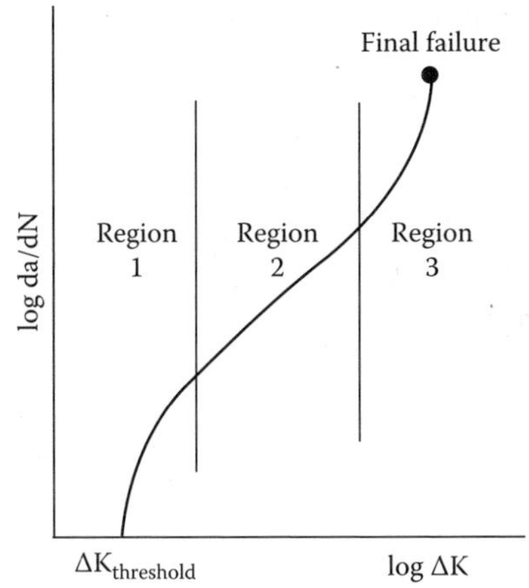

FIGURE 16.24
Fatigue crack growth rate as a function of stress intensity function.

regimes of behavior, and the central regime in which the Paris-Erdogan law applies is preceded and followed by the regimes in which m varies and takes much higher values (Figure 16.24).

Region 1 shows that there exists a value of ΔK below which the crack is nonpropagating; i.e., it merely opens and closes without growing forward. This is called the threshold for fatigue crack growth, ΔK_{th}. The rate of crack growth in the threshold region is much slower than calculation from the Paris-Erdogan law would predict.

If ΔK can be held below the value for ΔK_{TH} for the specific material, then an effective infinite fatigue life can be obtained despite the presence of cracks. This does tend to lead to very low design stresses even for small macrocracks.

In region 3, as K_{max} approaches the limiting fracture toughness of the material, K_{IC} or K_C, the Paris-Erdogan law underestimates the fatigue crack propagation rate. This acceleration of the logarithmic growth rate is associated with the presence of noncontinuum fracture modes, such as cleavage, intergranular, and fibrous fractures (these are activated at high levels of K). It is also found there will be a marked sensitivity to the mean stress.

16.8 Fatigue Design Philosophy

When designing a part or structure that will be subject to fatigue loading, consideration needs to be given to the design philosophy concept that the assembly has been designed against. There are two design philosophy concepts currently being used: fail-safe and safe-life.

16.8.1 Fail-Safe

A structure designed for a fail-safe philosophy will support designated loads with any single member failed or partially damaged. Sufficient stiffness shall remain to prevent divergence, sever vibrations, or other uncontrolled conditions within the normal design envelope.

Summarizing the fail-safe philosophy:

- Structure has the capability to contain fatigue or other types of damage
- Requires knowledge of:
  - Multiplicity of structural members
  - Load transfer capability between members
  - Tear-resistant material properties
  - Slow crack propagation properties
- Inspection controls
- Fatigue is a maintenance problem

16.8.2 Safe-Life

Safe-life components are those whose failure would result in a catastrophic failure with the potential for loss of life. These components must remain crack-free during their service life. The life of the component or structure in cycles is obtained by fatigue analysis or fatigue testing results divided by an appropriate life reduction factor, which will indicate that premature fracture is extremely remote.

The design philosophy will dictate the level of inspection that a component or structure is subjected to and the time when the part is withdrawn from service.

Summarizing the safe-life philosophy:

- Structure resists damage effects of variable load environment
- Requires knowledge of:
  - Environment
  - Fatigue performance
  - Fatigue damage accumulation
- Limit to service life
- Fatigue is a safety problem

16.9 Cycle Counting Methods

16.9.1 Introduction to Spectrum Cycle Counting

The actual service history of an aircraft, automotive, or rail vehicle is variable amplitude spectrum waveforms. To predict the durability of a structure, it is essential to decompose the complex variable cycles of the history data into events that can be processed with constant amplitude fatigue data, which are normally obtained from test or laboratory results. The procedure converts a complex irregular load-time history (Figure 16.25) into a series of simple peak-trough events containing individual cycles, equivalent to constant amplitude cycles. These are characterized by mean and amplitude values (or max/min values), called cycle counting.

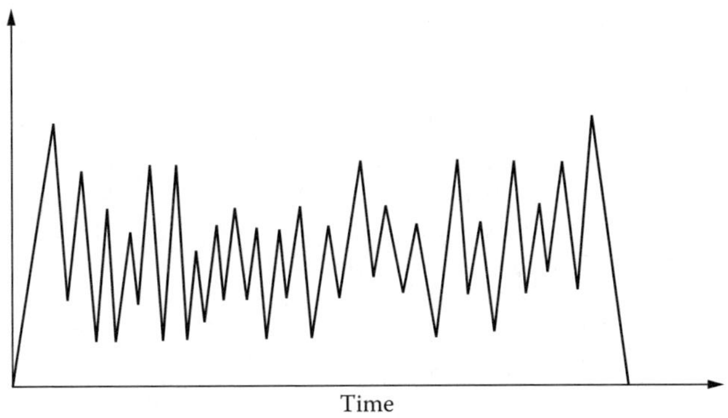

FIGURE 16.25
Schematic stress history of a detail subject to random variable amplitude loading.

Various methods for cycle counting of a load sequence can be found from literature. The most widely used counting methods are range-pair counting and the rain-flow counting. Both of these methods define individual cycles as closed stress-strain hysteresis loops.

There are a number of methods used for cycle counting; these include peak stress, level crossing (with or without reset levels), and range-mean. These methods have been shown to overestimate the fatigue life of a component subject to cumulative fatigue. Range-mean is found to most closely approach the true situation but is strongly influenced by the appropriate gate level or threshold used.

The rain-flow technique approaches closest to the true loading sequence, and in nearly all cases will give the most conservative fatigue life estimate.

It is this method that will be described here in more detail. The rain-flow or Pagoda roof method works in an analogy with raindrops falling and rolling on successive roofs. When one looks at the load history turned through 90°, with the time axis now as the vertical axis (Figure 16.27), it is clearly seen why it is called a Pagoda roof. This method requires one to keep track of all the wet parts of the roof.

Figure 16.26 shows the load history of a component that will be analyzed using the rain-flow method.

The rain-flow rules follow:

1. Rain-flow starts at the beginning of the test and again at the inside of every peak.
2. Rain flows down a pagoda roof and over the edge, where it falls vertically until it reaches a level opposite a maximum more positive (minimum more negative) than the maximum (minimum) from where it started.
3. Rain also stops when it is joined by rain from the pagoda roof above.
4. The beginning of the sequence is a minimum if the initial straining is in tension.
5. The horizontal length of each rain-flow is then counted as a half-cycle at the strain range.

In Figure 16.27 the rain-flow starts at 1, which is considered a minimum. It continues down to 2 and from there to 2′, and then down to 4, and finally stops opposite of 5. This is a (minimum) peak more negative than 1. Thus, a half-cycle from 1 to 4 is extracted.

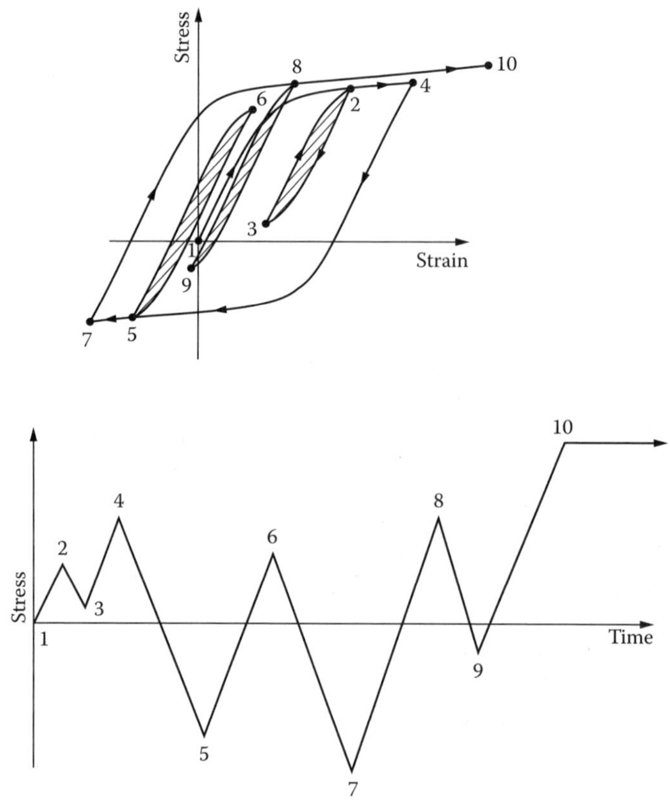

FIGURE 16.26
Stress-strain response for strain history shown in Figure 11.25.

The next rain-flow starts inside peak 2 and proceeds to 3, falling to a point opposite 4, which is more positive than the starting maximum 2. A half-cycle 2-3 is therefore extracted. A third flow starts at 3 but terminates at 2′, due to the rainfall from above 2. The half-cycle 3-2′ is extracted and paired with 2-3, since 3-2′ and 2-3 form a closed stress-strain loop 2′-3-2.

The next rain-flow begins inside peak 4 and proceeds to 5. The range 1-4 is now counted as a half-cycle. The rain-flow continues down to 5′ and then down to 7, where it falls to a point opposite of 10. This is a maximum greater than the maximum from which the rainfall started at 4. A half-cycle 4-5-7 is then extracted.

Rain now begins to flow from inside peak 5 and then continues down to 6, where it falls to a level opposite of 7, this being a greater minimum than 5. A half-cycle 5-6 is then extracted.

The flow from inside of peak 6 ends at 5′, which is where rain from above joins it. The half-cycle 6-5′ is paired with 5-6 to form a complete cycle 5′-6-5, which is extracted.

The next peak to be considered is 7, from where rain proceeds to 8 down to 8′ and then on to the final peak at 10. The range 4-5-7 is now counted as a half-cycle. The half-cycle 7-8-10 is then extracted. Rain starting from inside of 8 continues to 9 and then downwards until it is level with 10, which is more positive than 8. A half-cycle 8-9 is extracted and paired with the final half-cycle 9-8′. A full cycle 8′-9-8 is then counted.

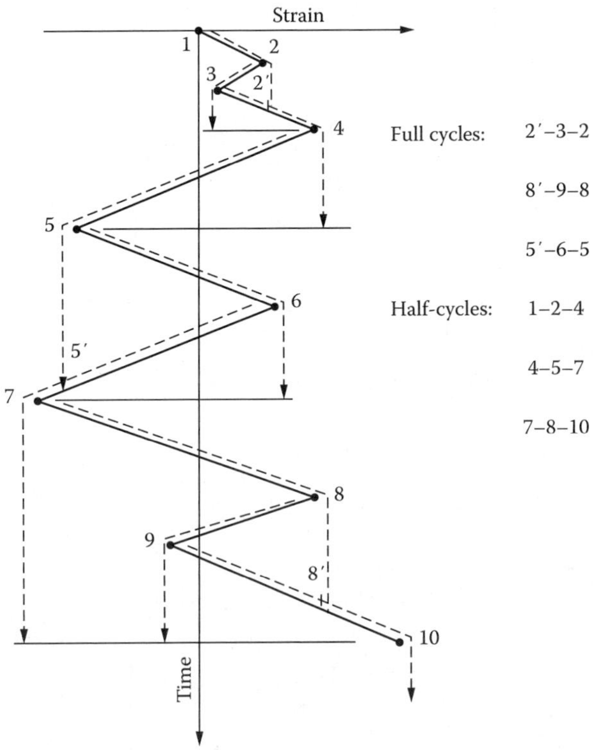

FIGURE 16.27
Rainflow diagram.

Thus the strain-time record shown contains three full cycles, 2′-3-2, 5′-6-5, and 8′-9-8, plus three half-cycles or reversals, 1-2-4, 4-5-7, and 7-8-10. Figure 16.27 shows that this combination of cycles and half-cycles is a true representation of the stress-strain behavior of the material exposed to the strain-time history. The three closed loops are clearly seen, as are the three half-cycles.

Note that each part of the strain-time record is counted once and once only. It is assumed that damage caused by a large event is not affected by its interruption to complete a small stress-strain loop. The damage of the interruption is simply added to that of the larger cycle or half-cycle.

Using this method a varied amplitude history can be reduced to a series of half-cycles whose maximum can be calculated and the cycles or half-cycles completely defined. Fatigue life can be calculated from constant amplitude data using a cumulative damage law.

Section III

Design of Mechanical Elements

17

Introduction to Material Selection

17.1 Introduction

One of the most difficult tasks encountered in engineering design is deciding the most appropriate material to use. With a wide choice of steels to choose from, it may seem daunting to a young engineer when faced with the decision to make the correct choice.

When a design is being modified or updated, the selection of material may have already been decided and so the problem will not arise. There may be a case where the material normally used has become obsolete and the material supplier has ceased to supply it. They may suggest a choice, and it will be up to the designer to select the most appropriate one. The design may have had a minimal modification, say including a change in cross section, and then the original choice can be retained. But where there has been a substantial change in the component section due to a redesign to, for example, reduce weight, then the engineer may have to consider carrying out a series of stress calculations to ensure that the life of the component is not compromised.

These notes are proposed as an aid for the young engineer when studying the important issues in material selection. Table 17.1 summarizes the material characteristics and properties to be considered for new and critical design applications.

The first objective in selecting the correct material is to carry out a simple stress analysis, as outlined in Chapter 14; this will then determine the approximate stress levels the component is being subject to. If the maximum stress is found to be in the order of, for example, 450 MPa, then there is no point in selecting a material such as aluminum, which has an ultimate strength of 450 MPa. In this case, there would be no Factor of Safety (FoS) if the loading were to be exceeded. A better choice would be a medium carbon steel with an ultimate strength of 620 MPa, giving a minimum FoS of 1.3.

17.2 Things to Consider

The following is a basic checklist that the student engineer will need to address as the design progresses. Some or all of the information may not be immediately available in the early phases of the design and must be acquired as the design phase progresses. It is important that the essential information is available before the initial design review.

TABLE 17.1

Summary of Material Characteristics and Properties to Be Considered for Critical Design Applications

1.1	**Static characteristics**
	Strength
	Ultimate strength
	Yield strength
	Shear strength
	Density
	Ductility
	Young's modulus
	Poisson's ratio
	Hardness
	Form
	Sectional dimensions (ruling section)
1.2	**Fatigue characteristics**
	Life required
	Vibration and shock
	Fatigue strength
	Spectrum loads
	Corrosion fatigue
1.3	**Fracture characteristics**
	Fracture toughness
	Flaw growth
	Crack stability
1.4	**Thermal properties**
	Coefficient of linear expansion
	Thermal shock resistance
1.5	**Manufacturability**
	Productability
	Availability
	Machinability
	Weldability
	Heat treatment
	Formability
	Spinning
	Deforming (forging)
1.6	**Corrosion environment**
	Seawater
	Galvanic corrosion
	Stress corrosion

17.2.1 Environment

The nature of the environment to which the component or structure will be exposed must be established at the earliest possible stage. If the atmosphere is in an engineering workshop at ambient temperature, then most corrodible materials will be suitable, but if the environment is in an offshore situation, then the choices will be limited to high-strength corrosion-resisting steels. If the component will be used in the aerospace environment, then obviously the choice will be limited to high-strength lightweight materials, such as aluminum, but as with the offshore industry, care must be taken to select a suitable corrosion protection package.

17.2.2 Strength

There are three principal uses of strength:

- Static strength: The ability to resist short-term steady loads at ambient temperature
- Dynamic strength: The ability to resist a fluctuating load
- Creep strength: The ability to resist a load at elevated temperatures to produce a progressive extension over an extended period of time

17.2.3 Durability

The dictionary definition of *durability* is "the ability to exist a long time without any significant deterioration." In the context of this chapter, this will include resistance to wear and abrasion and corrosion attacks.

17.2.4 Stiffness

Stiffness is the ability of a material to maintain its shape when subjected to a load or force. Consider Hooke's law, where a test material is incrementally loaded to produce an extension that is plotted against the load, and the resultant slope is used to demonstrate the relationship between stress and strain. Within the linear range of the extension, the material will return to its original size when the load or force is removed. If the test material is loaded so that this linear extension is reached, the material is said to have reached its *proportional limit*; any further extension past this limit will then result in the test material not returning to its original size when the applied load or force is removed. This phase is called the *nonproportional limit*.

A material that displays a steep stress–strain curve will be stiffer than one that has a shallow curve, and a component manufactured from the first material will deflect less for a given load or force than one manufactured from the second.

17.2.5 Weight

In the transport industry, including road and aerospace vehicles, it is paramount to keep weight to a minimum. In both cases, fuel is required to propel the vehicle, and with the current high cost of fuel it is essential to maximize the distance traveled per unit of fuel used. Substantial efforts are made in the design of lightweight structures.

The opposite is true in the case of a machine tool, where weight is important to minimize the effects of vibration and structural distortion when a tool is subject to cutting or forming forces.

17.2.6 Manufacturing

In the current atmosphere of high material costs, efforts are being made by manufacturers to minimize material wastage. In the nineteenth and early- to mid-twentieth centuries, it was not uncommon to generate significant amounts of waste from metal-cutting machines; in the current financial climate, efforts are being made to minimize this waste by making more use of forging, casting, and other metal-forming techniques, such as pressure die casting when the material properties allow its use. In the automotive and aerospace industries, significant use is made of fabrications manufactured from formed sections.

17.2.7 Cost

In any manufacturing enterprise, regardless of size, every effort is made to minimize cost. If the product manufactured is destined for the consumer market, then the selling price will be dictated by the buying public, and the manufacturing cost must be recovered from this; the difference will be the profit generated. When the manufacturer is selling through retailers, then the retailers will demand a percentage of the selling price, resulting in less profit for the manufacturer. The manufacturer will have to be cost conscious and scrutinize where cost savings can be made in the product manufacture.

17.2.8 Maintainability

A manufacturer has to make a decision whether to consider product maintenance. In the case of a low-cost product, such as a hair dryer, it may be argued that this will be a throw-away item at the end of the product's life and not to consider any maintenance. On the other hand, in a high capital cost item, such as a road vehicle, maintenance becomes a critical issue and efforts are made to ensure that the product is maintainable throughout its life span with the use of replaceable modules that can be returned either to the manufacturer or companies specializing in repairing and offering the items back for replacement.

17.3 A Model for Material Selection

The author has developed a material selection model similar to the one produced by Stuart Pugh and Bill Hollins for "total design." Figure 17.1 shows the basic model, and it is intended to put the decision making into a chronological order so that the important issues are dealt with in a timely manner and not considered out of order. A brief description of the main headings follows.

17.3.1 Geometry

The shape of the component is identified and all the boundary conditions defined. External and internal forces can then be evaluated.

17.3.2 Analysis

A detailed stress analysis must be carried out next to determine the stresses generated by the applied forces. In the initial stages of the design, this will be to try and establish the

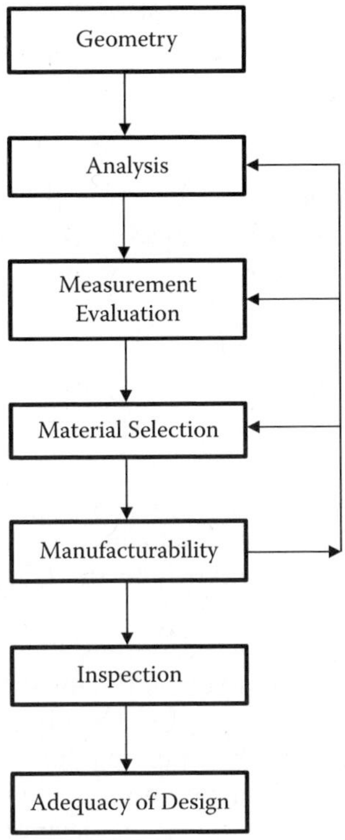

FIGURE 17.1
Model for material selection (core activities).

maximum and minimum stresses to enable a preliminary selection to be made. This activity may be revisited several times as better definition becomes available.

17.3.3 Measurement Evaluation

Where possible, a suitable prototype is constructed and tested to confirm the measured stresses and deflections agree with the calculated values. The model can be either a full-size physical model or a finite element model if there is sufficient confidence in the model's accuracy.

17.3.4 Material Selection

The stage has been reached where a suitable candidate material can now be considered. Each industry will have its own specific requirements; for example, the aerospace and automotive industry will have strength and density constraints as a high priority, whereas the agriculture industry will require cost and resistance to corrosion as their priorities.

17.3.5 Manufacturability

The manufacture of the component will have an important influence on the final selection of the material, depending on whether the component is machined, fabricated, cast, or forged. The production department will usually reach this decision depending on the range of machines at its disposal. The designer may have to revisit the design and make any suitable changes to the geometry and review the stress analysis.

17.3.6 Adequacy of Design

In keeping with the current total design philosophy, it is important that the final design is carefully reviewed to ensure that it will meet the original design specification and that any changes made do not compromise the design integrity.

17.4 A Material Database

There are a number of sources for material data, notably MIL-HDBK 5, which is published by the US government, and British Standards in the UK publishes a range of standards covering a wide range of materials.

The majority of manufacturing companies retain a stock of materials to support their manufacturing activity, and these tend to cover the most common materials used.

If a new material is used that is not in stock or is not a stock item at the local material suppliers, then a financial penalty may have to be paid if the material is unusual and has to be manufactured outside the normal range of sizes, etc. This will apply to rolled or extruded sections where special tooling has to be designed and manufactured. This will be an important consideration in the design of a suitable company-based database where only readily supplied material and sizes are included.

Databases can be held in two different forms:

- Paper based
- Computer based

17.4.1 Paper-Based Database

As outlined above, paper databases are available from government or commercial agencies. These will be comprehensive and cover a wide range of materials not required at the company level. They cover the majority of material properties and are a useful source of data. One disadvantage with this type of database is that they are soon out of date, and therefore need to be reissued periodically to include new materials and delete those materials that have become obsolete.

17.4.2 Computer-Based Database

A computer-based database can be available online from a subscription service; the advantage is that it is regularly updated as new materials are added. Like the paper-based database, it includes materials that the user does not necessarily require.

Young engineers should produce a personal database of material properties their own personal future reference. Figure 17.2 gives details of the template for database input used by the author for metallic materials. The template has space for more data than the reader may require, but gives room for expansion, as not every engineer will require the same

Template for Database Input

			Material specification No.
Specification			
Form		—	
Family		—	
Aircraft specification		—	
General Properties			
Atomic number		—	
Atomic weight		—	
Density	(ρ)	—	Mg/m^3
Basic alloying elements		—	
Price	(P)	—	£/kg
Mechanical Properties			
Ultimate strength	(f_t)	—	Mpa
Yield strength (0.2% proof)	(t_2)	—	Mpa
Yield strength (0.1% proof)	(t_1)	—	Mpa
Compressive stress (0.2% proof)	(c_2)		Mpa
(0.1% proof)	(c_1)		Mpa
Shear strength	(f_{so})	—	Mpa
Ultimate bearing stress	(f_b)		Mpa
Bearing stress	(b_{10})	—	Mpa
Ultimate torsional stress	(F_q)	—	Mpa
Torsional proof stress (0.1% proof)	(Q_1)	—	Mpa
Poisson's ratio	(v)	—	
Elongation	($5.65\sqrt{So}$)	—	% longitudinal
			% transverse
Ductility	(εf)	—	
Toughness	(K_{Ic})	—	MPA $m^{1/2}$
Hardness	(H)	—	HV
Modulus of elasticity	(E)	—	Gpa
Shear modulus	(G)	—	Gpa
Bulk modulus		—	

FIGURE 17.2
Template for database input.

(*Continued*)

Modulus of rupture		—	Mpa	
Fatigue ratio		(R)	—	
Impact resistance	(Izod)	—	J longitudinal	
			J transverse	
Loss coefficient	(η)	—		
Thermal Properties				
Melting point	(Tm)	—	°C	
Glass temperature	(Tg)	—	°C	
Minimum service temperature	(Tmin)	—	°C	
Maximum service temperature	(Tmax)	—	°C	
Specific heat	(Cp)	—	kJ/kg°C	
Thermal expansion		(α)	—	10^{-6}/°C (20–200°C)
Thermal shock reserve		—		
Latent heat of fusion		—		
Thermal conductivity	(λ)	—	W/m°Cat20°C	

Weldability	
Electron beam	Resistance to corrosion
Arc	Resistance to stress corrosion cracking
Resistance (overlap joint)	Resistance to exfoliation corrosion
Flame	Magnetic properties
Brazing	
Minimum welding properties	

Typical Uses

Remarks

Main References

FIGURE 17.2 (CONTINUED)
Template for database input.

type of information. The template can be modified to suit other purposes. The data generated can then be input into a suitable database program such as Microsoft Access or Excel, where a search facility exists, allowing for suitable candidate materials to be searched and extracted.

17.4.3 Material Classification and Coding

A suitable classification and coding scheme must be developed to enable the data to be stored within the database. The author uses the following description in the construction

TABLE 17.2

Performance Indices for Minimum Weight

Component Shape and Loading	Stress Limited	Stiffness Limited
Rods in tension	σ_{tu}/ρ	E/ρ
Short columns in compression	σ_{tu}/ρ	E/ρ
Thin wall pipes and pressure vessels under internal pressure	σ_{tu}/ρ	—
Helical springs for a specified length and load capacity	σ_s/ρ	—
Thin wall shafts in torsion	σ_s/ρ	G/ρ
Rods and pins in shear	σ_s/ρ	—
Beams with a fixed section shape, in bending	$\sigma_{tu}^{2/3}/\rho$	$E^{1/2}/\rho$
Solid shafts in torsion	$\sigma_s^{2/3}/\rho$	$G^{1/2}/\rho$
Solid shafts in bending	$\sigma_{tu}^{2/3}/\rho$	$E^{1/2}/\rho$
Rectangular beams with fixed width, in bending	$\sigma_{tu}^{1/2}/\rho$	$E^{1/3}/\rho$
Flat plates under pressure	$\sigma_{tu}^{1/2}/\rho$	$E^{1/3}/\rho$

Notation

Density	ρ
Modulus of elasticity	E
Shear modulus	G
Ultimate tensile strength	σ_{tu}
Ultimate shear stress	σ_s

TABLE 17.3

Performance Indices for Elastic Design

Component and Design Requirement	Maximize
Springs Specifies energy storage, volume to be minimized	σ_{tu}^2/E
Springs Specifies energy storage, mass to be minimized	$\sigma_{tu}^2/E\rho$
Compression seals and gaskets Maximum contact area with specified maximum contact pressure	σ_{tu}/E
Diaphragms Maximum deflection under specified pressure or force	$\sigma_{tu}^{3/2}/E$
Ties, columns Maximum longitudinal vibration frequencies	E/ρ
Beams Maximum longitudinal vibration frequencies	$E^{1/2}/\rho$
Plates Maximum flexural vibration frequencies	$E^{1/3}/\rho$
Ties, columns, beams, plates Maximum self-damping	η

Notation

Density	ρ
Modulus of elasticity	E
Ultimate tensile strength	σ_{tu}
Loss coefficient	η

of his own database. Tables 17.2 and 17.3 suggest a possible classification and coding scheme for use in the current exercise. Table 17.2 lists the possible material groups. Table 17.3 lists the subgroup numbers for the primary forms.

Table 17.4 shows a section of the material group table with the respective subgroup numbers against the material types. The respective database tables correlate the group and subgroup numbers together with their respective properties. For example, heat-resisting alloy forging would come under the Group 5000 and the Subgroup 170 for forging; therefore, the material code number will be 5170. MS is prefixed to the code number defining it as a material specification. The database would then be constructed using the listings outlined in Table 17.5. As this is based on using a relational database, the respective database tables are then related as per the table. Tables 17.6 through 17.11 are used.

Each table is to be populated with the respective details of the materials included in the database. As new materials are added, the database will become more comprehensive and useful.

When all the tables have been completed, it is then a simple matter to create a database report to identify candidate materials that meet the required specified criteria.

Table 17.12 is an example of a partly completed table that is covering material indices. Table 17.13 is a partly completed table of material properties that is used in the solution of the following two examples, which show how the database is used to identify candidate materials that meet a specified criterion.

Example 17.1

Consider a cantilever with a single static load applied at its free end.

From a previous calculation, it was established that the factored stress in the section was 400 MPa due to the load and an appropriate Safety Factor (SF). There is a further requirement that the component be as light as possible to minimize weight.

The designer puts together a simple list of criteria for the material to initially meet, from which a limited range of candidate materials was identified.

The list includes

1. Density (from database in Table 17.1): To be less than 2.8 kg/M^3.
2. Ultimate stress (from database in Table 17.2): To be greater than 400 MPa.
3. Material performance index: From Table 17.2 stress limited design for rectangular beam in bending. $\sigma_{tu}^{1/2} / \rho$ to be greater than 15.

The results of a search based on the above criteria are shown in Table 17.14.

From the above search, five candidate materials are identified as meeting the material requirements.

Maximizing the material performance index (σ_{tu}/ρ) defines MS 13201 as being the better performer. MS 13201 (L168) is supplied in the bar and extruded section, making it the ideal choice for manufacture.

Example 17.2

Consider an internally pressurized cylinder.

The requirement is for a thin-walled cylinder of 150 mm internal diameter with a minimum wall thickness and an internal pressure of 13.8 MPa (2000 lbf/in.^{2}). The material should exhibit a good ductility so that the component will not fail due to brittle fracture. Weight is not a problem.

The search criteria will be

TABLE 17.4

Part Proposed Classification System for Metallic Materials

Group	Material	Subgroup	Form
100000	**Aluminum alloys**		
	11000	**Low strength**	
		111000	Wire
		112000	Bar
		113000	Sheet and strip
		114000	Plate
		115000	Tube
		116000	Casting
		117000	Forging
	12000	**Medium strength**	
		121000	Wire
		122000	Bar
		123000	Sheet and strip
		124000	Plate
		125000	Tube
		126000	Casting
		127000	Forging
	13000	**High strength**	
		131000	Wire
		132000	Bar
		133000	Sheet and strip
		134000	Plate
		135000	Tube
		136000	Casting
		137000	Forging
200000	**Magnesium alloys**		
	21000	**Low strength**	

TABLE 17.5

Relationship between Database Tables

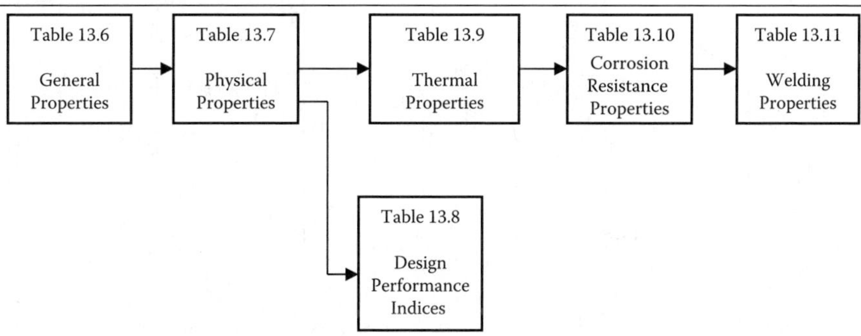

TABLE 17.6

General Properties

1	Atomic number
2	Atomic weight
3	Density
4	Basic alloying elements
5	Magnetic properties
6	Price per unit volume

TABLE 17.7

Physical Properties

1	Ultimate strength	(f_{tu})
2	Yield strength	(t_2)
3	Yield strength	(t_1)
4	Compressive strength	(c_2)
5	Compressive strength	(c_1)
6	Shear strength	(f_{so})
7	Ultimate bearing strength	(f_b)
8	Bearing stress	(b_{10})
9	Ultimate torsion stress	(F_q)
10	Torsional proof stress (0.1%)	(Q_1)
11	Poisson's ratio	(ν)
12	Elongation	($5.65\sqrt{So}$)
13	Ductility	(εF)
14	Toughness	(K_{1c})
15	Hardness	(H)
16	Modulus of elasticity	(E)
17	Shear modulus	(G)
18	Fatigue ratio	(R)
19	Impact resistance	(Charpy)
20	Impact resistance	(Izod)

TABLE 17.8

Design Performance Indices

1	E/ρ
2	$E^{1/2}/\rho$
3	$E^{1/3}/\rho$
4	$G^{1/2}/\rho$
5	σ/ρ
6	$\sigma^{2/3}/\rho$
7	$\sigma^{1/2}/\rho$
8	σ/E
9	σ^2/E
10	$\sigma^2/E\rho$
11	$\sigma^3/E2$
12	$\sigma^{2/3}/E$

TABLE 17.9

Thermal Properties

1	Melting point	(T_m)
2	Glass temperature	(T_g)
3	Minimum service temperature	(T_{min})
4	Maximum service temperature	(T_{max})
5	Specific heat	(C_p)
6	Coefficient of thermal expansion ($\times 10^{-6}$)	(α)
7	Thermal shock reserve	
8	Latent heat of fusion	
9	Thermal conductivity	(λ)

TABLE 17.10

Corrosion Resistance Properties

1	Resistance to corrosion
2	Resistance to stress corrosion cracking
3	Resistance to exfoliation cracking

TABLE 17.11

Welding Properties

1	Flame
2	Arc
3	Resistance
4	Brazing
5	Electron beam

1. Ultimate stress (from database in Table 17.7): To be greater than 500 MPa.
2. Elongation (from database in Table 17.7): To be greater than 25%.
3. Material performance index (from database in Table 17.8): From Table 17.2 stress limited design for thin wall. Pipes and pressure vessels under internal pressure. σ/ρ to be greater than 65.0.

The result of the search based on the above criteria is shown in Table 17.15.

MS 33002 is BS HR203, plate, sheet, and strip. It would obviously be unsuitable to manufacture a pressure vessel from this. The remaining material, MS 52002, is BS S129, a corrosion-resisting steel, and this is now obsolete. MS 52003 is BS S130, corrosion-resisting steel, and both are supplied in the form of bar and forgings.

Thus, MS 52003 would be the logical choice for this application. The elongation is better than specified, and the performance index is also very good. It will be easy to calculate the required minimum wall thickness using a factored ultimate stress in the thin wall stress calculation pd/2t.

TABLE 17.12
Part Table of Material Indices

	E	G	ρ	σ	σ/ρ	$\sigma^{1/2}/\rho$	$\sigma^{2/3}/\rho$	E/ρ	$E^{1/3}/\rho$	$E^{1/2}/\rho$	σ/E	σ^2/E	σ^3/E^2	$\sigma^{3/2}/E$	$\sigma^2/E.\rho$	$G^{1/2}/\rho$
Material	$\times10^9$	$\times10^9$	$\times10^0$	$\times10^6$	$\times10^6$	$\times10^6$	$\times10^6$	$\times10^9$	$\times10^9$	$\times10^9$	$\times10^{-3}$	$\times10^{-3}$	$\times10^{-3}$	$\times10^{15}$	$\times10^{-3}$	$\times10^0$
MS 11501	72	27	2.69	160	59.480	4.702	10.956	26.766	1.547	3.154	2.222	355.556	790.123	28.109	132.177	1.932
MS 11601	72	27	2.69	230	85.502	5.638	13.955	26.766	1.547	3.154	3.194	734.722	2347.029	48.446	273.131	1.932
MS 11602	71	27	2.71	215	79.336	5.411	13.243	26.199	1.528	3.109	3.028	651.056	1971.509	44.402	240.242	1.917
MS 12201	70	27	2.7	310	114.815	6.521	16.965	25.926	1.526	3.099	4.429	1372.857	6079.796	77.973	508.466	1.925
MS 12301	70	27	2.7	295	109.259	6.361	16.413	25.926	1.526	3.099	4.214	1243.214	5239.260	72.383	460.450	1.925
MS 12401	70	27	2.7	295	109.259	6.361	16.413	25.926	1.526	3.099	4.214	1243.214	5239.260	72.383	460.450	1.925
MS 12501	70	27	2.7	310	114.815	6.521	16.965	25.926	1.526	3.099	4.429	1372.857	6079.796	77.973	508.466	1.925
MS 12701	70	27	2.7	295	109.259	6.361	16.413	25.926	1.526	3.099	4.214	1243.214	5239.260	72.383	460.450	1.925
MS 13201	72	27	2.8	490	175.000	7.906	22.198	25.714	1.486	3.030	6.806	3334.722	22,694.637	150.647	1190.972	1.856
MS 13301	72	27	2.7	440	162.963	7.769	21.426	26.667	1.541	3.143	6.111	2688.889	16,432.099	128.188	995.885	1.925
MS 13401	72	27	2.8	450	160.714	7.576	20.973	25.714	1.486	3.030	6.250	2812.500	17,578.125	132.583	1004.464	1.856
MS 13501	72	27	2.8	400	142.857	7.143	19.389	25.714	1.486	3.030	5.556	2222.222	12,345.679	111.111	793.651	1.856
MS 13701	72	27	2.8	450	160.714	7.576	20.973	25.714	1.486	3.030	6.250	2812.500	17,578.125	132.583	1004.464	1.856
MS 32001	183	85	8.19	1000	122.100	3.861	12.210	22.344	0.693	1.652	5.464	5464.481	29,860.551	172.802	667.214	1.126
MS 32002	193	75	8.18	1080	132.029	4.018	12.869	23.594	0.706	1.698	5.596	6043.523	33,818.680	183.899	738.817	1.059
MS 33001	214		8.18	1080	132.029	4.018	12.869	26.161	0.731	1.788	5.047	5450.467	27,507.031	165.852	666.316	
MS 33002	210	80	8.37	650	77.658	3.046	8.965	25.090	0.710	1.731	3.095	2011.905	6227.324	78.913	240.371	1.069
MS 52001	193	77	7.72	880	113.990	3.843	11.895	25.000	0.749	1.800	4.560	4012.435	18,295.041	135.259	519.745	1.137
MS 52002	193	76	7.9	540	68.354	2.942	8.394	24.430	0.732	1.759	2.798	1510.881	4227.335	65.018	191.251	1.104
MS 52003	193	76	7.92	540	68.182	2.934	8.373	24.369	0.730	1.754	2.798	1510.881	4227.335	65.018	190.768	1.101
MS 52004	211	80	7.83	930	118.774	3.895	12.168	26.948	0.760	1.855	4.408	4099.052	18,066.912	134.413	523.506	1.142
MS 52005	211		7.82	1130	144.501	4.299	13.873	26.982	0.761	1.858	5.355	6051.659	32,409.357	180.026	773.869	
MS 52006	201	77	7.82	1270	162.404	4.557	14.997	25.703	0.749	1.813	6.318	8024.378	50,701.295	225.169	1026.135	1.122
MS 53001	170		7.9	772	97.722	3.517	10.652	21.519	0.701	1.650	4.541	3505.788	15,920.403	126.176	443.771	
MS 53002	170		7.92	772	97.475	3.508	10.626	21.465	0.699	1.646	4.541	3505.788	15,920.403	126.176	442.650	
MS 53003	193		7.92	540	68.182	2.934	8.373	24.369	0.730	1.754	2.798	1510.881	4227.335	65.018	190.768	
MS 54001	193	76	7.92	550	69.444	2.961	8.476	24.369	0.730	1.754	2.850	1567.358	4466.563	66.832	197.899	1.101

(*Continued*)

TABLE 17.12 (*Continued*)

Part Table of Material Indices

	E	G	ρ	σ	σ/ρ	$\sigma^{1/2}/\rho$	$\sigma^{2/3}/\rho$	E/ρ	$E^{1/3}/\rho$	$E^{1/2}/\rho$	σ/E	σ^2/E	σ^3/E^2	$\sigma^{3/2}/E$	$\sigma^2/E.\rho$	$G^{1/2}/\rho$
Material	$\times10^9$	$\times10^9$	$\times10^0$	$\times10^6$	$\times10^6$	$\times10^6$	$\times10^6$	$\times10^9$	$\times10^9$	$\times10^9$	$\times10^{-3}$	$\times10^{-3}$	$\times10^{-3}$	$\times10^{15}$	$\times10^{-3}$	$\times10^0$
MS 56001	193	78	7.7	950	123.377	4.003	12.550	25.065	0.751	1.804	4.922	4676.166	23,017.396	151.715	607.294	1.147
MS 56002	193	78	7.7	1200	155.844	4.499	14.665	25.065	0.751	1.804	6.218	7461.140	46,390.507	215.385	968.979	1.147
MS 56003	200	77	7.92	460	58.081	2.708	7.524	25.253	0.738	1.786	2.300	1058.000	2433.400	49.330	133.586	1.108
MS 62001	176	69	7.85	1320	168.153	4.628	15.329	22.420	0.714	1.690	7.500	9900.000	74,250.000	272.489	1261.146	1.058
MS 62002	204	80	7.84	930	118.622	3.890	12.153	26.020	0.751	1.822	4.559	4239.706	19,328.071	139.025	540.779	1.141
MS 62003	200	77	7.86	1080	137.405	4.181	13.392	25.445	0.744	1.799	5.400	5832.000	31,492.800	177.462	741.985	1.116
MS 62004	204	81	7.86	880	111.959	3.774	11.683	25.954	0.749	1.817	4.314	3796.078	16,375.240	127.966	482.962	1.145
MS 72001	110	42	4.42	900	203.620	6.787	21.090	24.887	1.084	2.373	8.182	7363.636	60,247.934	245.455	1665.981	1.466
MS 73001	106	45	4.51	390	86.475	4.379	11.836	23.503	1.049	2.283	3.679	1434.906	5279.370	72.659	318.161	1.487
MS 77001	110	42	4.42	900	203.620	6.787	21.090	24.887	1.084	2.373	8.182	7363.636	60,247.934	245.455	1665.981	1.466

TABLE 17.13

Part Table of Material Properties

Material	Form	Family	Ultimate Strength f_t MPa	Yield Strength 0.2% proof MPa	Yield Strength 0.1% proof MPa	Compressive Strength 0.2% proof MPa	Compressive Strength 0.1% proof MPa	Shear Strength MPa	Poisson's Ratio	Density Mg/m^3	Elongation % Longitude
MS 11501	Tube	Aluminum	160–200	60					0.33	2.69	18
MS 11601	Casting	Aluminum	230	185	175			180	0.33	2.68	2
MS 11602	Casting	Aluminum	215	190	170			172	0.33	2.71	1
MS 12201	Bar	Aluminum	310	270	263			172	0.33	2.7	8
MS 12301	Sheet/strip	Aluminum	295	240	232				0.33	2.7	8
MS 12401	Plate	Aluminum	295	240	231			171	0.33	2.7	8
MS 12501	Tube	Aluminum	310	240	229				0.33	2.7	9
MS 12701	Forging	Aluminum	295	255	248			177	0.33	2.7	8
MS 13201	Bar	Aluminum	490	440	432	391	403	260	0.33	2.8	7
MS 13301	Sheet/strip	Aluminum	440	380	367				0.33	2.8	8
MS 13401	Sheet/strip	Aluminum	460	410	402	421	435	270	0.33	2.8	8
MS 13501	Tube	Aluminum	400	290	275				0.33	2.8	10
MS 13701	Forging	Aluminum	450	395	375			270	0.33	2.8	6
MS 32001	Bar	Heat-resisting steel	1000	600				800	0.29	8.19	20
MS 32002	Bar	Heat-resisting steel	1080	695					0.29	8.18	20
MS 33001	Sheet/strip	Heat-resisting steel	1080	695					0.29	8.18	20
MS 33002	Plate	Heat-resisting steel	620	230					0.29	8.37	30
MS 52001	Bar	Corrosion-resisting steel	880–1080	650	649			810	0.3	7.72	12
MS 52002	Bar	Corrosion-resisting steel	540	210	200			417	0.29	7.9	35
MS 52003	Bar	Corrosion-resisting steel	540	210	200			417	0.29	7.92	35
MS 52004	Bar	Corrosion-resisting steel	930–1080	780	741			614	0.3	7.83	15
MS 52005	Bar	Corrosion-resisting steel	1130–1330	1030				746	0.3	7.82	12
MS 52006	Bar	Corrosion-resisting steel	1270–1470	1030				838	0.3	7.82	10
MS 53001	Sheet/strip	Corrosion-resisting steel	772	636	550	419	361		0.3	7.9	13

(*Continued*)

TABLE 17.13 (*Continued*)

Part Table of Material Properties

Material	Form	Family	Ultimate Strength f_t MPa	Yield Strength 0.2% proof MPa	Yield Strength 0.1% proof MPa	Compressive Strength 0.2% proof MPa	Compressive Strength 0.1% proof MPa	Shear Strength MPa	Poisson's Ratio	Density Mg/m^3	Elongation % Longitude
MS 53002	Sheet/strip	Corrosion-resisting steel	772	636	550	419	361		0.3	7.92	13
MS 53003	Sheet/strip	Corrosion-resisting steel	540	210	200				0.29	7.92	35
MS 54001	Tube	Corrosion-resisting steel	550–700	210–340					0.29	7.92	
MS 56001	Casting	Corrosion-resisting steel	950–1200	800					0.29	7.7	12
MS 56002	Casting	Corrosion-resisting steel	1200–1500	960					0.3	7.7	8
MS 56003	Casting	Corrosion-resisting steel	460	200					0.3	7.92	
MS 62001	Bar	Noncorrosion-resisting steel	1320–1520	1030				803	0.3	7.85	8
MS 62002	Bar	Noncorrosion-resisting steel	930–1080	740	695			614	0.3	7.84	13
MS 62003	Bar	Noncorrosion-resisting steel	1080–1280	880				713	0.3	7.86	10
MS 62004	Bar	Noncorrosion-resisting steel	880–1080	690				581	0.3	7.86	12
MS 72001	Bar	Titanium alloy	900–1160	830	806		869	495	0.31	4.42	8
MS 73001	Sheet/strip	Titanium alloy	390–540	290	274				0.31	4.51	22
MS 77001	Forging	Titanium alloy	900–1160	830	806			495	0.31	4.42	8

TABLE 17.14

Candidate Materials to Satisfy the Search Criteria for Cantilever: Example 17.1

Material Spec	Density kg/m^3	Ultimate Strength MPa	$\sigma_{tu}^{1/2}/\rho$ $\times 10^6$
MS13201	2.80	490	7.9057
MS13301	2.80	440	7.7689
MS13401	2.80	450	7.5761
MS13501	2.80	400	7.1429
MS13701	2.80	450	7.5761

TABLE 17.15

Candidate Materials for the Pressure Vessel: Example 17.2

Material Spec	Ultimate Strength MPa	Elongation %	σ/ρ $\times 10^6$
MS 33002	650	30	77.6583
MS 52002	540	35	68.3544
MS 52003	540	35	68.1818

17.5 Future Developments

In the immediate future any future development in material selection will be concentrating on a knowledge-based system specific to an engineering project.

17.5.1 Knowledge-Based Engineering

In the early 1990s, it became obvious that a parametric language could be developed for use in a CAD (computer-aided design) system where the operator would simply fill in a table and the system would then generate the appropriate model. Amendments to the design could then be implemented by making changes to the variables within the table. This approach will be able to save many man-hours where the design is modular.

A number of researchers have developed knowledge-based systems that can be integrated into the design process, allowing domain-specific knowledge to be stored regarding a part or process, together with other associated attributes.

Further work has been extended to develop the knowledge base using a material database together with a suitable solid modeling system that uses a rule-based technique, for example, using an "if-then" approach that is implemented to perform the material selection process. A material that satisfies all the constraints then becomes the most suitable candidate for a particular component operating within a set of specific conditions.

18

Beams

18.1 Introduction

A beam is considered to be a horizontal or near-horizontal structural member having a constant cross section loaded in the transverse direction to its length. (This is not to be confused with a strut which a vertical member that is loaded in compression normal to its neural axis and considered in Chapter 23.)

Beams can be classified into four basic groups:

1. Cantilever (built in at one end only)
2. Propped cantilever (built in at one end and simply supported at the free end)
3. Horizontal beam simply supported at each end
4. Horizontal beam in encastrè (built in at both ends)

Table 18.1 shows a number of standard loading cases for horizontal beams.

Table 18.2 also shows some common sectional properties.

The following is the minimum requirements for a beam:

- The beam is straight.
- The cross-sectional area is constant along the length of the beam (L).
- The beam has at least one longitudinal plane of symmetry.
- The beam is long in proportion to its depth (d).
  The ratio L/d is >8 for metals with compact sections.
  The ratio L/d is >15 for beams having thin webs.
- All the loads and reactions lie in the longitudinal plane of symmetry.
- The beam is not disproportionally wide.
- The maximum stress in the beam does not exceed the proportional limit (0.2%).

TABLE 18.1

Standard Loading Cases for Horizontal Beams

Case No.	Description	Graphic	Moment M Max	Deflection δ Max
1	Cantilever with concentrated load at the free end	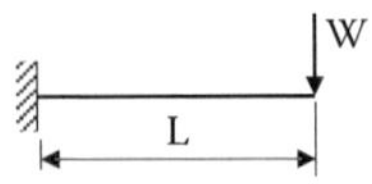	WL	$\frac{WL^3}{3EI}$
2	Cantilever with an intermediate load	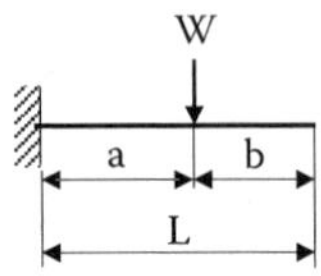	Wa	$\frac{Wa^3}{3EI}\left(1+\frac{3b}{2a}\right)$
3	Cantilever with uniformly distributed load	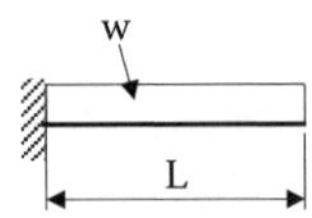	$\frac{wL^2}{2}$	$\frac{wL^4}{8EI}$
4	Propped cantilever with a central load	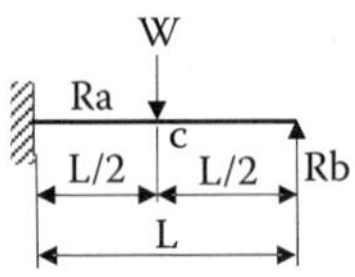	$Ma = \frac{Wb\left(L^2 - b^2\right)}{2.L^2}$ $Mc = \frac{Wb}{2} 2 - \left(2 - \frac{3b}{L} + \frac{b^3}{L^3}\right)$	$\frac{Wa^3b^2}{12EIL^3}(4L - a)$
5	Popped cantilever with an intermediate load.	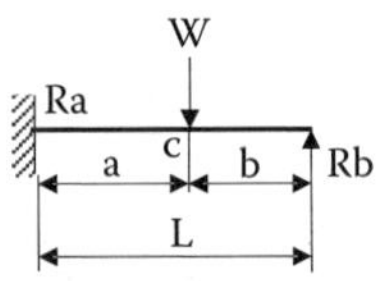	$Ma = \frac{3W^2}{16}$ $Mc = \frac{5WL}{32}$	$0.000932\frac{WL3}{3EI}$
6	Propped cantilever with a uniformly distributed load	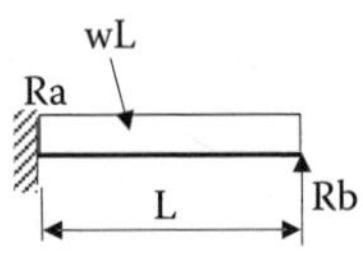	$\frac{WL}{8}$	$\frac{WL^3}{185EI}$
7	Beam supported at two points with single load at center of span	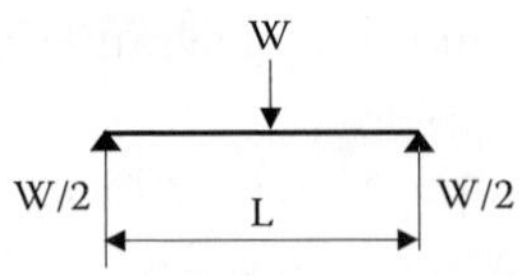	$\frac{WL}{4}$	$\frac{WL^3}{48EI}$
8	Beam supported at two points with an intermediate load	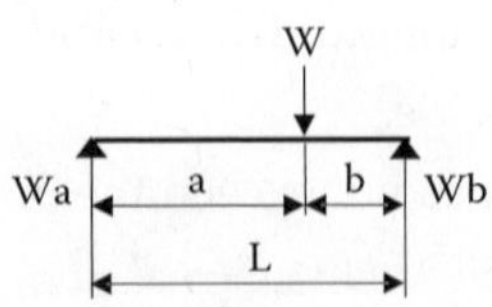	$\frac{Wab}{L}$	$\frac{WL^3}{48EI}$ $\left[\frac{3a}{L} - 4\left(\frac{a}{L}\right)^3\right]$
9	Beam supported at two points with uniformly distributed load along its length	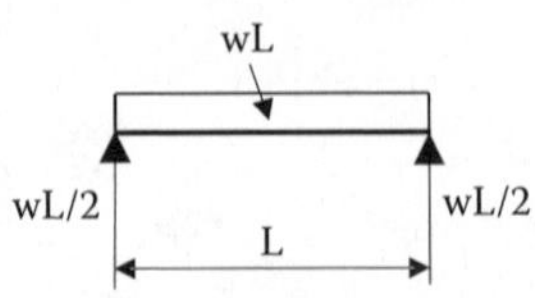	$\frac{wL^2}{8}$	$\frac{5wL^4}{384EI}$

(Continued)

TABLE 18.1 (*Continued*)

Standard Loading Cases for Horizontal Beams

Case No.	Description	Graphic	Moment M Max	Deflection δ Max
10	Beam in encastrè with a central load	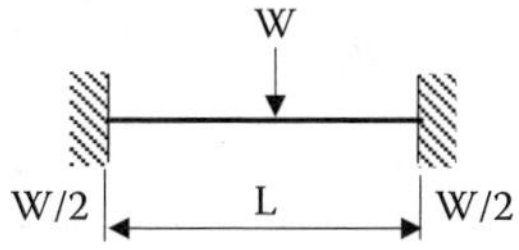	$\frac{WL}{8}$	$\frac{WL^3}{192EI}$
11	Beam in encastrè with an intermediate load	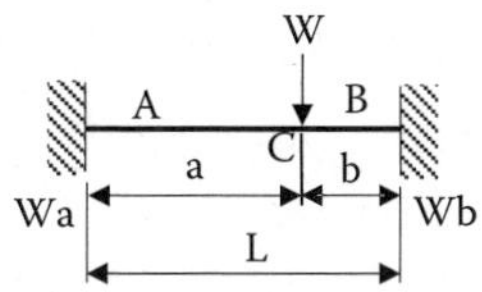	$M_A = -\frac{Wab^2}{L^2}$ $M_B = -\frac{Wba^2}{L^2}$ $M_C = \frac{2Wa^2b^2}{L^3}$	$\frac{2Wa^2b^2}{3EI(3L-2a)^2}$
12	Beam in encastrè with a uniformly distributed load	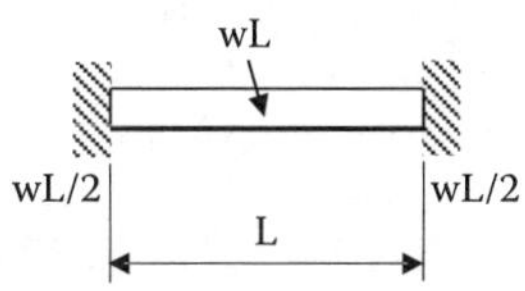	$\frac{wL^2}{12}$	$\frac{wL^4}{384EI}$

W = total load (N)
w = uniformly distributed load (N per m)
L = effective span (m)

TABLE 18.2

Some Common Sectional Properties

Case No.	Graphic	A	A_s	I_{xx}	Z_{xx}
1	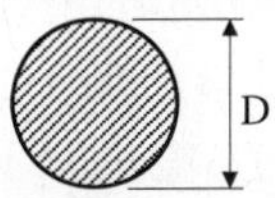	$\frac{\pi}{4}D^2$	$\frac{3}{4}A$	$\frac{\pi}{64}D^4$	$\frac{\pi}{32}D^3$
2	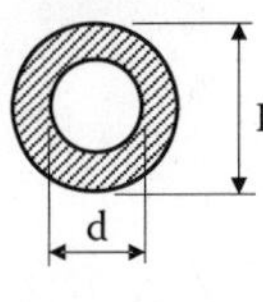	$\frac{\pi}{4}(D^2 - d^2)$	$\frac{D^2}{6}\left(B - b\left(t - \frac{2T}{D}\right)^3\right)$	$\frac{\pi}{64}(D^4 - d^4)$	$\frac{\pi}{32D}(D^4 - d^4)$
3	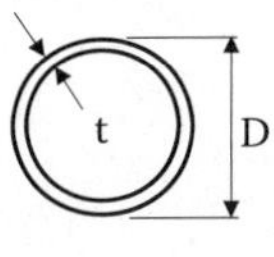	πDt	$\frac{A}{2}$	$\frac{\pi}{8}D^3t$	$\frac{D^2 \cdot t}{8}$
4	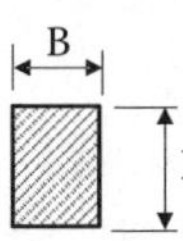	$B \cdot D$	$\frac{2}{3}A$	$\frac{1}{12}B \cdot D^3$	$\frac{1}{6}B \cdot D^2$

(*Continued*)

TABLE 18.2 (*Continued*)

Some Common Sectional Properties

Case No.	Graphic	A	A_s	I_{xx}	Z_{xx}
5	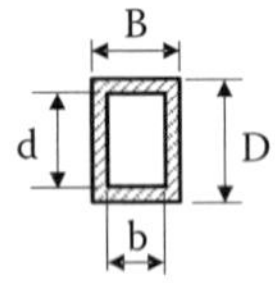	$B \cdot D - b \cdot d$	$\frac{2}{3}\frac{B \cdot D^3 - b \cdot d^3}{D^2}$	$\frac{1}{12}\left(B \cdot D^3 - b \cdot d^3\right)$	$\frac{1}{6D}\left(B \cdot D^3 - b \cdot d^3\right)$
6	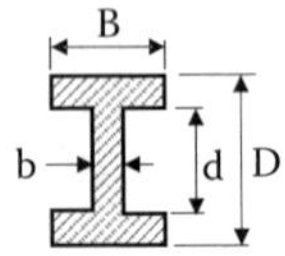	$B \cdot D - b \cdot d$	$d(B - b)$	$\frac{1}{12}\left(B \cdot D^3 - b \cdot d^3\right)$	$\frac{1}{6D}\left(B \cdot D^3 - b \cdot d^3\right)$
7	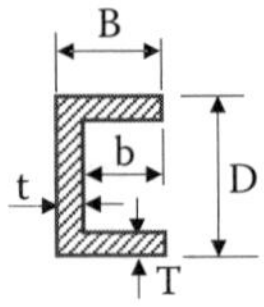	$D \cdot t + 2b \cdot T$	$t(D - 2T)$	$\frac{D^3}{12}\left(B - b\left(t - \frac{2T}{D}\right)^3\right)$	$\frac{D^2}{6}\left(B - b\left(t - \frac{2T}{D}\right)^3\right)$

A = cross sectional area
As = effective cross sectional area for the calculation of shear stress across the section
I_{xx} = second moment of area about x:x
$Z_{xx} = I_{xx}/y$ modulus of section about x:x

18.2 Basic Theory

Considering the stresses induced in the beam by the bending moment acting on the beam shown in Figure 18.1. Let the radius of curvature of the neutral axis at a particular section of the beam be R.

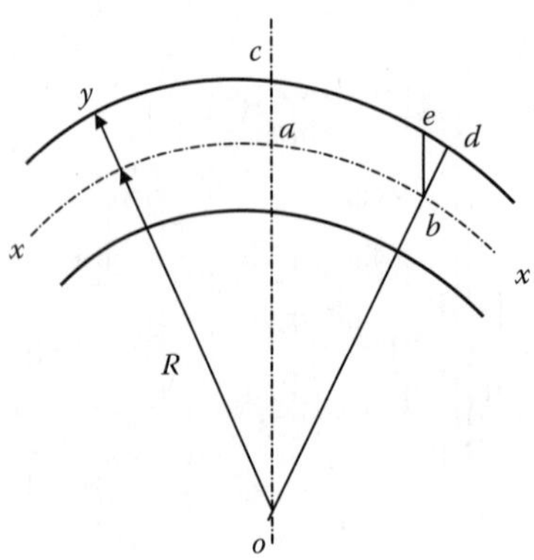

FIGURE 18.1
Stresses induced in a beam due to bending.

The layer originally of length ce has extended to cd:

$$\therefore \text{strain} = \frac{ed}{cd} = \frac{ed}{ab} = \frac{f_b}{R} \tag{18.1}$$

Using similar triangles,

$$\frac{ed}{bd} = \frac{ab}{ob} = \frac{y}{R} \tag{18.2}$$

From Equations 18.1 and 18.2,

$$\frac{f_b}{y} = \frac{E}{R} \tag{18.3}$$

Combining this with Equation 18.1 gives the general bending formula:

$$\frac{M}{I} = \frac{f_b}{y} = \frac{E}{R} \tag{18.4}$$

where:

M = the bending moment at the section being considered (Nm)
I = the moment of inertia (or second moment of area) of the section (m^4)
y = the distance from the neutral axis to the point at which the stress is required
R = the radius of curvature (m)
f_b = the bending stress (may be either tension or compression) (N/m^2)
E = the modulus of elasticity (N/m^2)

$$\text{Also } M = \sigma \cdot Z \tag{18.5}$$

where Z = I/y and is known as the *section modulus*.

Various values of I for a number of sections can be found in a number of steel manufactures or steel stocklist catalogues, there may also include other tables.

In a number of cases, it might be required to calculate the value for I, and this may require the section to be split up into various calculable parts using the *parallel axis theorem*, as discussed in Section 18.3.

The bending moment may be determined from the standard beam diagrams as shown in Table 18.1, where a list of several of these cases is shown alongside formulas for deflection and shear stresses.

Generally, there is no interest in knowing the bending stresses within the beam. Usually, the bending stresses at the outer fibers are needed, as this is the maximum value of stress. In unsymmetrical sections, the distance y must be taken in the correct direction across that portion of the section which is either in tension or compression. Ordinarily, only the maximum stress at the outer fibers which is at the greater distance y from the neutral axis is required.

18.3 Parallel Axis Theorem

The second moment of area of a section about an axis x:x parallel to an axis n:a passing through the centroid of the section is given by (see Figure 18.2)

$$I_{xx} = I_{na} + Ah^2 \tag{18.6}$$

where:

I_{xx} = the second moment about x:x axis (m^4)
I_{na} = the second moment of section about the center of gravity (CG) (m^4)
A = the area of section (m^2)
h = the perpendicular distance between x:x axis and CG (m)

Example 18.1

A mild steel shaft is loaded as shown in Figure 18.3a. Construct a bending moment diagram and determine the maximum diameter of the shaft required in the central portion of the shaft if the ultimate tensile strength (σ_t) of the material is limited to 460 MPa.

Consider if a safety factor (SF) of 5 is appropriate.

Solution

Taking moments about R_B.

Bending moments at positions 2, 3, and 4, respectively,

$$M_2 = (R_A \times 0.23\ \text{m})$$

$$M_3 = (R_A \times 0.88\ \text{m}) - (27\ \text{kN} \times 0.65\ \text{m})$$

$$M_4 = (R_A \times 1.16\ \text{m}) - (27\ \text{kN} \times 0.93\ \text{m}) - (35\ \text{kN} \times 0.28\ \text{m})$$

Substituting the value for R_A results in the following moments:

$$M_2 = (30.09 \times 10^3 \times 0.23) = 6.922\ \text{kNm}$$

$$M_3 = (30.095 \times 10^3 \times 0.88) - (27 \times 10^3 \times 0.65) = 8.934\ \text{Nm}$$

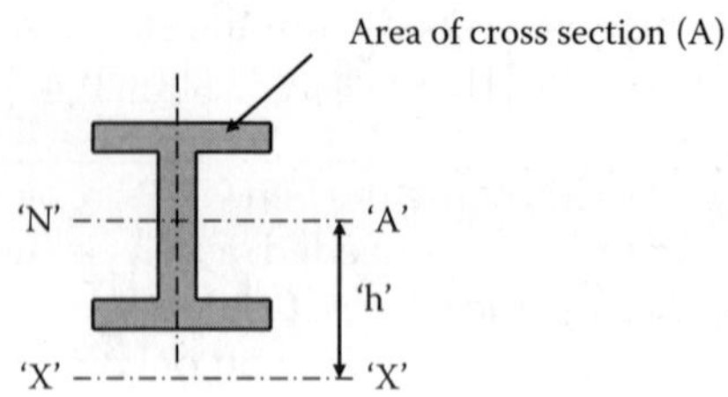

FIGURE 18.2
Parallel axis theorem.

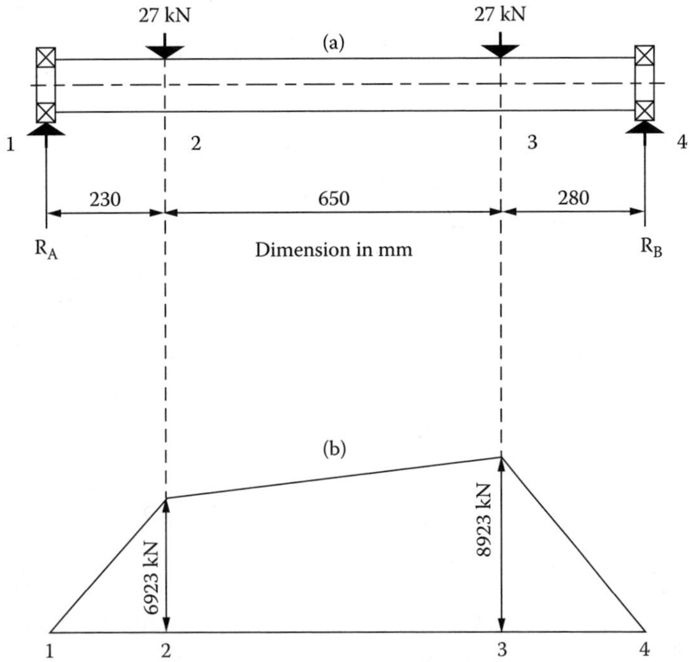

FIGURE 18.3
(a) Loaded beam. (b) Bending moment diagram.

$$M_4 = (30.095 \times 10^3 \times 1.16) - (27 \times 10^3 \times 0.93) - (35 \times 10^3 \times 0.28) = 0.00 \text{ kN m (check)}$$

The completed bending moment diagram is as shown in Figure 18.3b.

From the bending formula,

$$\frac{M}{I} = \frac{\sigma}{y} \quad \text{(from Equation 18.4)}$$

Rearranging the formula to solve for σ,

$$\sigma = \frac{M \cdot y}{I}$$

where the maximum bending moment M=8.933 kN·m

The second moment of area:

$$I = \frac{\pi \times d^4}{64} \quad \text{(Table 18.1, Case 1)}$$

$$y = \frac{d}{2}$$

Hence,

$$\sigma_t = 181.982 \times 10^3 \times d^3$$

Solving for d,

$$d = \sqrt[3]{\frac{181.982 \times 10^3 \ \text{N} \cdot \text{m}}{460 \ \text{MPa}}}$$

$$d = 73.41 \ \text{mm}$$

Now, with an SF of 5, the ultimate tensile stress is factored by the SF:

$$\sigma_{\text{allowable}} = \frac{460 \ \text{MPa}}{5 \ \text{(SF)}}$$

Therefore,

$$\sigma_{\text{allowable}} = 92 \ \text{MPa}$$

Hence, recalculating the revised diameter,

$$d = 125.5 \ \text{mm (Ans)}$$

Note:

The question was asked to consider whether an SF of 5 is appropriate. It is assumed that the loading on the shaft is steady, without any sudden impacts. In this instance, if the shaft rotates, then the stresses on either side of the shaft will alter from tensile to compressive, giving rise to a fluctuating stress that could eventually lead to a fatigue failure of the shaft. As a rule of thumb, providing that for *steel* components the maximum working stress is kept below the *fatigue limit*, the component should not suffer any long-term fatigue problems, and hence, this value would normally be below 40% of the ultimate tensile stress. Therefore, an SF of 5 in this instance is considered appropriate.

If the material of the shaft had been aluminum, then for this material there is no limiting stress, hence the number of cycles must be carefully considered. The SF will then have to be calculated based on the fatigue strength required.

Example 18.2

The cast-iron beam shown in Figure 18.4 carries a distributed load of 50 N/mm over its simply supported span of 3000 mm. Calculate the maximum and minimum stresses in the top and bottom flanges, together with the maximum deflection of the beam.

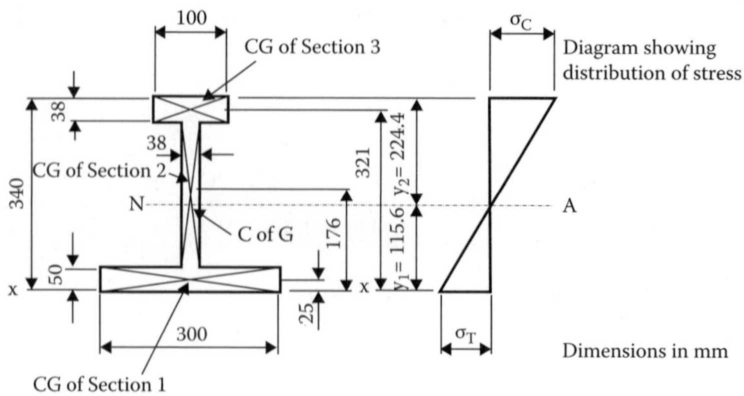

FIGURE 18.4
Example 18.2 I-beam section.

Solution

The first part of the answer is to establish the position of the neutral axis by taking moments about the edge x:x.

To answer this question, the section is broken down into smaller sections, as shown in Figure 18.4. The sectional properties of the individual sections are then calculated and summed, as in Table 18.2. From this, the position of the neutral axis of the individual sections can then be calculated.

The first part of the answer is to break the overall section into simple shapes sections, as depicted in Figure 18.4. A boundary edge is selected and used as a datum for the subsequent calculations. A table can be constructed as shown in Table 18.3; this will be used to identify the position of the neutral axis of the combined section.

The construction of the table should be achieved using hand calculations, although a spread sheet program such as Microsoft Excel® might be more convenient, particularly where there is a large number of sections. In this section, as it is symmetrical, the center of gravity will lie on the vertical axis; therefore, only the vertical position of the neutral axis will be considered. The elements of the table are considered self-explanatory.

When the vertical position of the neutral axis has been established (y_1) from the datum edge, the calculation for the moment of inertia (I) can now be completed. Table 18.4 shows the method in which the individual sections are calculated. In Table 18.5, columns 7, 8, and 9 are copied from Table 18.4 columns 2, 3, and 4. Column 9 contains the individual inertias for each section using the formulas from Table 18.3.

TABLE 18.3

Position of Neutral Axis for Example 18.2 Section

1	2	3	4	5	6
Section	**b**	**d**	**Area**	**y**	**Ay**
	mm	**mm**	**mm^2**	**mm**	**mm^3**
1	300	50	15,000	25	0.375×10^6
2	38	252	9,576	176	1.685×10^6
3	100	38	3,800	321	1.220×10^6
		ΣA=28,376		$\Sigma Ax=3.280\times10^6$	

$$y_1 = \frac{\Sigma Ax}{\Sigma A} = 115.60 \text{ mm}$$

$$y_2 = 340 - 115.6 = 224.4 \text{ mm}$$

TABLE 18.4

Combined Inertia Calculations for Example 18.2

1	2	3	4	5	6	7
Section	**b**	**d**	**$bd^3/12$**	**Area**	**h**	**$I_{xx}+(Ah^2)$**
	mm	**mm**	**mm^4**	**mm^2**	**mm**	**mm^4**
1	300	50	3.125×10^6	15,000	90.6	126.250×10^6
2	38	252	50.676×10^6	9,576	60.4	85.611×10^6
3	100	38	0.4573×10^6	3,800	205.4	160.78×10^6
						$I_{xx}=372.64\times10^6$

TABLE 18.5

Physical Properties for Example 18.3 Problem

1	2	3	4	5	6	7	8	9
Section	**b**	**d**	**A**	**x**	**Ax**	**I**	**h**	**I+(A×h²)**
	mm	**mm**	**mm²**	**mm**	**mm³**	**mm⁴**	**mm**	**mm⁴**
1			2654	228.0	605,112	1,340,000	102.945	29,466,224.2
2	12.00	250.00	3000	125.0	375,000	15,625,000	0.055	15,625,009.1
3	12.00	250.00	3000	125.0	375,000	1,562,5000	0.055	15,625,009.1
4			1310	20.9	27,379	572,000	104.155	14,783,225.9
5			1310	20.9	27,379	572,000	104.155	14,783,225.9
		Area=112,74 mm²			1,409,870 mm³			I_{NA}=90,282,694.1 mm⁴
Ybar=125.055 mm								

Column 8 (h) is the vertical distance of the c of g for each section to the vertical position of the neutral axis for the whole section, i.e., the modulus of (y_1−column 5 (y)) disregarding the negative sign when this occurs.

The parallel axis theorem is now invoked to complete the (column 9), that is $I_{na}+(Ah^2)$. The tabulations are then summed to give the I_{na} for the complete section.

The maximum tensile strength in the bottom flange:

$$\sigma_{tensile} = \frac{M \cdot y_1}{I_{xx}}$$

$$\sigma_{tensile} = \frac{56.25\ \text{MN} \cdot \text{mm} \times 115.6\ \text{mm}}{372.64 \times 10^6\ \text{mm}^4}$$

$$\sigma_{tensile} = 17.45\ \text{MPa}$$

The maximum compressive stress in the top flange:

$$\sigma_{compressive} = \frac{M \cdot y_2}{I_{xx}}$$

$$\sigma_{compressive} = \frac{56.25\ \text{MN} \cdot \text{mm} \times 224.4\ \text{mm}}{372.64 \times 10^6\ \text{mm}^4}$$

$$\sigma_{compressive} = 33.873\ \text{MPa}$$

The maximum deflection in the beam will be as follows.

From Table 18.1, Case 9, the maximum deflection can be calculated from the formula:

$$\delta_{max} = \frac{5wL^4}{384EI}$$

$$= \frac{5 \times 50\ \text{N/mm} \times (3000\ \text{mm})^4}{384 \times 208 \times 10^9\ \text{Pa} \times 372.64 \times 10^6\ \text{mm}^4}$$

$$\delta_{max} = 0.680\ \text{mm}$$

This deflection will occur at midspan of the beam, that is, 1500 mm from either end.

Example 18.3

A box girder 250 mm deep is fabricated as shown in Figure 18.5. The beam is uniformly loaded over a simple span of 6000 mm. The ultimate tensile stress of the material is 430 MPa. Use an SF of 5 to find the value of the uniformly distributed load the beam will carry. Also calculate the maximum deflection in the section.

Solution

Extracting data from structural steel manufacturer's catalogue.

178×76 Rolled steel channel:

$$\text{Area} = 26.54\ \text{cm}^2\ (2654.0\ \text{mm}^2)$$

$$\text{The second moment of area} = 134\ \text{cm}^4\ (1.34 \times 10^6\ \text{mm}^4)$$

70×70×10 Rolled steel equal angle:

$$\text{Area} = 13.1\ \text{cm}^2\ (1310.0\ \text{mm}^2)$$

$$\text{The second moment of area} = 57.2\ \text{cm}^4\ (0.5720 \times 10^6\ \text{mm}^4)$$

Side plates:

$$\text{Area} = 29.868\ \text{cm}^2\ (2986.8\ \text{mm}^2)$$

$$\text{The second moment of area} = 1562.5\ \text{cm}^4\ (15.625 \times 10^6\ \text{mm}^4)$$

All structural steel catalogues and steel designer manuals quote the physical properties of the sections in cm rather than mm. It will be found to be more convenient to convert to mm when constructing a table to calculate the positions of the neutral axes and second moment of areas.

First, identify the individual elements as shown in Figure 18.6, so that Table 18.5 can now be constructed to calculate the physical properties of the total section.

To find the maximum load (w) from the bending formula:

$$M = \frac{wL^2}{8} = \frac{I\sigma}{y}$$

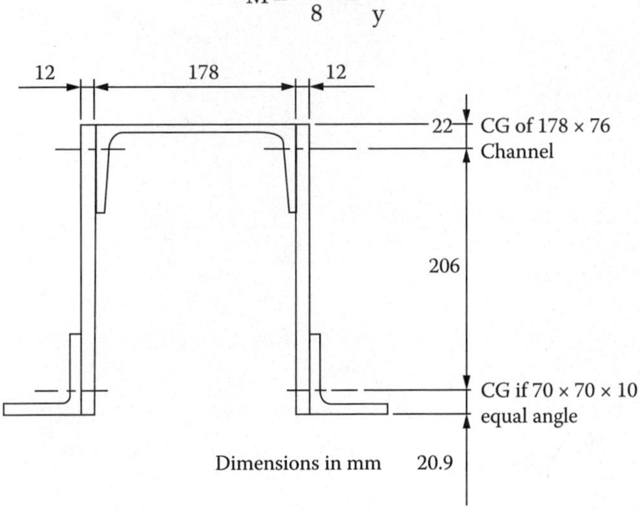

FIGURE 18.5
Example 18.3 beam section.

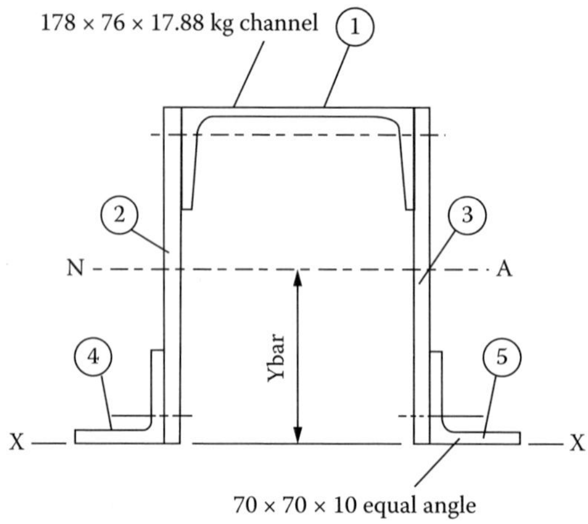

FIGURE 18.6
Identification of beam sections.

Rearranging to solve for w,

$$w = \frac{8 \cdot I \cdot \sigma}{L^2 \cdot y \cdot SF}$$

Inputting the following values,

$$I = 90.283 \times 10^6 \text{ mm}^4$$

$$\sigma = 430 \text{ MPa}$$

$$L = 6000 \text{ mm}$$

$$y = 125.055 \text{ mm}$$

$$SF = 5$$

gives

$$w = 13.797 \text{ N/m}$$

To calculate the maximum deflection in the section:
From Table 18.1, Case 9,

$$\delta_{max} = \frac{5wL^4}{384EI}$$

$$= \frac{5 \times 13.797 \text{ N/m} \times (6000 \text{ mm})^4}{384 \times 208 \times 10^9 \times \text{Pa} \times 90.283 \times 10^{-6} \text{ mm}^4}$$

$$\delta_{max} = 0.012 \text{ mm}$$

The completed physical properties for Example 18.3 are shown in Table 18.5.

Example 18.4

Consider the same uniformly loaded beam section as discussed in Example 18.3, but this time the beam is considered in an encastrè condition (Table 18.1, Case 12).

The equation for the calculation of the maximum bending deflection is

$$\delta_{max} = \frac{wL^4}{384EI} \quad (18.7)$$

Inputting the original values from Example 18.3 gives

$$\delta_{max} = 2.48 \times 10^{-3} \text{ mm}$$

From this, it can be clearly seen that the maximum deflection of the beam in encastrè is significantly less than the beam had it been simply supported.

Example 18.5

Shearing Force and Bending Moments (Figure 18.7)

The *shearing force* at a section of a beam is the algebraic sum of all the forces to one side of the section.

The *bending moment* at a section of a beam is the algebraic sum of the moments of all the forces to one side of the section.

At the point P in the cantilever shown in the Figure 18.7a, the shearing force is W_1+W_2, which is tending to shear the beam as shown in Figure 18.7b. This is opposed by the shearing resistance of the beam to the left of P.

The bending moment at the point P is $W_1x_1+W_2x_2$, which is tending to bend the beam as shown in Figure 18.7c. This is opposed by the bending resistance of the beam to the left of P.

In the case of the simply supported beam shown in Figure 18.8, the shearing force at P is either R_1-W_1 or R_2-R_2. These terms are equal since they equate upward and downward forces on the beam.

$$R_1 + R_2 = W_1 + W_2$$

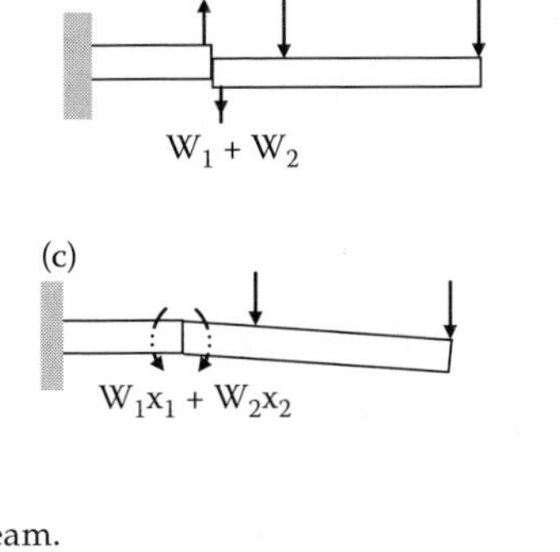

FIGURE 18.7
Shearing forces acting on a beam.

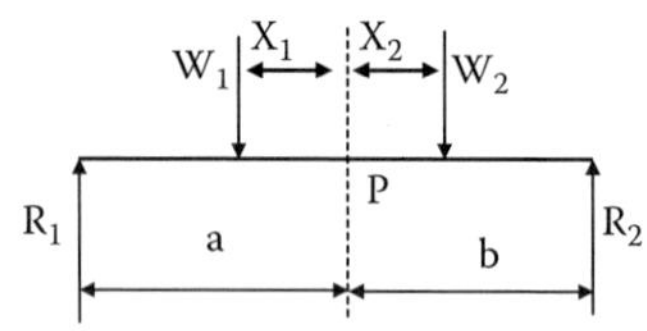

FIGURE 18.8
Simply supported beam.

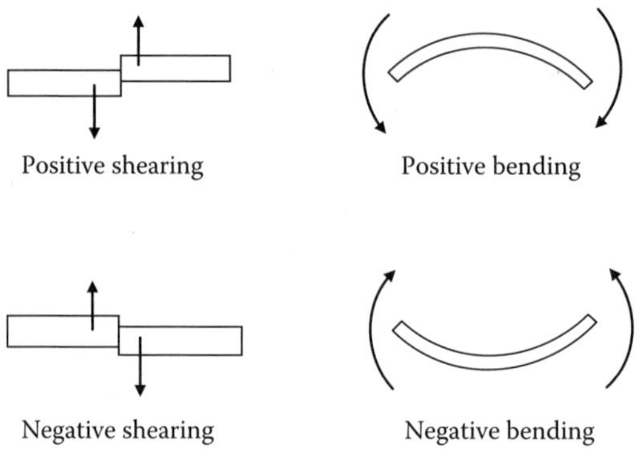

FIGURE 18.9
Conventions used in shearing and bending.

The bending moment at P is either $R_1a - W_1x_1$ or $R_2b - W_2x_2$. These terms are equal since, equating clockwise and counterclockwise moments about P.

$$R_1a + W_2x_2 = R_2b_2 + W_1x_1$$

It is immaterial which side of the section is chosen for the calculation of the shearing force or bending moment. When calculating the bending moment due to the distributed load, the part of the load to one side of the section may be considered as a concentrated load acting at the center of gravity of that part.

If the bending moment at a section of a beam changes sign, that point is called the *point of inflexion* or *contraflexure*.

If the resultant force to the right of a section is upward (or to the left is downward), then this is regarded as positive shearing force and the opposite kind of shearing will be regarded as negative, see Figure 18.9.

If the resultant bending moment to the right of the section is clockwise (or to the left is counterclockwise), then this will be regarded as a positive bending moment, and the opposite kind will be regarded as negative.

Thus, a positive bending moment causes the beam to bend upward (convex) and the negative bending moment bends it downward (concave).

18.4 Shearing Force and Bending Moment Diagrams

These are diagrams that show the results of the shearing force (SF) and bending moment (BM) along the entire length of the beam.

The four most common cases will now be considered.

18.4.1 Cantilever with a Concentrated End Load (Figure 18.10)

$$\text{SF at P} = -W$$

$$\text{BM at P} = Wx$$

$$\text{Maximum SF} = -W$$

$$\text{Maximum BM} = Wl$$

18.4.2 Cantilever with a Uniformly Distributed Load, w per Unit Length (Figure 18.11)

$$\text{SF at P} = -wx$$

$$\text{BM at P} = \frac{wx^2}{2}$$

$$\text{Maximum SF} = -wl$$

$$\text{Maximum BM} = \frac{wl^2}{2}$$

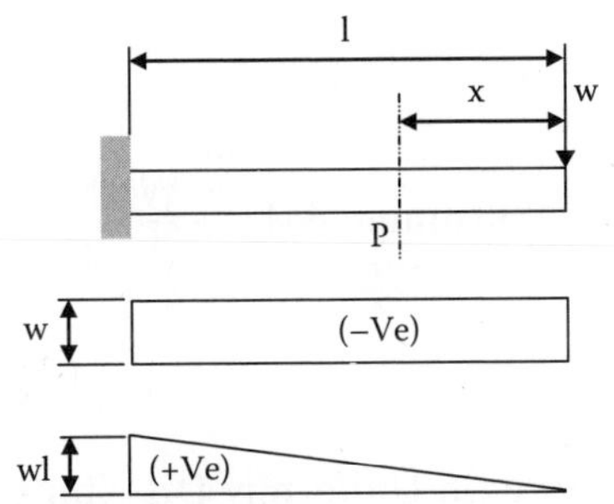

FIGURE 18.10
Cantilever with a concentrated load at the end.

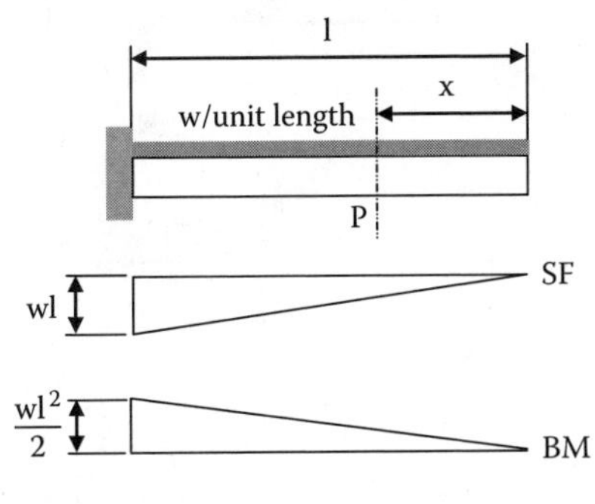

FIGURE 18.11
Cantilever with a uniformly distributed load.

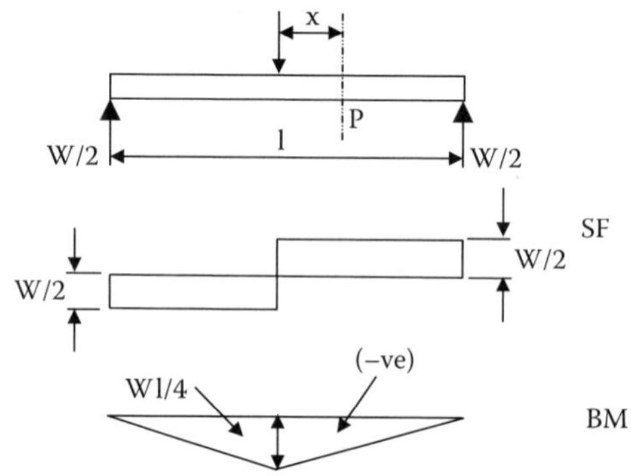

FIGURE 18.12
Simply supported beam with a central load.

18.4.3 Simply Supported Beam with Central Concentrated Load (Figure 18.12)

$$\text{SF at P} = \frac{W}{2}$$

$$\text{BM at P} = -\frac{W}{2}\left(\frac{l}{2} - x\right)$$

$$\text{Maximum SF} = \frac{W}{2}$$

$$\text{Maximum BM} = -\frac{W}{2}\cdot\frac{l}{2}$$

$$= -\frac{Wl}{4}$$

18.4.4 Simply-Supported Beam with Uniformly Distributed Load, w per Unit Length (Figure 18.13)

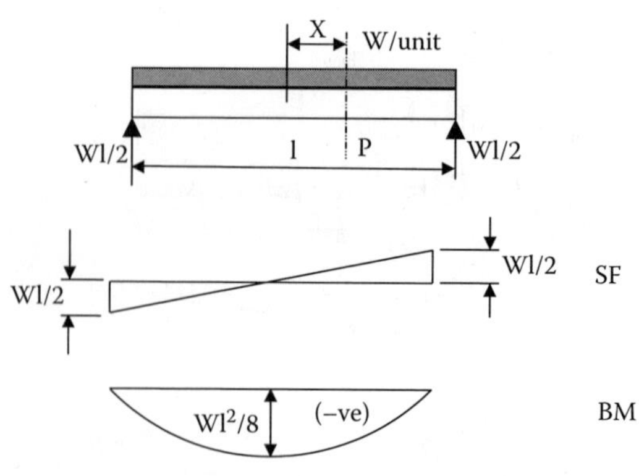

FIGURE 18.13
Simply supported beam with a uniformly distributed load.

$$\text{SF at P} = \frac{wl}{2} - w\left(\frac{l}{2} - x\right)$$

$$= wx$$

$$\text{BM at P} = \frac{wl}{2}\left(\frac{l}{2} - x\right) + \frac{w}{2}\left(\frac{l}{2} - x\right)^2$$

$$= -\frac{w}{2}\left(\frac{l^2}{4} - x^2\right)$$

$$\text{Maximum SF} = \frac{wl}{2}$$

$$\text{Maximum BM} = -\frac{wl^2}{8}$$

19

Torsion of Solid Sections

19.1 Introduction

If a shaft is subjected to a pure torque, i.e., without any bending, buckling, or axial thrust, every cross section is in a state of pure shear. The shearing stress induced in the shaft produces a moment of resistance, equal and opposite to the applied torque.

List of Symbols

T = Applied torsional moment

L = Length of member

G = Modulus of rigidity

Ip = Polar moment of inertia of section

J = Torsion constant of section

τ = Applied shear stress

q = Applied shear flow

θ = Angle of twist

A = Area of cross section

19.2 Basic Theory

In the theory of twisting, which relates the shear stress and angle of twist to the applied torque, it is assumed that:

1. The material is homogeneous, elastic, and obeys Hooke's law; i.e., the shear stress at any point is proportional to the shear strain at that point.
2. Stresses do not exceed the limits of proportionality.
3. Radial lines remain radial after twisting.
4. The plane cross section remains plane after twisting. (For noncircular shafts, the assumption that the plane sections remain plane after twisting is not justified and this theory ceases to apply.)

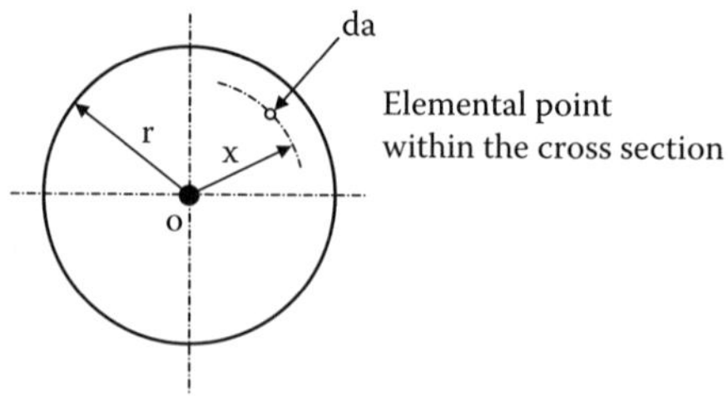

FIGURE 19.1
Cross section of a circular shaft subject to pure torsion.

From the third assumption above, it follows that the strain (and hence the stress) is directly proportional to the radius. Thus, if the shear stress at the surface of the shaft is τ, then the stress τ' on an element da at a distance x from the axis (see Figure 19.1).

$$\tau' = \frac{x}{r}\tau$$

$$\therefore \text{shear force on element} = \frac{x}{r}\tau\,da$$

$$\therefore \text{moment of force about O} = \frac{x}{r}\tau\,da\,x$$

$$\therefore \text{total moment of resistance} = \frac{\tau}{r}\int x^2 da$$

$$= \frac{\tau}{r}J$$

This is equal to the applied torque T, i.e.,

$$T = \frac{\tau}{r}J \qquad (19.1)$$

$$\text{or } \frac{T}{J} = \frac{\tau}{r}$$

This formula gives the shear stress at the surface in terms of T and J, but the stress at any other radius can be readily found since it is proportional to the radius at that point.

19.3 Modulus of Section

The maximum shear stress in a shaft is given by

$$\tau = \frac{T}{J}r$$

$$= \frac{T}{J/r}$$

The quantity J/r is known as the Modulus of Section and is denoted by Z.

Thus

$$\tau = \frac{T}{Z} \tag{19.2}$$

For a solid shaft

$$Z = \frac{\frac{\pi d^4}{32}}{\frac{d}{2}} \tag{19.3}$$

$$= \frac{\pi}{16} d^3$$

For a hollow shaft

$$Z = \frac{\frac{\pi(D^4 - d^4)}{32}}{\frac{D}{2}} \tag{19.4}$$

where D and d are the outer and inner diameters, respectively.

$$Z = \frac{\pi}{16}\left(\frac{D^4 - d^4}{D}\right) \tag{19.5}$$

19.4 Angle of Twist

Due to the shear strain in the shaft, the longitudinal line AB (see Figure 19.2) will rotate to the position AC, the end A being considered fixed.

The angle BAC is the shear strain, φ, and the angle BOC is the angle of twist, θ.

$$BC = L\varphi = r\theta$$

$$\therefore \varphi = \frac{r}{L}\theta = \frac{\tau}{G} \tag{19.6}$$

$$\therefore \frac{\tau}{r} = \frac{G\theta}{L}$$

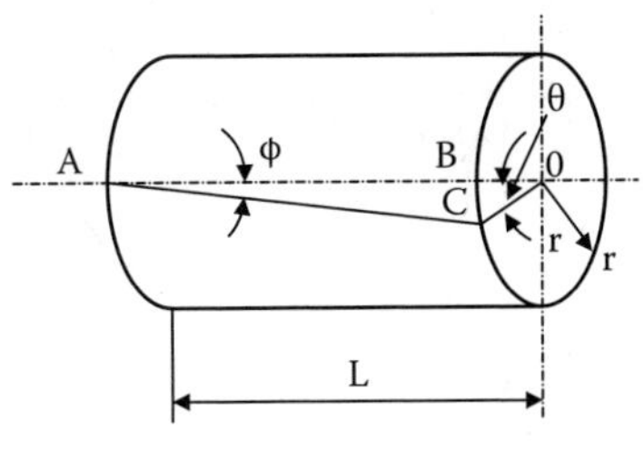

FIGURE 19.2
Angle of twist.

Combining this with Equation 19.1 gives the general twisting formula:

$$\frac{T}{J}=\frac{\tau}{r}=\frac{G\theta}{L} \tag{19.7}$$

19.5 Pure Torsion of Open Sections

19.5.1 Thick-Walled Open Sections

These types of sections have relatively thick rectangular elements having fillets at the junctions (Table 19.1).

Consider an L-shaped section as shown in Figure 19.3a. At the point where the curvature is concave or re-entrant is negative, the approximate maximum stress is given by:

$$f_s=\frac{T}{J} \quad \text{and occurs at point c}$$

$$(f_s = \text{shear stress})$$

J for various sections is given in Tables 19.2 and 19.3.

The parameter *n* is defined as follows:

For the concave radii as shown in Figure 19.3a

where

$$n=\frac{d}{1+\left(\frac{\pi^2\cdot d^4}{16A^2}\right)}\left\{1+\left[0.118\ln\left(1-\frac{d}{2r}\right)-0.238\frac{d}{2r}\right]\tanh\frac{2\theta}{\pi}\right\}$$

For convex curves (such as bulbs) as shown in Figure 19.2b

$$n=\frac{d}{1+\left(\frac{\pi^2\cdot d^4}{16A^2}\right)}\left[1+0.15\left(\frac{\pi^2 d^4}{16A^2}-\frac{d}{2r}\right)\right]$$

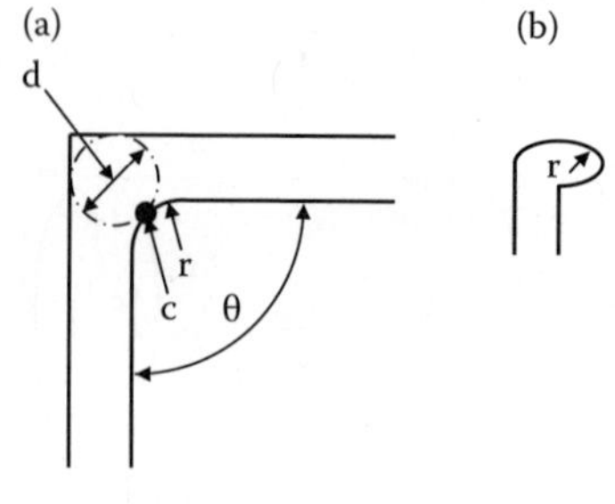

FIGURE 19.3
(a) L-shaped section. Concave curve. c=point of maximum stress; r=radius of fillet; θ=angle between elements (radians); d=diameter of largest inscribed circle; A=area of cross section. (b) Convex curve (i.e., at bulb).

TABLE 19.1

Torsional Properties of Solid Sections

Cross-Section	Torsional Constant J in $\theta = \frac{Tl}{GJ}$	Torsional Shear Stress f_s
Circle 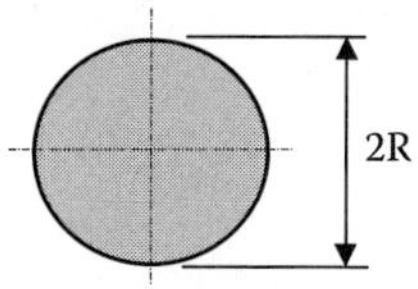	$J = \frac{\pi R^4}{2}$	Max $f_s = \frac{2T}{\pi R^3}$ at outer surface
Thick tube 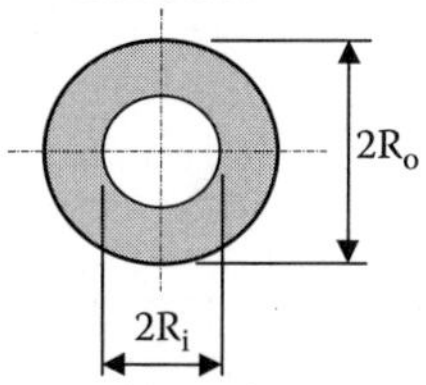	$J = \frac{\pi\left(R_o^4 - R_i^4\right)}{2}$	Max $f_s = \frac{2TR_o}{\pi\left(R_o^4 - R_i^4\right)}$ at outer surface
Solid ellipse 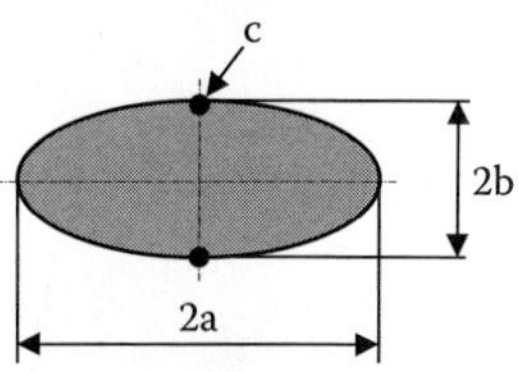	$J = \frac{\pi \cdot a^3 \cdot b^3}{a^2 + b^2}$	Max $f_s = \frac{2T}{\pi \cdot a \cdot b_2}$ at c
Square 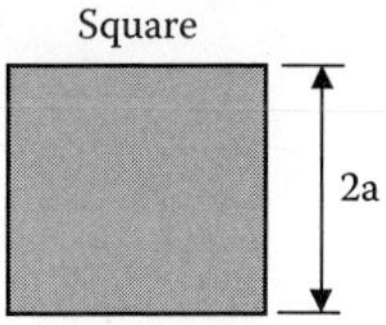	$J = 2.25 \cdot a^4$	Max $f_s = \frac{0.601 \cdot T}{a^3}$ At midpoint of each side
Rectangular 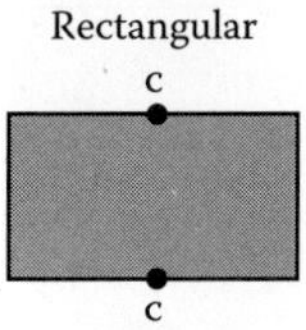	$J = ab^3\left[\frac{16}{3} - 3.36\frac{a}{b}\left(1 - \frac{b^4}{12a^4}\right)\right]$ For $a \geq b$	Max $f_s = T\left[\frac{3a + 1.8b}{8a^2b^2}\right]$ At midpoint of long sides
Circular segment 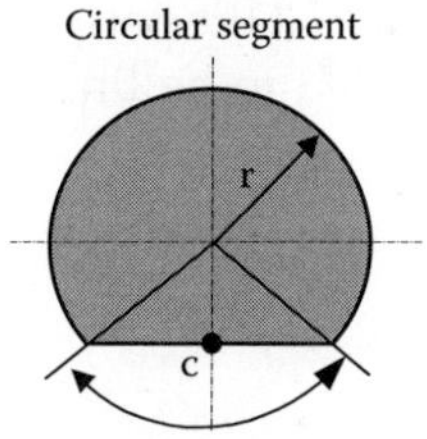	$J = 2cr^4$ $h = r(1 - \cos\alpha)$ for $0 \leq \frac{h}{4} \leq 1.0$ $c = 0.7854 - 0.3333\frac{h}{r} - 2.6183\left(\frac{h}{r}\right)^2 + 4.1595\left(\frac{h}{r}\right)^3 - 3.0769\left(\frac{h}{r}\right)^4 + 0.9299\left(\frac{h}{r}\right)^5$	Max $f_s = \frac{T\beta}{r^3}$ (at c) For $0 \leq \frac{h}{r} \leq 1.0$ $\beta = 0.6366 + 1.7398\frac{h}{r} - 5.4897\left(\frac{h}{r}\right)^2 + 14.062\left(\frac{h}{r}\right)^3 - 14.51\left(\frac{h}{r}\right)^4 + 6.434\left(\frac{h}{r}\right)^5$

TABLE 19.2

Torsional Constants for Various Sections (Part 1)

Section	**Torsional Constant** J as in $\frac{Tl}{GJ}$ and in $f_s = \frac{T}{J}n$
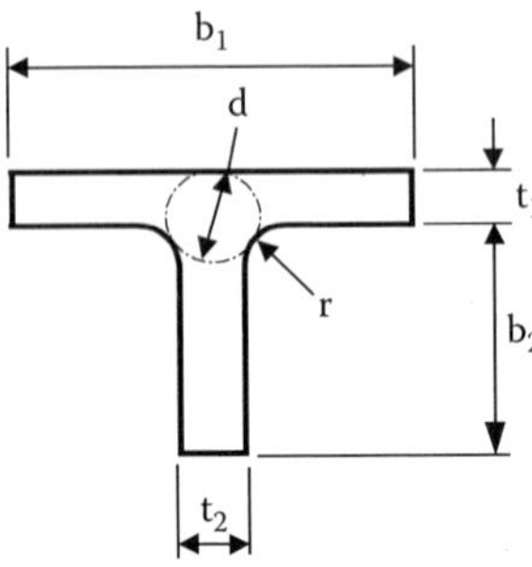	$J = K_1 + K_2 + ad^4$ if $t_1 \geq t_2$ $a = \frac{t_2}{t_1}\left(0.15 + 0.10\frac{r}{t_1}\right)$ if $t_2 \geq t_1$ $a = \frac{t_1}{t_2}\left(0.15 + 0.10\frac{r}{t_1}\right)$ for $t_2 < 2(t_1 + r)$ $d = 2\left[t_2 + t_1 + 3r - \sqrt{2(2r + t_1)(2r + t_2)}\right]$
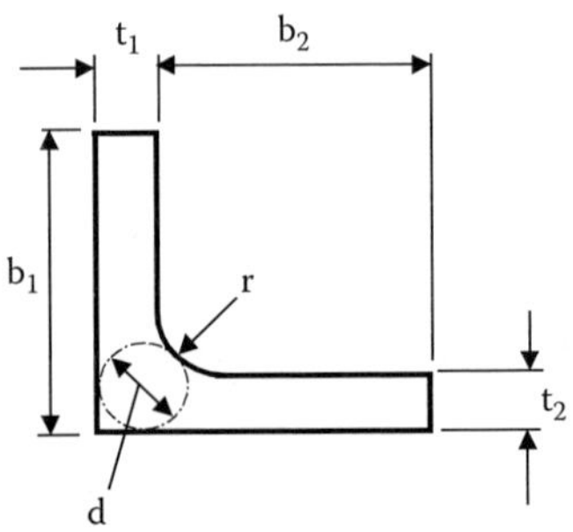	$J = K_1 + K_2 + ad^4$ if $t_1 \geq t_2$ $a = \frac{t_2}{t_1}\left(0.07 + 0.076\frac{r}{t_1}\right)$ if $t_2 \geq t_1$ $d = 2\left[t_2 + t_1 + 3r - \sqrt{2(2r + t_1)(2r + t_2)}\right]$

19.5.2 Thin-Walled Open Sections

The torsion in a thin-walled open section is balanced by the shear flow, q, about the periphery of the section. Consider Figure 19.4. The stress distribution is assumed to be linear across the thickness of the section as indicated in Figure 19.4.

$$\text{Torsional Constant } J = \frac{1}{3}\sum_{i=1}^{n} b_i t_i^3$$

$$\text{Maximum shear stress } f_s = \frac{T}{J} \cdot t$$

Note that this applies to pure torsion only; the effect of warping and associated differential bending loads in the flanges will be considered separately (see Table 19.4).

TABLE 19.3

Torsional Constants for Various Sections (Part 2)

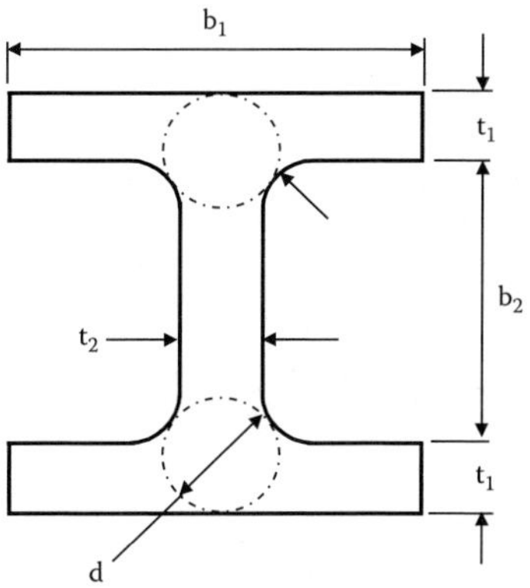

$$J = K_1 + K_2 + ad^4$$

$$a = \frac{t}{t'}\left(0.15 + 0.10\frac{r}{t_1}\right)$$

where

t = minimum of t_1 or t_2

t' = maximum of t_1 or t_2

For $t_2 < 2(t_1 + t_2)$

$$d = \frac{(t_1 + r)^2 + r \cdot t_2 + \frac{t_2^2}{4}}{(2r + t_1)}$$

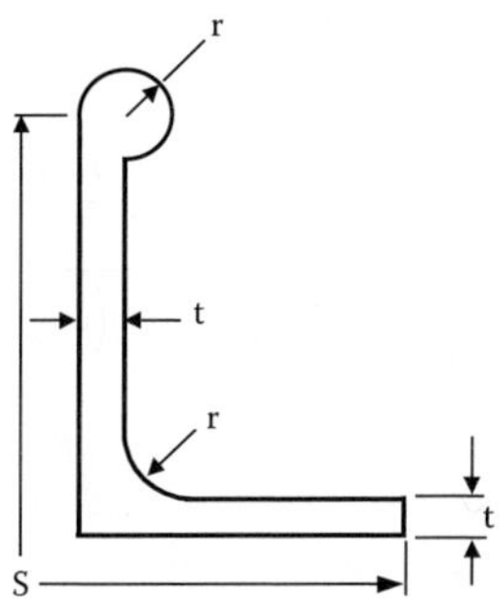

$$J = \frac{1}{3}st^3\left[1 + \left(\frac{4.14 + 4.71\left(\frac{r}{t}\right)^4}{\frac{s}{t}}\right)\right]$$

for 'n' at bulb or fillet see paragraph (2.1.6.1)

Sections composed of 3 or more rectangular elements

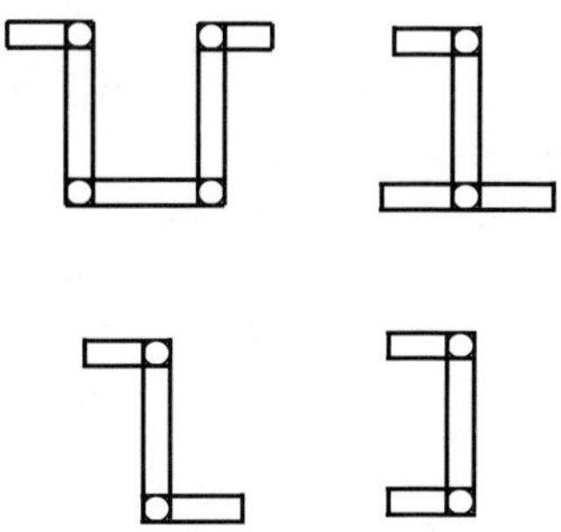

$J = \Sigma J$ of the constituant tee and angle sections:

$= \Sigma K_1 + \Sigma K_2 + \Sigma a \cdot d^4$

Where K_1 and K_2 are given as:

$$K_1 = b_1 t_1^3\left[\frac{1}{3} - 0.21\frac{t_1}{b_1}\left(1 - \frac{t_1^4}{12b_1^4}\right)\right]$$

$$K_2 = b_2 t_2^3\left[\frac{1}{3} - 0.105\frac{t_2}{b_2}\left(1 - \frac{t_2^4}{192b_2^4}\right)\right]$$

$$K_3 = \frac{1}{3}b_2 t_2^3$$

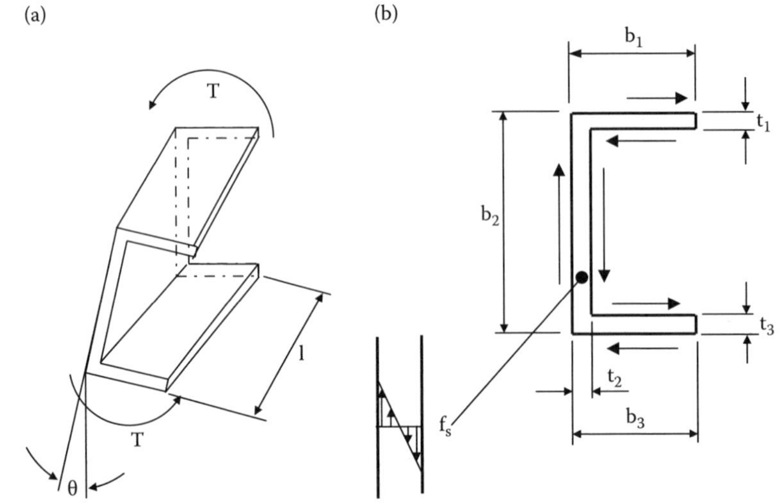

FIGURE 19.4
Torsion in a thin-walled section.

19.6 Thin-Walled Closed Sections

19.6.1 Single Cell Sections

The torsion of thin-walled closed sections is balanced at any section along a member by a uniform shear flow around the section. The stress is assumed to be uniform through the thickness of the section.

It will be obvious that the maximum stress will occur at the thinnest part of the section (see Figure 19.5).

$$\text{Shear flow q(load/mm)} = \frac{T}{2A} \tag{19.8}$$

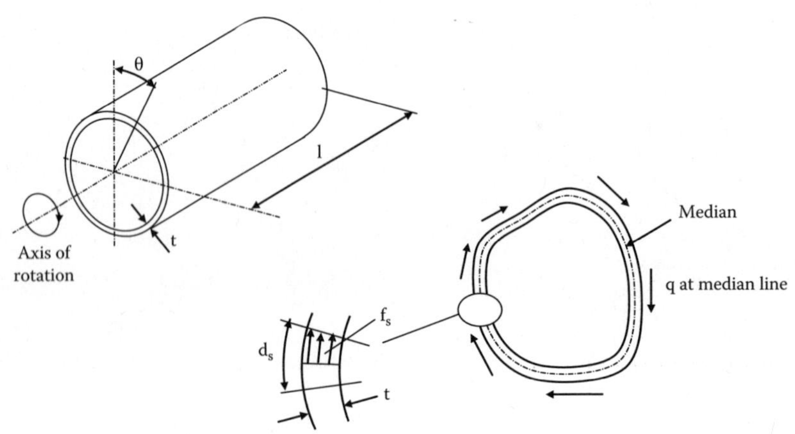

FIGURE 19.5
Torsion in a closed section. A = area enclosed by median contour; b = total length of median contour; d_s = length of differential element along median contour; t = local wall thickness.

TABLE 19.4

Torsional Constants for Thin-Walled Open Sections

Cross-Section	**Torsional Constant J in** $\theta = \frac{Tl}{GJ}$	**Torsional Shear Stress** $f_s = \frac{T}{J}t$
Thin rectangular 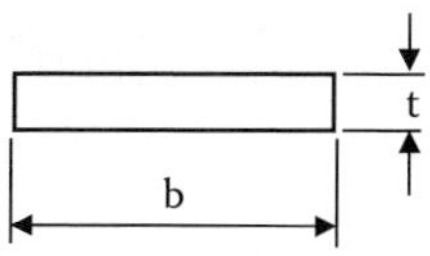	$J = \frac{1}{3} b \cdot t^3$	$f_s = \frac{3T}{bt^2}$
Thin circular 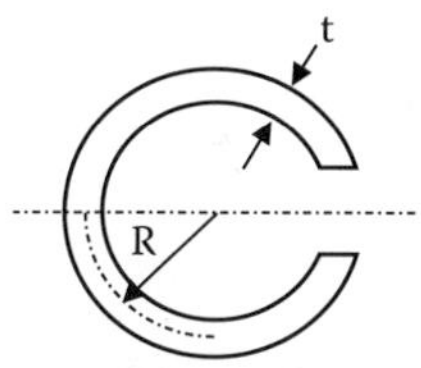 R = Mean radius	$J = \frac{2}{3} \pi R t^3$	$f_s = \frac{3T}{2\pi RT_2}$
Any formed section 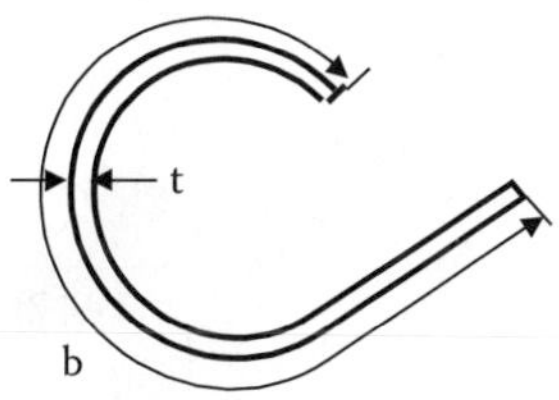	$J = \frac{1}{3} b \cdot t^3$	$f_s = \frac{3T}{bt^2}$
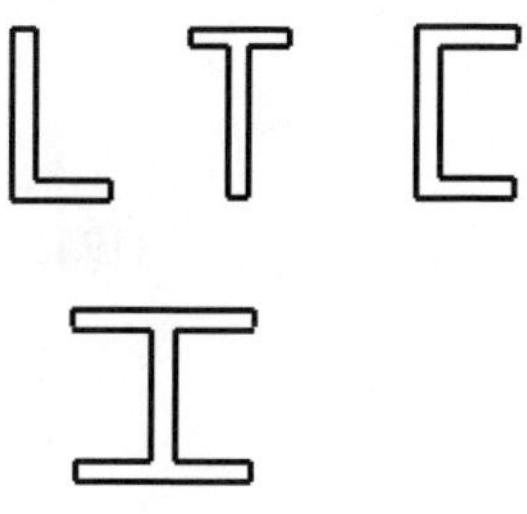	$J = \frac{1}{3} \sum_{i=1}^{n} bi \cdot ti^3$	$f_{si} = \frac{T}{J} rt_i$ $f_{smax} = \frac{T}{J} t_{max}$

$$\text{Shear stress } f_s = \frac{T}{2At} = \frac{q}{t} \tag{19.9}$$

The torsional constant J for thin walled closed sections subjected to a torque is given by:

$$J = \frac{4A^2}{\int_0^b \frac{d_s}{t}} = \frac{4A^2}{\sum_{i=1}^{n} \frac{b_i}{t_i}}$$

b_i = length of portion of the wall

t_i = thickness of portion of the wall

$$\text{angle of twist } q = \frac{Tl}{GJ} \tag{19.10}$$

$$\text{rate of twist } \frac{\theta}{l} = \frac{q}{2AG}\int \frac{d_s}{t} \tag{19.11}$$

19.7 Curved Members

19.7.1 Curved Torsion Members

In the previous section the maximum torsional shear stress on straight members was shown. This section will introduce an additional factor, k, to account for the effects of curvature on curved members.

If the member's radius of curvature is less than 10 times the member's depth, i.e., R/d < 10, then this factor will need to be applied. It is also assumed that the material properties will remain elastic (Table 19.5).

19.7.2 Circular Section

For a straight member:

$$f_{s_{max}} = \frac{T\cdot\left(\frac{d}{2}\right)}{J} = \frac{16\cdot T}{\pi\cdot d^3} \tag{19.12}$$

For a curved member:

$$f_{s_{max}} = k_1\cdot\left[\frac{16\cdot T}{\pi\cdot d^3}\right] \qquad k_1 = 1+\frac{5}{8}\left(\frac{d}{R}\right)+\frac{7}{32}\left(\frac{d}{R}\right)^2 \tag{19.13}$$

See Figure 19.6.

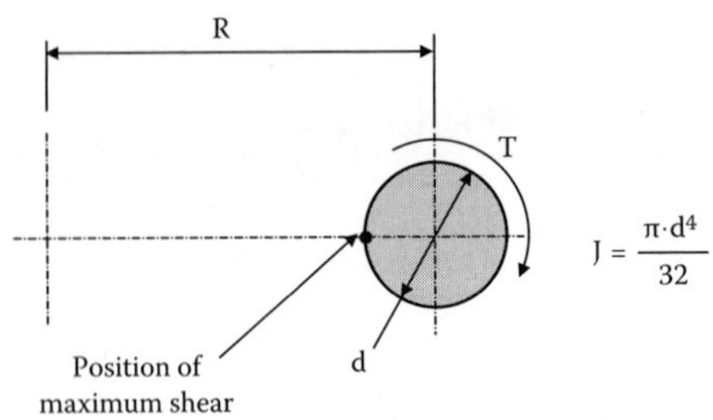

FIGURE 19.6
Torsion in a circular section.

TABLE 19.5

Torsional Constants for Thin-Walled Closed Sections

Cross-Section	Torsional Constant J in $\theta = \frac{Tl}{GJ}$	Torsional Shear Stress $f_s = \frac{T}{2at}$
Tube 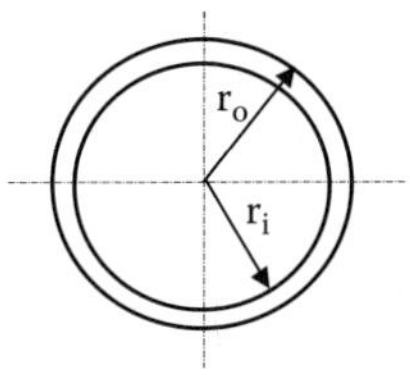	$J = \frac{1}{2}\pi\left(r_o^4 - r_i^4\right)$	$f_s = \frac{2Tr_o}{\pi\left(r_o^4 - r_i^4\right)}$
Ellipse 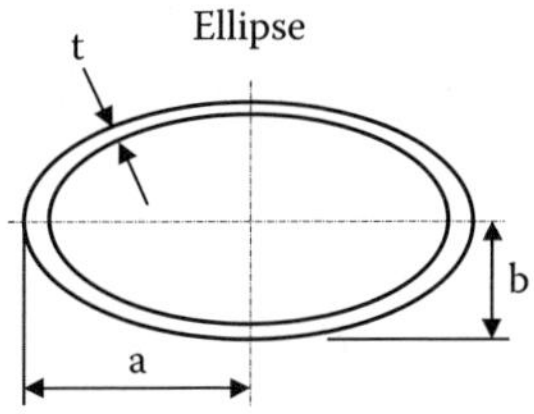	$J = \frac{4\pi^2 t\left[\left(a - \frac{1}{2}t\right)^2\left(b - \frac{1}{2}t\right)^2\right]}{U}$ U = length of ellipitical median boundary $U = \pi(a + b - t)\left[1 + 0.258\frac{(a-b)^2}{(a+b-t)^2}\right]$	$f_s = \frac{2T}{2\pi t\left(a - \frac{1}{2}t\right)\left(b - \frac{1}{2}t\right)}$
Rectangle 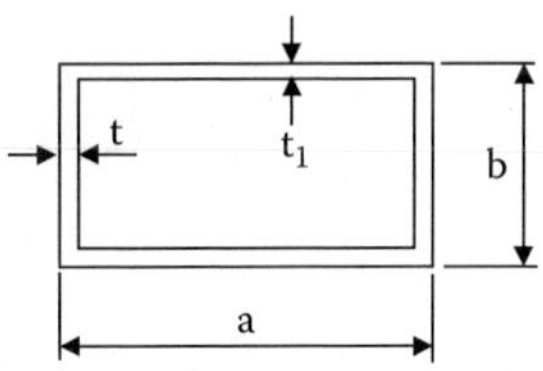	$J = \frac{2tt_1(a-t)^2(b-t_1)^2}{at + bt_1 - t^2 - t_1^2}$	$f_{s1} = \frac{T}{2t(a-t)(b-t_1)}$ Short sides $f_{s2} = \frac{T}{2t_1(a-t)(b-t_1)}$ Long sides
General 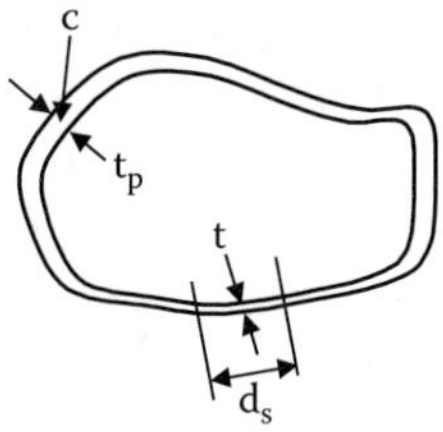	$J = \frac{4A^2}{\int_0^b \frac{d}{t}} = \frac{4A^2}{\sum_{i=1}^{n}\frac{b_i}{t_i}}$ Constant t: $J = 4A^2\frac{t}{b}$	at c $F_{sp} = \frac{T}{At_p}$ $f_s = \frac{T}{2At}$

A = area of enclosed cell; b = length of median line.

19.7.3 Square Section

For a straight member:

$$f_{s_{max}} = \frac{0.601 \cdot T}{b^3} \tag{19.14}$$

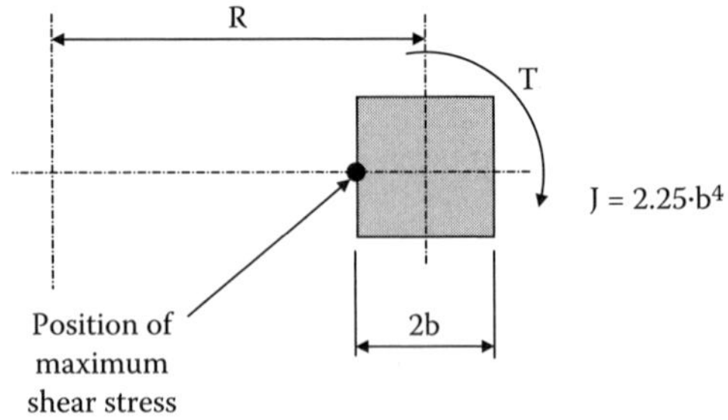

FIGURE 19.7
Torsion in a square section.

For a curved member:

$$f_{s_{max}} = k_2 \left[\frac{0.601 \cdot T}{b^3} \right] \tag{19.15}$$

$$k_2 = 1 + 1.2\left(\frac{b}{R}\right) + 0.56\left(\frac{b}{R}\right)^2 + 0.5\left(\frac{b}{R}\right)^3$$

See Figure 19.7.

19.7.4 Rectangular Sections

For a straight member:

$$f_{s_{max}} = T\left[\frac{3a + 1.8b}{8 \cdot a^2 \cdot b^2}\right] \tag{19.16}$$

For a curved member:

$$f_{s_{max}} = k_2 \cdot T\left[\frac{3a + 1.8b}{8 \cdot a^2 \cdot b^2}\right]$$

$$k_2 = 1 + 1.2\left(\frac{b}{R}\right) + 0.56\left(\frac{b}{R}\right)^2 + 0.5\left(\frac{b}{R}\right)^3 \tag{19.17}$$

The above formulas will apply as in Figures 19.8 and 19.9.

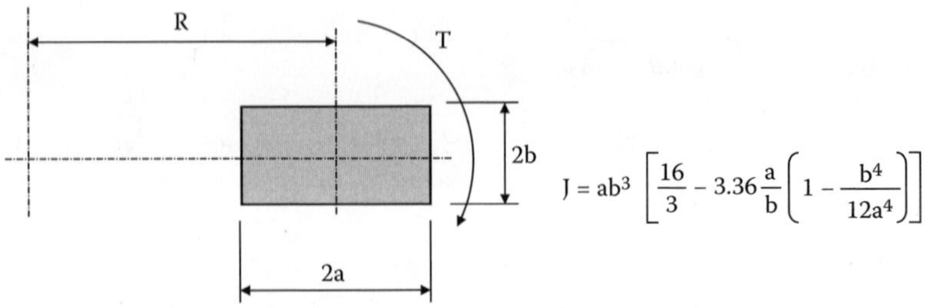

FIGURE 19.8
Torsion in a rectangular section.

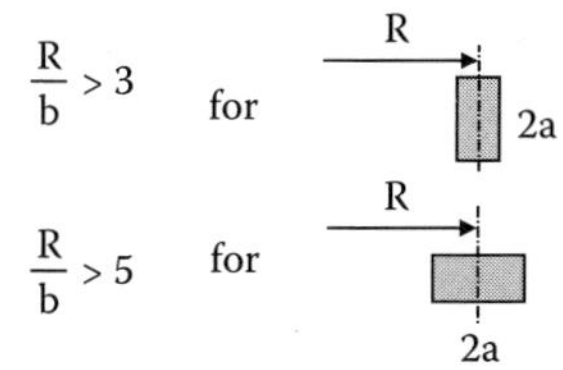

FIGURE 19.9
Ratio of R to b for the aspect ratio of the section.

19.7.5 Springs

Spring Nomenclature

α = Pitch angle of coils
n = Number of active coils in spring
δ = Axial deflection
G = Modulus of rigidity
P = Applied force
ν = Poisson's ratio

The axial deflections of springs are as follows (Figure 19.10):

$$\text{For circular sections: } \delta = \frac{64P \cdot R^3 \cdot n}{G \cdot d^4}\left[1 - \frac{3}{64}\left(\frac{d}{R}\right)^2 + \frac{3+\nu}{2(1+\nu)}(\tan a)^2\right] \tag{19.18}$$

$$\text{For square sections: } \delta = \frac{2.789P \cdot R^3 \cdot n}{G \cdot b^4} \quad \text{for } c > 3 \tag{19.19}$$

$$\text{For rectangular sections: } \delta = \frac{3 \cdot \pi \cdot P \cdot R^3 \cdot n}{8 \cdot G \cdot b^4}\left[\frac{a}{b} - 0.627\left(\tan h\left(\frac{\pi b}{2a}\right) + 0.004\right)\right]^{-1} \tag{19.20}$$

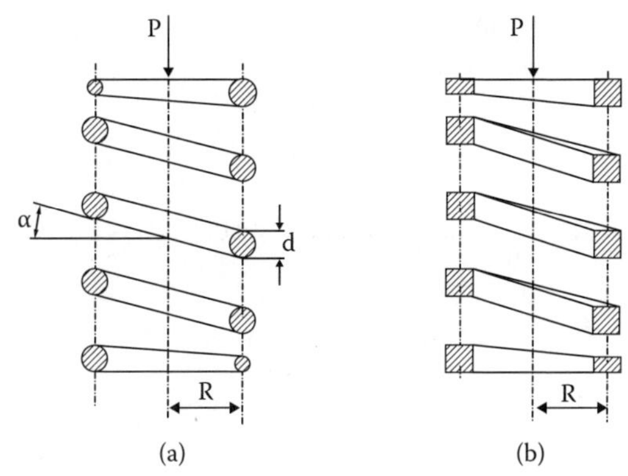

FIGURE 19.10
Spring forms. (a) Circular section and (b) square section.

19.8 Torsional Failure of Tubes

Failure in tubes due to torsion may be attributable to material failure or instability of the tube walls. The instability may be due to either:

d/t ratio may be too large

plastic regions existing in the wall of the tube (Figure 19.11)

r_m = radius to mean thickness

t = wall thickness

l = tube length

T = applied torque

$\overline{T}$ = allowable torque

J = torsional constant

f_s = applied shear stress

$\overline{f}_{sbe}$ = allowable elastic buckling stress at the mean radius(r_m)

$\overline{F}_{st}$ = allowable shear stress at r_o

$\overline{F}_{stb}$ = allowable buckling stress at the outside radius(r_o)

$\overline{F}_{stp}$ = allowable plastic stress at r_o

$\overline{F}_{su}$ = allowable ultimate shear stress for tube material

The shear stress curve is similar to that of the tensile stress curve with a $1/\sqrt{3}$ factor applied to the ordinates. Plastic torsion in tubes is treated similar to a bar in plastic bending (Figure 19.12 shows the elastic and plastic stress distribution in a solid shaft.)

The allowable shear stress, F_{st}, is also referred to as the *torsional modulus of rupture*. This is a fictitious stress acting at the surface of the tube and represents the same resisting moment as the true stress in the section.

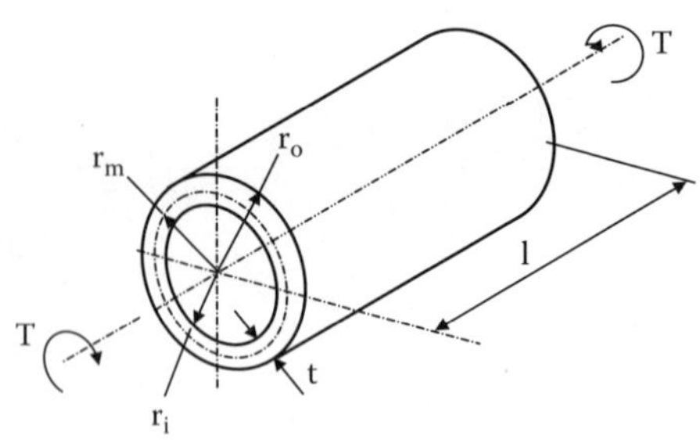

FIGURE 19.11
Notation.

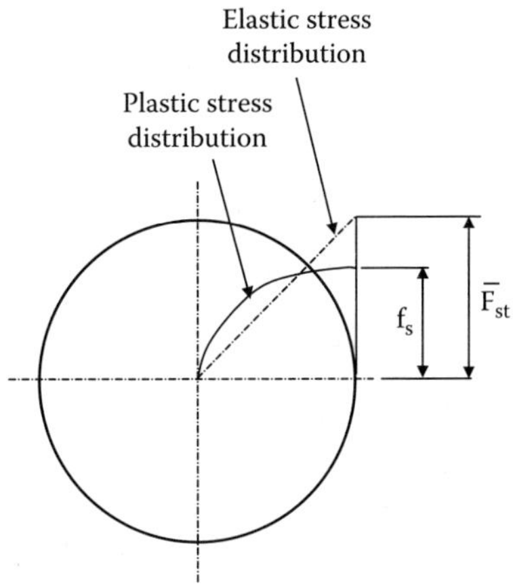

FIGURE 19.12
Elastic and plastic stress distribution.

Material property curves for the torsional modulus of rupture can be taken from MIL-HDBK 5* for a wide range of materials.

19.8.1 Modulus of Rupture: A Theoretical Approach

The first step to determine the modulus of rupture, F_{st}, is to calculate the stress that will lead to an instability failure, F_{stb}, and the stress to cause a material shear failure, F_{stp}.

The allowable F_{st} for a tube is the minimum of F_{stb} or F_{stp}.

The effects of any imperfections in the plastic range are considered small and can be disregarded.

19.8.1.1 Instability Failure

The elastic buckling stress (F_{stb}) is calculated by setting $E_{sec}=E$ and $M_r=1.0$ in the following equation:

Plastic Buckling Stress (F_{sb})

$$\bar{F}_{sb} = \frac{[3.2074H^4[A]+0.08213H^8[B+C]]\cdot E_{sec}}{D} \tag{19.21}$$

where

$$A = 37.3063K^4 + 7.0422\left(1+\frac{1}{M_r}\right)K^2 + 2$$

$$B = \frac{1}{(0.0276H^4 + 0.1660(3M_r - 1)H^2 + 1)}$$

$$C = \frac{1}{(0.9608H^4 + 0.9802(3M_r - 1)H^2 + 1)}$$

* MIL-HDBK 5 (Military Handbook 5): U.S. Government publication that covers a wide range of materials including steels, aluminium, copper, etc.

$$D = 10.773H\left[\frac{1}{t}\right]$$

$$H = \left[\frac{12}{t \cdot r_m}\right]^{\frac{1}{4}}$$

$$K = \frac{1}{H}$$

$$E_{sec} = \text{Secant Modulus}$$

The modulus ratio M_r is the ratio of the secant to the tangent modulus at a stress level equivalent to F_{sb}; hence an iteration is required to obtain a value.

$$M_r = \frac{1+\frac{3}{7}(m)\left(\sqrt{3}\frac{f_s}{F_{0.7}}\right)^{m-1}}{1+\frac{3}{7}\left(\sqrt{3}\frac{f_s}{F_{0.7}}\right)m^{-1}} = \frac{E_{sec}}{E_{tan}} \quad \text{at } \overline{F}_{sb} \tag{19.22}$$

E_{sec}, E_{tan}, Secant and Tangent moduli at $\overline{F}_{sb}$

m Stress-Strain curve shape factor

Allowable Buckling Shear Stress, $\overline{F}_{stb}$

$$\overline{F}_{stb} = \left(\frac{1+\nu-\nu^2-\nu^3}{1-\nu^4}\right)\cdot\overline{F}_{sb} \tag{19.23}$$

where

$$\nu = 1-\left(\frac{1}{r_o}\right)$$

The associated allowable torque is:

$$\overline{T}_b = \frac{\overline{F}_{stb}\cdot J}{r_o}$$

where

$$J = \frac{\pi(r_o^4 - r_i^4)}{2}$$

equivalent to F_{sb}, hence iteration is required to obtain a value.

19.8.2 Material Failure under Plastic Torsion

The allowable plastic stress at r_o

$$\overline{F}_{stp} = \frac{3}{4}\cdot\overline{F}_{su}\frac{[r_o^3 - r_i^3]r_o}{[r_o^4 - r_i^4]} \tag{19.24}$$

Assuming a constant shear stress, F_{su} through the thickness:

The allowable plastic torque:

$$\bar{T}_P = \frac{2\pi}{3} \cdot \bar{F}_{su}[r_o^3 - r_i^3] \tag{19.25}$$

19.9 Sand Heap Analogy for Torsional Strength

An experimental method for determining the torsional strength of noncircular sections has been developed by Bendix Aviation and McDonnell Aircraft Company and enables a fairly simple solution to complex shapes as found on aircraft landing components.

The procedure is presented here, as it has been helpful in the past and used by the author for checking complex shapes.

The following is a preamble to the method and is provided for information only.

A solid body, if subject to a severe twisting moment, will deform plastically when the intensity of the shear stress reaches the yield point in shear. The sand heap analogy method is an experimental representation of the stress distribution acting over the cross section of a component in a completely plastic state. The theory predicts that the maximum twisting moment a body can sustain before failure is equal to twice the volume enclosed by the surface, formed by the stress function. The slope of this surface will then be a maximum and therefore analogous to the natural sloping surface formed under gravity by heaping sand on a piece of cardboard having the proportional geometric configuration of the component's cross section.

Mass distribution and shape of the stress curve are important factors in limiting the maximum shear stress for calculation purposes. Experiments have verified the feasibility of using F_{su} as a conservative means of calculating the ultimate twisting moment.

19.9.1 Method (Solid Cross Section)

A horizontally positioned piece of cardboard shaped to the component's cross section is heaped with a fine-grade casting sand, such that a natural roof, defined by gravity, is formed (Figure 19.13).

In terms of the physical slope and sand volume, the following formula will apply.

$$T = \frac{(2F_{su})2q}{2kC^3} = \frac{q}{(2.54)^3 kC^3}(2F_{su}) = \frac{0.061}{kC^3}q(2F_{su}) \tag{19.26}$$

thus

$$T = \frac{0.083q}{C^3}(2F_{su})$$

where:

T = Allowable twisting moment (in.-lb)
q = Sand volume (cubic centimeters)
F_{su} = Ultimate shear stress
C = Scale factor, i.e., double scale, $C = 2$

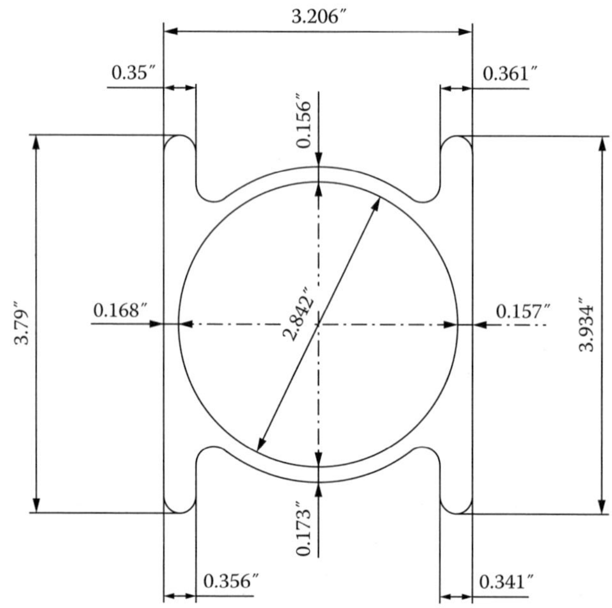

FIGURE 19.13
Sand heap analysis–example section.

Calculation of K and T is as follows (Figure 19.14):

K value:

$$\text{Plastic torsion capacity } T = \frac{2g\, f_{st}}{254^3 KC^3}$$

For a solid circular section:

$$\text{Plastic torsion capacity } T = \frac{4}{3}\frac{f_{st}}{r} J$$

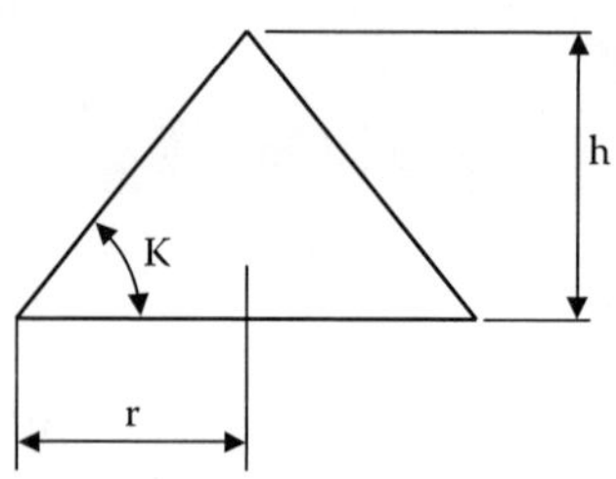

FIGURE 19.14
Angle of repose.

where

$$J = \frac{\pi d^4}{32}$$

$$T = \frac{2\pi f_{st} r^3}{3}$$

∴ For Circle:

$$\frac{2g\, f_{st}}{2.54^3 KC^3} = \frac{2}{3}\pi f_{st}$$

$$K = \frac{3g}{\pi r^3 \cdot 2.54^3 C^3}$$

where:

g = volume of circle (cm^3)
C = scale factor, or double scale c = 2.0
K = slope of repose angle, constant for same kind of sand

or:

$$g = \frac{\pi r^2 h(2.54^3 \cdot C^3)}{3} \text{ for a circle}$$

$$h = \frac{3g}{\pi r^2 \cdot 2.54^3 \cdot C^3}$$

$$K = \frac{h}{r} = \frac{3g}{\pi r^3 \cdot 2.54^3 \cdot C^3}$$

Calculation of T is as follows:

Measuring the sand volume of the desired section in cm^3

$$T = \frac{2g\, f_{st}}{2.54^3 KC^3}$$

where:

f_{st} = ultimate shear allowable (ksi)
g = sand volume of desired section (cm^3)
C = scale factor
K = repose angle (can be calculated from (I))

Note that this method requires the shear stress allowable to be measured in ksi (1000 lbf/in.^{2}).

Example 19.1

Consider a cone with a base of 1.50 and a height of 1.063.

$$\text{Total volume of cone} = 42 \text{ cm}^3$$

$$K = \frac{3.42}{\pi(1.5)^3 \cdot 2.54^3 (1)^3}$$

$$= 0.725$$

$$\text{let } h = 0.4 \text{ cm}$$

$$(R = 3.81 \text{ cm} \quad r = 3.231 \text{ cm})$$

$$\text{Volume of frustum} = 11\,\text{cm}^3$$

$$T_{\text{section}} = \frac{2 \cdot 11 \cdot 168}{2.54^3 \cdot 0.725 \cdot 1^3}$$

$$= 311.095 \text{ MPa}$$

Example 19.2

Figure 19.15 shows a section through a nose landing wheel strut. The original analysis used the sand heap method to establish the allowable torsion to be applied to the section.

Use a suitable CAD drawing program and setting the outer profile. The profile is then offset at the X dimension. The area is then calculated and tabulated.

By analysis of Figure 19.15, the following individual areas were obtained:

$A_1 = 1317.12\ \text{mm}^2$
$A_2 = 856.29\ \text{mm}^2$
$A_3 = 422.87\ \text{mm}^2$
$A_4 = 45.20\ \text{mm}^2$
$A_5 = 2.95\ \text{mm}^2$
$A_6 = 60.19\ \text{mm}^2$
$A_7 = 4.74\ \text{mm}^2$
$A_8 = A_4 + A_6 = 47.20\ \text{mm}^2 + 60.19\ \text{mm}^2 = 105.39\ \text{mm}^2$
$A_9 = A_5 + A_8 = 2.95\ \text{mm}^2 + 4.74\ \text{mm}^2 = 7.69\ \text{mm}^2$

$$V_{\text{total}} = \left[A_2 + \left(\frac{A_1 - A_2}{2}\right) + A_3 + \left(\frac{A_2 - A_3}{2}\right) + A_8 + \left(\frac{A_3 - A_8}{2}\right) + A_9 + \left(\frac{A_8 - A_9}{2}\right)\right]$$

$$= \left[856.29 + \left(\frac{1317.12 - 856.29}{2}\right) + 422.67 + \left(\frac{856.29 - 422.67}{2}\right) + 105.39\right.$$

$$\left. + \left(\frac{422.67 - 105.39}{2}\right) + 7.69 + \left(\frac{105.39 - 7.69}{2}\right)\right]$$

$$= 2.032 \tan\alpha[2046.98]$$

$$= 4159.47 \tan\alpha$$

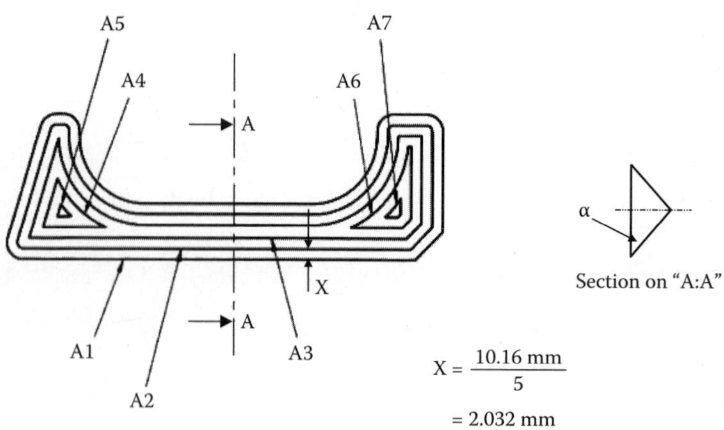

FIGURE 19.15
Section for Example 19.2.

Equivalent diameter circle:

$$\frac{1}{3}\pi r^3 \tan\alpha = 4159.47 \tan\alpha$$

$$r^3 = \frac{4159.47 \times 3}{\pi}$$

$$r = 15.84 \text{ mm}$$

$$\text{Dia}' = 31.67 \text{ mm}$$

$$\text{Torsion allowable} = \frac{p \cdot d^3}{16} \times 1158.32 \text{ MPa}$$

$$= 441.02 \text{ MPa}$$

20

Combined Torsion and Bending

Very often bending and torsion do not act alone when designing a component; they can often act simultaneously. As an example, a shaft that is supported as a cantilever with a drive pulley fitted on the free end will in this case be resisting a bending load due to the drive belt or chain tension and a torsion load again imposed due to the drive belt or chain. This example is not restricted to rotating shafts but can equally be applied to brackets or other nonrotating components.

In the first instance the bending moment (M) needs to be calculated and then the twisting moment (T) requires evaluation.

From the bending equation:

$$\frac{M}{I} = \frac{\sigma}{y} = \frac{E}{R} \tag{20.1}$$

from which:

$$\sigma = \frac{M \cdot y}{I} \tag{20.2}$$

and

$$\sigma = \frac{E \cdot y}{R} \tag{20.3}$$

The torsion equation:

$$\frac{T}{J} = \frac{\tau}{r} = \frac{G\theta}{L} \tag{20.4}$$

from which:

$$\tau = \frac{T \cdot r}{J} \tag{20.5}$$

and

$$\theta = \frac{\tau \cdot L}{G \cdot r} \tag{20.6}$$

For notation see Table 20.1.

When designing shafts or components subjected to combined torsion and bending for maximum strength, the material selected will determine the nature of the failure.

$$\text{i.e., when} \quad \frac{\sigma_x - \sigma_y}{2} = \frac{\sigma_E}{2} \tag{20.7}$$

TABLE 20.1

Notation

M	Maximum bending moment	Nm
I	Moment of inertia (second moment of area)	m^4
σ	Extreme fibre stress due to bending	Pa
y	Distance from the neutral axis to extreme fibre	m
E	Modulus of elasticity	N/m^2
R	Radius of bending on neutral axis	m
T	Twisting moment or torque	Nm
J	Polar moment of inertia	m^4
τ	Maximum shear stress due to twisting	Pa
r	Radius of circular section	m
G	Modulus of rigidity	N/m^2
θ	Angle of twist over length	Radians
L	Length	m
M_E	Equivalent bending moment due to combined bending and torque	Nm
T_E	Equivalent torque due to combined moment and torque	Nm

$$\text{or} \quad \sigma_x - \sigma_y = \sigma_E \tag{20.8}$$

For *ductile materials* such as mild steel, failure is considered to occur when the greatest shear stress reaches the maximum shear stress at the elastic limit in a simple tension test.

This is known as the *maximum-shear-stress theory* (*Guest's* or *Tresca criterion*) and gives good correlation with experimental results obtained with ductile materials.

It can be shown:

$$\tau = \pm\sqrt{(\sigma^2 + 4\tau^2)} \tag{20.9}$$

and the equivalent torque

$$\begin{aligned} T_E &= \sqrt{(M^2 + T^2)} \\ &= \frac{\pi \times d^2}{16} \times \tau \end{aligned} \tag{20.10}$$

When dealing with *brittle materials* such as cast iron that is subjected to combined torsion and bending, it is more important to determine the value of the principal stress, which in this case is a tensile stress and has a greater effect than shear stress. This approach is known as *Rankine's theory* (brittle materials will be generally weaker in tension than shear).

Rankine's theory for brittle materials is:

$$\sigma = \frac{\sigma}{2} \pm \frac{1}{2}\sqrt{(\sigma^2 + 4\tau^2)} \tag{20.11}$$

The equivalent bending moment:

$$M_E = \frac{1}{2}\left[M + \sqrt{(M^2 + T^2)}\right] \tag{20.12}$$

Example 20.1

Part of a spur reduction gear of 5:1 is shown in Figure 20.1. The pinion has 20 teeth with a module of 6 mm and an involute angle of 20°. The power transmitted is 45 kW at 13.5 rev/s.

Calculate the diameter of the pinion and wheel shafts taking the allowable shear stress as 55 MPa.

Solution

$$\text{Pitch diameter} = \text{module} \times \text{number of teeth}$$
$$= 6 \text{ module} \times 20 \text{ teeth}$$
$$= 120 \text{ mm}$$

and

$$\text{Torque} = \frac{\text{power}}{2\pi N}$$

Tangential tooth load (E):

$$E = \frac{\text{Torque}}{\text{radius of pitch circle}}$$
$$= 530.44 \text{ Nm}$$
$$= 8840.67 \text{ N}$$

Involute angle $\alpha = 20°$

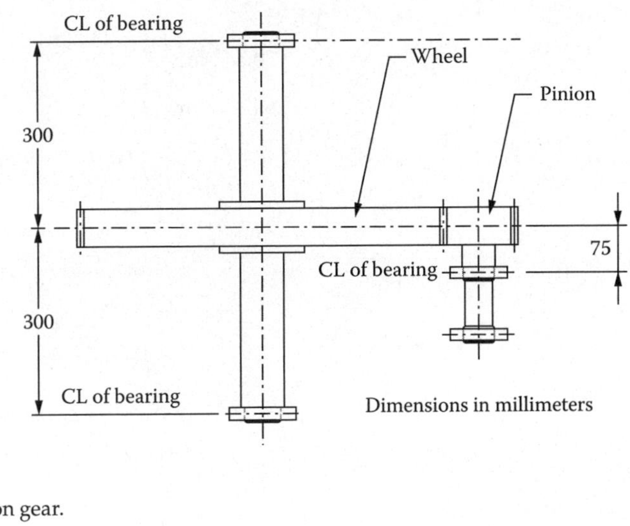

FIGURE 20.1
Spur reduction gear.

Maximum tooth load:

$$P = \frac{E}{\cos 20^\circ}$$

$$= \frac{8840.67}{0.9397}$$

$$= 9.408 \text{ kN}$$

The force E is the driving force and puts a torque on the shafts while P produces a bending effect (see Figure 20.2).

For the pinion shaft:

$$\text{Torsional moment } T = 530.44 \text{ Nm}$$

$$\text{Bending moment } M = 9408 \text{ N} \times 0.075 \text{ m}$$

$$\text{Equivalent torque } T_E = \sqrt{M^2 + T^2}$$

$$= \sqrt{705.6^2 + 530.44^2}$$

$$= \sqrt{779237}$$

$$= 705.6 \text{ Nm}$$

$$= 882.74 \text{ Nm}$$

Rearranging for d^3 and the allowable shear stress $\tau = 55\,\text{MPa}$.

$$882.74 \text{ Nm} = \frac{\pi}{16} d^3 \times 55 \times 10^6$$

$$d^3 = \frac{882.74 \times 16}{\pi \times 55 \times 10^6}$$

$$d = 43.4 \text{ mm, say.}$$

For the wheel shaft:

$$\text{Gear ratio} = 5:1$$

$$\text{Torsional moment} = 530.44 \text{ Nm} \times 5$$

$$= 2652.2 \text{ Nm}$$

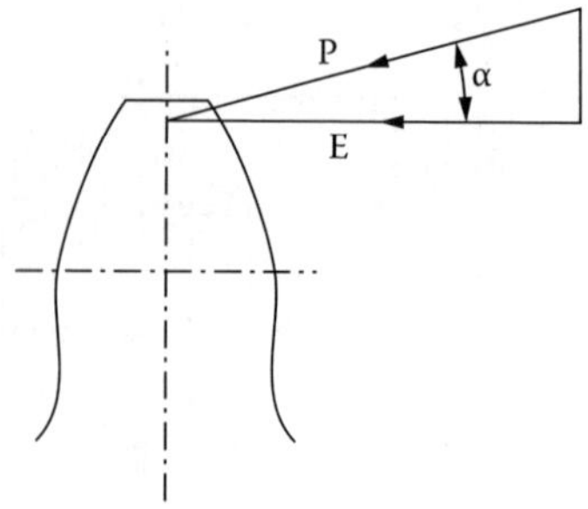

FIGURE 20.2
Forces acting on gear tooth.

$$\text{Bending moment} = \frac{WL}{4}$$

$$= 9408\ \text{N} \times \frac{0.6}{4}\ \text{m}$$

$$= 2652.2\ \text{Nm}$$

$$= 1411.2\ \text{Nm}$$

The reactions are assumed to act at the centre of the bearing.

$$= \sqrt{1411.2^2 + 2652.2^2}$$

Equivalent shear stress T_E or torque=3004.27 Nm

$$d^3 = 0.000278$$

$$d = 0.0653\ \text{m}$$

$$d = 65.3\ \text{mm}$$

If an allowance is made for the keyway:

d=65.3×1.1 (A factor of 1.1 is included to allow for the reduction in strength due to the presence of the keyway.)

$$d = 71.83\ \text{mm}$$

Example 20.2

Figure 20.3 shows a cast steel bracket subject to a force of 9 kN in the given directions. Calculate:

- The stress due to bending across the plane y:y.
- The stress due to the combined bending and torsion across the plane x:x.
- The stresses in the flange bolts resisting the torsional, sliding and overturning actions of the load.
- State the nature of the stresses in each case.

Solution

Considering y:y

$$M = \sigma Z \quad (\text{where } Z = \frac{\pi}{32^2} d^3 \text{ for circular section})$$

$$9 \times 0.28\ \text{kN} = \frac{\pi \times 0.100^3}{32} \times \sigma$$

$$\sigma = \frac{9 \times 0.28 \times 32}{\pi \times 0.100^3}$$

$$= 25668.5\ \text{kN/m}^2$$

$$= 25.67\ \text{MPa}$$

The bending moment (M) acting on this section is 9 kN×280 mm.

Considering section x:x

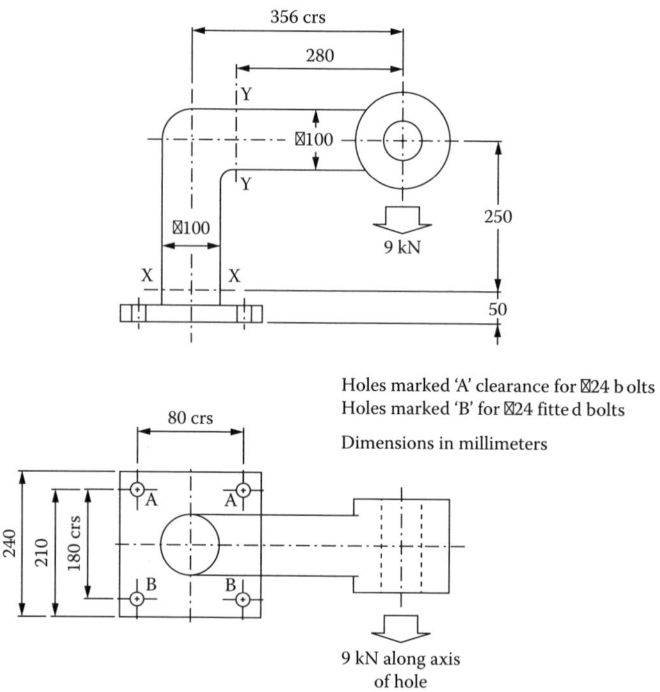

FIGURE 20.3
Steel bracket in Example 20.2.

$$M = 9\ kN \times 250\ mm$$

$$= 2.25\ kNm$$

The bending moment (M) acting on this section is 9 kN×250 mm.
Torsional moment (T):

$$T = 9\ kN \times 356\ mm$$

$$= 3.204\ kNm$$

$$Te = \sqrt{m^2 + T^2}$$

$$Te = \sqrt{2.25^2 + 3.204^2}$$

$$Te = 3.915\ kNm$$

$$T_{max} = \frac{16}{\pi d^3}\sqrt{m^2 + T^2}$$

$$T_{max} = \frac{16}{\pi \times 0.1^3\ m^3}\sqrt{2.25^2 + 3.204^2}$$

$$T_{max} = \frac{16}{\pi \times 0.1^3 m^3} \times 3.915\ kNm$$

$$T_{max} = 19938.9\ kN/m^2$$

$$T_{max} = 19.9389\ MPa$$

This is the maximum shear stress across section x:x due to the combined torsion and bending.

Consider the stresses acting on the bolts:
The twisting and sliding will generate a shear stress in the fasteners.
Consider the two fitted bolts only, as these will resist the sliding forces (see Figure 20.4).

$$= 20.1^2 \times \frac{\pi}{4}$$

$$= 317.310 \text{ mm}^2$$

Shear stress on bolt Y:

$$= \frac{13.3 \text{ kN}}{317.31 \text{ mm}^2}$$

$$= 0.0419 \text{ kN/mm}^2$$

$$= 0.0419 \text{ MPa}$$

Figure 20.5 shows the reactions acting on the bolts due to the load of 9 kN.

$$= \frac{22.3 \text{ kN}}{317.31 \text{ mm}^2}$$

$$= 0.0703 \text{ kN/mm}^2$$

$$= 0.0703 \text{ MPa}$$

Now the root area of Ø24 mm diameter bolt:

$$= 317.31 \text{ mm}^2$$

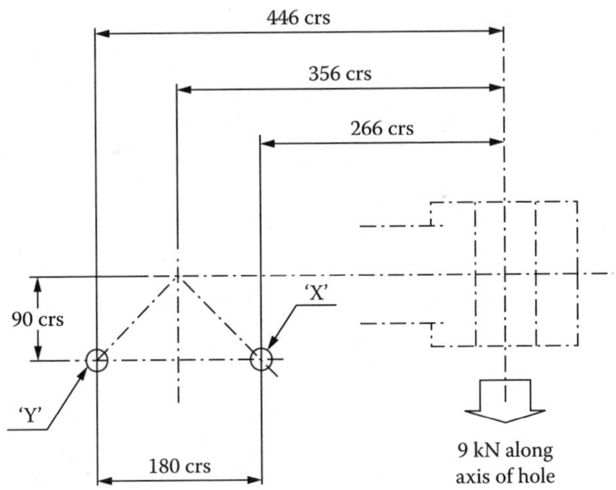

FIGURE 20.4
Bolts subject to shear load.

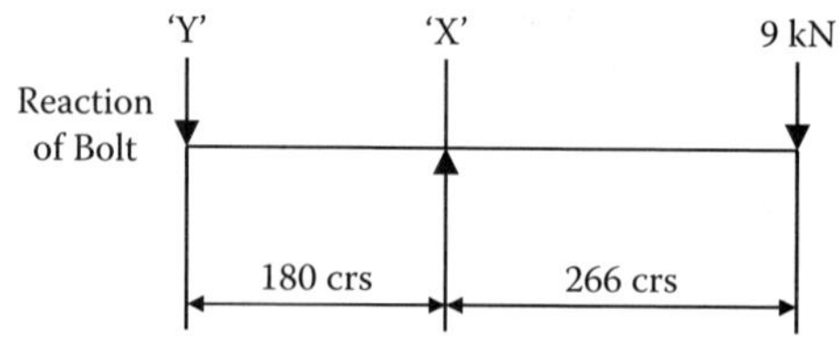

FIGURE 20.5
Reactions at fasteners.

The two bolts with clearance holes will resist the overturning effect about the edge O in Figure 20.6.

Shear stress acting on the bolt:

$$\text{Overturning moment} = 9\text{ kN} \times 300\text{ mm}$$

$$\text{Resisting moment} = \text{P} \times 210\text{ mm}$$

$$9 \times 300 = \text{P} \times 210$$

$$\text{P} = \frac{9 \times 300}{210}\text{ kN}$$

$$\text{P} = 12.86\text{ kN}$$

$$\text{Now load per bolt} = \frac{12.86}{2}\text{ kN}$$

$$= 6.43\text{ kN}$$

$$\text{Area at Root of } \varphi 24 = 317.31\text{ mm}^2$$

$$\text{Tensile stress in bolts} = \frac{6.43\text{ kN}}{317.31\text{ mm}^2}$$

$$= 20.3\text{ MPa}$$

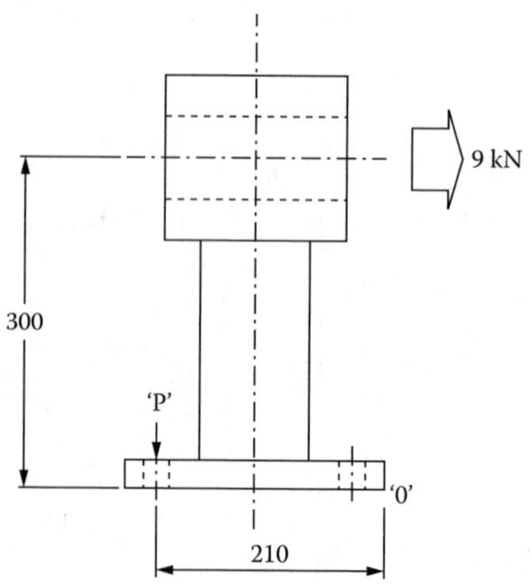

FIGURE 20.6
Overturning forces acting on fasteners.

Example 20.3

The overhead gear for an electric motorised lift is shown in Figure 20.7. The sheave is mounted and keyed centrally on the shaft, which is supported on bearing 510 mm between centres.

The ropes pass from the edge of the cage over the sheaves to the balancing mass.

The mass of the cage, the mass carried by the cage and the balancing mass are 1100, 910 and 1550 kg, respectively.

The maximum acceleration of the cage is 2 m/s^2, and assuming the shaft to be simply supported at the bearings, find the diameter of the shaft, using a maximum allowable shear stress of 40 MPa (neglect the inertia effect of the sheave, and friction in the bearings is negligible).

Solution

It is possible to apply shock and fatigue factors to the bending and twisting moments; values for these factors can be based on those given in Table 20.2, depending on the nature of the loading.

Tension in cable supporting the ascending cage:
To consider the tension in the cable, the dynamic situation is imagined to be static by applying d'Alembert's principle, which is stated briefly by putting an inertia force of $m \times a$ in the opposite direction to the acceleration.

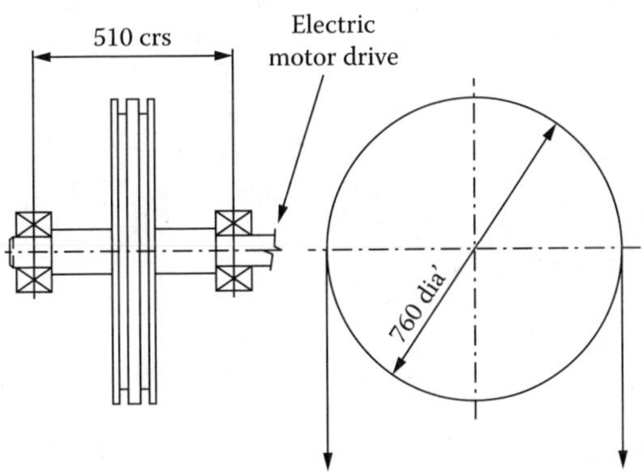

FIGURE 20.7
Overhead gear in Example 20.3.

TABLE 20.2
Factors of Safety for Various Applications

Stationary			Rotating		
	Factor k_b Bending	Factor k_t Torsion		Factor k_b Bending	Factor k_t Torsion
Load applied gradually	1.0	1.0	Gradually	1.5	1.0
Load applied suddenly	1.5–2.0	1.5–2.0	Minor sudden shock	1.5–2.0	1.0–1.5
			Sudden heavy shock	2.0–3.0	1.5–3.0

Tension in cable (T_1).
Gravitational force (mg).
Inertia force (ma).

See Figure 20.8a.

$$T_1 = mg + ma$$

$$T_1 = (1100+910)\,kg \times 9.81\,m/s^2 + (1100+910) \times 2\,kg \cdot m/s^2$$

$$T_1 = 19.72\,kN + 4.02\,kN$$

$$T_1 = 23.74\,kN$$

Tension in cable supporting descending balance mass:
See Figure 20.8b.

$$T_2 + ma = mg$$

$$T_2 = mg - ma$$

$$= 1550 \times 9.81 - 1550 \times 2$$

$$= 15.2 - 3.1\,kN$$

$$T_2 = 12.1\,kN$$

$$Torque = (T_1 - T_2) \times \text{radius of sheave}$$

$$T = (23.74 - 12.1) \times \frac{760}{2} \times 10^{-3}\,kNm$$

$$T = 11.64 \times 0.38\,kNm$$

$$T = 4.423\,kNm$$

Torque supplied by motor.

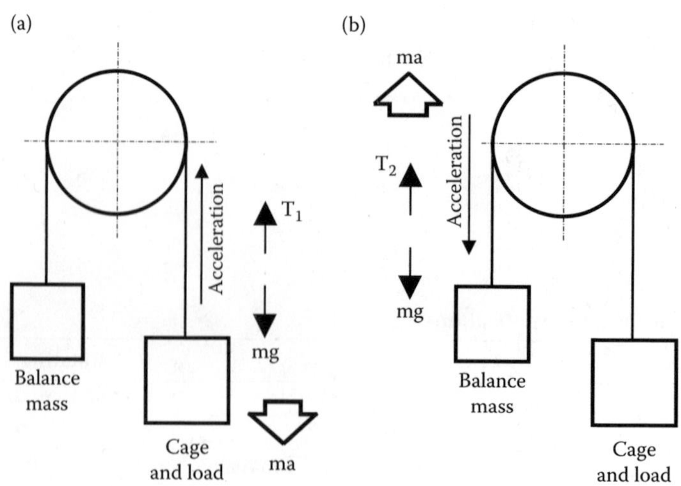

FIGURE 20.8
(a) Tension in cable due to ascending cage. (b) Tension in cable due to descending cage.

Bending of shaft:
Total bending force:

$$= 23.74 + 12.1 \text{ kN}$$

$$= 35.84 \text{ kN}$$

$$= \frac{F}{2} \times \frac{L}{2}$$

$$= \frac{35.83}{2} \times \frac{0.510}{2}$$

Maximum bending moment:

$$BM_{max} = 4.57 \text{ kNm}$$

To find the equivalent torsional moment and hence shaft diameter, consider the load as being applied as a sudden minor shock with a factor $(k_b)=1.5$ and a factor $(k_t)=1.1$.

$$d^3 = \frac{16}{\pi \times t} \times \sqrt{((k_b M)^2 + (k_t T)^2)}$$

$$d^3 = \frac{16}{\pi \times 40 \times 10^3} \times ((1.5 \times 4.57)^2 + (1.1 \times 4.423)^2)$$

$$d^3 = 0.0001273\sqrt{(47 + 23.7)}$$

$$= 0.0010738$$

$$d = 0.1022 \text{ m}$$

$$d = 102.2 \text{ mm}$$

Making an allowance for a keyway (as previously stated),

$$102.2 \text{ mm} \times 1.1 = 112.4 \text{ mm}$$

Now

$$\sigma_t = \frac{M \cdot y}{I}$$

where

$$M = 4.57 \times 10^3 \text{ Nm}$$

$$y = d/2 = 53.5 \text{ mm}$$

$$I = \frac{\pi \cdot D^4}{32}$$

$$I = \frac{\pi \times (107.31 \text{ mm})^4}{32}$$

$$= 13.018 \times 10^{-6}$$

$$\sigma_t = \frac{4.57 \times 10^3 \text{ Nm} \times 57.50 \text{ mm}}{13.018 \times 10^{-6} \text{ mm}^4}$$

$$= 18.78 \times 10^6 \text{ Pa}$$

It is suggested that the shaft may be manufactured from mild steel 220M07 (EN1A) where $\sigma_t = 480 \times 10^6$ Pa will give a Safety Factor (SF) > 6. Hence the shaft will be reasonably resistant to fatigue.

For a suitable stock size select a 110.0 mm diameter.

21

Shaft Design Basics

21.1 Introduction

Shafts are used in a multitude of engineering examples. There are two basic types of shaft:

- Stationary
- Rotating

Stationary shafts do not carry any torsional loading and are only subject to bending loads such as the king pin as used on the front suspension of an automobile.

Rotating shafts are components of a mechanical device that transmits rotational motion and power. It is integral to any mechanical system where power is being transmitted from a power source, such as an electric motor or an engine to other rotating parts within the system. Examples include gear type speed reducers, belt or chain drives, conveyors, pumps, etc.

Figure 21.1 depicts a typical shaft under discussion. The shaft can carry any number of components including gears, couplings, pulleys, cams, etc. and is generally supported on anti-friction bearings. The torque is transmitted to the shaft mounted components using keys, splines, taper pins, clamping bushes, press fits, etc.

21.2 Procedure for Design and Analysis of a Shaft

Figure 21.2 shows a "road map" for the design procedure used in the design of a rotating shaft. The following is a more detailed description of the headings from the road map.

21.2.1 Design Requirements for the Shaft

Before the design can begin it is important to define the end use of the shaft, i.e.:

- To connect a motor to a gearbox or a pump, etc.
- Used in an automotive drive line.
- What will the environment be?
- Will there be any special surface finish requirements?

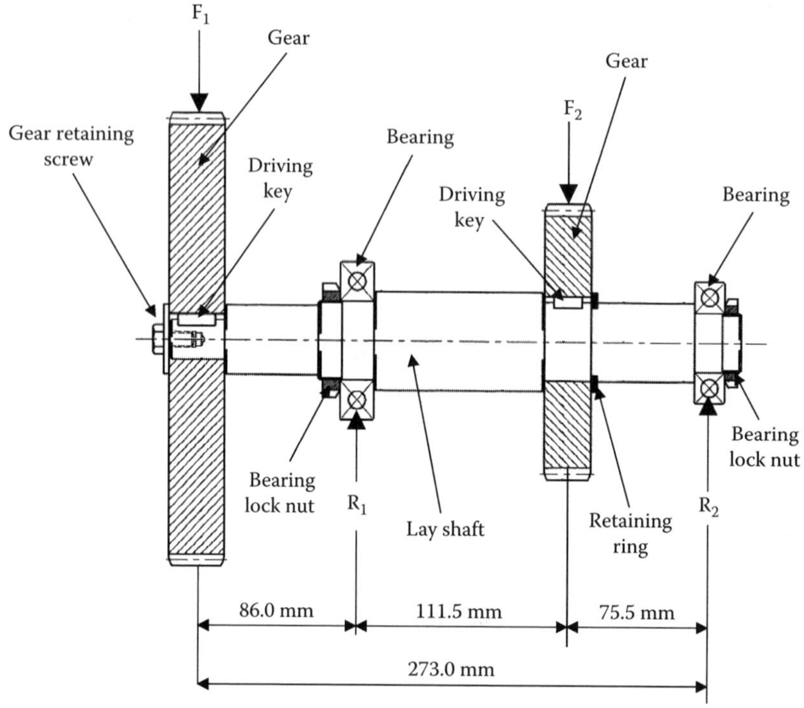

FIGURE 21.1
Typical layshaft.

21.2.2 Geometry of the Shaft

At the initial stage of the design the shaft will have been decided, as well as the mandatory task the shaft will be required to undertake, i.e.:

- Transfer power from a drive motor to a gearbox. In this case couplings will be required at each end of the shaft.
- Design a turning lathe spindle. Gears will be required to convey the power to the shaft and suitable bearings needed to support the spindle in the machine frame.
- Provide a driving force from an axle to a road wheel. This may be in an automotive situation, and here the shaft will be subject to a variable torque as the car or truck accelerates or decelerates during the course of its journey.

These are just some examples of the use of a shaft, and the next phase is to ensure that the shaft is "fit for function" and will not fail during its operating life.

21.2.3 Calculate the Forces Acting on the Shaft

Develop a free-body diagram by replacing the various machine elements mounted on the shaft by their statically equivalent load or torque components.

Specify the location of bearings to support the shaft. The reactions of the bearings supporting radial loads are generally assumed to act at the centre of the bearing.

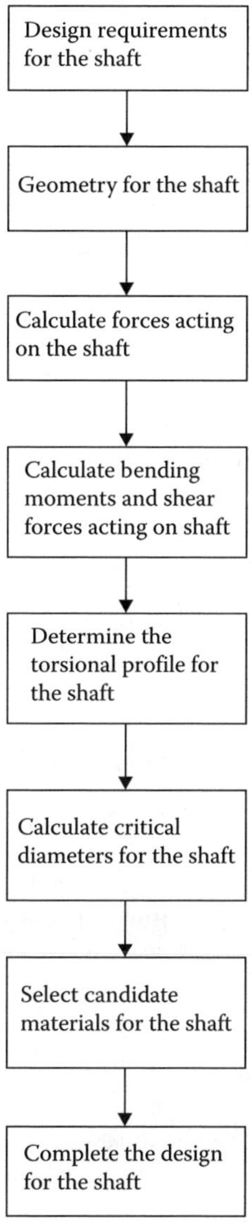

FIGURE 21.2
Road map for the design of a shaft.

An important point is that in a simple shaft there are only two bearings used to support the shaft, and they should, where possible, be placed on either side of the power transmitting elements to provide a stable support for the shaft and to produce reasonably balanced loading on the bearings. The bearings should be placed as close as possible to the power transmitting elements to minimise bending moments, and in addition, the shaft should be kept as short as possible to keep any shaft deflections at reasonable levels.

There are occasions where due to the length of the shaft (such as in "line shafting"), there may be more than two bearings needed to support the shaft. The analysis of this type of problem is outside the scope of this discussion.

21.2.4 Calculate the Bending Moments and Shear Forces Acting on the Shaft

Example 21.1

Consider Figure 21.1 and construct the bending moment and shear force diagrams for this figure assuming:

$$F_1 = 50\ N$$

$$F_2 = 20\ N$$

Solution

The first action is to calculate the reactions R_1 and R_2.

Taking moments about R_2:

$$(50.0\ N \times 273.0\ mm) + (20.0\ N \times 75.5\ mm) = 187.0R_1$$

Therefore $R_1 = 81.069\ N$.

Taking moments about R_1:

$$\left(50.0\ N \times 86.0\ mm\right) + (187.0R_2) = \left(20.0\ N \times 111.5\ mm\right)$$

Therefore $R_2 = -11.069\ N$.

The bending moment acting on the shaft will be calculated next.

From Figure 21.1, the positions of the respective force elements acting on the shaft are next identified and the individual moments then calculated as shown in Table 21.1.

The above results are plotted in Figure 21.3a and b; from the results it is obvious that the maximum bending moment occurs at position R_1.

This diagram is drawn for the x:y plane only and would need to be plotted for the y:z plane also if there were forces acting in the y:z plane.

The resultant internal moment at any section along the shaft may be expressed as:

$$M_x = \sqrt{M_{xy}^2 + M_{xz}^2}$$

There may be a number of force inputs into the shaft from gear, chain or pulley drives, and these will introduce variable bending moments in the shaft.

The calculated shear forces will also be considered when finalising the size of the shaft.

21.2.5 Determine the Torsional Profile of the Shaft

If the shaft is connecting a drive motor to, say, a hydraulic pump, in this situation normally the speed of the shaft will be constant. The torque the shaft needs to transmit may be variable depending upon the load being placed on the pump. In general the designer could determine the maximum torque the driving motor can produce and size the shaft based

TABLE 21.1

Bending Moments for Example 21.1

Position No.	Calculation	Bending Moment
1	0	= 0
2	$(50.0\ N \times 86.0\ mm)$	= 4300.0 N·mm
3	$(50.0\ N \times 197.5\ mm) - (81.069\ N \times 111.5\ mm)$	= 835.806 N·mm
4	$(50.0\ N \times 273.0\ mm) - (81.069\ N \times 187.0\ mm) + (20.0\ N \times 75.5\ mm)$	= 0.097 N·mm

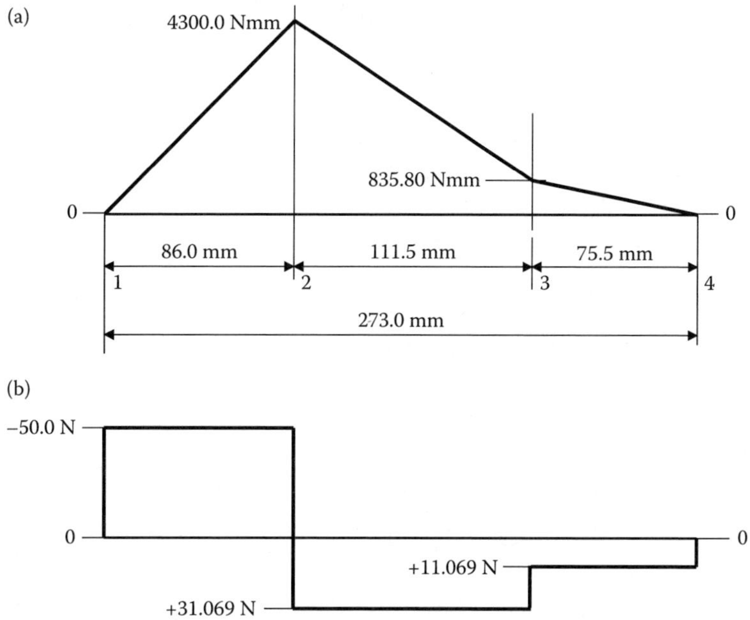

FIGURE 21.3
(a) Bending moment diagram for Example 21.1. (b) Shear force diagram for Example 21.1.

on this calculation, but this may result in a shaft size out of proportion to the duty it will be required to perform and be uneconomical to manufacture. It will be far better to size the shaft based on the maximum load the shaft would see and apply a Safety Factor (SF) to the load.

Questions to ask are: Will the driving load be gradually increased to its operating level? Or, will there be a sharp acceleration from the drive motor followed by a steady load? These factors will greatly influence the torque being applied to the shaft. As an example, in the first case the torsional profile will be a gentle increase followed by a stable period and at the completion of the duty allowed to come gently to rest.

In the second case with the sharply applied acceleration, this will introduce unwanted torsional vibrations in the shaft where the initial torque will be far larger than the shaft would experience when operating normally. In this situation a SF >3 will be needed to ensure a safe life for the shaft.

21.2.6 Calculate the Critical Diameters for the Shaft

If the shaft is "plain," i.e. no changes to the profile of the shaft, as in a connecting shaft, the calculation will be straightforward, as shown in the following.

Considering a shaft under pure torsion as shown in Figure 21.4.

$$\text{Power} = T\omega \tag{21.1}$$

where

T = torque in Nm

ω = angular velocity in rad/s

The unit of power is the watt; W = Nm/s.

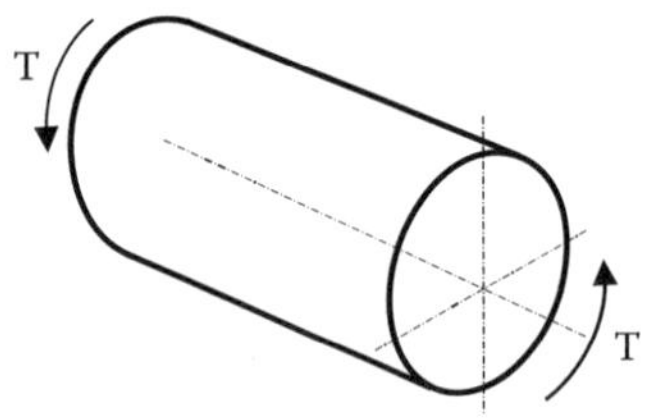

FIGURE 21.4
Basic shaft under torsion.

This formula can also be written as:

$$\text{Power} = 2\pi nT \tag{21.2}$$

where n = angular velocity in rev/s.

$$\frac{T}{J} = \frac{\tau}{r} = \frac{G\theta}{l} \tag{21.3}$$

where
- T = torque in Nm or Nmm
- J = polar second moment of area in m^4 or mm^4
- r = radius to extreme fibre under stress in m or mm
- G = modulus of rigidity in N/m^2 (Pa)
- τ = maximum shear stress in N/m^2 (Pa)
- l = length of shaft in m or mm

21.3 Section Modulus

$$Z_t = \frac{J}{r} \tag{21.4}$$

where Z_t is section modulus in m^3.
For solid circular sections:

$$T = \frac{\pi}{16} \times d^3 \times \tau \text{ or } T = Zt \times \tau \tag{21.5}$$

$$T = \frac{\pi}{16}\frac{D^4 d^4}{D}$$

For hollow sections:

$$T = \frac{\pi}{16} \cdot \left(\frac{D^4 - d^4}{D}\right)$$

where D is the outside diameter in m and d is the inside diameter in m.

For square sections:

$$T = 0.208\ S^3 \cdot \tau \qquad (21.6)$$

where S is the size of the square bar in m.

Values of J and Z_t for various shapes of solid and hollow sections can be found in books covering strength of materials, such as Timoshenko (see Bibliography). Tables 21.5 and 21.6 list properties of some common sections.

Example 21.2

A mild steel shaft is required to transmit 45 kW at 2.5 rev/s.

What diameter of shaft will be suitable using an ultimate shear stress value of 340 MPa? Also calculate the angle of twist over a length of 2.5 m if the modulus of rigidity is 86 GPa.

Solution

$$\text{Power} = 2 \cdot \pi \cdot n \cdot T$$

$$\text{therefore Torque(T)} = \frac{\text{Power}}{2 \cdot \pi \cdot n} = \frac{\pi}{16} d^3 \cdot \tau$$

Allowing for a SF of 4 will give an allowable shear stress (τ) of 85 MPa.

$$d^3 = \frac{\text{Power} \cdot 16}{2\pi^2 \cdot n \cdot t}$$

$$d^3 = \frac{45\ \text{kW} \times 16}{2\pi^2 \times 2.5\ \text{rad/s} \times 85\ \text{MPa}}$$

$$d^3 = 171.65 \times 10^{-6}\ \text{m}^3$$

$$d = 0.05556\ \text{m}$$

Say you have a 56 mm diameter; choose a suitable diameter available stock size.

21.3.1 Angle of Twist

$$\frac{T}{J} = \frac{G\theta}{l} = \frac{2\tau}{d} \qquad (21.7)$$

From the standard torsion equation:

$$\theta = \frac{\pi \times d^3 \times \tau \times 32 \times 2.5}{\pi \times d^4 \times 16 \times 86 \times 10^3}$$

$$\theta = \frac{85\ \text{MPa} \times 5\ \text{m}}{0.056\ \text{m} \times 86\ \text{GPa}}$$

$$\text{Angle of twist in radians} = 0.088$$

$$\text{Converting to degrees} = 0.088 \times 57.3 \text{ degrees per radian}$$

$$\theta = 5.0424 \text{ degrees}$$

Example 21.3

A hollow steel shaft with 220 mm inside diameter and 300 mm outside diameter transmits 2.24 MW at 2 rev/s.

What is the maximum shear stress generated in the shaft material?

Solution

$$D = 0.30 \text{ m}$$

$$d = 0.22 \text{ m}$$

$$P = 2.24 \text{ MW}$$

$$S = 2 \text{ rev/s}$$

$$T = \frac{2.24 \text{ MW}}{2\pi \times 2}$$

$$T = \frac{\pi}{16}\left(\frac{0.3 \text{ m}^4 - 0.22 \text{ m}^4}{0.3 \text{ m}}\right)\tau$$

Rearranging and solving for τ:

$$\tau = \frac{2.24 \text{ Mw}}{4\pi^2} \times \frac{16 \times 0.3 \text{ m}}{0.3 \text{ m}^4 - 0.22 \text{ m}^3}$$

$$\tau = 47.304 \text{ MPa}$$

This value is compared with the ultimate shear stress value for carbon steel of 480 MPa; hence the shaft is considered to be lightly loaded. (See Table 21.7 for the mechanical properties of a range of common materials.)

21.3.2 ASME Shaft Equations

In 1927 the ANSI/ASME standard for the design of transmission shafting B106-1M-1985 developed the following design:

$$D = \left[\frac{32 \cdot N}{\pi}\sqrt{\left[\frac{k_t \cdot M}{S'_n}\right]^2 + \frac{3}{4}\left[\frac{T}{S_y}\right]^2}\right]^{1/3} \tag{21.8}$$

where

k_t = stress concentration factor at shoulder; 1.5 to 2.5

M = bending moment; this is obtained from the bending moment diagram, thus creating a reversed bending moment on the shaft as it rotates (N·m)

T = torsion moment; this is usually uniform (N·m)

N = Factor of Safety (FoS)

D = diameter of shaft at the section being analysed

TABLE 21.2

Reliability Factors

Desired Reliability	Reliability Factor (C_r)
0.50	1.00
0.90	0.90
0.99	0.81
0.999	0.75

S'_n = modified endurance strength (this will depend upon the ultimate tensile strength (S_u)); $S'_n = S_n \cdot C_s \cdot C_r$

S_n = endurance strength (MPa)

S_y = torsional yield strength (MPa)

C_s = size factor (Larger diameter shafts tend to have lower fatigue strengths than smaller shafts for a number of reasons. Values of C_s range from about 0.9 for 50 mm diameter shafts to about 0.65 for shafts up to 250 mm diameter.)

C_r = reliability factor (Published fatigue data usually represent an average value of the endurance strength of a sample of test specimens. In the absence of any specific test data the failure distribution is often assumed to follow a normal or Gaussian distribution with a standard deviation of about 8% of the mean. For a 90% nominal reliability C_r is approximately 0.9, and for a 99% reliability C_r is about 0.8 (see Table 21.2).

The ASME equation can only be used subject to the following assumptions:

- Constant torque
- Fully reversed moment
- No axial load applied

Note: This equation can also be derived theoretically from the distortion energy failure theory as applied to fatigue loading.

Example 21.4

It is required to confirm that the critical diameter of a shaft that is manufactured from steel has a minimum diameter of 40.0 mm. The selected shaft material has a fatigue limit of 480 MPa, and the shaft is subject to a bending moment of 340 N·m. The desired reliability factor is 0.99.

A shoulder with a fillet radius of 0.50 mm is provided on the shaft to locate a bearing; a FoS of 2.5 is required.

Solution

The torque carried by the shaft:

$$\omega = 2\pi n$$

$$\omega = 2 \times p \times 4 \text{ rad/s}$$

$$= 25.13274 \text{ rad/s}$$

$$\text{Torque} = \frac{\text{power}}{\omega}$$

$$= 178 \text{ N} \cdot \text{m}$$

For Equation 21.8 the following values are to be inputs:

$$N = 2.5 \text{ (safety factor)}$$

$$k_t = 1.75$$

$$M = 340 \text{ N} \cdot \text{m}$$

$$T = 178 \text{ N} \cdot \text{m}$$

$$S_n = 480 \text{ MPa}$$

$$C_s = 0.9$$

$$C_r = 0.81$$

$$S'_n = 349.92$$

$$S_y = 125 \text{ MPa}$$

$$D = \left[\left[\frac{32 \times 2.5}{\pi} \right] \times \sqrt{ \left[\frac{1.75 \times 340 \text{ N} \cdot \text{m}}{349.92 \text{ MPa}} \right]^2 + \frac{3}{4} \times \left[\frac{178 \text{ N} \cdot \text{m}}{125 \text{ MPa}} \right]^{-2} } \right]^{1/3}$$

$$D = 37.678 \text{ mm}$$

From the above calculation it is considered that the critical diameter of 40.0 mm is satisfactory.

21.3.3 Fillet Radii and Stress Concentrations

When there is a change in the diameter of a shaft to create a shoulder against which to locate a machine element such as a gear or bearing, depending on the ratio of the two diameters and the radius in the fillet, a stress concentration will be generated. It is recommended that the fillet radius r be as large as possible to minimise the stress concentration. At times if the part is proprietary, the designer has no control over the radius and has to accommodate it in the design. For the purpose of design, fillets are classified into two categories: sharp and well rounded (see Figure 21.5).

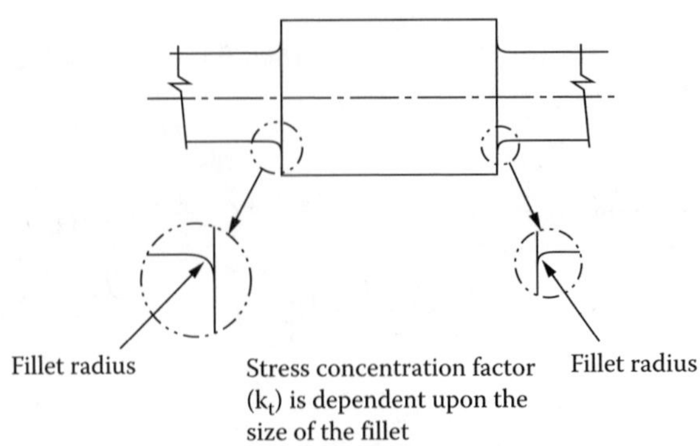

FIGURE 21.5
Fillets on a shaft.

Although the term *sharp* is used, this does not mean truly sharp and without any fillet radius at all. Such a shoulder configuration would have a very high stress concentration factor and should therefore be avoided. One such situation where it is likely to occur is where a ball or roller bearing is to be located. The inner race of the bearing has a factory produced radius, but it is small; therefore the fillet radius on the shaft will have to be smaller to allow the bearing to sit against the shoulder. Where the element has a large chamfer on its bore and is located against a shoulder, or where there is nothing locating against the shoulder, the fillet radius can be as large (well rounded) as possible and the corresponding stress concentration is smaller.

The symbol used for stress concentration is 'k'.

The following values are used for bending:

$$k_b = 2.5\ (\text{sharp fillet})$$

$$k_b = 1.5\ (\text{well rounded fillet})$$

It is possible to establish the stress concentration factor for a specific application. The following formula is used for elastic bending stress:

$$k = K_1 + K_2\left[\frac{2h}{D}\right] + K_3\left(\frac{2h}{D}\right)^2 + K_4\left(\frac{2h}{D}\right)^3 \cdots \tag{21.9}$$

where

$h = D - d$

D = large diameter on shaft

d = adjacent smaller diameter of shaft

Table 21.3 gives the coefficients for bending stress concentration factors, and there is a similar table covering elastic torsional stress which is shown in Table 21.4.

21.3.4 Undercuts

Most ball and roller bearings are manufactured with small fillet radii in their bores, and in these situations where the diameter of the shaft is stress critical, particularly in the vicinity of a bearing, it is possible to provide a local undercut in the shaft adjacent to the shoulder. Figure 21.6 provides details of such an undercut. This allows the bearing to be fitted against the abutment face and still provide a reasonable fillet radius. The undercut will allow the stress concentration local to the fillet radius to be reduced. Care will need

TABLE 21.3

Coefficients for Bending Stress Concentration Factors

	$0.25 \le \frac{h}{r} \le 2.0$	$\mathbf{2.0 \le \frac{h}{r} \le 20.0}$
K_1	$0.927 + 1.149\sqrt{h/r} - 0.08h/r$	$1.225 + 0.831\sqrt{h/r} - 0.010h/r$
K_2	$0.015 - 3.281\sqrt{h/r} + 0.837h/r$	$-3.790 + 0.958\sqrt{h/r} - 0.257h/r$
K_3	$0.847 + 1.716\sqrt{h/r} - 0.506h/r$	$7.374 - 4.834\sqrt{h/r} + 0.862h/r$
K_4	$-0.790 + 0.417\sqrt{h/r} - 0.246h/r$	$-3.809 + 3.046\sqrt{h/r} - 0.595h/r$

TABLE 21.4

Coefficients for Elastic Stress Concentration Factors from Table 21.3

	$0.25 \le \frac{h}{r} \le 4.0$
K_1	$-1.081 + 0.232\sqrt{h/r} + 0.065h/r$
K_2	$-0.493 - 1.820\sqrt{h/r} + 0.517h/r$
K_3	$1.621 + 0.908\sqrt{h/r} - 0.529h/r$
K_4	$-1.081 + 0.232\sqrt{h/r} + 0.065h/r$

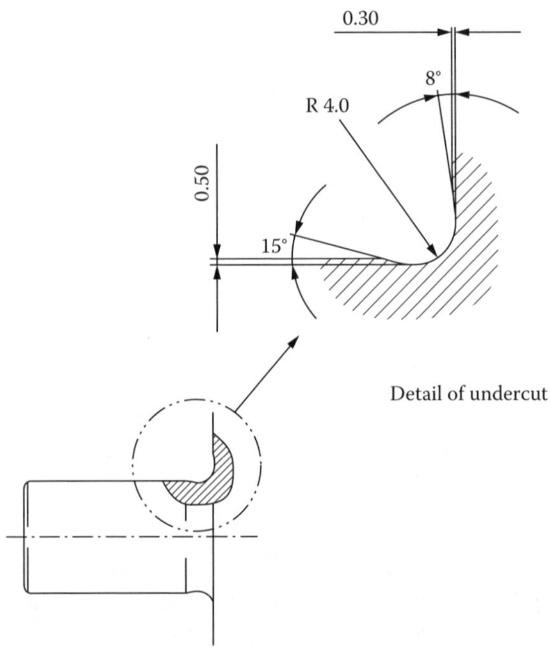

FIGURE 21.6
Detail of undercut in a shaft based on DIN 509.

to be exercised to ensure that the presence of the undercut does not compromise the local strength of the shaft as the diameter is reduced a little, and that the revised SF is still acceptable. Figure 21.7 shows a comparison between a roller bearing fillet radius and the undercut fillet radius, and it can be seen that the undercut allows a slightly larger fillet radius to be made, thereby reducing the stress concentration factor (k_t) by a small amount.

1. Establish the most efficient configuration for the shaft. Position the components along the shaft.
2. Select the most appropriate method for driving the shaft.
    - Determine load profile
    - Reactions/shear stresses/bending moments
    - Dimensions of shaft at critical points

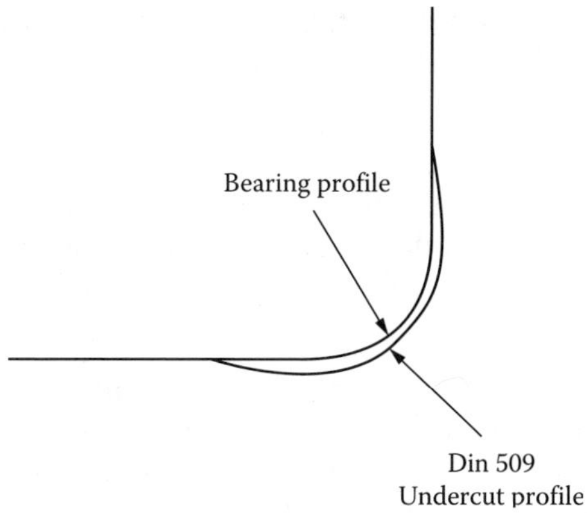

FIGURE 21.7
Comparison between bearing and undercut profiles.

3. Select the most appropriate bearings and components. The following rules are proposed to enable a suitable shaft to be derived to meet the design requirement.
   a. Develop a free-body diagram by replacing the various machine elements mounted on the shaft by their statically equivalent load or torque components. Specify the location of bearings to support the shaft. The reactions of the bearings supporting radial loads are assumed to act at the centre of the bearing. An important point is that generally there are only two bearings used to support the shaft, and they should, where possible, be placed on either side of the power transmitting elements to provide a stable support for the shaft and to produce reasonably balanced loading on the bearings. The bearings should be placed as close as possible to the power transmitting elements to minimise bending moments, and in addition, the shaft should be kept as short as possible to keep any shaft deflections at reasonable levels.
   b. Draw a bending moment diagram in the x–y and x–z planes as shown in Figure 21.3a and b. The resultant internal moment at any section along the shaft may be expressed as:

$$M_x = \sqrt{M_{xy}^2 + M_{xz}^2}$$

   c. Determine the magnitude of the power and torque to be transmitted by the shaft.
   d. Develop a torque diagram. Torque developed from one power transmitting element must balance the torque from other power transmitting elements.
   e. Establish the location of the critical cross section, or the x location where the torque and moment are the largest.
   f. Select the material which the shaft will be manufactured from, specifying:

      Ultimate tensile strength (σ_u)

      Yield strength (σ_y)

      Surface condition:

         Machined

         Ground

         Hot rolled

g. Consider the method of how each element on the shaft will be held in position axially and how the power transmission from each element to the shaft will take place. Design details such as fillet radii, shoulder heights and key seat dimensions should be specified.

h. Determine the forces being exerted on the shaft.

i. Resolve the radial forces into orthogonal components (vertical and horizontal forces).

j. Resolve the final reactions on the end supports (bearings) in each plane.

k. Analyse each critical point on the shaft to determine the minimum acceptable diameter to ensure adequate SFs for the loading at that point on the shaft.
   In general these critical points will include where a change in diameter will take place, where higher values of torque and bending moments will occur, together with stress concentrations that also occur.

l. Due to the nature of the shaft, as the shaft rotates, tensile and compression stresses will occur on either side of the shaft at each rotation. Carry out an endurance study to ensure the life of the shaft will not be compromised due to fatigue (Tables 21.5 through 21.7).

TABLE 21.5

Properties for Some Common Sections

	Circular Sections			**Non-Circular Sections**		
	Solid	**Tubular**		**Square**	**Rect'**	**Thin tube**
	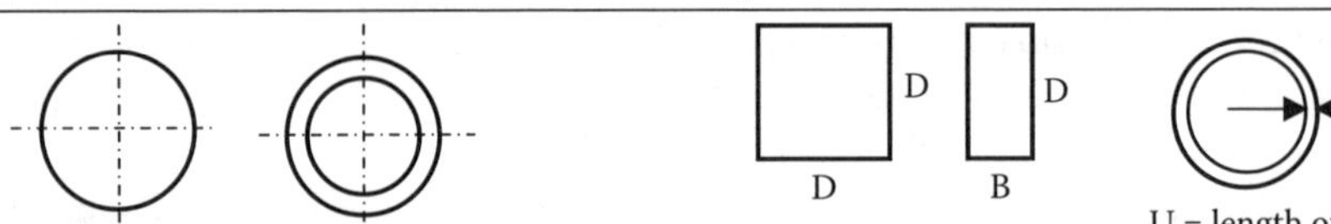 D = outer dia'	d = inner dia'		D	D, B	t = thickness at any point; U = length of median
A	$\frac{\pi}{4}\cdot D^4$	$\frac{\pi}{4}\cdot\left(D^2-d^2\right)$	A	D^2	BD	Mean of areas enclosed by outer and inner boundaries
I_p	$\frac{\pi}{32}\cdot D^4$	$\frac{\pi}{32}\cdot\left(D^4-d^4\right)$	K	0.1406 D^4	βBD^3	$\frac{4A^2}{\int (dU/t)}$
Z_p	$\frac{\pi}{16}\cdot D^3$	$\frac{\pi}{16\cdot D}\cdot\left(D^4-d^4\right)$	Z_p	0.2082 D^3	$\alpha\beta D^2$	2At t=minimum value

TABLE 21.6

Coefficients for Evaluating Rectangular Sections from Table 21.5

B/D	**1.0**	**1.5**	**2.0**	**2.5**	**3.0**	**4.0**	**6.0**	**8.0**	**≥10.0**
α	0.208	0.231	0.246	0.258	0.267	0.282	0.299	0.307	0.333
β	0.141	0.196	0.229	0.249	0.263	0.281	0.299	0.307	0.333

TABLE 21.7

Mechanical Properties for a Range of Common Materials

Mechanical Properties	Carbon Steel	Stainless Steel	Aluminium Alloy	Phosphor Bronze
Specification	220M07 (EN 1A)	304	BS 2014 T6 condition	BS 2874 PB-102
Ultimate strength (f_t)	480 N/mm^2	515 N/mm^2 (min)	483 N/mm^2	280 N/mm^2
Yield strength (0.2%) (t_2)	280 N/mm^2	205 N/mm^2 (min)	414 N/mm^2	120 N/mm^2
Shear strength (f_{so})	727 N/mm^2 (estimated)	340 N/mm^2 (estimated min)	290 M/mm^2	176 N/mm^2 (estimated)
Poisson's ratio (υ)	0.28	0.30	0.33	0.355
Elongation (%)	14	40	8	15
Modulus of elasticity (E)	195 kN/mm^2	193	73.1 kN/mm^2	110 kN/mm^2
Shear modulus (G)	77 kN/mm^2	82	27 kN/mm^2	41 kN/mm^2
Density (ρ)	7800 kg/m^3	8000 kg/m^3	2800 kg/m^2	8000 kg/m^3

22

Keys and Spline Calculations

22.1 Introduction

Keys and splines allow easy attachment and detachment of couplings and gears to shafts and are found in any number of examples in engineering where this feature is required. The types of keyways and splines shown in Figure 22.1 are considered in this chapter and include the following.

22.1.1 Feather Key

The feather key is considered the least strong connection and is restricted to the simpler connections where load reversals are minimal and where the shear strength of the key is considered sufficient.

22.1.2 Straight Spline

The straight sided spline is able to carry reversing loads and is reasonably simple to manufacture. The load carrying capacity of the spline is obviously better than the simple feather key, but due to pitch errors, not all the splines carry the full load, allowing other splines to become overloaded; therefore considerable care is required in the calculation to ensure there is adequate strength. The manufacture of the spline can be facilitated using fairly simple workshop machine tools using an indexing fixture. Most straight sided splines generally have six teeth, but this can be increased to eight teeth.

22.1.3 Involute Spline

The involute spline is judged to be the strongest connection method as the number of teeth is increased over that of the straight spline. Up to 24 teeth on the larger diameter splines is reasonable where the load carrying capacity is very high. They are also able to take high reversing loads. The manufacture is restricted to gear hobbing machines, which limits the types of manufacturing facilities to manufacture them.

Because the tooth form is essentially based on the gear tooth, pitch errors are minimised, ensuring a far greater number of teeth able to carry the torsional load. In any shaft calculation the strength of the shaft will dictate the load carrying capacity, and therefore this will be the first requirement to be calculated.

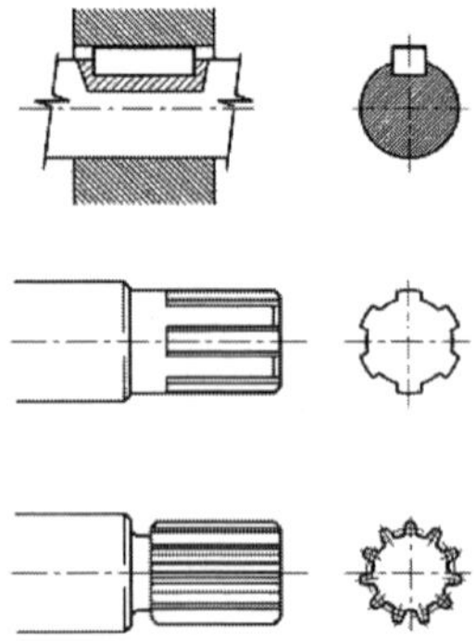

22.1.1 Feather key

22.1.2 Straight spline

22.1.3 Involute spline

FIGURE 22.1
Types of connections covered in this chapter.

22.2 Procedure for Estimating the Strength Capacity of Shaft

1. Estimate the capacity of the shaft for transmitting the required torque with a reduced shaft diameter that is the diameter of the shaft that does not include either the depth of the key or spline (see Figure 22.2). The general nomenclature for a key or spline is covered in Table 22.1. The nomenclature for a key connection is covered in Table 22.2, and for a straight sided spline, see Table 22.3.
2. Evaluate the shear and compressive stresses in the shaft and either the single key or spline.

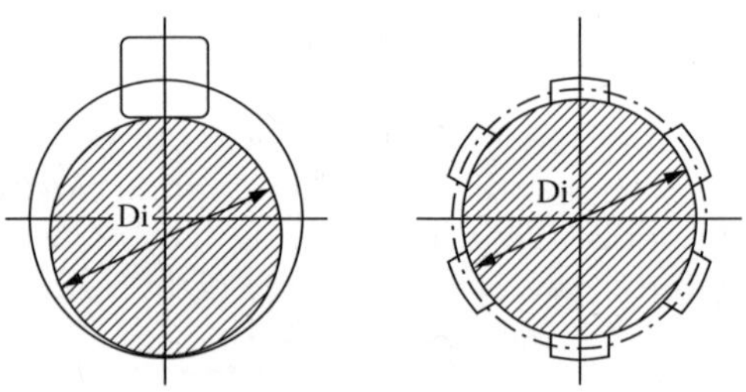

FIGURE 22.2
Reduced diameter for keys and splines.

TABLE 22.1
Nomenclature for Key and Spline

D_o	= Shaft outside diameter	m
D_i	= Reduced diameter	m
D_h	= Diameter of hole – hollow shaft	
T	= Applied torque	Nm
K_s	= Service factor	
σ_c	= Compressive stress of key	N/m^2
τ	= Shear stress in key	N/m^2

TABLE 22.2

Nomenclature for Key

D	= Nominal diameter of shaft	m
x	= Depth of keyface (key/shaft) taking side force	m
x_1	= Depth of keyface (key/hub) taking side force	m
r	= Radius of shaft (D_o/2)	m
L_e	= Effective length of key	m
c	= Key chamfer size	m
d	= Depth of key	m
w	= Width of key	m
t_1	= Depth of keyway	m
T	= Applied torque	Nm
F	= Force acting on key (T/r)	N
K_s	= Service factor	
	For fixed/close fit keyway, $K_s=K_a{\cdot}K_d/K_f$	
	For sliding fit keyway, $K_s=K_a{\cdot}K_d/K_w$	
σ_c	= Resulting compressive stress on key	N/m²
τ	= Resulting shear stress in key	N/m²

TABLE 22.3

Nomenclature for Straight Sided Spline

D_m	= Mean (pitch) diameter	m
r	= Mean radius of spline (D_m/2)	m
n	= Number of splines	
L_e	= Effective length of spline (straight length)	m
d	= Depth of spline	m
w	= Width of spline	m
T	= Applied torque	Nm
K_s	= Service factor	
	For a fixed/guided spline, $K_s=K_a{\cdot}K_d/K_f$	
	For a flexible/sliding fit spline, $K_s=K_a{\cdot}K_m{\cdot}K_d/K_w$	
σ_c	= Resulting compressive stress in spline	Nm²
τ	= Resulting shear stress in spline	Nm²

3. Consider the service factor relating to the application factor, design factor, fatigue life factor and wear factor from the tables for the particular design and application being considered.

Shear stress (τ) generated in a solid shaft due to an applied torque:

$$\tau = \frac{16 \cdot T \cdot K_s}{\pi \cdot D_i^3} \tag{22.1}$$

Shear stress (τ) generated in a hollow shaft due to an applied torque:

$$\tau = \frac{16 \cdot T \cdot D_i \cdot K_s}{\pi \cdot \left(D_i^4 - D_h^4\right)} \tag{22.2}$$

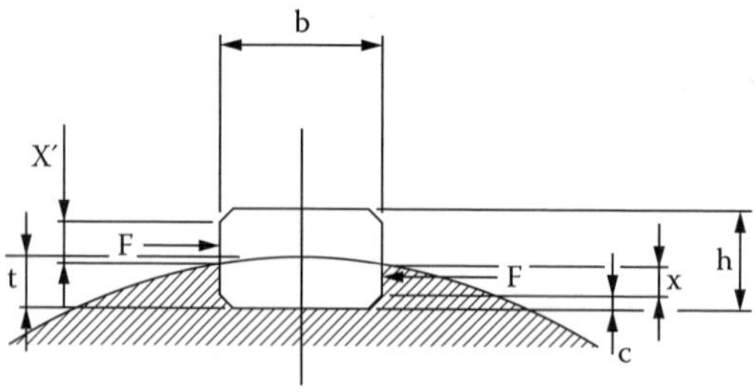

FIGURE 22.3
Nomenclature for key.

22.3 Strength Capacity of Key

$$x = t_1 - \left(r - \sqrt{r^2 - (W/2)^2}\right) - c \tag{22.3}$$

$$x_1 = D - x - 2 \cdot c$$

For a 25.000 mm diameter shaft fitted with a 8.0 mm × 7.0 mm key,

$$\begin{aligned} x &= 4 - (12.5 - \sqrt{12.2 - 42}) - 0.4 \\ &= 2.94\ \text{mm} \end{aligned} \tag{22.4}$$

$$\begin{aligned} x_1 &= 7 - 2.94 \\ &= 3.26\ \text{mm} \end{aligned}$$

where
K_s = Service factor
K_a = Application factor
K_d = Design factor
K_f = Fatigue life factor
K_w = Wear life factor
For values of these factors, see Tables 22.9 through 22.13.

Note: It is considered reasonable to use x = D/2; however, Figure 22.3 will provide a more accurate value of x.

22.4 Strength Capacity of an ISO Straight Sided Spline

For the nomenclature used for an ISO straight sided spline see Figure 22.4 and Table 22.3.
Compressive strength of straight sided spline:

$$\sigma = \frac{T \cdot K_s}{n \cdot d \cdot L_e \cdot r} \tag{22.5}$$

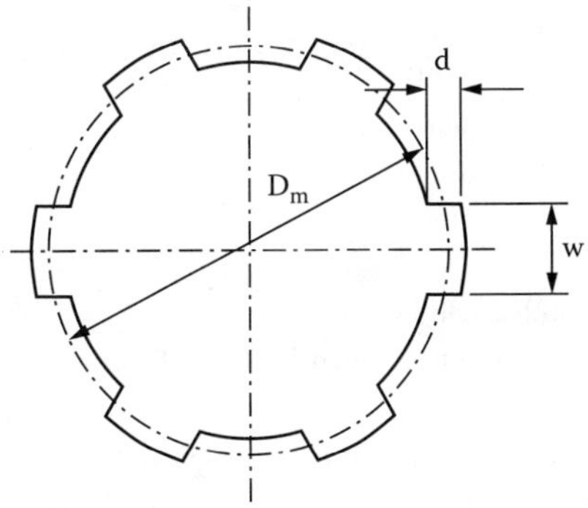

FIGURE 22.4
Nomenclature for straight sided spline.

22.5 Strength Capacity of ISO Involute Spline

For the nomenclature used for involute splines see Figure 22.5 and Table 22.4.

Note:

For 30° flat root splines, h = 0.9 m.

For fillet root splines, h = m.

Shear stress at mean diameter of spline resulting from the applied torque, including a factor of 2 (assuming only half the teeth are loaded due to pitch errors):

$$\tau = \frac{4 \cdot T \cdot K_s}{D_m \cdot n \cdot L_e \cdot h} \tag{22.6}$$

The compressive stress in teeth due to the applied torque:

$$\sigma_c = \frac{2 \cdot T \cdot K_s}{D_m \cdot n \cdot L_e \cdot h} \tag{22.7}$$

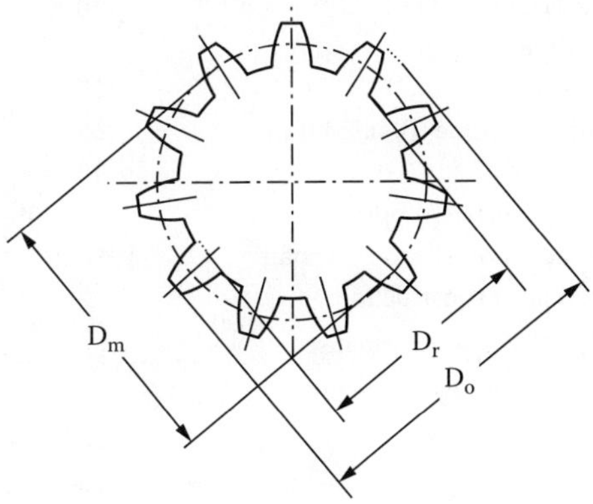

FIGURE 22.5
Nomenclature for involute spline.

TABLE 22.4

Nomenclature for Involute Splines

D_m	= Mean (pitch) diameter	m
D_o	= Outside diameter	m
D_r	= Root diameter	m
M	= Module	m
T	= Tooth thickness	m
H	= Depth of engagement of spline teeth	m
N	= No. of teeth	
L_e	= Length of engagement (straight length)	m
T	= Applied torque	Nm
K_s	= Service factor	
	For a fixed close fit/guided spline, $K_s = K_a/K_f$	
	For a flexible/sliding spline, $K_s = K_m \cdot K_d/k_w$	
σ_c	= Resulting compressive stress in shaft	N/m^2
τ	= Resulting shear stress in shaft	N/m^2

Tables 22.9 through 22.13 give the respective values for the factors K_d, K_a, K_m, K_f and K_w used in the calculations.

22.6 Example Calculations

The following example calculations compare the strength of a keyed shaft, a straight sided spline and an involute spline and are based on identical transmitted power and revolutions per second. The keys and splines are all considered to be close fitting and guided.

For input data for examples, see Table 22.5.

TABLE 22.5

Input Data for Example Calculations

1	Power transferred	10	kW
2	Rotational speed	25	rev/s
3	Shaft diameter	25	mm
4	Resulting torque (P/n.9549)	63.66	Nm
5	Design life	10×10^3	h
6	Number of start/stops	10×10^4	h
7	Total no. of revolutions of shaft	900×10^6	rev
8	Materials of construction:		
	BS EN 10083 C45 normalised		
	σ_t—ultimate tensile strength	580	MPa
	σ_y—yield strength	300	MPa
	σ_s—shear strength	200	MPa
	σ_c—compressive strength	130	MPa

TABLE 22.6

Factors Used for Example Calculations

K_d	Design factor	1.0
K_a	Application factor	1.0
K_m	Load distribution factor	1.0
K_f	Fatigue factor	0.5
K_w	Wear factor	0.7
K_s	Service factor:	
K_{shaft}	Shaft	2.0
K_k	Key (close fit)	2.0
K_{sp}	Spline (close fit)	2.0

For details of factors used in this example, see Table 22.6.

Now:

$$P = \frac{T\omega}{9549}$$

where

P is in kW

T is in Nm

ω is in rev/min

Solving for torque:

$$T = \frac{9549 \times P}{\omega}$$

$$= \frac{9549 \times 10\ \text{kW}}{1500\ \text{rev/min}}$$

$$T = 63.6\ \text{Nm}$$

22.6.1 Shaft Calculations

Assume a basic shaft diameter of 25.00 mm.

Shear stress in shaft (τ):

$$\tau = \frac{16 \cdot T \cdot K_s}{\pi \cdot D_o^3} \tag{22.8}$$

$$= \frac{16 \times 63.66\ \text{Nm} \times 2}{\pi \times (25.00\ \text{mm})^3}$$

$$= 41.5\ \text{MPa}$$

Factor of Safety (FoS):

$$FoS = \frac{\sigma_s}{\tau} \tag{22.9}$$

$$= \frac{200\ \text{MPa}}{41.5\ \text{MPa}}$$

$$FoS = 4.82$$

22.6.2 Key Calculations

Consider a key with dimensions 8.00 mm wide (w), 7.00 mm deep (d) and 70.00 mm long.

The effective length (L_e) will be 70.00 − (2×4.00 mm) = 62.00 mm.

From Figure 22.3:

$$x = 2.94\ \text{mm} \quad \text{By calculation.}$$

$$x_1 = 3.26\ \text{mm}$$

D_i (reduced shaft diameter)

$$= D_o - t_1$$

$$= 25.00 - 4.00$$

$$= 21.00\ \text{mm}$$

The shear stress in the reduced shaft diameter:

$$\tau = \frac{16 \cdot T \cdot K_s}{\pi \cdot D_i^3}$$

$$= \frac{16 \times 63.66\ \text{Nm} \cdot 2}{\pi \times (21.00\ \text{mm})^3}$$

$$= 70.02\ \text{MPa}$$

$$FoS = \frac{\sigma_s}{\tau}$$

$$= \frac{200\ \text{MPa}}{70.02\ \text{MPa}}$$

$$= 2.86$$

22.6.3 Straight Spline Calculations

Consider a straight sided spline with an outside diameter of 25.00 mm.

The nomenclature used for a straight sided spline is as shown in Figure 22.4.

For this example see Table 22.7 for input data for straight sided spline calculations.

TABLE 22.7

Input Data for Straight Sided Spline Calculation

D_i	= Reduced diameter	21.00	mm
n	= Number of teeth	6	
L_e	= Effective length	25.00	mm
r	= Mean radius of teeth	11.50	mm
c	= Chamfer/radius size at top and bottom of teeth	0.3 (assumed)	mm
y	= The effective depth of the teeth	1.40 ($(D-D_i)/2-2\cdot c$)	mm

The shear stress in the reduced diameter (D_i):

$$\tau = \frac{16 \cdot T \cdot K_s}{\pi \cdot D_i^3}$$

$$= \frac{16 \times 63.66\ \text{Nm} \cdot 2}{\pi \times (21.00\ \text{mm})^3}$$

$$= 70.02\ \text{MPa}$$

FoS:

$$\text{FoS} = \frac{\sigma_s}{\tau}$$

$$= \frac{200\ \text{MPa}}{70.02\ \text{MPa}}$$

$$= 2.86$$

The compressive stress in the spline and shaft:

$$\sigma_c = \frac{T \cdot K_s}{L_e \cdot n \cdot r \cdot y}$$

$$= \frac{63.66\ \text{Nm} \times 2}{25.0\ \text{mm} \times 6 \times 11.5\ \text{mm} \times 1.4\ \text{mm}}$$

$$= 52.72\ \text{MPa}$$

FoS:

$$= \frac{130\ \text{MPa}}{52.72\ \text{MPa}}$$

$$\text{FoS} = 2.47$$

22.6.4 Involute Spline Calculations

Considering an involute spline of 25.00×24 teeth.

The nomenclature used for an involute spline is covered in Figure 22.5.

TABLE 22.8

Input Data for Involute Spline Calculation

n	= Number of teeth	24	mm
m	= Module	1.00	mm
L	= Length of spline	10.0	mm
D	= Mean (pitch) diameter	24.00 (m·n)	mm
D_i	= Reduced diameter	22.50 (m·(n·1.5))	mm
D_o	= Outside diameter	25.00 (m·(n+1))	mm
p	= Pitch	3.1416 (m·π)	mm
t	= Tooth thickness	1.157 (p/2)	mm
h	= Tooth height	0.90 (0.9·m)	mm
c	= Chamfer/radius size at top and bottom of teeth	0.3 (assumed)	mm
y	= The effective depth of the teeth	1.40 ((D−D_i)/2−2·c)	mm

For this example see Table 22.8 for input data for involute sided spline calculations. The shear stress in the reduced shaft diameter (D_i):

$$t = \frac{16 \cdot T \cdot K_s}{\pi \cdot D_i^3}$$

$$= \frac{16 \times 63.66 \text{ Nm} \times 2}{\pi \times (2250 \text{ mm})^3}$$

$$= 56.927 \text{ MPa}$$

FoS:

$$= \frac{200 \text{ MPa}}{56.927 \text{ Pa}}$$

$$\text{FoS} = 3.513$$

The shear stress in the spline teeth:

$$\tau = \frac{4 \cdot T \cdot K_s}{L \cdot n \cdot t \cdot D}$$

$$= \frac{4 \times 63.66 \times 2}{10.00 \times 24 \times 57 \text{ mm} \times 24.00 \text{ mm}}$$

$$= 56.316 \text{ MPa}$$

FoS:

$$= \frac{200 \text{ MPa}}{56.316 \text{ MPa}}$$

$$\text{FoS} = 3.551$$

The compressive stress in the spline teeth:

$$\tau = \frac{2 \cdot T \cdot K_s}{L \cdot n \cdot h \cdot D_i}$$

$$= \frac{2.63 \times 66\ \text{Nm} \times 2}{10.00\ \text{mm} \times 24.0 \times 9\ \text{mm} \times 4.00\ \text{mm}}$$

$$= 49.12\ \text{MPa}$$

FoS:

$$= \frac{130\ \text{MPa}}{49.12\ \text{MPa}}$$

$$\text{FoS} = 2.647$$

Table 22.14 summarises the stresses calculated in the above examples.

Tables 22.9 through 22.13 give the respective values for the factors K_d, K_a, K_m, K_f and K_w used in the calculations.

TABLE 22.9

Showing Keyway/Spline Design Factor K_d

Coupling Design	Design Factor
Fixed (close fit) loaded.	1
No relative axial/radial movement of the hub and shaft occurs: the relative positions are fixed using a suitable construction design, e.g., threaded fasteners, press fits.	
Flexible/sliding—unloaded.	3
Relative position of the shaft and hub is not fixed. Axial movements of the hub along the shaft occur only when coupling is unloaded.	
The shaft is effectively fixed when loaded.	
Flexible/sliding when loaded.	9
Sliding coupling with loading relative positions of the shaft and hub is not fixed. Axial shifts of the hub along shaft occur in loaded couplings.	

TABLE 22.10

Showing Spline Application Factors K_a

	Type of Load			
	Uniform Generators, Fans	Light Shock Oscillating Pumps	Intermediate Shock Actuators	Heavy Shock Presses, Shears
Power Source	Application Factor (K_a)			
Uniform (turbine motor)	1.0	1.2	1.5	1.8
Light shock (hydraulic motor)	1.2	1.3	1.8	2.1
Medium shock (ICE)	2.0	2.2	2.4	2.8

TABLE 22.11

Showing the Spline Distribution Factors K_m

Misalignment (mm/mm)	Load Misalignment Factor (K_m)			
	12 mm Face Width	25 mm Face Width	50 mm Face Width	100 mm Face Width
0.001	1.0	1.0	1.0	1.5
0.002	1.0	1.0	1.5	2.0
0.004	1.0	1.5	2.0	2.5
0.008	1.5	2.0	2.5	3.0

Note: This factor relates to the reduction in strength as a result of the misalignment of the spline.

TABLE 22.12

Showing the Fatigue Life Factors for Splines K_f

Number of Torque Cycles (Start/Stop Cycles)	Fatigue Life Factor (K_f)	
	Unidirectional	Fully Reversed
1×10^3	1.8	1.8
1×10^4	1.0	1.0
1×10^5	0.5	0.4
1×10^6	0.4	0.3
1×10^7	0.3	0.2

TABLE 22.13

Showing Wear Life Factor for Splines K_w

No. of Revolutions of Spline	Life Wear Factor (K_w)
10×10^3	4.0
10×10^4	2.8
10×10^5	2.0
10×10^6	1.4
10×10^7	1.0
10×10^8	0.7
10×10^9	0.5

TABLE 22.14

Summary of Key, Straight Sided and Involute Spline Strengths from Examples

	Key	Straight Sided Spline	Involute Spline
Reduced shaft shear stress (MPa)	70.02	70.02	56.927
Factor of safety (FoS)	2.86	2.86	3.513
Compressive stress in spline teeth		52.72 MPa	49.12 MPa
Factor of safety (FoS)		2.47	2.647

23

Columns and Struts

23.1 Background

A column is vertical and supported at both ends. A strut may be inclined or even horizontal and have a variety of end fixings.

The notes below consider that the members are straight and manufactured from a homogeneous engineering material and are used within the elastic operating range of the material. It is further considered that the applied load or force is being applied along the centroid of the end features.

The column or strut will remain straight until the end force reaches a certain value and buckling begins. An increase in force will then result in the column further buckling, but a reduction in this force will then result in the column or strut returning to its original condition. The value of this critical force will depend upon the slenderness ratio and the end fixing conditions together with the material of construction.

The slenderness ratio (λ) is defined as:

$$\lambda = \frac{l}{k_{\min}} \tag{23.1}$$

where l is the effective length and $k_{\min}$ is the least radius of gyration of the section.

The principal end fixing conditions are as follows:

1. Pinned (hinged) at both ends
2. Fixed (built in) at both ends
3. Fixed at one end and free at the other end
4. Fixed at one end and pinned at the other end

Figure 23.1 depicts these end conditions.

The failure of a strut or column is a function of its length and will have a tendency to fail in pure bending; in this situation the Euler formula is suitable to analyse the condition.

$$\sigma_c = \frac{\pi^2 E}{\left(\dfrac{L}{k_{\min}}\right)^2} \tag{23.2}$$

Euler's theory takes no account of the compressive stress in the member. For a material with a compressive stress less than 300 MPa and a Young's modulus approximately 200 kPa, the strut will tend to fail in compression when the slenderness ratio (l/k) is less

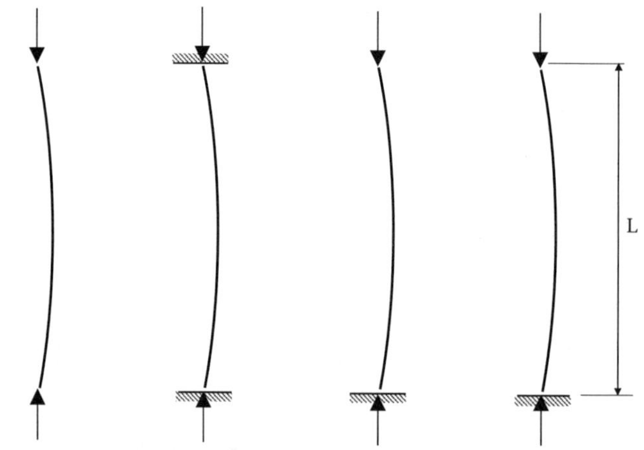

FIGURE 23.1
End fixing conditions.

than 80. It has been found that Euler's is not reliable for slenderness ratios less than 80 and should be avoided with slenderness ratios less than 120.

In many practical cases struts may have slenderness ratios below which the Euler formula is applicable. A number of empirical formulae have been developed to improve the prediction of the critical stress, and these include, among others:

- Rankine-Gordon
- Perry-Robertson
- Johnson-Euler
- Euler-Engesser

The first two theories (Rankine-Gordon and Perry-Robertson) will be considered in this chapter.

23.2 Rankine–Gordon Method

Gordon suggested an empirical formula be used (based on experimental data). Rankine modified this formula to the one used today.

Formula:

$$P_r = \frac{A\sigma_c}{1 + a\left(\frac{l}{k}\right)^2} \tag{23.3}$$

where

P_r = crushing or crippling load (Rankine-Gordon) (kN)
σ_c = direct crushing stress (MPa)

A = cross-sectional area of column or strut (m^2)
l = length of strut or column (m)
k = least radius of gyration of cross section (m)
a = constant depending upon end fixing

also $k = \sqrt{\left(\frac{I}{A}\right)}$ and I the least second moment of area of section

Typical values for σ_c and a are shown in Table 23.1; these will vary dependent upon materials and type of end fixings.

The safe load:

$$P = \frac{P_r}{\text{Factor of safety}} \tag{23.4}$$

For long columns, Euler's formula applies:

$$P_e = \frac{\pi^2 EI}{l^2} \text{ for hinged or rounded ends.}$$

where

$P_e = \frac{\pi^2 EI}{l^2}$ for one end fixed
E = modulus of elasticity (GPa)
P_e = crippling load (Euler)

For short columns, where buckling effects are absent and hence material is in direct compression, these equations reduce to:

$$P_r = P_e = A\sigma_c \tag{23.5}$$

Example 23.1

A strut in a framed structure is manufactured from a steel pipe 150 mm outside diameter and 12.5 mm wall thickness. The length is 3.05 m and the pipe is pin-jointed at both ends.

Using a Factor of Safety (FoS) of 5, what is the safe load?

TABLE 23.1

Typical Values of a for Use in the Rankine-Gordon Formula

		Value of a		
Material	**σ_c (MPa)**	**Fixed Ends**	**Hinged Ends**	**One End Fixed, the Other End Hinged**
Cast iron	560	1/6400	1/1600	1/3600
Low carbon steel	325	1/30,000	1/7500	1/16,875

Note: Since the above values of a are not exactly equal to the theoretical values, the Rankine-Gordon loads for long columns or struts will not be identical to those estimated by the Euler theory as estimated.

Solution

$$P_r = \frac{A\sigma_c}{1+\left(\frac{I^2}{k^2}\right)}$$

$$A = \frac{\pi}{4}\left(0.15^2 - (2 \text{x } 0.0125)^2\right)$$

$$= 0.00540 \text{ m}^2$$

$$\sigma_c = 325 \text{ MPa (from Table 23.1)}$$

$$a = \frac{1}{7500} \quad \text{(for hinged ends)}$$

$$l = 3.05 \text{ m}$$

$$k^2 = \frac{I}{A} \quad \text{(for a circular pipe)}$$

$$I = \frac{\pi}{64}\left(D^4 - d^4\right)$$

$$= \frac{\pi}{64}\left(0.15^4 - 0.125^4\right)$$

$$= \frac{\pi}{64}(0.000506 - 0.000244)$$

$$= 0.00001286 \text{ m}^4 \text{ or } 12.861 \times 10^{-6} \text{ m}^4$$

and

$$\text{FoS} = 5$$

$$\text{Safeload} = \frac{1.154}{5}$$

$$= 0.2308 \text{ MN}$$

$$\text{say} = 231 \text{ kN}$$

Example 23.2

A column for a crane gantry consists of two 406×152 universal beams, connected by 12.5 mm thick plates as shown in Figure 23.2.

The overall length is 8.2 m and the ends may be considered fixed.
Calculate the safe load using a Safety Factor (SF) of 5.

Solution

From the *Steel Designers' Manual*, for 406 mm×140 mm×46 kg universal beam the least second moment of area quoted:

$$I_{yy} = 539 \text{ cm}^4$$

$$A = 59 \text{ cm}^2$$

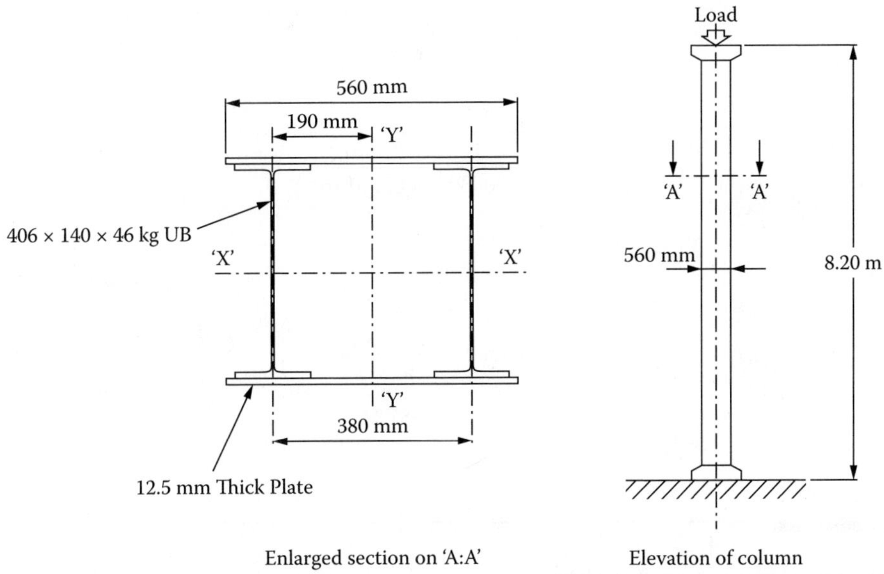

FIGURE 23.2
Graphic for Example 23.2.

For the whole column, it can be shown that the least value of I is about the Y:Y axis.

$$I_{yy} = (I_{cg} + Ah^2) \times 2$$

$$= (539 + 59 \times 19^2) \times 2$$

$$= 43676 \text{ cm}^4$$

To find I_{yy} for column, for two 406 mm × 140 mm sections:

$$I_{yy} \text{ for column} = 43676 \text{ cm}^4$$

$$\text{Area of cross section} = (2 \times 59) + (56 \times 2.5)$$

$$= 258 \text{ cm}^2$$

$$= 0.0258 \text{ m}^2$$

In this case the connecting plates can be disregarded and omitted from the second moment of area. Hence

$$k^2 = \frac{43676}{59 \times 2}$$

$$= 370.136 \text{ cm}^2$$

$$\text{and } l = 8.2 \text{ m}$$

To evaluate k^2 the area of the connecting rods is ignored.

From Table 23.1, case I mild steel – fixed ends,

$$a = \frac{1}{30000}$$

and

$$\sigma_c = 325 \text{ MPa}$$

$$P_r = \frac{0.0258 \text{ m}^2 \times 325 \text{ MPa}}{1 + \frac{1}{30000} \times \frac{(8.2\,\text{m})^2}{0.03701 \text{ m}^2}}$$

$$= 7.906 \text{ MN}$$

FoS of 5:

$$\text{Safe load} = \frac{7.906 \text{ MN}}{5 \text{ (SF)}}$$

$$= 1.581 \text{ MN}$$

23.3 Perry–Robertson Method

The Perry-Robertson formula more closely represents the behaviour of practical columns allowing for the uncertainty of the eccentricity of loading, initial curvature pre-existing in the column, local defects, etc.

Here the critical value for stress is calculated from:

$$\sigma_c = \frac{\sigma_y + (\eta + 1)\sigma_e}{2} - \sqrt{\left\{\left[\frac{\sigma_y + (\eta + 1)\sigma_e}{2}\right]^2 - \sigma_y \cdot \sigma_e\right\}} \tag{23.6}$$

where

$$\eta = 0.003 \frac{l}{k_{min}}$$

Figure 23.3 compares the results of the Perry-Robertson calculations against the Euler's predictions for steel and aluminium.

Notes on the use of the Perry-Robertson method:

1. In practice $\frac{l}{k_{min}} \leq 85$ (23.7)
2. When $\frac{l}{k_{min}} = 85$ all steels give a crippling stress (σ_c) of 210 MPa (approximately).

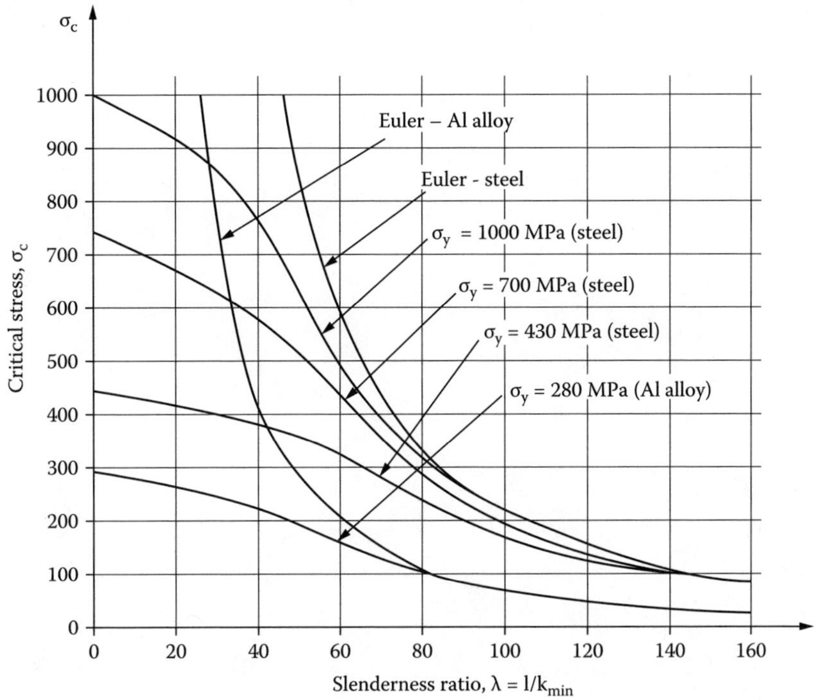

FIGURE 23.3
Comparison of Perry-Robertson against Euler's calculations.

24

Eccentric Loading

There are cases where the applied load acting on an element does not act on the neutral axis of a section. In this situation the principles of superposition will apply. Figure 24.1a through c—demonstrates the application. Figure 24.1a considers the element with an axial load acting directly on the neutral axis. Figure 24.1b then considers the effect of an eccentric load which is offset from the neutral axis; combining the loading of (a) and (b) will produce a load distribution as shown in (c).

It will be seen that the distribution of the load is offset where the loads are not equal about the neutral axis. The stresses in the element will reflect directly that of the load.

The general formulae for eccentric loading:

$$\sigma_1 = \frac{W_e}{A}\left(1 + \frac{X \cdot Y_1}{k^2}\right)$$

$$\sigma_2 = \frac{W_e}{A}\left(1 - \frac{X \cdot Y_2}{k^2}\right)$$

where

σ_1 = extreme fibre stress nearest load (MPa)
σ_2 = extreme fibre stress farthest from load (MPa)
W_e = actual eccentric load (kN)
A = area of section (m^2)
x = arm of eccentricity (m)
y_1 = perpendicular distance from neutral axis of section to outside edge nearest load (m)
y_2 = perpendicular distance from neutral axis of section to outside edge farthest from load (m)
k = radius of gyration (m)

The above may be demonstrated in the following example.

Example 24.1

Consider the cast iron shearing machine frame shown in Figure 24.2 which is subject to a force of 400 kN. A section of the frame is shown in Figure 24.3 on which is shown the position of the centroid, which has already been determined.

It is required to calculate the maximum tensile and compressive stresses in the section of the frame due to the given loading.

First find I_{xx} of the section; this will be the summation of parts (1), (2), (3) and (4). Use the parallel axis theorem for each part: This is determined as shown in Table 24.1. The next step is to determine the radius of gyration k^2.

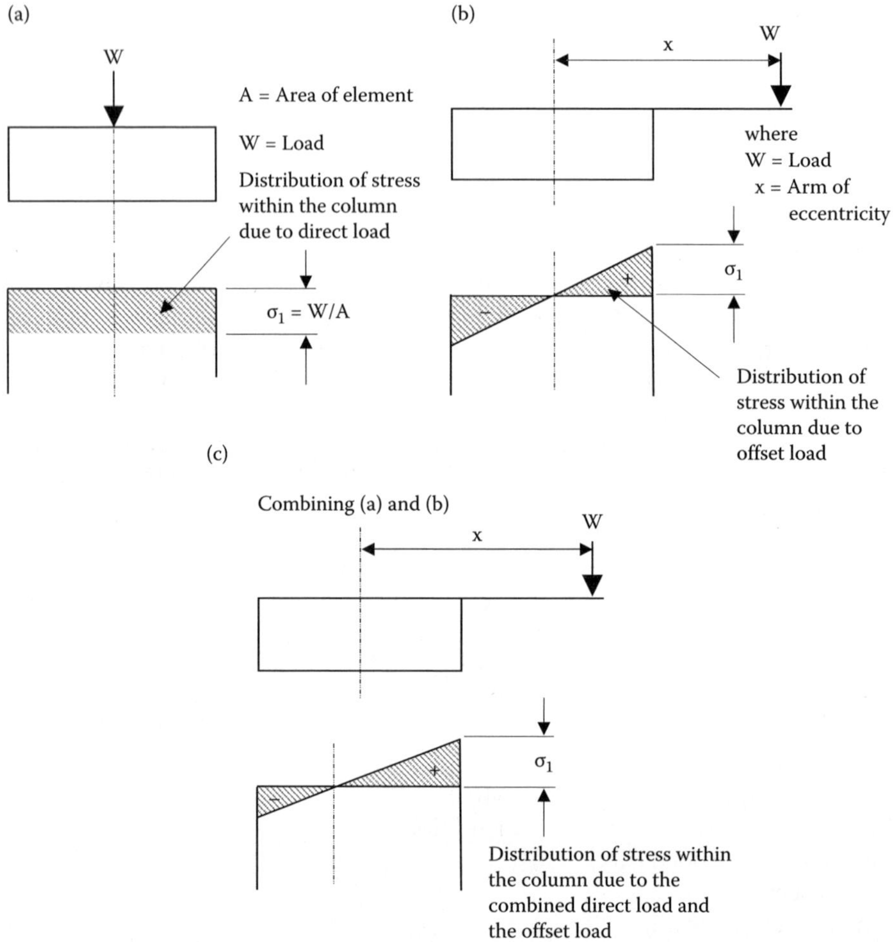

FIGURE 24.1
Principle of superposition.

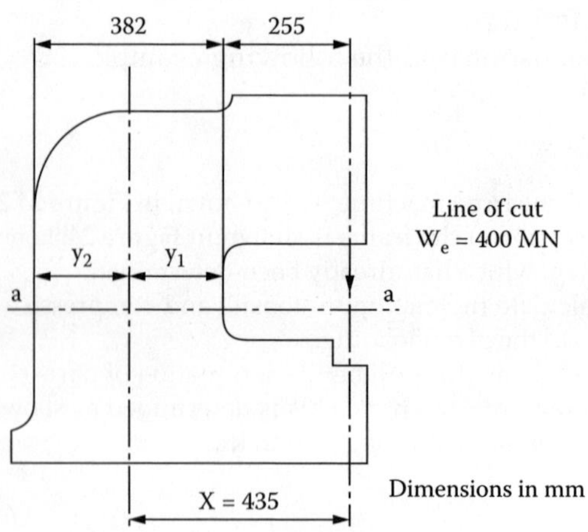

FIGURE 24.2
Outline elevation of frame.

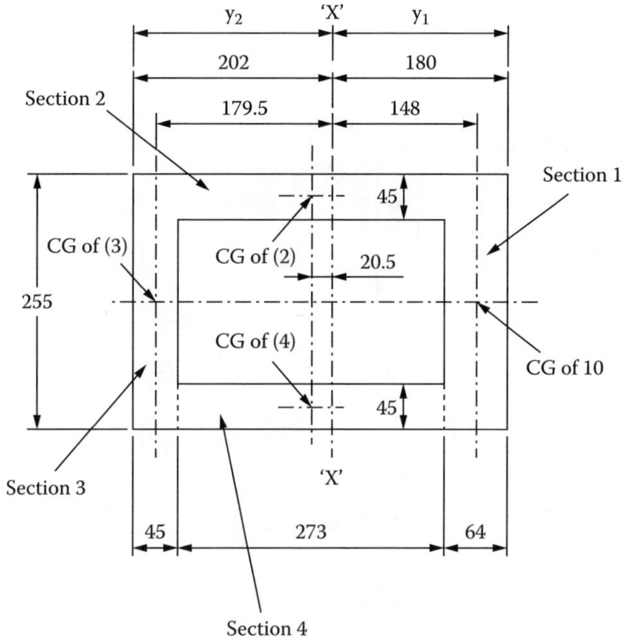

FIGURE 24.3
Enlarged section across section a:a.

Radius of gyration k^2:

$$k^2 = \frac{I}{A}$$

$$= \frac{897 \times 10^{-6}\ \text{m}^4}{52365 \times 10^{-6}\ \text{m}^2}$$

$$k^2 = 0.01713\ \text{m}^2$$

The radius of gyration is left in this form as it is used in the subsequent calculations.

The load W_e will tend to open the gap in the frame. Therefore the maximum fibre stress is tensile on the edge nearest the load (Section 1) and compressive on the edge farthest away from the load (Section 3).

From Figure 24.1, x = 0.435 m.

From Table 24.1:

$$\text{Total moment of inertia } (I_{xx}) = 897 \times 10^{-6}\ \text{m}^4$$

$$\text{Total area of the section} = 0.052365\ \text{m}^2$$

Maximum tensile stress:

$$\sigma_1 = \frac{W_e}{A}\left(1 + \frac{x \cdot y_1}{k^2}\right)$$

$$= \frac{400\ \text{kN}}{0.05237\ \text{m}^2}\left(1 + \frac{0.435\ \text{m} \times 0.180\ \text{m}}{0.01713\ \text{m}^2}\right)$$

$$\sigma_1 = 42.551\ \text{MPa}$$

TABLE 24.1

Moments of Inertia about Axis x:x for Example 24.1

Section	b	d	A	y	Ay	Ixx (b * d^3/12)	h	h^2	Ixx+Ah^2
1	0.255	0.064	0.0163	0.032	0.0005	5.57056E-06	0.1508	0.0227	0.000377
2	0.045	0.273	0.0123	0.2005	0.0025	7.62991E-05	0.0177	0.0003	8.01E-05
3	0.255	0.045	0.0115	0.3595	0.0041	1.93641E-06	0.1767	0.0312	0.00036
4	0.045	0.273	0.0123	0.2005	0.0025	7.62991E-05	0.0177	0.0003	8.01E-05
		S area =	0.0524 m^2	S Ay	0.0096 m^3		S Ixx =	0.000897 m^4	
	$\bar{X}$=	0.182828 m							
							Total Ixx=897.0×10^{-6} m^4		

Maximum compressive stress:

$$\sigma_1 = \frac{W_e}{A}\left(1 - \frac{x \cdot y_2}{k^2}\right)$$

$$= \frac{400 \text{ kN}}{0.05237 \text{ m}^2}\left(1 - \frac{0.435 \text{ m} \times 0.202 \text{ m}}{0.01713 \text{ m}^2}\right)$$

$$\sigma_2 = -31.541 \text{ MPa}$$

Note: If the direction of the loading was reversed, then obviously σ_1 would become compressive and σ_2 tensile.

25

Thick Cylinders

25.1 Introduction

When the wall thickness of a pressure vessel such as a cylinder or pipe is greater than one-tenth of the inside radius of the vessel, the meridional and circumferential (hoop) stresses cannot be considered uniform throughout the wall thickness, and the radial stress will not be negligible. At this point, the vessel will be considered to be a thick-walled cylinder, as in Figure 25.1.

$$(\sigma_r + d\sigma_r)(r + dr)d\theta = \sigma_r r d\theta + 2\sigma_c dr \frac{dr}{2}$$

i.e.

$$r d\theta_r + \sigma_r dr = \sigma_c dr$$

or

$$\sigma_r + r\frac{d\sigma_r}{dr} = \sigma_c \tag{25.1}$$

If the longitudinal stress and strain are denoted by σ_l and ε_l, respectively, then:

$$\varepsilon_l = \frac{\sigma_l}{E} - \nu\left(\frac{\sigma_r + \sigma_c}{E}\right)$$

It is assumed that ε_l is constant across the thickness, that is, a plane cross section of the cylinder remains plane after the application of pressure, and that σ_l is also of uniform thickness, both assumptions being reasonable on planes that are remote from the ends of the cylinder.

25.2 A Thick-Walled Cylinder Subject to Internal and External Pressures

Consider a thick-walled cylinder, as shown in Figure 25.1a, with internal and external radii r_1 and r_2, respectively, and subjected to internal and external pressures p_1 and p_2, respectively. Figure 25.1b shows the stresses acting on an element of radius r and thickness dr, subtending an angle dθ at the center. Table 25.1 gives the nomenclature used.

The radial and circumferential stresses, σ_r and σ_c, are assumed to be compressive, which in this context is considered positive.

Tensile stresses are positive; compressive stresses are negative.

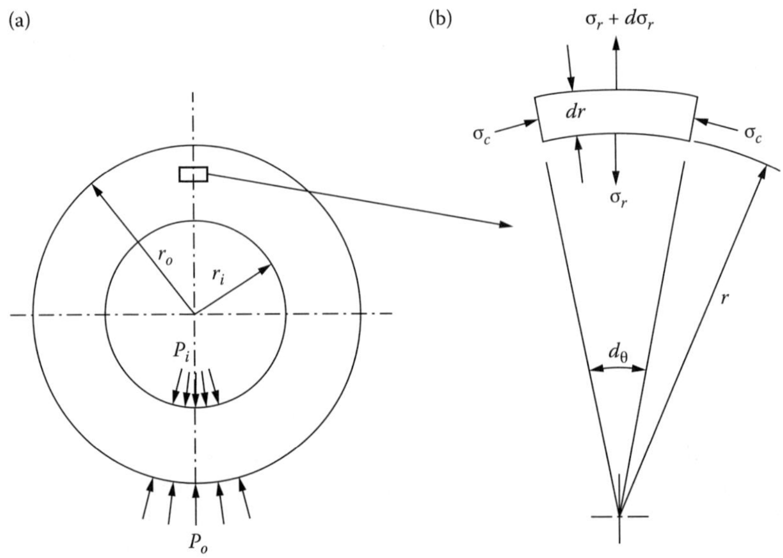

FIGURE 25.1
Elements of a thick-walled cylinder.

TABLE 25.1

Nomenclature
p_1 = Internal pressure (Pa)
p_2 = External pressure (Pa)
p_f = Interface pressure (Pa)
σ_r = Radial stress—compressive (Pa)
σ_c = Circumferential stress—compressive (Pa)
σ_a = Axial/longitudinal stress—tensile (Pa)
E = Modulus of elasticity (Pa)
ν = Poisson's ratio
ν_h = Poisson's ratio—hub
ν_s = Poisson's ratio—shaft
d = Diameter at point or analysis (m)
d_1 = Inside diameter of cylinder (m)
d_2 = Outside diameter of cylinder (m)
r = Radius at point of analysis (m)
r_1 = Inside radius of cylinder (m)
r_2 = Outside radius of cylinder (m)
r_f = Interface radius of cylinder (m)
ε_r = Radial strain
ε_c = Circumferential strain
ε_a = Axial/longitudinal strain
u = Radial deflection (m)
u_s = Radial deflection of shaft (m)
u_h = Radial deflection of hub (m)
u_t = Radial deflection of hub and shaft (m)
δr_h = Radial increase in hole (m)
δr_s = Radial decrease in hole (m)

The pressures p, p_1, p_2, and p_f are negative.

The following relationships are assumed for the strains ε_1, ε_2, and ε_3 associated with the stresses σ_1, σ_2, and σ_3. See Figure 25.2 for the orthogonal stress relationship.

$$\varepsilon_1 = \sigma_1 / E - \nu\sigma_2 / E - \nu\sigma_3 / E$$

$$\varepsilon_2 = \sigma_2 / E - \nu\sigma_1 / E - \nu\sigma_3 / E$$

$$\varepsilon_3 = \sigma_3 / E - \nu\sigma_1 / E - \nu\sigma_2 / E$$

It follows from these assumptions that $\sigma_r + \sigma_c$ is a constant; this will be denoted by 2a.

Thus,

$$\sigma_c = 2a - \sigma_r \tag{25.2}$$

Substituting Equation 25.2 into Equation 25.1,

$$\sigma_r + r\frac{d\sigma_r}{dr} = 2a - \sigma_r$$

or

$$\sigma_r + r\frac{d\sigma_r}{dr} = 2a - \sigma_r$$

Multiplying both sides by r,

$$\frac{d}{dr}\left(\sigma_r r^2 - ar^2\right) = 0$$

Therefore,

$$\sigma_r r^2 - ar^2 = b'$$

or

$$\sigma_r = a + \frac{b}{r^2} \tag{25.3}$$

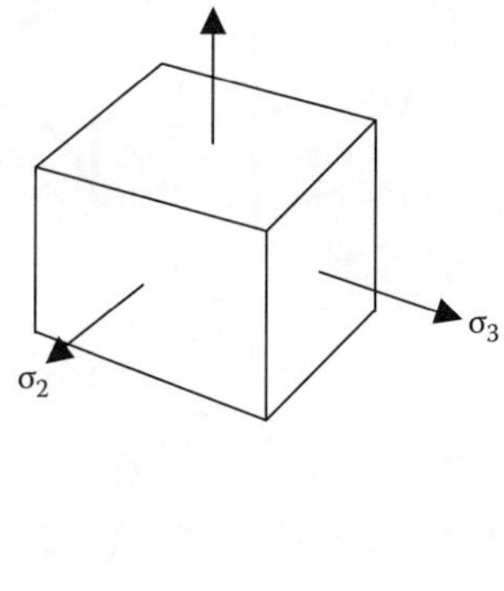

FIGURE 25.2
Stress relationships.

and from Equation 25.2,

$$\sigma_c = a - \frac{b}{r^2} \tag{25.4}$$

Equations 25.3 and 25.4 are known as *Lamé's equations*; in any given application, there will always be two conditions sufficient to solve for the constants a and b, together with the radial and circumferential stresses at any radius r, which can then be evaluated.

25.3 General Equations for a Thick-Walled Cylinder Subject to an Internal Pressure

Considering the common case of a cylinder subjected to an internal pressure only, see Figure 25.3.

$$\sigma_r = P_i \quad \text{when } r = r_i$$

and

$$\sigma_r = 0 \quad \text{when } r = r_o$$

$$P_i = a + \frac{b}{r_i^2}$$

and

$$0 = a + \frac{b}{r_o^2}$$

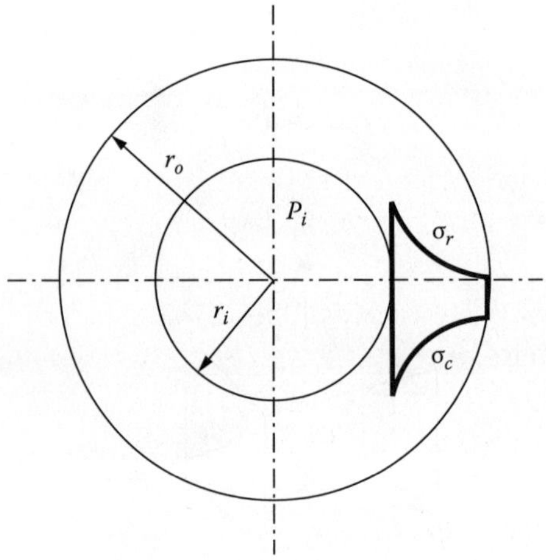

FIGURE 25.3
Thick-walled cylinder subject to an internal pressure.

from which

$$a = -P_i \frac{r_i^2}{r_o^2 - r_i^2}$$

and

$$b = P_i \frac{r_o^2 r_i^2}{r_o^2 - r_i^2}$$

Therefore,

$$\sigma_r = a + \frac{b}{r^2}$$
$$= -P_i \frac{r_i^2}{r_o^2 - r_i^2}\left(1 + \frac{r_o^2}{r^2}\right)$$

and

$$\sigma_c = a - \frac{b}{r^2}$$
$$= -P_i \frac{r_i^2}{r_o^2 - r_i^2}\left(1 + \frac{1}{r^2}\right)$$

The maximum radial and circumferential stresses occur at $r=r_i$, when $\sigma_r=P$:

$$\sigma_c = -P_i \frac{r_o^2 + r_i^2}{r_o^2 - r_i^2} \tag{25.5}$$

The negative sign indicates tension.

25.4 The General Equation for a Thick-Walled Cylinder Subject to Internal and External Pressures

The general equation for a thick-walled cylinder subject to internal and external pressures can easily be obtained from Equations 25.3 and 25.4.

Consider a cylinder with an internal diameter d_i, subject to an internal pressure P_i. The outside diameter d_o is subject to an external pressure P_o. The radial pressures at the surfaces are the same as the applied pressures; therefore,

$$\sigma_r = a + \frac{b}{r^2}$$
$$\sigma_c = a - \frac{b}{r^2} \tag{25.6}$$

as the radial pressures at the surfaces are the same as the applied pressures; therefore,

$$-P_i = a + \frac{b}{r_i^2}$$

$$-P_o = a + \frac{b}{r_o^2}$$

The general equation for a thick-walled cylinder is

$$\text{Subtracting } b = \frac{[(P_o - P_i) r_i^2 r_o^2]}{(r_o^2 - r_i^2)}$$

$$\text{Substituting this equation } a = \frac{(P_i \cdot r_i^2 - P_o \cdot r_o^2)}{(r_o^2 - r_i^2)}$$

and the resulting general equation is as follows:

$$\sigma_r = \frac{(P_i \cdot r_i^2 - P_o \cdot r_o^2)}{(r_o^2 - r_i^2)} + \frac{(P_o - P_i) \cdot r_i^2 \cdot r_o^2}{r^2 \cdot (r_o^2 - r_i^2)}$$

$$\sigma_c = \frac{(P_i \cdot r_i^2 - P_o \cdot r_o^2)}{(r_o^2 - r_i^2)} - \frac{(P_o - P_i) \cdot r_i^2 \cdot r_o^2}{r^2 \cdot (r_o^2 - r_i^2)}$$

If the external pressure is zero, the equation reduces to

$$\sigma_r = \frac{r_i^2 (r^2 - r_o^2)}{r^2 \cdot (r_o^2 - r_i^2)} P_i \text{ and } \sigma_c = \frac{r_i^2 \cdot (r^2 + r_o^2)}{r^2 \cdot (r_o^2 - r_i^2)} P_i$$

If the internal pressure is zero, the equation reduces to

$$\sigma_r = \frac{r_o^2 (r_i^2 - r^2)}{r^2 \cdot (r_o^2 - r_i^2)} P_o \text{ and } \sigma_c = -\left(\frac{r_o^2 \cdot (r_i^2 + r^2)}{r^2 \cdot (r_o^2 - r_i^2)} P_o \right)$$

If the cylinder has closed ends, then the axial stress can be found using the force equilibrium considerations only.

The pressure P_i acts on an area given by πr_i^2.
The pressure P_o acts on an area given by πr_0^2.
The axial stress σ_z acts on an area given by $\pi(r_0^2 - r_i^2)$.

Force equilibrium then gives

$$\sigma_z = \left(\frac{P_i \cdot r_i^2 - P_o \cdot r_o^2}{r_o^2 - r_i^2} \right)$$

Example 25.1: Interference Fit

When considering a force fit of a shaft into a hub, generally they are manufactured from the same material. This is a simple example of a compound cylinder.

Consider a press fit of a shaft into a hub. The compression of the shaft and the expansion of the hub result in a compressive pressure at the interface between the shaft and hub. The conditions are shown in Figure 25.4 and the individual components are shown in Figure 25.5.

The radial interference δr_f=sum of the shaft deflection δr_s and the hole deflection δr_h.

The longitudinal pressure, and hence σ_a, is assumed to be zero, and the internal pressure in the shaft hole and the external pressure outside the hub are also assumed to be zero:

$$E\varepsilon_c = E\frac{u}{r} = \sigma_c - \upsilon\sigma_a - \upsilon\sigma_r = \sigma_c - \upsilon\sigma_r \tag{25.7}$$

The radial increase in the hole diameter=u_h:

$$u_h = \left(\frac{r_f}{E_h}\right)(\sigma_c - \upsilon\sigma \cdot \sigma_r)$$

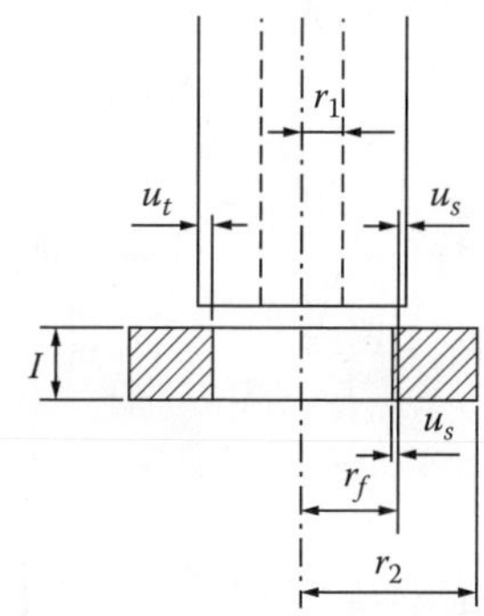

FIGURE 25.4
Interference fit.

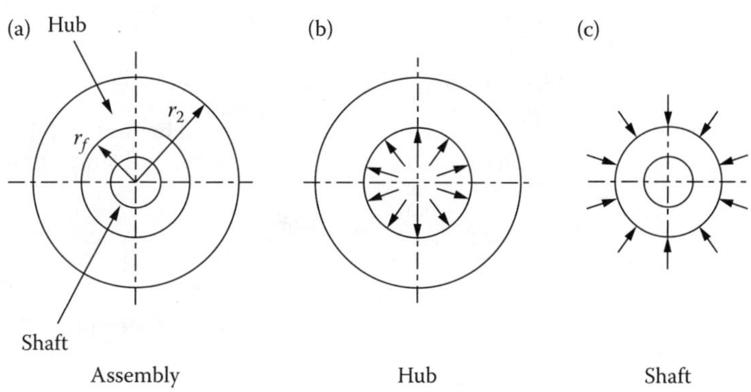

FIGURE 25.5
Components of shaft and hub.

The radial decrease in the shaft diameter=u_s:

$$u_s = \left(\frac{r_f}{E_s}\right)(\sigma_c - \upsilon\sigma \cdot \sigma_r)$$

The total interference $u_t = u_h + u_s$:
The equations for the hole are

$$\sigma_r = \frac{r_f^2 \cdot (r_f^2 - r_2^2)}{r_f^2 \cdot (r_2^2 - r_f^2)} P_f$$

$$\sigma_c = \frac{(r_2^2 + r_f^2)}{r_2^2 - r_f^2} P_f$$

The equations for the shaft are

$$\sigma_r = \frac{r_f^2 \cdot (r_1^2 - r_f^2)}{r^2 \cdot (r_f^2 - r_1^2)} P_f$$

$$\sigma_c = -\left(\frac{(r_f^2 + r_1^2)}{r_f^2 - r_1^2} P_f\right)$$

The interference equations are:

For the hole:

$$u_h = \frac{r_f \cdot P_f}{E_h}\left(\frac{r_2^2 + r_f^2}{r_2^2 - r_f^2} + \upsilon_h\right)$$

For the shaft:

$$u_s = -\frac{r_f \cdot P_f}{E_s}\left(\frac{r_f^2 + r_1^2}{r_f^2 - r_1^2} - \upsilon_h\right)$$

The total interference is therefore:

$$\begin{aligned} u_t &= u_h + (-u_s) \\ &= \frac{r_f \cdot P_f}{E_h}\left(\frac{r_2^2 + r_f^2}{r_2^2 - r_f^2} + \upsilon_h\right) + \frac{r_f \cdot P_f}{E_s}\left(\frac{r_f^2 + r_1^2}{r_f^2 - r_1^2} - \upsilon_h\right) \end{aligned}$$

If the hub and shaft are manufactured from the same material having an identical modulus of elasticity (E) and Poisson's ratio (ν), the above equation will simplify to

$$u_t = \frac{r_f \cdot P_f}{E}\left(\frac{2r_f^2(r_2^2 - r_1^2)}{(r_2^2 - r_f^2)(r_f^2 - r_1^2)}\right)$$

In a normal engineering application where the shaft is solid, for example, when $r_1 = 0$, the equation is further simplified to

$$u_t = \frac{2r_2^2 \cdot r_f \cdot P_f}{E(r_2^2 - r_f^2)}$$

There is often a requirement to calculate the interface pressure when the radial interference u_t is known (i.e., the shaft interference/2). When calculating the torque that can be transmitted or the force required to either make or separate the interference joint,

$$P_f = \frac{E(r_2^2 - r_f^2)\cdot u_t}{2\cdot r_2^2\cdot r_f}$$

By way of an example, consider a steel shaft 50 mm diameter pressed into a hub that has an outside diameter of 150 mm. The length of the hub is 50 mm and the interference is 0.05 mm. The assumed coefficient of friction is $\mu=0.15$.

The hub and shaft are both steel with the modulus of elasticity (E) $=210\times10^9$ Pa and the Poisson's ratio $\upsilon=0.3$:

$$\begin{aligned}P_f &= \frac{E(r_2^2 - r_f^2)\cdot u_t}{2\cdot r_2^2\cdot r_f}\\ &= \frac{210\times10^9\cdot(0.05^2-0.025^2)\cdot 0.5\times10^{-6}}{2\cdot 0.15^2\cdot 0.025}\\ &= 6.013\times10^6 \text{ Nm}\end{aligned}$$

$$\begin{aligned}\text{Torque } T &= \pi\cdot d\cdot L\cdot P_f\cdot\mu\cdot r_f\\ &= \pi\times 50\text{ mm}\times 50\text{ mm}\times 204.167\times10^6\text{ Pa}\times 0.15\times 25\text{ mm}\\ &= 6.013\times10^3 \text{ Nm}\end{aligned}$$

Example 25.2: Radial Distribution of Stress

A steel cylinder has an outside diameter of 200 mm and an inside diameter of 100 mm. The cylinder is subject to an internal pressure of 150 MPa. Determine the radial and circumferential stress distributions and show the results in the form of a spreadsheet. Assume the cylinder has closed ends.

The section is divided into incremental radii increasing from the inner radius of 50.0 mm at 2.5 mm intervals to the outside radius of 100.0 mm. The resultant stress is calculated at each incremental radius. The resultant stresses are tabulated in Table 25.2, and Figure 25.6 shows the results of the analysis.

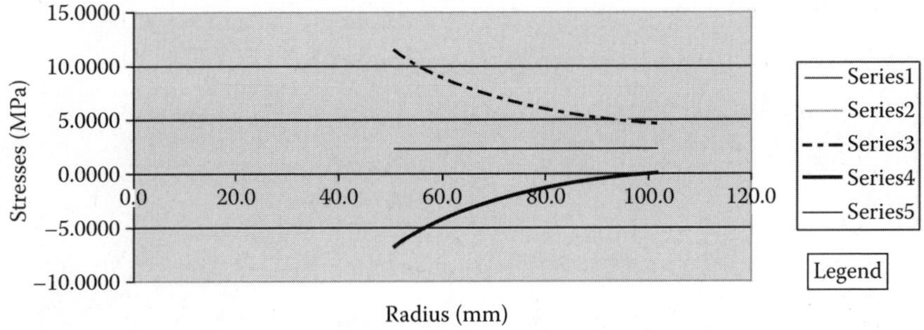

FIGURE 25.6
Thick-walled cylinder stresses.

TABLE 25.2

Spreadsheet Results for Example 25.2

Position	P_i	P_o	r_1	r_0	r	Radial Stress	Circumferential Stress	Axial Stress
	MPa	MPa	mm	mm	mm	MPa	MPa	MPa
1	150	0	50	100	50.0	150.000	250.000	50.000
2	150	0	50	100	52.5	131.406	231.406	50.000
3	150	0	50	100	55.0	115.289	215.289	50.000
4	150	0	50	100	57.5	101.229	201.229	50.000
5	150	0	50	100	60.0	88.889	188.889	50.000
6	150	0	50	100	62.5	78.000	178.000	50.000
7	150	0	50	100	65.0	68.343	168.343	50.000
8	150	0	50	100	67.5	59.739	159.739	50.000
9	150	0	50	100	70.0	52.041	152.041	50.000
10	150	0	50	100	72.5	45.125	145.125	50.000
11	150	0	50	100	75.0	38.889	138.889	50.000
12	150	0	50	100	77.5	33.247	133.247	50.000
13	150	0	50	100	80.0	28.125	128.125	50.000
14	150	0	50	100	82.5	23.462	123.462	50.000
15	150	0	50	100	85.0	19.204	119.204	50.000
16	150	0	50	100	87.5	15.306	115.306	50.000
17	150	0	50	100	90.0	11.728	111.728	50.000
18	150	0	50	100	92.5	8.437	108.437	50.000
19	150	0	50	100	95.0	5.402	105.402	50.000
20	150	0	50	100	97.5	2.597	102.597	50.000
21	150	0	50	100	100.0	0.000	100.000	50.000

26

Compound Cylinders

26.1 Introduction

A compound cylinder consists of two concentric cylinders, as shown in Figure 26.1, where the outer cylinder is shrunk onto an inner cylinder such that the latter is initially in compression prior to the application of any internal pressure.

The final stresses in the assembly will then be the result of those due to prestressing and those due to the internal pressure.

Compound cylinders are used to increase the pressure that can be contained in cylinders. Applications involving compound cylinders include gun barrels, etc.

If the radius of the common surface (or interface diameter) is r_o and the pressure at this surface before the application of any internal pressure is p_o, the initial stresses are determined by considering the two cylinders separately, the boundary conditions for the outer cylinder being $\sigma_r = P_o$ when $r = r_o$ and $\sigma_r = 0$ when $r = r_1$. For the inner cylinder $\sigma_r = P_o$ when $r = r_o$ and $\sigma_r = 0$ when $r = r_2$.

The stresses due to the internal pressure are obtained by considering the cylinder to be homogenous, with $\sigma_r = P$ at $r = r_2$ and $\sigma_r = 0$ at $r = r_1$.

The various stresses are then combined algebraically, as shown in Figure 26.2a, from which it is evident that the maximum resultant circumferential stress is less than for a homogeneous cylinder of the same cross section with the same internal pressure. Alternatively, for the same maximum stress, a thinner cylinder can be used if it is prestressed, the optimum conditions being when the resultant circumferential stress is the same at the inner surface of each cylinder as shown in Figure 26.2b.

26.2 Shrinkage Allowance

It is necessary for the inside diameter of the outer cylinder to be slightly smaller than the outside diameter of the inner cylinder to produce a desired initial pressure p_o at the common surface of the compound cylinder. It is common practice for the outer cylinder to be heated until it is possible to slide over the inner cylinder, and when cooled to the ambient temperature, it will generate the required pressure at the common surface. This procedure is termed the *shrinkage allowance*.

Let the circumferential stress at the outer surface of the inner cylinder due to p_o be σ_c'.

$$\text{The circumferential strain at the outer surface:} = \frac{\sigma_c'}{E} = \nu \frac{p_o}{E}$$

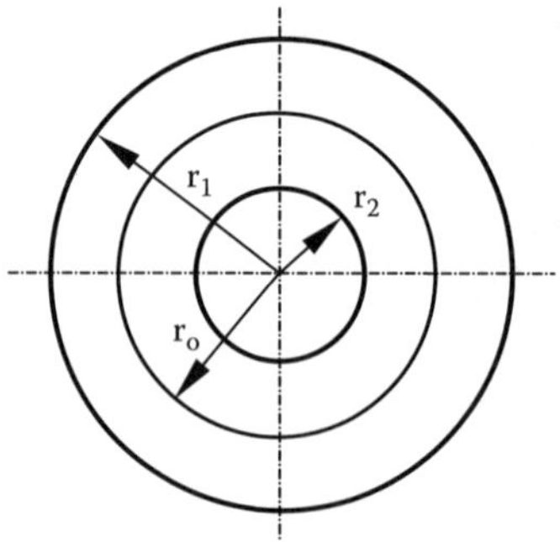

FIGURE 26.1
Compound cylinder.

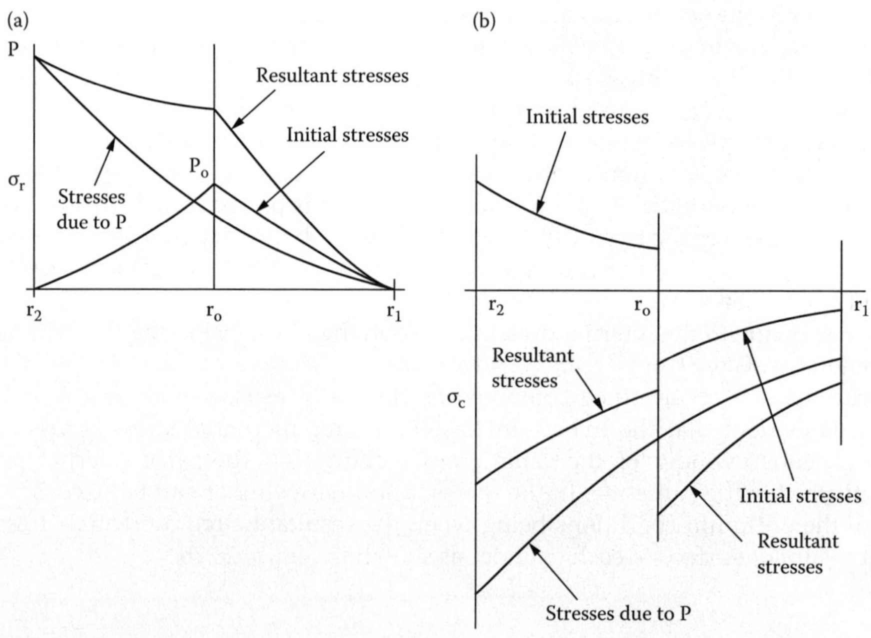

FIGURE 26.2
Algebraic combination of stresses. (a) Stress distribution in a homogeneous cylinder. (b) Stress distribution in a compound cylinder.

$$\text{Therefore: } = \frac{2r_o}{E}(\sigma_c' - \nu p_o)$$

For the outer cylinder, let the circumferential stress at the inner surface due to p_o be σ_c''.

$$\text{The circumferential strain at the inner surface: } = \frac{\sigma_c''}{E} - \nu\frac{p_o}{E}$$

$$\text{Therefore the decrease in diameter} = \frac{2r_o}{E}(\sigma_c'' - \nu p_o)$$

$$\text{Initial difference in diameter} = \frac{2r_o}{E}\{(\sigma_c' - \nu p_o) - (\sigma_c'' - \nu p_o)\}$$
$$= \frac{2r}{E}(\sigma_c' - \sigma_c'')$$

When the materials of the two cylinders have different modulii of elasticity, E_1 and E_2, together with the Poisson's ratio υ_1 and υ_2, respectively, for the inner and outer cylinders,

$$\text{Shrinkage allowance} = 2r_o\left\{\frac{\sigma_c' - \nu_1 p_o}{E_1} - \frac{\sigma_c'' - \nu_2 p_o}{E_2}\right\}$$

Example 26.1

Consider a compound cylinder subjected to an internal pressure of 150 MPa.

The radii of the inner, interface, and outer surfaces are, respectively, 100, 140, and 200 mm. Determine the shrinkage pressure necessary for the resultant maximum circumferential stress to be as small as possible and calculate this stress.

Initial stresses: $r_1 = 200$ mm (outside radius)

Let the shrinkage pressure=p $r_o = 140$ mm (interface radius)

$r_2 = 100$ mm (inside radius)

For the inner cylinder: $p_2 = 150$ MPa

$\sigma_r = p$ at $r = 140$ mm

$$p = a_1 + \frac{b_1}{140^2}$$
$$= a_1 + \frac{b_1}{19.60 \times 10^3}$$
$$\therefore\ p = a_1 + 51.02 \times 10^{-6} b_1$$
$$0 = a_1 + \frac{b_1}{100^2}$$
$$= a_1 + \frac{b_1}{10 \times 10^3}$$
$$\therefore\ 0 = a_1 + 100 \times 10^{-6} b_1 \tag{26.1}$$
$$p = a_1 + 51.02 \times 10^{-6} b_1 \tag{26.2}$$

Subtracting Equation 26.2 from 26.1

$$p = -48.98 \times 10^{-6} b_1$$

Therefore:

$$b_1 = -20.4167 \times 10^3 p \tag{26.3}$$

Substituting Equation 26.3 into 26.2

$$p = a_1 + 51.02 \times 10^{-6} \times (-20.4167 \times 10^3)p$$
$$p = a_1 - 1.0417p$$
$$\therefore\ a_1 = 2.0417p \tag{26.4}$$

Therefore at the inner surface:

$$\sigma_c = a_1 - \frac{b_1}{100^2}$$

$$\sigma_c = 2.0417p - \frac{(2.0417p)}{100^2}$$

$$\sigma_c = 4.0833p$$

For the outer cylinder:

$\sigma_r = p$ at $r = 140\,\text{mm}$

$$\therefore \quad p = a_2 + \frac{b_2}{140^2}$$

$$= a_2 + 51.02 \times 10^{-6} b_2 \qquad (26.5)$$

$\sigma_r = 0$ at $r = 200\,\text{mm}$

$$0 = a_2 + \frac{b_2}{200^2}$$

$$0 = a_2 + 25.0 \times 10^{-6} b_2 \qquad (26.6)$$

Subtracting Equation 26.6 from 26.5:

$$p = a_2 + 51.02 \times 10^{-6} b_2$$

$$0 = a_2 + 25.00 \times 10^{-6} b_2$$

$$p = \quad +26.02 \times 10^{-6} b_2$$

and

$$b_2 = 38.432 \times 10^3 p \qquad (26.7)$$

Substituting Equation 26.7 into 26.5:

$$p = a_2 + (51.2 \times 10^{-6} \times 38.432 \times 10^3)p$$

$$p = a_2 + 1.9608p$$

$$a_2 = -0.9608p$$

Therefore at the inner surface:

$$\sigma_c = -0.9608p - \frac{b_2}{140^2}$$

$$= -0.9608p - \frac{38.432 \times 10^3 p}{19.600 \times 10^3}$$

$$\sigma_c = -0.9608p - 1.9608p$$

$$\sigma_c = -2.922p$$

Stresses due to the internal pressure:

$\therefore \sigma_r = 100\,\text{MPa}$ at $r = 100\,\text{mm}$

$$100 = a + \frac{b}{100^2}$$

$$= a + 100 \times 10^{-6} b \qquad (26.8)$$

$\therefore \sigma_r = 0$ at $r = 200\,\text{mm}$

$$0 = a + \frac{b}{200^2}$$

$$= a + 25.00 \times 10^{-6} b \qquad (26.9)$$

Subtracting Equation 26.9 from 26.8

$$100 = a + 100 \times 10^{-6} b$$

$$0 = a + 25.00 \times 10^{-6} b$$

Hence:

$$100 = 75.00 \times 10^{-6} b$$

$$b = 1.333 \times 10^6$$

Substituting in Equation 26.8

$$100 = a + (100 \times 10^{-6} \times 1.333 \times 10^{6})$$
$$100 = a + 133.33$$
$$a = -33.33$$

Therefore at r = 100

$$\sigma_c = a - \frac{b}{100^2}$$
$$= -33.33 - \frac{1.333 \times 10^6}{100^2}$$
$$= -33.33 - 133.3$$
$$\sigma_c = -166.67 \text{ MPa}$$

and at r = 140 mm

$$\sigma_c = a - \frac{b}{140^2}$$
$$= -33.33 - \frac{1.333 \times 10^6}{140^2}$$
$$\sigma_c = -101.34 \text{ MPa}$$

For the resultant maximum circumferential stress to be as small as possible, the resultant stresses at the inner surfaces of each cylinder must be equal, as will be evident from Figure 26.2, i.e., $-166.67 + 4.0833p = -101.34 - 2.922p$.

From which:

$$-166.67 + 101.34 = 2.922p - 4.0833p$$
$$-65.33 = 7.0053p$$
$$\frac{66.67}{-7.0053} = p$$
$$9.326 \text{ MPa} = p$$

Resultant maximum stress: $-101.34 - (3 \times 9.326) = 129.32$ MPa.

Example 26.2

A compound steel cylinder has a bore of 75.0 mm with an external diameter of 175.0 mm; the diameter of the interface is 120.0 mm. Determine the radial pressure at the interface, which must be provided by shrinkage. If the resultant maximum circumferential stress in the inner cylinder under a superimposed internal pressure of 60 MPa is to be half the value of the maximum hoop tension, which would be produced in the inner cylinder if that cylinder alone were subjected to an internal pressure of 60 MPa.

Determine the final hoop tensions at the inner and outer surfaces of both cylinders under the internal pressure of 60 MPa and sketch a graph to show the circumferential stress varies across the cylinder wall.

1. Initial stresses in the inner cylinder.

Let the shrinkage pressure be p.

$$r_1 = 87.5 \text{ mm (outside radius)}$$
$$r_o = 60.0 \text{ mm (interface radius)}$$
$$r_2 = 37.5 \text{ mm (inside radius)}$$
$$P = 60.0 \text{ MPa}$$

$\sigma_r = p$ at $r = 60.0\,\text{mm}$ $$p = a_1 + \frac{b_1}{60.0^2}$$

hence: $$p = a_1 + 277.778 \times 10^{-6} b_1 \qquad (26.10)$$

and $\sigma_r = 0$ at $r = 37.5\,\text{mm}$ $$0 = a_1 + \frac{b_1}{37.5^2}$$

$$0 = a_1 + 711.111 \times 10^{-6} b_1 \qquad (26.11)$$

Subtracting Equation 26.10 from 26.11

$$\begin{aligned} p &= a_1 + 277.778 \times 10^{-6} b_1 \\ p &= a_1 + 711.111 \times 10^{-6} b_1 \\ p &= -433.33 \times 10^{-6} b_1 \\ b_1 &= -2307.705p \end{aligned}$$

Substituting in Equation 26.10

$$\begin{aligned} p &= a_1 + (277.78 \times 10^{-6} \times 2307.705p) \\ p &= a_1 - 0.64103p \\ 1.641p &= a_1 \end{aligned}$$

Therefore at the inner surface:

$$\begin{aligned} \sigma_c &= a_1 - \frac{b_1}{37.5^2} \\ &= 1.641p + \frac{2307.705p}{37.5^2} \\ &= 1.641p + 1.641p \\ \sigma_c &= 3.282p \end{aligned}$$

2. Stresses due to the internal pressure.

$\sigma_r = 60\,\text{MPa}$ at $r = 37.5\,\text{mm}$ $$60 = a_2 + \frac{b_2}{37.5^2}$$

$$60 = a_2 + 711.111 \times 10^{-6} b_2 \qquad (26.12)$$

and at $\sigma_r = 0$ at $r = 87.5\,\text{mm}$ $$0 = a_2 + \frac{b_2}{87.5^2}$$

$$0 = a_2 + 130.612 \times 10^{-6} b_2 \qquad (26.13)$$

Subtracting Equation 26.13 from 26.12

$$\begin{aligned} 60 &= a_2 + 711.111 \times 10^{-6} b_2 \\ 0 &= a_2 + 130.612 \times 10^{-6} b_2 \\ 60 &= 580.50 \times 10^{-6} b_2 \\ 103.359 \times 10^3 &= b_2 \end{aligned}$$

Substituting in Equation 26.12

$$60 = a_2 + (711.111 \times 10^{-6} \times 103.359 \times 10^3)$$
$$60 = a_2 + 73.50$$
$$60 - 73.50 = a_2$$
$$-13.50 = a_2$$

Therefore, at the inner surface: $\sigma_c = a_2 - \frac{b_2}{37.5^2}$

$$\sigma_c = a_2 - 711.111 \times 10^{-6} b_2$$
$$\sigma_c = -13.50 - (711.111 \times 10^{-6} \times 103.59 \times 10^3)$$
$$= -13.50 - 73.50$$
$$= -87.0 + 2.8928p$$

3. Inner cylinder alone with internal pressure.

σ_r=60 MPa at r=37.5 mm $\qquad 60 = a_3 + \frac{b_3}{37.5^2}$

$$60 = a_3 + 711.111 \times 10^{-6} b_3 \tag{26.14}$$

σ_r=0 at r=60 mm

$$0 = a_3 + \frac{b_3}{60.0^2} \tag{26.15}$$

Subtracting Equation 26.15 from 26.14

$$60 = a_3 + 711.111 \times 10^{-6} b_3$$
$$0 = a_3 + 277.78 \times 10^{-6} b_3$$
$$60 = +433.333 \times 10^{-6} b_3$$
$$138.462 \times 10^3 = b_3$$

Substituting in Equation 26.14

$$60 = a_3 + (711.111 \times 10^{-6} \times 138.462 \times 10^3)$$
$$a_3 = -38.4616$$

At the inner surface:

$$\sigma_c = a_3 - \frac{b_3}{37.5^2}$$
$$\sigma_c = -38.4616 - \frac{138.462 \times 10^3}{37.52}$$
$$\sigma_c = -136.923 \text{ MPa}$$
$$-87.0 + 2.8928\,p = \frac{-136.923}{2}$$
$$p = 6.4085 \text{ MPa}$$

4. Initial stresses in the outer cylinder:

$\sigma_r = p$ at $r = 60.0\,\text{mm}$

$$\therefore p = a_4 + \frac{b_4}{60^2}$$

$$p = a_4 + 277.78 \times 10^{-6} b_4 \tag{26.16}$$

$\sigma_r = 0$ at $r = 87.5\,\text{mm}$

$$\therefore 0 = a_4 + \frac{b_4}{87.5^2}$$

$$0 = a_4 + 130.612 \times 10^{-6} b_4 \tag{26.17}$$

Subtracting Equation 26.17 from 26.16

$$p = a_4 + 277.780 \times 10^{-6} b_4$$

$$0 = a_4 + 130.612 \times 10^{-6} b_4$$

$$p = +147.166 \times 10^{-6} b_4$$

$$6795.06p = b_4$$

Substituting in Equation 26.16

$$p = a_4 + (277.78 \times 10^{-6} \times 6795.06p)$$

$$p = a_4 + 1.8875p$$

$$-0.8875p = a_4$$

The resultant circumferential stresses in the compound cylinder are then as follows:
Inner cylinder, inner surface:

$$\sigma_{c1} = \frac{-136.923}{2}\ \text{MPa}$$

$$\sigma_{c1} = -68.4615\ \text{MPa}$$

Inner cylinder, outer surface:

$$\sigma_{c2} = \left(a_1 - \frac{b_1}{60.0^2}\right) + \left(a_2 - \frac{b_2}{60.0^2}\right)$$

$$= \left(1.6410p + \frac{-2307.705p}{60.0^2}\right) + \left(-13.50 - \frac{103.36 \times 10^3}{60.0^2}\right)$$

$$= 1.00 \times 6.4085 - 42.211$$

$$\sigma_{c2} = -35.8025\ \text{MPa}$$

Outer cylinder, inner surface:

$$\sigma_{c3} = \left(a_4 - \frac{b_4}{60.0^2}\right) + \left(a_2 - \frac{b_2}{60.0^2}\right)$$

$$\sigma_{c3} = \left(-0.8875p - \frac{6795.06p}{60.0^2}\right) + \left(-13.50 - \frac{103.359}{60.0^2}\right)$$

$$\sigma_{c3} = -59.995\ \text{MPa}$$

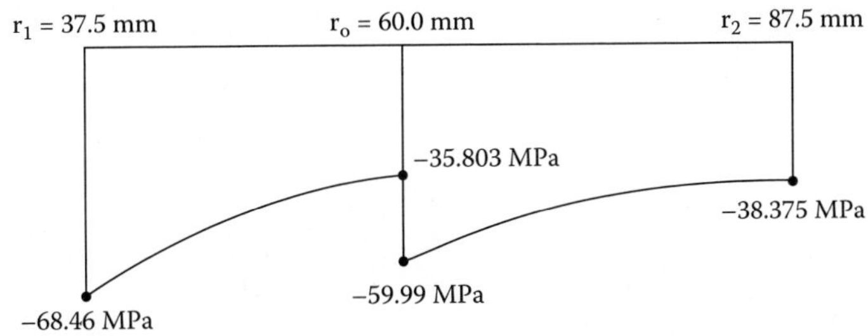

FIGURE 26.3
Variation in the circumferential stress across the cylinder.

Outer cylinder, outer surface:

$$\sigma_{c4} = \left(a_4 - \frac{b_4}{87.5^2}\right) + \left(a_2 - \frac{b_2}{87.5^2}\right)$$

$$\sigma_{c4} = \left(-0.8875p - \frac{6795.06p}{87.5^2}\right) + \left(-13.50 - \frac{103.359 \times 10^3}{87.5^2}\right)$$

$$\sigma_{c4} = -38.375 \text{ MPa}$$

The variation in the circumferential stress across the cylinder is shown in Figure 26.3.

Further Reading

Warren C. Young. *Roark's Formulas for Stress and Strain*. 6th ed. New York: McGraw-Hill, 1989.

27

Flat Plates

The theory of plates covers from circular plates to flat plates under a number of edge constraint conditions from simply supported to rigid edge restraints.

Table 27.1 gives descriptions of various restraints and loading conditions for some common circular plates covering simply supported and fixed edges. Table 27.2 also considers some common loadings on flat rectangular plates, including simply supported and fixed edges.

For the purposes of this section, a rectangular plate rigidly held at the edges and subject to a pressure on one side will be studied, as this is considered to be a more common design problem.

Example 27.1

A rectangular opening 406 mm × 252 mm in a pressure vessel is subjected to an internal pressure of 6 bar. It is closed by means of a cast iron flat plate cover (see Figure 27.1), which has a joint 19 mm wide and is assumed to withstand the same pressure. Design the flange and cover assuming the maximum allowable stress for the plate is 27.6 MPa and the stress in the studs is 42 MPa (this allows for the stress concentration factor at the root of the thread). For more information regarding stress concentration factors, see *Peterson's Stress Concentration Factors.*

Solution

The diameter and number of studs needs to be decided next. A provisional diameter of M16 will be assumed and this will give a core diameter of 13.546 mm.

$$\text{Area of the core} = 144.116 \times 10^{-6}\,\text{m}^2$$

$$\text{Load acting on each stud} = 42\ \text{MPa} \times 144.116 \times 10^{-6}\,\text{m}^2$$
$$= 6.053\ \text{kN}$$

The pressure is assumed to act at the centroid of the shaded triangular area and is resisted by the studs at the centroid of the pitch line. The distance S is the bending arm.

It was decided to use 14 studs to give a more symmetrical arrangement.

The dimensions of the flange will depend upon whether hexagonal or socket set screws are used.

In this example, hexagonal fasteners were chosen, as these would make the flange a little wider and increase the bending moment acting on the flange.

The cover thickness t is found by taking moments about the diagonal as shown in Figure 27.2.

TABLE 27.1

Circular Flat Plates (Constant Thickness)

Case	Description	Graphic	Stress & Deflection Equations
1	Concentrate central load with edge simply supported	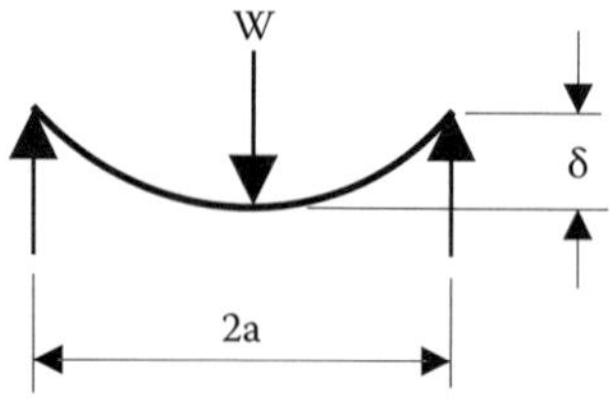	$\sigma = \frac{3W(1-v)}{2\pi t^2}\left(\frac{1}{v+1} + \log\frac{\sigma}{r_0} - \frac{1-v \cdot r_o^2}{v \cdot 4a^2}\right)$ $\delta max = \frac{3Wa^2(1-v)(3+v)}{4\pi Et^3}$
2	Uniformly distributed load with edge simply supported	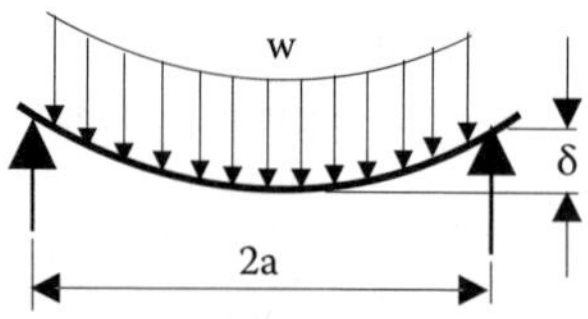	$\sigma = \frac{3w(3+v)}{8}\left(\frac{a^2}{t^2}\right)$ $\delta max = \frac{3wa^4(1-v)(5+v)}{16\pi Et^3}$
3	Concentrated central load with outer edge fixed	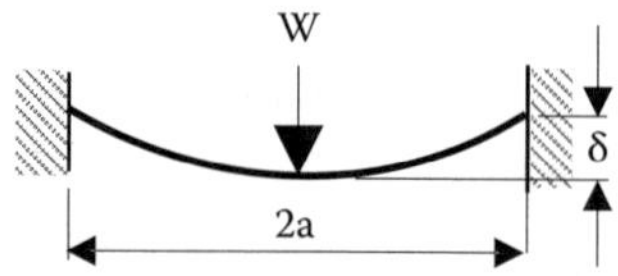	For $a > r_o$ $\sigma_{max} = \frac{3W(3+v)}{2\pi t^2}\left(\log\frac{a}{r_o} + \frac{r^2{}_o}{4a^2}\right)$ $\delta max = \frac{3Wa^2(1-v)}{4\pi Et^3}$
4	Uniformly distributed load with outer edge fixed,	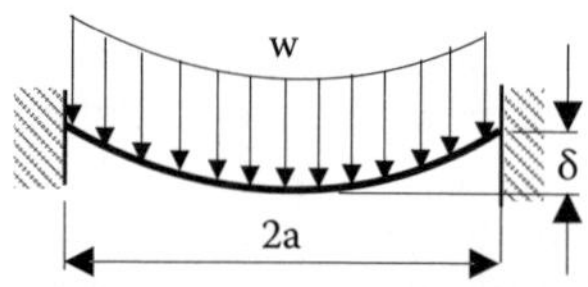	$\sigma_{max} = \frac{3Wa^2}{4t^2}$ $\delta_{max} = \frac{3wa^4(1-v^2)}{16Et^3}$ $\delta_{max} = H\frac{3wa^4(1-n^2)}{16Et^3}$ For thicker flat circular plates having $(t/a = 0.1,)$ multiply the deflection equation by the constant $H = 1 + 5.72\ (t/a)^2$

$$\text{Number of studs} = \frac{77.256}{6.053}$$

$$= 12.763$$

$$\text{Total pressure on triangular area} = \frac{77.256\text{ kN}}{2}$$

$$= 38.628\text{ kN}$$

$$\text{Length of the diagonal 'B'} = \sqrt{480^2 + 315^2}$$

$$= 574.13\text{ mm}$$

$$\text{Bending moment using bending arm } S = 53.72\text{ mm } \left(\text{see Figure 27.2}\right)$$

$$38.626\ \times\ 53.72 = 2074.99\text{ kN}$$

TABLE 27.2

Rectangular Plates – Common Loading

Case	Description	Graphic	Stress & Deflection Equations
1	Rectangular plate uniformly loaded with outer edges simply supported	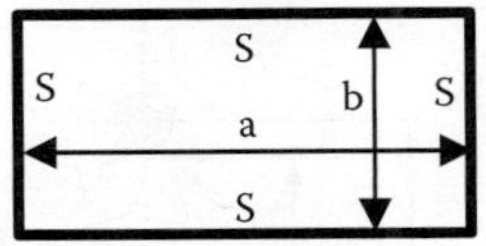 'S' indicates simply supported edges	$\sigma_{max} = \frac{\beta q b^2}{t^2}$ (at centre) q=load per unit area $\delta_{max} = \frac{-\alpha q b^4}{Et^3}$ (at centre)
2	Uniform over a small concentric circle of radius r_o, outer edges simply supported	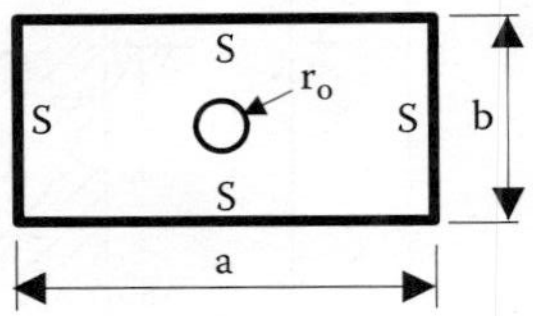	$\sigma_{max} = \frac{3W}{2\pi t^2}\left[(1+v)\ln\frac{2b}{\pi r'_o}+\beta\right]$ Note: $r' = \sqrt{1.6r_o^2 + t^2} - 0.675t$ if $r_o < t$ and $r'_o = r_o$ if $r_o > 0.5t$ $\delta_{max} = \frac{\alpha W b^2}{Et^3}$
3	Uniformly distributed load with outer edges fixed	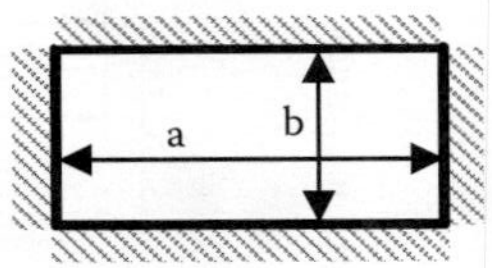	$\sigma_{max} = \frac{\beta q b^2}{t^2}$ (at centre of plate) q=load per unit area $\delta_{max} = \frac{\alpha q b^2}{Et^2}$ (at centre of plate)

Case 1:

a/b	1.0	1.2	1.6	2.0	3.0	4.0	5.0	∞
β	**0.2874**	**0.3762**	**0.5172**	**0.6102**	**0.7134**	**0.7410**	**0.7476**	**0.7500**
α	0.0444	0.0616	0.0906	0.1110	0.1335	0.1400	0.1417	0.1421

Case 3:

a/b	1.0	1.2	1.4	1.6	1.8	2.0	∞
β	**0.1386**	**0.1794**	**0.2094**	**0.2286**	**0.2406**	**0.2472**	**0.2500**
α	0.0138	0.0188	0.0226	0.0251	0.0267	0.0277	0.0284

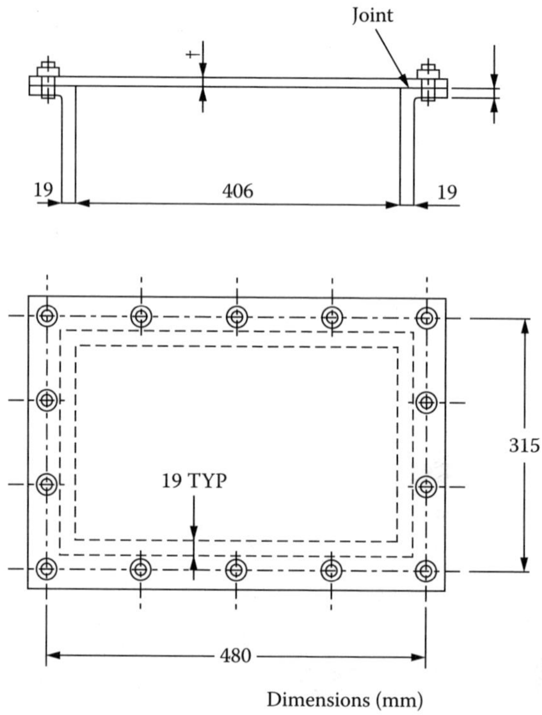

FIGURE 27.1
Flat plate cover for Example 27.1.

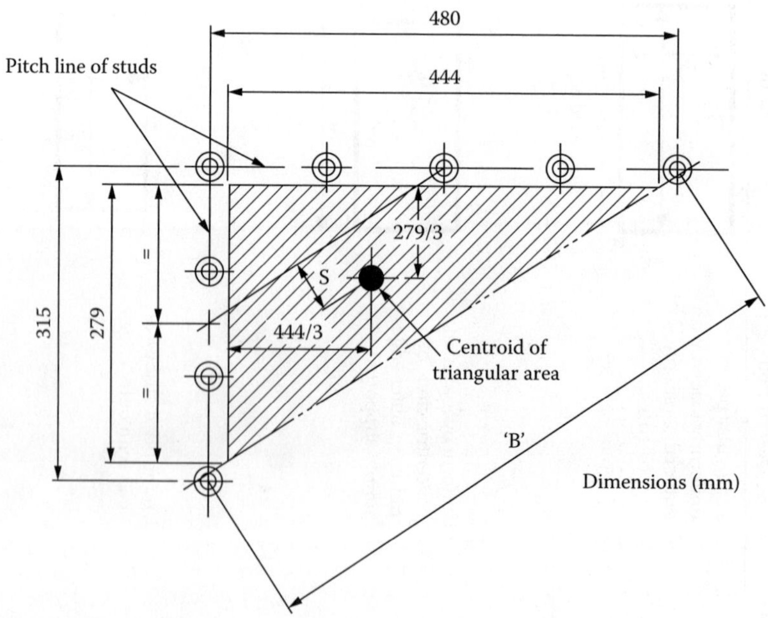

FIGURE 27.2
Part section of cover for Example 27.1.

$$\text{BM} = \sigma_z$$

$$2074.99 \text{ kN} = 27.6 \text{ x } 10^3 \text{ Z} \quad \text{where Z } = \frac{\text{Bt}^2}{6}$$

$$\frac{2074.99 \text{ x } 6}{0.57413 \text{ x } 27.6 \text{ x } 10^3} = \text{t}^2$$

$$0.0007854 \text{ m}^2 = \text{t}^2$$

$$\text{t} = 28.02 \text{ mm}$$

$$\text{S} = 53.72 \text{ mm by setting out to scale}$$

Use a 30 mm thick plate or the nearest larger stock standard size.

To find the flange thickness the method used here is to take the load acting on one stud producing bending on a section of flange of width of one pitch.

$$\text{Minimum pitch } = \frac{315 \text{ mm}}{3}$$

$$= 105 \text{ mm}$$

$$\text{Load on section} = \frac{77.3}{14} \text{ kN}$$

$$= 5.52 \text{ kN}$$

$$\text{Bending arm} = 18 \text{ mm}$$

$$\text{BM} = \sigma_2$$

$$5.52 \text{ kN x } 0.018 \text{ x } 10^3 \text{ m} = \frac{\sigma \text{bd}^2}{6}$$

$$99.36 \text{ Nm} = 27.6 \text{ x } 10^6 \text{ x } \frac{0.105}{6} \text{ d}^2$$

$$d = 0.0143 \text{ m}$$

To obtain sufficient depth of thread for the studs in the flange this should be 1.5 D=24 mm.

Example 27.2

A 200 mm steel tube 9.5 mm thick has the ends closed by a flat steel end cover which is held in place by a central bolt and hexagonal nut, as shown in Figure 27.3. The pressure in the tube is 14 bar and the pressure on the joint is twice this value.

Find the bolt diameter and the cover plate thickness allowing for a working stress of 70 MPa.

The bolt diameter may be neglected in this example.

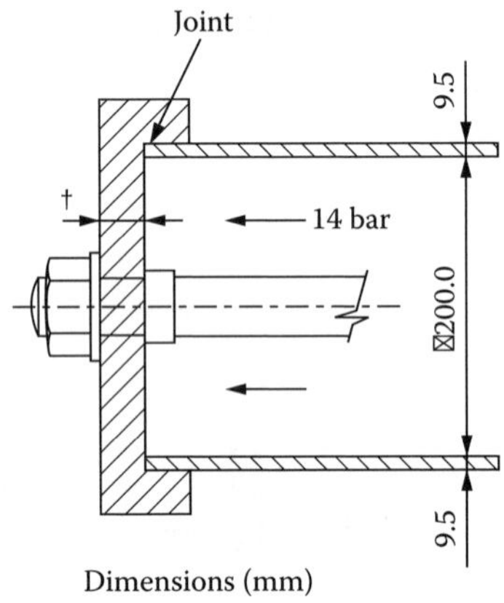

FIGURE 27.3
End plate for Example 27.1.

Solution

$$\text{Total bolt load for semi-circle} = \frac{61.5\text{ kN}}{2}$$

$$= 30.75\text{ kN}$$

$$\text{Load on Ø200 mm semi-circle area} = \frac{44\text{ kN}}{2}$$

$$= 22\text{ kN}$$

$$\text{Load due to joint pressure} = 2 \times 14\text{ bar} \times 10^5\text{ N/m}^2 \times (\pi \times 0.2095\text{ m} \times 0.0095\text{ m})$$

where 0.2095 m is the mean diameter of the tube.

$$\text{Load on end cover} = 14\text{ bar x }10^5\text{ N/m}^2\text{ x }\frac{\pi}{4}\text{ x }0.20^2\text{ m (N)}$$

$$= 43.982\text{ kN}$$

and 0.0095 m is the wall thickness.

$$= 17.507\text{ kN}$$

Hence

$$\text{Total load on bolt} = 43.982\text{ kN} + 17.507\text{ kN}$$

$$= 61.489\text{ kN}$$

Using the allowable stress of 70 MP:

$$= \frac{61.489\text{ kN}}{70\text{ MPa}}$$

$$= 0.0008784\text{ m}^2$$

Area at root of thread of bolt:

$$\text{Area} = \frac{\pi d^2}{4}$$

$$0.000878\ \text{m}^2 = 0.7854\ \text{d}^2$$

$$0.00119 = \text{d}^2$$

$$\text{d} = 0.03344\ \text{m}$$

$$= 33.44\ \text{mm root diameter}$$

Taking moments about the effective diameter of the end cover, the total load tending to bend the cover is the sum of half the load on Ø200 mm opening plus half the joint load, and acts at the centroids in each case.

These give a single resultant force, which is opposed by the central bolt, thus producing a couple balanced by the stress couple across the section of the cover.

$$\text{Total bolt load for semi-circle} = \frac{61.5\ \text{kN}}{2}$$

$$= 30.75\ \text{kN}$$

$$\text{Load on Ø200 mm semi-circle area} = \frac{44\ \text{kN}}{2}$$

$$= 22\ \text{kN}$$

$$\text{Load on joint for semi-circle} = \frac{17.5\ \text{kN}}{2}$$

$$= 8.75\ \text{kN}$$

Figure 27.4 illustrates the loading configurations:

For the semi-circle area centroid = 0.4244 × radius.
For the semi-circle of joint = 0.636 × radius.

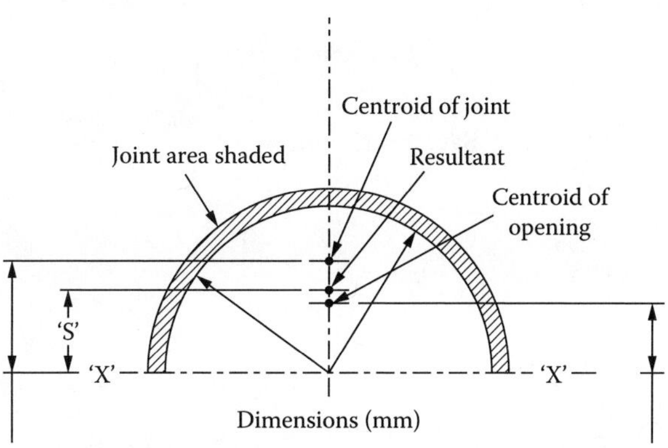

FIGURE 27.4
Loading arrangement for Example 27.2.

Centroid of Ø200 mm semi-circle area:

$$= 0.4244 \times 100 \text{ mm}$$
$$= 42.4 \text{ mm}$$

$$\text{Centroid of joint area} = 0.636 \times 104.75 \text{ mm}$$
$$= 66.621 \text{ mm}$$

To find the position of the resultant, take moments about axis X:X (Figure 27.4).

$$8.75 \text{ kN} \times 66.621 \text{ mm} + 22 \text{ kN} \times 42.4 \text{ mm}$$
$$= 30.75 \text{ kN} \times S$$

$$1515.733 \text{ kNmm} = 30.75 \times S$$

$$49.292 \text{ mm} = S$$

$$\text{Bending moment at X:X} = 30.75 \text{ kN} \times 0.0493 \text{ m}$$
$$= 1.516 \text{ kNm}$$

$$\text{Net width at X:X} = 219 \text{ mm} - 39 \text{ mm (bolt diameter)}$$
$$= 180 \text{ mm}$$
$$= 0.180 \text{ m}$$

$$1.516 \text{ kNm} = 70 \text{ MPa x} \frac{0.180 \text{ m x } t^2}{6} \quad \text{i.e.} \left(\frac{bd^3}{12}\right)$$

$$\frac{1.516 \text{ x } 6}{0.180 \text{ x } 70 \text{ x } 10^3} = t^2$$

$$t^2 = 0.0007219 \text{ m}^2$$

$$t = 0.02686 \text{ m}$$

$$t = 26.86 \text{ mm}$$

$$\text{Say, } t = 27 \text{ mm}$$

28

Methods of Attachments

The attachment of brackets to structural members requires special attention depending on the nature of the attachment and the type of loading on the bracket.

Three types of attachment are considered here. These represent the most common form of fitting:

- Bolts in tension
- Bolts in shear
- Welding

28.1 Bolts in Tension

Example 28.1

Consider a bracket depicted in Figure 28.1. The bracket is attached to the adjacent member using pairs of fasteners at positions a, b and c and is subject to vertical force P.

Solution

$$\text{The direct loading on the bolt} = \frac{P}{n}$$

where
P = total load
n = number of bolts

When the line of action does not pass through the central area of the bolt configuration, the following procedure is used.

28.1.1 Loading Producing a Tensile Load in Bolt

The bracket is considered to "heel" about point o under the influence of the load at P.

By inspection it will be apparent that the row of bolts farthest away from the centre of rotation will be the most heavily loaded.

$$m = \frac{P \cdot L \cdot l_c}{\Sigma l^2} \tag{28.1}$$

$$P = 150 \text{ kN}$$

$$L = 200 \text{ mm}$$

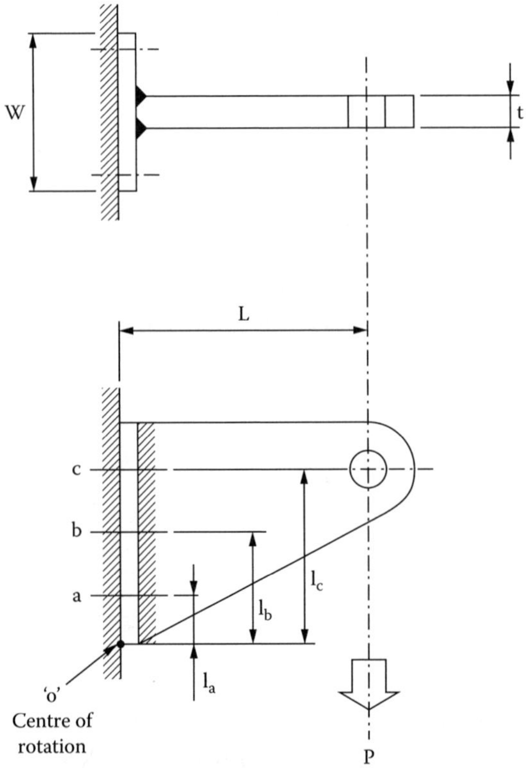

FIGURE 28.1
Detail of angle bracket: Example 28.1.

$$n = 6\ (2\text{ bolts per position})$$

$$l_a = 20\text{ mm}$$

$$l_b = 95\text{ mm}$$

$$l_c = 170\text{ mm}$$

$$\Sigma l^2 = \left(l_a{}^2 + l_b{}^2 + l_c{}^2\right)$$

$$\Sigma l^2 = \left(20^2 + 95^2 + 170^2\right)$$

$$\Sigma l^2 = 38325\text{ mm}^2$$

Maximum tension in bolts at position $l_c = F_{tension}$:

$$F_{tension} = \frac{150\text{ kN} \times 200\text{ mm}}{2 \times 225.44\text{ mm}}$$

$$F_{tension} = 66.536\text{ kN}$$

Total load on bolts = F_{total}

$$F_{total} = F_{tension} + \frac{P}{n}$$

$$F_{total} = 66.536 + \frac{150}{6}$$

$$= 91.536 \text{ kN}$$

$$\text{Considering fastener diameter} = 30 \text{ mm}$$

$$\text{Core diameter} = 25.706 \text{ mm}$$

$$\text{Maximum stress} = \sigma_{max}$$

$$\sigma_{max} = \frac{91.536 \times 10^3}{\frac{\pi \times 25.706^2}{4}}$$

$$\sigma_{max} = 176.37 \text{ MPa}$$

28.1.1.1 Permissible Stress

Using fasteners to BS 3692 strength grade 4.6, yield strength = 235 N/mm^2. Therefore use 6×30 mm diameter bolts.

$$\text{FoS} = \frac{\sigma_{yield}}{\sigma_{max}}$$

$$= 1.332$$

Note: It is generally recognised that under high loads the heel point O will suffer a degree of plastic deformation and the true heel point will migrate towards the first row of fasteners at position a. Some argue that this is the point where the moment should be measured.

It is considered that the former position should be adopted, as the second position will have the effect of reducing the load on the fasteners.

28.1.2 Load Producing a Tension and Shear Load in Bolt

Example 28.2

Consider the bracket shown in Figure 28.2 where the applied force is in the plane of the connections. In this case the load produces a tension and shear force on the bolts.

Solution

$$P_1 = P \times e$$

$$P_1 = 37.5 \times 10^3 \text{ kN/mm}^2$$

$$Y_1 = \frac{\Sigma y^2}{Y}$$

$$Y_1 = 900.321 \text{ mm}$$

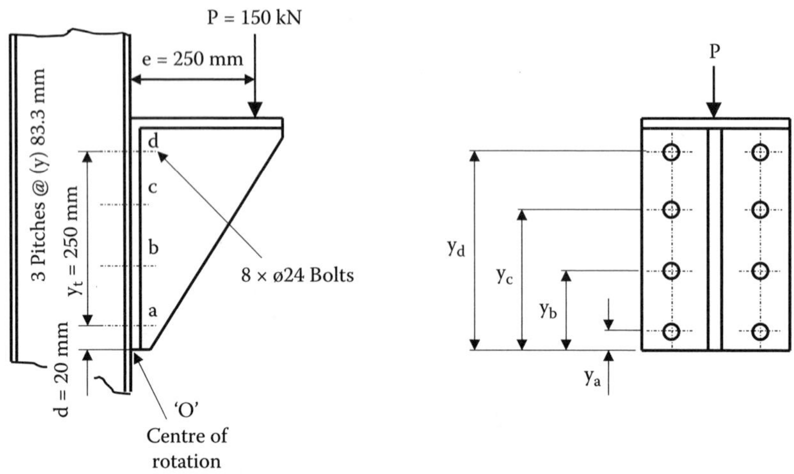

FIGURE 28.2
Detail of angle bracket: Example 28.2.

P =	150 kN	y_a =	20 mm+y
e =	250 mm moment arm	y_b =	20 mm+(2×y)
y =	83.3 mm pitch	y_c =	20 mm+(3×y)
Y =	250 mm	y_d =	20 mm+(4×y)
n =	8 bolts		
d =	20 mm		
A =	$\frac{\pi \cdot d^2}{4}$	Σy^2 =	$(y_a^2+y_b^2+y_c^2+y_d^2)$
∴ *A* =	314.16 mm²	Σy^2 =	243.087×10^3 mm²

The resultant stress on the top fastener is a vector quantity derived from the stress σ_t and the direct stress σ_s.

$$\sigma_r = (\sigma_{tension}{}^2 + \sigma_{shear}{}^2)^{0.5}$$

$$\sigma_r = \left(61.137^2 + 55.043^2\right)^{0.5}$$

$$\sigma_r = 82.265 \text{ MPa}$$

Tensile load:

$$F_t = \frac{P_1}{2\times Y_1}$$

$$F_t = 20.826 \text{ kN}$$

Direct load:

$$F_s = \frac{P}{n}$$

$$F_s = 18.75 \text{ kN}$$

$$\sigma_{shear} = \frac{P}{n \times A}$$

$$\sigma_{shear} = \frac{150 \text{ kN}}{8 \times 340.645 \text{ mm}^2}$$

$$= 55.043 \text{ MPa}$$

$$\sigma_{tension} = \frac{F_t}{\frac{\pi \times 20.319^2}{4}}$$

$$\sigma_{tension} = \frac{20.826 \text{ kN}}{340.645 \text{ mm}^2}$$

$$= 61.137 \text{ MPa}$$

Maximum stresses:

$$\sigma_{yield} = 235 \text{ kN/mm}^2$$

Permissible stress: Using fasteners to BS 3692 strength grade 4.6, yield strength = 235 kN/mm^2.

Factor of Safety (FoS):

$$\text{FoS} = \frac{\sigma_{allowable}}{F_r}$$

$$\text{FoS} = \frac{235 \text{ MPa}}{\sigma_r}$$

$$= 2.857$$

28.1.3 Bolts in Shear due to Eccentric Loading

Example 28.3

When the line of action is in the plane of the joint as shown in Figure 28.3, the fasteners will be in shear.

If the line of action of the load passes outside the centre of area, the bracket will tend to turn and the fasteners will be eccentrically loaded.

Solution

Let u_1=load/fastener per unit distance from O due to the couple WL.

$$P \cdot L = 4u_1 l_a^2 + 2u_1 l_b^2$$

Moments about O from which u_1 can be found:

$$P = 150 \text{ kN}$$

$$L = 225 \text{ mm}$$

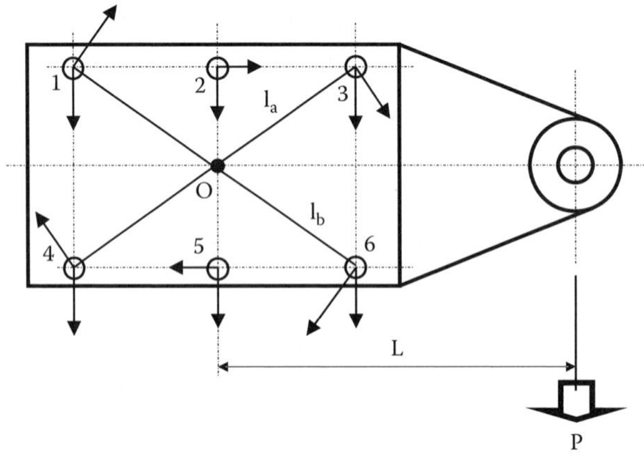

FIGURE 28.3
Bracket subject to eccentric loading.

$$L_a = 111.803 \text{ mm}$$

$$L_b = 111.803 \text{ mm}$$

$$n = 6$$

Re-arranging the above formulae to solve for u_1:

$$u_1 = \frac{P \cdot L}{\left[\left(4 \cdot L_a^2\right) + \left(2 \cdot L_b^2\right)\right]}$$

$$u_1 = 450.003 \text{ kN/m}$$

Load on fastener due to turning moment:

$$L_t = u_1 \cdot L_a$$

$$L_t = 50.312 \text{ kN}$$

Direct load on fastener due to load:

$$L_d = \frac{P}{n}$$

$$L_d = 25 \text{ kN}$$

Resultant load (solved using the sine and cosine rules):

$$c = \frac{L_t \times 0.894426}{0.6877428}$$

$$c = 65.432 \text{ kN} \left(\text{see Figure 28.4}\right)$$

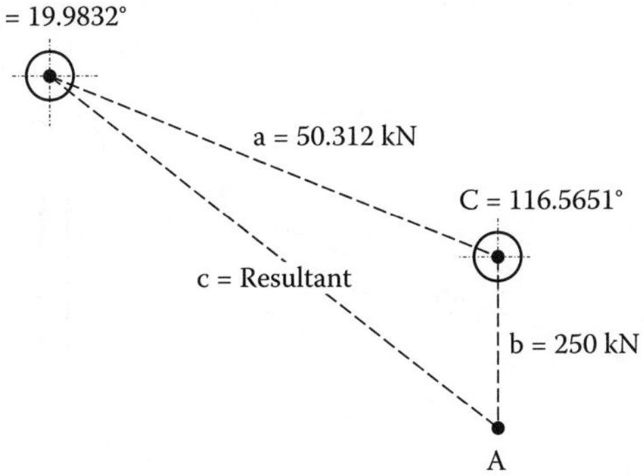

FIGURE 28.4
Resultant load for Example 28.3.

28.2 Welding (Permanent)

28.2.1 Strength of Welded Joints

Depending upon the angle between the fillets and line of action of the load, if reverse loads are experienced, these values must be modified and stress concentration factors can be used.

Table 28.1 gives the approximate strengths of some welded joints and Table 28.2 tabulates the stress concentration factors generated by different welds.

Example 28.4

Consider the welded bracket shown in Figure 28.5(a).

TABLE 28.1
Strength of Welded Joints

Type of Joint	Description of Stress Condition	Allowable Stress
Butt joints	Working stress for steel under tensile or compressive load	120–140 MPa
	Shear stress	80 MPa
Filler joints	Working stress–End fillets	95 MPa
	Side fillets	80 MPa
	Diagonal fillets	80–95 MPa

TABLE 28.2
Stress Concentration Factors of Various Welds

Type of Weld	Stress Concentration Factor
Reinforced butt	1.2
Toe of transverse fillet	1.5
T butt joint with sharp corners	2.0
End of parallel fillet	2.7

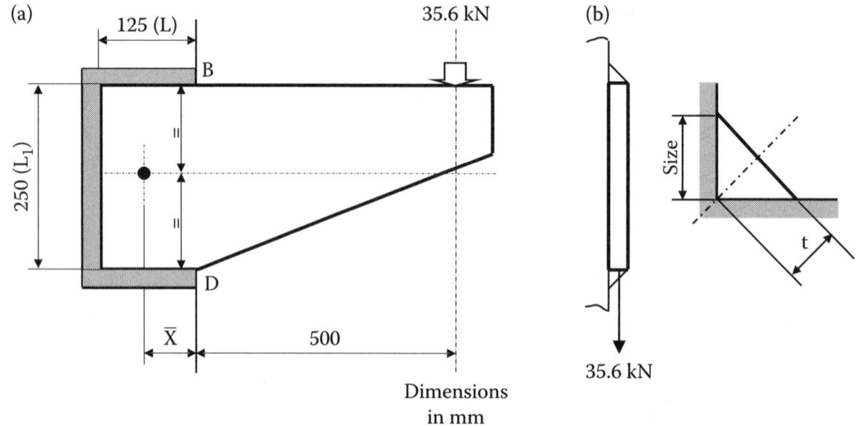

FIGURE 28.5
(a) Welded bracket for Example 28.4 and (b) weld throat dimensions.

The bracket is to be welded to a column and carries an eccentric load of 35.6 kN; find the size of weld required if the allowable weld stress is 77 MPa.

The weld is subjected to a direct shear force and an eccentric turning effect; note the 125 mm dimension is assumed to be the centroid of the vertical fillet weld.

Solution

To find the position of the centroid of the weld, take moments about BD to find X, and let t = throat thickness of weld as shown in Figure 28.5b and Figure 28.6 and considering t = 1 (unity).

Table 28.3 gives the calculated sectional properties for the example weld.

$$\frac{\Sigma Ay}{\Sigma A} = \frac{46875}{500}$$

$$\bar{X} = 93.75 \text{ mm}$$

$$\bar{Y} = 125 \text{ mm}$$

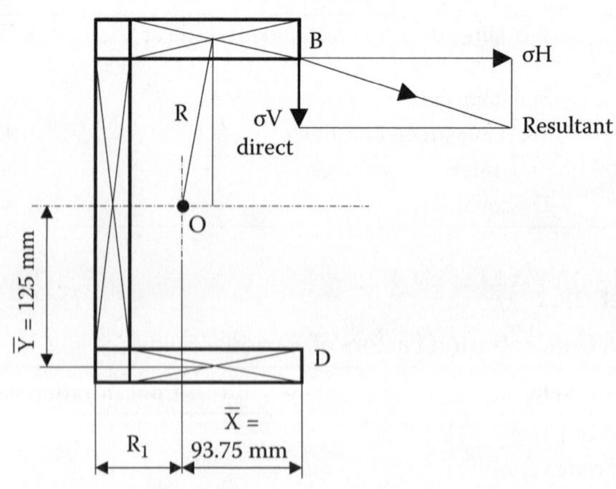

FIGURE 28.6
Centroid position of weld in Example 28.4.

TABLE 28.3

Sectional Properties of Weld in Example 28.4

Section	b	d	A	y	Ay
1	125	1	125	62.5	7812.5
2	1	250	250	125	31250
3	125	1	125	62.5	7812.5
		ΣA =	500	ΣAy =	46875

The larger resultant stresses at positions B and D are the vectors of the vertical and horizontal components generated by the eccentricity of the load together with the direct shear, as shown in Figure 28.7.

The polar second moment of area (J) of the weld arrangement:

$$J = 2a\left(\frac{1^2}{12} + R^2\right) + a_1\left(\frac{l_1^{\,2}}{12} + R_1^{\,2}\right)$$

$$= 2 \times 125t\left(\frac{125^2}{12} + 129^2\right) + 250t\left(\frac{250t}{12} + 31^2\right)$$

$$= 6.03 \times 10^6 t \text{ mm}^4$$

where

$R = 129$ mm

$R_1 = 31$ mm

Area $A = 125$ t

Area $A_1 = 250$ t

Stress at B and D:

$$\bar{x} = 93.75 \text{ mm}$$

$$= \frac{Q}{A} + \frac{M}{J} \cdot y$$

$$= \frac{35.6 \text{ kN}}{2 \times 125t + 250t} + \frac{35.6 \text{ kN} \times 593.75 \text{ mm}}{28.03 \times 10^6 t} \times 93.75 \text{ mm}$$

$$= \frac{0.3998}{t}$$

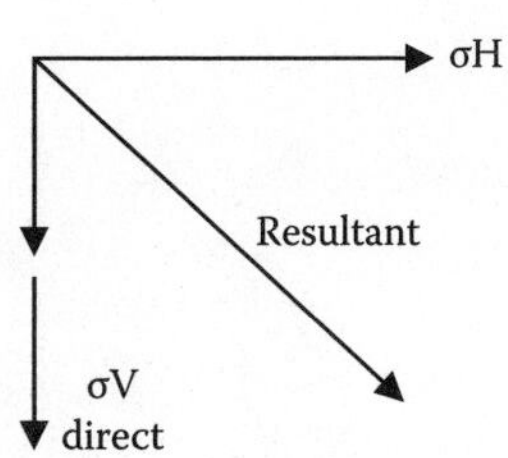

FIGURE 28.7
Resultant vector for Example 28.4.

Vertical σ=stress due to direct load+stress due to the eccentric load when

$$\sigma H = \frac{35.6\text{ kN} \times 593.75\text{ mm} \times 125\text{ mm}}{6.03 \times 10^6 t}$$

Horizontal σ when y=125 mm:

$$= \frac{0.438\text{ kN}}{t\text{ mm}^2}$$

The resultant:

$$\sigma = \sqrt{\left(\frac{0.3998}{t}\right)^2 + \left(\frac{0.438}{t}\right)^2}$$

$$= \frac{0.593\text{ kN}}{t\text{ mm}^2}$$

The allowable working stress=77 MPa=0.077 kN/mm^2.

To find throat thickness t mm:

$$\frac{0.593}{t} = 0.077$$

$$t = \frac{0.593}{0.077}$$

$$t = 7.701\text{ mm}$$

$$\text{Size of weld} = 1.414t$$

$$= 1.414 \times 7.701\text{ mr}$$

$$= 10.89\text{ mm}$$

29

The Design and Analysis of Helical Compression Springs Manufactured from Round Wire

29.1 Elastic Stresses and Deflections of Helical Compression Springs Manufactured from Round Wire

29.1.1 Introduction

This chapter covers the design data for helical compression springs manufactured from round wire and acting within the elastic range of the material. It is in four sections:

29.1. Elastic Stresses and Deflections of Helical Compression Springs Manufactured from Round Wire

29.2. Allowable Stresses for Helical Compression Springs Manufactured from Round Wire

29.3. Notes on the Design of Helical Compression Springs Made from Round Wire

29.4. Nested Helical Compression Springs

The design of a helical compression spring to meet specific requirements will require going through a number of iterations before a suitable spring design is finally established.

The procedure outlined in this chapter will enable a suitable spring design to be arrived at very quickly when all the constraints are clearly identified.

The design process outlined below identifies the use of each subsection in Section 29.1, although not all subs will be required for any one spring design. This process may have to be iterated a number of times before a suitable design will be established.

1. The type of end coil will need to be decided. These will influence both the stresses and the number of active coils in the spring. See Section 29.3, Notes on the Design of Helical Compression Springs Made from Round Wire.
2. Establish a suitable material and a value for the working stress. Guidance on this is given in Section 29.2, Allowable Stresses for Helical Compression Springs Manufactured from Round Wire.
3. The external diameter (*Do*) or internal diameter (*Di*) of the spring is often fixed by the design problem or can be easily established. Section 29.1.5 may be used to determine *d* and hence *c* and *D*.
4. Section 29.1.5 also covers the spring rate, deflection characteristics, and number of turns.

5. The spring will require to be checked that it conforms to the limitations of Section 29.1.3.
6. The closed length characteristic is checked using Section 29.1.7.
7. The buckling characteristic is checked using Section 29.1.8.
8. The outside diameter of the spring will increase slightly when loaded. This increase is generally less than 2%, but may be accurately determined, if necessary, by using Section 29.1.10.
9. Additional stresses may be incurred if the spring is subject to transverse loading (see Section 29.1.9).
10. Helix warping produces a stress increase over and above that which is predicted in Section 29.1.11. The increase is slight (being less than 5% for springs for which $N > 11$) and needs to be considered only in spring designs near their stress limit or where fatigue is the governing failure mode. See Section 29.1.11, for springs with ends closed and ground.

29.1.2 Notation

P = Axial load applied to spring

q = Shear stress due to P

δ = Deflection due to P

S = Rate or stiffness of spring

G = Modulus of rigidity

D = Mean diameter of coil

d = Diameter of wire

c = D/d spring index

n = Number of active coils

N = Total number of coils

L_0 = Free length

L_c = Solid length including end coils

$L_{c'}$ = Solid length omitting end coils (nd)

D_o = Outside diameter of coil

D_i = Inside diameter of end coils

C_1 to C_5 = Functions of c

29.1.3 Notes

The data in this chapter are subject to the following limitations.

1. The helix angle of a spring should not exceed 10°; therefore, for an unloaded compression spring,

$$L_0 < 0.55 n_o D + d' \tag{29.1}$$

   where values of d' are given in Table 29.1.
2. The deflection per coil under load should not be more than about a quarter of the coil diameter, D.

TABLE 29.1

Allowances for End Coil Forms

End coil formation	d′
One coil at each end closed and ground flat	1.5d
End coils unclosed but ground	–0.5d
End coils unclosed and unground	d
One coil at each end closed but unground	3.0d

Most springs will conform to conditions 1 and 2, but if these limitations are exceeded, then significant errors will arise; in this case it should be considered that the single spring should be replaced by a nest of two or more springs, each conforming to the conditions above (see Section 29.4).

29.1.4 Compression Spring Characteristics

29.1.4.1 Material Specifications

The materials covered in this chapter (but not limited to) are as shown in Table 29.2.

29.1.4.2 Wire Diameter

The wire diameter may be quoted in either imperial or metric diameters using readily available sizes.

Tables 29.7 and 29.8 gives approximately allowable stresses for standard wire sizes between 0.0076 and 0.300 in. and 0.250 and 12.00 mm.

29.1.4.3 Mean Diameter

The mean diameter is a basic characteristic of the helical spring. Although the outside or inside diameter of the spring may be quoted, the mean diameter will be found by either subtracting or adding the wire diameter to these diameters.

29.1.4.4 Spring Index c

The spring index is found by dividing the mean diameter D by the wire diameter d.

The preferred index is between 4 and 12. If the index is less than 4 the spring may be too difficult to manufacture and be too highly stressed; if the index is greater than 12 the spring may be too flimsy, tangle easily, or the coils may slip over each other as the spring is compressed to solid.

The larger the index the greater the deflection in comparison to the solid length.

TABLE 29.2

Material Specifications

Spring Material	Specification
Steel	S201 or S202
Stainless steel	S205
Nimonic 90	HR501HD
Nimonic 90	HR502
Phosphor bronze	BS 384

29.1.4.5 Spring Rate

This is another characteristic of the helical spring: in a helical spring design the deflection is proportional to the force acting on the spring. This is the same for most springs where the spring rate is linearly proportional.

29.1.4.6 Number of Active Coils

The number of active coils is those coils that actually deflect when the spring is either compressed or extended when subjected to the force.

29.1.4.7 Total Number of Coils

The total number of coils equals the number of active coils plus the number of end coils, if any, and may be taken as follows:

Type of Spring Ends

$$\text{Closed and ground flat} = N = n+2 \quad \text{where } N = \text{total number of coils}$$

$$\text{Closed but not ground} = N = n+2 \quad\quad n = \text{number of active coils}$$

$$\text{Open} = N = n$$

29.1.4.8 Solid Length

This is the length of the spring when all the coils are compressed and touching each other plus the type of spring end.

If the design calls for plating or an alternative coating, bear this in mind as this will increase the solid length.

The solid length should be specified as the maximum dimension. It should not be the same as the calculated nominal solid length.

29.1.4.9 Initial (Free) Length

This is the overall length of the spring in its fully unloaded condition.

If loads are specified then the free length should be a reference dimension. This will allow the spring maker to vary the free length to maintain the spring force relative to the spring deflection or specified length.

29.1.4.10 Clearance at Maximum Load

The free length of the spring should be chosen so that sufficient clearance is left between coils when the spring is compressed by the maximum working load.

This clearance should be not less than 15% of the deflection of the spring from the free length to the solid length.

29.1.4.11 Direction of Wind

A helical spring can be wound either right or left hand. If the direction of wind is not specified, it can be coiled either way.

If the springs are nested then the wind must be in opposite directions to eliminate coil binding.

When the spring is fitted over a screw thread then the coiling should be in the opposite direction of the screw thread.

29.1.4.12 Allowable Stresses

The allowable stresses will be determined by the material being specified and the operating temperature of the spring.

As a rule of thumb it is generally considered that if the maximum working stress in the spring is kept below 40% of the allowable stress for the material specified, then the spring will have an unlimited life. This will lead to a bulky spring, but if weight is not a problem, then this will be acceptable.

Where spring weight is critical, care will be required in calculating the maximum stress in the spring for the particular application, and the life of the spring will need to be calculated.

Tables 29.9 through 29.11 tabulate the temperature reduction factors and maximum temperatures required for the standard materials covered in this chapter.

29.1.4.13 Finish

Depending on the operating environment the spring will be working in, it may be desirable to specify a finish to the spring to protect against corrosion, etc. Care will be required in selecting the correct specification.

Some finishes or plating may have a detrimental effect on the fatigue life of the spring that could lead to premature failure.

29.1.5 Static Shear Stress

29.1.5.1 Basic Formulas

29.1.5.1.1 Spring Rate

The rate or stiffness of a spring is the load required to produce a unit of deflection.

$$S = \frac{P}{\delta} \quad \text{or} \quad S = \frac{8 \times P \times D}{\pi \times d^3}$$

Note that several authorities introduce a correction factor for variation in spring index, but this factor is small and can be safely ignored. The number of working coils cannot be precisely defined except by careful tests on individual springs, and the possibility of error from this source is greater than the effect of possible correction factors.

29.1.5.2 Maximum Shear Stress

This is the stress at the surface of the spring wire produced by a given load. The stress is higher on the inner surface of the coil.

Uncorrected for coil curvature:

$$q = \frac{8 \times P \times D}{\pi \times d^3}$$

With curvature correction:

$$q = \frac{8 \times P \times D}{\pi \times d^3} \times K \quad K = \frac{(c+0.2)}{(c-1)} \text{ (Stress concentration factor for curvature where } c = D / d)$$

29.1.5.2.1 Note A

This curvature correction factor was first proposed by National Physical Laboratory (NPL) in England in Spring Design Memorandum 1, published by the Ministry of Supply in 1943. It gives values very close to those obtained with Wahl's formulas:

$$K = \frac{4c-1}{4c-4} + \frac{0.615}{c}$$

The previous formula is much simpler to apply.

29.1.5.2.2 Note B

The correction factor should always be applied to springs subject to fatigue or compression to closure for extended periods of time.

The uncorrected formulas may be applied where fatigue or sustained compression (e.g., clamping, etc.) is not involved.

Alternatively:

$$q = \frac{G \times \delta}{\pi \times c^2 \times L_c}$$

29.1.5.3 Useful Relations

Note that imperial or Système International (SI) units may be used providing that they are consistent.

The problem of design is to find the dimensions of a spring for a specified performance. This is the inverse of the problem of finding rate and stress, when all the dimensions are known. The design problem can be solved from the basic formulas by a process of successive approximation, but the use of the following functions of the spring index gives the solution more readily:

$$D_o = (c+1) \times d \quad D_i = (c-1) \times d$$

$$d = \frac{D_o}{c+1} \quad d = \frac{D_i}{c-1} \quad c = \frac{D_o + D_i}{D_o - D_i} \quad c = \frac{D_o - d}{d}$$

D_o is used in preference to D, since the problem may be to find the remaining dimensions when the spring housing diameter is given. Values for the above functions are given in Tables 29.3 and 29.4 (covering imperial and metric (SI) springs, respectively), for the range of values of c from 2.0 to 12.0. Wherever possible, c should be restricted to the range from 3.0 to 10.0; greater than 10.0 may sometimes be necessary to avoid excessive lengths of springs.

A single spring of circular section is completely defined if any four of the following six parameters are known, P, δ, q, D_o, d, and n. (Note that S, L_{c1}, c, etc., are included in the forgoing since:

$$S = \frac{P}{\delta} \quad L_c = n \times d \quad c = \frac{D}{d}$$

TABLE 29.3

Values of Design Functions C_1, C_2, C_3, C_4, and C_5 (To Be Used for Imperial Springs)

c	D_i/D_o	C_1	C_2	C_3	C_4	C_5
2.00	0.333	0.084	10.00	0.000700	3.350	0.049
3.00	0.500	0.132	14.00	0.001240	7.000	0.093
3.10	0.512	0.137	14.40	0.001290	7.380	0.097
3.20	0.524	1.142	14.90	0.001350	7.770	0.102
3.30	0.535	0.147	15.40	0.001400	8.170	0.106
3.40	0.545	0.152	15.90	0.001450	8.580	0.111
3.50	0.555	0.157	16.30	0.001500	9.012	0.116
3.60	0.565	0.162	16.80	0.001560	9.451	0.121
3.70	0.574	0.167	17.30	0.001610	9.900	0.126
3.80	0.583	0.172	17.80	0.001660	10.359	0.132
3.90	0.592	0.177	18.40	0.001710	10.826	0.137
4.00	0.600	0.183	18.90	0.001770	11.300	0.142
4.10	0.608	0.188	19.40	0.001820	11.800	0.147
4.20	0.616	0.193	19.90	0.001870	12.300	0.152
4.30	0.623	0.199	20.50	0.001920	12.800	0.157
4.40	0.630	0.204	21.00	0.001980	13.300	0.162
4.50	0.637	0.209	21.60	0.002030	13.800	0.167
4.60	0.643	0.215	22.10	0.002080	14.300	0.172
4.70	0.649	0.220	22.70	0.002130	14.800	0.177
4.80	0.655	0.225	23.30	0.002190	15.300	0.182
4.90	0.661	0.231	23.80	0.002240	15.800	0.188
5.00	0.667	0.237	24.40	0.002290	16.300	0.193
5.10	0.672	0.242	25.00	0.002340	16.800	0.198
5.20	0.677	0.248	25.60	0.002400	17.300	0.204
5.30	0.682	0.253	26.20	0.002450	17.900	0.209
5.40	0.687	0.259	26.80	0.002500	18.400	0.215
5.50	0.692	0.264	27.40	0.002550	19.000	0.220
5.60	0.697	0.270	28.00	0.002610	19.500	0.226
5.70	0.702	0.276	28.60	0.002660	20.100	0.231
5.80	0.706	0.281	29.20	0.002710	20.600	0.236
5.90	0.710	0.287	29.80	0.002760	21.200	0.242
6.00	0.714	0.293	30.40	0.002820	21.700	0.247
6.10	0.718	0.299	31.10	0.002870	22.300	0.253
6.20	0.722	0.304	31.70	0.002920	22.900	0.258
6.30	0.726	0.310	32.40	0.002970	23.500	0.264
6.40	0.730	0.316	33.00	0.003030	24.100	0.207
6.50	0.733	0.322	33.70	0.003080	24.700	0.276
6.60	0.737	0.328	34.30	0.003130	25.300	0.282
6.70	0.740	0.344	35.00	0.003180	25.900	0.287
6.80	0.744	0.340	35.70	0.003240	26.500	0.293
6.90	0.747	0.346	36.30	0.003290	27.100	0.299
7.0	0.750	0.352	37.00	0.00334	27.70	0.305
7.1	0.753	0.358	37.70	0.00339	28.40	0.311
7.2	0.756	0.364	38.40	0.00344	29.00	0.316
7.3	0.759	0.370	39.00	0.00350	29.60	0.322

(Continued)

TABLE 29.3 (*Continued*)

Values of Design Functions C_1, C_2, C_3, C_4, and C_5 (To Be Used for Imperial Springs)

c	D_i/D_o	C_1	C_2	C_3	C_4	C_5
7.4	0.762	0.376	39.70	0.00355	30.20	0.328
7.5	0.765	0.382	40.40	0.00360	30.90	0.334
7.6	0.767	0.388	41.10	0.00365	31.50	0.340
7.7	0.770	0.394	41.80	0.00371	32.20	0.346
7.8	0.773	0.400	42.50	0.00376	32.80	0.351
7.9	0.775	0.406	43.20	0.00381	33.50	0.357
8.0	0.778	0.412	44.00	0.00386	34.20	0.363
8.1	0.780	0.418	44.70	0.00392	34.90	0.369
8.2	0.783	0.424	45.40	0.00397	35.50	0.375
8.3	0.785	0.431	46.10	0.00402	36.20	0.382
8.4	0.787	0.437	46.90	0.00407	36.90	0.388
8.5	0.789	0.443	47.60	0.00412	37.60	0.394
8.6	0.792	0.449	48.30	0.00418	38.30	0.400
8.7	0.794	0.456	49.10	0.00423	39.00	0.406
8.8	0.796	0.462	49.80	0.00428	39.70	0.412
8.9	0.798	0.468	50.60	0.00433	40.40	0.418
9.0	0.800	0.474	51.30	0.00439	41.10	0.424
9.1	0.802	0.481	52.10	0.00444	41.82	0.430
9.2	0.804	0.487	52.90	0.00449	42.55	0.437
9.3	0.806	0.494	53.60	0.00454	43.29	0.443
9.4	0.808	0.500	54.40	0.00460	44.03	0.449
9.5	0.810	0.506	55.20	0.00465	44.77	0.456
9.6	0.811	0.513	55.90	0.00470	45.51	0.462
9.7	0.813	0.519	56.70	0.00475	46.25	0.468
9.8	0.815	0.526	57.50	0.00480	46.99	0.475
9.9	0.817	0.532	58.30	0.00486	47.70	0.481
10.0	0.818	0.539	59.10	0.00491	48.40	0.487
10.1	0.820	0.545	59.90	0.00496	49.15	0.493
10.2	0.821	0.552	60.70	0.00501	49.90	0.499
10.3	0.823	0.558	61.50	0.00507	50.65	0.505
10.4	0.825	0.565	62.30	0.00512	51.41	0.512
10.5	0.826	0.571	63.10	0.00517	52.16	0.518
10.6	0.828	0.578	63.90	0.00522	52.92	0.524
10.7	0.829	0.584	64.70	0.00528	53.69	0.531
10.8	0.831	0.591	65.50	0.00533	54.45	0.537
10.9	0.832	0.598	66.40	0.00538	55.23	0.544
11.0	0.833	0.604	67.20	0.00543	56.00	0.551
12.0	0.846	0.671	75.70	0.00596	64.10	0.617

If only three parameters are specified, a value will need to be assumed for one of the unspecified parameters before the design can be completed. A satisfactory spring can be designed by taking a value of c in the middle of the range.

29.1.5.4 Relationships

Values of design functions, C_1, C_2, C_3, C_4, and C_5.

$$C_1 = \sqrt[4]{\frac{D_o^2}{L_c x S}} \quad C_2 = D_o\sqrt{\frac{q}{P}} \quad C_3 = \sqrt{\frac{\delta}{L_c q}} \quad C_4 = D_i\sqrt{\frac{q}{P}} \quad C_5 = \sqrt[4]{\frac{D_i^2}{L_c \times S}}$$

For imperial springs values of G other than 11.5×10^6 lbf/in^2 C_1 and C_2 should be multiplied by

$$\sqrt{\frac{11.5\times10^6}{G}}$$

and C_3 multiplied by

$$\sqrt[4]{\frac{11.5\times10^6}{G}}$$

TABLE 29.4

Values of Design Functions C_1, C_2, C_3, C_4, and C_5 (To Be Used for Metric Springs (SI Units))

c	D_i/D_o	C_1	C_2	C_3	C_4	C_5
2.00	0.333	0.292	10.00	0.0084	3.350	0.170
3.00	0.500	0.458	14.00	0.0149	7.000	0.323
3.10	0.512	0.475	14.40	0.0155	7.380	0.337
3.20	0.524	0.493	14.90	0.0163	7.770	0.354
3.30	0.535	0.510	15.40	0.0169	8.170	0.368
3.40	0.545	0.527	15.90	0.0175	8.580	0.385
3.50	0.555	0.545	16.30	0.0181	9.012	0.403
3.60	0.565	0.562	16.80	0.0188	9.451	0.420
3.70	0.574	0.580	17.30	0.0194	9.900	0.437
3.80	0.583	0.597	17.80	0.0200	10.359	0.458
3.90	0.592	0.614	18.40	0.0206	10.826	0.475
4.00	0.600	0.635	18.90	0.0213	11.300	0.493
4.10	0.608	0.652	19.40	0.0219	11.800	0.510
4.20	0.616	0.670	19.90	0.0225	12.300	0.527
4.30	0.623	0.691	20.50	0.0231	12.800	0.545
4.40	0.630	0.708	21.00	0.0238	13.300	0.562
4.50	0.637	0.725	21.60	0.0244	13.800	0.580
4.60	0.643	0.746	22.10	0.0250	14.300	0.597
4.70	0.649	0.763	22.70	0.0257	14.800	0.614
4.80	0.655	0.781	23.30	0.0264	15.300	0.632
4.90	0.661	0.802	23.80	0.0270	15.800	0.652
5.00	0.667	0.822	24.40	0.0276	16.300	0.670
5.10	0.672	0.840	25.00	0.0282	16.800	0.687
5.20	0.677	0.861	25.60	0.0289	17.300	0.708
5.30	0.682	0.878	26.20	0.0295	17.900	0.725

(Continued)

TABLE 29.4 (*Continued*)

Values of Design Functions C_1, C_2, C_3, C_4, and C_5 (To Be Used for Metric Springs (SI Units))

c	D_i/D_o	C_1	C_2	C_3	C_4	C_5
5.40	0.687	0.899	26.80	0.0301	18.400	0.746
5.50	0.692	0.916	27.40	0.0307	19.000	0.763
5.60	0.697	0.937	28.00	0.0314	19.500	0.784
5.70	0.702	0.958	28.60	0.0320	20.100	0.802
5.80	0.706	0.975	29.20	0.0326	20.600	0.819
5.90	0.710	0.996	29.80	0.332	21.200	0.840
6.00	0.714	1.017	30.40	0.0340	21.700	0.857
6.10	0.718	1.038	31.10	0.0346	22.300	0.878
6.20	0.722	1.055	31.70	0.0352	22.900	0.892
6.30	0.726	1.076	32.40	0.0358	23.500	0.916
6.40	0.730	1.097	33.00	0.0365	24.100	0.718
6.50	0.733	1.117	33.70	0.0371	24.700	0.958
6.60	0.737	1.138	34.30	0.0377	25.300	0.979
6.70	0.740	1.194	35.00	0.383	25.900	0.996
6.80	0.744	1.180	35.70	0.0390	26.500	1.017
6.90	0.747	1.201	36.30	0.0396	27.100	1.038
7.0	0.750	1.222	37.00	0.0402	27.70	1.058
7.1	0.753	1.242	37.70	0.0408	28.40	1.079
7.2	0.756	1.263	38.40	0.0414	29.00	1.097
7.3	0.759	1.284	39.00	0.0422	29.60	1.117
7.4	0.762	1.305	39.70	0.0428	30.20	1.138
7.5	0.765	1.326	40.40	0.0434	30.90	1.159
7.6	0.767	1.346	41.10	0.0440	31.50	1.180
7.7	0.770	1.367	41.80	0.0447	32.20	1.201
7.8	0.773	1.388	42.50	0.0453	32.80	1.218
7.9	0.775	1.409	43.20	0.0459	33.50	1.239
8.0	0.778	1.430	44.00	0.0465	34.20	1.260
8.1	0.780	1.451	44.70	0.0472	34.90	1.281
8.2	0.783	1.471	45.40	0.0478	35.50	1.301
8.3	0.785	1.496	46.10	0.0484	36.20	1.326
8.4	0.787	1.517	46.90	0.0490	36.90	1.346
8.5	0.789	1.537	47.60	0.0496	37.60	1.367
8.6	0.792	1.558	48.30	0.0503	38.30	1.388
8.7	0.794	1.582	49.10	0.0509	39.00	1.409
8.8	0.796	1.603	49.80	0.0515	39.70	1.430
8.9	0.798	1.624	50.60	0.0521	40.40	1.451
9.0	0.800	1.645	51.30	0.0529	41.10	1.471
9.1	0.802	1.669	52.10	0.0535	41.82	1.492
9.2	0.804	1.690	52.90	0.0541	42.55	1.517
9.3	0.806	1.714	53.60	0.0547	43.29	1.537
9.4	0.808	1.735	54.40	0.0554	44.03	1.558
9.5	0.810	1.756	55.20	0.0560	44.77	1.582
9.6	0.811	1.780	55.90	0.0566	45.51	1.603
9.7	0.813	1.801	56.70	0.0572	46.25	1.624
9.8	0.815	1.825	57.50	0.0578	46.99	1.648

(Continued)

TABLE 29.4 (*Continued*)

Values of Design Functions C_1, C_2, C_3, C_4, and C_5 (To Be Used for Metric Springs (SI Units))

c	D_i/D_o	C_1	C_2	C_3	C_4	C_5
9.9	0.817	1.846	58.30	0.0585	47.70	1.669
10.0	0.818	1.871	59.10	0.0591	48.40	1.690
10.1	0.820	1.891	59.90	0.0597	49.15	1.711
10.2	0.821	1.916	60.70	0.0603	49.90	1.732
10.3	0.823	1.936	61.50	0.0611	50.65	1.753
10.4	0.825	1.961	62.30	0.0617	51.41	1.777
10.5	0.826	1.982	63.10	0.0623	52.16	1.798
10.6	0.828	2.006	63.90	0.0629	52.92	1.818
10.7	0.829	2.027	64.70	0.0636	53.69	1.843
10.8	0.831	2.051	65.50	0.0642	54.45	1.864
10.9	0.832	2.075	66.40	0.0648	55.23	1.888
11.0	0.833	2.096	67.20	0.0654	56.00	1.912
12.0	0.846	2.329	75.70	0.0718	64.10	2.141

Values of design functions, C_1, C_2, C_3, C_4, and C_5.

$$C_1 = \sqrt[4]{\frac{D_o^2}{L_c \times S}} \quad C_2 = D_o\sqrt{\frac{q}{P}} \quad C_3 = \sqrt{\frac{\delta}{L_c \times q}} \quad C_4 = D_i\sqrt{\frac{q}{P}} \quad C_5 = \sqrt[4]{\frac{D_i^2}{L_c \times S}}$$

For metric springs values of G other than 79.3 MPa C_1 and C_2 should be multiplied by

$$\sqrt{\frac{79.3\text{ MPa}}{G}}$$

and C_3 multiplied by

$$\sqrt[4]{\frac{79.3\text{ MPa}}{G}}$$

29.1.6 Spring and Deflection Characteristics

The axial spring rate, S, of a compression spring may be calculated from Equation 29.2. Data on the transverse spring rate, S_t, is presented in Section 29.1.9.

$$S = \frac{G \times d}{8 \times n \times c^3} \tag{29.2}$$

The relationship between N and n is shown in Table 29.5. The value of N−n represents the total number of ineffective coils at both ends of the spring. In each range, the higher number corresponds to the compression under a large load and the lower number to compression under a small load.

TABLE 29.5
Number of Ineffective Coils for Various End Coil Forms

End Coil Formation	Total Number of Ineffective Coils (N–n)
One coil at each end closed and ground	1.5–2.0
End coils unclosed but ground	0.5–1.0
End coils unclosed and unground	0–0.5
One coil at each end closed but unground	1.5–2.0

29.1.7 Solid Length Characteristics

The nominal solid length of a spring is the length in which all coils are touching (see Figure 29.1). Some additional notes on the solid length are covered in Section 29.3.

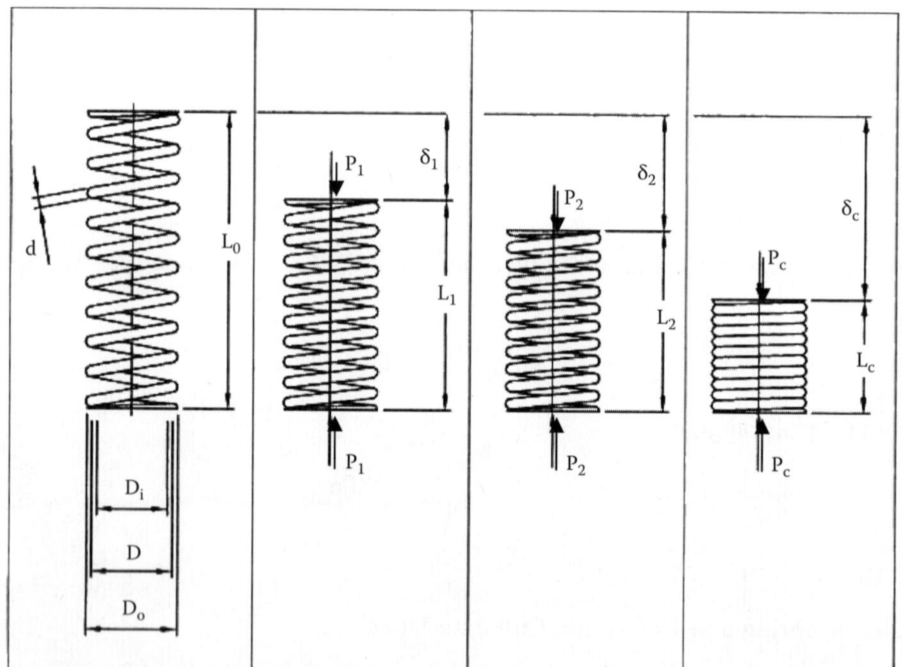

	1a	1b	1c	1d
Applied Load	0	Minimum working load P_1	Maximum working load P_2	Load at solid length P_c
Length	Free length L_0	Maximum working length L_1	Maximum working length L_2	Solid length L_c
Shear stress at intrados	0	q_1	q_2	q_c

FIGURE 29.1
Compression spring with ends closed and ground flat under 4 stages of loading.

TABLE 29.6
Effect of End Coil Forms on the Solid Length of the Spring

End Coil Formation	L_c
One coil at each end closed and ground	$d(N-0.5)$[a]
One coil at each end closed but unground	$d(N+1)$
End coils unclosed and unground	$d(N+1)$
End coils unclosed and ground	$d(N-0.5)$[a]

[a] This expression assumes that the tip of the end coil has been ground down to a thickness of 0.25d. That is, in the case of one coil at each end closed and ground, ¾ of the end coil is ground.

Most springs are scragged to introduce favorable residual stresses into the wire. If advantage of this process is not taken, then it will be necessary to ensure the spring will not be overstressed when compressed solid. At the minimum working length, clearance should be maintained between the coils. More detailed notes are contained in Section 29.3.

The nominal length of the spring will vary dependent upon the type of end coil formation and total number of coils in the spring, N. The relationship between n and N is presented in Table 29.5, and expressions for the nominal solid length are given in Table 29.6.

29.1.8 Buckling of Compression Springs

A compression spring will buckle if the deflection under a maximum force, as a proportion of the free length of the spring, exceeds a critical value of HD/L_0. A range of values of H for various end conditions is shown in Figure 29.2, and critical values of HD/L_0 are shown in Figure 29.3.

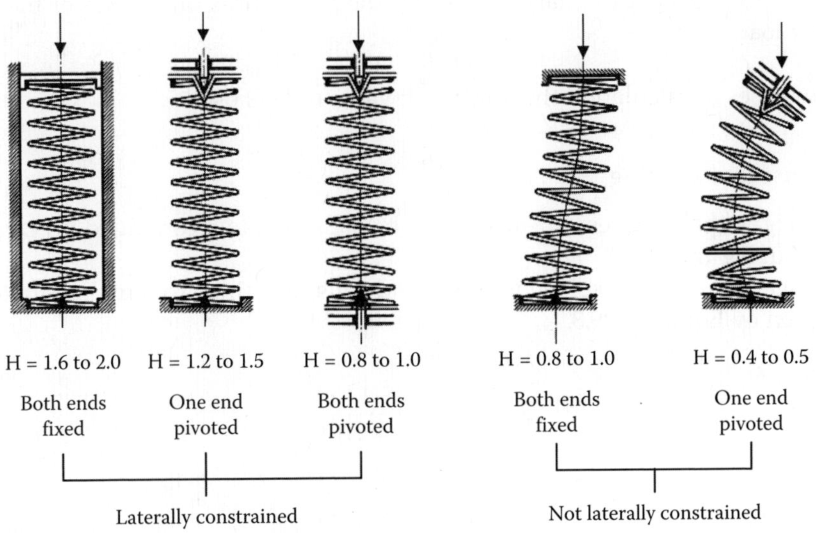

FIGURE 29.2
End function "H" for various end conditions.

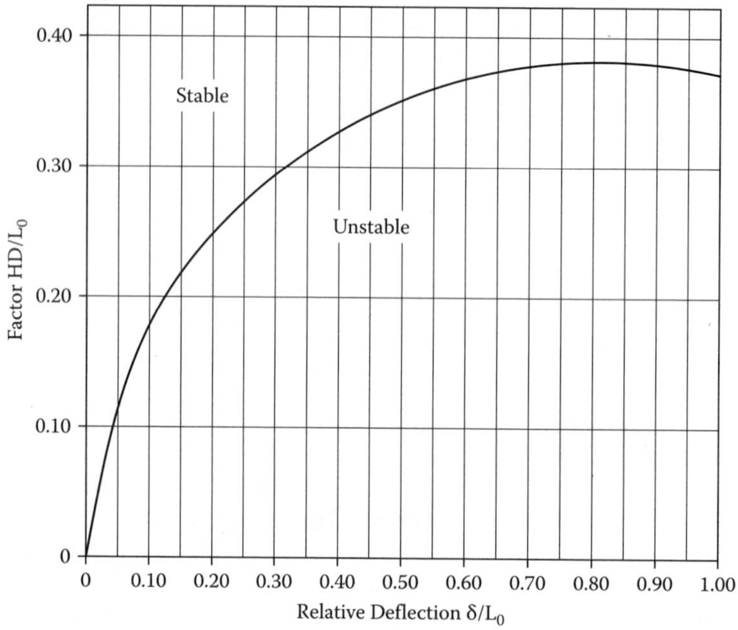

FIGURE 29.3
Critical values of HD/L_0.

There are two main reasons why a range of values for H has been given for each end condition:

1. The exact end condition as shown in Figure 29.2 is rarely met in actual practice.
2. Buckling is affected to a large extent by the squareness of the end of the spring under load.

For the purposes of calculation the minimum value of H is used.

29.1.9 Transverse Loading

In some designs the compressed spring may be subjected to a combined horizontal force as depicted in Figure 29.4.

The transverse rate of the spring with the end coils restrained to remain parallel to each other is given by Equation 29.3.

$$(S_t)_{p=0} = \frac{2S(1+\nu)}{1+\frac{(2+\nu)}{3}\left(\frac{L_0}{D}\right)^2} \tag{29.3}$$

In most instances, the spring may be subjected to an axial load and the transverse rate is then given by Equation 29.4.

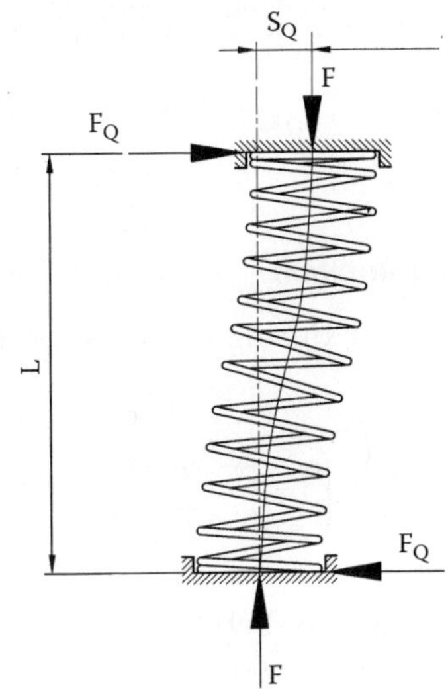

FIGURE 29.4
Compression spring under combined axial and transverse loading.

$$S_t = \frac{S\left(\frac{L_0}{L}-1\right)}{\left[\left(1+\frac{\left(\frac{L_0}{L}-1\right)}{2(1+\nu)}\right)\frac{\tan k}{k}-1\right]} \tag{29.4}$$

where

$$k = \frac{L}{2D}\left[2\left(\frac{L_0}{L}-1\right)\left(\frac{2+\nu}{1+\nu}\right)\left(1+\frac{\left(\frac{L_0}{L}-1\right)}{2(1+\nu)}\right)\right]^{0.5}$$

29.1.10 Increase of Spring Diameter under Compression

The coil diameter of a compression spring will increase when the spring is compressed. In general this expansion is within the region of 2%. The expansion will be dependent on whether the end coils are free to rotate about the helix angle.

The following derivation has been developed from Wahl.

29.1.10.1 Ends Free to Rotate

The relative change in the diameter of the spring, ΔD, during compression from the free length, L_0, to the solid length, L_c, is given by

$$\frac{\Delta D}{D} = \frac{1.3p^2 - 0.3d^2 - pd}{13D^2} \tag{29.5}$$

where the unloaded pitch,

$$p = \frac{L_0 - \bar{d}}{n_o} \tag{29.6}$$

29.1.10.2 Ends Restrained against Rotation

The relative change in the diameter during compression from free length, L_0, to solid length, L_c, is given by

$$\frac{\Delta D}{D} = \left(1 + \frac{p^2 - d^2}{\pi^2 \times D^2}\right)^{1/2} - 1 \tag{29.7}$$

In reality, the actual coil expansion will be between the values derived from Equations 29.5 and 29.7.

29.1.11 Helix Warping in Compression Springs

Various assumptions have been made in Section 29.1 about the deflection behavior of helical compression springs. Wahl (1953) gives a more detailed examination and shows that these assumptions lead to an underestimation of the stress, particularly in short springs.

Figure 29.5 gives the relationship between q_w/q and the stress increase due to helix warping and N for springs with one coil at each end closed and ground flat.

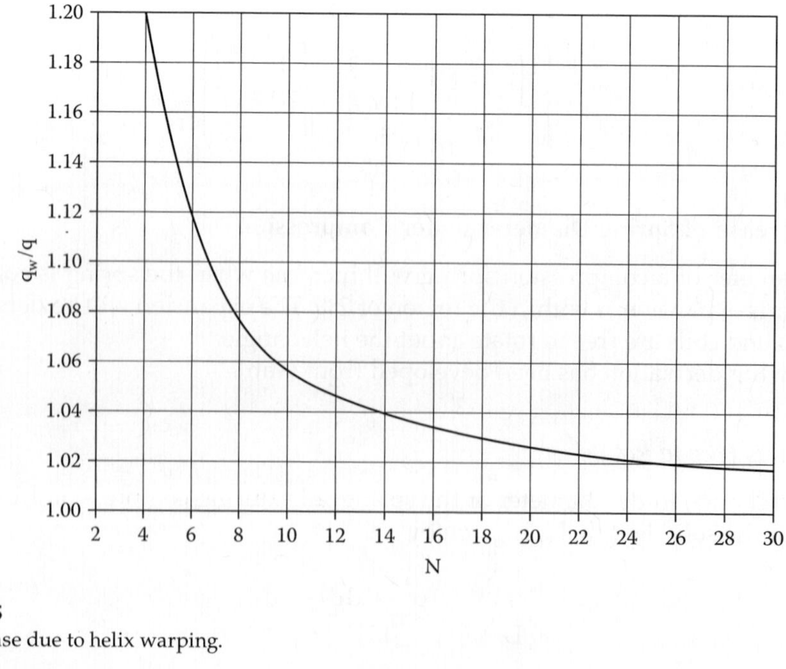

FIGURE 29.5
Stress increase due to helix warping.

29.1.12 Natural Frequency

The majority of helical compression springs are subject to dynamic applications such as:

Restraining a cam follower such as an inlet and exhaust valve
Absorbing shock or deaccelerating a load
Providing an accelerating force to a load

All springs have their own natural frequency, and it is the responsibility of the engineer to ensure that the spring will operate well under its first natural frequency; there will be a risk of damage, say in the case of an inlet or exhaust valve in an engine colliding with the piston if the spring cannot return the valve in time.

29.1.12.1 First Natural Frequency

A natural frequency curve is shown in Figure 29.6. The first natural frequency for a helical spring that is fixed at both ends is found to be

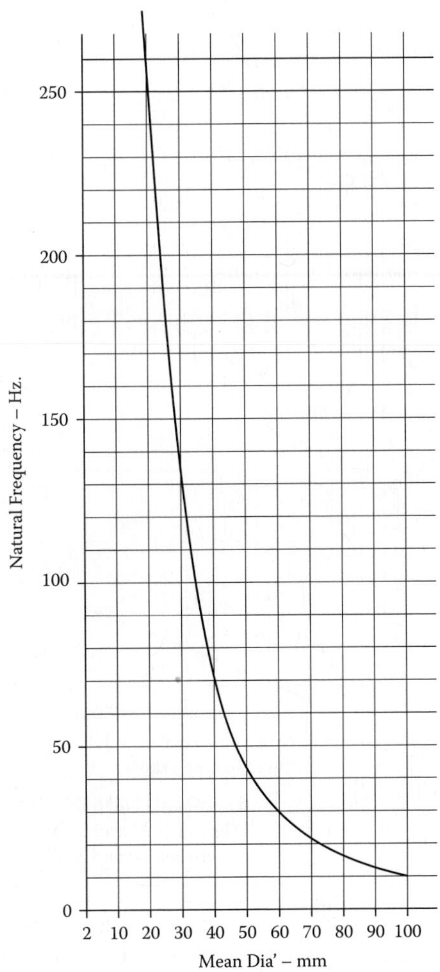

FIGURE 29.6
Natural frequency curve.

$$f_{nat} = \frac{d}{9D^2 n_t}\sqrt{\frac{G}{\rho}} \quad (29.8)$$

where

- d=wire diameter
- D=nominal coil diameter
- n_t=total number of coils
- G=shear modulus
- ρ=material density

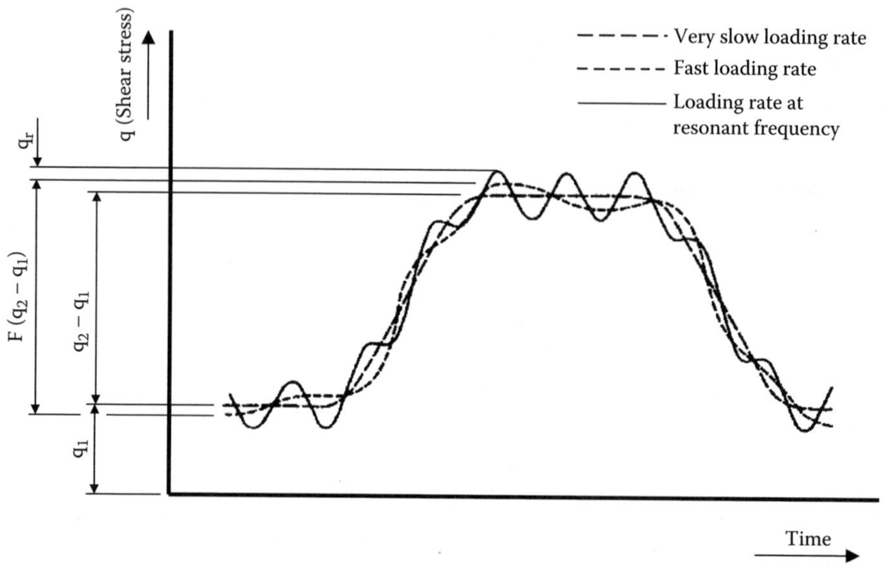

FIGURE 29.7
Stresses in spring subject to various cyclic loading.

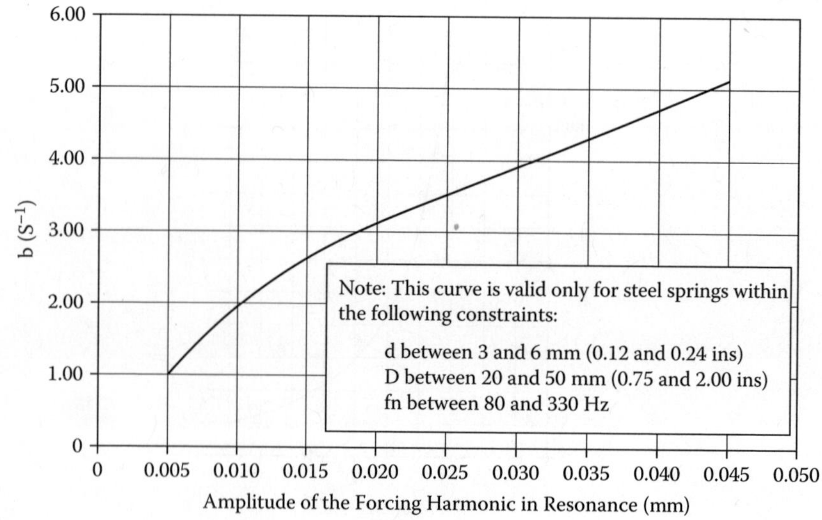

FIGURE 29.8
Damping coefficient.

Figure 29.7 shows typical stresses induced in a spring that is subjected to various cyclic loading.

Springs subjected to resonant forcing frequencies do exhibit a degree of damping. Figure 29.8 shows a typical damping curve for steel springs.

29.1.13 Example 1

29.1.13.1 Design of a Helical Compression Spring

A compression spring is to provide a load of 30 N when compressed to a length of 50 mm and a load of 85 N when compressed a further 20 mm. The outside diameter must not exceed 32 mm.

Both ends of the spring are fixed and laterally guided.

The spring shall have a fatigue life in excess of 10^7 cycles when driven by a cam rotating at 500 rev/min (8.33 Hz).

From the above information Figure 29.9 is produced.

29.1.13.2 End Coil Formulation

In this case it was considered that one coil at each end closed and ground flat would be appropriate.

29.1.13.3 Working Stress

A value of 500 MN/m² was chosen for the static design shear stress.

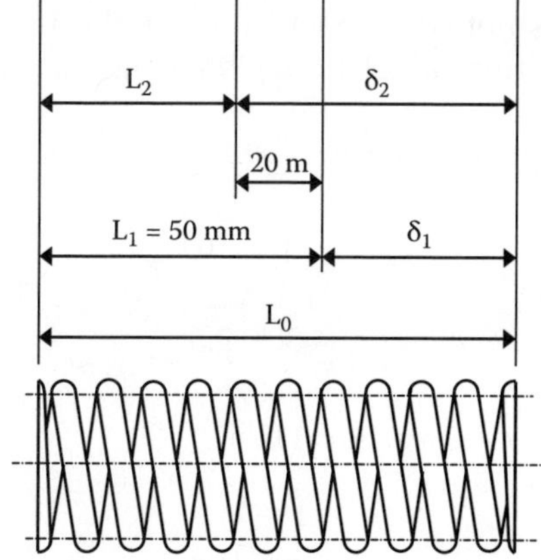

FIGURE 29.9
Compression spring diagram.

29.1.13.4 Determination of D, d, and c

The coil growth under compression is not considered to exceed 2% of the diameter, and allowing for the possible maximum tolerance on the outside diameter, a value of $D_o=30\,mm$ is considered appropriate.

$$\frac{P}{q} = \frac{85N}{500 \times 10^6 N/m^2}$$

At maximum working load:

$$= 0.17 \times 10^{-6} m^2$$

Considering case 2 from Table 29.3,

$$C_2 = D_o\sqrt{\frac{q}{P}}$$

$$= 30 \times 10^{-3}\sqrt{\frac{500 \times 10^6}{85}}$$

$$= 72.76$$

From Table 29.4 the spring index c corresponding to a value of $C_2=72.76$:
=11.654 (by interpolation)

Corresponding wire diameter d is calculated from

$$d = \frac{D_o}{c+1}$$

$$= \frac{30 \times 10^{-3}}{11.654 + 1}$$

$$= 2.371 \times 10^{-3}\ m$$

$$\text{say } 2.371\,mm$$

As this wire diameter does not correspond to a standard wire size, a wire diameter of 2.50 mm will be chosen for this calculation. Choosing a larger wire size will reduce the working stress.

Hence recalculating the spring index c:

$$c = \frac{(D_o - d)}{d}$$

$$= \frac{30 - 2.5}{2.5}$$

$$c = 11$$

and

$$D = 30 - 2.5$$

$$= 27.5\ mm$$

Recalculating the working stress in the spring at the maximum working load:

$$q = \frac{8PD}{\pi d^3} \times \frac{c+0.2}{c-1}$$
$$= \frac{8 \times 85 \times 27.5}{\pi \times 2.5^3}$$
$$= 426.67 \text{ MPa}$$

29.1.13.5 Determining Spring Rate, Deflection Characteristics, and Number of Coils

$$S = \frac{\text{Final load} - \text{initial load}}{\text{Deflection from initial load to final load}}$$
$$= \frac{85 \text{ N} - 30 \text{ N}}{(50-30) \times 10^{-3} \text{m}}$$
$$S = 2750 \text{ N/m}$$

Assuming a value of G for steel of 79.3GN/m^2,
Considering the number of working coils:

$$S = \frac{Gd}{8nc^3}$$
$$n = \frac{Gd}{8Sc^3}$$
$$= \frac{79.3 \times 10^9 \times 2.5 \times 10^{-3}}{8 \times 2750 \times 11^3}$$
$$= 6.77 \text{ coils}$$

From Table 29.5 for springs with ends closed and ground flat:

$$N - n = 1.75 \text{ (mid-range)}$$

Therefore,

$$N = 1.75 + 6.77$$
$$= 8.52 \text{ total coils}$$

The initial deflection (δ_1) produced by the initial load of 30 N is

$$\delta_1 = \frac{P}{S}$$
$$= \frac{30 \text{ N}}{2750 \text{ N/m}}$$
$$\delta_1 = 10.91 \text{ mm}$$

and the free length (L_0)

$$\begin{aligned} L_0 &= \delta_1 + L_1 \\ &= 10.91\text{ mm} + 50\text{ mm} \\ &= 60.91\text{ mm} \end{aligned}$$

29.1.13.6 Conformity with Limitations of Section 29.3

From Equation 29.1 and Table 29.1, together with the assumptions of Section 29.1.13.1 that one coil at each end is closed and ground,

$$n_o = N - 2$$

$$L_0 < 0.55 n_o D + \bar{d}$$

Since L_0=60.91 mm is less than 102.8 mm, the helix angle of the unloaded spring is less than 10°.

Furthermore, the deflection is given by

$$\begin{aligned} 0.55 n_o D + \bar{d} &= 0.55 \times (N-2) \times D + 1.5\,d \\ &= 0.55 \times 6.55 \times 27.5 + 1.5 \times 2.5 \\ &= 102.82\text{ mm} \end{aligned}$$

This is less than

$$\frac{D}{4}\left(\frac{27.5}{4}\right) = 6.88\text{ mm}$$

and so both criteria of Section 29.1.3 are satisfied.

29.1.13.7 Closed Length Characteristics

$$\begin{aligned} \frac{\delta_2}{n_o} &= \frac{10.91\text{ mm} + 20.0\text{ mm}}{6.55} \\ &= 4.72\text{ mm} \end{aligned}$$

Using Table 29.6, the nominal solid length is given by

$$\begin{aligned} L_c &= d(N - 0.5) \\ &= 2.5\text{ mm}(8.52 - 0.5) \\ &= 20.05\text{ mm} \end{aligned}$$

Since the minimum working length is 30 mm, the spring does not approach the solid length when working. If the working length was only slightly larger than the nominal solid length, it would be necessary to compare the dimensional tolerances on the working length with the maximum solid length.

The additional load necessary to close the spring to its solid length is given by

$$(L_2 - L_c) \times S = (30 \times 10^{-3} - 20.05 \times 10^{-3}) \times 2750$$
$$= 27.36 \text{ N}$$

Therefore, the total load needed to close the spring is

$$= 27.36 \text{ N} + 85 \text{ N}$$

$$= 112.36 \text{ N}$$

and the shear stress at closure:

$$= \frac{112.36}{\left(\frac{P}{q}\right)}$$

$$= \frac{112.36 \times 462.39}{85}$$

$$= 611.225 \text{ MN/m}^2$$

This stress would be permissible for a spring manufactured from BS 2803, which is low temperature, heat treated, and prestressed.

29.1.13.8 Buckling Characteristics

For a nonpivoted spring, as both ends of the spring are fixed and laterally constrained, from Figure 29.2 H has the values of 1.6 to 2.0.

In practice, however, H should be reduced by 20%, so H is taken to have a value of 1.6. Therefore,

$$\frac{HD}{L_0} = \frac{1.6 \times 27.5}{60.91}$$
$$= 0.722$$
$$\frac{Ha}{L_0} = \frac{1.6 \times 0}{60.91}$$
$$= 0$$

From Figure 29.3, where HD/L_0 is greater than 0.4, the spring will be stable for any deflection.

29.1.13.9 Determination of Coil Growth under Compression

From Equation 29.6,

$$p = \frac{L_0 - \bar{d}}{n_o}$$
$$= \frac{60.91 - 1.5 \times 2.5}{6.55}$$
$$= 8.727 \text{ mm}$$

For a worst case, that is, ends free to rotate, Equation 29.5 applies:

$$\frac{\Delta D}{D} = \frac{1.3p^2 - 0.3d^2 - pd}{13D^2}$$

$$= \frac{1.3 \times 8.727^2 - 0.3 \times 2.5^2 - 8.727 \times 2.5}{13 \times 27.5^2}$$

$$= 0.00766$$

Thus, the increase in the diameter of the spring is given by

$$\Delta D = 0.00766 \times 27.5 \text{ mm}$$

$$= 0.211 \text{ mm}$$

The maximum outside diameter of the spring will be 30.21 mm.

It is recommended to increase the outside diameter to 31.0 mm to cover any variations of manufacturing tolerances.

29.1.13.10 Natural Frequency of the Spring

From Equation 29.8 the natural frequency of the spring is calculated as follows:

$$\omega_n = \frac{d}{9D^2 n_t}\sqrt{\frac{G}{\rho}}$$

where:

$d = 2.5$ mm
$D = 27.5$ mm
$n_t = 8.52$
$G = 79.3 \times 10^9$ Pa
$\rho = 7,850$ kg/m^3

hence,

$\omega_n = 137$ Hz

In this example the operating frequency of the spring is 500 rev/min (8.33 Hz), which is well below the natural frequency of the spring. Therefore, in this example there will be no significant increase in the stress in the spring.

29.1.13.11 Stress Increase due to Helix Warping

From Figure 29.5 with N = 8.52, $q_w/q = 1.068$, this indicates that the stress in the spring previously calculated is some 6% low. Thus, the stress at static working load is

$$q = 425 \text{ MN/m}^2 \times 1.068$$

$$= 453.9 \text{ MN/m}^2$$

Similarly, the surge stress will be

$$= 561.77 \text{ MN/m}^2$$

The closure stress will not be affected by helix warping.

29.2 Allowable Stresses for Helical Compression Springs Manufactured from Round Wire

Section 29.2 provides data on both the static and fatigue strengths of various spring materials. The data relate to prestressed springs that have been heat treated after manufacture in cases where this has a beneficial effect on the fatigue life of the spring and operating in a noncorrosive ambient environment. For most carbon steel spring materials, an ambient temperature is within the range of −20 to +150°C.

In general the static and fatigue strengths should be checked in the design. Static criteria alone may be used for springs that are limited to a few hundred cycles in their working life.

29.2.1 Static Strength Data

Once the major parameters are established, the stress when the spring is compressed to solid may be calculated.

Usually, one of the first steps in spring design is to choose a value of stress that corresponds to the maximum working load. For a spring that only sustains a static load, this should be as high as reasonably possible, normally about 85% of the limiting static shear stress of the spring material.

This should give an adequate clearance between the coils when the spring is compressed to its minimum working length.

The shear stress q in Figures 29.10 and 29.11 refers to springs that have a degree of prestress. Prestressing allows springs to be loaded in service beyond the point that ordinarily produces plastic deformation and therefore a permanent set. Figures 29.10 and 29.11 show the stress distribution during the prestressing process. The compression spring is initially wound to be longer than required and is then stressed elastically under a compressive force from P to Q (Figure 29.10) and then plastically to closure from Q to R.

Releasing the force leads to a reverse residual stress to be generated, $-q_s$. Subsequently, the spring is able to accept a change of stress, $|q_s| + |q_r|$, while remaining elastic. In practice the spring is compressed to solid five or six times (S to T) to complete the process.

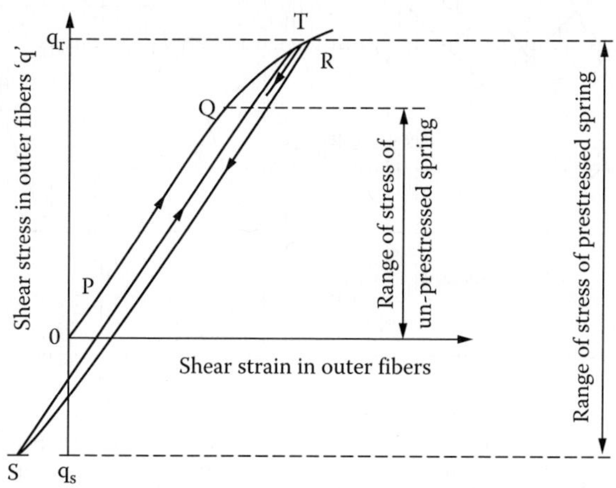

FIGURE 29.10
Shear stress in outer fibers of spring wire during prestressing.

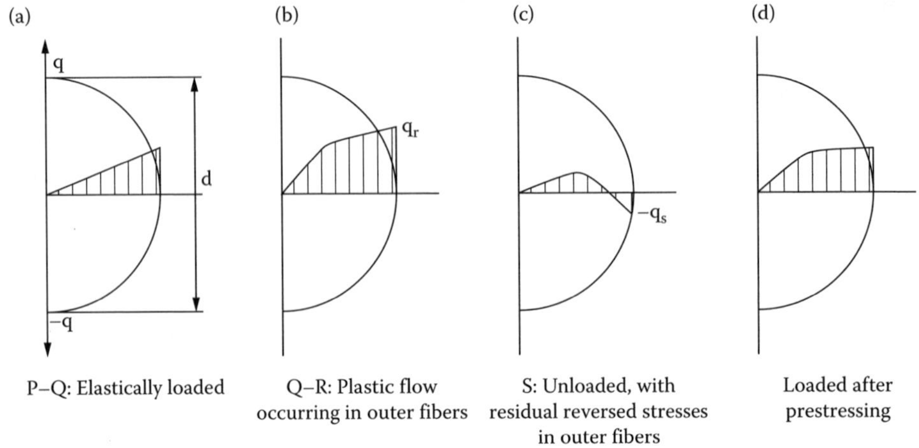

FIGURE 29.11
Shear stress distribution before, during, and after prestressing superimposed on wire diameter.

29.2.2 Fatigue Data

Fatigue data are generally presented on modified Goodman diagrams as in Figure 29.12a. This figure shows how these stresses are presented on the modified Goodman-type diagram and their relationship to the working stress cycle of the spring is shown in Figure 29.12b.

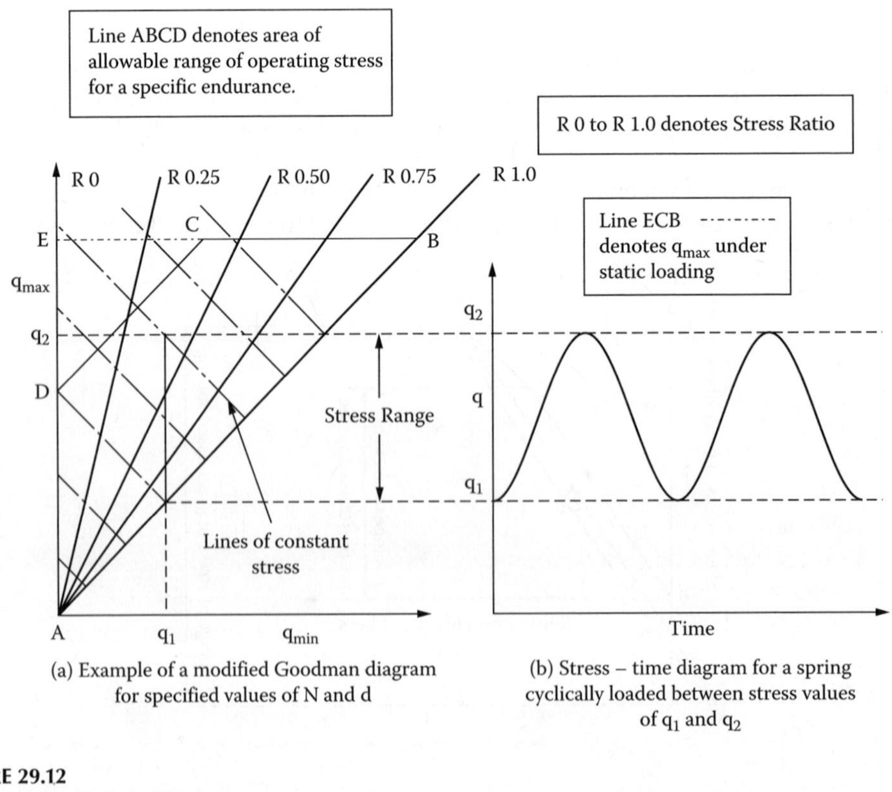

FIGURE 29.12
Presentation of fatigue data.

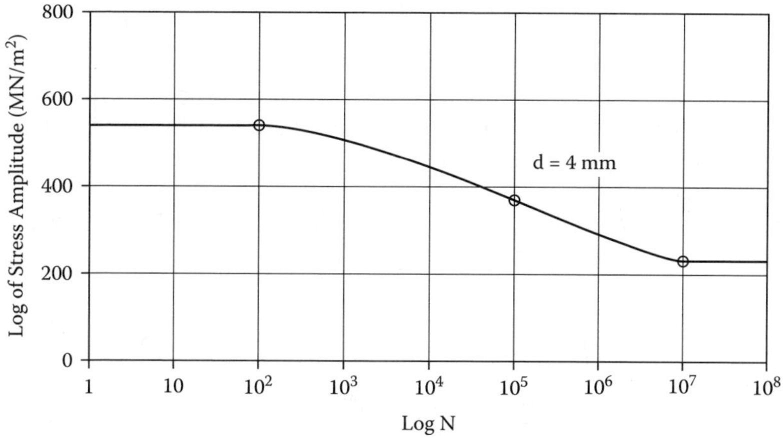

FIGURE 29.13
Estimated S-N curve for BS 1408 Range 3 at a constant mean stress: 500 MN/m².

These diagrams are specifically generated for springs of a particular wire diameter and material relating to springs that have been prestressed.

29.2.2.1 Estimation of an S-N Curve

Figure 29.13 shows a typical S-N curve; the stress amplitude (Sa) is plotted against the number of cycles to failure (N). The curve indicates the number of cycles to failure of the material, for a given specific stress amplitude, between 0 and 10^7 cycles. N is usually plotted on a logarithmic scale because the value may range between a few thousand to a few million cycles. The stress may also be plotted on a logarithmic scale. The line represents the best fit for the data points generated from the numerous fatigue tests undertaken. The fatigue test may have been a straight push-pull type of test or a bend test based on the Wohler rotating beam test.

The curve in Figure 29.13 is for a material manufactured from BS 1408 range 3 (unpeened carbon steel) subject to a constant mean stress of 500 MN/m².

29.2.3 Factors Affecting Spring Life

29.2.3.1 Spring Geometry

At small values of the spring index, c, the shear stress at the inner surface of the coil is significantly increased due to the curvature of the wire. This effect is accounted for in the curvature correction factor (see Section 29.1.5.2) when calculating q. Springs coiled from small-diameter wire can withstand higher operating stresses than larger wire diameters. There are two reasons for this:

1. A greater reduction in wire diameter by drawing during the manufacturing process improves the mechanical properties.
2. The springs may be coiled from cold wire. If hot coiled springs are subsequently heat treated, then decarburation problems may result.

29.2.3.2 Corrosion

The fatigue life of the spring may be significantly reduced if the spring is operating in a corrosive atmosphere. The fatigue life could reduce to as little as 10% of the fatigue life. Plating or other anticorrosion treatments may protect springs, but care will be required to ensure the treatment will not have a deleterious effect on the life of the spring due to, say, hydrogen embrittlement.

Corrosion-resistant materials such as titanium, stainless steel, or copper-nickel alloys would be a better choice of material.

29.2.3.3 Surface Finish

Machine marks or other surface damage on the surface of the spring may lead to early fatigue failure, as these will allow crack initiation to start much easier. A good surface finish can be obtained by specifying valve spring quality wire.

29.2.3.4 Elevated Temperatures

High operating temperatures tend to induce a permanent set or loss of load in springs at a fixed deflection and a reduction in the shear modulus. Temperatures as low as 120°C may cause a relaxation of, for instance, music wire (BS 1408M, ASTM A228, etc.), although other materials can withstand higher temperatures.

29.2.4 Treatments for Improving the Fatigue Life of Springs

29.2.4.1 Prestressing

Prestressing is a process where the compression spring is cycled between its operating range several times to improve the static and fatigue properties. It is described in more detail in Section 29.3.2.

29.2.4.2 Shot Peening

Shot peening is a process where small pellets are fired at the surface of the spring under very high pressure. It has the most beneficial effect where the torsional stress is greatest on the inside of the coils. Shot peening is difficult to accomplish with springs of small diameter wire and close coiled springs.

After shot peening the spring should be given a low-temperature heat treatment to further enhance its fatigue life. Peening leaves a very active surface, which should be protected against corrosion.

The process improves the fatigue life in three ways:

1. The process induces compressive stresses into the outer skin of the spring surface, thereby reducing the overall effects of the tensile stresses at the surface.
2. Part of the improvement in fatigue life due to shot peening may be attributed to increasing the surface hardness. It should be noted that shot peening has less effect on harder wires.
3. The process improves the surface finish and can remove small surface imperfections; the small indentations left by the shot will have no deleterious effects on crack propagation.

The process should not be relied on to improve very poor or decarburized surfaces.

TABLE 29.7

Spring Wire Data: Imperial Sizes and Allowables

Dimensions used:

Temperature: Degrees Centigrade

Strength: Lbf/in.^{2}

Modulus: Lbf/in.^{2}

Density: Lbf/in.^{3}

Material Wire Diameter	Stainless Steel S205	Steel S201 or 202	Steel Def. 106	Nimonic 90 HR501 HD	Nimonic 90 HR502	HD Phos. Brz. BS 384
0.3000	63,616	62,720	85,120	76,000	62,650	49,250
0.2760	63,616	62,720	85,120	76,000	62,650	49,250
0.2520	69,888	71,680	89,600	76,000	62,650	49,250
0.2320	69,888	71,680	89,600	76,000	62,650	49,250
0.2120	72,576	71,680	89,600	76,000	62,650	49,250
0.1920	75,266	80,640	94,080	80,640	62,650	49,250
0.1760	77,952	80,640	94,080	80,640	62,650	49,250
0.1600	80,640	80,640	94,080	80,640	62,650	49,250
0.1440	84,224	80,640	98,560	80,640	62,650	49,250
0.1280	86,910	89,600	98,560	80,640	62,650	49,250
0.1160	89,600	89,600	103,040	80,640	62,650	49,250
0.1040	89,600	89,600	107,520	80,640	62,650	52,000
0.0920	92,288	89,600	116,480	80,640	62,650	52,000
0.0800	95,872	98,560	116,480	80,640	62,650	52,000
0.0720	98,560	98,560	116,480	80,640	62,650	52,000
0.0640	98,560	98,560	125,440	80,640	62,650	52,000
0.0560	101,248	98,560	125,440	80,640	62,650	52,000
0.0480	103,936	107,520	125,440	80,640	62,650	52,000
0.0400	103,936	107,520	129,920	80,640	62,650	52,000
0.0360	107,520	107,520	134,400	89,340	62,650	52,000
0.0320	110,208	116,480	138,880	89,340	62,650	52,000
0.0280	112,896	116,480	143,360	89,340	62,650	52,000
0.0240	116,480	116,480	147,840	89,340	62,650	52,000
0.2200	116,480	125,440	147,840	89,340	62,650	52,000
0.0200	119,168	125,440	152,320	89,340	62,650	52,000
0.0180	119,168	125,440	152,320	89,340	62,650	0
0.0164	119,168	125,440	152,320	0	0	0
0.0148	121,856	125,440	152,320	0	0	0
0.0136	121,856	125,440	152,320	0	0	0
0.0124	121,856	125,440	152,320	0	0	0
0.0116	124,544	125,440	152,320	0	0	0
0.0108	124,544	134,400	152,320	0	0	0
0.0100	124,544	134,400	152,320	0	0	0
0.0092	124,544	134,400	0	0	0	0
0.0084	124,544	134,400	0	0	0	0
0.0076	124,544	134,400	0	0	0	0

Torsional modulus	10.00×10^6	12.00×10^6	12.00×10^6	12.00×10^6	12.00×10^6	12.00×10^6	6.50×10^6
Bending modulus	29.00×10^6	9.00×10^6	29.00×10^6	29.00×10^6	29.00×10^6	29.00×10^6	18.00×10^6
Density	0.283	0.283	0.283	0.283	0.300	0.300	0.320

TABLE 29.8

Spring Wire Data: Metric Sizes and Allowable Stresses

Dimensions used:

Temperature

Degrees Centigrade

Strength: N/m m²

Modulus: N/m m²

Density: N/m m³

Material Wire Diameter	Stainless Street S205	Steel S201 or 202	Steel Def. 106	Nimonic 90 HR501 HD	Nimonic 90 HR502	HD Phos. Brz. BS 384
12.000	400	431	567	400	432	340
9.000	400	431	567	400	432	340
8.000	440	431	567	400	432	340
7.100	440	431	567	400	432	340
6.300	480	494	618	400	432	340
5.600	500	494	618	400	432	340
5.000	520	557	618	556	432	340
4.500	540	557	649	556	432	340
4.000	560	557	649	556	432	340
3.550	580	557	680	556	432	340
3.150	600	616	680	556	432	340
2.800	620	616	710	556	432	340
2.500	640	616	741	556	432	358
2.240	640	616	803	556	432	358
2.000	660	679	803	556	432	358
1.800	680	679	803	556	432	358
1.600	700	679	865	556	432	358
1.400	700	679	865	556	432	358
1.250	720	741	865	556	432	358
1.120	720	741	865	556	432	358
1.000	740	741	895	616	432	358
0.900	740	741	926	616	432	358
0.800	760	804	957	616	432	358
0.710	760	804	988	616	432	358
0.630	780	804	988	616	432	358
0.560	800	863	1,019	616	432	358
0.500	820	863	1,050	616	432	358
0.450	820	863	1,050	616	432	358
0.400	840	863	1,050	616	432	358
0.355	840	918	1,050	616	432	358
0.315	840	918	1,050	616	432	358
0.280	860	918	1,050	616	432	358
0.250	860	918	1,050	616	432	358
Torsional modulus	68.95×10^3	82.74×10^3	82.74×10^3	82.74×10^3	82.74×10^3	44.82×10^3
Bending modulus	200×10^3	200×10^3	200×10^3	200×10^3	200×10^3	12.41×10^3
Density	76.81×10^6	76.81×10^6	76.81×10^6	81.43×10^6	81.43×10^6	86.85×10^6

TABLE 29.9

Spring Wire Data: Variation of Material Strength with Temperature

Material Temperature	Stainless Steel S205	Steel	Steel 201 Def. 106	Nimonic 90 HR501	Nimonic 90 HR502	HD Phos. Brz. BS 384
20	1.000	1.000	1.000	1.000	1.000	1.000
100	0.980	0.958	0.958	1.000	1.000	1.000
150	0.966	0.933	0.933	1.000	1.000	0.000
200	0.955	0.917	0.917	1.000	1.000	0.000
225	0.000	0.000	0.000	0.907	1.000	0.000
250	0.000	0.000	0.000	0.857	0.997	0.000
275	0.000	0.000	0.000	0.857	0.987	0.000
300	0.000	0.000	0.000	0.857	0.975	0.000
325	0.000	0.000	0.000	0.000	0.958	0.000
350	0.000	0.000	0.000	0.000	0.940	0.000
375	0.000	0.000	0.000	0.000	0.913	0.000
400	0.000	0.000	0.000	0.000	0.887	0.000
425	0.000	0.000	0.000	0.000	0.855	0.000
450	0.000	0.000	0.000	0.000	0.822	0.000
475	0.000	0.000	0.000	0.000	0.783	0.000
500	0.000	0.000	0.000	0.000	0.733	0.000
525	0.000	0.000	0.000	0.000	0.688	0.000
550	0.000	0.000	0.000	0.000	0.633	0.000
575	0.000	0.000	0.000	0.000	0.583	0.000
600	0.000	0.000	0.000	0.000	0.517	0.000

TABLE 29.10

Spring Wire Data: Maximum Allowable Temperatures and Variations of Bending and Torsional Modulus with Temperature

Material Type	Maximum Temperature
S205	200°C
S201	200°C
DEF 106	200°C
Nimonic 90 HR501	300°C
Nimonic 90 HR502	600°C
BS 384 HD wire	100°C

Material Temperature	Stainless Steel S205	Steel S201	Steel Def. 106	Mimonic 90 HR501	Nimonic 90 HR502	ID Phos. Brz. BS 384
100	1.000	1.000	1.000	1.000	1.000	1.000
200	0.980	0.958	0.958	1.000	1.000	1.000
300	0.966	0.933	0.933	0.992	0.992	0.000
600	0.955	0.917	0.917	0.979	0.979	0.000

29.2.4.3 Abrasive Cleaning

Abrasive cleaning improves the fatigue life by smoothing the surface of the spring, although it has the disadvantage of removing material from the wire. It is accomplished by blasting abrasive particles at the surface using compressed air or water. The surface should be immediately protected against corrosion.

29.3 Notes on the Design of Helical Compression Springs Made from Round Wire

29.3.1 General Notes

This section presents basic concepts and principles relating to the design and operation of helical compression springs manufactured from round wire.

Detailed treatment of their design is covered in Section 29.1, and the allowable stresses are covered in Section 29.2.

Springs having a constant wire diameter, coil diameter, and pitch are the essential concern of this chapter; deviations from these conditions are only permitted within the normal manufacturing tolerances.

There are many wide applications requiring compression springs, including:

To store and return energy as in a cam-operated system

To apply and maintain a constant force as in a relief valve

To attenuate vibration as in a machine structure

The energy stored in a spring is proportional to the square of the stress in the spring. Therefore, for the most efficient use of the volume of the spring material, the highest reasonable stress levels should be used as the basis of design.

Springs used in the most critical of applications, such as a pressure relief valve, require the most careful attention to every possible detail of the spring performance and manufacture. The cost of the spring should be viewed in relation to the cost of the associated equipment and replacement if a failure occurred.

Not all springs perform such critical duties, and since extra precision adds to the part cost, careful attention needs to be given to the function of each spring and tolerances relaxed according to its application.

It needs to be stressed that all the information resulting from the design considerations is passed on to the spring manufacturer in a clear and concise manner. A checklist for the manufacturing requirements is covered in Section 29.3.7.6.

29.3.2 Prestressing

In Section 29.2.4.1, the prestressing of the compression spring was considered. Prestressing (also known as scragging or presetting) induces beneficial residual stresses into the spring at the manufacturing stage. To carry out this process the spring is wound over length and is compressed to its solid length a number of times, usually five or six. This induces plastic deformation in the outer fibers of the spring coil. On release of all of the load, owing to the plastic flow that has occurred, the outer fibers of the coil, which in operation are the most highly stressed, are subjected to a residual shear stress by the inner fibers. This shear stress is in the opposite sense to that applied in the operation. The effect is to reduce the maximum stress, although the stress range remains unchanged. Therefore, the static strength and fatigue endurance of the spring are increased.

Figure 29.10 shows the spring load/deflection plot for a spring being prestressed. The first loading cycle is OPQR. Successive cycling of the load leads to a new load/deflection curve R_nQ parallel to OP, where R_n is the number of cycles the spring has been subject to.

The process also has the effect of increasing the apparent elastic limit of the material. The result of this prestressing means that in static loading the load carrying capacity of the spring is increased.

The improvements in the fatigue life and the static strength increase up to an optimum value with increasing overstrain. The amount of the increase and optimum overstrain will vary with the material. Discussions with the spring manufacturer will ensure the best possible results from the prestressing will be attained and the final details need not concern the designer.

29.3.3 Choice of Material

The choice of material is dictated by the following considerations.

29.3.3.1 Operation Reliability

Material of high quality and assured properties are required if the spring is to operate for a large number of cycles (i.e., in excess of 10^7 cycles) or at an unavoidably high stress (e.g., because of space restrictions), or because a failure will be particularly serious. Alternatively, materials of a lower quality may be acceptable where the spring is only loaded infrequently or the load is constant or is lightly stressed, or if replacement of the spring is simple and acceptable. Where the spring weight is to be kept to a minimum (as in aerospace applications) alternative materials such as titanium alloy may be acceptable.

29.3.3.2 Corrosion and Protection

Corrosion (including stress corrosion) is a common problem with many spring materials, and as a result, their fatigue strength may be seriously reduced. Particular care must be exercised in selecting a suitable material and surface finish depending up on the environment the spring will be operating in.

Usually it is best practice to use a nonreactive material rather than to attempt to protect the material that is prone to corrosion. If the corrosion protection were to break down, then the spring would be subject to corrosion attack.

If it is necessary to provide a surface protection, then it may be achieved by various methods. Plating should be done to an approved specification (such as BS 1706), although in arduous conditions pin holes are always likely to occur in the plating, leading to a general breakdown in the corrosion protection.

Painting or coating with a plastic film is feasible, but again, failure of the protection is always likely at the loading points or if the coils contact.

Immersion in oil or grease may be a means of providing the required protection. Titanium alloy springs give a greatly improved service if the cost will allow, but will require anodizing after manufacture.

If shot peening is carried out the surface of the spring is in a chemically active condition and must be protected immediately after peening. Protection should be an invariable sequel to shot peening, but it is advisable to make this clear when ordering.

29.3.3.3 Working Temperature

At high temperatures, oxidation and general reduction of material properties, such as stress relaxation, make carbon steel springs generally unacceptable. In such cases it may be necessary to consider using alloy steels or suitable nonferrous materials.

At low subzero (cryogenic) temperatures allotropic or embrittlement may be a problem, and again it may be necessary to avoid carbon steels and some alloy steels.

A guide to the suitability of a number of acceptable spring materials is listed in Tables 29.11 and 29.12. For more specific information on the material properties, reference should be made to the quoted standard or to the manufacturer.

29.3.3.4 Special Requirements

The choice of material may be dictated by specific physical properties that may be required in an application. As an example, instrument springs may require exceptional stability of the material properties or low hysteresis. A special alloy Ni-span C902 may be used when a constant modulus over a range of temperatures is required.

29.3.4 Loading

Depending how the spring is loaded will affect the permissible stress level and the magnitude of the induced stress.

It is essential to ensure that the spring is not stressed outside its elastic range during manufacture, assembly, or operation. The nominal working stress should be low enough to preclude overstressing. The additional loading factors described below should be taken into consideration when determining the nominal working stress, as these will tend to increase the stress values if not considered.

29.3.4.1 Cyclic Loading and Fatigue Properties

If the spring is subject to cyclical loading, consideration should be given to the fatigue life and endurance of the selected material. There are two factors that that will increase the working stress.

1. The nonuniform distribution of strain along the length of the spring
2. Natural frequency (see Section 29.1.12)

The actual size of these effects will depend upon the relationship of the loading frequency to the natural frequency of the spring. The natural frequency will only be known when the final dimensions of the spring are fixed.

29.3.4.2 Transverse Loading

When the spring is subject to a transverse load, this does not occur in isolation; it is generally combined with an axial load, as in a vibratory conveyor, for example.

The stress increase due to the transverse load will need to be determined to ensure that premature failure will not occur. The combined transverse and axial stress must not exceed the stress when the spring is compressed to its solid length due to the axial load alone. If there is a danger that this may occur, then the designer should provide a stop to prevent any excessive deflection occurring in both planes. The design procedure is covered in detail in Section 29.1.9. The designer should pay particular attention to the spring housing and end locations to ensure the spring remains in position under the influence of the transverse load.

TABLE 29.11
Alloy Steel Springs

Type	Material	Specification	Maximum Wire Diameter mm	Normal Working Temperature Range of Spring °C Minimum	Maximum
Carbon steels		BS 5216: HS2	13.25	−20	+130
		BS 5216: HS3	13.25	−20	+130
		BS 970: Pt. 5: 070A72	12.0	−20	+150
		BS 970: Pt. 5: 070A78	12.0	−20	+150
		BS 2803: 094A65 HS	12.5	−20	+150
Low-alloy steels	Silicon manganese	BS 970: Pt. 5: 250A58	40.0	−20	+150
	0.6–0.9% chromium	BS 970: Pt. 5: 250A61	40.0	−20	+150
		BS 970: Pt. 5: 527A60	80.0	−20	+150
	0.5% nickel-chromium	BS 970: Pt. 5: 805A60	80.0	−40	+150
	Molybdenum				
	1% chromium-vanadium	BS 970: Pt. 5: 735A50	40.0	−20	+175
Alloy steels	Austenitic chromium-nickel	BS 2056: EN58A	10.0	−200	+250
		BS 970: Pt. 4: 302S25	25.0	−200	+250
	Austenitic chromium-nickel molybdenum	BS 2056: EN58J	10.0	−200	+250
		BS 970: Pt. 4: 416S16	25.0	−200	+250
	17-7 chromium-nickel precipitation hardening	DTD 5086[a]	10.0	−90	+320
		FV 520B[b]	70.0	−90	+400

[a] UK Ministry of Defense specification.
[b] UK Trade description.

TABLE 29.12
Nonferrous Materials

Type	Specification	Maximum Wire Diameter mm	Normal Working Temperature Range of Spring °C Minimum	Maximum
Phosphor-bronze with 8% tin	BS 2873 ASTM B159-46	6.5	0	+100
Beryllium-copper with 2% beryllium	BS 2873 ASTM B197-51T	6.5	−50	+100
Ni-span C902[a]	—	3.0	−50	+100
Nickel-copper (Monel K-400)[b]	BS 3075-NA13	10.0	−40	+200
Nickel-copper-aluminum (Monel K-500)[b]	BS 3075-NA18	8.0	−200	+230
Nickel-chromium-iron (Inconel 600)[b]	BS 3075-NA14	10.0	−200	+370
Nickel-chromium-cobalt (Nimonic 90)[b]	BS 3075-2HR	8.0	−200	+550
Nickel-chromium P.H. (Inconel X-750)[b]	—	12.5	−200	+540
Titanium alloy (Ti-6A1-4V)	IMI 318b		−200	+125

[a] U.S. trade description.
[b] UK trade description.

29.3.4.3 Impact

In some designs the compression spring will be subject to a shock load (as in the case of an engine valve or pneumatic drill). The shock loading will create a surge wave of torsional stress along the wire of the spring. If the spring is subject to shock loading it will be necessary to reduce the design stress to minimize the spring being inadvertently overstressed.

29.3.4.4 Eccentric Loading

An eccentric load will have a significant effect on the spring behavior. The load/deflection characteristics will deviate from the axially loaded spring, and there will be a greater tendency to buckle where the maximum stresses will be higher.

A further problem with compression springs is that the eccentricity of the load in respect to the axis of the spring will induce uneven loading around each coil, causing the coil to warp, or take up a form other than the true helical shape. Helix warping will increase the maximum stress in the spring, but an increase in the number of coils will reduce the effects of the warping. Helix warping in compression springs is covered in more detail in Section 29.1.11.

29.3.4.5 Buckling

In Section 29.1.8 it was stated that buckling of an unrestrained compression spring will occur when the length/mean diameter ratio exceeds 4.

It is recommended that all spring designs be checked against buckling. The onset of buckling will be affected greatly by the type of seating, which may be represented by the theoretical end condition factor C. These factors are considered in Figure 29.2 for the various types of seating. It has been found that compression springs tend to buckle earlier than the predicted theory due to spring ends tending to lift off on a nonpivoted seating as the critical buckling deflection is approached; there may be some slight initial curvature of the spring axis and out-of-squareness of the ends.

If the spring as designed is found to be unstable and the dimensions cannot be altered to avoid this condition, it may be possible to restrict the lateral deflection of the spring by means of a close-fitting housing or rod. This may lead to fretting fatigue failure. When a long spring, say 10 to 20 times the mean diameter, is required, it may be preferable to stack several springs over a rod using spigotted washers between each spring, and this will reduce the tendency to buckle.

29.3.5 Design Features

29.3.5.1 End Forms

In a compression spring the center coils, remote from the ends, are the truly functional part of the spring; the shape and form of the end coils are extremely important. The end shape will govern the load carrying capacity. The type of end coil fabrication will affect both the free length of the unloaded spring and its solid length when fully compressed. Section 29.1.3 gives guidance on the estimation of the free length, and Section 29.1.7 covers the solid length calculations for each type of end coil formation.

It is good practice to locate the ends of the spring in either a recess or a spigot, particularly when there are significant lateral loads.

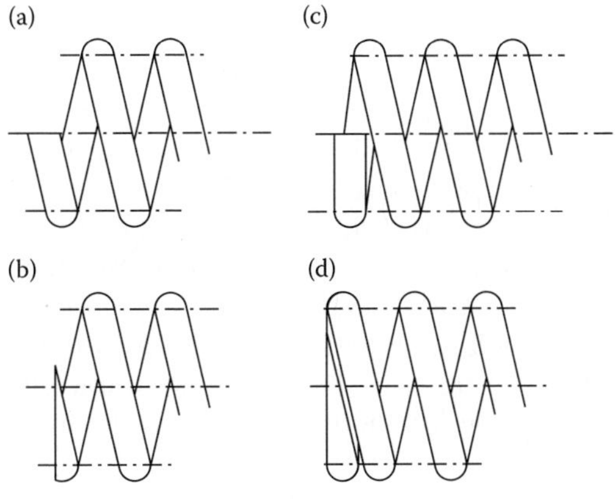

FIGURE 29.14
Spring end forms.

There are various forms used for the end coils of compression springs. The more common ones are those used in Figure 29.14:

a. Coils left unclosed and not ground.
b. Coils left unclosed, but ends are ground flat, square to the axis.
c. One or more coils at each end closed but not ground.
d. One or more coils at each end closed and ground flat, square to the axis.

It must be remembered that the more rigorous the requirements, the more expensive the spring will become. A spring of the form 29.14(d) could cost as much as three times the cost of 29.14(a).

Springs with end forms of the type 29.14(d) will be preferred for precision applications such as spring balances, etc. This form will give the minimum load tolerance and have the minimum tendency to buckle; the loading will be carried more uniformly through the spring. It will also fit in its seat without deformation of the end coils and provide good axiality of the load. The spring rate will also be more uniform over the whole loading range.

Form 29.14(c) is cheaper to produce than form 29.14(d), and while the end will settle into its seat, an offset load is produced by the deformation of the end coils as the load is applied. This will induce early buckling.

Form 29.14(b) will be cheaper again; the end coils will deform more under load, which means the spring rate is not constant and the spring is much less stable.

Form 29.14(a) is in the as coiled state and is therefore the cheapest form to produce. Its application is very limited and is consequently only used where the application is not critical.

29.3.5.2 Free Length

The nominal free length of a compression spring is one of the basic dimensional requirements required by the spring manufacturer. It is the length of the spring in the unloaded state and is measured between two parallel plates perpendicular to the axis of the spring and in contact with the end coils, or part of the end coils (see Figure 29.14).

Normally a good quality compression spring is made over the length, so that after prestressing the material is elastic, and the free length is as specified by the designer.

29.3.5.3 Solid Length

The solid length of a compression spring is the overall dimension along the axis when all the coils are touching.

Compression springs are normally designed and manufactured so that the yield stress is not exceeded when they are compressed from the free length to the solid length. This is to avoid changes in the free length should the spring be accidentally compressed solid during its operation or assembly. The nominal solid lengths for this calculation are given in the Table 29.13 and are also necessary to avoid the possibility for some coils contacting at the spring's minimum working length due to irregular pitching. If some do touch, others may be overstressed and the spring rate may alter. Experience has shown the minimum clearance between coils should be 15% of the wire diameter, and this will normally avoid the problem.

Table 29.13 relates to springs manufactured from wire less than 13 mm in diameter.

29.3.5.4 Tolerances

Care should be exercised when assigning dimensions and tolerances. Unnecessarily close tolerances will raise the cost of the spring substantially. Over-specification of the spring dimensions and tolerances should be avoided; examples of this are not uncommon, such as specifying the outside diameter, wire diameter, and inside diameter together. A spring performing a particularly critical duty will have to have minimum tolerances in order to achieve the required precision and consistency in manufacture.

The following features will require tolerances to be defined:

Final load/length requirements
Spring rate
Wire diameter
Coil diameter (outside or inside)
Free length
Straightness
End squareness and parallelism

A suitable dimensioning and tolerancing system that helps to avoid these difficulties is given in BS 1726 Part 1 on helical compression springs.

It should be recognized that even with these acceptable dimensional and load tolerances, the resulting shear stress may vary by as much as ±12%.

TABLE 29.13
Nominal Solid Length for End Coil Formations

End Coil Formation	Maximum Solid Length
One coil each end closed and ground flat	Nd
One coil each end closed but unground	(N+1.5)d
End coil unclosed and unground	(N+1)d

29.3.5.5 Surface Finish

Surface defects on the wire (such as scratch-like grooves left by the drawing process) can lead to premature failure. It is the responsibility of the spring manufacturer to avoid these through good material specification and quality control. It is good design practice to draw attention to the surface finish requirements in the design specification.

It is also good practice to avoid identification marking, but if it is essential, the identification mark should be etched rather than stamped, as again this could lead to premature failure of the spring. The marking should be applied in an area of minimum stress, such as the end coils.

The use of ground wire will avoid any decarburization of the surface. If the compression spring is subsequently hot formed, the manufacturer will need to ensure that no further decarburization takes place during coiling or heat treatment.

29.3.5.6 Surface Treatment

As discussed in Section 29.2.4.2, shot peening is the process of bombarding the surface of the spring with small hard pellets at high velocity. This is carried out to improve the fatigue life of the spring by inducing beneficial residual stress in the outer surface of the wire. It may not be possible to carry this out on small-diameter or closely coiled springs, as accessibility to the inner surface will be difficult. In this instance, the result of shot peening is to leave the surface very susceptible to corrosion, and it must be protected immediately. Other surface treatments, including plating, painting, anodizing, and plastic coating, are acceptable methods of protecting the surface to reduce corrosion. None of these coating methods will affect the strength of the spring apart from improving the fatigue life of the spring.

29.3.6 Design Procedures

29.3.6.1 Basic Design

The physical dimensions of the compression spring may be determined from the design working stress, loading requirements, and space limitations. The recommended sequence for preparing a design is given in Section 29.1.2, followed by details of the design procedure.

29.3.7 Manufacturing Requirements

Unlike the manufacture of machine components, spring manufacturers require as much latitude as possible compatible with the real design requirements. To a certain extent, the spring duty may dictate what information the manufacturer requires. From a previous example, a pressure relief valve will require an accurate spring rate or an engine valve specifying two precise points for force and deflections.

In the case of specific design requirements such as the end fixing arrangements, fit over a spigot or inside a recess will need to be supplied in full.

The manufacturer will normally stock material to the standard wire gauge sizes over a wide range. In the case of compression springs manufactured from nonstandard wire diameter this may be obtained by grinding to size from standard wire before coiling.

TABLE 29.14

Example of Spring Manufacturing Requirements

Designation of Spring:					
Reference drawing number/part number:					
Material:					
Geometric data	Diameter (or gauge) of wire:			Tolerance	Only one diameter to be specified
	Inside coil diameter:			Tolerance	
	Mean diameter of coil:			Tolerance	
	Outside coil diameter:			Tolerance	
	Number of coils:	Working			
		Total			
	Type of ends:				
	Free length:				
	Solid length:				
Performance	Initial tension:			Tolerance	Give sufficient information without duplication of data
	Load P_1:			Tolerance	
	Load P_2:			Tolerance	
	Length at load P_1:			Tolerance	
	Length at load P_2:			Tolerance	
	Spring rate:			Tolerance	
Treatments	Heat treatment:				
	Prestressing:				State yes or no
	Surface treatment:				
	Surface protection:				
Details to be finalized	Maximum load:				
	Deflection for assembly:				
	Wire length:				
	Weight:				
	Quantity required:				
Other information					

There may be a limit on the minimum diameter that can be obtained by this method, and a full discussion with the manufacturer will identify this.

It is essential that full descriptions of the material, geometric, surface finish, and end form requirements are provided if the spring is to meet the design requirements, with predictable and repeatable characteristics. Table 29.14 presents a suitable template showing the required spring details.

This data sheet is taken from BS 1726-1:2002.

It may not be possible for all the information to be identified due to the designer not always having the specialist knowledge and being able to prescribe all the requirements, but an experienced manufacturer will be able to supply this information, particularly for heat treatment, shot peening, and prestressing.

Table 29.15 should be completed with all the data required to manufacture in the spring. Table 29.16 will only be relevant to the designers and should be kept in the design file.

TABLE 29.15

Information Required by the Manufacturer

Designation of Spring:					
Reference drawing number/part number:					
Material:					
Geometric data	Diameter (or gauge) of wire:			Tolerance	Only one diameter to be specified
	Inside coil diameter:			Tolerance	
	Mean diameter of coil:			Tolerance	
	Outside coil diameter:			Tolerance	
	Number of coils:	Working			
		Total			
	Type of ends:				
	Free length:				
	Solid length:				
Performance	Initial tension:			Tolerance	Give sufficient information without duplication of data
	Load P_1:			Tolerance	
	Load P_2:			Tolerance	
	Length at load P_1:			Tolerance	
	Length at load P_2:			Tolerance	
	Spring rate:			Tolerance	
Treatments	Heat treatment:				
	Prestressing:				State yes or no
	Surface treatment:				
	Surface protection:				
Details to be finalized	Maximum load:				
	Deflection for assembly:				
	Wire length:				
	Weight:				
	Quantity required:				
Other information					

TABLE 29.16

Spring Specification Data Not Relevant to the Manufacturer

Shear stress in coils under load P_1:	
Shear stress in coils under load P_2:	
Shear stress in coils at solid length:	
Design life cycles:	
Maximum allowable temperature:	
Minimum allowable temperature:	
Environmental limitations:	
Natural frequency:	
Special notes:	

29.4 Nested Helical Compression Springs

29.4.1 General Notes to Section 29.4

When there is insufficient space available to fit a single spring without exceeding the maximum permissible stress, a design option may be to fit a nested spring, i.e., a spring within a spring.

The design of nested springs is treated here as a modification of the single spring, which satisfies the condition of loading, but not of space or stress.

The nest of two or three springs will give a higher volumetric efficiency, which may be used to either:

1. Reduce the outside diameter
2. Reduce the length
3. Reduce the maximum stress

Tables 29.18 through 29.20 enable a nest of springs to be derived from an equivalent single spring according to the requirements of either 1, 2, or 3.

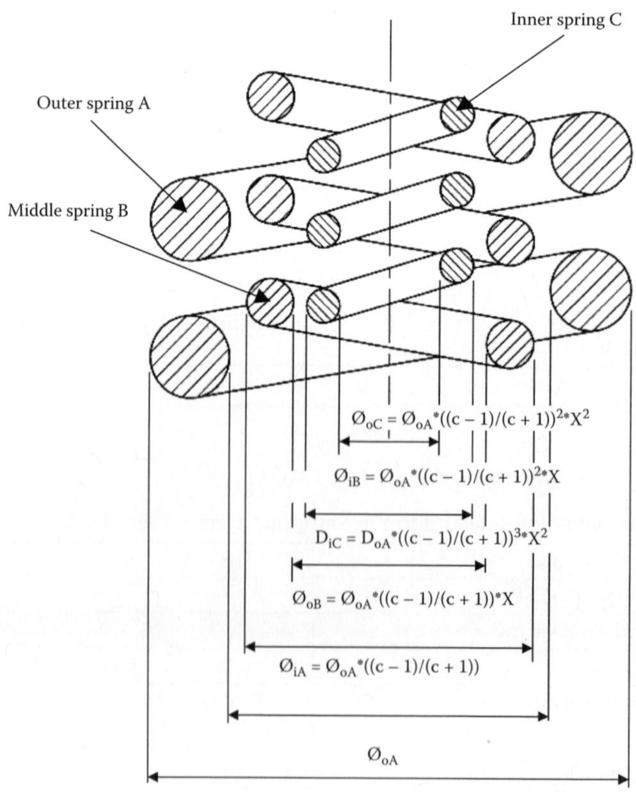

FIGURE 29.15
Nest of three springs.

TABLE 29.17

Recommended Values of Clearance Factor x

Outside Diameter of Inner Spring	Clearance	Inside Diameter of Outer Spring	Value of x
20.0	0.063	20.5	0.92
25.0	0.079	25.0	0.93
50.0	0.128	50.0	0.94
75.0	0.159	80.0	0.95
100.0	0.179	105.0	0.95
125.0	0.208	130.0	0.95
150.0	0.228	160.0	0.96
180.0	0.258	180.0	0.97

Clearance between the coils is provided by a factor x, illustrated in Figure 29.15. Clearance between the nest and the housing requires a separate allowance. Recommended values of the clearance factor are given in Table 29.17.

Each spring in a nest should have the same effective solid length and the same deflection for any given stress. It follows that each spring must have the same index c. Then if D_o of the outer spring and c are known, the coil and wire diameters can be written as in Figure 29.15.

Outside diameter of nests of two or three springs with clearance factors of 0.98, 0.94, and 0.90. The diameters are stated as fractions of the outside diameter of the single spring designed for the same load, stress, and effective solid length.

TABLE 29.18

To Reduce the Outside Diameter

	Nn=2			Nn=3		
	Clearance Factors					
Index c	0.98	0.94	0.90	0.98	0.94	0.90
2.0	0.951	0.954	0.958	0.944	0.950	0.955
3.0	0.898	0.905	0.912	0.878	0.888	0.897
4.0	0.862	0.871	0.880	0.826	0.839	0.852
4.1	0.859	0.869	0.877	0.822	0.836	0.849
4.2	0.856	0.866	0.875	0.818	0.832	0.845
4.3	0.853	0.863	0.872	0.814	0.828	0.841
4.4	0.851	0.861	0.870	0.810	0.824	0.838
4.5	0.848	0.858	0.868	0.806	0.821	0.835
4.6	0.846	0.856	0.865	0.802	0.817	0.832
4.7	0.844	0.854	0.863	0.798	0.814	0.829
4.8	0.842	0.852	0.861	0.796	0.810	0.825
4.9	0.840	0.850	0.859	0.792	0.807	0.822
5.0	0.837	0.847	0.858	0.788	0.804	0.819

(*Continued*)

TABLE 29.18 (*Continued*)

To Reduce the Outside Diameter

	Nn=2			Nn=3		
	Clearance Factors					
Index c	**0.98**	**0.94**	**0.90**	**0.98**	**0.94**	**0.90**
5.1	0.835	0.845	0.856	0.785	0.801	0.816
5.2	0.833	0.843	0.854	0.782	0.798	0.814
5.3	0.831	0.841	0.853	0.779	0.795	0.811
5.4	0.829	0.840	0.851	0.776	0.792	0.808
5.5	0.828	0.838	0.849	0.773	0.790	0.806
5.6	0.826	0.837	0.848	0.770	0.787	0.804
5.7	0.824	0.835	0.846	0.768	0.785	0.801
5.8	0.822	0.834	0.844	0.765	0.782	0.799
5.9	0.820	0.832	0.843	0.763	0.780	0.797
6.0	0.819	0.830	0.841	0.760	0.777	0.795
6.1	0.817	0.829	0.840	0.758	0.775	0.793
6.2	0.816	0.827	0.838	0.756	0.773	0.791
6.3	0.815	0.826	0.837	0.753	0.771	0.789
6.4	0.814	0.825	0.836	0.751	0.769	0.757
6.5	0.812	0.824	0.835	0.749	0.767	0.785
6.6	0.811	0.822	0.833	0.747	0.765	0.783
6.7	0.810	0.821	0.832	0.745	0.763	0.781
6.8	0.808	0.820	0.831	0.743	0.761	0.779
6.9	0.807	0.819	0.830	0.741	0.759	0.777
7.0	0.806	0.817	0.829	0.739	0.757	0.775
7.1	0.805	0.816	0.828	0.737	0.756	0.774
7.2	0.804	0.815	0.827	0.735	0.754	0.772
7.3	0.802	0.814	0.826	0.734	0.752	0.770
7.4	0.801	0.813	0.825	0.732	0.751	0.769
7.5	0.800	0.812	0.824	0.730	0.749	0.768
7.6	0.799	0.811	0.823	0.728	0.747	0.766
7.7	0.798	0.810	0.822	0.726	0.746	0.765
7.8	0.797	0.809	0.821	0.725	0.744	0.763
7.9	0.796	0.808	0.820	0.723	0.742	0.762
8.0	0.795	0.807	0.819	0.722	0.741	0.761
8.1	0.795	0.806	0.818	0.721	0.740	0.759
8.2	0.794	0.805	0.818	0.719	0.738	0.758
8.3	0.793	0.804	0.817	0.718	0.737	0.757
8.4	0.792	0.804	0.816	0.716	0.736	0.755
8.5	0.791	0.803	0.815	0.715	0.735	0.754
8.6	0.790	0.802	0.814	0.714	0.734	0.753
8.7	0.789	0.801	0.814	0.712	0.733	0.752
8.8	0.789	0.801	0.813	0.711	0.732	0.751
8.9	0.788	0.800	0.812	0.710	0.730	0.749
9.0	0.787	0.799	0.811	0.708	0.827	0.748
10.0	0.780	0.793	0.805	0.698	0.718	0.738

TABLE 29.19

To Reduce Stress

	Nn=2			Nn=3		
	Clearance Factors					
Index of Single Spring	**0.98**	**0.94**	**0.90**	**0.98**	**0.94**	**0.90**
2.0	2.05	2.04	2.04	2.06	2.05	2.04
3.0	3.16	3.15	3.13	3.20	3.18	3.16
4.0	4.28	4.26	4.24	4.38	4.34	4.31
4.1	4.40	4.37	4.35	4.50	4.46	4.43
4.2	4.51	4.48	4.46	4.62	4.58	4.55
4.3	4.62	4.59	4.57	4.74	4.70	4.66
4.4	4.74	4.71	4.69	4.86	4.82	4.78
4.5	4.85	4.83	4.80	4.98	4.94	4.90
4.6	4.97	4.94	4.91	5.10	5.06	5.02
4.7	5.08	5.05	5.02	5.28	5.18	5.13
4.8	5.19	5.17	5.14	5.34	5.30	5.25
4.9	5.31	5.28	5.25	5.46	5.42	5.37
5.0	5.42	5.39	5.36	5.58	5.53	5.48
5.1	5.53	5.50	5.47	5.70	5.65	5.60
5.2	5.64	5.61	5.58	5.82	5.77	5.71
5.3	5.75	5.72	5.69	5.95	5.89	5.83
5.4	5.87	5.83	5.80	6.07	6.01	5.95
5.5	5.98	5.95	5.91	6.19	6.13	6.07
5.6	6.10	6.06	6.03	6.32	6.25	6.19
5.7	6.22	6.18	6.15	6.45	6.38	6.31
5.8	6.33	6.30	6.26	6.57	6.50	6.43
5.9	6.44	6.41	6.37	6.69	6.62	6.54
6.0	6.56	6.52	6.48	6.81	6.76	6.66
6.1	6.67	6.63	6.60	6.93	6.86	6.78
6.2	6.78	6.74	6.71	7.05	6.98	6.90
6.3	6.90	6.86	6.82	7.17	7.10	7.02
6.4	7.02	6.97	6.93	7.30	7.22	7.14
6.5	7.13	7.08	7.04	7.42	7.34	7.26
6.6	7.24	7.20	7.15	7.54	7.46	7.38
6.7	7.36	7.31	7.27	7.66	7.57	7.49
6.8	7.47	7.42	7.38	7.78	7.69	7.61
6.9	7.58	7.53	7.49	7.90	7.81	7.73
7.0	7.69	7.64	7.60	8.03	7.93	7.84
7.1	7.81	7.76	7.71	8.15	8.05	7.96
7.2	7.92	7.87	7.82	8.27	8.17	8.08
7.3	8.04	7.99	7.93	3.39	8.29	8.20
7.4	8.15	8.10	8.04	8.52	8.41	8.32
7.5	8.27	8.22	8.16	8.65	8.54	8.44

(Continued)

TABLE 29.19 (*Continued*)

To Reduce Stress

	Nn=2			Nn=3		
	Clearance Factors					
Index of Single Spring	0.98	0.94	0.90	0.98	0.94	0.90
7.6	8.38	8.34	8.28	8.77	8.66	8.56
7.7	8.49	8.45	8.39	8.89	8.78	8.68
7.8	8.61	8.56	8.50	9.01	8.90	8.79
7.9	8.72	8.67	8.61	9.13	9.02	8.91
8.0	8.84	8.78	8.72	9.25	9.14	9.03
8.1	8.95	8.89	8.84	9.38	9.26	9.15
8.2	9.06	9.00	8.95	9.50	9.38	9.27
8.3	9.18	9.12	9.06	9.62	9.50	9.39
8.4	9.29	9.23	9.17	9.75	9.63	9.51
8.5	9.41	9.34	9.28	9.87	9.75	9.63
8.6	9.52	9.45	9.39	9.99	9.87	9.75
8.7	9.64	9.57	9.51	10.11	9.99	9.87
8.8	9.75	9.68	9.62	10.23	10.11	9.99
8.9	9.87	9.8	9.73	10.36	10.23	10.10
9.0	9.98	9.91	9.84	10.48	10.35	10.22
10.0	11.12	11.04	10.98	11.71	11.56	11.42

TABLE 29.20

To Reduce Length

	Nn=2			Nn=3		
	Clearance Factors					
Index of Single Spring	0.98	0.94	0.90	0.98	0.94	0.90
2.0	2.18	2.16	2.14	2.23	2.20	2.17
3.0	3.40	3.37	3.34	3.52	3.47	3.42
4.0	4.64	4.59	4.54	4.90	4.81	4.72
4.1	4.76	4.71	4.66	5.04	4.94	4.85
4.2	4.88	4.83	4.78	5.18	5.08	4.99
4.3	5.01	4.95	4.9	5.32	5.22	5.12
4.4	5.13	5.08	5.02	5.46	5.35	5.25
4.5	5.26	5.20	5.14	5.6	5.48	5.38
4.6	5.38	5.32	5.26	5.74	5.62	5.51
4.7	5.50	5.44	5.38	5.88	5.76	5.65
4.8	5.63	5.57	5.51	6.02	5.90	5.79
4.9	5.76	5.69	5.63	6.16	6.04	5.92
5.0	5.88	5.81	5.75	6.30	6.17	6.04

(Continued)

TABLE 29.20 (*Continued*)

To Reduce Length

	Nn=2			Nn=3		
	Clearance Factors					
Index of Single Spring	**0.98**	**0.94**	**0.90**	**0.98**	**0.94**	**0.90**
5.1	6.00	5.94	5.87	6.44	6.30	6.18
5.2	6.13	6.06	5.99	6.59	6.44	6.31
5.3	6.25	6.18	6.11	6.73	6.58	6.44
5.4	6.38	6.31	6.24	6.87	6.72	6.58
5.5	6.50	6.43	6.36	7.01	6.86	6.71
5.6	6.63	6.55	6.48	7.15	7.00	6.85
5.7	6.76	6.68	6.60	7.29	7.13	6.98
5.8	6.80	6.80	6.72	7.43	7.27	7.11
5.9	7.01	6.92	6.84	7.57	7.40	7.25
6.0	7.13	7.04	6.96	7.72	7.54	7.38
6.1	7.26	7.17	7.08	7.86	7.68	7.72
6.2	7.38	7.30	7.20	8.00	7.82	7.65
6.3	7.51	7.42	7.33	8.14	7.96	7.78
6.4	7.64	7.54	7.45	8.28	8.10	7.92
6.5	7.70	7.67	7.57	8.42	8.25	8.05
6.6	7.89	7.79	7.69	8.56	8.27	8.18
6.7	8.01	7.91	7.81	8.71	8.50	8.32
6.8	8.14	8.03	7.94	8.85	8.64	8.45
6.9	8.26	8.16	8.06	8.99	8.77	8.59
7.0	8.38	8.28	8.18	9.13	8.91	8.72
7.1	8.51	8.40	8.30	9.27	9.05	8.85
7.2	8.63	8.52	8.43	9.42	9.19	8.99
7.3	8.76	8.65	8.55	9.56	9.33	9.12
7.4	8.88	8.77	8.67	9.70	9.47	9.26
7.5	9.00	8.89	8.79	9.84	9.61	9.39
7.6	9.12	9.01	8.91	9.98	9.75	9.53
7.7	9.25	9.14	9.03	10.12	9.89	9.66
7.8	9.37	9.25	9.15	10.27	10.03	9.80
7.9	9.50	9.38	9.27	10.41	10.16	9.93
8.0	9.63	9.51	9.39	10.55	10.30	10.07
8.1	9.75	9.63	9.51	10.69	10.44	10.20
8.2	9.88	9.75	9.64	10.83	10.58	10.34
8.3	10.00	9.88	9.76	10.97	10.72	10.47
8.4	10.13	10.00	9.88	11.12	10.86	10.61
8.5	10.25	10.12	10.00	11.26	10.99	10.74
8.6	10.38	10.25	10.13	11.40	11.13	10.88
8.7	10.50	10.37	10.25	11.54	11.27	11.01
8.8	10.63	10.52	10.37	11.68	11.41	11.15
8.9	10.75	10.62	10.49	11.82	11.55	11.28
9.0	10.88	10.74	10.61	11.97	11.68	11.41
10.0	12.14	11.98	11.83	13.40	13.06	12.76

Values of index C for nests of two and three springs with clearance factors of 0.98, 0.94, and 0.90. The values of the index are tabulated against the index of the single spring, which has the same outside diameter, load, deflection, and effective solid length. The stress in each spring of the nest designed from this table will be lower than the stress in the corresponding single spring.

Values of the index c for nests of two or three springs with clearance factors of 0.98, 0.94, and 0.90. The values of the index are tabulated against the index of the single spring, which has the same outside diameter, load, deflection, and stress.

The length of each spring of a nest designed from this table will be less than the corresponding single spring.

29.4.2 Example 2

The following example considers replacing a single spring with a nest of three springs.

The single spring has the following characteristics:

$D_o = 120.0\,\text{mm}$

$d = 20.0\,\text{mm}$

$D = 100.0\,\text{mm}$

$C = 5.0$

$n = 8$

$L_{c1} = n.d = 160\,\text{mm}$

$L_c = (n - 1)d = 180$

$P = 62{,}560\,\text{N}$

$\delta = 86.36\,\text{mm}$

$S = 201.4\ \text{N/mm}$

$K = 1.3$

$q = 696.4\,\text{MPa}$

Let L_{c1} and q be retained. Find the outside diameter of a nest of three springs.

Section 29.1.5.2 shows that the maximum stress q remains unchanged only if c = 5.0, and therefore the stress concentration factor K is retained.

As the inside diameter of the coil of the single spring is $D_i = (D - d) = 80.0\,\text{mm}$, the clearance factor recommended in Table 29.17 is x = 0.95.

From Table 29.18 for Nn = 3, x = 0.95 and c = 5.0, the ratio D_o (nest)/D_o (single spring) = 0.79 is found by interpolation. Hence D_o (nest) = 0.79 × 120.0 = 94.8 mm.

The outside diameters of the other two springs will be obtained by means of the terms given in Figure 29.15.

Namely:

$$\left(\frac{c-1}{c+1}\right)x = \frac{4}{6}\cdot 0.95 = 0.6333$$

and

$$\left(\frac{c-1}{c+1}\right)^2 \cdot x^2 = 0.401$$

The outside diameter of the middle and inner spring is obtained from Figure 29.15:

$$D_{o\,middle} = D_o \cdot \left(\frac{c-1}{c+1}\right) \cdot x$$
$$= 94.8 \cdot \left(\frac{5-1}{5+1}\right) \cdot 0.95$$
$$= 60.04\,mm.$$

$$D_{o\,inner} = D_o \cdot \left(\frac{c-1}{c+1}\right)^2 \cdot x^2$$
$$= 94.8 \cdot \left(\frac{5-1}{5+1}\right)^2 \cdot 0.95^2$$
$$= 38.024\,mm.$$

Table 29.21 can now be constructed.

The reduction in the outside diameter (D_o) compared with the single spring:

It will also be noted that the solid length (Lc_1) is also shorter than the single spring.

This example of a spring nest is impracticable, as the inner spring is laterally unstable (see Section 29.3.4.5). The solution is to redesign the nest using only two springs. In this design the inner spring is stable and practicable.

From Table 29.22, considering N=2, x=0.95, and c=5.0 and interpolating:

$$D_o = 0.8395 \times 120.0\text{ mm}$$
$$= 100.74\text{ mm}$$

The reduction in the outside diameter:

$$\text{Reduction} = \left(1 - \frac{100.74}{120.00}\right) \cdot 100$$
$$= 16.05\%$$

TABLE 29.21

Summary of Nest Details for Example 2

D_o mm	94.8	60.04	38.03	
$d = \frac{D_o}{c+1}$ mm	15.8	10.07	6.338	
$D = D_o - d$ mm	79.0	49.97	31.69	
$D_i = D - 2d$ mm	63.2	39.90	25.35	
$n = \frac{L_{c1}}{d}$ mm	10.13	15.89	25.24	
$L_c = L_{c1} + d$ mm	175.8	170.07	166.34	
$S = \frac{Dd^4}{8nD^3}$ N/mm	118.54	49.27	19.08	$\Sigma S = 186.89$ N/mm
$L_c = L_{c1} + d$ mm	175.85	170.08	166.31	

TABLE 29.22
Summary of Nest Details for Example 2

D_o mm	100.74	63.80	
$d = \frac{D_o}{c+1}$ mm	16.79	10.63	
$D = D_o - d$ mm	83.95	53.17	
$D_i = D - 2d$ mm	67.16	42.54	
$n = \frac{L_{c1}}{d}$ mm	67.16	15.05	
$L_c = L_{c1} + d$ mm	176.79	170.63	
$S = \frac{Dd^4}{8nD^3}$ mm	176.79	53.70	$\Sigma S = 187.60$ N/mm

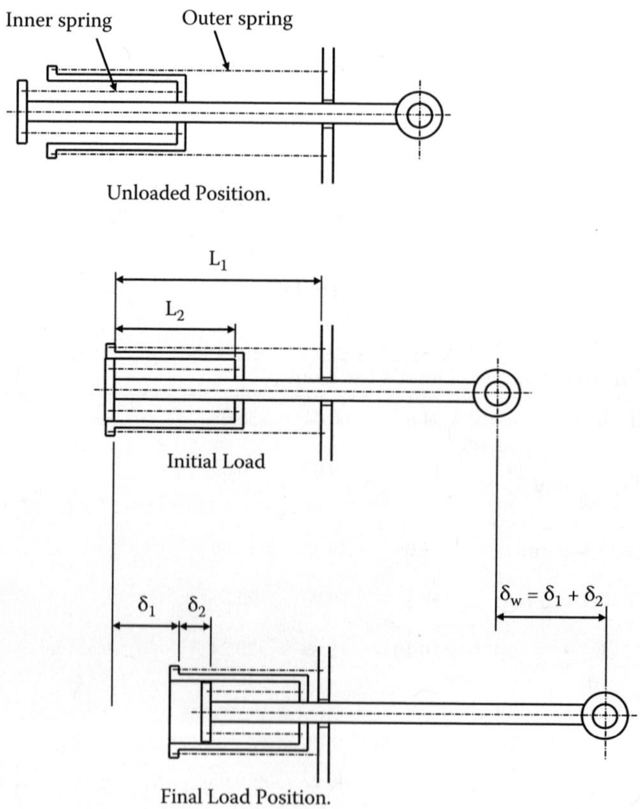

FIGURE 29.16
Nested springs in series.

29.4.3 Nested Springs in Series

This arrangement of spring assembly (see Figure 29.16) is generally used to vertically balance a mass such as a cover. It has the advantage that the deflection can be large but the solid length is small. In a housing of specific dimensions, the spring assembly will have a lower stress and also offer a greater resistance to buckling compared to that of a single spring.

The proportions in which the total working deflection (δ_w) comprises the deflections δ_1 and δ_2.

These deflections will depend upon the space required to accommodate the tube that transmits the load from the inner spring to the outer spring.

An approximate estimate of these deflections can be determined from the following formula:

$$\frac{d_1}{7L_1} = \frac{d_2}{3L_2} = \frac{d_w}{7L_1 + 3L_2}$$

For example if L_2=0.5 L_1, the proportion of the total deflection that is due to the outer spring will be:

$$\frac{d_1}{d_w} = \frac{7L_1}{7L_1 + 3L_2} = \frac{14L_2}{17L_2}$$
$$= 0.824$$

29.4.4 Example 3

A telescopic spring is required to balance an assembly to meet the following requirements:

Load range	6,900–12,200 N
Corresponding working deflection (δ_w)	320 mm
Maximum working stress (q)	827.4 MPa
Stress to closure limited to	956.3 MPa
Outside diameter of the nest not to exceed	140.0 mm

The inner spring is to be half the length of the outer spring in the initial loaded condition.

Considering the outer spring:

$$L_1 = 2L_2 \quad \text{and} \quad \frac{\delta_1}{\delta_w} = \frac{14}{17}$$
$$= 0.824$$
$$\delta_1 = 0.824\,\delta_w$$
$$= 0.824 \cdot 320\text{ mm.}$$
$$= 263.53\text{ mm.}$$

Spring rate:

$$= \frac{12200 - 6900}{263.53}$$
$$= 20.11\text{ N/mm}$$

Total deflection:

$$d = \frac{12200\,\text{N}}{20.11\,\text{N/mm}}$$
$$= 606.663\,\text{mm}$$

Initial deflection:

$$\delta_1 = \frac{6900\,\text{N}}{20.11\,\text{N/mm}}$$

$$C_2 = D_o\sqrt{\left(\frac{q}{P}\right)}$$
$$= 140.0\sqrt{\left(\frac{827.3\,\text{MPa}}{12200\,\text{N}}\right)}$$
$$= 36.457$$

Hence, c=6.922.

C_3=0.03974 (by interpolation from Table 29.4).

Now:

$$C_3 = \sqrt{\left(\frac{d}{L_{c1}.q}\right)}$$
$$0.03974 = \sqrt{\left(\frac{606.663}{L_{c1}\cdot 827.37}\right)}$$
$$\therefore L_{c1} = 464.29\text{ mm}$$

Deflection to closure:

$$\delta_c = 606.66\,\text{mm}\cdot\frac{956.27\text{ MPa}}{827.32\text{ MPa}}$$
$$= 701.22\,\text{mm}$$

Free length L_0':

$$= 701.22\,\text{mm} + 464\cdot 3\,\text{mm}$$
$$= 1165.50\,\text{mm}$$

Length at initial load:

$$= L' = L_0' - \delta_t$$
$$= 1165.5\text{ mm} - 343.113\text{ mm}$$
$$= 822.387\text{ mm}$$

Number of working coils:

$$
\begin{aligned}
n &= \frac{L_{c1}}{d} \\
&= \frac{464.3\ \text{mm}}{17.67\ \text{mm}} \\
&= 26.27\ \text{coils}
\end{aligned}
$$

Free length (L_0):

$$
\begin{aligned}
L_0 &= L_{c1} + 2d \\
&= 1165.5\,\text{mm} + (2 \cdot 17.67\,\text{mm}) \\
&= 1200.844\,\text{mm}
\end{aligned}
$$

Inner spring:

$$
\begin{aligned}
d_2 &= d_w - d_1 \\
d_2 &= 320.0\ \text{mm} - 263.53\ \text{mm} \\
&= 56.47\ \text{mm}
\end{aligned}
$$

Spring rate:

$$
\begin{aligned}
&= \frac{12200\ \text{N} - 6900\ \text{N}}{56.47\ \text{mm.}} \\
&= 93.86\ \text{N/mm.}
\end{aligned}
$$

Length of spring at initial load L_2:

$$
\begin{aligned}
L_2 &= 0.5\ L_1 \\
&= 0.5 \cdot 822.37\ \text{mm} \\
&= 411.19\ \text{mm}
\end{aligned}
$$

Initial deflection d_i:

$$
\begin{aligned}
d_i &= \frac{6900\ \text{N}}{93.855\ \text{N/mm}} \\
&= 73.52\ \text{mm}
\end{aligned}
$$

Deflection to closure:

$$
\begin{aligned}
\delta_c &= \frac{12200\ \text{N}}{93.855\ \text{N/mm}} \cdot \frac{956.27\ \text{MPa}}{827.32\ \text{MPa}} \\
&= 150.25\ \text{mm.}
\end{aligned}
$$

$$\begin{aligned}\therefore L_{c1} &= L_2 + \delta_i - \delta_c \\ &= 411.19\text{ mm} + 73.52\text{ mm} - 150.25\text{ mm} \\ &= 334.46\text{ mm}\end{aligned}$$

$$\begin{aligned}C_3 &= \sqrt{\left(\frac{\delta}{L_{c1}\cdot q}\right)} \\ &= \sqrt{\left(\frac{56.47\text{ mm} + 73.52\text{ mm}}{334.464\text{ mm}\cdot 827.37\text{ MPa}}\right)} \\ &= 0.02167\end{aligned}$$

By interpolation from Table 29.4:

$$\begin{aligned}c &= 4.0617 \\ C_2 &= 19.2085 \\ D_o &= C_2\sqrt{\left(\frac{P}{q}\right)} \\ &= 19.2085\sqrt{\left(\frac{12200\text{ N}}{827.37\text{ N/mm}^2}\right)} \\ &= 73.76\text{ mm}\end{aligned}$$

Number of working coils n:

$$\begin{aligned}n &= \frac{L_{c1}}{d} \\ &= \frac{334.464\text{ mm}}{14.572\text{ mm}} \\ &= 22.95\text{ coils}\end{aligned}$$

Further Reading

Ministry of Supply. Notes on the design of helical compression springs. In *Spring Design Memorandum* 1. London: HMSO, 1942.

Wahl, A.M. Diametral expansion of helical compression springs during deflection. *J. Appl. Mech.*, 20, 565–566, 1953.

30

Design and Analysis of Lugs and Shear Pins

30.1 Notation

The following notation is used in this design guide.

The dimensions a, c, d, t, and W are defined in the diagrams below.

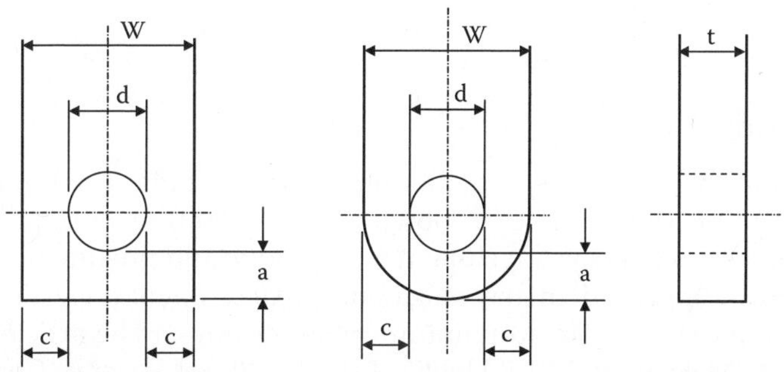

Notation.

σ_{tu} = Ultimate tensile strength of lug material (Pa)
σ_{tuw} = Ultimate tensile strength of lug material with the grain (Pa)
σ_{tux} = Ultimate tensile strength of lug material across grain (Pa)

When the plane of the lug contains both long- and short-transverse grain directions, σ_{tux} is the smaller of the two values.

σ_{ty} = Tensile yield strength of the lug material (Pa)
σ_{tyw} = Tensile yield strength of the lug material with grain (Pa)
σ_{tyx} = Tensile yield strength of the material across grain (Pa)

When the plane of the lug contains both long- and short-transverse grain directions, σ_{tyx} is the smaller of the two values.

σ_{cy} = Compressive yield strength of the bush material (Pa)
P_l = Applied limit load (N)
P_u = Applied ultimate load; this value is usually 1.5 P_l (Pa)
P_y = Applied yield load, this value is usually 1.15 P_l (Pa)
M = Applied bending moment on shear pin (Nm)
α = Angle between direction of loading on lug and the central plane of lug (degree)
P'_u = Allowable ultimate load (N)
P'_{bru} = Allowable ultimate load as determined by shear bearing (N)
P'_{bry} = Allowable yield-bearing load on bushing (N)

P'_{tu} = Allowable ultimate tension load (N)
P'_{tru} = Allowable ultimate transverse load (α = 90°) (N)
A_{br} = Projected bearing area (m^2)
A_t = Minimum net section for tension (m^2)
A_{av} = Weighted average area for transverse load (m^2)
K_{br} = Efficiency factor for shear bearing
K_t = Efficiency factor for tension
K_{tru} = Efficiency factor for transverse loading (ultimate)
K_{try} = Efficiency factor for transverse load (yield)
C = Yield factor
γ = Pin bending moment reduction factor for peaking
$r = (a - (D/2))/t$
R = Load ratio in interactive
R_α = Load ratio for axial load (α = 0°)
R_{tr} = Load ratio for transverse load (α = 90°)

30.2 Introduction

Lugs and shear pins are found in a number of aerospace applications, including landing gear attachment points (typical to that shown in Figure 30.1) together with general engineering components, such as the attachment of hydraulic and pneumatic cylinders, etc.

In 1950 an article appeared in the magazine *Product Engineering* titled "Analysis of Lugs and Shear Pins Made from Aluminum or Steel Alloys." The article was written by F.P. Cozzone, M.A. Melcon, and F.M. Hoblit of the Lockheed Aircraft Corporation. There was a follow-up article written in June 1953 in the same magazine entitled "Developments in the Analysis of Lugs and Shear Pins."

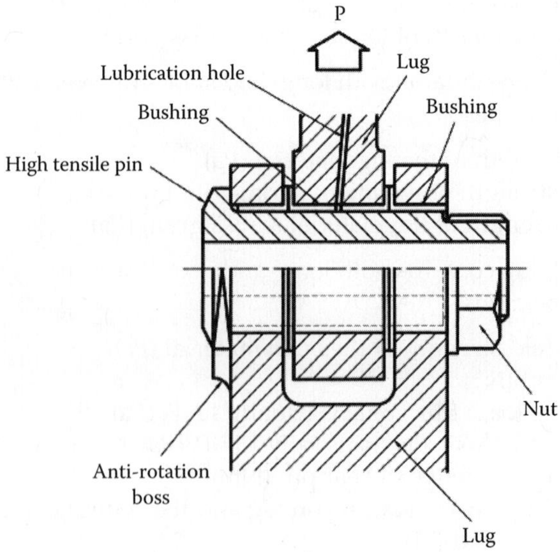

FIGURE 30.1
Typical lug assembly.

These two articles set the ground rules on how lugs and shear pins were to be analyzed and quickly became the standard used in the aerospace industry.

This guide will show the methods for analyzing lugs with either parallel or tapered sides with either round or square end forms, and deals with lug and shear pin failure modes associated with different load conditions covering axial, transverse, oblique, and out-of-plane loading.

This guide also shows how to analyze lugs with clearance fit shear pins and bushes. The stresses associated with the use of interference-fit pins or bushes are also considered.

The shear pin analyses investigate shear pin bending, and shear in double shear joints and pin shear in single shear joints will also be considered.

30.2.1 Method

The method described is semiempirical and is applicable to aluminum or steel alloy lugs.

A set of mutually perpendicular axes is applied to the lug, referred to as axial, transverse, and out-of-plane directions. The axial direction is parallel to the lug feature. The transverse direction lies perpendicular to the axial direction in the plane of the lug, and the out-of-plane direction lies perpendicular to the plane of the lug.

The axes intersect at the hole center (see Figure 30.2). The lug is independently analyzed for loads applied in the axial and transverse directions and for the combined oblique loading case. An out-of-plane check is also included.

30.2.2 Loading

The types of static loading considered in the analysis of lugs and shear pins follow:

Axial

Transverse

Oblique

Out-of-plane

30.2.3 Material Limitations

The calculation method may be applied to either steel or aluminum alloy materials.

In the case of axial failure modes it may be applied to lugs manufactured from any metallic material.

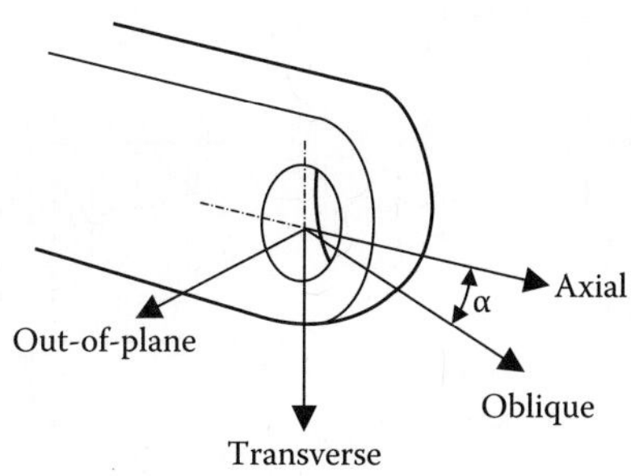

FIGURE 30.2
Lug loading.

30.2.4 Geometric Limitations

When the hole diameter (d) is less than the lug thickness (t) it is recommended that the thickness (t) should be set equal to the hole diameter (d) throughout the calculations, except when considering shear pin bending.

When the hole diameter (d) is greater that the lug thickness ($8 \times t$), lug stability should be checked.

The method for determining transverse areas is valid for the lug shapes shown in Figure 30.3.

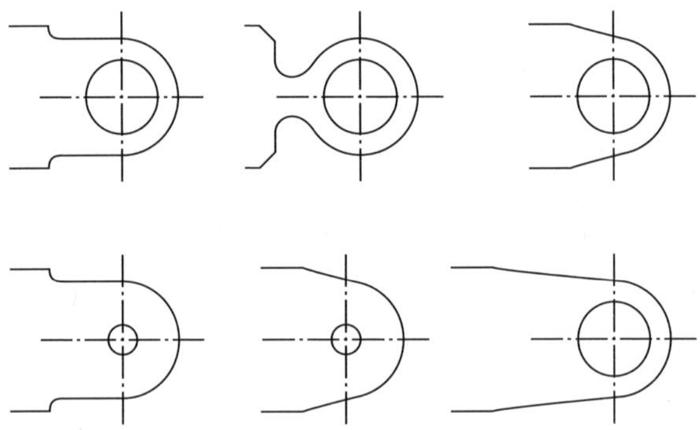

FIGURE 30.3
Types of lugs.

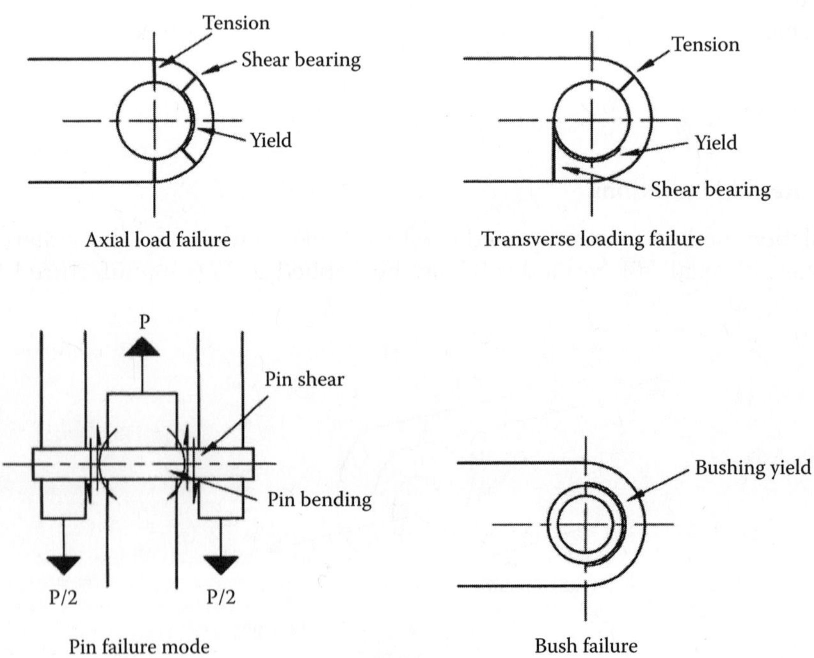

FIGURE 30.4
Lug failure modes.

30.2.5 Failure Modes

The failure modes (Figure 30.4) follow:

Tension
Shear tear out or bearing (shear bearing)
Yield

30.2.6 Notes

The margins of safety should be reduced by the relevant fitting or casting factors (see Section 30.4). If both fitting and casting factors are applicable, only the greater of the two need to be considered.

Margins of safety > 0.2 are preferable on lugs for use in aerospace vehicles. In general, in engineering projects where weight is not so important, the margin of safety (MoS) can be increased to 0.3 and above without any detrimental effects.

30.3 Analysis of Lugs with Axial Loading: Allowable Loads

A lug-shear pin combination under a tension load can fail in a variety of ways, each of which will require to be evaluated by the methods presented in this section.

Tension across the net section: Here stress concentrations will need to be considered.

Shear tear-out or bearing: These two conditions are closely related and are covered by a single calculation using empirical curves.

Shear acting on the pin: This will be analyzed in the usual way.

Bending of the shear pin: The ultimate strength of the pin is based on the Modulus of Rupture.

Excessive yielding of bushing (if used).

Yielding of the lug is considered to be excessive at a permanent set equal to 0.2 times the shear pin diameter. This condition must always be checked, as it is frequently reached at a lower load than would be anticipated from the ratio of the yield stress (F_{ty}) to the ultimate stress (F_{tu}) for the material.

Note the following:

1. Hoop tension at the tip of the lug is not a critical condition, as the shear-bearing condition precludes a hoop tension failure.
2. The lug should be checked for side loads (due to misalignment, etc.) by conventional beam formulas (see Section 30.4).

30.3.1 Analysis Procedure to Determine the Ultimate Axial Load

Calculate e/d, W/d, d/t, $A_{br}=dt$, $A_t=(W-d)/t$ (see Figures 30.5 and 30.6 for nomenclature).

Determine K_{br} using the coordinates e/d and d/t.

The ultimate load for shear-bearing failure, P'_{bru}, is

$$P'_{bru} = K_{br}F_{tu}A_{br} \tag{30.1}$$

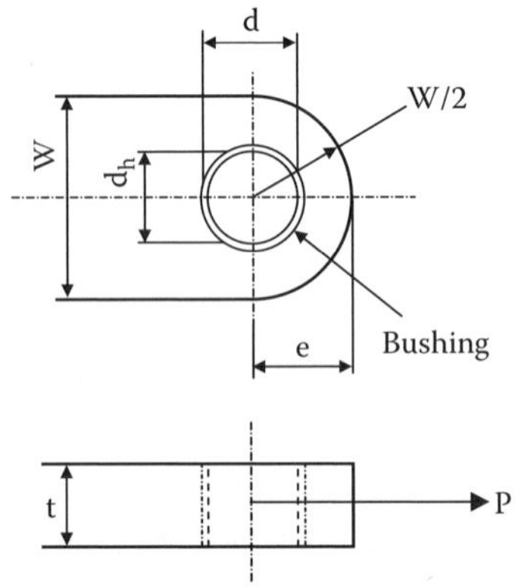

FIGURE 30.5
Detail of lug with bushing.

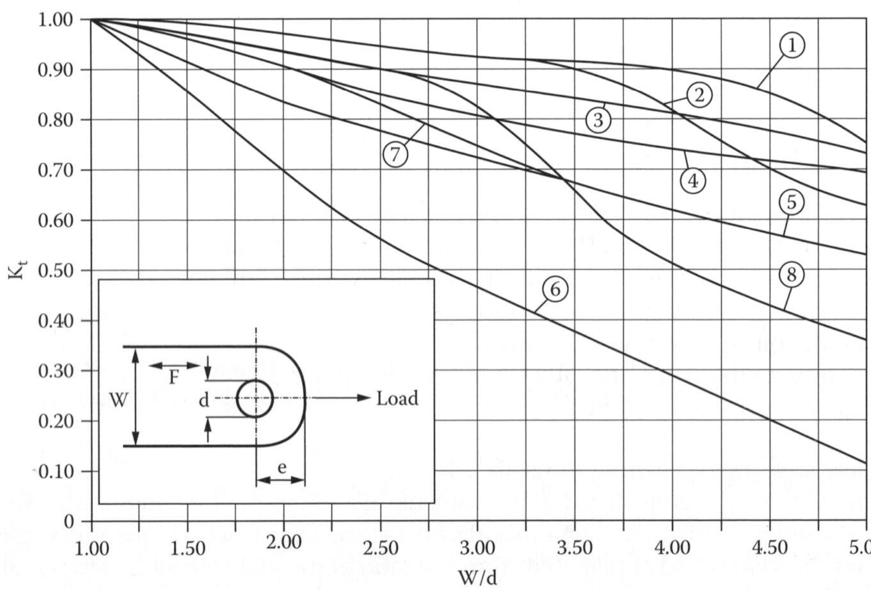

FIGURE 30.6
Effect of grain direction subject to axial load. L, T, N indicates the grain direction "F" in sketch; L = Longitudinal; T = Long transverse; N = Short transverse (normal).

Ultimate load for tension failure:

From Figure 30.6 using W/d, obtain K_t for the lug candidate material.

The ultimate load for tension failure P'_{tu} is

$$P'_{tu} = K_t F_{tu} A_t \tag{30.2}$$

where F_{tu} = ultimate tensile strength of lug material.

Load for yielding of the lug.

From Figure 30.8 with coordinates of e/d to obtain K_{bry}.

The yield load, P'_y, is

$$P'_y = K_{bry} A_{br} F_{ty} \quad (\text{for } K_{bry}, \text{ see Figure 30.8}) \tag{30.3}$$

Load for yielding of the bushing in bearing (if used).

$$P'_{bry} = 1.85\, F_{cy} A_{brb} \tag{30.4}$$

where F_{cy} = compressive yield stress of the bushing material.
For the shear pin bending stress see Section 30.7.

Curve 1.
4130, 4140, 4340, and 8630 steel.
2014-T6 and 7075-T6 plate ≤ 0.5 in. (12.7 mm) (L, T).
7075-T6 bar and extrusion (L).
2014-T6 hand-forging billet ≤ 144 in.^{2} (92903 mm^2) (L).
2014-T6 and 7075-T6 die forgings (L).

Curve 2.
2014-T6 and 7075-T6 plate > 0.5 in., ≤ 1.0 in. (>12.7 mm, ≤25.4 mm).
7075-T6 extrusion (T, N).
7075-T6 hand-forged billet ≤ 36 in.^{2} (23226 mm^2) (L).
2014-T6 hand-forged billet > 144 in.^{2} (92903 mm^2) (L).
2014-T6 hand-forged billet ≤ 36 in.^{2} (23226 mm^2) (T).
17-4 PH
17-7 PH-THD.

Curve 3.
2024-T6 plate (L, T).
2024-T4 and 2024-T42 extrusion (L, T, N).

Curve 4.
2024-T4 plate (L, T).
2024-T3 plate (L, T).
2014-T6 and 7075-T6 plate > 1.0 in. (25.4 mm).
2024-T4 bar (L, T).
7075-T6 hand-forged billet > 36 in.^{2} (23226 mm^2) (L).
7075-T6 hand-forged billet ≤ 16 in.^{2} (10323 mm^2) (T).

Curve 5.
195T6, 220T4, and 356T6 aluminum alloy casting.
7075-T6 hand-forged billet > 16 in.^{2} (10323 mm^2) (T).
2014-T6 hand-forged billet > 36 in.^{2} (10323 mm^2) (T).

Curve 6.
Aluminum alloy plate, bar, hand forged billet and die forging (N).
Note: for die forgings, N direction exists only at the parting plane.
7075-T6 bar (T).

Curve 7.
18-8 stainless steel, annealed.

Curve 8.
18-8 stainless steel, full hard.
Note: for ¼, ½, and 3/4 hard interpolate between curves 7 and 8.

In Figure 30.7, curve A is a cutoff to be used for all aluminum alloy hand-forged billets when the long-transverse grain direction has the general direction G. Curve B is to be used for all aluminum alloy plate, bar, and hand-forged billets when the short-transverse grain

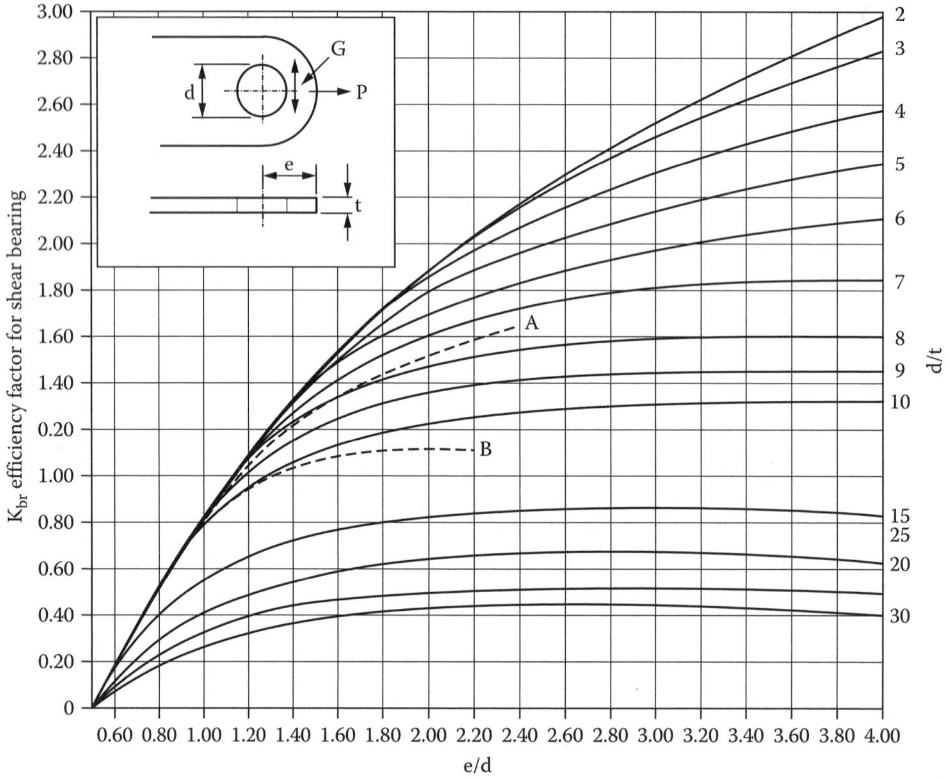

FIGURE 30.7
Shear efficiency factor K_{br}.

direction has the general direction G, and for die forgings when the lug contains the parting plane in a direction approximately normal to the direction G.

In addition to the limitations provided by curves A and B, in no event shall a K_{br} greater than 2.00 be used for lugs made from 12.7 mm thick or thicker aluminum alloy plate, bar, or hand-forged billets. (See Figure 30.8.)

30.4 Analysis of Lugs with Transverse Loading: Allowable Loads

In the case where the loading being applied to the lug is at $\alpha = 90°$, as in Figure 30.9, the loading is treated as transverse. In order to determine the ultimate and yield loads for the lug, the shape of the lug must be taken into account. This is accomplished by the use of a shape parameter given by

$$\text{Shape parameter} = \frac{A_{av}}{A_{br}}$$

where

A_{br} = the bearing area = dt

A_{vr} = the weighted average area given by:

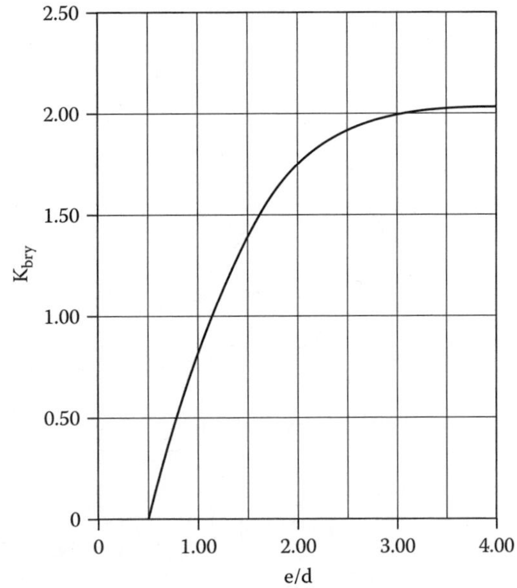

FIGURE 30.8
Factors for calculating yield axial loads attributable to shear bearing of lug and pin.

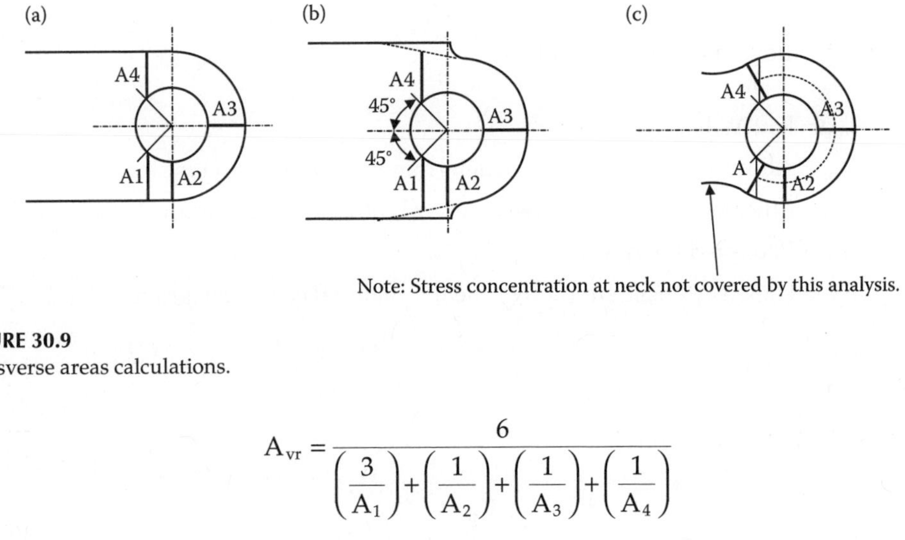

FIGURE 30.9
Transverse areas calculations.

$$A_{vr} = \frac{6}{\left(\frac{3}{A_1}\right)+\left(\frac{1}{A_2}\right)+\left(\frac{1}{A_3}\right)+\left(\frac{1}{A_4}\right)}$$

A_1, A_2, A_3, and A_4 are areas of the lug sections indicated in Figure 30.9.

Obtain the areas A_1, A_2, A_3, and A_4 as follows: A_1, A_2, A_3, and A_4 are measured on the planes indicated in Figure 30.9a (perpendicular to the axial center line), except that in a necked lug (Figure 30.9c). A_1 and A_4 should be measured perpendicular to the local centerline.

A_3 is the least area on any radial section around the hole. Since the choice of areas and the method of averaging has been substantiated only for the lugs of the shapes shown in Figure 30.10, thought should always be given to ensure that the areas A_1, A_2, A_3, and A_4 adequately reflect the strength of the lug.

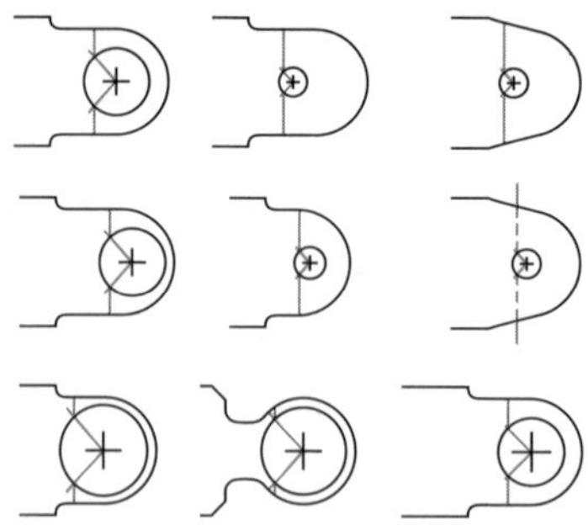

FIGURE 30.10
Substantiated lug shapes.

Obtain the weighted average of these areas using

$$A_{vr} = \frac{6}{\left(\frac{3}{A_1}\right)+\left(\frac{1}{A_2}\right)+\left(\frac{1}{A_3}\right)+\left(\frac{1}{A_4}\right)}$$

Compute:

$$A_{br} = dt, \quad \text{and} \quad \frac{A_{av}}{A_{br}}$$

Determine the allowable ultimate load:

$$P'_{tru} = K_{tru} A_{br} F_{tux}$$

P_{tru} = Allowable ultimate load as determined for the transverse load.
K_{tru} = Efficiency factor for transverse load (ultimate) (see Figure 30.11).
A_{br} = Pr ojected bearing area.
F_{tux} = Ultimate tensile strength of the lug material across the grain.

where:

Yield load P'_y of the lug:

Obtain K_{try} from Figure 30.11.

$$P'_y = K_{try} A_{br} F_{tyx}$$

where:

P'_y = Allowable yield load on the lug
K_{try} = Efficiency factor for transverse load (yield); see Figure 30.11
A_{br} = Projected bearing area
F_{tyx} = Tensile yield stress of lug material across the grain

Determine the allowable yield-bearing load on the bushing using

$$P_{bry} = 1.85\, F_{cy} A_{brb}$$

where:

P_{bry} = Allowable yield-bearing load on bushing

F_{cy} = Compressive yield stress of bushing material

A_{brb} = The smaller of the bearing areas of bushing on pin or bushing on lug (the latter may be smaller as a result of external chamfer on the bushing)

In Figure 30.11 all curves are for K_{tru} except for the one noted as K_{try}.

Note that the curve for 125, 000 HT steel agrees closely with test data. Curves for other materials have been obtained by the best available means of correcting for material properties and may possibly be very conservative in some places.

In no case should the ultimate transverse load be taken as less than that which can be carried by cantilever beam action of the portion of the lug under the load (Figure 30.12). The load that can be carried by cantilever beam action is indicated very approximately by curve A in Figure 30.7; should K_{tru} be below curve A, separate calculations, as a cantilever beam are warranted.

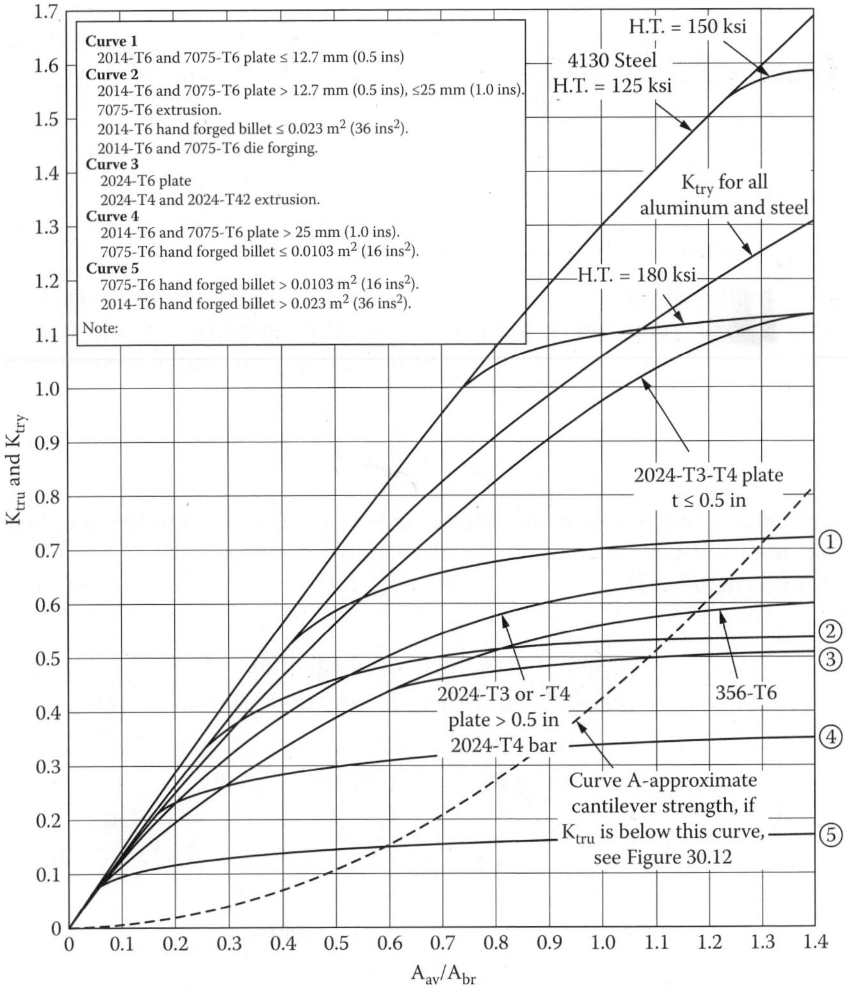

FIGURE 30.11
Efficiency factor for transverse load K_{tru}.

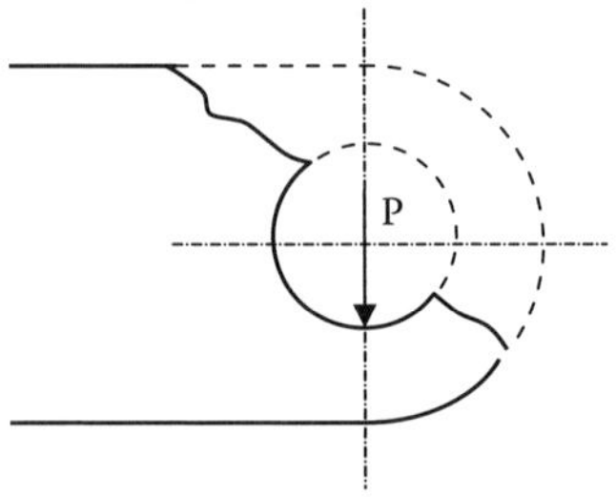

FIGURE 30.12
Lug as a cantilever carrying load.

Investigate pin bending as for axial load (see Section 30.6) with the following modifications: Take (P'_u) min$=P'_{tru}$.

In the equation:

$$r = \frac{\left[e - \left(\frac{d}{2}\right)\right]}{t} \text{ use for the } \left[e - \left(\frac{d}{2}\right)\right] \text{ term the edge distance at a} = 90^\circ$$

30.5 Oblique Loading: Allowable Loading

In the analysis of lugs subject to oblique loading it is convenient to resolve the loading into axial and transverse components; these are denoted by subscripts a and tr, respectively (see Figure 30.13). The two cases are analyzed separately and utilize the results by means of an interaction equation. The interaction equation:

$$R_a^{1.6} + R_{tr}^{1.6} = 1$$

where R_a and R_{tr} are ratios applied to critical loads in the indicated directions, is to be used for both ultimate and yield loads for both aluminum and steel alloys.

where, for ultimate loads:

$$R_a = \frac{\text{axial component of applied ultimate load}}{\text{smaller of } P'_{bru} \text{ and } P'_{tu} \text{ (from Equations 30.1 and 30.2)}}$$

$$R_{tr} = \frac{\text{transverse component of applied ultimate load}}{P'_{tru} \text{ (from analysis procedure for a} = 90^\circ\text{)}}$$

and for yield load:

$$R_a = \frac{\text{axial component of applied yield load}}{P'_y \text{ (fom Equation 30.3)}}$$

$$R_{tr} = \frac{\text{transverse component of applied yield load}}{P'_{try} \text{ (from analysis procedure for a} = 90^\circ\text{)}}$$

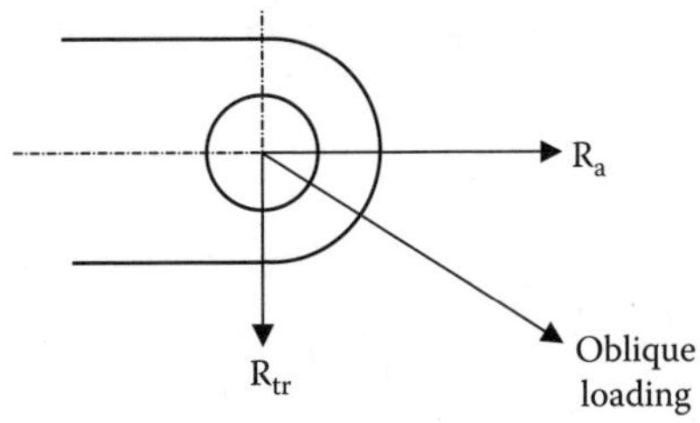

FIGURE 30.13
Oblique loading.

30.5.1 Analysis Procedure

Resolve the applied load into axial and transverse components and obtain the lug ultimate and yield margins of safety from the interaction.

$$\text{MoS.} = \frac{1}{(R_a^{1.6} + R_{tr}^{1.6} = 1)^{0.625}} - 1$$

Check pin shear and bushing yield as in Section 30.7.

Investigate pin bending using the procedure for axial load modified as follows:

$$(P_u')_{min} = \frac{P}{(R_a^{1.6} + R_{tr}^{1.6} = 1)^{0.625}}$$

In the equation

$$\frac{[e-(d/2)]}{t}$$

use for the [e–(d/2)] term the edge distance at the value of α corresponding to the direction of the load acting on the lug.

30.5.2 Out-of-Plane Loading

In general lugs should not be subject to out-of-plane loading; however, if there is a small component in the R_Z direction (Figure 30.14), then it is advisable to check the minimum

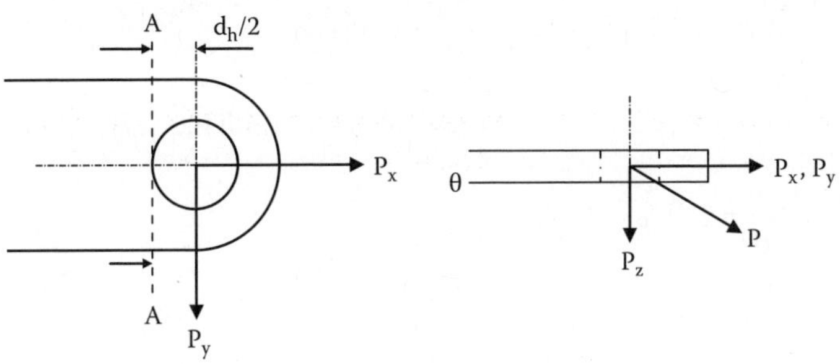

FIGURE 30.14
Out-of-plane loading, load direction, and bending section.

section for combined bending and shear. This check may be incorporated into the oblique load interaction formula detailed in Section 30.5.1.

The allowable load in the out-of-plane (R_Z) direction at section A:A in Figure 30.14 can be expressed as

$$\bar{P}_z = \frac{(W - d_h)\cdot t \cdot F_{tu}}{\sqrt{\left[\left(\frac{3d_h}{Kt}\right)^2 + \left(\frac{F_{tu}}{F_{su}}\right)^2\right]}}$$

where K is the form factor for bending. The allowable oblique load becomes

$$\bar{P} = \frac{1}{\left[\left(\frac{\mathrm{Cos}(\theta)\mathrm{Cos}(\alpha)}{\bar{P}_{ux\,min}}\right)^{1.6} + \left(\frac{\mathrm{Cos}(\theta)\mathrm{Sin}(\alpha)}{\bar{P}_{uy}}\right)^{1.6} + \left(\frac{\mathrm{Sin}(\theta)}{P_z}\right)^{1.6}\right]^{0.625}}$$

Estimate the oblique angle θ. Generally, if no design data are available, the oblique angle may be assumed to be at least 10° to allow for any misalignment, etc.

30.6 Bearing at Lug-to-Pin or -Bush Interface

The allowable bearing loads at the lug-to-pin interface are given below. A combined shear tear-out and bearing check has previously been performed for axial, transverse, and oblique loadings, respectively. However, this check may be a separate requirement, particularly for lugs subject to compression.

MIL HDBK 5 provides values of allowable ultimate and yield-bearing stress for (*a/d*) values of 1.5 and 2.0, valid for values of (*d/t*) up to 5.5. Bearing allowables for other values of (*a/d*) can be estimated by linear interpolation, as shown below.

$\frac{a}{d}$ range	Equation for bearing strength
$0.5 < \frac{a}{d} \le 1.5$	$\bar{F}_{br_n} = \left(\frac{a}{d} - 0.5\right)\bar{F}_{br1.5}$
$1.5 < \frac{a}{d} < 2.0$	$\bar{F}_{br_n} = \bar{F}_{br1.5} + 2\left(\frac{a}{d} - 1.5\right)\left(\bar{F}_{br2} - \bar{F}_{br1.5}\right)$
$\frac{a}{d} \ge 2.0$	$\bar{F}_{br_n} = \bar{F}_{br2}$

Ultimate bearing:

$$\bar{P}_{bru} = \bar{F}_{bru}\cdot d_{pin}\cdot t$$

Yield bearing:

$$\bar{P}_{bry} = \bar{F}_{bry} \cdot d_{pin} \cdot t$$

Note that the dimension a, the distance from the center of the hole to the edge of the lug is denoted as e in MIL HDBK 5.

30.7 Shear Pin Analysis

30.7.1 Shear Pin Bending in Double Shear Joint

In static tests of a single bolt fitting, failure of the shear pin due to bending failure will not be shown to be a factor in the failure of the lug. However, it is important to provide sufficient bending strength to ensure permanent bending deformation does not occur when subject to the limit loads so that the shear pin can be readily removed for inspection and maintenance operations.

Weakness in the shear pin can cause peaking up non-uniform-bearing loads on the lugs, influencing the lug tension and shear strength. The big unknown factor in shear pin bending is the true value of the bending moment acting on the pin because the moment arm to the resultant bearing forces is difficult to quantify.

An approximate method that is commonly used to determine the moment arm on the shear pin in a double shear joint is shown below:

$$\text{Moment arm (b)} = 0.5t_1 + 0.25t_2 + d$$

If there is no gap ($\delta = 0$) between the lug faces (see Figure 30.15) the moment arm is taken as

$$b = 0.5\,t_1 + 0.25\,t_2$$

The applied shear pin bending moment is given by

$$m = b \times \frac{P}{2}$$

where P is the minimum failure load for the center lug or twice that of the outer lugs, and b is the moment arm (see Figure 30.15). It is assumed that the maximum shear pin load is the minimum lug failure load.

Calculate bending stress from

$$\sigma_m = \frac{M}{Z}$$

where

$$Z = \frac{\pi}{32} d^3$$

$$M = \text{Bending moment}$$

If the pin strength is found to be inadequate by the above check, then use may be made of the tendency of the pin loads to peak near the shear faces, thereby reducing the bending moment arm.

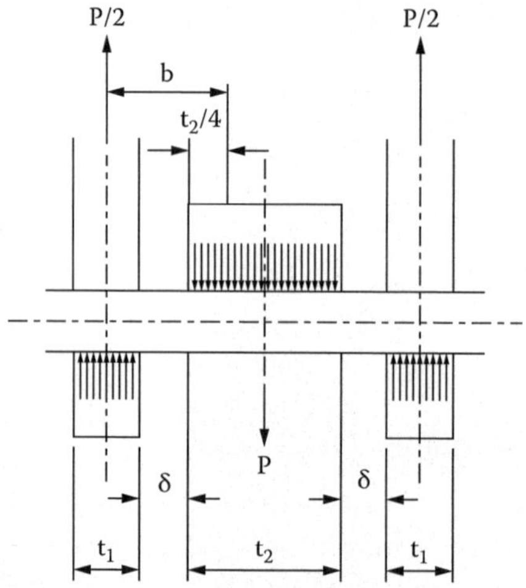

FIGURE 30.15
Pin moment arm.

No peaking in the outer lugs can occur unless a substantial head or nut bears firmly on the outer lugs. The reduction in the moment arm will apply only to the central lug.

A reduction in the shear pin bending up to 25% can be expected.

30.7.1.1 Shear Pin Bending: Load Peaking between Center Lug and Pin

If it is desired to take into account the reduction of pin bending that results from load peaking, as shown in Figure 30.16, the moment arm is obtained as follows: Calculate the inner lug r (use d_b instead of d, if bushing is used).

$$r = \frac{\left[\left(\frac{a}{D}\right) - \frac{1}{2}\right]D}{t_2}$$

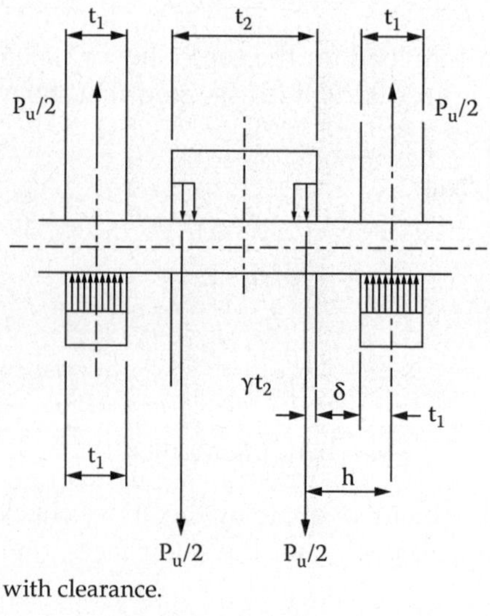

FIGURE 30.16 Pin moment arm with clearance.

Take the smaller of P_{bru} and P_{tu} for the inner lug as $(P_u)_{min}$ and calculate

$$\frac{(P_u)_{min}}{A_{br}F_{tux}}$$

From Figure 30.17 using $\frac{(P_u)_{min}}{A_{br}F_{tux}}$ and r obtain the reduction factor γ for peaking and calculate the moment arm.

The maximum bending moment in the shear pin:

$$h = \left(\frac{t_1}{2}\right) + \delta + \gamma\left(\frac{t_2}{4}\right)$$

and calculate the bending stress in the shear pin that results from M, assuming an MD/2I distribution (where I = moment of Inertia of the pin) and its Margin of Safety (MoS) as:

$$MoS = \frac{\left(\frac{MD}{2I}\right)}{F_b} - 1$$

where F_b = the Modulus of Rupture determined from Figure 30.18 or other sources.

$$M = \frac{P_u h}{2}$$

30.7.1.2 Shear Pin Bending, Including Excess Strength of Lug

This approach should only be used if there is an excess margin in the strength of the lug and the pin is still inadequate.

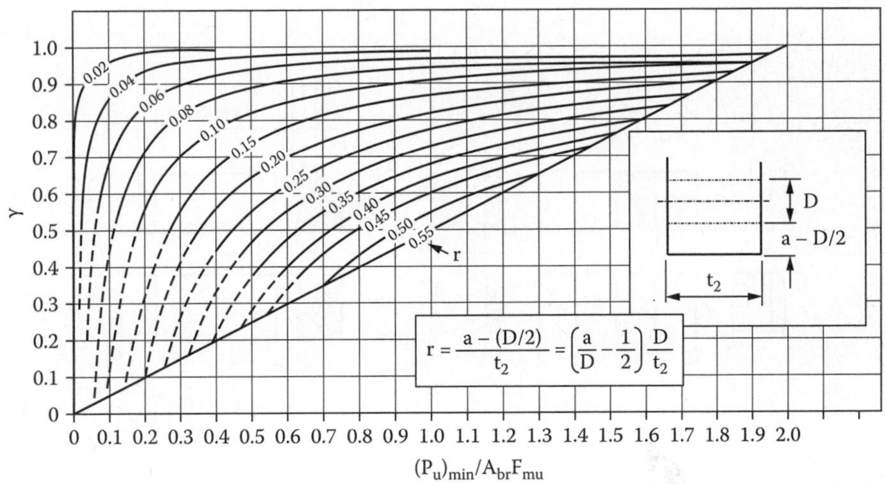

FIGURE 30.17
Peaking factors for pin bending. (Dashed lines indicate where the theoretical curves are not substantiated by test data.)

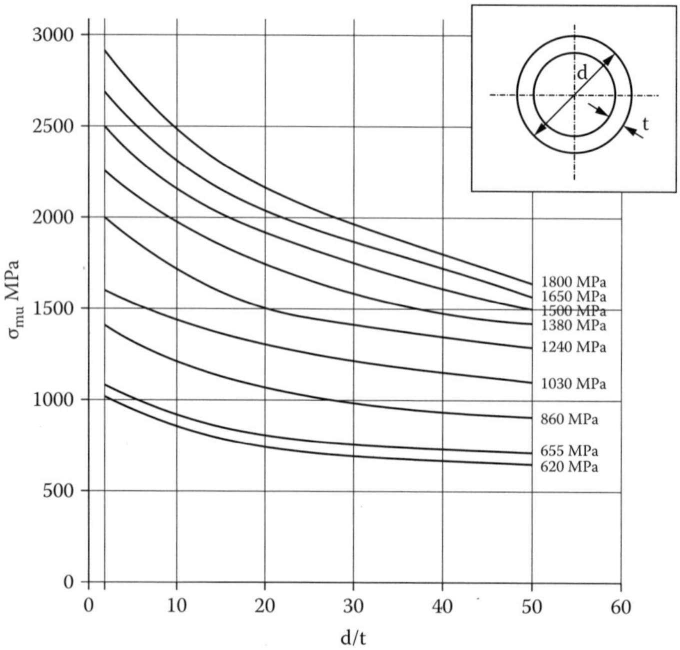

FIGURE 30.18
Bending Modulus of Rupture.

In this circumstance it may be assumed that a portion of the lug thickness in the joint is inactive (see Figure 30.19). The active thicknesses are chosen by trial and error to give approximately equal margins of safety (MoS) for the lug and the pin.

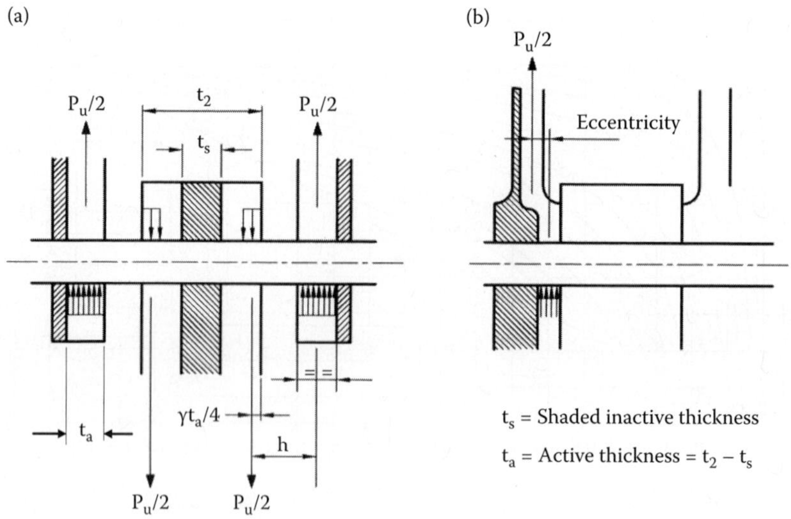

FIGURE 30.19
Load peaking.

Using the smaller of:

$$\frac{2c}{d}\frac{\lambda_t}{R_{tu}}$$

and

$$\frac{2a}{d}\frac{\lambda_s}{R_{tu}}$$

for the inner lug, obtain the reduction factor γ for peaking where:

$$r = \frac{a - d/2}{t_a}$$

and t_a is the active thickness shown in Figure 30.19a, noting that γ applies only to the inner lug. Calculate the moment arm h as indicated below:

$$\text{Maximum shear pin bending moment } M = \frac{P_u h}{2} \text{ as before.}$$

If the eccentricity occurs in an outer lug (see Figure 30.19b), the structure must be capable of withstanding the induced bending; otherwise, the shear pin must be strong enough for the analysis by the method given in Section 30.7.1.1.

30.7.2 Pin Shear

The allowable shear load for a solid shear pin in a double shear joint:

$$P_S = K_s \bar{F}_s A_{pin}$$

where

$$A_{pin} = \frac{\pi \times d_{pin}^2}{4}$$

In double shear, the applied load will be the load in the outer lugs or half the load in the center lug.

The allowable shear stress $\bar{F}_s$ is to be factored by K_s to allow for the variation in the ratio (d_h/t).

K_{su} can be determined from Figure 30.20. Limitations on the use of the curve are as follows:

$$\frac{d_h}{t} < 4.0 \quad t_o > 0.75 t_i$$

For a single shear joint the above equation will still apply, but in this case Ksu = 1.0.

30.8 Bush Analysis

The allowable yield load on bushing:

$$P_{bry} = 1.85 \bar{F}_{cy} A_{brb}$$

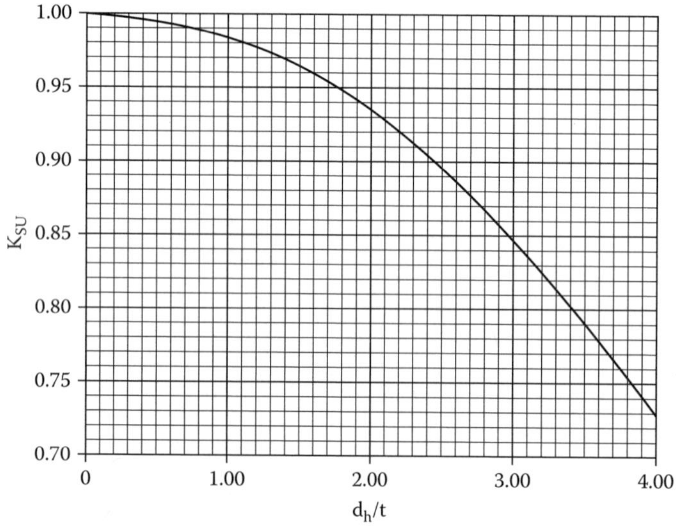

FIGURE 30.20
Allowable shear stress factor for pin in double shear.

where A_{brb} is the smaller bearing area of the busing on the shear pin or bushing on lug.

The bearing area on the lug may be less if the bushing has been chamfered. $\bar{F}_{cy}$ is the allowable compressive yield stress on the bush material.

30.9 Special Cases

30.9.1 Oil Holes

30.9.1.1 Axial Load

The calculation of P_{tux} or P_{qux}, respectively, should be modified depending on the location of the lubrication hole (see Figure 30.21). For the modification of these parameters the lug tension and bearing areas should be based on the lug thickness minus the lubrication hole diameter. The value of K_{qux} will need to be recomputed to account for the reduction in thickness.

30.9.1.2 Transverse Load

Obtain P_{uy} neglecting the lubrication hole and multiply by:

$$0.90\left(1-\frac{d_{\text{Lubrication hole}}}{t}\right)$$

30.9.1.3 Oblique Load

In this case proceed as normal and obtain T_{tux} and P_{qux} as per Section 30.9.1.1 and obtain P_{uy} as per Section 30.9.1.2. Disregard the lubrication hole in calculating the yield margin.

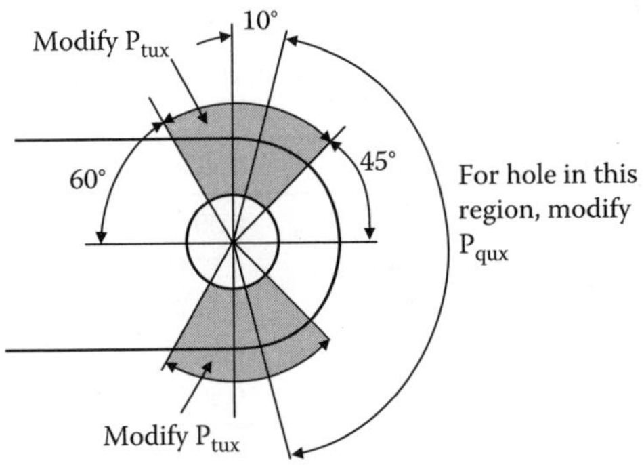

FIGURE 30.21
Modification of allowable load for the presence of a lubrication hole.

30.9.2 Eccentric Hole

If the hole is laterally located (see Figure 30.22a), i.e., e_1 is less than e_2, the ultimate and yield lug loads are determined by calculating P'_{bru}, P'_{tu}, and P'_y for the equivalent lug shown in Figure 30.22b and multiplying by the factor below:

$$\text{Factor} = \frac{e_1 + e_2 + 2d}{2e_2 + 2d}$$

30.9.3 Multiple Shear Connection

A shear pin-lug combination having a geometry shown in Figure 30.23 can be solved according to the following procedure:

1. The load carried by each individual lug can be determined by distributing the total applied load among the lugs, as shown in Figure 30.23. β is obtained from Table 30.1.
2. The maximum shear load acting on the shear pin is also given in Table 30.1.
3. The maximum bending moment on the shear pin is given by the formula

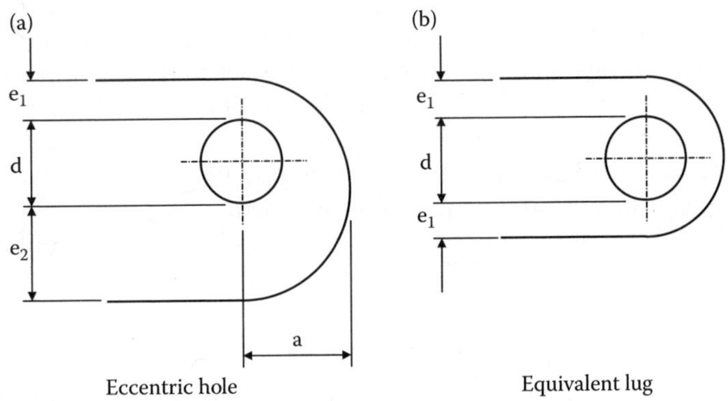

FIGURE 30.22
Equivalent lug for a lug with an eccentric hole.

TABLE 30.1
Pin Shear and Moment Arms in Multiple Shear Lugs

Number of Lugs Including Side Lugs	β	Pin Shear	Moment Arm
5	0.35	0.50 P′	0.28 (t′ + t″)/2
7	0.40	0.53 P′	0.33 (t′ + t″)/2
9	0.43	0.54 P′	0.37 (t′ + t″)/2
11	0.44	0.54 P′	0.39 (t′ + t″)/2
Infinity	0.50	0.50 P′	0.50 (t′ + t″)/2

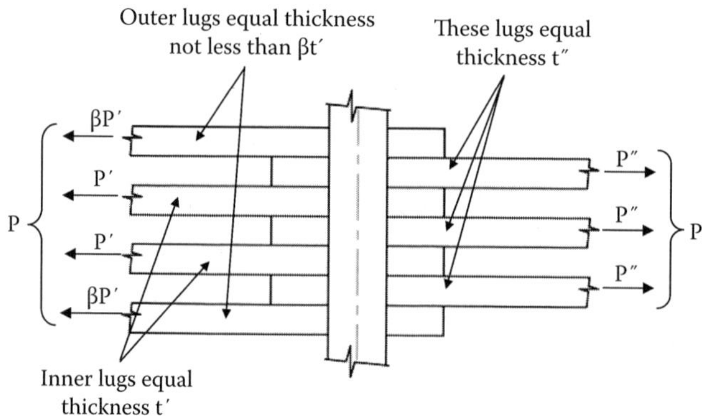

FIGURE 30.23
Multiple shear connections.

$$M = \frac{P \times \beta}{2}$$

where β is given in Table 30.1.

30.10 Stresses due to Interference-Fit Pins and Bushes

The insertion of circular interference-fit shear pins or bushes of similar or dissimilar materials in flat bars, strips, or lugs will introduce elastic stresses in these components. The analysis will be treated in two parts: solid shear pins (Section 30.10.1) and bushes (Section 30.10.2).

The method is based upon a two-dimensional analysis and will only give an average stress through the thickness of the lug. Local variations away from the average values will most likely be nonsignificant unless taper pins are used.

It is assumed that the material will be isotropic (obeys Hooke's law), but the method can be used for most engineering materials without any significant error.

30.10.1 Solid Circular Interference-Fit Shear Pins

In the case of an interference-fit shear pin fitted in an infinitely wide plate, i.e., $d_h/W=0$, the circumferential tensile, radial compressive, and shear stresses will be uniform around the hole.

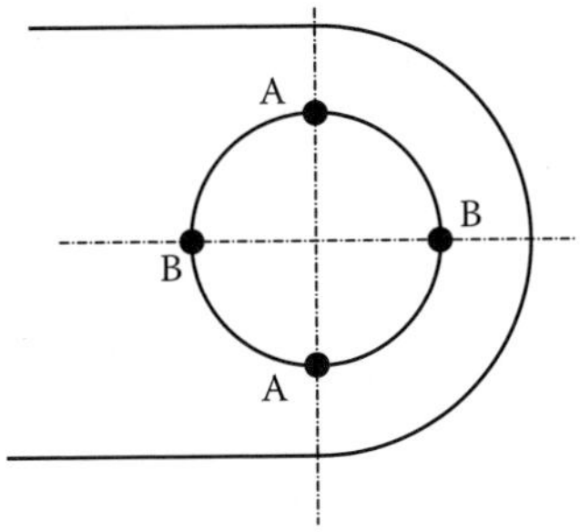

FIGURE 30.24
Points of high stress in bore.

For a plate of finite width, the stresses will vary around the hole, with the maximum value being dependent upon d_h/W.

Referring to Figure 30.24, at the points marked B the tensile stress σ, shear stress q, and interference pressure p will approach the values as determined for a shear pin in an infinite plate. The maximum tensile and shear stresses will occur at the points marked A, although the shear stress will only be slightly above its value as predicted at points B. The interference pressure will be at a maximum at points B and a minimum at points A.

At point A:

Maximum tensile and shear stresses

Minimum pressure

At point B:

Tensile and shear stresses will tend to reach values as predicted for an infinite plate

Maximum pressure

An estimate of the interface pressure at B is given by

$$\frac{p}{E_l e} = \left[1 - \left(\frac{d_h}{W}\right)^2\right]\left(\frac{q}{E_l \times e}\right)$$

Shear stress is given by:

$$\frac{q}{E_l e} = \frac{1}{2 + \left[\frac{E_l(1-\mu_p)}{E_p} - (1-\mu_l)\right]\left[1 - \left(\frac{d_h}{W}\right)^2\right]}$$

Tensile stress is given by:

$$\frac{\sigma}{E_l \cdot e} = \frac{q}{E_l \cdot e}\left[1 + \left(\frac{d_h}{W}\right)^2\right]$$

These formulas are derived from a pin in a ring, but they may be used with reasonable accuracy for a pin in a lug provided $(d_h/W) < 0.8$.

30.10.2 Interference-Fit Bushes

In the following analysis d_b is the bush internal diameter and K is the bush shape factor.

The maximum tensile stress occurs at the points marked A in Figure 30.24. The maximum shear stress is given by

$$q_{max} = \frac{(\sigma + p)}{2}$$

The form of the end of a bushed lug affects the interference pressure and maximum stress. Figure 30.25 provides a correction factor to (d_h/W) for the effect of round- and square-ended lugs.

The interference pressure is given by

$$\frac{p}{E_l e} = \frac{1}{\left[\frac{1+(K d_h/W)^2}{1-(K d_h/W)^2}\right] + \mu_l + \frac{E_l}{E_b}\left[\frac{1+(d_b/d_h)^2}{1-(d_b/d_h)^2} - \mu_b\right]}$$

and tensile stress is given by:

$$\frac{\sigma}{E_l e} = \frac{p}{E_l e}\left[\frac{1+(d_b/d_h)^2}{1-(d_b/d_h)^2}\right]$$

30.11 Stress Concentration Factor at Lug-to-Pin Interface

The method presented in this section is to determine the axial failure factor, K_{tux}.

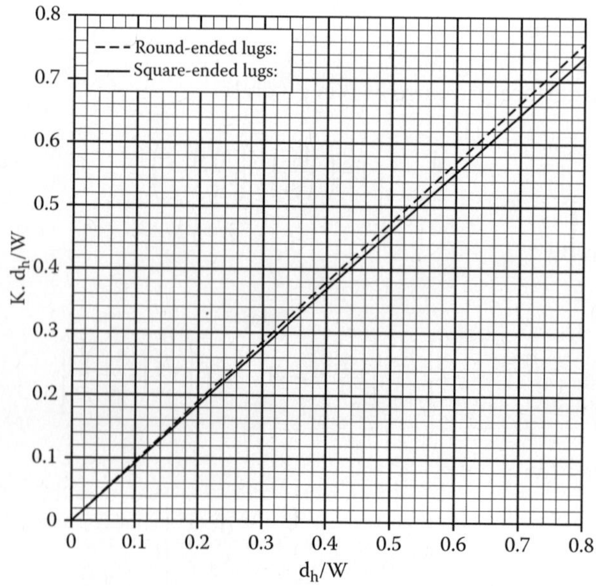

FIGURE 30.25
Bush shape factor.

For axially loaded lugs the stress concentration at the lug-to-pin interface is defined by the following equation:

$$K_e = (K_{0.2} + \eta_e(K_{100} - K_{0.2}))K_{th}$$

where:

K_e = Hole stress concentration factor for lugs with e percent clearance

$K_{0.2}$ = Stress concentration factor for 0.2% d_h clearance; refer to Figure 30.26 for round and square-ended lugs

K_{100} = Stress concentration factor for point loading; refer to Figures 30.26 and 30.27 for round and square-ended lugs ($e \leq 0.1\ d_h$)

η_e = Lug-to-pin clearance correction factor; refer to Figure 30.29

K_{th} = Lug thickness factor from Figure 30.30

In the case of square-ended lugs ($e < 0.1\%\ d_h$) K_e may be determined from Figure 30.28. This case represents a practical lower limit to the stress concentration.

30.12 Examples

In this example, a fitting (Figure 30.31) will be considered. This comprises of two lugs that connect to two struts. The fitting is used in an engine nacelle and is subject to engine heat.

Considering the dimensional and material properties of the lug:

Principal Lug Dimensions

a = 30.0 mm

c = 22.00 mm ($a - d_h/2$)

d_h = 17.50 mm (hole diameter with bush) (included for repairability)

d = 16.00 mm (hole diameter without a bush)

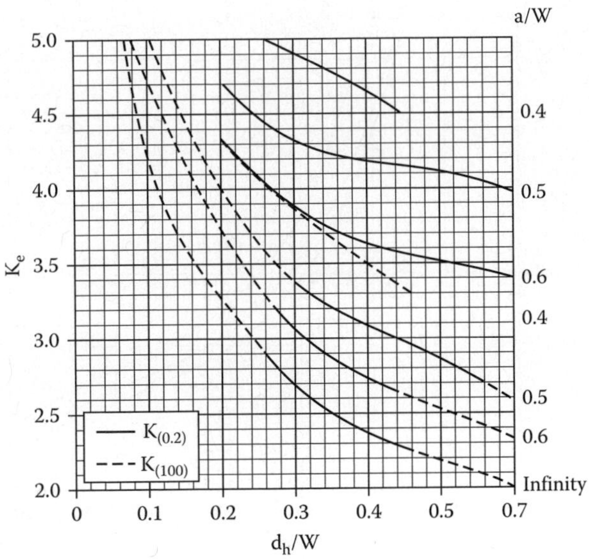

FIGURE 30.26
Stress concentration factor for round lugs.

W = 60.00 mm
R = 30.00 mm (W/2)
t = 7.50 mm
r = 8.75 mm ($d_h/2$)

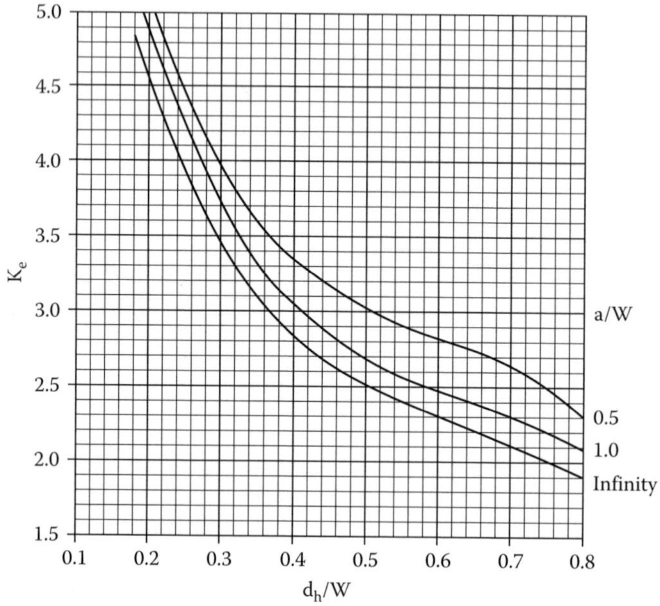

FIGURE 30.27
$K_{0.2}$ stress concentration factor for square lugs.

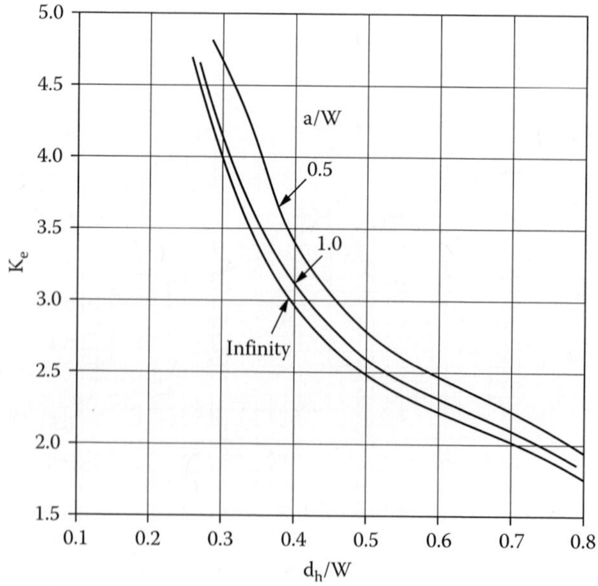

FIGURE 30.28
$K_{(e < 0.1)}$ stress concentration factor for square lugs.

Material properties and operating temperatures:

Material: 15-5PH Stainless Steel Bar/Forging (AMS 5659) Heat Treated to H1025

The in-flight operating temperature in the engine nacelle	T_{op} =	150°C
Ultimate tensile stress with thermal factor for 150°C	F_{tu} =	972.51 MPa
Yield tensile stress with thermal factor for 150°C	F_{ty} =	909.42 MPa
Yield compressive stress with thermal factor for 150°C	F_{cy} =	877.50 MPa
Ultimate shear stress with thermal factor for 150°C	F_{su} =	608.60 MPa
Ultimate bearing stress with thermal factor for 150°C	F_{bru} =	1, 650.12 MPa
Yield-bearing stress with thermal factor for 150°C	F_{bry} =	1, 323.86 MPa
Young's modulus at 150°C	E =	192.57 GPa

The table above summarizes the material properties used for the construction of the lug, and the allowables cover both longitudinal and transverse directions. Therefore, the analysis does not have to consider grain directions.

Note that the shear pin is made up of a nut-and-bolt assembly.
Summary of Allowables' Nut and Bolt Strengths

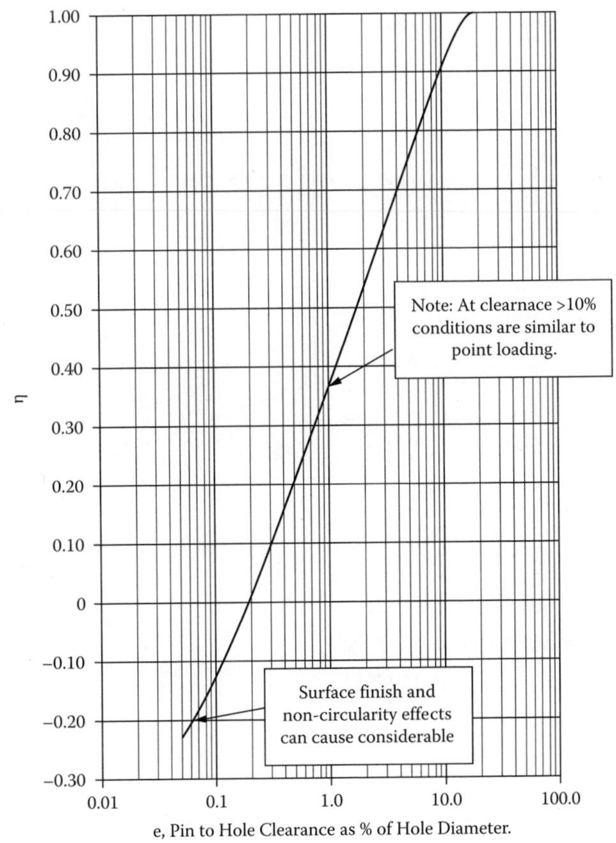

FIGURE 30.29
Correction factor for pin to hole clearance.

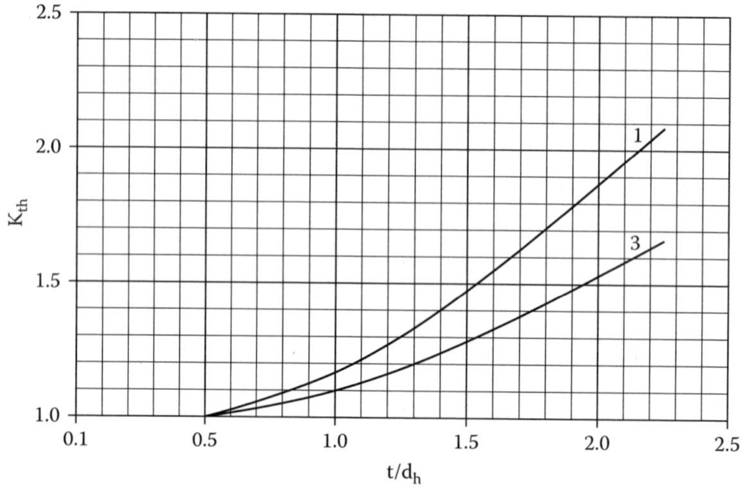

FIGURE 30.30
Effect of lug thickness.

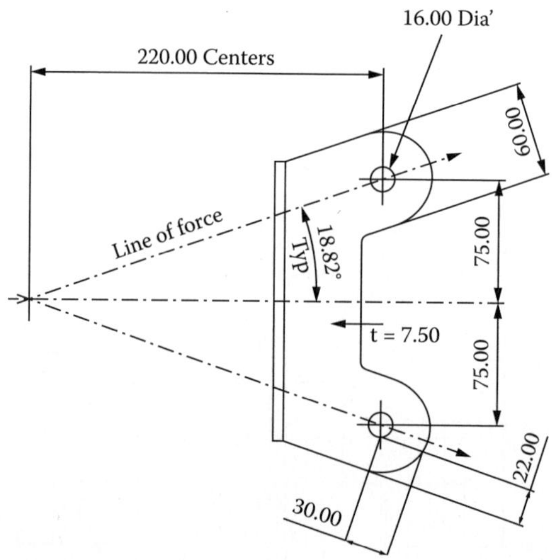

FIGURE 30.31
Detail of fitting.

Bolt maximum ultimate shear 32.47 N

Bolt maximum ultimate tension 45.37 N

Nut maximum ultimate tension 32.87 N

The bolt is manufactured from A286, and at the operating temperature of 150°C, there is no temperature knock-down factor to be applied.

Consider the axial tension acting on the lug. From Figure 30.31 it is clearly seen that the load acting on the lug is considered to be axial. The nominal load is

$$P_{tensile} = 76.33\,\text{kN tensile}$$

and

$$P_{compresssive} = 76.70\,kN \text{ compressive}$$

This analysis will consider the bore to be fitted with a bush, as this will give the minimum strength in the ligaments of the lug.

The fitting factor (ff) used throughout this analysis is 1.15.

From Figure 30.31,

$$a = 30.00 \text{ mm}$$
$$d_h = 17.50 \text{ mm}$$
$$t = 7.50 \text{ mm}$$

Hence:

$$a/d_h = 1.875$$
$$d_h/t = 2.133$$

The allowable axial load:

$$P_{tu} = K_t \times F_{tu} \times A_t$$
$$P_{tu} = 0.9231 \times 972.51 \text{ MPa} \times 318.75 \text{ mm}^2$$
$$P_{tu} = 285498.5 \text{ N}$$

$$RF = \frac{P_t}{F_a \times ff} = \frac{285498.5 \text{ N}}{76330 \text{ N} \times 1.15}$$
$$RF = 3.25$$

Consider the allowable shear-bearing load.

K_{br} is first deduced from Figure 30.7.

$$e/d = 1.87$$
$$d/t = 2.133$$

$$P_{qu} = K_{br} \times A_{br} \times F_{tu}$$
$$= 1.824 \times 131.25 \text{ mm}^2 \times 972.51 \text{ MPa}$$
$$P_{qu} = 232818.89 \text{ N}$$

$$RF = \frac{P_{qu}}{F_a \times ff} = \frac{232.819 \text{ N}}{76330 \text{ N} \times 1.15}$$
$$RF = 2.65$$

Consider the yield/proof of the lug. This section checks the minimum proof strength for either tension or shear bearing. (The critical bore size used is d_h, with bush included).

where:

P_{px} = Allowable yield load in the axial direction

K_{bry} = Axial yield factor

F_{ty} = Minimum allowable tensile yield stress, with respect to the grain direction in plane of the lug

F_{tu} = Minimum allowable ultimate tensile stress, with respect to the grain direction in plane of the lug

$$P_{px} = K_{bry} \times \frac{F_{ty}}{F_{tu}} \times (P_{tu}, P_{qu}) \text{ (This is checked for the larger of } P_{tu} \text{ and } P_{qu})$$

$$K_{bry} = 1.586 \text{ (from Figure 30.8)}$$

$$F_{tu} = 972.51 \text{ MPa}$$

$$F_{ty} = 909.42 \text{ MPa}$$

$$P_{tu} = 285498.5 \text{ N}$$

$$P_{qu} = 232818.89 \text{ N}$$

$$P_{px} = 1.586 \times \frac{909.42 \text{ MPa}}{972.51 \text{ MPa}} \times 285498.5 \text{ N} (P_{tu} \text{ is larger than } P_{qu})$$

$$= 423425.92 \text{ N}$$

Safety Factor (SF = 1.5(proof)

$$RF = \frac{P_{px} \times SF}{F_a \times ff}$$

$$= \frac{423425.92 \text{ N} \times 1.5}{76330 \text{ N} \times 1.15}$$

$$RF = 4.824$$

Further Reading

Cozzone, F.P., Melcon, M.A., and Hoblit, F.M. Analysis of lugs and shear pins made from aluminum or steel alloys. *Product Engineering*, 21, 113–117, 1950.

Melcon, M.A., and Hoblit, F.M. Developments in the analysis of lugs and shear pins. *Product Engineering*, 1953.

Niu, Michael C. *Airframe Structural Design*. Hong Kong: Conmilit Press Ltd., 1999.

31

Mechanical Fasteners

31.1 Threaded Fasteners

This chapter covers threaded fasteners. These are a very flexible method for attaching components as they allow future easy disassembly. Threaded fasteners are designed principally for tensile loads but will accept a small level of shear load. This chapter begins with a general description of threads and then progresses to cover pre-tension diagrams, and so forth.

31.2 Basic Types of Threaded Fasteners

Bolt

A bolt is a threaded fastener that passes through clearance holes in mating members and is secured with a nut at the end opposite the head of the bolt (Figure 31.1).

Screw

A screw is a threaded fastener that passes through a clearance hole in one member and into a threaded hole in the other mating member (Figure 31.2).

Threaded fasteners are available in a wide variety of head shapes (Figure 31.3), including some of the following.

31.3 Thread Standards

There are several national and international standards covering fastener threads. These include those listed in Table 31.1.

31.4 Thread Profiles

The pitch line or diameter is located at 0.5 the height of the theoretical sharp v-thread profile (Figure 31.4).

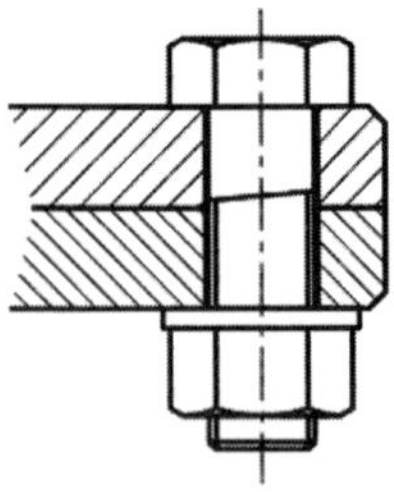

FIGURE 31.1
Bolt.

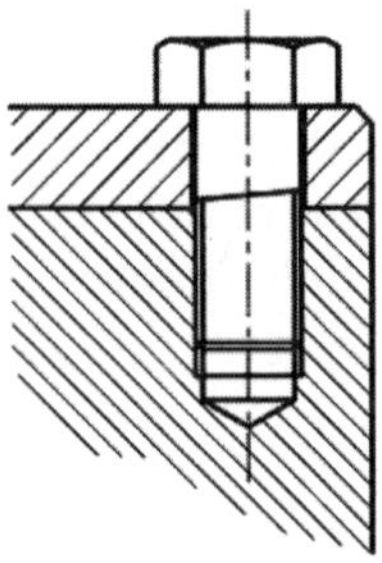

FIGURE 31.2
Screw.

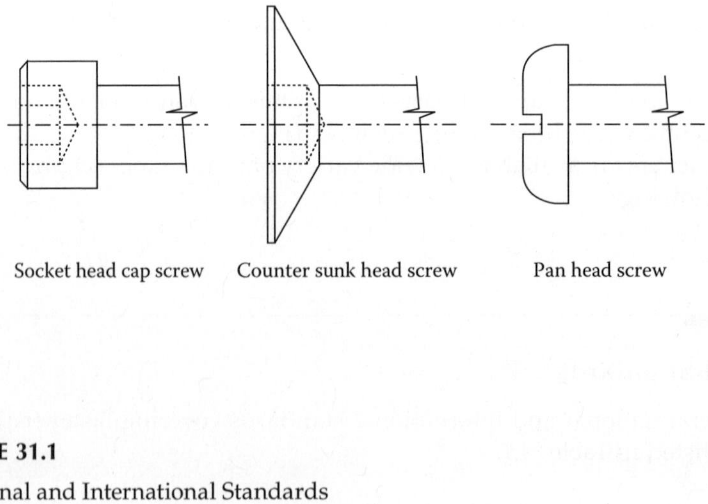

FIGURE 31.3
Types of heads.

TABLE 31.1
National and International Standards

Country	Standard Number	Description
UK	BS 3643	ISO metric screw threads
UK	BS 1500	Unified thread series
U.S.	B1.1-1949	Unified threads
U.S.	ANSI B1.1-1989/ASME B1.1-1989	Revision to B1.1-1949
U.S.	ANSI B1.13M-1983 (R1989)	Metric threads (M series)

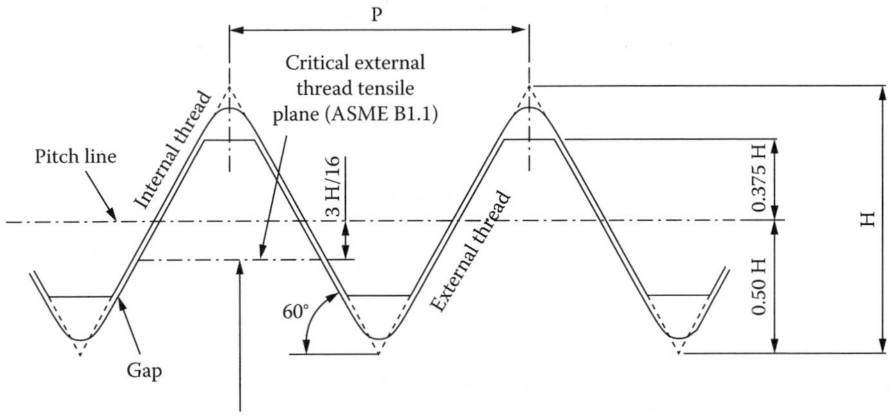

FIGURE 31.4
Typical thread profile.

31.5 Thread Series

The number of threads per unit length distinguishes groups of diameter-pitch combinations from each other when applied to a specific diameter.

ISO Metric Series. Metric system of diameters, pitches, and tolerance/allowances. This system is now the most common thread system in Europe and many parts of the world. There are two series of metric thread: metric coarse and metric fine. See Table 31.2 for ISO metric course thread details.

Unified Coarse Thread Series (UNC). Most commonly used in the majority of bolts, screws, and nuts in general engineering applications. This system is most commonly used in the United States and in Europe on old designs not yet converted to metric systems. See Table 31.3 for UNC thread details.

Unified Fine Thread Series (UNF). This system is used when there is a requirement for a finer pitch of thread, i.e., when the length of thread engagement is restricted.

Unified Fine J Series (UNJ). This series offer a finer pitch than that of the UNF series. Its most common application is in the aerospace industry.

31.6 Thread Designations

31.6.1 Metric Series

The following is an example of how a metric thread combination is described in a drawing or a specification.

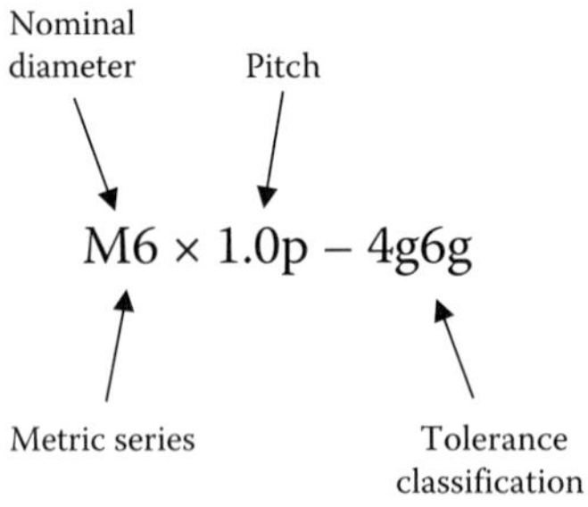

31.6.2 Imperial (Inch Series)

The following is how an imperial thread combination is called up on a drawing or a specification.

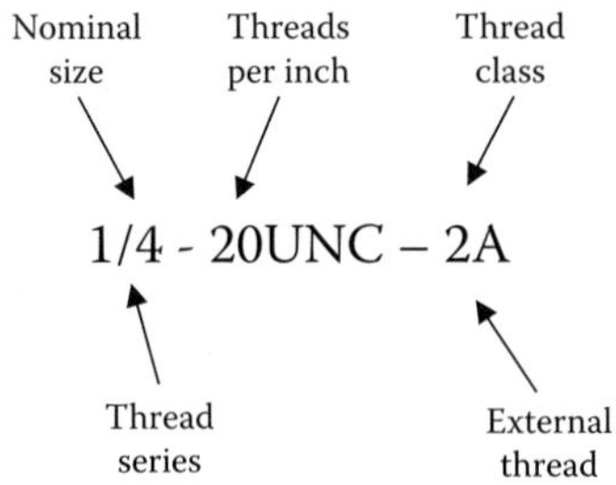

TABLE 31.2

ISO Metric Course Thread

Size Designation	Nominal Major Diameter (mm)	Pitch Diameter (mm)	Effective Diameter (mm)	Minor Diameter (mm)	Minor Diameter Area (mm²)	Depth (mm)	Shear Area External (mm²)	Shear Area Internal (mm²)
M3	3.00	0.50	2.675	2.387	4.473	0.307		
M3.5	3.50	0.60	3.110	2.764	6.000	0.368	4.674	6.606
M4	4.00	0.70	3.545	3.141	7.750	0.429	5.473	7.768
M5	5.00	0.80	4.480	4.018	12.683	0.491	7.073	9.997
M6	6.00	1.00	5.350	4.773	17.894	0.613	8.645	12.189
M8	8.00	1.25	7.188	6.466	32.841	0.767	12.160	16.826
M10	10.00	1.50	9.026	8.160	52.292	0.920	15.578	21.473
M12	12.00	1.75	10.863	9.853	76.247	1.074	18.975	26.113
M14	14.00	2.00	12.701	11.546	104.710	1.227	22.422	31.028
M16	16.00	2.00	14.701	13.546	144.120	1.227	26.095	35.563
M18	18.00	2.50	16.376	14.933	175.135	1.534	29.525	40.754
M20	20.00	2.50	18.376	16.933	225.189	1.534	33.276	45.379
M22	22.00	2.50	20.376	18.933	281.527	1.534	37.027	50.004
M24	24.00	3.00	22.051	20.319	324.274	1.840	40.458	54.999
M27	27.00	3.00	25.051	23.319	427.095	1.840	46.169	62.997
M30	30.00	3.50	27.727	25.706	518.990	2.147	51.633	69.537
M33	33.00	3.50	30.727	28.706	647.195	2.147	57.419	76.603
M36	36.00	4.00	33.402	31.092	759.274	2.454	63.091	84.043

Note: All dimensions in mm.

TABLE 31.3

UNC Threads

Designation	O Diameter	Core Diameter		Pitch	Depth	Effective Diameter
		Nut	Bolt			
1/4—20 UNC	0.2500	0.1959	0.1887	0.05000	0.03067	0.2175
5/16—18 UNC	0.3125	0.2524	0.2443	0.05556	0.03408	0.2764
3/8—16 UNC	0.3750	0.3073	0.2983	0.06250	0.03834	0.3344
7/16—14 UNC	0.4375	0.3602	0.3499	0.07143	0.04382	0.3911
1/2—13 UNC	0.5000	0.4167	0.4056	0.07692	0.04719	0.4500
9/16—12 UNC	0.5625	0.4723	0.4603	0.08333	0.05112	0.5084
5/8—11 UNC	0.6250	0.5266	0.5135	0.09091	0.05577	0.5660
3/4—10 UNC	0.7500	0.6417	0.6273	0.10000	0.06134	0.6850
7/8—9 UNC	0.8750	0.7547	0.7387	0.11111	0.06816	0.8028
1.0—8 UNC	1.0000	0.8647	0.8466	0.12500	0.07668	0.9188
1 1/8—7 UNC	1.1250	0.9704	0.9497	0.14286	0.08763	1.0322
1 1/4—7 UNC	1.2500	1.0954	1.0747	0.14286	0.08763	1.1572
1 3/8—6 UNC	1.3750	1.1946	1.1705	0.16667	0.10224	1.2667
1 1/2—6 UNC	1.5000	1.3196	1.2955	0.16667	0.10224	1.3917
1 3/4—5 UNC	1.7500	1.5335	1.5046	0.20000	0.12269	1.6201
2—4 1/2 UNC	2.0000	1.7594	1.7274	0.22222	0.13632	1.8557

Note: All dimensions in inches.

TABLE 31.4

Metric Grades of Steels for Fasteners

Grade	Fastener Size	Tensile Strength (MPa)	Yield Strength (MPa)	Proof Strength (MPa)
4.6	M5 to M36	400	240	225
4.8	M1.6 to M16	420	340[a]	310
5.8	M5 to M24	520	415[a]	380
8.8	M16 to M36	830	660	600
9.8	M1.6 to M16	900	720[a]	650
10.9	M6 to M36	1,040	940	830
12.9	M1.6 to M36	1,220	1,100	970

[a] Yield strength is approximate and is not included in the standard.

TABLE 31.5

SAE Grades of Steels for Fasteners

Grade Number	Fastener Size (in.)	Tensile Strength (ksi)	Yield Strength (ksi)	Proof Strength (ksi)	Head Marking
1	1/4 to 11/2	60	36	33	None
2	1/4 to 3/4	74	57	55	None
	>3/4 to 11/2	60	36	33	
4	1/4 to 11/2	115	100	65	None
5	1/4 to 1.0	120	92	85	3 points
	>1.0 to 11/2	105	81	74	
7	1/4 to 11/2	133	115	105	5 points
8	1/4 to 11/2	150	130	120	6 points

TABLE 31.6
ASTM Standard for Fastener Steels

ASTM Grade	Fastener Size (in.)	Tensile Strength (ksi)	Yield Strength (ksi)	Proof Strength (ksi)	Head Marking
A307	1/4 to 4	60	—	—	None
A325	1/2 to 1	120	92	85	A325
	>1.0 to 1 1/2	105	81	74	
A354-BC	1/4 to 2 1/2	125	109	105	BC
A354-BD	1/4 to 2 1/2	150	130	120	6 points
A449	1/4 to 1.0	120	92	85	
	>1.0 to 1 1/2	105	81	74	
	>1 1/2 to 3.0	90	58	55	
A574	0.060 to 1/2	180	—	140	(Socket head cap screws)
	5/8 to 4.0	170	—	135	

31.7 Material and Strength Designations

Table 31.4 covers the metric grades of steels for fasteners.

Table 31.5 covers the SAE grades of steels for fasteners.

Table 31.6 covers the ASTM standard for fastener steels.

31.8 Tensile and Shear Stress Areas

31.8.1 Tensile Stress Area

The average axial stress in a fastener is calculated using a tensile stress area (Figure 31.5).

$$f_{ave} = \frac{F}{A_t}$$

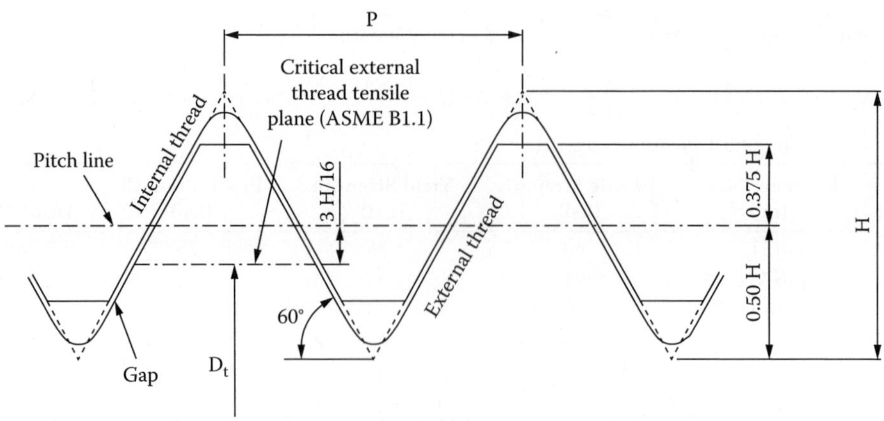

FIGURE 31.5
Tensile stress area.

$$A_t = \frac{\pi}{4}\left(\frac{D_r + D_p}{2}\right)^2$$

where

f_{ave} = Average axial stress
F = Axial force
D_r = Root diameter
D_p = Pitch diameter
A_t = Tensile stress area

An unthreaded rod having a diameter equal to the mean of the pitch diameter and minor diameter will have the same tensile strength as the threaded rod.

$$D_t = d_b - 2\left(\frac{3}{8}H + \frac{3}{16}H\right)$$

$$H = \frac{1}{2n} \cdot \tan(60°)$$

$$D_t = d_b - \frac{\tan(60°)}{n}\left(\frac{3}{8} + \frac{3}{16}\right)$$

$$D_t = d_b - \frac{9\sqrt{3}}{16n}$$

$$A_t = \frac{\pi}{4}D_t^2$$

$$A_t = \frac{\pi}{4}\left(d_b - \frac{0.9743}{n}\right)^2$$

where

D_t = Diameter at critical plane
d_b = Diameter of fastener
H = Theoretical height of thread
n = 1/p (threads per unit length)

This is the formula used by manufactures of imperial fasteners to publish the tensile area in their catalogs.

$$A_t = \frac{\pi}{4}\left(d_b - \frac{0.9328}{n}\right)^2$$

This formula is used to obtain a similar result for metric threads.

31.8.2 Shear Area of External Threads

The interaction between mating threads needs to be considered to establish the shear area of an external thread (Figure 31.6).

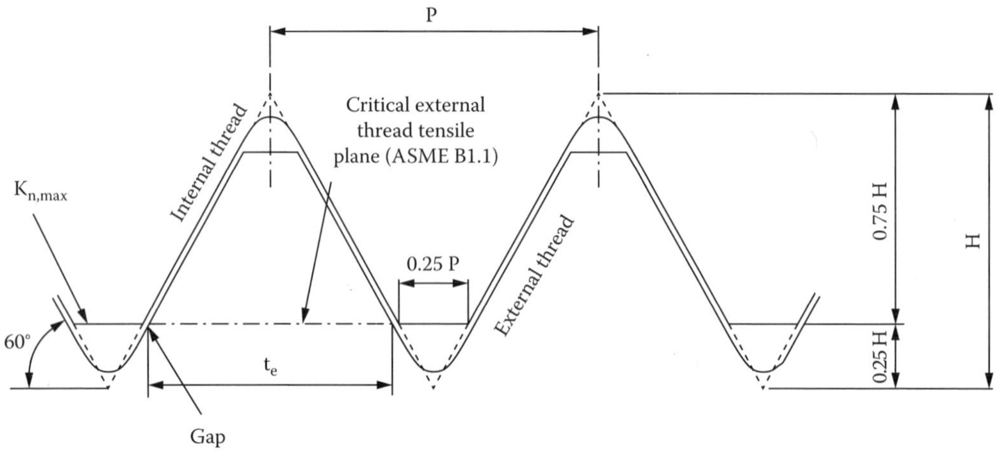

FIGURE 31.6
Shear area of external thread.

$$A_{s,c} = \pi \cdot K_{n,max} \cdot t_e \cdot n$$

$$\tan(30°) = \frac{0.5t_e}{0.75\,H - gap}$$

$$H = \frac{1}{2n}\tan(60°) = \frac{\sqrt{3}}{2n}$$

$$t_e = 2 \cdot \tan(30°)\left(0.75\frac{\sqrt{3}}{2n} - gap\right)$$

$$gap = \frac{1}{2}\left(K_{n,max} + \frac{1}{2}\frac{\sqrt{3}}{2n} - E_{s,min}\right)$$

where:

$A_{s,c}$ = Shear area of external thread
$K_{n,max}$ = Maximum minor diameter of internal thread
t_e = Thickness of external thread at critical shear plane
n = Threads per unit length

The gap equation is based on the tolerance data:
$E_{s,min}$ = Minimum pitch diameter of the external thread

$$t_e = 2 \cdot \frac{1}{\sqrt{3}}\left[\frac{3}{4}\frac{\sqrt{3}}{2n} - \frac{1}{2}\left(K_{n,max} + \frac{1}{2}\frac{\sqrt{3}}{2n} - E_{s,min}\right)\right]$$

$$t_e = \frac{1}{2n} + \frac{1}{\sqrt{3}}\left(E_{s,min} - K_{n,max}\right)$$

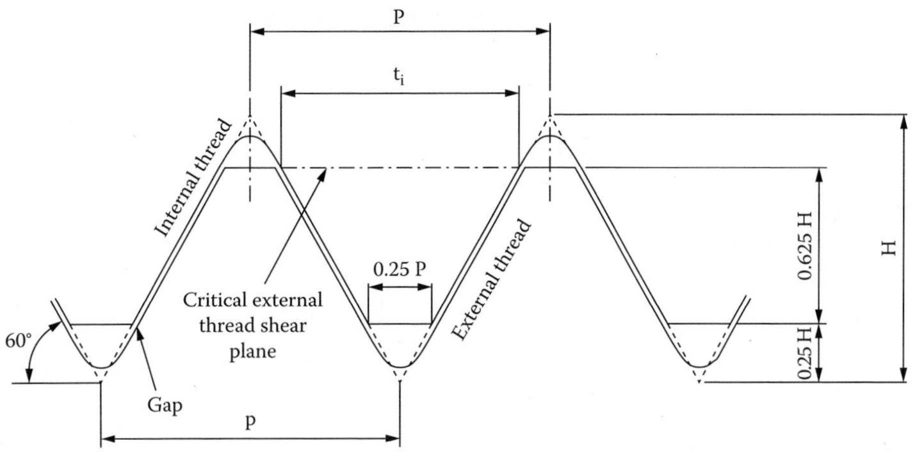

FIGURE 31.7
Shear area of internal thread.

This equation is in the British Standards (BS) and American National Standards Institute (ANSI) standards and gives the shear area per unit length of engagement, L_e, to obtain the actual shear area.

$$A_{s,c} = \pi \cdot n \cdot K_{n,max} \cdot \left[\frac{1}{2n} + \frac{1}{\sqrt{3}} (E_{s,min} - K_{n,max}) \right] L_e$$

31.8.3 Shear Area of Internal Threads

$$A_{s,i} = \pi \cdot D_{s,min} \cdot t_i \cdot n$$

where:

$D_{s,min}$ = Minimum major diameter (external thread)
t_i = Thickness of internal thread (critical plane)

Similar to the previous derivation, the shear area of an internal thread (Figure 31.7) can be derived by taking into account the tolerances of the thread system.

$$A_{s,i} = \pi \cdot D_{s,min} \cdot n \cdot \left[\frac{1}{2n} + \frac{1}{\sqrt{3}} (D_{s,min} - E_{n,max}) \right]$$

$E_{n,max}$ = Maximum pitch diameter of the internal thread

31.9 Length of Engagement

The length of engagement is considered the number of threads in engagement in either a nut or component multiplied by the thread pitch.

31.9.1 Length of Engagement Using Equal Strength Materials

If the external and internal thread materials have the same strengths, then:

Tensile strength (external thread) $$S_t = \frac{F_{max}}{A_t}$$

Shear strength (internal thread) $$0.5_{St} = \frac{F_{max}}{A_{s,i} \cdot L_e}$$

$$F_{max} = S_t A_t = 0.5 S_t \cdot A_{s,i}, L_e$$

Hence: $$L_e = \frac{2A_t}{A_{s,i}}$$

31.9.2 Length of Engagement Using Dissimilar Strength Materials

If the external and internal thread material does not have the same strengths, then:

Tensile strength (external thread) $$S_{t,e} = \frac{F_{max}}{A_t}$$

Shear strength (internal thread) $$0.5_{t,i} = \frac{F_{max}}{A_{s,i} \cdot L_e}$$

$$F_{max} = S_{te} A_t = 0.5 S_{t,i} \cdot A_{s,i}, L_e$$

Hence: $$L_e = \frac{2A_t \cdot S_{t,e}}{A_{s,i} \cdot S_{t,i}}$$

31.10 Fastener and Nut Design Philosophies

Standard fasteners and nuts of equal strengths are designed for the bolt to fail before the threads in the nut are stripped.

The design engineer has a responsibility to ensure that a failure of a machine element or other artifact will not endanger life. In this instance the length of thread engagement is an important consideration when designing machine elements using threaded fasteners.

31.11 Pitching of Fasteners

Considerable latitude exists for selecting the number of fasteners for a given application, as this will be influenced by the combination of pitch and size of fasteners in relation to the stiffness of the component being clamped.

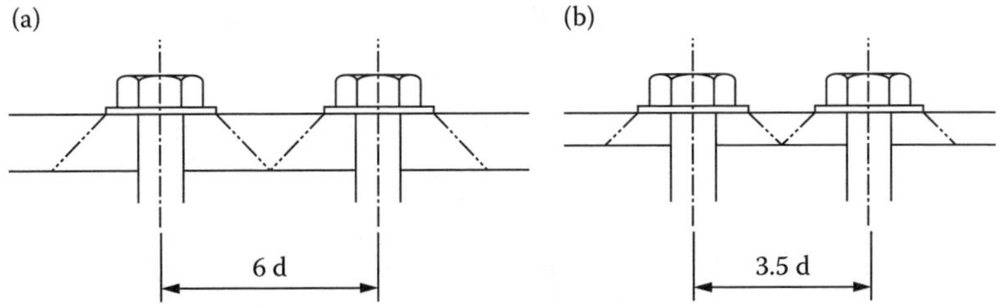

FIGURE 31.8
Effects of flange thickness on fastener pitch.

Minimum pitch will be determined by spanner or socket clearances.
Maximum pitch may be determined by:

- Fretting.
- Leakage aspect. This will be determined by the permissible degree of sealing required: the type of media involved (i.e., oil, fuel air, etc.) and the pressures involved.

31.11.1 Pressure Cone

In general it is assumed that a 90° frustum of a cone commences at the edges of a spreader washer or fastener head. The aim is to have the base of each cone at least coincidental at the joint face. The effect of the flange thickness on the fastener pitch is shown in Figure 31.8.

An increase in the number of small-diameter fasteners is preferable to a lesser number of large-diameter fasteners. Where a small number of fasteners will adequately deal with the externally applied load, it is still desirable to increase the number to give a reasonable pitch. This will reduce the frettage effects, especially in the case of flanged bearing outer races, where, in the interest of weight saving, a small number would otherwise be adopted.

31.12 Tension Connections

The purpose of a bolt is to join two elements together. In general most bolted static joints when torqued to a particular value will be sufficient to maintain adequate joint pressures and meet its requirements.

It is important for the design engineer to ensure that the joint pressures are maintained through the life of the component and not lead to premature failure of the fastener due to either:

1. A significant overload causing a joint separation resulting in an overload in the fastener
2. The joint being subject to fatigue loading

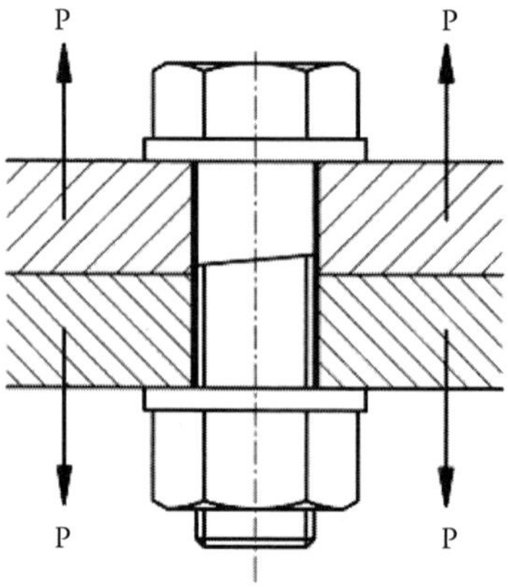

FIGURE 31.9
A typical bolted joint.

This section will describe the elements of the bolted joint and lead to an understanding of a pretension diagram.

It is important to realize that threaded fasteners are designed for static tension loading. They will tolerate a small amount of shear loading together with a degree of bending in the head if the clamping surfaces are not truly parallel.

If the fastener is expected to take a large amount of shear load then shear bolts should be considered.

Alternatively, dowel pins should be used to take out the shear component in the fittings.

In Figure 31.9 a typical bolted joint is depicted with an external force P applied trying to separate the joint faces.

31.13 Torque-Tension Relationship

The formula (Equation 31.1) is a good theoretical approximation for the torque required to tighten a nut onto a fastener, so to develop an end load P_i

$$T = P_i\left(\frac{p}{2\pi}\right) + \frac{E_s\mu_1}{2\cos\alpha} + \frac{d_o + d_i}{4}\mu_2 \tag{31.1}$$

The torque is in three parts: Pi(p/2π) represents that part of the torque absorbed in driving the mating thread helices over each other against the action of the axial load Pi to which they inclined, while $P_i(E_s\mu_1)/(2\cos\alpha)$ and $P_i((d_o - d_i)/4)\mu_2$ represent that absorbed in overcoming friction in the first case between the threads and in the second case at the bearing face under the nut. A typical distribution between the terms would be in the order of 10%, 40%, and 50% of the total. Friction conditions are therefore of predominating importance and unless known with a good degree of accuracy together with the bearing areas there is no point in using an expression more complicated than the simple empirical formula (Equation 31.2).

$$T = \frac{1}{5} P_i \cdot D \quad \text{or} \quad P_i = \frac{5T}{D} \tag{31.2}$$

where:

T = Tightening torque
P_i = Initial tensile load in the fastener developed by tightening
D = Basic major diameter (i.e., nominal diameter)

This applies to nuts and bolts of normal proportions and for threads that are lubricated with a thin film of grease or oil. This is generally accurate to about ±20%.

This does not apply to assemblies that are torque tightened from the bolt head.

31.14 Proof Load and Proof Stress

31.14.1 Fastener Preload

A preload in a bolt-nut combination is achieved by tightening the nut to a specified torque setting (Figure 31.10).

It should be borne in mind that not all the torque applied to the nut will be converted into a load in the bolt. A substantial amount of torque will be lost in friction between the contact faces of the bolt head and the nut, together with another loss due to the friction within the thread itself.

There will be an extension in the bolt due to the preload and a small compressive force in the abutment. In the analysis of the fastener preload the bolt and its associated abutments are treated as springs.

The bolt stiffness is treated as two separated parts.

L_d = Plain portion of the bolt
L_t = Threaded portion of the bolt

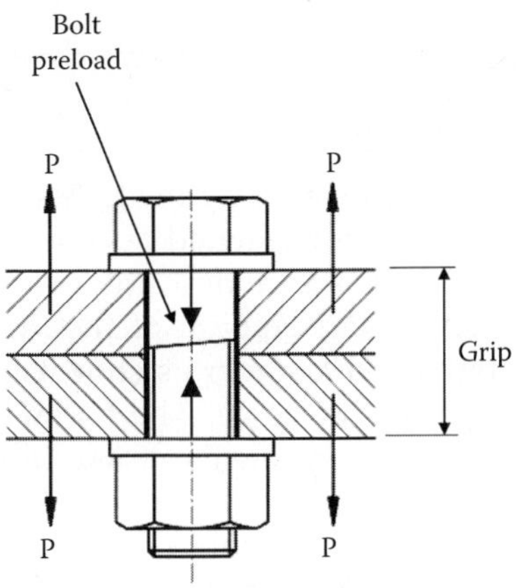

FIGURE 31.10
Bolt preload.

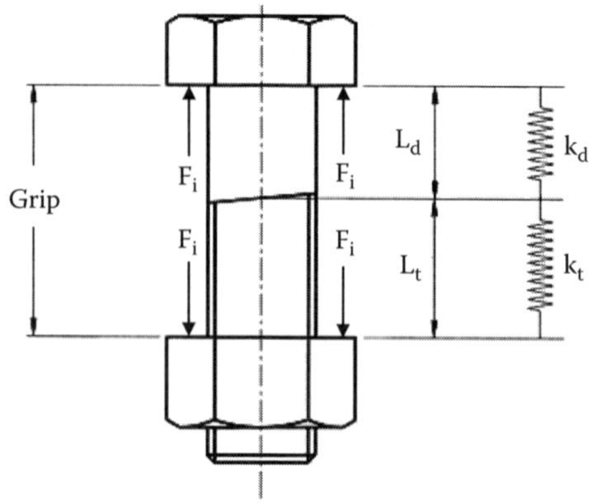

FIGURE 31.11
Bolt stiffness.

This latter part is for the part of the thread above the nut. The bolt stiffness can be described (see Figure 31.11):

$$\delta = \delta_d + \delta_t$$

$$\delta_d = \frac{F_i}{k_d} d$$

$$\delta_t = \frac{F_i}{k_t} \tag{31.3}$$

$$F_i = \frac{k_d k_t}{k_d + k_t} \delta$$

$$F_i = k_b \cdot n$$

31.15 Introduction to Pretension

It is not widely understood how a fastener in a fully tightened joint can survive in a situation where an untightened or loosely tightened fastener will fail within a very short space of time. When a force is applied to a joint fitted with a fully tightened fastener, the fastener is exposed to only a small amount of the force. When a force is applied to a joint fitted with a fully tightened fastener as shown in Figure 31.9, the fastener is exposed to only a small amount of the force. The following description will explain this. The bolt can be treated as a solid spring as shown in Figure 31.11 and when fitted through the connecting parts as shown in Figure 31.12, the bolt is extended under the tightening force. The connecting parts are subjected to a compressive force under the abutment faces of the bolt head and nut. If these forces are plotted in a tension diagram (Figure 31.13), an understanding of the loads involved becomes obvious. The following text explains this in more detail.

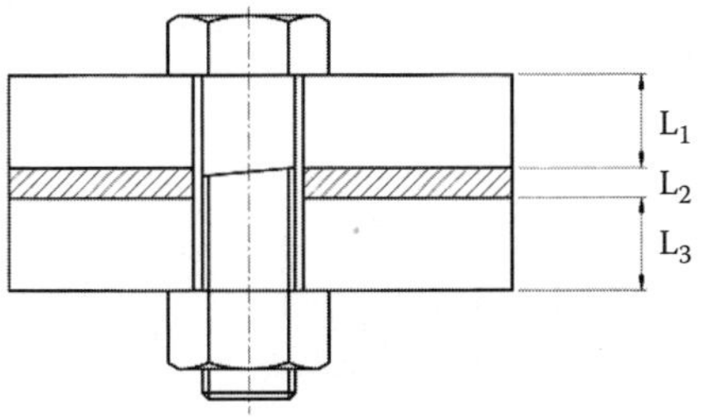

FIGURE 31.12
Abutment stiffness.

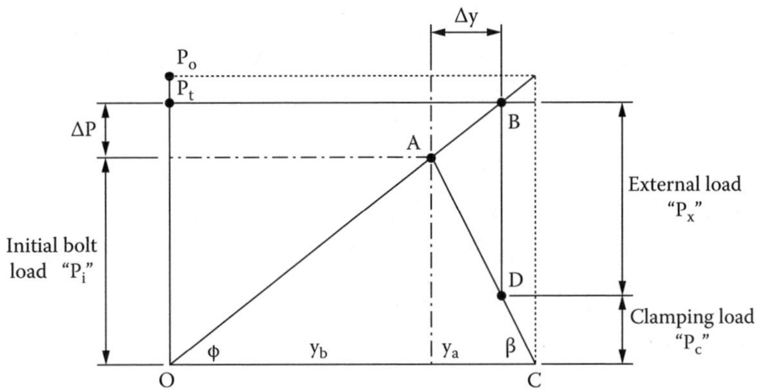

FIGURE 31.13
Tension diagram.

31.15.1 Why Preload?

Consider the arrangement shown in Figure 31.14a, where a fastener is located in a two-part joint and a compression spring is fitted under the head of the fastener; a pointer is also fitted to the head against a scale. The fastener has been partially tightened with a small load in the fastener and joint faces.

In Figure 31.14b the nut has been tightened to compress the spring so that the pointer indicates 1 on the scale. A working load has yet to be applied.

In the next figure (Figure 31.14c), with the same force in the fastener, a load of 1 unit is applied to the joint. This should have the effect of forcing the joint open, and it would be imagined that this additional force would increase the force in the fastener.

The initial reaction is to believe that the load in the fastener will increase and the joint faces will separate, producing a gap. But in fact, the fastener will keep its original value on the scale. What actually happens is the action of the force is to reduce the clamping force that exists between the end of the fastener to 0.5 unit, but the pointer on the scale will still register 1 unit. The fastener will not "feel" any of the applied force until it exceeds the clamping force.

Practically, when tightening takes place there will be an elongation of the fastener and a corresponding compression on the joint. This compression results in the fastener sustaining a proportion of the load. As the applied force reduces the clamping force that exists in the joint, an additional strain is felt by the fastener, which increases the force it sustains. The amount of the additional force the fastener sustains is less than the applied load acting

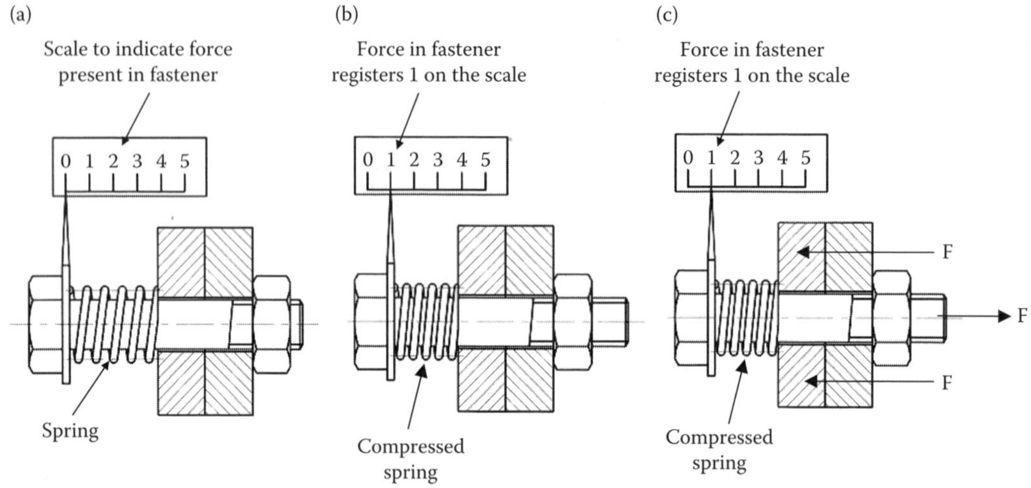

FIGURE 31.14
Description of preload.

on the joint. The actual amount of the force the fastener sustains will depend upon the ratio of stiffness of the fastener to the joint material.

31.16 Joint Diagrams

To help the design engineer visualize the loading within the bolted connection, joint diagrams have been developed. A joint diagram is a ready means of visualizing the load deflection characteristics of the fastener and the material it clamps. These diagrams assist in helping to understand how the joint sustains an external force and why the fastener does not sustain the whole of this force.

Figure 31.15 shows how a basic joint diagram is constructed. As the nut is tightened against the joint, the bolt will extend. Internal forces within the fastener resist this extension and a tension force or preload is generated. The reaction to this force is a clamping force that is the cause for the joint to be compressed. The force-extension diagram shows

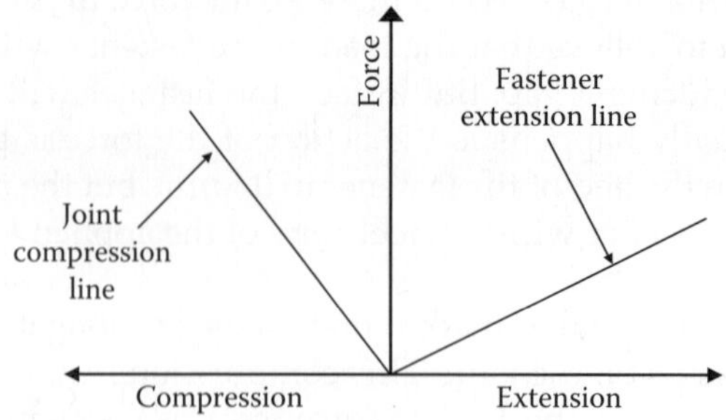

FIGURE 31.15
Basic joint diagram.

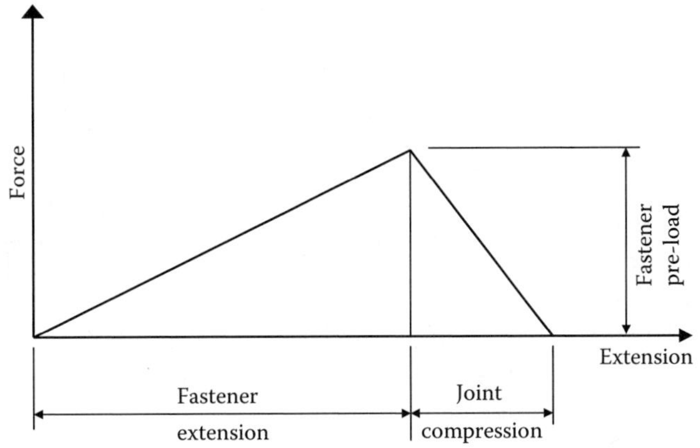

FIGURE 31.16
Basic pre-tension diagram.

the fastener extension and the joint compression. The slopes of the lines represent the stiffness of each component, the clamped joint being usually stiffer than the fastener.

Moving the compression line of the joint to the right forming a triangle, as it is recognized, forms the basic diagram shown in Figure 31.16, with the fastener and the joint in equilibrium. The clamping force tending to compress the joint is then seen to be equal to the preload in the fastener. Positive extension is to the right, such as that sustained by the fastener; negative extension (compression) is to the left and is sustained by the joint material.

31.16.1 Joint Diagrams with an External Load Applied

When an external force is applied to the joint it has the effect of reducing some of the clamping force caused by the fastener preload and applying an additional force to the fastener itself. This is illustrated in the joint diagram in Figure 31.17.

The external force acts through the joint material and then subsequently into the fastener.

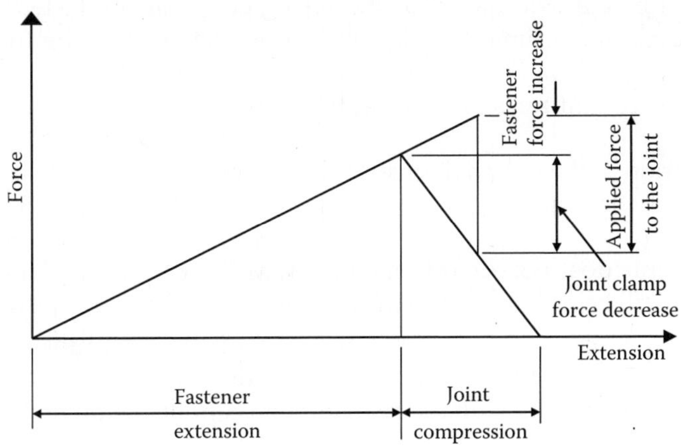

FIGURE 31.17
Pre-tension diagram with preload.

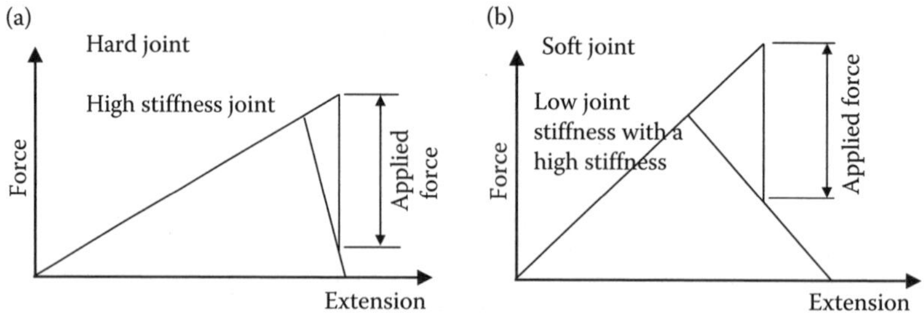

FIGURE 31.18
Effects of fastener stiffness.

It can be seen that the load on the fastener cannot be added without a subsequent reduction in the clamp force acting on the joint. As can be observed from a study of the diagram, the actual amount of increase in the fastener force is dependent upon the relative stiffness of the fastener to the joint.

As an illustration of the importance of the relative stiffness of the fastener to the joint, two figures are shown, Figure 31.18a presents a hard joint (low stiffness fastener with a high stiffness joint) and Figure 31.18b a soft joint (a high stiffness fastener with a low stiffness joint).

With a hard joint, because of the steep joint stiffness slope, in this case the fastener will only sustain a small proportion of the applied force.

A soft joint has a slope for the fastener greater than that of the joint. The fastener will therefore sustain the majority of the applied force.

High-performance fasteners generally have their shank diameters reduced down to the stress diameter of the thread. If the fastener is modified in this manner the stiffness of the fastener is reduced so that it will not support as much of any applied load that it would otherwise do. Provided that the shank diameter is not reduced below that of the stress diameter, the strength of the fastener will not normally be impaired.

31.16.2 Effects of a Large Increase in the External Load

With an increase in the external load, the force acting on the fastener proportionally increases, and at the same time there is a reduction in the clamping force acting on the joint.

If the external force continues to increase, then either:

1. The proportion of the external force acting on the fastener's preload would result in the yield of the fastener material being exceeded with the possibility of the fastener failure.

   Even if failure does not occur immediately, with the removal of the external load the preload will be reduced due to a permanent deformation of the fastener.

2. The clamping force acting on the joint will continue to reduce to a point when it becomes zero. Any further increase in the external force will result in a gap forming within the joint, with the fastener then supporting all the additional force (Figure 31.19).

   If a gap forms within the joint comprising the joint, then the fasteners are almost always subjected to nonlinear loadings from bending and shear forces. This will

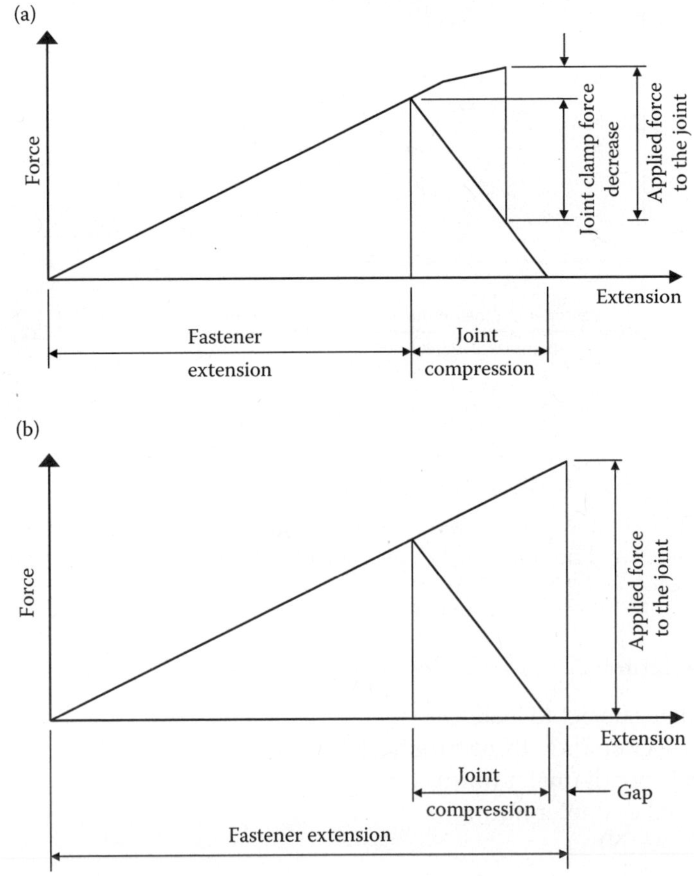

FIGURE 31.19
Effect of increase in the external load.

quickly result in the fastener failure; therefore, it is normal for the design criteria to include a statement that the applied forces must not under any circumstances result in a gap forming within the joint.

31.16.3 The Effect of a Compressive External Load

If a joint is subjected to a compressive external force (Figure 31.20), this has the effect of increasing the clamping force acting on the joint and decreasing the tension within the fastener. If the compressive external force is large enough, then:

1. The tension in the fastener can be reduced to a low value; if the joint is subjected to a cyclic load, the fastener could fail due to fatigue as it experiences tension variations under a compressive force. The fastener will be most susceptible to loosening due to vibration.
2. The yield limitations of the clamped material are exceeded, as the joint is supporting a compressive load in addition to that provided by the fasteners' preload. A loss of preload will result from permanent deformation when the external load is released.

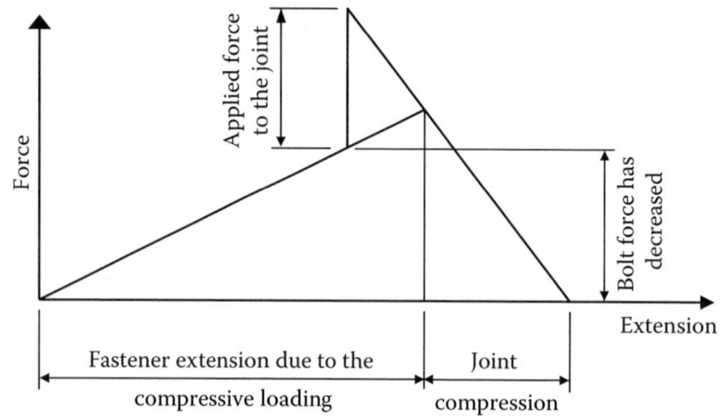

FIGURE 31.20
Effect of a compressive load.

Bolt pretension, also called preload or prestress is generated when a nut is tightened on a threaded bolt or screw. The load in the fastener increases and the deformation of the fastener also increases.

31.16.4 Nomenclature

A = Area of plate subject to load
A_b = Bolt cross section (bolt thread root area) (mm^2)
D_b = Bolt thread root diameter (mm)
d = Bolt nominal diameter (mm)
F_e = External load (N)
F_t = Thermal load (N)
F_p = Preload (N)
F = Total load on bolt (N)
E_z = Modulus of elasticity (N/mm^2)
L = Length of bolt joint (mm)
L_b = Length of bolt (mm)
L_j = Length of joint (mm)
k_z = Stiffness of component (N/mm)
k_b = Stiffness of bolt (N/mm)
k_j = Stiffness of joint (N/mm)
t_z = Thickness of plate (mm)
T = Bolt tightening torque (Nmm)
x_z = Deflection of item z/unit load (mm/N)
x_b = Deflection of bolt/unit load (mm/N)
α_z = Coefficient of thermal expansion of component z (mm/mm/deg.C)
δ = Deflection (mm)

31.16.5 Notes

A bolt tightened to a specific value is considered safer than a bolt simply tightened to an arbitrary value. A preload between 75% and 80% of the proof strength of the bolt material is normally used.

Consider a bolt being used to clamp a joint to a set preload value. The bolt has a low stiffness and the joint has a very high stiffness. An external load, when applied to the joint, will tend to separate the joint. Part of this load will cause a further extension in the bolt, and the remaining part of the load will result in a reduction of the compressive load on the joint.

It follows that an infinitely stiff bolt will result in no separation of the joint faces. The question is to determine a realistic stiffness for the bolt to withstand the influences of the external force.

Consider Figure 31.21, which shows an assembly bolted with a preload of P_i and an external load of P_x is applied.

For an infinitely stiff bolt, separation will not occur, as all the external load will be directly applied to the bolt with no resulting extension. For an infinitely stiff joint, separation will take place when the external load exceeds the preload.

The bolt loading diagrams (Figures 31.22 and 31.23) show the loading regime on the bolt and joint.

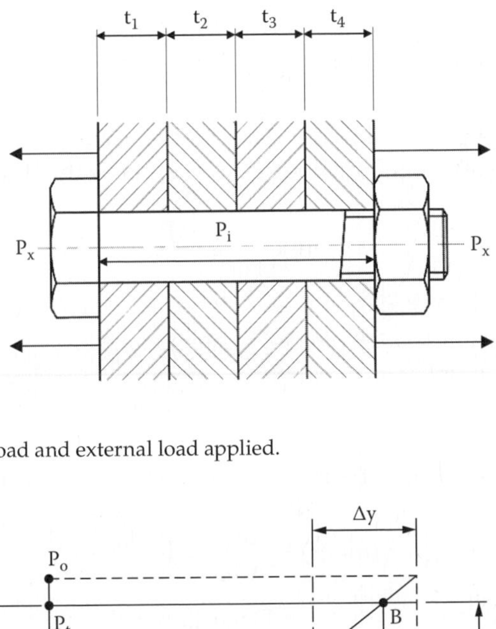

FIGURE 31.21
Bolted assembly with a preload and external load applied.

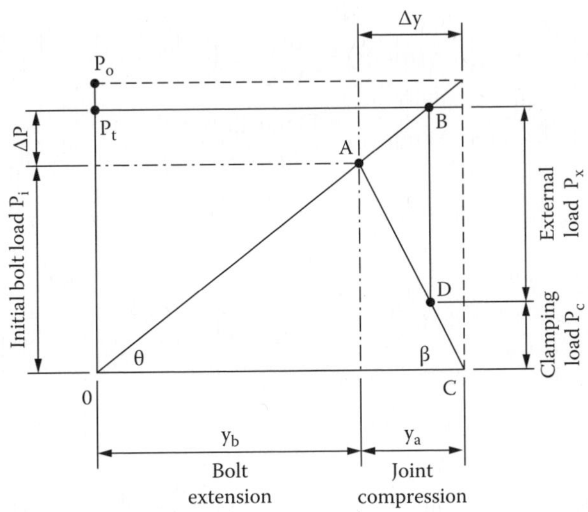

FIGURE 31.22
Bolt loading diagram with preload and external load applied.

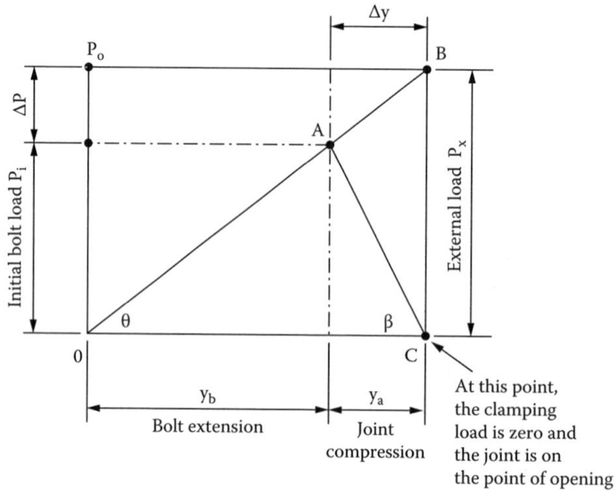

FIGURE 31.23
Bolt loading diagram with external load matching preload.

The calculation for the proportion of the load taken by the bolt and the joint using the component stiffness values uses the same basic formula as that for spring rate, i.e.,

$$\text{Stiffness} = k = \frac{P}{\delta}$$

31.17 Fastener Stiffness

The relationship E=stress/strain=f/ε is used to determine the stiffness of a bolt.

$$\text{Stress(f)} = \frac{\text{Force}}{\text{Area}} = \frac{P}{A}$$

$$\text{Strain(e)} = \frac{\text{Deflection}}{\text{Length}} = \frac{d}{L}$$

$$E = \frac{\frac{P}{A}}{\frac{\delta}{L}}$$

Therefore

$$\delta = \frac{P \cdot L}{E \cdot A}$$

$$k_b = \frac{P_b}{\delta_b} = \frac{P_b}{\frac{P_b \cdot L_b}{A_b \cdot E_b}} = \frac{A_b \cdot E_b}{L_b}$$

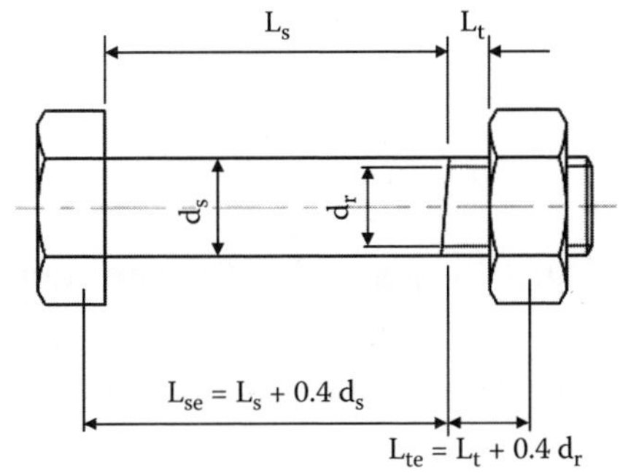

FIGURE 31.24
Stiffness characteristics of a bolt.

If the bolt length clamping the joint includes a number of different s, then the resulting stiffness is determined using the relationship

$$\frac{1}{k_{bt}} = \frac{1}{k_1} + \frac{1}{k_2} + \frac{1}{k_3} + \frac{1}{k_4}$$

To allow for a certain degree of elasticity of the bolt head and nut, a correction factor is often used to modify the length used in the stiffness calculations shown in Figure 31.24.

The stiffness of the bolt results from the stiffness of the bolt shank (diameter d_s) and the stiffness of the bolt thread (root diameter d_r).

The length used to calculate the shank stiffness $= L_{se} = L_s + 0.4d_s$.

The length used for the threaded length section $= L_{te} + L_t + 0.4d_r$.

31.18 Joint Stiffness

It is very difficult to calculate the stiffness of a joint that is based on holes drilled in a plate (see Figure 31.25). A rough estimation can be made by assuming the joint is an annulus with an outside diameter (OD) of 2.5 times the bolt diameter and an inside diameter (ID) equal to the bolt diameter.

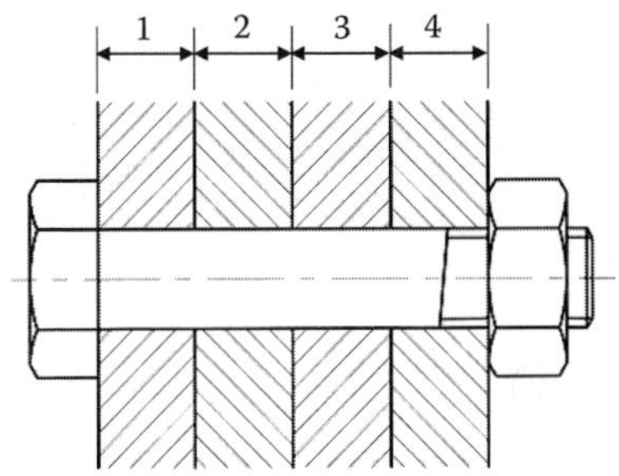

FIGURE 31.25
Multiple clamping surfaces.

The total joint stiffness is related to the individual stiffness values as follows:

$$\frac{1}{k_{jt}} = \frac{1}{k_1} + \frac{1}{k_2} + \frac{1}{k_3} + \frac{1}{k_4}$$

31.18.1 Calculation of Load Distribution Using Fastener/Joint Stiffness

A joint preload with a force P_x is then subject to an additional load P_e, which tends to separate the joint. The resulting deflections of the joint and bolt are the same providing that P_e is less than the separating force.

$$d = \frac{F_{be}}{k_b} = \frac{F_{je}}{k_j} \qquad F_e = F_{be} + F_{je}$$

It follows that

$$\begin{aligned} F_{be} &= F_{je}\frac{k_b}{k_j} \\ &= \frac{k_b}{k_j}(F_e - F_{be}) \\ &= \frac{k_b}{(k_b + k_j)}F_e \end{aligned}$$

Also

$$F_{je} = \frac{k_j}{(k_b + k_j)}F_e$$

Following the application of the external force, the resulting total force on the bolt is

$$\begin{aligned} F &= F_p + F_{be} \\ &= F_p + \frac{k_b}{(k_b + k_j)}F_e \end{aligned}$$

the total force on the joint

$$\begin{aligned} F_j &= F_p + F_{je} \\ &= \frac{k_j}{(k_b + k_j)}F_j - F_p \end{aligned}$$

31.19 Thermal Loading

If all the materials of the joint and the bolt are the same, then any changes in temperature will have minimal effects on the joint loadings. However, if the joint materials have thermal expansion coefficients different than the bolt material, changes in the joint loading will result from changes in temperature.

$$\text{Coefficient of thermal expansion of joint material} = \alpha_j$$

$$\text{Coefficient of thermal expansion of bolt} = \alpha_b$$

$$\text{Change of temperature} = \Delta T$$

$$\text{Length of joint} = \text{length of bolt} = (L_j = L_b = L)$$

$$\text{(as ambient temperature)}$$

$$\text{Expansion of joint} = \Delta L_j = \alpha_j \Delta T \cdot L$$

$$\text{Expansion of bolt} = \Delta L_b = \alpha_b \Delta T \cdot L$$

Overall stiffness of the joint is calculated as follows:

$$\frac{1}{k_t} = \frac{1}{k_j} + \frac{1}{k_b}$$

Therefore

$$k_t = \frac{k_b \cdot k_j}{(k_b + k_j)}$$

The resulting change in the joint load is calculated as follows:

$$F_t = \frac{k_b \cdot k_j}{(k_b + k_j)}(\Delta L_j - \Delta L_b)$$
$$= \frac{k_b \cdot k_j}{(k_b + k_j)} \Delta T \cdot L(\alpha_j - \alpha_b)$$

The total bolt load following temperature change:

Let: $F = F_p + F_t$

31.19.1 Initial Tension in Bolt

The initial tension in a bolt is roughly estimated for a bolt tightened by hand of an experienced mechanic as follows.

The tension resulting from this equation would be reasonably safe for M8 grade 8.8 bolts and above.

$$F_p = K \cdot d$$

where:

F_p=Preload (N)
K=Coefficient (varies between 1.75×106 N/m and 2.8×106 N/m)
d=Nominal diameter of bolt (mm)

For a bolt tightened with a torque wrench, the torque required to provide an initial bolt tension may be approximated by the formula

$$T = F_p \cdot K \cdot d$$

Typical K factors:

Steel thread condition	K
As received, stainless on mild or alloy	0.30
As received, mild or alloy on same	0.20
Cadmium plated	0.16
Molybdenum-disulfide grease	0.14
Polytetrafluoroethylene (PTFE) lubrication	0.12

A more accurate value can be obtained from the following:

$$T = \text{thread torque} + \text{collar torque}$$

$$= \frac{F_p \cdot d_m}{2}\left(\frac{p + \pi \cdot \mu \cdot d_m \cdot \sec\alpha}{\pi \cdot d_m - \mu \cdot p \cdot \sec\alpha}\right) + F_p \cdot \mu_c \cdot r_c$$

where

F_p = desired bolt preload (N)
p = thread pitch (mm)
d_m = mean diameter of thread (mm)
μ = coefficient of thread friction
μ_c = coefficient of collar friction
α = the thread angle/2 ($\alpha = 30°$ for standard metric threads and $\alpha = 29°/2$ for acme threads).
r_c = collar friction radius (mm)

It can be proved that the majority of the torque is required to overcome the thread and collar friction forces (approximately 90%). Therefore, any errors in the value of the friction coefficient will have a large variation on the bolt tensile load. The above formula is in essence not a lot more accurate than the approximate formulas above.

31.20 Fasteners Subject to Combined Shear and Tension

31.20.1 Interaction Curves: Load Ratios and Factors of Safety

Bolts and screws under combined tension stresses or loads may be predicted without using principal stresses or loads by using the interactive method.

The method represents applied and allowable stress or load conditions on the shank of a fastener by loads.

Load ratios are nondimensional coefficients R given by

$$R = \frac{\text{applied load}}{\text{allowable load}}$$

The method involves determining the allowable load for each separate failure mode such as tension, compression, bending, buckling, and shear. Load ratio R for each separate failure condition is calculated and combined using interaction equations if these loads act simultaneously on the fastener.

The equations generally take the following form:

$$R_1^x + R_2^y + R_3^z \cdots = 1.0$$

where ^{xyz} are exponents defining interaction relationship.

Failure is when the sum of the load ratios is greater than 1.0.

31.20.2 Interaction Curve

The margin of safety (MoS) assuming each load increases proportionately until failure occurs at a point B (see Figure 31.26):

$$MS = \frac{OB}{OA} - 1$$

The margin of safety assuming R_1 remains constant with R_2 increasing until failure occurs at point C:

$$MS = \frac{EC}{EA} - 1$$

The margin of safety assuming R_2 remains constant with R_1 increasing until failure occurs at point D:

$$MS = \frac{FD}{FA} - 1$$

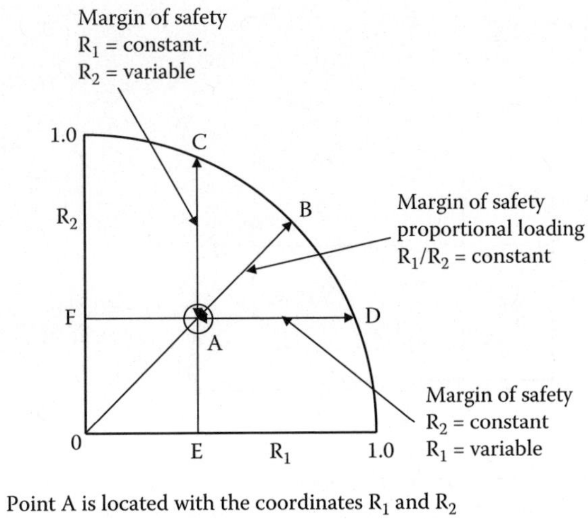

FIGURE 31.26
Interaction curve.

31.20.3 Interaction Equation

$$MS = \frac{1}{\sqrt{R_t^2 + R_s^2}} - 1$$

where:

$$R_t = \frac{\text{applied ultimate tension load}}{\text{allowable ultimate tension load}}$$

$$R_s = \frac{\text{applied ultimate shear load}}{\text{allowable ultimate shear load}}$$

Note that when calculating R_s, shear is applied to the bolt shank; this area should be used in calculating the allowable shear load.

31.21 Eccentric Loads

The design engineer is sometimes confronted with a problem requiring the design of a bracket to attach to an existing structural member, such as that depicted in Figure 31.27.

In this case the line of action passes outside the bolt group and as a consequence will create a shear load on the fasteners.

Figure 31.28 considers the loading seen by each fastener.

Let u_1=load/fastener per unit distance from O.

$$PL = 4u_1 I_a^2 + 2u_1 I_b^2 \tag{31.3}$$

Example 31.1

Consider a bracket similar in design to that depicted in Figure 31.28 with the principal load and dimensions as follows:

P=150 kN
L=225 mm

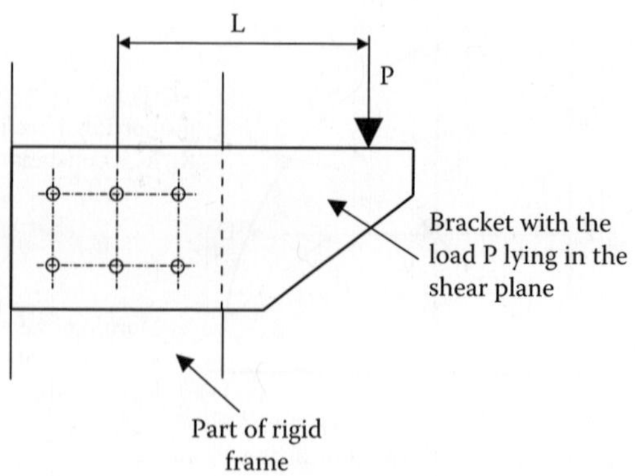

FIGURE 31.27
Bracket subject to shear loading.

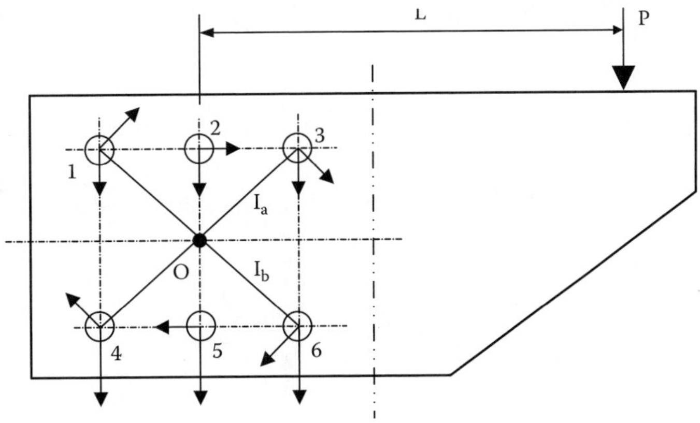

FIGURE 31.28
Centroid of bolt group.

$I_a = 111.803\,mm$
$I_b = 111.803\,mm$
$n = 6$

Determine the size of the fasteners if the allowable shear stress on the fasteners is to be limited to 400 MPa.

Rearranging Equation 31.3 to solve for u_1:

$$u_1 = \frac{P.L}{\left[\left(4.I_a^{\,2}\right)+\left(2.I_b^{\,2}\right)\right]}$$

$$u_1 = 450.0\,\frac{kN}{m}$$

Load on fastener due to turning moment:

$$L_t = u_1.I_a$$

$$L_t = 50.312\ kN$$

Now, direct load on fastener due to load:

$$L_d = \frac{P}{n}$$

$$L_d = 250\ kN$$

Resultant load (solve using the sine and cosine rules) (see Figure 31.29 for angle of forces):

$$c = \frac{L_t \cdot 0.894426}{0.6877428}$$

$$c = 65.432\ kN$$

This is the maximum shear load acting on fastener 6.

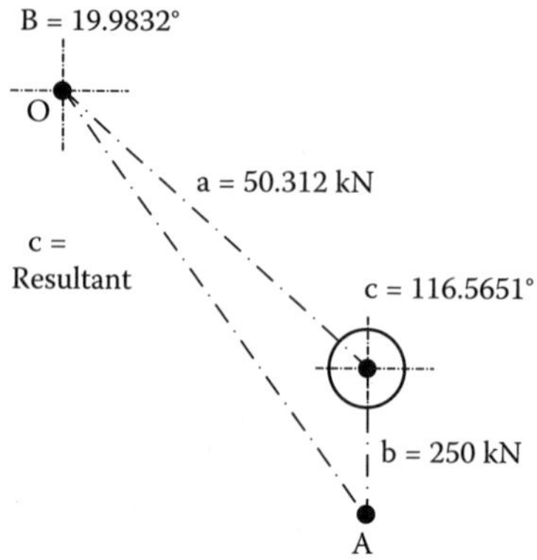

FIGURE 31.29
Angle of forces for Example 4.1.

31.21.1 Permissible Shear Stress

$$f_{fastener} = 400 \text{ MPa}$$

Diameter of fastener:

$$dia'(d) = \sqrt{\frac{4 \cdot f_s}{\pi \cdot f_{fastener}}}$$

$$d = \sqrt{\frac{4 \cdot 65.432 \text{ kN} / \text{mm}^2}{\pi \cdot 400 \text{ MPa}}}$$

$$= 14.432 \text{ mm}$$

Selecting a stock size fastener:

$$d = M16 \times 1.75 \text{ pitch}$$

31.22 Prying Forces

Example 31.2

Consider the bracket in Figure 31.30 where the applied force is in the plane of the connections. In this case the load produces a tension and shear force in the fasteners. (The fasteners have no preload applied in this example.)

Maximum stress

$$f_{tension} = \frac{F_t}{A}$$

$$f_{tension} = 61.38 \text{ MPa}$$

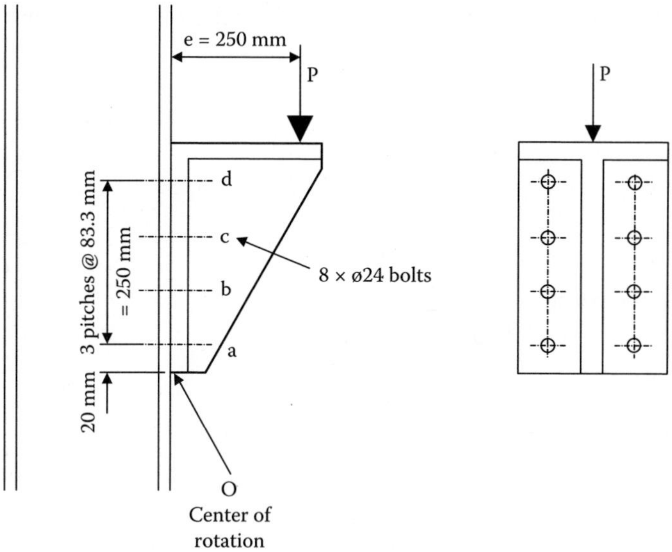

FIGURE 31.30
Bracket without preload.

$$f_{shear} = \frac{P}{n.A}$$

$$f_{shear} = 59.683 \text{ MPa}$$

$$P = 150 \text{ kN}$$
$$e = 250 \text{ mm(Moment arm)}$$
$$y = 83.3 \text{ mm pitch}$$
$$Y = 250 \text{ mm}$$
$$n = 8 \text{ bolts}$$
$$d = 20 \text{ mm}$$
$$A = \frac{\pi \cdot d^2}{4}$$
$$A = 314.16 \text{ mm}^2$$
$$P_1 = P \cdot e$$
$$P_1 = 37.5 \times 10^3 \text{ kN/mm}^2$$
$$Y_1 = \frac{\Sigma y^2}{Y}$$
$$Y_1 = 972.347 \text{ mm}$$

$$y_a = 20 \text{ mm} + y$$
$$y_b = 20 \text{ mm} + (2 \times y)$$
$$y_c = 20 \text{ mm} + (3 \times y)$$
$$y_d = 20 \text{ mm} + (4 \times y)$$
$$\Sigma y^2 = (y_a^2 + y_b^2 + y_c^2 + y_d^2)$$
$$\Sigma y^2 = 243.087 \times 10^3 \text{ mm}^2$$

The resultant stress on the top fastener is a vector quantity derived from the tensile stress $f_{tension}$ and the direct shear stress f_{shear}.

$$F_r = (f_{tension}{}^2 + f_{shear}{}^2)^{0.5}$$

$$F_r = 85.613 \text{ MPa}$$

Using fasteners to BS3692 strength grade 4.6, yield strength=235 kN/mm².

$$f_{allowable} = 235 \text{ MPa}$$

Factor of Safety (FoS)

$$FoS = \frac{f_{allowable}}{F_r}$$

$$FoS = 2.745$$

Tensile load on top fastener.

$$F_t = \frac{P \cdot e}{2 \cdot Y_1}$$

$$F_t = 19.283 \text{ kN}$$

Direct load

$$F_s = \frac{P}{n}$$

$$F_s = 18.75 \text{ kN}$$

Note that the above calculation is based on the fact that due to no preload on the fasteners the bracket will heel about the point of origin O. It is considered that if the prying load is high enough, plastic deformation will occur in the bracket at this point.

If this happens, then this will have the effect of reducing the loads, as the heel point will migrate upwards toward the first row of fasteners. Some argue that this is the point where the moment should be measured from. It is considered that the former position should be adopted, as the second position will have the effect of reducing the load on the fasteners.

In this second example of the above bracket (Figure 31.31), the fasteners have preload applied, so this will minimize any separation between the bracket and the supporting

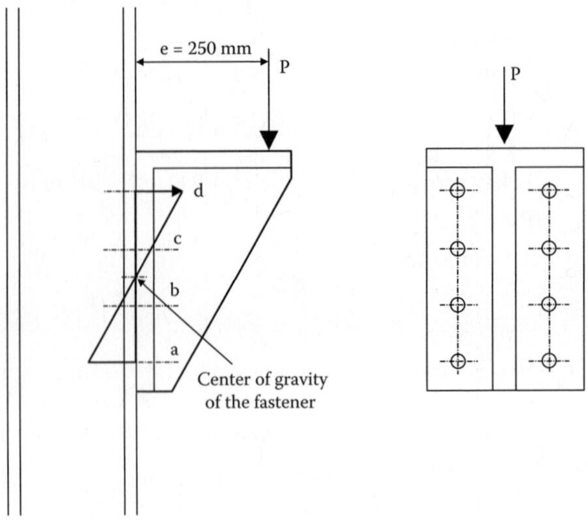

FIGURE 31.31
Bracket with preload.

steelwork. From Figure 31.30 it is clearly seen that the fastener deflections due to the applied moment are significantly reduced compared to Example 31.1.

31.23 Fasteners Subject to Alternating External Force

Consider a M12×1.75p grade 8.8 bolt used to clamp a fitting to a structural member.

The cross section of the part is (12×16 mm) = 192 mm^2 and is subject to an alternating load between 0 and 20,000 N.

Assume the stress concentration factor (SCF) for the thread is 3.0. Calculate the following:

1. The FoS for the bolt without any preload
2. The minimum preload required to prevent any loss of compression
3. The FoS for the bolt with a preload of 22,000 N
4. The minimum force in the part when the preload is 22,000 N

From BS 6104, Part 1, 1981, Table 6, the ultimate tensile load for this size and type of fastener is quoted as 67,400 N. The fastener core area is 84.267 mm^2.

The ultimate tensile stress (UTS) will be (67,400 N/84.267 mm^2) = 799.838 MPa. It is assumed that the yield strength is 500 MPa and the fatigue endurance limit for this material is 50% of the UTS = 400 MPa.

31.23.1 Factor of Safety (FoS) with No Preload

The factor of safety (FoS) is determined by multiplying the mean stress and the stress amplitude, multiplied by the SCF and plotted on a modified Goodman diagram.

The mean stress $$S_m = \frac{20000}{2 \times 84.3} = 118.6 \text{ MPa}$$

With no preload the entire load will be carried by the fastener.

For a load alternating between 0 and 200,000 N, the mean load is 10,000 N and the amplitude of the stress, multiplied by the SC (3.0), is (118.6 MPa × 3) = 355.8 MPa.

Plotting this on the diagram in Figure 31.32 shows that point A falls outside the safe zone, and therefore the fastener will be unsafe.

$$\frac{1}{\text{FoS}} = \left(\frac{\text{mean stress}}{S_y}\right) + \left(\frac{\text{stress amplitude} \times \text{scf}}{S_e}\right)$$

$$\frac{1}{\text{FoS}} = \left(\frac{118.6}{500}\right) + \left(\frac{118.6 \times 3.0}{400}\right)$$

hence:

$$\text{FoS} = 0.887$$

31.23.2 The Minimum Preload to Prevent Any Loss of Compression

For components of similar material and equal length, the spring constants are proportional to the cross-sectional areas of the part and fastener.

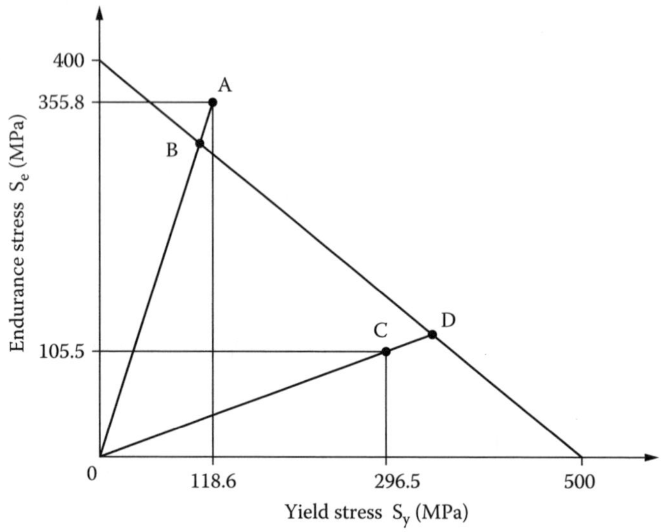

FIGURE 31.32
Modified Goodman diagram for Example 4.3.

Therefore, when the part has zero compression, the preload in the fastener is

$$P_{pre-load} = \frac{A_s \cdot P_{max}}{(A_s + A_p)}$$

$$= \frac{192.0 \text{ mm}^2 \times 20000 \text{ N}}{(84.267 \text{ mm}^2 + 192.0 \text{ mm}^2)}$$

$$= 13.90 \text{ kN}$$

31.23.3 Calculate the FoS for the Bolt with a Preload of 22,000 N

$$P_{b\,average} = \frac{P_{average} \times A_b}{(A_p + A_b)} + P_{pre-load}$$

$$P_{b\,average} = \frac{10000 \text{ N} \times 84.267 \text{ mm}^2}{192.0 \text{ mm}^2 + 84.267 \text{ mm}^2} + 22.00 \text{ kN}$$

$$= 25050.2 \text{ N}$$

$$\text{Stress}_{mean} = \frac{P_{b\,average}}{A_b}$$

$$= \frac{25050.2 \text{ N}}{84.267 \text{ mm}^2}$$

$$= 297.27 \text{ MPa}$$

$$P_{b\ amplitude} = \frac{A_b \times P_{b\ average}}{A_b + A_p}$$

$$= \frac{84.267\ mm^2 \times 10000\ N}{84.267\ N + 192.0\ N}$$

$$= 3.0502\ kN$$

$$Stress_{amplitude} = \frac{P_{b\ amplitude}}{A_b}$$

$$= \frac{3050.2\ N}{84.267\ mm^2}$$

$$= 36.197\ MPa$$

Point C (Figure 31.32) is inside the safe zone when plotted on the diagram; therefore, the bolt has a FoS when measured graphically from the diagram OD/OC:

$$\frac{1}{FoS} = \frac{OC}{OD} = \frac{314.29}{368.51}$$

$$\frac{1}{FoS} = 0.8529$$

$$FoS = 1.173$$

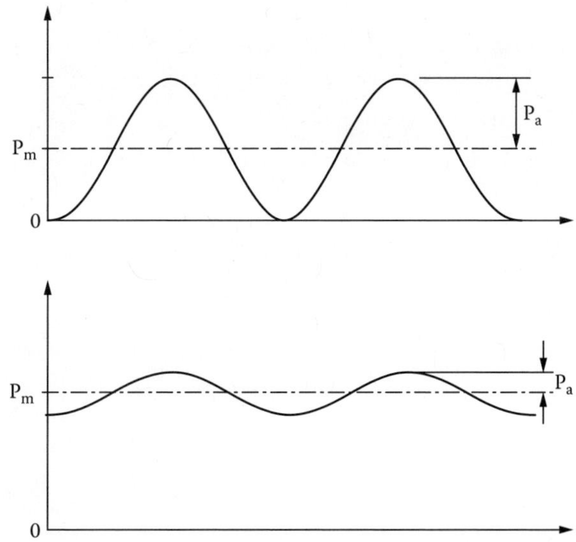

FIGURE 31.33
Effect of preload on the magnitude of the alternating force.

Algebraically:

$$\frac{1}{\text{FoS}} = \left(\frac{297.27\ \text{MPa}}{500.0\ \text{MPa}}\right) + \left(\frac{3.0 \times 36.197\ \text{MPa}}{400.0\ \text{MPa}}\right)$$

$$\text{FoS} = 1.155$$

The effect of the preload is to significantly reduce the magnitude of the alternating force and stress in the fastener (which is normally much more critical) while increasing the mean fastener force and stress (normally less critical). This is shown in the lower part of Figure 31.33.

31.23.4 The Minimum Force in the Part when the Preload Is 22,000 N

$$P_{\text{pre-load min}} = A_p \frac{P_{max}}{A_b + A_b} - P_{\text{pre-load}}$$

$$P_{\text{pre-load min}} = \left(\frac{192.0\ \text{mm}^2 \times 20000\ \text{N}}{192.0\ \text{mm}^2 + 84.267\ \text{mm}^2}\right) - 22.00\ \text{kN}$$

$$= -81.00\ \text{kN} \quad \text{Compression}$$

32

Limits and Fits

32.1 Introduction

The International Organization for Standardization (ISO) system (ISO 286-1988 and 2:1988) (on which the British Standards BS EN 20286-1993 and 2:1993 are based) provides for a wide range of hole and shaft tolerances to cater for a wide range of conditions.

However, it has been found for normal manufacturing processes that the range can be condensed to a small number of tolerance combinations.

The most commonly applied hole and shaft tolerances include:

Selected hole tolerances: H7, H8, H9, and H11

Selected shaft tolerances: c11, d10, e9, f7, g6, h6, k6, n6 p6, s6 and u6

The general standard adopted is the basic hole; the measurement of holes is generally controlled using plug gauges with "go no go" features, whereas a shaft is easily checked using micrometers or gap gauges.

In some instances a manufacturer may choose to use a shaft-based system, particularly where a single shaft may have to accommodate a variety of accessories, such as bearings, couplings or collars, etc.

Shaft-based systems may offer economies particularly where bar stock material may be available to standard shaft tolerances to the ISO system. For the nomenclature used in limits and fits, see Figure 32.1.

32.2 Tolerance Grade Numbers

32.2.1 Tolerance

Tolerance is the difference between the maximum and minimum size limits of a component.

32.2.2 International Tolerance Grade Numbers

Tolerance grade numbers are used to specify the size of a tolerance zone.

In the British Standard, the tolerance is the same for both the internal (hole) and external (shaft) parts having the same tolerance grade numbers.

Tolerance grades IT0 through IT16 are covered in the National and International standards.

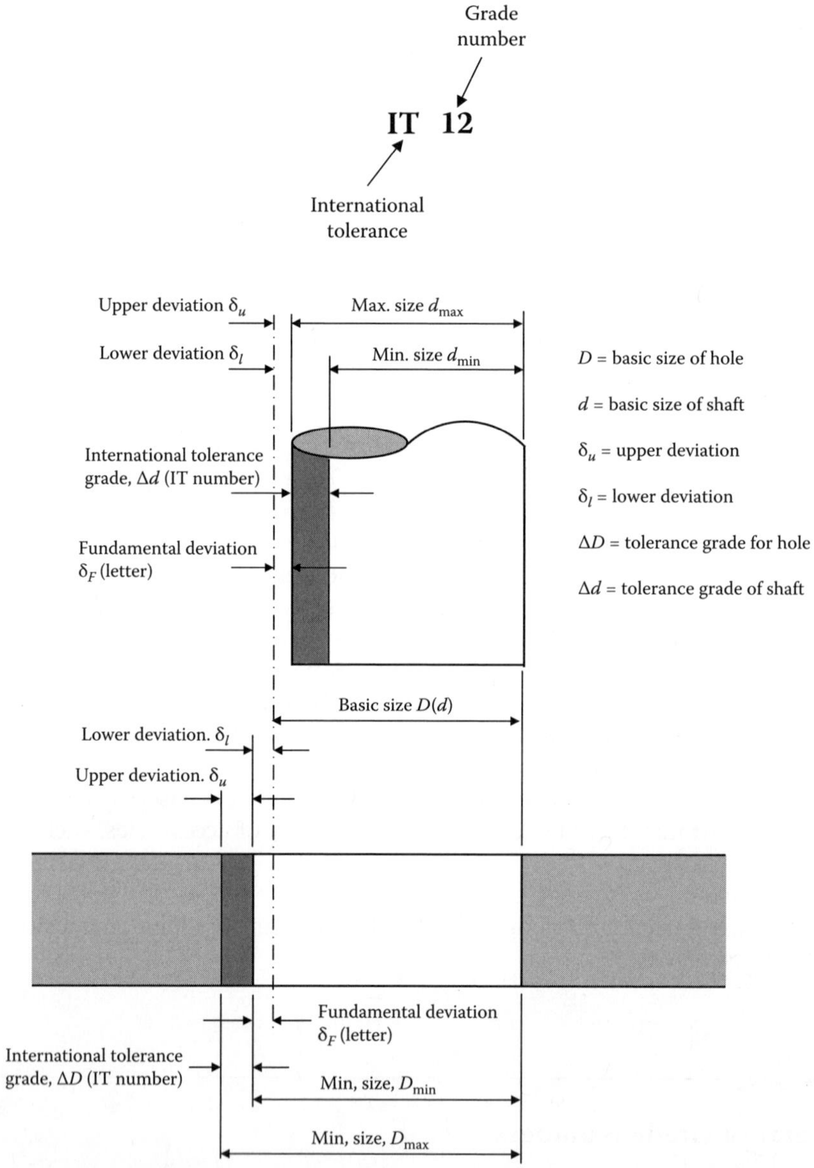

FIGURE 32.1
Nomenclature used in limits and fits.

32.3 Fundamental Deviations

32.3.1 Preamble

Consider a 25.000 mm nominal diameter hole and shaft.

An example of a fit specification:

25H7 hole

25g6 shaft

Note:

- All hole deviations are specified in uppercase.
- All shaft deviations are specified in lowercase.

32.3.2 Fundamental Deviation

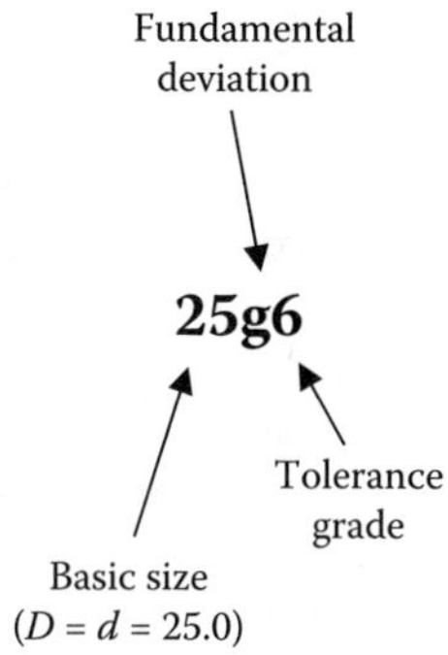

From the tolerance grade table for a basic size of 25.000 mm diameter, the variation between the upper and lower deviations for a grade IT6 will be 0.013 mm (see Table 32.1, 25.000 mm), and for grade IT7 will be 0.021 mm.

From Table 32.2, for ø25.000 g the upper deviation is −0.007 mm. The lower deviation will be −0.007 to 0.013 mm = 0.020 mm.

Hence the tolerance on a 25.000 mm diameter shaft machined to g6 is

$$ø25.000\text{ mm} - 0.007/-0.020\text{ mm}$$

TABLE 32.1

Tolerance Grades

	Tolerance Grades							
Basic Sizes	**IT5**	**IT6**	**IT7**	**IT8**	**IT9**	**IT10**	**IT11**	**IT12**
0–3	0.004	0.006	0.010	0.014	0.025	0.040	0.060	0.100
3–6	0.005	0.008	0.012	0.018	0.030	0.048	0.075	0.120
6–10	0.006	0.009	0.015	0.022	0.036	0.058	0.090	0.150
10–18	0.008	0.011	0.018	0.027	0.043	0.070	0.110	0.180
18–30	0.009	0.013	0.021	0.033	0.052	0.084	0.130	0.210
30–50	0.011	0.016	0.025	0.039	0.062	0.100	0.160	0.250
50–80	0.013	0.019	0.030	0.046	0.074	0.120	0.190	0.300
80–120	0.015	0.022	0.035	0.054	0.087	0.140	0.220	0.350
120–180	0.018	0.025	0.040	0.063	0.100	0.160	0.250	0.400
180–250	0.020	0.029	0.046	0.072	0.115	0.185	0.290	0.460
250–315	0.022	0.032	0.052	0.081	0.130	0.210	0.320	0.520
315–400	0.025	0.036	0.057	0.089	0.140	0.230	0.360	0.570
400–500	0.027	0.040	0.063	0.097	0.155	0.250	0.400	0.630

Source: BS EN 20286-2:1993.
Note: Values in mm.

32.3.3 Fundamental Deviations for Shafts

TABLE 32.2
Fundamental Deviation (Shaft)

Basic Size	Upper Deviation Letter					Lower Deviation Letter				
	c	d	f	g	h	k	n	p	s	u
0–3	–0.060	–0.020	–0.006	–0.002	0	0	0.004	0.006	0.014	0.018
3–6	–0.070	–0.030	–0.010	–0.004	0	0.001	0.008	0.012	0.019	0.023
6–10	–0.080	–0.040	–0.013	–0.005	0	0.001	0.010	0.015	0.023	0.028
10–14	–0.095	–0.050	–0.016	–0.006	0	0.001	0.012	0.018	0.028	0.033
14–18	–0.095	–0.050	–0.016	–0.006	0	0.001	0.012	0.018	0.028	0.033
18–24	–0.110	–0.065	–0.020	–0.007	0	0.002	0.015	0.022	0.035	0.041
24–30	–0.110	–0.065	–0.020	–0.007	0	0.002	0.015	0.022	0.035	0.048
30–40	–0.120	–0.080	–0.025	–0.009	0	0.002	0.017	0.026	0.043	0.060
40–50	–0.130	–0.080	–0.025	–0.009	0	0.002	0.017	0.026	0.043	0.070
50–65	–0.140	–0.100	–0.030	–0.010	0	0.002	0.020	0.032	0.053	0.087
65–80	–0.150	–0.100	–0.030	–0.010	0	0.002	0.020	0.032	0.059	0.102
80–100	–0.170	–0.120	–0.036	–0.012	0	0.003	0.023	0.037	0.071	0.124
100–120	–0.180	–0.120	–0.036	–0.012	0	0.003	0.023	0.037	0.079	0.144
120–140	–0.200	–0.145	–0.043	–0.014	0	0.003	0.027	0.043	0.092	0.170
140–160	–0.210	–0.145	–0.043	–0.014	0	0.003	0.027	0.043	0.100	0.190
160–180	–0.230	–0.145	–0.043	–0.014	0	0.003	0.027	0.043	0.108	0.210
180–200	–0.240	–0.170	–0.050	–0.015	0	0.004	0.031	0.050	0.122	0.236
200–225	–0.260	–0.170	–0.050	–0.015	0	0.004	0.031	0.050	0.130	0.258
225–250	–0.280	–0.170	–0.050	–0.015	0	0.004	0.031	0.050	0.140	0.284
250–280	–0.300	–0.190	–0.056	–0.017	0	0.004	0.034	0.056	0.158	0.315
280–315	–0.330	–0.190	–0.056	–0.017	0	0.004	0.034	0.056	0.170	0.350
315–355	–0.360	–0.210	–0.062	–0.018	0	0.004	0.037	0.062	0.190	0.390
355–400	–0.400	–0.210	–0.062	–0.018	0	0.004	0.037	0.062	0.208	0.435

Source: BS EN20286-2:1993.
Note: Values in mm.

32.3.4 Fundamental Deviations for Holes

Table 32.3 gives the lower deviation for the basic size of holes.

TABLE 32.3
Fundamental Deviation (Hole)

Basic Size	Lower Deviation
	H
0–3	0.000
3–6	0.000
6–10	0.000

(Continued)

TABLE 32.3 (*Continued*)

Fundamental Deviation (Hole)

Basic Size	Lower Deviation
	H
10–14	0.000
14–18	0.000
18–24	0.000
24–30	0.000
30–40	0.000
40–50	0.000
50–65	0.000
65–80	0.000
80–100	0.000
100–120	0.000
120–140	0.000
140–160	0.000
160–180	0.000
180–200	0.000
200–225	0.000
225–250	0.000
250–280	0.000
280–315	0.000
315–355	0.000
355–400	0.000

Note: Values in mm

32.3.5 Upper and Lower Deviations

32.3.5.1 Shaft Letter Codes c, d, f, g, and h

Upper deviation is equivalent to the fundamental deviation.

Lower deviation is equivalent to:

$$\text{Upper deviation} - \text{tolerance grade}$$

32.3.5.2 Shaft Letter Codes k, n, p, s, and u

Lower deviation is equivalent to the fundamental deviation.

Upper deviation is equivalent to:

$$\text{Lower deviation} + \text{tolerance grade}$$

32.3.5.3 Hole Letter Code H

$$\text{Lower deviation} = 0$$

$$\text{Upper deviation} = \text{tolerance grade}$$

32.4 Preferred Fits Using the Basic Hole System

32.4.1 Loose Running Fit (Example)

Determine the loose running fit tolerance for a hole and shaft that has a basic diameter of 25.000 mm.

From Table 32.4, the specification for a loose running fit is 25H11/25c11, as detailed in Table 32.5.

TABLE 32.4

Preferred Fits Using the Basic Hole

Type of Fit	Description	Symbol
Clearance	**Loose running fit:** For wide commercial tolerances or allowances on external members.	H11/c11
	Free running fit: Not for use where accuracy is essential, but good for large temperature variations, high running speeds, or heavy journal pressures.	H9/d9
	Close running fit: For running on accurate machines and for accurate location at moderate speeds and journal pressures.	H8/f7
	Sliding fit: Where parts are not intended to run freely, but must move and turn freely and locate accurately.	H7/g6
	Locational clearance fit: Provides snug fit for location of stationary parts, but can be freely assembled and disassembled.	H7/h6
Transition	**Locational transition fit:** For accurate location, a compromise between clearance and interference.	H7/k6
	Locational transition fit: For more accurate location where greater interference is permissible.	H7/n6
Interference	**Locational interference fit:** For parts requiring rigidity and alignment with prime accuracy of location but without special bore pressure requirements.	H7/p6
	Medium drive fit: For ordinary steel parts or shrink fits on light sections, the tighter fit usable with cast iron.	H7/s6
	Force fit: Sui for parts that can be highly stressed or for shrink fits where the heavy pressing force required is impractical.	H7/u6

TABLE 32.5

Hole and Shaft Sizes for Loose Running Fit (H11/c11)

	Hole	Shaft
Tolerance grade	0.130 mm	0.130 mm
Upper deviation	+0.130 mm	–0.110 mm
Lower deviation	0	–0.240 mm
Maximum diameter	25.130 mm	24.890 mm
Minimum diameter	25.000 mm	24.760 mm
Average diameter	25.065 mm	24.825 mm
Maximum clearance	$C_{max} = D_{max} - d_{min} = (25.130 - 24.760) = 0.370$ mm	
Minimum clearance	$C_{min} = D_{min} - d_{max} = (25.000 - 24.890) = 0.110$ mm	

There are two forms of showing the tolerances on a drawing:

The first method:

Hole	Shaft
Ø25.000 mm + 0.130 / –0 mm	Ø25.000 mm – 0.110 / –0.240 mm

The second method:

Hole	Shaft
Ø25.000 mm	Ø24.825 mm
Ø25.130 mm	Ø24.760 mm

This second method is preferred from a machining point of view. It gives the machinist a specific dimension to aim for; if the mean dimension is calculated and if he or she misses it there is still a small margin to recover the machined feature and still remain within the tolerance zone. This method is sometimes referred to as maximum metal conditions.

32.4.2 Location Clearance Fit (Example)

Determine the location clearance fit tolerance for a hole and shaft that has a basic diameter of 25.000 mm.

From Table 32.4, the specification for a location clearance fit is 25H7/25h6, as detailed in Table 32.6.

As with the example in Section 32.4.1, there are two methods for showing the tolerances.

The first method:

Hole	Shaft
Ø25.000 mm + 0.021 / –0 mm	Ø25.000 mm + 0 / +0.013 mm

The second method:

Hole	Shaft
Ø25.000 mm	Ø24.987 mm
Ø25.021 mm	Ø25.000 mm

TABLE 32.6

Hole and Shaft Sizes for Location Clearance Fit (H7/h6)

	Hole	Shaft
Tolerance grade	0.021 mm	0.013 mm
Upper deviation	0.021 mm	0.000 mm
Lower deviation	0.000 mm	–0.013 mm
Maximum diameter	25.021 mm	25.000 mm
Minimum diameter	25.000 mm	24.987 mm
Average diameter	25.011 mm	24.994 mm
Maximum clearance	$C_{max} = D_{max} - d_{min} = (25.021 - 24.987) = 0.034$ mm	
Minimum clearance	$C_{min} = D_{min} - d_{max} = (25.000 - 25.000) = 0.000$ mm	

32.5 Surface Finish

Any discussion about limits and fits invariably brings up the question of surface finish. It is impractical to apply a close tolerance such as 25g6 (0.013) to a machined part that has a surface roughness in the order of, say, 12.5 microns.

Table 32.7 gives an indication of the surface roughness obtained by the various manufacturing processes.

TABLE 32.7

Surface Roughness Values Obtainable by Standard Manufacturing Processes

	Manufacturing Process	Obtainable with Difficulty	Normally Obtainable	Roughing
Casting	Sand casting	0.8–1.6	6.3–12.5	
	Permanent mold		1.6–6.3	12.5–25
	Die casting		0.8–3.2	
Manual	Forging	1.6–3.2	3.2–25	
	Extrusion	0.4–0.8	0.8–6.3	
	Rolling	0.4–0.8	0.8–3.2	
Surface process	Flame cut		25–50	
	Hack saw		6.3–50	
	Bandsaw, chipping		3.2–50	
	Filing	0.8–1.6	1.6–12.5	
	Emery polish	0.1–0.4	0.4–1.6	1.6–3.2
Machining	Shell milling	1.6–3.2	3.2–25	25–50
	Drilling	3.2–6.3	6.3–25	
	Planning and shaping		1.6–12.5	
	Face milling	0.8–1.6	1.6–12.5	12.5–50
	Turning	0.2–1.6	1.6–6.3	6.3–50
	Boring	0.2–1.6	1.6–6.3	6.3–50
	Reaming	0.4–0.8	0.8–6.3	6.3–12.5
	Cylindrical grinding	0.025–0.4	0.4–3.2	3.2–6.3
	Centerless grinding	0.05–0.4	0.4–3.2	
	Internal grinding	0.024–0.4	0.4–3.2	3.2–6.3
	Surface grinding	0.025–0.4	0.4–3.2	3.2–6.3
	Broaching	0.2–0.8	0.8–3.2	3.2–6.3
	Super finishing	0.025–0.1	0.1–0.4	
	Honing	0.025–0.1	0.1–0.4	
	Lapping	0.006–0.05	0.05–0.4	
Gear Manufacturing	Milling, spiral bevel	1.6–3.2	3.2–12.5	12.5–25
	Milling, with form cutter	1.6–3.2	3.2–12.5	12.5–50
	Hobbing	0.8–3.2	3.2–12.5	12.5–50
	Shaping	0.4–1.6	1.6–12.5	12.5–250
	Planning	0.4–1.6	1.6–18.5	12.5–50
	Shaving	0.4–0.8	0.8–3.2	
	Grinding (crisscross)	0.4–0.8	0.8–1.6	
	Grinding	0.1–0.4	0.4–0.8	
	Lapping	0.05–0.2	0.2–0.8	

33

Energy Systems (Flywheels)

33.1 Flywheels Basics

When a circular mass that is at rest is rotated, it gains kinetic energy. Slowing the mass down or bringing it to rest is achieved by removing kinetic energy from the mass. A rotating mass that is designed to store kinetic energy as a result of its rotational motion is called a *flywheel*.

An energy storage system that involves the use of a flywheel is called a "flywheel energy storage (FES) system."

FES systems are used in internal combustion engines to even out the fluctuating power strokes of the engine, and in shearing presses to provide an energy intensification during a cutting stroke. They are also found in a number of high-tech energy conservation systems used on passenger road vehicles as a means of storing energy recovered from braking. A flywheel is designed to have the majority of its mass concentrated in the rim to store the maximum kinetic energy. The downside is that the stresses in the rim of the flywheel will be increased, and if the flywheel exceeds its critical safe speed, then a catastrophic failure will result.

Basic energy formulae:

Consider the simple flywheel depicted in Figure 33.1.

$$\text{Rotational kinetic energy} = \tfrac{1}{2} I \omega^2 \tag{33.1}$$

$$\text{Moment of inertia} = mk^2 \tag{33.2}$$

$$\text{Change in energy} = \tfrac{1}{2} I\left(\omega_1^2 - \omega_2^2\right) \tag{33.3}$$

$$\text{Torque} = I\alpha \tag{33.4}$$

$$\text{Power} = T\omega \tag{33.5}$$

where:

I = moment of inertia (kg m^2) about the axis of rotation
m = mass of flywheel (kg)
k = radius of gyration (m) about the axis of rotation
T = torque applied (Nm)
α = angular acceleration (rad/s^2)

Table 33.1 gives some basic equations used in the design of flywheel systems.

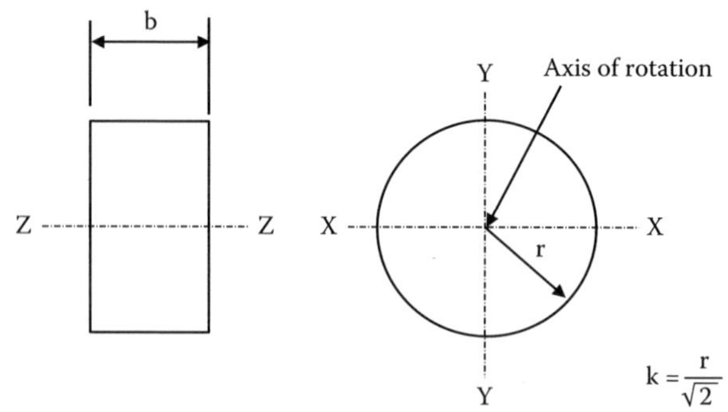

FIGURE 33.1
Simple flywheel.

TABLE 33.1
Basic Equations for Flywheel Systems

Case	Description	Symbol	Equation	Units
1	Angular velocity	ω	$= \frac{2\pi n}{60}$	$\left(\frac{rad}{s}\right)$
2	Angular acceleration	α	$= \frac{(\omega_1 - \omega_2)}{t}$	$\left(\frac{rad}{s^2}\right)$
3	Angular acceleration	α	$= \frac{T}{I_z}$	$\left(\frac{rad}{s^2}\right)$
4	Energy stored	E	$= \frac{I_z \cdot w^2}{2}$	N·m (Joules)
5	Change in kinetic energy when angular velocity changes from ω_1 to ω_2	$E_2 - E_1$	$= \frac{I_z\left(w_2^2 - w_1^2\right)}{2}$	N·m (Joules)
Assuming a flywheel is accelerated by application of a driving torque T_1 between angles θ_1 and θ_2 and decelerated by resisting its motion with a torque T_o between angles θ_3 and θ_4.				
6	The energy input to the flywheel (work input) between angles θ_1 and θ_2	U_1	$= T_1(\theta_2 - \theta_1)$	
7	The energy from a flywheel (work output) between angles θ_1 and θ_2	U_o	$= T_o(\theta_4 - \theta_3)$	

33.2 Torque—Angle Diagrams

As an example, a typical four-stroke engine will exhibit a variation of torque against a crank angle as shown in Figure 33.2

The complete cycle of the engine will take two complete revolutions, which divides into four equal strokes:

1. Induction
2. Compression

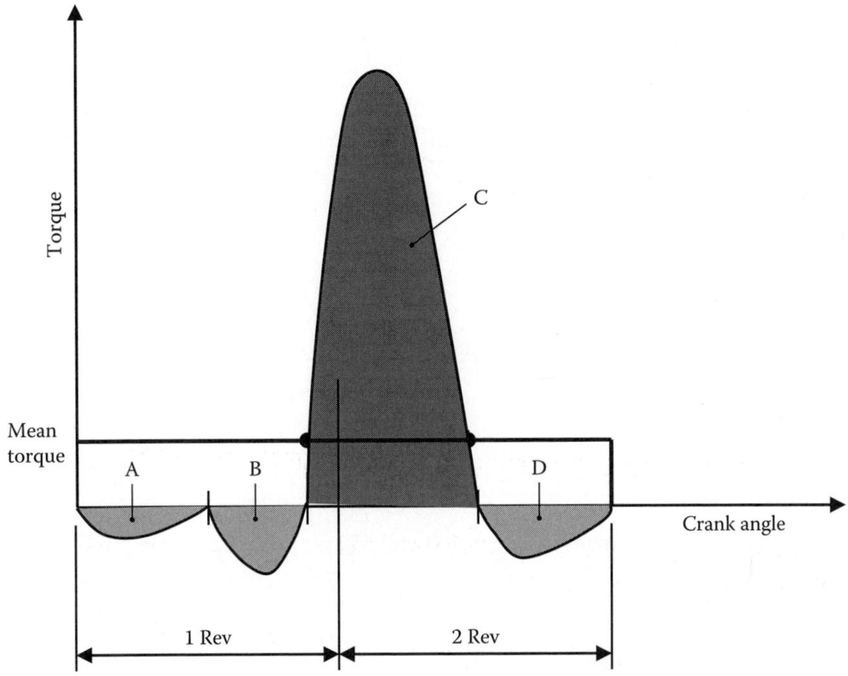

FIGURE 33.2
Four-stroke cycle.

3. Ignition
4. Exhaust

One complete revolution of the crankshaft is 2π radians, hence it will take 4π radians to complete the cycle. In Figure 33.2, the area C represents the energy generated (the power stroke). Areas A, B, and D represent the induction, compression, and exhaust strokes, respectively, and these require negative torque, as energy is required to complete this operation.

For the engine to be effective, area C must be larger than the other areas to give a positive mean torque. By definition, *mean torque* requires the areas to be equal both above and below the mean line. The area of a rectangle that is formed between the mean line and the zero line will be the work output available of the engine per work cycle.

The total work produced will be angle × torque:

$$W = T_m \times 4\pi = \text{net area } (A_{net}) \tag{33.6}$$

where

$$T_m = \frac{A_{net}}{4\pi}$$

Beware that the angle 4π is only applicable to four-stroke engines. A two-stroke engine will have an angle of 2π.

33.3 Coefficient of Speed Fluctuation (C_s)

The difference between the maximum and minimum speeds during a cycle is called the maximum fluctuation of speed. The ratio of the maximum fluctuation of speed is called the coefficient of speed fluctuation.

$$C_s = \left(\frac{\omega_2 - \omega_1}{\omega} \right) \tag{33.7}$$

where:

ω_1 = smallest angular velocity
ω_2 = largest angular velocity
ω = mean angular velocity

Example 33.1

The speed of a shaft fluctuates between 400 and 550 rev/min. Calculate the C_s:

Solution

$$\omega_1 = 2\pi N_1 = 2\pi \left(\frac{400}{60} \right) = 41.89 \text{ rad / s}$$

$$\omega_2 = 2\pi N_2 = 2\pi \left(\frac{550}{60} \right) = 57.60 \text{ rad / s}$$

$$\omega = \left(\frac{\omega_2 + \omega_1}{2} \right) = 49.75 \text{ rad / s}$$

$$C_s = \left(\frac{\omega_2 - \omega_1}{\omega} \right) = 0.316$$

In this instance, it would have been possible to use the speed in rev/min to calculate C_s:

$$\text{Mean speed} = N_m = \left(\frac{400 + 550}{2} \right) = 475 \text{ rev / min}$$

$$C_s = \left(\frac{N_2 - N_1}{N_m} \right) = \left(\frac{150}{475} \right) = 0.316$$

33.4 Coefficient of Energy Fluctuation (C_f)

To determine the size of a flywheel (I_z) necessary to ensure that the speed is maintained within a specified range (i.e., the angular velocity be limited to a range ω_{min} to ω_{max} when the kinetic energy level varies between E_{min} and E_{max}):

$$C_f = \frac{\text{Largest fluctuation in kinetic energy}}{\text{Work done}} \tag{33.8}$$

$$C_f = \frac{\frac{I\omega_2^2}{2} - \frac{I\omega_1^2}{2}}{W} = \frac{1}{2}\left(\omega_2^2 - \omega_1^2\right) \tag{33.9}$$

Factorizing the parentheses,

$$C_f = \frac{1}{2W}\left(\omega_2 + \omega_1\right)\left(\omega_2 - \omega_1\right) \tag{33.10}$$

Now,

$$\left(\omega_2 + \omega_1\right) = 2\omega \text{ and } \left(\omega_2 - \omega_1\right) = C_s \cdot \omega \tag{33.11}$$

Then,

$$C_f = \frac{1}{2W}(2\omega)(\omega \cdot C_s) \tag{33.12}$$

$$C_f = \frac{1}{W}\omega^2 C_s \tag{33.13}$$

Rearranging Equation 33.13 making I the subject:

$$I = \frac{C_f \cdot W}{\omega^2 C_s} \tag{33.14}$$

Reconsidering the cycle depicted in Figure 33.2, the points A and B indicate where the diagram exceeds the mean torque. For the purposes of discussion, it may be considered convenient to redraw the diagram with point A at the start of the cycle, as shown in Figure 33.3.

As stated, by definition, *mean torque* requires the areas to be equal both above and below the mean line. Hence, area A_1 must equal area A_2.

Now, let the energy at point A in Figure 33.3 be E_A.

At point B, the energy will have increased by an amount equal to A_1.

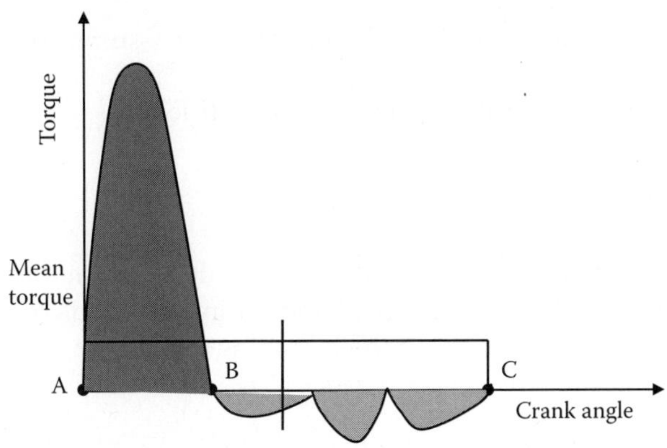

FIGURE 33.3
Modified four-stroke cycle.

The energy at point B will be

$$E_B = E_A + A_1 \tag{33.15}$$

At point C, the energy will have reduced by an amount equal to A_1.

The energy at point C will be

$$E_C = E_B - A_2 \tag{33.16}$$

$$= E_A + A_1 - A_2 \tag{33.17}$$

$$= E_A \tag{33.18}$$

On completion of one cycle, the energy must be returned to the starting value; therefore, points A and C are the same point.

From the figures, the maximum fluctuation in energy can be deduced. In this case, the maximum energy will be at point B and the minimum energy will occur at points A and C, being the same point.

Hence, the fluctuation will be

$$E_A - E_B = A_1 \tag{33.19}$$

Therefore, in this case,

$$C_f \cdot W = A_1 \tag{33.20}$$

33.5 Stresses Generated within Flywheel Section due to Speed of Rotation

A flywheel section can be simplified to that of a rotating ring, enabling the determination of the stresses due to the rotation. The theory involving the stresses in a thick-walled cylinder may be used; the primary difference will be due to the inertial forces acting on the ring section.

The following conditions will apply in using the above assumption:

- The outside radius of the ring is greater than its thickness.
- The section is constant.
- The stresses are constant across its width.

The tangential and radial stresses at radius r resulting from the rotation of a ring (see Figure 33.4) with an outside radius of r_2 and an inside radius of r_1 and rotating at an angular velocity (ω) will be found using the following equations:

Tangential tensile inertial stress:

$$\sigma_t = \frac{\rho\omega^2}{8}\left[(3+\nu)\left(r_1^2 + r_2^2 + \frac{r_1^2 \cdot r_2^2}{r^2}\right) - (1+3\nu)r^2\right] \tag{33.21}$$

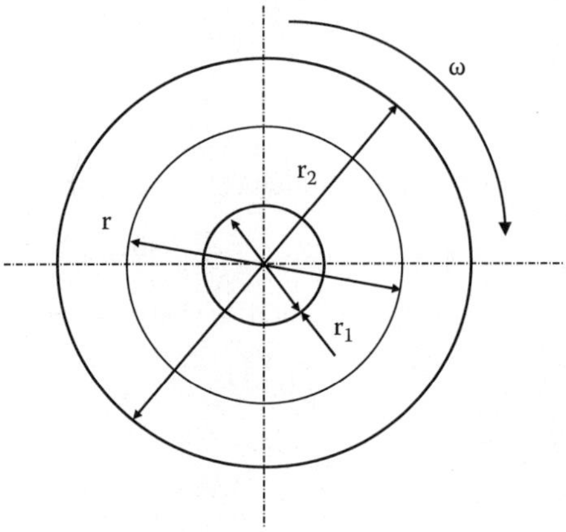

FIGURE 33.4
Tangential and radial stresses in a flywheel.

Radial tensile inertial stress:

$$\sigma_r = \rho\omega^2\left(\frac{(3+\nu)}{8}\right)\left(r_1^2 + r_2^2 - \frac{r_1^2 \cdot r_2^2}{r^2} - r^2\right) \quad (33.22)$$

The maximum tangential stress occurs in the bore and is equal to

$$\text{Max } \sigma_t = \frac{\rho\omega^2}{4}\left[(1-\nu)r_1^2 + (3+\nu)r_2^2\right] \quad (33.23)$$

The maximum radial stress will occur at a radius $r = \sqrt{(r_1 \cdot r_o)}$ and is equal to

$$\text{Max } \sigma_r = \frac{\rho\omega^2}{8}(3+\nu)(r_2 - r_1)^2 \quad (33.24)$$

Example 33.2

A single-cylinder engine working on the four-stroke cycle develops 9 kW at 5.8 rev/s. Find a suitable size of a cast iron flywheel if the energy absorbed is to be 0.8 of the energy developed per cycle.

Take the speed fluctuation as 0.5% above and below the mean. The average rim stress should not exceed 7 MPa.

Solution

$$\text{The density of cast iron} = 7200 \text{ kg/m}^3$$

$$\text{Work done per rev} = \frac{\text{Power}}{\text{Working cycles/s}}$$

$$\text{Four-stroke working cycles per second} = \frac{5.8 \text{ rev/s}}{2}$$

$$\text{Work done per working cycle} = \frac{9000\ \text{W}}{5.8/2\ \text{per s}}$$

$$= 3103.44\ \text{Nm}$$

$$\text{Energy to be stored in flywheel} = 0.8 \times 3103.44\ \text{Nm}$$

$$= 2482.76\ \text{Nm}$$

To find the permissible rim velocity to generate a stress of 7 MPa,

$$\sigma = \rho\omega^2 r^2$$

where
- σ = centrifugal stress (N/m²)
- ρ = density of material (kg/m³)
- ω = angular velocity (rad/s)
- r = radius of wheel (m)

$$7 \times 10^6 \frac{\text{N}}{\text{m}^2} = 7200 \frac{\text{kg}}{\text{m}^3} \times 36.442 \frac{\text{rad}^2}{\text{s}^2} \times \text{r}^2\text{m}^2$$

$$\frac{7 \times 10^6}{7200 \times 36.442^2} = \text{r}^2 \quad \left(\text{where} \frac{\text{kg} \times \text{m}}{\text{s}^2}\right) = \text{N}$$

$$0.7321 = \text{r}^2 \quad \text{and} \quad \omega = \text{rev}/\text{s} \times 2\pi$$

$$\text{r} = 0.8556\ \text{m}\ (855.6\ \text{mm})$$

$$\text{Diameter of flywheel} = 1.71\ \text{m}$$

$$\text{Change in energy} = \frac{\text{I}}{2}\left(\omega_1^2 - \omega_2^2\right)$$

$$2482.76\ \text{Nm} = \frac{\text{I}}{2} \times \left(36.6242^2 - 36.2598^2\right)$$

So

$$0.5\%\ \text{up on}\ 36.442\ \text{rad/s} = 36.6242\ \text{rad/s}$$

$$0.5\%\ \text{down on}\ 36.442\ \text{rad/s} = 36.2598\ \text{rad/s}$$

$$2482.76\ \text{Nm} = \frac{\text{I}}{2}(36.6242 + 36.2598) \times (36.6242 - 36.2598)$$

$$2482.76\ \text{Nm} = \frac{\text{I}}{2}(26.559)$$

$$\frac{2 \times 2482.76}{26.559} = \text{I}$$

$$186.962\ \text{Nm s}^2 = \text{I}$$

$$\text{or } 186.962\ \text{kg m}^2 = \text{I}$$

Now,

$$I = Mk^2$$

where $k = \frac{D}{2}$ for a solid disc.

Hence, $k = \frac{1.71}{2}$

$$k = 0.855 \text{ m}$$

$$186.962 \text{ kg m}^2 = M \times 0.855^2 \text{m}^2$$

Therefore, M (mass) = 255.75 kg

$$\text{Now, volume} = \frac{\text{mass}}{\text{density}}$$

$$= \frac{255.75 \text{ kg}}{7200 \text{ kg/m}^3}$$

$$= 0.0355 \text{ m}^3$$

$$\text{Mean circumference} = \pi D$$

$$= 1.71\,\pi$$

$$= 5.372 \text{ m}$$

$$\text{Cross-sectional area of rim} = \frac{0.0355 \text{ m}^3}{5.372 \text{ m}}$$

$$= 0.0066 \text{ m}^2$$

$$\text{Area} = b \times t \text{ (breadth} \times \text{thickness)}$$

Some proportion should be considered between the breadth and the thickness:

$$\text{say } b = 1.2\,t$$

$$0.0066\text{m}^2 = 1.2\,t^2$$

$$0.0055 = t^2$$

$$t = 0.0742 \text{ m}$$

$$t = 74.16 \text{ mm}$$

$$\text{and } b = 1.2\,t$$

$$b = 89.0 \text{ mm}$$

Example 33.3

A motor generator set for a rolling mill, running at 8.5 rev/s, has a cast-steel flywheel designed to take up the fluctuations of demand. The rotating parts of the motor and generator have a combined mass of 20 tonnes and a radius of gyration of 460 mm; the flywheel has a mass of 50 tonnes and a radius of gyration of 1.6 m. If an excess of energy of 11 MNm is required when the mill is running over that supplied by the motor, calculate the decrease in speed.

Solution

Total moment of inertia:

$$\begin{aligned} I &= mk^2 \text{ for motor and generator} + mk^2 \text{ for flywheel} \\ &= 20\times10^3 \text{ kg}\times0.460^2 \text{ m}^2 + 50\times10^3\text{kg}\times1.6^2 \text{ m}^2 \\ &= 4232 \text{ kg m}^2 + 128{,}000 \text{ kg m}^2 \\ &= 132{,}232 \text{ kg m}^2 \end{aligned}$$

$$\text{Change in energy} = \frac{I}{2}\omega_1^2 - \frac{I}{2}\omega_2^2$$

If R = the new speed (rev/s),

$$\text{then} \quad 11\times10^6 \text{ Nm} = \frac{1}{2}\times132{,}232\times\left[8.5^2\times(2\pi)^2 - R^2(2\pi)^2\right]$$

$$11\times10^6 \text{ Nm} = \frac{1}{2}\times132{,}232\times4\pi^2\left(8.5^2 - R^2\right)$$

$$4.213 = 8.5^2 - R^2$$

$$R^2 = 72.25 - 4.213$$

$$R = 8.248 \text{ rev/s}$$

$$\begin{aligned} \text{Decrease in speed} &= 8.5 - 8.248 \text{ rev/s} \\ &= 0.252 \text{ rev/s} \end{aligned}$$

Example 33.4

A flywheel has a mass of 2.03 tonnes and a radius of gyration of 1.22 m. It is keyed to the crankshaft of an engine that develops 15 kW running at 2.5 rev/s. Assuming the mean torque is constant at all speeds, and neglecting all masses but the flywheel, find the time required to get the engine to full speed.

Solution

Average torque:

$$\begin{aligned} T &= \frac{\text{power}}{2\pi N} \\ &= \frac{15\times10^3}{2\pi\times2.5}\frac{\text{Nm}}{\text{s}}\times\frac{\text{s}}{\text{rev}} \\ &= 954.93 \text{ Nm} \end{aligned}$$

$$T = I\alpha \text{ where } \alpha = \omega/t \text{ and}$$

$$I = mk^2$$

$$= 2.03 \times 10^3 \times 1.22^2 \text{ kg m}^2$$

$$= 3021.45 \text{ kg m}^2$$

$$T = I\alpha$$

$$954.93 \text{ Nm} = 3021.45 \text{ kg m}^2 \times \frac{2.5 \times 2\pi}{t} \frac{\text{kg m}^2}{\text{s}}$$

$$t = 49.72 \text{ s}$$

Example 33.5

A cast-iron flywheel for a shearing machine has a rim section of 240 mm wide. The outside diameter is 1.7 m and the inside diameter is 1.22 m. The flywheel has a v-belt drive from an electric motor running at 10 rev/s. The motor pulley is 480 mm diameter. The energy absorbed during the effective cutting stroke of 75 mm is 80% of the energy supplied. When cutting a section, the speed of the flywheel is reduced by 0.25 rev/s. Determine the average cutting force exerted at the blade. Taking the shearing strength of the steel as 350 MPa, find the area of the steel being sheared. The inertia of the flywheel spokes and the rotating parts of the motor may be neglected for the purposes of this example. The density of cast iron is 7200 kg/m³.

Solution

$$\text{Mass of rim} = \frac{\pi}{4}(1.7^2 - 1.22^2) \times 0.24 \times 7200 \text{ m}^2 \times \text{ m} \times \frac{\text{kg}}{\text{m}^3}$$

$$= 1902.21 \text{ kg}$$

$$\frac{\text{Maximum speed of flywheel}}{\text{Motor speed}} = \frac{\text{Motor pulley diameter}}{\text{Flywheel diameter}}$$

$$\text{Maximum speed of flywheel} = \frac{0.48 \text{ m} \times 10 \text{ rev/s}}{1.7 \text{ m}}$$

$$= 2.824 \text{ rev/s}$$

$$\text{Radius of gyration } k^2 = \frac{R_1^2 + R_2^2}{2}$$

$$k^2 = \frac{1.7^2 + 1.22^2}{2}$$

$$= 2.189 \text{ m}^2$$

$$\text{at } 2.824 \text{ rev/s} \times \omega_1 = 2.824 \times 2 \times 2\pi$$

$$= 17.744 \text{ rad/s}$$

$$\text{at } (2.824 - 0.25 \text{ rev/s}) = 2.574 \times 2$$

$$= 16.173 \text{rad/s}$$

$$\text{Moment of inertia} = mk^2$$

$$= 1902.21 \times 2.189 \text{ kg m}^2$$

$$= 4163.94 \text{ kg m}^2$$

$$\text{Kinetic energy at 2.824 rev/s} = \frac{I}{2}\omega^2$$

$$= \frac{1}{2} \times 4163.94 \text{ kg m}^2 \times 17.744^2 \text{rad/s}$$

$$= 0.6555 \text{ MNm}$$

$$\text{Kinetic energy at 2.574 rev/s} = \frac{I}{2} \times \omega_2{}^2$$

$$= \frac{1}{2} \times 4163.94 \text{ kg m}^2 \times 16.173^2 \text{ rad/s}$$

$$= 0.5446 \text{ MNm}$$

$$\text{Energy given out} = 0.6555 - 0.5446 \text{ MNm}$$
$$= 0.1109 \text{ MNm}$$

$$\text{Energy absorbed or work done at blade} = 0.1109 \text{ MNm} \times 0.8$$
$$= 0.08872 \text{ MNm}$$

$$\text{Work done} = \text{average cutting force} \times \text{distance moved}$$

$$\text{Average cutting force} = \frac{0.08872 \text{ MNm}}{0.075 \text{ m}} \times 10^3$$
$$= 1182.9 \text{ kN}$$

$$\text{Area of section being sheared} = \frac{\text{Force}}{\text{Shear stress}}$$

$$= \frac{1182.9 \text{ kN}}{350 \times 10^3 \text{ kN/m}^2}$$

$$= 0.00338 \text{ m}^2$$

$$= 3379.8 \text{ mm}^2$$

33.6 Geared Systems

In some designs, the flywheel mass is connected to the drive motor through a speed-changing device, such as a gear box. Here, the inertia is significantly modified. Consider the layout depicted in Figure 33.5, in this case the damping and the gear inertias will be ignored.

Let θ_m be the motor rotation and θ_o be the output rotation.

The gear ratio will be

$$G_r = \frac{\theta_o}{\theta_m} \tag{33.25}$$

This ratio is fixed and is therefore not a function of time; the angular speed and accelerations will also be in the same ratio:

$$G_r = \frac{\omega_o}{\omega_m} \tag{33.26}$$

where ω is the angular velocity.

$$G_r = \frac{\alpha_o}{\alpha_m} \tag{33.27}$$

where α is the angular acceleration.

The power transmitted through the shaft is

$$\text{Power} = \omega T \text{ (assuming there is no power lost)} \tag{33.28}$$

The input and output powers must be equal:

$$\text{Hence,} \quad \omega_m T_m = \omega_o T_o \tag{33.29}$$

$$\text{Therefore,} \quad T_m = \frac{\omega_o T_o}{\omega_m} = G_r T_o \tag{33.30}$$

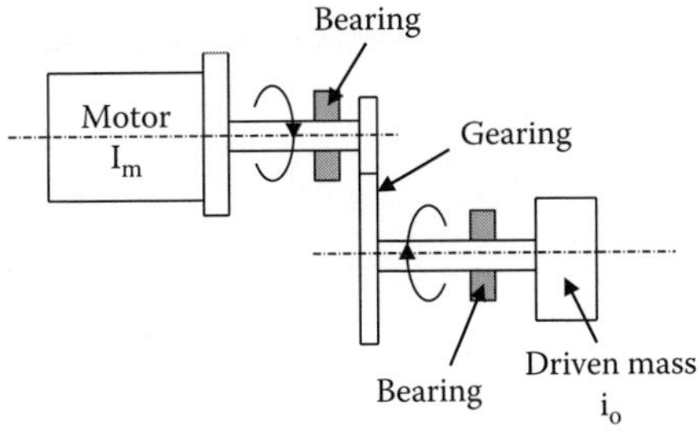

FIGURE 33.5
Gear drive.

This, therefore, is the output torque referred to the motor. In addition to this, the inertia torque of the motor ($I_m\ \alpha_m$) must also be added:

$$T_m = I_m\alpha_m + I_o\alpha_m \times G_r^{\ 2} \tag{33.31}$$

$$T_m = \alpha_m\left(I_m + G_r^2 I_o\right) \text{and} \left(I_m + G_r^2 I_o\right) \tag{33.32}$$

This is the effective moment of inertia I_e referred back to the motor shaft.

Example 33.6

A load is driven by a motor through a reduction gear having a ratio of 3.5. The motor characteristic has a variable torque of 300 sin θ Nm per half revolution. The motor has a moment of inertia of 0.4 kg m^2 and the load has a moment of inertia of 0.5 kg m^2. It is a requirement that the speed of the load is maintained at 350 rev/min, being limited to a variation of ±1.0%. Calculate the moment of inertia of a flywheel to be added to the motor shaft maintaining this requirement.

Solution

The coefficient of speed fluctuation (C_s) is

$$C_s = \frac{\Delta\omega}{\omega} = 3.5\% \text{ or } 0.35$$

The coefficient of energy fluctuation (C_f) is

$$C_f = \frac{\text{Maximum fluctuation of energy}}{\text{Work done per cycle}}$$

$$C_f \cdot W = \text{maximum fluctuation in energy}$$

$$C_f \cdot W = I\omega^2 C_s \ (\omega = \text{mean angular velocity})$$

Motor speed = 1225 rev/min

$$\omega = 2\pi \times \frac{1225}{60} = 128.28 \text{ rad / s}$$

W = work done per cycle

In this case, the half-revolution will be chosen as it is repeated every half-revolution (refer to Figure 33.5).

Now,

$$W = \int_0^\pi 300 \ \sin\theta \ d\theta$$

$$W = 300\left[\cos q\right]_0^\pi$$

$$W = 600 \text{ J}$$

Mean torque:

$$\text{Torque}_{\text{mean}} = \frac{W}{\text{base}}$$

$$= \frac{600}{\pi}$$

$$= 190.98 \text{ Nm}$$

The angle where the graph cuts the mean torque is $\sin^{-1}$ (190.98/300) = 39.57 (0.69 rad) and 180° − 39.57° = 140.43° (2.451 rad).

The area represented by B is found from

$$W = \int_{0.69}^{2.45} 300 \sin\theta \; d\theta$$

$$W = 300\big[\cos\theta\big]_{0.69}^{2.45}$$

$$W = 462.44 \text{ J}$$

Now the individual areas A and C are equal to 462.44 J (i.e., 231.22 J).

The maximum fluctuation of energy will be from (mean −231.22) to (mean +232.22):

$$\text{So,} \quad C_f \cdot W = 462.44$$

$$C_f = \frac{I\omega^2 C_s}{W}$$

Rearranging for I to be the subject,

$$I = \frac{C_f \cdot W}{\omega^2 C_s}$$

$$I = \frac{462.44}{128.28^2 \times 0.035}$$

$$I = 0.803 \text{ kg m}^2$$

$$I_T = I_m + G_r{}^2 I_o + I$$

$$0.803 = 0.4 + (0.3)^2 \times 0.5 + I$$

$$0.803 = 0.445 + I$$

$$I = 0.358 \text{ kg m}^2$$

34

Introduction to Linkages

34.1 Introduction

The use of linkages may be found in many walks of life. An example in the home is of the rear suspension on a mountain bike. In the building industry, a further example will be of the hydraulically operated buckets fitted to wheeled loaders; in the aerospace industry, extensive use of linkages is found for extending and retracting the aircraft landing gear units including wing flaps and so on. In the automotive industry, the vehicle suspension systems rely heavily on the use of linkages for maintaining road wheel positions.

In manufacturing, robotic manipulators rely wholly on the science of linkages optimising their sections for strength, mass and deflections to enable them to carry heavy loads at high speeds from position to position.

The definition of a link is that of a rigid body having two or more pairing elements which connect it to other bodies with the prime purpose of transmitting force or motion.

This chapter will briefly cover planar, spatial and spherical mechanisms but focus on four bar linkages as these are the most commonly found and will give details of determining velocities and accelerations of linkages.

Linkages can be constructed from either open or closed chains or a combination of both. Each link in a chain is connected by a joint to one or more other links. These create a kinematic chain that can then be modelled using graphical methods in which the links are considered as paths and the joints are the vertices.

The movement of the ideal joint can be considered as a sub-group of the group of Euclidean displacements where the number of parameters within the sub-group is called the 'degrees of freedom' (DOF) of the joint.

One of the prime functions of a mechanical linkage is to transform a given input force or movement into a desired output force or movement. The ratio of the output force to the input force is referred to as the 'mechanical advantage' of the linkage, while the ratio of the input movement or speed to the output movement or speed is referred to as the 'velocity ratio'.

The velocity ratio and the mechanical advantage are defined so that they produce the same number in an ideal linkage.

By definition, a kinematic chain in which one of the links is fixed and held stationary is called a mechanism. Conversely, a linkage that is designed to be stationary is called a structure.

34.2 Brief History

The use of levers predates Western civilisation for moving heavy objects with only manpower available. Examples here include the construction of the pyramids during ancient Egyptian times. Archimedes studied the geometry of the lever and together with Hero of Alexandria became the prime sources of machine theory. This work lasted up to the fifteenth century when Leonardo da Vinci (1452–1519) in his Codex gave information about the use of linkages which he incorporated into his designs.

In the mid-1700s, several designs of steam engines were being proposed amongst them. Thomas Newcomen (1664–1729) designed his successful atmospheric steam engine, originally used in the tin mines of Cornwall for pumping water out of the mines. James Watt (1736–1819) recognised that the efficiency of these early engines could be significantly improved by using different cylinders for expansion and condensation of steam. His search for a linkage that could generate approximately straight lines inspired the mathematician J. J. Sylvester (1814–1897) who was an authority on the Peaucellier linkage which generated a straight line from a rotating crank. This work, in turn, inspired A. B. Kemp (1849–1922) who showed that linkages could be used for addition and multiplication and also for tracing a given algebraic curve.

Around this period, Charles Babbage (1791–1871) proposed his analytical and difference engine which made considerable use of cams and linkages, giving an indication of the degree of workmanship that was being achieved at this time.

In the late 1800s, F. Reuleaux (1829–1905), A. B. W Kennedy (1847–1928) and L. Burmester (1840–1927) formalised the analysis and synthesis of linkages using descriptive geometry. P. L. Chebyshev (1821–1894) developed analytical techniques for the study and design of linkages.

In recent years, G. N. Sandor (1913–1997) and F. Freudenstein (1926–2006), using the newly developed digital computer for solving the loop equations of a linkage, were able to establish the dimensions required to generate a desired function and begin the revolution for computer-aided-design of linkages. These techniques are now fundamental in the analysis of complex machine systems and the control of robotic manipulators.

By combining the computer's ability to compute the roots of polynomial equations, R. F. Kaufman united Freudenstein's techniques with the geometric methods of Reuleaux and Burmester to implement in *KINYN* an interactive computer program.

34.3 Kinematic Definitions

The following are some definitions that are commonly used in the analysis of linkages.

34.3.1 Kinematic Chain

An assemblage of links and bodies, together with joints, interconnected in such a way as to provide an output motion that is in response to an input motion.

34.3.2 Mechanism

A kinematic chain in which at least one link or body is fixed in relation to the frame of reference (in which it itself may be in motion).

34.3.3 Machine

A combination of resistant bodies was arranged in such a way as to compel the mechanical forces of nature to do work accompanied by determinate motions.

A kinematic link or bar is a rigid body or assembly which constitutes the parts of a mechanism and is the smallest element of the mechanism that transmits motion to other links. A rigid body is one which does not deform or change shape due to the application of a force. In a mechanism, three types of links, rigid link, flexible link and fluid link, are the most widely used.

34.3.4 DOF

DOF is defined as the number of input parameters which must be independently controlled in order to bring the mechanism into useful engineering purposes. It also defines the number of independent relative motions, both translational and rotational, that a pair can have. DOF= 6−number of restraints.

To find the number of DOF for planar mechanisms, an equation referred to as Grubler's equation is used where

$$F = 3(n-1) - 2j_1 - j_2 \tag{34.1}$$

where

F = mobility or number of DOF
n = number of links including frame
j_1 = joints with a single degree of freedom
j_2 = joints with two DOF
$F > 0$, results in a mechanism with 'F' DOF
$F = 0$, results in a statically determinate structure
$F < 0$, results in a statically indeterminate structure

34.3.5 Rigid Links

A rigid link possesses at least two nodes, where a node is an attachment point to other links or bodied via joints.

Links can be classified into the following three categories as depicted in Figure 34.1a through c.

1. Binary link: This is a link that has two nodes.
2. Ternary link: A link that forms three nodes.
3. Quaternary link: A link that has four nodes.

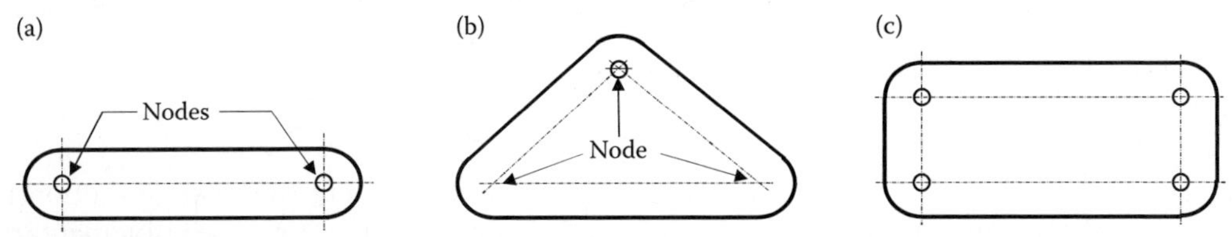

FIGURE 34.1
Classification of links. (a) Binary, (b) ternary and (c) quaternary links.

The ternary and quaternary links do not have any relative movement between the nodes within the link.

34.3.6 Order of a Link

The order of a link indicates the number of nodes per link as indicated above; there are binary (two), ternary (three) and quaternary (four) nodes. Links or bodies can be of any shape and are not restricted to those shown in Figure 34.1a through c.

Link order is equal to the number of nodes.

34.3.7 Joints

A joint is a connection between two or more links at their nodes allowing motion to occur between the links.

- A pivot is a joint that allows rotary motion at the joint.
- A slider is a joint that only allows a linear motion.

34.3.8 Kinematic Pairs

Kinematic pairs refer to the type of connection that exists between adjacent links. See Section 34.4 for a more detailed description of kinematic pairs and the definition in the differences between lower and higher pairs.

34.3.9 Mobility

This is the number of input parameters that have to be controlled independently to bring the mechanism to a set position.

See Section 34.6 for a fuller description.

34.4 Kinematic Pairs

The part of a link that makes contact with another link is called an 'element', examples being either the bore of a cylinder or a gear tooth or a rotating connection with another link.

The connections between links are made by pairs of elements coming into contact; these connections are known as 'kinematic pairs'. The kinematic pairs fall into two categories:

1. Lower pairs
2. Higher pairs

Lower pairs can be defined as those having surface contact where the relative motion is either purely turning or sliding. The contact surfaces of a lower pair are complementary to each other, such as a journal or plain bearing or a piston within a cylinder. Lower pairs may be defined as those allowing one degree of freedom.

Higher pairs may be defined as those having line or point contact, such as gear teeth or a follower on a cam. The relative motion in higher pairs is a combination of sliding and turning. The contact surfaces are dissimilar. Kinematic pairs formed between gears, cams and followers and ball and roller bearings belong to the higher pair category. Higher pairs usually allow two DOFs.

34.4.1 Relative Motion between Kinematic Pairs

The nature of the relative motion between kinematic pairs can be further classified into the following types:

34.4.1.1 Lower Pairs

1. *Sliding pair:* A kinematic pair is said to be a 'sliding pair' when two links are so connected that one element is constrained to have a sliding motion relative to the element. Figure 34.2a illustrates a prismatic bar sliding in a rectangular hole. The connection between a piston within a cylinder of a slider-crank mechanism is a further example of a sliding pair.
2. *Revolute pair:* When two links are so connected that only a rotation motion is possible between them, the kinematic pair is called a 'turning pair'. A circular shaft that is retained by two collars as shown in Figure 34.2b depicts an example of a turning pair.
3. *Cylindrical pair:* Where two links are so connected that rotational motion together with linear motion is allowed, the kinematic pair is referred to as a cylindrical pair as shown in Figure 34.2c.
4. *Helical pair:* Figure 34.2d shows a screw pair which has rotation as well as translation but these two movements are related to each other. Therefore, a screw pair has one degree of freedom because the relative movement between the screw and nut can be expressed by a single coordinate 'θ' or 'x'. These two coordinates are related by

$$\frac{\Delta\theta}{2\pi} = \frac{\Delta x}{L}$$

   where L is the lead of the screw.
5. *Spherical pair:* When one link is in the form of a sphere that is allowed to turn within a fixed link, such a pair is referred to as a 'spherical pair', such an example being a ball and socket joint (Figure 34.2e).
6. *Planar pair:* Where the relative motion between items 1 and 2 can be described by x and y coordinates in an x–y plane. The x and y coordinates describe relative translation and 'θ' describes relative rotation about the z-axis. This pair has three DOFs. This is shown in Figure 34.2f.

34.4.1.2 Higher Pairs

7. *Gear pair:* The contact between the involute curves that form the meshing teeth of two gears is known as a gear pair. The contact between the teeth is considered a point contact. Other examples in this category include cams and followers. Figure 34.2g shows a gear pair.

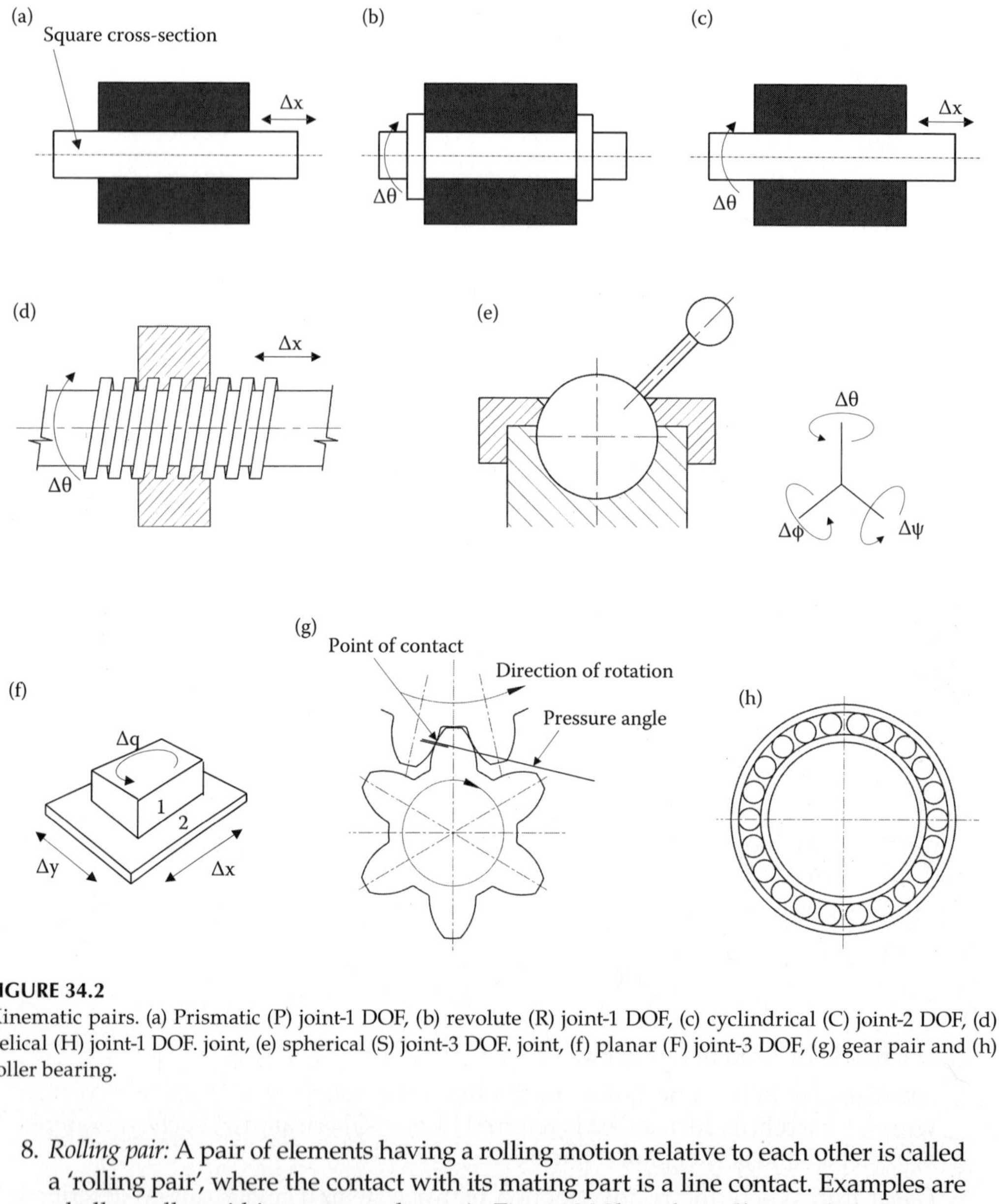

FIGURE 34.2
Kinematic pairs. (a) Prismatic (P) joint-1 DOF, (b) revolute (R) joint-1 DOF, (c) cyclindrical (C) joint-2 DOF, (d) helical (H) joint-1 DOF. joint, (e) spherical (S) joint-3 DOF. joint, (f) planar (F) joint-3 DOF, (g) gear pair and (h) roller bearing.

8. *Rolling pair:* A pair of elements having a rolling motion relative to each other is called a 'rolling pair', where the contact with its mating part is a line contact. Examples are a ball or roller within a cage as shown in Figure 34.2h and a pulley in a belt drive.

34.4.2 Nature of Kinematic Constraints

Subject to the nature of the mechanical constraints, kinematic pairs can be classified into the following two categories.

34.4.3 Closed Pair

When two links that form a kinematic pair (Figure 34.3) are restrained mechanically, such a pair is called a 'closed pair'. Examples include screw pairs and turning pairs.

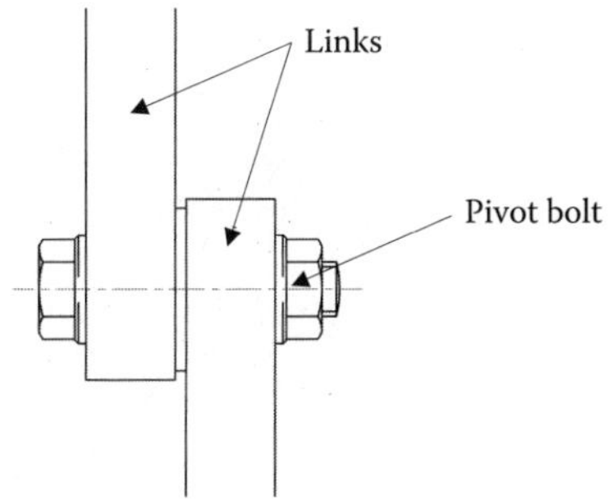

FIGURE 34.3
Closed pairs.

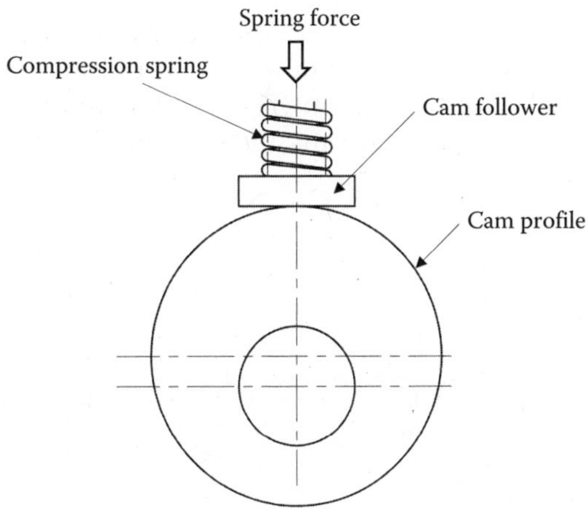

FIGURE 34.4
Open pairs.

34.4.4 Open Pair

Two links that form a pair (Figure 34.4) that is held together by means of an external force such as gravity or spring force will constitute an 'open pair'. The cam follower used in an IC engine valve operating mechanism is an example of an open pair.

34.5 Planar, Spherical and Spatial Mechanisms

Linkages may be classified into three fundamental groups:

1. Planar
2. Spherical
3. Spatial

34.5.1 Planar Mechanism

A planar motion is defined as that, when all the particles that make up the body move in parallel planes. If the pivot points on a planar mechanism are projected, they will meet at infinity.

An example of a planar link is shown in Figure 34.5a. Planar links will be considered extensively in this introduction.

34.5.2 Spherical Mechanism

A body is considered to have spherical motion when all its particles move on the surface of concentric spheres. That is, if the pivot points are projected, they will all meet at the centre of a sphere. Figure 34.5b depicts a universal joint that moves in a spherical motion. Other examples include some types of robotic arms.

34.5.3 Spatial Mechanism

In a spatial mechanism, particles that make up the body move in paths that do not always remain in a plane. Figure 34.5c illustrates one such spatial mechanism. The mechanism shown is referred to as a *'Stewart'* platform (Figure 34.6) and is used extensively as the base mechanism in flight simulator platforms. Other examples can also be seen in modern folding pushchair frames.

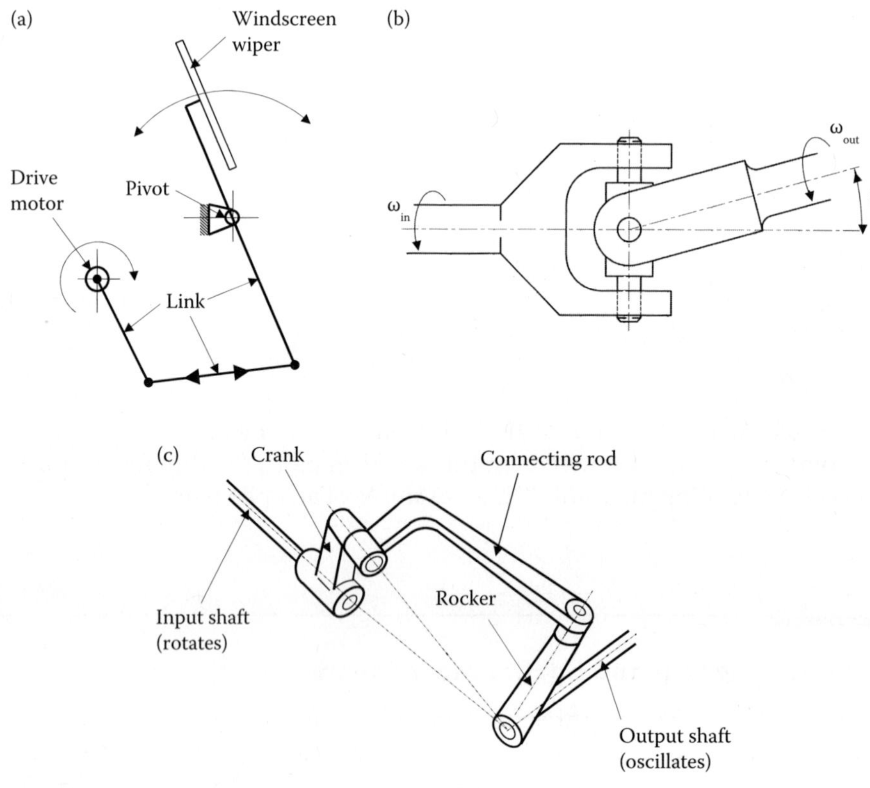

FIGURE 34.5
Planar, spherical and spatial mechanisms. (a) Example of a planar linkage, (b) example of a spherical mechanism (universal joint) and (c) example of a spatial mechanism.

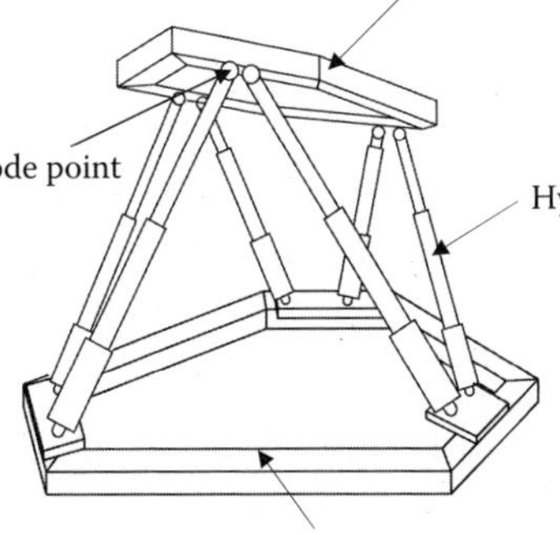

FIGURE 34.6
Stewart platform.

Both planar and spherical mechanisms can be considered as subsets of spatial mechanisms.

Mechanical linkages are used extensively in mechanical engineering where they are designed to transform a given input force and movement into a desired output force and movement.

The ratio of the output force to the input force is referred to as the mechanical advantage of the linkage. The ratio of the input velocity to the output velocity is known as the velocity ratio. The mechanical advantage and the velocity ratio are defined so that they yield the same number in an ideal linkage.

The definition of a link is that of a rigid body having two or more pairing elements that connect with other bodies for the specific purpose of transmitting force and/or motion.

In a linkage mechanism, at least one link occupies a fixed position and is referred to as either the ground or fixed link.

Linkages come in all sorts of arrangements covering a three-bar linkage to six-bar linkages, although they are not restricted to six bars. The more complicated linkages may have more, but these are outside the scope of this introduction. Stevenson and Watt linkages usually found on steam locomotives are generally six-bar linkages.

Fixed linkages that have only three bars and have no movement between adjacent links are usually called a truss which is used in structures.

One of the most commonly used linkages is that of the four-bar linkage (usually referred to as a four-bar) and is the simplest closed chain linkage. Four-bar consists of three moveable linkages connected to a fixed link and connected in a loop connected by four rotating joints. The joints are configured so that the links all move in parallel planes and the assemblage is called a 'planar four-bar linkage'.

Planar four-bar linkages are important mechanisms in machine design and examples include

- Double wishbone suspension
- Watt and Chebyshev linkages (these linkages give an approximately straight line motion)
- Crank slider (four-bar and one degree of freedom)
- Pantograph (four-bar and two DOF, i.e. only one pivot joint is fixed)

34.6 Mobility

When considering a linkage, an important consideration is the number of DOFs the mechanism will have. The mobility of a linkage is the number of input parameters which have to be controlled independently to bring the mechanism to a set position. This is, generally, the number of independent parameters including joint angles and slide distances that are needed to specify the configuration of the linkages disregarding any deformation of the links.

The mobility of a system is the sum of the unconstrained DOFs for the links in the system minus the constraints imposed by the joints.

A free link usually has three DOFs (x, y, θ). One link is always fixed and is referred to as either the fixed or ground link. Before any joints are attached, the number of DOFs of a linkage assembly with 'n' links is

$$\text{DOF} = 3(n-1) \tag{34.2}$$

Connecting two links using a joint (which has one degree of freedom) adds two constraints. Connecting two links that have a joint with two DOFs includes one restraint to the system. The number of 1 DOF joint=j_1 and the number of joints with 2 DOF joints=j_2; the 'mobility' of a system can be expressed as

$$\text{Mobility} = m = 3(n-1) - 2j_1 - j_2 \tag{34.3}$$

Figure 34.7 gives examples of the mobility of some planar linkages.

Note: In the case of a linkage that has a mobility of '0', this is essentially a structure. This discussion will be developed further in Sections 34.7 and 34.8.

34.7 Chebyshev–Gruber–Kutzbach Criterion

In Section 34.6, the discussion introduced the concept of mobility and how it was arrived at. In this and the following sections, a more formalised method of establishing mobility is discussed.

The Chebyshev–Gruber–Kutzbach criterion is used to determine the DOF within a kinematic chain of linkages, that is, a coupling of rigid bodies by means of mechanical constraints and it computes the number of parameters that define the configuration of a

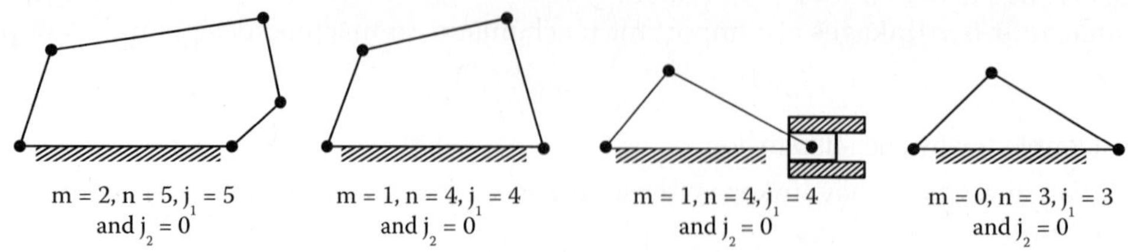

FIGURE 34.7
Mobility of planar linkages.

linkage that consists of a number of individual links and joints together with the DOF that exists at each joint.

Gruber's equation is written as

$$F = 3(n-1) - 2I - h \tag{34.4}$$

where

F = total DOFs of the mechanism
n = number of links (including the frame)
I = number of lower pairs (one degree of freedom)
h = number of higher pairs (two DOFs)

Example 34.1

Consider a transom that is fitted above a door as shown in Figure 34.8 where the transom is opened and closed by means of the link (1).

It is required to calculate its degree of freedom.

Solution

$$n = 4(\text{link 1, 5, 6 and frame 2}),\ I = 4(\text{at A, B, C and D}),\ h = 0$$

$$F = 3(4-1) - 2\times 4 - 1\times 0$$

$$F = 1$$

Note: In this example, 'D' and 'E' as a prismatic pair count only as one lower pair.

The number of DOFs of a mechanism is also called the mobility of the mechanism. The mobility is the number of input parameters (usually paired variables) that have to be individually controlled to bring the mechanism into a particular desired position.

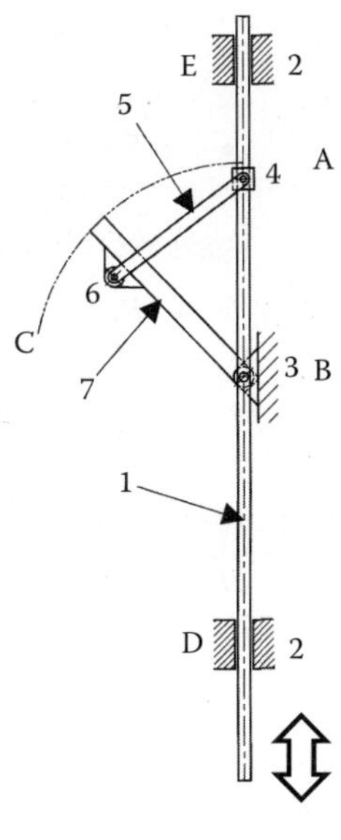

FIGURE 34.8
Example 34.1.

In order to control a mechanism, the number of independent input motions must equal the DOFs of the mechanism. As an example, the transom in Example 34.1 has only a single degree of freedom; therefore, only one independent input motion is required to open and close the transom, that is, push and pull on the link (1) opens and closes the window.

34.8 Grashof's Law

Prior to discussing Grashof's Law, it will be useful to extend the definitions as covered in Sections 34.3 and 34.4.

In a planar mechanism, the simplest group of lower pair mechanisms will be the four-bar chain or linkage and comprises four bar-shaped links and turning pairs as depicted in Figure 34.9a through e.

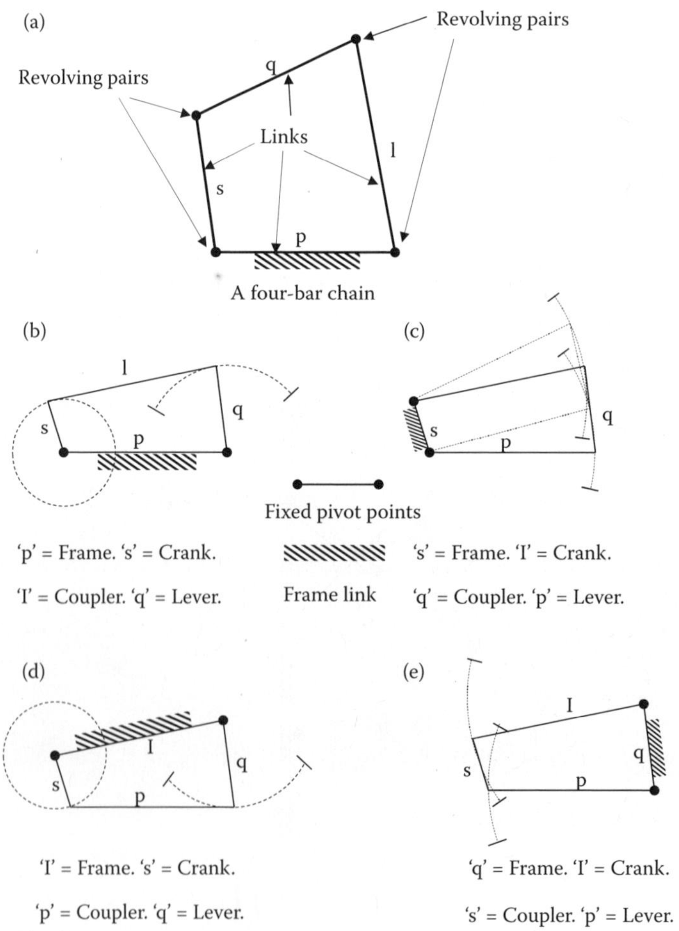

FIGURE 34.9
(a)–(e): Inversions of various linkages.

The fixed link is referred to as either the 'fixed link' or the 'frame'. The link that is opposite the frame is called the 'coupler link'. The links that are hinged to the frame are called 'side links'. A link that is free to rotate through 360° with respect to a second link is said to revolve relative to the second link (not necessary the frame). Where it is possible for all four links to become simultaneously aligned, such a state is called a 'change point'.

Some important concepts in link mechanisms are

1. *Crank:* A side frame that revolves relative to the frame is called a 'crank'.
2. *Rocker:* Any link that does not revolve is referred to as a 'rocker'.
3. *Crank–rocker mechanism:* In a four-bar chain, if the shorter side link revolves and the other one rocks (that is oscillates), this is called a 'double-crank' mechanism.
4. *Double-crank mechanism:* In a situation where both side links revolve, this is called a 'double-crank' mechanism.
5. *Double-rocker mechanism:* In a four-bar chain where both side links rock, this is called a 'double-rocker' mechanism.

34.8.1 Classification

Referring to Figure 34.9a, the following notation is defined:

In a four-bar chain, the line segment between hinges on a given link is called a 'bar' where 'b' is the length of the shortest bar, 'c' is the length of the longest bar and 'a' and 'd' are the lengths of intermediate bars.

In certain cases, it may be required to design a four-bar mechanism where the input link is driven and rotates continuously. It is essential that the input link is able to rotate through a complete revolution without any binding. Grashof provides a law which provides a simple test for this condition.

Grashof's law states that the sum of the longest and shortest links cannot be greater than the sum of the remaining links if there is to be continuous relative rotation between two members.

From Figure 34.9a through e, it can be seen that one of the links (usually the shortest link) will be able to rotate continuously if the following conditions are met

$$\text{s(shortest link)} + \text{l(longest link)} \le p + q \tag{34.5}$$

This being so will satisfy Grashof's law and different inversions will be possible by fixing the different links within the chain.

The mechanism satisfying Grashof's criterion is called Grashof's linkage.

For a non-Grashof linkage, only a rocker–rocker mechanism will occur as in the case where

$$\text{s(shortest link)} + \text{l(longest link)} > p + q \tag{34.6}$$

In this case, all three mobile links will rock.

Example 34.2

In a four-bar mechanism, the lengths of the driver-crank, coupler and follower link are 150, 250 and 300 mm, respectively. The fixed link length is 'L_a'. Determine the range of values for 'L_a' so as to make the mechanism:

1. Crank–crank mechanism
2. Crank–rocker mechanism

Solution

For the crank–crank mechanism, the conditions to be satisfied are

$$150\ \text{mm} + 250\ \text{mm} \leq L_a + 300\ \text{mm}$$

$$400\ \text{mm} \leq L_a + 300\ \text{mm}$$

$$L_a \leq 100\ \text{mm (Ans)}$$

For the crank–rocker mechanism, the conditions to be satisfied are

- The link adjacent to the fixed link must be the smallest link.
- $s+l \leq p+q$.

Two possibilities have to be considered:

1. When L_o is the longest link.
2. When L_o is not the longest link.

When L_o is the longest link, from Grashof's criterion:

$$L_o + 150 \leq 250 + 300$$

or

$$L_o \leq 400$$

When L_o is not the longest link, from Grashof's criterion:

$$300 + 150 \leq L_o + 250$$

or

$$L_o \geq 200$$

Hence, the crank–rocker mechanism, range of values for L_o is

$$200 \leq L_o \leq 400\ \text{mm (Ans)}$$

34.9 Four-Bar Linkage

34.9.1 Planar Four-Bar Linkages

The four-bar linkage or four-bar chain is fundamental in a wide range of mechanisms and takes many different shapes and sizes. As a mechanism, it is fairly simple to manufacture, inexpensive, easy to maintain and robust. It has the ability to operate at low and high speeds, transmit small or large loads and operate in a wide range of environments with suitable protection. It is possible to design planar four-bar linkages to cover a wide range of movements including complex path patterns. Examples of planar four-bar linkages can be found in a wide range of industries including earth-moving machines, landing gear systems in aviation, suspension units in the automotive industry and automatic assembly machines in the manufacturing industry, to name but a few.

34.9.2 Inversion

A mechanism is defined as a kinematic chain where one of the links is fixed; thus, by fixing the individual links one at a time, a number of variants of the mechanisms are obtained. These variants are dependent upon the number of links. These variants of the kinematic chain are called 'inversions of kinematic chain' or 'inversions of mechanism'.

It should be noted that the relative motion between the various links is not changed in any manner through the process of inversion, but their absolute motions (measured in respect of the fixed link) may be altered drastically.

Most mechanisms are reversible where in one configuration the 'driver' link transmits motion to a 'follower'; in another configuration of the same mechanism, the role is reversed and the driver link then becomes the follower and the link that initially was the follower then becomes the driver. An example of this is when a reciprocating steam engine is considered where the piston is the driver and the flywheel is the follower, while in a reciprocating air compressor, the flywheel becomes the driver.

Consider the four-bar linkage depicted in Figure 34.10a. Figure 34.10b through e shows the various inversions possible with the basic linkage. The four inversions look identical to each other but their mobility can be significantly varied by altering the proportions of lengths of the various links.

Planar four-bar linkages are constructed from links connected in a loop by four one degree of freedom joints. A joint may be either hinged (revolute), denoted by 'R', or sliding (prismatic), denoted by 'P'. A planar quadrilateral linkage formed of four links and four revolute joints is designated 'RRRR'. A slider–crank linkage constructed from four links

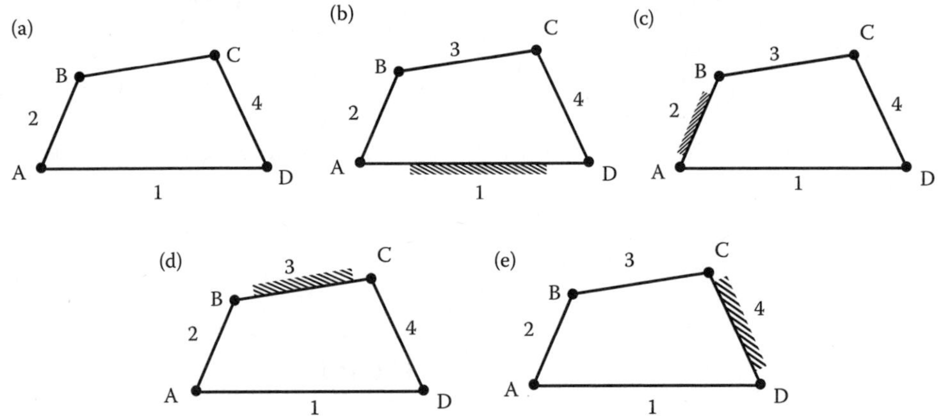

FIGURE 34.10
Variants of a four-bar chain. (a) Basic kinematic chain, (b) variant 'b', (c) variant 'c', (d) variant 'd' and (e) variant 'e'.

and connected by three revolute joints and one prismatic joint is designated 'RRRP'. If a double slider is used, this will be a 'PRRP' linkage (see Figure 34.11).

As described earlier in the chapter, one link of the chain is usually fixed and is called the 'ground' link, fixed link or the frame. The two links that are connected to the frame are referred to as the 'grounded links' and these are usually the input and output links of the system. The remaining link is called the 'floating link' and this is also called a 'coupler' or 'connecting link' as it connects the input to the output.

Section 34.8 describes Grashof's law for a planar quadrilateral four-bar linkage where the sum of the shortest link + longest link should be either shorter or equal to the sum of the remaining links. This is the requirement for the shortest link (usually referred to as the crank) to rotate fully with respect to a neighbouring link.

34.9.3 Slider–Crank Linkage

A slider–crank linkage or chain is a specialised form of a four-bar linkage where one turning pair is replaced by a sliding pair. Various inversions of slider–crank linkages are obtained by fixing different links and these are discussed further below with reference to Figure 34.12.

34.9.3.1 Link 1 Fixed

If link 1 of Figure 34.12 is fixed and link 2 is allowed to act as a crank, the resulting mechanism is called a slider–crank mechanism. This converts reciprocating motion into rotary motion or vice versa.

34.9.3.2 Link 2 Fixed

When link 2 of the slider–crank linkage is fixed and link 3 acts as a crank, link 1 will rotate about centre 'O' along the slider 4, which in turn will reciprocate along the link as in Figure 34.13. This inversion produces two popular forms of mechanisms:

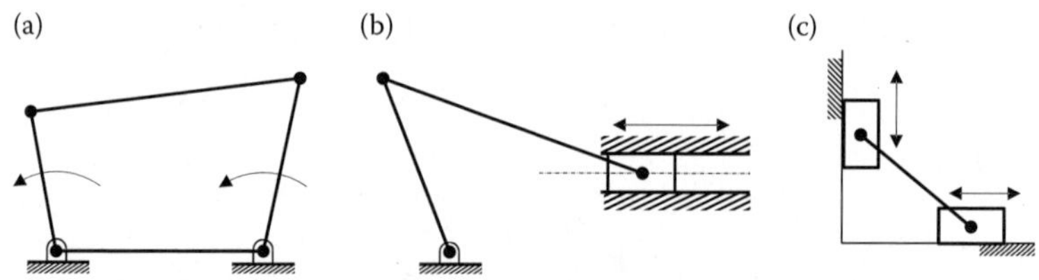

FIGURE 34.11
Examples of revolute and prismatic joints. (a) Planar linkage (RRRR), (b) slider-crank linkage (RRRP) and (c) double-slider linkage (PRRP).

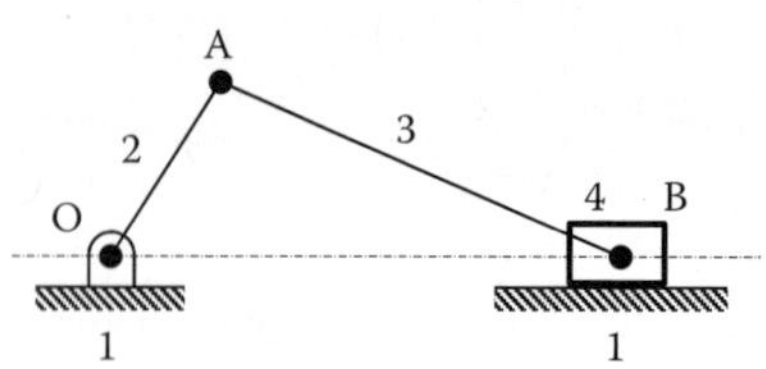

FIGURE 34.12
Slider–crank mechanism.

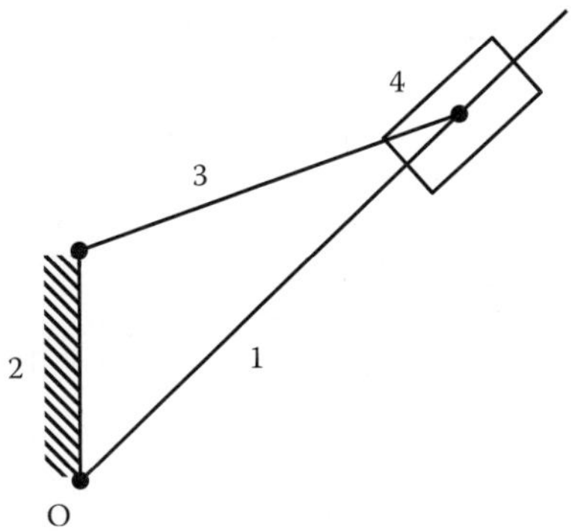

FIGURE 34.13
Inversion of a slider–crank chain.

1. Whitworth quick return mechanism
2. Rotary engine configuration

Figure 34.14 depicts the Whitworth quick return mechanism that is used in a number of machine tools including the 'shaper' in which a fixed tool is passed over a work piece at the correct cutting speed and returned to its starting point in a shorter time, thereby reducing the overall cycle time.

In this mechanism, the link 'A:C' (link 3) that forms the turning pair is fixed. The driving crank 'C:B' revolves at a constant angular speed around the centre 'C'. A sliding block which is attached to the crank pin at 'B' is fitted in and slides in the slotted bar 'A:P'. This causes the slotted bar to oscillate about the pivot point 'A'. The slotted link is connected to

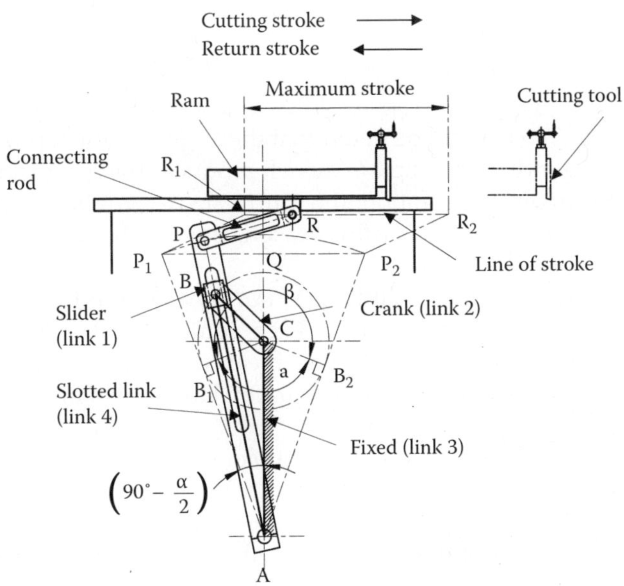

FIGURE 34.14
Whitworth quick return mechanism.

the shaper ram that carries the cutting tool via a connecting link 'P:R' and this reciprocates along the line of stroke 'R_1:R_2'.

Positions 'A:P_1' and 'A:P_2' are tangential to the circle 'A:B_1' and 'A:B_2'. The forward-cutting action occurs when the crank rotates from the position 'C:B_1' to 'C:B_2' (or through the angle 'β') when rotating in the clockwise direction. The return stroke occurs when the crank rotates from position 'C:B_2' to 'C:B_1' (or through the angle 'α' when rotating in the clockwise direction).

As the crank is rotating at a uniform angular speed, it follows that

$$\frac{\text{Time of cutting stroke}}{\text{Time of return stroke}} = \frac{\beta}{\alpha} = \frac{\beta}{360° - \beta} \tag{34.7}$$

or

$$\frac{360° - \alpha}{\alpha} \tag{34.8}$$

As the cutting tool travels the distance of 'R_1:R_2' during the total cutting cycle, the length of stroke will be

$$= R_1R_2 = P_1P_2 = 2 \cdot P_1Q = 2P_1 \sin \angle P_1 \cdot AQ \tag{34.9}$$

$$= 2AP_1 \sin\left(90° - \frac{\alpha}{2}\right) = 2AP \cos\frac{\alpha}{2} \quad \text{where } AP_1 = AP \tag{34.10}$$

$$2AP \times \frac{CB_1}{AC} \qquad \text{where } \cos\frac{\alpha}{2} = \frac{CB_1}{AC} \tag{34.11}$$

$$2AP \times \frac{CB}{AC} \qquad \text{where } CB_1 = CB \tag{34.12}$$

From a study of Figure 34.14, it will be noted that the angle 'β' which produces the forward stroke of the ram is larger than the angle 'α' which produces the return stroke. As the crank is rotating at a uniform angular rotation, the return stroke will be completed in a shorter time to that of the forward stroke. For this reason, this mechanism is known as a quick return mechanism.

The output from a planar four-bar linkage is not restricted to the output link; in some applications, the output of the four-bar linkage could be taken from the coupler. In Figure 34.9 where the motion is required to be linear, such an application may be an automatic assembly machine such as a 'pick and place'. This is also an example of a double-rocker mechanism.

Extensive use is made of loci generated by coupler curves in a wide range of industries from automatic assembly machines as described to industrial manipulators.

There are a number of 'atlases' containing a wide range of curves generated by points on the coupler link or extensions of the coupler in four-bar linkages. One such atlas has been produced by Hrones and Nelson which covers thousands of curves, but these are now being overtaken by modern simulation packages such as MATLAB® reducing the number of iterations to arrive at a more optimised curve.

34.10 Mechanical Advantage of a Four-Bar Linkage

One of the principal criteria for the designer to be aware of is the ability of a particular mechanism to be able to transmit force or torque. In some mechanisms including gear trains, these mechanisms transmit a constant torque ratio between the input and the output. In a linkage, this is not the case, as the geometric nature of the linkage will be varying all the time through the mechanism's motion, and as such there will be a variation between the input and output ratios.

If it is considered that the mechanism is a conservative system (i.e. energy losses due to friction, heat sound, etc. are negligible when compared with the total energy that is transmitted by the system as shown in Figure 34.15), it also assumes that the effects of inertial forces are also negligible, that is, 'power in' will equal 'power out' ($P_{in}=P_{out}$).

Hence, the 'torque in' (T_{in}) times the 'angular velocity in' (ω_{in}) will equal the 'torque out' (T_{out}) times the 'angular velocity out' (ω_{out}).

That is

$$P_{in} = T_{in} \cdot \omega_{in} = T_{out} \cdot \omega_{out} = P_{out} \tag{34.13}$$

Alternatively:

$$P_{in} = F_{in} \cdot V_{in} = F_{out} \cdot V_{out} = P_{out} \tag{34.14}$$

where T_{in} and F_{in} are torques applied to the linkage chain and T_{out} and F_{out} are those that are exerted by the chain, and V_{in} and V_{out} are the velocities of the points through which F_{in} and F_{out}, respectively, act.

In vector form:

$$V \cdot F = VF\cos(\arg F - \arg V) \tag{34.15}$$

Also,

$$V \cdot F = V_x F_x + V_y F_y \tag{34.16}$$

It should be noticed that the units of torque times angular velocity together with the scalar products of force times velocity both represent power as in Equation 34.13.

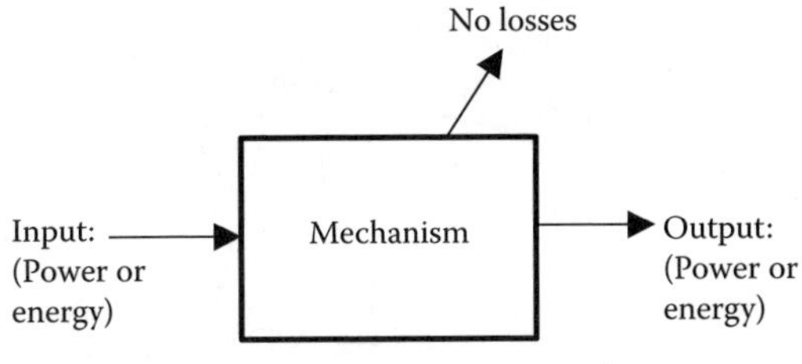

FIGURE 34.15
Conservation of power and energy through a mechanism.

That is

$$\frac{T_{out}}{T_{in}} = \frac{\omega_{in}}{\omega_{out}} \tag{34.17}$$

These relationships are dimensionless.

In cases where the input force generates an output torque (or vice versa):

By definition, the 'mechanical advantage (MA) of a system is the ratio of the magnitude of 'F_{out}' over 'F_{in}', that is

$$MA = \frac{F_{out}}{F_{in}} \tag{34.18}$$

where

$$F = | F |$$

Torque is the product of force 'F' times radius 'r'. Combining Equations 34.13 and 34.18 and solving for MA, the following equation will result

$$MA = \left(\frac{T_{out}}{r_{out}}\right)\left(\frac{r_{in}}{T_{in}}\right) = \left(\frac{r_{in}}{r_{out}}\right)\left(\frac{T_{out}}{T_{in}}\right) \tag{34.19}$$

and

$$MA = \left(\frac{r_{in}}{r_{out}}\right)\left(\frac{\omega_{in}}{\omega_{out}}\right) \tag{34.20}$$

In SI units, these dimensions are rad/s to m/s and N/Nm (or the inverse).

A power, force or torque requirement can be converted into a velocity ratio and velocity analysis can then be used to design a mechanism for a required mechanical advantage.

Bear in mind that velocity ratio is normally a function of the mechanism position, and therefore the mechanical advantage will also be a function of the mechanism position.

To summarise, the mechanical advantage of a planar four-bar linkage is defined as the ratio of the output power or torque exerted by the driven link or follower link to the required input power or torque at the driver link.

The mechanical advantage of the four-bar linkage is directly proportional to the sine of the angle 'γ' between the coupler and the follower and inversely proportional to the sine of the angle 'β' between the coupler and the driver. For one complete cycle of the mechanism, both these angles 'γ' and 'β' are continuously changing and as a result the mechanical advantage also changes as the linkage moves (see Figure 34.16).

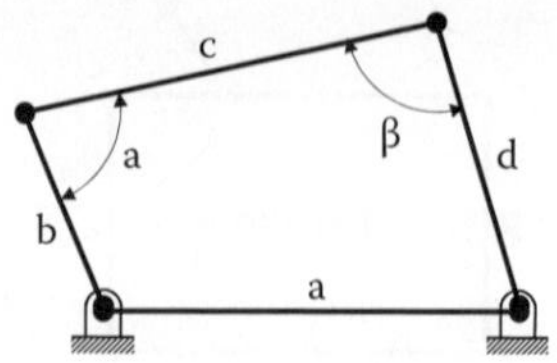

FIGURE 34.16
Mechanical advantage.

34.11 Freudenstein's Equation

Freudenstein developed a simple algebraic method for determining the position of an output link knowing the length of the four links and the position of the input link.

Consider the four-bar linkage chain shown in Figure 34.17.

The position vector of the links has the following relationship:

$$l_1 + l_2 + l_3 + l_4 = 0 \tag{34.21}$$

Equating the horizontal distances:

$$l_1 \cos\theta_1 + l_2 \cos\theta_2 + l_3 \cos\theta_3 + l_4 \cos\theta_4 = 0 \tag{34.22}$$

Equating the vertical distances:

$$l_1 \sin\theta_1 + l_2 \sin\theta_2 + l_3 \sin\theta_3 + l_4 \sin\theta_4 = 0 \tag{34.23}$$

Making the assumption that $\theta_1 = 180°$, hence, $\sin\theta_1 = 0$, $\cos\theta_1 = -1$.

Rearranging and squaring both sides in respect to l_3:

$$l_3^2 \cos^2\theta_3 = (l_1 - l_2 \cos\theta_2 - l_4 \cos\theta_4)^2 \tag{34.24}$$

$$l_3^2 \sin^2\theta_3 = (-l_2 \sin\theta_2 - l_4 \sin\theta_4)^2 \tag{34.25}$$

Summing Equations 34.24 and 34.25 and using the following relationships:

$$\cos(\theta_2 - \theta_4) = \cos\theta_2 \cos\theta_4 + \sin\theta_2 \sin\theta_4 \text{ and } \sin^2\theta + \cos^2\theta = 1 \tag{34.26}$$

The following relationship will result:

$$\frac{l_3^2 - l_1^2 - l_2^2 - l_4^2}{2 \cdot l_2 - l_4} + \frac{l_1}{l_4}\cos\theta_2 + \frac{l_1}{l_2}\cos\theta_4 = \cos(\theta_2 - \theta_4) \tag{34.27}$$

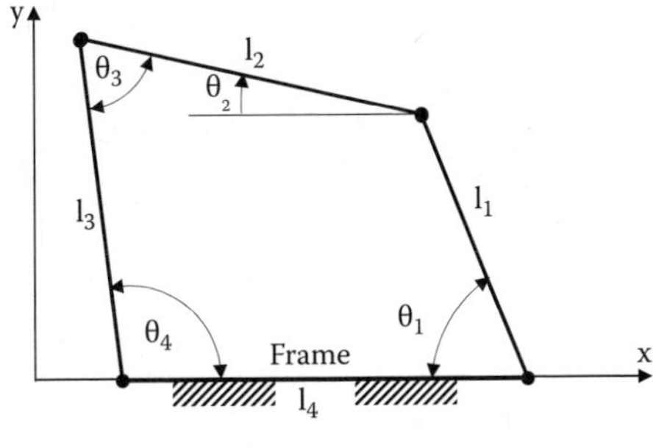

FIGURE 34.17
Four-bar linkage vector representation.

The following will result in Freudenstein's equation:

$$K_1 \cos\theta_2 + K_2 \cos\theta_4 + K_3 = \cos(\theta_2 - \theta_4) \tag{34.28}$$

where

$$K_1 = \frac{l_1}{l_4} \tag{34.29}$$

$$K_2 = \frac{l_1}{l_2} \tag{34.30}$$

$$K_3 = \frac{(l_3^2 - l_1^2 - l_2^2 - l_4^2)}{2 \cdot l_2 \cdot l_4} \tag{34.31}$$

The equation can be used to design a mechanism from a required relationship between input to output; in this case, the unknowns are the lengths of the links or more frequently the ratios of the links l_2/l_1, l_4/l_1 and l_3/l_1. If three desired values of the output are fixed for three desired values of the input (co-ordinating input and output positions), three equations in three unknowns are produced and hence the ratio of the links relating to the length of the fixed link l_1.

To obtain an expression for the output θ_4, it will be found more convenient to write Equation 34.28 as follows:

$$K_1 \cos\theta_2 + K_2 \cos\theta_1 - K_3 = \cos\theta_2 \cos\theta_1 + \sin\theta_2 \sin\theta_1 \tag{34.32}$$

Collecting terms containing θ_1 gives

$$\begin{gathered} \sin\theta_1 \sin\theta_2 + \cos\theta_1(\cos\theta_2 - K_2) = K_1 \cos\theta_2 - K_3 \\ A \sin\theta_1 + B \cos\theta_1 = C \end{gathered} \tag{34.33}$$

$$\left.\begin{aligned} A &= \sin\theta_2 \\ B &= (\cos\theta_2 - K_2) \\ C &= (K_1 \cos\theta_2 - K_3) \end{aligned}\right\}$$

Solving for l_1, recall from trigonometry that

$$\sin\theta_1 = \frac{2t}{1+t^2}$$

where

$$t = \tan\frac{1}{2}\theta_1$$

and

$$\cos\theta_1 = \frac{1-t^2}{1+t^2}$$

Equation 34.33 becomes

$$(B + C)t^2 - 2At + (C - B) = 0 \tag{34.34}$$

This equation is a quadratic in 't' and the solution will be

$$t = \frac{A \pm \sqrt{A^2 + B^2 - C^2}}{B + C} \tag{34.35}$$

There being two solutions:

$$\theta_1^+ = 2\tan^{-1}\frac{A + \sqrt{A^2 + B^2 - C^2}}{B + C} \tag{34.36}$$

Corresponding to the positive sign, and

$$\theta_1^- = 2\tan^{-1}\frac{A - \sqrt{A^2 + B^2 - C^2}}{B + C} \tag{34.37}$$

These two values relate to the two possible ways that the linkage can be assembled as depicted in Figure 34.18 for the given values of l_1, l_2, l_3, l_4 and ABCD will be referred to as an 'open' linkage and ABCD as a 'crossed' linkage.

Example 34.3

A four-bar linkage has the following parameters:

Crank	= 100 mm
Coupler	= 200 mm
Follower	= 200 mm
Fixed link	= 200 mm

Consider the quadrilateral chain shown in Figure 34.19. Determine the Freudenstein equation when the input link (l_1) is indexed every 15° between 0° and 360°.

The Freudenstein equation is given as

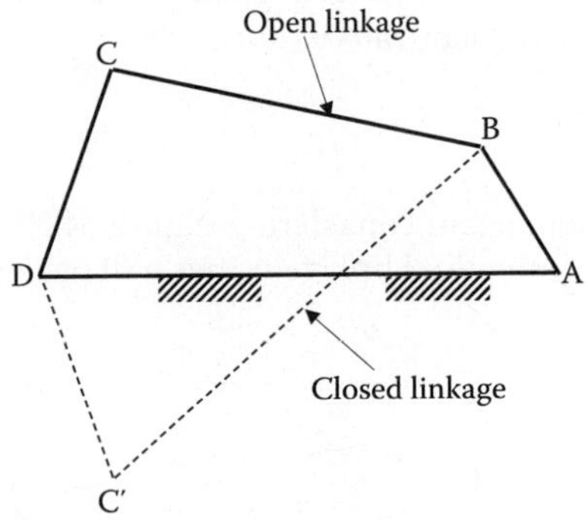

FIGURE 34.18
Open and closed linkages.

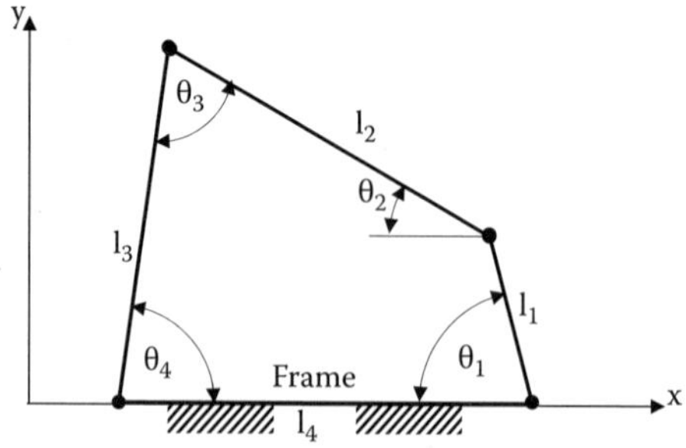

FIGURE 34.19
Example 34.3, kinematic chain.

$$K_1 - K_2 \cdot \cos\theta_4 + K_3 - \cos(\theta_4 - \theta_1) = 0$$

where

$$K_1 = \frac{l_4}{l_3} = \frac{200}{100} = 1.00$$

$$K_2 = \frac{l_4}{l_1}$$

$$K_3 = \frac{(l_1^2 - l_2^2 + l_3^2 + l_4^2)}{2 \cdot l_1 \cdot l_3}$$

Table 34.1 tabulates the results from the above equations.

34.12 Drawing Velocity Vectors for Linkages

In the slider–crank linkage example shown in Figure 34.20, link 'OA' is driven and connects to a further link 'AB'. This, in turn, connects to a slider block at 'B', which is allowed to reciprocate in a horizontal plane.

This section will concentrate on determining the velocity of the links using a graphical method for evaluating the link velocities. This is considered easier than using vector arithmetic and is capable of achieving accurate values.

Example 34.4

To demonstrate the method, again considering Figure 34.20, the following values are applied to the figure. The individual link velocities will be determined for the position shown:

'OA' = 75 mm
'AB' = 200 mm
'x' = 75 mm
'θ' = 45°
'N' = 2000 rev/min

TABLE 34.1

Example of Freudenstein's Equation, Four-Bar Linkage—Freudenstein's Equation

Link Lengths (mm)				
a	100			
b	200			
c	200			
d	200			
Ratio 1	1			
Ratio 2	2			
Ratio 3	1.25			
Input Angle, θ_1		Output Angle, θ_2		
Degree	Radian	Degree	Radian	Freudenstein's Equation
0	0.0000	41	0.7156	5.2120
15	0.2618	47	0.8203	3.4004
30	0.5236	53	0.9250	4.6194
45	0.7854	60	1.0472	4.9145
60	1.0472	68	1.1868	1.5152
75	1.3090	75	1.3090	−0.5935
90	1.5708	83	1.4486	0.9970
105	1.8326	89	1.5533	2.1873
120	2.0944	94	1.6406	−0.3358
135	2.3562	97	1.6930	3.1452
150	2.6180	96	1.6755	3.4402
165	2.8798	89	1.5533	0.4053
180	3.1416	76	1.3264	1.5482
195	3.4034	60	1.0472	5.1509
210	3.6652	48	0.8378	3.3238
225	3.9270	40	0.6981	4.5219
240	4.1888	34	0.5934	3.7234
255	4.4506	31	0.5411	1.0047
270	4.7124	29	0.5061	4.3657
285	4.9742	29	0.5061	37.859
300	5.2360	29	0.5061	30.661
315	5.4978	31	0.5411	0.1115
330	5.7596	33	0.5760	2.3958
345	6.0214	37	0.6458	−0.2732
360	6.2832	41	0.7156	4.0967

Note: Freudenstein's equation: $'F' = \text{Ratio } 1 - \text{Ratio } 2 \times \cos\theta_4 + \text{Ratio } 3 - \cos(\theta_1 - \theta_4) = 0$.

In this example, there are eight steps required to find the absolute velocity of the slider 'B' as well as the relative velocity of the slider with respect to the joint 'A':

Step 1: Draw a space diagram of the mechanism. This will generally be a scaled representation of the mechanism at the desired orientation of the links. The scale chosen will be determined on the method of drawing. If the drawing is being sketched, the scale will be dictated by the size of the paper being used.

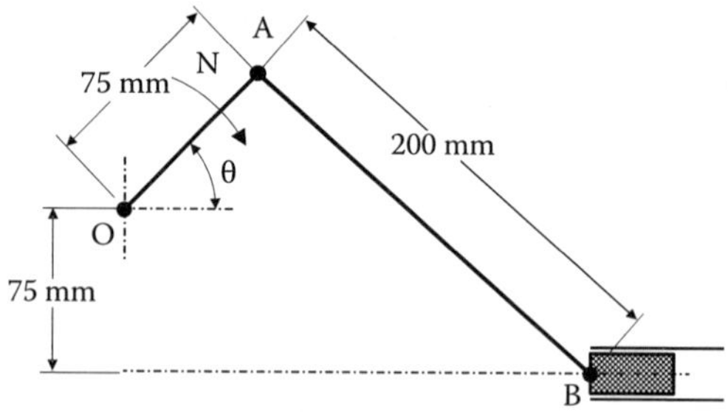

FIGURE 34.20
Space diagram for a slider–crank mechanism.

Alternatively, if the sketch is being drawn using a suitable CAD type program, then the sizes will be the actual dimensions.

Step 2: The first vector 'OA' is drawn as in Figure 34.21a with a length of 75 mm inclined at an angle of 45° to the horizontal.

Step 3: The diagram is continued by drawing in the vector 'AB' having a length of 200 mm and intersecting with the axis, which is 75 mm below the fixed point 'O' at 'B'. At the completion of step 3, a diagram will consist of the two lines 'OA' and 'AB' as shown in Figure 34.21b.

Step 4: Calculation of the linear velocity of joint 'A'.

The rotational speed 'N' of the link 'OA'=2000 rev/min; this is equivalent to 209.44 rad/s.

$$\text{The linear velocity of the link 'OA'} = \text{angular velocity} \times \text{length of the link 'OA'}$$

$$\text{Linear velocity of link OA} = 209.44\ \text{rad/s} \times 75\ \text{mm}$$

$$= 15{,}708\ \text{mm/s}$$

Step 5: The velocity vector diagram can now be progressed by drawing the velocity vector for the link 'OA'. This is drawn perpendicular to the link having a length equivalent to 15,708 mm, depending on the scale being drawn. This will represent the velocity vector $V_{oa'}$ with a value of 15,708 mm/s as shown in Figure 34.21c.

Step 6: The following step is to draw the velocity vector for the link 'AB'. There are two points known about this vector:

- It will start from the point 'A_1' of the velocity vector diagram.
- It will be perpendicular to the orientation of the link 'AB' as shown in Figure 34.20.

At this point, the value of the 'A_1B_1' is not known (the linear velocity of the joint 'B' with respect to the joint 'A'). The diagram will look similar to Figure 34.21d.

Step 7: To find the location of point 'B_1', the use of the property of point 'B' will be used. Point 'B' will always have the motion only in the horizontal direction; hence, it will pass through the point 'O' (Joint 'B' is connected to the slider and is only allowed to reciprocate in the horizontal direction). Connecting the horizontal line with point 'O' gives the velocity of the joint 'B' with respect to joint 'O'. This now completes the final velocity vector of the mechanism. The final velocity diagram will look like Figure 34.21e.

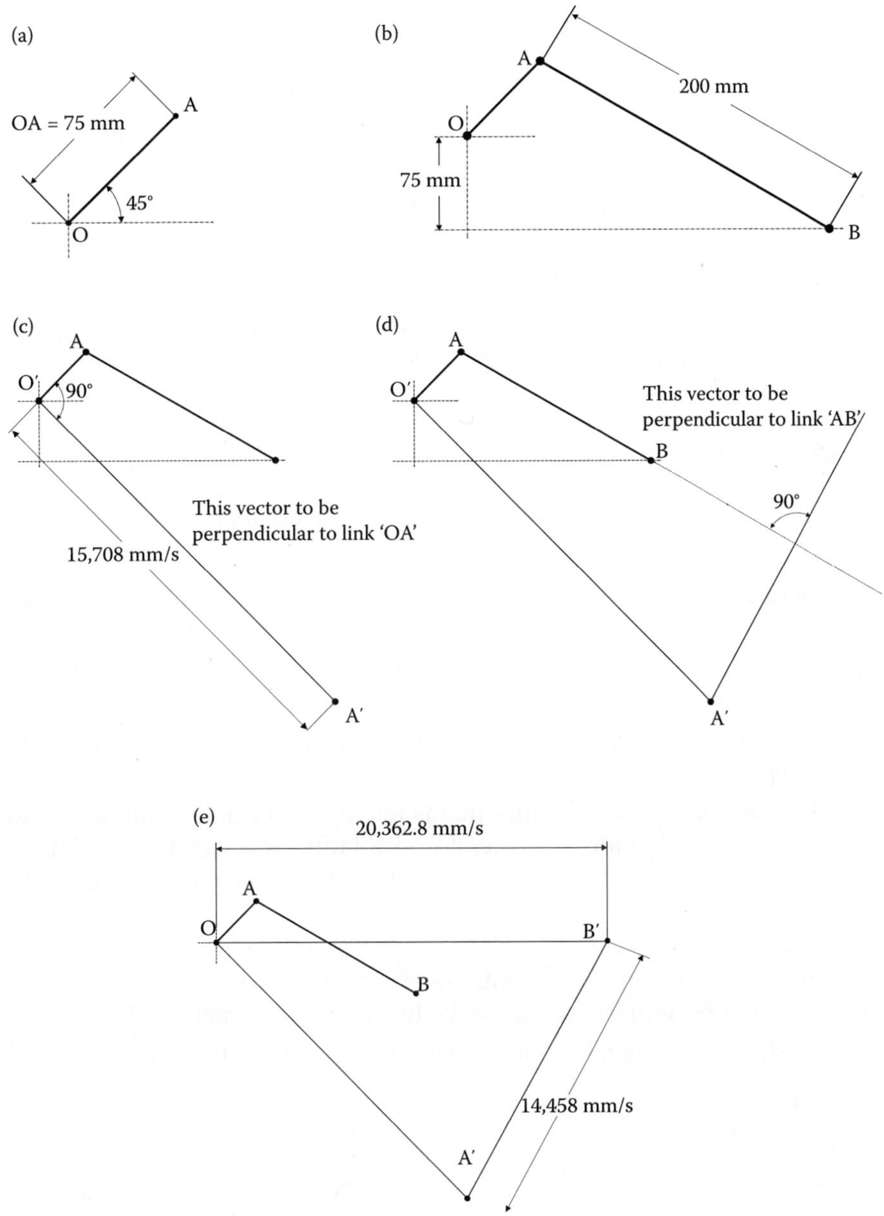

FIGURE 34.21
Velocity diagrams for Example 34.4. (a) Link OA step 2, (b) space diagram at step 3, (c) velocity diagram at step 5, (d) velocity diagram at step 6 and (e) velocity diagram at step 7.

Step 8: All that remains is to measure the length of the vectors to obtain the velocity of joint 'B' with respect to 'A' from the length of the line 'A_1B_1' which measures 14,458 mm/s and the velocity of point 'B' with respect to 'O' (or the ground) measuring 20,362.8 mm/s.

This procedure can be repeated for the mechanism with varying values of 'θ' and the velocity profile of the mechanism can be resolved.

The method outlined above can be used to determine velocity vectors for other mechanisms without difficulty. Its prime advantage is that the method is easily understood.

Section 34.13 will focus on drawing the acceleration vector diagram.

34.13 Drawing Acceleration Vectors for Linkages

In Section 34.12, a velocity diagram for a slider–crank mechanism was developed. In this section, an acceleration vector diagram for the same example will be discussed.

Reiterating, the space diagram for the slider–crank mechanism is shown in Figure 34.20.

Before proceeding, there are a number of points that need to be borne in mind regarding the acceleration diagrams for a mechanism.

Item 1: Typically, there will be two components of acceleration:

- Radial
- Tangential

Item 2: The direction of the radial component of acceleration will always be parallel to the orientation of the link.

Item 3: The direction of the tangential component will always be perpendicular to the orientation of the link.

Item 4: In the case of the end of a link that is moving at a constant angular velocity (i.e., a link that is pivoting at one end and rotating at a constant rev/min with respect to the pivot point), the rotating end will only have a radial acceleration component with no tangential component.

Item 5: Where a joint moves in a straight line (such as a slide or a cylinder), there will be no radial acceleration component. The direction of the total acceleration of the joint, in such cases, will be parallel to the line of the movement of the joint.

Item 6: The radial acceleration for a link can be calculated from the following equation:

$$A_{xy} = xy \times W_{xy}^2 = \frac{V_{xy}^2}{xy}$$

where

A_{xy} is the radial acceleration of joint 'x' with respect to joint 'y'
W_{xy} is the angular velocity of joint 'x' with respect to joint 'y'
V_{xy} is the linear velocity of joint 'x' with respect to joint 'y'
xy is the length of link 'xy'

Example 34.5

The following steps are required to draw an acceleration vector diagram for the slider–crank mechanism shown in Figure 34.20.

Step 1: Orientation of link 'OA'. The joint 'A' is rotating at a constant angular velocity (2000 rev/min) about joint 'O'; from item 4, joint 'A' will only have a radial component with no tangential component of acceleration. From item 2, the radial component in the acceleration vector diagram will be orientated parallel to the link 'OA'.

Step 2: Value of the radial acceleration component for the link 'OA'. From the Example statement 34.12, the angular velocity of the link 'OA' (Woa) is 2000 rev/min=209.44 rad/s and the length of the link 'OA' is 75 mm.

From Item 6, the radial acceleration component for the link 'OA' is calculated as

$$\text{Aroa} = 75\ \text{mm} \times 209.44^2 = 3,289,884\ \text{mm}/\text{s}^2$$

Step 3: Representing Aroa in the acceleration diagram. Due to the very large numbers encountered with the acceleration values, it will be sensible to give them a scale factor to make the drawing more manageable. In this instance, a scale factor of 10^4 will give the length of the vector Aroa of 328.99 mm in the acceleration vector diagram. As there is no tangential component for the link, 'Arao' will represent Aao; this is shown by the vector 'O_2A_2' in Figure 34.22a.

Step 4: Drawing the radial component of acceleration for the joint 'B' with respect to 'A', Arba for the link 'AB'.

Start by drawing a line from point 'A_2' to the point 'z' as shown in Figure 34.22b and orientated along the link 'AB' of the space diagram.

Using item 6 to calculate the value of Arba:

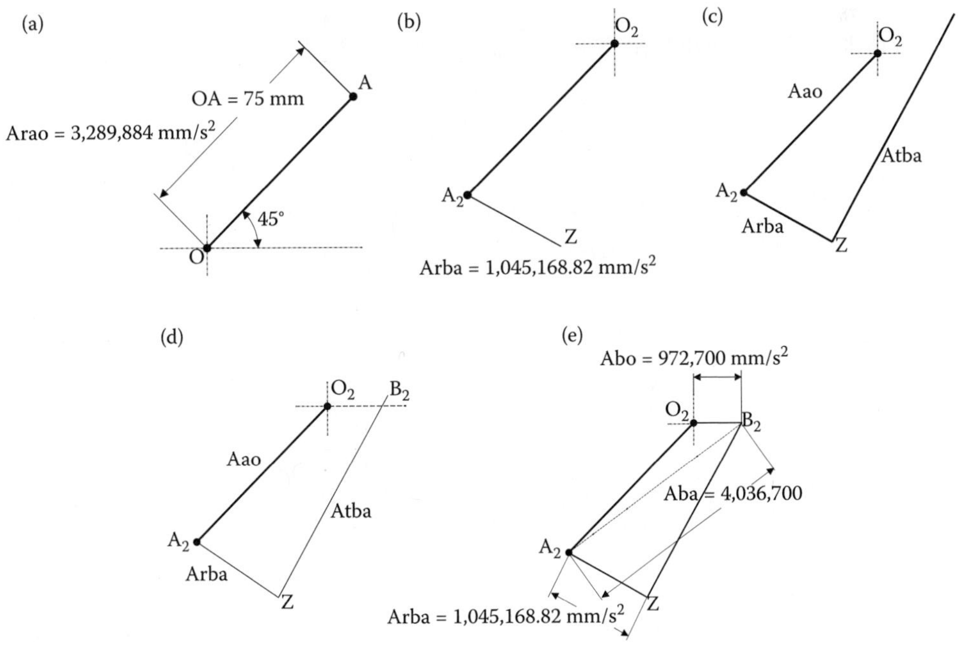

FIGURE 34.22
Acceleration diagrams for Example 34.5. (a) Link OA step 1, (b and c) acceleration diagram Step 5 and (d and e) acceleration diagram Step 6.

$$\text{Arba} = \frac{(\text{Vba})^2}{\text{BA}}$$

$$= \frac{14,458^2}{200}$$

$$= 1,045,168.82 \text{ mm / s}^2$$

The values of 'Vba' and 'BA' are obtained from Example 34.4.

The scaled value of Arba = 1, 045, 168.82/10^4 = 104.517 mm

Step 5: Drawing the tangential component (Atba) for the link AB. From item 3, the tangential component of the acceleration will be perpendicular to the radial component. At the moment, the value of the tangential component (Atba) of acceleration is unknown.

Referring to Figure 34.22c, a line is drawn perpendicular to the line A_2z.

Step 6: Drawing the acceleration vector for the slider 'B'. From item 5, the total acceleration of the joint 'B' with respect to the joint 'O' is represented by a horizontal straight line starting from 'O_2' until it intersects the line representing the vector 'Atba'.

In Figure 34.22d, the line O_2 represents the total linear acceleration for the slider with respect to joint 'O'.

Step 7: Calculating the output acceleration values. The length of the lines 'A_2B_2' and 'O_2B_2' represents the linear acceleration of the joint 'B' with respect to the joint 'O'. This is shown in Figure 34.22e.

Measuring the respective lengths and multiplying by the scale factor, the following values are obtained:

Linear acceleration of joint 'B' with respect to joint 'A'

$$\text{Aba} = 4,036,700 \text{ mm / s}^2$$

Linear acceleration of joint 'B' with respect to the joint 'O'

$$\text{Abo} = 972,700 \text{ mm / s}^2$$

35

Introduction to Cams and Followers

35.1 Introduction

Modern-day automated production machinery relies on servo-driven motion control systems, and this has unquestionably provided considerable flexibility in the design of machine units. This has led to a revolution in machine design where motion control can be accomplished using servo-motors that are connected to its control system by cables instead of drives and linkages.

An added advantage is that the "cam profile" is stored electronically and can be amended within the control unit program without the need to modify or machine a new cam profile.

It is important for the design engineer to be aware of the history of the early motion control and the reasons for its evolution.

Cams are a very versatile mechanism with many applications. They generate motion, including rectilinear, oscillatory, or rotary motion to a mechanical element nominally referred to as a follower. This is in contact with the cam profile.

Many types of motion are possible for the follower, as the cam face can have any desired profile within certain limitations, including acceptable dynamic forces.

The cam is generally attached to a rotating shaft. In some cases the rotation may not be continuous but can be oscillatory. The cam profile will determine the type of motion the follower will have, such as a slow rise at a particular acceleration to a given value followed by a dwell for a limited period and a rapid return to the start point of the cycle.

35.2 Background

In the early and mid twentieth century, high production automatic machinery was exclusively driven by cams and gears. In the late twentieth century the electric servo and hydraulic servo systems began to be developed with a high degree of reliability and offered a more versatile operating system. Although this was thought to be the death knell of the use of cams in machinery, this has not occurred across the board. There are numerous examples, including packaging machinery, labeling, etc., where repetitious and accurate movements are required to position or provide a stop within a process line. These are usually provided by a cam-operated mechanism, as it generally offers a cost-effective and reliable method. Most people's thoughts generally turn to the automotive engine, where cams are used to operate the valve gear for the induction and exhaust process. This is undertaken under the most adverse conditions of heat and vibration and provides a very reliable and cheap system that has not found any alternative to replace it.

This section will provide the design engineer with some advice on the design of a suitable cam-operated system using either a plate cam or a cylindrical cam. An imaginary project is considered such that the various elements can be demonstrated and the effect on the overall design is considered.

35.3 Requirements of a Cam Mechanism

The basic requirements for a cam-operated mechanism should consider the following points:

1. The mechanism should be "stiff" and have minimal deflections in any linkages, etc.
2. The cam follower must be able to maintain contact with the cam profile under all operating conditions.
3. The mechanism must be able to operate and maintain positional tolerances.
4. The mechanism must be able to operate at the design speed and maintain clearances with other mechanisms operating within the same space envelope.
5. The output from the follower must meet the design requirements.

35.4 Terminology

There are two principal designs of rotary cams: plate cams and cylindrical cams.

35.4.1 Plate Cams

Plate-type cams are the most common and recognizable form of a cam, where examples can be found in automotive engines for the operation of the inlet and exhaust valves where the cam is machined onto the "cam shaft." Other examples will be found in automatic machine tools, packaging machinery, and some quick-release work vices. In general, the profiles can easily be machined using standard machine tools.

The cam followers for plate-type cams come in two forms: sliding follower and rotary or swinging arm follower.

The cam followers move in a plane perpendicular to the cam face.

Figures 35.1 and 35.2 show both forms of follower and also show the terminology used in the cam designations.

35.4.2 Cylindrical Cams

In certain instances it may not be possible to fit a plate-type cam due to space limitations. In this instance a cylindrical cam (Figure 35.3) may be a more convenient solution.

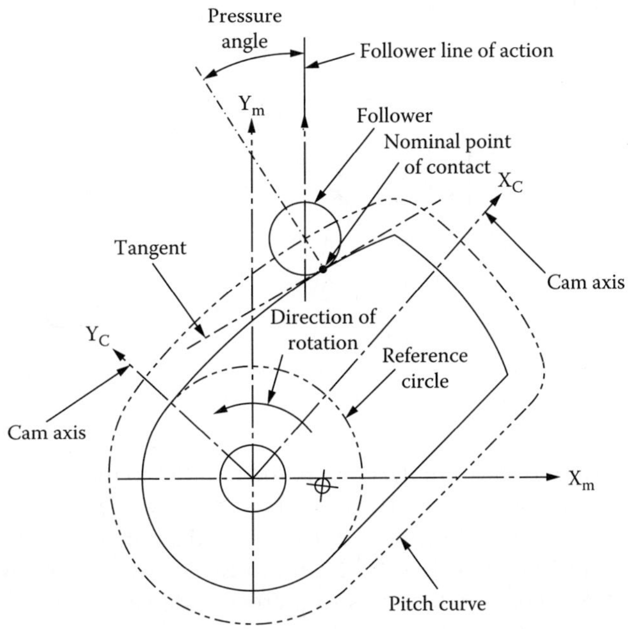

FIGURE 35.1
Cam driving an offset translating follower.

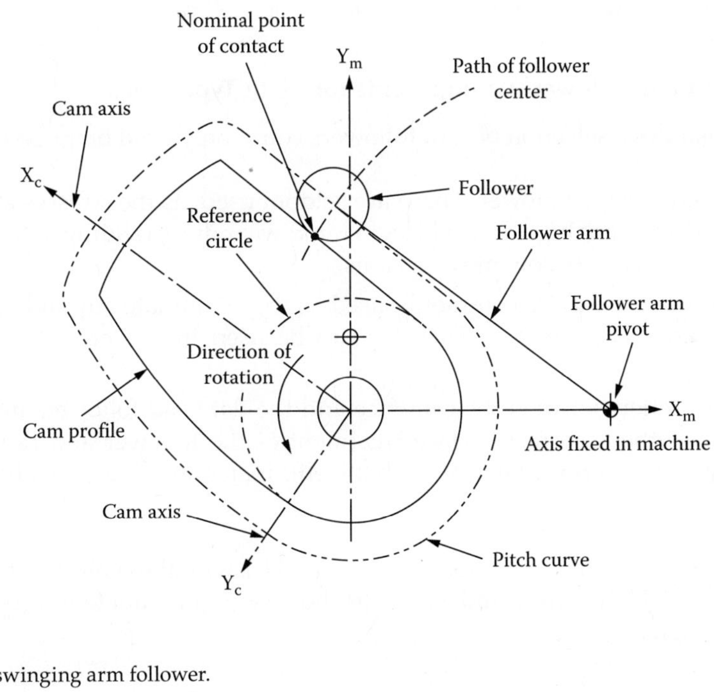

FIGURE 35.2
Cam driving a swinging arm follower.

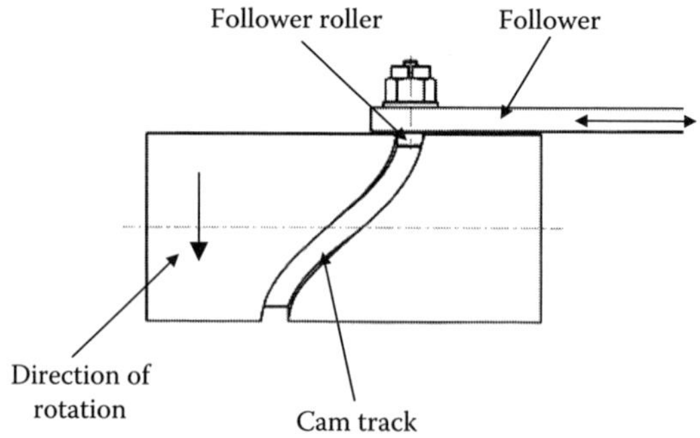

FIGURE 35.3
Cylindrical cam and follower.

In this design the cam follower moves in a plane parallel to the axis of the cam.

Depending on the size of the cam, the cam track may be made up of segments allowing for the track to be modified to meet particular requirements. This type of cam and follower arrangement is more likely to be found in earlier versions of metal-cutting automatic lathes where the follower connects to a cross-slide carrying the cutting tool. The track will advance the follower at a specific linear rate for either form cutting or even cross-drilling, and on completion this will be followed by a short dwell and then a rapid return to its starting position.

An advantage to the use of the cylindrical cam is the roller follower is essentially trapped in the cam track and therefore follows the cam profile accurately. The mechanism is sometimes fitted with a light spring to keep the roller against one side of the track when returning to the start position.

A further variation to the cylindrical cam is where the cam profile is machined onto the edge of the cylinder. These are referred to as a face or shell cams.

35.4.3 Typical Cam Follower Arrangements for Plate-Type Cams

Figure 35.4 illustrates a selection of cam followers commonly used on plate-type cams:

(a) Depicts a knife edge follower. These are seldom used, as the stresses at the tip of the follower are very high and subject to rapid wear due to sliding at the cam face. They are only used in slow-moving cams.
(b) Roller followers are more commonly used as they eliminate any sliding between the roller and the cam face. They tend to be used in moderate to high-speed systems.
(c) Although friction between the cam face and the flat head followers may present a problem without correct surface treatment of the follower and cam face, the advantage of this type of follower is that it eliminates any side thrust between the contact surfaces.

Figure 35.4d through f shows variations from (b) and (c), with the centerline position of the roller position in (d) being offset, and (e) and (f) show swinging cam followers for the roller and flathead designs.

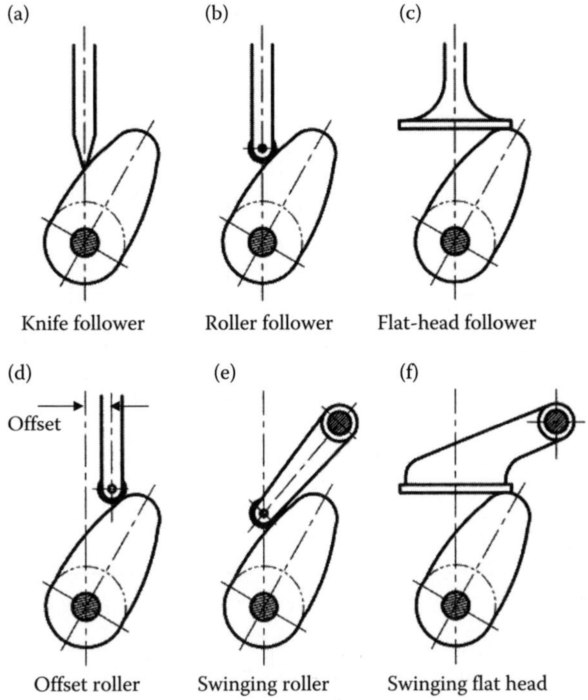

FIGURE 35.4
Typical cam follower designs.

The advantage of using this type of cam followers in (e) and (f) is the pressure angle of the cam system can be increased. This will be discussed in Section 35.7.

35.5 The Timing Diagram

The timing diagram is an important feature in the cam design. As the name suggests, this covers all aspects of the mechanism's operating envelope, including the cam rise time, dwell, return, and final dwell, usually the base circle of the cam.

Once a preliminary design has been selected, consideration should be given to its operation using the timing diagram (Figure 35.5). This will investigate the feasibility of the design and highlight any potential problems, including any possible crash points in the design.

The timing diagram is only concerned with the angular rotation of the cam or the timescale the cam operation is based upon. The shape of the cam profile is irrelevant, as the timing diagram is only concerned with the start of the lift period and its completion. Other machine activities may be in operation at the same time, and the diagram will help to synchronize these operations.

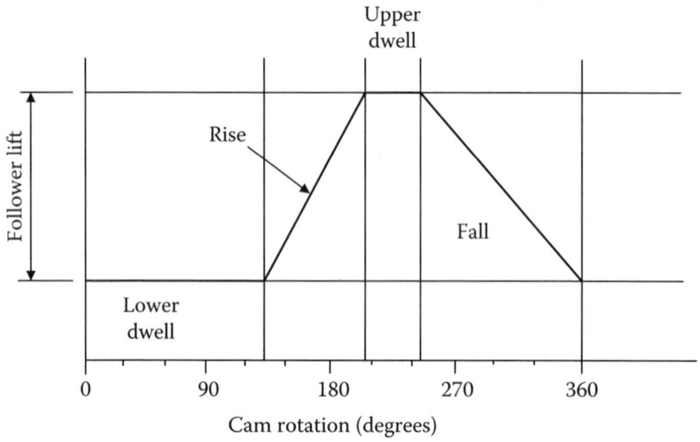

FIGURE 35.5
Typical timing diagram.

35.6 Cam Laws

The cam profile is important as it dictates the following:

1. The speed of rotation of the cam
2. How the follower behaves (will the follower leave the profile?)
3. The life of the follower

The profile can be classified by examination of the follower motion requirements, investigating the velocity, acceleration, and jerk generated due to the profile.

The cam laws cover and include the following:

Constant velocity
Parabolic (constant) acceleration
Simple harmonic motion
Cycloidal (sine) acceleration

The first step is to identify the cam rotation required to lift (or drop) the follower from the lower dwell to the upper dwell point on the cam profile. This is best carried out using the timing diagram (Figure 35.6). In constructing the diagram, a straight line is drawn between the lower and upper points between the dwell points, without giving consideration to the profile requirements at this point.

The cam duration can be expressed in either degrees (θ) or time in seconds (t).

The following expression can describe the follower displacement y due to a rotation θ of the cam:

$$y = f(\theta) \quad \text{or} \quad y = f(t) \tag{35.1}$$

depending on the units of the x axis used.

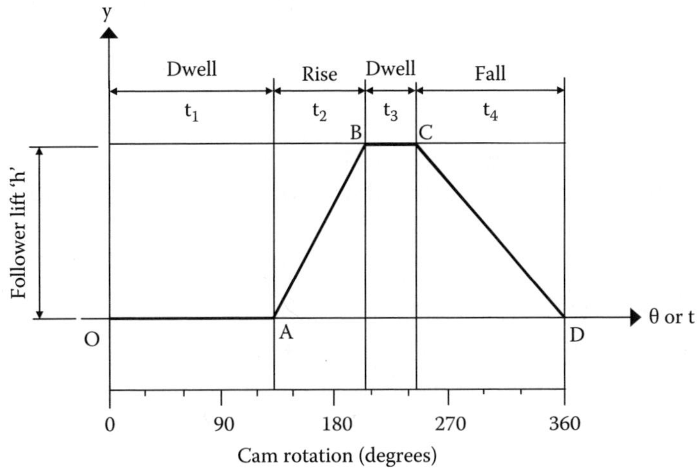

FIGURE 35.6
Timing diagram.

The first differential of y with respect to t gives the velocity of the follower:

$$v = \frac{dy}{dt} \quad \dot{y} = \frac{df(t)}{dt} = \frac{df(\theta)}{d\theta}\frac{d\theta}{dt} = \omega\frac{df}{d\theta} \tag{35.2}$$

where ω = angular velocity of the cam.

The second differential of y with respect to t will give the acceleration of the follower:

$$a = \frac{d^2y}{dt} \quad \ddot{y} = \omega^2\frac{d^2f}{dq^2} + \dot{\omega}\frac{df}{dq} \tag{35.3}$$

The third differential of y with respect to t will give the jerk* of the follower:

$$J = \frac{d^3y}{dt^3} = \dddot{y} = \omega\frac{df}{dq} + 3\omega\dot{\omega}\frac{d^2f}{dq^2} + \omega^3\frac{d^3f}{dq^3}$$

In the case of the cam rotating at a constant rotational speed the above equations for acceleration can be reduced to the following:

Acceleration: $$a = \omega^2\frac{d^2f}{d\theta^2}$$

Jerk: $$J = \omega^3\frac{d^3f}{d\theta^3}$$

* Jerk is an undesirable feature, as it can result in the follower leaving the cam profile momentarily, or at its worst, resulting in substantial damage to the follower.

Let T = time taken for the cam to rotate through an angle of α degrees

$$= \frac{\alpha}{\omega}$$

where α is the rotation of the cam in degrees for a follower rise of h.

If a cam is designed to have a rise or fall at a constant velocity, it will experience very large changes in acceleration and jerk.

35.6.1 Constant Velocity of the Follower

In this first case, the follower moves at a constant velocity from the base circle O to the maximum lift of the follower A (Figure 35.7).

Consider a nominal cam profile with the following characteristics:

$h=25.0$ mm	Full lift
$\alpha=65°$	Angular rotation to full lift
$\theta=6.5°$	Interval spacing
$\omega=360°/s$	Angular velocity

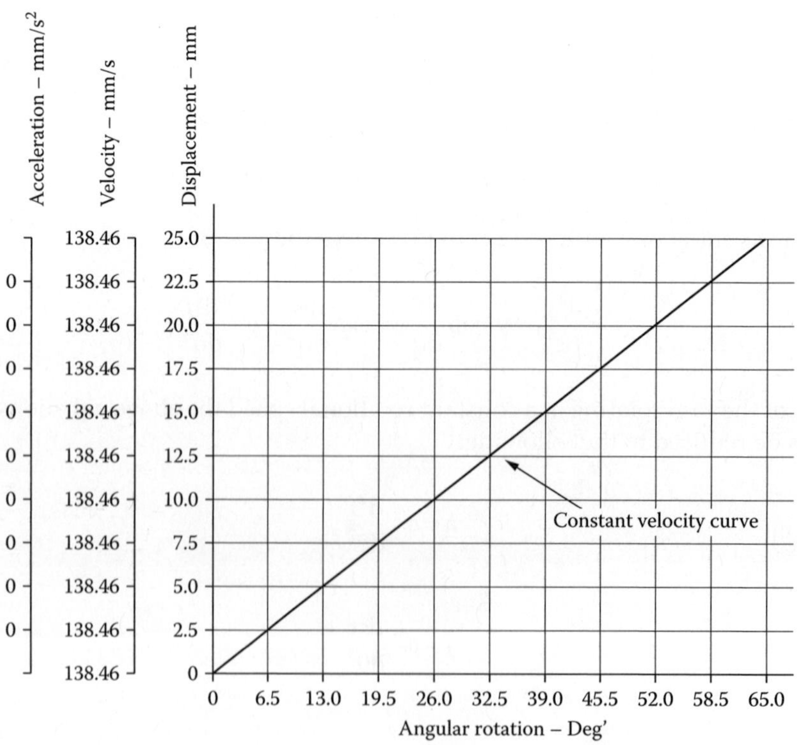

FIGURE 35.7
Constant velocity curve.

TABLE 35.1

Constant Velocity Values

y=25 mm
α=65°
θ=2.5°
w=360°/s

y (mm)	θ°	v (mm/s)	a (mm/s²)
0.0	0.0	138.46	Infinity
2.5	6.5	138.46	0
5.0	13.0	138.46	0
7.5	19.5	138.46	0
10.0	26.0	138.46	0
12.5	32.5	138.46	0
15.0	39.0	138.46	0
17.5	45.5	138.46	0
20.0	52.0	138.46	0
22.5	58.5	138.46	0
25.0	65.0	138.46	Infinity

Displacement: $y = \frac{ht}{T}$ or $y = \frac{h\theta}{\alpha}$

Velocity: $v = \frac{h}{T}$ or $v = \frac{h\omega}{\alpha}$

Acceleration: $a = 0$

At positions O and A where the acceleration will be infinite. Tabulating the above using a set of notional values.

Table 35.1 shows the above using a set of notional values.

This type of motion is unacceptable except at very low speeds.

35.6.2 Parabolic Motion

In this second case (Figure 35.8), the profile is also known as a constant acceleration cam law. This law has been widely used in the past, as it produces the lowest maximum follower acceleration. It does produce an infinite jerk at both the beginning and the end of the motion cycle and also at the point of inflection. This makes this a very poor choice where dynamic considerations are important, and for this reason it is only suitable for low-speed systems.

Mathematically we have:

Displacement: $y = 2h\left(\frac{t}{T}\right)^2$ or $y = 2h\left(\frac{\theta}{\alpha}\right)^2$

Velocity: $v = 4h\frac{t}{T2}$ or $v = \frac{4h\omega\theta}{\alpha^2}$

Acceleration: $a = \frac{4h}{T}$ or $a = \frac{4h\omega^2}{\alpha^2}$

Jerk: $J = 0$ Except at o, B and A where it is infinite.

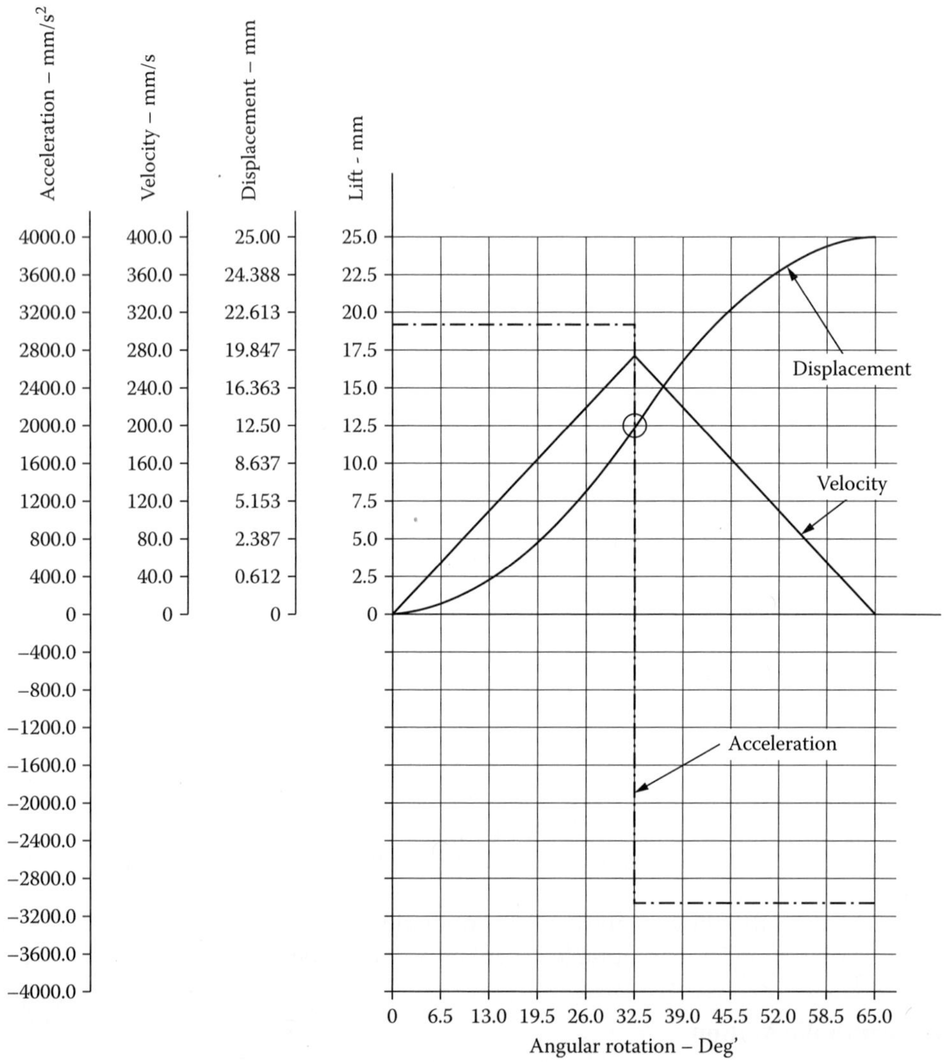

FIGURE 35.8
Parabolic motion.

The tabulated values of the notional cam are shown in Table 35.2.

Due to the sharp increase and decrease in the acceleration curve, jerk will be extremely high, resulting in the possibility of the follower leaving the cam profile. For this reason this kind of profile is only suitable for low-speed systems.

35.6.3 Simple Harmonic Motion

This third case of cam laws shows an improvement on the velocity and acceleration curves compared that of the parabolic cam law in that they are smoother without any discontinuities.

See Table 35.3 to consider the mathematics and Figure 35.9 to compare the displacement, velocity, acceleration, and jerk curves for this type of motion.

TABLE 35.2
Parabolic Motion Values

y=25 mm
α=65°
θ=2.5°
ω=360°/s

y (mm)	θ°	Disp. (mm)	v (mm/s)	a (mm/s²)	J (mm/s³)
0.0	0.0	0.0	0.0	3,067.46	0
2.5	6.5	0.5	55.4	3,067.46	0
5.0	13.0	2.0	110.8	3,067.46	0
7.5	19.5	4.5	166.2	3,067.46	0
10.0	26.0	8.0	221.5	3,067.46	0
12.5	32.5	12.5	276.9	0.00	0
15.0	39.0	17.0	221.5	−3,067.46	0
17.5	45.5	20.5	166.2	−3,067.46	0
20.0	52.0	23.0	110.8	−3,067.46	0
22.5	58.5	24.5	55.4	−3,067.46	0
25.0	65.0	25.0	0.0	−3,067.46	0

TABLE 35.3
Simple Harmonic Motion Values

y=25 mm
α=65°
θ=2.5°
ω=360°/s

y (mm)	θ°	Disp. (mm)	v (mm/s)	a (mm/s²)	Jerk (mm/s³)
0.0	0.0	0.000	0.00	3,784.32	0.0
2.5	6.5	0.612	67.21	3,599.10	−20,347.4
5.0	13.0	2.387	127.84	3,061.58	−38,703.1
7.5	19.5	5.153	175.96	2,224.37	−53,270.2
10.0	26.0	8.637	206.85	1,169.42	−62,622.9
12.5	32.5	12.500	217.49	0.00	−65,845.6
15.0	39.0	16.363	206.85	−1,169.42	−62,622.9
17.5	45.5	19.847	175.96	−2,224.37	−53,270.2
20.0	52.0	22.613	127.84	−3,061.58	−38,703.1
22.5	58.5	24.388	67.21	−3,599.10	−20,347.4
25.0	65.0	25.000	0.00	−3,784.32	0.0

Displacement: $$y=\frac{h}{2}\left(1-\cos\frac{\pi t}{T}\right) \quad \text{or} \quad y=\frac{h}{2}\left(1-\cos\frac{\pi\theta}{\alpha}\right)$$

Velocity: $$v=\frac{\pi h}{2T}\sin\frac{\pi t}{T} \quad \text{or} \quad v=\frac{\pi h\omega}{2\alpha}\sin\frac{\pi\theta}{\alpha}$$

Acceleration: $$a=\frac{\pi^2 h}{2T^2}\cos\frac{\pi t}{T} \quad \text{or} \quad a=\frac{\pi^2 h\omega^2}{2\alpha^2}\cos\frac{\pi\theta}{\alpha}$$

Jerk: $$J=-\frac{\pi^3 h}{2T^3}\sin\frac{\pi t}{T} \quad \text{or} \quad J=-\frac{\pi^3 h\omega^3}{2\alpha^3}\sin\frac{\pi\theta}{\alpha}$$

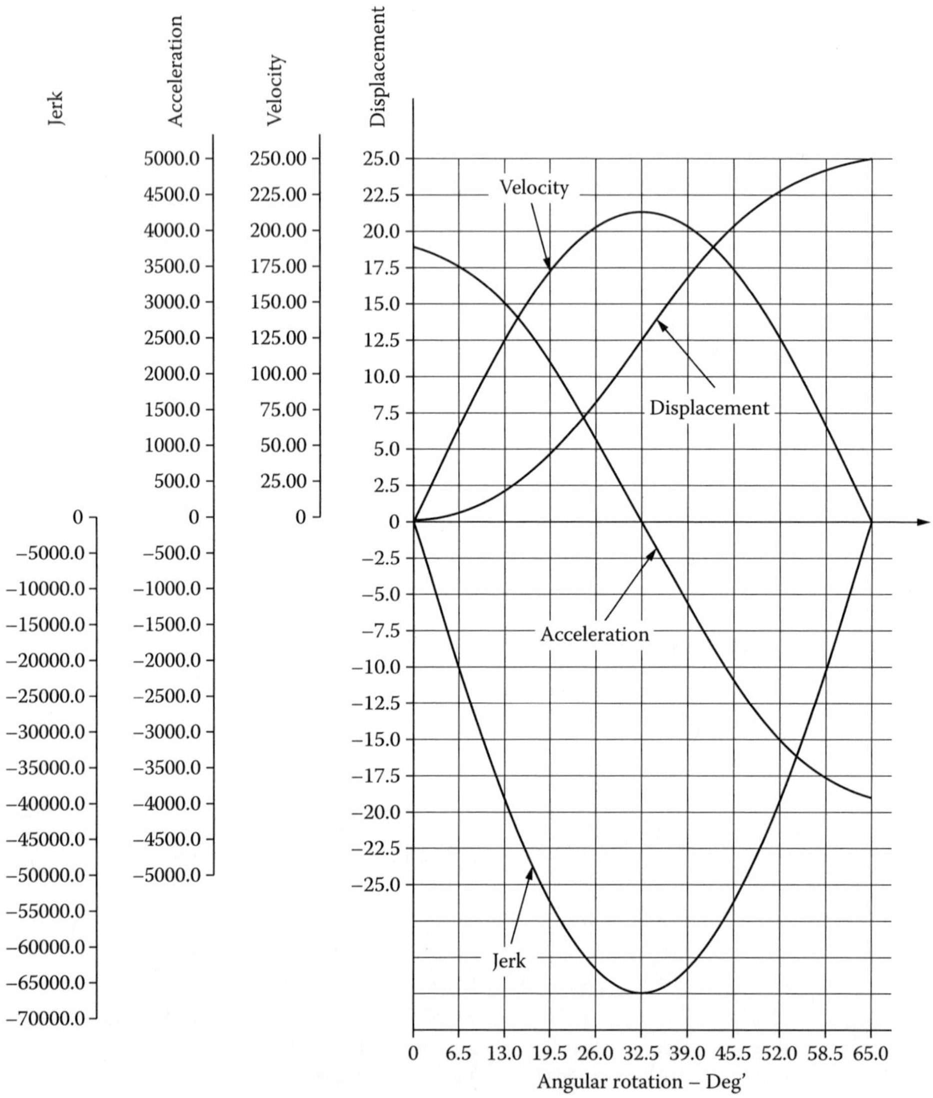

FIGURE 35.9
Simple harmonic curve.

35.6.4 Cycloidal Motion

This fourth case (Figure 35.10) has the highest nominal acceleration of the cam laws reviewed so far. This case also has a finite jerk.

Although the acceleration has a greater value than that of the simple harmonic motion for the same cam parameters, this curve is widely used in high-speed applications. Since the curve has zero acceleration at the beginning and the end of the motion segment, it is easily coupled to dwells at these points.

Coupling two cycloidal curves together without a dwell period in between is to be avoided, as the pressure angle will tend to be high in this instance. See Table 35.4 for the tabulated values.

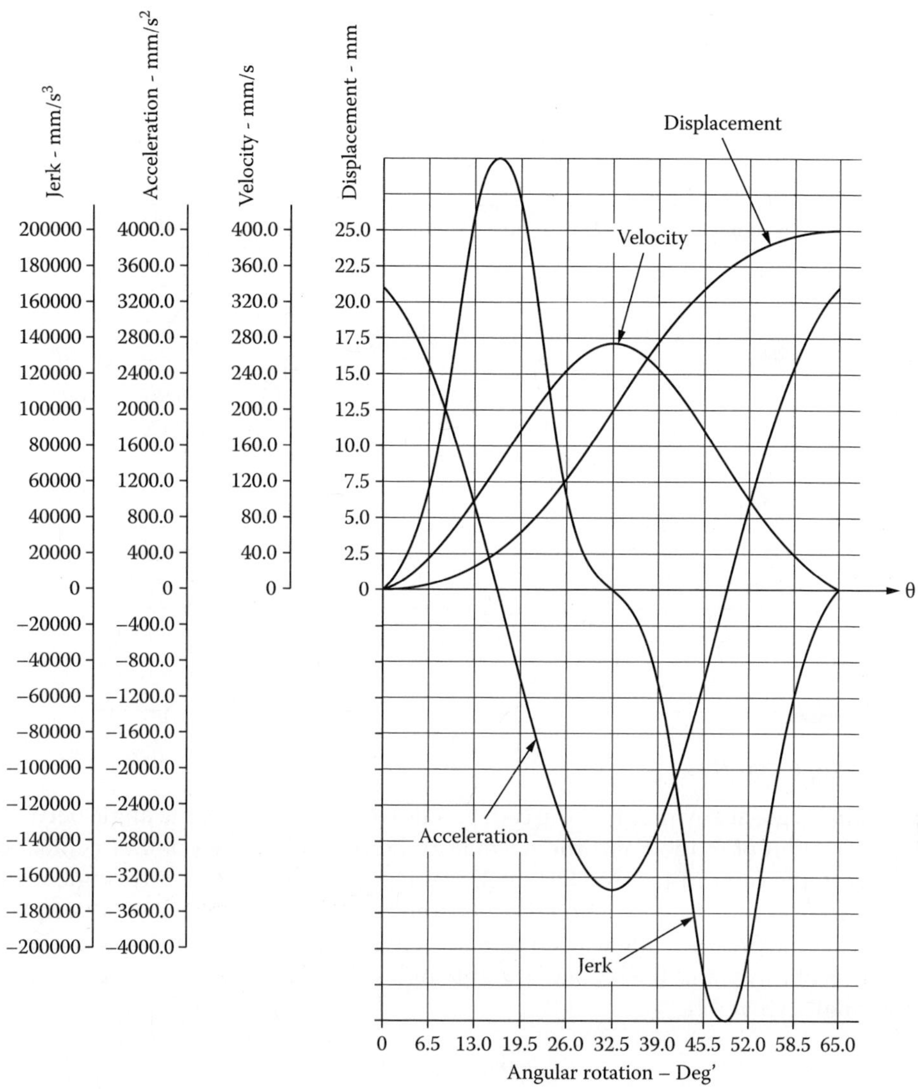

FIGURE 35.10
Cycloidal motion curve.

The mathematics for the cycloidal curve:

Displacement: $y = h\left(\frac{t}{T} - \frac{1}{2\pi}\sin\frac{2\pi t}{T}\right)$ or $y = h\left(\frac{\theta}{\alpha} - \frac{1}{2\pi}\sin\frac{2\pi\theta}{\alpha}\right)$

Velocity: $v = \frac{h}{T}\left(1 - \cos\frac{2\pi t}{T}\right)$ or $v = \frac{h\omega}{\alpha}\left(1 - \cos\frac{2\pi\theta}{\alpha}\right)$

Acceleration: $a = \frac{2\pi h}{T^2}\sin\frac{2\pi t}{T}$ or $a = \frac{2\pi h\omega^2}{\alpha^2}\sin\frac{2\pi\theta}{\alpha}$

Jerk: $J = -\frac{4^3 h}{T^3}\cos\frac{2\pi t}{T}$ or $J = -\frac{4\pi^2 h\omega^3}{\alpha^3}\cos\frac{2\pi\theta}{\alpha}$

TABLE 35.4

Cycloidal Motion Values

$y = 25\,mm$
$\alpha = 65°$
$\theta = 2.5°$
$\omega = 360°/s$

y (mm)	θ°	Disp. (mm)	v (mm/s)	a (mm/s²)	J (mm/s³)
0.0	0.0	0.0	0.0	0.00	167,675
2.5	6.5	0.2	26.4	2,832.15	135,652
3.5	13.0	1.2	95.7	4,582.52	51,814
7.5	19.5	3.7	181.2	4,582.52	−51,814
10.0	26.0	7.7	250.5	2,832.15	−135,652
12.5	32.5	12.5	276.9	0.00	−167,675
15.0	39.0	17.3	250.5	−2,832.15	−135,652
16.0	45.5	21.3	181.2	−4,582.52	−51,814
20.0	52.0	23.8	95.7	−4,582.52	51,814
22.5	58.5	24.8	26.4	−2,832.15	135,652
25.0	65.0	25.0	0.0	0.00	167,675

35.7 Pressure Angle

The pressure angle at any point on the cam profile can be defined as the angle between the follower direction of motion and the normal to the tangent at the point of contact of the follower with the cam profile (see Figure 35.11). The value of the pressure angle is important, as it dictates the forces acting on the follower. From Figure 35.11 it can be seen that the force acting at the pressure angle α can be resolved into two components, one tangential to the direction of travel and the second at right angles to it. This angle α is referred to as the pressure angle. Therefore,

$$P = P_t + P_n$$

P_n is undesirable as it exerts a side thrust on the follower guides or bearings, and if this is large enough, it will cause the follower to bind and increase the friction.

It is common practice for conventional translating-type followers to keep the pressure angle below 30° and at or below 45° for swinging-type followers. These values are on the conservative side and in many cases may be increased, but beyond these limits trouble could develop and further analysis will be required.

The pressure angle is important for the following reasons:

- As stated above, the pressure angle increases the side thrust acting on the follower, and this in turn increases the forces between the follower and cam profile, resulting in undesirable wear.
- Reducing the pressure angle increases the size of the cam, and often this is undesirable where space is at a premium. Furthermore, large cams require greater care in manufacture.

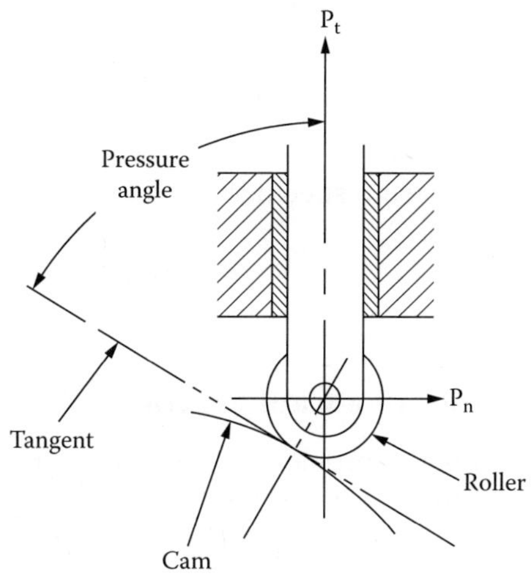

FIGURE 35.11
Pressure angle.

- Larger cams also mean more mass, and hence higher inertia forces. This in turn will lead to increased vibration in high-speed machines.
- The inertia of the larger cam may impede with quick starting and stopping.

There are ways to reduce the pressure angle; these include:

- Increase the base circle diameter.
- Reduce the total follower travel.
- Increase the amount of cam rotation for a given follower displacement.
- Change the follower motion type.
- Change the follower offset; this may generally be reduced.

35.8 Design Procedure

The following is a recommended design procedure to establish an optimum cam-operated mechanism.

1. Determine a timing diagram highlighting the key requirements for displacements, velocities, and accelerations.
2. Determine the space constraints in which the mechanism will be required to fit.
3. Select a suitable cam mechanism such as a plate-type cam or cylindrical cam that will fit best in the available space.
4. Consider the cam laws the mechanism will be required to meet.
5. Calculate the minimum size of cam that will meet the pressure angle and profile constraints.
6. Calculate the maximum contact stress and dynamic load on the follower.

7. Do the dimensions of the cam mechanism and materials meet the design requirements?
8. Will an external supplier manufacture the cam mechanism? If so, ensure it is in the consultation loop.
9. Finally, prepare a product design specification (PDS).

35.9 Graphical Construction of a Cam Profile

In the graphical design of cams the problem of construction is simplified by employing inversion, that is, to imagine the cam being held stationary and the follower then being rotated in the opposite direction to the cam rotation. This preserves the correct sequence when the motion takes place.

To illustrate the point, Figures 35.12 and 35.13 are shown; the profile of the cam and the follower is then placed in equal increments around the periphery.

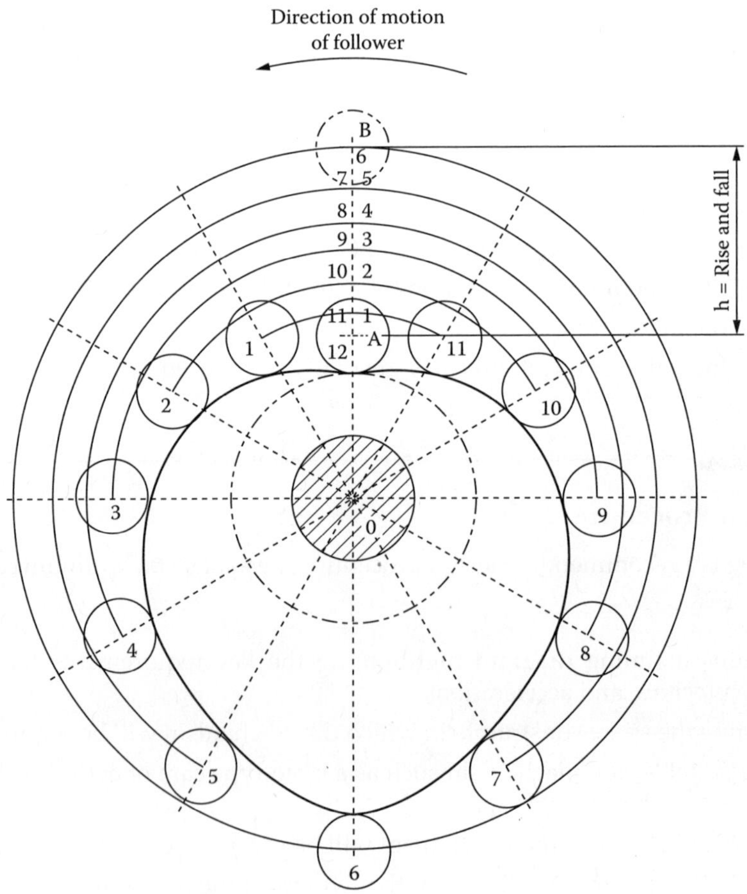

FIGURE 35.12
Constant velocity of follower–graphical construction.

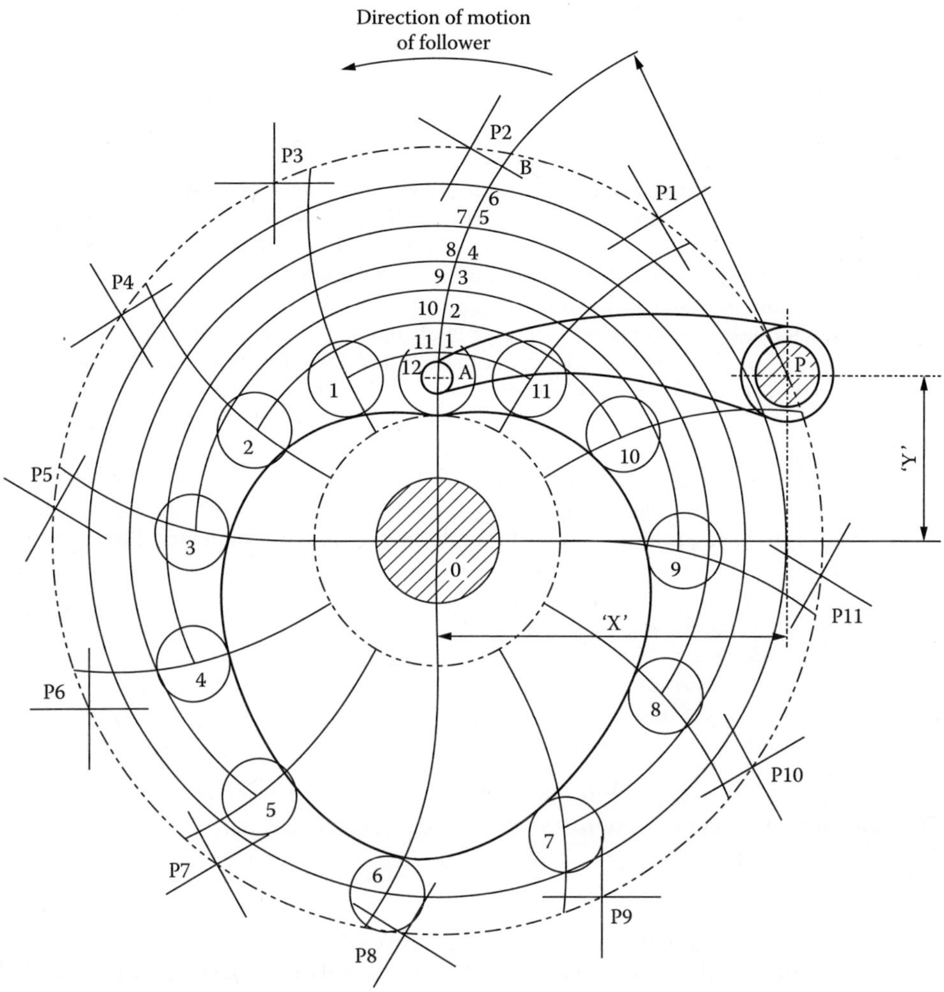

FIGURE 35.13
Swinging link follower having a uniform angular velocity–graphical construction.

The first construction (Figure 35.12) shows a cam designed for imparting a constant velocity on a translating-type roller that is on the centerline of the cam. For the purposes of the construction, one complete revolution of the cam is to give one rise and one fall to the follower.

Example 35.1

A plate cam fitted with a translating-type follower (Figure 35.12) with the following details is to be designed:

Shaft diameter = 10.0 mm
Base circle diameter = 23.0 mm
Roller diameter = 7.0 mm
Rise and fall of follower = 15.0 mm
Rotation of the cam = clockwise

Step 1. Draw the shaft diameter and base circle at center O and extend the centerline upwards to represent the follower position. The original drawing was produced using AutoCad; therefore it was drawn full size.

Step 2. Offset the base circle diameter plus half the radius of the follower diameter by the rise dimension (15.0 mm) to position B. Considering the cam is held stationary, the follower will rotate in a counterclockwise direction.

Step 3. Subdivide the line A:B into six equal parts, numbering them 1, 2, 3, …, and produce a series of arcs continuing around the circle as shown in Figure 35.12.

Step 4. Radiate from the center O a series of lines at 30° angular rotation.

Step 5. Locate positions 1, 2, 3, …, and on the intersections draw the rollers.

Step 6. If you are using a drafting program such as AutoCad, the next step is to draw straight lines between the points of intersection, following this with a "PEDIT" command. The lines are joined and followed by the "SPLINE-FIT." This will generate a suitable curve.

Example 35.2

It is required to design a cam using a swinging link that is fitted with a roller follower. The profile is to have uniform velocity for both the rise and fall segments of the cam.

Cam data:

Shaft diameter = 10.0 mm
Base circle diameter = 23.0 mm
Roller diameter = 7.0 mm
Position of the pivot from the cam centerline:
X = 30.0 mm
Y = 15.0 mm
Total angular movement of the link = 30°
Rotation of the cam = clockwise

Refer to Figure 35.13.

Step 1. Draw the shaft diameter and base circle at center O and position the swinging link at position P with a length P:A. The original drawing was produced in AutoCad; therefore, it was drawn full size.

Step 2. Draw a circle with a radius of O:P in order for the successive positions of the link to be positioned. Subdivide this circle into 12 equal parts and number them as P1, P2, P3, …, P12. Considering the cam is held stationary, the follower will rotate in a counterclockwise direction.

Step 3. From the position P construct an arc of a circle A:B with a radius P:A such that the angle A:P:B = 30°. Divide this arc into six equal parts, numbering them 1, 2, 3, …, and produce a series of arcs continuing around the circle.

Step 4. From each position of P1, P2, P3, …, construct an arc as in step 2, originating from P2, P3, etc.

Step 5. Locate points 1, 2, 3, …, on these lines and with these points as centers; then draw the rollers.

Step 6. As in the previous, if you are using AutoCad, "PEDIT" and "SPLINE-FIT" will produce a suitable curve.

36

Geared Systems

36.1 Introduction

Rotary transmission between shafts is accomplished using different methods:

1. Pulleys, using v-belts, timing belts, chains
2. Hydraulic pumps and motors, servo-electro motors
3. Gearing

This third group will be discussed in this chapter. Very simplistically, a gearbox is a device for mechanically transmitting power from one shaft to another and either maintaining, increasing, or decreasing the speed N of the second shaft to the first. Because of the speed change, there will be a change in the torque T.

The gearbox designer's prime task is to design a system that will provide the required characteristics to match the demands of the drive system, whether it be a metal-cutting machine tool drive, automotive gearbox, winch drive, etc.

Gears are wheels that mesh with each other through interlocking teeth. Rotation of one wheel will cause the rotation of the other, albeit in the opposite direction, without any slip between them. There are a number of gear teeth designs; the most common is the cycloidal and involute form. The fundamental basis of the involute form is there is no sliding between the tooth surfaces, the action is purely rolling, and, hence, there is minimum damage between the teeth, provided the tooth size is adequate for the power being transmitted and there is adequate lubrication in the gearbox.

36.2 Types of Gears

There are many different types of gears, and the classification or identification depends largely on how the gears are used. The most common combination is where the two shafts are parallel with each other; here, the gears will be either spur or helical and the shafts will rotate in the opposite direction to each other. When the two shafts are not parallel with each other, the connecting gears can be either skew or spiral or even bevel gears.

This chapter will identify the more common configurations.

36.2.1 Spur Gears

This type of gear (Figure 36.1) is most commonly used to connect two parallel shafts that rotate in opposite directions. The teeth are parallel to the axis of the gear and are identical in profile. Because there is no axial force generated due to the tooth loading, these are used as sliding gears for change-speed mechanisms in gearboxes.

The *spur pinion* (i.e., the gear having the smaller number of teeth in a pair of mating gears) is also used to mesh with an internal spur gear. The principal problem with straight-cut spur gears is that they are inherently noisy.

36.2.2 Internal Spur Gears

In internal spur gears, the teeth are cut on the internal diameter of an annulus or ring (Figure 36.2). The teeth are parallel to the axis and the same profile meshing in a similar manner to the spur gear. The gears are used to connect two parallel shafts usually opposed, and when engaged, the shafts run in the same direction.

This type of gear is used in epicyclic transmissions, reduction gears for machine tools, and for clutches and couplings. In the latter case, however, the external spur pinion has the same number of teeth as the internal wheel and, therefore, rotates as a single body in the same direction.

FIGURE 36.1
Spur gear.

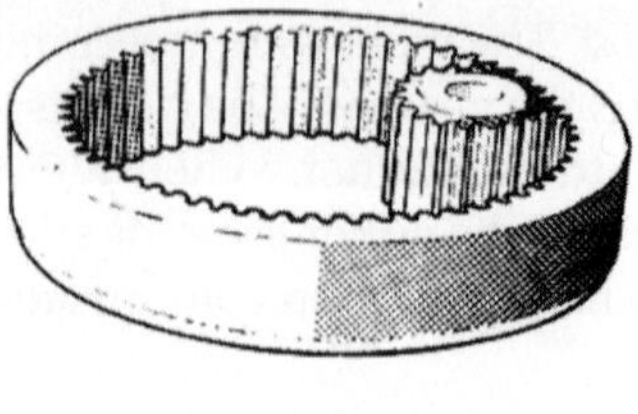

FIGURE 36.2
Internal spur gear.

36.2.3 Rack and Pinion

The pitch line of a spur rack moves in a longitudinal direction and the rack may be considered a spur gear of infinite radius or zero curvature (Figure 36.3).

A spur pinion (as in Figure 36.1) is most commonly used to transmit the longitudinal motion to the rack, and the most common application can be found on lathes, when in this case the rack is used to transmit motion to the saddle. The rack tooth is of fundamental importance to the study of involute teeth.

36.2.4 Helical Gears

Helical gears (Figure 36.4) are used in the same way as spur gears, but differ from them in that the teeth are cut on the periphery of the gear blank and are of helical or screw form.

The pinion cannot be moved longitudinally along its axis without imparting angular motion to its mating wheel, and for this reason helical gears are used in automotive gearboxes as constant mesh on syncro-mesh applications.

The transverse section of a helical gear is identical with a spur gear insofar as the tooth profile is concerned and is involute on this section. Mating gears must, however, be made with the same helix angle but opposite hand, and the tooth loading produces an axial thrust.

Helical gears are considered quieter than spur gears, and where noise will be a problem, these will be used in those applications where possible.

36.2.5 Double Helical Gears

As the name implies, double helical gears (Figure 36.5) are the equivalent of two helical gears secured together, but in actual practice they are manufactured as one piece. The teeth may be continuous or separated by a gap, although the two helices forming the double helix are opposite hand and meet at a common apex. They are produced by a variety of processes and are distinguished by the name of the manufacturing process by which the teeth are cut.

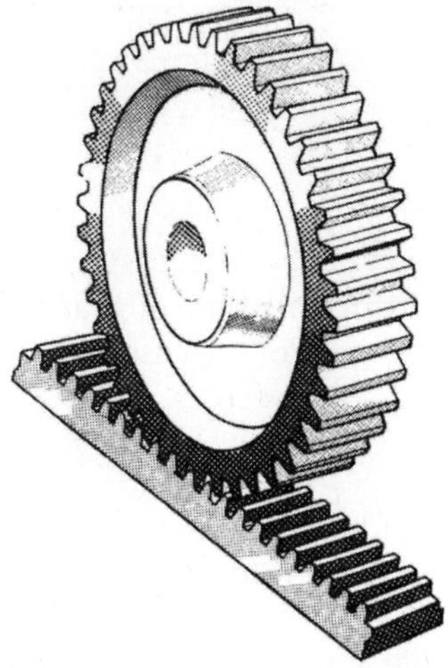

FIGURE 36.3
Rack and pinion.

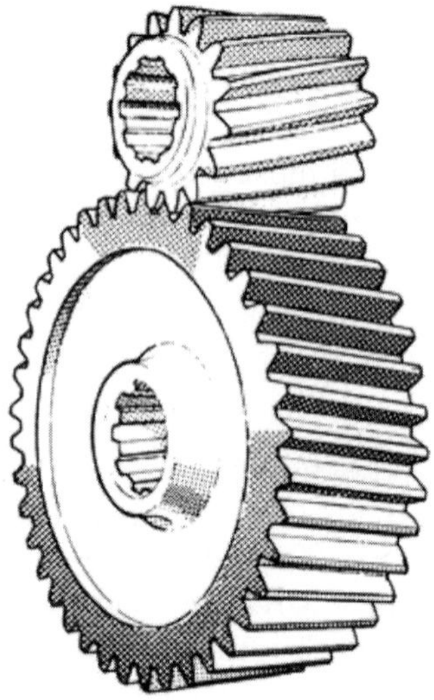

FIGURE 36.4
Helical gears.

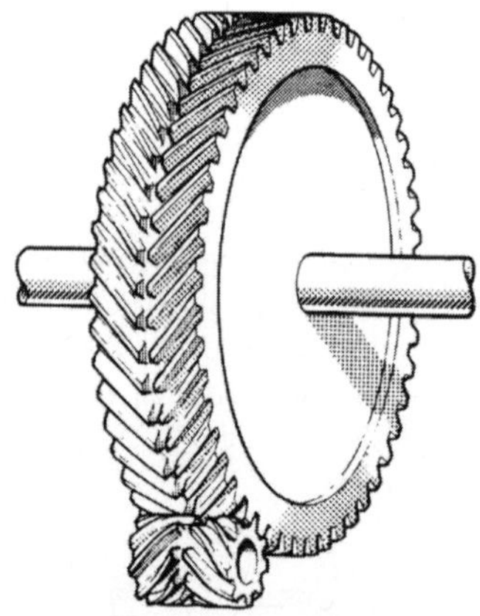

FIGURE 36.5
Double helical gears.

It is the usual practice when cutting this type of gear to know the direction of the teeth in relation to the driving end, especially in the case of shaft gears.

36.2.6 Spiral Bevel Gears

Spiral bevel gears (Figure 36.6) are identical to helical gears; the only reason they are called spiral gears is that the two gears are used to connect two shafts that are not parallel with each other.

Spiral gears used in this way give only a theoretical point contact instead of a line contact, as is the case for the previous gear types, and a longitudinal sliding motion is introduced between the teeth.

FIGURE 36.6
Spiral bevel gear.

36.2.7 Bevel Gears

In the case of two intersecting shafts, bevel gears (Figure 36.7) would be used in this application. The most common type is that in which the teeth are radial to the point of intersection of the shaft axes or apex. Such gears are called *straight bevel gears.*

The teeth are similar to spur gears in that they make a line contact across the face of the teeth. The teeth are proportionally smaller at the front of the gear than the tooth form at the back of the gear in the ratio of

1. The distance from the apex to the front end.
2. The distance from the apex to the rear face.

The shaft angle most commonly encountered is 90°, and the bevel gears with this angle are often referred to as *miter gears.*

36.2.8 Spiral Gears

Spiral gears (Figure 36.8) are similar to helical gears; the only reason they are called spiral gears is that the two gears are used to connect two shafts that are not parallel with each other.

Spiral gears used in this way give only a theoretical point contact instead of a line contact, as is the case for the previous gear types, and a longitudinal sliding motion is introduced between the teeth.

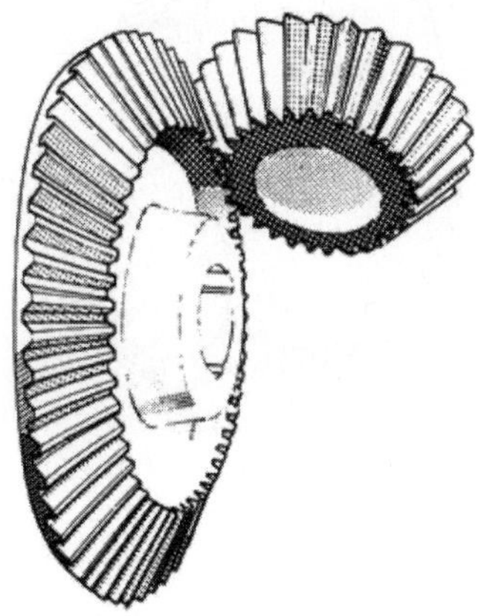

FIGURE 36.7
Bevel gears.

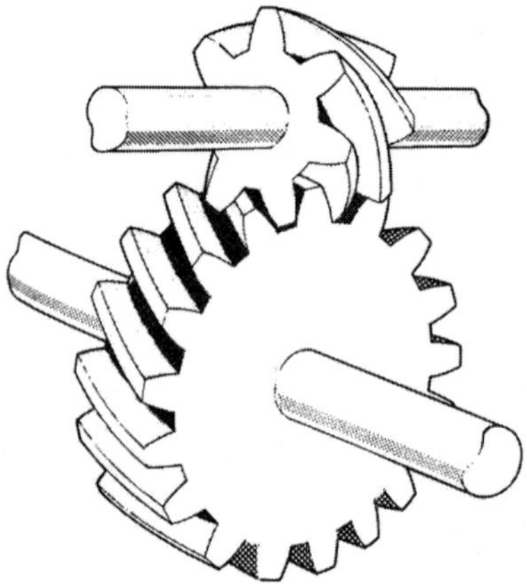

FIGURE 36.8
Spiral gears.

36.2.9 Worm and Worm Wheels

Worm gears (Figure 36.9) are used in the same manner as spiral gears, that is, to connect skew shafts, but not necessarily at right angles.

The *worm gear* exactly resembles one of a pair of spiral gears, but the wheel is throated and has concave teeth. While resembling spiral gears in purpose and general dimensions, worm gears differ from spiral gears in that they give line contact instead of point contact and will carry greater loads. A further difference in spiral gears is that the center distance is important in relation to the helix angle. This problem does not arise when calculating the worm gears.

The sliding velocity is high with worm gears when compared with other types of gears, and very special attention is necessary with the choice of materials and lubrication.

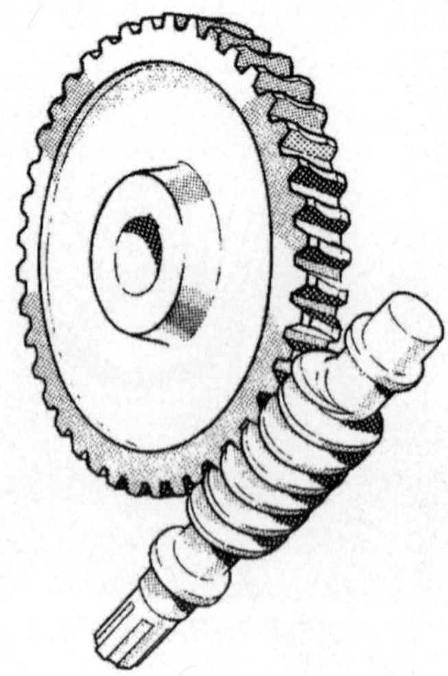

FIGURE 36.9
Worm and worm wheel.

36.3 Form of Tooth

The function of a pair of gears is to transmit motion from one shaft to another with a uniform velocity ratio. Either the cycloidal or the involute form of tooth will satisfy this requirement, but the cycloidal has the disadvantage that a slight variation in the theoretical center distance destroys the uniformity of angular velocity.

It is difficult to ensure that the exact theoretical center distances can be maintained under all circumstances. The involute tooth form is much more tolerant to the small errors that can occur between the theoretical and manufactured center distances. This has ensured the universal adoption of the involute tooth form.

A further factor for the adoption of the involute tooth form is that it can be produced by a straight-sided rack cutter, and, in fact, every type of involute gear was originally produced by means of this basic principle. It can, therefore, be stated that all the types of gears produced in Section 36.2 are produced with an involute tooth form with the exception of bevel gears (Figure 36.7) and spiral bevel gears (Figure 36.8).

In the case of bevel gears, the teeth are generated by a straight-sided cutter, representing the side of a tooth of an imaginary crown gear, and the form of tooth produced is termed *octoid*.

A *crown gear* is in effect a bevel gear having a pitch angle of 90°, and the teeth of a true involute crown gear should have very slightly curved tooth profiles.

The difference between the tooth shapes on a gear being cut by an imaginary crown gear having a straight-sided cutter and one with a tooth profile that is slightly curved is so small that it can be ignored.

Both forms give theoretically correct action, and for practical purposes it can be safely assumed that the form of tooth produced on the transverse plane of both straight and spiral bevel gears is of involute form.

Spur gears (Figure 36.1), spur racks (Figure 36.3), helical gears (Figure 36.4), spiral gears (Figure 36.8), and worm and worm wheels (Figure 36.9) are all of involute form on their transverse planes, that is, on the plane of rotation.

This point should be kept in mind when undertaking any calculations or layouts regarding the form of the tooth, especially in the case of helical gears, spiral gears, or worms.

For these types of gears, which in any case are closely related to each other, the form of tooth on the normal and axial planes is such that the form is entirely dependent on the involute form on the transverse plane and the spiral or helix angle.

36.4 Layout of Involute Curves

The involute tooth form has been constructed on the transverse plane of the spur gear pair shown in Figure 36.1, and it will be particularly noted that the origin of the wheel and pinion tooth shapes is based on the base circle diameters D_o and d_o.

The base circle diameter should be considered as a disc around which a line of tape is wrapped around the diameter, the two ends of the tape commencing and finishing at point A, as shown in Figure 36.10.

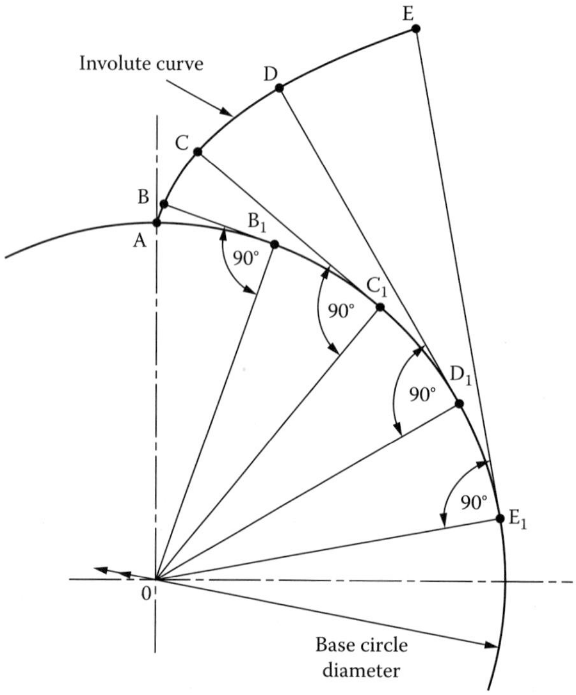

FIGURE 36.10
Development of the involute curve.

By securing one end of the string or tape at position A, unwinding the free end of the tape will describe the path of the involute curve, through points B, C, D, and E, when the tape is tangential to the base disc at the corresponding points B_1, C_1, D_1, and E_1.

Before the involute form can be applied to a gear tooth, some considerations should be given to the question of the pitch diameter and pressure angle.

The pitch diameter and pressure angle together with the addendum and dedendum define which part of the involute curve is being utilized on a gear tooth, and before any tooth layout can made it is necessary to have this information.

The pitch diameter is always greater than the base circle, and where the involute bisects the pitch diameter circle is called *the pitch point*. From this point, the pressure angle is obtained in conjunction with the base circle and the center of the gear.

Figure 36.11 shows the construction of a spur gear tooth in relation to the pressure angle and the pitch diameter; it should be noted that a right-angle triangle is formed by the three sides a, b, and c when

$$a = \frac{\text{Base diameter}}{2} \tag{36.1}$$

$$b = \frac{\text{Pitch diameter}}{2} \tag{36.2}$$

$$\psi = \text{Pressure angle of gear} \tag{36.3}$$

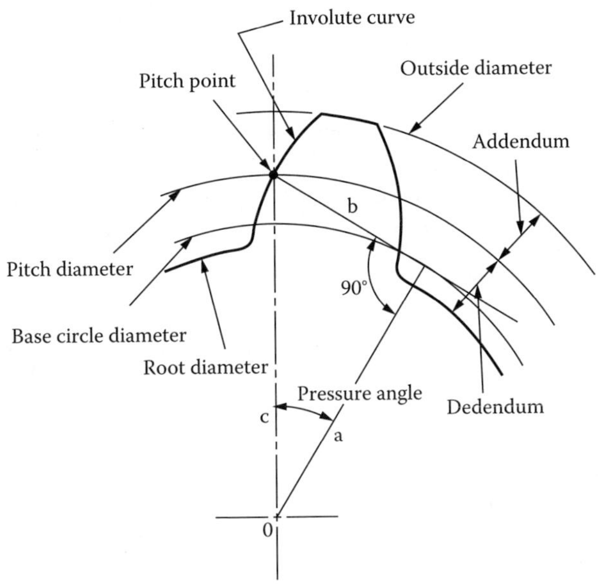

FIGURE 36.11
Description of the involute tooth.

Therefore, given the pitch diameter D and the pressure angle ψ of a gear, the base circle diameter D_o is equal to D cos ψ when $D=2\times c$ and $D_o=2\times a$.

Alternatively, given D and D_o, the pressure angle can be obtained as follows:

$$\cos\psi = \frac{D_o}{D} \tag{36.4}$$

There are several methods of setting out the involute tooth form, and the method selected is dependent on the accuracy required for the purpose of the proposed tooth shape.

For example, if an accurate projection drawing is required, it is recommended that the tooth shape is generated, especially if a gear has a small number of teeth, as the shape changes very rapidly on gears of this nature.

For a gear having a large number of teeth, the shape does not change, and in fact it will be found that the shape is practically equivalent to the arc of a circle.

For setting out tooth shapes when the accuracy is not the first essential, the following procedure can be followed. This method is suitable for helical, spiral, worm, and bevel gears on their transverse planes, as well as for spur gears.

Referring to Figure 36.12, the first step is to draw the pitch diameter D together with the outside and root diameters. The outside diameter is larger than the pitch circle diameter by an amount equal to twice the addendum. The root diameter is smaller than the pitch diameter by an amount equal to twice the dedendum of the gear. The pressure angle ψ is then set out from the center line of the gear and from the gear center O, while another line from the pitch point C at right angles to the line OC_1 is drawn to form a triangle OCC_1. The base circle is drawn equal to a radius of OC_1.

An arc is drawn from C to D measuring C_1C. A further arc is then drawn from D to E, measuring D_1D. This process is repeated until the involute curve reaches the outside

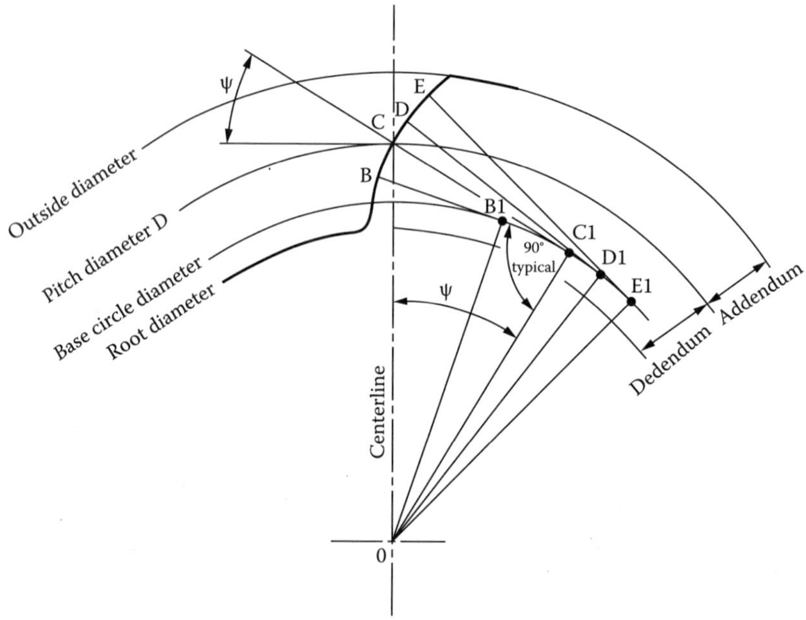

FIGURE 36.12
Layout of the involute tooth.

diameter and a similar procedure is adopted for the curve below the pitch diameter to the base circle.

In the example above, only four points have been used in the construction of the involute curve in the figure. More points could be used, and in fact greater accuracy will be obtained, but all this will be dependent on the size of the tooth being set out.

The other side of the tooth is drawn in an identical way after setting off the tooth thickness on the pitch line. The arcs used to construct the curve are exactly the same as were used for the first side.

36.5 Involute Functions

In addition to being able to generate gear tooth profiles on their transverse planes in the manner described, it is also possible to plot the tooth form using coordinates previously calculated by means of involute trigonometry, in conjunction with which involute functions are used.

To define an involute function, an involute tooth form is shown in Figure 36.13, where

Ψ = Pressure angle of gear on transverse plane at radius D/2
D = Pitch circle diameter
D_0 = Base circle diameter
J = Outside diameter
Ψ_1 = Involute function of angle ψ

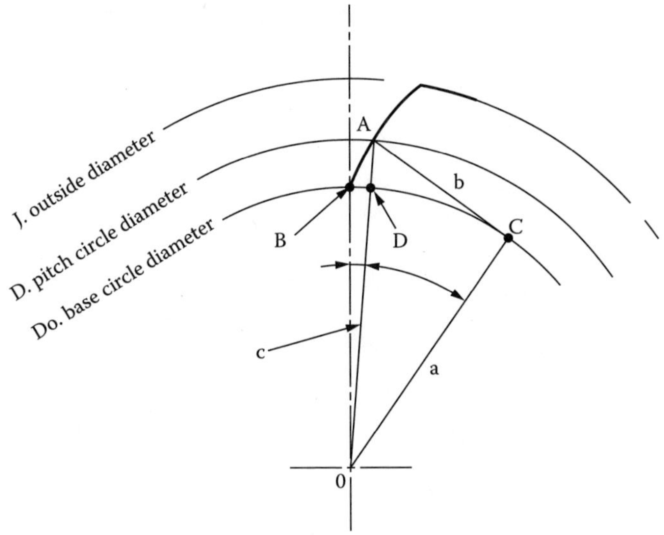

FIGURE 36.13
Involute function.

The involute function is the vectorial angle (in radians) of an involute curve from its origin and is also a function of the pressure angle at the radius at which the vectorial angle is taken. The pressure angle at the point of origin B is zero.

The vectorial angle at A on the involute curve is therefore ψ_1, which is the angle subtended by OB and OA, the pressure angle at A at radius D/2 being ψ.

Therefore, the angle ψ_1 (in radians) is known as the involute function of the angle ψ and is written as inv Ψ.

$$\frac{b}{a} = \tan\psi \tag{36.5}$$

$$b = a \cdot \tan\psi \tag{36.6}$$

Length b is then rewound on the base circle diameter and is then equal to a $(\psi+\psi_1)$.

Therefore,

$$a \cdot (\psi + \psi_1) = a \cdot \tan\psi$$

$$(\psi + \psi_1) = \tan\psi$$

$$\psi_1 = \tan\psi \quad (\psi \text{ radians})$$

$$\psi_1 = \text{inv} \cdot \psi$$

When the number of teeth are not an integer, this will require one of the given variables to be modified, keeping as close to the theoretical values as possible. For spur gear dimensional terms, see Figure 36.14, and for basic spur gear design formula, see Table 36.1.

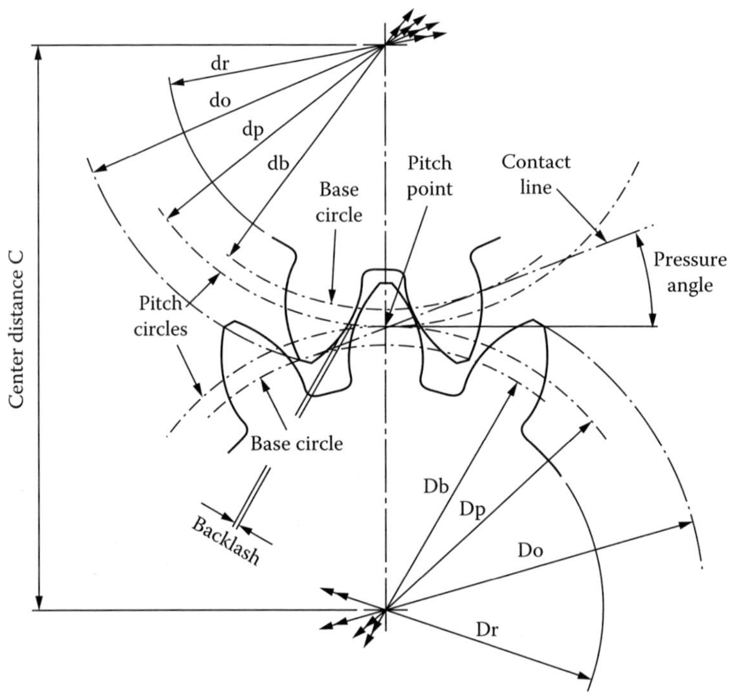

FIGURE 36.14
Spur gear dimensional terms.

TABLE 36.1
Spur Gear Design Formulae

Description	Notation	Formula
Center distance	C	Given
Outside diameter	D_o	Given
Pressure angle	Ψ	Given
Pitch diameter	D_p	N×m
Number of teeth	N	D/m
Module	m	Given
Addendum	A	1.00×m
Dedendum	B	1.25×m
Circular pitch	P_c	$p \cdot D_p/N$, pm
Base circle diameter	D_b	$D_p \times \cos$ (C)
Root diameter	D_r	$D_p - (2.5\times m)$

Example 36.1

To find the involute function of 20°, the procedure is as follows:
Let:

$$\psi = 20° = 0.3490658 \text{ radians}$$

$$\text{inv } \psi_1 = \tan \psi - \text{arc } \psi$$

where
tan ψ = natural tan of given angle
arc ψ = numerical value of given angle in radians
Therefore,

$$\text{inv } \psi = \tan 20° - 20°(\text{converted to radians})$$

$$\text{inv } \psi = 0.3639702 - (20 * 0.0174533)$$

Note:

$$1° = 0.0174533 \text{ radians}$$

$$\text{inv } \psi = 0.3639701 - 0.3490659$$

Hence,

$$\text{inv } 20° = 0.0149043$$

The involute function for 20° is 0.0149043, accurate to seven decimal places.

Using this procedure with the modern programmable calculators, it is safer to calculate the involute function rather than use the functions from a table of involute functions, as these may contain errors.

36.6 Spur Gearing: Basic Design Principles

The following is the general formulae for the design of spur gearing:

$$E = \frac{T}{r} = \frac{\text{Power}}{\omega r} = \frac{\text{Power}}{V} \tag{36.7}$$

Lewis formula for strength (American Gear Manufacturers Association):

$$\sigma = \frac{W \cos\phi}{K_v \cdot F \cdot m \cdot y} \tag{36.8}$$

where

$$W = \frac{19.1\,P}{N \cdot D}$$

$$y = 0.484 - \frac{4.24}{t + 6}$$

$$K_v = \frac{3.54}{3.54 + \sqrt{V}}$$

$$\phi = \text{Pressure angle } (20°)$$

36.6.1 Notation

For gear notation, see Table 36.2.

See Figure 36.15 for a general description of a gear set.

36.6.2 Working Stress σ_w

Allowable values for σ_w are given in Table 36.3. These cover the more common materials used in gear manufacture.

36.6.3 Width of Teeth

Using the Lewis factor to find the wheel proportions involves fixing the width in terms of the circular pitch and substituting the appropriate formula. Generally, for slow speeds and where shafts are inaccurately adjusted, the face width may be 1.25–2.5 times the pitch (the average works out 3–4 times the pitch); for high speeds, smooth engagement and high wear, the width may be 6–times. High ratios result in finer tooth pitches.

TABLE 36.2
Gear Notation

E	Tangential load on teeth	(N)
T	Torque transmitted	(Nm)
r	Pitch circle radius	(m)
V	Pitch line velocity	(m/s)
y	Lewis form factor	(See Table 36.3)
b	Width of teeth	(m)
P	Circular pitch	(m)
σ_w	Safe working stress	(MPa)
D	Pitch circle diameter for wheel	(m)
d	Pitch circle diameter for pinion	(m)
T	Number of teeth for wheel	(m)
t	Number of teeth for pinion	(m)
m	Module	(m)
A	Addendum = module	(m)
B	Dedendum = 1.25 × module	(m)

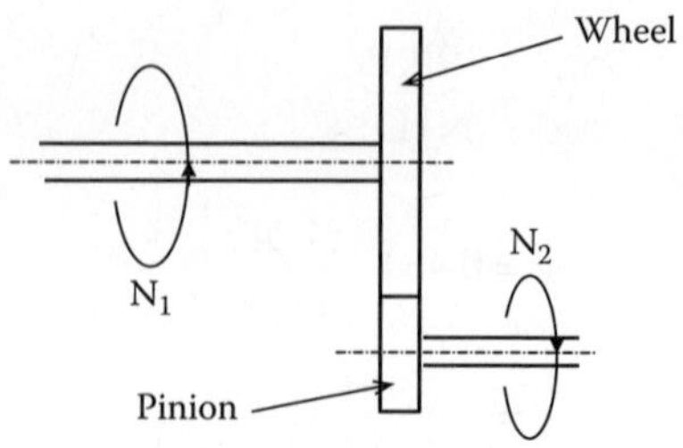

FIGURE 36.15
Simple gear drive.

TABLE 36.3

Lewis Factor Stresses (σ_w) in MPa for Various Materials

	Pitch Circle Velocity (m/min)							
Material	**30**	**60**	**120**	**180**	**270**	**370**	**550**	**730**
Cast iron	54.4	40.9	32.7	27.2	20.4	16.3	13.9	11.7
Cast steel	136	102	81.7	68.0	51.0	40.8	34.7	29.7
Forged steel	163	122.5	95.3	81.7	68.0	51.0	44.2	37.2
Phosphor bronze	71.5	71.5	44.2	37.4	32.6	27.2	20.8	16.6
Nickel chrome steel (1544 MPa Ult)	—	—	224.6	187.0	150.0	126.0	93.6	75.9

Example 36.2

Two spur gears are to have a ratio of 4:3 and a 6 mm module. If the center distance is to be approximately 125 mm, calculate the following:

- The circular pitch
- The number of teeth on each gear
- The pitch circle diameters
- The exact center distance

Solution

1. The circular pitch:

$$\begin{aligned} \rho &= \pi m \\ &= 6\pi \\ &= 18.849 \text{ mm} \end{aligned} \tag{36.9}$$

2. The number of teeth on each gear:

$$\text{From } 0.5(D + d) = \text{center distance}$$

$$D + d = 125 \times 2$$

Also:

$$m = \frac{D}{T} = \frac{d}{t}$$

$$6 = \frac{D}{T} \text{ and}$$

$$6T = D \quad \text{and} \quad 6t = d$$

Substituting in Equation 36.2:

$$6T + 6t = 250$$

Also, T/t=4/3 and substitute in Equation 36.9 for T:

$$6 \times \frac{4}{3} t + 6t = 250$$

$$\text{Hence, } t = 17.86$$

Both gears must contain a whole number of teeth; therefore, to make T a whole number, t must be divisible by 3.

Hence, the nearest multiple of 3 is 18.

$$\text{Therefore,}\quad T = \frac{4}{3} \times 18$$

$$T = 24 \text{ teeth}$$

$$t = 18 \text{ teeth}$$

3. The pitch circle diameters:

$$\begin{aligned}\text{Wheel pitch circle diameter } D &= T \times \text{module}\\ &= 24 \times 6\\ &= 144 \text{ mm}\end{aligned}$$

$$\begin{aligned}\text{Pinion pitch circle diameter } d &= t \times \text{module}\\ &= 18 \times 6\\ &= 108 \text{ mm}\end{aligned}$$

4. Exact center distance:

$$\begin{aligned}c &= 0.5 \times (D + d)\\ &= 0.5 \times (144 + 108)\\ &= 126.0 \text{ mm}\end{aligned}$$

Example 36.3

Design a pair of involute spur gears to transmit power of 30 kW at 8.5 rev/s. Gear ratio of 4:1. Assume a 20-tooth pinion of 6 mm module and a ratio of face width to pitch = 4. The pinion is forged steel and the wheel is cast steel.

Find the tooth proportions of wheel and pinion.

Solution

$$\begin{aligned}\text{Torque} &= \frac{\text{power}}{2\pi N}\\ &= \frac{30 \times 10^3}{2\pi \times 8.5}\,\frac{\text{Nm}}{\text{s}} \times \text{s}\\ &= 561.72 \text{ Nm}\end{aligned}$$

Also, torque = E × r, where E is the tangential load and

$$E \times \frac{\text{pcd}}{2} = 561.72 \text{ Nm}$$

Also

$$\text{pcd} = \frac{t \cdot p}{\pi}$$

$$E \times \frac{tp}{2\pi} = 561.72\ \text{Nm}$$

$$E = \frac{2\pi \times 561.72}{t \times p}$$

Assuming 20° involute, full depth:

$$y = 0.154 - \frac{0.912}{t} \quad \text{and} \quad t = 20$$

$$y = 0.1084$$

Pitch circle diameter:

$$\begin{aligned} \text{pcd of pinion (d)} &= \text{module} \times t \\ &= 6 \times 20\ \text{mm} \\ &= 120\ \text{mm}. \end{aligned}$$

and circular pitch:

$$= \frac{\pi d}{t} = \frac{\pi \times 120\ \text{mm}}{20} \text{ or } (\pi \times m)$$

$$= 18.85\ \text{mm}$$

Given b=4×circular pitch:

$$\begin{aligned} b &= 4 \times 18.85 \\ &= 75.40\ (\text{say } 75.0\ \text{mm}) \end{aligned}$$

To find allowable working stress where

$$\sigma_w = \frac{E}{ybp}$$

$$\sigma_w = \frac{2\pi \cdot 562\ \text{Nm}}{t \cdot p^2 \cdot y \cdot b}$$

$$\sigma_w = \frac{2\pi \times 562\ \text{Nm}}{20 \times 0.0189^2\ \text{m} \times 0.1084 \times 0.075\ \text{m}}$$

$$= 60.796\ \text{MPa}$$

To find the pitch line velocity:

$$\begin{aligned} &= \pi d \cdot \text{rev/min} \\ &= \pi \times 0.12\ \text{m} \times 8.5\ \text{rev/s} \times 60 \\ &= 192.265\ \text{m/min} \end{aligned}$$

From Table 36.3 for a forged steel pinion:
180 m/min gives 81.7 MPa and 270 m/min gives 68 MPa.
By interpolation 192.3 m/min is equivalent to 79.83 MPa.
The calculated stress value σ_w of 60.8 MPa would, therefore, be satisfactory.

Blank outside diameter for pinion:

$$= d + (2 \times \text{module})$$

$$= 120\text{ mm} + (2 \times 6\text{ mm})$$

$$= 132.0\text{ mm}$$

Tangential load at pitch line:

$$E = \frac{562\text{ Nm}}{\text{radius of pinion}}$$

$$= \frac{562\text{ Nm}}{0.06\text{ m}}$$

$$= 9366.7\text{ N}$$

Load per mm of tooth width:

$$= \frac{9366.7\text{ N}}{75\text{ mm}}$$

$$= 124.9\text{ N/mm}$$

Data for Cast-Steel Wheel

The gear ratio is 4:1 and the number of teeth in the pinion is 20. Therefore, the number of teeth in the wheel = 20 × 4 = 80.

Blank outside diameter for wheel:

$$= D + 2\text{ m}$$

$$= 4 \times 120\text{ m} + 2 \times 6\text{ m}$$

$$= 492\text{ mm}$$

Stress in wheel:

$$\sigma_w = \frac{E}{ybp}$$

$$= \frac{9367\text{ N}}{y \times 0.075\text{ m} \times 0.0189\text{ m}}$$

where

$$y = 0.154 - \frac{0.912}{80\text{ t}}$$

$$= 0.1426$$

$$\sigma_w = \frac{9366\text{ N}}{0.1426 \times 0.075\text{ m} \times 0.0189\text{ m}}$$

$$= 46.335\text{ MPa}$$

σ_w for cast steel from Table 36.3, 180 m/min gives 68 MPa and 270 m/min yields 51 MPa. By interpolation, 192 m/min is equivalent to 65.73 MPa; therefore, cast-steel wheel stress is satisfactory.

Wheel and pinion shafts would be sized using the procedure outlined in Example 36.1. It is calculated that a shaft diameter of 66 mm including keyway allowance will be acceptable.

Table 36.4 summarizes the gear details for this example.

Wheel Arms

Because of the large diameter of the wheel and as it is cast, it would be considered desirable to design the wheel with spokes:

a. To reduce the amount of cast metal in the melt and reduce any shrinkage problems
b. To reduce the mass and therefore the inertia in the wheel

For the wheel arms (or spokes), it is considered that four arms would be reasonable and be elliptical in cross section (see Figure 36.16).

$$\text{Boss diameter} = 2 \times 66$$
$$= 132 \text{ mm}$$

TABLE 36.4
Summarized Particulars for Example 36.3

Item	Number of Teeth	Module (mm)	PC Diameter (mm)	Teeth Width (mm)	Blank Diameter (mm)
Forged steel pinion	20	6	120	75	132
Cast-steel wheel	80	6	480	75	492

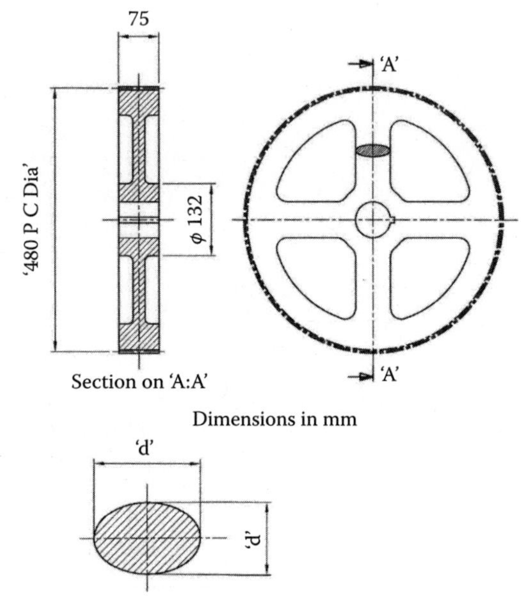

FIGURE 36.16
Detail of spur gear wheel used in Example 36.3.

$$\text{Bending arm radius} = \frac{480\text{ mm}}{2} - \frac{132\text{ mm}}{2}$$

$$= 174\text{ mm}$$

and

$$b = 22\text{ mm}$$

$$\text{Tangential load} = 9366\text{ N}$$

$$\text{Load per wheel arm} = 2341.5\text{ N}$$

$$\text{Bending moment} = 2341.5\text{ N} \times 0.174\text{ m}$$

$$= 407.42\text{ Nm}$$

$$\text{Bending moment} = \sigma Z$$

For elliptical section:

$$Z = \frac{\pi b d^2}{32}$$

Say 3b = d and allowable bending stress for cast steel allowing for load reversal:

$$\sigma = 44\text{ MPa}$$

$$407.42\text{ Nm} = 4 \times 10^6 \times \frac{\pi}{32} \times \frac{d}{3} \times d^2$$

$$\frac{407.42 \times 32}{44 \times 10^6 \times \pi} = \frac{d}{3} \times d^2$$

$$94.324 \times 10^{-6} \times 3 = d^3$$

$$d = 65.65\text{ mm}$$

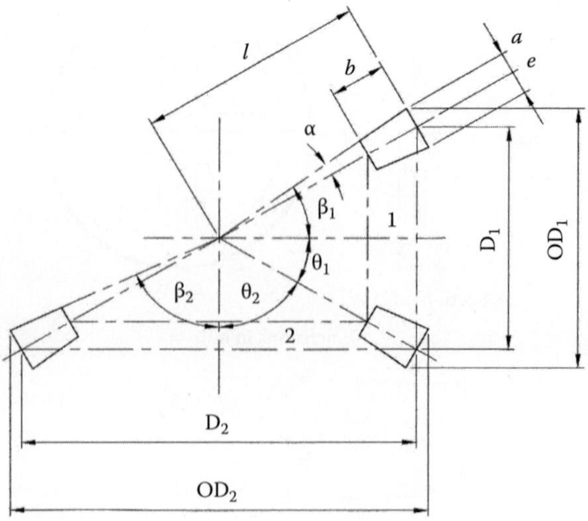

FIGURE 36.17
Bevel gear set used in Example 36.4.

36.7 Bevel Gearing

A skeletal layout of a pair of bevel gears is shown in Figure 36.17, with the more important features used in the calculation of the gear proportions required for the manufacture identified.

Since the bevel tooth is only a full tooth at the pitch cone diameter, the method for calculating the tooth proportions differ slightly from that used for the ordinary spur wheel and pinion.

The following modified form of the Lewis formula has given satisfactory results. The tooth load is assumed to act at the mean radius of the tooth width, and the formula has a correction to allow for this.

36.7.1 Modified Lewis Formula for Bevel Gears

$$E = r^2 ybp\sigma_w \tag{36.10}$$

where:

r = ratio of mean pitch radius at pitch radius
y = form factor based on the number of teeth in the equivalent spur gear
E = tangential load at pitch circle (N)
b = width of tooth (m)
p = circular pitch (m)
σ_w = allowable working stress (N/m²)

The width of the tooth will usually vary from one-fourth to one-third the length of the pitch cone surface.

Note: Where the formula is common to both the pinion and the wheel, the suffix is omitted.

The definitions and basic formulae used in the calculation of bevel gears is covered in Table 36.5.

TABLE 36.5

Basic Formulae for Bevel Gears

Symbol	Definition	Formula
t	Number of teeth in pinion	$t=\pi d/p$
T	Number of teeth in wheel	$T=\pi D/p$
D	Pitch circle diameter	$D=mT$ or $d=mt$
OD	Outside or blank diameter	$OD=D+2a\cos\theta$
a	Addendum	a = module
e	Dedendum	$e=1.25\,m$
θ_1	Pinion pitch cone angle	$\theta_1=\tan^{-1} t/T$
θ_2	Wheel pitch cone angle	$\theta_2=\tan^{-1} T/t$
β	Face angle	$\beta=\theta+\tan^{-1} a/l$
P	Circular pitch	$p=\pi m$
m	Module	$D/T=m=d/t$
n_1	Number of teeth in equivalent spur wheel	$n_1=t/\cos\theta$
l	Length of pitch cone surface	$l=D/(2\sin\theta)$

Example 36.4

Design a pair of bevel gears to transmit 25 kW at 8.5 rev/s of the driving shaft and have a reduction ratio of 2.5:1 with the shafts at right angles to each other. See Figure 36.4 for details.

Solution

Both gears are to be cast steel. For this problem, assume the pinion has 24 teeth, and an involute angle of 20°, together with a module of 6 mm:

$$\text{Mean torque} = \frac{\text{power}}{2\pi \times N}$$

$$= \frac{25 \times 10^3 \text{ Nm}}{2\pi \times 8.5 \text{ rev/s}}$$

$$= 468.10 \text{ Nm}$$

Pitch cone angles:

$$t = \text{number of teeth in pinion}$$

$$= 24 \text{ teeth}$$

$$T = \text{number of teeth in wheel}$$

$$= 24 \times 2.5$$

$$= 60 \text{ teeth}$$

Pinion pitch cone angle:

$$\theta_1 = \tan^{-1}\frac{t}{T}$$

$$= \tan^{-1} \times \frac{24}{60}$$

$$= \theta_1 = 21.80°$$

For shafts at right angles:
 Wheel pitch cone angle:

$$\theta_2 = 90° - 21.48°$$

$$= 68.20°$$

Design of teeth:

$$E = r^2 ybp\sigma_w$$

To find the ratio (r):

Let the width of the teeth = 1/3 the height of pitch cone surface

$$h = 1/3 \times 1 \text{ (see Figure 36.18)}$$

$$r_1 = \text{pitch radius of pinion}$$

$$r_m = \text{mean pitch radius of pinion}$$

$$\frac{r_m}{r_1} = \frac{5}{6} = r$$

By similar triangles

To find b in terms of the circular pitch (p):

$$1 = \frac{D_1}{2\sin\theta_1}$$

$$= \frac{tp}{2\pi}\sin\theta_1$$

where θ_1=21.80° and t=24.

$$b = \frac{tp}{6\pi\sin\theta_1}$$

$$= \frac{24 \times p}{6\pi \times 0.3714}$$

$$= 3.428p$$

To find the value of y:

First, find the number of the teeth in the equivalent spur wheel for pinion:

$$n^1 = \frac{t}{\cos\theta_1}$$

$$= \frac{24}{0.9305}$$

$$= 25.79 \text{ (say 26)}$$

For 20° involute full tooth:

$$y = 0.154 - \frac{0.912}{26}$$

$$= 0.1189$$

Now, torque=468.10 Nm and also

$$\text{torque} = E \times \frac{D_1}{2} \text{ or } 468.10 = \frac{tp}{2\pi}$$

For a tangential load:

$$E = \frac{2\pi \times 468.10}{24P}$$

$$= \frac{122.55}{P}$$

Working stress:

$$\frac{122.55}{P} = r^2 ybp\sigma_w \times \frac{122.55 \times 36}{p^3 \times 3.428 \times 25 \times 0.119} = \sigma_w$$

where

$$\begin{aligned} p &= \pi m \\ &= \pi \times 6 \\ &= 18.85 \text{ mm} \\ &= 0.01885 \text{ m} \end{aligned}$$

and then

$$\sigma_w = 64.3 \text{ MPa}$$

This value is compared with Table 36.2 for pitch circle velocity of $\pi n \times d$, where d is found as follows:

For pinion pitch diameter:

$$\begin{aligned} D_1 &= m \times t \\ &= 6 \text{ mm} \times 24 \text{ t} \\ &= 144 \text{ mm} \end{aligned}$$

For wheel pitch diameter:

$$\begin{aligned} D_2 &= m \times T \\ &= 6 \text{ mm} \times 60 \text{ t} \\ &= 360 \text{ mm} \end{aligned}$$

Therefore,

$$\begin{aligned} \text{Pitch circle velocity} &= \pi \times 510 \text{ rev/min} \times 0.144 \text{ m} \\ &= 230.719 \text{ m/min (say 231 m/min)} \end{aligned}$$

From Table 36.3 for cast steel:

180 m/min gives 68MPa.

270 m/min gives 51MPa.

By interpolation,

231 MPa gives $\sigma_w = 60.63$ MPa

Therefore, stress by calculation is considered satisfactory.

To find the width of a tooth:

$$\begin{aligned} &= 3.428 \times p \\ &= 3.428 \times 18.9 \\ &= 64.79 \text{ mm} \end{aligned}$$

Blank or outside diameters:

For pinion:

$$\begin{aligned} OD_1 &= D_1 + 2a\cos\theta_1 \\ &= 144 \text{ mm} + 2 \times 6 \times \cos 21.8° \\ &= 155.142 \text{ mm} \end{aligned}$$

For wheel:

$$OD_2 = D_2 + 2a\cos\theta_2$$
$$= 360\text{ mm} + 2\times 6\times \cos 68.2°$$
$$= 364.456\text{ mm}$$

Face angles:

$$\beta = \theta + \tan^{-1}\frac{a}{1}$$

$$1 = \frac{D_1}{2\sin\theta_1}$$
$$= \frac{144}{2\times \sin 21.8°}$$
$$= 193.877\text{ mm}$$

$$\gamma = \tan^{-1}\frac{a}{1}$$
$$= \tan^{-1}\left(\frac{6}{193.877}\right)$$
$$= \tan^{-1}(0.03095)$$

This is common for both pinion and wheel.
For pinion:

$$\text{Face angle } \beta_1 = 21.8° + \tan^{-1} 0.03095$$
$$= 23.57°$$

For wheel:

$$\text{Face angle } \beta_2 = 68.2° + \tan^{-1} 0.03095$$
$$= 69.97°$$

Wheel Arms

The load on the arms is calculated from the torque at the mean pitch radius, and the tangential tooth load at the center of the teeth would be

$$E = \frac{\text{Mean torque}}{\text{Mean pitch radius}}$$

From the example above,

$$E = \frac{468\text{ Nm}}{(5/6)\times 0.144\text{ m}}$$
$$= 3900\text{ N}$$

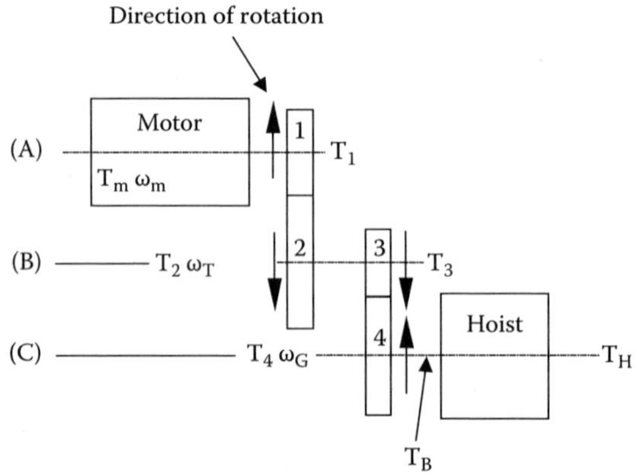

FIGURE 36.18
Diagram of motor/hoist gear drive.

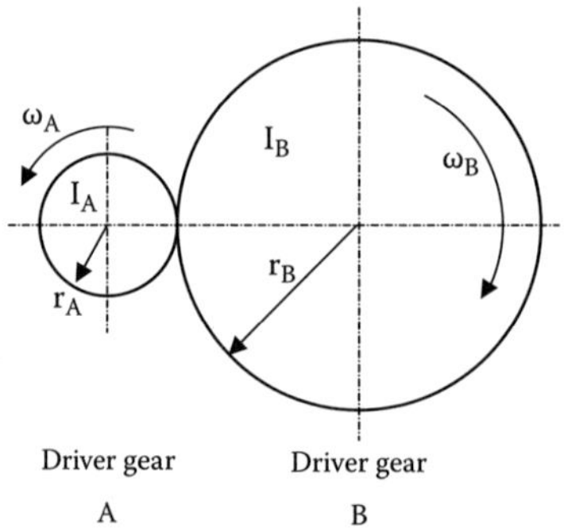

FIGURE 36.19
Referred inertia for a simple gear.

TABLE 36.6
Summarized Particulars for Example 36.4

Item	Number of Teeth	Module (mm)	PC Diameter (mm)	Teeth Width (mm)	Blank Diameter (mm)	Pitch Cone Angle (°)	Face Angle (°)
Forged steel pinion	24	6	144	64.8	155.14	21.8	23.56
Cast-steel wheel	60	6	360	364.46	492	68.2	69.97

The arms would then be calculated to resist this load (Figure 36.18).

The final details of the bevel gear pair are as shown in Figure 36.19 and summarized in Table 36.6.

36.8 Basic Gear Transmission Theory

Consider a simple gear drive with input and output shafts (Figure 36.20).

In this simple case, the gear ratio (n) is defined as

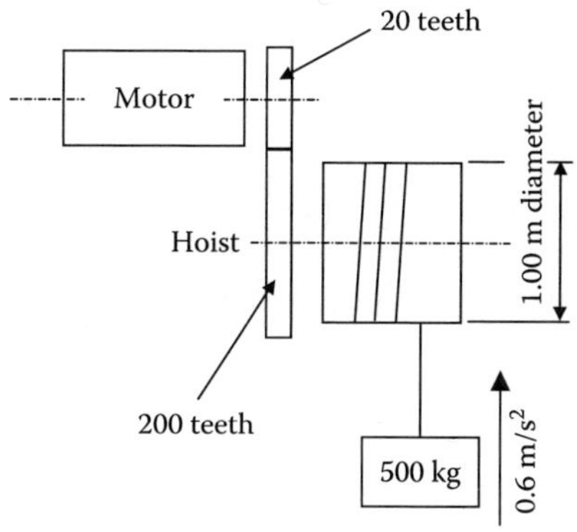

FIGURE 36.20
Diagram of hoist.

$$\eta = \frac{\text{Input speed}}{\text{Output speed}} \tag{36.11}$$

Usually, the speed of the shafts is described as rev/min, but the ratio will be the same if the units of angular velocity (ω) are used, that is,

$$\eta = \frac{\text{Input speed}}{\text{Output speed}} = \frac{\omega_1}{\omega_2} \tag{36.12}$$

36.8.1 Torque and Efficiency

The shaft power (SP) transmitted by a torque T Nm applied to a shaft that is rotating at N rev/min is given by

$$S_P = \frac{2\pi NT}{60} \tag{36.13}$$

In an ideal gear drive, the input and output powers are the same:

$$S_P = \frac{2\pi N_1 T_1}{60} = \frac{2\pi N_2 T_2}{60} \quad N_1 T_1 = N_2 T_2 \tag{36.14}$$

$$n = \frac{T_2}{T_1} = \frac{N_1}{N_2} \tag{36.15}$$

It follows then that as the input shaft speed is reduced, the torque will increase, and vice versa.

In a real drive or gearbox, power is lost due to heat and friction and the power output will be reduced.

The efficiency of the gear drive is defined by

$$\eta = \frac{\text{power out}}{\text{power in}} = \frac{2\pi N_2 T_2 \times 60}{2\pi N_1 T_1 \times 60} = \frac{N_2 T_2}{N_1 T_1} \tag{36.16}$$

As the input and output torques are different, the gearbox will require restraining to prevent it from rotating. A holding torque T_3 must be applied to the gearbox through its attachments.

The total torque must equate to zero, that is, $T_1+T_2+T_3=0$.

Using the convention that anticlockwise rotation is positive and clockwise is negative, the holding torque, together with its direction of rotation, can be determined. The direction of rotation of the output shaft will be dependent upon the internal configuration of the gears within the box.

36.9 Types of Gear Trains

36.9.1 Simple Gear Train

Figure 36.21 shows a typical spur gears drive chain. The direction of rotation is reversed from one gear to another. The only function of the idler gear is to change the direction of rotation; it has no effect on the gear ratio.

t=number of teeth on the specific gear
D=pitch circle diameter (PCD)
m=modem=D/t
$D_A=m\ t_A$
$D_B=m\ t_B$
$D_C=m\ t_C$
ω=angular velocity
v=linear velocity on the pitch circle diameter; $v=\omega\ D/2$
Considering the velocities of the gears at the pitch circles, we have

$$\text{Linear velocity at point: } V_{P_1}=\frac{\omega_1}{D_1/2} \tag{36.17}$$

$$\text{Linear velocity at point: } V_{P_2}=\frac{\omega_2}{D_2/2} \tag{36.18}$$

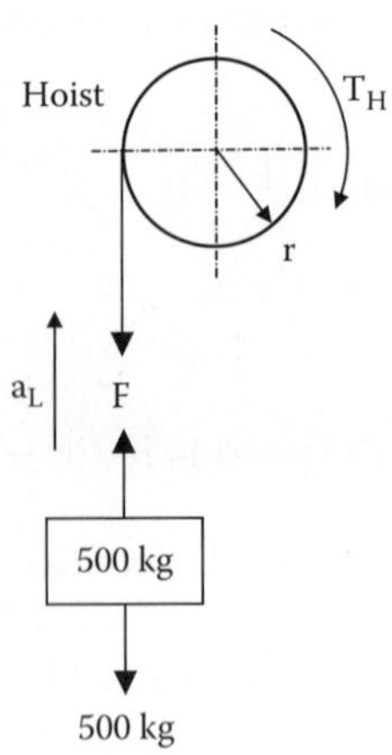

FIGURE 36.21
Torque required by hoist.

$$\text{Linear velocity at point: } V_{P3} = \frac{\omega_3}{D_3/2} \tag{36.19}$$

The velocity v of any point on the PCD must be the same for all the gears; otherwise they would be slipping. It follows that

$$\frac{\omega_A \cdot D_A}{2} = \frac{\omega_B \cdot D_B}{2} = \frac{\omega_C \cdot D_C}{2}$$

$$\omega_A \cdot D_A = \omega_B \cdot D_B = \omega_C \cdot D_C$$

$$\omega_A m t_A = \omega_B m t_B = \omega_C t_C$$

$$\omega_A t_A = \omega_B t_B = \omega_C t_C \tag{36.20}$$

$$N_A \cdot t_A = N_B \cdot t_B = N_C \cdot t_C$$

In terms of rev/min, $N_A{\cdot}t_A = N_B{\cdot}t_B = N_C{\cdot}t_C$

If gear A is the input and gear C the output, then

$$n = \frac{N_A}{N_C} = \frac{t_C}{t_A} \tag{36.21}$$

36.9.2 Compound Gears

A compound gear train has one or more gears fitted on one shaft, as illustrated in Figure 36.22.

In this example, gears A and C are driving gears, and gears B and D are driven gears. Gears B and C are locked on the same shaft and therefore revolve at the same speed.

The velocity of each tooth on gears A and B is the same, so that $\omega_A{\cdot}t_A = \omega_B{\cdot}t_B$, as they are simple gears. Likewise, for gears C and D, $\omega_C{\cdot}t_C = \omega_D{\cdot}t_D$:

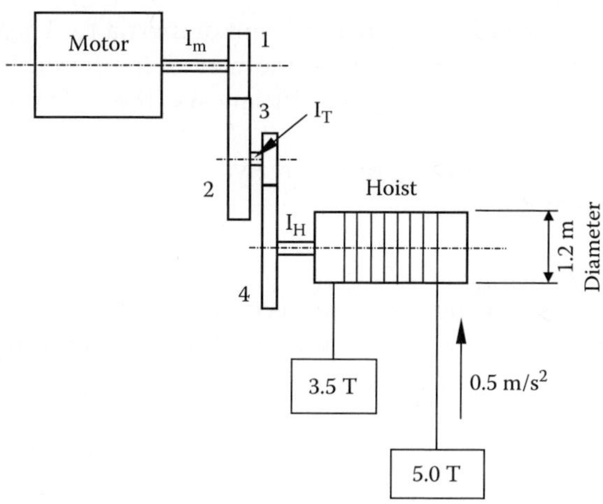

FIGURE 36.22
Diagram of hoist mechanism for Example 36.3.

$$\frac{\omega_A}{t_B} = \frac{\omega_B}{t_A} \text{ and } \frac{\omega_C}{t_D} = \frac{\omega_D}{t_C}$$

$$\omega_A = \frac{t_B \cdot \omega_B}{t_A} \text{ and } \omega_C = \frac{t_D \cdot \omega_D}{t_C}$$

$$\omega_A \cdot \omega_C = \frac{t_B \cdot \omega_B}{t_A} \times \frac{t_D \cdot \omega_D}{t_C} = \frac{t_B \cdot t_D}{t_A \cdot t_C} \times \omega_B \cdot \omega_D$$

$$\frac{\omega_A \cdot \omega_C}{\omega_B \cdot \omega_D} = \frac{t_B \cdot t_D}{t_A \cdot t_C}$$

As gears B and C are fitted on the same shaft, $\omega_B = \omega_C$.

$$\frac{\omega_A}{\omega_B} = \frac{t_B \cdot t_D}{t_A \cdot t_C} = n.$$

and since $\omega = 2\pi N$, the gear ratio can be written as

$$\frac{N(in)}{N(out)} = \frac{t_B \cdot t_D}{T_a \cdot t_C} = n \tag{36.22}$$

36.10 Power Transmission in a Gear Train

In a gear train, power is lost between the teeth due to friction resulting from loads imposed on them and in the bearings. Power loss due to overcoming shaft inertia also contributes to the reduction in efficiency.

Consider the gear train for a hoist driven by an electric motor. The drive chain consists of two sets of reducing gears.

A motor is attached to the system with a moment of inertia I_m. The moment of inertia of the middle shaft is I_T, and I_H is the combined moment of inertia of the hoist; this acts as a load to the system. The gear ratio and gear efficiency of the gear set 1–2 are $n_{1/2}$ and $\eta_{1/2}$, respectively.

Gear set 3–4 consists of $n_{3/4}$ and $\eta_{3/4}$, respectively.

Let: T_m = torque at the motor

T_H = torque of the hoist

T_B = friction torque of bearings

Draw a free body diagram using Newton's second law, $\Sigma T = I\alpha$.

In this instance, consider rotation in a clockwise direction as positive.

For shaft (A):

$$T_m - T_1 = I_m \cdot \alpha_m \tag{36.23}$$

For shaft (B):

$$T_2 - T_3 = I_T \cdot \alpha_T \tag{36.24}$$

Note: As there is a gear mating between gear 1 and gear 2, the analysis has to include its own gear ratio and gear efficiency and relate it to the transfer shaft, I_T.

Previously,

$$\eta_{1/2} = \frac{T_2}{T_1}\eta_{1/2}$$

It follows that

$$T_2 = \frac{h_{1/2} \cdot T_1}{\eta_{1/2}} \tag{36.25}$$

For shaft (C):

$$T_4 - T_B - T_H = I_H \cdot \alpha_H$$

$$\text{Also,} \quad T_4 = T_3\eta_{3/4} \tag{36.26}$$

Using power, $P = T_m\ \omega_m$, the power transfer to each gear component is as follows:

1. Power transfer by the motor:

$$P_M = T_m \cdot \omega_m \tag{36.27}$$

2. Power at gear 1:

$$P_1 = T_1 \cdot \omega_m = (T_m - I_m \cdot \alpha_m)\omega_m \tag{36.28}$$

3. Power at gear 2:

$$P_2 = P_1 \cdot \eta_{1/2} \tag{36.29}$$

4. Power at gear 3:

$$P_3 = T_3 \cdot \omega_T = (T_2 - I_T \cdot \alpha_T)\omega_T \tag{36.30}$$

5. Power at gear 4:

$$P_4 = P_3 \cdot \eta_{3/4}$$

6. Power at hoist:

$$P_H = T_H \cdot \omega_H = (T_4 - T_B - I_H \cdot \alpha_H)\omega_H \tag{36.31}$$

7. Overall power transfer efficiency, η_o:

$$\eta_o = \frac{P_H}{P_m} \tag{36.32}$$

Thus, if friction torque T_B effect is neglected:

$$\eta_o = \frac{P_H}{P_m} = \left|\frac{P_T}{P_m}\right| \times \left|\frac{P_H}{P_m}\right| = \eta_{1/2} \times \eta_{3/4} \tag{36.33}$$

$$\text{Also,}\quad \eta_o = \frac{T_H}{T_m}\eta_{1/2} \cdot \eta_{3/4} \tag{36.34}$$

36.11 Referred Moment of Inertia ($I_{referred}$)

Consider a simple gear system. In order for the driver gear A to begin to rotate, it must have enough torque available to overcome its own inertia, IA, and then additional torque to begin to accelerate the driven gear B. The torque required to accelerate B must be initially calculated.

1. Torque at B to overcome I_B:

$$T_B = I_B \cdot \alpha_B \tag{36.35}$$

To refer α_B to gear A, the gear ratio is used:

$$\eta = \frac{\omega_B}{\omega_A} = \frac{\alpha_B}{\alpha_A} \tag{36.36}$$

Thus, $T_B = I_B\, n\alpha_A$.

2. The gear efficiency is related to power and torque of the mating gears:

$$\eta_G = \frac{P_B}{P_A} = \frac{T_B \cdot n}{T_A} \tag{36.37}$$

where η_G is the efficiency of the gear set.

3. Torque required at gear A to accelerate I_B:

$$T_A = \frac{T_B \cdot n}{\eta_G} = \frac{I_B\alpha_B n}{\eta_G} = \frac{(I_B \cdot n\alpha_A)n}{\eta_G} = \frac{I_B n^2 \alpha_A}{\eta_G} \tag{36.38}$$

4. Total torque required at gear A to accelerate I_A and I_B:

$$T_{Total} = I_A \cdot \alpha_A + T_A \tag{36.39}$$

$$T_{Total} = \left[I_A + \frac{I_B n^2}{\eta_G} \right] \alpha_A \tag{36.40}$$

This can be expressed in a general form $T_{Total} = I_{equiv} \cdot \alpha_A$ when referred to gear A:

$$\text{Hence,} \quad I_{equiv} = \left(I_A + \frac{I_B n^2}{\eta_G} \right) \tag{36.41}$$

This derivation of I_{equiv} for the simple gear system can be extended to cover for a double set of reducing gears, as discussed in Section 36.8, by neglecting the friction torque effect T_B.

$$I_{equiv} = I_m + I_T \frac{(n_{1/2})^2}{n_{1/2}} + \frac{I_G (n_{1/2})^2 (n_{3/4})^2}{(n_{1/2})(n_{3/4})} \tag{36.42}$$

36.12 Gear Train Applications

Example 36.5

An electric motor accelerates a 500 kg load with an acceleration of 0.6 m/s² using a simple gear system (shown in Figure 36.20). The load is carried by a rope that encircles a hoist pulley that has a diameter of 1.0 m. The gear connected to the hoist has 200 teeth, and the gear connected to the motor shaft has 20 teeth. Assume the gear efficiency is 90%.

The masses and the radius of gyrations for each shaft are tabulated as follows:

	Mass (kg)	Radius of Gyration (mm)
Motor shaft	250	100
Hoist shaft	1100	300

Calculate the torque required by the motor to raise the load with an acceleration of 0.6 m/s². Friction losses may be neglected.

Total torque required by motor to raise load:

$$T_{total} = T_m + T_{equiv} \tag{36.43}$$

where

T_m = torque to accelerate load through the gear system

T_{equiv} = torque to overcome equivalent inertia (referred to the motor)

From

$$I_{equiv} = I_m + \frac{I_H \cdot n^2}{\eta_g}$$

$$I_m = \text{motor shaft inertia}$$

$$I_m = m \cdot r^2$$

$$= 250\ \text{kg} \times (0.1\ \text{m})^2$$

$$= 250\ \text{kg m}^2$$

$$= 99.0\ \text{kg m}^2$$

Motor shaft:

$$I_{equiv} = 2.5\ \text{kg m}^2 + \left(\frac{99.0\ \text{kg m}^2 \times 0.12}{0.9}\right)$$

$$= 3.6\ \text{kg m}^2$$

Gear ratio (n):

$$n = \frac{N_1}{N_2}$$

$$= \frac{20}{200}$$

$$n = 0.1$$

Acceleration of hoist:

$$a_H = \alpha_H \cdot r_H$$

$$\alpha_H = \frac{a_H}{r_H}$$

$$= \frac{0.6\ \text{m/s}^2}{0.5\ \text{m}}$$

$$= 1.2\ \text{rad/s}$$

From the gear ratio, calculate the angular acceleration of the motor:

$$a_m = \frac{a_H}{n}$$

$$= \frac{1.2\ \text{rad/s}}{0.1}$$

$$= 12\ \text{rad/s}$$

Torque due to equivalent inertia:

$$T_m = I_{equiv} \times a_m$$

$$= 3.6 \times 12\ \text{rad/s}$$

$$= 43.2\ \text{Nm}$$

Torque required at hoist (see Figure 36.21):
From Newton's second law:

$$\Sigma F = ma$$

$$F - 500g = 500a_H F = 500(g + a)$$

$$F = 500 \times (9.81 + 0.6)$$

$$F = 5205\ \text{N}$$

Torque at hoist:

$$\begin{aligned} T_H &= F \cdot r \\ &= 5205\,\text{N} \times 0.5\,\text{m} \\ &= 2602\,\text{Nm} \end{aligned}$$

Due to the gear efficiency (as the hoist is connected to the gear system),
Torque required to accelerate the load:

$$\begin{aligned} T_{equiv} &= \frac{T_H \cdot n_{1/2}}{\eta_{1/2}} \\ &= \frac{2602.5 \times 0.1}{0.9} \\ &= 289.17\ \text{Nm} \end{aligned}$$

Total torque referred to motor:

$$\begin{aligned} T_{total} &= 43.2 + 289.17\,\text{Nm} \\ &= 332.37\,\text{Nm} \end{aligned}$$

Example 36.6

An electric motor is used to raise and accelerate a load using a hoist and driving through two sets of reducing gears, as in Figure 36.22.

	Moment of Inertia (kg m²)	Gear Ratio
Motor shaft, I_m	5	Set ½=1:3.5
Transfer shaft, I_T	40	
Hoist shaft, I_H	500	Set ¾=1:4.5

Gear efficiency for both sets of gears is 90%. Neglecting friction, determine the total torque required by the motor to accelerate a load of 5 tonnes at an acceleration of $0.5\,\text{m/s}^2$.

Given that

$$\begin{aligned} I_m &= 5\,\text{kg m}^2 \\ I_T &= 40\,\text{kg m}^2 \\ I_H &= 500\,\text{kg m}^2 \\ n_{1/2} &= 1:3.5 \\ n_{3/4} &= 1:4.5 \\ \eta_G &= 0.9 \end{aligned}$$

Neglecting friction effects:
Total torque required by the motor is

$$T_{total} = T_{m1} + T_{m2}$$

where

T_{m1} = torque to overcome equivalent inertia (referred to motor)
T_{m2} = torque to accelerate the load through the gear system
Consider T_{m1}:

$$T_{m1} = I_{equiv} \cdot \alpha_m$$

For a compound gear set neglecting friction:

$$I_{equiv} = I_m + \frac{I_T.(n_{1/2})^2}{\eta_{G1/2}} + \frac{I_H.(n_{1/2})^2(n_{3/4})}{(\eta_{G1/2}).(\eta_{G3/4})}$$

$$I_{equiv} = 5 + \frac{40(1/3.5)^2}{0.9} + \frac{500(1/3.5)^2(1/4.5)^2}{(0.9).(0.9)}$$

$$= 11.116 \text{ Nm}$$

Given that the linear acceleration of the load $a_H = 0.5\,\text{m/s}^2$:

$$\alpha_H = \frac{a_H}{r_H}$$

$$= \frac{0.5 \text{ m/s}^2}{0.5 \text{ m}}$$

$$= 1.0 \text{ rad/s}^2$$

From the gear ratio:

$$\frac{\alpha_H}{\alpha_m} = \frac{\alpha_H}{\alpha_T} \times \frac{\alpha_T}{\alpha_m}$$

$$= \frac{1}{4.5} \times \frac{1}{3.5}$$

$$= \frac{1}{15.75}$$

Thus,

$$a_m = 15.75 \cdot \alpha_H$$

$$= 15.75 \cdot (1.0)$$

$$= 15.75 \text{ rad/s}^2$$

Therefore,

$$T_{m1} = I_{equiv} \cdot \alpha_m$$

$$= 11.116 \times 15.75$$

$$= 175.08 \text{ Nm}$$

Determine T_{m2}.

From Newton's second law, $\Sigma F = ma$:

$$3500 - F_1 = 3500 \cdot a$$

$$F_1 = 3500(g - a)$$

$$= 3500(9.31)$$

$$= 32,585 \text{ N}$$

$$F_2 - 5000g = 5000a$$

$$F_2 = 5000(g - a)$$

$$= 46,550 \text{ N}$$

Resultant force at hoist:

$$F_r = F_2 - F_1$$

$$= 13,965 \text{ N}$$

Torque at hoist:

$$T_H = F_r \times r_H$$

$$= 13,965 \times 0.5$$

$$= 6982.5 \text{ Nm}$$

$T_{m2} = T_H$ referred to the motor:

$$T_{m2} = \frac{T_H \; n_{1/2} n_{3/4}}{h_{1/2} \cdot h_{3/4}}$$

$$T_{m2} = \frac{6982.5(1/3.5)(1/4.5)}{(0.9)(0.9)}$$

$$= 547.33 \text{ Nm}$$

Total torque at the motor is

$$T_{total} = T_{m1} + T_{m2}$$

$$= 175.08 + 547.33$$

$$T_{total} = 722.41 \text{ Nm}$$

37

Power Screws

37.1 Power Screws

37.1.1 Introduction

Power screws (or translation screws) are used to convert rotary motion into a linear motion, that is, to convert a rotational torque and power into a linear force and power. The reverse is also possible, with a power screw that is used to convert linear motion to rotary motion, though seldom used in this application.

The main applications of power screws include the following:

- To raise a load, for example, screw jack
- To obtain a smooth and accurate motion in a machine tool, for example, leadscrew on a lathe
- To clamp a workpiece, for example, vice
- To load a component, for example, universal test machine

There are three essential parts of a power screw (screw, nut, and a frame or similar to hold the screw and nut in their correct and respective positions).

Power screws operate in two different ways:

- The screw rotates in its bearings and the nut has an axial motion. The lead screw of a lathe is an example of this category.
- The nut is kept stationary and the screw moves in an axial direction. Screw jacks and machine vices are examples of this category.

The fundamental difference between a power screw and a fastening screw is that the efficiency of a power screw is high where friction is low, whereas fastening screws have a low efficiency and there is a high degree of fiction which is necessary so that the screw will not undo. In the limiting case of zero friction, the power screw would have an efficiency of 100%, whereas the fastening screw would not fasten at all.

Due to the sliding friction that is present in a screw, the efficiency is generally low unless there is a way the friction can be minimized. One way to minimize friction is to use recirculating ball screws, but this tends to be an expensive option. They are used extensively in machine leadscrews for activating machine slides used in combination with servo or stepper motor drives where extra precision is required. There are other advantages of using basic power screws that override the disadvantage of low efficiency.

Power screws offer the following advantages:

1. Power screws have a large load-carrying capacity.
2. Power screws are simple in design.
3. The manufacture of a power screw is easy and does not require any specialized equipment.
4. Power screws provide a large mechanical advantage. A load of 20 kN can be raised applying an effort of 300 N. Most power screws are used in applications such as screw jacks, vices, clamps, and valves.
5. Power screws are able to provide accurate linear motion, such as that required in machine tool applications.
6. There are few parts in a screw jack. This reduces cost and improves reliability and reduces maintenance.
7. Power screws can be designed to be self-locking. In a screw jack, this property will prevent the load from descending on its own.

The disadvantages of power screws can be summarized as follows:

1. Power screws have poor efficiencies, as low as 40% in some cases. This makes them unsuitable for continuous operations other than as a lead screw in a machine tool.
2. High friction within the threads can cause rapid wear of the screw. Where square threads are used, the nut is manufactured from a soft material and replaced when worn.

In a trapezoidal thread, the nut is of a split form and able to compensates for wear.

The efficiency of a power screw is improved where the sliding friction is replaced by a rolling friction. This principle is used in recirculating ball screws.

37.2 Terminology

The following terminology is used in describing thread forms such as those used in power screws and fastening screws (Figure 37.1).

1. Pitch p: This is the axial distance between two adjacent threads.
2. Lead L: The distance that a screw or nut will advance for one complete revolution. Generally, the lead will equal the pitch for a single-start thread, but for a multi-start thread as found on some metal cutting center lathes, the lead will equal the pitch multiplied by the number of starts.
3. Nominal Diameter D: The nominal diameter is the outside diameter of the thread. For example, an M25.0 metric thread has an outside diameter of 25.0 mm.

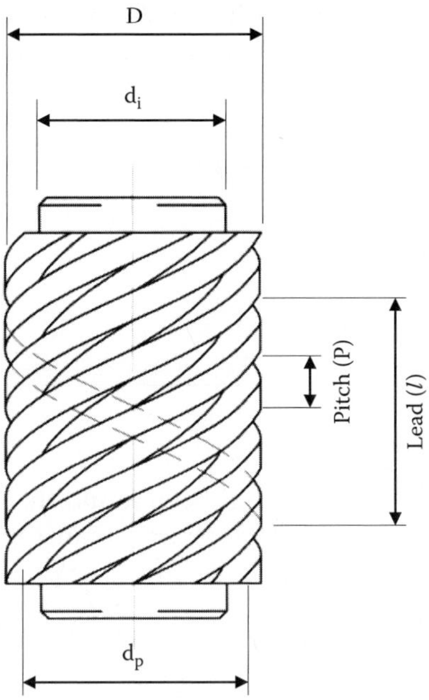

FIGURE 37.1
Terminology.

4. Root Diameter d_i: The root diameter is the inside diameter of the thread. This is also known as the minor or core diameter.
5. Pitch Diameter d: The pitch diameter is a diameter that is midpoint between the outside and root diameters, that is,

$$d = \frac{D + d_i}{2} \tag{37.1}$$

As there will be a small clearance between the tips and roots of the thread and nut, the pitch diameter will be a little larger than the mean diameter. For design purposes, there will be little error in using the mean diameter as the pitch diameter.

6. Helix Angle θ: The helix angle is the angle that is made by the thread with a line drawn perpendicular to the centerline of the thread. It is also the slope of the inclined plane if the helix were unwrapped and laid on a flat surface. It is given by:

$$\tan\theta = \frac{L}{\pi d} \tag{37.2}$$

7. Friction Angle φ: The friction angle is the angle that is made by the maximum friction force and the normal reaction at the surface of the thread. It is also the

angle to the horizontal made by an inclined plane of identical material when a block on the plane is on the point of sliding down the plane without acceleration. It is given by

$$\tan\varphi = \mu \tag{37.3}$$

where μ is the coefficient of friction.

37.3 Coefficient of Friction

Returning to the coefficient of friction, it has been found that the coefficient of friction (μ) in a screw thread is dependent on three factors: workmanship, the type of lubrication, and the number of cycles.

37.3.1 Workmanship

Workmanship is the accuracy of the thread form and the surface finish achieved. This depends on the method used to generate the screw thread, this can be hobbing, screw cutting, and thread rolling where the thread diameter is small.

37.3.2 The Type of Lubrication

The coefficient of friction is practically independent on the load, rubbing velocity, or materials used in the construction. An average value of 0.15 can be taken for the coefficient at the thread surface when lubricated with mineral oil.

The values of coefficient of friction for the thrust collar with a sliding contact will vary from 0.17 for soft steel running on cast iron, down to 0.08 for hardened steel in contact with bronze for starting, and will drop to 0.12 for the soft steel/cast iron combination and 0.06 for the hardened steel/bronze combination when running. If a thrust ball bearing is used at the collar surface, then the friction will significantly drop to 0.01/0.02.

37.3.3 The Number of Cycles

When first installed, the coefficient of friction of the screw thread will be at its maximum. As the revolutions of the screw increase with use, the contact surfaces will "bed down" and the coefficient of friction will reduce over time and reach the average values.

Bear in mind that the values that are quoted above are for normal running or operating circumstances. The start-up or initial friction levels will be higher, and for start-up the values should be increased by one-third.

As the thread is assumed to be lubricated, the materials of construction for the screw and nut will not affect the coefficient of friction.

Table 37.1 tabulates some basic coefficient values of friction of materials used in the manufacture of power screws.

TABLE 37.1
Typical Friction Factors for Power Screws

	Nut Material			
Screw Material	**Steel**	**Brass**	**Bronze**	**Cast Iron**
Steel (dry)	0.15–0.25	0.15–0.23	0.15–0.19	0.15–0.25
Steel (lubricated)	0.11–0.17	0.10–0.16	0.10–0.15	0.11–0.17
Bronze (lubricated)	0.08–0.12	0.04–0.06	—	0.06–0.09

37.4 Thread Forms

The threads used for fastening purposes such as 60° vee forms are unsuitable for power screws. The purpose of fastening threads is to have high friction between the thread and nut or the part the screw is fastening. As discussed previously, the purpose of a power screw is to reduce friction between the screw and the nut. Threads with smaller angles, such as trapezoidal threads, are preferred for power transmission. Figure 37.2a through d shows the various thread forms suitable for power screws.

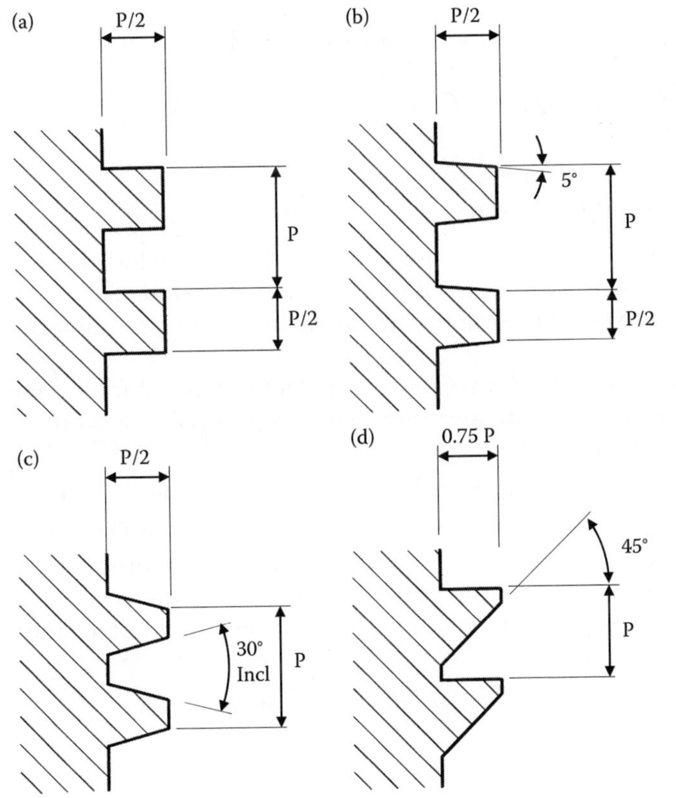

FIGURE 37.2
(a) Power screw thread forms—square, (b) power screw thread forms—modified square, (c) power screw thread forms—trapezoidal, and (d) power screw thread forms—buttress.

37.4.1 Square Thread Form

As seen from Figure 37.2a, the form has faces that are perpendicular to the axis of the thread. It can be cut on a center lathe but not on a thread-hobbing machine or thread-rolling machine. This is the reason that square threads are not widely used, only in special circumstances. The thread form is usually replaced with a modified square section that has the thread faces inclined at 5°, as shown in Figure 37.2b. The modified square can be manufactured using any of the following methods:

- The square thread is the most efficient of all the thread forms.
- There is no radial pressure or side thrust generated on the nut. This radial pressure is known as the bursting pressure on the nut. As there is no side thrust, the motion of the nut will be uniform and the life of the nut is increased.

Square threads have a number of disadvantages:

- Square threads are notoriously difficult to manufacture.
- The strength of the thread depends on the thread thickness at the core diameter and, compared with the trapezoidal thread form, the square form has less load-carrying capacity.
- The wear of the square thread surface can become a serious problem in the service life of the thread. It is not possible to be able to compensate for wear, therefore when worn, either the nut or screw needs replacing.

37.4.2 Trapezoidal Metric Form

The trapezoidal metric form has a face angle of 15° (with an included angle of 30°) and is the metric equivalent of the ACME thread form that has an included angle of 29° (Figure 37.2c).

The advantages of using the trapezoidal thread form can be summarized as follows:

- The trapezoidal thread form can be manufactured using multipoint cutting tools and is a more economical operation compared to single-point cutting tools for the square thread form.
- The trapezoidal thread form has more thickness at the core diameter compared with the square thread. Such screws have a large load-carrying capacity.
- The axial wear on the surface of a trapezoidal thread form can be compensated using a "split"-type nut design and can only be used with the trapezoidal form. This arrangement will be found mostly on the leadscrew of a lathe.

The disadvantages of the trapezoidal thread form are as follows:

- The efficiency is less than the square thread form.
- Trapezoidal thread forms generate side thrust or radial pressures on the nut. The radial pressure (or bursting pressure) on the nut affects the performance.
- When using fine pitch screws, care will be needed to prevent thread jumping.

37.4.3 Buttress Thread Form

The thread form combines the advantages of the square thread, and the trapezoidal is used where heavy axial loads act along the axis of the thread in one direction only (Figure 37.2d).

The advantages of the buttress thread form are as follows:

- It has higher efficiency than trapezoidal threads.
- The thread is stronger than comparable square or trapezoidal thread forms due to the greater thickness at the thread's base.
- It can be economically machined using thread-milling machines.
- The axial wear can be compensated using the "split-type" nut.

The one disadvantage of the buttress thread is that it can only transmit power in one direction only. It is used extensively on vice threads. It is also used in the armament's field for connecting the barrel to the housing of antiaircraft guns.

37.5 Thread Pitch/Diameter Relationship

There are no standardized metric pitch sizes and thread depths for power screws, each thread is designed for the purpose for which it is going to be used, and the main criterion will be thread strength.

Table 37.2 is presented as a guide only for the proportions of the square and trapezoidal metric thread forms.

There is a particular method for the designation of square and trapezoidal threads. A power screw having a single-start square thread is designated by the letters Sq followed by the nominal diameter and the pitch expressed in millimeters separated by the sign x. As an example,

TABLE 37.2

Thread Pitch/Diameter Relationship

Diameter (mm)		Pitch (mm)
From	Up to and including	
10	12	3
15		4
20		5
25		6
30	35	8
4	45	10
50	55	12
60	65	13
70	75	14
80	85	15
90	95	16
100		17

$$\text{Sq } 30 \times 8$$

This indicates a single-start square thread having a nominal diameter of 30 and 8 mm pitch.

Similarly, a single-start ISO metric trapezoidal thread will be designated by the letters Tr followed by the nominal diameter and the pitch expressed in millimeters and separated by the sign ×. As an example,

$$\text{Tr } 40 \times 10$$

This indicates a single-start trapezoidal thread with a nominal diameter of 40 mm and a pitch of 10 mm.

Multiple-start trapezoidal threads are designated by the letters Tr followed by the nominal diameter and the lead separated by the sign x, and in brackets the letter p followed by the pitch expressed in millimeters. As an example,

$$\text{Tr } 40 \times 20 \text{ (P10)}$$

In the above designation,

$$\text{Lead} = 20 \text{ mm}$$
$$\text{Pitch} = 10 \text{ mm}$$

Therefore, the number of starts = 20 / 10 = 2

The above designation indicates a two-start trapezoidal thread having a nominal diameter of 40 and a 10 mm pitch. In the case of a left-hand thread, the letters LH are added to the thread designation, as follows:

$$\text{Tr } 40 \times 20 \text{ (P10) LH}$$

37.6 Multiple-Threaded Screws

When higher traveling speeds are required, an option is to use multi-start screws. They can be manufactured in double-start or triple-start screws.

These screws have the following examples:

- It provides larger axial movement of the sliding member compared to the single-start screw.
- The efficiency of the multi-start screw is more than the single-start screw due to the increase in the helix angle.

The disadvantages of a multi-start screw are as follows:

- The mechanical advantage of using a multi-start thread is lower than that for a single-start thread. The torque required to raise a load will be significantly increased for the multi-start screw.

- The self-locking property in a multi-start screw may be lost. This could be dangerous in the case of a screw jack, where the load may descend on its own.

37.7 Screw Torque and Thrust

37.7.1 Torque Required for Raising the Load

The screw thread can be considered as an inclined plane with an angle α, as shown in Figure 37.3.

In this case, the screw is considered to be vertical where the load is being raised. The following forces act on the inclined plane:

1. Load (W): This always acts parallel to the screw axis, in this case acting downward.
2. Normal reaction (N): This acts perpendicular (normal) to the plane.
3. Frictional force (μN): This acts opposite to the motion of the load. As the load is moving up the inclined plane in Figure 37.3, frictional force acts downward on the inclined plane.
4. Effort (P): This acts perpendicular to the load W. In this case, it will act toward the right to overcome the friction and raise the load.

Consider the equilibrium of the horizontal forces acting on the load as depicted in Figure 37.3:

$$P = \mu N \cos\alpha + N \sin\alpha \tag{37.4}$$

Consider the equilibrium of the vertical forces:

$$P = N \cos\alpha - \mu N \sin\alpha \tag{37.5}$$

Dividing the expression (37.4) by Equation 37.5,

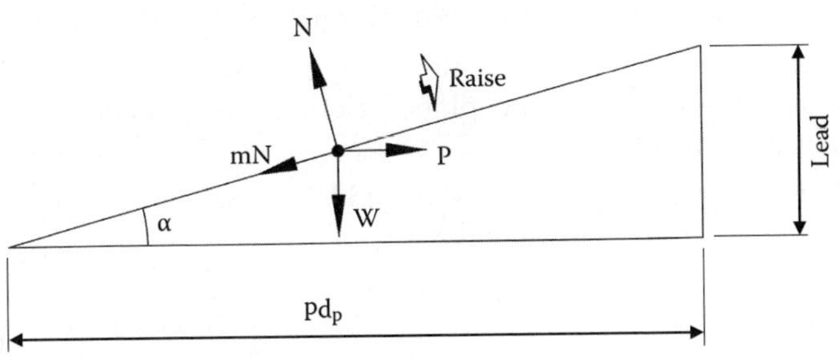

FIGURE 37.3
Raising the load.

$$P = \frac{W(\mu \cos\alpha + \sin\alpha)}{(\cos\alpha - \mu \sin\alpha)} \quad (37.6)$$

Dividing the numerator and the denominator of the right-hand side by cos α,

$$P = \frac{W(\mu + \tan\alpha)}{(1 - \mu \tan\alpha)} \quad (37.7)$$

The coefficient of friction μ is expressed as

$$\mu = \tan\varphi \quad (37.8)$$

where φ is the friction angle.

Substituting μ=tan φ in Equation 37.7:

$$P = \frac{W(\tan\phi + \tan\alpha)}{(1 - \tan\phi \tan\alpha)}$$

or

$$P = W\tan(\varphi - \alpha) \quad (37.9)$$

The torque T required to raise the load can be expressed as

$$T = \frac{Pd_P}{2}$$

$$T = \frac{Wd_P}{2}\tan(\varphi + \alpha) \quad (37.10)$$

37.7.2 Torque Required for Lowering the Load

When the load is being lowered, the system of forces acting at a point on the inclined plane is shown in Figure 37.4.

1. Load (W): This always acts parallel to the screw axis, in this case acting downward.
2. Normal reaction (N): This acts perpendicular (normal) to the plane.
3. Frictional force (μN): This acts opposite to the motion of the load. In this case, the load is moving down the inclined plane, frictional force acts along the inclined plane in an upward direction.
4. Effort (P): This acts perpendicular to the load W. In this case, it will act toward the left to overcome the friction lowering the load.

Considering the equilibrium of the vertical and horizontal forces:

$$P = \mu N \cos\alpha - N \sin\alpha \quad (37.11)$$

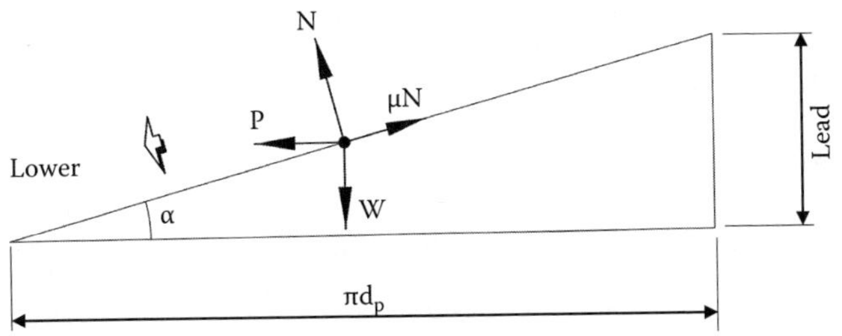

FIGURE 37.4
Lowering the load.

$$W = N\cos\alpha + \mu N\sin\alpha \tag{37.12}$$

Dividing expression 37.11 by Equation 37.12,

$$P = \frac{W(\mu\cos\alpha - \sin\alpha)}{(\cos\alpha + \mu\sin\alpha)} \tag{37.13}$$

Dividing the numerator and denominator of the right-hand side by cos α,

$$P = \frac{W(\mu - \tan\alpha)}{(1 + \mu\tan\alpha)} \tag{37.14}$$

Substituting $\mu = \tan\varphi$ in Equation 37.14,

$$P = \frac{W(\tan\varphi - \tan\alpha)}{(1 + \tan\varphi\tan\alpha)} \tag{37.15}$$

or

$$P = W\tan(\varphi - \alpha) \tag{37.16}$$

and

$$T = \frac{wd_p}{2}\tan(\varphi - \alpha) \tag{37.17}$$

Example 37.1

By way of an example, consider a trapezoidal single-start metric screw orientated in a vertical plane and having a nominal diameter of 40 mm with a pitch of 10 mm. It carries an axial load of 6 kN.

Assuming a coefficient of friction of 0.15 between the thread contact faces, calculate the torque required to continuously raise or lower the load.

Solution

Thread details:

$D = 40\,\text{mm}$

$p = 10\,mm$
$L = 10\,mm$ (Lead for single-start thread)
$F = 6\,N$
$\mu = 0.15$
d_p = Pitch diameter

$$d_p = D - \frac{p}{2} = 35\text{ mm}$$

Coefficient of friction:
cof = 0.15
Face angle:
$\alpha = 15°$

From Equation 37.4,

$$\tan\varphi = \frac{\mu}{\cos\alpha}$$

$$\tan\varphi = \frac{0.15}{\cos 15°}$$

$$= 0.1553$$

$$= 8.827°$$

Helix angle:
From Equation 37.2,

$$\tan\theta = \frac{L}{\pi d}$$

$$\tan\theta = \frac{10\text{ mm}}{\pi * 35\text{ mm}}$$

$$= 0.0909$$

$$= 5.1965°$$

Raising the load:

$$T = \frac{Wd_p}{2}\tan(\varphi + \alpha)$$

$$T = \frac{6000 * 35\text{mm}}{2}\tan(8.827° + 5.196°)$$

$$= 26.224\text{ Nm}$$

Lowering the load:

$$T = \frac{Wd_p}{2}\tan(\varphi - \alpha)$$

$$T = \frac{6000 * 35\text{mm}}{2}\tan(8.827° - 5.196°)$$

$$= 6.663\text{ Nm}$$

37.8 Self-Locking

Example 37.1 shows that the torque required to lower the load is 6.663 Nm. If the applied torque had been reduced to zero, then the load would have been held in position.

This feature is known as *self-locking* and is seen as a built-in safety feature that is important in many applications, such car jacks and other lifting devices.

In situations where the load will self-lower without any application of torque, the thread is no longer considered self-locking. This is also known as *overhauling*, and will occur when

$$\varphi - \theta = 0 \text{ or } \varphi = \theta \ (\text{where } \varphi \text{ is a non-square load face})$$

Simply stated, when the helix angle of the thread is equal to or greater than the friction angle, the thread will no longer be self-locking. As seen earlier in Section 37.6, multi-start threads usually have large helix angles and there is a possibility of overhauling. Therefore, if overhauling is a requirement, then then a multi-start thread will be required.

37.9 Efficiency of a Screw Thread

Drawing a force diagram for lifting the load in Example 37.1, it would look like Figure 37.4. Suppose for the purposes of illustration, the load W moves from the lower end to the upper end of the inclined plane. The output would consist of raising the load. Therefore,

$$\text{Work output} = \text{force} \times \text{distance moved in the direction of the force} = (\text{Wl})$$

The input would consist of rotating the screw by means of an effort P:

$$\text{Work input} = \text{force} \times \text{distance moved in the direction of the force} = \text{P}(\pi \text{d}_\text{p})$$

The efficiency η of the screw is given by:

$$\eta = \frac{\text{work output}}{\text{work input}} = \frac{\text{Wl}}{\text{P}(\pi \text{d}_\text{p})} = \frac{\text{W}}{\text{P}} \tan\alpha \tag{37.18}$$

Substituting Equation 37.9 in Equation 37.18:

$$\eta = \frac{\tan\alpha}{\tan(\varphi - \alpha)} \tag{37.19}$$

In the case of a square threaded power screw, its efficiency will be dependent on the angles of (α) and (φ), and α is also dependent on the thread's lead (l) and the pitch diameter (d_p) since:

$$\tan\alpha = \frac{1}{\pi \text{d}_\text{p}} \tag{37.20}$$

The above shows that the efficiency of a square thread can be dependent upon the following factors:

- Pitch diameter of the screw
- Lead of the screw
- Coefficient of friction

The above expression 37.19 for the efficiency of a square thread has certain limitations; it only considers the friction loss at the contact surfaces of the thread. It does not consider any other frictional losses, such as the collar friction loss. This will be considered later in the chapter.

37.10 Efficiency of a Self-Locking Thread

From Equation 37.19, the efficiency of a square threaded screw is given as

$$\eta = \frac{\tan\alpha}{\tan(\varphi - \alpha)}$$

For a self-locking screw,

$$\varphi \geq \alpha$$

Substituting the limiting value into Equation 37.19 gives

$$\eta \leq \frac{\tan\varphi}{\tan(\varphi + \alpha)} \quad \text{or} \quad \eta \leq \frac{\tan\varphi}{\tan(2\varphi)} \tag{37.21}$$

Substituting

$$\tan(2\varphi) = \frac{2\tan\varphi}{(1 - \tan^2\varphi)} \tag{37.22}$$

In the above expression 37.21,

$$\eta \leq \frac{\tan\varphi(1 - \tan^2\varphi)}{2\tan\varphi} \tag{37.23}$$

or

$$\eta \leq \left[\frac{1}{2} - \frac{\tan^2\varphi}{2}\right]$$

The preceding equation clearly shows that the efficiency of a self-locking square thread power screw is less than 50%.

37.11 Efficiency of a Trapezoidal Thread

The forces acting on a trapezoidal thread is illustrated in Figure 37.5.

The thread angle is 2θ:

- For ISO metric trapezoidal thread, 2θ=30°, inclusive
- For an ACME thread form, 2θ=29°, inclusive

In the case of the square thread, where W is the axial load being raised by the screw. It is also the normal force that is acting on the screw flank.

In the trapezoidal or ACME thread form, the force acting normal to the flank of the thread form is

$$W_\theta = \frac{W}{\cos\theta} \quad \text{or} \quad \frac{W}{\sec\theta}$$

The frictional force is dependent on the normal force. As a result, the effect of the thread angle is to increase the fictional force by the term (sec θ). To account for this, the coefficient of friction is taken as (μ sec θ) instead of (μ).

In the case of the trapezoidal thread form, the equation derived for the square threaded screws is amended by making this change.

Considering Section 37.7.1

Torque required for raising the load:

Modifying Equation 37.7,

$$P = \frac{W(\mu\sec\theta + \tan\alpha)}{(1+\mu\sec\theta\tan\alpha)} \tag{37.24}$$

and

$$T = P\frac{d_p}{2} = \frac{Wd_p}{2}\cdot\frac{(\mu\sec\theta + \tan\alpha)}{(1-\mu\sec\theta\tan\alpha)} \tag{37.25}$$

Torque required for lowering the load:

Modifying Equation 37.14,

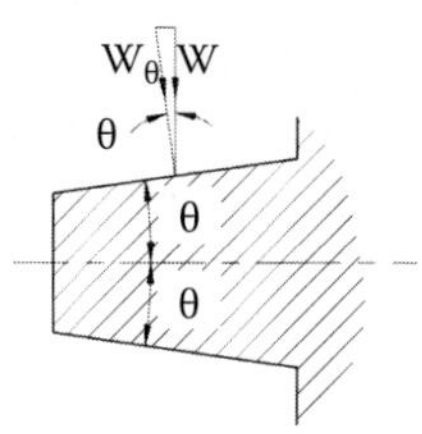

FIGURE 37.5
Forces acting on ACME or trapezoidal thread.

$$P = \frac{W(\mu \sec\theta - \tan\alpha)}{(1 + \mu \sec\theta \tan\alpha)} \tag{37.26}$$

and

$$T = P\frac{d_p}{2} = \frac{Wd_p}{2} \cdot \frac{(\mu \sec\theta - \tan\alpha)}{(1 + \mu \sec\theta \tan\alpha)} \tag{37.27}$$

The efficiency of a square thread is determined by Equation 37.18.

$$\eta = \frac{W}{P}\tan\alpha \tag{37.18}$$

Substituting Equation 37.7 into Equation 37.18,

$$\eta = \frac{\tan\alpha(1 - \mu\tan\alpha)}{(\mu + \tan\alpha)} \tag{37.28}$$

Now, substituting the expression (μ sec θ) for μ, the efficiency of a trapezoidal thread is given by the expression

$$\eta = \frac{\tan\alpha(1 - \mu \sec\theta \tan\alpha)}{(\mu \sec\theta + \tan\alpha)} \tag{37.29}$$

37.12 Collar Friction

The thrust from the power screw is generally taken by a thrust collar, where the rotary action of the power screw is converted into a linear motion. If a rolling element bearing is used, then the friction will be reasonably small enough to be disregarded. If, on the other hand, there is a rotating thrust face rather than a bearing, then the friction could be significant and should therefore be taken into account in the torque calculations.

Assuming that the thrust washer has been "bedded" in, then the frictional collar torque is given by

$$T = \mu\, F\, r_m \tag{37.30}$$

where r_m is the mean radius of the thrust face.

The torque calculated for the screw must be increased to cover for the additional torque required by the thrust washer. Hence, the efficiency will be lower.

This can be illustrated by Example 37.2.

Example 37.2

Consider the screw thread given in Example 37.1, where it transmits the thrust load by means of a collar of an outside diameter of 60 mm and an inside diameter of 30 mm.

Recalculate the torque now required to raise and lower the load, and the efficiency when raising the load.

Solution

The mean radius of the thrust collar is

$$r_m = \frac{(30\text{ mm} + 15\text{ mm})}{2} = 22.5\text{ mm}$$

The collar friction torque will be

$$T = \mu F r_m$$

$$T = 0.15 \times 6000 \times 0.0225\text{ Nm}$$

$$T = 20.25\text{ Nm}$$

The screw torque required to raise the load from Example 37.1 = 26.224 Nm:

$$\therefore \text{Total torque} = 26.224 + 20.25\text{ Nm}$$

Total torque = 46.47 Nm

Lowering the load, the screw torque from Example 37.1 = 6.663 Nm:

$$\therefore \text{Total torque} = 6.663 + 20.25\text{ Nm}$$

Total torque = 26.91 Nm

The efficiency when raising the load will be

$$\eta = \frac{6000 \times 0.01}{2\pi \times 46.47}$$

$$= 0.205\ (20.5\%)$$

37.13 Stress Analysis

It has been found that due to pitch errors and thread deflection, the majority of the load is carried on the first one or two threads. Clearances, surface finishes, fillet radii, and the quality of machining will have a significant effect on the local stresses within the power screw components.

A simplified form of stress analysis is shown in this section where the effects above are not considered. It is assumed that the threads are evenly loaded and all the threads are in intimate contact with the threads in the nut.

The simplified analysis is still useful to undertake, for it gives the following:

- A fundamental understanding of the stresses that is present within the power screw and its associated parts.
- Appropriate design factors can be used to minimize any factors of uncertainties that may exist.

Consider a power screw thread and nut that is loaded with an axial force F, as shown in Figure 37.6.

37.13.1 Nomenclature

a = thickness of nut
h = thickness of the thread at the core diameter
t = depth of thread
l = length of thread in contact with the load

37.13.2 Assumptions

1. The load is evenly distributed and considered acting at the pitch diameter of the thread.
2. The nut thickness is within the range of 0.75–1.5 D.
3. For square, modified square, and trapezoidal threads, the depth of thread t = 0.5 p, for buttress threads, t = 0.75 p.

Ignoring any clearances, calculating the thread height h using the geometry derived from the thread face angles, Table 37.3 is constructed.

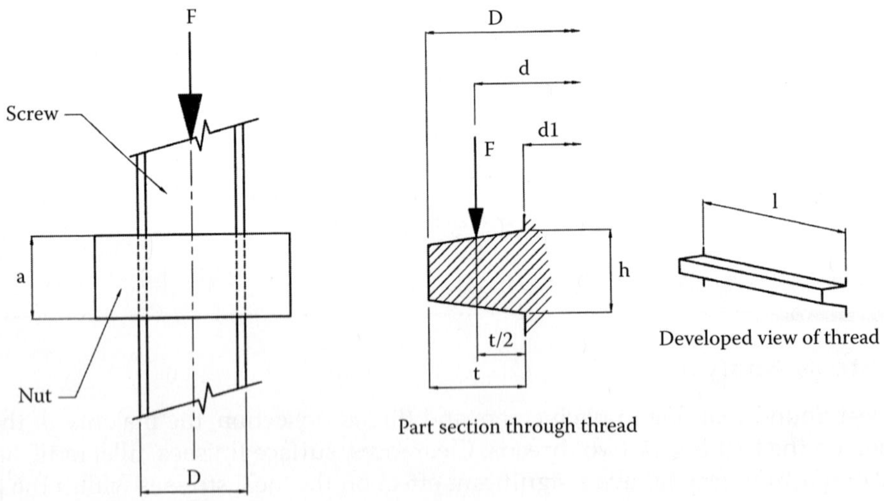

FIGURE 37.6
Thread and nut dimensions.

TABLE 37.3
Thread Thicknesses

Type of Thread	Square	Modified Square	Trapezoidal Metric	Buttress
Thread height (h)	0.5 p	0.5437 p	0.634 p	0.875 p

37.13.3 Number of Threads in Nut

Number of threads in the nut n:

$$n = \frac{a}{p} \quad (37.31)$$

where
a = thickness of nut
p = pitch of screw
This formula is true for all thread forms and will apply to both single-start and multi-start threads.

37.13.4 Length of Thread

Thread length l that carries the load:

$$'l' = \pi \, d_p n \quad (37.32)$$

where d_p is the pitch diameter of the thread.

37.13.5 Bearing Pressure in the Thread

The bearing pressure (stress) is given by

$$\sigma_b = \frac{F}{t\,l} \quad (37.32)$$

where
σ_b = bearing stress
F = force acting on thread
t = depth of thread
l = total length of thread

In a well-designed power screw, failure will not occur due to either crushing or bearing stress in the thread. The bearing pressure must be kept reasonably low to allow adequate lubrication of the threads, thus preventing excessive wear.

The allowable pressure will be dependent on a couple of factors, namely, the rubbing speed and the lubricant properties.

The higher the rubbing speed, the lower the allowable pressure. Typical maximum rubbing speed values based on the assumption that all the threads are in contact with the nut is given in Table 37.4. The figures quoted are for a general lubricant.

TABLE 37.4
Maximum Rubbing Speeds and Allowable Bearing Pressures

Maximum Rubbing Speed (m/s)	Allowable Bearing Pressure (MPa)
<0.05	20
0.05–0.10	10
–0.20	5
>0.20	2.5

37.13.6 Bending Stress in the Thread (σ_b)

Considering the thread as a cantilever, the bending stress can be expressed as

$$\sigma_b = \frac{My}{I} \tag{37.34}$$

where

σ_b = bending stress
M = bending moment due to the action of the force
y = distance from the neutral axis to the outermost fiber
I = second moment of area of section

Thread properties:

$$M = \frac{F\,t}{2} \tag{37.35}$$

$$y = \frac{h}{2} \tag{37.36}$$

$$I = \frac{l\,h^3}{12} \tag{37.37}$$

Substituting the above values in Equation 37.35,

$$\sigma_b = \frac{3\,F\,t}{l\,h^2} \tag{37.38}$$

37.13.7 Shear Stress in the Thread (σ_s)

The average shear stress in the thread:

$$\sigma_s = \frac{F}{l\,h} \tag{37.39}$$

As bending is combined with the shear stress, the shear stress distribution will not be uniform.

The maximum shear stress will occur on the neutral axis of the thread and will be approximately 50% higher than the average shear stress for a rectangular section.

Hence, the maximum shear stress:

$$\sigma_s = \frac{1.5\,F}{l\,h} \tag{37.40}$$

In this case, there is no need to combine the bending stress with the shear stress in the thread, as the maximum bending stress occurs at the outer fibers of the thread, whereas the maximum shear stress occurs on the neutral axis where the bending stress will be zero.

37.13.8 Axial Stress in Root of the Thread (σ)

There will be an axial stress at the root of the thread, which will be either tensile or compressive depending on the direction of the axial force. The stress area is slightly larger than the root area and is generally the area of a circle having a diameter that is equal to the mean of the root diameter and the pitch diameter.

$$d_t = \frac{d_i + d_p}{2} \tag{37.41}$$

where

d_t = tensile diameter
d_i = root diameter
d_p = pitch diameter

The stress area (tensile area):

$$A_t = \frac{\pi\, d_t^2}{4} \tag{37.42}$$

The tensile (or compressive) stress:

$$\sigma_t = \frac{F}{A_t} \tag{37.43}$$

For a conservative design, the root area of the thread should be used in preference to the tensile area.

37.13.9 Torsional Shear in the Root (σ_{st})

The torsional shear stress in the root is

$$\sigma_{st} = \frac{16\, T}{\pi\, d_i^3} \tag{37.44}$$

In this formula, the root diameter is used in preference to the tensile diameter, the threads contribute very little to the strength of the screw in torsion.

37.13.10 Combined Shear Stress in the Root of the Thread (σ_{smax})

Both axial stress and torsional shear stress occur in the root of the thread and they are required to be combined. The maximum combined stress is given by the general combined stress formula:

$$\sigma_{smax} = \sqrt{\left(\frac{\sigma}{2}\right)^2 + \sigma_{st}^2} \tag{37.45}$$

The combined axial stress is usually not as significant as the combined shear stress and is often disregarded.

TABLE 37.5

Buckling End Condition Constants

End Condition	C
Fixed—pivoted	0.25
Pivoted—pivoted	1.0
Fixed—pivoted	2.0
Fixed—fixed	4.0

37.14 Buckling

Unsupported power screws may be subject to buckling if the applied load is compressive; as a rule of thumb, when the exposed length of screw is greater than eight times the root diameter, then the buckling stress must be considered.

General buckling of columns and struts is discussed in greater detail in Chapter 23.

In this section, buckling in relation with the power screw is considered briefly. The Euler formula will be used to estimate the critical load (F_c) at which buckling will occur for relatively long lengths of screw (L_c), and the second moment of area (I).

The Rankine formula can also be used:

$$F_c = \frac{C\pi^2 E}{L_c^2}\left(\frac{\pi d_r^4}{64}\right) \tag{37.46}$$

where

C is the end condition constant for various cases, as shown in Table 37.5

F_c is the critical (buckling) load

37.14.1 Stability Factor

The stability factor is defined as

$$\frac{\text{Buckling load}}{\text{Applied load}} \tag{37.47}$$

If the above factor is greater than 2.5, then buckling will not occur in the unsupported length of the screw.

38

Clutches

38.1 Introduction

A clutch can be described as a device to connect or disconnect the power transmission system from the source of power at the will of either the operator or an overload safety device.

The flow of mechanical power is controlled via the clutch, unless the system is a direct drive.

38.2 Positive Clutches

With this type of clutch, the engaging features mesh to produce a rigid connection between two adjacent shafts, usually the driving shaft and the driven shaft. They are mostly used in situations requiring a simple and rapid disconnection. This type of clutch will require to be connected when the shafts are stationary and unloaded.

The merits of positive clutches are as follows:

- Simple
- No slip
- No heat generated
- Compact
- Low cost

Their disadvantage is that they cannot be easily connected while the shafts are rotating and under load.

38.2.1 Square Jaw (Dog) Clutch

The engaging surfaces are usually of a square design, as shown in Figure 38.1a.

38.2.2 Spiral Jaw Clutch

In this design, the engaging surfaces may be a spiral jaw design as depicted in Figure 38.1b.

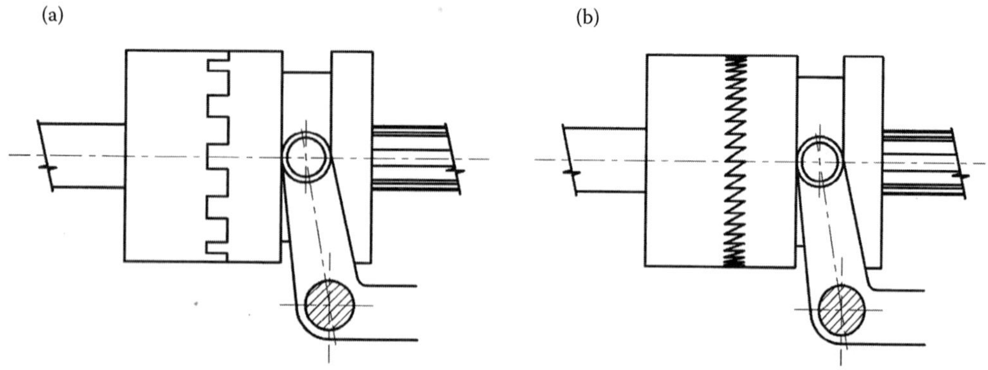

FIGURE 38.1
(a) Square jaw (dog) clutch and (b) spiral jaw clutch.

38.3 Friction Clutches

Friction clutches work on the basis that frictional forces are generated between two contacting surfaces. Friction clutches have the advantage over positive clutches in that:

- The friction surfaces are able to slip during engagement; this enables the driver of a vehicle to pick up and accelerate the vehicle with a minimum of shock without stalling the engine.

Friction clutches can be characterized as follows:

- Single-plate clutch
- Multiplate clutch
- Cone clutch

Multiplate clutches are considered outside the scope of this discussion.

38.3.1 Single-Plate Clutch

In this design, the clutch consists of two flanges as shown in Figure 38.2.

One flange is rigidly connected to one shaft (usually the driving shaft) and the other flange is free to move axially on the driven shaft, along a splined joint. The actuating force is provided by a spring that forces the driven flange to connect with the driving flange. The face of the driving flanged is lined with a high-friction material, such as cork, leather, or ferodo. In the case of automotive applications, the friction material is held in a separate disc that is allowed to move axially but is retained by the driven shaft. This disc is sandwiched between the two flanges providing the friction for the clutch to function correctly. This method allows the friction disc to be replaced easily.

There are two basic assumptions used in the design for disc clutches:

- Uniform pressure distribution between the mating surfaces
- Uniform rate of wear at the mating surfaces

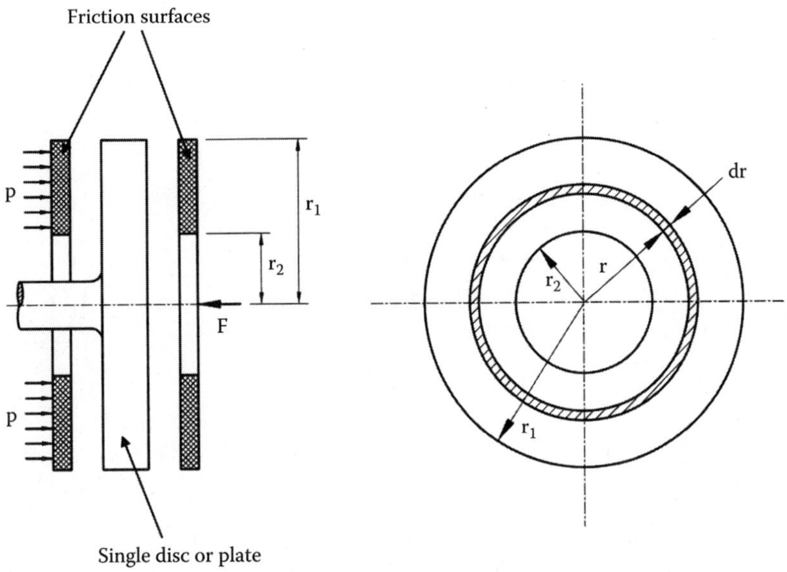

FIGURE 38.2
Forces acting on a single-disc clutch

These assumptions are only valid for an unworn, accurately machined clutch with rigid outer discs.

38.3.2 Nomenclature

Table 38.1 depicts the nomenclature used in the description of friction disc clutches.

38.4 Considering Uniform Pressure

When the pressure is uniformly distributed over the entire area of the friction face, the intensity of pressure is given by

$$p = \frac{F}{\pi\left(r_o^2 - r_i^2\right)} \tag{38.1}$$

The elemental force acting on an elemental annular ring is given by

$$dF = pdA \tag{38.2}$$

The area of an elemental annular ring on the disc clutch, as shown in Figure 38.2:

$$dA = 2\pi\ r\ dr \tag{38.3}$$

and

TABLE 38.1

Nomenclature Used in the Description of a Friction Disc Clutch

Symbol	Description
F	Applied force (N)
P	Transmitted power (kW)
T	Torque (Nm)
T_f	Frictional torque
μ	Coefficient of fiction
r	Elemental radius (m)
dr	Elemental thickness (m)
r_o	Outer radius of clutch plate (m)
r_i	Inner radius of clutch plate (m)
n	Rotational speed (rev/min)
p	Pressure between friction surfaces (N/m²)
p_{max}	Maximum pressure between friction surfaces (N/m²)

$$dF = 2\pi \ r \ p \ dr \tag{38.4}$$

Integrating Equation 38.4 to give the total axial force,

$$F = \int_{r_i}^{r_o} 2\pi \ r \ p \ dr \tag{38.5}$$

The frictional torque dTf that can be developed on an elemental ring will be the product of the elemental normal force, given by μ dF and the radius f.

$$dT = \mu \ dF \ r \tag{38.6}$$

$$dT = \mu \ (2\pi \ r \ p \ dr) \ r \tag{38.7}$$

$$dT = 2\pi \ p \ \mu \ r^2 \ dr \tag{38.8}$$

where μ is the coefficient of friction between the two mating surfaces.

Integrating Equation 38.8 between the limits of the annular ring, r_o and r_i will give the total torque (T):

$$T = \int_{r_i}^{r_o} 2\pi \ r^2 \mu \ p \ dr \tag{38.9}$$

Assuming uniform pressure distribution,

$$F = 2\pi \ p \int_{r_i}^{r_o} r \ dr \tag{38.10}$$

$$F = 2\pi \ p \left[\frac{r^2}{2}\right]_{r_i}^{r_o} \tag{38.11}$$

$$F = \pi \ p \ \left(r_o^2 - r_i^2\right) \tag{38.12}$$

F is the necessary force that is required to clamp the clutch discs together.

Integrating Equation 38.9 to obtain T.

As p is a constant, the integration becomes

$$T = 2\,\pi\,\mu\,p \int_{r_i}^{r_o} r^2\,dr \tag{38.13}$$

$$T = 2\,\pi\,\mu\,p \left[\frac{r^3}{3}\right]_{r_i}^{r_o} \tag{38.14}$$

$$T = \frac{2}{3}\,\pi\,\mu\,p\left(r_o^3 - r_i^3\right) \tag{38.15}$$

Equation 38.15 is the torque capacity of a clutch having a single frictional interface. In reality, clutches will have an even number of frictional surfaces. For a clutch having n frictional surfaces, the torque capacity is given by

$$T = \frac{2}{3}\,n\,\pi\,\mu\,p\left(r_o^3 - r_i^3\right) \tag{38.16}$$

where n is the number of frictional surfaces.

From Equations 38.5 and 38.12 and substituting in Equation 38.16 for the pressure p, this gives an expression for the torque capacity as a function of the axial clamping force:

$$T = \frac{2}{3}\,n\,\mu\,F\left(\frac{r_o^3 - r_i^3}{r_o^2 - r_i^2}\right) = n\,\mu\,F\,R_m \tag{38.17}$$

where R_m is the mean radius of friction surfaces:

$$R_m = \frac{2}{3}\left(\frac{r_o^3 - r_i^3}{r_o^2 - r_i^2}\right) \tag{38.18}$$

38.5 Considering Uniform Wear

The wear on the friction surfaces is assumed to be proportional to the product of pressure and velocity:

$$pv = p\,\omega\,r = \text{constant} \tag{38.19}$$

where ω is the angular velocity.

At a constant angular velocity, the maximum pressure will occur at the inner radius (r_i), see Figure 38.3.

$$p_{max}\,\omega\,r_i = p\,\omega\,r \tag{38.20}$$

Deleting the angular velocity ω from both sides of the equation, the relationship for the pressure as a function of the radius is

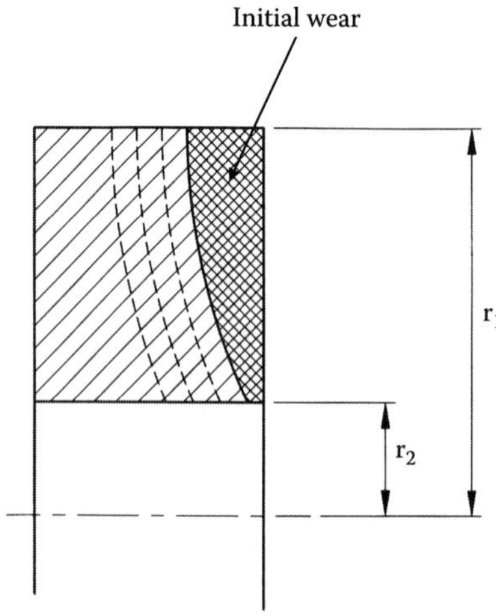

FIGURE 38.3
Uniform axial wear.

$$p = p_{max} \frac{r_i}{r} \tag{38.21}$$

Integrating Equation 38.5 to obtain F,

$$F = \int_{r_i}^{r_o} 2\pi\, r\, p\, dr$$

$$= \int_{r_i}^{r_o} 2\pi\, r\, p_{max} \frac{r_i}{r}\, dr$$

$$= 2\pi\, p_{max}\, r_i \int_{ri}^{ro} dr$$

$$= 2\pi\, p_{max}\, r_i\, [ri]_{r_i}^{r_o}$$

$$F = 2\pi\, p_{max}\, r_i\, (r_o - r_i) \tag{38.22}$$

Rearranging Equation 38.22 and solving for p_{max},

$$p_{max} = \frac{F}{2\pi\, r_i\, (r_o - r_i)} \tag{38.23}$$

Substituting the value for p from Equation 38.21 into Equation 38.9 and integrating to obtain T,

$$T = \int_{r_i}^{r_o} 2\pi\, \mu\, p_{max}\, r_i\, r\, dr$$

$$= 2\pi\, \mu\, p_{max}\, r_i \int_{r_i}^{r_o} r\, dr$$

$$= 2\pi\, \mu\, p_{max}\, r_i \left[\frac{r^2}{2}\right]_{r_i}^{r_o}$$

$$T = \pi\, \mu\, p_{max}\, r_i\, (r_o^2 - r_i^2) \tag{38.24}$$

Substituting p_{max} obtained from Equation 38.23 into Equation 38.24 to obtain T,

$$T = \mu\, F \left(\frac{r_o - r_i}{2} \right) \tag{38.25}$$

The mean radius will be

$$r_{mean} = \frac{(r_o + r_i)}{2} \tag{38.26}$$

When comparing Equation 38.16 with Equation 38.24, note that a clutch designed for uniform wear will have a lower torque capacity than one designed for uniform pressure. This will be due to the higher initial wear at the outer radius moving the center of pressure toward the inner radius.

Clutches are usually designed based on uniform wear assumption, because the uniform wear assumption will give a lower torque capacity clutch compared to one designed on the uniform pressure assumption. The clutch designed using the uniform wear will have larger clutch dimensions and will still be able to carry a larger torque when the clutch is old. Designing a clutch for the uniform pressure assumption will result in clutch slip as the clutch becomes older.

38.6 Calculating Frictional Clutch Dimensions Using the Uniform Wear Assumption

The design will depend upon the following:

- Maximum torque
- Friction material properties (μ, p_{max})
- Clutch plate dimensions

From Equation 38.24,

$$T = \pi\, \mu\, p_{max}\, r_i\, (r_o^2 - r_i^2)$$

Knowing the required torque capacity and frictional material properties, there remain two unknowns: r_o and r_i.

Differentiating Equation 38.24 with respect to r_i and equating the result to zero,

$$\frac{dT}{dr_i} = \pi\, \mu\, p_{max} \left[(r_o^2 - r_i^2) + r_i\, (-2r_i) \right] = 0$$

$$\left[(r_o^2 - r_i^2) + r_i\, (-2r_i) \right] = 0$$

$$r_o^2 - 3r_i^2 = 0 \tag{38.27}$$

The maximum torque for any outer radius r_o is found to occur at

$$r_i = \sqrt{\frac{1}{3}}\, r_o \tag{38.28}$$

$$r_i = 0.577\, r_o \tag{38.29}$$

To ensure that the torque is fully transmitted under all working conditions without slip occurring, a service factor is applied to obtain the clutch torque capacity:

$$T_c = \beta\, T_{max} \tag{38.30}$$

where

T_c=clutch torque capacity
β=service factor
T_{max}=maximum input torque

Table 38.2 lists representative values for service factors. Table 38.3 lists some typical values for friction coefficients linings for dry and immersed in oil, permissible contact pressures and allowable temperature limits.

TABLE 38.2

Some Representative Values for Service Factors

		Type of Drive			
Description of Drive Systems	**Description of Drive Systems**	**Small Electric Motors, Turbines**	**IC Engine, 4–6 Cylinders**	**IC Engine, 2–3 Cylinders**	**IC Engine, Single Cylinder**
Steady power source, steady load, no shock or overload	Belt drives, small generators, centrifugal pumps, fans, machine tools	1.5	1.7	1.9	2.2
Steady power source with some irregularity of load up to 1.5 times normal power	Light machinery for wood, metal and textiles, conveyor belts	1.8	2.0	2.4	2.7
	Large conveyor belts, larger machinery, reciprocating pumps	2.0	2.2	2.4	2.7
Frequent stop-starts, overloads, High inertia starts, large pulsating power source	Presses, punches, piston pumps, cranes, hoists	2.5	2.7	2.9	3.2
	Rolling mills, stone crushers, heavy machinery, single-cylinder compressors	3.0	3.2	3.4	3.7

TABLE 38.3

Typical Frictional Properties for Various Materials

Material	μ Dry	μ Oil	p_{max} (MPa)	t (°C)
Cork	0.3–0.5	0.15–0.25	0.055–0.1	80
Woven material	0.25–0.45	0.08–0.10	0.35–0.70	200–260
Wood	0.20–0.45	0.12–0.16	0.35–0.62	90
Sintered metal	0.15–0.45	0.05–0.08	1.04–2.10	230–680
Cast iron	0.15–0.25	0.03–0.06	0.69–1.73	260

38.7 Cone Clutches

Figure 38.4 shows a simple form of a cone clutch.

The cone clutch consists of a cup or driver, and a cone or follower. The cup is restrained on the drive shaft by either a spline or key, and its internal face fits the external face of the cone, which, in turn, is retained by either a spline or key. The external face of the cone generally has friction material fitted to it. The cone is moved axially on the driven shaft to bring the two conical surfaces together.

The cone clutch is able to transfer a higher torque than can plate or disk clutches of the same size. Due to the wedging action and increased surface area, cone clutches are generally used in low-peripheral-speed applications.

They have the following advantages and disadvantages over a plate or disc clutch:

- Advantages:
  The clutch is simple in design.
  Less axial force is required to engage the clutch.

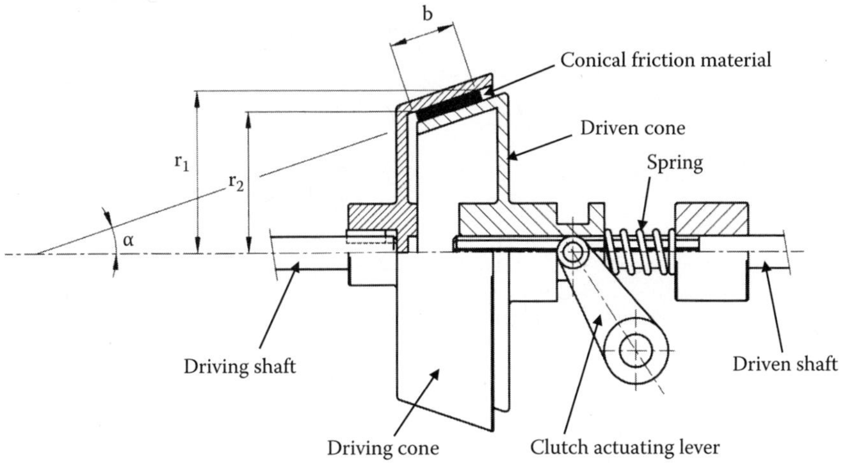

FIGURE 38.4
Cone clutch.

- Disadvantages:
  There is a tendency to grab.
  There is some reluctance to disengage.

The cone angle (α) is always greater than 8° and is normally between 12° and 15°. If the angle is less than this value, then there is a strong possibility that the clutch will jam in engagement.

38.7.1 Torque Capacity of a Cone Clutch

Consider an elemental ring of radius r and thickness dr, as shown in Figure 38.5 and Table 38.4.

$$\text{Sloping length} = \frac{dr}{\sin\alpha} \tag{38.31}$$

$$\text{Area of elementary ring} = 2\pi\, r \frac{dr}{\sin\alpha} \tag{38.32}$$

$$\text{Normal force (F) acting on elementary ring} = p\ 2\pi\ r \frac{dr}{\sin\alpha} \tag{38.33}$$

$$\text{Axial component of the above normal force} = \frac{p\ 2\pi\ r\ dr}{\sin\alpha} \times \sin\alpha$$

$$= 2\pi\ p\ r\ dr \tag{38.34}$$

$$\text{Total axial force } F_a = \int_{ri}^{ro} 2\pi\ p\ r\ dr \tag{38.35}$$

$$\text{Frictional force acting on elemental ring} = \mu\ p\ 2\pi\ r \frac{dr}{\sin\alpha} \tag{38.36}$$

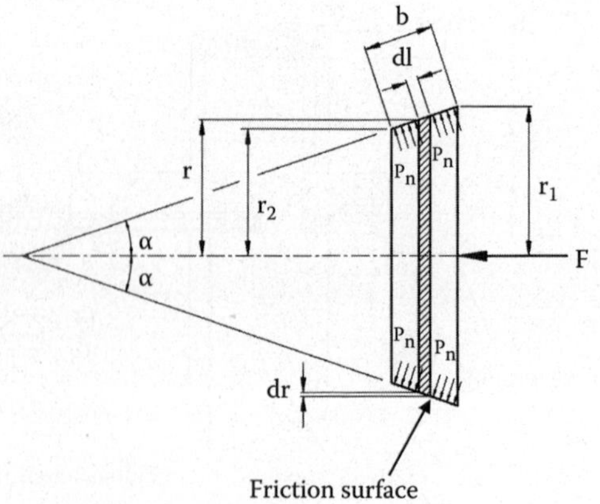

FIGURE 38.5
Cone clutch analysis.

TABLE 38.4

Nomenclature for a Cone Clutch

Symbol	Description
r_o	Outside radius of cone
r_i	Inside radius of cone
r_m	Mean radius of cone
l	Sloping length of cone face
α	Semi-angle of cone
P	Intensity of normal pressure at the contact surfaces
μ	Coefficient of friction
F	Normal force
F_a	Axial force

$$\text{Moment of friction force about the axial} = \mu\ p\ 2\pi\ r \frac{dr}{\sin\alpha} \times r$$

$$= 2\pi\ p\ \mu\ r^2 \frac{dr}{\sin\alpha} \tag{38.37}$$

$$\text{Total torque (T)} = \int_{r_i}^{r_o} 2\pi\ \mu\ p\ r^2 \frac{dr}{\sin\alpha} \tag{38.38}$$

Uniform pressure theory: p constant.

Equation 38.35 becomes

$$F_a = \int_{r_i}^{r_o} 2\pi\ p\ r\ dr$$

$$= 2\pi\ p \int_{r_i}^{r_o} r\ dr$$

$$= \pi\ p\ \left(r_o^2 - r_i^2\right) \tag{38.39}$$

Rearranging Equation 38.38 for p to be the subject,

$$p = \frac{F_a}{\pi\left(r_o^2 - r_i^2\right)} \tag{38.40}$$

Equation 38.38 becomes

$$T = \int_{r_i}^{r_o} 2\pi\tau p\ \mu\ r^2 \left(r_o^2 - r_i^2\right) \frac{dr}{\sin\alpha}$$

$$= \frac{2\pi\ p\ \mu}{\sin\alpha} \int_{r_i}^{r_o} r^2\ dr \tag{38.41}$$

$$T = \frac{2\pi\ \mu\ p}{\sin\alpha} = \frac{r_o^3 - r_i^3}{3} \tag{38.42}$$

$$T = \frac{2\pi \mu p}{\sin\alpha} = \frac{F_a}{\mu \left(r_o^2 - r_i^2\right)} \times \frac{r_o^3 - r_i^3}{3}$$

$$T = \frac{2\pi \mu p}{3 \sin\alpha} = \frac{F_a}{\mu \left(r_o^2 - r_i^2\right)} \times \frac{r_o^3 - r_i^3}{\left(r_o^2 - r_i^2\right)}$$

$$T = \frac{2\pi \mu p}{3 \sin\alpha} \times \frac{2}{3} \times \frac{D_o^3 - D_i^3}{\left(D_o^2 - D_i^2\right)}$$

$$= \frac{\mu F_a}{2 \sin\alpha}$$

where

$$D_m = \frac{2}{3}\left(\frac{D_o^3 - D_i^3}{D_o^2 - D_i^2}\right)$$

$$\text{Axial force } F_a = \pi p \left(D_o^2 - D_i^2\right) \tag{38.43}$$

Consider uniform wear assumption
For uniform wear condition,
$p \cdot r = C$ (constant)
Equation 38.35 can be rewritten as

$$F_a = \int_{r_i}^{r_o} 2 \pi p r \, dr$$

$$= 2\pi p \int_{r_i}^{r_o} dr$$

$$F_a = 2\pi c \left(r_o - r_i\right)$$

or

$$c = \frac{f_a}{2\pi \left(r_o - r_i\right)}$$

Equation 38.38 can be rewritten as

$$T = \int_{r_i}^{r_o} 2\pi \mu p r^2 \frac{dr}{\sin\alpha}$$

$$= \frac{2\pi \mu c}{\sin\alpha} \int_{r_i}^{r_o} r \, dr$$

$$T = \frac{2\pi \mu c}{\sin\alpha} = \frac{r_o^2 - r_i^2}{2}$$

Substituting for c

$$T = \frac{2\pi\mu\left(r_o^2 - r_i^2\right)}{\sin\alpha} \times \frac{F_a}{2\pi\left(r_o - r_i\right)}$$

$$T = \frac{\mu\ Fa}{2\sin\alpha} \times \frac{(D_o - D_i)}{2} \times \frac{\mu\ f_a\ D_m}{2\ \sin\alpha} \tag{38.44}$$

where

$$D_m = \text{mean diameter} = \left(\frac{D_o - D_i}{2}\right).$$

Example 38.1

A cone clutch is to transmit 10 kW at 800 rev/min. The face width is 75 mm and the mean diameter is 350 mm, with the face angle at 15°. Assuming the coefficient of friction is 0.2, calculate the axial force required to hold the clutch in full contact and the normal pressure on the cone surface.

Solution

Given: $P = 10\,kW$
$N = 800$ rev/min
$\beta = 75\,mm$
$D_{mean} = 350\,mm$
$\alpha = 15°$
$\mu = 0.2$

$$T = \frac{60\ P}{2\pi\ N}$$

$$T = \frac{60 \times 10 \times 10^3}{2\ \pi \times 800}$$

$$T = 119.366\ Nm$$

Torque transmitted:

$$T = \frac{\mu\ F_a\ D_{mean}}{2\ \sin\alpha}$$

$$119.366 = \frac{0.2 \times F_a \times 350}{2 \times \sin 15}$$

$$F_a = 882.69\,N$$

$$\text{Now}: F_a = \pi \cdot D_{mean}\ p \cdot \beta \cdot \sin\alpha$$

Rearranging to solve for p (normal pressure on the cone surface),

$$p = \frac{F_a}{\pi \cdot D_{mean} \cdot \beta \cdot \sin\alpha}$$

$$p = \frac{882.69}{\pi \cdot 350 \cdot 75 \cdot \sin 15°}$$

Normal pressure on the cone surface

$$p = 0.041\ N/mm^2.$$

39

Design of Brakes

39.1 Introduction

The purpose of a brake is to absorb kinetic energy from a rotating object and convert the energy into heat.

The primary requirements for a brake are as follows:

- Producing a force that is capable of reducing the speed of a rotating shaft or a slide
- Absorbing the generated heat and dissipating it in the process

There are three basic types of brake design:

- Mechanical brakes
- Electric brakes
- Hydraulic brakes

This discussion is limited to mechanical brakes only.

39.2 Mechanical Brakes

All mechanical brakes work by generating frictional forces as two surfaces rub against each other during the braking process. The stopping power or the capacity of a brake is dependent on the surface area of the frictional surface, together with the actuation force applied. The durability of a brake or its service life will depend on the type of braking material used to form the brake lining and the surface finish of the object being braked.

Mechanical brakes are primarily of three types:

- Shoe or block brakes, where the braking force is applied radially
- Band brakes, where the force is applied tangentially
- Disc brakes, where the force is applied axially

39.2.1 Shoe or Block Brake

In this form of brake, a brake shoe is brought into contact with a rotating drum or wheel using a suitable force (see Figure 39.1). The contacting surface of the shoe is fitted with a frictional material. A good example of this type of brake is seen on horse-drawn trailers.

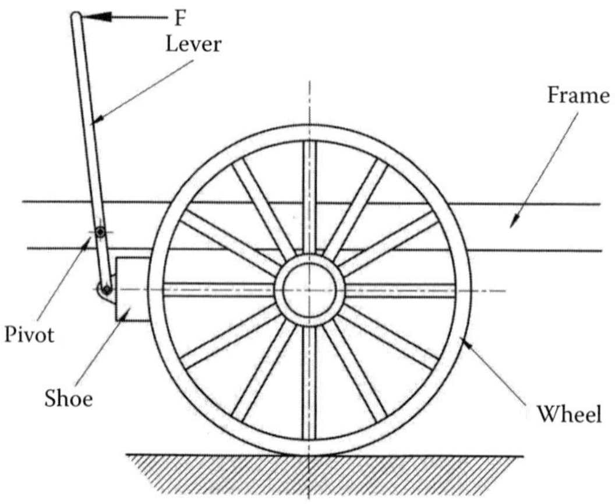

FIGURE 39.1
Shoe or block brake.

39.2.2 Double-Shoe Brake

This type of brake design is most common on automotive applications in which the double brake shoes expand against a circular drum within the wheel (see Figure 39.2). The brake is activated by a hydraulic cylinder, as shown in Figure 39.2.

39.2.3 Band Brakes

Band brakes consist of a flexible band that surrounds a drum connected to a driven member and the band is made to contract onto the drum, creating the braking effect (see Figure 39.3). This design is rarely used in modern automotive braking systems, although some examples may be found on vintage military trucks such as the Diamond T heavy-duty truck that used a band brake arrangement acting directly on the prop shaft and used as a parking brake.

These days, band brakes may be found in automatic transmission gearboxes using planetary gear trains, band brakes act directly on the planetary drum's circumference and being applied to sequence the gear ratios. The bands are actuated using hydraulic cylinders.

The principles of operation are the same as that for flat belt drives. The formulae in Equation 39.1 relate the band tensions (T_1 and T_2), the lap angle θ_t, and the coefficient of friction, μ.

$$e^{\mu\theta_t} = \frac{(T_1 - T_c)}{(T_2 - T_c)} \tag{39.1}$$

where

θ_t = total band lap angle (rad)
μ = coefficient of friction
T_1 = maximum band tension (N)
T_2 = minimum band tension (N)
T_c = band tension associated with centrifugal force (N)

The tension due to the band rotational inertia for $T_c = 0$ for the band is not moving for the brake application; hence, the formula is

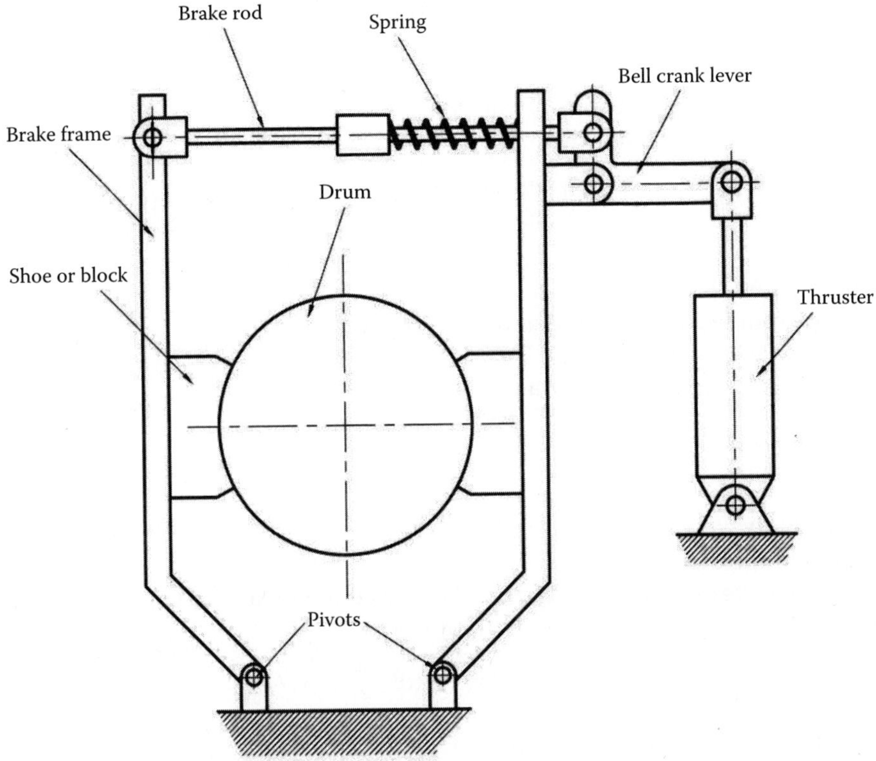

FIGURE 39.2
Double-shoe brake.

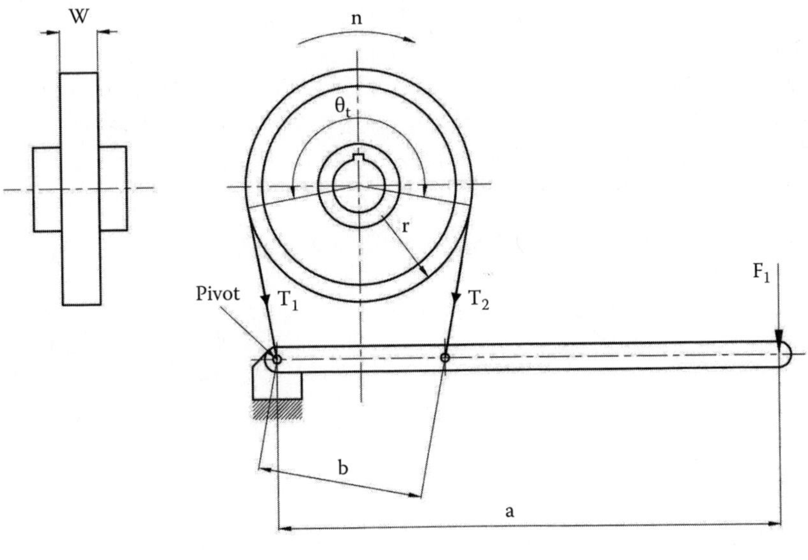

FIGURE 39.3
Band brake.

$$e^{\mu\theta_t} = \frac{T_1}{T_2} \tag{39.2}$$

For the normal band brake arrangement shown the tension between the actuating force and the slack side band tension is

$$F_i = T_2 \cdot \frac{b}{a} \tag{39.3}$$

where

F_i=actuating force (N)

b=distance from tensioning band and pivot point

a=pivot actuating force radius (m)

The torque capacity of the band brake is

$$M = (T_1 - T_2)r \tag{39.4}$$

where

M=torque (Nm)

r=radius of brake drum

Figure 39.4 depicts an element of the band brake acting on a drum of radius r. The force (R) acting on an element of width W is

$$R = pwrd\theta \tag{39.5}$$

Resolving forces in the vertical direction,

$$R = (T + dT + T)\sin\left(\frac{d\theta}{2}\right) \tag{39.6}$$

Ignoring the dθ·dT term for a small angle $\sin\theta \rightarrow \theta$, therefore,

$$R = T \cdot d\theta \tag{39.7}$$

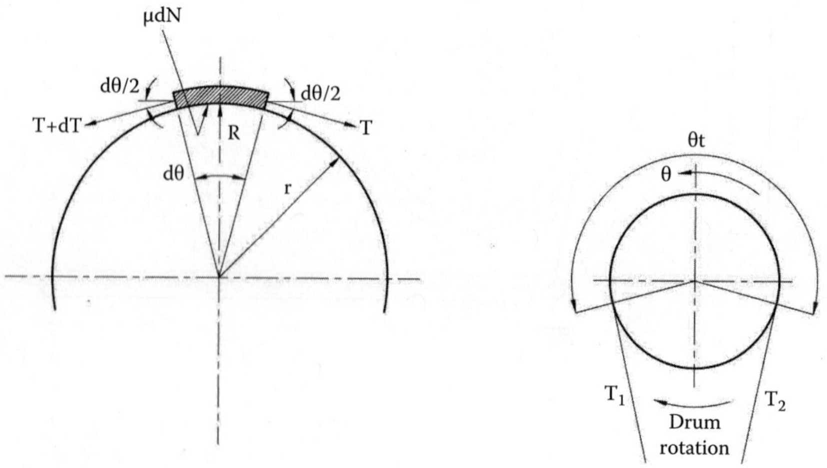

FIGURE 39.4
Maximum band tension.

Substituting for R,

$$T \cdot d\theta = pWrd\theta \tag{39.8}$$

Therefore,

$$T = pWr \tag{39.9}$$

The maximum tension in the belt can be evaluated from

$$T_{1-max} = p_{max} \cdot W \cdot r \tag{39.10}$$

39.2.4 Disc Brakes

Disc brakes came into prominence in the early 1950s to overcome the problem with brake fade that occurred mostly in competition cars and was used successfully by Jaguar in the 1953 Le Mans automotive race in which they won convincingly against conventional braked race cars (see Figure 39.5).

Disc brakes where originally invented in 1902 by Frederick W. Lanchester in his factory located in Birmingham, UK. Unfortunately, the choice of braking materials available to him at that time was very limited and, as a result, excessive wear was experienced during use (see Figure 39.6 for a facsimile of his patent drawing).

The basic design of a disc brake uses a pad that is squeezed against a disc using a caliper to create the braking effect, as shown in Figure 39.5. The heat generated within the disc is easily dissipated and had little effect on the overall braking efficiency. *Hydraulic*-powered disc brakes are the most commonly used form of brake for motor vehicles today, but the principles of a disc brake are applicable to almost any rotating shaft.

39.2.5 Frictional Linings

Mechanical brakes use some types of frictional material lining. Table 39.1 tabulates a range and properties of material combinations used in braking applications. At one time, asbestos was the accepted friction material but this has been replaced by non-asbestos materials

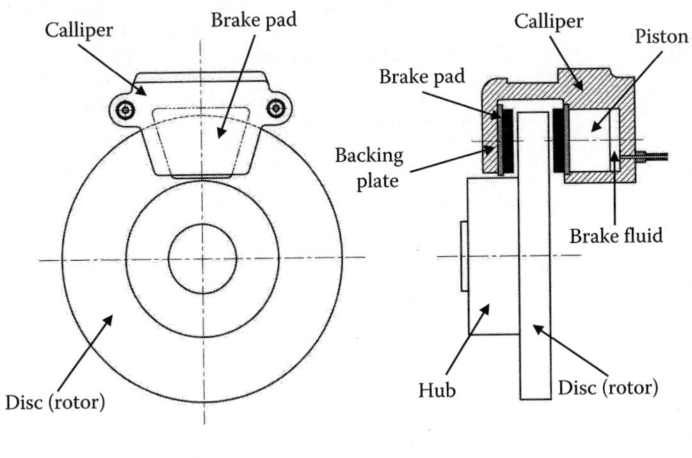

FIGURE 39.5
Terminology of a disc brake.

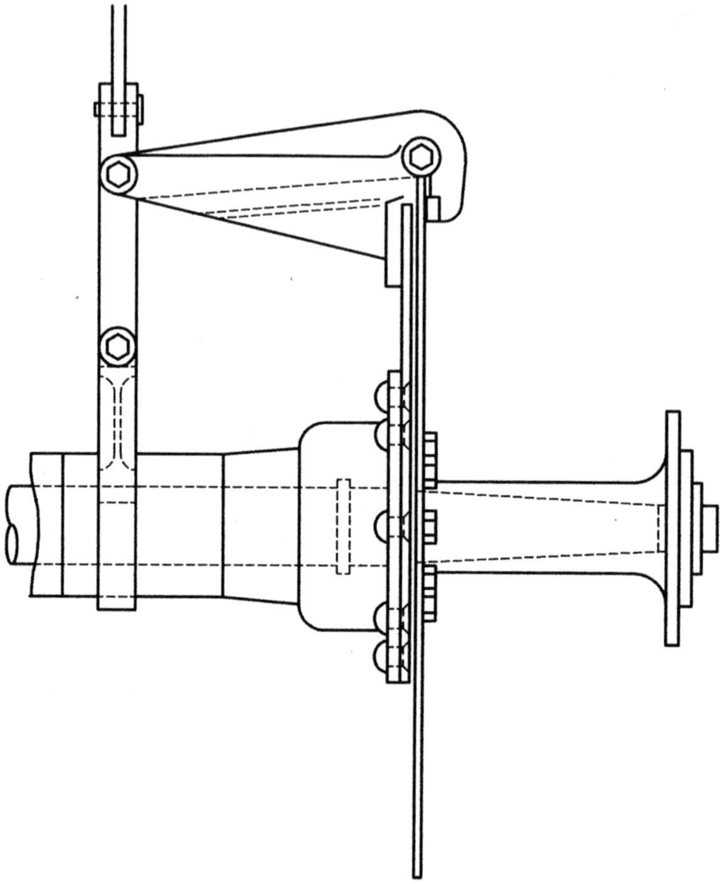

FIGURE 39.6
A facsimile of F.W. Lanchester's patent drawing (1902) for a disc brake.

TABLE 39.1
Brake Material Properties

Material Combination	Coefficient of friction μ		Maximum Pressure	Maximum Temperature
	Wet	Dry	MPa	°C
Cast iron/cast iron	0.05	0.15–0.20	0.8	300
Cast iron/steel	0.06	0.15–0.2	0.8–1.3	300
Cast iron/hard steel	0.06	0.15–0.25	0.7–1.72	260
Wood/cast iron or steel	0.16	0.2–0.35	0.6	150
Leather/cast iron or steel	0.12–0.15	0.3–0.5	0.25	100
Cork/cast iron or steel	0.15–0.25	0.3–0.5	0.1	100
Felt/cast iron or steel	0.18	0.22	0.06	140
Carbon-graphite/cast iron or steel	0.05–0.1	0.25	2.1	500
Kevlar/cast iron or steel	0.05–0.1	0.35	3.0	325

Notes: The properties tabulated above are offered as a guide only. For design work, practical test results or suppliers' information should be used. The use of asbestos is banned within the European Union and must not be used for new or replacement equipment.

due to environmental and human health problems associated with the use of asbestos. Cotton-based linings are on occasion used for situations involving light-duty applications. Copper is sometimes included in the material to improve the heat dissipation.

Cermet, a combination of ceramic and metallic materials, is currently used in situations involving extremely rigorous braking duties. It has high temperature resistance and hardness characteristics.

Brake linings may be attached to the brake shoes of mechanical brakes by either rivets or bonding. Riveting is simpler than bonding, but the usable lining is dictated by the depth of the rivet head countersink and there is a danger that the drum surface will be damaged by the rivet heads if the lining is not replaced in time.

40

Couplings

40.0 Introduction

The basic function of a coupling is to connect two shafts together that are used in a variety of applications, such as connecting a motor to either a pump or a generator. Couplings may be either rigid or flexible and be able to compensate for shaft misalignment; there may be a requirement to reduce shock loading and vibration. There are a wide variety of commercially available shaft couplings, ranging from simple keyed rigid coupling to complicated flexible designs requiring a complicated procedure to select and size. The types of shaft misalignments generally encountered are shown in Figure 40.1.

40.1 Rigid Couplings

These types of coupling cannot accommodate any shaft misalignment and must have good lateral and angular alignment. There are three types of rigid couplings:

1. Keyed sleeve coupling
2. Split clamp coupling
3. Rigid coupling
   a. Unprotected flange
   b. Protected flange

40.1.1 Keyed Sleeve Coupling

This is one of the simplest types of rigid coupling (often referred to as a "muff" coupling) and consists of a cylindrical sleeve (Figure 40.2) that is keyed to the shafts to be connected. Square keys are used to transmit the torque, and it is important to correctly size the coupling and the key for the torque to be transmitted safely.

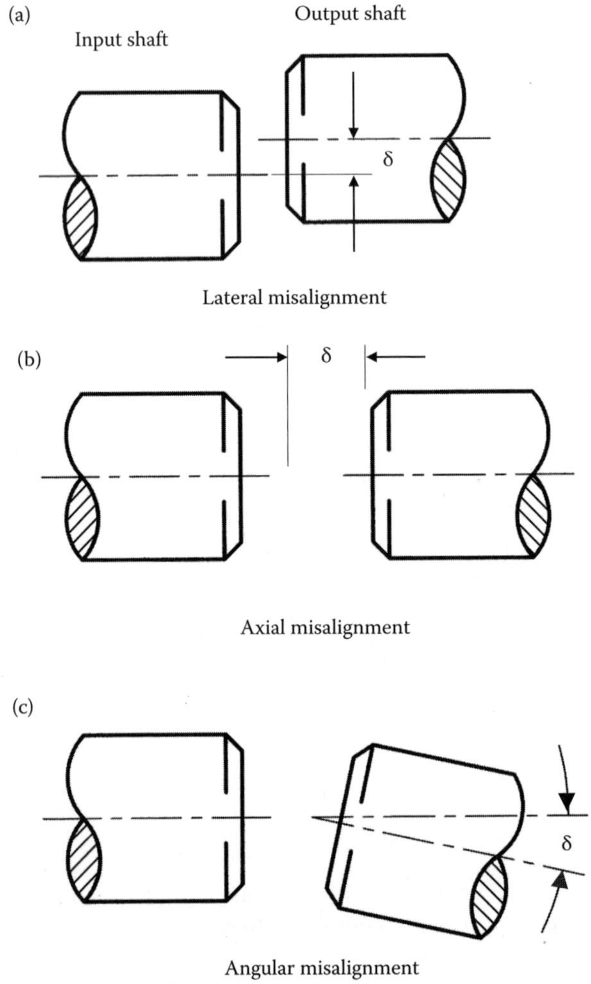

FIGURE 40.1
Types of shaft misalignments.

The empirical proportions that have been developed over the years for sizing the sleeve are as follows:

The outer diameter of the sleeve is

$$D_{outside} = 2d + 13\ \text{mm} \tag{40.1}$$

The length of the sleeve is

$$L = 3.5\ d \tag{40.2}$$

where d is the diameter of the shaft.

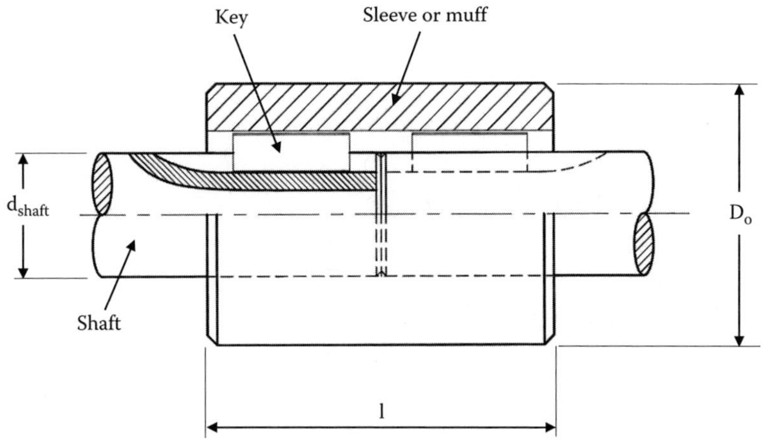

FIGURE 40.2
Keyed sleeve coupling.

1. **Design of the sleeve**

   Consider the sleeve as a hollow shaft.

   Let

   T = torque to be transmitted by the coupling

   τ_s = permissible shear stress in the sleeve material

   The sleeve material is usually gray cast iron and the safe value for shear stress may be taken as 14 MPa.

   The torque transmitted by a hollow section is

$$T = \frac{\pi}{16} \times \tau_s \left(\frac{D^4 - d^4}{D} \right) \tag{40.3}$$

$$T = \frac{\pi}{16} \times \tau_s \times D^3 (1 - k^4) \tag{40.4}$$

   where k = d/D.

   This expression allows the induced shear stress in the sleeve to be checked.

2. **Design of the key**

   The connecting key is usually made in two parts, therefore the length of the key in each shaft will be

$$l = \frac{L}{2} = \frac{3.5\ d}{2} \tag{40.5}$$

   After the length of each key has been established, the induced shearing and crushing stresses can be calculated from the torque to be transmitted.

Considering the shearing stress in the key,

$$\tau = \frac{2\,T}{lwd} \tag{40.6}$$

where

τ = permissible shear stress in the key

l = length of the key

w = width of the key

d = diameter of the connecting shafts.

Considering the crushing stress in the key,

$$\sigma_c = \frac{4\,T}{ltd} \tag{40.7}$$

where

t = thickness of the key

σ_c = permissible compressive stress for the material

Note: The depth of the keyway in each shaft as well as the diameter of each shaft should be identical.

Example 40.1

A keyed sleeve coupling is used to connect two steel shafts transmitting 60 kW at 500 rev/min. The material for the shafts and key is plain carbon steel, having an allowable shear and crushing stresses of 40 and 80 MPa, respectively. The material of the coupling is gray cast iron, for which the allowable shear stress may be taken as 15 MPa.

Calculate the proportions of the coupling and shaft diameters together with the size of the keys.

Solution

From the question,

$$P = 60\ \text{kW}$$

$$N = 500\ \text{rev/min}$$

$$\tau_s = 40\ \text{MPa}$$

$$\sigma_c = 80\ \text{MPa}$$

$$\tau_c = 15\ \text{MPa}$$

Diameter of Shafts

Torque to be transmitted:

$$T = \frac{P \times 60}{2\pi N}$$

$$T = \frac{60 \times 10^3 \times 60}{2\pi\ 500}$$

$$= 1146\ \text{Nm}$$

From Equation 40.3,

$$T = \frac{\pi}{16} \times \tau_s \times d^3$$

$$1146 \times 10^3 = \frac{\pi}{16} \times 40 \times d^3$$

$$1146 \times 10^3 = 11.78\, d^3$$

$$\therefore\ d = \sqrt[3]{\left(\frac{1146 \times 10^3}{7.85}\right)}$$

$$d = 52.65\,\text{mm, say } 55.0\,\text{mm (ans).}$$

Diameter of the Sleeve

From Equation 40.1,

The sleeve diameter will be

$$D = 2d + 13\ \text{mm}$$

Therefore, the outside diameter of the sleeve is

$D_{sleeve} = 2 \times 55.0 + 13$

$D_{sleeve} = 122.0\,\text{mm}$, say 125.0 mm (ans).

Length of the Sleeve

$l = 3.5\ d$

$l = 192.5\,\text{mm}$, say 195.0 mm (ans).

Consider the induced shear stress (τ_c) in the sleeve.

As the sleeve is made from grey cast iron and can be considered to be a hollow shaft, the torque (T) transmitted will be

$$1146 \times 10^3 = \frac{\pi}{16} \times \tau_c \left(\frac{D^4 - d^4}{D}\right)$$

$$1146 \times 10^3 = \frac{\pi}{16} \times \tau_c \left(\frac{125^4 - 55^4}{125}\right)$$

$$= 369.12 \times 10^3\, \tau_c$$

Therefore,

$$\tau_c = \frac{1146 \times 10^3}{369.12 \times 10^3}\ \text{MPa}$$

$$\tau_c = 3.105\ \text{MPa}$$

Factor of safety (FoS):

$$\text{FoS} = \frac{15\ \text{MPa}}{3.105\ \text{MPa}}$$

FoS = 4.83

As the induced shear stress in the sleeve (3.105 MPa) is significantly less than the permissible shear stress for gray cast iron (15 MPa), the sleeve can be considered safe.

TABLE 40.1

Sizes and Proportions of Square and Rectangular Keys

Shaft		Key	Keyway											
			Width b						Depth					
				Tolerance Class					Shaft t_1		Hub t_2			
Nominal Diameter				Free Fit		Normal Fit		Close Fit					Radius r	
Over	Up to and Including	Size b×h	Nominal	Shaft (H9)	Hub (D10)	Shaft (N9)	Hub (Js9)	Shaft/ Hub P9	Nominal	Tolerance	Nominal	Tolerance	Max	Min
6	8	2×2	2	+0.025	+0.06	−0.004	+0.012	−0.006	1.2	+0.1	1.0	+0.1	0.16	0.08
8	10	3×3	3	0	+0.02	−0.029	−0.012	−0.031	1.8	0	1.4	+0	0.16	0.08
10	12	4×4	4	+0.03	+0.078	0	+0.015	−0.012	2.5		1.8		0.16	0.08
12	17	5×5	5	0	+0.030	−0.030	−0.015	−0.042	3.0		2.3		0.25	0.16
17	22	6×6	6						3.5		2.8		0.25	0.16
22	30	8×7	8	+0.036	+0.098	0	+0.018	−0.015	4.0	+0.2	3.3	+0.2	0.25	0.16
30	38	10×8	10	0	+0.040	−0.036	−0.018	−0.051	5.0	0	3.3	0	0.40	0.25
38	44	12×8	12	+0.043	+0.120	0	+0.021	−0.018	5.0	+0.2	3.3	+0.2	0.40	0.25
44	50	14×9	14	0	+0.050	−0.043	−0.021	−0.061	5.5	0	3.8	0	0.40	0.25
50	58	16×10	16						6.0		4.3		0.40	0.25
58	65	18×11	18						7.0		4.4		0.40	0.25
65	75	20×12	20	+0.052	+0.149	0	+0.026	−0.022	7.5		4.9		0.60	0.40
75	85	22×14	22	0	+0.065	−0.052	−0.026	−0.074	9.0		5.4		0.60	0.40
85	95	25×14	25						9.0		5.4		0.60	0.40
95	110	28×16	28						10.0		6.4		0.60	0.40
110	130	32×18	32	+0.062	+0.180	0	+0.031	−0.026	11.0		7.4		0.60	0.40
130	150	36×20	36	0	+0.080	−0.062	−0.031	−0.088	12.0	+0.3	8.4	+0.3	1.00	0.70
150	170	40×22	40						13.0	0	9.4	0	1.00	0.70
170	200	45×25	45						15.0		10.4		1.00	0.70
200	230	50×28	50						17.0		11.4		1.00	0.70
230	260	56×32	56	+0.074	+0.220	0	+0.037	−0.032	20.0		12.4		1.60	1.20
260	290	63×32	63	0	+0.100	−0.074	−0.037	−0.106	20.0		12.4		1.60	1.20
290	330	70×36	70						22.0		14.4		1.60	1.20
330	380	80×40	80						25.0		15.4		2.50	2.00
380	440	90×45	90	+0.087	+0.260	0	+0.043	−0.037	28.0		17.4		2.50	2.00
440	500	100×50	100	0	+0.120	−0.087	−0.043	−0.124	31.0		19.4		2.50	2.00

Design for the Key

From Table 40.1 for a shaft of 55.0 mm diameter,

The width of the keyw is 18.0 mm.

The crushing stress (80 MPa) for the key is twice the shear stress (40 MPa) an 18 mm square key will be acceptable.

The thickness of the key t = 18.0 mm.

The length of the key in each shaft:

From the calculation of the length of the sleeve l = 195.0 mm, the length of engagement for each shaft in the sleeve (L) is

$$L = \frac{195.0}{2} = 97.5 \text{ mm}$$

Considering the shear stress in the key,

$$T = L \times w \times \tau_s \times \frac{d}{2}$$

$$1145 \times 10^3 = 97.5 \times 18 \times \tau_s \times \frac{55}{2}$$

Therefore, $\tau_s = 23.72$ MPa.

FoS:

FoS = 1.69

Considering the crushing stress of the key,

$$T = L \times \frac{t}{2} = \sigma_c \times \frac{d}{2}$$

$$1145 \times 10^3 = 97.5 \times \frac{18}{2} \times \sigma_{cs} \times \frac{55}{2}$$

Therefore, $\sigma_{cs} = 47.45$ MPa.

FoS = 1.69

As the induced shear and crushing stresses are less than the permissible stresses, the key is considered satisfactory.

40.1.2 Split Clamp Coupling

The split clamp coupling (also referred to as a split sleeve or a split muff coupling) is a rigid type of coupling where the sleeve is made of two halves that is split along a plane that passes through the axis of the shafts, as shown in Figure 40.3. In the construction of the coupling, the two halves of the sleeve are clamped together by the use of bolted fasteners. The fasteners are placed in recesses that are formed in the sleeve halves. A small clearance is provided on the parting plane between the two halves such that when the fasteners are tightened the sleeve is clamped rigidly onto the surface of the shafts. Torque is transmitted by means of the friction on the surface of the shaft. A key is incorporated between the sleeve and the shafts, which will also transmit the torque.

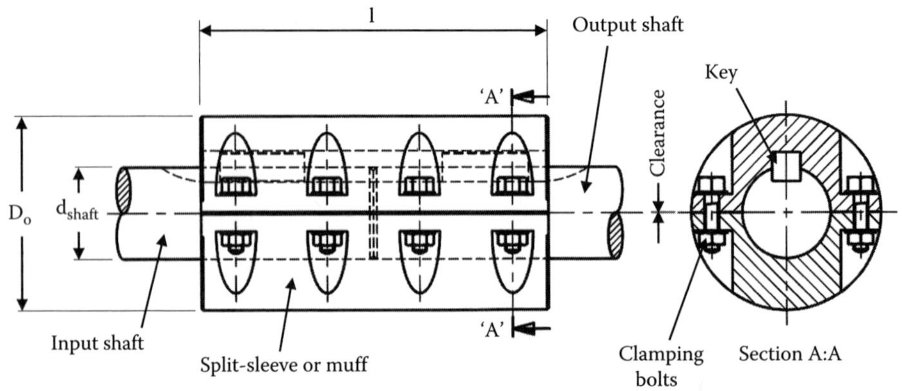

FIGURE 40.3
Split clamp coupling.

The advantages of the clamp coupling are as follows:

- The coupling is easy to assemble and dismantle.
- It is easily removed without disturbing the shaft in an axial direction unlike the sleeve coupling.
- Compared to a flanged coupling, the clamp coupling has small diametral dimensions.

The disadvantages of the clamp coupling are as follows:

- There is difficulty in dynamic balancing of the coupling and as a result it is limited to low speed applications.
- The coupling is unsuitable when subject to shock loading.
- It is essential to provide a suitable guard around the coupling to comply with factory safety regulations.

It is difficult to apportion the percentage of torque transmitted between friction or by the key; therefore, it is usual to design the clamp coupling assuming the total torque is transmitted by either the friction alone or the total torque is transmitted by the key alone. Therefore, it is usual practice to design the coupling by either friction or shear resistance of the key.

The initial design of the clamp coupling is based on the following empirical proportions:

1. **Design of sleeve halves**

$$D = 2.0\,d + 13\text{ mm} \tag{40.8}$$

$$L = 3.5\,d \tag{40.9}$$

where

D = outer diameter of the sleeve halves (mm)

L = length of the sleeve (mm)

d = diameter of the shaft (mm)

The design of the split coupling follows in a similar way to that covered in Section 40.1.1.

2. Design of clamping bolts:

Nomenclature:

T	Torque transmitted by the shaft
d	Diameter of the shaft
d_e	Effective (or root) diameter of the bolt
n	Number of bolts
σ_t	Permissible tensile stress for bolt material
μ	Coefficient of friction between the sleeve and the shaft
L	Length of the sleeve
p	Pressure on the shaft due to bolting force

The force exerted by each bolt:

$$F = \frac{\pi}{4} \times d_e^2 \times \sigma_t \tag{40.10}$$

The force exerted on each shaft by the bolts:

$$F = \frac{\pi}{4} \times d_e^2 \times \sigma_t \times \frac{n}{2} \tag{40.11}$$

Consider a uniform pressure distribution over the surface of the shaft and coupling:

$$p = \frac{\text{Force}}{\text{Projected area}} \tag{40.12}$$

$$p = \frac{\frac{\pi}{4} \times d_c^2 \times \sigma_t \times \frac{n}{2}}{\frac{1}{2} L \times d} \tag{40.13}$$

Frictional force between the shaft and the sleeve:

$$F = \mu \times \text{pressure} \times \text{area}$$

$$F = \mu \times p \times \frac{1}{2} \times \pi d \times L \tag{40.14}$$

$$= \mu \times \frac{\frac{\pi}{4} \times d_c^2 \times \sigma_t \times \frac{n}{2}}{\frac{1}{2} L \times d} \times \frac{1}{2} \pi d \times L$$

$$= \mu \times \frac{\pi}{4} \times d_e^2 \times \sigma_t \times \frac{n}{2} \times p$$

$$= \mu \times \frac{\pi^2}{8} \times d_e^2 \times \sigma_t \times n \tag{40.15}$$

Torque that can be transmitted by the coupling by friction alone:

$$T = F \times \frac{d}{2} \tag{40.16}$$

$$= \mu \times \frac{\pi^2}{8} \times d_e^2 \times \sigma_t \times n \times \frac{d}{2}$$

$$= \frac{\pi^2}{16} \times \mu \times d_e^2 \times \sigma_t \times n \times d \tag{40.17}$$

The effective (root) diameter of the bolt can now be calculated as

$$d_e = \sqrt{\frac{T}{\frac{\pi^2}{16} \times \mu \times \sigma_t \times n \times d}} \tag{40.18}$$

Example 40.2

A keyed split sleeve coupling is to be designed to replace the sleeved coupling used in Example 40.1 for connecting the two steel shafts and transmitting 60 kW at 500 rev/min.

The material for the shafts and key is plain carbon steel having an allowable shear and crushing stresses of 40 and 80 MPa, respectively. The material of the coupling is gray cast iron for which the allowable shear stress may be taken as 15 MPa. The number of bolts connecting the two halves together is eight, and the permissible tensile stress for the bolts is 70 MPa.

The coefficient of friction between the sleeve and the shaft surface may be taken as 0.25.

Calculate the proportions of the coupling and shaft diameters together with the size of the keys and bolt sizes.

Solution

From the question,

$$P = 60 \text{ kW}$$

$$N = 500 \text{ rev/min}$$

$$\tau_s = 40 \text{ MPa}$$

$$\sigma_c = 80 \text{ MPa}$$

$$\tau_c = 15 \text{ MPa}$$

$$\sigma_{bolt} = 70 \text{ MPa}$$

$$n = 8$$

$$\mu = 0.25$$

1. **Design for the shaft:**
   From Example 40.1, the size of the shafts was calculated as 55.0 mm diameter.

2. **Design of the split sleeve:**
   The diameter and length of the sleeve were as calculated in Example 40.1:

$$D_{sleeve} = 125 \text{ mm}$$

$$L = 195 \text{ mm}$$

3. **Design for the key:**
   The width and thickness of the key will be as calculated in Example 40.1: 18.0 mm square key

4. **Design of the bolts:**
   Let

$$d_c = \text{core or root diameter of bolt}$$

From Example 40.1,
The torque transmitted (T):

$$T = \frac{P \times 60}{2\pi\, N}$$

$$= \frac{60 \times 10^3 \times 60}{2\pi\, 500}$$

$$T = 1146\ \text{Nm}$$

From Equation 40.18,

$$d_c = \sqrt{\frac{T}{\frac{\pi^2}{16} \times \mu \times \sigma_t \times n \times d}}$$

$$= \sqrt{\frac{1146 \times 10^3}{\frac{\pi^2}{16} \times 0.25 \times 70.0 \times 8 \times 55.0}}$$

Therefore: $d_c = 15.53\ \text{mm}.$

From Table 40.5, the standard core (minor) diameter of the bolt for course series is found to be 16.933 mm diameter and the nominal diameter of the bolt is 20.0 mm (M20).

40.1.3 Rigid Flange Coupling

Flange couplings consist of two separate components, as shown in Figure 40.4a and b; each flange is fitted to the shaft end with keys. The two flanges are coupled together using bolts and nuts. One flange has a circular projection that fits into a circular relief in the mating flange; this helps to keep the shafts in line and to maintain alignment.

There are two types of rigid flange couplings:

- Unprotected rigid coupling
- Protected rigid coupling

1. **Unprotected flange coupling:**
   In this type of coupling (Figure 40.4a), each boss of the flange is connected to the shaft by means of a key, and the flanges are connected together using bolts. The number of bolts used in the flange is dependent on its diameter. Table 40.2 indicates the number of bolts for a range of shaft sizes.

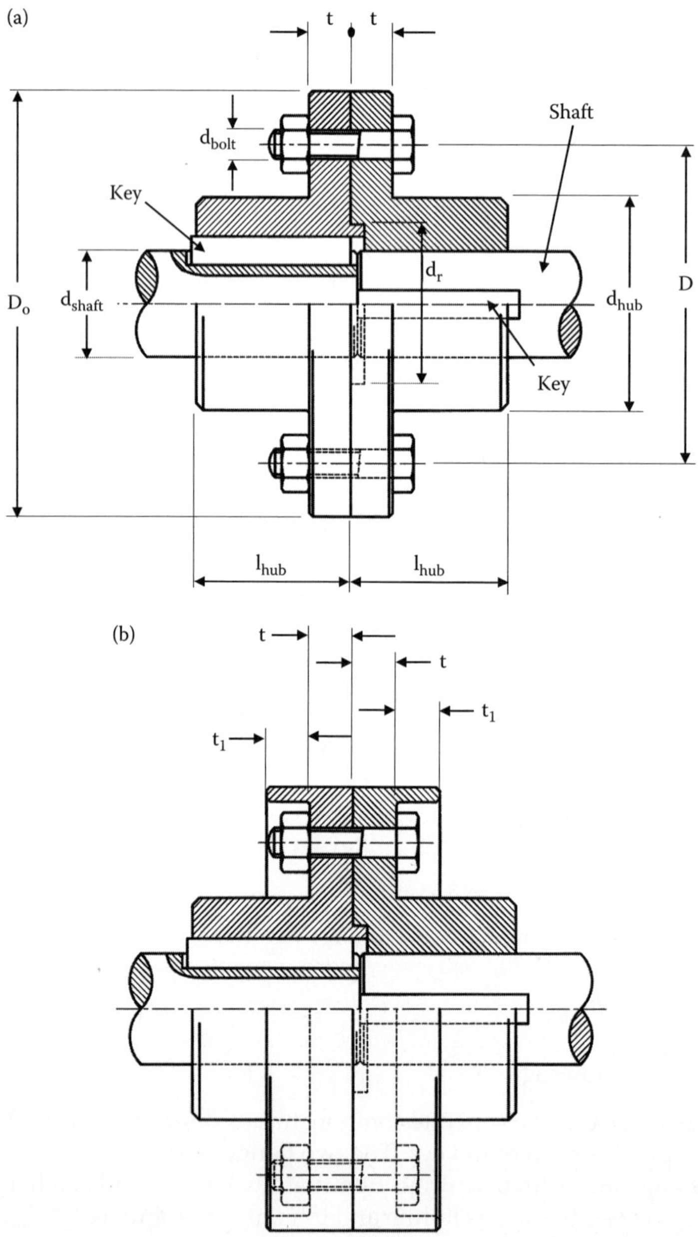

FIGURE 40.4
(a) Unprotected rigid flange coupling. (b) Protected rigid flange coupling.

The keys are staggered at right angles to each other such that they are not in line with each other. The reason for this is to divide the weakening effect created by the keyways.

In the unprotected design, the bolt heads and nuts are left exposed. In situations where the flanges are accessible to hands or fingers, a separate method of guarding will be required to comply with the Factories Act and Health and Safety legislation.

TABLE 40.2

Number of Bolts in Flange

	Shaft Diameter	
No. of Bolts 'N'	**From (mm)**	**Up to and Including (mm)**
3	—	40.0
4	40.0	100.0
6	100.0	180.0

2. **Protected flange coupling:**

In the protected design, the flange is provided with a shroud that shelters the bolt heads and nuts, as shown in Figure 40.4b.

The rigid flange couplings have the following advantages:

- The coupling has a high torque capacity.
- They are easy to assemble and disassemble.
- The design and manufacture of the couplings is straightforward due to their simple construction.

Their use has the following notable disadvantages:

- The coupling cannot tolerate any misalignment between the axes of the shafts.
- The coupling can only be used in situations that are free from shock and vibration
- This design requires more radial space.

Over the years, shop floor engineers have developed a set of empirical proportions that have been used, resulting in the design of robust flanges.

These proportions have been tabulated in Table 40.3 and illustrated in Figure 40.4b. These proportions are based upon the shaft diameter.

TABLE 40.3

Nomenclature and Proportions of Flange

Feature	Description	Proportion
d	Diameter of shafts	Reference dimension
d_{hub}	Outside diameter of the hub	$2d$
l_{hub}	Length of the hub	$1.5d$
d_r	Diameter of spigot and recess	$1.5d$
D	Pitch circle diameter for bolts	$3d$
D_o	Outside diameter of flanges	$(4d+2t_1)$
d_{bolt}	Diameter of the bolt	Calculated
t	Thickness of flanges	$0.5d$
t_1	Thickness of the protecting rim	$0.25d$
L	Length of the key	

TABLE 40.4

Nomenclature for Pin Flexible Coupling

Feature	Description
d_{bush}	Diameter of the bush
l_{bush}	Length of the bush in contact with the flange
P_{bush}	Bearing pressure on the bush
F_{bush}	Tangential Force acting on the bush due to torque
N	Number of bushes
D	Pitch circle diameter of bushes (and pins)

40.1.3.1 Size of the Hub

To size the hub, consider it as a hollow shaft that is transmitting the same torque (T) as a solid shaft:

$$T = \frac{\pi}{16} \times \tau_c \left(\frac{D^4 - d^4}{D} \right) \tag{40.19}$$

From Table 40.4, the hub diameter (D) is twice the diameter of the shaft diameter (d), and from this relationship, the induced shear stress in the hub can be evaluated.

The length of the hub (l_{hub}) can be taken as 1.5 d.

40.1.3.2 Key Dimensions

The key is designed using standard proportions and is assessed for shearing and crushing stresses.

The material for the key would usually be the same as the shaft material, and its length is generally taken as the length of the hub.

40.1.3.3 Flange Diameter

The torque is transmitted to the flange through the junction of the flange with the hub, and this torque induces a shear stress at the junction.

The torque transmitted is calculated as follows:

$$T = \pi \times \frac{d_{hub}^2}{2} \times \tau_s \times t \tag{40.20}$$

From Table 40.3, the thickness of the flange can be taken as d/2.

From Equation 40.20, the shear stress in the flange can be calculated.

40.1.3.4 Design of the Bolts

The bolts will be subjected to a shear stress (τ_s) when transmitting the torque (T). Table 40.2 indicates the number of bolts used in the flange based on the shaft diameter (d), and the pitch circle diameter of the bolts (D) is taken as 3 d.

The shear stress in each bolt:

$$\text{Load (L)} = \frac{\pi}{4} d_{\text{bolt}}^2 \times \tau_s \tag{40.21}$$

Therefore, the total load on all the bolts:

$$\text{Total load (L)} = \frac{\pi}{4} d_{\text{bolt}}^2 \times \tau_s \times N \tag{40.22}$$

The torque transmitted:

$$T = \frac{P}{4} \times d_{\text{bolt}}^2 \times \tau_s \times N \times \frac{D}{2} \tag{40.23}$$

From Equation 40.23, the diameter of the bolt (d_{bolt}) can be calculated:

$$d_{\text{bolt}} = \sqrt{\left(\frac{8T}{\pi \times \tau_s \times N \times D}\right)} \tag{40.24}$$

Example 40.3

A protective type gray cast iron flange is required to transmit 25 kW at 1500 rev/min driving a compressor using an electric motor. A service factor of 1.25 may be assumed.

The following are the permissible stresses that may be used:

Shear stress for gray cast iron = 8 MPa

Shear stress (t_s) for shaft, bolts, and key materials = 40 MPa

Crushing stress for bolts and keys = 80<Pa

Solution

From the question,

$$P = 25 \text{ kW}$$

$$N = 1500 \text{ rev/min}$$

$$\tau_s, \tau_k, \tau_b = 40 \text{ MPa}$$

$$\tau_f = 8 \text{ MPa}$$

$$\sigma_b, \sigma_k = 80 \text{ MPa}$$

Shaft diameter:

$$T = \frac{P \times 60}{2\pi\, N}$$

$$= \frac{25 \times 10^3 \times 60}{2\pi \times 1500}$$

$$T = 159.15 \text{ Nm}$$

A service factor of 1.25 is to be applied to the calculated torque; therefore, the maximum torque that is transmitted by the shaft:

$$T_{max} = 159.15 \times 1.25 \text{ Nm}$$
$$= 198.94 \text{ Nm}$$

The Diameter of the Shaft

$$198.94 = \frac{\pi}{16} \times \tau_s \times d^3$$

$$198.94 = \frac{\pi}{16} \times 40 \times d^3$$

Rearranging and solving for d,

Hence: d = 29.368 mm.

Say 35.0 mm diameter stock size.

Diameter of the Hub

From Table 40.3, the diameter of the hub is twice the diameter of the shaft.

$$d_{hub} = 2d$$
$$d_{hub} = 2 \times 35.0 \text{ mm}$$
$$= 70.0 \text{ mm diameter.}$$

Length of the Hub

$$l_{hub} = 1.5d$$
$$l_{hub} = 1.5 \times 35.0 \text{ mm}$$
$$= 52.5 \text{ mm length}$$

Check the hub for the induced shear stress. The hub material is gray cast iron.

Considering the hub as a hollow shaft, the maximum torque that can be transmitted (T_{max}):

$$198.94 \times 10^3 = \frac{\pi}{16} \times \tau_c \left[\frac{d_{hub}^4 - d^4}{d_{hub}}\right]$$

$$198.94 \times 10^3 = \frac{\pi}{16} \times \tau_c \left[\frac{70^4 - 35^4}{70}\right]$$

Rearranging and solving for τ_c:

$$\tau_c = 3.151 \text{ MPa}$$

FoS:

$$\text{FoS} = \frac{\text{Permissible shear stress}}{\text{Induced shear stress}}$$

$$\text{FoS} = \frac{8 \text{ MPa}}{3.151 \text{ MPa}}$$
$$= 2.54$$

With a FoS of 2.54, the design of the hub is considered safe.

The Strength of the Key

Because the crushing stress for the key material is twice its shear stress ($\sigma_{ck} = 2\tau_k$) in this application, a square key may be used. From Table 40.1, for a shaft of 35.0 mm diameter:

$$\text{Width of key (w)} = 10.0 \text{ mm}$$

$$\text{Thickness of key (t)} = 10.0 \text{ mm}$$

The length of the key (L) will be equal to the length of the hub (l_{hub}).

$$L = l_{hub} = 52.5 \text{ mm.}$$

Considering the shearing and crushing stresses in the key:

1. The shear stress in the key due to the maximum torque (T_{max}):

$$198.94 \times 10^3 = L \times w \times \tau_k \times \frac{d}{2}$$

$$198.94 \times 10^3 = 52.5 \times 12 \times \tau_k \times \frac{35}{2}$$

Rearranging and solving for τ_k:

$$\tau_k = 18.04 \text{ MPa}$$

FoS:

$$\text{FoS} = \frac{\text{Permissible shear stress}}{\text{Induced shear stress}}$$

$$\text{FoS} = \frac{40 \text{ MPa}}{18.04 \text{ MPa}}$$

$$= 2.22$$

2. The crushing stress in the key due to the maximum torque (T_{max}):

$$198.94 \times 10^3 = L \times \frac{t}{2} \times \sigma_{ck} \times \frac{d}{2}$$

$$198.94 \times 10^3 = 52.5 \times \frac{12}{2} \times \sigma_{ck} \times \frac{35.5}{2}$$

Rearranging and solving for σ_{ck}:

$$\sigma_{ck} = 35.6 \text{ MPa}$$

$$\text{FoS} = \frac{\text{Permissible crushing stress}}{\text{Induced crushing stress}}$$

$$\text{FoS} = \frac{80 \text{ MPa}}{35.6 \text{ MPa}}$$

$$= 2.25$$

With a FoS of 2.22 for the shear and crushing stress, the key is considered safe.

The Design of the Flange

The thickness of the flange (t) is taken as t/2.
Therefore,

$$t = \frac{35}{2}$$

$$t = 17.5 \text{ mm}$$

Considering the induced shear stress between the flange and the hub due to the maximum torque (T_{max}):

$$198.94 \times 10^3 = \frac{\pi D_o^2}{2} \times \tau_c \times t$$

$$198.94 \times 10^3 = \frac{\pi \times 70^2}{2} \times \tau_c \times 17.5$$

Rearranging and solving for τ_c,

$$\tau_c = 1.477\,\text{MPa}$$

$$\text{FoS} = \frac{\text{Permissible shear stress}}{\text{Induced shear stress}}$$

$$\text{FoS} = \frac{8\text{ MPa}}{1.477\text{ MPa}}$$

$$= 5.42$$

As the FoS is 5.42, the design of the flange is considered safe.

Size of the Bolts

From Table 40.3, the nominal diameter of the bolt $= d_{bolt}$.

From Table 40.2, the number of bolts in the flange for a 35 mm diameter shaft is 3 and from Table 40.3 the pitch circle diameter (D) is

$$D = 3\,d$$

Therefore,

$$D = 3 \times 35.0\text{ mm}$$

$$= 105.0\text{ mm}$$

The bolts will be subjected to a shear load due to the torque being transmitted (T_{max}):

$$198.94 \times 10^3 = \frac{\pi}{4} \times d_{bolt}^2 \times \tau_b \times N \times \frac{D}{2}$$

$$198.94 \times 10^3 = \frac{\pi}{4} \times d_{bolt}^2 \times 40 \times 3 \times \frac{105}{2}$$

Rearranging and solving for d_{bolt}:

$$d_{bolt} = 6.34\,\text{mm diameter}$$

Considering the ISO metric course series of thread, the nearest bolt size will be M8 (Table 40.5).

The remaining proportions of the flange:
Outside diameter of flange

$$D_o = (4\,d + 2\,t_1)$$

$$D_o = (4 \times 35.0) + (2 \times 0.25 \times 35)$$

$$= 148.75\text{ mm}$$

TABLE 40.5

ISO Metric Course Thread Proportions

Size Designation	Nominal Major Diameter (mm)	Pitch Diameter (mm)	Effective Diameter (mm)	Minor Diameter (mm)	Minor Diameter Area (mm)	Depth of Thread (mm)	Shear Area External (mm^2)	Shear Area Internal (mm^2)
M3	3.00	0.50	2.675	2.387	4.473	0.307		
M3.5	3.50	0.60	3.110	2.764	6.000	0.368	4.674	6.606
M4	4.00	0.70	3.545	3.141	7.750	0.429	5.473	7.768
M5	5.00	0.80	4.480	4.018	12.683	0.491	7.073	9.997
M6	6.00	1.00	5.350	4.773	17.894	0.613	8.645	12.189
M8	8.00	1.25	7.188	6.466	32.841	0.767	12.160	16.826
M10	10.00	1.50	9.026	8.160	52.292	0.920	15.578	21.473
M12	12.00	1.75	10.863	9.853	76.247	1.074	18.975	26.113
M14	14.00	2.00	12.701	11.546	104.710	1.227	22.422	31.028
M16	16.00	2.00	14.701	13.546	144.120	1.227	26.095	35.563
M18	18.00	2.50	16.376	14.933	175.135	1.534	29.525	40.754
M20	20.00	2.50	18.376	16.933	225.189	1.534	33.276	45.379
M22	22.00	2.50	20.376	18.933	281.527	1.534	37.027	50.004
M24	24.00	3.00	22.051	20.319	324.274	1.840	40.458	54.999
M27	27.00	3.00	25.051	23.319	427.095	1.840	46.169	62.997
M30	30.00	3.50	27.727	25.706	518.990	2.147	51.633	69.537
M33	33.00	3.50	30.727	28.706	647.195	2.147	57.419	76.603
M36	36.00	4.00	33.400	31.092	759.274	2.454	63.091	84.043

Say 150 mm outside diameter.

Thickness of the protective circumferential flange:

$$t_1 = 0.25\,d$$

$$t_1 = 0.25 \times 35.0$$

$$= 8.75\text{ mm}$$

Say 10 mm thickness.

40.2 A Template for Designing a Rigid Flange Coupling

The basic procedure for establishing the dimensions for a rigid flange coupling comprises the following steps.

Step 1 Determine the shaft diameter

The shaft diameter is calculated using the following two equations:

$$M_t = \frac{60 \times 10^6 \times P}{2\pi n} \tag{40.25}$$

and

$$\tau = \frac{16\ M_t}{\pi d^3} \tag{40.26}$$

Step 2 Dimensions of the flanges

The dimensions of the flanges are calculated using the following empirical equations:

$$d_{hub} = 2\ d$$

$$l_{hub} = 1.5\ d$$

$$D = 3\ d$$

$$t = 0.5\ d$$

$$t_1 = 0.25\ d$$

$$d_r = 1.5\ d$$

$$D_o = (4\ d + t_1)$$

The torsional shear stress in the hub is calculated by treating it as a hollow shaft subjected to a torsional moment M_t.

The outer and inner diameters of the hub are d_{hub} and d, respectively. The torsional shear stress in the hub is given by the following equations:

$$\tau = \frac{M_t r}{J} \tag{40.27}$$

$$J = \frac{\pi\left(d_{hub}^4 - d^4\right)}{32} \tag{40.28}$$

$$r = \frac{d_{hub}}{2} \tag{40.29}$$

The junction of the flange and the hub will be subject to the shearing stress while transmitting the torsional moment M_t.

From Figure 40.4a,

$$\text{Shear area} = (\pi d_{hub}) \times t \tag{40.30}$$

$$\text{Shear force} = \text{area} \times \text{shear stress}$$

$$= \pi d_{hub} t\tau \tag{40.31}$$

$$\text{Re sisting torque} = \text{shear force} \times \left(\frac{d_{hub}}{2}\right) \tag{40.32}$$

Therefore,

$$M_t = (\pi d_{hub} t\tau) \times \frac{d_{hub}}{2} \tag{40.33}$$

or:

$$M_t = \frac{1}{2}\left(\pi d_{hub}^2 t\tau\right) \quad (40.34)$$

Step 3 Diameter of bolts

The number of bolts used in the flange is based on the following empirical guidelines:

$$N = 3 \quad \text{for } d < 40 \text{ mm}$$

$$N = 4 \quad \text{for } 40 < d < 100 \text{ mm}$$

$$N = 6 \quad \text{for } 100 < d < 180 \text{ mm}$$

The diameter of the bolt is determined from the following expression:

$$\tau = \frac{8\,M_t}{\pi D N d_1{}^2} \quad (40.35)$$

This expression can be rearranged to solve for d_1^2:

$$d_1^2 = \frac{8\,M_t}{\pi D N \tau} \quad (40.36)$$

where τ is the permissible shear stress for the bolt material.

The compressive stress generated by the bolt:

$$\text{Crushing area of each bolt} = d_1 t$$

$$\text{Crushing area of all bolts} = N d_1 t$$

$$\text{Compressive force} = N d_1 t \sigma_c$$

$$\text{Torque} = (N d_1 t\, \sigma_c) \times D/2$$

or

$$\sigma_c = \frac{2\,M_t}{N d_1 t D} \quad (40.37)$$

This equation is used to confirm the tensile stress in the bolt.

Step 4 Dimensions of the key

The cross-sectional dimensions of the key are determined from Table 40.1. The length of the key in each flange is l_{hub}.

Therefore,

$$l = l_{hub}$$

These dimensions of the key are used to check the shear and compressive stresses in the key using equations:

$$\tau = \frac{2\,M_t}{dbl} \quad (40.38)$$

$$\sigma_c = \frac{4\,M_t}{dhl} \tag{40.39}$$

where b and h are the characteristic dimensions of the standard square or rectangular key.

40.3 Pin-Type Flexible Coupling

Figure 40.5a and b shows two styles of a pin-type flexible coupling. This is a modification of the rigid flange coupling described in the previous section. The previous bolts are replaced with shouldered screws, known as pins. Elastomeric bushes (fitted in brass sleeves) are fitted over the pins, and these then are fitted into one-half of the flange, whereas the screws are retained in the other flange.

There is no recess or spigot as in the rigid coupling; it is usual for a clearance of 5 mm to be allowed for between the flanges to ensure there is no rigid contact between them other than through the bushes. The drive then takes place through the compressible elastomeric bushes.

Couplings required for a lower duty are manufactured from cast iron, as with the rigid flange couplings.

For couplings involving higher duty strengths, these are manufactured from forged alloy steel.

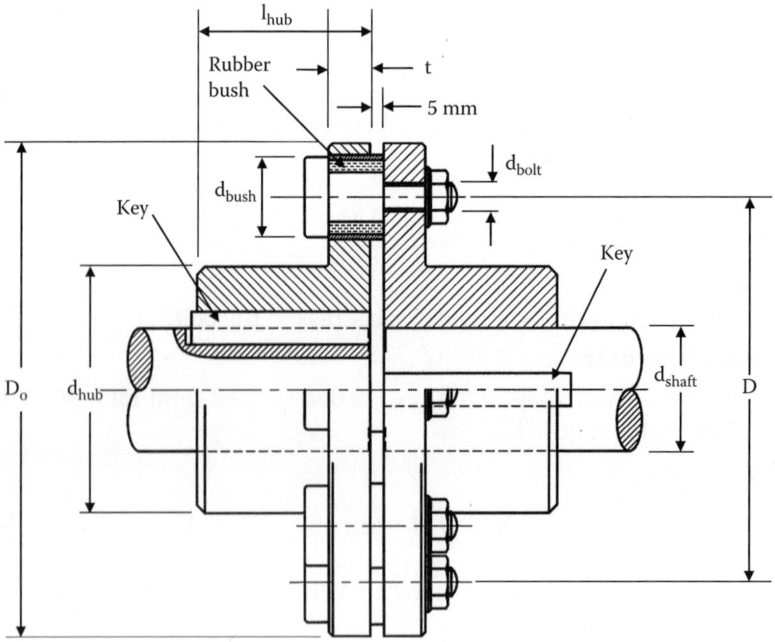

FIGURE 40.5
Pin flexible coupling.

In some proprietary couplings, alternate bushes are fitted in both flanges, enabling additional pins and bushes to be incorporated, thereby reducing the resisting forces acting on the bushes, as shown in Figure 40.5b. This style of coupling is generally for a heavy-duty coupling and these flanges are manufactured from forged alloy steel.

This analysis will be confined to bushes being fitted into one flange only.

The proportions of the rigid flange coupling are modified for the design of the flexible bush pin coupling. The problem is to reduce the bearing pressure on the elastomeric bushes, which should not exceed 0.5 MPa. In order to maintain this low bearing pressure, the pitch circle diameter is increased together with the pin size.

Table 40.4 presents the nomenclature used for the pin flexible coupling.

It can be deduced that the bearing load acting on each bearing (and pin):

$$F_{bush} = P_{bush} \times d_{bush} \times l_{bush} \tag{40.40}$$

Therefore, the total bearing load acting on each bush (and pin):

$$= F_{bush} \times N$$

$$= P_{bush} \times d_{bush} \times l_{bush} \times N \tag{40.41}$$

The torque transmitted by the coupling:

$$T = F_{bush} \times N \times \left(\frac{D}{2}\right)$$

$$T = P_{bush} \times d_{bush} \times l_{bush} \times N \times \left(\frac{D}{2}\right) \tag{40.42}$$

The pin should be a light press fit into the right-hand flange; this is to minimize any bending stresses in the threaded portion of the pin.

The direct shear stress acting on the pin due to the torsion (see Figure 40.6):

$$\tau = \frac{P_{bush}}{\frac{\pi}{4}(d_1)^2} \tag{40.43}$$

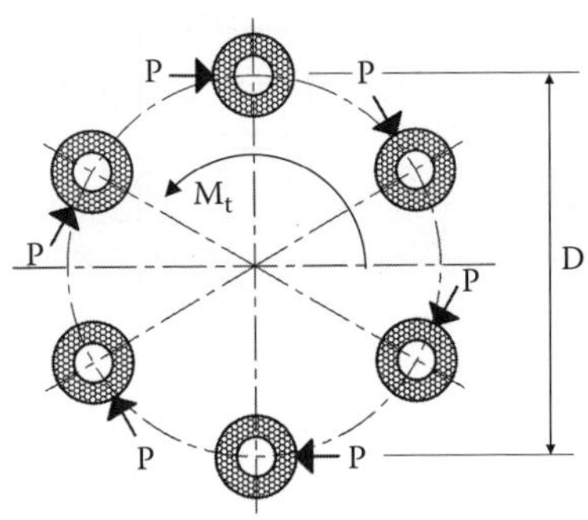

FIGURE 40.6
Forces acting on elastomeric bushes.

As the pin is not rigidly held in the elastomeric bush (in the left-hand flange), the tangential force P acting on the pin will exert a bending moment on the pin as shown in Figure 40.7. The large diameter of the pin can be treated as a cantilever beam of length L.

Assuming a uniform distribution along the bush of the load P, the maximum bending moment acting on the pin will be

$$M = P\left(\frac{L}{2} + 5\text{ mm}\right) \tag{40.44}$$

The bending stress can be calculated from

$$\sigma = \frac{M}{Z} \tag{40.45}$$

$$\sigma = \frac{P\left(\frac{L}{2} + 5\text{mm}\right)}{\frac{\pi}{32}(d_1)^3} \tag{40.46}$$

The pin is subject to a combination of shear and bending stresses; hence, the pin should be checked for either the maximum principle stress or the maximum shear stress.

Maximum principal stress:

$$= \frac{1}{2}\left[\sigma + \sqrt{\sigma^2 + 4\tau^2}\right] \tag{40.47}$$

Maximum shear stress acting on the pin:

$$= \frac{1}{2}\sqrt{\sigma^2 + 4\tau^2} \tag{40.48}$$

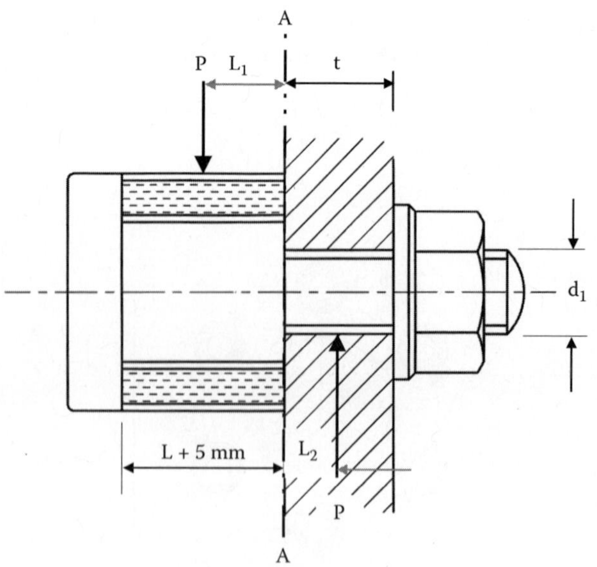

FIGURE 40.7
Bending moments acting on elastomeric bush and pin.

In analyzing the flexible coupling, the assumption is made that the power is transmitted by the shear resistance of the pins. It is seen that as the flange rotates a force P is exerted on each of the elastomeric bushes, as shown in Figure 40.6. Equating the couples due to the resisting torque with the torque gives

$$M_t = P \times \frac{D}{2} \times N \tag{40.49}$$

where

M_t = torque transmitted by the coupling (N mm)
P = force acting on each of the elastomeric bushes (N)
D = pitch circle diameter of the pins and bushes (mm)
N = number of pins and bushes

The force P is equal to the product of the projected area of the bush normal to its centerline and the intensity of the allowable pressure acting on the bush.

Therefore,
$$P = D_{bush} \times L \times I_p \tag{40.50}$$

where

D_{bush} = outer diameter of the bush (mm)
L = effective length of the bush in direct contact with the input flange (mm)
I_p = permissible intensity of pressure acting on the bush (N/mm^2)

Combining Equations 40.49 and 40.50,

$$M_t = \frac{1}{2} \times I_p \times D_{bush} \times L \times d \times N \tag{40.51}$$

The permissible intensity of pressure between the bush and the flange is usually restricted to 1 N/mm^2 and the ratio of the outer diameter of the bush to the length of the bush is generally assumed to be 1.0.

Therefore,

$$\frac{l_{bush}}{d_{bush}} = 1.0 \tag{40.52}$$

$$\text{and } Ip = 1\ N/mm^2 \tag{40.53}$$

Substituting Equation 40.52 into Equation 40.51,

$$M_t = \frac{D_{bust}^2 \times D \times N}{2} \tag{40.54}$$

The outer diameter of the bush can be obtained from Equation 40.54.

Rearranging and substituting for D_{bush},

$$D_{bush} = \sqrt{\frac{2\ M_t}{dN}} \tag{40.55}$$

Equation 40.43 identifies the direct shear stress due to the force P, and combining Equation 40.43 with Equation 40.49 gives

$$\tau = \frac{8\,M_t}{\pi \times d_1^2 \times D \times N} \tag{40.56}$$

In Figure 40.5a and b, a 5.0 mm clearance is shown between the driving and driven flanges. This clearance is important for it accommodates any angular misalignment that may exist between the two shafts.

The most satisfactory position of a shaft coupling is to be located as close to a shaft bearing as possible. This is to minimize any bending moments acting on the shaft due to coupling forces.

40.3.1 A Template for Designing a Pin-Type Flexible Coupling

A template for establishing the basic dimensions of a bushed pin flexible coupling is covered in the following steps.

1. Shaft diameter

The calculation for the shaft diameter uses the following two equations:

$$M_t = \frac{60 \times 10^6\,(kW)}{2\pi N} \tag{40.57}$$

$$\tau = \frac{16\,M_t}{\pi d^3} \tag{40.58}$$

2. Dimensions of the flanges

The dimensions for the flanges are calculated using the following empirical relationships:

$$d_{hub} = 2\,d$$

$$l_{hub} = 1.5\,d$$

$$D = 3\,d \text{ to } 4\,d$$

$$t = 0.5\,d$$

$$t_1 = 0.25\,d$$

To calculate the shear stress in the hub, consider it as a hollow shaft subjected to a torsional moment M_t. The inner and outer diameters of the hub are d and d_{hub}, respectively.

The torsional shear stress in the hub is given by the following expressions:

$$\tau = \frac{M_t r}{J} \tag{40.59}$$

$$J = \frac{\pi\left(d_{hub}^4 - d^4\right)}{32} \tag{40.60}$$

$$r = \frac{d_{hub}}{2} \tag{40.61}$$

The shear stress in the flange at the junction with the hub is calculated from the following equation:

$$M_t = \frac{1}{2}\pi d_{hub}^2 t\tau \quad (40.62)$$

3. Diameter of the pins

In the pin flexible coupling, the number of pins used can be either four or six, depending on the size of the coupling and its torque rating.

The diameter of the pins can be determined from the following empirical formula:

$$d_{pin} = \frac{d}{2\sqrt{n}} \quad (40.63)$$

where n is the number of pins.

The shear stress in the pin can be calculated from Equation 40.56:

$$\tau = \frac{8\,M_t}{\pi \times d_1^2 \times D \times N}$$

Assuming the pins will be manufactured from a medium carbon steel such as 080M40, the permissible shear stress should be less than say 35 MPa. This allows for a FoS of 6.

The bending stresses in the pin can also be calculated and ensure that it is in the permissible limits for the material.

4. Outer diameter of the elastomeric bushes

The outer diameter of the bushes can be calculated from Equation 40.54:

$$M_t = \frac{1}{2}D_{bush}^2 \times D \times N$$

5. Length of elastomeric bush

The effective length of the elastomeric bush is given by Equation 40.52:

$$l_{bush} = d_{bush}$$

6. Dimensions of the keys

Table 40.1 gives the cross section of a range of standard square and rectangular keys. The length of the key in each shaft will be

$$l_{key} = l_{hub} \quad (40.64)$$

Therefore, the length of the key will be identical to the length of the hub.

When the key dimensions have been established, the next step will be to check for shear and compressive stresses in the key using the following equations:

$$\tau = \frac{2\,M_t}{dbl} \quad (40.65)$$

where b = width of the key

$$\sigma_c = \frac{4\,M_t}{dhl} \tag{40.66}$$

where h = height of the key

Example 40.4

A pin flexible coupling is required to connect the output shaft of an electric motor to the shaft of a centrifugal fluid pump. The motor will deliver 30 kW at 750 rev/min. It can be assumed that the starting torque to be 150% of the rated torque.

Calculate the respective sizes of the individual components making up the coupling.

The yield strength of the key material in compression can be assumed to be equal to the yield strength in tension at 460 N/mm².

Solution

Given:

Input = 30 kW
Speed (N) = 750 rev/min
Design torque = 1.5 (rated torque)

The design procedure for finding the dimensions of the pin bushed flexible coupling will be shown using the following steps.

Step 1 Shaft

1.1 Material for the shaft

Considering a grade of medium carbon steel (British Standard 080M40) having a yield stress of 460 MPa. The FoS for the shaft is assumed to be 2:

$$\tau = \frac{\sigma_y}{(FoS)} = \frac{0.5\,\sigma_y}{(FoS)} = \frac{0.5(460)}{(2)} = 115\text{ MPa}$$

1.2 Diameter of the shaft

The starting torque of the motor is 150 percent of the rated torque. Hence,

$$\begin{aligned} M_t &= \frac{60\times 10^6(\text{Power})}{2\pi N} \\ &= \frac{60\times 10^6(30)(1.5)}{2\pi(750)} \\ &= 572,957.8\text{ N mm} \end{aligned}$$

Now

$$\tau = \frac{16\,M_t}{\pi d^3}$$

or

$$115 = \frac{16(572,957.8)}{\pi d^3}$$

Rearranging and solving for the diameter (d),
d=29.385 mm say 30.0 mm (ans).

Step 2 Dimensions of the flange
The dimensions of the flanges are as follows:

$$d_{hub} = 2 \cdot d = 2(30) = 60.0 \text{ mm}$$

$$l_{hub} = 1.5 \cdot d = 1.5(30) = 45.0 \text{ mm}$$

$$D_o = 4 \cdot d = 4(30) = 120.0 \text{ mm}$$

$$t = 0.5 \cdot d = 0.5(30) = 15.0 \text{ mm}$$

$$t_1 = 0.25 \cdot d = 0.25(30) = 7.5 \text{ say } 8.0 \text{ mm}$$

The hub is treated as a hollow shaft subject to a torsional moment. From Equation 40.60,

$$J = \frac{\pi\left(d_{hub}^4 - d^4\right)}{32}$$

$$= \frac{\pi\left(60.0^4 - 30.0^4\right)}{32}$$

$$= 1,192,823.46 \text{ mm}^4$$

$$r = \frac{d_{hub}}{2} = \frac{60.0}{2} = 30.0 \text{ mm}$$

The torsional stress in the hub is given by

$$\tau = \frac{M_t r}{J}$$

$$= \frac{(572,957.8)(30)}{(1,192,823.46)}$$

$$= 14.41 \text{ MPa}$$

Permissible shear stress in flange:

$$\tau = \frac{S_{su}}{(FoS)} = \frac{0.5\ S_{ut}}{(FoS)} = \frac{0.5(200)}{6} = 16.67 \text{ MPa}$$

Therefore,

$$\tau < 16.67 \text{ MPa}$$

The shear stress in the flange is within the acceptable limit.

Step 3 Dimensions of the pins
The number of pins to carry the torque is 6.

$$n = 6$$

The diameter of the pin is calculated from

$$d_{pin} = \frac{d}{2\sqrt{n}}$$

$$= \frac{30.0}{2 \times \sqrt{6}}$$

$$= 6.12 \text{ mm, say } 8.0 \text{ mm}$$

Shear stress in the pin:

$$\tau = \frac{8\,M_t}{\pi d_1^2 Dn}$$

$$= \frac{8(572{,}957.8)}{\pi(8^2)(120)(6)}$$

$$\tau = 31.66 \text{ MPa} < 35.0 \text{ MPa}$$

The pin design is considered safe for the shear strength.

The bending stress in the pin will be considered when the design of the elastomeric bushes is determined.

Step 4 Dimensions of the elastomeric bushes

The allowable intensity of pressure between the elastomeric bush and the flange (if manufactured from cast iron) is taken as 1 MPa.

The ratio of the effective length to the diameter for the elastomeric bushes is assumed to be 1.0.

From Equation 40.55,

$$d_{bush} = \sqrt{\frac{2\,M_t}{DN}}$$

$$= \sqrt{\frac{2(572{,}957.8)}{(120)(6)}}$$

$$= 39.89 \text{ mm, say } 40.0 \text{ mm}$$

Therefore, $d_{bush} = l_{bush} = 40.0$ mm.

Step 5 Bending moment acting on the pin

From Figure 40.7, the force P acting on each pin will depend on the applied torque and is given in the relationship:

$$M_t = P \times \frac{D}{2} \times N$$

Therefore,

$$P = \frac{2\,M_t}{DN}$$

$$= \frac{2(572{,}957.8)}{(120)(6)}$$

$$= 1591.55 \text{ N}$$

Figure 40.8 shows the subassembly of the bush and pin. It is assumed that the force P is evenly distributed over the effective length of the bush (40.0 mm).

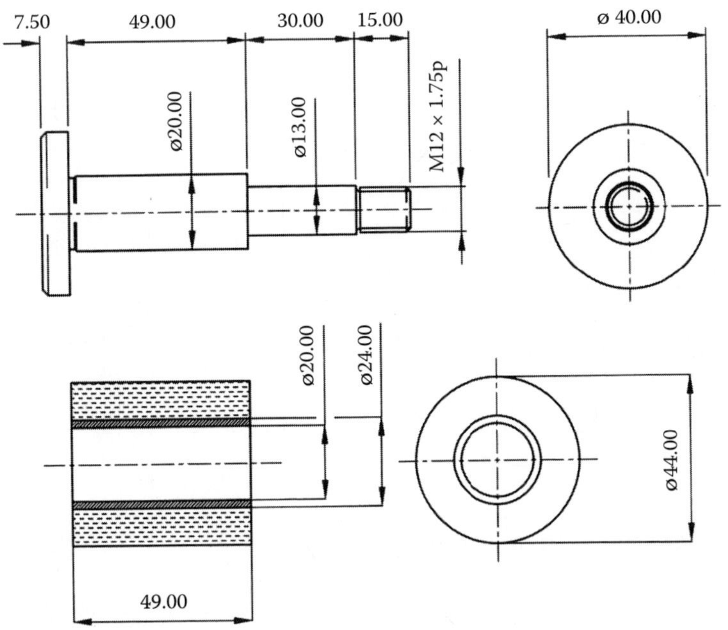

FIGURE 40.8
Details of elastomeric bush and pin.

$$M_{pin} = P\left(\frac{40}{2} + 5\right)$$

$$= 1591.55 \times \left(\frac{40}{2} + 5\right)$$

$$= 39{,}775 \text{ Nmm}$$

$$\sigma_{pin} = \frac{32\ M_{pin}}{\pi d_{pin}^3}$$

$$\frac{460}{2} = \frac{32(39{,}775)}{\pi\left(d_{pin}^3\right)}$$

$$d_{pin} = 12.08 \text{ mm, say } 13.0 \text{ mm}$$

In step 3 of this procedure, the initial calculation of the pin diameter, considering the shear stress only, established the diameter as 8.0 mm using an empirical formula. This subsequent calculation which considers the bending stress shows that the pin diameter should be 13.0 mm to keep the bending stress within 230 MPa. In order to provide a shoulder for the pin to abut to the flange, the enlarged diameter is taken as 13.0 + 6.0 or 19.0 mm. This would be increased to 20.0 mm to provide a rational diameter for a standard bush to be selected. The inner diameter of the elastomeric bush will be 24.0 mm; this allows for a 2.0 mm-thick brass lining.

The outer diameter of the bush will be 44.0 mm. The minimum thickness of the bush is usually 10.0 mm.

Figure 40.9 gives the dimensions of the driving flange.

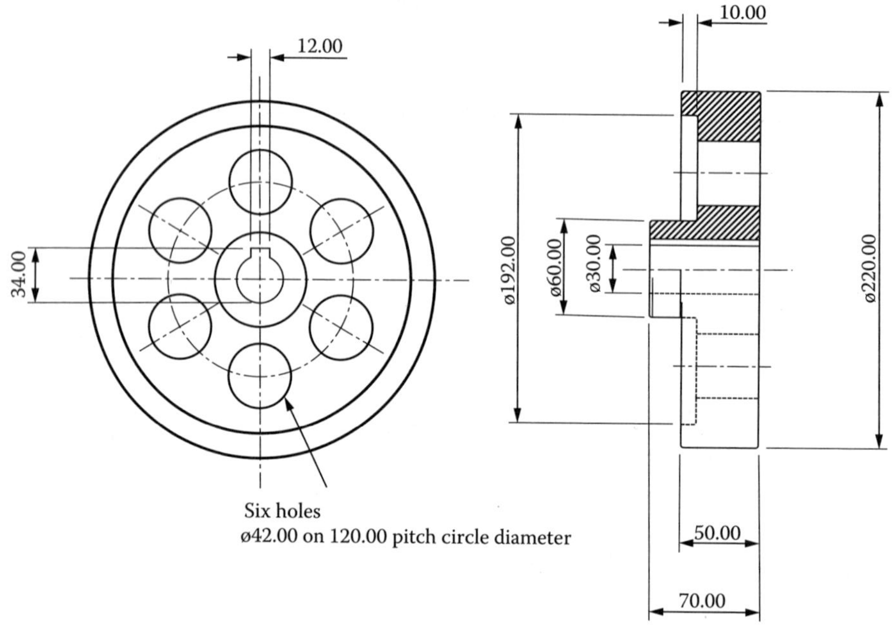

FIGURE 40.9
Finalized dimensions of example driving flange.

Step 6 Dimensions for keys

From Table 40.1, the standard cross section of a flat sunk key for a 35.0 mm diameter shaft is 10×8 mm. The length of the key is equal to l_{hub}:

$$l = l_{hub} = 45.0\ \text{mm}$$

The final dimensions of the flat key will be 10×8×45 mm.

The shear stress in the key when subject to a torque of 572,957.8 N mm will be

$$\tau = \frac{2\,T}{lhd}\ \text{(from Equation 40.6)}$$

$$\tau = \frac{2(572,957.8)}{(45)(8)(35)}$$

$$= 90.95\ \text{N/mm}^2$$

$$\text{FoS} = \frac{460\ \text{N/mm}^2}{90.95\ \text{N/mm}^2}$$

$$= 5.06$$

Therefore, the shear strength of the key is acceptable.

The compressive stress in the key will be

$$\sigma_c = \frac{4\,T}{lhd}$$

$$\sigma_c = \frac{4(572,957.8)}{(45)(8)(35)}$$

$$= 181.90\ \text{N/mm}^2$$

$$\text{FoS} = \frac{460\ \text{N/mm}^2}{181.90\ \text{N/mm}^2}$$

$$= 2.53$$

Therefore, the compressive stress of the key is acceptable.

40.4 Universal Joint

The universal joint invented in the sixteenth century by Gerolamo Cardano is one of the earliest method of transmitting power between two shafts making an angle with each other. It was recognized that there was variable velocity of the output shaft to that of the input shaft. This was recognized in the seventeenth century by Robert Hook, who proposed joining two Cardan shafts that were offset by 90°. This arrangement cancelled out the velocity variations of the single joint, thereby making it the first constant velocity (CV) joint.

An application of a universal joint may be found in the transmission from a vehicles gearbox connecting to the differential in the rear axle. Other examples will be found on multi-spindle drilling machines.

Figure 40.10 illustrates the basic design of a universal joint. In this design, a spider is assembled into the ends of the joint such that they can pivot about two axes normal to each other. Rotation is transmitted from one side of the joint to the other. The axis direction can also change during rotation.

It can be shown, without proof, that if the angular velocity in the input shaft is ω_1 and the output shaft is ω_2, the relationship between ω_1 and ω_2 is

$$\frac{\omega_2}{\omega_1} = \frac{\cos\phi}{1 - \sin^2\phi\cos^2\theta} \tag{40.67}$$

FIGURE 40.10
Universal joint.

where ϕ is the angle between the two axes, and θ is the angle rotated by the input shaft.

Example 40.5

Determine the largest and smallest values of the velocity ratio of two shafts connected with a universal joint with the axis are at 45°.

$$\frac{\omega_2}{\omega_1} = \frac{\cos\phi}{1-\sin^2\phi\cos^2\theta}$$

$$= \frac{\cos 45}{1-\sin^2 45\ \cos^2\theta}$$

$$\text{When } \cos\theta = 1 \quad \frac{\omega_2}{\omega_1} = \frac{2}{\sqrt{2}} = 1.414 \text{ (maximum)}$$

$$\text{When } \cos\theta = 0 \quad \frac{\omega_2}{\omega_1} = \frac{1}{\sqrt{2}} = 0.707 \text{ (minimum)}$$

40.5 Oldham Coupling

In cases where there is an axial misalignment between the input and output shafts, an Oldham coupling would be a solution to the problem. In 1821 John Oldham invented the coupling to solve a paddle placement problem concerning a paddle steamer design.

An Oldham coupling consists of three discs (see Figure 40.11). One disc is coupled to the input shaft and another coupled to the output shaft. These two discs are fitted with slots.

A third disc that has a tenon machined on either side sits in the middle between the inner and outer discs.

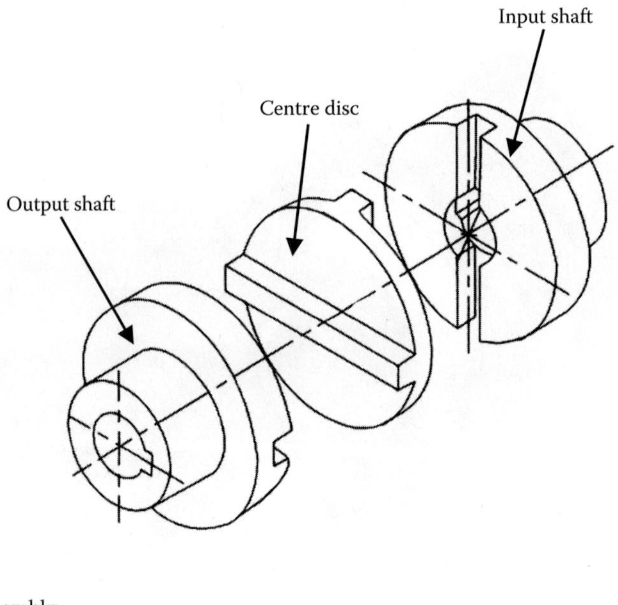

FIGURE 40.11
Oldham coupling assembly.

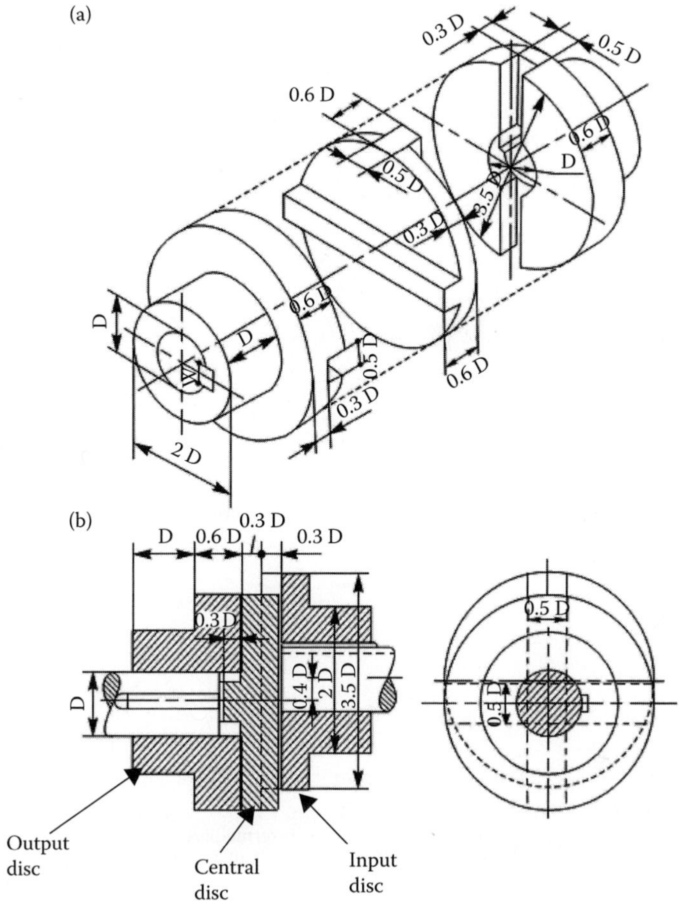

FIGURE 40.12
Oldham coupling proportions.

The principle of operation is the center disc rotates at the same speed as the input and the output shafts and traces a circular locus, twice per revolution, around the mid-position between the input and output shafts (Figure 40.12).

The principle advantage of this type of coupling as compared to using two universal joints is its compact size.

40.6 CV Joints

40.6.1 Rzeppa Joint

The first practical CV joint was invented by Alfred Hans Rzeppa (pronounced Sheppa) in 1926. The joint was able to transmit an even torque over a wide range of angles. A later version used six balls as intermediate members, kept in a plane that bisected the angle between the input and output shafts (see Figure 40.13) A controlled guide ball cage

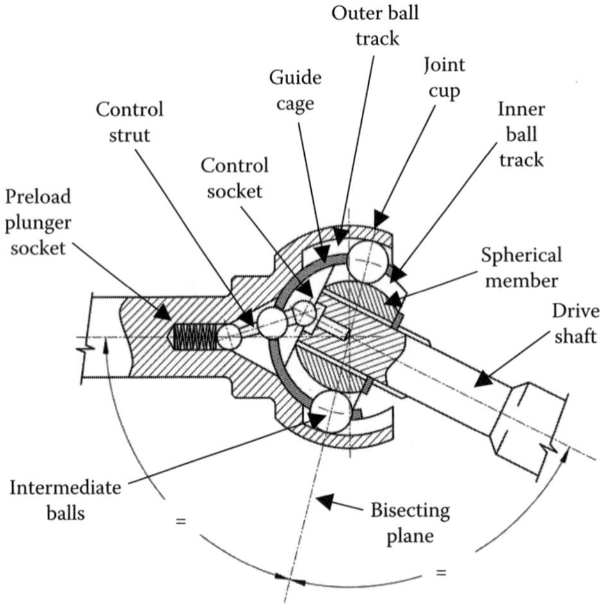

FIGURE 40.13
Rzeppa CV joint.

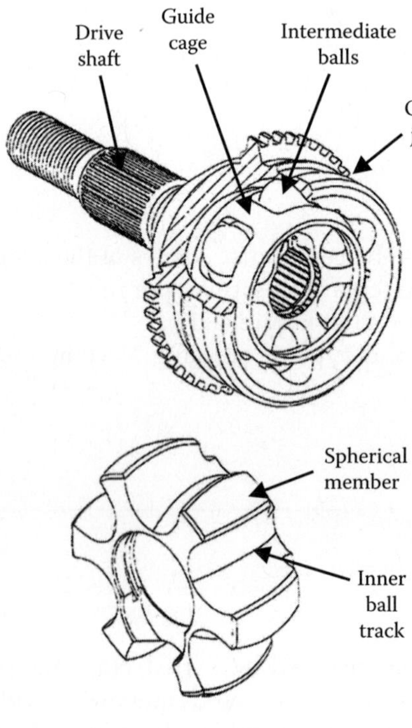

FIGURE 40.14
Birfield CV joint.

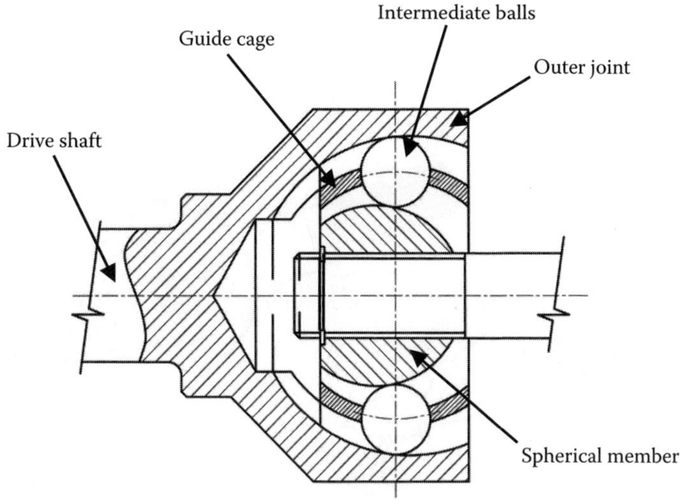

FIGURE 40.15
Section through Birfield joint.

maintained the balls at the bisecting angle using a pivoted control strut that swiveled the cage at the correct angle.

The Rzeppa joint had a tendency to jam, and suffered mechanical wear as a result.

40.6.2 Birfield Joint

The Birfield CV joint manufactured by the GKN Birfield Transmissions group; this included GKN Hardy Spicer in Birmingham and GKN Laycock in Sheffield, and Salisbury Transmissions in Witton Birmingham (the name Birfield came from the contraction of **Bi**rmingham and Shef**field**). The joint design is based upon the Rzeppa principle and was used in the Alec Issigonis-designed BMC Mini that used the joint within its front-wheel-drive transmission (see Figures 40.14 and 40.15).

The company developed and improved the original Rzeppa design by incorporating converging ball tracks that did not rely on a controlled ball cage to maintain the intermediate ball members on the median plane.

40.6.3 Carl Weiss CV Joint

Carl W. Weiss of New York patented a successful constant velocity joint in 1925 (as shown in Figure 40.16) that was developed and manufactured by the Bendix Products Corporation.

40.6.4 Tripode (Tripot)-Type CV Joint

Figure 40.17 shows this type of joint which was developed by Michel Orain of Glaenzer Spicer in France. The joint has a three-pointed yoke that is part of the drive shaft, and has barrel-shaped roller bearings on the ends. These bearings fit into a cup having three matching grooves. Due to the limited range of movement, a typical tripod joint has approximately 26° of angular articulation together, with up to 50 mm plunge travel.

They are generally used on the inboard side of front-wheel-drive vehicles and on rear-wheel-drive vehicles where the range of motion is substantially lower.

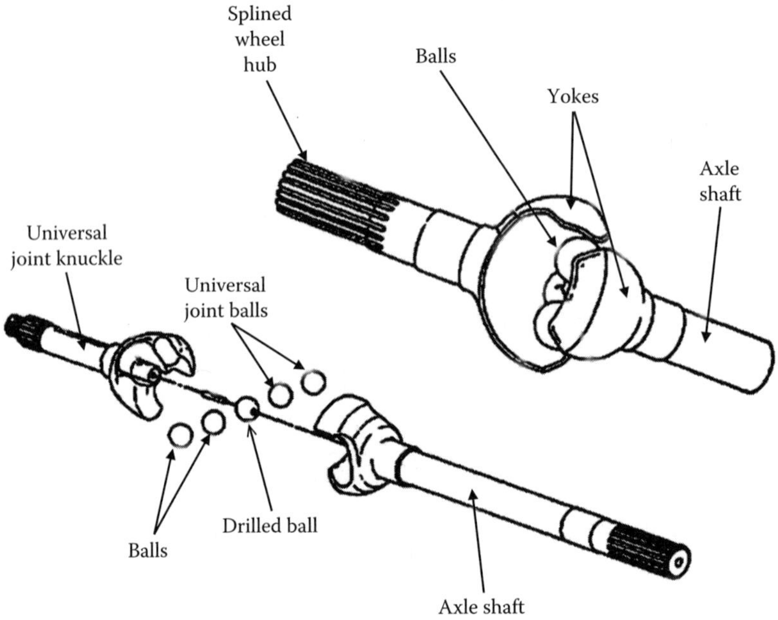

FIGURE 40.16
Bendix Weiss CV joint.

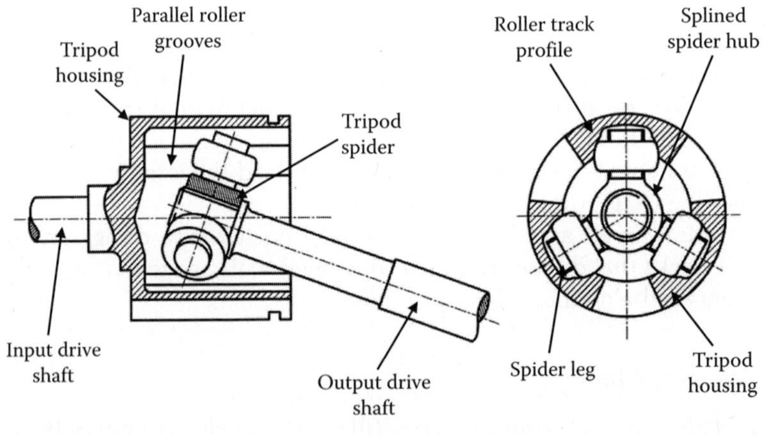

FIGURE 40.17
Tripode-type CV joint.

41

Bearings

41.0 Introduction

A bearing is a machine element that allows one part to bear on another part, and allows easy movement between the two parts. A bearing constrains relative motion and reduces friction between moving parts. The design of the bearing may provide axial movement of the part or rotational around a fixed axis.

Bearing classification:
Bearings are broadly classified into two groups:

- Fluid film
- Rolling contact

Fluid film bearings:
In a fluid film bearing, the entire load acting on the shaft is carried by a thin film of fluid that exists between the rotating and stationary elements of the bearing. There are basically four types of fluid film bearings (Figure 41.1):

- Sliding-contact bearing
- Journal bearing
- Thrust bearing
- Sliding bearing

Rolling contact bearings:
With rolling contact bearings, the load on the rotating shaft is carried by a series of rolling elements, either balls or rollers, placed between the rotating and stationary elements. The rolling contact types of bearings are known as

- Ball bearing
- Roller bearing

Consideration of bearing friction between fluid film and rolling bearings:
The fluid film bearing has a high starting torque before it passes through the boundary lubrication before the hydrodynamic oil wedge becomes established. Once running under a hydrodynamic film, the fluid film bearing exhibits friction characteristics comparable to a rolling-element bearing.

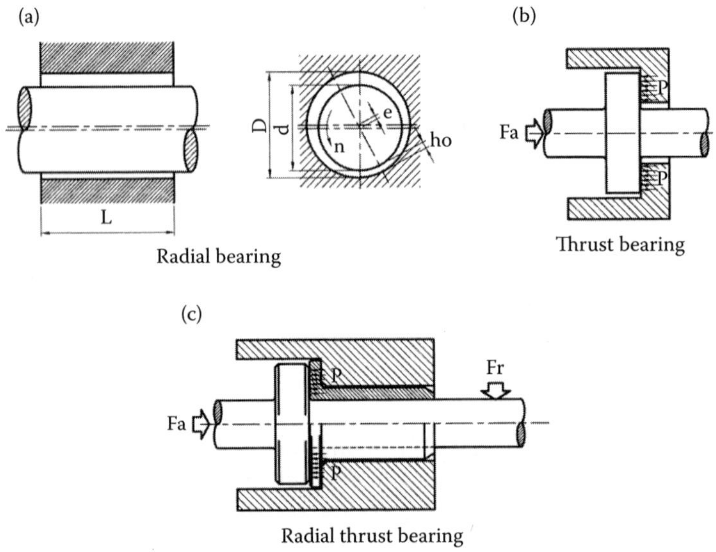

FIGURE 41.1
Fluid film bearings.

Running friction for a rolling-element bearing is lower than its starting friction. If torque characteristics are critical to a bearing design, then starting and running frictional characteristics should be measured experimentally.

When a bearing must be started repeatedly under heavy load, rolling-element bearings offer a better choice than do sleeve bearings. When the increased complexity is acceptable, an externally pressurized (hydrostatic) bearing is the better choice. When starting load is light and load increases gradually with speed, the conventional hydrodynamic sleeve bearing usually is preferred.

41.1 Fluid Film Bearings

41.1.1 Advantages of Fluid Film Bearings

Fluid film bearings have certain advantages over rolling contact bearings:

- The bearing and housing design is simple.
- They are able to occupy less radial space.
- They operate more silently.
- The design of the shaft is simpler.
- Fluid film bearings are ideally suited for medium- and high-speed operations.
- They have good shock loading capacity.

Fluid film bearings have the following disadvantages:

- They require careful attention to lubrication.

- These bearings are designed to carry either radial load or axial load only.
- The power loss due to friction is greater than in rolling contact bearings.
- The finish of the rubbing contact surfaces has to be very good compared to that for fitting a rolling contact bearing.

41.1.2 Journal Bearings

In this type of bearing, the bearing pressure is exerted at right angles to the axis of the shaft. The portion of the shaft that lies within the bearing is known as the *journal*. The shaft is generally manufactured from mild steel in low-stress applications (Figure 41.2).

41.1.3 Pivot (or Footstep) Bearing

In this design, the bearing pressure is exerted parallel to the shaft where the axis is vertical. Note: The end of the shaft rests within the bearing (Figure 41.3).

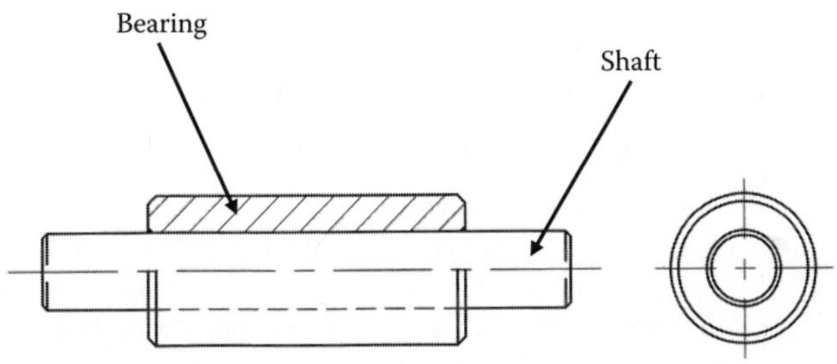

FIGURE 41.2
Journal bearing.

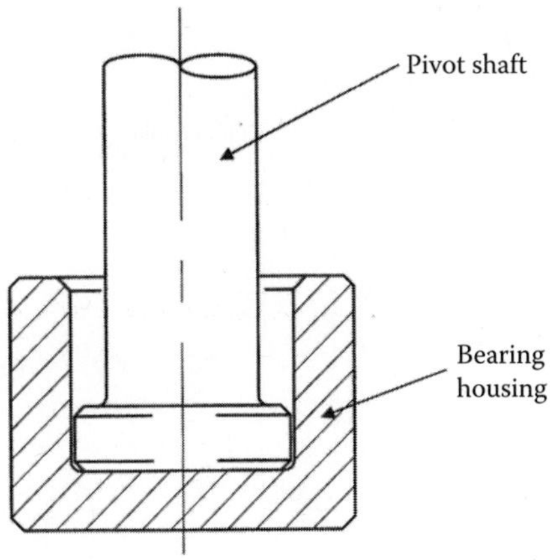

FIGURE 41.3
Pivot (or footstep) bearing.

41.1.4 Flexural Bearing

This is a special type of bearing in which motion occurs by the bending of a metal strip. The range of motion is small but it is quite accurate. This application is usually found in scientific instruments (Figure 41.4).

41.1.5 Thrust Bearing

This is also known as a collar bearing. In this design, the shaft has a stepped feature that allows the load that is parallel to the axis of the shaft to be carried by the thrust bearing. Unlike the pivot or footstep bearing, the shaft can pass through and beyond the thrust bearing. Thrust bearings are most likely found in bevel mountings, propeller drives, turbines, and so on, where the shaft carries an axial load (Figure 41.5).

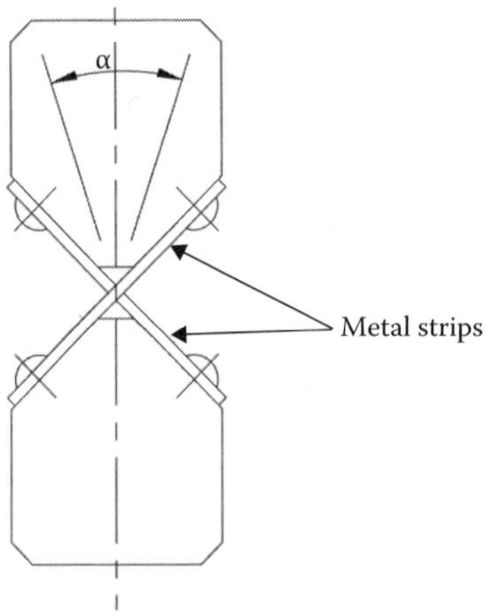

FIGURE 41.4
Flexural bearing.

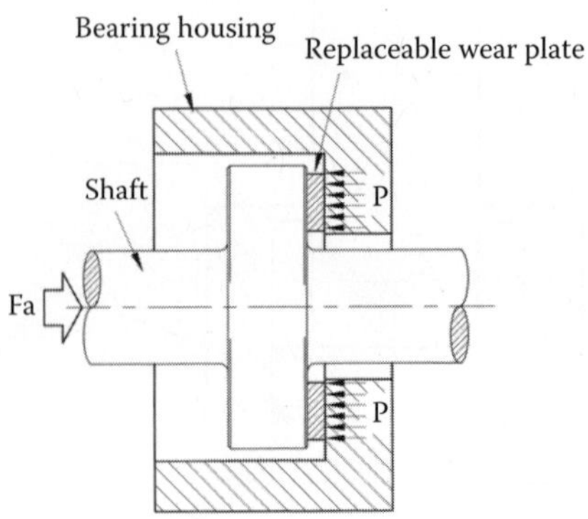

FIGURE 41.5
Thrust bearing.

41.1.6 Linear Bearings

Sliding-contact linear bearings are essentially just sliding-contact rotary bearings with a very large radius of curvature. There are three basic configurations for linear fluid film bearings (Figure 41.6): boxway, dove tail, and twin rail.

41.1.6.1 Boxway

Boxway bearings support the largest loads in all directions (Figure 41.7).

41.1.6.2 Dovetail

Dovetail bearings have less load capacity, but have fewer required precision surfaces and are easy to adjust (Figure 41.8).

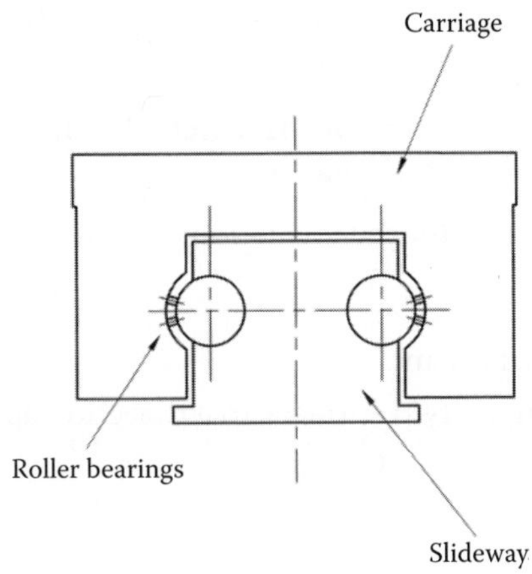

FIGURE 41.6
Linear bearing.

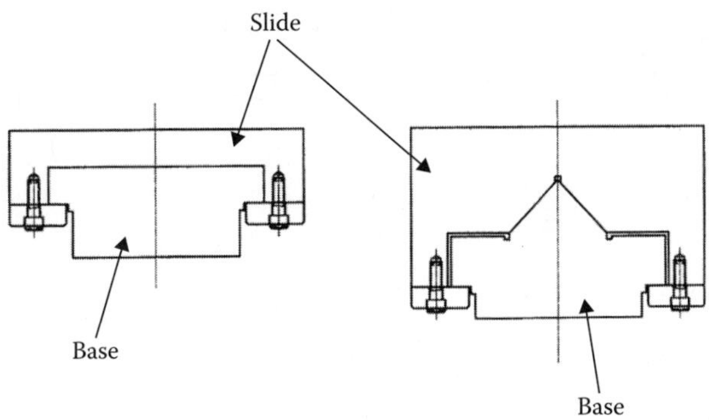

FIGURE 41.7
Boxway slideway.

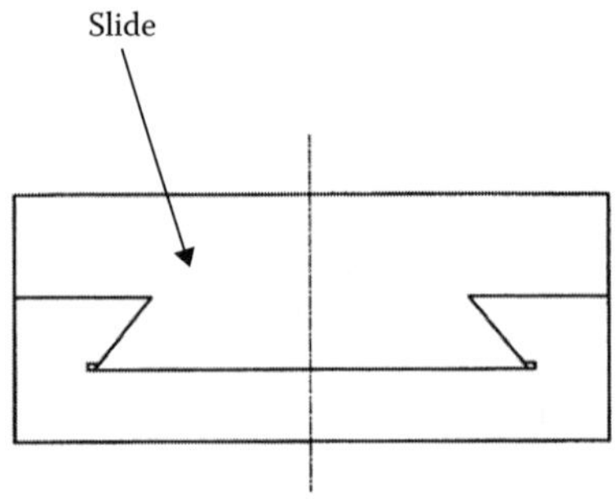

FIGURE 41.8
Dovetail slide.

41.1.6.3 Twin Rail

Twin rails are the easiest to build, but the rails (only supported at the ends) are subject to bending deformations unless more complex supports are used (Figure 41.9).

41.1.7 Factors Covering the Formation of the Thick Oil Film in a Hydrodynamic Lubricated Bearing

The following factors are essential for the formation of a thick film of lubricant in a hydrodynamic lubricated bearing:

- A continuous supply of lubricant
- Relative motion between the two surfaces in a direction approximately tangential to the surfaces

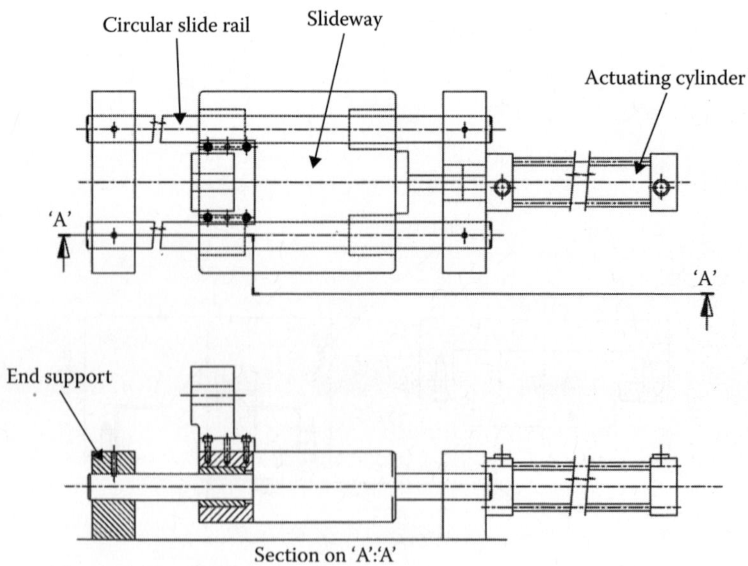

FIGURE 41.9
Twin-rail slideway.

- The ability of one of the surfaces to take a small inclination to the other surface in the direction of the relative motion
- The line of action of the resultant lubricant pressure must coincide with the line of action of the external load between the surfaces

The Reynolds equation in its simplified form:

$$\frac{\partial}{\partial x}\left(\frac{h^3}{12\mu}\frac{\partial p}{\partial x}\right)+\frac{\partial}{\partial z}\left(\frac{h^3}{12\mu}\frac{\partial p}{\partial z}\right)=\frac{U}{2}\frac{\partial h}{\partial x} \tag{41.1}$$

where
U: surface speed of the wedge in x-direction
p: pressure at any point (x, z) in the film
μ: absolute viscosity of the lubricant
h: film thickness, measured in the y-direction

The left-hand side of the equation represents the flow under the pressure gradient. The corresponding right-hand side represents a pressure-generation mechanism. In this equation, it is assumed that the lubricant is incompressible and Newtonian. The wedge shape is assumed to be a linear profile, as shown in Figure 41.10. The bearing is very long in the Z-direction and the variation of pressure is in the X and Z directions.

Reviewing the right-hand term in more detail:

$$\frac{\partial}{\partial x}\left(\rho\frac{u_1+u_2}{2}h\right)+\frac{\partial}{\partial y}\left(\rho\frac{v_1+v_2}{2}h\right)+\rho\frac{\partial h}{\partial t}+h\frac{\partial \rho}{\partial t} \tag{41.2}$$

Squeeze film   Compression

$$\rho\frac{U}{2}\frac{\partial h}{\partial x}+\frac{1}{2}\rho h\frac{\partial U}{\partial x}+\frac{1}{2}Uh\frac{\partial \rho}{\partial x} \tag{41.3}$$

Physical wedge   Stretch

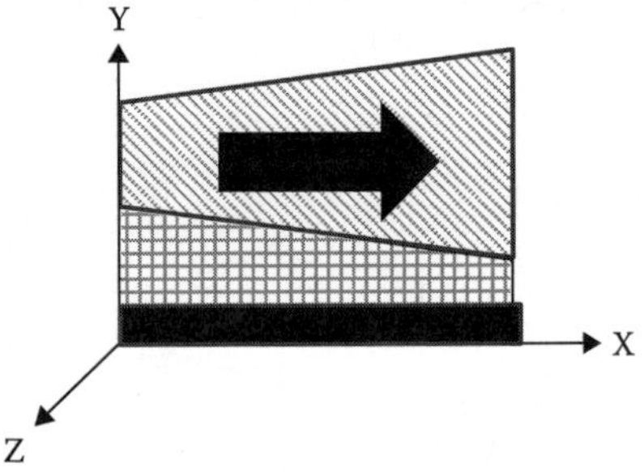

FIGURE 41.10
Wedge-shaped profile.

Consider two moving surfaces 1 and 2, as shown in Figure 41.11. For surface 1, the velocities are u_1, v_1, and w_1 along the three coordinate axes X, Y, and Z, respectively. For surface 2, similarly, the velocities are u_2, v_2, and w_2, respectively. Equation 41.2 represents the full form of the right-hand side of Reynolds equation. For the purpose of explanation, partial derivative of only the first term of Equation 41.2 is written in Equation 41.3, in this equation u_1 and u_2 have been replaced by U.

In Equation 41.2, the first term $\rho\frac{U}{2}\frac{\partial h}{\partial x}$ represents a physical wedge. The second term $\frac{1}{2}(\rho h)\frac{\partial U}{\partial x}$ is known as the *stretch*. All three terms of Equation 41.3 contribute in the pressure-generation mechanism.

The third term in Equation 41.2 $\rho\frac{\partial h}{\partial t}$ is referred to as the *squeeze film*; this term represents how the film thickness changes in respect to time.

The final term $h\frac{\partial \rho}{\partial t}$ is the compressibility of the fluid with time and is termed as *compression*.

The simplified form of the Reynolds equation (41.3) has only the physical wedge term, $\rho\frac{U}{2}\frac{\partial h}{\partial x}$.

41.1.8 Wedge Film Journal Bearings

The load-carrying capability of a wedge film bearing results when the journal and/or the bearing rotates relative to the load. The most common case is that of a steady load, a fixed (nonrotating) bearing and a rotating journal. Figure 41.12a is that of a journal at rest with metal-to-metal contact at point A on the line of action of the supported load.

When the journal starts to rotate slowly in an anticlockwise direction as depicted in Figure 41.12b, the point of contact will move to position B so that the angle AOB is the angle of sliding friction of the surface in contact at B. In the absence of a lubricant, there will be a dry metal-to-metal friction. If a lubricant is present in the clearance space of the bearing and journal, then a thin absorbed film of lubricant may partly separate the surfaces, but a continuous fluid film completely separating the surfaces will not exist because of slow speed.

When the speed of the journal is increased, a continuous fluid film will be established, as shown in Figure 41.12c. The center of the journal will have moved so that the minimum film thickness is at C. It may also be noted that the direction of rotation will be from D to C, the film is continuously narrowing, and hence is a converging film. The curved converging can be considered as a wedge-shaped film of a slipper bearing wrapped around the journal. A small consideration will show that from C to D in the direction of rotation, as

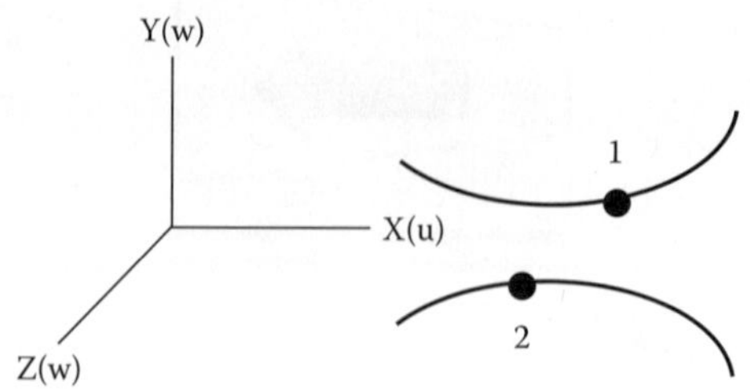

FIGURE 41.11
Reynolds equation.

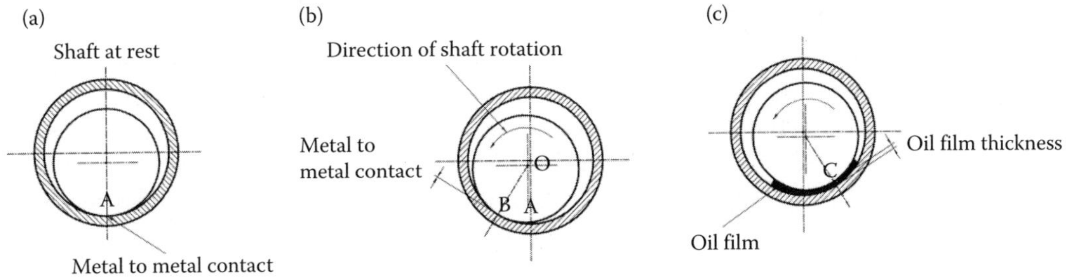

FIGURE 41.12
Wedge-shaped film bearing.

shown in Figure 41.12c. The film is now diverging and is unable to give rise to a positive pressure or a supporting action.

Figure 41.12 gives two views of the bearing that is shown in Figure 41.12c, with the variation of pressure within the converging film. Due to the side leakage at the edges of the bearing, the angle of contact on which the pressure acts will be less than 180°.

41.1.9 Squeeze Film Journal Bearings

It has been shown in the previous sections that in a wedge film journal bearing, the bearing carries a steady load and the journal rotates relative to the bearing. In certain cases, the bearings either revolve slowly or oscillate so that the wedge film is not able to provide a satisfactory film thickness. If the load is uniform or varying in magnitude while acting in a constant direction, this becomes a thin film. If the load reverses in direction, the squeeze film may be sufficient to carry the dynamic loads without contact being made between the journal and the bearing. Such bearings are known as *squeeze film journal bearings*.

41.1.10 Materials for Bearings

The most common materials used in bearing construction include the following:

- *Lead-based babbits*: Approximately 85% lead (Pb); remainder includes tin (Sn), antimony (Sb), and copper (Cu). (Pressure rating not exceeding 14 MPa.)
- *Tin-based babbits*: Approximately 90% tin (Sn); remainder includes copper (Cu), antimony (Sb), and lead (Pb). (Pressure rating not to exceed 14 MPa.)
- *Phosphor bronze*: Major composition copper (Cu); remainder includes tin (Sn), lead (Pb), and phosphorus (P). (Pressure rating not to exceed 14 MPa.)
- *Gun metal*: Major composition copper (Cu); remainder includes tin (Sn) and zinc (Zn). (Pressure rating not to exceed 10 MPa.)
- *Cast iron*: (Pressure rating not to exceed 3.5 MPa.)

Other materials that are commonly used also include silver; carbon-graphite, Teflon (PTFE), and so on.

41.1.11 Bearing Lubrication

Chapter 45 covers the more general elements of lubrication.

In this section, only the elements relating to bearing lubrication will be considered.

41.1.11.1 Stribeck Curve

Professor Richard Stribeck (1861–1950) at the Prussian Technical Testing Institute studied the variation of friction between liquid lubricated surfaces and produced the curve known as the Stribeck curve. This curve plots the variation of the coefficient of friction against a dimensionless lubrication $\eta N/P$ where η is the dynamic viscosity, N the speed (e.g., revolutions per minute of a bearing), and P the load projected on to the geometrical surface (see Figure 41.13).

The dimensionless parameter is known under a number of names (i.e., Stribeck number, lubrication parameter, and others).

41.1.11.2 Classification Based on the Lubrication Mechanism

The Stribeck curve identifies three friction regimes for sliding and rolling contact bearings and these have been broadly categorized into the following areas:

1. Solid and boundary friction
2. Thin film or mixed friction
3. Thick film or fluid friction

41.1.11.3 Solid or Boundary Lubrication

42.1.11.3(a) Solid Film Lubricated Bearings

In extreme operating temperature applications, ordinary mineral lubricants will not be satisfactory and in these situations solid film lubricants, including molybdenum disulfide and graphite, are used which are able to withstand high operating temperatures. Examples include furnace equipment or trunnion bearings in hot drawing mills.

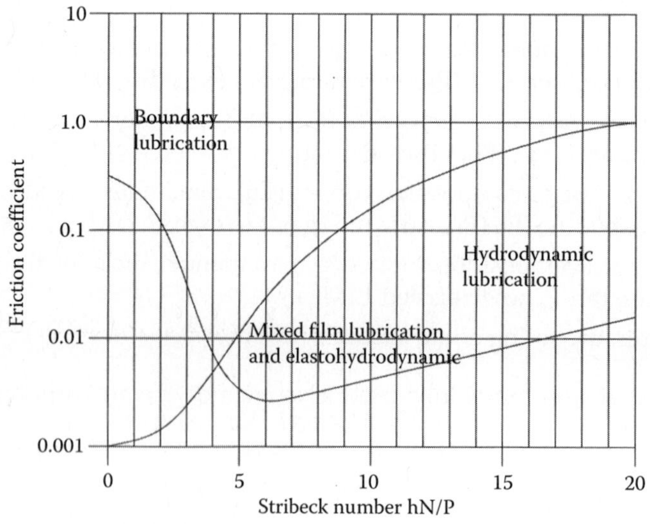

FIGURE 41.13
Stribeck curve.

41.1.11.3(b) Boundary Lubricated Bearings

If the bearing surfaces are in direct contact with each other there will be a significant amount of friction generated, together with heat when power is applied to move the bearing. There will also be rapid wear between the surfaces. When a thin film of lubricant is introduced between the rubbing surfaces, this provides a continuously renewed surface film that separates the surfaces a small amount sufficient to lift the surface roughness clear of each other and reduces friction, as shown in Figure 41.14a. The typical coefficient of friction is 0.05–0.20. It is possible for some part of the surface to make contact, resulting in a degree of wear to take place.

41.1.11.3(c) Thin Film or Mixed Film Lubrication

Even though the surfaces are separated by a thin film of lubricant, some high spots will still make metal-to-metal contact (Figure 41.14b). Surface wear will be mild and the coefficient of friction will be in the range of 0.004–0.10.

41.1.11.4 Thick Film or Hydrodynamic Lubrication

In this type of bearing, the film pressure is developed by the lubricant being forced through a wedge-shaped zone, as shown in Figure 41.14c. The rubbing surfaces are separated by a film of lubricant between 8 and 20 μm, and as a result there will be no surface-to-surface contact; hence, wear will be minimal. Coefficient of friction will range within 0.002–0.10. Hydrodynamic lubrication is in this category.

41.1.11.5 Elastohydrodynamic Lubricated Bearings

In the Stribeck curve (Figure 41.15), within the mixed film lubrication and hydrodynamic lubrication regions is an area referred to as elastohydrodynamic lubrication.

In rolling contact bearings within the Hertzian contact area between the ball or roller and the outer race, because the area of contact is very small, the contact pressures are extremely high. Roller bearing loads pressure may reach 34,450 kPa and ball bearing load pressures may reach 689,000 kPa. Under these pressures, it would appear that the lubricant would be entirely squeezed from between the contact surfaces. However, viscosity increases under extremely high pressure, preventing the lubricant from being entirely squeezed out and,

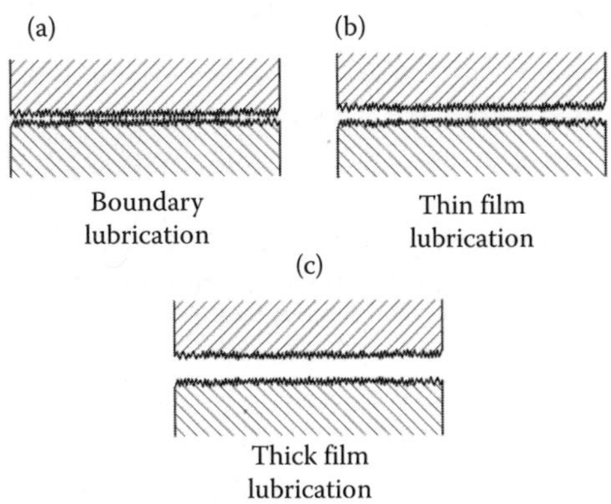

FIGURE 41.14
Oil film generation.

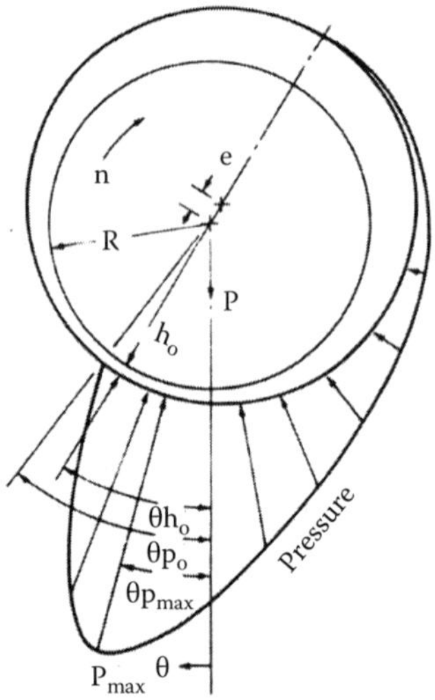

FIGURE 41.15
Elastohydrodynamic lubricated bearing.

as a consequence, a thin film of lubricant is maintained. This type of lubrication is seen occurring between gear tooth contact areas and cam operating sliders and rollers.

41.1.11.6 Hydrostatic Lubricated Bearings

Hydrostatic lubricated bearings are externally pressurized where the lubricant is fed into the bearing cavity to separate the surfaces with a thick film of lubricant. This type of bearing does not require any motion between the surfaces to generate the lubricant film; therefore, they are able to operate from very low speeds to high speeds. This is depicted in Figure 41.16.

41.1.12 Lubricant Properties

41.1.12.1 Viscosity

The viscosity of a fluid is its resistance to shear or flow, and is a measure of the fluids adhesive/cohesive or frictional properties. The viscosity will arise due to internal molecular friction within a fluid producing the frictional drag effect. There are two related measures of fluid viscosity: dynamic (or absolute) and kinematic viscosity.

Some common viscosities for various materials are tabulated in Table 41.1.

41.1.12.2 Dynamic (Absolute) Viscosity

Dynamic (absolute) viscosity—coefficient of absolute viscosity—is a measure of internal resistance. Dynamic viscosity is the tangential force per unit area required to move one horizontal plane with respect to another plane—at a unit velocity—when maintaining a unit distance apart in the fluid.

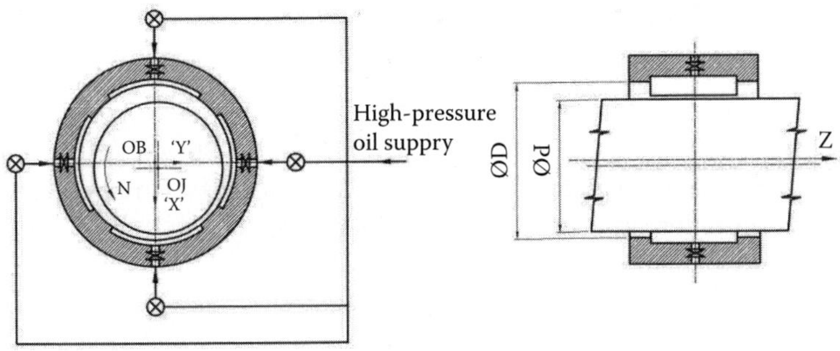

FIGURE 41.16
Hydrostatic lubricated bearing.

TABLE 41.1
Viscosities for Some Common Materials

Approximate Viscosities of Common Materials (at ambient temperature 21°C)	
Material	**Viscosity in Centipoise**
Water	1
Milk	3
SAE 10 motor oil	85–140
SAE 20 motor oil	140–420
SAE 30 motor oil	420–650
SAE 40 motor oil	650–900
Castor oil	1,000
Syrup	5,000
Honey	10,000
Chocolate	25,000
Ketchup	50,000
Mustard	70,000
Sour cream	100,000
Peanut butter	250,000

The shearing stress between the layers of a non-turbulent fluid moving in straight parallel lines can be defined for a Newtonian fluid as that whose viscosity does not change with rate of flow.

The dynamic viscosity can be expressed as

$$\tau = \mu\, d_c / d_y \tag{41.4}$$

where

τ = shearing stress (N/m^2)
μ = dynamic viscosity (N s/m^2)
d_c = unit velocity (m/s)
d_y = unit distance between layers (m)

Equation 41.4 is known as *Newton's law of friction*.

In the SI system, the dynamic viscosity units are N s/m^2, Pa s, or kg/(m s), where

- 1 Pa s = 1 N s/m^2 = 1 kg/(m s)

  Dynamic viscosity may also be expressed in the metric CGS (centimeter-gram-second) system as *g/(cm s), dyne s/cm^2*, or *poise (p)*, where

- 1 poise = 1 dyne s/cm^2 = 1 g/(cm s) = 1/10 Pa s = 1/10 N s/m^2

  For practical use, the *poise* is normally too large and the unit is often divided by 100—into the smaller unit *centipoise (cP)*, where

- 1 p = 100 cP
- 1 cP = 0.01 poise = 0.01 gram per cm second = 0.001 pascal second = 1 millipascal second = 0.001 N s/m^2

  Water at 20.2°C (68.4°F) has the absolute viscosity of 1 cP.

41.1.12.3 Kinematic Viscosity

Kinematic viscosity is the ratio of *absolute (or dynamic) viscosity to density*—a quantity in which no force is involved. It can be obtained by dividing the absolute viscosity of a fluid with the fluid's mass density:

$$\nu = \mu / \rho \tag{41.5}$$

where
ν = kinematic viscosity (m^2/s)
μ = absolute or dynamic viscosity (N s/m^2)
ρ = density (kg/m^3)

In the SI system, the theoretical unit of kinematic viscosity is m^2/s—or stoke (St), where

- 1 St (stokes) = 10^{-4} m^2/s = 1 cm^2/s

Since the stoke is a large unit, it is often divided by 100 into the smaller unit centistoke (cSt), where

- 1 St = 100 cSt
- 1 cSt (centistoke) = 10^{-6} m^2/s = 1 mm^2/s

The specific gravity for water at 20.2°C (68.4°F) is almost one, and the kinematic viscosity for water at 20.2°C (68.4°F) is for practical purpose 1.0 mm^2/s (cSt). A more exact kinematic viscosity for water at 20.2°C (68.4°F) is 1.0038 mm^2/s (cSt) (Figures 41.17 and 41.18).

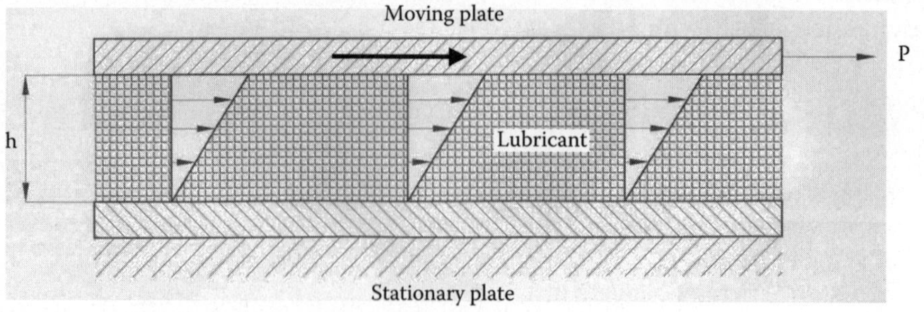

FIGURE 41.17
Moving flat-plate analogy.

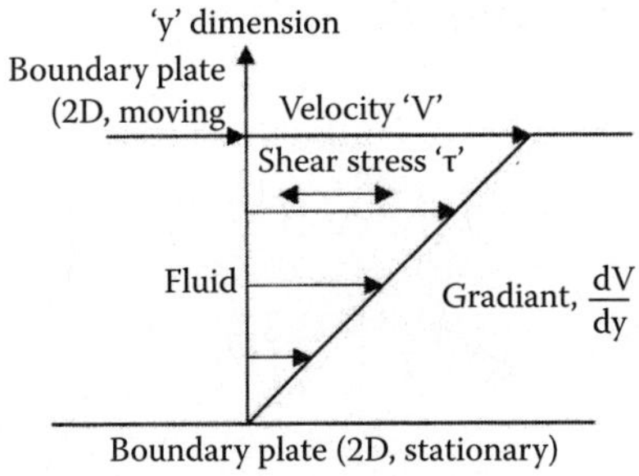

FIGURE 41.18
Viscosity gradient in an oil film.

- Kinematic Viscosity of Some Common Liquids and Fluids
  A conversion from absolute to kinematic viscosity in imperial units can be expressed as

$$\nu = 6.7197\ 10^{-4}\mu/\gamma \tag{41.6}$$

where

ν = kinematic viscosity (ft^2/s)

μ = absolute or dynamic viscosity (cP)

γ = specific weight (lb/ft^3)

Viscosity is the measurement of a fluid's internal resistance to flow. This is typically designated in units of centipoise or poise but can be expressed in other acceptable measurements as well. Some conversion factors are as follows:

100 centipoise = 1 poise

1 centipoise = 1 mPa s (milliPascal second)

1 poise = 0.1 Pa s (pascal second)

Centipoise = centistoke × specific gravity

41.1.13 Terms Used in Hydrodynamic Journal Bearings

Figure 41.19 shows the elements of a hydrodynamic journal bearing, in which O is the center of the journal and O′ is the center of the bearing.

where
D = Diameter of the bearing
d = Diameter of the journal
l = Length of the bearing

The following terms are important in the description of a hydrodynamic bearing.

41.1.13.1 Diametral Clearance c

This is the difference between the inner diameter of the bearing and the external diameter of the journal and is expressed as

$$c = D - d \tag{41.7}$$

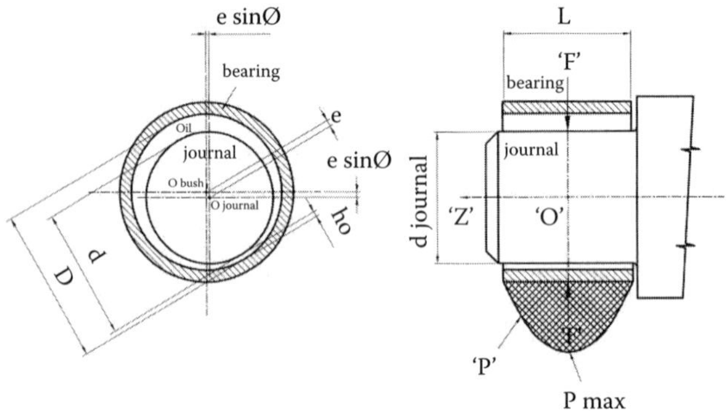

FIGURE 41.19
Terms used in hydrodynamic journal bearings.

Note: The diametral clearance (c) in a bearing should be kept as small as possible to produce the necessary velocity gradient, enabling the pressure build up that will support the load. Small clearances have the advantage of reducing the side leakage. However, the allowance must be made for manufacturing tolerances within the bearing and journal. A common "rule of thumb" is clearance in industrial machines is 0.025 mm/cm of journal diameter (Figure 41.20).

41.1.13.2 Radial Clearance c_1

The radial clearance is the difference between the radii of the bearing and the journal, and is expressed as

$$c_1 = R - r = \frac{D - d}{2} = \frac{c}{2} \tag{41.8}$$

41.1.13.3 Diametral Clearance Ratio

This is the ratio of the diametral clearance to the diameter of the journal, and is expressed as

$$= \frac{c}{d} = \frac{D - d}{d}$$

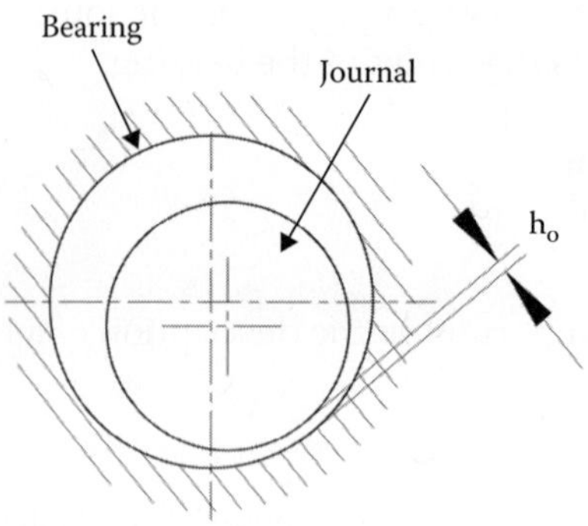

FIGURE 41.20
Minimum oil film thickness.

41.1.13.4 Eccentricity e

This is the radial distance between the center of the bearing (O) and the displaced center of the journal (O′) under load, and is denoted by e.

41.1.13.5 Minimum oil Film Thickness

This is the minimum distance between the bearing and the journal under complete lubrication conditions. It is denoted by h_o and occurs at the line of contact as shown in Figure 41.21. Its value can be assumed to be c/4.

41.1.13.6 Attitude or Eccentricity Ratio

This is the ratio of the eccentricity to the radial clearance, and may be expressed as

$$e = \frac{e}{c_1} = \frac{c_1 - h_o}{c1} = 1 - \frac{h_o}{c_1} = 1 - \frac{2h_o}{c} \tag{41.9}$$

41.1.13.7 Short and Long Bearings

If the ratio of length to the diameter of the journal (i.e., *l*/d) is less than 1, then the bearing is considered to be a *short bearing*. If the ratio *l*/d is greater than 1, then the bearing is known as a *long bearing*.

Notes:

1. If the length of the journal (*l*) is equal to the diameter (d), then the bearing is called a *square bearing*.
2. Due to the side leakage of the lubricant from the bearing, the pressure in the film will be atmospheric. The average pressure will be higher for a long bearing than for a short or square bearing. Therefore, from the point of view of limiting side leakage, a bearing having a large *l*/d ratio is preferable. However, space requirements, manufacturing, tolerances, and shaft deflections are better met with a short bearing. The value of *l*/d may be taken as 1 or 2 for general industrial machines. In crankshaft bearings, the *l*/d ratio is frequently less than 1.

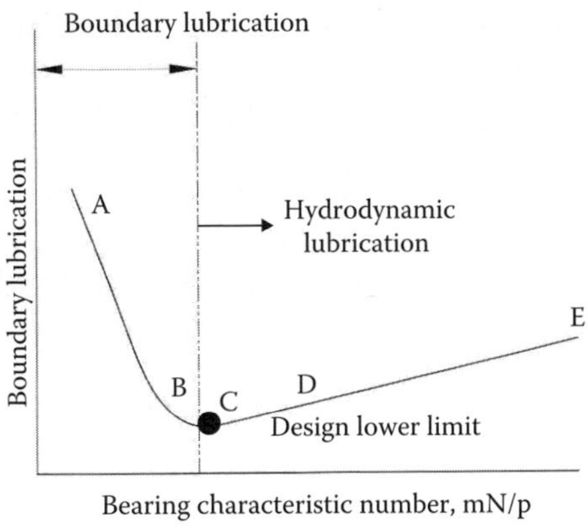

FIGURE 41.21
Coefficient of friction.

41.1.14 Bearing Characteristic Number and Bearing Modulus for Journal Bearings

The coefficient of friction in the design of a bearing is of major importance, as it affords a means of determining the loss of power due to bearing friction. It has been shown through experiments that the coefficient of friction for a fully lubricated journal bearing is a function of three variables:

$$\frac{ZN}{p} \tag{41.10}$$

$$\frac{d}{c} \tag{41.11}$$

$$\frac{l}{d} \tag{41.12}$$

Hence, the coefficient of friction can be expressed as

$$\mu = \phi\left(\frac{ZN}{p}, \frac{d}{c}, \frac{l}{d}\right) \tag{41.13}$$

where

μ = Coefficient of friction
ϕ = A functional relationship
Z = Absolute viscosity of lubricant (kg·m/s)
N = Speed of the journal (rev/min)
p = Bearing pressure on the projected bearing area (N/mm^2)
= Load on the journal = $l \times d$
d = Diameter of the journal (mm)
l = Length of the bearing (mm)
c = Diametral clearance (mm)

The factor ZN/p is referred to as a *bearing characteristic number* and is dimensionless.

41.1.15 Coefficient of Friction

The variation of the coefficient of friction with the operating values of bearing characteristic number (ZN/p) is determined from experiments carried out by the McKee Brothers during actual tests of friction (refer to Figure 41.21). The factor ZN/p is a predictor of bearing performance.

The curve shows that from point A to point B with an increase in the bearing characteristic number (ZN/p), the coefficient of friction (μ) decreases till point B is reached. The coefficient of friction continues to decrease (at a lower rate) until a minimum point C is reached. From this point onward to point D, the coefficient increases, and from D to E, it increases at a constant rate.

The zone from A to C is known as *boundary lubrication* (or imperfect lubrication) *zone,* where there is a potential for metal-to-metal contact occurring. It is possible for some oiliness to be present in this region. The portion of the curve from C to D is known as the *hydrodynamic lubrication zone.* The calculated bearing characteristic number should lie somewhere in the portion of the curve D to E. This zone is known as the *design zone.*

Referring back to point C: at this position on the curve, the value of ZN/p is known as the *bearing modulus*, denoted as K. The bearing should not be operated at this value of bearing modulus, because any slight decrease in speed or slight increase in pressure will break the oil film and the journal will operate with metal-to-metal contact. This results in high friction, increased wear, and heating within the bearing. In order to prevent such conditions, the bearing should be designed for the value of ZN/p to be at least three times the minimum value of the bearing modulus (K). If the bearing is subject to large variations of load and heavy impacts, then the value of ZN/p should be greater than 15 K.

For a well-lubricated journal bearing, the coefficient of friction can be calculated using the empirical relationship established by the McKee Brothers based on their experimental data:

$$\mu = \frac{33}{10^8}\left(\frac{ZN}{p}\right)\left(\frac{d}{c}\right)+k \tag{41.14}$$

where the values for Z, N, p, d, and c have the same meaning as discussed in Section 41.1.8 and k is the factor to correct for side leakage.

This will depend upon the ratio of the length to the diameter of the bearing (l/d), for example, k=0.002 for l/d ratios between 0.75 and 2.8.

41.1.16 Sommerfeld Number

In the design of fluid bearings, the Sommerfeld number (S) is a dimensionless quantity used extensively in hydrodynamic lubrication. It is also referred to as a bearing characteristic number. The Sommerfeld number is important in lubrication analysis because it contains all the variables normally specified by the designer.

$$\text{Sommerfield number 'S'} = \frac{ZN}{p}\left(\frac{d}{c}\right)^2 = 14.3 \times 10^6 \tag{41.15}$$

where

S=Sommerfeld Number —
d=Shaft diameter (mm)
c=Radial clearance (mm)
Z=Absolute viscosity of lubricant (kg/m^2)
N=Speed of rotating shaft (rev/min)
p=Load per unit of projected bearing area (N/mm^2)

41.1.17 Design Procedure

There are two basic methods of designing fluid film bearings:

- Based on dimensional analysis attributed to M.D. Hersey, based on the analysis by the McKee Brothers
- This method developed by Raimondi and Boyd (1958) based on hydrodynamic theory

41.1.17.1 Hersey Method of Fluid Film Bearing Design

This method is based on the dimensional analysis applied to an infinitely long bearing. The analysis incorporates a side flow correction factor developed from experiments undertaken by the McKee Brothers.

The McKee equation for coefficient of friction, for a full bearing is given by

$$\mu = K_1 \frac{KN}{p} \frac{d}{c} + K_2 \tag{41.16}$$

where

p=Pressure on bearing (projected area) (P/ld)
l=Length of the bearing
d=Diameter of the journal
P=Load acting on bearing
N=Speed of the journal
K=Absolute viscosity of the lubricant
c=Difference between bearing L/dia' and journal diameter
K_1=Constant
K_2=Side flow factor=0.002 for (l/d) 0.75–2.8

The constant K_1 is dependent on the system of units employed, for example,

$$K_1 = \frac{473}{10^{10}} \tag{41.17}$$

when ν is in centipoise, p is in lb/in.^{2}, and d and c are in inches.

The steps to the procedure should consider the following:

The basic design parameters are provided by the designer from the operating conditions:

- Bearing load (P)
- Journal diameter (d)
- Journal speed (N)

Depending upon the type of application, selected design parameters may be obtained from a design handbook, typically:

l/d ratio
Bearing pressure (p)
c/d ratio
Type of lubricant and an operating temperature

The heat generation in the bearing is given by

$$H_g = \mu P v$$

where v is the rubbing velocity.

The heat dissipation is given by

$$H_d = \lambda A(t_b - t_a) \tag{41.18}$$

where

A=Projected bearing area
λ=Heat dissipation coefficient
t_b=Bearing surface temperature
t_a =Ambient temperature

The next steps are as follows:
The value of KN/P should be within the design zone.

- Heat generation and heat dissipation is calculated to check for thermal equilibrium.
- Iteration with the selected parameters are required if thermal equilibrium is not established.
- Provision for external cooling is required if stable temperatures are to be achieved.

The method described here is relatively old but still applicable.

41.1.17.2 Method Developed by Raimondi and Boyd

This method is based on the hydrodynamic theory. The Reynolds equation 41.1 has no general solution. Therefore, assuming no side flow, Sommerfeld proposed a solution and defined a parameter, known as the Sommerfeld number, given by

$$\frac{d}{c}\mu = \phi\left[\left(\frac{d}{c}\right)^2 \frac{KN}{p}\right] \tag{41.19}$$

The Summerfeld number is important to designers because it includes design parameters, bearing dimensions d and c, friction μ, speed of rotation N, and bearing pressure p. What it does not include is the bearing arc. Therefore, the functional relationship can be obtained for bearings having different arcs, say 360°, 60°, and so on.

Raimondi and Boyd (1958 provided a methodology for computer-aided solution of Reynolds equation using an iterative technique. For l/d ratios of 1, 1:2, and 1:4, and for bearing angles of 360° to 60°, extensive design data is available.

Charts have been prepared by Raimondi and Boyd covering various design parameters, in dimensionless form and plotted with respect to the Sommerfeld number.

This detailed work is given in "A Solution of Finite Journal Bearing and Its Application to Analysis and Design Part 1, II and III," Trans ASLE, Vol. 1 No. 1, *Lubrication Science and Technology*, Pergamon, New York, 1958, pp 159–202.

The design parameters which are given by Raimondi and Boyd are as follows:
Design parameters:

h_o/c	Minimum film thickness
$(d/c)\mu$	Coefficient of friction
$Q/(dcNl)$	Flow
Qs/Q	Flow ratio
p/p_{max}	Pressure ratio
θp_o (°)	Terminating position of film
θh_o (°)	Minimum film thickness position

The above design parameters are defined in Figure 41.22. The pressure profile shown is only for the positive part of the bearing where the converging zone is present. The negative part has not been shown because it serves no useful purpose in the design.

41.2 Rolling Contact Bearings

Rolling bearings use rolling elements (balls or rollers) to reduce friction between fixed and moving surfaces. These bearings are able to accommodate high axial thrust and moment loads, depending on the design of the bearing and how they are mounted. The rolling elements are constrained between an inner race and outer race. Generally, a separator (cage) keep the rolling elements equally spaced apart so that they do not rub against each other and result in galling or skidding.

41.2.1 Types of Rolling Contact Bearings

There are a wide range of types and sizes of rolling-element bearings that meet the majority of requirements. They range from subminiature bearings for instruments, to giant bearings used in the bases of large cranes.

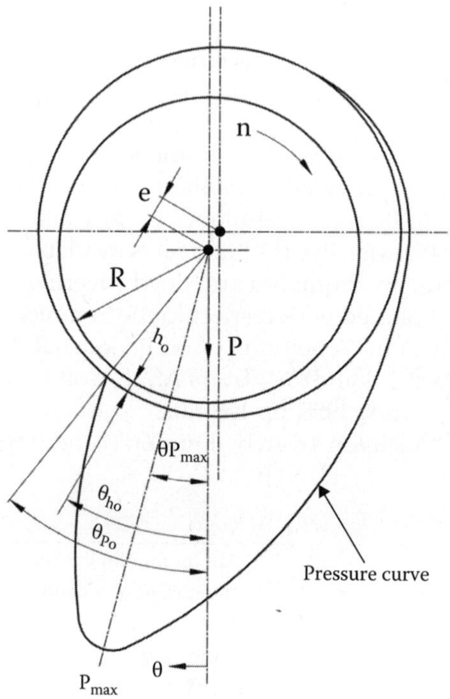

FIGURE 41.22
Nomenclature for a journal bearing.

41.2.1.1 Deep-Grooved Ball

These are a low-cost standard geometry bearing with the balls retained in an annulus between the outer and inner race/rings. Their advantage is that they have a low starting friction. These bearings are also supplied with a range of seals/shields and filled with grease. Rolling bearings eliminate the need for sleeves to protect the shaft from wear or damage. A relatively large radial space is needed, but quite small axial space is required. These bearings can provide both axial and radial location of shafts in housings (Figure 41.23).

41.2.1.2 Cylindrical Rolling

These bearings operate on the same principle as the ball bearing, but the balls are replaced with cylindrical rolling elements. This type of bearing is able to take much higher radial loads. They are not able to resist any axial loads. If axial loads are present, then an alternative bearing such as a thrust bearing is required. These bearings require separate lubrication.

Caution: Cylindrical roller bearings have no tolerance to angular deflection of shafts (Figure 41.24).

41.2.1.3 Double-Row Spherical

These bearings have a spherical contoured bore and are able to accept large angular displacements of shafts. They have a large load capacity. As with the cylindrical roller bearings, they require lubrication (Figure 41.25).

41.2.1.4 Taper (Rolling)

Taper roller bearings have similar properties to cylindrical bearings but are able to accept a wider range of loads. They have applications in machine tool work spindles on lathes where they can accept the cutting forces. They require a good flow of lubrication (Figure 41.26).

FIGURE 41.23
Deep-grooved ball bearing.

FIGURE 41.24
Cylindrical roller bearing.

FIGURE 41.25
Double-row spherical bearing.

FIGURE 41.26
Taper roller bearing.

41.2.1.5 Needle (Rolling)

These are rolling bearings based on small diameter needle bearings. A very small radial space is required and they have a high radial load capacity. They are subject to lower speed limits relative to ball bearings and they require lubrication. These bearings are able to run directly on a suitably hardened shaft eliminating the need for an inner race (Figure 41.27).

FIGURE 41.27
Needle roller bearing.

41.2.2 Rolling Contact Bearing Terminology

Some of the important terminologies that apply to the selection of rolling contact bearings are described in the following sections.

41.2.2.1 Bearing Rating Life Calculations

The *rating life* is the bearing life calculated for 90% reliability. This is duration that a group of apparently identical bearings will complete or exceed before the formation of surface damage or spalling. The basic formula for calculating bearing L_{10} rating life is

$$L_{10} = \frac{(c/P)^e \times 10^6}{N \times 60} \tag{41.20}$$

where

c = Dynamic capacity (N)
P = Equivalent bearing load (N)
N = Bearing rotation speed (rev/min)
e = 3.0 for ball bearings
10/3 for roller bearings

41.2.2.2 Equivalent Radial Load

The load rating of a bearing is given for radial loads only. Therefore, if a bearing is subjected to both a axial and radial load, then the equivalent radial load is to be calculated using the following expressions:

$$P_e = V P_r, \text{ or}$$

$$P_e = XV P_r + YP_a \tag{41.21}$$

where

P_e = Equivalent radial load
P_r = Given radial load
P_a = Given axial load
V = Rotation factor

1.0, inner race rotating
1.0, for self-aligning bearings when inner race is stationary
1.2, outer race rotating

X = A radial factor
Y = An axial factor

The values for radial load factor X and axial load factor Y for a dynamically loaded bearing may be obtained from Table 41.2.

The factor C_o is obtained from the bearing manufactures catalogues.

TABLE 41.2

Values of X and Y for Dynamically Loaded Bearings

		$\frac{P_a}{P_r} \leq e$		$\frac{P_a}{P_r} > e$		
Type of Bearing	**Specifications**	**X**	**Y**	**X**	**Y**	**e**
Deep-groove ball bearing	$\frac{P_a}{C_o} = 0.025$				2.0	0.22
	= 0.04	1.0	0	0.56	1.8	0.24
	= 0.07				1.6	0.27
	= 0.13				1.4	0.31
	= 0.25				1.2	0.37
	= 0.50				1.0	0.44
Angular contact ball bearings	Single row	1.0	0	0.35	0.57	1.14
	Two rows in tandem		0	0.35	0.57	1.14
	Two rows back to back		0.55	0.57	0.93	1.14
	Double row		0.73	0.62	1.17	0.86
Spherical roller bearings	For bores:	1.0		0.67		
	25–35 mm		2.1		3.1	0.32
	40–45 mm		2.5		3.7	0.27
	50–100 mm		2.9		4.4	0.23
	100–200 mm		2.6		3.9	0.26
Taper roller bearings	For bores:	1.0	0	0.4		
	30–40 mm				1.60	0.37
	45–110 mm				1.45	0.44
	120–150 mm				1.35	0.41

41.2.3 The Bearing Selection Procedure

Given the shaft diameter together with the magnitude of the radial and axial load, a suitable type of bearing can be selected from the bearing manufacturer catalogue. The choice of bearing will be either a ball bearing or roller bearing, dependent on the choice of bearings available. It is not unusual for a number of alternative manufacturer catalogues to be reviewed to determine the availability of bearings to meet the design requirements. The equivalent radial load is to be determined using Equation 41.24. If the bearing is a tapered roller bearing, then the manufacturer catalogue should be consulted for the equation that will give the equivalent radial load. The value of dynamic load rating C is calculated for the given bearing life and equivalent radial load. From the known value of C, a suitable bearing of the size that conforms to the required shaft size should be chosen. However, some amendment to the shaft size may be required to accommodate a suitable bearing.

42

Engineering Seals

42.0 Introduction

Engineering seals in mechanical engineering is of primary importance in preventing the loss of hydraulic fluid from a fluid pressure system, preventing the ingress of foreign particles entering bearing systems and slideways. This chapter will demonstrate the more important aspects of sealing between contacting surfaces.

42.1 Classification of Engineering Seals

There are essentially two types of engineering seals:

- Static seals
- Dynamic seals

Static seals cover mechanical joints that remain stationary and covers packing or gaskets. In certain situations, dynamic seals are sometimes used in a static application, as an example "O"-rings and labyrinth seals are used in static applications where a rotating shaft exits an outer cover and is fitted with a sealing flange, as depicted in Figure 42.1.

Dynamic seals will be found where the shaft is rotating in a bearing and passing through the external face of a housing, and the shaft extends through a sealing flange as in Figure 42.2.

This chapter considers other sealing methods.

42.2 O-Rings

An *O-ring* is a simple form of a seal. It closes off a gap between two surfaces so that liquids or gases cannot pass through in any direction. The shape of an O-ring is like a doughnut—or torus, in its correct mathematical description—from which it derives its name.

O-rings have a number of advantages, including the following:

1. Low cost
2. Suitable for static and dynamic applications
3. Space efficient

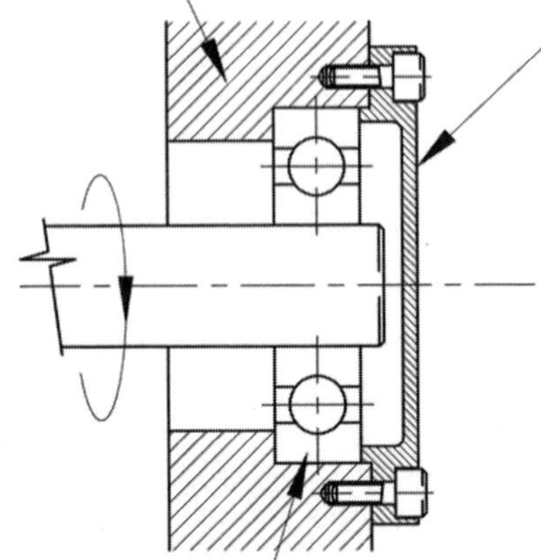

FIGURE 42.1
Static seal.

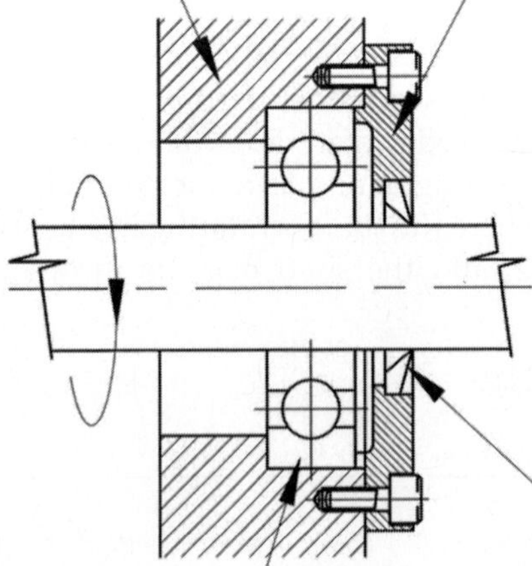

FIGURE 42.2
Dynamic seal.

4. Seals in both directions
5. Fluid pressure assists sealing
6. Suitable for all fluids using appropriate elastomers

Typical Engineering Applications

O-rings are used in a variety of applications covering domestic and industrial products, including hydraulic and pneumatic piping systems together with hydraulic valves, vehicle transmissions and braking systems, aerospace systems, and special-purpose electromechanical devices.

Definition

A toroidal-shaped seal manufactured from a resilient material that prevents ingress or egress of fluids under static or dynamic situations.

Notes on Selection and Use

O-rings are one of the most widely used seals due to their relatively low cost, ease of use, and high performance in service. O-rings rely on a confined housing that elastically deforms the cross section of the ring, thereby creating an initial seal, and then, as fluid pressure is applied, the ring deforms against the housing, thereby increasing its effectiveness. They can be guaranteed to be completely leakproof when used as a static seal up to extremely high pressures, and are satisfactorily leakproof under dynamic conditions. Commonly available O-rings in various materials can operate up to 10 MPa in standard form and up to 35 MPa when backup (anti-extrusion) washers are incorporated to prevent the extrusion of the O-ring through the clearance in the housing. When used in dynamic environments, for instance when used as a rotary seal, peripheral speeds of up to 7.5 m/s can easily be accommodated. If clearances are large, higher speeds could cause extrusion and localized *nibbling* damage to the ring surface. It should be noted that as dynamic seals, O-rings are best suited for small-diameter, short-stroke, medium-pressure applications, but if there is very slow reciprocating motion, the higher friction can cause premature failure. As a general rule, seals should be arranged so that the direction of pressure and any rubbing friction should be in opposition.

Although O-rings of suitable materials can operate over a temperature range of −40°CC to + 250°C, contraction or expansion of metal housing can make significant changes to the original machined sizes, and corrections may be necessary to avoid the seal extruding. In static housing, for instance a pressure vessel end cap, stretching of assembly bolts may also change ring clearances.

Groove finish is important because this will affect the wear life of the seal, although for some static uses a rougher finish than that needed for compression gaskets can be tolerated as long as all the edges are de-burred. Great care is needed in handling and assembling the O-ring to ensure that no premature damage occurs. Protective cones or tubes are often used, especially if the O-ring has to pass over a threaded part to be fitted into its housing. Lubricants are recommended for rings and their housings, and greases with a wide temperature ranges with a wide temperature range are usually specified for this purpose.

A Note of Caution

Always wear protective gloves when handling O-rings that have perished due to overheating. In some circumstances, highly toxic chemicals may be released from the elastomer.

In the UK, there are two current standards covering the design and manufacture of O-rings.

BS 1806:1989

This standard specifies the dimensions for toroidal sealing rings and their housings. Section sizes according to the standard covers 1.02, 1.27, 1.52, 1.78, 2.62, 3.53, and 6.99 mm.

BS 4518:1982

This standard covers the specification for metric dimensions of toroidal sealing rings and their housings. Section sizes according to the standard include 1.6, 2.4, 3.0, 4.1, 5.7, and 8.4 mm. Tables 42.1 through 42.8 specify the standard dimensions for this range of O-rings.

TABLE 42.1

Dimensions for Dynamic, Pneumatic, and Hydraulic Applications

O-ring size (mm)	2.4	3.0	5.7	8.4
Width "E"	3.2/3.4	4.0/4.2	7.5/7.7	11.0/11.2
Radial depth "F" "pneumatic"	2.13/2.20	2.70/2.77	5.22/5.38	7.57/7.96
Radial depth "hydraulic"	1.97/2.09	2.50/2.65	4.95/5.18	7.50/7.96
Maximum clearance "G"	0.14	0.15	0.18	0.2
Lead in "C"	0.6	0.7	1.0	1.5
Maximum rad "r"	0.5	1.0	1.0	1.0

TABLE 42.2

Dimensions for Static Applications

O-ring size (mm)	1.6	2.4	3.0	5.7	8.4
Width "E"	2.3/2.5	3.1/3.3	3.7/3.9	6.4/6.8	9.0/9.2
Radial depth "F"	1.18/1.25	1.84/1.97	2.35/2.50	4.70/4.95	7.20/7.50
Maximum clearance "G"	0.12	0.14	0.15	0.18	0.2
Lead in "C"	0.6	0.7	0.8	1.2	1.5
Maximum rad "r"	0.5	0.5	1.0	1.0	1.0

TABLE 42.3

Dimensions for Face Seal Static Applications

O-ring size (mm)	1.6	2.4	3.0	5.7	8.4
Depth "H"	1.2/1.3	1.7/1.8	2.2/2.3	4.4/4.5	6.6/6.7
Maximum rad "r"	0.2	0.5	1.0	1.0	1.0
Maximum rad "T"	0.8	1.3	2.0	3.0	4.0
Chamfer "M"	2.20/2.32	3.30.3.42	4.20/4.32	7.80/7.92	11.50/11.62
Minimum length "L"	4.0	5.0	6.0	10.0	14.0

TABLE 42.4

O-Ring 1.6 mm Diameter

1.6mm± 0.08mm Diameter O-Ring						
	Diameters		**Flange Groove Diameters**			
			Internal Pressure		**External Pressure**	
Ref′ No.	**Shaft Diameter**	**Hub (Cylinder) Diameter**	**V max**	**W (H11)**	**V (h11)**	**W min**
0031-16	3.5	6	1	6.3	3.5	7.5
0041-16	4.5	7	2.3	7.3	4.5	8.5
0051-16	5.5	8	3.3	8.3	5.5	9.5
0061-16	6.5	9	4.3	9.3	6.5	10.5
0071-16	7.5	10	5.8	10.3	7.5	11.5
0081-16	8.5	11	6.8	11.3	8.5	12.5
0091-16	9.5	12	7.8	12.3	9.5	13.5
0101-16	10.5	13	8.8	13.3	10.5	14.5
0111-16	11.5	14	9.8	14.3	11.5	15.5
0121-16	12.5	15	10.8	15.3	12.5	16.5
0131-16	13.5	16	11.8	16.3	13.5	17.5
0141-16	14.5	17	12.8	17.3	14.5	18.5
0151-16	15.5	18	14	18.3	15.5	19.5
0161-16	16.5	19	15	19.3	16.5	20.5
0171-16	17.5	20	16	20.3	17.5	21.5
0181-16	18.5	21	17	21.3	18.5	22.5
0191-16	19.5	22	18	22.3	19.5	23.5
0221-16	22.5	25	21	25.3	22.5	26.5
0251-16	25.5	28	24	28.3	25.5	29.5
0271-16	17.5	30	26	30.3	27.5	31.5
0291-16	29.5	32	28	32.3	29.5	33.5
0321-16	32.5	35	31	35.3	32.5	36.5
0351-16	35.5	38	34	38.3	35.5	39.5
0371-16	37.5	40	36	40.3	37.5	41.5

TABLE 42.5

O-Ring 2.4mm Diameter

2.4 ± 0.08mm Diameter O-Ring						
	Diameters		**Flange Groove Diameters**			
			Internal Pressure		**External Pressure**	
Reference No.	**Shaft Diameter**	**Hub (Cylinder) Diameter**	**V max**	**W (H11)**	**V (h11)**	**W min**
0036-24	4	8	—	8.4	4	10
0046-24	5	9	1	9.4	5	11
0056-24	6	10	2.5	10.4	6	12
0066-24	7	11	4	11.4	7	13
0076-24	8	12	5	12.4	8	14

(Continued)

TABLE 42.5 (*Continued*)

O-Ring 2.4 mm Diameter

2.4 ± 0.08 mm Diameter O-Ring						
	Diameters		Flange Groove Diameters			
			Internal Pressure		External Pressure	
Reference No.	Shaft Diameter	Hub (Cylinder) Diameter	V max	W (H11)	V (h11)	W min
0086-24	9	13	6.4	13.4	9	15
0096-24	10	14	7.4	14.4	10	16
0106-24	11	15	8.4	15.4	11	17
0116-24	12	16	9.5	16.4	12	18
0126-24	13	17	10.5	17.4	13	19
0136-24	14	18	11.5	18.4	14	20
0146-24	15	19	12.5	19.4	15	21
0156-24	16	20	13.5	20.4	16	22
0166-24	17	21	14.5	21.4	17	23
0176-24	18	22	15.5	22.4	18	24
0186-24	19	23	16.5	23.4	19	25
0196-24	20	24	17.5	24.4	20	26
0206-24	21	25	18.5	25.4	21	27
0216-24	22	26	19.5	26.4	22	28
0246-24	25	29	22.5	29.4	25	31
0276-24	28	32	25.5	32.4	28	34
0296-24	30	34	27.5	34.4	30	36
0316-24	32	36	29.5	36.4	32	38
0346-24	35	39	32.5	39.4	35	41
0356-24	36	40	33.5	40.4	36	42
0376-24	38	42	35.5	42.4	38	44
0396-24	40	44	37.5	44.4	40	46
0416-24	42	46	39.5	46.4	42	48
0446-24	45	49	42.5	49.4	45	51
0456-24	46	50	43.5	50.4	46	52
0476-24	48	52	45.5	52.4	48	54
0496-24	50	54	47.5	54.4	50	56
0516-24	52	56	49.5	56.4	52	58
0546-24	55	59	52.5	59.4	55	61
0556-24	56	60	53.5	60.4	56	62
0576-24	58	62	55.5	62.4	58	64
0586-24	59	63	56.5	63.4	59	65
0596-24	60	64	57.5	64.4	60	66
0616-24	62	66	59.5	66.4	62	68
0626-24	63	67	60.5	67.4	63	69
0646-24	65	69	6205	69.4	65	71
0676-24	68	72	65.5	72.4	68	74
0696-24	70	74	67.5	74.4	70	76

TABLE 42.6

O-Ring 3.0 mm Diameter

3.0 ± 0.10 mm Diameter "O"-Ring

	Diameters		Flange Groove Diameters			
			Internal Pressure		External Pressure	
Reference No.	**Shaft Diameter**	**Hub (Cylinder) Diameter**	**V max**	**W (H11)**	**V (h11)**	**W min**
0195-30	20	25	17	25	20	28
0215-30	22	27	19	27	22	30
0225-30	23	28	20	28	23	31
0245-30	25	30	22	30	25	33
0255-30	26	31	23	31	26	34
0265-30	27	32	24	32	27	35
0275-30	28	33	25	33	28	36
0295-30	30	35	27	35	30	38
0315-30	32	37	29	37	32	40
0325-30	33	38	30	38	33	41
0345-30	35	40	32	40	35	43
0355-30	36	41	33	41	36	44
0365-30	37	42	34	42	37	45
0375-30	38	43	35	43	38	46
0395-30	40	45	37	45	40	48
0415-30	42	47	39	47	42	50
0425-30	43	48	40	48	43	51
0445-30	45	50	42	50	45	53
0495-30	50	55	47	55	50	58
0545-30	55	60	52	60	55	63
0555-30	56	61	53	61	56	64
0575-30	58	63	55	63	58	66
0595-30	60	65	57	65	60	68
0625-30	63	68	60	68	63	71
0645-30	65	70	62	70	65	73
0695-30	70	75	67	75	70	78
0745-30	75	80	72	80	75	83
0795-30	80	85	77	85	80	88
0845-30	85	90	82	90	85	93
0895-30	90	95	87	95	90	98
0945-30	95	100	92	100	95	103
0995-30	100	105	97	105	100	108
1045-30	105	110	102	110	105	113
1095-30	110	115	107	115	110	118
1145-30	115	120	112	120	115	123
1195-30	120	125	117	125	120	128
1245-30	125	130	122	130	125	133

(*Continued*)

TABLE 42.6 (*Continued*)

O-Ring 3.0 mm Diameter

3.0 ± 0.10 mm Diameter "O"-Ring						
	Diameters		Flange Groove Diameters			
			Internal Pressure		External Pressure	
Reference No.	Shaft Diameter	Hub (Cylinder) Diameter	V max	W (H11)	V (h11)	W min
1295-30	130	135	127	135	130	138
1345-30	135	140	132	140	135	143
1395-30	140	145	137	145	140	148
1445-30	145	150	142	150	145	153
1495-30	150	155	147	156	150	158
1545-30	155	160	152	160	155	163
1595-30	160	165	157	165	160	168
1645-30	165	170	162	170	165	173
1695-30	170	175	167	175	170	178
1745-30	175	180	172	180	175	183
1795-30	180	185	177	185	180	188
1845-30	185	190	182	190	185	193
1895-30	190	195	187	195	190	198
1945-30	195	200	192	200	195	203
1995-30	200	205	197	205	200	208
2095-30	210	215	207	215	210	218
2195-30	220	225	217	225	220	228
2295-30	230	235	227	235	230	238
2395-30	240	245	237	245	240	248
2445-30	245	250	242	250	245	253
2495-30	250	255	247	255	250	258

TABLE 42.7

O-Ring 5.7 mm Diameter

5.7 ± 0.12 mm Diameter O-Ring						
	Diameters		Flange Groove Diameters			
			Internal Pressure		External Pressure	
Reference No.	Shaft Diameter	Hub (Cylinder) Diameter	V max	W (H11)	V (h11)	W min
0443-57	45	55	41	55	45	59
0453-57	46	56	42	56	46	60
0493-57	50	60	46	60	50	64
0523-57	53	63	49	63	53	67
0543-57	55	65	51	65	55	69
0553-57	56	66	52	66	56	70
0593-57	60	70	56	70	60	74
0623-57	63	73	59	73	63	77
0643-57	65	75	61	75	65	79

(Continued)

TABLE 42.7 (*Continued*)

O-Ring 5.7 mm Diameter

5.7 ± 0.12 mm Diameter O-Ring						
	Diameters		Flange Groove Diameters			
			Internal Pressure		External Pressure	
Reference No.	Shaft Diameter	Hub (Cylinder) Diameter	V max	W (H11)	V (h11)	W min
0693-57	70	80	66	80	70	84
0743-57	75	85	71	85	75	89
0793-57	80	90	76	90	80	94
0843-57	85	95	81	95	85	99
0893-57	90	100	86	100	90	104
0943-57	95	105	91	105	95	109
0993-57	100	110	96	110	100	114
1043-57	105	115	101	115	105	119
1093-57	110	120	106	120	110	124
1143-57	115	125	111	125	115	129
1193-57	120	130	116	130	120	134
1243-57	125	135	121	135	125	139
1293-57	130	140	126	140	130	144
1343-57	135	145	131	145	135	149
1393-57	140	150	136	150	140	154
1443-57	145	155	141	155	145	159
1493-57	150	160	146	160	150	164
1543-57	155	165	151	165	155	169
1593-57	160	170	156	170	160	174
1643-57	165	175	161	175	165	179
1693-57	170	180	166	180	170	184
1743-57	175	185	171	185	175	189
1793-57	180	190	176	190	180	194
1843-57	185	195	181	195	185	199
1893-57	190	200	186	199	190	204
1943-57	195	205	191	204	195	209
1993-57	200	210	195	209	200	214
2093-57	210	220	205	219	210	224
2193-57	220	230	215	229	220	234
2293-57	230	240	225	239	230	244
2393-57	240	250	235	249	240	254
2493-57	250	260	245	259	250	264
2593-57	260	270	255	269	261	275
2693-57	270	280	265	279	271	285
2793-57	280	290	275	289	281	295
2893-57	290	300	285	290	291	305
2993-57	300	310	295	309	301	315
3093-57	310	320	305	319	311	325
3193-57	320	330	315	329	321	335
3393-57	340	350	335	349	341	355

(*Continued*)

TABLE 42.7 (*Continued*)

O-Ring 5.7 mm Diameter

5.7 ± 0.12 mm Diameter O-Ring						
	Diameters		Flange Groove Diameters			
			Internal Pressure		External Pressure	
Reference No.	Shaft Diameter	Hub (Cylinder) Diameter	V max	W (H11)	V (h11)	W min
3593-57	360	370	355	369	361	375
3793-57	380	390	375	389	381	395
3893-57	390	400	385	399	391	405
3993-57	400	410	395	409	401	415
4193-57	420	430	415	429	422	436
4393-57	440	460	435	449	442	456
4593-57	460	470	455	469	462	476
4793-57	480	490	475	489	482	496
4893-57	490	500	485	499	492	506
4993-57	500	510	495	509	502	516

TABLE 42.8

O-Ring 8.4 mm Diameter

8.4 ± 0.15 mm Diameter O-Ring						
	Diameters		Flange Groove Diameters			
			Internal Pressure		External Pressure	
Reference No.	Shaft Diameter	Hub (Cylinder) Diameter	V max	W (H11)	V (h11)	W min
1441-84	145	160	140	160	145	165
1491-84	150	165	145	165	150	170
1541-84	155	170	150	170	155	175
1591-84	160	175	156	175	160	180
1641-84	165	180	160	180	165	185
1691-84	170	185	165	185	170	190
1741-84	175	190	170	190	175	195
1791-84	180	195	175	195	180	200
1841-84	185	200	180	200	185	205
1891-84	190	205	185	205	190	210
1941-84	195	210	190	210	195	215
1991-84	200	215	195	215	200	220
2041-84	205	220	200	220	205	225
2091-84	210	225	205	225	210	230
2191-84	220	235	215	235	220	240
2291-84	230	245	225	245	230	250
2341-84	235	250	230	250	235	255
2391-84	240	255	235	255	240	260
2491-84	250	265	245	265	250	270

42.2.1 O-Ring Reference Number

In the standard BS 4518, the O-ring reference number comprises its inside diameter (mm—four digits—one decimal place) and O-ring section (mm—two digits—one decimal place); the decimal point is excluded.

For example, a reference number 0545-30 is an O-ring having an inside diameter of 54.5 mm and a section of 3.0 mm.

42.2.2 Surface Texture

The surface texture of the surfaces that is in direct contact with the O-ring should not exceed the following values:

1.	For nonsealing surfaces	1.6 μm CLA
2.	For static applications	0.8 μm CLA
3.	For dynamic applications	0.4 μm CLA

Notes

The sealing surfaces include rod, hole, and bottom of the groove.

In dynamic applications, it is important that the surface finish is greater than 0.2 μm, because a dry surface would result from the seal-wiping action.

42.2.3 BS 4518 Internal and External Groove Dimensions for Dynamic and Hydraulic Applications

Figures 42.3 and 42.4 give the nomenclature for internal and external groove, respectively.

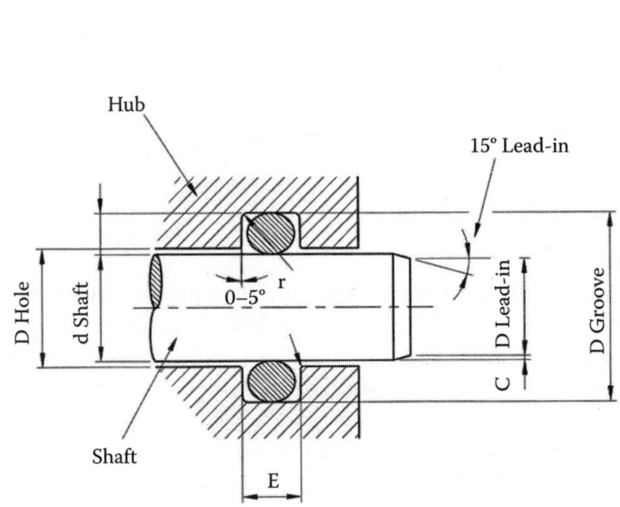

FIGURE 42.3
Internal groove details for an O-ring.

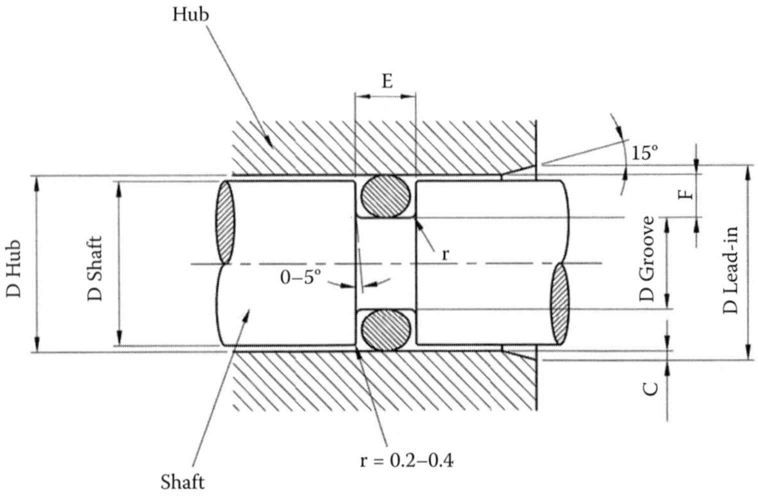

FIGURE 42.4
External groove details for an O-ring.

42.3 Rotary Seals

Rotary seals provide a solution of sealing between a rotating shaft and a fixed housing. There are four basic seals:

1. Labyrinth seals
2. Felt seals
3. Lip seals
4. Ferrofluid seals

42.3.1 Labyrinth Seals

The definition of a labyrinth is that of a complicated network of passages that prevents an easy passage between the entry and exit.

A *labyrinth seal* is a type of mechanical seal that provides a tortuous path to help prevent leakage. An example of such a seal is sometimes found within an *axle's bearing* to help prevent leakage of the oil lubricating the bearing. It is a noncontact seal and therefore has zero friction. It has virtually unlimited life and is unaffected by temperature and can be used for high shaft-rotating speeds.

This type of seal does not work efficiently if the shaft is not rotating and also not effective if there is a high-pressure differential across the seal.

Sealing is dependent on the form of the labyrinth gap, together with the length of the leakage path. Rings on the shaft and grooves in the housing provide the basic labyrinth. It is considered that at least three grooves should be used to provide adequate sealing. Clearances can vary between 0.25 mm and 1.0 mm, depending on the speed of rotation and temperature the seal is operating in.

Sealing on gas turbines utilizes a screw thread on either the compressor or turbine shaft within a close-fitting plain bore. The lead on the screw thread draws any lubricant back into the casing as the shaft rotates.

Figures 42.5a and b and 42.6 depict three types of labyrinth seals.

Figure 42.5a shows an axial labyrinth seal. In this design, the seal has a labyrinth on the axial side of the housing.

Figure 42.5b shows a radial labyrinth seal where the passageway is attached to the radial side of the housing. This design requires a split housing and offers a better sealing efficiency than do axial labyrinth seals.

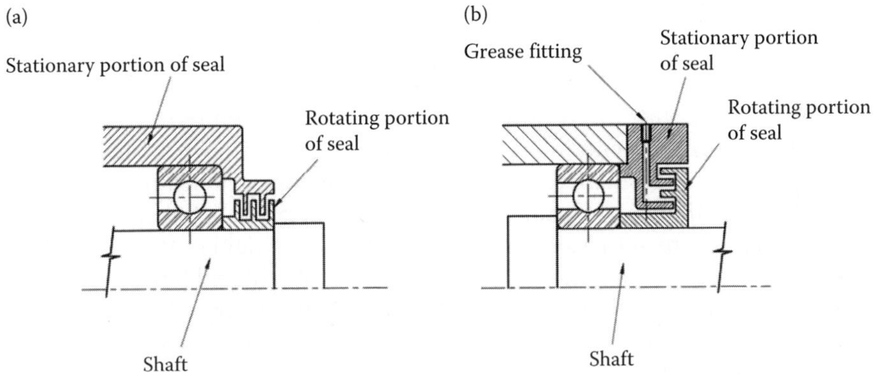

FIGURE 42.5
(a) Axial labyrinth seal and (b) radial labyrinth seal.

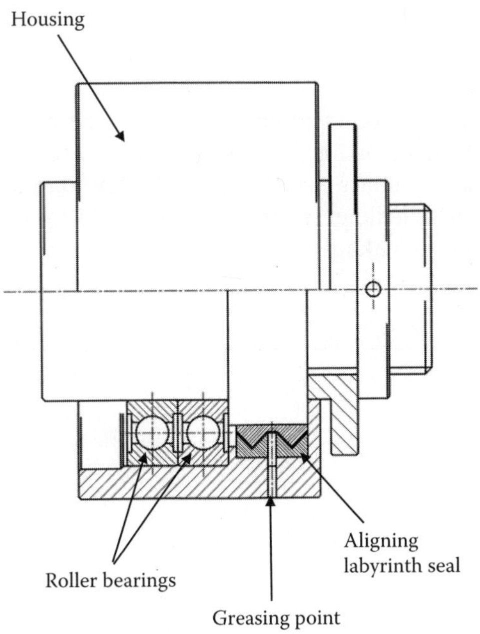

FIGURE 42.6
Aligning labyrinth seal.

Figure 42.6 illustrates an aligning labyrinth seal. In this design, the labyrinth passageway is slanted and has sufficient clearance to prevent contact between the housing projections and the shaft.

The sealing efficiency can be further improved by filling the labyrinth passageway with grease having a consistency grade between 150 and 200.

42.3.2 Felt Seals

Felt seals are commonly used near ball and roller bearings as lubricant retainers and preventing dirt or dust from entering the bearing.

Felt seals have long been used for sealing due to a number of favorable properties, such as wicking where the capillary properties of the seal ensures lubrication after long idle periods, and oil absorption in that the storage capacity is largely a function of density. However, approximately 78% of the volume of a felt seal serves as oil storage. Filtration: wool felt seals in a dry state are highly effective as filters in removing particles of 0.7 μm and larger. When saturated with lubricant, they trap and retain even smaller particles.

Felt seals have a good resilience, as they maintain a constant sealing pressure despite any wear, end play, minor misalignment, or deformity of metallic assemblies.

Felt seals are usually pre-saturated with lubricant having a higher viscosity, and this offers the bearing positive protection. If the seal runs dry, then it will still tend to protect and polish the shaft rather than cause any damage. Felt seals are not recommended for applications where the oils have an extremely low viscosity or where the lubrication is pressurized.

When the seal is correctly installed, the seal will be effective over a wide range of operating conditions and a wide range of speeds. Normal maximum operating speed is 10 m/s but may be as high as 20 m/s if the rubbing surfaces are highly polished and lubricant is always available.

Figure 42.7 shows a typical felt seal application.

42.3.3 Lip Seals

A *lip seal* assembly is comprised of a rubbing elastomeric ring seal element within a metal cup and pre-tensioned using a spring. The seal friction is reduced as an oil film is generated between the seal and the shaft (see Figure 42.8). Any damage to the shaft where the seal makes contact will cause leakage, as the optimum oil film thickness will be exceeded locally. Shaft finish is vital, as leakage will occur if there is any irregular surface present.

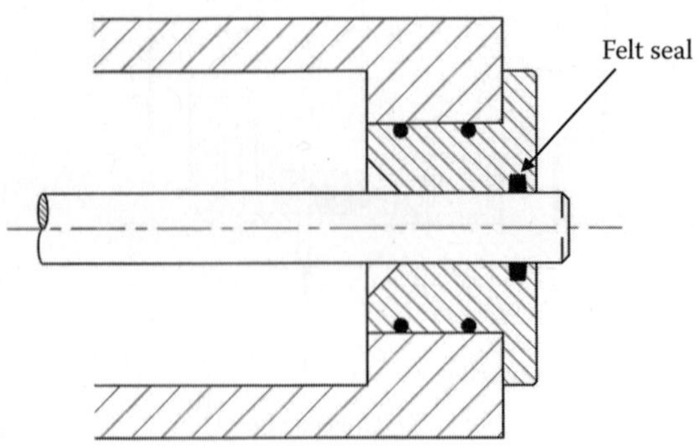

FIGURE 42.7
Typical felt seal application.

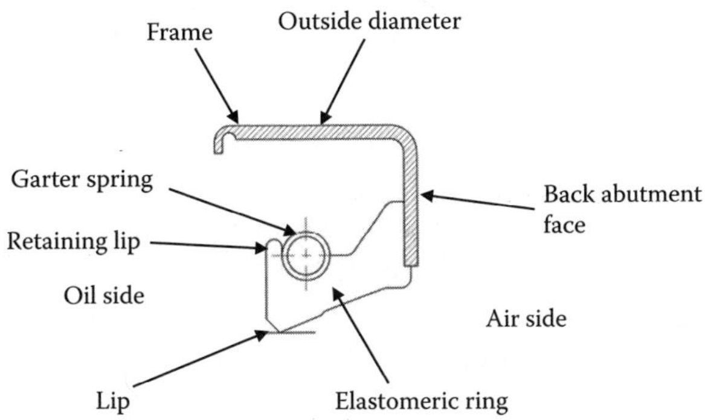

FIGURE 42.8
Lip seal.

42.3.4 Ferrofluid Seals

A *ferrofluidic seal* takes advantage of the response of a magnetic fluid to an applied magnetic field.

The ferrofluid seal (FS) consists of at least four essential elements: ferrofluid, a permanent magnet, two pole pieces, and a magnetically permeable shaft. The ferrofluid occupies the annular gap between the pole pieces and the shaft at the place where the largest magnetic field intensity exists, and holds it firmly. The ferrofluid ring preserves the tightness with zero leakage.

When fluid is applied to this gap, it assumes the shape of a liquid O-ring and provides hermetic sealing, long life, virtually frictionless sealing, and smooth operation. This type of seal is non-contaminating, reliable, and is able to operate at high speeds and operate over a wide range of temperatures that may be extended by the use of cooling or heating circuits.

These seals must be maintained regularly, as the fluid properties deteriorate over time.

Figure 42.9 illustrates a typical FS.

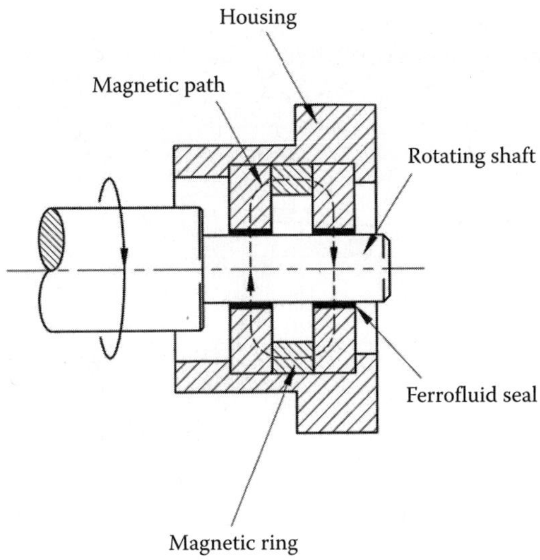

FIGURE 42.9
A typical ferrofluid seal.

42.4 Mechanical Seals

42.4.1 Introduction

Prior to examining how mechanical seals function, here is a review of previous methods used to form the seal between the shaft and the pump housing that can contain a pressurized fluid and withstand the friction caused by the rotating shaft.

One such method still widely used is *gland packing*. This is a braided ropelike material that is packed around the shaft—physically filling the gap between the shaft and the pump housing.

The principal problems found using the gland packing are as follows:

- The friction of the shaft rotating wears away at the packing over time, leading to increased leakage until the packing is adjusted or repacked.
- The friction of the shaft also means that packing also must be flushed with large volumes of water in order to keep it cool.
- Packing must press against the shaft in order to reduce leakage—the pump requires more power to overcome the turning resistance, thereby wasting energy.
- Because the packing must contact the shaft, it will eventually wear a groove in the shaft, resulting in additional cost in repairing or replacing the shaft.

Mechanical seals were designed to overcome the problems experienced using gland packing.

42.4.2 Main Elements of a Mechanical Seal

Mechanical seals are being used to replace packed gland and lip seals on fluid pumps (Figure 42.10). Pumps with mechanical seals are able to perform more efficiently and more reliably for extended periods of time.

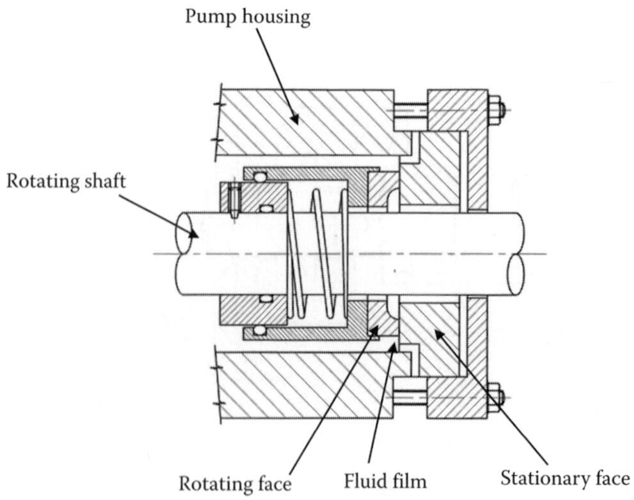

FIGURE 42.10
An example of a mechanical seal.

Mechanical seals are fitted to prevent pumped fluid from leaking along the drive shaft. The controlled leakage path is between two flat surfaces that are associated with the rotating shaft and the housing. This leakage path gap varies as the faces are subject to fluctuating external loads that tend to move the faces relative to each other.

The design of the seal housing must be redesigned to accommodate the mechanical seal and, unlike the packing gland seal arrangement, it does not provide any support to the shaft.

To enable the mechanical seal to perform over an extended period of time with low friction, the seal faces are generally hydrodynamically lubricated. The fluid film will need to carry a substantial load. When the load becomes too high for the film surface, contact will occur, with subsequent bearing failure taking place. The thickness of the lubricating film is in the order of 3 μm thick or less. This thickness is critical for the bearing function. Mechanical seals often have one face of a suitable solid lubricant, such that the seal can still function for a limited period of time without the fluid film.

42.4.3 Design Features

The following are the basic set of parts comprising a mechanical seal:

- A set of machined and lapped primary sealing faces—the close contact between these two flat mating surfaces that are perpendicular to the shaft—minimizes leakage. Dissimilar materials used for the sealing faces one hard and the other softer, in order to prevent adhesion of the two faces. One of the faces is usually a non-galling material such as *carbon graphite* and the other usually a relatively hard material, such as *silicon-carbide* or *ceramic*:
    1. One sealing face is held stationary in the housing.
    2. The other face is fixed, and rotates with the shaft.
- A set of secondary static seals, typically O-rings, wedges, and/or V-rings:
    1. One static seal, seals stationary component(s) to the housing.
    2. The other seal, seals the rotating component(s) to the shaft (it will normally move axially on the shaft or shaft sleeves).
- A spring member to maintain the faces in contact, such as a single spring, multiple springs, or metal bellows
- Other mechanical seal hardware, including shaft seals, gland rings, collars, compression rings, etc.

All these components must be compatible with the fluid being contained and the associated environment.

42.4.4 Assembly Options

There are a number of mechanical seal options:

- External seal: This design is installed on the outside of the stuffing box with the sealed pressure inside. This design provides a good access, allowing the seal components to be cleaned.

- Internal seal: Generally, mechanical seals are mounted inside the stuffing box with the sealed pressure outside the seal.
- Double seals: In this option, double mechanical seals are mounted in pairs and used for sealing hazardous, toxic, or abrasive fluids, and are often provided with clean flushing fluid between the seals. Double seals also provide an additional degree of safety where the pressure differentials are likely to reverse and there is a high risk of the sealing failing.

42.5 Packing

42.5.1 Introduction

Throughout the history of engineering, the solution to seal a valve shaft (rotary and axial movement) and process pump shafts (rotary movement) is to fit a packed gland between the housing and shaft. In modern times, the packed gland is being replaced by new solutions, including mechanical seals.

42.5.2 Pumps

New pump applications rarely use packed glands, mainly due to high pump maintenance requirements. Specific operating conditions require different types of packed gland seals. They require regular checks and maintenance adjustments. Proper lubrication of the gland packing requires a specific leakage rate. Manufactures recommendations are to be observed closely.

A service life between 1 and 2 years can be expected and when subject to favorable conditions this can be extended to several years. The service life can be drastically cut when the seal is exposed to extremely bad fluid conditions such as sediments, additives, overheating. Packed glands should preferably be used in conjunction with shaft sleeves in order to avoid any damage to the shaft by aggressive fluids or improper treatment of the packed gland.

42.5.3 Valves

Various types of valves use packed glands, including gate, globe, and ball valves. The packed gland offers a cost-effective option having the capability of sealing under a wide range of pressures and temperatures with a wide range of fluids when selecting appropriate packing materials.

Packing glands generally creep over a period of time, requiring regular inspection and tightening to maintain the seal. Optional seals such as O-rings are often used in conjunction with packing glands to reinforce the seal. The packing may be continuously spring loaded, thereby extending the maintenance period. Figure 42.11 shows a typical valve stem packing.

42.5.4 Properties

The properties of a packing gland system require good sealing and minimal maintenance, including the following:

- Compatibility with the working fluid at the operating temperature
- Ability to conform to the shaft due to the influence of the gland force

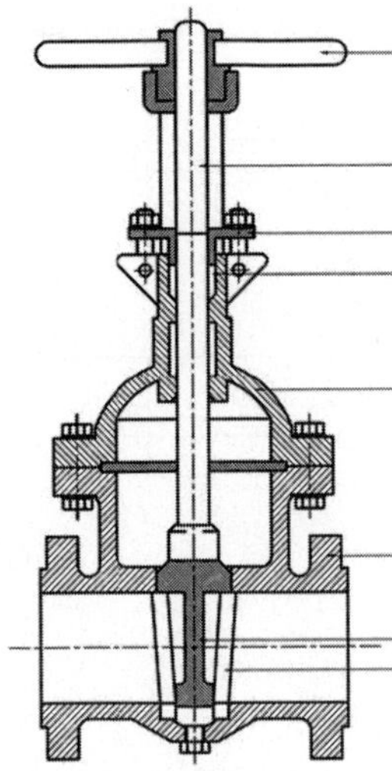

FIGURE 42.11
An example of a packing gland for a valve stem.

- Use of a lubricant that is insoluble and immiscible with the working fluid
- Nonabrasive to avoid damage to either the shaft or the housing
- Noncorrosive to avoid damage to both the shaft and the housing
- Able to retain its properties over time

42.5.5 Materials

There are a variety of packing materials available for seals operating at medium temperatures and pressures. These use "soft packing" such as the following:

- Plaited
- Braided
- Twisted
- Plastic

When the seals operate at higher temperatures and pressures, reinforcing wires can be added to the plaited and braided packing. For high temperatures, woven types of copper and aluminum packing are available.

42.5.6 The Lantern Ring and Its Purpose

A *lantern ring* is an annular ring with channels for the passage of water that is inserted between the gland packing (Figure 42.12). Fluid is fed from an external pipe to the lantern ring for lubrication and the cooling of the packing. It is used in applications were there are a number of packing rings within the stuffing box. The lantern rings distribute cooling water to all packing rings. It is also used where the fluid to be sealed is contaminated with abrasives or is corrosive and a clean fluid is required for lubrication. During installation, it is crucial that the holes in the lantern ring meet or are aligned with the holes in the pipes. It is important to inspect and align the lantern ring during each overhaul of the pumps.

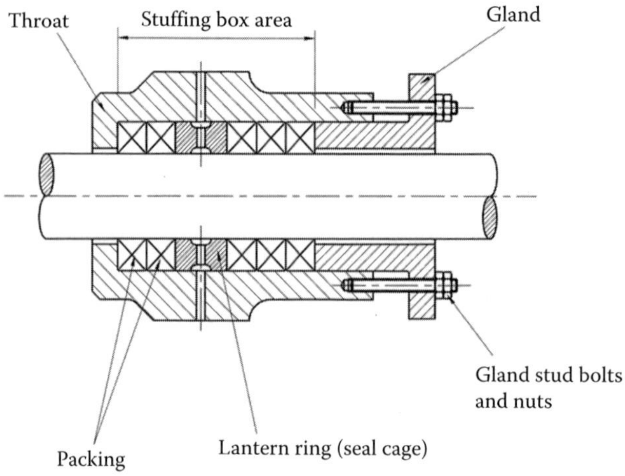

FIGURE 42.12
Example showing lantern rings.

42.5.7 Shaft Design

The shaft surfaces in contact with the seal should have a surface finish of 0.4 μm or better. The maximum operating clearances between the packing, housing, and shaft should be such to prevent extrusion of the packing material along the shaft. The minimum clearance should be sufficient to prevent the packing gland contacting the static housing and rotating shaft.

42.5.8 Temperature Range for Various Packing Materials

Table 42.9 identifies the range of temperatures various packing materials can accommodate and is provided for guidance only.

All the gland packing are a combination of the above-mentioned basic materials and great research is being carried out by the manufacturers to develop new and more efficient types of packing. The packing made of a combination of graphite, carbon, glass, and so on are used for harsh applications, whereas gland packing made of materials such as

TABLE 42.9
Operating Temperatures for a Range of Packing Materials

Packing Materials	Temperature Range (°C)
Jute, flax, hemp	0–60
Cotton	0–70
Rubberized cotton	0–80
PTFE	–250 to 220
Aramid	–200 to 280
Aluminum mesh	0–420
Wrapped metal foils—various types	0–450
Carbon graphite	–150 to 350
Graphite fiber	–200 to 600
Copper mesh	0–800
Aluminum silica filament/inconel reinforced	0–1200

PTFE, polytetrafluoroethylene.

TABLE 42.10

Recommended Surface Speeds for a Range of Packing Materials

Packing Materials	Surface Speed (m/s)
Copper mesh	0–0.5
Jute	0–4
Hemp and cotton	0–6
Flax	0–10
PTFE	0–12
Wrapped metal foils—various types	0–20
Aramid	0–22
Aluminum mesh	0–25
Carbon graphite	0–30
Graphite fiber	0–50

PTFE, polytetrafluoroethylene.

jute, flax, hemp, and cotton are used for light applications, such as seawater, potable water, and so on. The compound polytetrafluoroethylene is versatile and can be used for a wide variety of applications.

42.5.9 Surface Velocities for Packing Materials

Table 42.10 gives the recommended surface speeds for a range of packing materials and is provided for guidance only.

42.6 Gaskets

42.6.1 Introduction

A gasket is used to provide a seal between two flat mating surfaces, it provides a means of preventing leakage either from or into the joined features. The gasket may be formed using the following:

- Paper
- O-ring
- Fiber
- Anaerobic
- Copper

The gasket is formed under compression and fills any irregularities between the mating surfaces.

Originally, asbestos was a common material used in the manufacture of heat-resisting gaskets as found on high-pressure steam systems and automotive engines when sealing between the cylinder head and cylinder block, but in recent years, due to health hazards associated with exposure to asbestos particles, non-asbestos gasket materials are now used, when practical.

42.6.2 Flange Design

The design of the flange is important when designing a joint sealed by a gasket. The following points must be considered:

- Flange surface finish
- Area of the seal surfaces to ensure high clamping pressure
- Fastener placement to ensure even clamping pressure across the flange
- Fastener size to ensure adequate strength and force is provided
- Flange material strength
- Flange material to be compatible with the gasket material

The circular flange is the most common shape; however, gaskets can be used to seal a wide variety of faces that have a large number of separate pressurized cavities and complex shapes.

42.6.3 Gasket Design

Gaskets are available in a wide variation of design based on industrial use, budget, and chemical and physical parameters.

Sheet Gaskets
This is considered a simple straightforward gasket in that it is cut out of a sheet of fibrous material, such as high-temperature graphite. This type of gasket can fill various chemical requirements, depending on the inertness of the materials used.

Spiral-Wound Gaskets
Spiral-wound gaskets are a combination of metallic and filler material. The metallic element of the gasket is made up of a thin profiled strip that is wound in the form of a spiral and filled with a sealant, usually flexible graphite. This results in alternating layers of metal and filler. The filler material acts as the sealing element and the metal provides the structural support.

Kammprofile Gaskets
Kammprofile gaskets are used to replace older seals, as they have both a flexible nature and reliable performance. Kammprofiles are essentially a covered metal gaskets consisting of a serrated metallic core with a soft facing material. The Kammprofile style gasket is mainly used in vessel applications where often the gasket width is limited. These gaskets are generally reusable.

The depth and profile of the groove is specially design to facilitate densification of the facing material and create a leak tight seal with low gasket stresses.

Properties of Kammprofile Gaskets

- Low sealing stress required for installation
- Suitable for a range of chemicals
- Suitable for high-pressure and high-temperature applications
- Can be refurbished and reused
- Are able to be manufactured to suit standard flange applications and vessels
- Can be manufactured for use with narrow flanges

42.6.4 Possible Failures in Gaskets

The principal failures in the gasket arise due to a number of reasons:

- Uneven distribution of face loading. This may be due to incorrect torqueing of the fasteners due to human error or using uncalibrated torque tools.
- Relaxation of the tightening torque due to vibration, temperature increase or decrease, and other factors.
- The mating surfaces of the joint are not flat or parallel, or due to incorrect surface finish.

43

Belt Drives

43.0 Introduction

There are three basic categories of belt drives used in power transmission:

- Flat belts
- V-belts
- Synchronous belts (aka timing belts)

Belts in the first two categories are collectively known as *"friction drives."* The third category comes under the heading *"synchronous drives."*

43.1 Flat Belt Drive

Flat belt drives became common after the "Industrial Revolution," when the early machine shops were driven by overhead line shafting originating from a central power source, such as a water wheel or steam engine. They were also used in agriculture and mining applications. The early machine tools were directly linked to and driven from an overhead pulley that ran at a continuous speed and was always running. The machine usually had three different pulley sizes attached to the work spindle for slow, medium, or high speed. When the machine operator required changing spindle speed using a piece of wood, he would have to flip the belt from one pulley to anther (a procedure not without its dangers).

The flat belts had a few advantages in that if the cutting load was exceeded, the belt would slip until the load was reduced. Belt fracture was also a common occurrence, but belt repair was easily facilitated using a steel comb stapled through the leather, allowing a separate piece of belt to be grafted in, getting the machine back in production in a short time.

Because flat belts have a tendency to track across a wide pulley, pulleys were made with a slight convex or "crowned" surface (rather than flat) to allow the belt to self-center.

Traditionally, flat belts were manufactured from leather or felt. Modern flat belts are manufactured from rubber or polymers. These materials also allow lower shaft and bearing loads to transmit significant power. The strongest flat belts now transmit over 2935.83 kW/m. of belt width.

Flat belt drives can be configured as open-drives or cross-belt drives.

With open belt drives, the driven pulley rotates in the same direction as the drive pulley.

In the cross-belt configuration, the driven pulley rotates in the opposite direction.

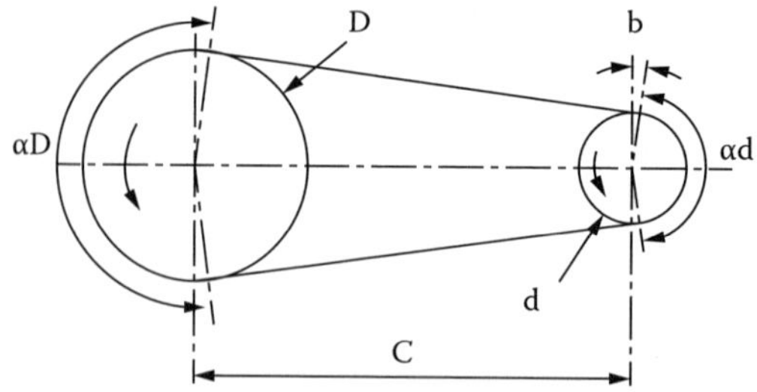

FIGURE 43.1
Open belt drive.

43.1.1 Nomenclature for an Open Flat Belt Drive

See Figure 43.1.

D	Diameter of large pulley
d	Diameter of small pulley
C	Center distance
α_D	Angle of wrap of the larger pulley
α_d	Angle of wrap for the small pulley

43.1.2 Basic Formulas for an Open Belt Drive

$$\alpha_D = 180^o + 2\beta$$

$$\alpha_d = 180^o - 2\beta$$

Where angle β is

$$\beta = \sin^{-1}\left(\frac{D-d}{2C}\right)$$

$$L_o = \text{Length of open belt}$$

$$L_o = \frac{\pi}{2}(D+d) + 2C + \frac{1}{4C}(D-d)^2$$

This formula may be verified by simple geometry.

43.1.3 Nomenclature for a Cross Belt Drive

See Figure 43.2.

D	Diameter of large pulley
d	Diameter of small pulley
C	Centre distance
α_D	Angle of wrap of the larger pulley
α_d	Angle of wrap for the small pulley

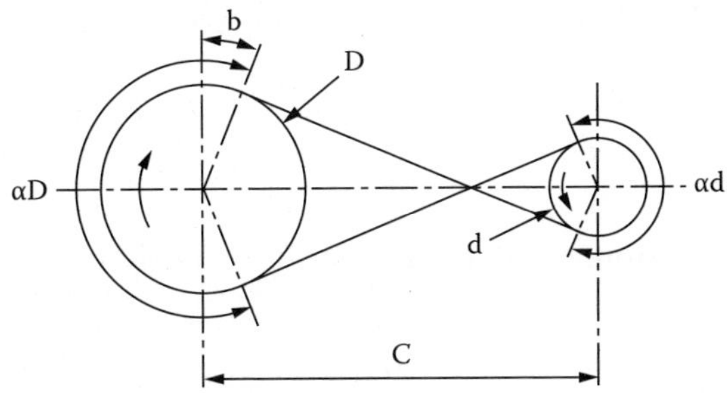

FIGURE 43.2
Cross belt drive.

43.1.4 Basic Formulas for a Cross Belt Drive

$$\alpha_D = \alpha_d = 180^\circ + 2\beta$$

where angle β is

$$\beta = \sin^{-1}\left(\frac{D-d}{2C}\right)$$

$$L_o = \text{Length of cross belt}$$

$$L_o = \frac{\pi}{2}(D+d) + 2C + \frac{1}{4C}(D+d)^2$$

43.1.5 Flat Belt Tensions

Belt drives primarily operates on the principles of friction, that is the friction existing between the belt and the surface of the pulley is responsible for the power transmitted from one pulley to the other.

Due to the presence of friction between the pulley and the belt surface, tensions on both sides of the belt will not be equal. It is therefore important that the high and low tensions are identified as shown in Figure 43.3.

When the driving pulley rotates in the direction shown in the Figure 43.3, from the fundamental concept of friction, the belt will oppose the motion of the pulley. Thereby, friction f on the belt will be opposite to the motion of the pulley. Friction in the belt acts in

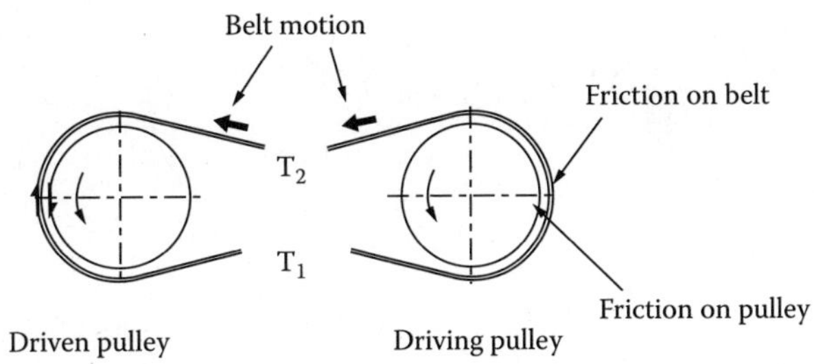

FIGURE 43.3
Belt tensions.

the direction as shown in Figure 43.3, and it will impart a motion on the belt in the same direction. The friction f acts in the same direction as T_2. Equilibrium of the belt segment suggests that T_1 will be higher than T_2. T_1 is referred to as the *tight side* and T_2 as the *slack side*, hence T_1 is the higher-tension side and T_2 is the lower-tension side.

43.1.6 Derivation of Relationship between Flat Belt Tensions

Figure 43.4 depicts a free body diagram for a belt segment.

The belt segment subtends an angle $d\phi$ at the center. Hence, the length of the belt segment

$$dl = f \cdot d\phi \tag{43.1}$$

At the impending condition, that is when the belt is just in motion with respect to the pulley, the forces acting on the belt segment are as shown in Figure 43.4. This belt segment is subjected to a normal force acting from the pulley on the belt segment, and due to the impending motion, the frictional force will be acting in the direction as shown in the figure:

$$f = \mu dl \tag{43.2}$$

where μ is the coefficient of friction between the belt and the pulley.

The centrifugal force acting on the belt segment due to the motion of the belt is denoted as CF, and its magnitude is calculated as

$$CF = \frac{\left[m(rd\phi) \times v^2\right]}{r} = mv^2 d\phi \tag{43.3}$$

where v is the peripheral velocity of the pulley, and m is the mass of the belt of unit length:

$$m = bt\rho \tag{43.4}$$

where b is the width, t is the thickness, and ρ is the density of the belt material.

From the equation of equilibrium in the tangential and normal direction,

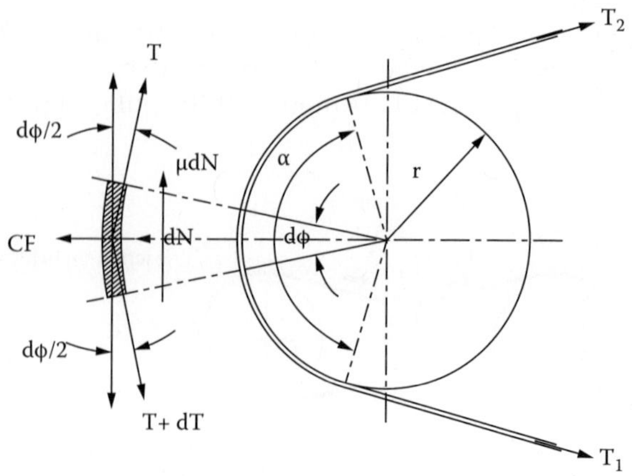

FIGURE 43.4
Free body diagram of forces acting on a belt segment.

$$\Sigma F_t = 0$$

$$T\cos\frac{d\phi}{2} - (T + dT)\cos\frac{d\phi}{2} + \mu dN = 0 \tag{43.5}$$

$$\Sigma F_n = 0$$

$$mv^2 d\phi + dN + T\sin\frac{d\phi}{2} - (T + dT)\left(\sin\frac{d\phi}{2}\right) = 0 \tag{43.6}$$

For small angle $d\phi$,

$$\cos\frac{d\phi}{2} \approx 1 \quad \text{and} \quad \sin\frac{d\phi}{2} \approx \frac{d\phi}{2} \tag{43.7}$$

A simplified form of Equation 43.5

$$dN = \frac{dT}{\mu} \tag{43.8}$$

From Equation (43.6) and using Equation (43.8),

$$mv^2 d\phi + \frac{dT}{\mu} - T d\phi = 0$$

or

$$\frac{dT}{T - mv^2} = \mu d\phi \tag{43.9}$$

Consider entire angle wrap:

$$\int_{T_1}^{T_2} \frac{dT}{T - mv^2} = \int_0^{\alpha} \mu \, d\phi \tag{43.10}$$

Determination of relationship between belt tensions is

$$\frac{T_1 - mv^2}{T_2 - mv^2} = e^{\mu\alpha} \tag{43.11}$$

In Equation 43.11, α is expressed in radians.

The product $\mu\alpha$ should be a minimum to determine the tension ratio for the driver or driven pulley.

43.1.7 Elastic Creep and Initial Tension

The presence of friction between the pulley and belt causes a differential tension in the belt. This differential tension is responsible for the belt to either elongate or contract and causes a relative motion between the belt and the pulley surface. This relative motion is created due to the phenomena known as *elastic creep*.

The belt will always have an initial tension when fitted over the drive and driven pulleys. This initial tension will be the same throughout the belt length when there is no motion. During the rotation of the drive, the tight side tension will be higher than the initial tension and the slack side tension will be lower than the initial tension.

Consider an elementary length of the belt on the drive pulley (see Figure 43.4). The magnitude of the initial tension in the belt is as follows:

$$\text{Tight side elongation} \propto (T_1 - T_i)$$

$$\text{Slack side elongation} \propto (T_i - T_2)$$

where T_i is the initial tension in the belt.

As the belt length remains unchained, the elongation in the belt will be the same as the contraction in the belt, that is,

$$T_i = \frac{T_1 + T_2}{2} \tag{43.12}$$

It should be noted that with the increase in the initial tension, the power transmission can be increased. If the initial tension is gradually increased, then T_1 will also increase and T_2 will decrease at the same time. Therefore, if it happens that T_2 is equal to zero, then $T_1 = 2T_i$ and maximum power transfer will be achieved.

43.1.8 Velocity Ratio of Flat Belt Drive

The velocity ratio of a belt drive is defined as the ratio of the speeds of the driver and driven pulley:

$$\text{Let } N_1, D = \text{speed (rpm) and diameter of driving pulley}$$

$$N_2, d = \text{speed (rpm) and diameter of driven pulley}$$

$$t = \text{thickness of belt}$$

Velocity ratio:

$$\frac{N_2}{N_1} = \frac{D}{d}$$

If the thickness of the belt is considered,

Velocity ratio:

$$\frac{N_2}{N_1} = \frac{D + t}{d + t}$$

The slip of the belt over the pulleys reduces the velocity ratio. Therefore, let

$$sD = \text{percentage slip between belt and driving pulley}$$

$$sd = \text{percentage slip between belt and driven pulley}$$

$$s = \text{total percentage slip.}$$

Velocity ratio with slip:

$$\frac{N_2}{N_1} = \frac{D+t}{d+t}\left(1 - \frac{s}{100}\right)$$

43.1.9 Power Transmission of a Flat Belt Drive

Consider the maximum power to be transmitted without any belt slip.

The power transmitted by a pulley is shown to be

$$P = \pi ND(T_1 - T_2) = v(T_1 - T_2)$$

When the belt starts to slip, the tension ratio is

$$\left(\frac{T_1}{T_2}\right) = e^{\mu\theta}$$

It may be considered easier to use the belt velocity instead of the pulley speed: $v = \pi ND$

$$P = vT_1\left(1 - e^{\mu\theta}\right)$$

This is the maximum power that can be transmitted at a given speed with no belt slip occurring.

Example 43.1

The tension in a pulley belt is 110 N when stationary. Calculate the tension in each side and the power transmitted when the belt is on the point of slipping on the smaller wheel. The wheel is 240 mm diameter and the coefficient of friction between the belt and wheel is 0.32. The angle of lap is 165° and the wheel speed is 1500 rev/min.

Solution

Belt velocity $v = \pi = \pi \times (1500/60) \times 0.24 = 18.85\ \text{m/s}$

Lap angle $\theta = (165/180) \times \pi = 2.88$

Initial tensions $T_1 = T_2 = 110\ \text{N}$ when stationary hence $T_1 + T_2 = 220\ \text{N}$

$T_2 = 220 - T_1$

Tension in belt:

$$\frac{T_1}{T_2} = e^{\mu\theta} = e^{0.32\times 2.88} = 2.513$$

$$T_1 = 2.513T_2 = 2.513(220 - T_1) = 552.9 - 2.513T_1$$

$$T_1 + 2.513T_1 = 552.9$$

$$3.513T_1 = 552.9$$

$$T_1 = 157.39$$

$$T_1 = \frac{157.39}{2.513} = 62.6\ \text{N}$$

$$P = v(T_1 - T_2) = 18.85(157.4 - 62.6)$$

$$P = 1.786\ \text{kW}$$

Power check:

$$P = vT_1\left(1 - e^{\mu\theta}\right) = 18.85 \times 157.4 \times (1 - 0.398)$$

$$P = 18.85 \times 157.4 \times 0.6$$

$$P = 1.786\ \text{kW}$$

43.2 Wedge and V-Belt Drives

Flat belt drives have been replaced by V-drives for many applications, as they are able to transmit more power with more compact drive arrangements. A V-drive is able to use a number of belts, whereas a flat belt uses only one belt. The design of the V-belt allows the individual belt to wedge itself into the groove in the drive and driven pulleys reducing the chance of slip. V-belts do slip only when the subject to high loads, increasing the number of belts will reduce the chance of slip. In this instance, V-belts used in higher-power duties generally must be matched to ensure that the drive power is shared.

Considering a flat belt drive, the tangential force at the point of slipping is μR, where μ is the coefficient of friction and R is the radial force between the belt and the pulley.

For a V-belt drive, the equivalent friction force at the point of slipping is $\mu R/\sin\beta$, where β is half the included angle of the V-groove. Hence a V-drive has a maximum friction of $(1/\sin\beta) \times$ the friction in a flat belt, for example, a V-belt having an inclusive angle of $(2{\cdot}\beta) = 20°$, the maximum friction force will be $1/\sin(10) = 5.75 \times$ the maximum flat belt friction.

When designed and installed correctly, a V-belt system can be expected to have a continuous service life of 25,000 h and achieve a drive efficiency of approximately 95%.

43.2.1 Nomenclature of a Wedge and V-Belt

Figure 43.5 shows typical wedge and V-belt sections. The geometrical features of the belt are indicated in the figure. The pitch line of the section can also be referred to as the neutral axis and is often shown as N-A on calculations and drawings. The design calculations for a wedge and V-belt drive are based on the pitch line or neutral axis. These belts are available in various sections, dependent on the power rating (see Tables 43.1 through 43.3).

43.2.2 Standard Wedge and V-Belt Sections

There are two standards: wedge belts and V-belts. The wedge belts have a deeper cross section compared to the standard V-belt (see Table 43.1), with the nominal top width of both belts being identical.

Figure 43.6 shows the terminology for pulley groove dimensions for a wedge belt.

The angle of contact factor is derived from Figure 43.7.

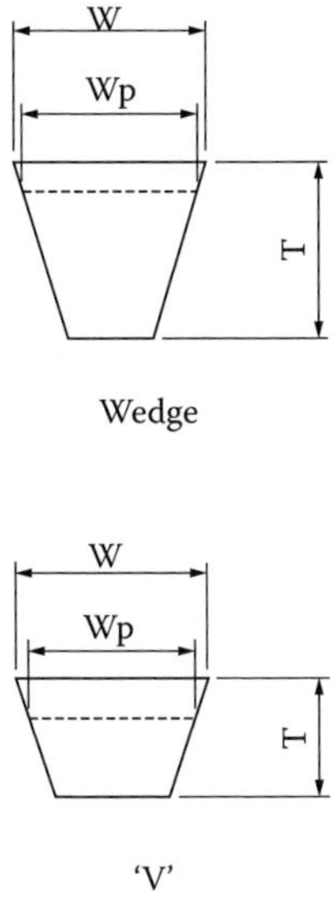

FIGURE 43.5
Typical wedge and vee belt sections.

TABLE 43.1
Specification for Wedge and V-belts

Section	Pitch Width (mm)	Nom' Top Width (mm)	Nom' Height (mm)
Wedge Belts			
SPZ	8.5	9.5	8.0
SPA	11.5	13.0	10.0
SPB	14.0	16.0	14.0
SPC	19.0	22.0	18.0
V-Belts			
Z	8.5	10.0	6.0
A	11.0	13.0	8.0
B	14.0	17.0	11.0

TABLE 43.2
Specification for Pulley Dimensions for Wedge Belts

Belt Section	ip (mm)	b (mm)	h (mm)	e (mm)	f (mm)	α (°)	g (mm)
SPZ	8.5	2.5	9.0	12.0	8.0	34 (dp < 80) 38 (dp > 80)	10.0 10.2
SPA	11.0	3.3	11.0	15.0	10.0	34 (dp < 118) 38 (dp > 118)	13.1 13.5
SPB	14.0	5.7	19.0	25.5	12.5	34 (dp < 190) 38 (dp > 190)	16.6 16.9
SPC	19.0	5.7	19.0	25.5	17.0	34 (dp < 315) 38 (dp > 315)	22.5 22.9

TABLE 43.3
Typical Power Rating

Belt Section	Pulley Dia' (mm)	Power (kW) at Motor Speed		
		2880 (rev/min)	1440 (rev/min)	960 (rev/min)
SPZ	67	1.75	1.05	0.77
	140	6.97	4.02	2.84
SPA	100	4.12	2.61	1.92
	200	15.1	9.4	6.72
SPB	160	12.1	7.85	5.73
	400	25.7	29.8	22.0
SPC	224	18.5	16.6	12.7
	560	–	62.7	53.3

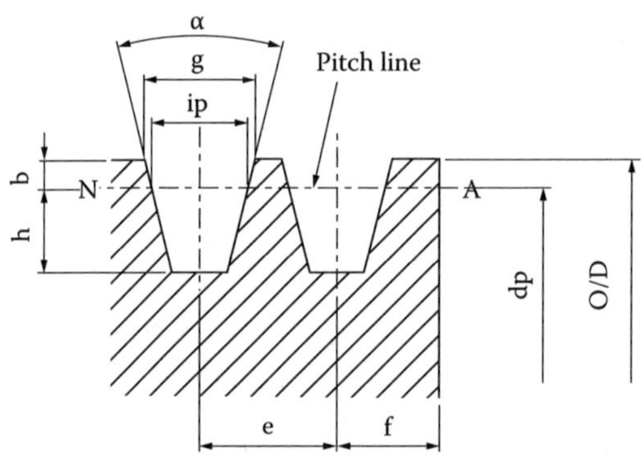

FIGURE 43.6
Pulley groove dimensions.

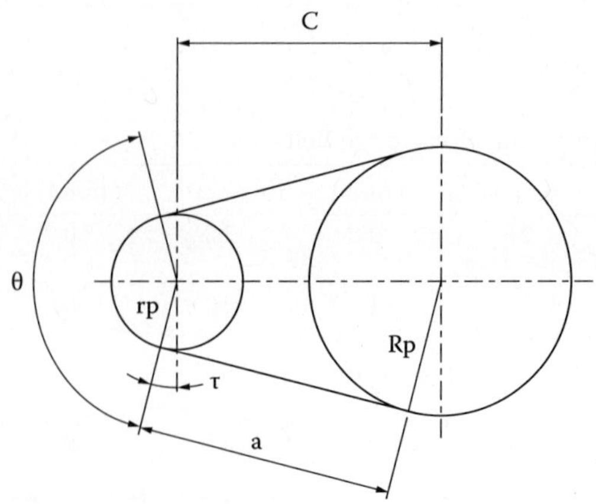

FIGURE 43.7 Angle of contact factor.

43.2.3 V-Belt Design

Basic theory:

$$\text{Length of the curved element} = rd\theta$$

$$\text{Density of belt material} = \rho$$

$$\text{Cross-sectional area of belt} = A$$

$$\text{Volume} = Ard\theta$$

$$\text{mass in dm} = \rho Ard\theta$$

The equation for centrifugal force (CF) is:

$$CF = mv^2r.$$

Note: The CF acting on the tiny mass is written in calculus form as d(CF):

$$d(CF) = dm\, v^2/r$$

$$d(CF) = \rho Ard\theta v^2/r$$

$$d(CF) = \rho\, A\, d\theta\, v^2$$

The normal force R pressing the element to the pulley

$$R = Td\theta$$

This is reduced by the centrifugal force, so

$$R = Td\theta - \rho Ad\theta v^2$$

$$R = d\theta(T - \rho\, Av^2)$$

The change in force from one side of the element to the other is dT=μR. Substituting for R,

$$dT = \mu d\theta\left(T - \rho Av^2\right)$$

$$\frac{dT}{\left(T - \rho Av^2\right)} = \mu d\theta$$

Integrating this expression,

$$\frac{\left(T_1 - \rho Av^2\right)}{\left(T_2 - \rho Av^2\right)} = e\mu\theta$$

Let ρAv2=Tc, which is the centrifugal force term:

$$\frac{(T_1 - T_c)}{(T_2 - T_c)} = e\mu\theta$$

As the angle of lap will be smallest on the small pulley, the belt will always slip first on this pulley, providing that the coefficient of friction remains the same.

43.2.4 Power Transmission of a V-Belt Drive

The equation for power transmitted

$$P = \pi ND(T_1 - T_2)$$

Subtracting T_c from each term makes no difference:

$$P = \pi ND(T_1 - T_c) - (T_2 - T_c)$$

But,

$$(T_2 - T_c) = (T_1 - T_c)^{-\mu\theta}$$

Substituting for $(T_2 - T_c)$ gives

$$P = \pi ND(T_1 - T_c)(1 - e^{-\mu\theta})$$

or

$$P = v(T_1 - T_c)(1 - e^{-\mu\theta})$$

43.3 Synchronous Belts

43.3.1 Introduction

Synchronous belts (also timing belts) have a toothed profile that mate with corresponding grooves in the pulleys, providing the same positive engagement as gears or chains. They are used in applications where indexing, positioning, or a constant speed ratio is required.

The first tooth profile used on synchronous belts was the trapezoidal shape (Figure 43.8a). It is still recognized as the standard. Recent modifications to tooth profiles have improved on the original curvilinear shape (Figure 43.8b). The full-rounded profile distributes tooth loads better to the belt tension members. It also provides greater tooth shear strength for improved load capacity.

A modified curvilinear tooth design (Figure 43.8c) has a different pressure angle, tooth depth, and materials for improved load/life capacity and nonratcheting resistance.

Synchronous belts can wear rapidly if pulleys are not aligned properly, especially in long-center-distance drives, where belts tend to rub against pulley flanges. To prevent the belt from riding off the pulleys, one of the pulleys is usually flanged.

Undertensioning will cause performance problems. The drive may be noisy because belt teeth do not mate properly with pulley grooves, or the belt may prematurely wear from ratcheting. High forces generated during belt ratcheting are transmitted directly to shafts and bearings can cause severe damage.

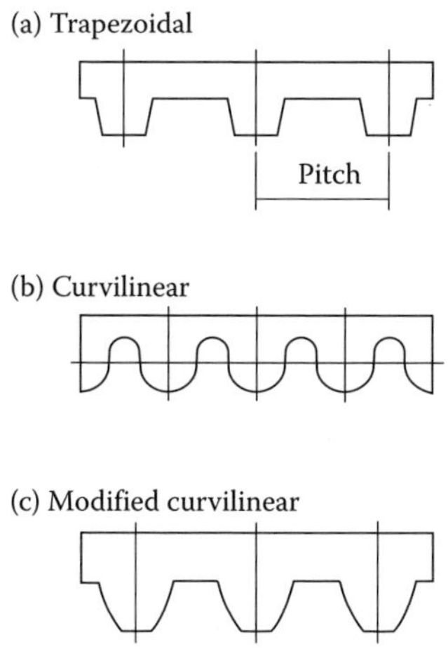

FIGURE 43.8
Basic synchronous tooth profiles.

43.3.2 Relevant Standards

BS 4548:1987: Specification for synchronous belt drives for industrial applications

This standard has been declared obsolescent but is still current.

ISO 5296:2012 Synchronous Belt Drives:belts with pitch codes MXL, XXL, XL, L, H, XH, and XXH—Metric and inch dimensions

This is the current ISO standard. *Note:* the British Standard will not be compatible with this standard.

ISO 5294:2012: Synchronous belt drives—pulleys

43.3.3 Materials

43.3.3.1 Belts

Synchronous or synchronous belts are constructed using elastomer, for example, natural rubber, neoprene, polyurethane, and polychloroprene.

The core is reinforced to provide increased tensile strength. Early versions were originally reinforced with steel to achieve the strength. In modern belts, the most common reinforcement is glass fiber. Aramid is used when maximum capacity is required. The belts are often offered with a nylon facing to provide the necessary wear resistance, and the coating can also be provided with a conductive coating.

43.3.3.2 Pulleys

The pulleys are often made from cast or ductile iron although aluminum is becoming more the choice due to its low weight. These applications can include high-speed drives with low inertia. Steel or stainless steel is preferred to iron when the drive is expected to exceed the safe operating limits for cast iron (2000 m/min) or ductile iron (2500–3000 m/min). The use of plastic gears may be appropriate in some domestic power tools (including garden equipment) were the belt service life may be short.

TABLE 43.4

Classical Synchronous Belts

Section	Description	Pitch (mm)	Width (Available) (mm)
MXL	Extra light	2.032	3.05, 4.826, 6.35
XL	Extra light	5.08	6.35, 9.652
L	Light	9.525	12.7, 19.05, 25.4
H	Heavy	12.7	19.05, 25.4, 38.1, 50.8, 76.2
XH	Extra heavy	22.225	50.8, 76.2, 50.8, 76.2, 101.6, 127.0
XXH	Double extra heavy	31.75	50.8, 76.2, 101.6, 127.0

TABLE 43.5

Curvilinear Synchronous Belts

Section	Description	Pitch (mm)	Width (Available) (mm)
3M	3 mm high torque drive	3	6.0, 9.0, 15
5M	5 mm high torque drive	5	9.0, 15.0, 25.0
8M	8 mm high torque drive	8	20.0, 30.0, 50.0, 85.0
14M	14 mm high torque drive	14	40.0, 55.0, 85.0, 115.0, 170.0
20M	20 mm high torque drive	20	115.0, 170.0, 230.0, 290.0, 340

TABLE 43.6

GT Curvilinear Synchronous Belts

Section	Description	Pitch (mm)	Width (Available) (mm)
2MR (gates)	2 mm high torque belt	2	3.0, 6.0, 9.0
3MR (gates)	3 mm high torque belt	3	6.0, 9.0, 15.0
5 mm (gates)	5 mm high torque belt	5	9.0, 15.0, 25.0

Pulleys may be mounted on shafts using the conventional methods such as keyways, pins, or proprietary shaft-locking bushes (such as Taper-Lock®). Pulleys can have one or two flanges to ensure that the belt retains its position. For drives with horizontal axes, it is recommended that one pulley has two flanges (usually the drive pulley) and the other pulley has one flange to retain the belt. On drives having vertical axes, the lower face on each pulley should include a flange, and one pulley should include two flanges.

43.3.4 Basic Synchronous Belt Specifications

There are three classes of synchronous belts:

- Classical synchronous belts (see Table 43.4)
- HTD—Curvilinear belts (see Table 43.5)
- GT—Curvilinear belts (see Table 43.6)

43.3.5 Design Methods

There are two basic methods for designing a synchronized belt drive system.

43.3.5.1 Power Method

In the power method, the following points are considered:

- The drive speed and the maximum driven torque is required (this includes any inertia load, shock loading, friction or resistance are used in the calculations for the required drive power.
- Table 43.7 provides a range of service factors for a selection of equipment using synchronous drives. It is recommended that the information provided by the supplier of the belts selected is used for the final detail design work.
- Judgment is required when applying service factors when load-isolation measures are incorporated in the design such as flywheels, slipping clutches, fluid couplings, etc. These provide relatively smooth drive conditions that smooth out the load, particularly with relatively high shock loadings.
- The required design power is obtained based on the product of the driven power required and the service factor.
- The belt section may be determined from Figure 43.9 when the rotational speed of the small pulley and the product of the design power×service factor is established.
- The final drive geometry will be defined together with suitable pulley sizes and belt center. See Table 43.8 for some typical pulley sizes.
- A basic power rating for the belt is calculated (see Table 43.9).
- Finally a suitable belt width can now be selected from Table 43.10 (See also Section 43.3.11).

43.3.5.2 Torque Method

The classical synchronous belt, such as the MXL, together with the more advanced curvilinear belts are designed on the torque levels.

The MXL belts generally operate at low belt speeds, and therefore, the torque levels are similar to the normal range of pulley rotational speed.

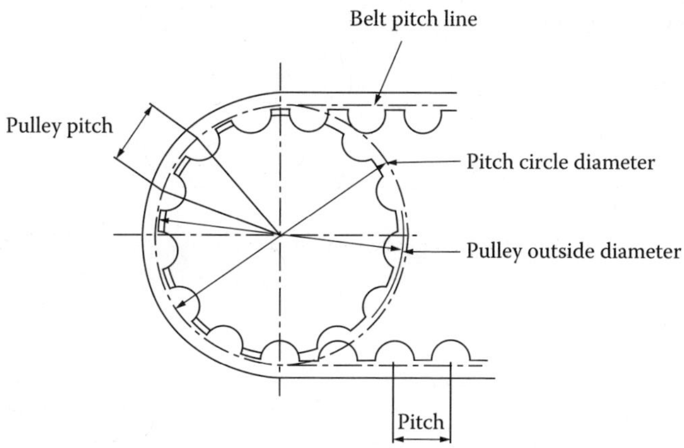

FIGURE 43.9
Synchronous belt in mesh.

TABLE 43.7

Service (Load) Factors—Synchronous Drives

	Operating Period/Day								
	<8 h	<16 h	>16 h	<8 h	<16 h	>16 h	<8 h	<16 h	>16 h
	Driving—Smooth Starting Torque 100%–150% Full Load			Driving—Moderate Vibration Starting Torque 150%–250% Full Load			Driving—Heavy Vibration Starting Torque 250%–400% Full Load		
Description	Electric Motors with Slow Starting Torque (Up to 1.5 Times the Rated Torque)			Electric Motors with Medium Starting Torque (1.5–2.5 Times the Rated Torque)			Electric Motors with High Starting Torque and Braking Torque (More than 2.5 Times the Rated Torque)		
Instrumentation Office equipment Medical equipment	1.0	1.2	1.4	1.1	1.3	1.5	1.2	1.4	1.6
Conveyor system for light materials Oven screens Woodworking drills	1.1	1.3	1.6	1.2	1.4	1.6	1.3	1.5	1.7
Stirrers, mixers, liquid Paper-making: agitators Pumps: centrifugal Heavy woodworking tools	1.2	1.4	1.6	1.4	1.6	1.8	1.6	1.8	2.0
Stirrers, mixers, semi-liquid Machine tools for drilling, grinding, milling, shaping Textile machinery: spinning	1.3	1.5	1.7	1.5	1.7	1.9	1.6	1.8	2.0
Conveyor system: belt conveyors for heavy materials, Elevators, screw conveyors	1.4	1.6	1.8	1.6	1.8	2.0	1.8	2.0	2.2
Centrifuges, Paper pulpers Hammer mills	1.5	1.7	1.9	1.7	1.9	2.1	1.9	2.1	2.3
Brickworks machinery Loam mills, fans Blowers: positive displacement Pulverizers	1.6	1.8	2.0	1.8	2.0	2.2	2.0	2.2	2.4
Mills: ball, roller, and gravel Reciprocating compressors Reciprocating pumps Steelmaking machines	1.7	1.9	2.1	1.9	2.1	2.3	2.1	2.3	2.5

43.3.6 Preliminary Selection

When the design power has been established (Power × Service Factor), a synchronous belt can be initially selected from a chart similar to Figure 43.11 (Timing Belt Selection Chart). This figure is provided for guidance only and is similar to other published material offered by various belt manufacturers.

TABLE 43.8

Classical Synchronous Pulley Sizes

MXL			XL			L			H			XH			XXH		
Teeth	Pitch Circle Dia′	O/D	Teeth	Pitch Circle Dia′	O/D	Teeth	Pitch Circle Dia′	O/D	Teeth	Pitch Circle Dia′	O/D	Teeth	Pitch Circle Dia′	O/D	Teeth	Pitch Circle Dia′	O/D
10	6.47	5.96	10	16.17	15.67	10	30.32	29.56	10	40.43	39.08	18	127.34	124.54	18	181.91	178.87
11	7.11	6.61	11	17.79	17.29	11	33.35	32.59	11	44.47	43.12	20	141.49	138.68	20	202.13	199.09
12	7.76	7.25	12	19.40	18.90	12	36.38	35.62	12	48.51	47.16	22	155.64	152.83	23	232.45	219.30
14	9.06	8.55	13	21.02	20.52	13	39.41	38.65	13	52.56	51.20	24	169.79	167.01	25	252.66	239.50
16	10.35	9.84	14	22.64	22.14	14	42.45	41.68	14	5.60	55.25	26	183.94	181.15	26	262.76	259.72
18	11.64	11.13	15	24.26	23.76	16	48.51	44.72	15	60.64	59.29	28	198.08	195.30	30	303.19	300.15
20	12.94	11.78	16	25.87	25.37	17	51.54	47.75	16	64.68	63.33	30	212.23	209.45	34	343.62	340.56
21	13.58	13.07	17	27.49	26.99	18	54.57	50.78	17	68.72	67.37	32	226.38	223.60	40	404.25	401.19
22	14.23	13.72	18	29.11	28.61	19	57.61	56.84	18	72.77	71.42	40	282.98	280.19	48	485.10	482.07
24	15.52	15.02	20	32.34	31.84	20	60.64	59.88	19	76.81	75.46	48	339.57	336.78	60	606.38	603.32
28	18.11	17.60	21	33.96	33.46	21	63.67	62.91	20	80.85	79.50	60	424.47	421.67	72	727.66	648.41
30	19.40	18.90	22	35.57	35.07	22	66.70	65.94	21	84.99	83.54	72	509.36	506.58	90	909.67	906.53
32	20.70	20.19	24	38.81	38.31	24	72.77	72.00	23	92.98	91.63	84	594.25	591.46			
36	23.29	22.78	25	40.43	39.93	25	75.80	75.04	25	101.06	99.71	90	636.70	633.91			
40	25.87	25.37	26	42.04	41.54	26	78.83	78.07	26	105.11	103.76	96	679.15	676.35			
42	27.17	26.67	28	45.28	44.78	28	84.89	84.13	28	113.19	111.84	120	848.93	846.15			
44	28.46	27.94	30	48.51	48.01	30	90.96	90.19	30	121.28	119.93						
48	31.05	30.53	32	51.74	51.24	32	97.02	96.26	32	129.36	128.01						
60	38.81	38.30	36	58.21	57.71	36	109.15	108.39	33	133.40	132.05						
72	46.57	46.05	40	64.68	64.18	40	121.28	120.51	34	137.45	136.10						
			42	67.91	67.41	42	127.34	126.58	35	141	140.14						
			44	71.15	70.65	44	133.40	132.64	36	49	144.18						
			48	77.62	77.12	48	145.53	144.77	38	145.53	152.27						
			50	80.85	80.35	50	151.60	150.83	40	153.62	160.35						
			54	87.32	86.82	54	163.72	162.96	42	161.70	168.44						
			60	97.02	90.52	60	181.91	181.15	44	177.87	176.52						
			72	116.43	115.93	72	218.30	220.57	48	194.04	192.69						

TABLE 43.9
Power Rating for Synchronous Belts

Belt Designation	Width (mm)	Power Rating (P_r) (kW)
XL	9.652	$P_r = 0.746 \cdot Z\left(0.0916 - 7.07 \times 10^{-5} \times Z^2\right)$
L	25.4	$P_r = 0.746 \cdot Z\left(0.436 - 3.01 \times 10^{-4} \times Z^2\right)$
H	76.2	$P_r = 0.746 \cdot Z\left(3.73 - 1.41 \times 10^{-3} \times Z^2\right)$
XH	101.6	$P_r = 0.746 \cdot Z\left(7.21 - 4.68 \times 10^{-3} \times Z^2\right)$
XXH	127.0	$P_r = 0.746 \cdot Z\left(11.4 - 7.81 \times 10^{-3} \times Z^2\right)$

TABLE 43.10
Belt Width Factors

	Belt Width											
Belt Section	**3.05**	**4.826**	**6.35**	**9.625**	**12.7**	**19.05**	**25.4**	**38.1**	**50.8**	**76.2**	**101.6**	**127.0**
MXL	0.43	0.73	1.00									
XL			0.62	1.00								
L					0.45	0.72	1.00					
H						0.21	0.29	0.45	0.63	1.00		
XH									0.45	0.72	1.00	
XXH`									0.35	0.56	0.78	1.00

43.3.7 Designation of Synchronous Belts

Synchronous belts are identified by a standard number. The first digits specify the belt length in increments of 0.1 in., followed by a letter (or letters) indicating the belt section (Table 43.10). The belt width (following the belt section) is indicated in one-hundredths of an inch. As an example, an L section belt having a pitch length of 30.00 and 0.75 in. wide would be specified as 300L075 Synchronous Belt. A similar method is used for defining metric belt designations.

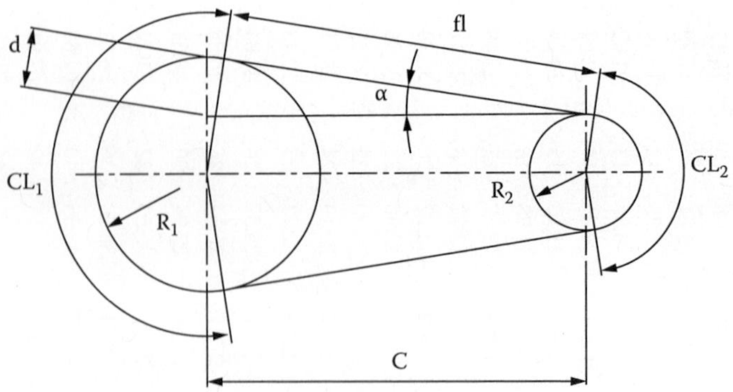

FIGURE 43.10
Calculating synchronous belt length.

TABLE 43.11

Belt Length and Center Distance

Ident	Description	Formula	Units
1.	Pitch dia' of pulley—"P"		mm
2.	Drop distance—"d"	$[P_1 - P_2]/2$	mm
3.	Center distance "C"	C	mm
4.	Belt contact angle "a"	arcsin(d/C)	°
5.	Belt fall length "fl"	d/tan a	mm
6.	Contact length—small pulley "CL1"	$P_1 \cdot \pi \cdot (90+a)/180$ (°)	mm
7.	Contact length—large pulley "CL2"	$P_2 \cdot \pi \cdot (90-a)/180$ (°)	mm
8.	Belt length "L"	$2.fl + CL1 + CL_2$	mm
9.	Total no. of teeth on belt "N"	L/pitch	
10.	No. of teeth in mesh (small pulley)	CL_2/pitch (rounded to nearest integer)	

43.3.8 Power Rating of Synchronous Belts

This method is based on the method shown in Machinery's handbook. It is preferable to use the calculation tool provided by the belt manufacturers to size the belts for detail design.

The power ratings of belts for the basic belt widths (in brackets) are as identified in Table 43.11.

The following is the nomenclature for Table 43.11:

$$P_r = \text{Power rating in kW}$$

$$r = \frac{\text{Speed of faster shaft (rev/min)}}{1000}$$

$$Z = \frac{P_2 \times r}{25.4}$$

$$P_2 = \text{Pitch diameter of smallest pulley (mm)}$$

43.3.9 Belt Length and Center Distance

Figure 43.10 and Table 43.11 tabulate the formulas used for calculating the length of the synchronous belt. Like a chain drive, the belt has fixed pitches and this may have an impact on the center distance, particularly if the centers are fixed. This may result in the belt having a degree of slackness that increases the chance of tooth "jumping." To overcome this possibility, it is usual to fit a jockey pulley to maintain tension in the belt.

An approximation of the center distance as a function of belt length is given by

$$C = \frac{Y + \sqrt{Y^2 - 2(d_2 - d_1)^2}}{4}$$

where

$$Y = L - \frac{\pi(d_2 + d_1)}{2}$$

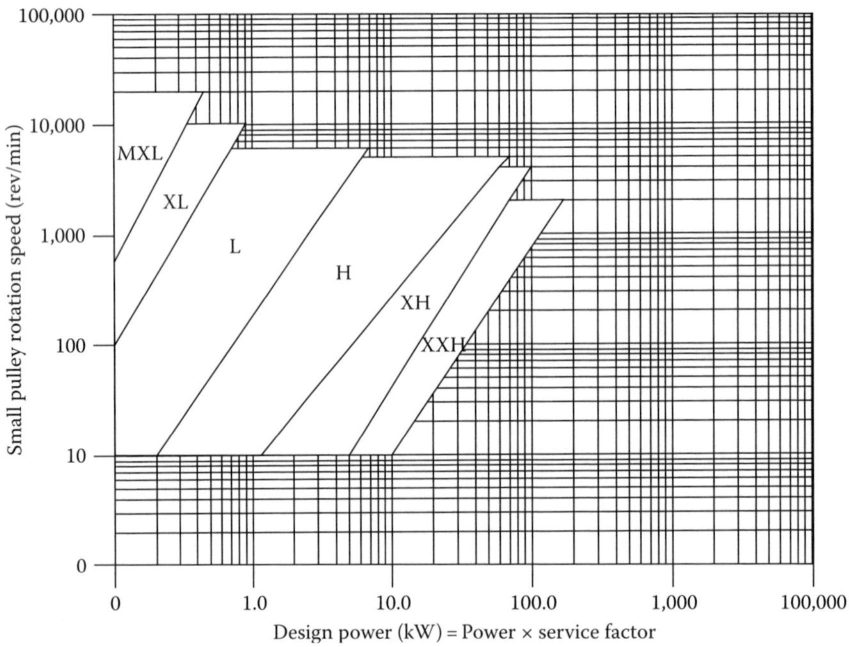

FIGURE 43.11
Synchronous belt selection chart.

43.3.10 Synchronous Belt Width Determination

The following points must be considered when establishing the appropriate width of the synchronous belt for a specific application:

- Determine the design power to be transferred in kW; power×service factor.
- Select the most suitable belt and calculate the basic power using the belt size, smaller pulley speed, and smaller pulley size.
- If the basic power is less than the design power, modify one or more of belt size, pulley size, or pulley speed.
- Divide the basic power by the design power to obtain the belt width factor.
- Use Table 43.11 to select a width with a width factor greater than that calculated.

43.3.11 Synchronous Pulley Sizes

Pulley dimensions (pitch circle and outside diameters) covering a range of pulleys for the classical timing belts is given in Table 43.8. In practice the range of pulleys available is much broader than covered here. Consulting the individual supplier catalogues is recommended. Pulleys for the higher-specification belts will also be covered in the catalogues.

43.3.12 Advantages of Using Synchronous Timing Belt

The following is the advantages of using synchronous timing belts.

- Precision registration and timing without any loss of high torque carrying capacity
- Minimal vibration
- Reduced noise
- Positive slip-proof engagement
- Wide speed range
- Virtually no elongation (stretching) due to wear
- High mechanical efficiency (up to 98%)
- Clean operation, no need for lubrication
- Long, dependable trouble-free service
- Excellent abrasion resistance
- Rust resistance
- Resists chemicals and contaminants
- Increased drive design options
- Weight savings
- Reduced safety issues

43.3.13 Disadvantages of Using Synchronous Timing Belts Compared to Chain Drives

The following are the disadvantages of using synchronous timing belts.

- Availability of numerous chain sizes allows for smaller pitch diameters and/or number of teeth while achieving the desired ratio
- Higher speeds and power capacity using chain drives
- Detachable chain links allow easier installations in certain instances
- Need for accurate initial alignment of drive components for synchronous drives
- Higher drive ratios at shorter center distances for chain drives
- Lower bearing loads using chain drives
- Chain drives are less affected by temperature and humidity
- Chain drives are less affected by grease and oil

44

Roller Chain Drives

44.1 Introduction

Roller chain drives is a type of segmented link drive most commonly used for transmission of mechanical power of many types of domestic, industrial and agricultural machinery. It consists of short cylindrical rollers held together by side links and driven by a toothed wheel called a sprocket.

Though Hans Renold is credited with inventing the roller chain in 1880, Leonardo da Vinci showed sketches of a chain with roller bearings in a sixteenth century Codex.

44.2 Advantages and Disadvantages of Using a Roller Chain

Roller chain drives have characteristics that in certain applications are more favorable than belt or gear drives for providing power transmission between two or more shafts. Figure 44.1 gives the salient features of an individual roller chain link, and Figure 44.2 shows a typical chain sprocket.

The advantages of a roller chain drive over a belt drive system are as follows:

- There is no slip or creep (no power loss resulting from slippage).
- Chain drives are more compact for a given capacity.
- There are lower loads on the drive/driven shafts (high tension is not required as with a belt drive).
- Chain drives are easier to install.
- They are unaffected by environmental issues, such as ultraviolet rays or high ambient temperatures.
- They will not deteriorate with age.
- Chain drives are more effective at lower speeds.
- They require little adjustment.

The advantages of a roller chain drive over gear drive systems are as follows:

- Chain drives are more flexible with center distances.
- They are more economical.

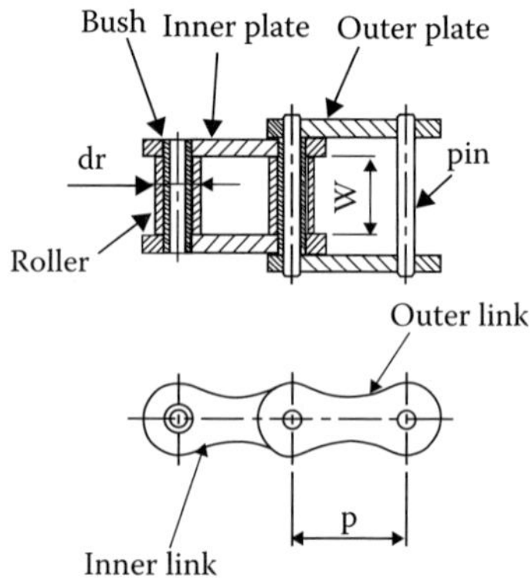

FIGURE 44.1
Detail of individual roller chain link.

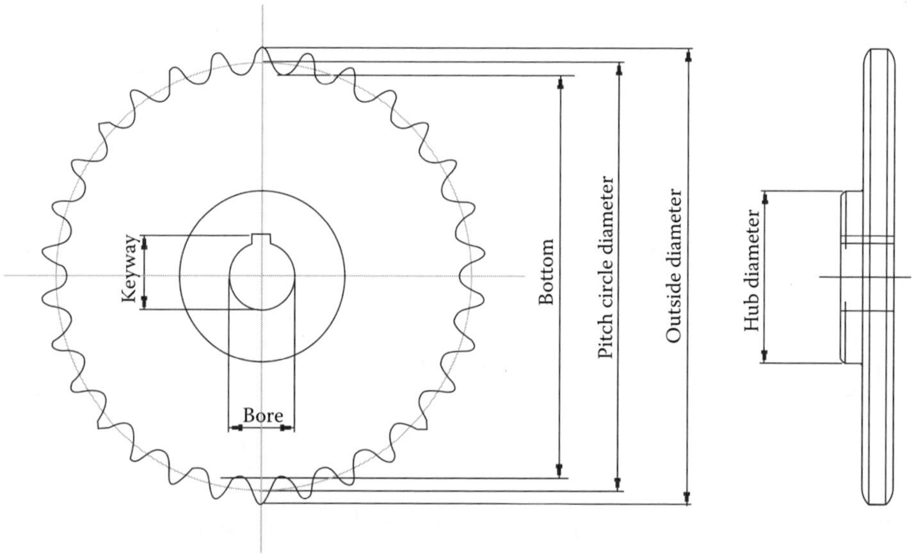

FIGURE 44.2
Typical roller chain sprocket.

- They are simpler to install.
- Chain drives absorb shock better than gear drives.

The disadvantages of a roller chain drive over a belt drive system are as follows:

- Chain drives generally require lubrication.
- Chain drives are noisier than belt drives.
- They are more expensive than belt drives.
- Impracticable for extremely long center-to-center distances where flat belts could be used more effectively.
- They are less efficient than flat belts at extremely high-speed ranges.

- They must be used on parallel shafts; they are unable to twist like belts.
- Chordal action—slight pulsation in the output drive; this becomes less pronounced as the number of sprocket teeth is increased

44.3 Geometry Design Proportions of Roller Chains

Figure 44.1 and Table 44.1 list the main geometric properties of a roller chain link and nomenclature.

Main roller chain properties are based on national standard recommendations. Specific sizes of chains also define the corresponding dimensions for the sprockets to engage correctly with the chain.

TABLE 44.1

Nomenclature of Roller Chain Properties

α	Roller seating angle
B	Maximum width over bearing pins
b_1	Minimum width between the inner plates
B_a	Tooth side relief
b_{axf}	Tooth side relief factor
b_r	Tooth width
b_s	Minimum shroud diameter
d_1	Maximum roller diameter
d_2	Maximum bearing pin body diameter
D_a	Tip diameter
D_g	Measuring pin diameter
D_p	Pitch diameter
D_r	Root diameter
d_r	Maximum bush or roller diameter
D_s	Maximum shroud diameter
h_1	Maximum inner plate depth
h_2	Maximum outer or intermediate plate depth
h_a	Height of the tooth above pitch line
h_{max}	Maximum plate depth
K	Number of strands
M_r	Measurement over pins or direct measurement
P	Pitch
p_1	Transverse pitch
Pc	Chordal pitch equals to chain pitch
r_a	Shroud fillet radius
r_e	Tooth flank radius
r_i	Roller seating radius
r_x	Tooth side radius
SC	Seating clearance
t_1	Thickness of inner plates
t_2	Thickness of outer plates
Z	Number of teeth

Toothed Sprocket Properties

Sprocket dimensional properties are based on a specific chain size, together with the national standard recommendations. For more details on exact sprocket dimensions, please refer to the corresponding chain standards.

Figure 44.2 illustrates a typical roller chain sprocket.

In this chapter, two types of tooth form are considered:

- Theoretical tooth form
- Simplify ISO tooth form

The theoretical tooth form is designed so that the chain rollers ride out toward the tips of the sprocket teeth as the chain wears and elongates. The final tooth form may not exactly match the theoretical tooth form.

Simplified ISO tooth form is determined by the minimum and maximum tooth gap forms and shall have tooth flanks of a form that lies between maximum and minimum flank radii and blending smoothly with the roller seating profile subtending the respective angles.

For maximum service life, smooth operation and optimum performance, the following points should be considered when determining the optimum number of teeth in the pinion:

1. As most drives have an even number of pitches in the chain, the use of a pinion with an odd number of teeth ensures even distribution of chain and wheel tooth wear.
2. Pinions for normal, steady drives should generally not have fewer than 17 teeth, the reason being that a chain forms a polygon around the pinion. When the pinion speed is constant, the chain speed is subject to regular cyclic variation. The percentage of cyclic variation becomes less marked as the number of teeth increases—and, in fact, becomes insignificant for the majority of applications when the number of teeth in the pinion exceeds 17.
3. A minimum of 23 teeth is recommended on moderate shock drives where the speed of the pinion exceeds 50% of the maximum rated speed, and for heavy shock drives where the speed of the pinion exceeds 25% of the maximum rated speed.
4. The pinion should be heat treated to HV 10–550 for smooth drives where the pinion speeds exceed 70% of the maximum speed and operates under full horsepower rating.

The following equations describe the profiles of the sprocket tooth form (Figure 44.3):

$$D_p = \frac{\pi}{\sin\frac{p}{z}} \tag{44.1}$$

$$D_a = D_p \cdot \cos\frac{\pi}{z} + 2 \cdot h_a \tag{44.2}$$

$$D_f = D_p - 2r_i \tag{44.3}$$

$$b_s = p_t(k-1) + b_f \tag{44.4}$$

$$b_a = b_{ax} p \tag{44.5}$$

$$r_i = \frac{d_r}{2} + SC \tag{44.6}$$

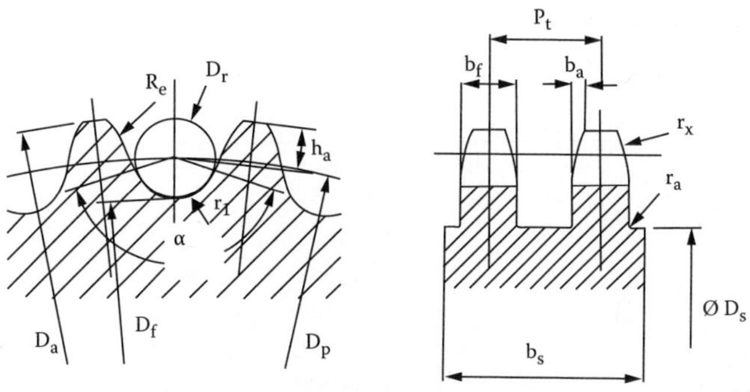

FIGURE 44.3
Roller chain sprocket proportions.

$$D_z = D_p \cdot \cos\frac{\pi}{z} - h_{max} - 2 \cdot r_a \tag{44.7}$$

Measuring Toothed Sprocket

The roller seating radius is a critical dimension when measuring the roller chain sprocket. It controls the position of the chain bushing relative to the tooth profile. If wear has taken place and the chain bushing is sitting lower than the root diameter, then there will be a possibility for the chain to catch the tip of the sprocket, causing eventual damage and shock to the chain, resulting in a reduced life.

One method for checking roller chain sprockets is to take a measurement over pins that have an identical diameter as the chain roller. Figure 44.4a and b shows the sprocket profile for a duplex chain considering even and uneven teeth.

Figure 44.4a and b depicts the method for measuring the sprocket over pins for even and uneven toothed sprockets, respectively, and Table 44.2 shows the terminology used in this method.

Measuring a toothed sprocket:

When measuring an even-toothed sprocket over pins (see Figure 44.4a),

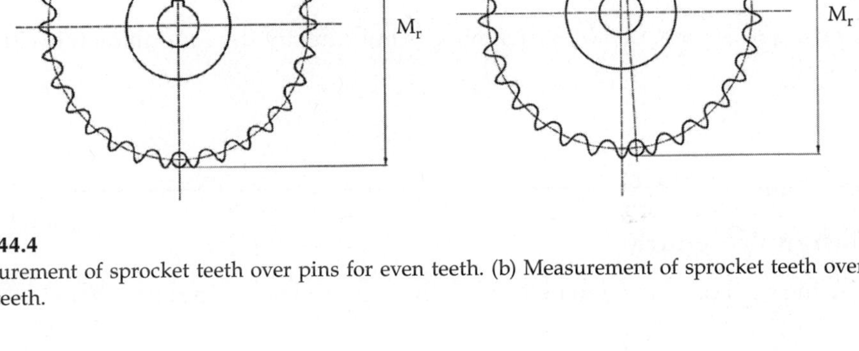

FIGURE 44.4
(a) Measurement of sprocket teeth over pins for even teeth. (b) Measurement of sprocket teeth over pins for uneven teeth.

TABLE 44.2

Nomenclature for Measuring Sprocket Properties

Symbol	Description
D_p	Pitch diameter
D_a	Tip diameter
D_f	Root diameter
d_r	Maximum bush or roller diameter
Z	Number of sprocket teeth
P	Chordal pitch equals to chain pitch
p_t	Strand transverse pitch
k	Number of strands
SC	Seating clearance
r_1	Roller seating radius
r_e	Tooth flank radius
α	Roller-seating angle
h_a	Height of tooth above pitch polygon
b_f	Tooth width
b_a	Tooth side relief
$b_{ax\,f}$	Tooth side relief factor
r_x	Tooth side radius
r_a	Shroud fillet radius
b_s	Minimum shroud width
D_s	Maximum shroud diameter
h_{max}	Maximum plate depth h_{max}=max(h2, h3)

$$M_f = D_P + 2 \cdot D_g - d_r \tag{44.8}$$

When measuring an odd number of teeth over pins (see Figure 44.4b),

$$M_f = D_P \cdot \cos\frac{\pi}{z} + 2 \cdot D_g - d_r \tag{44.9}$$

where:

D_P = Pitch diameter
D_g = Measuring pin diameter
M_f = Measurement over pins or direct measurement
Z = Number of sprocket teeth
d_r = Maximum chain roller diameter

Note:
For measuring over pins, $D_g = d_r$. When measuring directly the root diameter without the use of pins, $D_g = 0$.

44.4 Design Procedure

The following method is used to determine the type and size of a roller chain.

44.4.1 Calculate the Design Power to Be Transmitted (kW)

The design power is the specified input power to the chain drive multiplied by the service factor obtained from Table 44.2.

$$\text{Design power} = \text{specified input power x service factor} \tag{44.10}$$

44.4.2 Determine the Chain Service Factor

From Table 44.3, select the most appropriate service factor covering the chain drive application.

The service factor takes into consideration the conditions the chain drive will be operating under. Shock loads created by the driving machine and driven load require a more robust design. From Table 44.3 determine the type of the driven machinery, and from the correct column for the driving machinery determine the selection factor for the number of teeth on the smallest sprocket.

44.4.3 Determine the Drive Ratio

Use Table 44.4 to select a ratio based on the standard sprocket sizes available. It is preferred to select one pulley with an odd number of teeth, combined with an even number of chain pitches.

Ideally, chain sprockets with a minimum of 19 teeth should be selected. If the chain is operating at high speed or is subjected to impulsive loading, then the smaller sprocket ideally should have a minimum of 25 teeth and should also be hardened.

It is recommended that larger chain sprockets should have a maximum of 114 teeth.

TABLE 44.3
Service Factor

		Source of Power		
			Internal Combustion Engine	
Type of Impact	**Machines**	**Electric Motor or Turbine**	**With Hydraulic Drive**	**Without Hydraulic Drive**
Smooth	Belt conveyors with small load fluctuation, chain conveyors, centrifugal blowers, general textile machines with small load fluctuations	1.0	1.0	1.2
Some impact	Centrifugal compressors, marine engines, conveyors with some load fluctuations, automatic furnaces, dryers, pulverizers, general machine tools, compressors, general paper mills	1.3	1.2	1.4
Large impact	Press, construction or mining machines, vibration machines, oil well rigs, rubber mixers, general machines with reverse or large impact loads	1.5	1.4	1.7

TABLE 44.4

Chain Reduction Ratios to One Using Preferred Sprockets

Number of Teeth	Number of Teeth					
Driven Sprocket (Z_2)	Drive Sprocket (Z_1)					
	15	17	19	21	23	25
25	—	—	—	—	—	1.00
38	2.53	2.23	2.00	1.80	1.65	1.52
57	3.80	3.35	3.00	2.71	2.48	2.28
76	5.07	4.47	4.00	3.62	3.30	3.04
95	6.33	5.59	5.00	4.52	4.13	3.80
114	7.60	6.70	6.00	5.43	4.96	4.56

Calculate the drive ratio R (velocity ratio) given the speed of the input drive and the output drive from the following formula:

$$\text{Drive ratio R} = \frac{\text{Speed of input (rev / min)}}{\text{Speed of output (rev / min)}} = \frac{\text{Number of teeth on input sprocket}}{\text{Number of teeth on driven sprocket}} \tag{44.11}$$

For larger ratio drives, the angle of lap on the smallest sprocket (Z1) should not be less than 120° (see Figure 44.5).

44.4.4 Select Chain Size

From the power rating chart (Figures 44.6 and 44.7), select the most appropriate chain size. The design power rating for simplex, duplex, and triplex chains are shown in the three columns on the left-hand side of the chart. Wherever possible, due to cost factors, use a simplex chain. The design power on the vertical axis is referenced with the speed of the smaller sprocket on the horizontal axis to obtain a chain size and lubrication requirements.

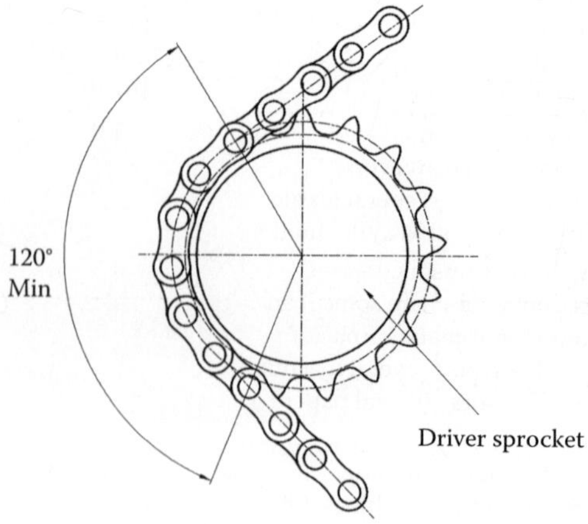

FIGURE 44.5
Angle of lap.

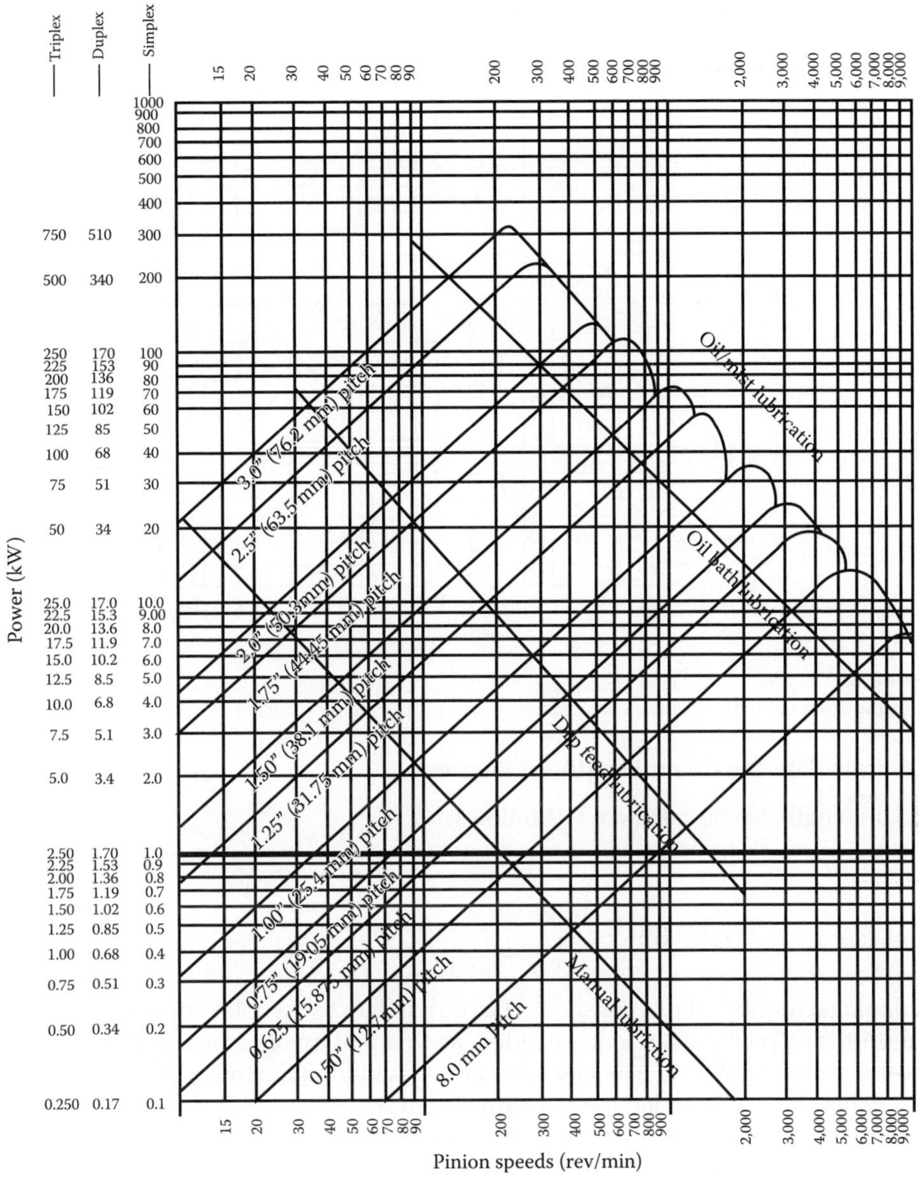

FIGURE 44.6
Power rating chart (BS/ISO). *Note*: Rating chart based on using 19T driving sprocket.

44.4.5 Determine the Sprocket Size

The sprocket pitch diameters can be calculated as follows:

$$\text{Pitch circle diameter} = \frac{\text{Pitch of chain (p)}}{\sin\left(\dfrac{180}{\text{No. of teeth in sprocket}}\right)} \tag{44.12}$$

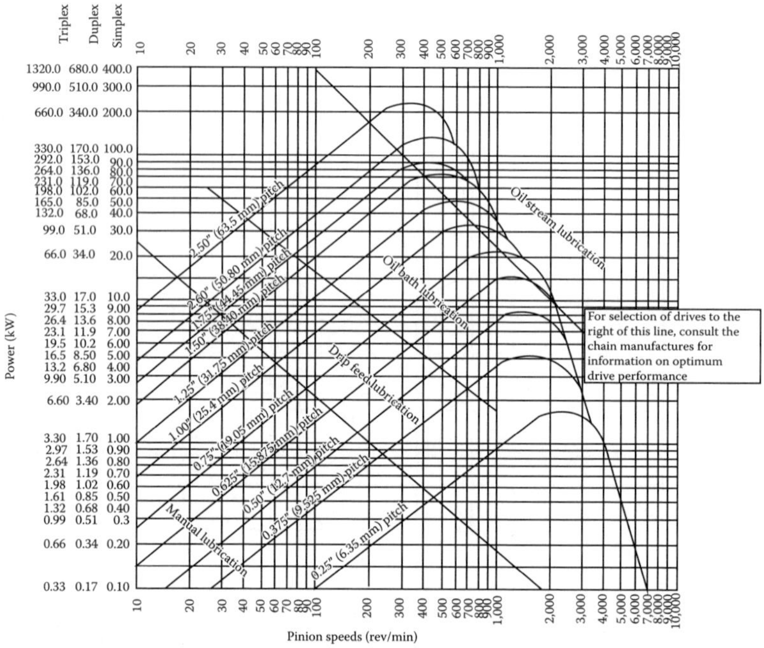

FIGURE 44.7
Power rating chart (US). *Note*: Rating chart based on using 19T driving sprocket.

44.4.6 Calculate the Chain Length (in Pitches)

The chain length can be calculated as follows:

$$\text{Length (L)} = \frac{Z_1 + Z_2}{2} + \frac{2C}{p} + \frac{\left(\frac{Z_2 - Z_1}{2\pi}\right)^2 p}{C} \tag{44.13}$$

The calculated number of pitches should be rounded up to a whole number of even pitches. Odd numbers of pitches should be avoided, as this will involve the use of a cranked link that is not recommended. If a jockey sprocket is used for adjustment purposes, two additional pitches should be added to the chain length (L).

C is the proposed center distance in mm and should generally be between 30 and 50 pitches.

For example,

$$\text{For a 38.1 mm pitch chain, } C = 38.1 \times 40 = 1524 \text{ mm.}$$

44.4.7 Calculate the Actual Center-to-Center Distance (mm)

The actual center distance for the chain length (L) that has been calculated by the method in Section 44.4.8 will generally be greater than that originally considered. The revised center distance can be calculated from Equation 44.12:

$$C = \frac{p}{8}\left[2L - Z_2 - Z_1 + \sqrt{(2L - Z_2 - Z_1)^2 - \left(\frac{\pi}{3.88}(Z_2 - Z_1)^2\right)}\right] \tag{44.14}$$

44.4.8 Lubrication

Chain life will vary extensively depending on the method used to lubricate the drive. A correctly lubricated chain can last as long as 100 times more than the same chain incorrectly lubricated. It is recommended to use a good grade of clean petroleum oil without additives and free flowing at the prevailing temperature. Some additives leave a varnish or gum deposits that prevent the oil entering the chain joints. Heavy oils or greases are generally too stiff to enter the chain joints and should not be used unless the oil is indirectly heated and the chain is immersed in the molten oil.

With correct lubrication, a separate wedge of lubricant is formed between the bushings and pins in the chain joints, similar to that formed between the mating surfaces of a journal bearing. The viscosity of the oil will greatly affect its separating force and the ability to become a wedge between the moving surfaces. The highest viscosity oil that will flow between the chain link plates and fill the pin bushing will provide the best wear protection. This being essential to minimize metal-to-metal contact, and if sufficient volume of lubricant is supplied this will also provide effective cooling and also improve damping at higher chain speeds.

Figure 44.8 is a lubrication chart recommending types of lubrication methods against chain pitch and chain speeds.

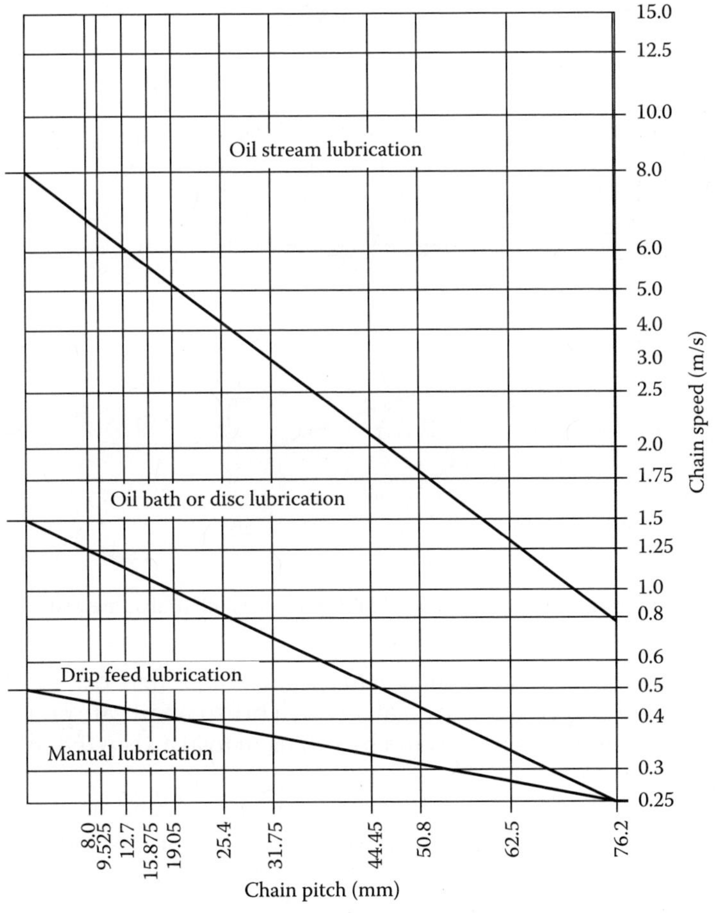

FIGURE 44.8
Lubrication chart.

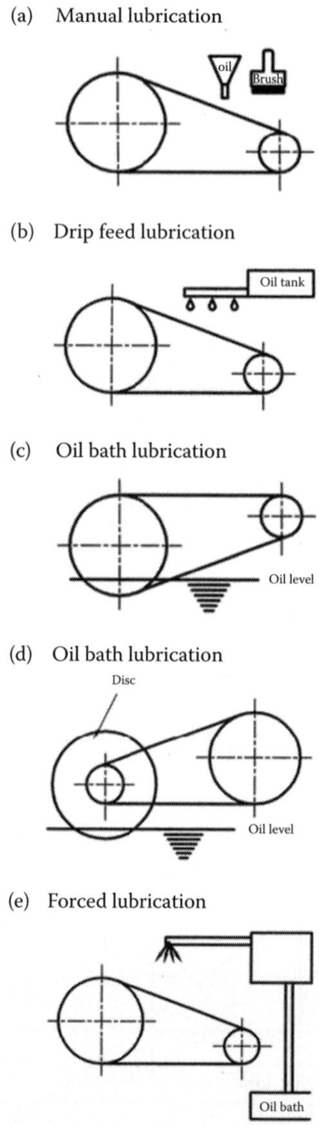

FIGURE 44.9
(a) Manual method of lubrication. (b) Drip method of lubrication. (c) Bath or disc method of lubrication. (d) Oil stream method of lubrication. (e) Forced lubrication.

Figure 44.9a depicts the manual method of lubrication for a chain drive that is low power and low speed. The oil is applied periodically with a brush or oil can, preferably every 8 h of operation. Volume and frequency should be sufficient to just keep the chain wet with oil and allow penetration with clean fluid into the chain joints.

Figure 44.9b describes the drip lubrication method in which oil drips are directed between the link plate edges from the drip lubricator. The volume and frequency should be sufficient to allow the penetration of the oil into the chain pins.

The bath or disc lubrication method is shown in Figure 44.9c in which the strand of the chain runs through a sump of oil in the chain housing where the oil level covers the chain strand at its lowest point.

In some designs, a disc is attached to the lower chain sprocket drive and this runs through the oil and flings the oil onto the chain strand. The oil is deposited onto the chain via deflector plates within the housing. The peripheral speed of the disc should be a minimum of 180 and 2440 m/min.

The fourth method, "oil stream lubrication" (Figure 44.9d), uses a continuously circulating stream of oil directed onto the chain from a circulating pump. It is essential that the spray holes are in line with the chain plate edges. The spray holes should be positioned such that the oil is delivered onto the chain just before it engages with the drive sprocket. This ensures that the oil is centrifuged through the chain and assists in cushioning the roller impact on the sprocket wheel (Table 44.5).

Example 44.1

The task is to provide a suitable roller chain drive for a small rotary kiln (Figure 44.10) to transmit 3.0 kW from a geared electric motor running at 200 rev/min to a rotary kiln at 100 rev/min. Assume moderate shock loads. Use a center distance that is twice the pitch diameter of the wheel sprocket. Figure 44.11 shows the proposed drive schematic of the kiln.

Calculate:

1. The length of the chain in pitches.
2. The exact center distance.

TABLE 44.5

Combinations for Input and Output Sprockets for a Drive Ratio of 2.00

Z_1	30	38	47	57
Z_2	57	76	95	114
R	1.90	2.00	2.02	2.00
Z_2 (PCD) mm	288.18	384.15	480.14	576.13

UNIQUE PERENY RTC Series rotary kiln

FIGURE 44.10
A small rotating kiln.

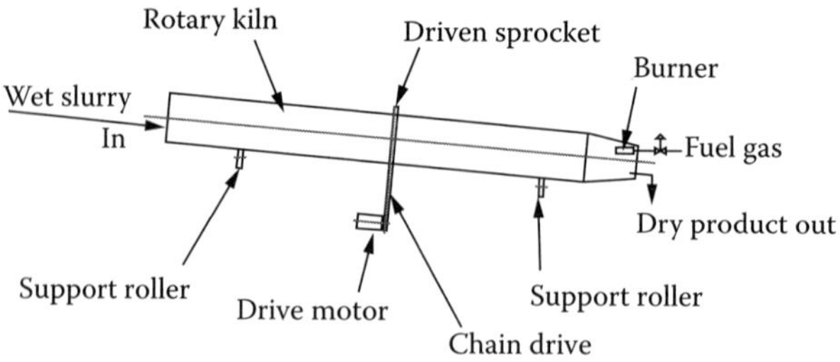

FIGURE 44.11
A schematic of a rotary kiln.

Rotary kilns (or rotary dryers) are used extensively in industry for drying solids or slurries.

The kiln is a cylindrical vessel, inclined slightly to the horizontal, that is rotated slowly about its axis. The material to be processed is fed into the upper end of the cylinder. As the kiln rotates, material gradually moves down toward the lower end, and may undergo a certain amount of stirring and mixing. Hot gases pass along the kiln, sometimes in the same direction as the process material (co-current), but usually in the opposite direction (countercurrent). The hot gases may be generated in an external furnace, or may be generated by a flame inside the kiln. Such a flame is projected from a burner pipe (or "firing pipe") which acts like a large Bunsen burner. The fuel for this can be gas, oil, pulverized petroleum coke, or pulverized coal.

Kilns come in a wide range of sizes from small to very large, several meters in diameter.

Solution

Known data:

$$\text{Power transmitted } (P) = 3.0 \text{ kW}$$

Drive motor — geared electric motor — speed 200 rev / min

Driven machine — rotary kiln — speed 80 rev / min — medium impulse load

Estimated center distance (C) — 2 x pitch circle diameter wheel sprocket

1. Calculate the drive ratio:

$$\text{Drive ratio (R)} = \frac{\text{Speed of driver (rev / min)}}{\text{Speed of driven (rev / min)}}$$

$$= \frac{200}{100}$$

$$(R) = 2.0/1$$

2. Select the sprocket size:

$$Z_1 = \text{number of teeth in pinion (driver)}$$

$$Z_2 = \text{number of teeth in wheel (driven)}$$

$$Z_2 = R \times Z_1 \tag{44.15}$$

Consider chain pitch = 15.875 mm

Note:

The size of the driven sprocket will be dictated by the throat of the rotary kiln. Assuming the design allowed for a throat of 500.0 mm, the appropriate combination would be $Z_1 = 57$ teeth and $Z_2 = 114$ teeth, having a pitch circle diameter (PCD) of 576.13 mm and giving a drive ratio of 2.0 (see Table 44.5).

The most suitable combination is

$$Z_1 = 57,\ Z_2 = 114$$

Exact drive ratio (R) = 114/57

= 2.0 (required drive ratio)

Determine the service factor:

From Table 44.3,

Medium impact load − service load = 1.3.

3. Calculate the design power:

$$P = \text{Theoretical power x service factor}$$
$$P = 3.0\ \text{kW x } 1.3$$
$$P = 3.9\ \text{kW}$$

4. Select the chain size: from Figure 44.8 together with an approximate chain speed.

Chain selection:

15.875 mm pitch, manual lubrication

19.05 mm pitch, drip lubrication

5. Determine the center distance:

$$\text{Given: } C = 2 \text{ x PCD wheel (44.16)}$$
$$= 2 \text{ x } 114 \text{ mm}$$
$$C = 228 \text{ mm}$$

6. Length of chain in pitches:

$$L = \frac{Z_1 + Z_2}{2} + \frac{2C}{p} + \frac{\left(\frac{Z_2 - Z_1}{2\pi}\right)^2 p}{C} \tag{44.17}$$

$$L = \frac{57 + 114}{2} + \frac{2 \times 228}{15.875} + \left[\left(\frac{\frac{114 - 57}{2\pi}}{228}\right)^2 \times 15.875\right]$$

$$L = 158 \text{ pitches}$$

TABLE 44.6

Nomenclature for Figure 44.14

Curve	Description
A	Chain drive power capacity limited by link plate fatigue
B	Chain drive power capacity limited by roller and bush impact fatigue
C	Chain drive power capacity limited by pin-bush galling

7. Actual center distance:

$$CD = \frac{p}{8}\left[2 \cdot L - Z_2 - Z_1 + \sqrt{\left[(2 \cdot L - Z_2 - Z_1)^2 - \left[\frac{\pi}{3.88}(Z_2 - Z_1)^2\right]\right]}\right] \quad (44.18)$$

$$CD = \frac{15.875}{8}\left[2 \times 113 - 57 - 23 + \sqrt{\left[(2 \times 158 - 114 - 57)^2 - \left[\frac{\pi}{3.88}(114 - 57)^2\right]\right]}\right]$$

$$CD = 163 \text{ mm}$$

44.5 Roller Chain Calculation Basics

The following notes cover the basic issues to be considered when designing a roller chain drive system.

44.5.1 Wear

Wear in the roller chain drive is an important consideration. Roller chains are normally most affected by chain joint wear and sprocket wear.

Chain joint wear will result in the roller chain increasing in length. Sprockets are designed to accept an increase in length up to 3% from wear. When the chain has elongated beyond this value, it no longer fits the sprockets and the chain drive will not operate correctly. If the worn chain is to be replaced, it is recommended that the sprockets are either replaced as well or the sprocket (if possible) is reversed on the shaft; this sometimes extends the life of the worn sprocket.

Sprocket wear will result in the modification of the tooth shape. The teeth will begin to take on a hooked shape, as shown in Figure 44.12. For idler sprockets, these usually wear at the bottom of the tooth shape (see Figure 44.13). When the tooth shape is worn deeply enough, the roller chain may bind against the tooth tips as the link enters and leaves the idler sprocket. This type of wear can be the source of shock loads in the chain, resulting in reduced chain life.

See Section 44.4.9 on considerations for lubrication.

Normally at the beginning of the chain drive service, the wear progresses rapidly as the chain beds in. Initial wear can be minimized by preloading the chain, a practice some chain manufacturers undertake to increase the service life of the chain.

The chain joint wear continues at a reduced rate and becomes what is known as normal wear. If the chain is adequately lubricated, then the chain joint wear continues to exhibit normal wear to a point when the chain reaches the end of its useful life. When this point is reached, the chain joint wear begins to rapidly increase again.

44.5.2 Ultimate Tensile Strength of the Chain F_u

The ultimate tensile strength of the roller chain is the highest value of load that the chain can withstand in a single application of load before breaking. A minimum value of the ultimate tensile strength is used as a quality-control criterion to ensure that the chain has been manufactured and assembled correctly.

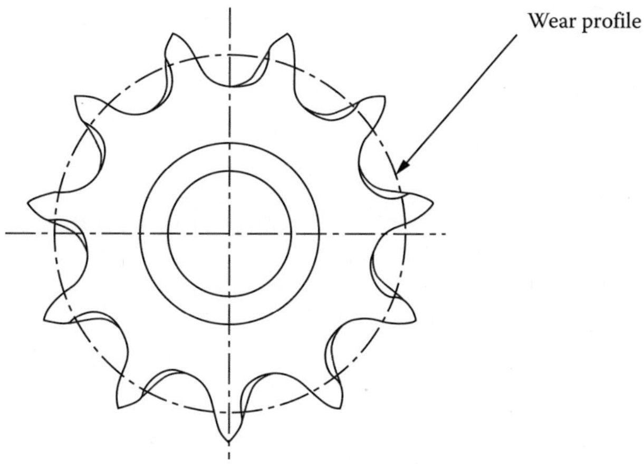

FIGURE 44.12
Wear profile on a driver sprocket.

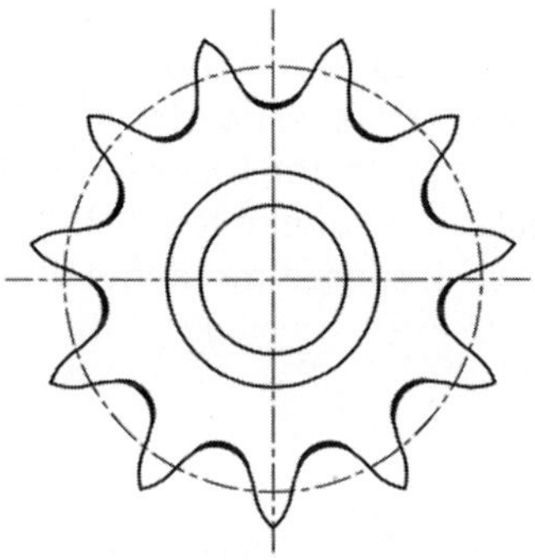

FIGURE 44.13
Wear profile on an idler sprocket.

The default value of the ultimate strength is arrived from standard recommendations for a given chain size. This tensile strength may differ for the same size of chain between different chain manufacturers as well as different materials.

44.5.3 Specific Chain Mass m

The specific mass of the chain (m) is used for computing the centrifugal force acting on the chain, and is also used in vibration analysis in the chain drive. The default value is taken from standard recommendations or the closest value of a steel chain produced by the chain manufacturers.

44.5.4 Chain Construction Factor

The chain construction factor defines the actual quality of the chain. It has a direct impact on the chain power rating as well as the chain's permissible bearing area pressure. The default factor is equal to unity. If the chain is manufactured from material that has improved strength to the default chain, then the factor will be greater than unit, or if the quality of the chain is better than that mentioned in national standards.

44.5.5 Chain Power Rating P_R

The chain power rating (P_R) represents the chain capacity rating for specific operating conditions. Normally, the chain capacity is limited by link plate fatigue, roller and bushing impact fatigue and galling between the pin and bushing. See Figure 44.14 and Table 44.6.

44.5.6 Power Correction Factors

Chain power rating equations provide valid power capacity for chain drives that work under normal operating conditions.

The criteria for a roller chain drive operating under normal conditions are covered in Table 44.7.

44.5.7 Shock Factor F_s

The service factor (F_S) takes into account the dynamic dependency to overloads on the chain drive operating conditions, driver, and driven characteristics. The shock factor determines the degree of the service factor together with the dynamic factor of safety (FoS). The peak loads can be caused by unexpected shock, and these peak overloads can significantly increase when there are large moments of inertia within the drive system (see Table 44.8).

44.5.8 Service Factor S_f

The service factor (S_f) considers any dynamic overloads occurring in the chain drive due to the driving conditions. The values of the FoS are listed in Table 44.9.

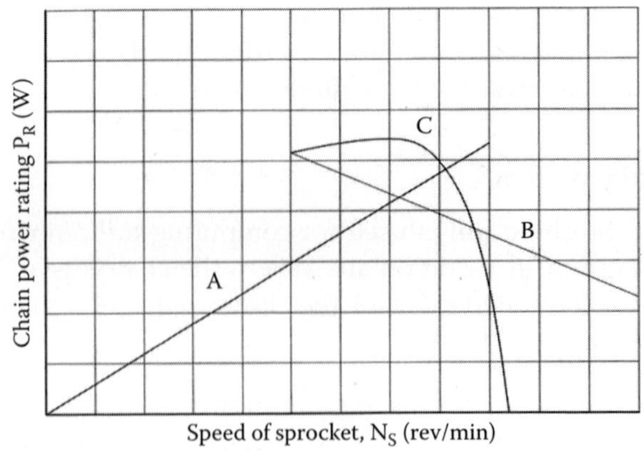

FIGURE 44.14
Chain power rating.

TABLE 44.7

Criteria for Normal Operating Conditions

1	Chain drive with two sprockets on parallel horizontal shafts
2	Small sprocket with 19 teeth
3	Simplex chain without a cranked link
4	Chain length:
	120 pitches of ISO chain
	100 pitches of ANSI and CSN chains
5	Speed ratio of 1:3 or 3:1
6	Expected service life:
	15,000h for ISO, ANSI, and Din chains
7	Operating temperature between 5°C and +70°C
8	Sprocket correctly aligned and chain maintained with correct adjustment
9	Uniform operation without overload, shocks, or frequent starts
10	Clean and adequate lubrication throughout the working life of the chain

TABLE 44.8

Shock Factors in Chain Drives

Shock Factor F_s	Application
1.0	Smooth running
1.5	Smooth running with occasional shocks
2.0	Slight shocks with moderate temporary overloads
3.0	Moderate shocks with heavy temporary peak overloads
4.0	Heavy shocks with moderate constant peak overloads
5.0	Heavy shocks with heavy constant peak overloads

TABLE 44.9

Values of FOS

Driven Machine Characteristics	Driver Machine Characteristics		
	Smooth Running	Slight Shocks	Moderate Shocks
Smooth running	1.0	1.1	1.3
Moderate shocks	1.4	1.5	1.7
Heavy shocks	1.8	1.9	2.1

44.5.9 Definitions of Characteristics of Driver Machines

Table 44.10 defines the characteristics of driver machines used in roller chain drive systems. Table 44.11 gives examples of driven machine types used in roller chain drive systems.

44.5.10 Sprocket Size Factor (f_2)

The sprocket size factor (f_2) takes into account the number of teeth on the smallest sprocket. The factor is equal to one if the power rating equations consider the number of teeth on the smallest sprocket in the drive. The size of the smallest sprocket may have a specific impact on each portion of the chain power rating.

TABLE 44.10

Characteristics of Driver Machines

Driver Machine Characteristics	Examples of Drive Machine Types
Smooth running	Electric motors, steam and gas turbines, internal combustion engines fitted with a hydraulic coupling
Slight shocks	Six-cylinder or more internal combustion engines fitted with mechanical coupling. Electric motors subject to frequent stop and starts greater than twice a day
Moderate shocks	Internal engines having fewer than six cylinders fitted with mechanical coupling

TABLE 44.11

Characteristics of Driven Machines

Driven Machine Characteristics	Examples of Driven Machine Types
Smooth running	Centrifugal pumps and compressors, printing machines, uniformly loaded belt conveyors, escalators, liquid mixers and agitators, rotary driers
Moderate shocks	Reciprocating pumps and compressors having three or more cylinders, concrete mixers, non-uniformly loaded conveyors, solid agitators and mixers
Heavy shocks	Excavators, roll and ball mills, rubber-processing machines, planers, presses, compressors having more than one or two cylinders, oil drilling rigs

The relationship between the sprocket size factor (f_2) and the number of teeth on the smallest sprocket is

$$f_2 = \left(\frac{19}{Z_s}\right)^{1.08} \tag{44.20}$$

where Z_s is the number of teeth on the smallest sprocket.

44.5.11 Lubrication Factor (f_4)

The lubrication factor (f_4) determines the amount of lubrication the chain is receiving, and has a direct bearing on the chain power capacity as well as service life. If the lubrication is adequate, then the factor will be equal to 1. If the lubrication is inadequate, then the factor will decrease the chain power rating, which will be limited by pin-bush galling. Table 44.12 shows the effect on the degree of lubrication received.

44.5.12 Center Distance Factor (f_5)

The minimum center distance factor (f_5) is one half the sum of the outside sprocket diameters, to avoid tooth interference. To ensure adequate wrap on the small sprocket (approximately 120°), it is suggested to have minimum center distance of the sum of the outside diameter of the large sprocket plus one-half the outside diameter of the small sprocket.

TABLE 44.12

Effect of Lubrication with Respect to Chain Speed

	Lubrication Factor			
	Lubrication			No Lubrication
		Insufficient		
Chain Speed (m/s)	**Recommended**	**Clean Environment**	**Soiled Environment**	
Up to 4	1.0	0.6	0.3	0.15
Up to 7		0.3	0.15	Inadmissible
Up to 12		Inadmissible		
More than 12				

It is good practice to set the çenter distance at 30–50 times the chain pitch. The longest practical center distance is about 80 times the chain pitch, because chain sag and catenary tension become very large.

The center distance factor corrects the design power and takes into account the difference of the actual center distance from the normal. The reason for the center distance factor comes from modification of load tension distribution and its affect to chain fatigue. The center distance factor is determined as follows:

$$f_5 = \left(\frac{X_B}{X}\right)^{0.4} \tag{44.21}$$

where

f_5 = Center distance factor

X_B = Number of chain links for normal operating conditions

X = Actual number of chain links in the drive

44.5.13 Ratio Factor (f_6)

The design power is corrected by the ratio factor (f_6) and it takes into account the difference between the transmission ratio from the normal. The reason for this factor comes from the modification of load tension distribution and its effect on the chain fatigue. Figure 44.15 is a chart showing the ratio factor with respect to the actual transmission ratio.

Transmission ratio is given from the number of teeth on the driver and driven sprockets:

$$\text{For } z_1 < z_2,\ i = Z_2 / Z_1 \tag{44.21}$$

$$\text{For } z_1 > z_2,\ i = Z_1 / Z_2 \tag{44.22}$$

where

f_6 = Ratio factor

i = Transmission ratio

Z_1 = Number of teeth on the driver sprocket

Z_2 = Number of teeth on the driven sprocket

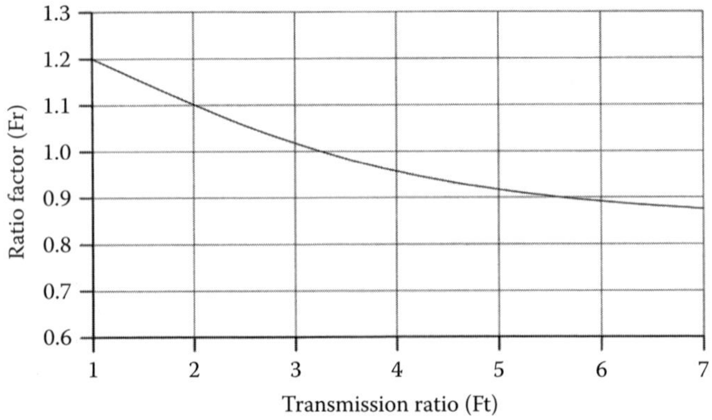

FIGURE 44.15
Relationship between the ratio factor and the transmission ratio.

44.5.14 Service Life Factor (f_7)

The service life factor (f_7) corrects the design power and takes into account the difference of the required service life from the normal. By default, the ratio factor is determined as follows:

$$f_7 = \left(\frac{L_h}{L_{hB}}\right)^{0.4} \tag{44.23}$$

where
f_7 = Service life factor
L_h = Required service life (h)
L_{hB} = Normal expected service life (h)

44.5.15 Pressure in Chain Bearing Area Analysis

The fluctuating tensile load acting in the roller chain during service acts on the contact surfaces between the pin and bushing, and it is this that causes the level of pressure between the chain bearing area. If this pressure exceeds the permissible pressure, then the chain service life will be significantly reduced. The level of the actual bearing pressure in the chain bearing area is computed from Equation 44.22 and is based on the maximum tension in the chain span:

$$P_B = \left(\frac{F_{T\max}}{A}\right) \tag{44.24}$$

where
P_B = Actual pressure in chain bearing area (Pa)
$F_{T\max}$ = Maximum tension in taught section of chain span (N)
A = Chain bearing area (m^2)

44.5.16 Chain Bearing Area (A)

The chain bearing area (A) is defined by the width of the internal chain link and the pin diameter, as shown in Table 44.13.

TABLE 44.13

Area of Pin and Bush

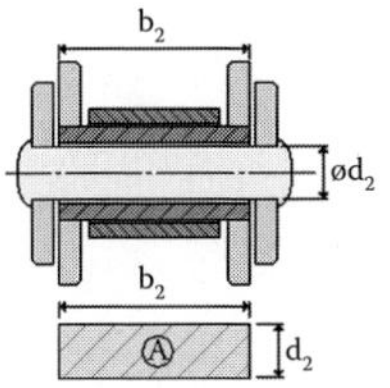

$A = b_2 d_2$

where

A = Chain bearing area (m^2)

b_2 = Width of internal chain link (m)

d_2 = Chain pin diameter (m)

44.5.17 Permissible Pressure in Chain Bearing Area (P_o)

The permissible pressure obtained from Figure 44.16 is also corrected by the chain construction factor, so the amount of permissible pressure in the chain-bearing area (P_o) is defined as follows:

$$P_o = P_{Bo}\ \varphi \tag{44.25}$$

where

P_o Permissible pressure in chain-bearing area for normal conditions (Pa)

P_{Bo} Specific permissible pressure in chain-bearing area for normal operating conditions (Pa)

φ Chain construction factor

44.5.18 Specific Friction Factor (λ)

The specific friction factor corrects the permissible pressure in the chain bearing area. The size of the factor is dependent on the difference between the actual operating conditions and the normal, and is defined as follows:

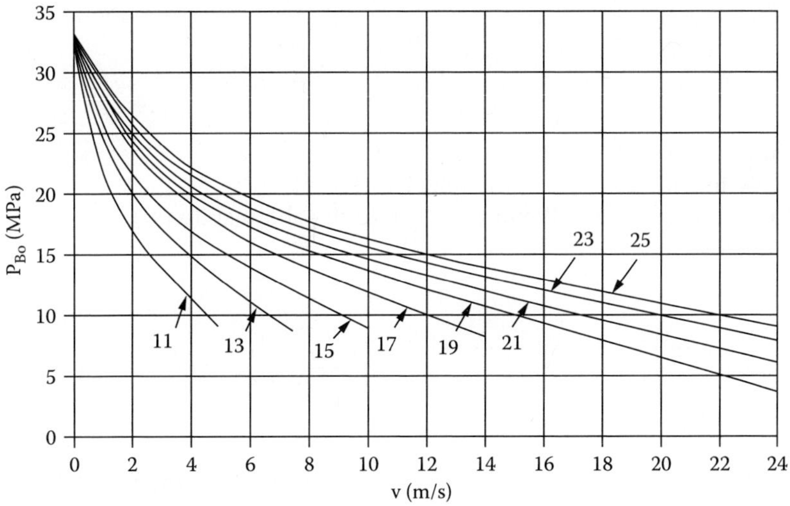

FIGURE 44.16
Permissible pressure in chain bearing.

$$\lambda = \frac{f_4}{f_1, f_5, f_6, f_7} \tag{44.26}$$

where

λ = Specific friction factor
f_1 = Service factor
f_4 = Lubrication factor
f_5 = Center distance factor
f_6 = Ratio factor
f_7 = Service life factor

44.6 Static Safety Factor

The static FoS from the chain breaking under a constant load is determined as follows:

$$S_s = \frac{F_u}{F_{tmax}} \geq S_{smin} \tag{44.27}$$

where

S_{smin} = Minimum allowed static safety factor
F_u = Ultimate tensile strength of the selected chain (N)
F_{tmax} = Maximum tension in the taut section of the chain span (N)

44.7 Dynamic Safety Factor

The dynamic FOS from selected chain breaking from peak load is determined as follows:

$$S_D = \frac{F_u}{F_{tmax}.Y} \geq S_{Dmin} \tag{44.28}$$

where

S_{Dmin} = Minimum allowed dynamic safety factor
F_u = Ultimate tensile strength of the selected chain (N)
F_{tmax} = Maximum tension in the taut section of the chain span (N)
Y = Shock factor

44.8 Design Power and Chain Power Rating

The chain power rating P_R is compared to the design power P_D. The chain power rating must be greater than the design power for a satisfactory life:

$$P_D > P_R$$

where:

$$P_D = P \cdot f_1 \cdot f_2 \cdot f_5 \cdot f_6 \cdot f_7 \cdot \frac{1}{f_4} \tag{44.29}$$

$P_R = P_{RN} \cdot f_3 \cdot C_f$
P = Power to transmit (W)
P_{RN} = Single-strand chain power rating for normal operating conditions (W)
f_1 = Service factor
f_2 = Sprocket size factor
f_3 = Chain strand factor
f_4 = Lubrication factor
f_5 = Center distance factor
f_6 = Ratio factor
f_7 = Service life factor
C_f = Chain construction factor

45

Lubrication

45.1 Introduction

Lubrication is the process or technique of reducing friction between two components and wear on one or both surfaces in contact with each other and moving relative to each other. The lubricant may be either a liquid (requiring continuous replacement) or a solid, in the form of molybdenum disulfide (MoS_2) or graphite, or a combination of both.

Lubrication is required for the correct operation of machine systems including cams, bearings, pumps, cutting tools, etc. Without lubrication, the pressure between the contacting surfaces would generate enough heat to lead to rapid surface damage, which could lead to the surfaces literally welding themselves resulting in seizure.

45.2 Types of Lubrication

There are five distinct forms of lubrication:

1. Hydrodynamic
2. Hydrostatic
3. Elastohydrodynamic
4. Boundary
5. Solid film

Each type of lubricant has its own special properties, which will be examined further.

45.2.1 Hydrodynamic Lubrication

Hydrodynamic lubrication is where the load-carrying surfaces of a bearing are separated by a relatively thick film of lubricant. Hydrodynamic lubrication is also referred to as "full-film" or "fluid lubrication."

45.2.2 Hydrostatic Lubrication

Hydrostatic lubrication is obtained by introducing the lubricant under pressure into the load-bearing area. The pressure is high enough to separate the two surfaces with a relatively thick film of lubricant, which can be air or water, as well as oil. Unlike either hydrostatic or elastohydrodynamic lubrication, this type of lubrication does not require any motion between one surface and another.

45.2.3 Elastohydrodynamic Lubrication

Elastohydrodynamic lubrication is the phenomenon that occurs when a lubricant is introduced between surfaces that are moving relative to each other, such as rolling contact with mating gears or in a rotating plain bearing. Stresses generated between two circular bodies of different diameters is explained using Hertzian theory of contact stresses.

45.2.4 Boundary Lubrication

Boundary lubrication is handy in situations where there is insufficient surface area, a reduction in the velocity of the moving surface, a lessening in the quantity of lubricant that is delivered to the bearing surface or an increase in the bearing load, as well as an increase in the lubricant temperature resulting in decrease in the viscosity, any of which factors may result in the prevention of a film thick enough for full-film lubrication. When this occurs, the highest asperities may be separated by lubricant films only several molecular dimensions in thickness.

45.2.5 Solid–Dry Lubricants

Solid-dry lubrication is used for conditions when conventional liquid lubricants are inadequate and also the easiest and most convenient method of lubricating contact surfaces.

Typical applications for using dry lubrication include the following:

- *Reciprocating motion.* In applications that include sliding or reciprocating motion requiring lubrication to minimize wear such as in gear and chain drives, liquid lubrication will be squeezed out where solid lubricants are retained. Dry lubrication helps prevent fretting corrosion and galling.
- *Coating fasteners.* A solid lubricant applied to fasteners helps to reduce thread seizure and enables release even after a long stay and exposure to high temperatures.
- *High temperature.* Graphite and MoS_2 exhibit high temperature properties and are able to survive in oxidizing atmospheres.
- *Extreme contact pressures.* Due to the lamellar structure that orients parallel to the sliding surface results in a high bearing-load combined with a low shear stress, most applications in metal forming involving plastic deformation will utilize solid lubricants.

45.2.5.1 Graphite

Graphite is structurally composed of planes of polycyclic carbon atoms that are hexagonal in orientation. It is best suited for lubrication in a regular atmosphere. Water vapor is a necessary component for graphite lubrication, where the absorption of water reduces the bonding energy between the hexagonal planes of graphite to a lower level than the adhesion energy between a substrate and the graphite. Because water vapor is a requirement for lubrication, graphite is not suitable for use in a vacuum or low-pressure atmosphere.

In cases in an oxidative atmosphere at temperatures up to a continuous 450°C, graphite is effective and can in fact withstand much higher temperature peaks. The thermal conductivity is generally low ~1.3 W/mK at 40°C.

There are two main groups of graphite: natural and synthetic. Synthetic graphite is a high-temperature sintered product and is characterized by its high purity of carbon (99.5%–99.9%). The primary grade synthetic graphite can approach the good lubricity of quality natural graphite.

45.2.5.2 Molybdenum Disulfide (MoS_2)

MoS_2 is a mined material found within thin veins of granite and is highly refined in order to achieve a purity suitable for use as a lubricant. Just like graphite, MoS_2 has a hexagonal crystal structure and has the intrinsic property of easy shear. MoS_2 lubrication performance often exceeds that of graphite and is effective in a vacuum where graphite is not. The temperature limitation of MoS_2 at 400°C is restricted by oxidation. The particle and film thickness are important parameters of MoS_2 and should be matched to the surface roughness of the substrate. Large particles may result in excessive wear by abrasion caused by the impurities in the MoS_2. Small particles may result in accelerated oxidation.

45.2.5.3 Boron Nitride

Boron nitride is a ceramic powder lubricant. Its principle feature is its high temperature resistance of 1200°C in an oxidizing atmosphere. Boron has a high thermal conductivity.

Boron nitride is available in two chemical structures:

- Hexagonal, which is the lubricating version
- Cubic, which is used as an abrasive and cutting tool component

45.2.5.4 PTFE

Polytetrafluoroethylene (PTFE) is widely used as an additive in lubricating oils and greases. Due to the low surface energy of PTFE, stable unflocculated dispersions of PTFE in oil or water can be produced. Contrary to the other solid lubricants, PTFE does not have a layered structure. The macromolecules of PTFE slip easily along each other, similar to lamellar structures. PTFE shows one of the lowest coefficient of static and dynamic friction, down to 0.04. Operating temperature is limited to approximately 260°C.

45.2.5.5 Tungsten Disulfide (WS_2)

Tungsten disulfide (WS_2) is one of the most lubricous materials currently known to science. With a coefficient of friction at 0.03, it offers excellent dry lubricity unmatched by any other substance.

WS_2 can be used instead of MoS_2 and graphite in almost all applications. Molybdenum and tungsten are from the same chemical family. Tungsten is heavier and more stable and is used extensively by NASA, and the aerospace and automotive industries.

In recent years, the price of tungsten was almost 10 times the cost of molybdenum disulfide, but with the steady price increase of molybdenum, the prices are now within a comparative range, making the use of tungsten disulfide commercially attractive.

45.2.5.6 Self-Lubricating Composites

Solid lubricants such as graphite, MoS_2, PTFE, and some other antifriction and antiwear additives are often compounded in polymers and sintered materials. MoS_2, for example, is compounded in materials for sleeve bearings, elastomeric O-rings, carbon brushes, etc. Solid lubricants are compounded in plastics to form a self-lubricating or internally lubricated thermoplastic composite. PTFE particles, for example, when compounded in plastic, form a PTFE film over the mating surface, resulting in a reduction of friction and wear. MoS_2 compounded in nylon reduces wear, friction, and stick-slip. The primary use of graphite-lubricated thermoplastics find use in applications operating in aqueous environments.

45.3 Viscosity

The term *viscosity* describes the property of a fluid that resists the force tending to cause the fluid to flow, and is the measure of the extent to which a fluid possesses this property.

Consider a plate having an area A moving under a force F having a velocity of V over a second fixed plate separated by a film of fluid having a thickness h. The shear stress on the fluid will be F/A. In a Newtonian fluid, the shear stress is directly proportional to the velocity gradient in the fluid: $\frac{dV}{dh}$

Thus,

$$\frac{P}{A} = m \cdot \frac{dV}{dy} \tag{45.1}$$

The proportionality constant m is the viscosity and has units of $\frac{N}{m^2}$.

45.3.1 Dynamic Viscosity

The SI unit for dynamic viscosity is the pascal second (Pa·s). This equates to a Newton second per square meter $\left(\frac{N \cdot s}{m^2}\right)$. In cgs units, the poise is the corresponding unit, which has dimensions of dyne seconds per cm^2. The centipoise (cP = 10^2 poise) is the most common unit used in tables and handbooks—1 Pa·s = 1000 cP. The viscosity of water at ambient temperature is approximately 1 cP.

45.3.2 Kinematic Viscosity

Dynamic viscosity is the ratio and fluid density $\rho \left(\frac{kg}{m^3}\right)$. The SI unit of kinematic viscosity is the $\left(\frac{m^2}{s}\right)$. The equivalent cgs unit is the stoke (St), which has the unit of $\left(\frac{cm^2}{s}\right)$. The unit commonly used is the centistoke (cSt) $-1\,\frac{m^2}{s} = 10^6\,\text{cSt}$.

$$\text{Absolute kinematic vis cos ity}\left(\frac{m^2}{s}\right) = \frac{m}{\rho}\left(\frac{Pa \cdot s}{kg/m^3}\right)$$

$$\text{Kinematic viscosity } (\text{cS}) = \frac{\text{m}}{\rho} \; \frac{(\text{cP})}{(\text{g / cm}^3)}$$

For water with a density of approximately 1000 kg/m^3 (the conversion is simply 1 cSt = 1 cP).

45.3.3 Saybolt Universal Viscosity

The Saybolt universal second (SUS or SSU) is a measure of the *kinematic viscosity* used in *classical mechanics*. It is the measure of time that 60 cm^3 of oil takes to flow through a calibrated tube at a controlled temperature of 38°C. The rate is measured in seconds; the greater the number of seconds the more viscous is the fluid.

This viscosity (seconds) t = Saybolt universal viscosity—seconds. The dynamic viscosity relates to the Saybolt universal (t) by the following formula:

$$\text{Vis cos ity m (P)} = \left(0.00022 \cdot \text{t} - \frac{0.18}{\text{t}}\right)$$

45.3.4 SAE Viscosity Grading

The Society of Automotive Engineers (SAE) is the society of engineers of the American automobile industry, taking independent responsibility for standardization and classification in American automobile construction. The SAE has also taken the responsibility for the motor oils and gear oils into viscosity grades that is used worldwide.

There are nine grades of single-grade motor oils covering six grades of winter oils (SAE 0W to SAE 25W) and three single grades of summer oils. In addition, there are five grades of single-grade gear oils—SAE 75W-90 to SAE 85W-140 (see Table 45.1). Table 45.2 gives a conversion table for a range of viscosity grades.

TABLE 45.1

Comparison between SAE Grades

	Viscosity (cP) at Temperature (0°C) max		Viscosity mm^2/s (cSt) at 100°C	
SAE Viscosity Grade	**Cranking**	**Pumpability**	**Min**	**Max**
0 W	3250 at −30°C	30,000 at −35°C	3.8	
5 W	3600 at −25°C	30,000 at −30°C	3.8	
10 W	3500 at −20°C	30,000 at −25°C	4.1	
15 W	3500 at −15°C	30,000 at −20°C	5.6	
20 W	4500 at −10°C	30,000 at −15°C	5.6	
25 W	6000 at −5°C	30,000 at −10°C	9.3	
20 W			5.6	9.3
30 W			9.3	12.5
40 W			12.5	16.3
50 W			16.3	21.9
60 W			21.9	26.1

TABLE 45.2
Viscosity Conversion Table

Centipoise (CPS)	Poise	CentiStokes	Stokes	Saybolt Seconds Universal
Millipascal Seconds (mPa.s)	(P)	(cSt)	(S)	(SSU)
1	0.01	1.0	0.01	31
2	0.02	2.0	0.02	34
4	0.04	4.0	0.04	38
7	0.07	7.0	0.07	47
10	0.10	10.0	0.10	60
15	0.15	15.0	0.15	80
20	0.20	20.0	0.20	100
25	0.24	25.0	0.24	130
30	0.30	30.0	0.30	160
40	0.40	40.0	0.40	210
50	0.50	50.0	0.50	260
60	0.60	60.0	0.60	320
70	0.70	70.0	0.70	370
80	0.80	80.0	0.80	430
90	0.90	90.0	0.90	480
100	1.00	100.0	1.00	530
120	1.20	120.0	1.20	580
140	1.40	140.0	1.40	690
160	1.60	160.0	1.60	790
180	1.80	180.0	1.80	900
200	2.00	200.0	2.00	1000
220	2.20	220.0	2.20	1100
240	2.40	240.0	2.40	1200
260	2.60	260.0	2.60	1280
280	2.80	280.0	2.80	1380
300	3.00	300.0	3.00	1475
320	3.20	320.0	3.20	1530
340	3.40	340.0	3.40	1630
360	3.60	360.0	3.60	1730
380	3.80	380.0	3.80	1850
400	4.00	400.0	4.00	1950
420	4.20	420.0	4.20	2050
440	4.40	440.0	4.40	2160
460	4.60	460.0	4.60	2270
480	4.80	480.0	4.80	2380
500	5.00	500.0	5.00	2480
550	5.50	550.0	5.50	2660
600	6.00	600.0	6.00	2900
700	7.00	700.0	7.00	3380
800	8.00	800.0	8.00	3880
900	9.00	900.0	9.00	4300
1,000	10.00	1000.0	10.00	4600

(*Continued*)

TABLE 45.2 (*Continued*)
Viscosity Conversion Table

Centipoise (CPS) Millipascal Seconds (mPa.s)	Poise (P)	CentiStokes (cSt)	Stokes (S)	Saybolt Seconds Universal (SSU)
1,100	11.00	1100.0	11.00	5200
1,200	12	1200	12	5620
1,300	13	1300	13	6100
1,400	14	1400	14	6480
1,500	15	1500	15	7000
1,600	16	1600	16	7500
1,700	17	1700	17	8000
1,800	18	1800	18	8500
1,900	19	1900	19	9000
2,000	20	2000	20	9400
2,100	21	2100	21	9850
2,200	22	2200	22	10300
2,300	23	2300	23	10750
2,400	24	2400	24	11200
2,500	25	2500	25	11600
3,000	30	3000	30	14500
3,500	35	3500	35	16500
4,000	40	4000	40	18500
4,500	45	4500	45	2100
5,000	50	5000	50	23500
5,500	55	5500	55	26000
6,000	60	6000	60	28000
6,500	65	6500	65	30000
7,000	70	7000	70	32500
7,500	75	7500	75	35000
8,000	80	8000	80	37000
8,500	85	8500	85	39500
9,000	90	9000	90	41080
9,500	95	9500	95	43000
15,000	150	15000	150	69400
20,000	200	20000	200	92500
30,000	300	30000	300	138500
40,000	400	40000	400	185000
50,000	500	50000	500	231000
60,000	600	60000	600	277500
70,000	700	70000	700	323500
80,000	800	80000	800	370000
90,000	900	90000	900	415500
100,000	1000	100000	1000	462000
125,000	1250	125000	1250	578000
150,000	1500	150000	1500	694000
175,000	1750	175000	1750	810000
200,000	2000	200000	2000	925000

45.3.5 ISO Viscosity Classification

Here, you will find kinematic viscosity for industrial oils according to ISO 3448.

In the 1970s, the International Standards Organization (ISO) together with ASTM (American Society for Testing and Materials), DIN (Deutsches Institut für Normung), and others started an attempt to create a common viscosity classification. The result is known as the International Standards Organization Viscosity Grade. This classification applies mainly for use with industrial lubrication. The viscosity of each grade ranges ±10% from the mid-point. The step-in between two subsequent viscosity grades is approximately 50% of the lower one. The temperature of 40°C is related to the operating temperature in machinery. Table 45.3 tabulates the grades.

TABLE 45.3
ISO Viscosity Grading

		Kinematic Viscosity Limits (cSt)	
Viscosity Grade	**Midpoint Viscosity at 40°C**	**Min**	**Max**
IS VG 2	2.2	1.98	2.42
IS VG 3	3.2	2.88	3.52
IS VG 5	4.6	4.14	5.06
IS VG 7	6.8	6.12	7.48
IS VG 10	10	9.00	11.00
IS VG 15	15	13.50	16.50
IS VG 22	22	19.80	24.20
IS VG 32	32	28.80	35.20
IS VG 46	46	41.40	50.60
IS VG 68	68	61.20	74.80
IS VG 100	100	90.00	110.00
IS VG 150	150	135.00	165.00
IS VG 220	220	198.00	242.00
IS VG 320	320	288.00	352.00
IS VG 460	460	414.00	506.00
IS VG 680	680	612.00	748.00
IS VG 1000	1000	900.00	1100.00
IS VG 1500	1500	1350.00	1650.00

46

Introduction to Stepper Motors and Their Drives

46.0 Introduction

Modern automatic machine tools and robotics make extensive use of stepper motors in their control systems.

Basically, a stepper motor is a brushless DC electric motor that indexes through a full rotation in a number of fixed equal steps. It can be considered as a digital motor where each step is defined as a step angle.

In the example shown in Figure 46.1, there are six distinct steps for the rotor to complete a 360° rotation. This defines the step at 60°. Because the motor moves in a discrete manner, it can be considered that a stepper motor is a digital motor, making the motor a very suitable digital interface when used with a microcontroller.

For many years, stepper motors have been used in applications such as printers and machine equipment. This type of motor will firmly hold its position at a given step, providing a relatively high holding torque. Other benefits include a higher torque at lower speeds than a typical DC motor, including no need for any mechanical braking.

The stepper motor's position can be controlled to move through an angular position and hold that position until directed to move to a further position. In its simplest form, the motor operates in open-loop and normally does not require a feedback encoder. Provided that the motor is correctly sized and has sufficient torque, it will be sufficient for the safe operation of the drive system. In cases where the size is insufficient to match the torque required, there is a danger that the motor may stall and the discrete counts will be lost, leading to errors. In critical operations, the stepper motor can be fitted with a feedback encoder to monitor the step signals sent and the actual angular rotations made.

46.1 The Basic Construction

For the purposes of this discussion, Figure 46.2 shows a basic stepper motor.

46.1.1 Main Components

A stepper motor has some basic components; the first component is a soft iron stator. Each stator is wrapped with multiple windings that will be energized using a voltage source that initiates a current flow in the stator to produce a polarity at each end (or pole) of the stator.

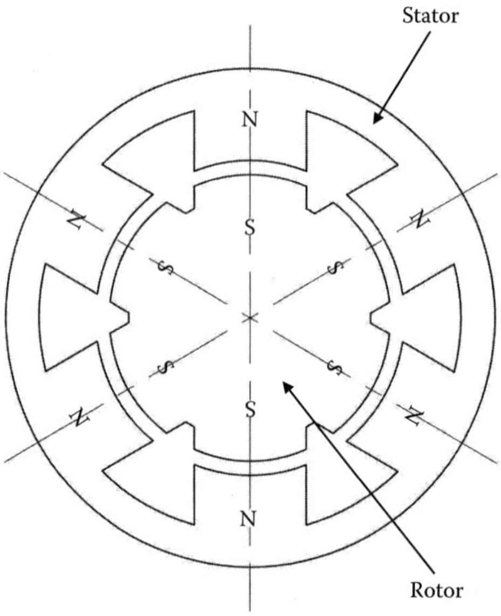

FIGURE 46.1
Stepper motor.

The second component is a soft iron rotor; this is the actual rotating part of the stepper motor and is located between the poles of the stators. The rotor can be either magnetized (as shown here) or nonmagnetized, depending on the type of motor selected. The various types of motor will be discussed later in this chapter.

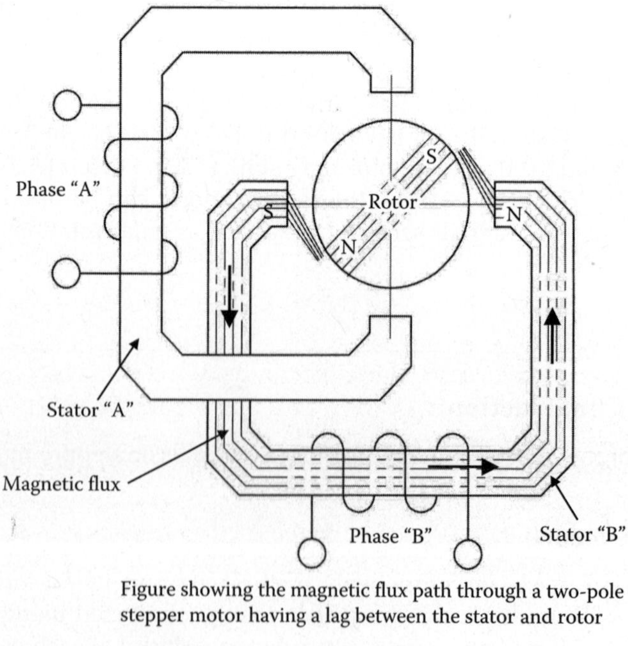

FIGURE 46.2
Basic stepper drive.

When a voltage is applied across the windings around the stator, a current will be induced through the winding. (Remember the right-hand rule). It will be seen that each end of the stator will be magnetized to opposite poles. Magnetic flux will flow from north to south, thereby continuing through the magnetic rotor to the opposite stator pole. The flux will want to travel the path of least resistance.

As the rotor rotates, it will position itself to minimize the flux path (or reluctance). By increasing the number of stators and phases and increasing the charge on a winding will attract the rotor poles accordingly and then removing the applied voltage will allow other stators to attract the rotor poles sequentially.

46.1.2 Types of Stepper Motors

These come in three different forms.

- Permanent magnet
  Magnetic rotor
- Variable reluctance
  Nonmagnetic, geared motor
- Hybrid
  Combines the characteristics of the permanent magnet and variable reluctance motors

It will be shown that the hybrid motor offers high rotational resolutions, together with an improved torque.

Permanent Magnetic Stepping Motor
As stated in the opening introduction, there are three types of stepper motors. The first type to be described is the permanent magnet stepping motor in which the rotor is magnetized. The rotor can be rotated in a particular direction by sequentially energizing the windings to the pairs of corresponding stators. The current direction dictates the direction of rotation of the rotor. To hold the rotor at a particular position, step the rotor to the required angle and stop the sequence, maintaining voltage on the appropriate phase. In subsequent sections, it will be seen that this is not the end of the story. Current limiting will need to be implemented in certain applications. For the purposes of this discussion, this stopping is to be considered the fundamental operational characteristics of this type of motor (Figure 46.3).

Variable Reluctance Stepping Motor
The variable reluctance type of stepping motor uses a nonmagnetic, soft iron rotor (Figure 46.4). The rotor has teeth that are carefully offset from the stator poles to accomplish rotation. Rotation of the rotor is accomplished by sequentially energizing the stator poles. The rotor rotates to minimize the reluctance of the magnetic flux path. The variable reluctance motor normally has three or five stator windings that are energized sequentially to create rotation of the rotor. It should be noted that the motor step angle is actually half of that for a corresponding permanent magnet motor with the same number of stator/windings.

Note that the individual stator windings are configured differently to that of the permanent magnet stepper motor. All the windings have a common terminal that is connected to a voltage source. The opposite ends of each winding are kept separate from the adjacent windings. More teeth can be added to increase the resolution of this type of motor.

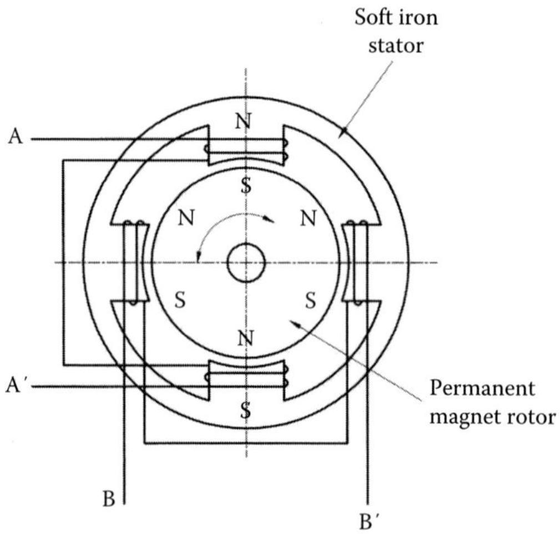

FIGURE 46.3
Permanent magnet stepping motor.

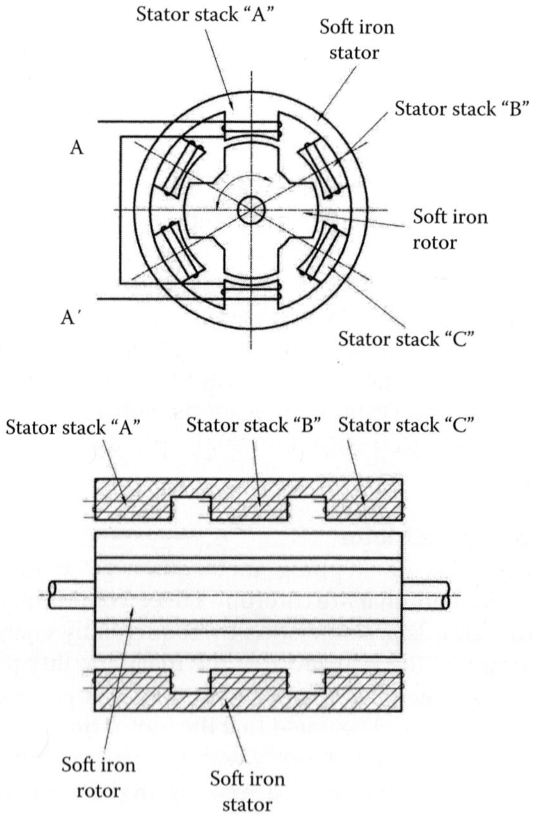

FIGURE 46.4
Variable reluctance stepping motor.

Variable reluctance motors have three or five stator windings, where each is energized one at a time to create a polarity on the appropriate stator poles. The rotor rotating to minimize the reluctance of the magnetic flux path. To rotate the rotor in a particular direction, the stator pole/winding energizing sequence is the reverse of that used on the permanent magnet motor. An advantage with the variable reluctance motor is that the step angle is actually half of that of the permanent magnet motor having the same number of stator/windings.

Hybrid Motor

The hybrid stepping motor borrows characteristics from both the permanent magnet and the variable reluctance stepping motors. Figure 46.5a shows the basic design of the hybrid stepper motor. With this design of motor, the rotor is magnetized and is also fitted with

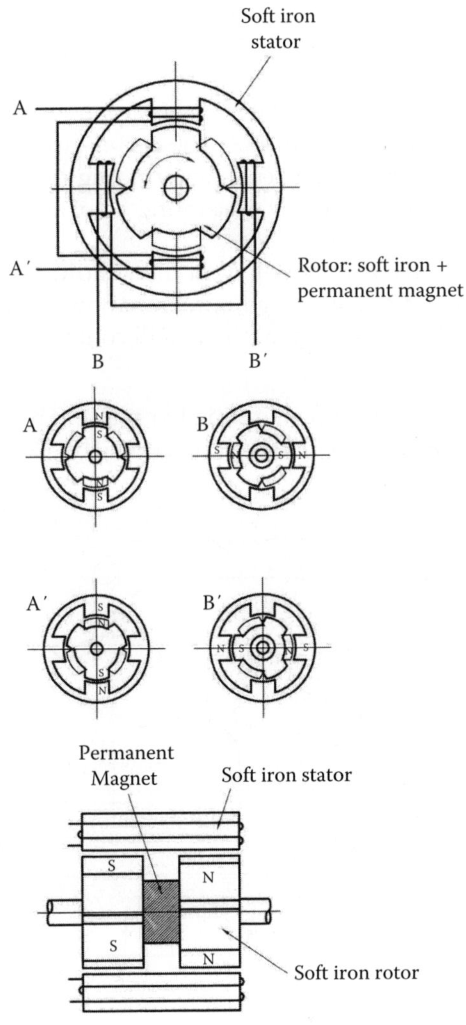

FIGURE 46.5
Hybrid stepping motor.

teeth, and each stator is also fitted with teeth or poles. The figure illustrates that the rotor is in two sections, and the two rotors are opposite in polarity. Also note that the teeth on one rotor are offset by one tooth compared to those on the other rotor, as shown in Figure 46.5b.

46.1.3 Unipolar and Bipolar Stepper Motors

There are basic two types of stepper motors available in market:

- Unipolar stepper motor
- Bipolar stepper motor

Unipolar Stepper Motor

The unipolar stepper motor has five or six wires and four coils (actually two coils divided by center connections on each coil). The center connections of the coils are tied together and used as the power connection. They are called unipolar steppers because power always comes in on this one pole.

Taking a closer look to how the unipolar type of stepping motor works, consider the simplified stepping motor shown in Figure 46.6. The center-tap lead is connected to a positive voltage source. Connecting one of the leads on winding A to earth allows the current to flow in one half of the winding, thereby generating a polarity on the stator poles, and the rotor will rotate accordingly.

Next, the grounding source is removed from the winding A lead, and one of the winding B leads is now connected to earth. Again, current flows in one half of the winding and the appropriate stator poles are energized. This will continue to rotate the rotor 360°.

Bipolar Stepper Motor

Bipolar stepping motors have a single winding per phase. The current in a winding needs to be reversed in order to reverse a magnetic pole. The driving circuit becomes more

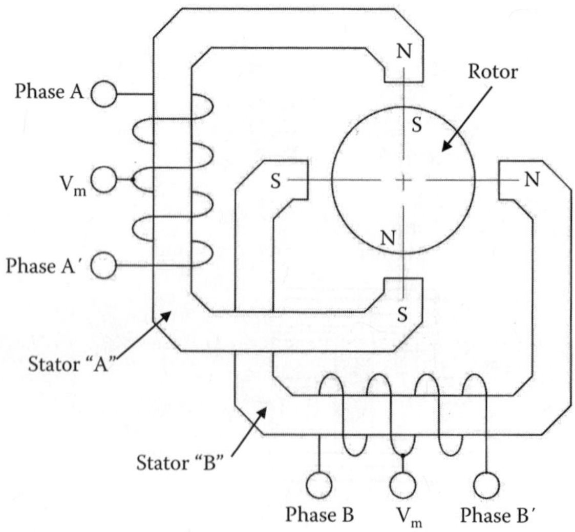

FIGURE 46.6
Unipolar stepper motor.

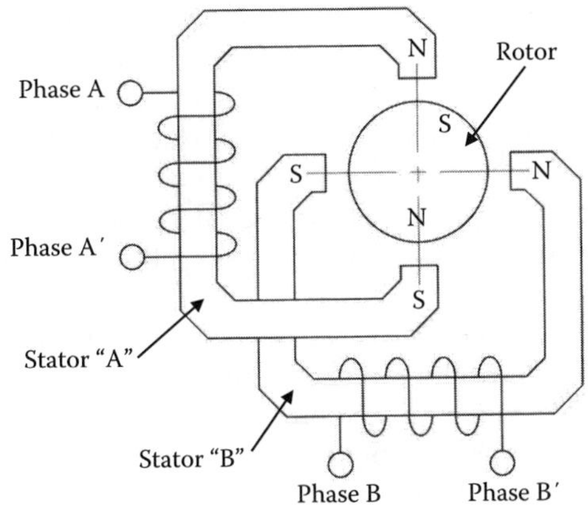

FIGURE 46.7
Bipolar stepper motor.

complicated, typically with an H-bridge arrangement. In this arrangement, there are two leads per phase, in which none are common.

The bipolar motor allows current flow in both directions through each winding. Consider Figure 46.7 applying a voltage to lead A′ and earthing lead A generates current flow, resulting in the stator polarities shown in the figure. Removing the voltage from winding A and applying a positive voltage to lead B′ on winding B and driving lead B to ground generates current flow and stator polarities. This will continue to rotate the rotor 360°.

Summary of Differences between Unipolar and Bipolar Motors
The following differences between unipolar and bipolar motors increase the cost of both the motor with the additional circuitry and controllers, with increased component counts and more complex circuit boards:

1. Bipolar uses the entire winding.
2. Unipolar uses half the windings at any one time.
3. Unipolar windings wire is thinner than bipolar.
   More wire is required, which equates to increased resistance and increased windings temperatures.
4. Bipolar requires more complex circuitry.

46.1.4 Half-Stepping and Micro-Stepping

Up to this point, the discussion has been centered on full-step stepping motors where the rotor moves through the basic step angle (i.e., a 1.8° step motor takes 200 steps per one motor revolution). There are two types of full-step excitation modes, also known as "one-phase on, full-step" excitation. The motor is operated with only one phase energized at a time. This mode requires the least amount of power from the driver of any of the excitation modes.

In dual-phase mode, also known as "two-phase on, full-step" excitation, the motor is operated with both phases energized at the same time. This mode provides improved torque and speed performance. Dual-phase excitation provides approximately 30%–40%

more torque than does a single-phase excitation, but requires twice as much power from the driver.

Half-step excitation is an alternating single- and dual-phase operation that results in steps that are half the basic step angle. This mode, due to the smaller step angle, provides twice the resolution and also results in a smoother operation. The problem is that half-stepping produces 15% less torque than does dual-phase full-stepping. Modified half-stepping eliminates this torque decrease by increasing the current applied to the motor when a single phase is energized.

Micro-stepping is a technique that increases the motor resolution by controlling both the direction and amplitude of current flow in each winding. Current is proportioned in each winding according to sine and cosine functions. Micro-stepping can divide a motor's basic step up to 256 times and also improves low-speed smoothness and minimizes low-speed resonance effects. The down side of micro-stepping is that it produces around 30% less torque than does dual-phase full-stepping.

46.1.5 Resonance

Stepper motors have a natural resonant frequency as a result of the motor being a spring-mass system. When the step rate equals the motor's natural frequency, there may be an audible change in noise made by the motor, as well as an increase in vibration. The resonant point will vary with the application and load, but typically occurs somewhere between 70 and 120 steps per second. In severe cases, the motor may lose steps at the resonant frequency. Changing the step rate is the simplest means of avoiding many problems related to resonance in a system. Also, half-stepping or micro-stepping usually reduces resonance problems. When accelerating to speed, the resonance zone should be passed through as quickly as possible.

46.1.6 Fatigue/Life

In determining the life of a stepper motor drive chain, there are certain issues that could reduce the operating life of the system that be given consideration.

As the stepper motor has no brushes to wear out, the life of the motor is dictated by the life of the bearings and generally exceeds the life of other mechanical components in the drive chain.

If the motor is run at or near its rated torque, life will be affected. The motor's life will be increased exponentially with reduced operating loads. In general, motors should be sized to run at 40%–60% of their maximum load capability. Allowances should be made for environmental factors such as high humidity, exposure to harsh chemicals and aggressive environments.

46.1.7 Summary

This section has reviewed the basic design of stepper motors, and by nature they can be considered as a digital version of a motor moving in distinct steps as defined by their step angle. Three distinctly different types of motors were discussed. The first motor was a basic permanent magnet stepping motor with a magnetized rotor. The second motor was a variable reluctance motor whose characteristics offer a high resolution resulting from the geared nature of the nonmagnetic nature of the rotor. The third motor was the hybrid, which offers the best of both worlds using a geared bipolar rotor to generate optimal torque and resolution.

46.2 Selecting the Size of a Stepping Motor

46.2.1 Overview

When selecting a stepper motor for a particular application, consider these eight parameters:

Ident	Description	Units
1	Operating speed	Steps/s
2	Torque	Nm
3	Rotary inertia	kg·m^2
4	Required step angle	
5	Time to accelerate	ms
6	Time to decelerate	ms
7	Type of drive to be used	
8	Size and weight considerations	

Some of the information will be provided from application specifications from the manufactures, such as size and weight considerations, step angles, and operating speeds. Other information will be required to be calculated.

46.2.2 Calculating Torque

- Torque (Nm):

  $$T = Fr$$

  where

  F = force in N

  r = radius in m

- Rotary inertia (I = moment of inertia kg·m^2):

$$I\left(Nm^2\right) = \frac{mr^2}{2} \text{ (for a disc)}$$

$$I\left(Nm^2\right) = \frac{mr^2}{2}\left(r_1^2 + r_2^2\right) \text{ (for a hollow cylinder)}$$

where

m = mass in kg

r = radius in m of a solid disc

r_1 = inner radius of hollow cylinder

r_2 = outer radius of hollow cylinder

Calculating load:

- The following formulae is for the equivalent inertia to overcome friction in the system together with sufficient torque to start or stop all inertia loads:

$$T = I\alpha$$

where

T = torque in Nm

I = moment of inertia in Nm

α = angular acceleration in rad/s^2

Frictional and Rotational Acceleration Considerations

The formula for calculating the torque required to rotationally accelerate an inertial load is

$$T = 2 \times I_o \frac{\omega'}{t} \times \frac{\pi\phi}{180}$$

where

T = torque in Nm

I_o = inertial load in kg·m^2

π = 3.1416

ϕ = step angle in degrees

ω' = step rate in steps per second

t = time in s

Comparison between Stepping Motors and Servo Motors

For the completeness of this chapter, a comparison between the two categories of stepper motors and servo motors is considered.

Advantages/Disadvantages of Stepper Motors

Advantages

- Low cost for control achieved
- High torque at start-up and low speeds
- Ruggedness
- Simplicity of construction
- Able to operate in an open-loop control systems
- Low maintenance
- Reduced chance of stalling or slipping
- Able to work in any environment
- Can be used in robotics on a wide scale
- High reliability
- The rotation angle is proportional to the input pulse
- The motor has full torque at standstill (provided the windings are energized).
- Precise positioning and repeatability of movement since good stepper motors have an accuracy of between 3%–5% of a step and the error is noncumulative from one step to the next
- Excellent response to starting/stopping/reversing
- Very reliable as there are no contact brushes in the motor. Therefore, the life of the motor is simply dependent on the life of the bearings

- The motor response to digital input pulses provides open-loop control, making the motor simpler and less costly to control
- It is possible to achieve very-low-speed synchronous rotation with the load coupled directly to the drive shaft
- A wide range of rotational speeds can be achieved as the speed is proportional to the frequency of the input pulses

Disadvantages

- Requires a dedicated control system
- Uses more current than a DC motor
- Torque reduces at higher speeds
- Resonances can occur if not properly controlled
- Not easy to operate at extremely high speeds

Servo Motors

One of the main differences between a stepping motor and a servo motor is that servo motors, by definition, run using a control loop and require a feedback of some description. There are a number of different types of control loops. Generally, the PID (proportional, integral, derivative) control loop is mainly used for servo motors.

When using a control loop such as PID, the motor may need to be tuned. Tuning is the procedure of making the motor respond in a desired manner. Tuning a motor can be a difficult and tedious procedure but it has the major advantage that it lets the user have more control over the behavior of the motor.

Because servo motors have a control loop to continuously check the state of the motor, they are generally more reliable than a stepper motor. When a stepper motor misses a step for any reason, there is no control loop to check or compensate for the error. The control loop in the servo motor is constantly checking to see if the motor is on the correct path and, if it is not, then it makes the necessary adjustments.

In general, servo motors run more smoothly than stepper motors except when microstepping is used. Also as speed increases, the torque of the servo motor remains constant, making it a better proposition for when high speeds are required.

Some advantages of servo motors over stepper motors are as follows:

- High intermittent torque
- High torque-to-inertia ratio
- High speeds
- Work well with velocity control
- Available in all sizes
- Quiet in operation

Some disadvantages of servo motors are as follows:

- More expensive than stepper motors
- Cannot work open loop—feedback is always required
- Require tuning of control loop parameters
- More maintenance required due to brushes on brushed DC motors

47

Squirrel-Cage Motors

47.0 Introduction

In the early days of mass manufacture and the widespread of industrial use of electricity, individual machines were powered by flat leather belts driven from roof-mounted network of rotating shafts and pulleys, which in turn were driven by a centralized power source.

In later years, machines were converted to run on individual AC motors, which revolutionized machine shops by removing the need for a centralized power source using either a steam engine or waterwheel.

The General Electric Company (GE) began developing three-phase induction motors in 1891. By 1896, General Electric and Westinghouse had signed a cross-licensing agreement for the bar-winding-rotor design, later called the *squirrel-cage rotor.*

This chapter discusses the mechanical aspects of the motor, where the interest is in selecting a motor to drive a given mechanical load. It will be essential also to appreciate in broad terms the thermal and electrical behavior of motors when selecting a suitable motor.

47.1 Operating Characteristics

The squirrel-cage induction motor is available in single-phase and three-phase forms. The single-phase motor is not self-starting, whereas the three-phase motor is self-starting. Therefore, only the three-phase motor will be considered in this chapter.

The three-phase, squirrel-cage induction motor is relatively small in physical size for a given power rating compared to other types of motors. The squirrel-cage induction motor has a good speed regulation under varying load conditions, and because of its rugged construction and reliable operation, this type of motor is widely used in a wide variety of industrial applications.

47.2 Construction Details

Figure 47.1 shows a cutaway drawing of a squirrel-cage induction motor.

The three-phase, squirrel-cage induction motor consists of four components: a stator, a rotor, and two end shields housing the bearings supporting the rotor shaft. The principle

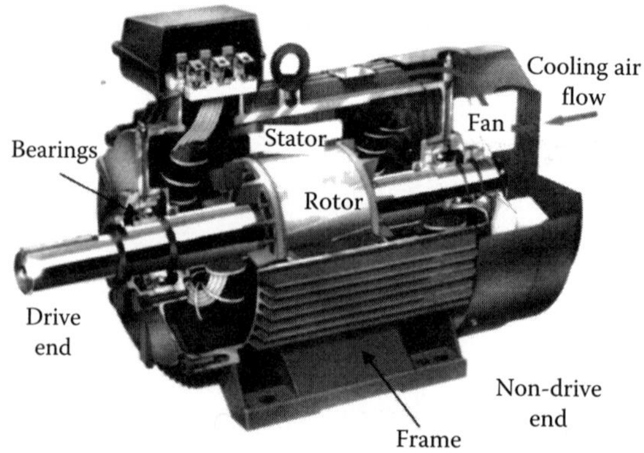

FIGURE 47.1
Component details of a squirrel-cage motor.

advantage of this type of motor is that it requires minimal maintenance. There are no commutator or slip rings to require servicing (compared to a synchronous motor), and hence there are no brushes that require replacing.

The motor frame is generally manufactured from cast steel. The stator core is pressed directly into the frame. The two end shields housing the support bearings are bolted to the cast-steel frame. The bearings supporting the rotor can be either sleeve bearings or ball bearings.

47.2.1 Stator

Figure 47.2 illustrates a typical stator.

A typical stator consists of a three-phase winding that is mounted in the slots of a laminated silicon steel core. The winding itself consists of preformed coils of wire connected so that there are three single-phase windings spaced at 120 electrical degrees apart. The three separate single-phase windings are then connected, usually internally, in a wye or delta configuration. Three or nine leads from the three-phase stator windings are brought out to a terminal box mounted on the external surface of the frame for either single- or dual-voltage connections.

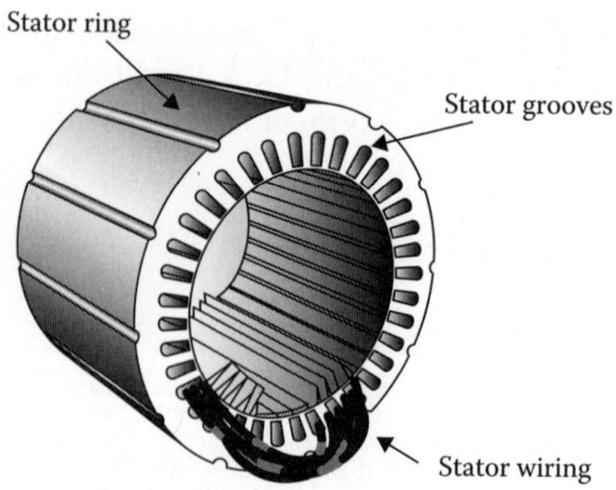

FIGURE 47.2
Typical stator design.

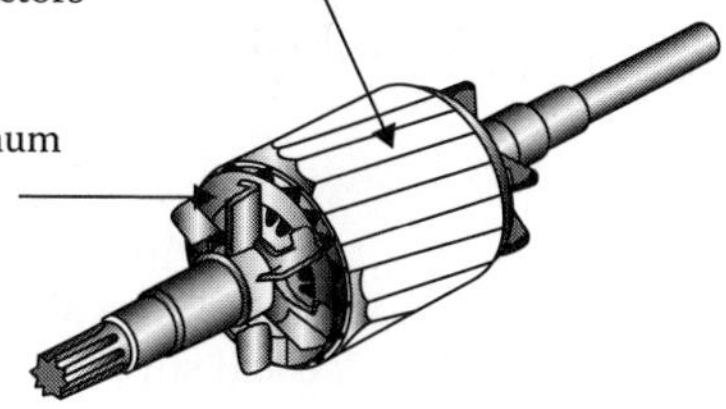

FIGURE 47.3
A typical rotor design.

47.2.2 Rotor

The revolving part of the motor consists of steel laminations arranged in a cylindrical core, and copper or aluminum bars are mounted near the surface of the rotor. These bars are brazed or welded to two copper end rings. On some smaller squirrel-cage induction motors, these end rings are cast in aluminum. Figure 47.3 illustrates a typical rotor. The end rings also circulate cooling air around the rotor when it is running. Note also that the rotor bars between the end rings are skewed at an angle to the end rings. This is to help the motor to run more quietly and more smoothly in operation. The rotor shaft has a feature to allow a pulley or coupling to be attached to the rotor shaft.

47.2.3 Shaft Bearings

Figure 48.4 depicts a typical sleeve bearing, fitted into the end shields made from a Babbitt material, which provides a smooth and polished long-wearing surface for the rotor shaft. An oil slinger fits loosely around the rotor shaft and slings lubricating oil onto the rotating shaft and bearing surfaces. An oil inspection cup at each end enables a maintenance fitter to check the level of the oil in the sleeve bearings.

In some motor designs, ball bearing units are used instead of sleeve bearings. Instead of lubricating oil, grease is used to lubricate the ball bearings. Fittings are provided to allow grease to be injected into the bearings at periodic intervals.

47.3 Principles of Operation

In the paragraph regarding the construction of the stator construction, it was stated that slots in the stator core contain three separate single-phase windings. When three currents, 120 electrical degrees apart, pass through these windings, a rotating magnetic field results. This field travels around the inside of the stator core. The speed of the rotating magnetic field is dependent on the number of stator poles and the frequency of the power source. This speed is called the *synchronous speed* and is determined by the formula:

$$S = \frac{120 \times f}{p} \tag{47.1}$$

where
S = synchronous speed
f = frequency (Hz)
p = number of poles per phase

Example 47.1

A three-phase squirrel-cage induction motor has six poles on the stator winding and is connected to a three-phase, 50 Hz electrical supply. Calculate the synchronous speed of the revolving magnetic field.

Solution

From Equation 47.1,

$$S = \frac{120 \times 50}{6}$$

$$S = 1000 \text{ rev/min (ans)}$$

As this magnetic field rotates at synchronous speed, it cuts the copper bars of the rotor and induces voltages in the bars of the squirrel-cage windings. These induced voltages set up currents in the rotor bars, which in turn create a field in the rotor core. This rotor field reacts with the stator field to create a torque in the rotor. The rotor will always turn at a speed slightly less than the synchronous speed of the stator field. This ensures that the stator field will always cut the rotor bars. If the rotor turns at the same speed as the stator field, then the stator field will not cut the rotor bars and, as a result, there will be no induced voltage or torque.

47.4 Speed Regulation and Percentage Slip

The squirrel-cage induction motor has good speed regulation characteristics (the ratio of difference in speed from low load to full load). Speed performance is measured in terms of percent slip. The synchronous speed of the rotating magnetic field of the stator is used as a reference point.

Because the synchronous speed is dependent on on the number of stator poles and the voltage frequency, these two quantities remain constant. The synchronous speed also remains constant.

If the speed of the rotor at full load is deducted from the synchronous speed of the stator, then the difference is the number of revolutions per minute that the rotor slips behind the rotating field of the stator.

Example 47.2

If the three-phase squirrel-cage induction motor in Example 48.1 has a synchronous speed of 1000 rev/min and a full load speed of 950 rev/min, calculate the percentage slip of the motor.

Solution

$$\text{Synchronous speed} = 1000 \text{ rev/min}$$

$$\text{Full-load rotor speed} = 950 \text{ rev/min}$$

$$\text{Percentage slip} = \left[\frac{(\text{synchronous speed} - \text{rotor speed})}{\text{synchronous speed}}\right] \times 100$$

$$\text{Percentage slip} = 5\% \text{ (ans)}$$

As the value of percentage slip decreases toward 0%, the motor speed performance is improved. The average range of slip for a squirrel-cage induction motor is between 2% and 6%.

Figure 47.4 shows a speed curve together with a percentage slip for a squirrel-cage induction motor that is operating between no load and full load. Rotor friction and windage losses generate sufficient torque to cause the motor to slip behind the synchronous speed. As a mechanical load is applied to the motor shaft, the rotor will tend to slow down. With the stator field turning at a fixed speed, the magnetic field cuts the rotor bars a greater number of times within a given period. The induced voltages within the rotor bars increase, resulting in more current generated in the rotor bars, resulting in a stronger rotor field. There is a larger magnetic reaction between the stator and rotor fields causing a stronger torque being generated. This will increase the stator current being taken from the line. The motor is able to handle the increased mechanical load with very little decrease in the speed of the rotor.

Figure 47.5 illustrates a typical set of slip/torque curves for a squirrel-cage induction motor. The torque output of the motor (measured in Nm) increases linearly with an increase in the value of percentage slip as the mechanical load is progressively increased to a point where full load is reached. Beyond full load, the curve reaches a

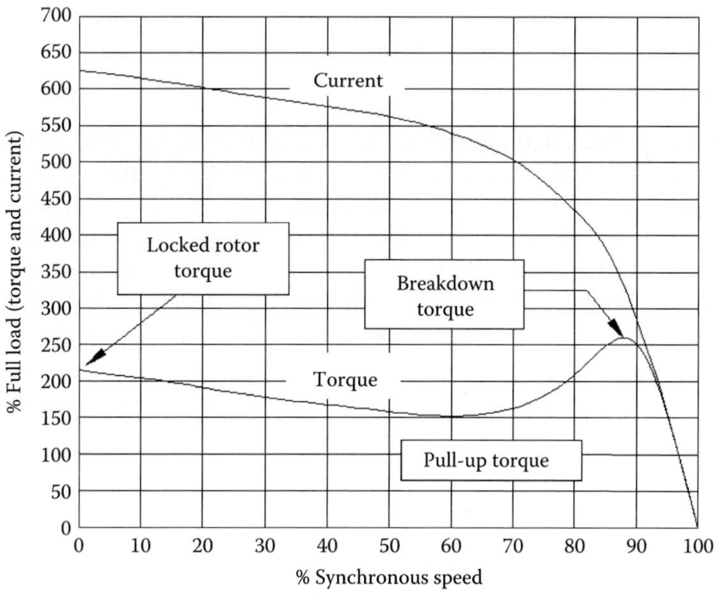

FIGURE 47.4
Sample torque and current vs speed curve.

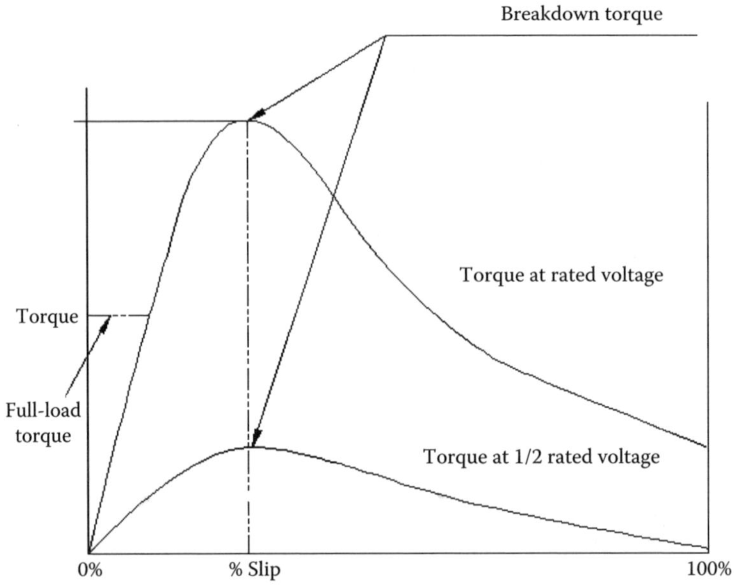

FIGURE 47.5
Typical slip/torque curves.

maximum point called the *breakdown torque*. If the motor continues to be loaded beyond this point, there will be a corresponding reduction in torque until a point is reached when the motor will stall. Starting torque is not shown on the figure but is generally approximately 300% of the running torque.

47.5 Starting Current

When a squirrel-cage induction motor is connected across the full-line voltage, the starting surge of current momentarily reaches as high as 400%–600% or more of the rated full-load current. At the instance the motor starts, the rotor is at a standstill. The stator field cuts the rotor bars at a faster rate than when the rotor is rotating. There will be relatively high induced voltages within the rotor, which will result in a heavy current draw. The resulting input current to the stator windings will be high at the instance of starting. Because of this high starting current, starting protection rated as high as 300% should be provided.

47.6 Power Factor

In AC circuits, the power factor is the ratio of the real power that is used to do work and the apparent power that is supplied to the circuit. The power factor can get values in the range from 0 to 1. When all the power is reactive power with no real power (usually inductive load), the power factor is 0.

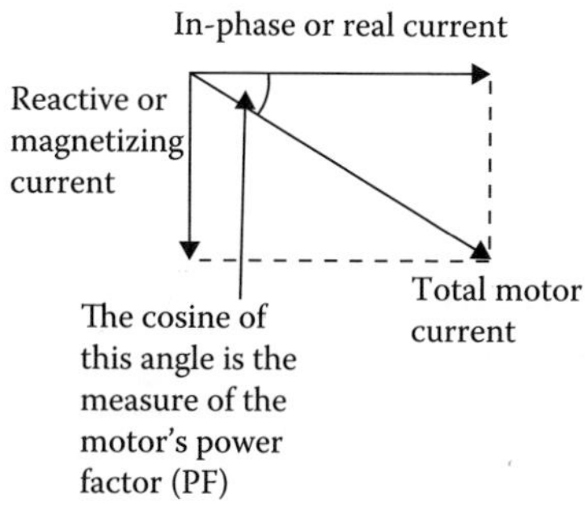

FIGURE 47.6
Power factor.

The power factor of a squirrel-cage induction motor is poor at no-load and low-load conditions. At no-load, the power factor can be as low as 15% lagging. However, as load is applied to the motor, the power factor increases. At the motor's rated load, the power factor can be as high as 85%–90% lagging.

The reason for the power factor being low at no-load is because the magnetizing component of the input current is a large part of the total input current of the motor. When the load on the motor is increased, the in-phase current supplied to the motor increases, but the magnetizing component of the current remains practically the same. This means that the resultant line current is more nearly in phase with the voltage and, therefore, the power factor is improved when the motor is loaded, compared with an unloaded motor that has the magnetizing current as a major component of the input current.

The increase in the power factor from a no-load condition to full-load is shown in Figure 47.6. In the no-load diagram, the in-phase current (*I*w) is small, when compared to the magnetizing current (*I*m) thus, the power factor is poor at no load. In the full-load diagram, the in-phase current has increased, while the magnetizing current remains the same. As a result, the angle of lag of the line current decreases and the power factor increases.

47.7 Revering Rotation

The direction of rotation of a three-phase squirrel-cage induction motor can be easily reversed. The motor will rotate in the opposite direction if any two of the three line leads are reversed, as shown in Figure 47.7. The leads are reversed at the motor.

47.8 Speed Control

A squirrel-cage induction motor has almost no speed variations without external controls. Recall from Section 47.3 that the speed of the motor is dependent on:

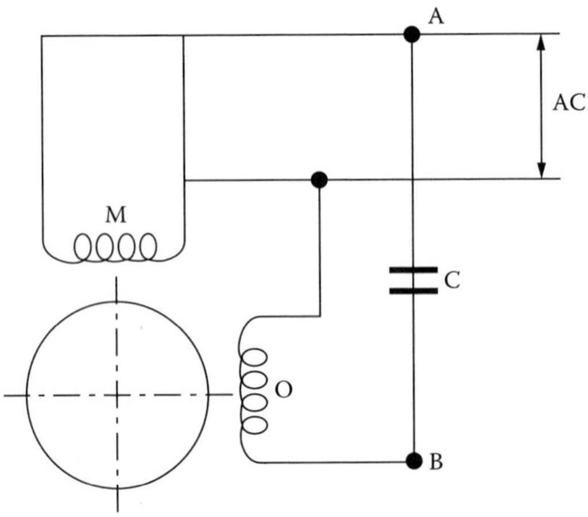

FIGURE 47.7
Reversing rotation.

- The frequency of the three-phase source (50 Hz in the UK)
- The number of poles of the stator winding

The frequency is maintained nationally by the generating board. Since the number of poles in the motor is also fixed, the synchronous speed remains constant. As a result, it is not possible to obtain a range of speeds without changing the applied frequency. It is possible using a variable-frequency AC electronic drive system to change the frequency, or by changing the number of poles using external controllers.

47.9 Electrical Connections

There are two basic line connections:

- Star (Y) connection
- Delta (Δ) connection

Figure 47.8 illustrates the comparison between star and delta connections. Table 47.1 compares the differences between the two methods.

47.10 Motor Nameplates

Motor nameplates are normally located on all electric motors and are vital to the proper selection and installation of the motor. The most useful information given on the nameplate refers to the characteristics of the motor. In most countries, it is a requirement for manufacturers to display all information on the motor's nameplate, but unfortunately this is not always the case.

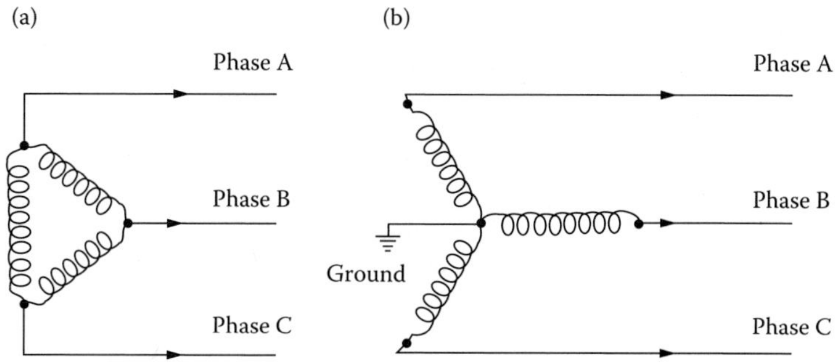

FIGURE 47.8
Comparison between star and delta connections.

TABLE 47.1
Comparison between Star (Y) and Delta (Δ) Connections

Star (Y) Connection	Delta (Δ) Connection
In star connections, the starting or finishing ends (similar ends) of three coils are connected together to form the neutral point. A common wire is taken out from the neutral point called "neutral."	In delta connections, the opposite ends of three coils are connected together, that is, the end of each coil is connected with the start of another coil and the three wires are taken out from the coil joints.
There is a *neutral* or *star point* in Star connections	There is no neutral point in delta connections
Three-phase four-wire system is derived from star connections (*three-phase, four-wire system*)	Three-phase three-wire system is derived from star connections (*three-phase, three-wire system*)
Line current is equal to phase current: Line current = phase current $I_L = I_{PH}$	Line voltage is equal to phase voltage: Line voltage = phase voltage $V_L = V_{PH}$
Line voltage is √3 times phase voltage: $V_L = \sqrt{3}V_{PH}$	Line current is √3 times of phase current: $I_L = \sqrt{3}I_{PH}$
The total power of three phases could be found by $P = \sqrt{3} \times V_L \times I_L \times \cos\Phi$ or $P = 3 \times V_{PH} \times I_{PH} \times \cos\Phi$	The total power of three phases could be found by $P = \sqrt{3} \times V_L \times I_L \times \cos\Phi$ or $P = 3 \times V_{PH} \times I_{PH} \times \cos\Phi$
The speeds of star-connected motors are slow because they receive 1/√3 voltage	The speed s of delta-connected motors are high because each phase receives the total line voltage.
In star connection, the phase voltage is low as 1/√3 of the line voltage, so it needs low numbers of turns, hence saving copper.	In delta connection, the phase voltage is equal to the line voltage; hence, it needs more turns.
Low insulation is required as phase voltage is low.	Heavy insulation is required as phase voltage is equal to the line voltage.
In power transmission, a star-connection system is general and typical to be used.	In power distribution and industries, delta connection is general and typical to be used.

Source: © WWW.electricaltechnology.org.

TABLE 47.2

Information on Motor Nameplate

Ident	Description
47.1	Voltage
47.2	Frequency
47.3	Phase
47.4	Current
47.5	Type
47.6	Power factor
47.7	kW or horsepower
47.8	Full-load speed
47.9	Efficiency
47.10	Duty
47.11	Insulation class
47.12	Maximum ambient temperature
47.13	Altitude
47.14	Enclosure
47.15	Frame
47.16	Bearings
47.17	NEMA: letter code
47.18	NEMA: design letter
47.19	NEMA: service factor

The essential information that can be found on a motor's nameplate is shown in Table 47.2. The principal headings are listed in the following sections.

47.10.1 Voltage

This data states the voltage the motor is designed for. Nameplate-defined parameters for the motor, such as power factor, efficiency, torque, and current, are at rated voltage and frequency. When the motor is used outside these parameters, its performance will be affected.

47.10.2 Frequency

Standard motors are designed for an input frequency of 50 or 60 Hz. When more than one frequency is marked on the nameplate, the appropriate parameters will also have to be indicated on the nameplate.

47.10.3 Phase

This parameter identifies the number of power lines that supply the motor. The standard covers either single-phase or three-phase.

47.10.4 Current

The current stated on the nameplate corresponds to the rated power output, together with the voltage and frequency. If the phases are unbalanced, or if the voltage turns out to be lower than indicated, there will be a current deviation.

47.10.5 Type

Type indicates the type of enclosure. This usually uses the manufacturer's catalogue identification number.

47.10.6 Power Factor

The power factor is indicated on the nameplate as either "PF" or "P.F or cosΦ." Power factor is an expression of the ratio of active power (W) to apparent power (VA) expressed as a percentage.

Remember: The power factor is numerically expressed as equal to the cosine of the angle of the input current with respect to its voltage.

47.10.7 Mechanical Output

The output rating of the motor is expressed as either kW or horsepower (HP). That is its ability to deliver the torque needed for the load at rated speed.

47.10.8 Full-Load Speed

Full-load speed is the speed of the motor at which rated full-load torque is delivered at the rated power output. Normally, the full-load speed is given in either RPM or rev/min. This speed is sometimes referred to as the *slip speed* or *actual rotor speed.*

47.10.9 Efficiency

Efficiency is the motor's output power divided by its input power multiplied by 100. Efficiency is expressed as a percentage. Efficiency is guaranteed by the manufacturer to be within a certain tolerance band, this varies depending on the design standard used, for example, IEC or NEMA.

47.10.10 Duty

Duty is the cycle of operation that a motor can safely operate. This is indicated by either Cont or S1 on the name plate. This indicates that the motor can operate fully loaded 24 h a day. If interval intermediate is shown, a time will also appear. This indicates that the motor can operate at full load for the specified period. The motor should then be stopped and allowed to cool down before running is resumed.

47.10.11 Insulation

The class of insulation refers to the insulation material used in winding the motor stator. It is an expression of the standard classification of the thermal tolerance of the motor's winding. The insulation class is a letter designation such as "B" or "F" depending on the windings ability to survive a given operating temperature for a given life. For example, in a class "B" system, the maximum operating temperature is 130°C. For class "F," it is 155°C, and for class "H," it will be 180°C. The further in the alphabet, the better the performance. For example, a class "F" insulation has a longer nominal life at a given operating temperature than does a class "B" insulation.

47.10.12 Maximum Ambient Temperature

The maximum ambient temperature at which the motor can operate is sometimes shown on the nameplate. This specifies the maximum surrounding air temperature at which the motor can operate to deliver its rated power.

47.10.13 Altitude

This indicates the maximum height above sea level at which the motor will remain within its design temperature rise, meeting all other nameplate data.

Manufacturers guarantee the standard motor ratings at any altitude up to 1000 m (300 feet) above sea level. Motors suitable for operation at altitudes are of a special design and/or have a different insulation class. For example, standard motors with a service factor of 1.15 may be able to operate up to an altitude of 3300 m by utilizing the service factor. At an altitude of 3300 m, the service factor would be 1.00. It may therefore to derate the motor or increase the frame size.

If the altitude is not indicated on the nameplate, then the maximum height above sea level should be restricted to 1000 m.

47.10.14 Enclosure

Enclosure classifies a motor as to the degree of protection from its environment and its method of cooling. Enclosure is shown as IP or ENCL on the nameplate.

IP ratings (known as Ingress Protection or International Protection ratings) are defined to the international standard EN 60529 (British Standard IEC 60509:1989). These standards define the levels of sealing effectiveness of electrical enclosures against intrusion from foreign bodies, such as dirt and water. The standard that the motor is designed to is indicated on the nameplate.

The rating consists of the letters IP followed by two digits. The first digit identifies the level of protection that the enclosure provides against solid bodies, and the second digit describes the degree of protection of the equipment inside the enclosure against water. Table 47.3 is a reference chart showing the ratings and their definitions.

47.10.15 Frame

There are three basic frames of motors:

- Foot-mounted
- Flange-mounted
- Face-mounted

By far the most common are foot-mounted motors. The frame size on the nameplate defines the motor mounting dimensions, such as foot-hole mounting pattern and shaft height. The standard also defines the size of the drive shaft and the dimensions of the keyway for the attachment of pulleys, etc.

TABLE 47.3

IP Rating Reference Chart

IP Rating	First Digit Solids	Second Digit Liquids
IP 54	Protected from limited dust ingress	Protected from water spray from any direction, limited ingress protection
IP 55	Protected from limited dust ingress	Protected from low-pressure water jets from any direction, limited ingress protection
IP 56	Protected from limited dust ingress	Protected from high-pressure water jets from any direction, limited ingress protection
IP 57	Protected from limited dust ingress	Protected from immersion between 15 cm and 1 m in-depth, limited ingress protection
IP 58	Protected from limited dust ingress	Protected from long-term immersion up to a specified pressure, limited ingress protection
IP 60	Protected total limited dust ingress	Not protected from liquids, limited ingress protection
IP 61	Protected total limited dust ingress	Protected from condensation, limited ingress protection
IP 62	Protected total limited dust ingress	Protected from water spray <15° from vertical, limited ingress protection
IP 63	Protected total limited dust ingress	Protected from water spray <60° from vertical limited ingress protection
IP 64	Protected total limited dust ingress	Protected from water spray from any direction, limited ingress protection
IP 65	Protected total limited dust ingress	Protection from low-pressure water jets from any direction, limited ingress protection
IP 66	Protected total limited dust ingress	Protected from high-pressure water jets from any direction, limited ingress protection
IP 67	Protected total limited dust ingress	Protection from immersion between 15 cm and 1 m in depth, limited ingress protection
IP 68	Protected total limited dust ingress	Protected from long-term immersion up to a specified depth, limited ingress protection
IP 69K	Protected total limited dust ingress	Protected from steam jet cleaning, limited ingress protection

47.10.16 Bearings

The bearings used in any AC motor is identified on the nameplate. These require the most maintenance. The data given is for the drive end (DE) bearing and the nondrive end (NDE) and identifies the bearings fitted, together with the type of grease required.

47.10.17 NEMA

Besides the above information, NEMA nameplates have additional supplementary information. The most important are as follows:

- Letter code
- Design letter
- Service factor

TABLE 47.4
NEMA Code Letters

NEMA Code Letter	Locked Rotor (kVA/HP)
A	0–3.15
B	3.15–3.55
C	3.55–4.0
D	4.0–4.5
E	4.5–5.0
F	5.0–5.6
G	5.6–6.3
H	6.3–7.1
I	Not used
J	7.1–8.0
K	8.0–9.0
L	9.0–10.0
M	10.0–11.2
N	11.2–12.5
O	Not used
P	12.5–14.0
Q	Not used
R	14.0–16.0
S	16.0–18.0
T	18.0–20.0
U	20.2–22.4
V	22.4 and upward

47.10.17.1 NEMA Letter Code

A letter code defines the locked rotor current kVA on a per-horsepower basis. The letter code consists of letters from A to V. The further away from the letter code A, the higher the inrush current per horsepower (see Table 47.4).

47.10.17.2 Design Letter

The design letter covers the operating characteristics of torque and current of the motor. Design letters A, B, C, and D define the various categories. The majority of motors are design A or B.

A design A motor has the torque characteristic of motor B, but there is no limit in the starting inrush current. For a design B motor, the manufacturer is limited to the inrush current to ensure that users can apply their own motor starting devices were applicable.

Caution: When replacing a motor in an application, it is important to check the design letter.

Some manufacturers assign letters to their products that do not meet the industry standard and this may lead to starting problems.

47.10.17.3 Service Factor

A motor that is designed to operate at its nameplate power rating will have a service factor of 1.0. With this service factor, the motor is able to operate at 100% of its rated power.

In some applications, a motor may on occasion be required to exceed the rated power. In these cases, a motor having a service factor of 1.15 can be applied to the rated power. A 1.15 service factor motor can operate at 15% higher than the motor's nameplate power.

It must be expected that any motor operating continuously at a service factor that exceeds 1.00 will have a reduced life expectancy compared to one that is operating within its rated power.

Example 47.3

A three-phase 50 Hz synchronous generator runs at 187.5 rev/min. Determine the number of poles of the machine. What type of prime mover would be expected from this machine?

Solution

$$\text{Frequency, f} = 50\ \text{Hz}$$

$$\text{Speed, N} = 187.5\ \text{rev/min}$$

Number of poles of the machine:

$$\begin{aligned} P &= \frac{120f}{N_s} \\ &= \frac{120 \times 50}{187.5} \\ &= 32 \end{aligned}$$

Because the speed of the synchronous generator is very low, the prime mover could be a water turbine (hydraulic turbine). For such a large number of poles, the machine would be a salient-pole type.

NOTE TO EXAMPLE 47.3

In *salient pole* type of rotor consist of large number of projected *poles* (*salient poles*) mounted on a magnetic wheel. Construction of a *salient pole* rotor is as shown in Figure 47.9. The projected *poles* are made up from laminations of steel and are suitable for lower rotor speeds, as opposed to medium- and high-speed rotors where the poles are cylindrical (see Figure 47.10a and b)

- *Salient pole rotors* have large diameter and shorter axial length.
- They are generally used in lower speed electrical machines, say 100–1500 RPM.
- As the rotor speed is lower, more number of poles are required to attain the required frequency. (N_s=120f/p; therefore, f=N_s*p/120; that is, frequency is proportional to number of poles). Typically, the number of salient poles is between 4 and 60.
- Flux distribution is relatively poor than for non-salient pole rotor; hence, the generated emf waveform is not as good as cylindrical rotor.
- Salient pole rotors generally need damper windings to prevent rotor oscillations during operation.

Non-salient pole rotors are cylindrical in shape having parallel slots on it to place rotor windings. It is made up of solid steel.

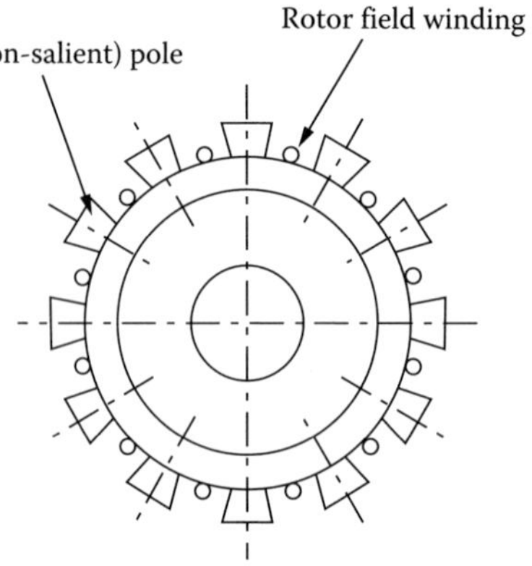

FIGURE 47.9
Non-salient pole rotor.

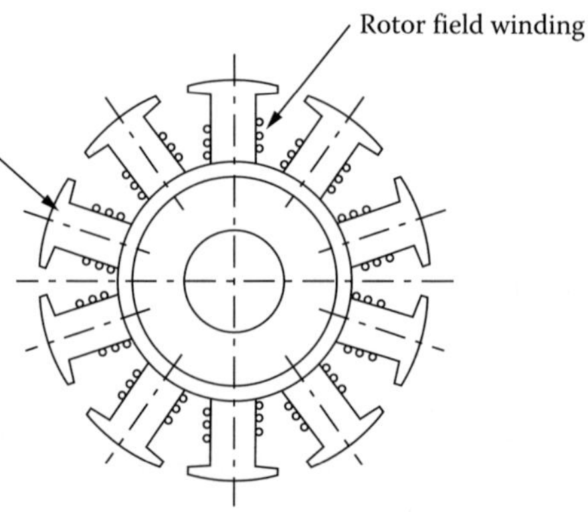

FIGURE 47.10
Salient pole rotor.

The *construction of non-salient pole rotor* (*cylindrical rotor*) is as shown in Figure 47.10a and b. Sometimes, they are also referred to as a drum rotor.

- They are smaller in diameter but having longer axial length.
- *Cylindrical rotors* are used in high-speed electrical machines, usually 1500–3000 RPM.
- Windage loss as well as noise is less compared to salient pole rotors.
- Their construction is robust compared to salient pole rotors.
- Number of poles is usually 2 or 4.
- Damper windings are not needed in *non-salient pole rotors*.
- Flux distribution is sinusoidal and hence gives better emf waveform.
- *Non-salient pole rotors* are used in nuclear, gas, and thermal power plants.

STAR-CONNECTED SYSTEM

Star connection is used where a neutral terminal is required to obtain phase voltage like above image.

In a star-connected system, $VL = \sqrt{3}\ V_{PH}$; mean phase voltage is root three times less than line voltage.

In a star-connected system, IL=I phase.
Star-connected systems require less insulation level.
Star-connected systems are used where low starting current is required.

DELTA-CONNECTED SYSTEM

In a delta-connected system, line voltage is equal to phase voltage, while phase current is $\sqrt{3}$ times less than line current.
Insulation level is high because line voltage=phase voltage.
Delta-connected systems are generally used where high starting torque is required.

48

Synchronous Motors

48.0 Introduction

Synchronous motors and induction motors are the most widely used types of AC motors used in industry. With a synchronous electric motor running at a steady state, the rotation of the rotor is synchronized with the frequency of the supply current. Synchronous motors are available in a wide range of sizes, from sub-fractional self-excited to large high-horsepower industrial sizes, rated up to 15 mW and speeds ranging from 150 to 1800 rev/min.

48.1 Construction of the Synchronous Motor

Figure 48.1 illustrates the significant points of a synchronous motor; the principal components of the motor are the stator, the rotor, and the frame. The stator core is built up using thin silicon laminations and insulated by a surface coating, to minimize any eddy current and hysteresis losses. Axial slots within the stator carry three-phase windings and are wound for a specific number of poles equal to the rotor poles.

The rotor in a synchronous motor is generally of the salient pole type and is excited by a DC field, with slip rings and brushes conducting the current to the rotor. The rotor poles connect to each other and move at the same speed, hence the name synchronous motor.

The direct current excites the rotor windings and creates electromagnetic poles. In certain cases, permanent magnets can also be used.

48.2 Working of a Synchronous Motor

A synchronous motor is a doubly-excited machine, that is, two electrical inputs are provided to it. Its stator winding, which consists of a three-phase winding, is provided with three-phase supply and rotor is provided with DC supply. The three-phase stator winding carrying three-phase currents produces three-phase rotating magnetic flux. The rotor carrying DC supply also produces a constant flux. Considering the frequency to be 50 Hz, from the above relation we can see that the three-phase rotating flux rotates about 3000 revolutions in 1 min, or 50 revolutions in 1 sec. At a particular instant, rotor and stator poles might be of same polarity (N–N or S–S) (see Figure 48.2), causing a repulsive

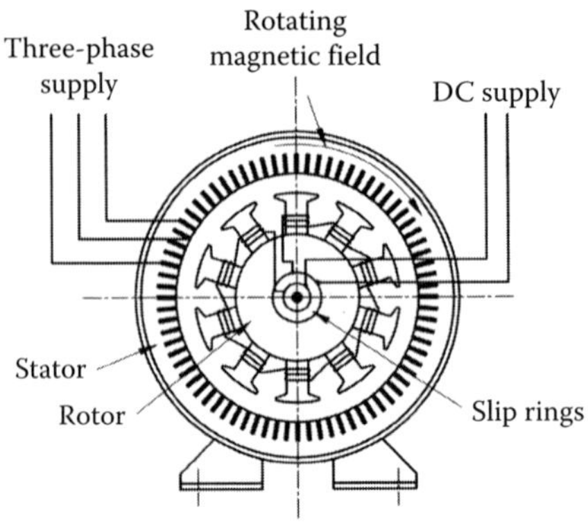

FIGURE 48.1
Construction of a synchronous motor.

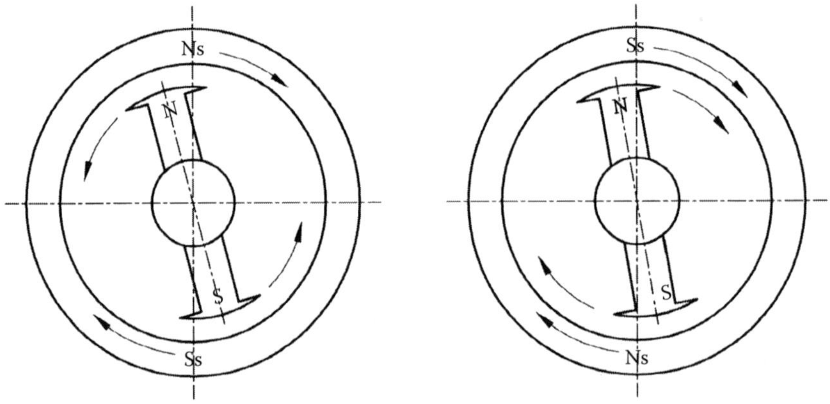

FIGURE 48.2
Operation of a synchronous motor.

force on the rotor, and the very next second it will be N–S, causing an attractive force. But due to inertia of the rotor, it is unable to rotate in any direction due to attractive or repulsive force and remains in standstill condition. Hence, it is not self-starting. To overcome this inertia, rotor is initially fed some mechanical input that rotates it in the same direction as the magnetic field to a speed very close to synchronous speed. After some time, magnetic locking occurs and the synchronous motor rotates in synchronism with the frequency.

48.3 Synchronous Speed

Synchronous motors are so called because the speed of the rotor of this motor is same as the rotating magnetic field. It is basically a fixed-speed motor because it has only one speed, which is synchronous speed and therefore has no intermediate speed, or in other words it is in synchronism with the supply frequency.

The synchronous speed of the motor in rev/min is given by the expression:

$$N_s = 120\frac{f}{p} \tag{48.1}$$

and when expressed in rad/s, by

$$\omega_s = 2\pi\frac{f}{P}$$

where
f is the frequency of the AC supply current in Hz
p is the number of poles per phase
P is the pair number of poles (P=p/2)

Example 48.1

A three-phase, 12-pole (6-pole-pair) synchronous motor is operating from an AC supply having a frequency of 50 Hz. The number of poles per phase is 12/3=4; hence, the synchronous speed will be

$$N_s = 120 \times \frac{50}{4} = 1500 \text{ rev/min}$$

48.4 Operation

As briefly discussed in Section 48.2, synchronous motors are not self-starting and require an external means of speeding the rotor close to synchronism before they can be synchronized. The following are two methods employed to achieve this.

48.4.1 Starting the Motor with an External Power Source

To overcome this inertia, the rotor is initially rotated in the desired direction by a mechanical input, the DC excitation is not initially fed to the rotor. The rotor is rotated in the required direction and as a magnetic field is developed within the stator at a speed close to the synchronous speed. After a period of time, magnetic locking occurs and the synchronous motor rotates in synchronism with the line frequency, and the external motor is cut off.

48.4.2 Damper Winding

Synchronous motors are not self-starting machines. These machines can be made self-starting by providing a special winding in the rotor poles, known as damper winding or squirrel-cage windings. The damper winding consists of short-circuited copper bars embedded in the face of the rotor poles.

When an AC supply is provided to stator of a three-phase synchronous motor, stator winding produces a rotating magnetic field. Due to the damper winding present in the rotor winding of the synchronous motor, the machine starts as an induction motor (induction machines work on the principle of induction). Damper windings in synchronous motors will carry out the same task of induction motor rotor windings. Therefore, due to damper windings, the synchronous motor starts as an induction motor and continues to accelerate. The exciter for the synchronous motor moves along with the rotor. When the motor attains about 95% of the synchronous speed, the rotor windings is connected to exciter terminals and the rotor is magnetically locked by the rotating magnetic field of stator and it then runs as a synchronous motor.

48.5 Difference between Synchronous and Induction Motors

Synchronous motors and induction motors are the most widely used types of AC powered motors.

A synchronous motor run in exact synchronous with the line frequency, whereas an induction motor requires "slip" between the line frequency and motor speed. The rotor must rotate slightly slower than the AC current alterations. Timing motors utilize the advantages of the accuracy of the synchronous motors in timers in appliances and precision servomechanisms where the motor must operate at a precise speed. The speed accuracy is dictated by the accuracy of the line frequency.

48.6 Applications of Synchronous Motors

The following list covers applications of the use of synchronous motors in industry.

1. Because synchronous motors have a higher efficiency compared to induction motors, they are usually employed where loads require a constant speed. Other typical applications include fans, blowers, centrifugal pumps, compressors, reciprocating pumps, and paper and rolling mills.
2. Synchronous motors are used to regulate the voltage at the end of transmission lines.
3. In textile and paper industries, synchronous motors are employed to attain wide range of speeds with a variable-frequency drive system.
4. Because a synchronous motor is capable of operating under either leading or lagging power factor, it can be used for power factor improvement. A synchronous motor under no-load with leading power factor is connected in power system where static capacitors cannot be used.
5. Synchronous motors are available in a wide range of sizes including sub-fractional self-excited sizes. These are commonly used in devices such as where correct time is required, including microwave ovens, timers, and analogue electric clocks.

48.7 Other Types of Synchronous Motors

Synchronous machines are the general category that covers synchronous motors. These also include synchronous generators. Generator action will occur when the field poles are driven ahead of the resultant air-gap flux due to the forward action of the prime mover. Motor action will occur when the field poles are trailed behind the resultant air-gap flux due to the retarding action of the motor shaft load.

There are two main types of synchronous motors depending on the manner the rotor is magnetized:

- Non-excited motors
- Reluctance motors

48.7.1 Non-Excited Motors

Non-excited motors have their rotors manufactured from steel. At synchronous speed, it rotates in step with the rotating magnetic field of the stator, so that it has an almost constant magnetic field through it. The external stator field magnetizes the rotor, inducing the magnetic poles required to turn it. The material of the rotor is a high-retentivity steel, such as cobalt. These are manufactured in permanent magnet, reluctance, and hysteresis designs.

48.7.2 Reluctance Motors

These motors have rotors consisting of a solid steel casting having projecting (salient) toothed poles. In this design, there are fewer rotor than stator poles, to minimize torque ripple and also to prevent the poles from aligning simultaneously with each other, thereby preventing a position occurring which cannot generate torque. The size of the air gap in the magnetic circuit, and therefore the reluctance, will be a minimum when the poles are aligned with the magnetic field of the stator, and this increases with the angle between them. A torque is created that pulls the rotor into alignment with the nearest pole of the stator field. Hence, at the synchronous speed, the rotor is "locked" to the rotating magnetic field of the stator. This in itself cannot start the motor, so the rotor poles generally have a squirrel-cage winding embedded within them that provides a torque at a speed below the synchronous speed. Therefore, the motor starts as an induction motor until its speed approaches synchronous speed until the rotor locks into the rotating magnetic stator field.

48.8 Permanent Magnet Synchronous Motors

A further variation of this type of motor uses permanent magnets embedded within the steel rotor creating a constant magnetic field. In this design, the stator carries windings that are connected to an AC supply, thus producing a rotating magnetic field. At synchronous speed, the rotor poles lock into the rotating magnetic field in the stator.

Permanent magnet synchronous motors have similarities to brushless DC motors.

This type of motor requires a variable-frequency power source to enable the motor to be started, due to the constant magnetic field in the rotor having no induction windings.

49

Introduction to Reverse Engineering

49.1 Introduction

Engineering is the branch of science and technology concerned with the design, building, and use of engines, machines, systems, and structures. At a higher level, there are two types of engineering: forward engineering and reverse engineering.

Forward engineering is the traditional process of moving from a high-level abstraction and logical design to the physical embodiment of the design. In certain situations, there may be a physical part that has no technical details available, such as drawings, parts lists or bills of materials, or having no engineering data such as thermal or electrical properties.

This process of replicating an existing part or assembly or product, without the aid of drawings, documentation, or computer models, is known as *reverse engineering*.

There may be a number of reasons why a part may be reverse engineered, some of which are as follows:

- The original manufacture has gone out of business and replacement parts may not be available.
- There is inadequate design documentation for the original design.
- The original documentation has been lost or may never have existed.
- To create adequate understanding of the original system in another form.
- To recreate the physical representation of that system.
- To update obsolete materials or outdated manufacturing processes with more up to date and less expensive technologies.

Reverse engineering has now become a common technique in a wide range of industries, including software engineering, automotive, consumer products, electronics, and mechanical design.

49.2 History

Wartime situations are the largest users of reverse engineering; it is often used to copy other nations' technology, devices, or information, or parts that have been obtained in the field or intelligence operations. History will recount any number of examples where reverse engineering has changed a military strength when a country was on its back foot. Consider the following historical examples:

Egyptian chariot: The Assyrian army dominated Egypt with their chariots. The Egyptians managed to capture one and, after reverse engineering, put the replica into mass production. Using their improved design, the Egyptians went on to dominate the region for the next thousand years.

Carthaginian quinquereme (multi-oared war galley): The Carthaginians were superior to the Romans during the First Punic War. The Romans had no answer to the feared Carthaginian quinquireme (multi-oared war galley). A number of quinquiremes were wrecked on the Italian coast following a storm. Roman engineers copied the design and constructed several of their own ships to bolster the Roman navy. In 241 B.C., the Roman fleet sank 50 Carthaginian ships during the battle of Aegeus.

Tupolev TU-4: During World War II, three American Boeing B29 bombers on a mission over Japan were forced to land in the USSR. At that time, the USSR had no long-range heavy bombers. Within a few years, they had copied the B29, which they developed into the TU-4.

49.3 Reverse Engineering

Reverse engineering is the process of very carefully dissembling a product down to its individual component parts and meticulously measuring every feature of each component. This was the only method employed for centuries, and it was a time-consuming procedure. The problems came if there was any curvature in the shape of the component.

Today, with the advent of noncontact laser scanners and computer-controlled measurement machines linked to a CAD/CAM design package, the procedure of data collection from the surface feature is much easier. There is software linked into CAD packages currently available that make the procedure of curve fitting much easier and straightforward.

One aspect of reverse engineering was to provide spares for replacement broken or worn-out parts for which either replacement parts was not available from the original supplier or no technical data was available. Reengineering, or reverse engineering, such parts can in most cases a less-expensive option compared to reimporting a replacement part. This option is able to allow the creation of additional spares to maintain the product over a longer period.

49.4 Reinvention

Up to now, the emphasis has been to obtain the best possible geometric representation of the part or assembly, such that it may be copied accurately. One important feature of reverse engineering is that when copying a competitor's product, in trying to gain a competitive advantage, the engineer will dissect the parts carefully and try to step back into the design using classical design methodology. It is desirable that a product design specification (PDS) should be started as if starting the design from scratch, and if the company currently manufactures a similar product, it will have some if not most of the elements in a PDS of their own product.

The company will have carried out a customer market survey based on the positioning of their own product within the market place and will also have carried out comparative surveys on their competitor products; therefore, these comments will be readily available on the company's file.

Company's attempting to enter a particular market but do not have a suitable product, may attempt to consider a market leader's product and determine what improvements can be made to that product that can be the basis of their new design.

As the investigation into the redesign progresses, other elements will be considered, including backward morphological analysis where alternative options are investigated to see if there is a better way of achieving similar or improved results over the market leader (see Table 49.1).

Before attempting reverse engineering, consider a well-planned life cycle analysis and cost/benefit analysis to justify the reverse-engineered project. Reverse engineering is typically cost effective only if the parts to be reverse engineered reflect a high investment or will be reproduced in significant quantities. If the component or part is absolutely required and is considered mission-critical, then the cost of reverse engineering would be undertaken regardless of the cost.

49.5 "S" Curves

At this point, it may be appropriate to discuss "S" curves in the context of the design life of a product. Figure 49.1 depicts a typical S curve as would be expected when charting the life of a product from infancy to decline. The figure shows the four life stages of a project: infancy, rapid growth, maturity, and decline.

49.5.1 Infancy Stage

Generally, a new product starts with a promising idea and will use as many components as were used on the old product. The timing of the start of the new product is important; it would be ideal to reach the stage of rapid growth as the product it is replacing enters the decline stage.

49.5.2 Rapid Growth Stage

Depending how well marketing performs and the spread of advertising, the replacement product will start with a small advantage if the product it is replacing had a good slice of the market and is well placed in its market segment. If the customer's perception is that the new product is a good replacement for the original, the new product should have an advantage. Plotting sales against time will show how well the sales are progressing, and these sales data can be compared against the sales figures of the previous product (Figures 49.2 and 49.3).

49.5.3 Maturity Stage

When the maturity stage is reached, the sales will generally plateau and remain fairly consistent over a period. In some cases, the maturity period may be quite short, depending

TABLE 49.1

Reverse Engineering and Redesign Methodology

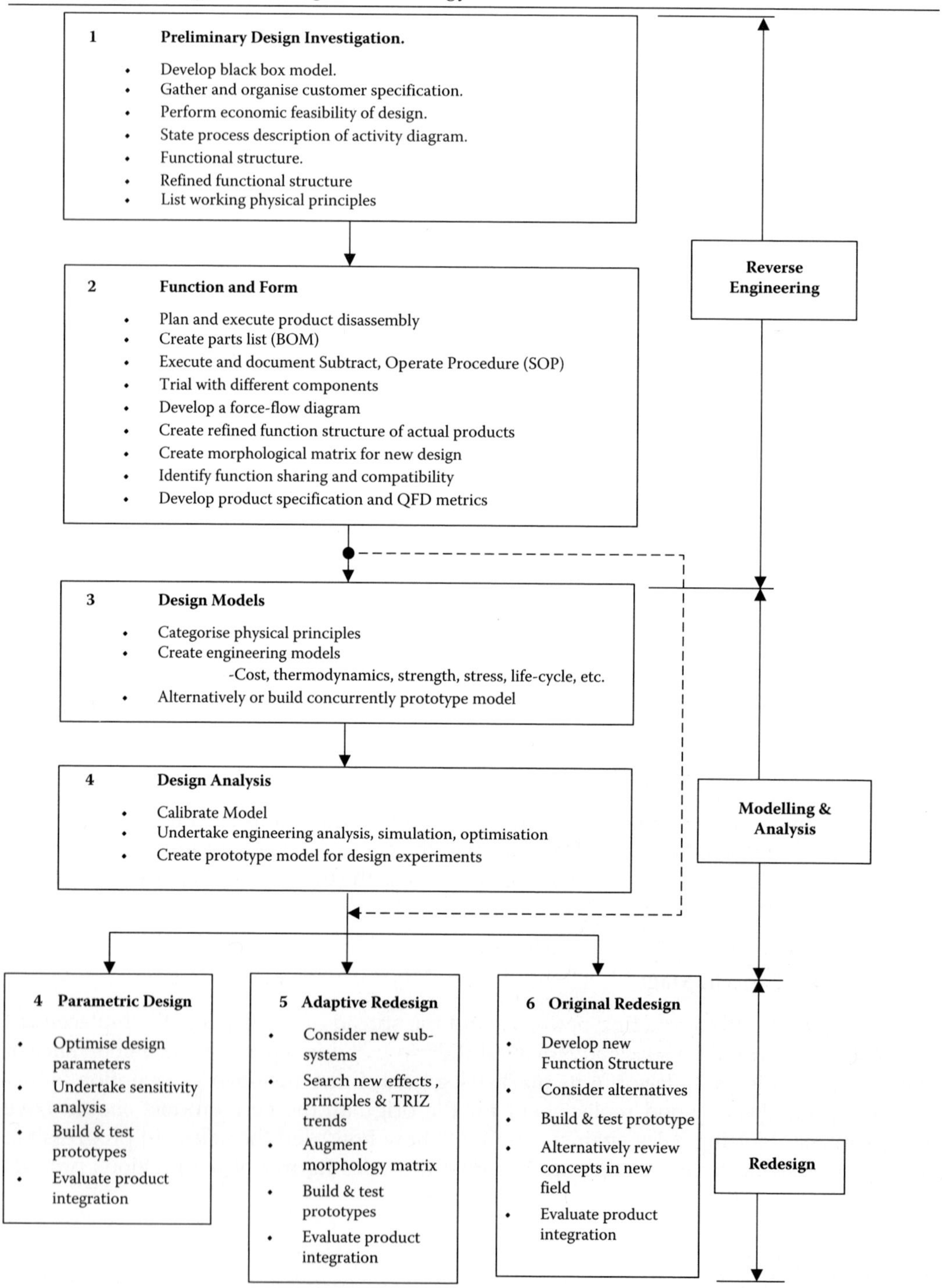

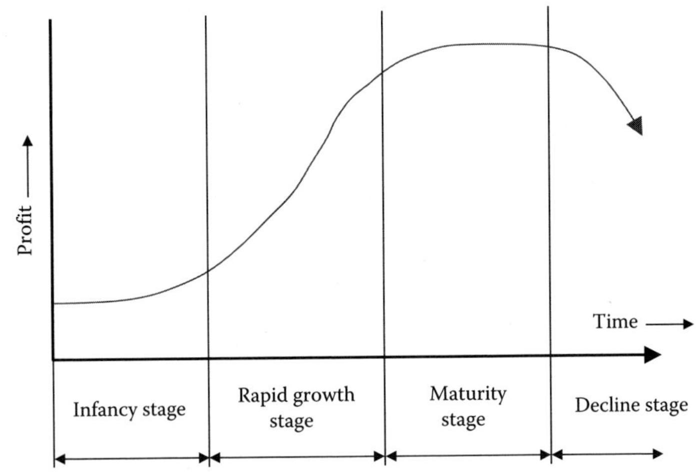

FIGURE 49.1
The S curve of evolution for the life of a product.

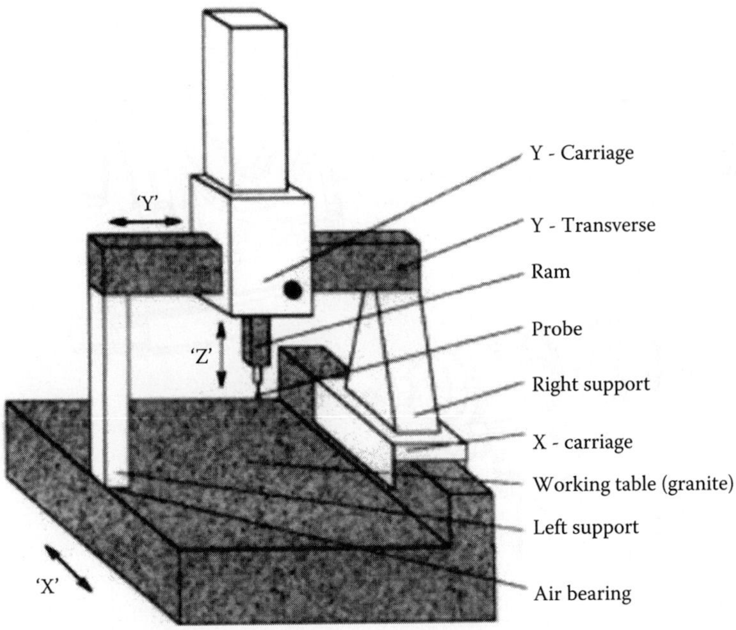

FIGURE 49.2
A typical coordination measuring machine (CMM).

on how well the product is received in the market place. In other cases, particularly in the branded goods market, some brands acquire a loyal following and the maturity stage will be extended. An example is one brand of toothpaste where the maturity phase has lasted more than 20 years and is still selling.

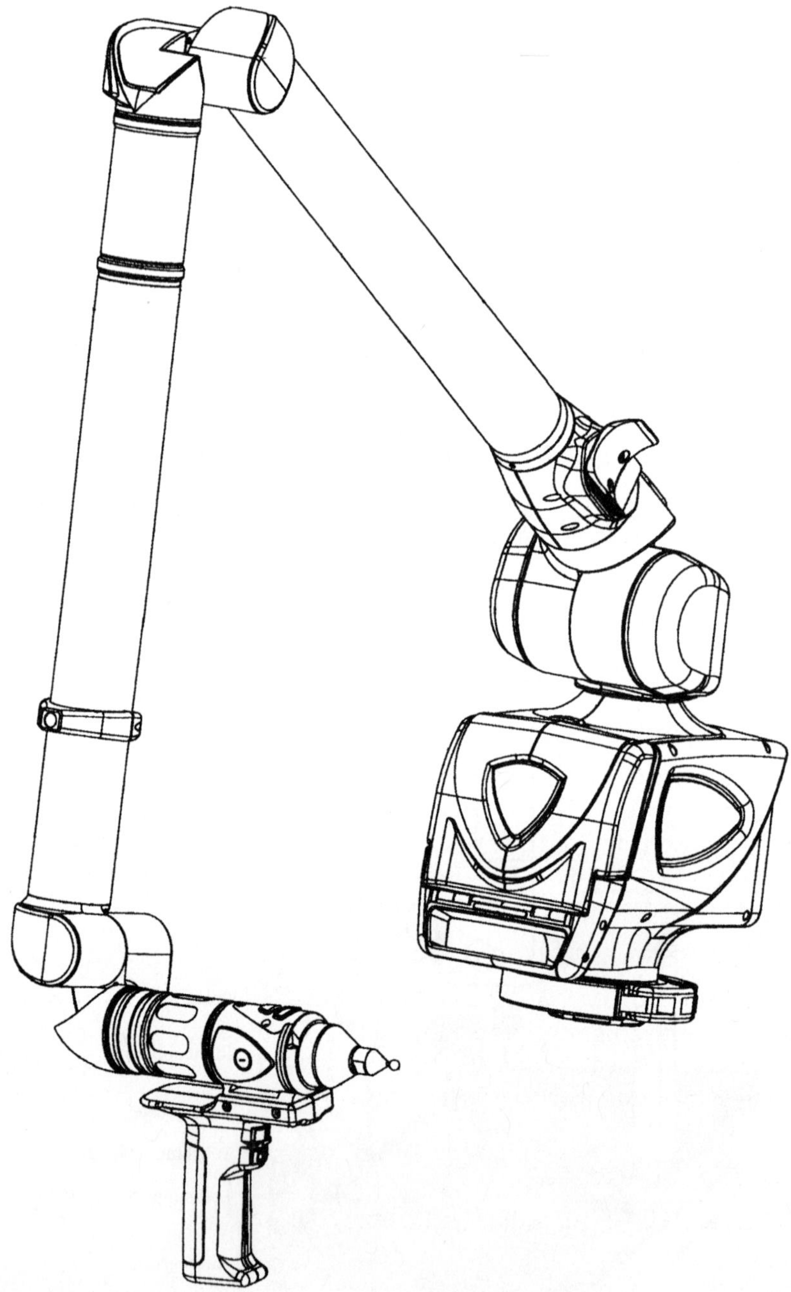

FIGURE 49.3
A typical coordinate measuring arm.

49.5.4 Decline Stage

This stage is marked by a gradual decline in sales. The forward-thinking company will have instigated the next-generation product range prior to the maturity phase of the existing product. The new product range would incorporate the latest in technology and improved manufacturing practice and roll out of the new product range coinciding when sales have reached a figure that the marketing considers irreversible.

When CD records were first introduced onto the market, it was considered they would sign the death knell of vinyl records, and as such vinyl record sales slumped till a point was reached when sales were marked in the tens to a few hundred. Recently, there has been a sudden surge in the sales of vinyl records where today their sales are considered significant. Enthusiasts claim the sound from a vinyl record is superior to that of the CD disc.

49.6 Measurement Technologies

49.6.1 Contact or Noncontact Measurements

There is a clear split in engineering inspection whether the method of surface measurement should be undertaken using either a mechanical contact or an optical method:

- Contact techniques use a mechanical probe of some description; this either touches onto the surface or is dragged across it.
- Noncontact methods use some form of radiation (generally light) to perform the measurement.

Contact techniques (usually coordinate measuring machine (CMM) or mechanical arm) are not affected by the optical characteristics of the surface being measured, but can affect the surface being measured (e.g., soft materials such as silicones or rubbers that will distort when the probe touches the surface while materials such as aluminum and titanium can actually be scored by the effects of the tip of the probe). Contact methods tend to be slow.

By contrast, noncontact techniques (white light or laser scanners/fringe projection) are generally fast and should not affect the surface being measured, as only light is used in the measurement. However, light itself can be affected by the surface, and this, in turn, will affect the quality of the measurement (e.g., clear/translucent materials such as glass can be very challenging, as are shiny surfaces including mirrors or highly polished metal).

49.6.2 Point, Line, or Area

The contact and noncontact methods can be further subdivided:

- *Point scanners* acquire a single measurement point at a time. Each point tends to be very accurate, but very slow, generally than 10 measurements per second. Examples include CMMs or laser trackers.
- *Line scanners* (usually *laser scanners*) are considerably faster, but tend to space measurements along a line. Measurements are spaced regularly along that line and the space between lines depends on how quickly the operator moves the scanner. Line scanners are significantly faster than point scanners, achieving oin the order of 30,000 measurements per second, but their regular measurement grid reduces the X–Y resolution available.

- *Area scanners* (also known as *full-field scanners*) are often viewed as the successor to line scanners. Rather than scanning a single line at a time, they scan thousands of lines—giving a much higher acquisition rate and reducing the challenges of aligned scanned lines. Area scanners gather more information about the object they scan and can, therefore, be more accurate.

49.6.3 Area or Full-Field Scanners

An area or full-field scanner operates by taking a digital photograph of an object to be measured and calculating the range from the scanner to each pixel of the image. From this, the x–y–z coordinates of each pixel are determined.

49.6.4 Difference between Single-Point and Area Scanners

There is a fundamental difference between full-field noncontact scanners and mechanical touch probes as used on CMMs and mechanical arms: A full-field scanner is able to capture millions of measurements that reveal intricacies of freeform shapes that the point-based measurement system is simply too slow to acquire. However, when inspecting a regular feature such as a hole or circular projection, a CMM's x-y resolution only requires either three or four points to measure a hole diameter or projection, whereas a full-field system will have to capture many more points to equal this precision.

49.7 Methods of Acquiring Surface Geometries

49.7.1 Coordinate Measuring Machines

CMMs consist of a bridge supported on a gantry consisting of two side columns straddling a table, this carrying the component to be measured. The table can be fixed and the gantry slides the full length of the table. The second option is for the table to slide between the fixed gantry columns.

The bridge carries the measuring equipment and is able to move in the X and Y axes; the third axis (Z) is provided by the addition of a vertical quill or spindle that moves up and down through the center of the carriage. A touch probe forms the sensing device on the end of the quill.

The movement of the X, Y, and Z axes fully describes the measuring envelope.

It has been long established that the CMM 3D accuracy is on the order of 1 or 2 μm, which is required for the detailed inspection of precision engine or aerospace components. Unfortunately, their measurement speeds are slow. A white-light scanner will make millions of measurements within seconds, whereas a CMM would only manage perhaps a dozen measurements within the same time.

In many applications, the accuracy of the CMM is simply not required. An accuracy of tens of microns measured in seconds is usually more than adequate for many products, especially if this means the measurement can be carried out on every item.

Generating a *point cloud* on a complex surface, such as a body panel, gas turbine blade, or wing section can be a slow undertaking. In many cases, an ideal solution would be to use a combination of using a CMM to measure the cardinal points on the surface and then white light to generate the free-form shape.

In summary, the CMM is ideal where the application requires a small number of extremely accurate measurements or is on a difficult surface such as glass or mirrors, whereas white-light scanners will speed up the freeform measurements.

49.7.2 Mechanical Arm

Mechanical arms as shown in Figure 49.3 is finding a niche market in small manufacturing enterprises or local inspection stations within larger manufacturing facilities. As with the CMM, they use a point-touch probe obtaining 3D measurements. Compared to the white-light scanner, it may be quicker to take 3D measurements with the arm of specific features on the component and establish a go/no-go acceptance for the component. The accuracy will be generally good enough for most workshop purposes, but the arm is essentially a very manual piece of equipment and it must be ensured that the measuring envelope will cover the component perimeters, otherwise may have to incrementally move around the part obtaining surface data.

As discussed, the arm (as with the CMM) is good for measuring a few critical measurements, but when it comes to measuring freeform surfaces the white-light scanners will outclass the mechanical arm. In the time it takes an arm to take one or two measurements, the white-light scanner will have taken millions of measurements and digitized them.

The white-light scanner can be automated either as an integral inspection station within a production line or a fixed position within the inspection department.

49.7.3 Laser Tracker

Laser trackers are instruments that accurately measure large objects by determining the positions of optical targets held against those objects. The accuracy of laser trackers is on the order of 0.025 mm over a distance of several meters. Some examples of laser tracker applications are to align aircraft wings during assembly, and to align and collimate machine tools.

To use the laser tracker, the technician first sets up a laser tracker on a tripod with an unobstructed view of the object to be measured. The technician removes a target, such as a spherically mounted retro-reflector, from the base of the laser tracker and carries it to the

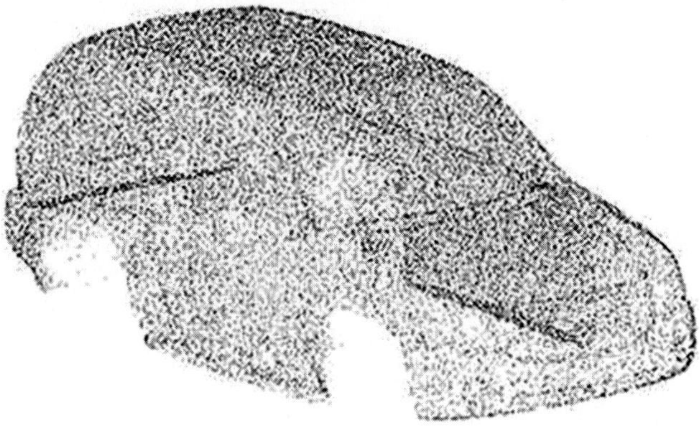

FIGURE 49.4
An example of a point cloud.

object to be measured, moving smoothly to allow the laser tracker to follow the movement of the target. The technician places the target against the object and triggers measurements to be taken at selected points, usually with a remote control device. Measurements can be imported into different types of software to plot the points or to calculate deviation from the correct position.

The targets are known as *retro-reflective* because they reflect the laser beam back in the same direction it came from (in this case, back to the laser tracker).

49.7.4 Structured-Light 3D Scanner

This scanner is a 3D device that measures the 3D object or shape using a projected light pattern and using a camera system.

The principle is that a narrow band of light is projected onto a 3D surface that produces a line of illumination that appears distorted when viewed from a different perspective from the projector. This can be used for an exact geometric reconstruction of the surface shape.

Although many other variants of structured-light projection are possible, patterns of parallel stripes are widely used. The displacement of the stripes allows for an exact retrieval of the 3D coordinates of any details on the object's surface.

49.7.5 Computed Tomography Scanning

This technology is fairly new to metrology and is particularly useful in reverse engineering applications, allowing the inspection of part interiors using X-rays without any subsequent damage or destruction to the parts themselves.

A CAD file can be generated from the computed tomography (CT) data set. Exported CAD file formats are recognized by many software programs, such as CAD, FEA, and CFD. The CAD file created by CT scanning can show not only the external components, but the internal features as well. This allows for first-time rapid prototyping of internal components without the task of creating an entirely new CAD file by hand.

Image-based finite element method (IBFEM) converts the 3D image data from X-ray CT directly into meshes for FEM. Benefits of this method include modelling complex geometries (e.g., composite materials) or accurately modelling "as manufactured" components at the microscale.

49.8 Point Cloud

Point clouds are created by 3D scanners where a large number of points on the object's surface are generated and output as a data file. The point cloud represents the set of points that the scanner has measured (Figure 49.4 is an example of a point cloud).

The point cloud can be used for a variety of reasons. Although not directly usable in most 3D applications, they are usually converted to a polygon mesh or triangular mesh models, on-uniform rational basis spline surface models, or CAD models.

49.9 Summary

It may be said that reverse engineering begins with the product and works through the design process in the opposite direction to arrive at a product statement, and in doing so, it uncovers information about the design ideas that were used in the production of the original product.

50

General Tables

The following tables are given for reference:

TABLE 50.1

Using a Safety Factor of 5—Allowable Working Stress for a Steady or Permanent Working Load

	Type of Stress			
Material	**Tension (t) (MPa)**	**Compression (c) (MPa)**	**Bending (b) (MPa)**	**Shear (MPa)**
Cast iron	29	83	41–55	27.5
Mild steel	89.5–117	89.5–117	89.5–117	69–117
Cast steel	117–145	117–145	117–145	89.5–117
Steel casting	55–69	83–110	69–96.5	16.5
Rolled copper	41.5	—	—	—
Brass	21	—	—	—

TABLE 50.2

Allowable Working Stress for a Load Varying from Zero to Maximum Value

Material	Tension (t) (MPa)	Compression (c) (MPa)	Bending (b) (MPa)	Shear (MPa)
Cast iron	19.3	58.6	27.5	24
Mild steel	59.3–78.6	59.3–83	59.3–78.6	45–59.3
Cast steel	78.6–96.5	78.6–96.5	78.6–96.5	59.3–78.6
Steel casting	36.5–55	55–73	45.5–65	32.4–55
Rolled copper	20.7	—	—	11.0
Brass	13.8	—	—	—

TABLE 50.3

Allowable Working Stresses for a Fluctuating Load Producing Equal Stresses in Opposite Directions

Material	Tension and Compression (MPa)	Bending (MPa)	Shear (MPa)
Cast iron	9.7	13.8	12.0
Mild steel	29.6–39.6	29.6–39.4	22.8–29.6
Cast steel	39.3–48.3	39.3–48.3	29.6–39.3
Steel casting	18.6–27.6	22.8–32.4	15.9–27.6

TABLE 50.4

Specific Gravity of Engineering Materials

Material	Specific Gravity
Aluminum—Cast.	2.56
Aluminum—Wrought.	2.67
Brass—Cast.	8.10
Brass—Sheet 75% Cu	8.45
Copper—Cast.	8.79
Copper—Sheet.	8.81
Copper—Wire.	8.91
Gunmetal 83% Cu.	8.46
Gunmetal 79% Cu.	8.73
Iron—Cast.	6.90–7.50
Iron—Wrought (Average).	7.70
Lead.	11.40
Monel.	8.87
Phosphor Bronze—Cast.	8.60
Steel.	7.73–7.90
Tin—Hammered.	7.39
Tin—Pure.	7.29
Zinc—Cast.	6.86
Zinc—Rolled.	7.19

Giving the specific gravity of a range of metals.
Giving the basic SI units used in engineering.

TABLE 50.5

Basic SI Units

Quantity	Name of Unit	Symbol	Definition
Mass	Kilogram	kg	
Length	Meter	m	
Time	Second	s	
Energy	Joule	J	= Nm = kg m^2/s^2
Power	Watt	W	= J/s = kg m^2/s^3
Pressure	Pascal	Pa	= N/m^2 = kg/m s^2
Force	Newton	N	= m a = kg m/s^2
Frequency	Hertz	Hz	= s^{-1}

Note: a = acceleration.

Specifying the thread details for metric coarse thread properties ranging from 1.6 to 68.0 mm.

TABLE 50.6

I.S.O. Metric Coarse Threads

Outside Dia'	Core Dia'	Pitch	Depth	Effective Dia'	Tapping Drill	Clearance Dia'
1.6	1.1706	0.35	0.2147	1.373	1.25	1.65
1.8	1.3706	0.35	0.2147	1.573	1.45	1.85
2.0	1.5092	0.40	0.2454	1.740	1.60	2.05
2.2	1.6480	0.45	0.2760	1.908	1.75	2.25
2.5	1.9480	0.45	0.2760	2.208	2.05	2.60
3.0	2.3866	0.50	0.3067	2.675	2.50	3.10
3.5	2.7638	0.60	0.3681	3.110	2.90	3.60
4.0	3.1412	0.70	0.4294	3.545	3.30	4.10
4.5	3.5798	0.75	0.4601	4.013	3.80	4.60
5.0	4.0184	0.80	0.4908	4.480	4.20	5.10
6.0	4.7732	1.00	0.6134	5.350	5.00	6.10
7.0	5.7732	1.00	0.6134	6.350	6.00	7.20
8.0	6.4664	1.25	0.7668	7.188	6.80	8.20
10.0	8.1596	1.50	0.9202	9.026	8.50	10.20
12.0	9.8530	1.75	1.0735	10.863	10.20	12.20
14.0	11.5462	2.00	1.2269	12.701	12.00	14.25
16.0	13.5462	2.00	1.2269	14.701	14.00	16.25
18.0	14.9328	2.50	1.5336	16.376	15.50	18.25
20.0	16.9328	2.50	1.5336	18.376	17.50	20.25
22.0	18.9328	2.50	1.5336	20.376	19.50	22.25
24.0	20.3194	3.00	1.8403	22.051	21.00	24.25
27.0	23.3194	3.00	1.8403	25.051	24.00	27.25
30.0	25.7060	3.50	2.1470	27.727	26.50	30.50
33.0	28.7060	3.50	2.1470	30.727	29.50	33.50
36.0	31.0924	4.00	2.4538	33.402	32.00	36.50
39.0	34.0924	4.00	2.4538	36.402	35.00	39.50
42.0	36.4790	4.50	2.7605	39.077	37.50	42.50
45.0	39.4790	4.50	2.7605	42.077	40.50	45.50
48.0	41.8646	5.00	3.0672	44.752	43.00	48.75
52.0	45.8646	5.00	3.0672	48.752	47.00	52.75
56.0	49.2522	5.50	3.3739	52.428	50.50	56.75
60.0	53.2522	5.50	3.3739	56.428	54.50	60.75
64.0	56.6388	6.00	3.6806	60.103	58.00	64.75
68.0	60.6388	6.00	3.6806	64.103	62.00	68.75

Note: All dimensions in mm.

TABLE 50.7

General Units

Dimensions:
Mass = M, Length = L, Time = T

Heat energy:
Unit of heat energy is the Joule (J).
Mechanical equivalent of heat energy = 4.186 J/calorie.
1 calorie (cal) = quantity of heat required to raise 1 g of water through 1°CC

Angular measure:
1 Radian = $360/(2\pi) = 57.296°$

A range of general units related to engineering.
The Greek alphabet and the Roman equivalents.

TABLE 50.8

Greek Alphabet

alpha	A	a
beta	B	β
gamma	Γ	γ
delta	Δ	δ
epsilon	E	ε
zeta	Z	ζ
eta	H	η
theta	Θ	θ
iota	I	t
kappa	K	κ
lambda	Λ	λ
mu	M	M
nu	N	ν
xi	Ξ	ξ
omicron	O	o
pi	Π	π
rho	P	ρ
sigma	Σ	σ
tau	T	τ
upsilon	Υ	υ
phi	Φ	φ
chi	X	χ
psi	Ψ	ψ
omega	Ω	ω

TABLE 50.9

Mathematical Expressions	
y ∞ x:	y is proportional to x
y = f(x):	y is a function of x
y ≈ x:	y is approximately equal to x
y ≠ x:	y is not equal to x
y < x:	y is less than x
y > x	y is greater than x

Some general mathematical expressions that may be encountered.
Definition of some SI units including Newton, Joule, Pascal and Watt.

TABLE 50.10

Other Important Units

Force:
One Newton (1 N) is the force required to give a mass of one kilogram (1 kg) an acceleration of one meter per second per second.
$1\,N = 1\,kg \times m/s^2$

Energy:
One Joule (1 J) is the work done by a force of one Newton (1 N) when moving through a distance of one meter (1 m).
$1\,J = 1\,N \times 1\,m$

Pressure:
One Pascal (1 P) is the pressure exerted by a force of one Newton acting on an area of one square meter (1 m^2)
$1\,Pa = 1\,N/m^2$

Power:
One Watt (1 W) is the power due to a force of one Newton moving at a speed of one m/s or due to a torque of one Nm when rotating at a speed of one radian per second.
$1\,W = 1\,N \times 1\,m/s$

or $1\,W = 1\,Nm \times 1\,rad/s$

The effects of galvanic corrosion of a range of metals in sea water.

TABLE 50.11

Galvanic Corrosion

Galvanic series in flowing sea water.

Alloy	Voltage range of Alloy vs. Reference Voltage[a]	
		Corroded
Anodic or Active End		↑
Magnesium	–1.60 to –1.63	
Zinc	–0.98 to –1.03	
Aluminum Alloys	–0.70 to –1.90	
Cadmium	–0.70 to –0.76	
Cast Irons	–0.60 to –0.72	
Steel	–0.60 to –0.70	
Aluminum Bronze	–0.30 to –0.40	
Naval Brass	–0.30 to –0.40	
Copper	–0.28 to –0.36	
Lead-Tin Solder (50/50)	–0.26 to –0.35	
Admiralty Brass	–0.25 to –0.34	
Manganese Bronze	–0.25 to –0.33	
Silicon Bronze	–0.24 to –0.27	**Direction of Attack**
Stainless Steel (400 Series)[b)]	–0.20 to –0.35	
Copper-Nickel (90-10)	–0.21 to –0.28	
Lead	–0.19 to –0.25	
Copper Nickel (70-30)	–0.13 to –0.22	
Stainless Steel (17-4 PH)[c]	–0.10 to –0.20	
Silver	–0.09 to –0.14	
Monel	–0.04 to –0.14	
Stainless Steel (300 Series)[b,c]	0.00 to –0.15	
Titanium and Titanium Alloys[c)]	+0.06 to –0.05	
Inconel 625[c)]	+0.10 to –0.04	
Hastelloy C-276[c]	+0.10 to –0.04	
Platinum[c)]	+0.25 to +0.18	
Graphite	+0.30 to +0.20	
Cathodic or Noble End		↓
		Protected

[a] These numbers refer to Saturated Calomel Electrodes.

[b] In poorly aerated or low velocity water, or inside crevices, these alloys may start to corrode and exhibit potentials closer to –0.5V.

[c] When covered with slime films of marine bacteria, these alloys may exhibit potentials between +0.3 and +0.4 V.

TABLE 50.12

I.S.O Metric Fasteners

A	C	D	E	F	G	H	J	K	L	M	N	P (Min)	P (Intsn)	Q	T
M2	4	3.38 3.50		5.0	0.3	1.6		3.8				3.50	4.0		1.3
M2.2*	4.6	4.38 4.00		5.0	0.3	1.8		4.0				3.85	4.4		1.5
M2.5	5.2	4.38 4.50		6.5	0.5	2.0		4.5	1.5	5.0	1.25	4.38	5.0		1.6
M3	6.4	5.38 5.50	2.0	7.0	0.5	2.4		5.5	1.8	6.0	1.50	5.25	6.0	5.5	2.0
M3.5*	6.9	5.88 6.00				2.8		6.0	2.1	7.0	1.75	6.10	7.0		2.4
M4	8.1	6.85 7.00	2.8	9.0	0.8	3.2		7.0	2.4	8.0	2.00	7.00	8.0	7.0	2.6
M4.5*	8.1	6.85 7.00				3.5		8.0	2.7	9.0	2.25	7.85	9.0		3.1
M5	9.2	7.85 8.00	3.5	10.0	1.0	4.0		8.5	3.0	10.0	2.50	8.75	10.0	8.5	3.9
M6	11.5	9.78 10.00	4.0	12.5	1.6	5.0	3.0	10.0	3.6	12.0	3.00	10.50	12.0	10.0	3.3
M8	15.0	12.73 13.00	5.5	17.0	1.6	6.5	5.0	13.0	4.8	16.0	4.00	14.00	16.0	13.0	5.0
M10	19.6	16.73 17.00	7.0	21.0	2.0	8.0	6.0	16.0	6.0	20.0	5.00	17.50	20.0	16.0	6.0
M12	21.9	18.67 19.00	8.0	24.0	2.5	10.0	7.0	18.0			6.00	21.00	24.0	18.0	7.0
M14*	25.4	21.67 22.00	9.0	28.0s	2.5	11.0	8.0	21.0							8.0
M16	27.7	23.67 24.00	10.0	30.0	3.0	13.0	8.0	24.0						24.0	9.0
M20	34.6	29.67 30.00	13.0	37.0	3.0	16.0	9.0	30.0						30.0	11.0
M24	41.6	35.38 36.00	15.0	44.0	4.0	19.0	10.0							36.0	

Dimensions are given of a range of common metric fasteners.

Bibliography

The following book titles, although not referenced directly will be found useful to Students of design.

CONSTRADO. *Steel Designer's Manual*, 4th ed., Oxford, UK: BSP Professional Books, 1989.

MIL.HNBK-5J. Metallic Materials & Elements for Aerospace Vehicle Structures, U.S. Department of Defense, 2003.

Pilkey, W.D. *Peterson's Stress Concentration Factors*, 2nd ed., New York: John Wiley & Sons, 2007.

Pugh, S. *Total Design Integrated Methods for Successful Product Engineering*, Wokingham, UK: Addison-Wesley, 1990.

Radzevich, S.P. *Dudley's Handbook of Practical Gear Design & Manufacture*, 2nd ed., Boca Raton, FL: CRC Press, 2012.

Richards, K.L. *Design Engineer's Handbook*, Boca Raton, FL: CRC Press, 2012.

Timoshenco, S. and Goodier, J.N. *Theory of Elasticity*, 2nd ed., New York: McGraw-Hill, 1951.

Young, W.C. *Roark's Formulas for Stress and Strain*, 6th ed., New York: McGraw-Hill, 1989.

Index

C

M

Q

R

T

U